LENS	Laser Engineered Net Shaping		PROM	Programmable Read-Only Memory
LFG	low force grove		PS	Production System
LOM	Laminated Object Manufacturing		P/M	Powder Metallurgy
LP	Lean Production		PVD	Physical Vapor Deposition
LSI	Large Scale Integration		QC	Quality Control
MAP	Manufacturing Automation Protocol		QMS	Quality Management System
MCU	Machine Control Unit		RAM	Random Access Memory
MDI	Manual Data Input		RIM	Reaction Injection Molding
MIG	Metal-Inert Gas		ROM	Read-Only Memory
MIM	Metal Injection Molding		RSW	Resistance Spot Welding
MMFA	mixed model final assembly		SAW	Submerged Arc Welding
MPS	Manufacturing Production System		SCA	Single Cycle Automatic
mrp	Material Requirements Planning		SEM	scanning electron microscope
MRPII	Manufacturing Resources Planning		SGC	Solid Ground Curing
MRR	metal removal rate		SLA	Stereolithography Apparatus
MSD	Manufacturing System Design		SLM	Selective Laser Melting
NC	Numerically Control		SLS	Selective Laser Sintering
NDT	NonDestructive Testing *(NDE = Evaluation) (NDI = Inspection)*		SMAW	Shielded Metal Arc Welding
			SMED	single minute exchange of dies
OAW	Oxyacetylene Welding		SPC	Statistical Process Control
OCR	Optical Character Recognition		SPF	Single Piece Flow
OM	Orthogonal Machining		SQC	Statistical Quality Control
OPM	Orthogonal Plate Machining		TCM	Thermochemical Machining
OS	Operating System		TIG	Tungsten-Inert Gas
OTT	Orthogonal Tube Turning		TIR	Total Indicator Readout
PAW	Plasma Arc Welding *(PAC = Cutting) (PAM = Machining)*		TPS	Toyota Production System
			TQC	Total Quality Control
PC	process capability		UHSMC	Ultra-high-speed machining center
PCB	Printed Circuit Board		USM	Ultrasonic Machining *(USW = Welding)*
PCN	polycrystalline cubic boron nitride		VA	Value Analysis
PD	Pitch Diameter		WAN	Wide Area Network
PDES	Product Design Exchange Specification		WIP	Work-In-Progress (or Process)
PIM	Powder Injection Molding		WJM	Water Jet Machining
PLC	Programmable Logic Controller		WLK	Withdrawl Kanban
POK	Production Ordering Kanban		YAG	Yttrium-Aluminum Garnet
POUS	point of use storage			

Degarmo's Materials and Processes in Manufacturing

12th Edition

J T. BLACK

Auburn University-Emeritus

RONALD A. KOHSER

Missouri University of Science & Technology

WILEY

VP AND EDITORIAL DIRECTOR	Laurie Rosatone
SENIOR DIRECTOR	Don Fowley
ACQUISITIONS EDITOR	Linda Ratts
EDITORIAL MANAGER	Gladys Soto
DEVELOPMENT EDITOR	Chris Nelson
CONTENT MANAGEMENT DIRECTOR	Lisa Wojcik
CONTENT MANAGER	Nichole Urban
SENIOR CONTENT SPECIALIST	Nicole Repasky
PRODUCTION EDITOR	Padmapriya Soundararajan
PHOTO RESEARCHER	Billy Ray
COVER PHOTO CREDIT	(top to bottom) © Senohrabek/iStockphoto; © Nerthuz/Shutterstock; © DigtialStorm/iStockphoto; © Jon Patton/iStockphoto

This book was set in Source Sans Pro 9.5/12.5 by SPi Global and printed and bound by LSC Communications Kendallville.

This book is printed on acid free paper. ∞

Founded in 1807, John Wiley & Sons, Inc. has been a valued source of knowledge and understanding for more than 200 years, helping people around the world meet their needs and fulfill their aspirations. Our company is built on a foundation of principles that include responsibility to the communities we serve and where we live and work. In 2008, we launched a Corporate Citizenship Initiative, a global effort to address the environmental, social, economic, and ethical challenges we face in our business. Among the issues we are addressing are carbon impact, paper specifications and procurement, ethical conduct within our business and among our vendors, and community and charitable support. For more information, please visit our website: www.wiley.com/go/citizenship.

Evaluation copies are provided to qualified academics and professionals for review purposes only, for use in their courses during the next academic year. These copies are licensed and may not be sold or transferred to a third party. Upon completion of the review period, please return the evaluation copy to Wiley. Return instructions and a free of charge return shipping label are available at: www.wiley.com/go/returnlabel. If you have chosen to adopt this textbook for use in your course, please accept this book as your complimentary desk copy. Outside of the United States, please contact your local sales representative.

ISBN: 978-1-118-98767-4 (PBK)

Library of Congress Cataloging in Publication Data:

Names: DeGarmo, E. Paul (Ernest Paul), 1907- author. | Black, J. Temple, author. | Kohser, Ronald A. author.
Title: Degarmo's materials and processes in manufacturing / by J.T. Black, Auburn University-Emeritus, Ronald A. Kohser, Missouri University of Science & Technology.
Other titles: Materials and processes in manufacturing
Description: 12th edition. | Hoboken, NJ : Wiley, 2017. | Includes bibliographical references and index. |
Identifiers: LCCN 2017014996 (print) | LCCN 2017016331 (ebook) | ISBN 9781119299158 (pdf) | ISBN 9781119299585 (epub) | ISBN 9781118987674 (pbk. : acid-free paper)
Subjects: LCSH: Manufacturing processes. | Materials.
Classification: LCC TS183 (ebook) | LCC TS183 .D4 2017 (print) | DDC 670—dc23
LC record available at https://lccn.loc.gov/2017014996

The inside back cover will contain printing identification and country of origin if omitted from this page. In addition, if the ISBN on the back cover differs from the ISBN on this page, the one on the back cover is correct.

10 9 8 7 6 5 4 3 2 1

Preface

It's a world of manufactured goods. Whether we like it or not, we all live in a technological society. Every day we come in contact with hundreds of manufactured items, made from every possible material. From the bedroom to the kitchen to the workplace, we use appliances, phones, cars, trains, and planes, TVs, cell phones, VCRs, DVDs, furniture, clothing, sports equipment, books and more. These goods are manufactured in factories all over the world using manufacturing processes.

Basically, manufacturing is a value-adding activity, where the conversion of materials into products adds value to the original material. Thus, the objective of a company engaged in manufacturing is to add value and to do so in the most efficient manner, with the least amount of waste in terms of time, material, money, space, and labor. To minimize waste and increase productivity, the processes and operations need to be properly selected and arranged to permit smooth and controlled flow of material through the factory and provide for product variety. Meeting these goals requires an engineer who can design and operate an efficient manufacturing system. Here are some trends that are having impacts on the manufacturing world.

- **Manufacturing is a global activity**

  Manufacturing is a global activity with work often being performed at locations based on proximity to materials, labor, or marketplace. US firms often have plants in other countries, and foreign companies operate plants in the United States. Final product assembly often involves components made at a variety of locations.

- **It's a digital world**

  Information technology and computers are growing exponentially, with usage in virtually every aspect of manufacturing. Design and material selection are performed on computers, and this information is then transmitted to manufacture, where machines are often operated and controlled by computers. Computerized inspection processes ensure product quality.

- **Lean manufacturing is widely practiced**

  Most manufacturing companies have restructured their factories (their manufacturing systems) to become lean producers—making goods of superior quality, cheaper and faster, in a flexible way (i.e., they are more responsive to the customers). Almost every plant is doing something to become leaner. Many have adopted some version of the Toyota Production System. More importantly, these manufacturing factories are also designed with the internal customer (the workforce) in mind, so things such as ergonomics and safety are key design requirements. While this book is all about materials and processes for making products, the design of the factory cannot be ignored when it comes to making the external customer happy with the product and the internal customer satisfied with the employer.

- **New products and materials need new processes**

  The number and variety of products and the materials from which they are made continues to proliferate, while production quantities (lot sizes) have become smaller. Existing processes must be modified to be more flexible, and new processes must be developed.

- **Customers expect great quality**

  Consumers want better quality and reliability, so the methods, processes, and people responsible for that quality must improve continually. Reducing the number and magnitude of flaws and defects often requires continual changes to the manufacturing system.

- **Rapid product development is required**

  Being competitive often requires reducing the time to market for new products. Many companies are taking holistic or systemwide perspectives, including concurrent engineering efforts to bring product design and manufacturing closer to the customer. Products are being designed to be easier to manufacture and assemble (design for manufacture/assembly). Manufacturing systems are becoming more flexible (able to rapidly adapt to and assimilate new products).

- **3-D printing and additive manufacturing is exploding**

  New and improved processes, new materials, and expanded capability machines and equipment are entering the market on an almost weekly basis. Technology that once produced lookalike prototype parts is now producing fully functional products from the full range of materials, including metals, ceramics, polymers, and biomaterials.

History of the Text

E. Paul DeGarmo was a mechanical engineering professor at the University of California, Berkley, when he wrote the first edition of *Materials and Processes in Manufacturing*, published by Macmillan in 1957. The book quickly became the emulated standard for introductory texts in manufacturing. Second, third, and fourth editions followed in 1962, 1969, and 1974. DeGarmo began teaching at Berkeley in 1937, after earning his master's of science degree in mechanical engineering from California Institute of Technology. He was a founder of the Department of Industrial Engineering (now Industrial Engineering and Operations Research) and served as its chairman from 1956–1960. He was also assistant dean of the College of Engineering for three years while continuing his teaching responsibilities.

Paul DeGarmo observed that engineering education had begun to place more emphasis on the underlying sciences at the expense of hands-on experience. Most of his students were coming to college with little familiarity with materials, machine tools, and manufacturing methods that their predecessors had acquired through their former "shop" classes. If these engineers and technicians were to successfully convert their ideas into reality, they needed a foundation in materials and processes, with emphasis on capabilities and limitations. Paul sought to provide a text that could be used in either a one- or two-semester course designed to meet these objectives. The materials sections were written with an emphasis on use and application. Processes and machine tools were described in terms of how they worked, what they could do, and their relative advantages and limitations, including economic considerations. The text was written for students who would be encountering the material for the first time, providing clear descriptions and numerous visual illustrations.

Paul's efforts were well-received, and the book quickly became the standard text in many schools and curricula. As materials and processes evolved, the advances were incorporated into subsequent editions. Computer usage, quality control, and automation were added to the text, along with other topics, so that it continued to provide state-of-the-art instruction in both materials and processes. As competing books entered the market, their subject material and organization tended to mimic the DeGarmo text.

Paul DeGarmo retired from active teaching in 1971, but he continued his research, writing, and consulting for many years. In 1977, after the publication of the fourth edition of *Materials and Processes in Manufacturing*, he received a letter from Ron Kohser, then an assistant professor at the University of Missouri-Rolla, containing numerous suggestions regarding the materials chapters. Paul DeGarmo asked Dr. Kohser to rewrite those chapters for the upcoming fifth edition. After that edition, Paul decided he was really going to retire and, after a national search, recruited J T. Black, then a professor at Ohio State, to co-author the book with Dr. Kohser.

For the sixth through 11th editions (published in 1984 and 1988 by Macmillan, 1997 by Prentice Hall, and 2003, 2008 and 2012 by John Wiley & Sons), Dr. Kohser and Dr. Black have shared the responsibility for the text. The chapters about engineering materials, casting, forming, powder metallurgy, additive manufacturing, joining and nondestructive testing have been written or revised by Dr. Kohser. Dr. Black has responsibility for the introduction and chapters about material removal, metrology, surface finishing, quality control, manufacturing systems design, and lean engineering.

Paul DeGarmo died in 2000, three weeks short of his 93rd birthday. For the 10th edition, which coincided with the 50th anniversary of the text, Dr. Black and Dr. Kohser honored their mentor with a change in the title to include his name—*DeGarmo's Materials and Processes in Manufacturing*. We recognize Paul DeGarmo for his insight and leadership and are forever indebted to him for selecting us to carry on the tradition of his book for this, the 12th edition.

Purpose of the Book

The purpose of this book is to provide basic information on materials, manufacturing processes and systems to students of engineering and technology. The materials section focuses on properties and behavior. Aspects of smelting, refining, or other material production processes are presented only as they affect subsequent use and application. In terms of the processes used to manufacture items (converting materials into useful shapes with desired properties), this text seeks to provide a descriptive introduction to a wide variety of options, emphasizing how each process works and its relative advantages and limitations. The goal is to present this material in a way that can be understood by individuals seeing it for the very first time. This is not a graduate text where the objective is to thoroughly understand and optimize manufacturing processes. Mathematical models and analytical equations are used only when they enhance the basic understanding of the material. Although the text is introductory in nature, new and emerging technologies, such as direct-digital and micro- and nano-manufacturing processes, are included as they transition into manufacturing usage.

Organization of the Book

E. Paul DeGarmo wanted a book that explained to engineers how the things they designed could be made. *DeGarmo's Materials and Processes in Manufacturing* is still being written to provide a broad, basic introduction to the fundamentals of manufacturing. The text begins with a survey of engineering materials, the "stuff" that manufacturing begins with, and seeks to provide the basic information that could be used to match the properties of a material to the service requirements of a component. A variety of engineering materials are presented, along with their properties and means of modifying them. The materials section can be used in curricula that lack preparatory courses in metallurgy, materials science, or strength of materials, or where the student has not yet been exposed to those topics. In addition, various chapters in this section can be used as supplements to a basic materials course, providing additional information about topics such as heat treatment, plastics, composites, and material selection.

Following the materials chapters are sections about casting, powder metallurgy, forming, material removal, and joining. Each section begins with a presentation of the fundamentals on which those processes are based. These introductions are followed by a discussion about the various process alternatives, which can be selected to operate individually or be combined into an integrated system.

Reflecting the many recent developments and extreme interest in additive manufacturing (often called 3-D printing), the chapter about this technology has been significantly updated to present the various technologies in place at the time of textbook printing. Uses and applications are summarized, including prototype manufacture, rapid tooling, and direct-digital manufacture. The advantages and limitations of additive manufacturing are summarized, along with a description of current and future trends.

Manufacturing processes are often designed to accommodate specific materials. A separate chapter presents those processes that are somewhat unique to plastics, ceramics, and composites.

Chapters have been included to provide information about surface engineering, measurements, nondestructive testing, and quality control. Engineers need to know how to determine process capability and, if they get involved in Six Sigma projects, to know what sigma really measures. There is also introductory material about surface integrity, since so many processes produce the finished surface and impart residual stresses to the components.

Many of the advances in manufacturing relate to the way the various processes are implemented and integrated in a production plant or on the shop floor—the design of manufacturing systems. Aspects of automation, numerical control, and robotics are presented in a separate chapter. In addition, there is expanded coverage of lean engineering, in which the mass production system is converted into a lean production system, capable of rapidly manufacturing variations of a product, small quantities of a product, or even one-of-a-kind items on a very flexible and continual basis.

With each new edition, new and emerging technology is incorporated, and existing technologies are updated to accurately reflect current capabilities. Through its nearly 60-year history and 11 previous editions, the DeGarmo text was often the first introductory book to incorporate processes such as friction-stir welding, microwave heating and sintering, and machining dynamics.

Each chapter closes with a listing of *Review Questions*, designed to assess a student's understanding of the material presented in the text. The *Problems* section further applies this understanding, with a bit of focus on application. Somewhat open-ended case studies are provided in the Internet supplements that accompany this text. These have been designed to make students aware of the great importance of properly integrating design, material selection, and manufacturing to produce cost competitive, reliable products.

The DeGarmo text is intended for use by engineering (mechanical, lean, manufacturing, industrial, and materials) and engineering technology students, in both two- and four-year undergraduate degree programs. In addition, the book is also used by engineers and technologists in other disciplines concerned with design and manufacturing (such as aerospace and electronics). Factory personnel find this book to be a valuable reference that concisely presents the various production alternatives and the advantages and limitations of each. Additional or more in-depth information about specific materials or processes can be found in an expanded list of supplemental references that is organized by topic.

Supplements

An instructor solutions manual for instructors adopting the text for use in their courses is available on a companion website: www.wiley.com/college/black. A collection of manufacturing process videos also is available to both instructors and students in the Fundamental Manufacturing Processes Sampler also on that site.

Acknowledgments

The authors wish to acknowledge the multitude of assistance, information, and illustrations that have been provided by a variety of industries, professional organizations, and trade associations. The text has become known for the large number of clear and helpful photos and illustrations that have been provided graciously by a variety of sources. In some cases, equipment is photographed or depicted without safety guards, so as to show important details, and personnel are not wearing certain items of safety apparel that would be worn during normal operation.

Over the many editions, hundreds of reviewers, user faculty, and students have submitted suggestions and corrections to the text. We continue to be grateful for their input.

The authors also would like to acknowledge the contributions of Dr. Elliot Stern for the dynamics of machining section in Chapter 21, Dr. Brian Paul for writing the micromanufacturing chapter, Dr. Kavit Antani for his contributions in lean engineering and system design, Dr. Andres Carrano for his work in measurements/metrology, Prof. Julia Morse for her contributions to NC and CNC, and Mr. Kevin Slattery of the Boeing Company for his review of the chapter on additive manufacturing.

The authors want to thank Linda Pitchford at Auburn University for her assistance during the preparation of the book.

The authors thank the John Wiley & Sons team that worked on the book, including Padmapriya Soundararajan, Jen Devine, and Chris Nelson.

Both Dr. Black and Dr. Kohser lost their wives to COPD and cancer, respectively, since the publication of the 11th edition. Carol Black was a great editor, and Barb Kohser endured being a "textbook widow" through the preparation of seven editions. They will be dearly missed.

About the Authors

J T. Black received his PhD from Mechanical and Industrial Engineering, University of Illinois, Urbana, in 1969, a master's of science degree in industrial engineering from West Virginia University in 1963, and his bachelor of science degree in industrial engineering, Lehigh University, in 1960. J T. is professor emeritus from Industrial and Systems Engineering at

Auburn University. He was the chairman and a professor of Industrial and Systems Engineering at University of Alabama-Huntsville. He also taught at Ohio State University, University of Rhode Island, University of Vermont, and University of Illinois. He taught his first processes class in 1960 at West Virginia University. J T. is a Fellow in the American Society of Mechanical Engineers, the Institute of Industrial Engineering and the Society of Manufacturing Engineers. J loves to write music (mostly down-home country) and poetry, play tennis in the backyard, and show his champion pug dogs.

Ron Kohser received his PhD from Lehigh University Institute for Metal Forming in 1975. He then joined the faculty of the University of Missouri-Rolla, now the Missouri University of Science & Technology, where he held positions of professor of metallurgical engineering, dean's teaching scholar, department chairman and associate dean for undergraduate instruction. Ron consistently carried a full teaching load, including metallurgy for engineers; introduction to manufacturing processes; material selection, fabrication, and failure analysis; materials processing; powder metallurgy; and metal deformation processes. In 2013, he retired as professor emeritus, and moved to the Lake of the Ozarks, where he and his wife helped build their retirement home. He looks forward to some time of fishing on the lake.

About the Cover

Although there are some similarities to function and design, the four airplanes on the cover represent a spectrum of manufacturing and operating conditions, and meeting those conditions often requires different "materials" and "processes." The 1903 Wright Flyer was a hand-built one-of-a-kind airplane with a wooden propeller and an airframe made of ash and spruce wood, stiffened by bracing wires, and covered with linen cloth. As airplane technology developed, the materials changed, along with the methods to convert the new materials to the desired shapes. Desired material properties usually included light weight and high strength, coupled with features that might include fracture resistance, corrosion resistance, and others.

The four airplanes on the cover are a large commercial passenger airliner or freight hauler, a small private light aircraft,

the F35 Lightning II military stealth fighter, and the SR-71 Blackbird supersonic, stealth spy plane. Most manufacturing decisions for the passenger airliner or freight hauler are driven by economics, with the bottom line being the resulting cost per mile of travel for a passenger or ton of cargo. For the small private airplane, objectives might include purchase and operational affordability, as well as ease of maintenance at diverse local facilities. Desired features for the military fighter are usually performance driven, with cost being secondary. The fighter must be versatile and able to perform a range of operations under various combat conditions. Supersonic speed, agility, and stealth might come at justifiable additional cost. Common materials include aluminum, titanium, and carbon-fiber composites. The SR-71 Blackbird was designed to operate under never-before-attained conditions, namely Mach 3 supersonic speed and altitudes as high as 80,000 feet (more than 15 miles). Prolonged supersonic flight would raise surface temperatures to more than 1000°F, bringing a new and critical demand to the usual material properties of light weight and high strength. Body sheet and other components had to be fabricated from titanium, and thermal expansion was a critical design property. Until spy planes were replaced by satellite surveillance, the SR-71 was the ultimate airplane.

These four planes are also powered by different types of engines with widely different designs, horsepower, operating temperatures, and speeds of internal components. The landing gear must support different weights and endure different shocks and impacts. The production quantity also varies. Only a few Blackbirds were made, while passenger planes are usually made in the hundreds, and several thousand of a particular model private plane might be produced. A $5000 die, mold, or tool adds $500 per part if the production run is only 10 planes, but only $5 per plane if the production run is 1000. Some processes favor small quantities, while others favor large runs.

Nearly all of the materials and processes described in this book find themselves employed in one or more of the four airplanes. Each of the material-process combinations was selected because it offered the best match to the needs of the specific product and component. However, as new materials and processes are developed, the current "best" solutions will be constantly changing. We invite the reader to open the text and explore this fascinating area of engineering and technology.

Contents

CHAPTER 1

Introduction to DeGarmo's Materials and Processes in Manufacturing

1.1 Materials, Manufacturing, and the Standard of Living

Manufacturing is critical to a country's economic welfare and standard of living because the standard of living in any society is determined, primarily, by the *goods* and *services* that are available to its people. Manufacturing companies contribute about 20% of the GNP, employ about 18% of the workforce, and account for 40% of the exports of the United States. In most cases, materials are utilized in the form of manufactured goods. **Manufacturing** and **assembly** represent the organized activities that convert raw materials into salable goods. The manufactured goods are typically divided into two classes: producer goods and consumer goods. **Producer goods** are those goods manufactured for other companies to use to manufacture either producer or consumer goods. **Consumer goods** are those purchased directly by the consumer or the general public. For example, someone has to build the machine tool (a lathe) that produces (using machining processes) the large rolls that are sold to the rolling mill factory to be used to roll the sheets of steel that are then formed (using dies) into body panels of your car. Similarly, many service industries depend heavily on the use of manufactured products, just as the agricultural industry is heavily dependent on the use of large farming machines for efficient production.

Processes convert materials from one form to another adding value to them. The more efficiently materials can be produced and converted into the desired products that function with the prescribed quality, the greater will be the companies' productivity and the better will be the standard of living of the employees.

The history of mankind has been linked to our ability to work with tools and materials, beginning with the Stone Age and ranging through the eras of copper and bronze, the Iron Age, and recently the age of steel. Although ferrous materials still dominate the manufacturing world, we have entered the age of tailor-made plastics, composite materials, and exotic alloys.

A good example of this progression is shown in **Figure 1.1**. The goal of the manufacturer of any product or service is to continually improve. For a given product or service, this improvement process usually follows an S-shaped curve, as shown in Figure 1.1a, often called a product life-cycle curve. After the initial invention/creation and development, a period of rapid growth in performance occurs, with relatively few resources required. However, each improvement becomes progressively more difficult. For a significant gain, more money and time and innovation are required. Finally, the product or service enters the maturity phase, during which additional performance gains become very costly.

For example, in the automobile tire industry, Figure 1.1b shows the evolution of radial tire performance from its birth in 1946 to the present. Growth in performance is actually the superposition of many different improvements in material, processes, and design.

These innovations, known as **sustaining technology**, serve to continually bring more value to the consumer of existing products and services. In general, sustaining manufacturing technology is the backbone of American industry and the ever-increasing productivity metric.

Although materials are no longer used only in their natural state, there is obviously an absolute limit to the amounts of many materials available here on earth. Therefore, as the variety of man-made materials continues to increase, resources must be used efficiently and recycled whenever possible. Of course, recycling only postpones the exhaustion date.

Like materials, processes have also proliferated greatly in the past 50 years, with new processes being developed to handle the new materials more efficiently and with less waste. A good example is the laser, invented around 1960, which now finds many uses in machining, measurement, inspection, heat treating, welding, additive manufacturing, surgery, and many

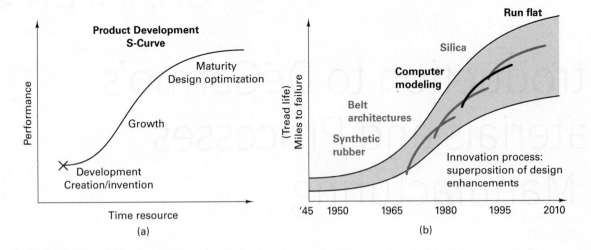

FIGURE 1.1 (a) A product development curve usually has an "S"-shape. (b) Example of the S-curve for the radial tire. (Courtesy of Bart Thomas, Michelin)

more. New developments in manufacturing technology often account for improvements in productivity. Even when the technology is proprietary, the competition often gains access to it, usually quite quickly.

Starting with the product design, materials, labor, and equipment are interactive factors in manufacturing that must be combined properly (integrated) to achieve low cost, superior quality, and on-time delivery. **Figure 1.2** shows a breakdown of costs for a product (like a car). Typically about 40% of the selling price of a product is the **manufacturing cost**. Because the selling price determines how much the customer is willing to pay, maintaining the profit often depends on reducing manufacturing cost. The internal customers who really make the product are called direct labor. They are usually the targets of automation, but typically they account for only about 10% of the manufacturing cost, even though they are the main element in increasing productivity. In Chapters 42 and 43, a new manufacturing strategy is presented that attacks the materials cost, indirect costs, and general administration costs, in addition to labor costs. The materials costs include the cost

of storing and handling the materials within the plant. The strategy depends on a new factory design and is called **lean production**.

Referring again to the total expenses shown in Figure 1.2 (selling price less profit), about 68% of dollars are spent on people, but only 5 to 10% on director labor, the breakdown for the rest being about 15% for engineers and 25% for marketing, sales, and general management people. The average labor cost in manufacturing in the United States is $10 to $25 per hour for hourly workers. Reductions in direct labor will have only marginal effects on the total people costs. The optimal combination of factors for producing a small quantity of a given product may be very inefficient for a larger quantity of the same product. Consequently, a systems approach, taking all the factors into account, must be used. This requires a sound and broad understanding on the part of the decision makers on the value of materials, processes, and equipment to the company, and their customers, accompanied by an understanding of the manufacturing systems. Materials, processes, and manufacturing systems are what this book is all about.

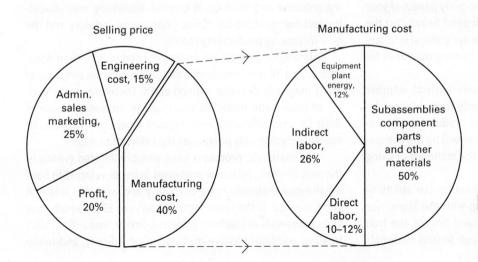

FIGURE 1.2 Manufacturing cost is the largest part of the selling price, usually around 40%. The largest part of the manufacturing cost is materials, usually 50%.

1.2 Manufacturing and Production Systems

Manufacturing is the economic term for making goods and services available to satisfy human wants. Manufacturing implies creating value by applying useful mental or physical labor. The *manufacturing processes* are arranged in the factory to form a *manufacturing system* (MS). The manufacturing system is a complex arrangement of physical elements characterized by measurable parameters. The manufacturing system takes inputs and produces products for the external customer, as shown in **Figure 1.3**.

The inputs to the manufacturing system includes materials, information, and energy. The system is a complex set of elements that includes machines (or machine tools), people, materials-handling equipment, and tooling. Workers are the internal customers. They process materials within the system, which gain value as the material progresses from process to machine. Manufacturing system outputs may be finished or semifinished goods. Semifinished goods serve as inputs to some other process at other locations. Manufacturing systems are dynamic, meaning that they must be designed to adapt constantly to change. Many of the inputs cannot be fully controlled by management, and the effect of disturbances must be counteracted by manipulating the controllable inputs or the system itself. Controlling the input material availability and/or predicting demand fluctuations may

be difficult. A national economic decline or recession can cause shifts in the business environment that can seriously change any of these inputs. In manufacturing systems, not all inputs are fully controllable. To understand how manufacturing systems work and be able to design manufacturing systems, computer modeling (simulation) and analysis are used. However, modeling and analysis are difficult because

1. In the absence of a system design, the manufacturing systems can be very complex, be difficult to define, and have conflicting goals.

2. The data or information may be difficult to secure, inaccurate, conflicting, missing, or even too abundant to digest and analyze.

3. Relationships may be awkward to express in analytical terms, and interactions may be nonlinear; thus, many analytical tools cannot be applied with accuracy. System size may inhibit analysis.

4. Systems are always dynamic and change during analysis. The environment can change the system, and vice versa.

5. All systems analyses are subject to errors of omission (missing information) and commission (extra information). Some of these are related to breakdowns or delays in feedback elements.

Because of these difficulties, digital simulation has become an important technique for manufacturing systems modeling and analysis as well as for manufacturing system design.

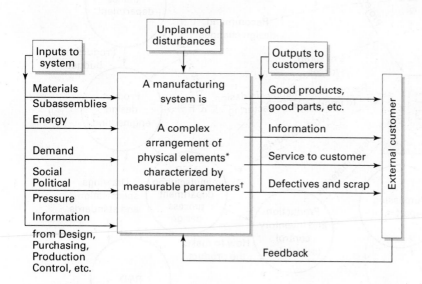

* Physical elements:
  • Machines for processing
  • Tooling (fixtures, dies, cutting tools)
  • Material handling equipment (which includes all transportation and storage)
  • People (internal customers) operators, workers, associates

† Measurable system parameters:
  • Throughput time (TPT)
  • Production rate (PR)
  • Work-in-process inventory
  • % on-time delivery
  • % defective
  • Daily/weekly/monthly volume
  • Cycle time or takt time (TT)
  • Total cost or unit cost

FIGURE 1.3 Here is our definition of a manufacturing system with its inputs and outputs. (From Design of the Factory with a Future, 1991, McGraw-Hill, by J T. Black)

The entire company is often referred to as the enterprise or the production system. The production system services the manufacturing system, as shown in **Figure 1.4**. In this book, a production system will refer to the total company and will include within it the manufacturing system. The production system includes the manufacturing system plus all the other functional areas of the plant for information, design, analysis, and control. These subsystems are connected by various means to each other to produce either goods or services or both.

Goods refers to material things. **Services** are nonmaterial things that we buy to satisfy our wants, needs, or desires. Service production systems include transportation, banking, finance, savings and loan, insurance, utilities, health care, education, communication, entertainment, sporting events, and so forth. They are useful labors that do not directly produce a product. Manufacturing has the responsibility for designing

processes (sequences of operations and processes) and systems to create (make or manufacture) the product as designed. The system must exhibit flexibility to meet customer demand (volumes and mixes of products) as well as changes in product design.

As shown in **Table 1.1**, production terms have a definite rank of importance, somewhat like rank in the army. Confusing *system* with *section* is similar to mistaking a colonel for a corporal. In either case, knowledge of rank is necessary. The terms tend to overlap because of the inconsistencies of popular usage.

An obvious problem exists here in the terminology of manufacturing and production. The same term can refer to different things. For example, *drill* can refer to the machine tool that does these kinds of operations; the operation itself, which can be done on many different kinds of machines; or the cutting tool,

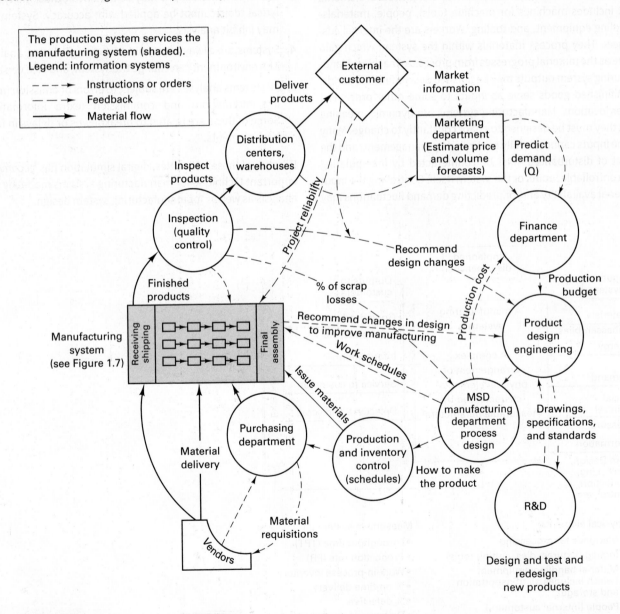

FIGURE 1.4 The production system includes and services the manufacturing system. The functional departments are connected by formal and informal information systems, designed to service the manufacturing that produces the goods.

TABLE 1.1   Production Terms for Manufacturing Production Systems

Term	Meaning	Examples
Production system; the enterprise	All aspects of workers, machines, and information, considered collectively, needed to manufacture parts or products; integration of all units of the system is critical.	Company that makes engines, assembly plant, glassmaking factory, foundry; sometimes called the enterprise or the business.
Manufacturing system (sequence of operations, collection of processes) or factory	The collection of manufacturing processes and operations resulting in specific end products; an arrangement or layout of many processes, materials-handling equipment, and operators.	Rolling steel plates, manufacturing of automobiles, series of connected operations or processes, a job shop, a flow shop, a continuous process.
Machine or machine tool or manufacturing process	A specific piece of equipment designed to accomplish specific processes, often called a *machine tool*; machine tools linked together to make a manufacturing system.	Spot welding, milling machine, lathe, drill press, forge, drop hammer, die caster, punch press, grinder, etc.
Job (sometimes called a *station*; a collection of tasks)	A collection of operations done on machines or a collection of tasks performed by one worker at one location on the assembly line.	Operation of machines, inspection, final assembly; e.g., forklift driver has the job of moving materials.
Operation (sometimes called a *process*)	A specific action or treatment, often done on a machine, the collection of which makes up the job of a worker.	Drill, ream, bend, solder, turn, face, mill extrude, inspect, load.
Tools or tooling	Refers to the implements used to hold, cut, shape, or deform the work materials; called *cutting tools* if referring to machining; can refer to *jigs* and *fixtures* in workholding and *punches* and *dies* in metal forming.	Grinding wheel, drill bit, end milling cutter, die, mold, clamp, three-jaw chuck, fixture.

which exists in many different forms. It is therefore important to use modifiers whenever possible: "Use the *radial* drill *press* to drill a hole using a 1-in.-diameter spade drill." The emphasis of this book will be directed toward the understanding of the processes, machines, and tools required for manufacturing and how they interact with the materials being processed. In the last chapters of the book, an introduction to systems aspects is presented.

Production System—The Enterprise

The highest-ranking term in the hierarchy is **production system**. A production system includes people, money, equipment, materials and supplies, markets, management, and the manufacturing system. In fact, all aspects of commerce (manufacturing, sales, advertising, profit, and distribution) are involved. **Table 1.2** provides a partial list of production systems. Another term for them is "industries" as in the "aerospace industry." Further discussion on the enterprise is found in Chapter 42.

Much of the information given for manufacturing systems is relevant to the service system. Most require a service production system (SPS) for proper product sales. This is particularly true in industries, such as the food (restaurant) industry, in which customer service is as important as quality and on-time delivery. **Table 1.3** provides a short list of service industries.

Manufacturing Systems

A collection of operations and processes used to obtain a desired product(s) or component(s) is called a **manufacturing system**. The manufacturing system design is therefore the arrangement of the manufacturing processes in the factory. Control of a system applies to overall control of the whole, not

TABLE 1.2   Partial List of Production Systems for Producer and Consumer Goods

Aerospace and airplanes	Foods (canned, dairy, meats, etc.)
Appliances	Footwear
Automotive (cars, trucks, vans, wagons, etc.)	Furniture
Beverages	Glass
Building supplies (hardware)	Hospital suppliers
Cement and asphalt	Leather and fur goods
Ceramics	Machines
Chemicals and allied industries	Marine engineering
Clothing (garments)	Metals (steel, aluminum, etc.)
Construction	Natural resources (oil, coal, forest, pulp and paper)
Construction materials (brick, block, panels)	Publishing and printing (books, CDs, newspapers)
Drugs, soaps, cosmetics	Restaurants
Electrical and microelectronics	Retail (food, department stores, etc.)
Energy (power, gas, electric)	Ship building
Engineering	Textiles
Equipment and machinery (agricultural, construction and electrical products, electronics, household products, industrial machine tools, office equipment, computers, power generators)	Tire and rubber Tobacco Transportation vehicles (railroad, airline, truck, bus) Vehicles (bikes, cycles, ATVs, snowmobiles)

merely of the individual processes or equipment. The entire manufacturing system must be controlled to schedule and control the factory—all its inputs, inventory levels, product quality, output rates, and so forth.

TABLE 1.3 **Types of Service Industries**

Advertising and marketing
Communication (telephone, computer networks)
Education
Entertainment (radio, TV, movies, plays)
Equipment and furniture rental
Financial (banks, investment companies, loan companies)
Health care
Insurance
Transportation and car rental
Travel (hotel, motel, cruise lines)

Manufacturing Processes

A **manufacturing process** converts unfinished materials to finished products, often using machines or machine tools. For example, injection molding, die casting, progressive stamping, milling, arc welding, painting, assembling, testing, pasteurizing, homogenizing, and annealing are commonly called processes or manufacturing processes. The term *process* can also refer to a sequence of steps, processes, or operations for production of goods and services, as shown in **Figure 1.5**, which shows the processes to manufacture an Olympic-type medal.

A **machine tool** is an assembly of related mechanisms on a frame or bed that together produce a desired result. Generally, motors, controls, and auxiliary devices are included. Cutting tools and workholding devices are considered separately.

A machine tool may do a single process (e.g., cutoff saw) or multiple processes, or it may manufacture an entire component. Machine sizes vary from a tabletop drill press to a 1000-ton forging press.

Job and Station

In the classical manufacturing system, a **job** is the total of the work or duties a worker performs. A **station** is a location or area where a production worker performs tasks or a job.

A job is a group of related operations and tasks performed at one station or series of stations in cells. For example, the job at a final assembly station may consist of four tasks:

1. Attach carburetor.
2. Connect gas line.
3. Connect vacuum line.
4. Connect accelerator rod.

The job of an operator of a turret lathe (a semiautomatic machine tool) may include the following operations and tasks: load, start, index and stop, unload, inspect. The operator's job may also include setting up the machine (i.e., getting ready for manufacturing). Other machine operations include drilling, reaming, facing, turning, chamfering, and knurling. The operator can run more than one machine or service at more than one station.

The terms *job* and *station* have been carried over to unmanned machines. A job is a group of related operations generally performed at one station, and a station is a position or location in a machine (or process) where specific operations are performed. A simple machine may have only one station. Complex machines can be composed of many stations. The job at a station often includes many simultaneous operations, such as "drill all five holes" by multiple spindle drills. In the planning of a job, a process plan is often developed (by the engineer) to describe how a component is made using a sequence of operations. The engineer begins with a part drawing and selects the raw material. Follow in **Figure 1.6** the sequence of machining operations that transforms the cylinder in a pinion shaft.

Operation

An **operation** is a distinct action performed to produce a desired result or effect. Typical manual machine operations are loading and unloading. Operations can be divided into suboperational elements. For example, loading is made up of picking up a part, placing part in jig, and closing jig. However, suboperational elements will not be discussed here.

Operations categorized by function are

1. *Materials handling and transporting:* change in the location or position of the product.
2. *Processing:* change in volume and quality, including assembly and disassembly; can include packaging.
3. *Packaging:* special processing; may be temporary or permanent for shipping.
4. *Inspecting and testing:* comparison to the standard or check of process behavior.
5. *Storing:* time lapses without further operations.

These basic operations may occur more than once in some processes, or they may sometimes be omitted. *Remember, it is the manufacturing processes that add value and quality to the materials.* Defective processes produce poor quality or scrap. Other operations may be necessary but do not, in general, add value, whereas operations performed by machine tools that do material processing usually do add value.

Treatments

Treatments operate continuously on the workpiece. They usually alter or modify the product-in-process without tool contact. Heat treating, curing, galvanizing, plating, finishing, (chemical) cleaning, and painting are examples of treatments. Treatments usually add value to the part.

These processes are difficult to include in manufacturing cells because they often have long cycle times, are hazardous to the workers' health, or are unpleasant to be around because of high heat or chemicals. They are often done in large tanks

How an olympic medal is made using the CAD/CAM process

Artist's model of medal

(1) An oversized 3D plaster model is made from the artist's conceptual drawings.

Laser scan model

Model of medal

(2) The model is scanned with a laser to produce a digital computer called a computer-aided design (CAD).

CAD

CAM

Top

Die set

Cavity in die forms medal

CAD

Computer

CNC machine tool

Bottom

(3) The computer has software to produce a program to drive numerical control machine to cut a die set.

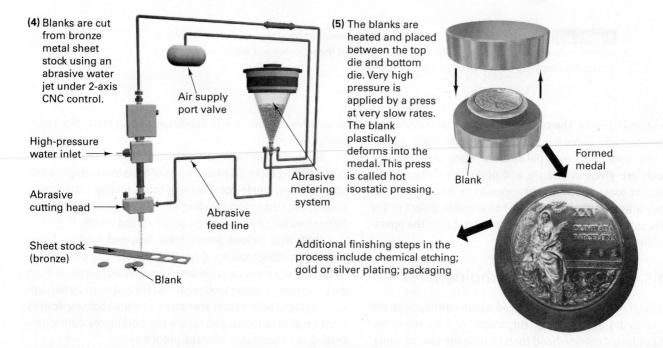

(4) Blanks are cut from bronze metal sheet stock using an abrasive water jet under 2-axis CNC control.

Air supply port valve

High-pressure water inlet

Abrasive cutting head

Sheet stock (bronze)

Blank

Abrasive metering system

Abrasive feed line

(5) The blanks are heated and placed between the top die and bottom die. Very high pressure is applied by a press at very slow rates. The blank plastically deforms into the medal. This press is called hot isostatic pressing.

Blank

Formed medal

Additional finishing steps in the process include chemical etching; gold or silver plating; packaging

FIGURE 1.5   The manufacturing process for making Olympic medals has many steps or operations, beginning with design and including die making. (Courtesy J T. Black)

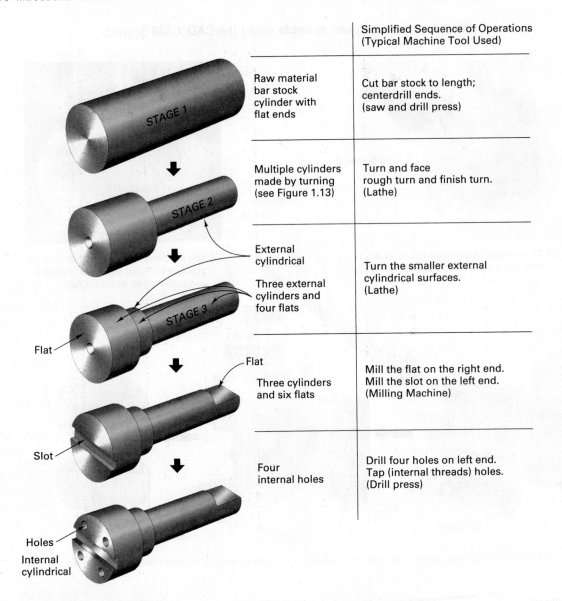

	Simplified Sequence of Operations (Typical Machine Tool Used)
Raw material bar stock cylinder with flat ends	Cut bar stock to length; centerdrill ends. (saw and drill press)
Multiple cylinders made by turning (see Figure 1.13)	Turn and face rough turn and finish turn. (Lathe)
External cylindrical / Three external cylinders and four flats	Turn the smaller external cylindrical surfaces. (Lathe)
Three cylinders and six flats	Mill the flat on the right end. Mill the slot on the left end. (Milling Machine)
Four internal holes	Drill four holes on left end. Tap (internal threads) holes. (Drill press)

FIGURE 1.6 The component called a pinion shaft is manufactured by a "sequence of operations" to produce various geometric surfaces. The engineer determines the sequence and selects the processes and tooling needed to make the component.

or furnaces or rooms. The cycle time for these processes may dictate the cycle times for the entire system. These operations also tend to be material specific. Many manufactured products are given decorative and protective surface treatments that control the finished appearance. A customer may not buy a new vehicle because it has a visible defect in the chrome bumper, although this defect will not alter the operation of the car.

Tools, Tooling, and Workholders

The lowest mechanism in the production term rank is the **tool**. Tools are used to hold, cut, shape, or form the unfinished product. Common hand tools include the saw, hammer, screwdriver, chisel, punch, sandpaper, drill, clamp, file, torch, and grindstone.

Basically, mechanized versions of such hand tools are called cutting tools. Some examples of tools for cutting are drill bits, reamers, single-point turning tools, milling cutters, saw blades, broaches, and grinding wheels. Noncutting tools for forming include extrusion dies, punches, and molds.

Tools also include workholders, jigs, and fixtures. These tools and cutting tools are generally referred to as the **tooling**, which usually must be considered (purchased) separate from machine tools. Cutting tools wear and fail and must be periodically replaced before parts are ruined. The workholding devices must be able to locate and secure the workpieces during processing in a repeatable, mistake-proof way.

Tooling for Measurement and Inspection

Measuring tools and instruments are also important for manufacturing. Common examples of measuring tools are rulers, calipers, micrometers, and gages. Precision devices that use laser optics or vision systems coupled with sophisticated electronics are becoming commonplace. Vision systems and coordinate measuring machines are becoming critical elements for achieving superior quality.

Integrating Inspection into the Process

The integration of the **inspection** process into the manufacturing process or the manufacturing system is a critical step toward building products of superior quality. An example will help. Compare an electric typewriter with a computer that does word processing. The electric typewriter is flexible. It types whatever words are wanted in whatever order. It types a specific font and type size. The computer can do all of this but can also, through its software, change font or type size, set italics; set bold, dark type; vary the spacing to justify the right margin; plus many other functions. It checks immediately for incorrect spelling and other defects like repeated words. The software system provides a signal to the hardware to flash the word so that the operator will know something is wrong and can make an immediate correction. If the system were designed to prevent the typist from typing repeated words, then this would be a *poka-yoke,* a term meaning defect prevention. Defect prevention is better than immediate defect detection and correction. Ultimately, the system should be able to forecast the probability of a defect, correcting the problem at the source. This means that the typist would have to be removed from the process loop, perhaps by having the system type out what it is told (convert oral to written directly). Poka-yoke devices and source inspection techniques are keys to designing manufacturing systems that produce superior-quality products at low cost.

Products and Fabrications

In manufacturing, material things (goods) are made to satisfy human wants. **Products** result from manufacturing, which also includes conversion processes such as refining, smelting, and mining.

Products can be manufactured by fabricating or by processing. **Fabricating** is the manufacture of a product from pieces such as parts, components, or assemblies. Individual products or parts can also be fabricated. Separable discrete items such as tires, nails, spoons, screws, refrigerators, or hinges are fabricated.

Processing is also used to refer to the manufacture of a product by continuous means, or by a continuous series of operations, for a specific purpose. Continuous items such as steel strip, beverages, breakfast foods, tubing, chemicals, and petroleum are "processed." Many processed products are marketed as discrete items, such as bottles of beer, bolts of cloth, spools of wire, and sacks of flour.

Separable discrete products, both piece parts and assemblies, are fabricated in a plant, factory, or mill, for instance, a textile or rolling mill. Products that flow (liquids, gases, grains, or powders) are processed in a *plant* or *refinery*. The *continuous-process industries* such as petroleum and chemical plants are sometimes called processing industries or flow industries.

To a lesser extent, the terms *fabricating industries* and *manufacturing industries* are used when referring to fabricators or manufacturers of large products composed of many parts, such as a car, a plane, or a tractor. Manufacturing often includes continuous-process treatments such as electroplating, heating, demagnetizing, and extrusion forming.

Construction or building is making goods by means other than manufacturing or processing in factories. Construction is a form of project manufacturing of useful goods like houses, highways, and buildings. The public may not consider construction as manufacturing because the work is not usually done in a plant or factory, but it can be. Companies can now build a custom house of any design in a factory, truck it to the building site, and assemble it on a foundation in two or three weeks.

Agriculture, fisheries, and commercial fishing produce real goods from useful labor. Lumbering is similar to both agriculture and mining in some respects, and mining should be considered processing. Processes that convert the raw materials from agriculture, fishing, lumbering, and mining into other usable and consumable products are also forms of manufacturing.

Workpiece and Its Configuration

In the manufacturing of goods, the primary objective is to produce a component having a desired geometry, size, and finish. Every component has a shape that is bounded by various types of surfaces of certain sizes that are spaced and arranged relative to each other. Consequently, a component is manufactured by producing the surfaces that bound the shape. Surfaces may be:

1. Plane or flat.
2. Cylindrical (external or internal).
3. Conical (external or internal).
4. Irregular (curved or warped).

Figure 1.6 illustrated how a shape can be analyzed and broken up into these basic bounding surfaces. Parts are manufactured by using a set or sequence of processes that will either (1) remove portions of a rough block of material (bar stock, casting, forging) to produce and leave the desired bounding surface (2) add portions of material (welding, additive manufacturing) or (3) cause material to form into a stable configuration that has the required bounding surfaces (casting, forging). Consequently, in designing an object, the designer specifies the shape, size, and arrangement of the bounding surface. The part design must be analyzed to determine what materials will provide the desired properties, including mating to other components, and what processes can best be employed to obtain the end product at the most reasonable cost. This is often the job of the engineer.

Roles of Engineers in Manufacturing

Many engineers have as their function the designing of products. The products are brought into reality through the processing or fabrication of materials. In this capacity designers are a key factor in the material selection and manufacturing procedure. A **design engineer**, better than any other person, should know what the design is to accomplish, what assumptions can be made about service loads and requirements, what service environment the product must withstand, and what appearance the final product is to have. To meet these requirements, the material(s) to be used must be selected and specified. In most cases, to utilize the material and to enable the product to have the desired form, the designer knows that certain manufacturing processes will have to be employed. In many instances, the selection of a specific material may dictate what processing must be used. On the other hand, when certain processes must be used, the design may have to be modified for the process to be utilized effectively and economically. Certain dimensional sizes can dictate the processing, and some processes require certain sizes for the parts going into them. In converting the design into reality, many decisions must be made. In most instances, they can be made most effectively at the design stage. It is thus apparent that design engineers are a vital factor in the manufacturing process, and it is indeed a blessing to the company if they can *design for manufacturing,* that is, design the product so that it can be manufactured and/or assembled economically (i.e., at low unit cost). Design for manufacturing uses the knowledge of manufacturing processes, and so the design and manufacturing engineers should work together to integrate design and manufacturing activities.

Manufacturing engineers select and coordinate specific processes and equipment to be used or supervise and manage their use. Some design special tooling so that standard machines can be utilized in producing specific products. These engineers must have a broad knowledge of manufacturing processes and material behavior so that desired operations can be done effectively and efficiently without overloading or damaging machines and without adversely affecting the materials being processed. Although it is not obvious, the most hostile environment the material may ever encounter in its lifetime is the processing environment.

Industrial and **lean engineers** are responsible for manufacturing systems design (or layout) of factories. They must take into account the interrelationships of the factory design and the properties of the materials that the machines are going to process as well as the interreaction of the materials

and processes. The choice of machines and equipment used in manufacturing and their arrangement in the factory are key design tasks.

The **lean engineer** has expertise in cell design, setup reduction (tool design), integrated quality control devices (poka-yokes and decouplers) and reliability (maintenance of machines and people) for the lean production system. See Chapters 42–43 for a discussion of cell design and lean engineering.

Materials engineers devote their major efforts in developing new and better materials. They, too, must be concerned with how these materials can be processed and with the effects that the processing will have on the properties of the materials. Although their roles may be quite different, it is apparent that a large proportion of engineers must concern themselves with the interrelationships of materials and manufacturing processes.

As an example of the close interrelationship of design, materials selection, and the selection and use of manufacturing processes, consider the common desk stapler. Suppose that this item is sold at the retail store for $20. The wholesale outlet sold the stapler for $16, and the manufacturer probably received about $10 for it. Staplers typically consist of 10 to 12 parts and some rivets and pins. Thus, the manufacturer had to produce and assemble the 10 parts for about $1 per part. Only by giving a great deal of attention to design, selection of materials, selection of processes, selection of equipment used for manufacturing (tooling), and utilization of personnel could such a result be achieved.

The stapler is a relatively simple product, yet the problems involved in its manufacture are typical of those that manufacturing industries must deal with. The elements of design, materials, and processes are all closely related, each having its effect on the performance of the device and the other elements. For example, suppose the designer calls for the component that holds the staples to be a metal part. Will it be a machined part rather than a formed part? Entirely different processes and materials need to be specified depending on the choice. Or, if a part is to be changed from metal to plastic, then a whole new set of fundamentally different materials and processes would need to come into play. Such changes would also have a significant impact on cost as well as the service (useful life) of the product.

Changing World Competition

In recent years, major changes in the world of goods manufacturing have taken place. Three of these are:

1. Worldwide competition for global products and their manufacture.

2. High-tech manufacturing advanced technology, computerization.

3. New manufacturing systems designs, strategies, and management.

Worldwide (global) competition is a fact of manufacturing life, and it will get stronger in the future. The goods you buy today may have been made anywhere in the world. For many U.S. companies, suppliers in China, India, and Mexico are not uncommon.

The second aspect, advanced manufacturing technology, usually refers to new machine tools or processes controlled by computers. Additive manufacturing is an example of a new computer controlled process. Companies that produce such machine tools, though small, can have an enormous impact on factory productivity. Improved processes lead to better components and more durable goods. However, the new technology is often purchased from companies that have developed the technology, so this approach is important but may not provide a unique competitive advantage if your competitors can also buy the technology, provided that they have the capital. Some companies develop their own unique process technology and try to keep it proprietary as long as they can.

The third change and perhaps the real key to success in manufacturing is to implement lean manufacturing system design that can deliver, on time to the customer, super-quality goods at the lowest possible cost in a flexible way. Lean production is an effort to reduce waste and improve markedly the methodology by which goods are produced rather than simply upgrading the manufacturing process technology.

Manufacturing system design is discussed extensively in this book, and it is strongly recommended that students examine this material closely after they have gained a working knowledge of materials and processes. The next section provides a brief discussion of manufacturing system designs (factory designs).

Manufacturing System Designs

Five manufacturing system designs can be identified: the job shop, the flow shop, the linked-cell shop, the project shop, and the continuous process. See **Figure 1.7**. The continuous process deals with liquids and/or gases (such as an oil refinery) rather than solids or discrete parts and is used mostly by the chemical engineer.

The most common of these layouts is the **job shop**, characterized by large varieties of components, general-purpose machines, and a functional layout (**Figure 1.8**). This means that machines are collected by function (all lathes together, all mills together, all grinding machines together), and the parts are routed around the shop in small lots to the various machines. The layout of the factory shows the multiple paths through the shop. The material is moved from machine to machine in carts or containers (called the lot or batch).

Flow shops are characterized by larger volumes of the same part or assembly, special-purpose machines and equipment, less variety, less flexibility, and more mechanization. Flow shop layouts are typically either continuous or interrupted and can be for manufacturing or assembly, as shown

in **Figure 1.9**. If continuous, a production line is built that basically runs one large-volume complex item in great quantity and nothing else. The common light bulb is made this way. A transfer line producing an engine block is another typical example. If interrupted, the line manufactures large lots but is periodically "changed over" to run a similar but different component.

The **linked-cell manufacturing system (L-CMS)** is composed of manufacturing and subassembly cells connected to final assembly (linked) using a unique form of inventory and information control called *kanban*. The L-CMS is used in lean production systems where manufacturing processes and subassemblies are restructured into U-shaped cells so that they can operate on a one-piece-flow basis, like final assembly.

As shown in **Figure 1.10**, the lean production factory is laid out (designed) very differently than the mass production system. More than 70% of all manufacturing industries have adopted lean production. Hundreds of manufacturing companies have dismantled their conveyor-based flow lines and replaced them with U-shaped subassembly cells, providing flexibility while eliminating the need for line balancing. Chapters 42–43 discuss subassembly cells and manufacturing cells.

The **project shop** is characterized by the immobility of the item being manufactured. In the construction industry, bridges and roads are good examples. In the manufacture of goods, large airplanes, ships, large machine tools, and locomotives are manufactured in project shops. It is necessary that the workers, machines, and materials come to the site. The number of end items is not very large, and therefore the lot sizes of the components going into the end item are not large. Thus, the job shop usually supplies parts and subassemblies to the project shop in small lots.

Continuous processes are used to manufacture liquids, oils, gases, and powders. These manufacturing systems are usually large plants producing goods for other producers or mass-producing canned or bottled goods for consumers. The manufacturing engineer in these factories is often a chemical engineer.

Naturally, there are many hybrid forms of these manufacturing systems, but the job shop is a very common system. Because of its design, the job shop has been shown to be the least cost-efficient of all the systems. Component parts in a typical job shop spend only 5% of their time in machines and the rest of the time waiting or being moved from one functional area to the next. Once the part is on the machine, it is actually being processed (i.e., having value added to it by the changing of its shape) only about 30 to 40% of the time. The rest of the time parts are being loaded, unloaded, inspected, and so on. The advent of **numerical control** machines increased the percentage of time that the machine is making chips because tool movements are programmed and the machines can automatically change tools or load or unload parts.

However, there are a number of trends that are forcing manufacturing management to consider means by which the job shop system itself can be redesigned to improve its overall

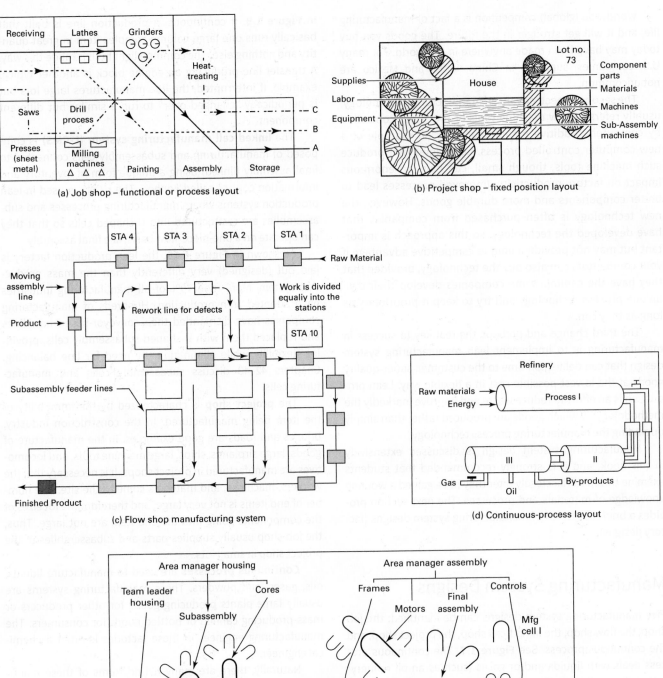

efficiency. These trends have forced manufacturing companies to convert their batch-oriented job shops into linked-cell manufacturing systems, with the manufacturing and subassembly cells structured around specific products.

A technique called Value Stream Mapping (VSM) is widely employed as a means to examine the product flow and the associated information flow. In a value stream map, all of the steps or processes required to bring a product (or a family

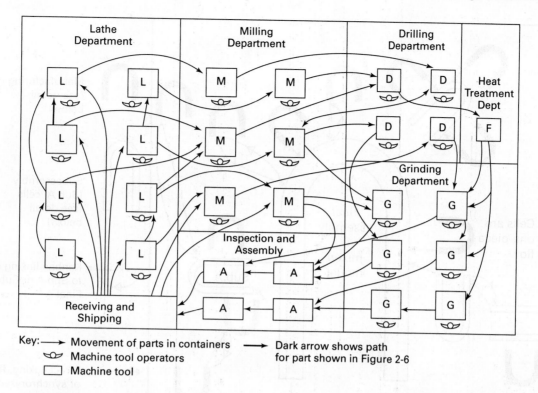

FIGURE 1.8 Schematic layout of a job shop where processes are gathered fuctionally into areas or departments. Each square block represents a manufacturing process or machine tool. Sometimes called the "spaghetti design."

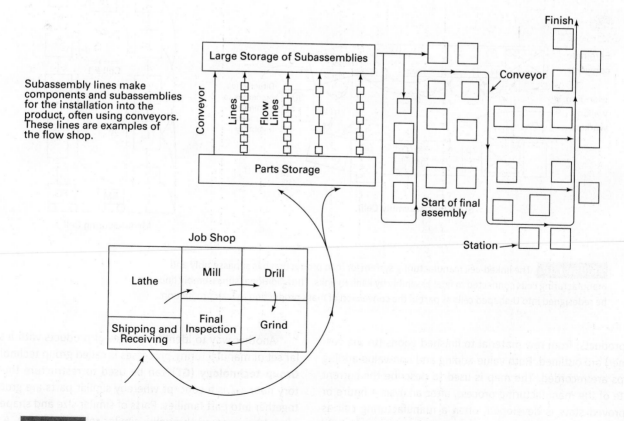

FIGURE 1.9 Flow shops and lines are common in the mass-production system. Final assembly is usually a moving assembly line. The product travels through stations in a specific amount of time. The work needed to assemble the product is distributed into the stations, called division of labor. The moving assembly line for cars is an example of the flow shop.

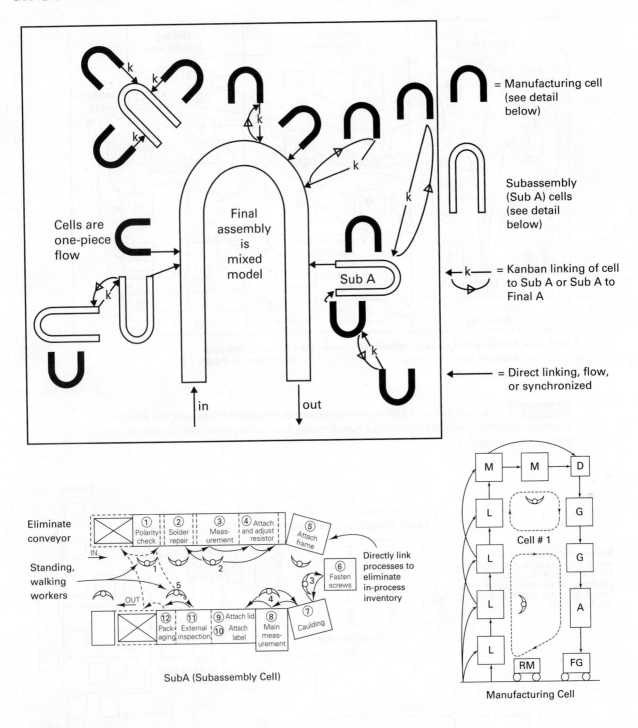

FIGURE 1.10 The linked-cell manufacturing system for lean production has subassembly and manufacturing cells connected to final assembly by kanban links. The traditional subassembly lines can be redesigned into U-shaped cells as part of the conversion of mass production to lean production.

of products) from raw material to finished goods (to the customer) are outlined. Both value-adding and non-value-adding steps are recorded. The map is used to describe the current state of the manufacturing process, after analysis a future or improved state is developed, often a manufacturing cell as shown in Figure 1.10. VSM is discussed more extensively in Chapters 43.

Another way to identify families of products with a similar set of manufacturing processes is called group technology. **Group technology (GT)** can be used to restructure the factory floor. GT is a concept whereby similar parts are grouped together into part families. Parts of similar size and shape can often be processed through a similar set of processes. A part family based on manufacturing would have the same set or sequences of manufacturing processes. The machine tools needed to process the part family are gathered into a cell. Thus, with GT, job shops can be restructured into cells, each cell specializing in a particular family of parts. The parts are handled less, machine setup time is shorter, in-process inventory is lower, and the time needed for parts to get through the manufacturing system (called the throughput time) is greatly reduced.

Basic Manufacturing Processes

It is the manufacturing processes that create or add value to a product. The manufacturing processes can be classified as:

- Casting, foundry, or molding processes
- Forming or metalworking processes
- Machining (material removal) processes
- Nano, micro, and nontraditional processes
- Joining and assembly
- Surface treatments (finishing)
- Additive manufacturing or 3D printing
- Heat treating
- Other

These classifications are not mutually exclusive. For example, some finishing processes involve a small amount of metal removal or metal forming. A laser can be used either for joining metal removal, heat treating, or additive manufacturing. Occasionally, we have a process such as shearing, which is really metal cutting but is viewed as a (sheet) metal-forming process. Assembly may involve processes other than joining. The categories of process types are far from perfect.

Casting and **molding** processes are widely used to produce parts that often require other follow-up processes, such as **machining**. Casting uses molten metal to fill a cavity. The metal retains the desired shape of the mold cavity after solidification. An important advantage of casting and molding is that, in a single step, materials can be converted from a crude form into a desired shape. In most cases, a secondary advantage is that excess or scrap material can easily be recycled. **Figure 1.11** illustrates schematically some of the basic steps in the *lost-wax or investment casting process,* one of many processes used in the foundry industry.

Casting processes are commonly classified into two types: permanent mold (a mold can be used repeatedly) or nonpermanent mold (a new mold must be prepared for each casting made). Molding processes for plastics and composites are included in the chapters on forming processes.

Forming and **shearing** operations typically utilize material (metal or plastics) that has been previously cast or molded. In many cases, the materials pass through a series of forming or shearing operations, so the form of the material for a specific operation may be the result of all the prior operations. The basic purpose of forming and shearing is to modify the shape and size and/or physical properties of the material.

Metal-forming and shearing operations are done both "hot" and "cold, " a reference to the temperature of the material at the time it is being processed with respect to the temperature at which this material can recrystallize (i.e., grow new grain structure). **Figure 1.12** shows the process by which the fender of a car is made using a series of metal-forming processes.

Metal cutting, machining, or metal removal processes refer to the removal of certain selected areas from a part to obtain a desired shape or finish. Chips are formed by interaction of a cutting tool with the material being machined. **Figure 1.13** shows a chip being formed by a single-point cutting tool in a machine tool called a lathe. The manufacturing engineer may be called on to specify the cutting parameters such as cutting speed, feed, or depth of cut (DOC). The engineer may also have to select the cutting tools for the job.

Cutting tools used to perform the basic turning on the lathe are shown in Figure 1.13. The cutting tools are mounted in machine tools, which provide the required movements of the tool with respect to the work (or vice versa) to accomplish the process desired. In recent years many new machining processes have been developed.

The seven basic machining processes are shaping, drilling, turning, milling, sawing, broaching, and abrasive machining. Each of these basic processes is extensively discussed. Historically, eight basic types of machine tools have been developed to accomplish the basic processes. These machine tools are called shapers (and planers), drill presses, lathes, boring machines, milling machines, saws, broaches, and grinders. Today, most machine tools are capable of performing more than one of the basic machining processes. Shortly after numerical control was invented in the mid 1950s, machining centers were developed that could combine many of the basic processes, plus other related processes, into a single machine tool with a single workpiece setup.

Aside from the chip-making processes, there are processes wherein metal is removed by chemical, electrical, electrochemical, or thermal sources. Generally speaking, these nontraditional processes have evolved to fill a specific need when conventional processes were too expensive or too slow when machining very hard materials. One of the first uses of a laser was to machine holes in ultra-high-strength metals. Lasers are being used today to drill tiny holes in turbine blades for jet engines. Because of its ability to produce components with great precision and accuracy, metal cutting, using machine tools, is recognized as having great value-adding capability.

In recent years a new family of processes has emerged called additive manufacturing, or 3D printing in Chapter 33. These additive-type processes produce components directly from the software using specialized machines driven by computer-aided design packages. Prototypes and parts can

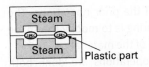
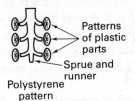

To make the foam parts, metal molds are used. Beads of polystyrene are heated and expanded in the mold to get parts.

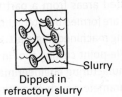

A pattern containing a sprue, runners, risers, and parts is made from single or multiple pieces of foamed polystyrene plastic.

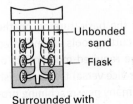

The polystyrene pattern is dipped in a ceramic slurry, which wets the surface and forms a coating about 0.005 in. thick.

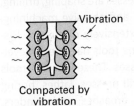

The coated pattern is placed in a flask and surrounded with loose, unbonded sand.

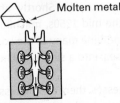

The flask is vibrated so that the loose sand is compacted around the pattern.

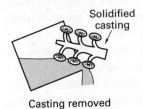

During the pouring of molten metal, the hot metal vaporizes the pattern and fills the resulting cavity.

The solidified casting is removed from flask and the loose sand reclaimed.

FIGURE 1.11   Schematic of the lost-foam casting process.

be quickly made, field tested, and modified for use. Early versions of these machines produced only nonmetallic components, but modern machines can make metal parts, like that shown in **Figure 1.14**, an antenna bracket for a space satellite. Additive manufacturing enabled a 40% reduction in weight compared to the previous design (a major consideration for launch cost) while providing the necessary strength and rigidity. In contrast to other processes additive manufacturing has the ability to produce components with good precision and unique geometry. Companies have sprung up where you can send your CAD drawing over the Internet and a unique part is made in hours.

Metalforming Process for Automobile Fender

Sheet metal bending/forming

Single draw punch and die

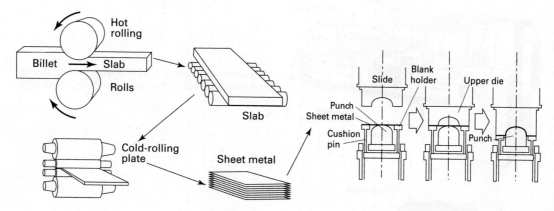

(a) Cast billets of metal are passed through successive rollers to produce sheets of steel rolled stock.

(b) The flat sheet metal is "formed" into a fender, using sets of dies mounted on stands of large presses.

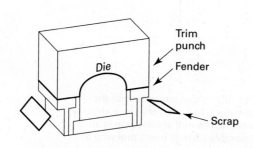

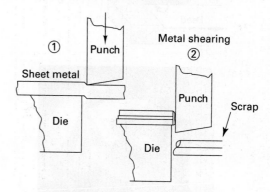

(c) The fender is cut out of the sheet metal in the last stage using shearing processes.

(d) Sheet metal shearing processes use punches and dies to cut metal.

FIGURE 1.12 The forming processes used to make a fender for a car start with hot rolling and finish with sheet metal shearing.

Perhaps the largest collection of processes, in terms of both diversity and quantity, are the **joining processes**, which include the following:

1. Mechanical fastening
2. Soldering and brazing
3. Welding
4. Press, shrink, or snap fittings
5. Adhesive bonding
6. Assembly processes

Many of these joining processes are often found in the assembly area of the plant. **Figure 1.15** provides one example where all but welding are used in the sequence of operations to produce a computer. Starting in the upper left corner, microelectronic

fabrication methods produce entire integrated circuits (ICs) of solid-state (no moving parts) components, with wiring and connections, on a single piece of semiconductor material, usually single-crystalline silicon. Arrays of ICs are produced on thin, round disks of semiconductor material called wafers. Once the semiconductor on the wafer has been fabricated, the finished wafer is cut up into individual ICs, or chips. Next, at level 2, these chips are individually housed with connectors or leads making up "dies" that are placed into "packages" using adhesives. The packages provide protection from the elements and a connection between the die and another subassembly called the printed circuit boards (PCBs). At level 3, IC packages, along with other discrete components (e.g., resistors, capacitors, etc.), are soldered onto PCBs and then assembled with even larger circuits on PCBs. This is sometimes referred to as electronic assembly. Electronic packages

The Machining Process
(turning on a lathe)

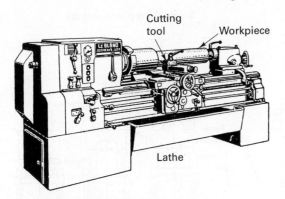

Cutting tool   Workpiece

Lathe

The workpiece is mounted in a workholding device in a machine tool (lathe) and is cut (machined) with a cutting tool.

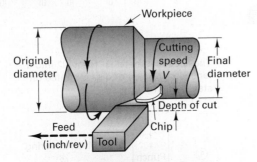

Workpiece

Original diameter

Cutting speed *V*

Final diameter

Depth of cut

Feed (inch/rev)   Tool   Chip

The workpiece is rotated while the tool is fed at some feed rate (inches per revolution). The desired cutting speed *V* determines the rpm of the workpiece. This process is called turning.

The cutting tool interacts with the workpiece to form a chip by a shearing process. The tool shown here is an indexable carbide insert tool with a chip-breaking groove.

FIGURE 1.13 Single-point metal-cutting process (turning) produces a chip while creating a new surface on the workpiece. (Courtesy J T. Black)

at this level are called cards or printed wiring assemblies (PWAs). Next, series of cards are combined on a back-panel PCB, also known as a motherboard or simply a board. This level of packaging is sometimes referred to as card-on-board packaging. Ultimately, card-on-board assemblies are put into housings using mechanical fasteners and snap fitting and finally integrated with power supplies and other electronic peripherals through the use of cables to produce final commercial products.

Finishing processes are yet another class of processes typically employed for cleaning, removing burrs left by machining, or providing protective and/or decorative surfaces on workpieces. Surface treatments include chemical and mechanical cleaning, deburring, painting, plating, buffing, galvanizing, and anodizing.

Heat treatment is the heating and cooling of a metal for the specific purpose of altering its metallurgical and mechanical properties. Because changing and controlling these properties

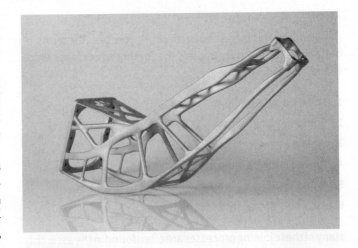

FIGURE 1.14 High-strength aluminum antenna bracket for a space satellite produced by additive manufacturing. The part is approximately 40 cm (16 in.) in length and weighs less than one kilogram (2.2. pounds). (Courtesy of EOS GmbH)

is so important in the processing and performance of metals, heat treatment is a very important manufacturing process. Each type of metal reacts differently to heat treatment. Consequently, a designer should know not only how a selected metal can be altered by heat treatment but, equally important, *how a selected metal will react, favorably or unfavorably, to any heating or cooling that may be incidental to the manufacturing processes.*

Other Manufacturing Operations

In addition to the processes already described, there are many other fundamental manufacturing operations that must be considered. Inspection determines whether the desired objectives stated by the designer in the specifications have been achieved. This activity provides feedback to design and manufacturing with regard to the process capability. Essential to this inspection function are measurement activities. In the factory, measurements by attributes or variables (Chapter 10) inspect the outcomes from the process and determine how they compare to the specifications. The many aspects of quality control are presented in Chapter 12. Chapter 11 covers testing, where a product is tried by actual function or operation or by subjection to external effects. Although a test is a form of inspection, it is often not viewed that way. In manufacturing, parts and

materials are inspected for conformance to the dimensional and physical specifications, while testing may simulate the environmental or usage demands to be made on a product after it is placed in service. Complex processes may require many tests and inspections. Testing includes life-cycle tests, destructive tests, nondestructive testing to check for processing defects, wind-tunnel tests, road tests, and overload tests.

Transportation of goods in the factory is often referred to as *material handling* or *conveyance* of the goods and refers to the transporting of unfinished goods (work-in-process) in the plant and supplies to and from, between, and during manufacturing operations. Loading, positioning, and unloading are also material-handling operations. Transportation, by truck or train, is material handling between factories. Proper manufacturing system design and mechanization can reduce material handling in countless ways.

Automatic material handling is a critical part of continuous automatic manufacturing. The word *automation* is derived from automatic material handling. Material handling, a fundamental operation done by people and by conveyors and loaders, often includes positioning the workpiece within the machine by indexing, shuttle bars, slides, and clamps. In recent years, wire-guided automated guided vehicles (AGVs) and automatic storage and retrieval systems (AS/RSs) have been developed in an attempt to replace forklift trucks on the factory floor. Another form of material handling, the mechanized

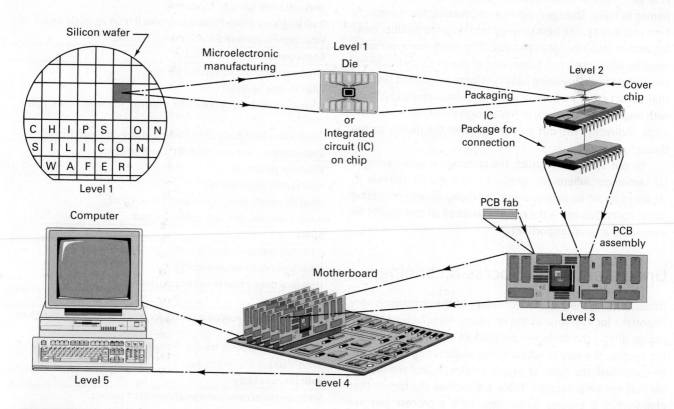

FIGURE 1.15 How an electronic product is made.

removal of waste (chips, trimming, and cutoffs), can be more difficult than handling the product. Chip removal must be done before a tangle of scrap chips damages tooling or creates defective workpieces.

Most texts on manufacturing processes do not mention **packaging**, yet the packaging is often the first thing the customer sees. Also, packaging often maintains the product's quality between completion and use. (The term *packaging* is also used in electronics manufacturing to refer to placing microelectronic chips in containers for mounting on circuit boards.) Packaging can also prepare the product for delivery to the user. It varies from filling ampules with antibiotics to steel-strapping aluminum ingots into palletized loads. A product may require several packaging operations. For example, Hershey Kisses are (1) individually wrapped in foil, (2) placed in bags, (3) put into boxes, and (4) placed in shipping cartons.

Weighing, filling, sealing, and labeling are packaging operations that are highly automated in many industries. When possible, the cartons or wrappings are formed from material on rolls in the packaging machine. Packaging is a specialty combining elements of product design (styling), material handling, and quality control. Some packages cost more than their contents (e.g., cosmetics and razor blades).

During **storage**, nothing happens intentionally to the product or part except the passage of time. Part or product deterioration on the shelf is called **shelf life**, meaning that items can rust, age, rot, spoil, embrittle, corrode, creep, and otherwise change in state or structure, while supposedly nothing is happening to them. Storage is detrimental, wasting the company's time and money. The best strategy is to keep the product moving with as little storage as possible. Storage during processing must be *eliminated*, not automated or computerized. Companies should avoid investing heavily in large automated systems that do not alter the bottom line. Have the outputs improved with respect to the inputs, or has storage simply increased the costs (indirectly) without improving either the quality or the throughput time?

By not storing a product, the company avoids having to (1) remember where the product is stored, (2) retrieve it, (3) worry about its deteriorating, or (4) pay storage (including labor) costs. Storage is the biggest waste of all and should be eliminated at every opportunity.

Understand Your Process Technology

Understanding the process technology of the company is very important for everyone in the company. Manufacturing technology affects the design of the product and the manufacturing system, the way in which the manufacturing system can be controlled, the types of people employed, and the materials that can be processed. **Table 1.4** outlines the factors that characterize a process technology. Take a process you are familiar with and think about these factors. One valid criticism of American companies is that their managers seem to have an aversion to understanding their companies' manufacturing

technologies. Failure to understand the company business (i.e., its fundamental process technology) can lead to the failure of the company.

The way to overcome technological aversion is to run the process and study the technology. Only someone who has run a drill press can understand the sensitive relationship between feed rate and drill torque and thrust. All processes have these "know-how" features. Those who run the processes must be part of the decision-making for the factory. The CEO who takes a vacation working on the plant floor and learning the processes will be well on the way to being the head of a successful company.

TABLE 1.4   Characterizing a Process Technology

Mechanics (statics and dynamics of the process)
How does the process work?
What are the process mechanics (statics, dynamics, friction)?
What physically happens, and what makes it happen? (Understand the physics.)
Economics or costs
What are the tooling costs, the engineering costs?
Which costs are short term, which long term?
What are the setup costs?
Time spans
How long does it take to set up the process initially?
What is the throughput time?
How can these times be shortened?
How long does it take to run a part once it is set up (cycle time)?
What process parameters affect the cycle time?
Constraints
What are the process limits?
What cannot be done?
What constrains this process (sizes, speeds, forces, volumes, power, cost)?
What is very hard to do within an acceptable time/cost frame?
Uncertainties, process reliability, and safety
What can go wrong?
How can this machine fail?
What do people worry about with this process?
Is this a reliable, safe, and stable process?
Skills
What operator skills are critical?
What is not done automatically?
How long does it take to learn to do this process?
Flexibility
Can this process be adapted easily for new parts of a new design or material?
How does the process react to changes in part design and demand?
What changes are easy to do?
Process capability
What are the accuracy and precision of the process?
What tolerances does the process meet? (What is the process capability?)
How repeatable are those tolerances?

Product Life Cycle and Life-Cycle Cost

Manufacturing systems are dynamic and change with time. There is a general, traditional relationship between a product's life cycle and the kind of manufacturing system used to make the product. **Figure 1.16** simplifies the **product life cycle** into these steps, again using an S-shaped curve.

1. *Startup.* New product or new company, low volume, small company.

2. *Rapid growth.* Products become standardized and volume increases rapidly. Company's ability to meet demand stresses its capacity.

3. *Maturation.* Standard designs emerge. Process development is very important.

4. *Commodity.* Long-life, standard-of-the-industry type of product.

5. *Decline.* Product is slowly replaced by improved products.

The maturation of a product in the marketplace generally leads to fewer competitors, with competition based more on price and on-time delivery than on unique product features. As the competitive focus shifts during the different stages of the product life cycle, the requirements placed on manufacturing—cost, quality, flexibility, and delivery dependability—also change. The stage of the product life cycle affects the product design stability, the length of the product development cycle, the frequency of engineering change orders, and the commonality of components—all of which have implications for manufacturing process technology.

During the design phase of the product, much of the cost of manufacturing and assembly is determined. Assembly of the product is inherently integrative as it focuses on groups of parts and the supply chain.

It is crucial to achieve this integration during the design phase because about 70% of the life-cycle cost of a product is determined when it is designed. Design choices determine materials; fabrication methods; assembly methods; and, to

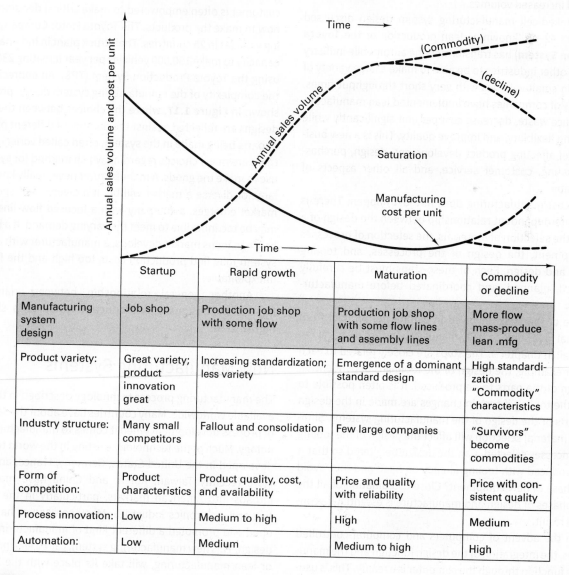

	Startup	Rapid growth	Maturation	Commodity or decline
Manufacturing system design	Job shop	Production job shop with some flow	Production job shop with some flow lines and assembly lines	More flow mass-produce lean .mfg
Product variety:	Great variety; product innovation great	Increasing standardization; less variety	Emergence of a dominant standard design	High standardization "Commodity" characteristics
Industry structure:	Many small competitors	Fallout and consolidation	Few large companies	"Survivors" become commodities
Form of competition:	Product characteristics	Product quality, cost, and availability	Price and quality with reliability	Price with consistent quality
Process innovation:	Low	Medium to high	High	Medium
Automation:	Low	Medium	Medium to high	High

FIGURE 1.16 Product life-cycle costs change with the classic manufacturing system designs.

a lesser degree, material-handling options, inspection techniques, and other aspects of the production system. Manufacturing engineers and internal customers can influence only a small part of the overall cost if they are presented with a finished design that does not reflect their concerns. Therefore, all aspects of production should be included if product designs are to result in real functional integration.

Life-cycle costs include the costs of all the materials, manufacture, use, repair, and disposal of a product. Early design decisions determine about 60% of the cost, and all activities up to the start of full-scale development determine about 75%. Later decisions can make only minor changes to the ultimate total unless the design of the manufacturing system is changed.

In short, the concept of product life cycle provides a framework for thinking about the product's evolution through time and the kind of market segments that are likely to develop at various times. Analysis of life-cycle costs shows that the design of the manufacturing system determines the cost per unit, which generally decreases over time with process improvements and increased volumes.

The linked-cell manufacturing system design discussed in Chapters 42–43 (known as lean production or the **Toyota Production System**) has transformed the automobile industry and many other industries to be able to make a large variety of products in small volumes with very short throughput times. Thousands of companies have implemented lean manufacturing to reduce waste, decrease cost per unit significantly while maintaining flexibility, and improve quality. This is a new business model affecting product development, design, purchasing, marketing, customer service, and all other aspects of the company.

Low-cost manufacturing does not just happen. There is a close, interdependent relationship between the design of a product, the selection of materials, the selection of processes and equipment, the design of the processes, and tooling selection and design. Each of these steps must be carefully considered, planned, and coordinated before manufacturing starts.

Some of the steps involved in getting the product from the original idea stage to daily manufacturing are discussed in more detail in Chapter 9. The steps are closely related to each other. For example, the design of the tooling is dependent on the design of the parts to be produced. It is often possible to simplify the tooling if certain changes are made in the design of the parts or the design of the manufacturing systems. Similarly, the material selection will affect the design of the tooling or the processes selected. Can the design be altered so that it can be produced with tooling already on hand and thus avoid the purchase of new equipment? Close coordination of all the various phases of design and manufacture is essential if economy is to result.

With the advent of computers and computer-controlled machines, the integration of the design function and the manufacturing function through the computer is a reality. This is usually called CAD/CAM (computed-aided design/computer-aided

manufacturing). The key is a common database from which detailed drawings can be made for the designer and the manufacturer and from which programs can be generated to make all the tooling. In addition, extensive computer-aided testing and inspection (CATI) of the manufactured parts is taking place. There is no doubt that this trend will continue at ever-accelerating rates as computers become cheaper and smarter, but at this time, the computers necessary to accomplish complete computer-integrated manufacturing (CIM) are expensive and the software very complex. Implementing CIM requires a lot of manpower as well.

Comparisons of Manufacturing System Design

When designing a manufacturing system, two customers must be taken into consideration: the external customer who buys the product and the internal customer who makes the product. The external customer is likely to be global and demand greater variety with superior quality and reliability. The internal customer is often empowered to make critical decisions about how to make the products. The Toyota Motor Company is making vehicles in 25 countries. Their truck plant in Indiana has the capacity to make 150,000 vehicles per year (creating 2300 jobs), using the Toyota Production System (TPS). An appreciation of the complexity of the manufacturing system design problem is shown in **Figure 1.17**, where the choices between the system designs are reflected against the number of different products, or parts being made in the system, often called *variety*. Clearly, there are many choices regarding which method (or system) to use to make the goods. A manufacturer never really knows how large or diverse a market will be. If a diverse and specialized market emerges, a company with a focused flow-line system may be too inflexible to meet the varying demand. If a large but homogeneous market develops, a manufacturer with a flexible system may find production costs too high and the flexibility unexploitable.

Another general relationship between manufacturing system designs and production volumes is shown in **Figure 1.18**.

New Manufacturing Systems

The manufacturing process technology described in this text is available worldwide. Many countries have about the same level of process development when it comes to manufacturing technology. Much of the technology existing in the world today was developed in the United States, Germany, France, and Japan. More recently Taiwan, Korea, and China have made great inroads into American markets, particularly in the automotive and electronics industries. Many companies have developed and promoted a different kind of manufacturing system design. This new manufacturing system, called lean production or lean manufacturing, will take its place with the American Armory System and the Ford System for mass production. This

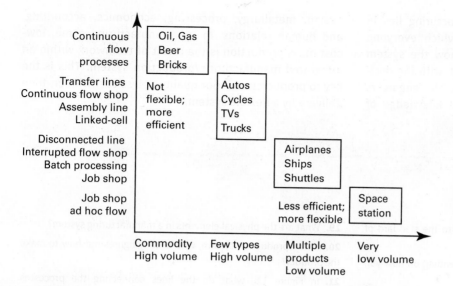

The figure shows in a general way the relationship between manufacturing systems and production volumes. The upper left represents systems with low flexibility but high efficiency compared to the lower right, where volumes are low and so is efficiency. Where a particular company lies in this matrix is determined by many forces, not all of which are controllable. The job of manufacturing lean and industrial engineers is to design and implement a system that can achieve low unit cost, superior quality, with on-time delivery in a flexible way.

FIGURE 1.17   This figure shows in a general way the relationship between manufacturing systems and production volumes.

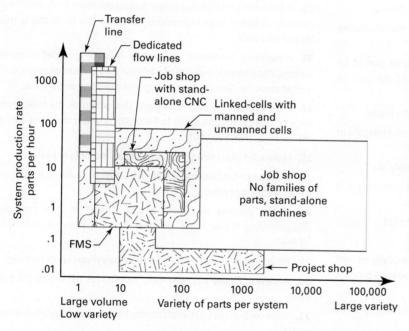

This part variety-production rate matrix shows examples of particular manufacturing system designs. This matrix was developed by Black based on real factory data. Notice there is a large amount of overlap in the middle of the matrix, so the manufacturing engineer has many choices regarding which method or system to use to make the goods. This book will show the connection between the process and the manufacturing system used to produce the products, turning raw materials into finished goods.

FIGURE 1.18   Different manufacturing system designs produce goods at different production rates.

new manufacturing system, developed by the Toyota Motor Company, has been successfully adopted by many American companies.

For lean production to work, units with no defects (100% good) must flow rhythmically to subsequent processes without interruption. To accomplish this, an integrated quality control program has to be developed. The responsibility for quality has been given to manufacturing, and the internal customer. All the employees are inspectors and are empowered to make it right the first time. There is a companywide attitude toward constant quality improvement. Make quality easy to see, stop the line when something goes wrong, and inspect things 100% if necessary to prevent defects from occurring. The results of this system are astonishing in terms of quality, low cost, and on-time delivery of goods to the customer.

The most important factor in economical and successful manufacturing is the manner in which the resources—labor, materials, and capital—are organized and managed so as to provide effective coordination, responsibility, and control. Part of the success of lean production can be attributed to a different management approach. This approach is characterized by a holistic attitude (i.e., respect) toward people.

The real secret of successful manufacturing lies in designing a manufacturing system in which everyone who works in the system understands how the system works how the flow of goods is controlled, with the decision making placed at the correct level. The engineers also must possess a broad fundamental knowledge of design, metallurgy, processing, economics, accounting, and human relations. In the manufacturing game, low-cost mass production is the result of teamwork within an integrated manufacturing/production system. This is the key to producing superior quality at less cost with on-time delivery by a flexible system.

Review Questions

1. What role does manufacturing play relative to the standard of living of a country?

2. Aren't all goods really consumer goods, depending on how you define the customer? Discuss.

3. The Subway sandwich shop is an example of a job shop, a flow shop, or a project shop, which?

4. How does a system differ from a process? From a machine tool? From a job? From an operation?

5. Is a cutting tool the same thing as a machine tool? Discuss.

6. What are the major classifications of basic manufacturing processes?

7. Casting is often used to produce a complex-shaped part to be made from a hard-to-machine metal. How else could the part be made?

8. In the lost-wax casting process, what happens to the foam?

9. In making a gold medal, what do we mean by a "relief image" cut into the die?

10. How is a railroad station like a station on an assembly line?

11. Because no work is being done on a part when it is in storage, it does not cost you anything. True or false? Explain.

12. What forming processes are used to make a paper clip?

13. What is tooling in a manufacturing system?

14. It is acknowledged that chip-type machining is basically an inefficient process. Yet it is probably used more than any other to produce desired shapes. Why?

15. Compare Figure 1.1 and Figure 1.16. What are the stages of the product life cycle for a computer?

16. In a modern safety razor with three or four blades that sells for $1, what do you think the cost of the blades might be?

17. List three purposes of packaging operations.

18. *Assembly* is defined as "the putting together of all the different parts to make a complete machine." Think of (and describe) an assembly process. Is making a club sandwich an assembly process? What about carving a turkey? Is this an assembly process?

19. What are the physical elements in a manufacturing system?

20. In the production system, who usually figures out how to make the product?

21. In Figure 1.8, what do the lines connecting the processes represent?

22. Characterize the process of squeezing toothpaste from a tube (extrusion of toothpaste) using Table 1.4 as a guideline. See the index for help on extrusion.

23. It has been said that low-cost products are more likely to be more carefully designed than high-priced items. Do you think this is true? Why or why not?

24. Proprietary processes are closely held or guarded company secrets. The chemical makeup of a lubricant for an extrusion process is a good example. Give another example of a proprietary process.

25. If the rolls for the cold-rolling mill that produces the sheet metal used in your car cost $300,000 to $400,000, how is it that your car can still cost less than $20,000?

26. Make a list of service systems, giving an example of each.

27. What is the fundamental difference between a service system and a manufacturing system?

28. In the process of buying a calf, raising it to a cow, and disassembling it into "cuts" of meat for sale, where is the "value added"?

29. What kind of process is powder metallurgy: casting or forming?

30. In view of Figure 1.2, who really determines the selling price per unit?

31. What costs make up manufacturing cost (sometimes called factory cost)?

32. What are major phases of a product life cycle?

33. How many different manufacturing systems might be used to make a component with annual projected sales of 16,000 parts per year with 10 to 12 different models (varieties)?

34. In general, as the annual volume for a product increases, the unit cost decreases. Explain.

Problems

1. The Toyota truck plant in Indiana produces 150,000 trucks per year. The plant runs one eight-hour shift and makes 400 trucks per day. About 1300 people work on the final assembly line. Each truck has about 20 direct labor hours per car in it.

> **a.** Assuming the truck sells for $26,000 and workers earn $50 per hour in wages and benefits, what percentage of the cost of the truck is in direct labor?
>
> **b.** What is the production rate of the final assembly line?

2. A company is considering making automobile bumpers from aluminum instead of from steel. List some of the factors it would have to consider in arriving at its decision.

3. Many companies are critically examining the relationship of product design to manufacturing and assembly. Why do they call this concurrent engineering?

4. We can analogize your university to a manufacturing system that produces graduates. Assuming that it takes four years to get a college degree and that each course really adds value to the student's knowledge base, what percentage of the four years is "value adding" (percentage of time in class plus two hours of preparation for each hour in class)?

5. What kind of manufacturing system (design) is your university?

6. What are the major process steps in the assembly of a subway sandwich?

7. What is the relationship between Figure 1.2 and Figure 1.4?

CHAPTER 2

Properties of Materials

2.1 Introduction

The history of civilization has been intimately linked to the materials that have shaped this world, so much so that we have associated periods of time with the dominant material, such as the Stone Age, Bronze Age, and Iron Age. Stone was used in its natural state, but bronze and iron were made possible by advances in processing. Each contributed to the comfort, productivity, safety, and security of everyday living. One replaced the other when advantages and new capabilities were realized, iron being lighter and stronger than bronze, for example. The close of the 20th century has been called the Silicon Age, and the multitude of devices that have been made possible by the transistor and computer chip (fast and compact computers, cell phones, GPS units, etc.) have revolutionized virtually every aspect of our lives. From the manufacturing perspective, the current era lacks a single material designation. We now use a wide array of materials, coming from the categories of metals, ceramics, polymers, and combinations known as composites.

With each of these materials, the ultimate desire is to convert it into some form of useful product. **Manufacturing** has been described as the various activities that are performed to convert "stuff" into "things." Successful products begin with appropriate materials. You wouldn't build an airplane out of lead, or an automobile out of concrete—you need to start with the right stuff. But "stuff" rarely comes in the right shape, size, and quantity for the desired use. Parts and components must be produced by subjecting materials to one or more processes (often a series of operations) that alter their shape, properties, or both. Another definition of manufacturing is those activities by which a material is altered in shape, form, or properties to add value.

Much of a manufacturing education relates to understanding: (1) the **structure** of materials, (2) the **properties** of materials, (3) the **processing** of materials, and (4) the **performance** of materials, as well as the interrelations between these four factors, as illustrated in **Figure 2.1**. This chapter will begin to address the properties of engineering materials. Chapters 3 and 4 will develop the subject of structure and begin to provide the reasons behind the various properties. Chapter 5 introduces the possibility of controlling and modifying structure to produce desired properties. Many engineering materials do not have a single set of properties, but instead offer a range

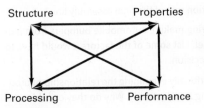

FIGURE 2.1 The interdependent relationships between structure, properties, processing, and performance.

or spectrum of possibilities. Properties can be tailored to specific needs or even altered during the manufacturing process. As one example, a material may be made weak and ductile for easy shaping (enabling low forming loads, extended tool life, and prevention of cracking or fracture), and then, once the shape has been produced, made strong for enhanced performance when in use.

When selecting a material for a product or application, it is important to ensure that its properties will be adequate for the anticipated operating conditions. The various requirements of each component must first be estimated or determined. These requirements typically include mechanical characteristics (strength, rigidity, resistance to fracture, and the ability to withstand vibrations or impacts) and physical characteristics (weight, electrical properties, and appearance), as well as features relating to the service environment (ability to operate under extremes of temperature or resist corrosion). Candidate materials must possess the desired properties within their range of possibilities.

To help evaluate the properties of engineering materials, a variety of standard tests have been developed, and data from these tests have been tabulated and made readily available. Proper use of this data, however, requires sound engineering judgment. It is important to consider which of the evaluated properties are significant, under what conditions the test values were determined, and what cautions or restrictions should be placed on their use. Only by being familiar with the various test procedures, their capabilities, and their limitations can one determine if the resulting data are applicable to a particular problem.

Metallic and Nonmetallic Materials

Whereas engineering materials are often grouped as metals, ceramics, polymers, and composites, a more simplistic distinction might be to separate into metallic and nonmetallic.

The common **metals** and metallic materials include iron, copper, aluminum, magnesium, nickel, titanium, lead, tin, and zinc, as well as their many alloys, including steel, brass, and bronze. They possess the metallic properties of luster, high thermal conductivity, and high electrical conductivity; they are relatively ductile; and some have good magnetic properties. Some common **nonmetals** are wood, brick, concrete, glass, rubber, and plastics. Their properties vary widely, but they generally tend to be weaker, less ductile, and less dense than the metals, with poor electrical and thermal conductivities.

Although metals have traditionally been the more important of the two groups, the nonmetallic materials have become increasingly important in modern manufacturing. Advanced ceramics, composite materials, and engineered plastics have emerged in a number of applications. In many cases, metals and nonmetals are viewed as competing materials, with selection being based on how well each is capable of providing the required properties. Where both perform adequately, the total cost often becomes the deciding factor, where the total cost includes both the cost of the material and the cost of fabricating the desired component. Factors such as product lifetime, environmental impact, energy requirements, and recyclability are also considered.

Physical and Mechanical Properties

A common means of distinguishing one material from another is through their **physical properties**. These include such features as density (weight); melting point; optical characteristics (transparency, opaqueness, or color); the thermal properties of specific heat, coefficient of thermal expansion, and thermal conductivity; electrical conductivity; and **magnetic properties**. In some cases, physical properties are of prime importance when selecting a material, and several will be discussed in more detail near the end of this chapter.

More often, however, material selection is dominated by the properties that describe how a material responds to applied loads or forces. These **mechanical properties** are usually determined by subjecting prepared specimens to standard test conditions. When using the obtained results, however, it is important to remember that they apply only to the specific conditions that were employed in the test. The actual service conditions of engineered products rarely duplicate the conditions of laboratory testing, so considerable caution should be exercised.

Stress and Strain

When a force or load is applied to a material, it deforms or distorts (becomes *strained*), and internal reactive forces (*stresses*) are transmitted through the solid. For example, if a weight, W, is suspended from a bar of uniform cross section and length L, as in **Figure 2.2**, the bar will elongate by an amount ΔL. For a given weight, the magnitude of the **elongation**, ΔL, depends

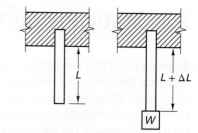

FIGURE 2.2 Tension loading and the resultant elongation.

on the original length of the bar. The amount of elongation per unit length, expressed as $e = \Delta L/L$, is called the **unit strain**. Although the ratio is that of a length to another length and is therefore dimensionless, **strain** is usually expressed in terms of millimeters per meter, inches per inch, or simply as a percentage.

Application of the force also produces reactive stresses, which serves to transmit the load through the bar and on to its supports. **Stress** is defined as the force or load being transmitted divided by the cross-sectional area transmitting the load. Thus, in Figure 2.2, the stress is $S = W/A$, where A is the cross-sectional area of the supporting bar. Stress is normally expressed in megapascals in SI units (where a Pascal is one Newton per square meter) or pounds per square inch in the English system.

In Figure 2.2, the weight tends to stretch or lengthen the bar, so the strain is known as a *tensile strain* and the stress as a *tensile stress*. Other types of loadings produce other types of stresses and strains (**Figure 2.3**). Compressive forces tend to shorten the material and produce *compressive stresses and strains*. **Shear** *stresses and strains* result when two opposing forces acting on a body are offset with respect to one another.

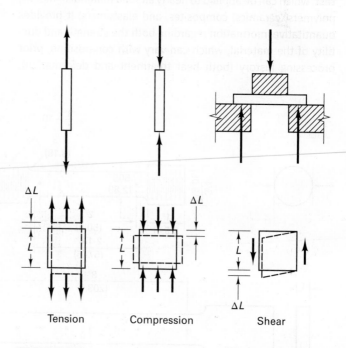

Tension   Compression   Shear

FIGURE 2.3 Examples of tension, compression, and shear loading—and their response.

2.2 Static Properties

When the forces that are applied to a material are constant, or nearly so, they are said to be *static*. Because static or steady loadings are observed in many applications, it is important to characterize the behavior of materials under these conditions. For design engineers, the strength of a material may be of primary concern, along with the amount of elastic stretching or deflection that may be experienced when the product is under load. Manufacturing engineers, wanting to shape products with mechanical forces, need to know the stresses necessary to produce permanent deformation. At the same time, they want to perform this deformation without inducing cracking or fracture.

As a result, a number of standardized tests have been developed to evaluate the **static properties** of engineering materials. Individual test results can be used to determine if a given material or batch of material has the necessary properties to meet specified requirements. The results of multiple tests can provide the characterization information that is used when selecting materials for various applications. In all cases, it is important to determine that the conditions for the product being considered are indeed similar to those of the standard testing. Even when the service conditions differ, however, the results of standard tests may still be helpful in qualitatively rating and comparing various materials.

Tensile Test

The most common of the static tests is the **uniaxial tensile test**, which can be applied to nearly all solid materials—metals, polymers, ceramics, composites, and elastomers. It provides quantitative information regarding both the strength and ductility of the material, which can vary with composition, prior processing history (both heat treatment and deformation),

speed of load application, and temperature. Room temperature conditions are generally assumed, unless otherwise specified.

The test begins with the preparation of a standard specimen with prescribed geometry, like the round and flat (rectangular) specimens described in **Figure 2.4**. The standard specimens ensure meaningful and reproducible results and have been designed to produce uniform uniaxial tension in the central portion while ensuring reduced stresses in the enlarged ends or shoulders that are placed in moving grips.

Strength Properties. The standard specimen is then inserted into a testing machine like the one shown in **Figure 2.5**. A tensile force or load, W, is applied and measured by the testing machine, while the elongation or stretch (ΔL) of a specified length (**gage length**) is simultaneously monitored. A plot of the coordinated load–elongation data produces a curve similar to that of **Figure 2.6**. Because the loads will differ for different-size specimens and the amount of elongation will vary with different gage lengths, it is important to remove these geometric or size effects if we are to produce data that are characteristic of a given material and not a particular specimen. If the load is divided by the *original* cross-sectional area, A_o, and the elongation is divided by the *original* gage length, L_o, the size effects are eliminated, and the resulting plot becomes known as an **engineering stress–engineering strain curve** (see Figure 2.6). This is simply a load–elongation plot with the scales of both axes modified to remove the effects of specimen size.

In Figure 2.6 it can be noted that the initial response is linear. Up to a certain point, the stress and strain are directly proportional to one another. The stress at which this proportionality ceases is known as the **proportional limit**. Below this value, the material obeys **Hooke's law**, which states that the strain is directly proportional to the stress. The proportionality constant, or ratio of stress to strain, is known as **Young's modulus** or the **modulus of elasticity**. This is an inherent property

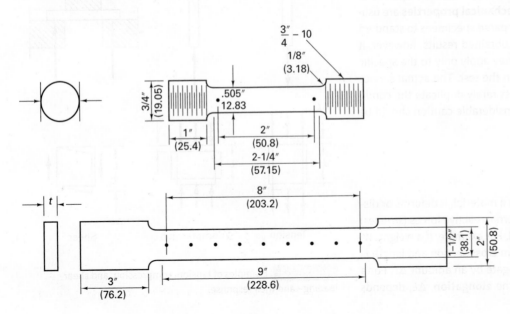

FIGURE 2.4 Two common types of standard tensile test specimens: (a) round; (b) flat. Dimensions are in inches, with millimeters in parentheses.

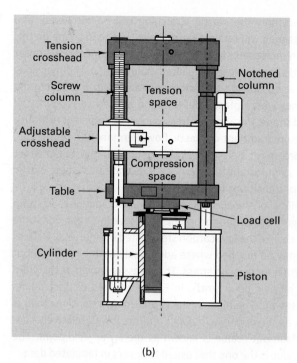

(a)

(b)

FIGURE 2.5 (a) Universal (tension and compression) testing machine; (b) Schematic of the load frame showing how motion of the darkened yoke can produce tension or compression with respect to the stationary (white) crosspiece. [(a)Courtesy of Tinius Olsen, Inc., Horsham, PA; (b) Courtesy of Satec Systems, Inc., Grove City, PA]

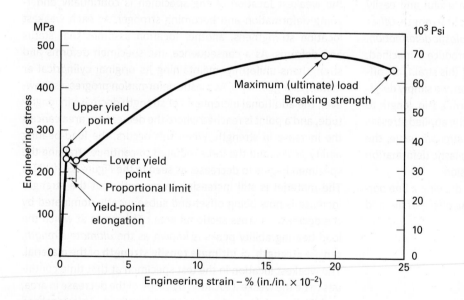

FIGURE 2.6 Engineering stress—strain diagram for a low-carbon steel.

of a given material[1] and is of considerable engineering importance. As a measure of **stiffness**, it indicates the ability of a material to resist deflection or stretching when loaded and is commonly designated by the symbol E.

Up to a certain stress, if the load is removed, the specimen will return to its original length. The response is elastic or recoverable, like the stretching and relaxation of a rubber band. The uppermost stress for which this behavior is observed is known as the **elastic limit**. For most materials the elastic limit and proportional limit are almost identical, with the elastic limit being slightly higher. Neither quantity should be assigned great engineering significance, however, because the determined values are often dependent on the sensitivity and precision of the test equipment.

The amount of energy that a material can absorb while in the elastic range is called the **resilience**. The area under a load–elongation curve is the product of a force and a distance and is therefore a measure of the energy absorbed by the specimen. If the area is determined up to the elastic limit, the absorbed energy will be elastic (or potential) energy and is regained when the specimen is unloaded. If the same determination is performed on an engineering stress–engineering strain diagram, the area beneath the elastic region corresponds to an energy per unit volume and is known as the **modulus of resilience**.

Elongation beyond the elastic limit becomes unrecoverable and is known as **plastic deformation**. When the load is

---

[1] The modulus of elasticity is determined by the binding forces between the atoms. Because these forces cannot be changed, the elastic modulus is characteristic of a specific material and is not alterable by the structure modifications that can be induced by processing.

removed, only the elastic stretching will be recovered, and the specimen will retain a permanent change in shape (in this case, an increase in length). For most components, the onset of plastic flow represents failure because the part dimensions will now be outside allowable tolerances. In manufacturing processes where plastic deformation is used to produce the desired shape, the applied stresses must be sufficient to induce the required amount of plastic flow. Permanent deformation, therefore, may be either desirable or undesirable, but in either case, it is important to determine the conditions where elastic behavior transitions to plastic flow.

Whenever the elastic limit is exceeded, increases in strain no longer require proportionate increases in stress. For some materials, like the low-carbon steel tested in Figure 2.6, a stress value may be reached where additional strain occurs without any further increase in stress. This stress is known as the **yield point**, or *yield-point stress*. In Figure 2.6, two distinct points are observed. The highest stress preceding extensive strain is known as the *upper yield point*, and the lower, relatively constant, "run-out" value is known as the *lower yield point*. The lower value is the one that usually appears in tabulated data.

Most materials, however, do not have a well-defined yield point and exhibit stress–strain curves more like that shown in **Figure 2.7**. For these materials, the elastic-to-plastic transition is not distinct, and detection of plastic deformation would be dependent on machine sensitivity or operator interpretation. To solve this dilemma, we simply define a useful and easily determined property known as the **offset yield strength**. *Offset yield strength does not describe the onset of plastic deformation*, but instead defines the stress required to produce a specified, acceptable, amount of permanent strain. If this strain, or "offset," is specified to be 0.2% (a common value), we simply determine the stress required to plastically deform a 1-in. length to a final length of 1.002 in. (a 0.2% strain). If the applied stresses are then kept below this *0.2% offset yield strength* value, the user can be guaranteed that any resulting plastic deformation will be less than 0.2% of the original dimension.

Offset yield strength is determined by drawing a line parallel to the elastic line, but displaced by the offset strain, and reporting the stress where the constructed line intersects the actual stress–strain curve. Figure 2.7 shows the determination of both 0.1% offset and 0.2% offset yield strength values, S_1 and S_2, respectively. The intersection values are reproducible and are independent of equipment sensitivity. It should be noted that the offset yield strength values become meaningful only when they are reported in conjunction with the amount of offset strain used in their determination. While 0.2% is a common offset for many mechanical products (and is generally assumed unless another number is specified), applications that cannot tolerate that amount of deformation may specify offset values of 0.1% or even 0.02%. It is important, therefore, to verify that any tabulated data being used was determined under the desired conditions.

As shown in Figure 2.6, the load (or engineering stress) required to produce additional plastic deformation continues to increase. This load is the product of the material strength times the cross-sectional area. During tensile deformation, the specimen is continually increasing in length. The cross-sectional area, therefore, must be decreasing. For the overall load-bearing ability of the specimen to increase, the material must be getting stronger. The mechanism for this phenomenon will be discussed in Chapter 3, where we will learn that the strength of a metal continues to increase with increased plastic deformation.

During the plastic deformation portion of a **tensile test**, the weakest location of the specimen is continually undergoing deformation and becoming stronger. As each weakest location strengthens, another location assumes that status and deforms. As a consequence, the specimen deforms and strengthens uniformly, maintaining its original cylindrical or rectangular geometry. As plastic deformation progresses, however, the additional increments of strength decrease in magnitude, and a point is reached where the decrease in area cancels the increase in strength. When this occurs, the load-bearing ability peaks, and the force required to continue straining the specimen begins to decrease, as seen in the Figure 2.6. (NOTE: The material is still increasing in strength, but the strength increase is now being offset and subsequently dominated by the decrease in cross-sectional area.) The stress at which the load-bearing ability peaks is known as the *ultimate strength,* **tensile strength**, or **ultimate tensile strength** of the material. The weakest location in the test specimen at that time continues to be the weakest location by virtue of the decrease in area, and further deformation becomes localized. This localized reduction in cross-sectional area is known as **necking** and is shown in **Figure 2.8**.

If the straining is continued, necking becomes intensified, and the tensile specimen will ultimately fracture. The stress at which fracture occurs is known as the **breaking strength** or **fracture strength**. For ductile materials, necking precedes fracture, and the breaking strength is less than the ultimate tensile strength. For a brittle material, fracture usually terminates the stress–strain curve before necking and often before the onset of plastic flow.

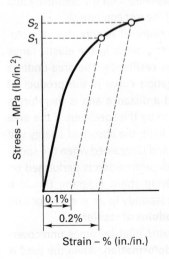

FIGURE 2.7 Stress–strain diagram for a material not having a well-defined yield point, showing the offset method for determining yield strength. S_1 is the 0.1% offset yield strength; S_2 is the 0.2% offset yield strength.

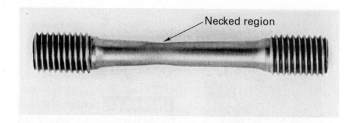

FIGURE 2.8 A standard 0.505-in.-diameter tensile specimen showing a necked region that has developed prior to failure. (E. Paul DeGarmo)

Ductility and Brittleness.

When evaluating the suitability of a material for certain manufacturing processes or its appropriateness for a given application, the amount of plasticity that precedes fracture, or the **ductility**, can often be a significant property. For metal deformation processes, the greater the ductility, the more a material can be deformed without fracture. Ductility also plays a key role in toughness, a property that will be described shortly.

One of the simplest ways to evaluate ductility is to determine the **percent elongation** of a tensile test specimen at the time of fracture. As shown in **Figure 2.9**, ductile materials do not elongate uniformly when loaded beyond necking. If the percent change of the entire 8-in. gage length were computed, the elongation would be 31%. However, if only the center 2-in. segment is considered, the elongation of that portion is 60%. A valid comparison of material behavior, therefore, requires similar specimens with the same standard gage length.

In many cases, material "failure" is defined as the onset of localized deformation or necking. Consider a sheet of metal being formed into an automobile body panel. If we are to ensure uniform strength and corrosion resistance in the final panel, the operation must be performed in a way that maintains uniform sheet thickness. For this application, a more meaningful

measure of material ductility would be the **uniform elongation** or the *percent elongation prior to the onset of necking*. This value can be determined by constructing a line parallel to the elastic portion of the diagram, passing through the point of highest force or stress. The intercept where the line crosses the strain axis denotes the available uniform elongation. Because the additional deformation that occurs after necking is not considered, uniform elongation is always less than the total elongation at fracture (the generally reported elongation value).

Another measure of ductility is the **percent reduction in area** that occurs in the necked region of the specimen. This can be computed as

$$\text{R.A.} = \frac{A_o - A_f}{A_o} \times 100\%$$

where A_o is the original cross-sectional area and A_f is the smallest area in the necked region. Percent reduction in area, therefore, can range from 0% (for a brittle glass specimen that breaks with no change in area) to 100% (for extremely plastic soft bubble gum that pinches down to a point before fracture).

When materials fail with little or no ductility, they are said to be **brittle**. Brittleness, however, is simply the lack of ductility and should not be confused with a lack of strength. Strong materials can be brittle, and brittle materials can be strong.

Toughness.

Toughness, or *modulus of toughness*, is the work per unit volume required to fracture a material. The tensile test can provide one measure of this property because toughness corresponds to the total area under the stress–strain curve from test initiation to fracture and thereby encompasses both strength and ductility. A high-strength, low-ductility material can have the same toughness or fracture resistance as a low-strength, high-ductility one. Caution should be exercised when using toughness data, however, because the work or energy to fracture can vary markedly with different conditions of testing. Variations in the temperature or the speed of loading can significantly alter both the stress–strain curve and the toughness.

In most cases, toughness is associated with impact or shock loadings, and the values obtained from high-speed (dynamic) impact tests often fail to correlate with those obtained from the relatively slow-speed (static) tensile test.

True Stress–True Strain Curves.

The stress–strain curve in Figure 2.6 is a plot of **engineering stress**, S, versus **engineering strain**, e, where S is computed as the applied load divided by the original cross-sectional area, A_o, and e is the elongation, ΔL, divided by the original gage length, L_o. As the test progresses, the cross section of the test specimen is continually changing, first in a uniform manner and then nonuniformly after necking begins. The actual stress should be computed based on the instantaneous cross-sectional area, A, not the original, A_o. Because the area is decreasing, the actual or true stress will be greater than the engineering stress plotted in Figure 2.6. **True stress**, σ, can be computed by taking

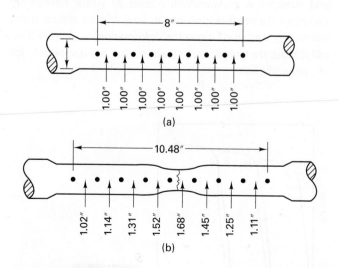

(a)

(b)

FIGURE 2.9 Final elongation in various segments of a tensile test specimen: (a) original geometry; (b) shape after fracture.

simultaneous readings of the load, W, and the minimum specimen diameter. The actual area can then be computed, and true stress can be determined as

$$\sigma = \frac{W}{A}$$

The determination of **true strain** is a bit more complex. In place of the change in length divided by the original length that was used to compute engineering strain, true strain is defined as the summation of the incremental strains that occur throughout the test. For a specimen that has been stretched from length L_o to length L, the *true*, *natural*, or *logarithmic strain*, would be

$$\varepsilon = \int_{L_o}^{L} \frac{d\ell}{\ell} = \ln \frac{L}{L_o} = \frac{D_o^2}{D^2} = 2\ln \frac{D_o}{D}$$

The preceding equalities make use of the following relationships for cylindrical specimens that maintain constant volume (i.e., $V_o = L_o A_o = V = LA$)

$$\frac{L}{L_o} = \frac{A_o}{A} = \frac{D_o^2}{D^2}$$

NOTE: Because these relations are based on cylindrical geometry, they apply only up to the onset of necking.

Figure 2.10 depicts the type of curve that results when the data from a uniaxial tensile test are converted to the form of true stress versus true strain. Because the true stress is a measure of the material strength at any point during the test, it will continue to rise even after necking. Data beyond the onset of necking should be used with extreme caution, however, because the geometry of the neck transforms the stress state from uniaxial tension (stretching in one direction with compensating contractions in the other two) to triaxial tension, in which the material is stretched or restrained in all three directions. Because of the triaxial tension, voids or cracks (**Figure 2.11**) tend to form in the necked region and serve as a precursor to final fracture. Measurements of the external diameter no longer reflect the true load-bearing area, and the data are further distorted.

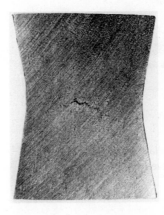

FIGURE 2.11 Section of a tensile test specimen stopped just prior to failure, showing a crack already started in the necked region, which is experiencing tri-axial tension. (Photo by E. R. Parker, courtesy E. Paul DeGarmo)

Strain Hardening and the Strain-Hardening Exponent. **Figure 2.12** is a true stress–true strain diagram, which has been modified to show how a ductile metal (such as steel) will behave when subjected to slow loading and unloading. Loading and unloading within the elastic region will result in simply cycling up and down the linear portion of the curve between points O and A. However, if the initial loading is carried through point B (in the plastic region), unloading will follow the path BeC, which is approximately parallel to the line OA, and the specimen will exhibit a permanent elongation of the amount OC. Upon reloading from point C, elastic behavior is again observed as the stress follows the line CfD, a slightly different path from that of unloading. Point D is now the yield point or yield stress for the material in its partially deformed state. A comparison of points A and D reveals that plastic deformation has made the material stronger. If the test were again interrupted at point E, we would find a new, even higher-yield stress. Thus, within the region of plastic deformation, each of the points along the true stress–true strain curve represents the yield stress for the material at the corresponding value of strain.

When metals are plastically deformed, they become harder and stronger, a phenomenon known as **strain hardening**. Therefore, if a stress induces plastic flow, an even greater stress will be required to continue the deformation. In Chapter 3 we will discuss the atomic-scale features that are responsible for this phenomenon.

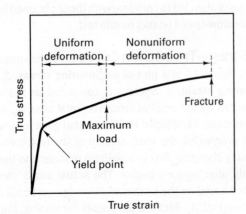

FIGURE 2.10 True stress–true strain curve for an engineering metal, showing true stress continually increasing throughout the test.

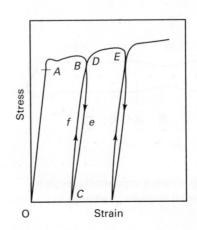

FIGURE 2.12 Stress–strain diagram obtained by unloading and reloading a specimen.

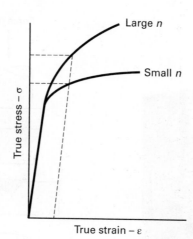

FIGURE 2.13 True stress–true strain curves for metals with large and small strain hardening. Metals with larger *n* values experience larger amounts of strengthening for a given strain.

Various materials strain harden at different rates; that is, for a given amount of deformation different materials will exhibit different increases in strength. One method of describing this behavior is to mathematically fit the plastic region of the true stress–true strain curve to the equation

$$\sigma = K\varepsilon^n$$

and determine the best-fit value of *n*, the **strain-hardening exponent**.[2] As shown in **Figure 2.13**, a material with a high value of *n* will have a significant increase in strength with a small amount of deformation. A material with a small *n* value will show little change in strength with plastic deformation.

Damping Capacity.

In Figure 2.12 the unloading and reloading of the specimen follow slightly different paths. The area between the two curves is proportional to the amount of energy that is converted from mechanical form to heat and is therefore absorbed by the material. When this area is large, the material is said to exhibit good **damping capacity** and is able to absorb mechanical vibrations or damp them out quickly. This is an important property in applications such as crankshafts and machinery bases. Gray cast iron is used in many applications because of its high damping capacity. Materials with low damping capacity, such as brass and steel, readily transmit both sound and vibrations. Bronze, another low-damping material, is often used in the manufacture of bells.

Rate Considerations.

The rate or speed at which a tensile test is conducted can have a significant effect on the various properties. **Strain rate** sensitivity varies widely for the engineering materials. Plastics and polymers are very sensitive to testing speed, as are metals with low melting points, such as lead and zinc. Those materials that are sensitive to

speed variations exhibit higher strengths and lower ductility when speed is increased. It is important to recognize that standard testing selects a standard speed, which may or may not correlate with the conditions of product application. When tensile tests are conducted over a range of speeds, the data can be used to generate a **strain-rate-hardening exponent**, designated as *m*, which describes the change in yield strength as testing speed is increased.

Compression and Bending Tests

When a material is subjected to compressive loadings, the relationships between stress and strain are similar to those for a **tension** test. Up to a certain value of stress, the material behaves elastically. Beyond this value, plastic flow occurs. In general, however, a **compression** test is more difficult to conduct than a standard tensile test. Test specimens must have larger cross-sectional areas to resist bending or buckling. As deformation proceeds, the material strengthens by strain hardening, and the cross section of the specimen also increases, combining to produce a substantial increase in required load. Friction between the testing machine surfaces and the ends of the test specimen will alter the results if not properly considered. The type of service for which the material is intended, however, should be the primary factor in determining whether the testing should be performed in tension or compression.

Hard, brittle materials, such as glass and other ceramics, are often evaluated through three-point and four-point bending tests, as shown in **Figure 2.14**. The material is elastically flexed, compressing the upper surface and stretching the bottom, with failure occurring in the form of sudden fracture. Key properties that are determined include the **flexural modulus** or modulus of elasticity in bending and the **flexural strength** or modulus of rupture.

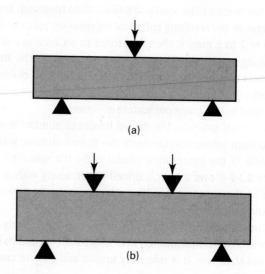

FIGURE 2.14 Schematic of the (a) three-point and (b) four-point bending tests that are commonly applied to brittle materials.

---

[2] Taking the logarithm of both sides of the equation yields $\log \sigma = \log K + n \log \varepsilon$, which has the same form as the equation $y = mx + b$ if y is $\log \sigma$ and x is $\log \varepsilon$. This is the equation of a straight line with slope m and intercept b. Therefore, if the true stress—true strain data were plotted on a log-log scale with stress (σ) on the *y*-axis and strain (ε) on the *x*, the slope of the data in the plastic region would be *n*.

Hardness Testing

The wear resistance and strength of a material can also be evaluated by assessing its "**hardness**," a somewhat hard-to-define property of engineering materials. A number of easy-to-perform and typically nondestructive tests have been developed using various phenomena. The most common of the hardness tests are based on resistance to permanent deformation in the form of penetration or indentation. Other tests evaluate resistance to scratching, wear resistance, resistance to cutting or drilling, or elastic rebound (energy absorption under impact loading). Because these phenomena are not the same, the results of the various tests often do not correlate with one another. Although hardness tests are among the easiest to perform on the shop floor, caution should be exercised to ensure that the selected test clearly evaluates the phenomena of interest. The various ASTM specifications[3] provide details regarding sample preparation, selection of loads and penetrators, minimum sample thicknesses, spacing and near-edge considerations, and conversions between scales.

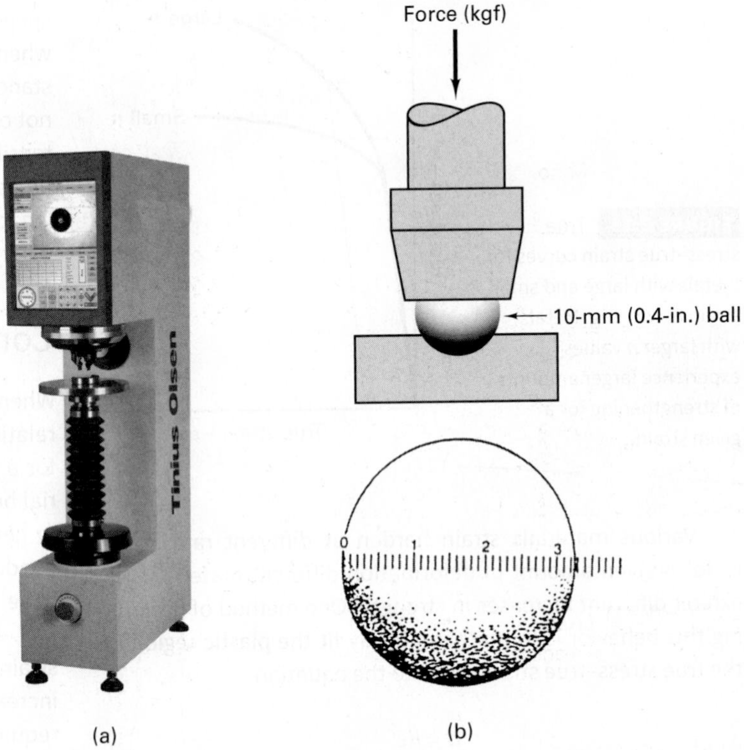

Brinell Hardness Test.

The **Brinell hardness test** was one of the earliest accepted methods of measuring hardness. A tungsten carbide or hardened steel ball is pressed into the flat surface of a material by a standard load, which is then maintained for a period of time to permit sufficient plastic deformation to occur to support the load. (A 10-mm-diameter ball and 3000-kg load are typically used when testing irons and steel, with smaller diameters and lighter loads being used with softer metals. A hold time of 10–15 seconds is recommended for iron or steel and up to 30 seconds for softer metals.) The load and ball are then removed, and the diameter of the resulting spherical indentation (usually in the range of 2 to 5 mm) is then measured to an accuracy of 0.05 mm using a special grid or traveling microscope. The **Brinell hardness number** (BHN or HB) is equal to the load divided by the surface area of the spherical indentation when the units are expressed as kilograms per square millimeter.

In actual practice, the Brinell hardness number is determined from tables that correlate the Brinell number with the diameter of the indentation produced by the specified load. **Figure 2.15** shows a typical Brinell tester, along with a schematic of the testing procedure, which is actually a two-step operation—load then measure.

The Brinell test measures hardness over a relatively large area and is somewhat indifferent to small-scale variations in the material structure. It is relatively simple and easy to conduct

FIGURE 2.15 (a) Brinell hardness tester; (b) Brinell test sequence showing loading and measurement of the indentation under magnification with a scale calibrated in millimeters. [(a) Courtesy of Tinius Olsen, Inc., Horsham, PA; (b) Courtesy of Wilson Hardness, an Instron Company., Norwood, MA]

and is used extensively on irons and steels. On the negative side, however, the Brinell test has the following limitations:

1. It cannot be used on very hard or very soft materials.
2. The results may not be valid for thin specimens. It is best if the thickness of material is at least 10 times the depth of the indentation. Some standards specify the minimum hardnesses for which the tests on thin specimens will be considered valid.
3. The test is not valid for case-hardened surfaces.
4. The test must be conducted far enough from the edge of the material so that no edge bulging occurs.
5. The substantial indentation may be objectionable on finished parts.
6. The edge or rim of the indentation may not be clearly defined or may be difficult to see.

The Rockwell Test.

The **Rockwell hardness test** is the most widely used hardness test and is similar to the Brinell test, with the hardness value again being determined by an indentation or penetration produced by a static load. **Figure 2.16**(a) shows the key features of the Rockwell test. A small indenter, either a hardened steel ball of 1/16, 1/8, 1/4, or 1/2-in. in diameter, or a 120-degree diamond-tipped cone called a **brale**, is first seated firmly against the material by the application of a 10-kilogram "minor" load. This causes a slight

---

[3] ASTM hardness testing specifications include E3, E10, E18, E103, E140, and E384.

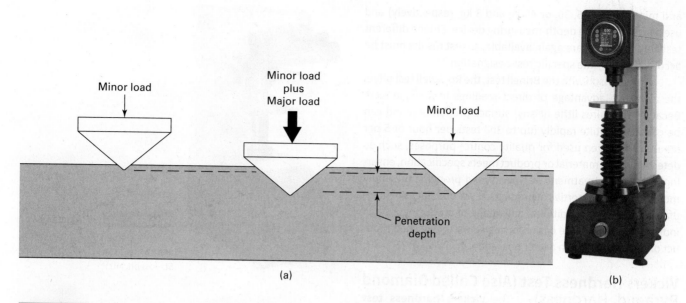

Minor load

Minor load plus Major load

Minor load

Penetration depth

(a)

(b)

FIGURE 2.16 (a) Operating principle of the Rockwell hardness tester; (b) Typical Rockwell hardness tester with digital readout. (Courtesy of Tinius Olsen, Inc., Horsham, PA)

elastic penetration into the surface and removes the effects of any surface irregularities. The position of the indenter is noted, and "major" load of 60, 100, or 150 kg is then applied to the indenter to produce a deeper penetration by inducing plastic deformation. When the indenter ceases to move, the major load is removed. With the minor load still applied to hold the indenter firmly in place, the testing machine, like the one shown in **Figure 2.16(b)**, now displays a numerical reading. This Rockwell hardness number is really an indication of the distance of indenter travel or the *depth* of the plastic or permanent penetration that was produced by the major load (as opposed to the diameter that was used in the Brinell test), with each unit representing a penetration depth of 2 μm.

To accommodate a wide range of materials with a wide range of strength, there are 15 different Rockwell test scales, each having a specified major load and indenter geometry. **Table 2.1** provides a partial listing of Rockwell scales, which are designated by letters, and some typical materials for which they are used. Because of the different scales, a Rockwell hardness number must be accompanied by the letter corresponding to the particular combination of load and indenter used in its determination, typically appearing as a subscript following the capital R and preceding the test result number. The notation R_C60 (also Rockwell C 60 or 60 HRC), for example, indicates that a Rockwell test was conducted with the brale indenter in combination with a major load of 150 kg, and a reading of 60 was obtained. The B and C scales are the most common, with B being used for copper and aluminum and C for steels.[4] ASTM standard E18 provides guidance on the proper Rockwell scale to use.

Rockwell tests should not be conducted on thin materials (typically less than 1.5 mm or ¹⁄₁₆-in.), on rough surfaces, or on materials that are not homogeneous, such as gray cast iron. Because of the small size of the indentation, variations in roughness, composition, or structure can greatly influence test results. For thin materials, or where a very shallow indentation is desired (as in the evaluation of surface-hardening treatments such as nitriding or carburizing or the testing of very soft materials), one of the **Rockwell superficial hardness tests** would be preferred. Operating on the same Rockwell principle, these tests employ significantly reduced major

TABLE 2.1   Some Common Rockwell Hardness Tests

Scale Symbol	Penetrator	Load (kg)	Typical Materials
A	Brale	60	Cemented carbides, thin steel, shallow case-hardened steel
B	¹⁄₁₆-in. ball	100	Copper alloys, soft steels, aluminum alloys, malleable iron
C	Brale	150	Steel, hard cast irons, titanium, deep case-hardened steel
D	Brale	100	Thin steel, medium case-hardened steel
E	⅛-in. ball	100	Cast iron, aluminum, magnesium
F	¹⁄₁₆-in. ball	60	Annealed coppers, thin soft sheet metals
G	¹⁄₁₆-in. ball	150	Hard copper alloys, malleable irons
H	⅛-in. ball	60	Aluminum, zinc, lead

[4] The Rockwell C number is computed as 100 – (depth of penetration in μm / 2 μm), while the Rockwell B number is 130 – (depth of penetration in μm / 2 μm).

and minor loads (15, 30, or 45 kg and 3 kg, respectively) and use a more sensitive depth-measuring device. Fifteen different test configurations are again available, so test results must be accompanied by the specific test designation.

In comparison with the Brinell test, the Rockwell test offers the attractive advantage of direct readings in a single step. Because it requires little (if any) surface preparation and can be conducted quite rapidly (up to 300 tests per hour or 5 per minute), it is often used for quality control purposes, such as determining if a material or product meets specification, ensuring that a heat treatment was performed properly, or simply monitoring the properties of products at various stages of manufacture. It has the additional advantage of producing a small indentation that can be easily concealed on the finished product or easily removed in a later operation.

Vickers Hardness Test (Also Called Diamond Pyramid Hardness).

The **Vickers hardness test** is similar to the Brinell test, but uses a 136° square-based diamond-tipped pyramid as the indenter and loads between 1 and 120 kg. Like the Brinell value, the Vickers hardness number (HV) or diamond pyramid hardness (DPH) is also defined as load divided by the surface area of the indentation expressed in units of kilograms per square millimeter. The advantages of the Vickers approach include the increased accuracy in determining the diagonal of a square impression as opposed to the diameter of a circle and the assurance that even light loads will produce some plastic deformation. The use of diamond as the indenter material enables the test to evaluate any material, including those too hard for the Brinell test, and effectively places the hardness of all materials on a single scale.

Like the other indentation or penetration methods, the Vickers test has a number of attractive features: (1) it is simple to conduct, (2) little time is involved, (3) little surface preparation is required, (4) the marks are quite small and are easily hidden or removed, (5) the test can be done on location, (6) it is relatively inexpensive, and (7) it provides results that can be used to evaluate material strength or assess product quality.

Microindentation Hardness.

Hardness tests have also been developed for applications where the testing involves a very precise area of material, or where the material or modified surface layer is exceptionally thin. These tests were previously called **microhardness tests**, but the newer **microindentation** term is more appropriate because it is the size of the indentation that is extremely small, not the measured value of hardness.[5] Special machines, such as the one shown in **Figure 2.17**, have been constructed for this type of testing, which must be performed on specimens with a polished metallographic surface. The location for the test is selected under high magnification. A predetermined

FIGURE 2.17 Microindentation hardness tester. (Image used with permission from LECO Corporation, St. Joseph, MI)

load between 1 and 1000 grams is then applied through a small diamond-tipped penetrator. The standard Vickers indenter can be used, providing a continuous micro through macro scale. In the **Knoop test**, the diamond-tipped indenter is modified to have a long diagonal seven times the length of the short diagonal, and the longer length of the indentation is measured under a magnification between 200 and 1000X. **Figure 2.18** compares the indenters for the Vickers and Knoop tests and shows a series of Knoop indentations progressing left-to-right across a surface-hardened steel specimen, from the hardened surface to the unhardened core. The hardness value, known as the **Knoop hardness number**, is again obtained by dividing the load in kilograms by the projected area of the indentation, expressed in square millimeters.

Other Hardness Determinations.

When testing soft, elastic materials, such as rubbers and nonrigid plastics, a **durometer** is often be used. This instrument, shown in **Figure 2.19**, measures the resistance of a material to elastic penetration by a spring-loaded conical steel indenter. No permanent deformation occurs. A similar test is used to evaluate the strength of molding sands used in the foundry industry and will be described in Chapter 14.

In the **scleroscope** test, hardness is measured by the rebound of a small diamond-tipped "hammer" that is dropped from a fixed height onto the surface of the material to be tested. This test evaluates the resilience of a material, and the surface on which the test is conducted must have a fairly high polish to yield good results. Because the test is based on resilience, scleroscope hardness numbers should only be used to compare similar materials. A comparison between steel and rubber, for example, would not be valid.

The **LEEB test** is another impact method. A carbide-tipped impact body is driven into the test surface by a spring force, with the impact and rebound creating a deformation induced indentation. This method is primarily used for portable hardness testing, where the specimen cannot be conveniently brought to a testing machine.

---

[5] The ASTM Standard E384 has been renamed "Standard Test Method for Microindentation Hardness of Materials" to reflect this change in nomenclature.

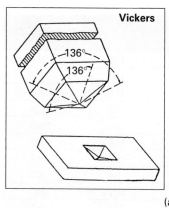 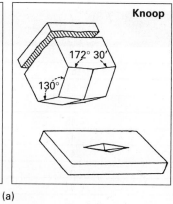

(a)

(b)

FIGURE 2.18 (a) Comparison of the diamond-tipped indenters used in the Vickers and Knoop hardness tests. (b) Series of Knoop hardness indentations progressing left-to-right across a surface-hardened steel specimen (hardened surface to unhardened core). (Courtesy of Buehler Ltd., Lake Bluff, IL)

Another definition of hardness is the ability of a material to resist being scratched. A crude but useful test that employs this principle is the **file test**, where one determines if a material can be cut by a simple metalworking file. The test can be either a pass–fail test using a single file, or a semiquantitative evaluation using a series of files that have been pretreated to various levels of known hardness. This approach is commonly used by geologists where ten selected materials are used to create a scale that enables the hardness of rocks and minerals to be classified from 0 to 10.

FIGURE 2.19 Durometer hardness tester. (Courtesy of New Age Testing Instruments, an AMETEK Company, www.hardnesstesters.com, Southampton, PA)

Relationships Among the Various Hardness Tests.
Because the various hardness tests often evaluate different phenomena, there are no simple relationships between the different types of hardness numbers. Approximate relationships have been developed, however, by testing the same material on a variety of devices. **Table 2.2** presents a correlation of hardness values for plain carbon and low-alloy steels. It may be noted that for Rockwell C numbers above 20, the Brinell values are approximately 10 times the Rockwell number. Also, for Brinell values below 320, the Vickers and Brinell values agree quite closely. Because the relationships among the various tests will differ with material, mechanical processing, and heat treatment, correlations such as Table 2.2 should be used with caution.

Relationship of Hardness to Tensile Strength.
Table 2.2 and **Figure 2.20** show a definite relationship between tensile strength and hardness. For plain carbon and low-alloy steels, the tensile strength (in pounds per square inch) can be

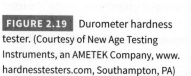

TABLE 2.2 **Hardness Conversion Table for Steels**

| Brinell Number | Vickers Number | Rockwell Number | | Scleroscope Number | Tensile Strength | |
		C	B		ksi	MPa
	940	68		97	368	2537
757 [a]	860	66		92	352	2427
722 [a]	800	64		88	337	2324
686 [a]	745	62		84	324	2234
660 [a]	700	60		81	311	2144
615 [a]	655	58		78	298	2055
559 [a]	595	55		73	276	1903
500	545	52		69	256	1765
475	510	50		67	247	1703
452	485	48		65	238	1641
431	459	46		62	212	1462
410	435	44		58	204	1407

(continued)

TABLE 2.2   (continued)

Brinell Number	Vickers Number	Rockwell Number		Scleroscope Number	Tensile Strength	
		C	B		ksi	MPa
390	412	42		56	196	1351
370	392	40		53	189	1303
350	370	38	110	51	176	1213
341	350	36	109	48	165	1138
321	327	34	108	45	155	1069
302	305	32	107	43	146	1007
285	287	30	105	40	138	951
277	279	28	104	39	34	924
262	263	26	103	37	128	883
248	248	24	102	36	122	841
228	240	20	98	34	116	800
210	222	17	96	32	107	738
202	213	14	94	30	99	683
192	202	12	92	29	95	655
183	192	9	90	28	91	627
174	182	7	88	26	87	600
166	175	4	86	25	83	572
159	167	2	84	24	80	552
153	162		82	23	76	524
148	156		80	22	74	510
140	148		78	22	71	490
135	142		76	21	68	469
131	137		74	20	66	455
126	132		72	20	64	441
121	121		70		62	427
112	114		66		58	

[a] Tungsten, carbide ball; others, standard ball.

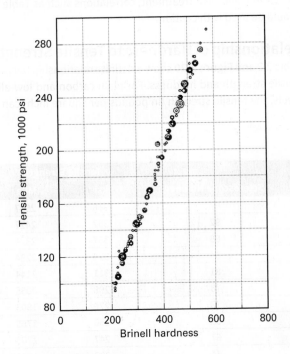

FIGURE 2.20 Relationship of hardness and tensile strength for a group of standard alloy steels. (Courtesy of ASM International, Materials Park, OH)

estimated by multiplying the Brinell hardness number by 500. In this way, an inexpensive and quick hardness test can be used to provide a close approximation of the tensile strength of the steel. For other materials, however, the relationship is different and may even exhibit too much variation to be dependable. The multiplying factor for age-hardened aluminum is about 600, whereas for soft brass it is around 800.

2.3 Dynamic Properties

Products or components can also be subjected to a wide variety of **dynamic loadings**. These may include (1) sudden impacts or loads that change rapidly in magnitude, (2) repeated cycles of loading and unloading, or (3) frequent changes in the mode of loading, such as from tension to compression. To select materials and design for these conditions, we must be able to characterize the mechanical properties of engineering materials under dynamic loadings.

Most dynamic tests subject standard specimens to a well-controlled set of test conditions. The conditions experienced by actual parts, however, rarely duplicate the controlled conditions of the standardized tests. Whereas identical tests on different materials can indeed provide a comparison of material behavior, the assumption that similar results can be expected for similar conditions may not always be true. Because dynamic conditions can vary greatly, the quantitative results of standardized tests should be used with extreme caution, and one should always be aware of test limitations.

Impact Test

Several tests have been developed to evaluate the **toughness** or fracture resistance of a material when it is subjected to a rapidly applied load, or impact. Of the tests that have become common, two basic types have emerged: (1) bending impacts, which include the standard Charpy and Izod tests, and (2) tension impacts.

The bending **impact tests** utilize specimens that are supported as beams. In the **Charpy test**, shown schematically in **Figure 2.21**, the standard specimen is a square bar with a one centimeter by one centimeter cross section, containing a V-, keyhole-, or U-shaped notch. The test specimen is positioned horizontally, supported on the ends, and an impact is applied to the center, behind the notch, to complete a three-point bending. The **Izod test** specimen, although somewhat similar in size and appearance, is supported vertically as a cantilever beam and is impacted on the unsupported end, striking from the side of the notch (**Figure 2.22**). Impact testers, like the one shown in **Figure 2.23**, supply a predetermined impact energy in the form of a heavy pendulum swinging from a fixed starting height. After breaking or deforming the specimen at the bottom of its swing, the pendulum continues its upward swing with an energy equal to its original minus that absorbed by the impacted specimen. The loss of energy is measured by the

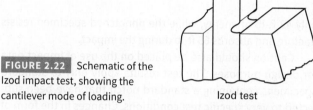

FIGURE 2.21 Schematic of the standard Charpy impact test specimen showing the three-point bending type of impact loading.

FIGURE 2.22 Schematic of the Izod impact test, showing the cantilever mode of loading.

angle that the pendulum attains during its upward swing. Additional information can be obtained by measuring the lateral expansion and appearance of the fracture surface.

The test specimens for bending impacts must be prepared with geometric precision to ensure consistent and reproducible results. Notch profile is extremely critical, for the test measures the energy required to both initiate and propagate a fracture. The effect of notch profile is shown dramatically in **Figure 2.24**. Here two specimens have been made from the same piece of steel with the same reduced cross-sectional area. The one with the keyhole notch fractures and absorbs

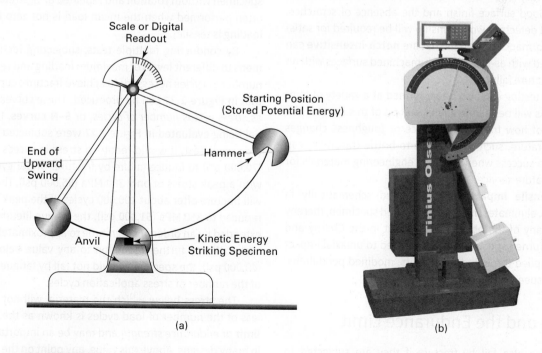

(a)

(b)

FIGURE 2.23 (a) Basic principle of a pendulum impact test, (b) Typical impact testing machine. (Courtesy of Tinius Olsen Inc., Horsham, PA)

(a)   (b)

FIGURE 2.24   (a) Notched and (b) unnotched impact specimens before and after testing. Both specimens had the same cross-sectional area, but the notched specimen fractures while the other doesn't. (Courtesy of Tinius Olsen Inc.)

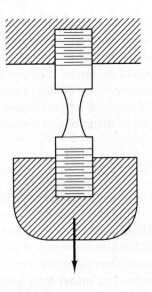

FIGURE 2.25   Tensile impact test.

only 43 ft-lb of energy, while the unnotched specimen resists fracture and absorbs 65 ft-lb during the impact.

Caution should also be placed on the use of impact data for design purposes. The test results apply only to standard specimens containing a standard notch that have been subjected to very specific test conditions. Changes in the form of the notch, minor variations in the overall specimen geometry, or faster or slower rates of loading (speed of the pendulum) can all produce significant changes in the results. Under conditions of sharp notches, wide specimens, and rapid loading, many ductile materials lose their energy-absorbing capability and fail in a brittle manner. (e.g., the standard impact tests should not be used to evaluate materials for bullet-proof armor because the velocities of loading are extremely different.)

The results of standard tests, however, can be quite valuable in assessing a material's sensitivity to notches and the multiaxial stresses that exist around a notch. Materials whose properties vary with notch geometry are termed **notch-sensitive**. Good surface finish and the absence of scratches, gouges, and defects in workmanship will be required for satisfactory performance. Materials that are **notch-insensitive** can often be used with as-cast or rough-machined surfaces with no risk of premature failure.

Impact testing can also be performed at a variety of temperatures. As will be seen in a later section of this chapter, the evaluation of how fracture resistance or toughness changes with temperature, such as a ductile-to-brittle transition, can be crucial to success when selecting engineering materials for low-temperature service.

The **tensile impact test**, illustrated schematically in **Figure 2.25**, eliminates the use of a notched specimen, thereby avoiding many of the objections inherent in the Charpy and Izod tests. Turned specimens are subjected to uniaxial impact loadings applied through drop weights, modified pendulums, or variable-speed flywheels.

Fatigue and the Endurance Limit

Materials can also fail by fracture if they are subjected to repeated applications of stress, even though the peak stresses have magnitudes less than the ultimate tensile strength and usually less than the yield strength. This phenomenon, known as **fatigue**, can result from either the cyclic repetition of a particular loading cycle or entirely random variations in stress. Almost 90% of all metallic fractures are in some degree attributed to fatigue.

For experimental simplicity, a periodic, sinusoidal loading is often utilized, and conditions of equal-magnitude tension–compression reversals provide further simplification. These conditions can be achieved by placing a cylindrical specimen in a rotating drive and hanging a weight to produce elastic bending along the axis, as shown in the schematics of **Figure 2.26**. With each rotation, the surface of the specimen experiences a sinusoidal application of tension and compression. Alternative methods of testing involve the repeated bending or flexing of a specimen without rotation and repeated axial stretching (a test often performed when the mean load is not zero because all loading is tensile).

By conducting multiple tests, subjecting identical specimens to different levels of maximum loading and recording the number of cycles necessary to achieve fracture, curves such as that in **Figure 2.27** can be produced. These curves are known as *stress versus number of cycles*, or **S–N curves**. If the material being evaluated in Figure 2.27 were subjected to a standard tensile test, it would require a stress in excess of 480 MPa (70,000 psi) to induce failure by fracture. Under cyclic loading with a peak stress of only 380 MPa (55,000 psi), the specimen will fracture after about 100,000 cycles. If the peak stress were reduced to 350 MPa (51,000 psi), the fatigue lifetime would be extended by an order of magnitude to approximately 1,000,000 cycles. With a further reduction to any value below 340 MPa (49,000 psi), the specimen would not fail by fatigue, regardless of the number of stress application cycles.

The stress below which the material will not fail regardless of the number of load cycles is known as the **endurance limit** or *endurance strength* and may be an important criterion in many designs. Above this value, any point on the curve is the **fatigue strength**, the maximum stress that can be sustained for a specified number of loading cycles.

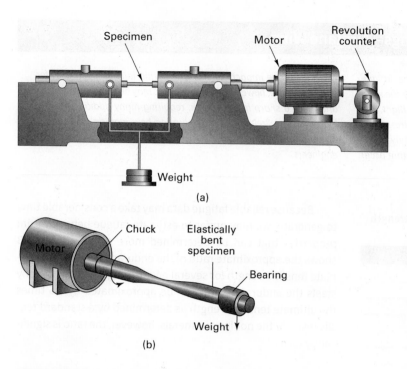

FIGURE 2.26 Schematic diagrams of (a) a Moore rotating-beam fatigue machine (Adapted from Hayden et. al., *The Structure and Properties of Materials*, Vol. 3, p. 15, Wiley, 1965) and (b) an alternate rotating beam configuration.

A different number of loading cycles is generally required to determine the endurance limit for different materials. For steels, 10 million cycles are usually sufficient. For several of the nonferrous metals, 500 million cycles may be required. For aluminum, the curve continues to drop such that, if aluminum has an endurance limit, it is at such a low value that a cheaper and much weaker material could be used. In essence, if aluminum is used under realistic stresses and cyclic loading, it will fail by fatigue after a finite lifetime.

The fatigue resistance of an actual product is sensitive to a number of additional factors. One of the most important of these is the presence of stress raisers (or stress concentrators), such as sharp corners, small surface cracks, machining marks, or surface gouges. Data for the *S–N* curves are obtained from polished-surface, "flaw-free" specimens, and the reported

lifetime is the cumulative number of cycles required to initiate a fatigue crack and then grow or propagate it to failure. Precracked specimens may be tested to study the rate of crack growth under various test conditions. If a part already contains a surface crack or flaw, the number of cycles required for crack initiation and subsequent fracture can be reduced significantly. In addition, the stress concentrator magnifies the stress experienced at the tip of the crack, accelerating the rate of subsequent crack growth. Great care should be taken to eliminate stress raisers and surface flaws on parts that will be subjected to cyclic loadings. Proper design and good manufacturing practices are often more important than material selection and heat treatment.

Operating temperature can also affect the fatigue performance of a material. **Figure 2.28** shows *S–N* curves for Inconel 625 (a high-temperature Ni–Cr–Fe alloy) determined over a range of temperatures. As temperature is increased, the fatigue strength drops significantly. Because most test data are generated at room temperature, caution should be exercised when the product application involves elevated service temperatures.

Fatigue lifetime can also be affected by changes in the environment. When metals are subjected to corrosion during the cyclic loadings, the condition is known as **corrosion fatigue**, and both specimen lifetime and the endurance limit can be significantly reduced. Moreover, the nature of the environmental attack need not be severe. For some materials, tests conducted in air have been shown to have shorter lifetimes than those run in a vacuum, and further lifetime reductions have been observed with increasing levels of humidity. The test results can also be dependent on the frequency of the loading cycles. For slower frequencies, the environment has a longer time to act between loadings. At high frequencies, the environmental effects may be somewhat masked. The application of test data to actual products, therefore, requires considerable caution.

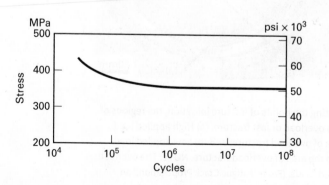

FIGURE 2.27 Typical *S–N* curve for steel showing an endurance limit. Specific numbers will vary with the type of steel and treatment.

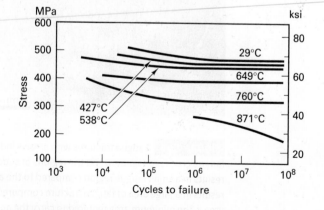

FIGURE 2.28 Fatigue strength of Inconel alloy 625 at various temperatures. (Courtesy of Huntington Alloy Products Division, The International Nickel Company, Inc., Toronto, Canada)

Fatigue Response to Complex Loadings—Airplane Wing

If the magnitude of the load varies during service, the fatigue response can be extremely complex. For example, consider the wing of a commercial airplane. As the wing vibrates during flight, the wing-fuselage joint is subjected to a large number of low-stress loadings. Although large in number, these in-flight loadings may be far less damaging than a few high-stress loadings, like those that occur

when the plane impacts the runway during landing. From a different perspective, however, the heavy loads may be sufficient to stretch and blunt a sharp fatigue crack, requiring many additional small-load cycles to "reinitiate" it. Evaluating how materials respond to complex patterns of loading is an area of great importance to design engineers.

TABLE 2.3   Ratio of Endurance Limit to Tensile Strength for Various Materials

Material	Ratio
Aluminum	0.38
Beryllium copper (heat-treated)	0.29
Copper, hard	0.33
Magnesium	0.38
Steel	
AISI 1035	0.46
Screw stock	0.44
AISI 4140 normalized	0.54
Wrought iron	0.63

Residual stresses can also alter fatigue behavior. If the specimen surface is in a state of compression, such as that produced from shot peening, carburizing, or burnishing, it is more difficult to initiate a fatigue crack, and lifetime is extended. Conversely, processes that produce residual tension on the surface, such as welding or machining, can significantly reduce the fatigue lifetime of a product.

Because reliable fatigue data may take a considerable time to generate, we may prefer to estimate fatigue behavior from properties that can be determined more quickly. **Table 2.3** shows the approximate ratio of the endurance limit to the ultimate tensile strength for several engineering metals. For many steels the endurance limit can be approximated by 0.5 times the ultimate tensile strength as determined by a standard tensile test. For the nonferrous metals, however, the ratio is significantly lower.

Fatigue Failures

Components that fail as a result of repeated or cyclic loadings are called **fatigue failures**. These fractures form a major part of a larger group known as progressive fractures. Consider the fracture surfaces shown in **Figure 2.29**. Arrows identify the points of fracture initiation, which often correspond to discontinuities in the form of surface cracks, sharp corners, machining marks, or even "metallurgical notches," such as an abrupt change in metal structure. When the load is applied, the stress at the tip of the crack exceeds the strength of the material, and

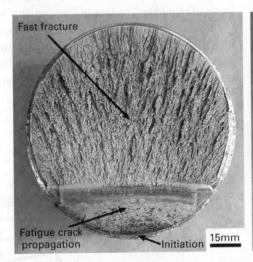
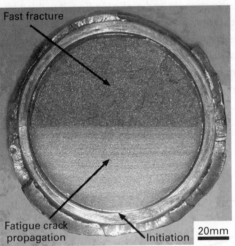

FIGURE 2.29 Fatigue fractures with arrows indicating the points of fracture initiation, the regions of fatigue crack propagation, and the regions of sudden overload or fast fracture. (a) High applied load results in a small fatigue region compared to the area of overload fracture; and (b) Low applied load results in a large area of fatigue fracture compared to the area of overload fracture. NOTE: The overload area is the minimum area required to carry the applied loads. (From "Fatigue Crack Propagation," an article published in the May 2008 issue of *Advanced Materials & Processes* magazine, Reprinted with permission of ASM International, Materials Park, OH. All rights reserved. www.asminternational.org.)

the crack grows a very small amount. Crack growth continues with each successive application of load until the remaining cross section is no longer sufficient to withstand the peak stresses. Sudden overload fracture then occurs without warning through the remainder of the material.

The overall fracture surface tends to exhibit two distinct regions: a smooth, relatively flat region where the crack was propagating by cyclic fatigue, and a fibrous, irregular, or ragged region that corresponds to the sudden overload tearing. The size of the overload region reflects the area that must be intact to support the highest applied load. In **Figure 2.29a**, the applied load is high, and only a small fatigue area is necessary to reduce the specimen to the point of overload fracture. In the part b of the figure, where the load is reduced, the fatigue fracture propagates about halfway through before sudden overload occurs.

The smooth areas of the fracture often contain a series of parallel ridges radiating outward from the origin of the crack. These ridges may not be visible under normal examination, however. They may be extremely fine; they may have been obliterated by a rubbing action during the compressive stage of repeated loading; or they may be very few in number if the failure occurred after only a few cycles of loading ("low-cycle fatigue"). Electron microscopy may be required to reveal these ridges, or **fatigue striations**, that are characteristic of fatigue failure and are caused by the growth associated with each successive application of the cyclic load. **Figure 2.30** shows an example of these markings at high magnification. Larger marks, known as "**beach marks**" may appear on the fatigued surface, lying parallel to the striations. These can be caused by interruptions to the cyclic loadings, changes in the magnitude of the applied load, and isolated overloads (not sufficient to cause ultimate fracture). **Ratchet marks**, or offset steps, can appear on the fracture surface if multiple fatigue cracks nucleate at different points and grow together.

For some fatigue failures, the overload area may exhibit a crystalline appearance, and the failure is sometimes attributed to the metal having "crystallized." As will be noted in Chapter 3, engineering metals are almost always crystalline materials. The final overload fracture simply propagated along the intercrystalline surfaces (grain boundaries), revealing the already-existing crystalline nature of the material. The conclusion that the material failed because it crystallized is totally erroneous, and the term is a definite misnomer.

Another common error is to classify all progressive-type failures as fatigue failures. Other progressive failure

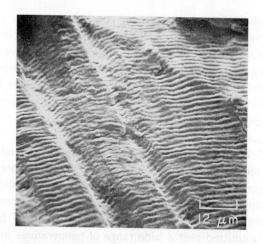

FIGURE 2.30 Fatigue fracture of AISI type 304 stainless steel viewed in a scanning electron microscope at 810X. Well-defined striations are visible. (From "Interpretation of SEM Fractographs," *Metals Handbook*, Vol. 9, 8th ed., p.70. Reprinted with permission of ASM International, Materials Park, OH, All rights reserved. www.asminternational.org.)

mechanisms, such as creep failure and stress–corrosion cracking, will also produce a two-region fracture. In addition, the same mechanism can produce fractures with different appearances depending on the magnitude of the load, type of loading (torsion, bending, or tension), temperature, and operating environment. Correct interpretation of a metal failure generally requires far more information than that acquired by a visual examination of the fracture surface.

2.4 Temperature Effects (Both High and Low)

The test data used in design and engineering decisions should always be obtained under conditions that simulate those of actual service. A number of engineered products, such as jet engines, space vehicles, gas turbines, and nuclear power plants, are required to operate under temperatures as low as −130°C (−200°F) or as high as 1250°C (2300°F). To cover these extremes, the designer must consider both the short- and long-range effects of temperature on the mechanical and physical

Are Fatigue Failures "Time Dependent?"

The failure of materials under repeated loads below their static strength is primarily a function of both the magnitude and number of loading cycles. If the frequency of loading is increased, the time to failure should decrease proportionately, maintaining the same overall number of cycles. If the time does not change, the failure is dominated by one or more environmental factors, and fatigue is a secondary component.

properties of the material being considered. From a manu-facturing viewpoint, the effects of temperature are equally important. Numerous manufacturing processes involve heat, and the elevated temperature and processing may alter the material properties in both favorable and unfavorable ways. A material can often be processed successfully, or economi-cally, only because heating or cooling can be used to change its properties.

Elevated temperatures can be quite useful in modifying the strength and ductility of a material, and important data can be obtained through **hot hardness** and **hot tensile** test-ing (conducting the standard tests with elevated temperature specimens). **Figure 2.31** summarizes the results of tensile tests conducted over a wide range of temperatures using a medium-carbon steel. Similar effects are presented for magne-sium in **Figure 2.32**. As expected, an increase in temperature will typically induce a decrease in strength and hardness and an increase in elongation. For manufacturing operations such as metal forming, heating to elevated temperature may be extremely attractive because the material is now both weaker

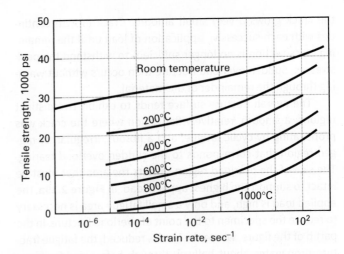

FIGURE 2.33 The effects of temperature and strain rate on the tensile strength of copper. (From A. Nadai and M. J. Manjoine, *Journal of Applied Mechanics*, Vol. 8, 1941, p. A82, courtesy of ASME)

and more ductile. The tensile and hardness test-ing of heated specimens can provide extremely useful data.

Figure 2.33 shows the combined effects of temperature and strain rate (speed of testing) on the ultimate tensile strength of copper. For a given temperature, the **rate of deformation** can also have a strong influence on mechanical properties. Room-temperature standard-rate tensile test data will be of little value if the application involves a material being hot-rolled at speeds of 1300 m/min (5000 ft/ min).

The effect of temperature on material tough-ness became the subject of intense study in the 1940s when the increased use of welded-steel construction led to catastrophic failures of ships and other structures experiencing impact while operating in cold environments. Welding produces a monolithic (single-piece) product where cracks can propagate through a joint and continue on to other sections of the structure! Extremely large fractures were being created. **Figure 2.34** shows the effect of decreasing temperature on the frac-ture resistance of two low-carbon steels. Although similar in form, the two curves are significantly displaced. The steel indicated by the solid line becomes brittle (requires very little energy to frac-ture) at temperatures below −4°C (25°F), whereas the other steel retains good fracture resistance down to −26°C (−15°F). The temperature at which the toughness goes from high energy absorption to low energy absorption is known as the **ductile-to-brittle transition temperature (DBTT)**. As the temperature crosses this transition, the fracture appearance also changes. At high temperatures, ductile fracture occurs, and the specimen deforms

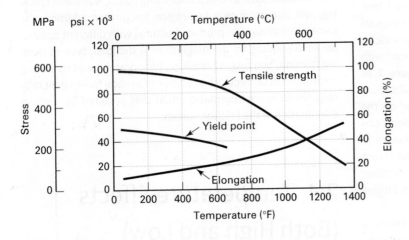

FIGURE 2.31 The effects of temperature on the tensile properties of a medium-carbon steel.

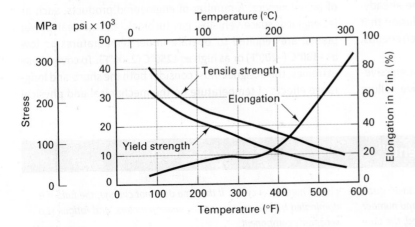

FIGURE 2.32 The effects of temperature on the tensile properties of magnesium.

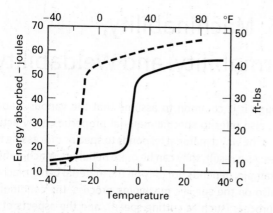

FIGURE 2.34 The effect of temperature on the impact properties of two low-carbon steels.

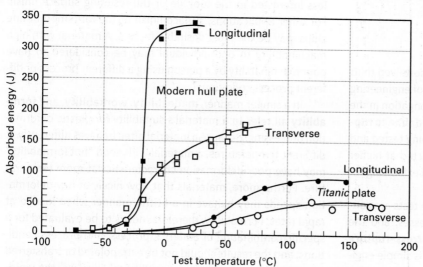

FIGURE 2.35 Notch toughness impact data: steel from the *Titanic* versus modern steel plate for both longitudinal and transverse specimens. (Courtesy I&SM, Sept. 1999, p. 33, Iron and Steel Society, Warrendale, PA)

prior to the ultimate fracture. At low temperatures, the fracture is brittle in nature, and little or no deformation is observed. A plot of percent shear in the fracture surface (relative amount of ductile fracture) versus temperature results in a curve similar to that for toughness, and the 50% point is reported as the **fracture appearance transition temperature (FATT)**.

Figure 2.35 shows the ductile-to-brittle transition temperature for steel salvaged from the *RMS Titanic* along with data from currently used ship plate material. Although both are quality materials for their era, the *Titanic* steel has a much higher **transition temperature** and is generally more brittle. The *Titanic* struck an iceberg in salt water. The water temperature at the time of the accident was –2°C, and the results show that the steel would have been quite brittle. All steels tend to exhibit a ductile-to-brittle transition, but the temperature at which it occurs varies with carbon content, alloy type and amount, and other metallurgical features. Metals such as aluminum, copper, and some types of stainless steel do not have

a ductile-to-brittle transition and can be used at low temperatures with no significant loss of toughness.

Two separate curves are provided for each of the steels in Figure 2.35, reflecting test specimens cut in different orientation with respect to the direction of product rolling. Here we see that processing features can further affect the properties and performance of a material. Because the performance properties can vary widely with the type of material, chemistry variations within the class of material, and prior processing, special cautions should be taken when selecting materials for low-temperature applications.

Creep

Long-term exposure to elevated temperatures can also lead to failure by a phenomenon known as **creep**. If a tensile-type specimen is subjected to a constant load at elevated temperature, it will elongate continuously until rupture occurs, even though the applied stress is below the yield strength of the material at the temperature of testing. Although the rate of elongation is often quite small, creep can be an important consideration when designing equipment such as steam or gas turbines, power plant boilers, and other devices that operate under loads or pressures for long periods of time at high temperature.

If a test specimen is subjected to conditions of fixed load and fixed elevated temperature, an elongation-versus-time plot can be generated, similar to the one shown in **Figure 2.36**. The curve contains three distinct stages: a short-lived initial stage, a rather long second stage where

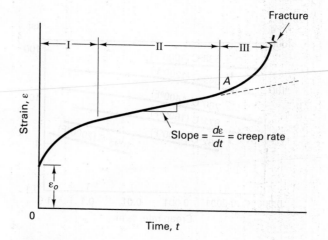

FIGURE 2.36 Creep curve for a single specimen at a fixed elevated temperature, showing the three stages of creep and reported creep rate. Note the nonzero strain at time zero due to the initial application of the load.

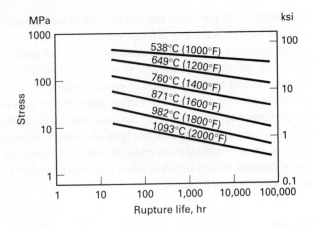

FIGURE 2.37 Stress–rupture diagram of solution-annealed Incoloy alloy 800 (Fe–Ni–Cr alloy). (Courtesy of Huntington Alloy Products Division, The International Nickel Company, Inc., Toronto, Canada)

the elongation rate is somewhat linear, and a short-lived third stage leading to fracture. Two significant pieces of engineering data are obtained from this curve: the rate of elongation in the second stage, or **creep rate**, and the total elapsed **time to rupture**. These results are unique to the material being tested and the specific conditions of the test. Tests conducted at higher temperatures or with higher applied loads would exhibit higher creep rates and shorter rupture times.

When creep behavior is a concern, multiple tests are conducted over a range of temperatures and stresses, and the **rupture time** data are collected into a single **stress–rupture diagram**, like the one shown in **Figure 2.37**. This simple engineering tool provides an overall picture of material performance at elevated temperature. In a similar manner, creep rate data can also be plotted to show the effects of temperature and stress. **Figure 2.38** presents a creep-rate diagram for the same high-temperature nickel-base alloy.

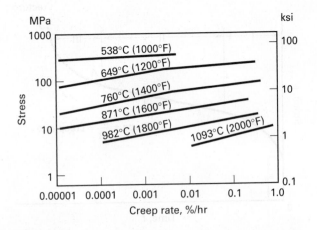

FIGURE 2.38 Creep-rate properties of solution-annealed Incoloy alloy 800. (Courtesy of Huntington Alloy Products Division, The International Nickel Company, Inc., Toronto, Canada)

2.5 Machinability, Formability, and Weldability

Although it is common to assume that the various "-ability" terms also refer to specific material properties, they actually refer to the way a material responds to specific processing techniques. As a result, they can be quite nebulous. **Machinability**, for example, depends not only on the material being machined but also on the specific machining process, the conditions of that process (such as cutting speed), and the aspects of that process that are of greatest interest. Machinability ratings are frequently based on relative tool life. In certain applications, however, we may be more interested in how easy a metal is to cut, or how it performs under high machining speeds, and less interested in the tool life or the resulting surface finish. For other applications, surface finish or the formation of fine chips may be the most desirable feature. A material with high machinability to one individual may be considered to have poor machinability by a person using a different process or different process conditions.

In a similar manner, **malleability**, **workability**, and **formability** all refer to a material's suitability for plastic deformation processing. Because a material often behaves differently at different temperatures, a material with good "hot formability" may have poor deformation characteristics at room temperature. Furthermore, materials that flow nicely at low deformation speeds may behave in a brittle manner when loaded at rapid rates. Formability, therefore, needs to be evaluated for a specific combination of material, process, and process conditions, and the results should not be extrapolated or transferred to other processes or process conditions. Likewise, the **weldability** and **castability** of a material will also depend on the specific process and the specific process parameters.

2.6 Fracture Toughness and the Fracture Mechanics Approach

A discussion of the mechanical properties of materials would not be complete without mention of the many tests and design concepts based on the fracture mechanics approach. Instead of treating test specimens as flaw-free materials, fracture mechanics begins with the premise that *all materials contain flaws or defects of some given size*. These may be **material defects**, such as pores, cracks, or inclusions; **manufacturing defects**, in the form of machining marks, arc strikes, or contact damage to external surfaces; or **design defects**, such as abrupt section changes, excessively small fillet radii, and holes. When the specimen is subjected to loads, the applied stresses are amplified or intensified in the vicinity of these defects, potentially causing accelerated failure or failure under unexpected conditions.

Fracture mechanics seeks to identify the conditions under which a defect will grow or propagate to failure and, if possible, the rate of crack or defect growth. The methods concentrate on three principal quantities: (1) the size of the largest or most critical flaw, usually denoted as a; (2) the applied stress, denoted by σ and (3) the fracture toughness, a quantity that describes the resistance of a material to fracture or crack growth, which is usually denoted by K with subscripts to signify the conditions of testing. Equations have been developed that relate these three quantities at the onset of crack growth or propagation for various specimen geometries, flaw locations, and flaw orientations. If nondestructive testing or quality control methods have been applied, the size of the largest flaw that could go undetected is often known. By mathematically placing this worst possible flaw in the worst possible location and orientation, and coupling this with the largest applied stress for that location, a designer can determine the value of fracture toughness necessary to prevent that flaw from propagating during service. Specifying any two of the three parameters allows the computation of the third. If the material and stress conditions were defined, the size of the maximum permissible flaw could be computed. Inspection conditions could then be selected to ensure that flaws greater than this magnitude are cause for product rejection. Finally, if a component is found to have a significant flaw and the material is known, the maximum operating stress can be determined that will ensure no further growth of that flaw.

In the past, detection of a flaw or defect was usually cause for rejection of the part (Detection = Rejection). With enhanced methods and sensitivities of inspection, almost every product can now be shown to contain flaws. Fracture mechanics comes to the rescue. According to the philosophy of fracture mechanics, each of the flaws or defects in a material can be either **dormant** or **dynamic**. Dormant defects are those whose size remains unchanged through the lifetime of the part and are indeed permissible. A major goal of fracture mechanics, therefore, is to define the distinction between dormant and dynamic for the specific conditions of material, part geometry, and applied loading. The basic equation of fracture mechanics assumes the form of $K \geq \alpha\,\sigma\,\sqrt{\pi a}$, where K is the **fracture toughness** of the material (a material property that depends on composition and microstructure, temperature, and the rate of load application), σ is the maximum applied tensile stress, a is the size of the largest or most critical flaw, and α is a dimensionless factor that considers the location, orientation, and shape of the flaw. The left side of the equation considers the material, and

the right side describes the usage condition (a combination of flaw and loading). The relationship is usually described as a greater than or equal. When the material number, K, is greater than the usage condition, the flaw is dormant. When equality is reached, the flaw becomes dynamic, and crack growth or fracture occurs. Alternative efforts to prevent material fracture generally involve overdesign, excessive inspection, or the use of premium-quality materials—all of which increase cost and possibly compromise performance.

Fracture mechanics can also be applied to fatigue, which has already been cited as causing as much as 90% of all dynamic failures. The standard method of fatigue testing applies cyclic loads to polished, "flaw-free" specimens, and the reported lifetime includes both crack initiation and crack propagation. In contrast, fracture mechanics focuses on the growth of an already existing flaw. **Figure 2.39** shows the **crack**

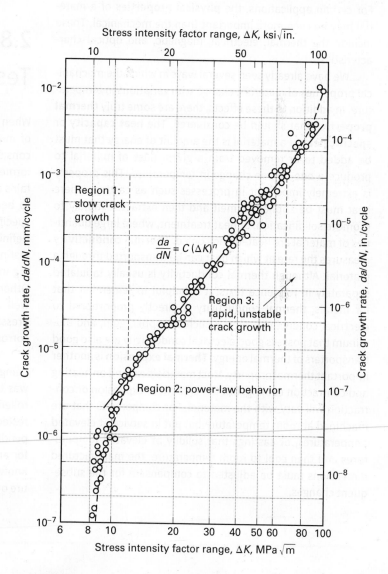

FIGURE 2.39 Plot of the fatigue crack growth rate vs. ΔK for a typical steel—the fracture mechanics approach. Similar shape curves are obtained for most engineering metals. (Courtesy of ASM International, Materials Park, OH)

growth rate (change in size per loading cycle denoted as da/dN) plotted as a function of the fracture mechanics parameter, ΔK (where ΔK increases with an increase in either the flaw size and/or the magnitude of applied stress). Because the fracture mechanics approach begins with an existing flaw, it provides a far more realistic guarantee of minimum service life.

Fracture mechanics is a truly integrated blend of design (applied stresses), inspection (flaw-size determination), and materials (fracture toughness). The approach has proven valuable in many areas where fractures could be catastrophic.

2.7   Physical Properties

For certain applications, the **physical properties** of a material may be even more important than the mechanical. These include the thermal, electrical, magnetic, and optical characteristics.

We have already seen several ways in which the mechanical properties of materials change with variations in temperature. In addition to these effects, there are some truly **thermal properties** that should be considered. The **heat capacity** or **specific heat** of a material is the amount of energy that must be added to or removed from a given mass of material to produce a one-degree change in temperature. This property is extremely important in processes such as casting, where heat must be applied to melt and then extracted rapidly to promote solidification, or heat treatment, where large quantities of material are heated and cooled. **Thermal conductivity** measures the rate at which heat can be transported through a material. Although thermal conductivity is usually tabulated separately in reference texts, it is helpful to remember that for metals, thermal conductivity is directly proportional to electrical conductivity. Metals such as copper, gold, and aluminum that possess good electrical conductivity are also good transporters of thermal energy. **Thermal expansion** is another important thermal property. Most materials expand on heating and contract on cooling, but the amount of expansion or contraction will vary with the material. For components that are machined at room temperature but put in service at elevated temperatures, or castings that solidify at elevated temperatures and then cool to room temperature, the manufactured dimensions must be adjusted to compensate for the subsequent changes.

Electrical conductivity or **electrical resistivity** may be another important design consideration. These properties will vary not only with the material, but also with the temperature and the way the material has been processed.

The **magnetic response** of materials can be classified as diamagnetic, paramagnetic, ferromagnetic, antiferromagnetic, or ferrimagnetic. These terms refer to the way in which the material responds to an applied magnetic field. Material properties, such as saturation strength, remanence, and magnetic hardness or softness, describe the strength, duration, and nature of this response.

Still other physical properties that may assume importance include *weight* or **density**, **melting**, and **boiling points**, and the various **optical properties**, such as the ability to transmit, absorb, or reflect light or other electromagnetic radiation.

2.8   Testing Standards and Testing Concerns

When evaluating the mechanical and physical properties of materials, it is important that testing be conducted in a consistent and reproducible manner. **ASTM International**, formerly the American Society of Testing and Materials, maintains and updates many testing standards, and it is important to become familiar with their contents. For example, ASTM specification E370 describes the "Standard Test Methods and Definitions for Mechanical Testing of Steel Products." Tensile testing is described in specifications E8 and E83, impact testing in E23, creep in E139, and penetration hardness in E10. Other specifications describe fracture mechanics testing, as well as procedures to evaluate corrosion resistance, compressive strength, shear strength, torsional properties, and corrosion-fatigue.

In addition, it is important to note not only the material being tested but also the location from which the specimen was taken and its orientation. Rolled sheet, rolled plate, and rolled bars, for example, will have different properties when tested parallel to the direction of rolling (longitudinal) and perpendicular to the rolling direction (transverse). See Figure 2.35 for an example. This variation of properties with direction is known as **anisotropy** and may be crucial to the success or failure of a product.

Review Questions

1. What eras in the history of mankind have been linked to materials?
2. Provide two definitions of the term *manufacturing*.
3. Knowledge of what four aspects and their interrelations is critical to the successful application of a material in an engineering design?
4. Give an example of how we might take advantage of a material that has a range of properties.
5. What are some of the possible property requirements for a product or component?

6. What are some properties commonly associated with metallic materials?

7. What are some of the more common nonmetallic engineering materials?

8. What are some of the important physical properties of materials?

9. Why should caution be exercised when applying the results from any of the standard mechanical property tests?

10. What are the standard units used to report stress and strain in the English system? In the metric or SI system?

11. What are static properties?

12. What is the most common static test to determine mechanical properties?

13. What is engineering stress? Engineering strain? Why is it important to present data in this form?

14. What is Young's modulus or stiffness, and why might it be an important material property?

15. What are some of the tensile test properties that are used to describe or define the elastic-to-plastic transition in a material?

16. Why is it important to specify the "offset" when providing yield strength data?

17. How is the offset yield strength determined?

18. During the plastic deformation portion of a tensile test, a cylindrical specimen first maintains its cylindrical shape (increasing in length and decreasing in diameter) then transitions into a state called "necking." What is the explanation for this behavior?

19. What are the test conditions associated with tensile strength or ultimate tensile strength?

20. How would the tensile test curves differ for a ductile material and a brittle material?

21. What are two tensile test properties that can be used to describe the ductility of a material?

22. What is uniform elongation, and when might it be preferred to the normal elongation at fracture?

23. Is a brittle material a weak material? What does "brittleness" mean?

24. What is the toughness of a material, and how might the tensile test provide insight?

25. What is the difference between true stress and engineering stress? True strain and engineering strain?

26. Explain how the plastic portion of a true stress–true strain curve can be viewed as a continuous series of yield strength values.

27. What is strain hardening or work hardening? How might this phenomenon be measured or reported? How might strain hardening be used when manufacturing products?

28. Give examples of applications utilizing high damping capacity. Low damping capacity.

29. How might tensile test data be misleading for a "strain rate sensitive" material?

30. What type of tests can be used to determine the mechanical properties of brittle materials?

31. What are some of the different material characteristics or responses that have been associated with the term *hardness*?

32. What units could be applied to the Brinell hardness number?

33. Although the Brinell hardness test is simple and easy to conduct, what are some of its negatives or limitations?

34. What are the similarities and differences between the Brinell and Rockwell hardness tests?

35. Why are there different Rockwell hardness scales?

36. How might hardness tests be used for quality control purposes?

37. What are the attractive features of the Vickers hardness test?

38. When might a microhardness test be preferred over the more-standard Brinell or Rockwell tests?

39. What is the attractive feature of the Knoop microhardness indenter compared to the Vickers?

40. Why might the various types of hardness tests fail to agree with one another?

41. What is the relationship between penetration hardness and the ultimate tensile strength for steel?

42. Describe several types of dynamic loading.

43. Why should the results of standardized dynamic tests be applied with considerable caution?

44. What are the two most common types of bending impact tests? How are the specimens supported and loaded in each?

45. What aspects or features can significantly alter impact data?

46. What is "notch-sensitivity," and how might it be important in the manufacture and performance of a product?

47. Which type of dynamic condition accounts for almost 90% of metal failures?

48. Are the stresses applied during a fatigue test above or below the yield strength (as determined in a tensile test)?

49. Is a fatigue S–N curve determined from a single test specimen or a series of identical specimens?

50. What is the endurance limit? What occurs when stresses are above it? Below it?

51. What features may significantly alter the fatigue lifetime or fatigue behavior of a material?

52. What relationship can be used to estimate the endurance limit of a steel?

53. Describe the growth of a fatigue crack.

54. What material, design, or manufacturing features can contribute to the initiation of a fatigue crack?

55. How might the relative sizes of the fatigue region and the overload region provide useful information about the design of the product?

56. What are fatigue striations, and why do they form?

57. Why is it important for a designer or engineer to know a material's properties at all possible temperatures of operation?

58. What mechanical property changes are typically observed when temperature is increased?

59. Why should one use caution when using steel at low (below zero Fahrenheit) temperature?

60. What metals exhibit a ductile-to-brittle transition on cooling? Which do not?

61. How might the orientation of a piece of metal (with respect to its rolling direction) affect properties such as fracture resistance?

62. How might we evaluate the long-term effect of elevated temperature on an engineering material?

63. What two pieces of significant data are obtained from a single creep test?

64. What is a stress–rupture diagram, and how is one developed?

65. Why are terms such as machinability, formability, and weldability considered to be poorly defined and therefore quite nebulous?

66. What is the basic premise of the fracture mechanics approach to testing and design?

67. What are some of the types of flaws or defects that might be present in a material?

68. What three principal quantities does fracture mechanics attempt to relate?

69. What is a dormant flaw? A dynamic flaw? How do these features relate to the former "detection = rejection" criteria for product inspection?

70. How is fracture mechanics applied to fatigue conditions?

71. What are the three most common thermal properties of a material, and what do they measure?

72. Describe an engineering application where the density of the selected material would be an important material consideration.

73. Why is it important that property testing be performed in a standardized and reproducible manner?

74. Why is it important to consider the orientation of a test specimen with respect to the overall piece of material?

Problems

1. Select a product or component for which physical properties are more important than mechanical properties.

    a. Describe the product or component and its function.

    b. What are the most important properties or characteristics?

    c. What are the secondary properties or characteristics that would also be desirable?

2. Repeat Problem 1 for a product or component whose dominant required properties are of a static mechanical nature.

3. Repeat Problem 1 for a product or component whose dominant requirements are dynamic mechanical properties.

4. A fuel tanker or railroad tanker car has been involved in an accident beneath a bridge or highway overpass, resulting in an intense heat fire. Concern has been expressed that the exposure to extreme elevated temperature has weakened the steel structure of the bridge or overpass. Has the steel been sufficiently weakened that the bridge or overpass should be closed or the flow of traffic restricted? Describe several ways in which hardness tests could be conducted (or used in a modified form) to provide a quick, on-site, determination of the current strength of the steel.

5. One of the important considerations when selecting a material for an application is to determine the highest and lowest operating temperature along with the companion properties that must be present at each extreme. The ductile-to-brittle transition temperature, discussed in Section 2.4, has been an important factor in a number of failures. An article that summarized the features of 56 catastrophic brittle fractures that made headline news between 1888 and 1956 noted that low temperatures were present in nearly every case. The water temperature at the time of the sinking of the *Titanic* was above the freezing point for salt water, but below the transition point for the steel and rivets used in construction of the hull of the ship.

    a. Which of the common engineering materials exhibits a ductile-to-brittle transition?

    b. For plain carbon and low-alloy steels, what is a typical value (or range of values) for the transition temperature?

    c. What type of material would you recommend for construction of a small vessel to transport liquid nitrogen within a building or laboratory?

    d. Figure 2.35 summarizes the results of impact testing performed on hull plate from the *RMS Titanic* and similar material produced for modern steel-hulled ships. Why should there be a difference between specimens cut longitudinally (along the rolling direction) and transversely (across the rolling direction)? What advances in steel making have led to the significant improvement in low-temperature impact properties?

6. Several of the property tests described in this chapter produce results that are quite sensitive to the presence or absence of notches or other flaws. The fracture mechanics approach to materials testing incorporates flaws into the tests and evaluates their performance. The review article mentioned in question 4 earlier, cites the key role of a flaw or defect in nearly all of the headline-news fractures.

    a. What are some of the various "flaws or defects" that might be present in a product? Consider flaws that might be present in the starting material, flaws that might be introduced during manufacture, and flaws that might occur due to shipping, handling, use, maintenance, or repair.

    b. What particular properties might be most sensitive to flaws or defects?

    c. Discuss the relationship of flaws to the various types of loading (tension vs. compression, torsion, shear).

    d. Fracture mechanics considers both surface and interior flaws and assigns terms such as "crack initiator," "crack propagator," and "crack arrestor." Briefly discuss why location and orientation may be as important as the physical size of a flaw.

7. a. Would elastic flexing (without permanent deformation) be an asset or a liability to a golf club shaft?

    b. If an asset, how might the amount of flexing be controlled through material and/or design?

    c. Repeat a and b for another piece of sporting equipment, such as: baseball bat, tennis racket, hockey stick, or snow ski.

CHAPTER 3

Nature of Materials

3.1 Structure—Property—Processing—Performance Relationships

The success of many manufactured products depends on the selection of materials whose properties meet the requirements of the application. Primitive cultures were often limited to the naturally occurring materials in their environment. As civilizations developed, the spectrum of construction materials expanded. Materials could now be processed and their properties altered and improved. The alloying or heat treatment of metals and the firing of ceramics are examples of techniques that can substantially alter the properties of a material. Fewer compromises were required, and enhanced design possibilities emerged. Products, in turn, became more sophisticated. Although the early successes in altering materials were largely the result of trial and error, we now recognize that the **properties** and **performance** of a material are a direct result of its **structure** and **processing**. If we want to change the properties, we will most likely have to induce changes in the material structure.

Because all materials are composed of the same basic components—particles that include *protons*, *neutrons*, and *electrons*—it is amazing that so many different materials exist with such widely varying properties. This variation can be explained, however, by the many possible combinations these units can assume in a macroscopic assembly. The subatomic particles combine in different arrangements to form the various elemental *atoms*, each having a nucleus of protons and neutrons surrounded by the proper number of electrons to maintain charge neutrality. The specific arrangement of the electrons surrounding the nucleus affects the electrical, magnetic, thermal, and optical properties as well as the way the atoms bond to one another. Atomic bonding then produces a higher level of structure, which may be in the form of a *molecule, crystal,* or *amorphous aggregate*. This structure, along with the imperfections that may be present, has a profound effect on the mechanical properties. The size, shape, and arrangement of multiple crystals, or the mixture of two or more different structures within a material, produce a higher level of structure, known as **microstructure**. Variations in microstructure further affect the material properties.

Because of the ability to control structures through processing, and the ability to develop new structures through techniques such as composite materials, engineers now have at their disposal a wide variety of materials with an almost unlimited range of properties. The specific properties of these materials depend on all levels of structure, from subatomic to macroscopic (**Figure 3.1**). This chapter will attempt to develop an understanding of the basic structure of engineering materials and how changes in that structure affect their properties and performance.

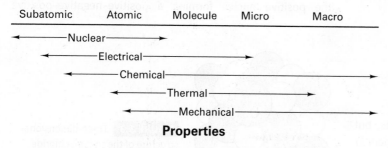

Structure Levels

Properties

FIGURE 3.1 General relationships between structural level and the various types of engineering properties.

3.2 The Structure of Atoms

Experiments have revealed that atoms consist of a relatively dense nucleus composed of positively charged protons and neutral particles of nearly identical mass, known as neutrons. Surrounding the nucleus are the negatively charged electrons, which appear in numbers equal to the protons to maintain a neutral charge balance. Distinct groupings of these basic particles produce the known elements, ranging from the relatively simple hydrogen atom to the unstable transuranium atoms over 250 times as heavy. Except for density and specific heat, however, the weight of atoms has very little influence on their engineering properties.

The light electrons that surround the nucleus, however, play an extremely significant role. These electrons are arranged in a characteristic structure consisting of shells and subshells, each of which can contain only a limited number of electrons. The first shell, nearest the nucleus, can contain only 2. The second shell can contain 8, the third, 18, and the fourth, 32. Each shell and subshell is most stable when it is completely filled. For atoms containing electrons in the third shell and beyond, however, relative stability is achieved with eight electrons in the outermost layer or subshell.

If an atom has slightly less than the number of outer-layer electrons required for stability, it will readily accept electrons from another source. It will then have more electrons than protons and becomes a negatively charged atom, or **negative ion**. Depending on the number of additional electrons, **ions** can have negative charges of 1, 2, 3, or more. Conversely, if an atom has a slight excess of electrons beyond the number required for stability (such as sodium, with one electron in the third shell), it will readily give up the excess and become a **positive ion**. The remaining electrons become more strongly attached, so further removal of electrons becomes progressively more difficult.

The number of electrons surrounding the nucleus of a neutral atom is called the **atomic number**. More important, however, are those electrons in the outermost shell or subshell, which are known as **valence electrons**. These are influential in determining chemical properties, electrical conductivity, some mechanical properties, the nature of interatomic bonding, the atom size, and optical characteristics. Elements with similar electron configurations in their outer shells tend to have similar properties.

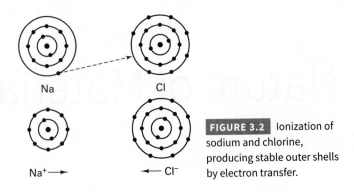

FIGURE 3.2 Ionization of sodium and chlorine, producing stable outer shells by electron transfer.

Three types of **primary bonds** are generally recognized, the simplest of which is the **ionic bond**. If more than one type of atom is present, the outermost electrons can break free from atoms with excesses in their valence shell, transforming them into positive ions. These electrons then transfer to atoms with deficiencies in their outer shell, converting them into negative ions. The positive and negative ions have an electrostatic attraction for each other, resulting in a strong bonding force. **Figure 3.2** presents a crude schematic of the ionic bonding process for sodium and chlorine. Ionized atoms do not usually unite in simple pairs, however. All positively charged atoms attract all negatively charged atoms. Therefore, each sodium ion will attempt to surround itself with negative chlorine ions, and each chlorine ion will attempt to surround itself with positive sodium ions. Because the attraction is equal in all directions, the result will be a three-dimensional structure, like the one shown in **Figure 3.3**. Because charge neutrality must be maintained within the structure, equal numbers of positive and negative charges must be present in each neighborhood. General characteristics of materials joined by ionic bonds include high density, moderate to high strength, high hardness, brittleness, high melting point, and low electrical and thermal conductivities (because all electrons are captive to specific atoms, movement of electrical charge requires movement of entire atoms or ions).

A second type of primary bond is the **covalent bond**. Here the atoms in the assembly find it impossible to produce completed shells by electron transfer but achieve the same goal through electron sharing. Adjacent atoms share outer-shell electrons so that each achieves a stable electron configuration. The shared (negatively charged) electrons locate between the positive nuclei, forming a positive–negative–positive

3.3 Atomic Bonding

Atoms are rarely found as free and independent units, but are usually linked or bonded to other atoms in some manner as a result of interatomic attraction. The electron structure of the atoms plays a dominant role in determining the nature of the bond.

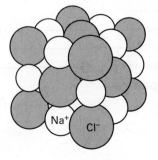

FIGURE 3.3 Three-dimensional structure of the sodium chloride crystal. Note how the various ions are surrounded by ions of the opposite charge.

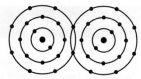

FIGURE 3.4 Formation of a chlorine molecule by the electron sharing of a covalent bond.

bonding link. **Figure 3.4** illustrates this type of bond for a pair of chlorine atoms, each of which contains seven electrons in the valence shell. The result is a stable two-atom molecule, Cl_2. Stable molecules can also form from the sharing of more than one electron from each atom, as in the case of nitrogen (**Figure 3.5a**). The atoms in the assembly need not be identical (as in HF, **Figure 3.5b**), the sharing does not have to be equal, and a single atom can share electrons with more than one other atom. For atoms such as carbon and silicon, with four electrons in the valence shell, one atom may share its valence electrons with each of four neighboring atoms. The resulting structure is a three-dimensional network of bonded atoms, like the one shown in **Figure 3.5c**. Each atom is the center of a four-atom tetrahedron formed by its four neighbors as shown in **Figure 3.5d**. Because each atom wants only four neighbors, carbon and silicon materials, such as polymers or plastics and computer chips, tend to be light in weight. The covalent bond tends to produce materials with high strength and high melting point. Because atom movement within the three-dimensional structure (plastic deformation) requires the breaking of discrete bonds, covalent materials are characteristically brittle. Electrical conductivity depends on bond strength, ranging from conductive tin (weak covalent bonding), through semiconductive silicon and germanium, to insulating diamond (carbon). Ionic or covalent bonds are commonly found in ceramic and polymeric materials.

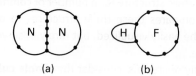

(a)   (b)

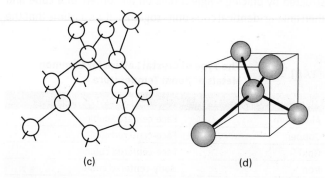

(c)   (d)

FIGURE 3.5 Examples of covalent bonding in (a) nitrogen molecule, (b) HF, and (c) silicon. Part (d) shows the tetrahedron formed by a silicon atom and its four neighbors.

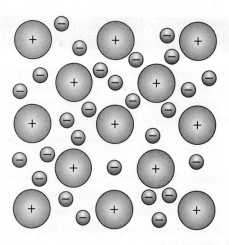

FIGURE 3.6 Schematic of the metallic bond showing the positive ions and free electrons.

A third type of primary bond is possible when a complete outer shell cannot be formed by either electron transfer or electron sharing. This bond is known as the **metallic bond** (**Figure 3.6**). If each of the atoms in an aggregate has only a few valence electrons (one, two, or three), these electrons can be easily removed to produce "stable" ions. The positive ions (nucleus and inner nonvalence electrons) then arrange in a three-dimensional periodic array and are surrounded by wandering, universally shared, valence electrons, sometimes referred to as an electron cloud or electron gas. These highly mobile, free electrons account for the high electrical and thermal conductivity values as well as the opaque (nontransparent) characteristic observed in metals (the free electrons are able to absorb the various discrete energies of light radiation). They also provide the "cement" required for the positive–negative–positive attractions that result in bonding. Bond strength, and therefore material strength and melting temperature, varies over a wide range. More significant, however, is the observation that the positive ions can now move within the structure without the breaking of discrete bonds. Materials bonded by metallic bonds can be deformed by atom-movement mechanisms and produce an altered-shape that is every bit as strong as the original. This phenomenon is the basis of metal plasticity, enabling the wide variety of forming processes used in the fabrication of metal products.

3.4 Secondary Bonds

Weak or **secondary bonds**, known as **van der Waals forces**, can form between molecules that possess a nonsymmetrical distribution of electrical charge. Some molecules, such as hydrogen fluoride and water, can be viewed as electric dipoles. Certain portions of the molecule tend to be more positive or more negative than others (an effect referred to

Why Do Water Molecules Have an Attraction for One Another?

The H_2O molecule can be viewed as a 109° boomerang or elbow with oxygen in the middle and the two hydrogen atoms on the extending arms. The eight valence electrons (six from oxygen and two from hydrogen) associate with oxygen, giving it a negative charge. The *hydrogen arms have lost their valence electron and are positive. Therefore, when two or more water molecules are present, the positive hydrogen locations of one molecule are attracted to the negative oxygen location of an adjacent molecule.*

as **polarization**). The negative part of one molecule tends to attract the positive region of another, forming a weak bond. Van der Waals forces contribute to the mechanical properties of a number of molecular polymers, such as polyethylene and polyvinyl chloride (PVC).

3.5 Atom Arrangements in Materials

As atoms bond together to form larger aggregates, we find that the particular arrangement of the atoms has a significant effect on the material properties. Depending on the manner of atomic grouping, materials are classified as having **molecular structures**, **crystal structures**, or **amorphous structures**.

Molecular structures have a distinct number of atoms that are held together by primary bonds. There is only a weak attraction, however, between a given molecule and other similar groupings. Typical examples of molecules include O_2, H_2O, and C_2H_4 (ethylene). Each molecule is free to act more or less independently, so these materials exhibit relatively low melting and boiling points. Molecular materials tend to be weak because the molecules can move easily with respect to one another. Upon changes of state from solid to liquid or liquid to gas, the molecules remain as distinct entities.

Solid metals and most minerals have a crystalline structure. Here the atoms are arranged in a three-dimensional geometric array known as a **lattice**. Lattices are describable through a unit building block, or **unit cell**, that is essentially repeated throughout space. Crystalline structures will be discussed in detail in the following section.

In an amorphous structure, such as glass, the atoms have a certain degree of local order (arrangement with respect to neighboring atoms), but when viewed as an aggregate, they lack the periodically ordered arrangement that is characteristic of a crystalline solid.

3.6 Crystal Structures

From a manufacturing viewpoint, metals are an extremely important class of materials. They are frequently the materials being processed and often form both the tool and the machinery performing the processing. They are characterized by the metallic bond and possess the distinguishing characteristics of strength, good electrical and thermal conductivity, luster, the ability to be plastically deformed to a fair degree without fracturing, and a relatively high specific gravity or density compared to nonmetals. The fact that some metals possess properties different from the general pattern simply expands their engineering utility.

When metals solidify, the atoms assume a crystalline structure; that is, they arrange themselves in a geometric lattice. Many metals exist in only one lattice form. Some, however, can exist in the solid state in two or more lattice forms, with the particular form depending on the conditions of temperature and pressure. These metals are said to be **allotropic** or **polymorphic** (poly means "more than one"; morph means "structure"), and the change from one lattice form to another is called an **allotropic transformation**. The most notable example of such a metal is iron, where the allotropic change makes it possible for heat-treating procedures that yield a wide range of final properties. It is largely because of its allotropy that iron has become the basis of our most important alloys.

There are 14 basic types of crystal structures or lattices. Fortunately, however, nearly all of the commercially important metals solidify into one of three lattice types: body-centered cubic, face-centered cubic, or hexagonal close-packed. **Table 3.1** lists the room temperature structure for a number of common metals. **Figure 3.7** compares these three structures to one another, along with the easily visualized, but rarely observed, simple cubic structure.

To begin our study of crystals, consider the **simple cubic** structure illustrated in **Figure 3.7a**. This crystal can be constructed by placing single atoms on all corners of a cube and then linking identical cube units together. If we assume that the

TABLE 3.1	The Type of Crystal Lattice for Common Metals at Room Temperature
Metal	**Lattice Type**
Aluminum	Face-centered cubic
Copper	Face-centered cubic
Gold	Face-centered cubic
Iron	Body-centered cubic
Lead	Face-centered cubic
Magnesium	Hexagonal
Silver	Face-centered cubic
Tin	Body-centered tetragonal
Titanium	Hexagonal

	Lattice structure	Unit cell schematic	Unit cells	Number of nearest neighbors	Packing efficiency	Typical metals
a	Simple cubic	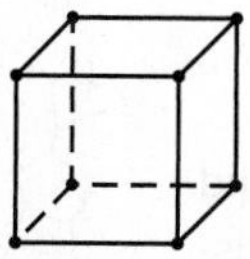	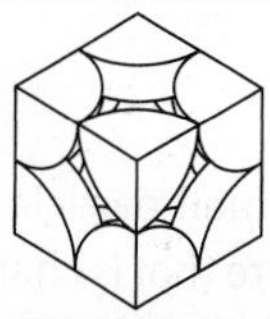	6	52%	None
b	Body-centered cubic	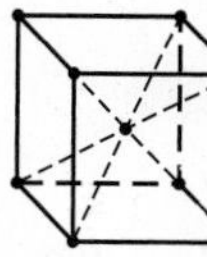		8	68%	Fe, Cr, Mn, Cb, W, Ta, Ti, V, Na, K
c	Face-centered cubic	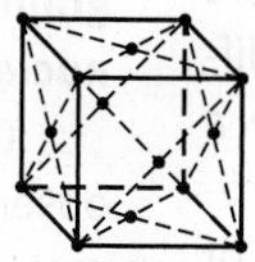		12	74%	Fe, Al, Cu, Ni, Ca, Au, Ag, Pb, Pt
d	Hexagonal close-packed	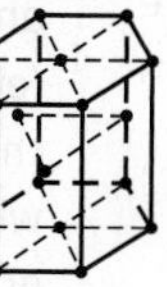	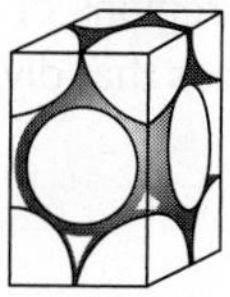	12	74%	Be, Cd, Mg, Zn, Zr

FIGURE 3.7 Comparison of crystal structures: simple cubic, body-centered cubic, face-centered cubic, and hexagonal close-packed.

atoms are rigid spheres with atomic radii touching one another, computation reveals that only 52% of available space is occupied. Each atom is in direct contact with only six neighbors (plus and minus in each of the x, y, and z cube edge directions). Both of these observations are unfavorable to the metallic bond, where atoms desire both a high number of immediate neighbors and high-efficiency packing.

The largest region of unoccupied space is in the geometric center of the cube, where a sphere of 0.732 times the atom diameter could be inserted.[1] If the cube is expanded to permit the insertion of an entire atom, the **body-centered-cubic (BCC)** structure results (**Figure 3.7b**). Each atom now has eight nearest neighbors, and 68% of the space is occupied. This structure is more favorable to metals and is observed in room temperature iron, chromium, manganese, and the other metals listed in Figure 3.7b. Compared to materials with other structures, body-centered-cubic metals tend to be high strength.

In seeking efficient packing and a large number of adjacent neighbors, consider maximizing the number of spheres in a single layer and then stacking those layers. The layer of maximized packing is known as a **close-packed plane** and exhibits the hexagonal symmetry shown in **Figure 3.8**. The next layer is positioned with its spheres occupying either the "point-up" or "point-down" triangular recesses in the original layer. Depending on the sequence in which the various layers are stacked, two distinctly different structures can be produced. Both have twelve nearest neighbors (six within the original plane and

three from each of the layers above and below) and a 74% efficiency of occupying space.

If the layers are stacked in sets of three (original location, point up recess of the original layer, and point down recess of the original layer), rotation of the resulting structure yields cubic symmetry where an atom has been inserted into the center of each of the six cube faces, like a dice with the number five on each of its sides. (The close-packed planes are those formed by connecting any three cube-face diagonals.) This is the **face-centered-cubic (FCC)** structure shown in **Figure 3.7c**. It is the observed structure for many engineering metals and tends to provide the exceptionally high ductility (ability to be plastically deformed without fracture) that is characteristic of aluminum, copper, silver, gold, and elevated temperature iron.

A stacking sequence alternating any two of the close-packed layers results in a structure known as **hexagonal close-packed (HCP)**, where the individual close-packed planes can be clearly identified (**Figure 3.7d**). Metals having this structure, such as magnesium and zinc, tend to have poor ductility, fail in a brittle manner, and often require special processing procedures.

---

[1] The diagonal of a cube is equal to the square root of three times the length of the cube edge, and the cube edge is here equal to two atomic radii or one atomic diameter. Thus the diagonal is equal to 1.732 times the atom diameter and is made up of an atomic radius, open space, and another atomic radius. Because two radii equals one diameter, the open space must be equal in size to 0.732 times the atomic diameter.

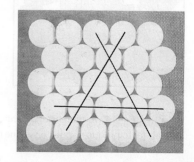

FIGURE 3.8 Close-packed atomic plane showing three directions of atom touching or close-packing. (Courtesy Ronald Kohser)

3.7 Development of a Grain Structure

When a crystalline material solidifies, a small particle of solid forms from the liquid with a lattice structure that is characteristic of the material. This particle then acts like a seed or nucleus and grows as other atoms attach themselves. The basic crystalline unit, or unit cell, is repeated, as illustrated in **Figure 3.9**.

In actual solidification, many nuclei form independently throughout the liquid and have random orientations with respect to one another. Each then grows until it encounters its neighbors. Because the adjacent lattice structures have different alignments or orientations, growth cannot produce a single continuous structure, and a polycrystalline solid is produced. **Figure 3.10** provides a two-dimensional illustration of this phenomenon. The small, continuous regions of solid are known as crystals or **grains**, and the surfaces that divide them

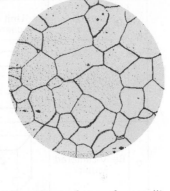

FIGURE 3.11 Photomicrograph of polycrystalline iron showing grains and grain boundaries. (Courtesy Ronald Kohser)

(i.e., the surfaces of crystalline discontinuity) are known as **grain boundaries**. The process of solidification is one of crystal **nucleation and growth**.

Grains are the smallest unit of structure that can be observed with an ordinary light microscope. If a piece of solid material is polished to mirror finish with a series of abrasives and then exposed to an attacking chemical for a short time (etched), the grain structure can be revealed. The atoms along the grain boundaries are more loosely bonded and tend to react with the chemical more readily than those that are part of the grain interior. When viewed under reflected light, the attacked boundaries scatter light and appear dark compared to the relatively unaffected (still flat) grains, as shown in **Figure 3.11**. In some cases, the individual grains may be large enough to be seen by the unaided eye, as with some galvanized steels, but magnification is usually required.

The number and size of the grains vary with the rate of nucleation and the rate of growth. The greater the nucleation rate, the smaller the resulting grains. The greater the rate of growth, the larger the grains. Because the resulting grain structure will influence certain mechanical and physical properties, it is an important property to control and specify. One means of specification is through the **ASTM grain size number**, defined in ASTM specification E112 as

$$N = 2^{n-1}$$

where N is the number of grains per square inch visible in a prepared specimen at 100X magnification, and n is the ASTM grain size number. Low ASTM numbers mean a few massive grains, whereas high numbers refer to materials with many small grains.

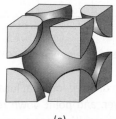

(a)

(b)

FIGURE 3.9 Growth of crystals to produce an extended lattice: (a) unit cell; (b) multicell aggregate. (Courtesy Ronald Kohser)

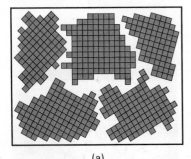

(a)

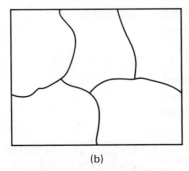

(b)

FIGURE 3.10 Two-dimensional schematic representation of the growth of crystals (a) to produce a polycrystalline material (b).

3.8 Elastic Deformation

The mechanical properties of a material are strongly dependent on its crystal structure. An understanding of mechanical behavior, therefore, begins with an understanding of the way crystals react to mechanical loads. Most studies begin with carefully prepared single crystals. Through them, we learn that the mechanical behavior depends on (1) the

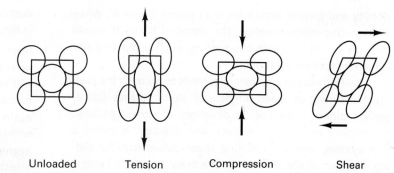

FIGURE 3.12 Distortion of a crystal lattice in response to various elastic loadings.

Unloaded   Tension   Compression   Shear

type of lattice, (2) the interatomic forces (i.e., bond strength), (3) the spacing between adjacent planes of atoms, and (4) the density of the atoms on the various planes.

If the applied loads are relatively low, the crystals respond by simply stretching or compressing the distance between adjacent atoms (**Figure 3.12**). The basic lattice does not change, and all the atoms remain in their original positions relative to one another. The applied load serves only to alter the force balance of the atomic bonds, and the atoms assume new equilibrium positions with the applied load as an additional component of force. If the load is removed, the atoms return to their original positions, and the crystal resumes its original size and shape. The mechanical response is **elastic** in nature, and the amount of stretch or compression is directly proportional to the applied load or stress.

Elongation or compression in the direction of loading results in an opposite change of dimensions at right angles to that direction. The ratio of lateral contraction to axial stretching is known as **Poisson's ratio**. This value is always less than 0.5 and is usually about 0.3.

3.9 Plastic Deformation

As the magnitude of applied load becomes greater, distortion (or elastic strain) continues to increase, and a point is reached where the atoms either (1) break bonds to produce a fracture, or (2) slide over one another in a way that would reduce the load. For metallic materials, the second phenomenon generally requires lower loads and occurs preferentially. The atomic planes shear over one another to produce a net displacement or permanent shift of atom positions, known as **plastic deformation**. Conceptually, this is similar to the distortion of a deck of playing cards from rectangular to parallelogram as the cards slide over each other. Because the freely moving electrons of the metallic bond continue to cement the structure together, the result is a permanent change in shape that occurs without a concurrent deterioration in properties.

Consider the close-packed plane of Figure 3.8, with the three directions of atom touching (the close-packed directions) identified by bold lines. If we were to look across the top surface of this plane along one of the three close-packed

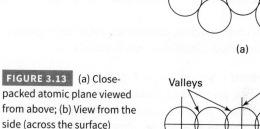

(a)

FIGURE 3.13 (a) Close-packed atomic plane viewed from above; (b) View from the side (across the surface) showing the ridges and valleys that lie in directions of close-packing.

Valleys   Ridges

(b)

directions, as in **Figure 3.13**, we would see a series of parallel ridges. The point-up or point-down depressions, which became the sites for the next layer of atoms, lie along valleys that parallel the ridges. If the upper layer were to slide in one of the ridge directions, its atoms would simply traverse the valleys and would encounter little resistance. Movement in any other direction would require atoms to climb over the ridges, requiring a greater applied force. Deformation, therefore, prefers to occur by movement along close-packed planes in directions of atom touching. If close-packed planes are not available within the crystal structure, plastic deformation tends to occur along planes having the highest atomic density and greatest separation. The rationale for this can be seen in the simplified two-dimensional array of **Figure 3.14**. Planes A and A' have higher

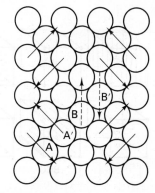

FIGURE 3.14 Simple schematic illustrating the lower deformation resistance of planes with higher atomic density and larger interplanar spacing.

density and greater separation than planes B and B'. When experiencing relative motion, the atoms of B and B' would interfere significantly with one another, whereas planes A and A' do not experience this difficulty.

Plastic deformation, therefore, tends to occur by the preferential sliding of maximum-density atomic planes (close-packed planes if present) in directions of closest packing. A specific combination of plane and direction is called a **slip system**, and the resulting shear deformation or sliding is known as **slip**. The ability of a metal to deform along a given slip system depends on the ease of shearing along that system and the orientation of the plane with respect to the applied load. Consider a deck of playing cards. The deck will not "deform" when laid flat on the table and pressed from the top, or when stacked on edge and pressed uniformly. The cards will slide over one another, however, if the deck is skewed with respect to the applied load so as to induce a shear stress along the plane of sliding.

With this understanding, consider the deformation properties of the three most common crystal structures:

1. *Body-centered cubic.* In the BCC structure, there are no close-packed planes. Slip occurs on the most favorable alternatives, which are those planes with the greatest interplanar spacing (six of which are illustrated in **Figure 3.15**). Within these planes, slip occurs along the directions of closest packing, which are the diagonals through the body of the cube. If each specific combination of plane and direction is considered as a separate slip system, we find that the BCC materials contain 48 attractive ways to slip (plastically deform). The probability that one or more of these systems will be oriented in a favorable manner is great, but the force required to produce deformation is extremely large because there are no close-packed planes. As a result, materials with this structure generally possess high strength with moderate ductility. (Refer to the typical BCC metals in Figure 3.7.)

2. *Face-centered cubic.* In the FCC structure, each unit cell contains four close-packed planes, as illustrated in Figure 3.15. Each of those planes contains three close-packed directions, the diagonals along the cube faces, giving 12 possible means of slip. The probability that one or more of these will be favorably oriented is great, and for this structure, the force required to induce slip is quite low. Metals with the FCC structure are relatively weak and possess excellent

ductility, as can be confirmed by a check of the metals listed in Figure 3.7.

3. *Hexagonal close-packed.* The hexagonal lattice also contains close-packed planes, but only one such plane exists within the lattice. Although this plane contains three close-packed directions and the force required to produce slip is again rather low, the probability of favorable orientation with respect to the applied load is small. In a polycrystalline aggregate, the majority of the crystals will be unfavorably oriented. As a result, metals with the HCP structure tend to have low ductility and are often classified as brittle.

3.10 Dislocation Theory of Slippage

A theoretical calculation of the strength of metals based on the sliding of entire atomic planes over one another predicts yield strengths on the order of 3 million pounds per square inch or 20,000 MPa. The observed strengths in actual testing are typically 100 to 150 times lower than this value. Extremely small laboratory-grown crystals, however, have been shown to exhibit the full theoretical strength.

An explanation can be provided by the fact that plastic deformation does not occur by all of the atoms in one plane slipping simultaneously over all the atoms of an adjacent plane. Instead, deformation is the result of the progressive slippage of a localized disruption known as a **dislocation**. Consider a simple analogy. A carpet has been rolled onto a floor, and we now want to move it a short distance in a given direction. One approach would be to pull on one end and try to "shear the carpet across the floor," simultaneously overcoming the frictional resistance of the entire area of contact. This would require a large force acting over a small distance. An alternative approach might be to form a wrinkle at one end of the carpet and walk the wrinkle across the floor to produce a net shift in the carpet as a whole—a low-force-over-large distance approach to the same task. In the region of the wrinkle, there is an excess of carpet with respect to the floor beneath it, and the movement of this excess is relatively easy.

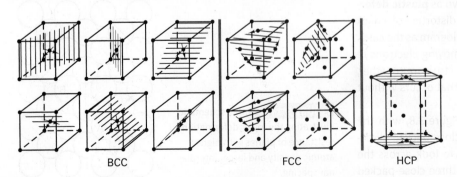

BCC   FCC   HCP

FIGURE 3.15 Slip planes within the BCC, FCC, and HCP crystal structures.

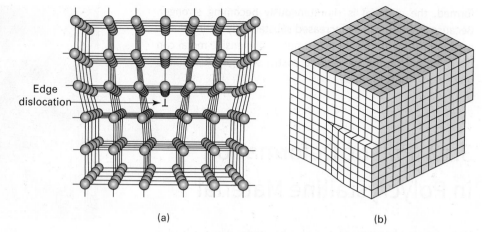

FIGURE 3.16 Schematic representation of (a) edge and (b) screw dislocations. [(a) From *Elements of Physical Metallurgy*, by A.G. Guy, Addison-Wesley Publishing Co., Inc., Reading, MA, 1959; (b) Adapted from *Materials Science and Engineering*, 7th Ed., by William D. Callister Jr., John Wiley & Sons, Inc., 2007]

(a)   (b)

Electron microscopes have revealed that crystals of material do not have all of their atoms in perfect arrangement, but rather contain a variety of localized imperfections. Two such imperfections are the **edge dislocation** and **screw dislocation** (**Figure 3.16**). Edge dislocations are the edges of extra half-planes of atoms. Screw dislocations correspond to partial tearing along a crystal plane. In each case the dislocation is a disruption to the regular, periodic arrangement of atoms and can be moved about with a rather low applied force. It is the motion of these atomic-scale dislocations under applied load that is responsible for the observed macroscopic plastic deformation.

Crystalline materials contain dislocations, usually in abundant quantities. The ease of deformation depends on the ease of making them move. Barriers to dislocation motion, therefore, would tend to increase the overall strength of a metal. These barriers take the form of other crystal imperfections and may be of the point type (missing atoms or **vacancies**, extra atoms or **interstitials**, or **substitutional atoms** of a different variety, as shown in **Figure 3.17**), line type (other *dislocations*), or surface type (*crystal grain boundaries* or *free surfaces*). To increase the strength of a material, we can either remove all

defects to create a perfect crystal (nearly impossible) or work to impede the movement of existing dislocations by adding other crystalline defects (the basis of a variety of strengthening mechanisms that will be discussed later).

3.11 Strain Hardening or Work Hardening

As noted in our discussion of the tensile test in Chapter 2, most metals become stronger when they are plastically deformed, a phenomenon known as **strain hardening** or **work hardening**. Understanding of this phenomenon can now come from our knowledge of dislocations and a further extension of the carpet analogy. Suppose that our goal this time is to move the carpet diagonally. The best way would be to move a wrinkle in one direction, and then move a second one perpendicular to the first. But suppose that both wrinkles were started simultaneously. We would find that wrinkle 1 would impede the motion of wrinkle 2, and vice versa. In essence, the feature that makes deformation easy can also serve to impede the motion of other, similar dislocations.

In metals, plastic deformation occurs through dislocation movement. As dislocations move, they are more likely to encounter and interact with other dislocations or other crystalline defects, thereby producing resistance to further motion. In addition, mechanisms exist that markedly increase the number of dislocations in a metal during deformation (usually by several orders of magnitude), thereby enhancing the probability of interaction. Transmission electron microscope studies have confirmed the existence of dislocations and the slippage theory of deformation. By observing the images of individual dislocations in a thin metal section, we can see the increase in the number of dislocations and their interactions during deformation.

The effects of strain hardening become attractive when one considers that mechanical deformation is frequently used in the shaping of metal products. As the product shape is being

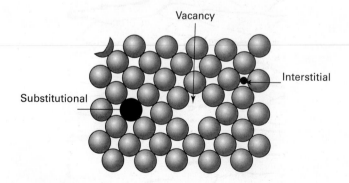

FIGURE 3.17 Two-dimensional schematic showing the various types of point defects: vacancy, interstitial, and substitutional (which can be larger or smaller than the host atoms). Because of the presence or absence of both physical size and electrical charge, local distortion will occur around each of the defects.

formed, the material is simultaneously becoming stronger. Because strength can be increased substantially during deformation, a strain-hardened (deformed), inexpensive metal can often be substituted for a more costly, stronger one that is machined or cast to shape.

3.12 Plastic Deformation in Polycrystalline Material

Macroscopic observations can also be made. When a load is applied to a single metal crystal, deformation begins on the slip system that is most favorably oriented. The net result is often an observable slip and rotation, like that of a skewed deck of cards (**Figure 3.18**). Dislocation motion becomes more difficult as strain hardening produces increased resistance, and rotation makes the slip system orientation less favorable. Further deformation may then occur on alternative systems that now offer less resistance, a phenomenon known as **cross slip**.

Commercial materials, however, are not single crystals, but usually take the form of polycrystalline aggregates. Within each crystal, deformation proceeds in the manner previously described. Because the various grains have different orientations, an applied load will produce different deformations within each of the crystals. This can be seen in **Figure 3.19**, where a metal has been polished and then deformed. The relief of the polished surface reveals the different slip planes for each of the grains.

One should note that the slip lines do not cross from one grain to another. The grain boundaries act as barriers to the dislocation motion (i.e., the defect is confined to the crystal in which it occurs). As a result, metals with a finer grain structure—more grains per unit area—tend to exhibit greater strength and hardness, coupled with increased impact resistance. This near-universal enhancement of properties is an attractive motivation for grain size control during processing.

3.13 Grain Shape and Anisotropic Properties

When a material is deformed, the grains tend to elongate in the direction of flow (**Figure 3.20**). Directionally varying properties accompany this nonsymmetric structure. Mechanical properties (such as strength and ductility), as well as physical properties (such as electrical and magnetic characteristics), may all exhibit directional differences. Properties that vary with direction are said to be **anisotropic**. Properties that are uniform in all directions are **isotropic**.

The directional variation of properties can be harmful or beneficial. By controlling metal flow in processes such as forging, enhanced strength, or fracture resistance can be imparted

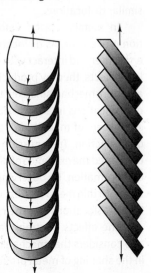

FIGURE 3.18 Schematic representation of slip and crystal rotation resulting from deformation. (From Richard Hertzberg, *Deformation and Fracture Mechanics of Engineering Materials*. Reprinted with permission of John Wiley & Sons, Inc.)

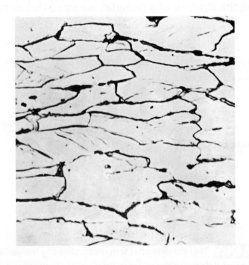

to certain locations or directions. Caution should be exercised, however, because an improvement in one direction is generally accompanied by a decline in another. Moreover, directional variation in properties may create problems during subsequent processing, such as the further forming of rolled metal sheets. For these and other reasons, both the part designer and the part manufacturer should consider the effects of directional property variations.

3.14 Fracture

When metals are deformed, strength and hardness increase, and ductility decreases. If too much plastic deformation is attempted, fracture may occur. If plastic deformation precedes the break, the fracture is known as a **ductile fracture**. Fractures can also occur when the load is below that required for plastic deformation. These sudden, catastrophic failures, known as **brittle fractures**, are characteristic of ceramic materials and are also observed in metals having the BCC or HCP crystal structure. Whether the fracture is ductile or brittle, however, often depends on the specific conditions of material, temperature, state of stress, and rate of loading.

3.15 Cold Working, Recrystallization, and Hot Working

During plastic deformation, a portion of the deformation energy is stored within the material in the form of additional dislocations and increased grain boundary surface area.[2] If a deformed polycrystalline metal is subsequently heated to a high enough temperature, the material will seek to lower its energy. Atom movement within the material (atomic diffusion) enables the nucleation and growth of new crystals that consume and replace the original structure (**Figure 3.21**). This process of reducing the internal energy through the formation of new crystals is known as **recrystallization**. The temperature at which recrystallization occurs is different for each metal and also varies with the amount of prior deformation. The greater

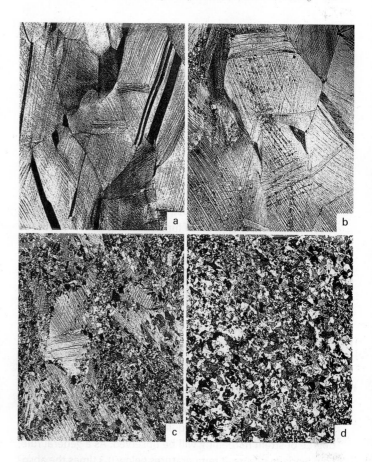

FIGURE 3.21 Recrystallization of 70–30 cartridge brass: (a) cold-worked 33%; (b) heated at 580°C (1075°F) for 3 seconds, (c) 4 seconds, and (d) 8 seconds; 45X. (Courtesy of J. E. Burke, General Electric Company, Fairfield, CT)

the amount of prior deformation, the more stored energy, and the lower the recrystallization temperature. There is a lower limit, however, below which recrystallization will not take place in a reasonable amount of time. **Table 3.2** gives the lowest practical recrystallization temperatures for several materials. This is the temperature above which atomic diffusion becomes significant, and can often be estimated by taking 0.4 times the

TABLE 3.2	The Lowest Recrystallization Temperature of Common Metals
Metal	**Temperature [°F(°C)]**
Aluminum	300 (150)
Copper	390 (200)
Gold	390 (200)
Iron	840 (450)
Lead	Below room temperature
Magnesium	300 (150)
Nickel	1100 (590)
Silver	390 (200)
Tin	Below room temperature
Zinc	Room temperature

[2] A sphere has the least amount of surface area of any shape to contain a given volume of material. When the shape becomes altered from that of a sphere, the surface area must increase. Consider a round balloon filled with air. If the balloon is stretched or flattened into another shape, the rubber balloon is stretched further. When the applied load is removed, the balloon snaps back to its original shape, the one involving the least surface energy. Metals behave in an analogous manner. During deformation, the distortion of the crystals increases the energy of the material. Given the opportunity, the material will try to lower its energy by returning to spherical grains.

melting point of the metal when the melting point is expressed as an absolute temperature (Kelvin or Rankine).

When metals are plastically deformed at temperatures below their recrystallization temperature, the process is called **cold working**. The metal strengthens by strain hardening, and the resultant structure consists of distorted grains. As deformation continues, the metal decreases in ductility and may ultimately fracture. It is a common practice, therefore, to recrystallize the material after a certain amount of cold work. Through this recrystallization anneal, the structure is replaced by one of new crystals that have never experienced deformation. All strain hardening is lost, but ductility is restored, and the material is now capable of further deformation without the danger of fracture.

If the temperature of deformation is sufficiently above the recrystallization temperature, the deformation process becomes **hot working**. Recrystallization begins as soon as sufficient driving energy is created, (i.e., deformation and recrystallization take place simultaneously), and extremely large deformations are now possible. Because a recrystallized grain structure is constantly forming, the final product will not exhibit the increased strength of strain hardening.

An intermediate condition, known as **warm working** may also be specified. Temperatures that are lower than hot working reduce the energy required for the process, but some degree of strain hardening will be observed. As a general rule, cold working occurs at temperatures below 0.3 times the absolute melting temperature; warm working between 0.3 and 0.5, and hot working above 0.5.

Recrystallization can also be used to control or improve the grain structure of a material. A coarse grain structure can be converted to a more attractive fine grain structure through recrystallization. The material must first be plastically deformed to store sufficient energy to provide the driving force. Subsequent control of the recrystallization process then establishes the more desirable final **grain size**.

3.16 Grain Growth

Recrystallization is a continuous process in which a material seeks to lower its overall energy. Ideally, recrystallization will result in a structure of uniform crystals with a comparatively small grain size. If a metal is held at or above its recrystallization temperature for any appreciable time, however, the grains in the recrystallized structure can continue to increase in size. In effect, some of the grains become larger at the expense of their smaller neighbors as the material seeks to further lower its energy by decreasing the amount of grain boundary surface area. Because engineering properties tend to diminish as the size of the grains increase, control of recrystallization is of prime importance. A deformed material should be held at elevated temperature just long enough to complete the recrystallization process. The temperature should then be decreased to stop the process and avoid the property changes that accompany **grain growth**.

3.17 Alloys and Alloy Types

Our discussion thus far has been directed toward the nature and behavior of pure metals and materials. For most manufacturing applications, however, metals are not used in their pure form. Instead, engineering metals tend to be **alloys**, materials composed of two or more different elements, and they tend to exhibit their own characteristic properties.

There are three ways in which a metal might respond to the addition of another element. The first response, and probably the simplest, occurs when the *two materials are insoluble in one another in the solid state*. In this case the base metal and the alloying addition each maintain their individual identities, structures, and properties. The alloy in effect becomes a composite structure, consisting of two types of building blocks in an intimate mechanical mixture.

The second possibility occurs when the *two elements exhibit some degree of solubility in the solid state*. The two materials can form a **solid solution**, where the alloy element dissolves in the base metal. The solutions can be: (1) *substitutional* or (2) *interstitial*. In the substitutional solution, atoms of the alloy element occupy lattice sites normally filled by atoms of the base metal. In an interstitial solution, the alloy element atoms squeeze into the open spaces between the atoms of the base metal lattice.

A third possibility exists where the *elements combine to form* **intermetallic compounds**. In this case, the atoms of the alloying element interact with the atoms of the base metal in definite proportions and in definite geometric relationships. The bonding is primarily of the nonmetallic variety (i.e., ionic or covalent), and the lattice structures are often quite complex. Because of the type of bonding, intermetallic compounds tend to be hard, but brittle, high-strength materials.

Even though alloys are composed of more than one type of atom, their structure is still one of crystalline lattices and grains. Their behavior in response to applied loadings is similar to that of pure metals, with some features reflecting the increased level of structural complexity. Dislocation movement can be further impeded by the presence of unlike atoms. If neighboring grains have different chemistries and/or structures, they may respond differently to the same type and magnitude of load.

3.18 Atomic Structure and Electrical Properties

In addition to mechanical properties, the structure of a material also influences its physical properties, such as its electrical behavior. **Electrical conductivity** refers to the net movement of charge through a material. In metals, the charge carriers are the valence electrons. The more perfect the atomic arrangement, the greater the freedom of electron movement, and the higher the electrical conductivity. Lattice imperfections or irregularities act as impediments to electron transport, and lower conductivity.

The electrical resistance of a metal, therefore, depends largely on two factors: (1) lattice imperfections and (2) temperature. Vacant atomic sites, interstitial atoms, substitutional atoms, dislocations, and grain boundaries all act as disruptions to the regularity of a crystalline lattice. Thermal energy causes the atoms to vibrate about their equilibrium position. These vibrations cause the atoms to be out of position, which further interferes with electron travel. For a metal, electrical conductivity will decrease with an increase in temperature. As the temperature drops, the number and type of crystalline imperfections becomes more of a factor. The best metallic conductors, therefore, are those with fewer defects (such as pure metals with large grain size) at low temperature.

The electrical conductivity of a metal is due to the movement of the free electrons in the metallic bond. For covalently bonded materials, however, bonds must be broken to provide the electrons required for charge transport. Therefore, the electrical properties of these materials are a function of bond strength. Diamond, for instance, has strong bonds and is a strong insulator. Silicon and germanium have weaker bonds that are more easily broken by thermal energy. These materials are known as **intrinsic semiconductors** because moderate amounts of thermal energy enable them to conduct small amounts of electricity. Continuing down Group IV of the periodic table of elements, we find that tin has such weak bonding that a high number of bonds are broken at room temperature, and the electrical behavior resembles that of a metal.

The electrical conductivity of intrinsic semiconductors can be substantially improved by a process known as **doping**. Silicon and germanium each have four valence electrons and form four covalent bonds. If one of the bonding atoms is replaced with an atom containing five valence electrons, such as phosphorus or arsenic, the four covalent bonds would form, leaving an additional valence electron that is not involved in the bonding process. This extra electron would be free to move about and provide additional conductivity. Materials doped in this manner are known as ***n*-type extrinsic semiconductors**.

A similar effect can be created by inserting an atom with only three valence electrons, such as aluminum. An electron will be missing from one of the bonds, creating an **electron hole**. When a voltage is applied, a nearby electron can jump into this hole, creating a hole in the location that it vacated. Movement of electron holes is equivalent to a countermovement of electrons and thus provides additional conductivity. Materials containing dopants with three valence electrons are known as ***p*-type extrinsic semiconductors**. The ability to control the electrical conductivity of semiconductor material is the functional basis of solid-state electronics and circuitry.

In ionically bonded materials, all electrons are captive to atoms (ions). Charge transport, therefore, requires the movement of entire atoms, not electrons. Consider a large block of salt (sodium chloride). It is a good electrical insulator, until it becomes wet, whereupon the ions are free to move in the liquid solution and conductivity is observed.

Review Questions

1. What enables us to control the properties and performance of engineering materials?

2. What are the next levels of structure that are greater than the atom?

3. What is meant by the term *microstructure*?

4. What is the most stable configuration for an electron shell or subshell?

5. What is an ion, and what are the two varieties?

6. What properties or characteristics of a material are influenced by the valence electrons?

7. What are the three types of primary bonds, and what types of atoms do they unite?

8. What are some general characteristics of ionic-bonded materials?

9. Where are the bonding electrons located in a covalent bond?

10. What are some general properties and characteristics of covalent-bonded materials?

11. Why are the covalently bonded hydrocarbon polymers light in weight?

12. Ionic and covalent bonds are most commonly found in what types of engineering materials?

13. Where are the bonding electrons located in a metallic-bonded material?

14. What are some unique property features of materials bonded by metallic bonds?

15. For what common engineering materials are van der Waals forces important?

16. What is the difference between a crystalline material and one with an amorphous structure?

17. What is a lattice? A unit cell?

18. What are some of the general characteristics of metallic materials?

19. What is an allotropic or polymorphic material?

20. Why did we elect to focus on only three of the fourteen basic crystal structures or lattices? What are those three structures?

21. Why is the simple cubic crystal structure not observed in the engineering metals?

22. What is the efficiency of filling space with spheres in the simple cubic structure? Body-centered cubic structure? Face-centered cubic structure? Hexagonal close-packed structure?

23. What is the dominant characteristic of body-centered cubic metals? Face-centered cubic metals? Hexagonal close-packed metals?

24. What is a close-packed plane? Which of the common crystal structures contain close-packed planes?

25. What is a grain? A grain boundary?

26. What is the most common means of describing or quantifying the grain size of a solid metal?

27. What is implied by a low ASTM grain-size number? A large ASTM grain-size number?

28. How does a metallic crystal respond to low applied loads?

29. What is Poisson's ratio, and under what conditions is it determined?

30. What is plastic deformation?

31. Why do metals retain their strength during plastic deformation?

32. What is a slip system in a material? What types of planes and directions tend to be preferred?

33. What structural features account for each of the dominant crystal structure properties cited in Question 23?

34. What is a dislocation? Using the carpet analogy, describe how dislocations account for the lower-than-predicted strength of metals.

35. What is the difference between an edge dislocation and a screw dislocation?

36. What are some of the common barriers to dislocation movement that can be used to strengthen metals?

37. What are the three major types of point defects in crystalline materials?

38. What is the mechanism (or mechanisms) responsible for the observed deformation strengthening or strain hardening of a metal?

39. Why is a fine grain size often desired in an engineering metal?

40. What is an anisotropic property? Why might anisotropy be a concern?

41. What is the difference between brittle fracture and ductile fracture?

42. How does a metal increase its internal energy during plastic deformation?

43. What is required in order to drive the recrystallization of a cold worked or deformed material?

44. How might the lowest recrystallization temperature of a metal be estimated?

45. In what ways can recrystallization be used to enable large amounts of deformation without fear of fracture?

46. What is the major distinguishing feature between hot and cold working?

47. What is warm working?

48. How can deformation and recrystallization improve the grain structure of a metal?

49. Why is grain growth usually undesirable?

50. What types of structures can be produced when an alloy element is added to a base metal?

51. As a result of their ionic or covalent bonding, what types of mechanical properties are characteristic of intermetallic compounds?

52. How is electrical charge transported in a metal (electrical conductivity)?

53. What features in a metal structure tend to impede or reduce electrical conductivity? How might we improve conductivity?

54. What is the difference between an intrinsic semiconductor and an extrinsic semiconductor?

55. What is required for electrical conductivity in ionic-bonded materials?

Problems

1. A prepared sample of metal reveals a structure with 32 grains per square inch at 100X magnification.

    a. What is its ASTM grain-size number?

    b. Would this material be weaker or stronger than the same metal with an ASTM grain-size number of 5? Why?

2. Brass is an alloy of copper with a certain amount of zinc dissolved and dispersed throughout the structure. Based on the material presented in this chapter:

    a. Would you expect brass to be stronger or weaker than pure copper? Why?

    b. Low brass (Copper alloy 240) contains 20% zinc. Cartridge brass (Copper alloy 260) contains 30% dissolved zinc. Which would you expect to be stronger? Why?

3. It is not uncommon for processing operations to expose manufactured products to extreme elevated temperature. Zinc coatings can be applied to steel by immersion into a bath of molten zinc (hot-dip galvanizing). Welding actually melts and resolidifies the crystalline metals. Brazing deposits molten filler metal. How might each of the following materials and their properties be altered by an exposure to elevated temperature?

    a. A recrystallized polycrystalline metal

    b. A cold-worked metal

    c. A solid-solution alloy, such as brass where zinc atoms are dissolved and dispersed throughout copper

4. Polyethylene consists of fibrous molecules of covalently bonded atoms tangled and interacting like the fibers of a cotton ball. Weaker van der Waals forces act between the molecules with a strength that is inversely related to separation distance.

    a. What properties of polyethylene can be attributed to the covalent bonding?

    b. What properties are most likely the result of the weaker van der Waals forces?

    c. If we pull on the ends of a cotton ball, the cotton fibers go from a random arrangement to an array of somewhat aligned fibers. Assuming we get a similar response from deformed polyethylene, how might properties change? Why?

    d. Would the properties of the deformed polyethylene be isotropic or anisotropic?

5. The following metals are deformed at a temperature of 540°C or 1000°F. For each of them, determine if the operation would be classified as cold working, warm working, or hot working. (NOTE: Convert to absolute temperature before evaluating!)

    a. Aluminum (Melting point of 660°C)

    b. Copper (Melting point of 1085°C)

    c. Iron (Melting point of 1538°C)

    d. Tungsten (Melting point of 3422°C)

CHAPTER 4

Equilibrium Phase Diagrams and the Iron–Carbon System

4.1 Introduction

As our study of engineering materials becomes more focused on specific metals and alloys, it is increasingly important that we acquire an understanding of their natural characteristics and properties. What is the basic structure of the material? Is the material uniform throughout, or is it a mixture of two or more distinct components? If there are multiple components, how much of each is present, and what are the different chemistries? Is there a component that may impart undesired properties or characteristics? What will happen if temperature is increased or decreased, pressure is changed, or chemistry is varied? The answers to these and other important questions can be obtained through the use of equilibrium phase diagrams.

4.2 Phases

Before we move to a discussion of equilibrium phase diagrams, it is important that we first develop a working definition of the term **phase**. As a starting definition, a phase is simply a form of material possessing a characteristic structure and characteristic properties. Uniformity of chemistry, structure, and properties is assumed throughout a phase. More rigorously, a phase has *a definable structure, a uniform and identifiable chemistry* (also known as **composition**), and distinct *boundaries* or **interfaces** that separate it from other different phases.

A phase can be continuous (like the air in a room) or discontinuous (like grains of salt in a shaker). A phase can be solid, liquid, or gas. In addition, a phase can be a pure substance or a solution, provided that the structure and composition are uniform throughout. Alcohol and water mix in all proportions and will therefore form a single phase when combined. There

are no boundaries across which structure and/or chemistry changes. Oil and water, on the other hand, tend to separate into regions with distinct boundaries and must be regarded as two distinct phases. Ice cubes in water are another two-phase system because there are two distinct structures with interfaces between them.

4.3 Equilibrium Phase Diagrams

An **equilibrium phase diagram** is a graphic mapping of the natural tendencies of a material or a material system, assuming that equilibrium has been attained for all possible conditions. There are three primary variables to be considered: *temperature, pressure,* and *composition*. The simplest phase diagram is a **pressure–temperature (*P–T*) diagram** for a fixed-composition material. Areas of the diagram are assigned to the various phases, with the boundaries indicating the equilibrium conditions of transition.

As an introduction, consider the *pressure–temperature* diagram for water, presented as **Figure 4.1**. With the composition fixed as H_2O, the diagram maps the stable form of water for various conditions of temperature and pressure. If the pressure is held constant and temperature is varied, the region boundaries denote the melting and boiling points. For example, at 1 atmosphere pressure, the diagram shows that water melts at 0°C and boils at 100°C. Still other uses are possible. Locate a temperature where the stable phase is liquid at atmospheric pressure. Maintaining the pressure at one atmosphere, drop the temperature until the material goes from liquid to solid (i.e., ice). Now, maintain that new temperature and begin to decrease the pressure. A transition will be encountered where solid goes directly to gas without melting (sublimation). The combined process just described,

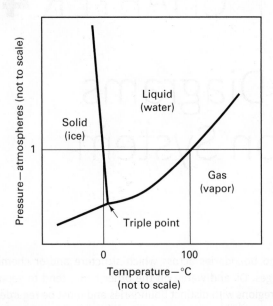

FIGURE 4.1   Pressure–temperature equilibrium phase diagram for water.

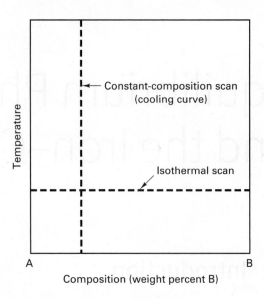

FIGURE 4.2   Mapping axes for a temperature–composition equilibrium phase diagram.

known as **freeze-drying**, is employed in the manufacture of numerous dehydrated products. With an appropriate phase diagram, process conditions can be determined that might reduce the amount of required cooling and the magnitude of pressure drop required for sublimation. A process operating about the **triple point** (where the three phases come together) would be most efficient.

Temperature–Composition Diagrams

Although the *P–T* diagram for water is an excellent introduction to phase diagrams, *P–T* phase diagrams are rarely used for engineering applications. Most engineering processes are conducted at atmospheric pressure, and variations are more likely to occur in temperature and composition as we consider both alloys and impurities. The most useful mapping, therefore, is usually a **temperature–composition phase diagram** at 1 atmosphere pressure. For the remainder of the chapter, this will be the form of phase diagram that will be considered.

For mapping purposes, temperature is placed on the vertical axis and composition on the horizontal. **Figure 4.2** shows the axes for mapping the *A–B* system, where the left-hand vertical corresponds to pure material *A,* and the percentage of *B* (usually expressed in weight percent) increases as we move toward pure material *B* at the right side of the diagram. The temperature range often includes only solids and liquids because few processes involve engineering materials in the gaseous state. Experimental investigations that provide the details of the diagram take the form of either vertical or horizontal scans that seek to locate the various phase transitions.

Cooling Curves

Considerable information can be obtained from vertical scans through the diagram where a fixed composition material is heated and slowly cooled. By plotting the cooling history in the form of a temperature-versus-time plot, known as a **cooling curve**, the transitions in structure will appear as characteristic points, such as slope changes or isothermal (constant-temperature) holds.

Consider the system composed of sodium chloride (common table salt) and water. Five different cooling curves are presented in **Figure 4.3**. Curve (a) is for pure water being cooled from the liquid state. A decreasing-temperature line is observed for the liquid where the removal of heat produces a concurrent drop in temperature. When the freezing point of 0°C is reached (point *a*), the material begins to change state and releases heat energy as part of the liquid-to-solid transition. Heat is being continuously extracted from the system, but because its source is now the change in state, there is no companion decrease in temperature. An **isothermal** or constant-temperature **hold** (*a–b*) is observed until the solidification is complete. From this point, as heat extraction continues, the newly formed solid experiences a steady drop in temperature. This type of curve is characteristic of pure metals and other substances with a distinct melting point.

Curve (b) in Figure 4.3 presents the cooling curve for a solution of 10% salt in water. The liquid region undergoes continuous cooling down to point *c*, where the slope abruptly decreases. At this temperature, small particles of ice (i.e., solid) begin to form, and the reduced slope is attributed to the energy released in this transition. The formation of these ice particles leaves the remaining solution richer in salt and imparts a lower freezing temperature. Further cooling is required to form additional ice, which continues to enrich the remaining liquid and

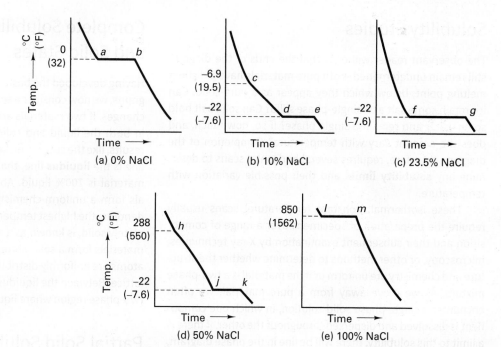

FIGURE 4.3 Cooling curves for five different solutions of salt and water: (a) 0% NaCl; (b) 10% NaCl; (c) 23.5% NaCl; (d) 50% NaCl; and (e) 100% NaCl.

further lowers its freezing point. Instead of possessing a distinct melting point or freezing point, this material is said to have a freezing range. When the temperature of point d is reached, the remaining liquid undergoes an abrupt reaction and solidifies into an intimate mixture of solid salt and solid water (discussed later), and an isothermal hold is observed. Further extraction of heat produces a drop in the temperature of the fully solidified material.

For a solution of 23.5% salt in water, a distinct freezing point is again observed, as shown in curve (c). Compositions with richer salt concentration, such as curve (d), show phenomena similar to those in curve (b), but with salt being the first

solid to form from the liquid. Finally, the curve for pure salt, curve (e), exhibits behavior similar to that of pure water.

If the observed transition points are now transferred to a temperature–composition diagram, such as **Figure 4.4**, we have part of a map that summarizes the behavior of the system. Line a–c–f–h–l denotes the lowest temperature at which the material is totally liquid and is known as the **liquidus** line. Line d–f–j denotes a particular three-phase reaction and will be discussed later. Between the lines, two phases coexist, one being a liquid and the other a solid. The equilibrium phase diagram, therefore, can be viewed as a collective presentation of cooling curve data for an entire range of alloy compositions.

Salt on Highways in Winter

The cooling curve studies just presented have provided some key information regarding the salt–water system, including some insight into the use of salt on highways in the winter. With the addition of

salt, the freezing point of water can be lowered from 0°C (32°F) to as low as −22°C (−7.6°F).

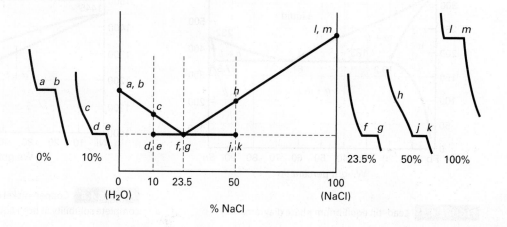

FIGURE 4.4 Partial equilibrium diagram for NaCl and H₂O derived from cooling-curve information.

Solubility Studies

The observant reader will note that the ends of the diagram still remain undetermined. Both pure materials have a distinct melting point, below which they appear as a pure solid. Can ice retain some salt as a single-phase solid? Can solid salt hold some water and remain a single phase? If so, how much, and does the amount vary with temperature? Completion of the diagram, therefore, requires several horizontal scans to determine any **solubility limits** and their possible variation with temperature.

These isothermal (constant temperature) scans usually require the preparation of specimens over a range of composition and their subsequent examination by X-ray techniques, microscopy, or other methods to determine whether the structure and chemistry are uniform or if the material is a two-phase mixture. As we move away from a pure material, we often encounter a single-phase solid solution, in which one component is dissolved and dispersed throughout the other. If there is a limit to this solubility, there will be line in the phase diagram, known as a **solvus** line, denoting the conditions of saturation where the single-phase solid solution becomes a two-phase mixture. **Figure 4.5** presents the equilibrium phase diagram for the lead–tin system, using the conventional notation in which Greek letters are used to denote the various single-phase solids. The upper portion of the diagram closely resembles the salt–water diagram, but the partial solubility of one material in the other can be observed on both ends of the diagram.

Complete Solubility in Both Liquid and Solid States

Having developed the basic concepts of equilibrium phase diagrams, we now consider a series of examples in which solubility changes. If two materials are completely soluble in each other in both the liquid and solid states, a rather simple diagram results, like the copper–nickel diagram of **Figure 4.6**. The upper line is the **liquidus** line, the lowest temperature for which the material is 100% liquid. Above the liquidus, the two materials form a uniform-chemistry liquid solution. The lower line, denoting the highest temperature at which the material is completely solid, is known as a **solidus** line. Below the solidus, the materials form a solid-state solution in which the two types of atoms are uniformly distributed throughout a single crystalline lattice. Between the liquidus and solidus is a **freezing range**, a two-phase region where liquid and solid solutions coexist.

Partial Solid Solubility

Many materials do not exhibit **complete solubility** in the solid state. Each is often soluble in the other up to a certain limit or saturation point, which varies with temperature. Such a diagram has already been observed for the lead–tin system in Figure 4.5.

At the point of maximum solubility, 183°C, lead can hold up to 19.2 wt% tin in a single-phase solution, and tin can hold

---

Lead–Tin Solders

Lead-tin solders have had a long history in joining electronic components. With the miniaturization of components, and the evolution of the circuit board or chip to ever-smaller features, exposure to the potentially damaging temperatures of the soldering operation *became an increasing concern. The lead-tin diagram of Figure 4.5 reveals why 60–40 solder (60 wt% tin) became the primary joining material in the lead-tin system. Of all possible alloys, it has the lowest (all liquid) melting temperature.*

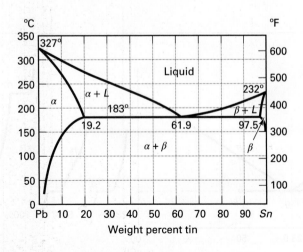

FIGURE 4.5 Lead–tin equilibrium phase diagram.

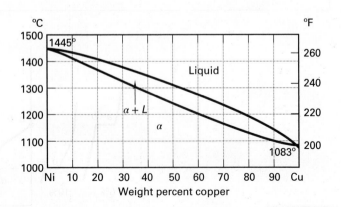

FIGURE 4.6 Copper–nickel equilibrium phase diagram, showing complete solubility in both liquid and solid states.

up to 2.5% lead within its structure and still be a single phase. If the temperature is decreased, however, the amount of **solute** that can be held in solution decreases in a continuous manner. If a saturated solution of tin in lead (19.2 wt% tin) is cooled from 183°C, the material will go from a single-phase solution to a two-phase mixture as a tin-rich second phase precipitates from solution. This change in structure can be used to alter and control the properties in a number of engineering alloys.

Insolubility

If one or both of the components is totally insoluble in the other, the diagrams will also reflect this phenomenon. In Figure 4.5, tin is only slightly soluble in lead. If it were completely insoluble, the left-hand regions of that diagram would extend to a vertical line at 100% tin.

Utilization of Diagrams

Before moving to more complex diagrams, let us first return to a simple phase diagram, such as the one in **Figure 4.7**, and develop several useful tools. For each condition of temperature and composition (i.e., for each point in the diagram), we would like to obtain three pieces of information:

1. *The phases present.* The stable phases can be determined by simply locating the point of consideration on the temperature–composition mapping and identifying the region of the diagram in which the point appears.

2. *The composition of each phase.* If the point lies in a single-phase region, there is only one component present, and the composition (or chemistry) of the phase is simply the composition of the alloy being considered. If the point lies in a two-phase region, a **tie-line** must be constructed. A tie-line is simply an isothermal (constant-temperature) line drawn through the point of consideration, terminating at the boundaries of the single-phase regions on either side. The compositions where the tie-line intersects the neighboring single-phase regions will be the compositions of those respective phases in the two-phase mixture. For example, consider point a in Figure 4.7. The tie-line for this temperature runs from S_2 to L_2. The tie-line intersects the solid phase region at point S_2. Therefore, the solid in the two-phase mixture at point a has the composition of point S_2. The other end of the tie-line intersects the liquid region at L_2, so the liquid phase that is present at point a will have the composition of point L_2.

3. *The amount of each phase present.* If the point lies in a single-phase region, all of the material, or 100%, must be of that phase. If the point lies in a two-phase region, the relative amounts of the two components can be determined by a **lever law** calculation using the previously drawn tie-line. Consider the cooling of alloy X in Figure 4.7 in a manner sufficiently slow so as to preserve equilibrium at all temperatures. For temperatures above t_1, the material is a single-phase liquid. Temperature t_1 is the lowest temperature for which the alloy is 100% liquid. If we draw a tie-line at this temperature, it runs from S_1 to L_1 and lies entirely to the left of composition X. At temperature t_3, the alloy is completely solid, and the tie-line lies completely to the right of composition X. As the alloy cools from temperature t_1 to temperature t_3, the amount of solid goes from zero to 100% while the segment of the tie-line that lies to the right of composition X also goes from 0 to 100%. Similarly, the amount of liquid goes from 100% to 0 as the segment of the tie-line lying to the left of composition X also goes from 100% to 0. Extrapolating these observations to intermediate temperatures, such as temperature t_2, we predict that the fraction of the tie-line that lies to the left of point a corresponds to the fraction of the material that is liquid. This fraction can be computed as:

$$\% \text{ Liquid} = \frac{a - S_2}{L_2 - S_2} \times 100\%$$

where the values of a, S_2, and L_2 are their composition values in weight percent B read from the bottom scale of the diagram. In a similar manner, the fraction of solid corresponds to the fraction of the tie-line that lies to the right of point a.

$$\% \text{ Solid} = \frac{L_2 - a}{L_2 - S_2} \times 100\%$$

Each of these mathematical relations could be rigorously derived from the conservation of either A or B atoms, as the material divides into the two different compositions of S_2 and L_2. Because the calculations consider the tie-line as a lever with the fulcrum at the composition line and the component phases at either end, they are called *lever-law* calculations.

Equilibrium phase diagrams can also be used to provide an overall picture of an alloy system, or to identify the transition points for phase changes in a given alloy. For example, the

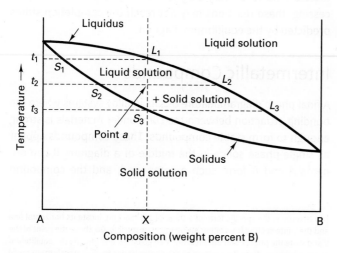

Equilibrium diagram showing the changes that occur during the cooling of alloy X.

temperature required to redissolve a second phase or melt an alloy can be easily determined. The various changes that will occur during the slow heating or slow cooling of a material can now be predicted. In fact, most of the questions posed at the beginning of this chapter can now be answered.

Solidification of Alloy X

Let us now apply the tools that we have just developed—tie-lines and lever-laws—to follow the solidification of alloy X in Figure 4.7. At temperature t_1, the first minute amount of solid forms with the chemistry of point S_1. As the temperature drops, more solid forms, but the chemistries of both the solid and liquid phases shift to follow the tie-line endpoints. The chemistry of the liquid follows the liquidus line, and the chemistry of the solid follows the solidus. Finally, at temperature t_3, solidification is complete, and the composition of the single-phase solid is now that of alloy X.

The composition of the first solid to form is different from that of the final solid. If the cooling is sufficiently slow, such that equilibrium is maintained or approximated, the composition of the solid changes during cooling and follows the endpoint of the tie-line. These chemistry changes are made possible by **diffusion**, the process by which atoms migrate through the crystal lattice given sufficient time at elevated temperature. If the cooling rate is too rapid, however, the temperature may drop before sufficient diffusion occurs. The resultant material will have a nonuniform chemistry. The initial solid that formed will retain a chemistry that is different from the solid regions that form later. When these nonequilibrium variations occur on a microscopic level, the resultant structure is referred to as being **cored**. Variation on a larger scale is called **macrosegregation**.

Three-Phase Reactions

Several of the phase diagrams that were presented earlier contain a feature in which phase regions are separated by a horizontal (or constant temperature) line. These lines are further characterized by either a V intersecting from above or an inverted-V intersecting from below. The intersection of the V and the line denotes the location of a **three-phase reaction**.

One common type of three-phase reaction, known as a **eutectic**, has already been observed in Figures 4.4 and 4.5. It is possible to understand these reactions through use of the tie-line and lever-law concepts that have been developed. Refer to the lead–tin diagram of Figure 4.5, and consider any alloy containing between 19.2 and 97.5 wt% tin at a temperature just above the 183°C horizontal line. Tie-line and lever-law computations reveal that the material contains either a lead-rich or tin-rich solid and remaining liquid. At this temperature, any liquid that is present will have a composition of 61.9 wt% tin, regardless of the overall composition of the alloy. If we now focus on this liquid and allow it to cool to just below 183°C, a transition occurs in which the liquid of composition 61.9% tin

transforms to a mixture of lead-rich solid with 19.2% tin and tin-rich solid containing 97.5% tin. The three-phase reaction that occurs on cooling through 183°C can be written as:

$$\text{Liquid}_{61.9\%\ Sn} \xrightarrow{183°C} \alpha_{19.2\%\ Sn} + \beta_{97.5\%\ Sn}$$

Note the similarity to the very simple chemical reaction in which water dissociates or separates into hydrogen and oxygen: $H_2O \rightarrow H_2 + \frac{1}{2}O_2$. Because the two solids in the lead–tin eutectic reaction have chemistries on either side of the original liquid, a similar separation must have occurred. The chemical separation requires atom movement, but the distances cannot be great. The resulting structure, known as **eutectic structure**, will be an intimate mixture of the two single-phase solids, with a multitude of interphase boundaries.

In the preceding example, the eutectic structure always forms from the same chemistry at the same temperature and has its own characteristic set of physical and mechanical properties. Alloys with the eutectic composition have the lowest melting point of all neighboring alloys and generally possess relatively high strength. For these reasons, they are often used as casting alloys or as filler material in soldering or brazing operations.

The eutectic reaction can be written in the general form of:

$$\text{liquid} \rightarrow \text{solid}_1 + \text{solid}_2$$

Figure 4.8 summarizes the various types of three-phase reactions that may occur in equilibrium phase diagrams, along with the generic form of the reaction shown below the figures.[1] These include the **eutectic**, **peritectic**, **monotectic**, and **syntectic** reactions, where the suffix -ic denotes that at least one of the three phases in the reaction is a liquid. If the same prefix appears with an -oid suffix, the reaction is of a similar form, but all phases involved are solids. Two all-solid reactions are the **eutectoid** and the **peritectoid**. The separation that occurs in the eutectoid reaction produces an extremely fine two-phase mixture. The combination reactions of the peritectic and peritectoid tend to be very sluggish. During actual material processing, these reactions may not reach the completion states predicted by the equilibrium diagram.

Intermetallic Compounds

A final phase diagram feature occurs in alloy systems where the bonding attraction between the component materials is strong enough to form stable compounds. These compounds appear as single-phase solids in the middle of a diagram. If components A and B form such a compound, and the compound

---

[1] To determine the specific form of a three-phase reaction, locate its horizontal line and the V intersecting from either above or below the line. Go above the point of the V and write the phases that are present. Then go below and identify the equilibrium phase or phases. Write the reaction as the phases above the line transform to those below. Apply this method to the diagrams in Figure 4.8 to identify the specific reactions, and compare them to their generic forms presented below the figures, remembering that the Greek letters denote single-phase solids.

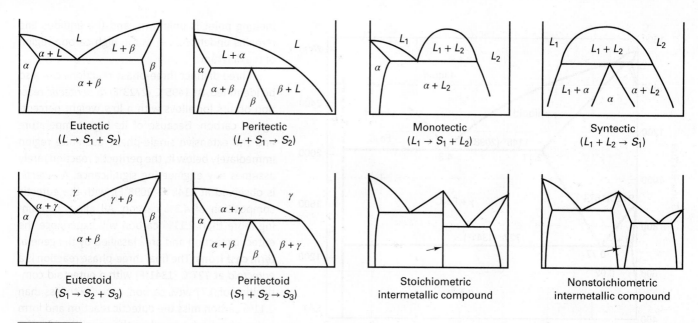

FIGURE 4.8 Schematic summary of three-phase reactions and intermetallic compounds.

cannot tolerate any deviation from its fixed atomic ratio, the product is known as a **stoichiometric intermetallic compound**, and it appears as a single vertical line in the A–B phase diagram. (Note: An example of this is the Fe_3C or iron carbide that appears at 6.67 wt% carbon in the upcoming iron–carbon equilibrium diagram.) If some degree of chemical deviation is tolerable, the vertical line expands into a single-phase region, and the compound is known as a **nonstoichiometric intermetallic compound**. Figure 4.8 shows schematic representations of both stoichiometric and nonstoichiometric compounds.

The single-phase intermetallic compounds appear in the middle of equilibrium diagrams, with locations consistent with whole number atomic ratios, such as AB, A_2B, AB_2, A_3B, AB_3, etc.[2] In general, they tend to be hard, brittle materials because these properties are a consequence of their ionic or covalent bonding. If they are present in large quantities or lie along grain boundaries in the form of a continuous film, the overall alloy can be extremely brittle. If the same compound is dispersed throughout the alloy in the form of small discrete particles, the result can be a considerable strengthening of the base metal.

Complex Diagrams

The equilibrium diagrams for actual alloy systems may be one of the basic types just discussed or some combination of them. In some cases the diagrams appear to be quite complex and formidable. However, by focusing on a particular composition and analyzing specific points using the tie-line and lever-law

concepts, even the most complex diagram can be interpreted and understood. If the properties of the various components are known, phase diagrams can then be used to predict the behavior of the resultant structures.

All of the preceding material has been for **binary phase diagrams**, where the mapping is for combinations of two elements or components. Three component systems can be mapped by **ternary phase diagrams**, which take the form of three-dimensional plots where chemistry is an equilateral triangle base, and temperature ascends upward. Because of the complexity of these diagrams, one of the three components is often fixed and the remaining two are varied. The resulting plot is simply a vertical slice through the ternary diagram and can be presented on a flat sheet like a binary diagram. Ternary diagrams will not be considered in this text.

4.4 Iron–Carbon Equilibrium Diagram

Steel, composed primarily of iron and carbon, is clearly the most important of the engineering metals. For this reason, the **iron–carbon equilibrium diagram** assumes special importance. The diagram most frequently encountered, however, is not the full iron–carbon diagram but the iron–iron carbide diagram shown in **Figure 4.9**. Here, a stoichiometric intermetallic compound, Fe_3C, is used to terminate the carbon range at 6.67 wt% carbon. The names of key phases and structures, and the specific notations used in the diagram, have evolved historically and will be used in their generally accepted form.

There are four single-phase solids within the diagram. Three of these occur in pure iron, and the fourth is the iron

[2] The use of "weight percent" along the horizontal axis tends to mask the whole number atomic ratio of intermetallic compounds. Many equilibrium phase diagrams now include a second horizontal scale to reflect "atomic percent." Intermetallic compounds then appear at atomic percents of 25, 33, 50, 67, 75, and similar values that reflect whole number atomic ratios.

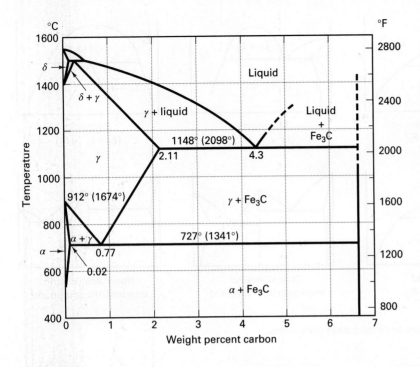

FIGURE 4.9 The iron–carbon equilibrium phase diagram. Single phases are α, ferrite; γ, austenite; δ, δ–ferrite; and Fe₃C, cementite.

carbide intermetallic that forms at 6.67% carbon. Upon cooling, pure iron solidifies into a body-centered-cubic solid that is stable down to 1394°C (2541°F). Known as **delta-ferrite**, this phase is present only at extremely elevated temperatures and has little engineering importance. From 1394 to 912°C (2541 to 1674°F) pure iron assumes a face-centered-cubic structure known as **austenite** in honor of the famed metallurgist Roberts-Austen of England. Designated by the Greek letter γ, austenite exhibits the high formability that is characteristic of the face-centered-cubic structure and is capable of dissolving over 2% carbon in single-phase solid solution. Hot forming of steel takes advantage of the low strength, high ductility, and chemical uniformity of austenite. Most of the heat treatments of steel begin by forming the high-temperature austenite structure. *Alpha ferrite*, or more commonly just **ferrite**, is the stable form of iron at temperatures below 912°C (1674°F). This body-centered-cubic structure can hold only 0.02 wt% carbon in solid solution and forces the creation of a two-phase mixture in most steels. Upon further cooling to 770°C (1418°F), iron undergoes a transition from nonmagnetic-to-magnetic. The temperature of this transition is known as the **Curie point**, but because it is not associated with any change in phase (but is an atomic-level transition), it does not appear on the equilibrium phase diagram.

The fourth single phase is the stoichiometric intermetallic compound, Fe₃C or iron carbide, and goes by the name **cementite**. Like most intermetallics, it is quite hard and brittle, and care should be exercised in controlling the structures in which it occurs. Alloys with excessive amounts of cementite, or cementite in undesirable form, tend to have brittle characteristics. Because cementite dissociates prior to melting, its exact

melting point is unknown, and the liquidus line remains undetermined in the high carbon region of the diagram.

Three distinct three-phase reactions can also be identified. At 1495°C (2723°F), a *peritectic* reaction occurs for alloys with a low weight percentage of carbon. Because of its high temperature and the extensive single-phase austenite region immediately below it, the peritectic reaction rarely assumes any engineering significance. A *eutectic* is observed at 1148°C (2098°F), with the eutectic composition of 4.3% carbon. All alloys containing more than 2.11% carbon will experience the eutectic reaction and are classified by the general term **cast irons**. The final three-phase reaction is a *eutectoid* at 727°C (1341°F) with a eutectoid composition of 0.77 wt% carbon. Alloys with less than 2.11% carbon miss the eutectic reaction and form a two-phase mixture when they cool through the eutectoid. These alloys are known as **steels**. The point of maximum solubility of carbon in iron, 2.11 wt%, therefore, forms an arbitrary separation between steels and cast irons.

4.5 Steels and the Simplified Iron–Carbon Diagram

If we focus on the materials normally known as steel, the phase diagram of Figure 4.9 can be simplified considerably. Those portions near the delta phase (or peritectic) region are of little significance, and the higher carbon region of the eutectic reaction only applies to cast irons. By deleting these segments and focusing on the eutectoid reaction, we can use the simplified diagram of **Figure 4.10** to provide an understanding of the properties and processing of steel.

Rather than beginning with liquid, our considerations generally begin with high-temperature, face-centered-cubic, single-phase austenite. The key transition will be the conversion of austenite to the two-phase ferrite plus carbide mixture as the temperature drops. Control of this reaction, which arises as a result of the drastically different carbon solubilities of the face-centered and body-centered structures, enables a wide range of properties to be achieved through heat treatment.

To begin to understand these processes, consider a steel of the eutectoid composition, 0.77% carbon, being slow cooled along line *x–x′* in Figure 4.10. At the upper temperatures, only austenite is present, with the 0.77% carbon being dissolved in solid solution within the face-centered structure. When the steel cools through 727°C (1341°F), several changes occur simultaneously. The iron wants to change crystal structure from the face-centered-cubic austenite to the

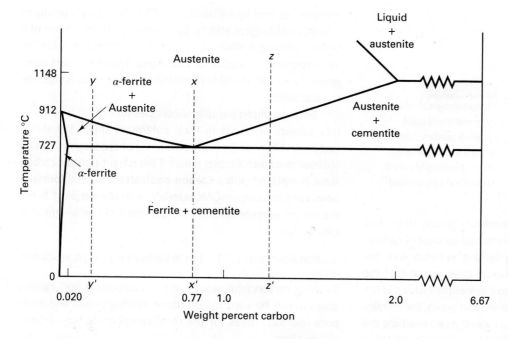

FIGURE 4.10 Simplified iron–carbon phase diagram with labeled regions. Figure 4.9 shows the more-standard Greek letter notation.

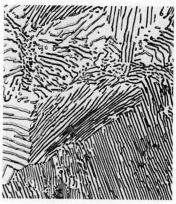

FIGURE 4.11 Pearlite; 1000X. (Courtesy of United States Steel Corporation)

FIGURE 4.12

Photomicrograph of a hypoeutectoid steel showing regions of primary ferrite (white) and pearlite; 500x. (Courtesy of United States Steel Corporation)

body-centered-cubic ferrite, but the ferrite can only contain 0.02% carbon in solid solution. The excess carbon is rejected and forms the carbon-rich intermetallic (Fe$_3$C) known as cementite. The net reaction at the eutectoid composition and temperature is

$$\text{Austenite}_{0.77\% \text{ C; FCC}} \rightarrow \text{Ferrite}_{0.02\% \text{ C; BCC}} + \text{Cementite}_{6.67\% \text{ C}}$$

Because the chemical separation occurs entirely within crystalline solids, the resultant structure is a fine mixture of ferrite and cementite. Specimens prepared by cutting, polishing, and etching in a weak solution of nitric acid and alcohol reveal a lamellar structure composed of alternating layers or plates, as shown in **Figure 4.11**. Because it always forms from a fixed composition at a fixed temperature, this structure has its own set of characteristic properties (even though it is composed of two distinct phases) and goes by the name **pearlite** because of its metallic luster and resemblance to mother-of-pearl when viewed at low magnification.

Steels having less than the eutectoid amount of carbon (less than 0.77%) are called **hypoeutectoid steels** (*hypo* means "less than"). Consider the cooling of a typical hypoeutectoid alloy along line *y–y′* in Figure 4.10. At high temperature the

material is entirely austenite. Upon cooling, however, it enters a region where the stable phases are ferrite and austenite. Tie-line and lever-law calculations show that the low-carbon ferrite nucleates and grows, leaving the remaining austenite richer in carbon. At 727°C (1341°F), the remaining austenite will have assumed the eutectoid composition (0.77% carbon), and further cooling transforms it to pearlite. The resulting structure, therefore, is a mixture of *primary* or *proeutectoid ferrite* (the **primary phase** is the one that forms before the reaction) and regions of pearlite as shown in **Figure 4.12**.

Pearlite

If sheets of black and white construction paper were stacked in alternating sequence and the pile then cut at an arbitrary angle with a machete, the cut surface would consist of black and white lines, like those appearing in Figure 4.11. The lines in the figure are the cut

edges of alternating layers of ferrite and cementite. Because ferrite is a metal and cementite is an intermetallic ceramic, pearlite is actually a naturally occurring composite material.

FIGURE 4.13
Photomicrograph of a hypereutectoid steel showing primary cementite along grain boundaries; 500X. (Courtesy of United States Steel Corporation)

Hypereutectoid steels (*hyper* means "greater than") are those that contain more than the eutectoid amount of carbon. When such a steel cools, as along line *z–z'* in Figure 4.10, the process is similar to the hypoeutectoid case, except that the primary or proeutectoid phase is now cementite instead of ferrite. As the carbon-rich phase nucleates and grows, the remaining austenite decreases in carbon content, again reaching the eutectoid composition at 727°C (1341°F). This austenite then transforms to pearlite upon slow cooling through the eutectoid temperature. **Figure 4.13** is a photomicrograph of the resulting structure, which consists of primary cementite and pearlite. In this case the continuous network of primary cementite (a brittle intermetallic compound) will cause the overall material to be extremely brittle.

It should be noted that the transitions just described are for equilibrium conditions, which can be approximated by slow cooling. Upon slow heating, the transitions will occur in the reverse manner. When the alloys are cooled rapidly, however, entirely different results may be obtained because sufficient time may not be provided for the normal phase reactions to occur. In these cases, the equilibrium phase diagram is no longer a valid tool for engineering analysis. Because the rapid-cool processes are important in the heat treatment of steels and other metals, their characteristics will be discussed in Chapter 5, and new tools will be introduced to aid our understanding.

4.6 Cast Irons

As shown in Figure 4.9, iron–carbon alloys with more than 2.11% carbon experience the eutectic reaction during cooling. These alloys are known as *cast irons*. Being relatively inexpensive, with good fluidity and rather low liquidus (full-melting) temperatures, they are readily cast and occupy an important place in engineering applications.

We should note, however, that most commercial cast irons also contain a significant amount of silicon. Cast irons typically contain 2.0–4.0% carbon, 0.5–3.0% silicon, less than 1.0%

manganese, and less than 0.2% sulfur. The silicon produces several metallurgical effects. By promoting the formation of a tightly adhering surface oxide, silicon enhances the oxidation and corrosion resistance of cast irons. Therefore, cast irons generally exhibit a level of corrosion resistance that is superior to most steels.

Because silicon partially substitutes for carbon (both have four valence electrons in their outermost shell), the three-component iron–carbon–silicon ternary phase diagram can be reduced to a much simpler binary if the weight-percent-carbon scale is replaced with a **carbon equivalent**. Several formulations exist to compute this number, with the simplest being the weight percent carbon plus one-third of the weight percent silicon:

$$\text{Carbon Equivalent (CE)} = (\text{wt \% Carbon}) + 1/3 (\text{wt \% Silicon})$$

By using carbon equivalent, the two-component iron–carbon diagram can be used to determine melting points and compute microstructures for the three-component iron–carbon–silicon alloys.

Silicon also tends to promote the formation of **graphite** as the carbon-rich phase instead of the Fe_3C intermetallic. The eutectic reaction, therefore, now has two distinct possibilities, as shown in the modified phase diagram of **Figure 4.14**:

$$\text{Liquid} \rightarrow \text{Austenite} + Fe_3C$$
$$\text{Liquid} \rightarrow \text{Austenite} + \text{Graphite}$$

The final microstructure of cast iron, therefore, will contain either the carbon-rich intermetallic compound, Fe_3C, or

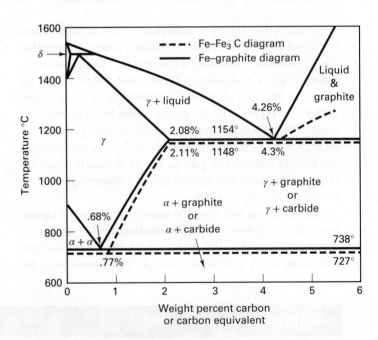

FIGURE 4.14 An iron–carbon diagram showing two possible high-carbon phases. Solid lines denote the iron–graphite system; dashed lines denote iron–cementite (or iron–carbide).

pure carbon in the form of graphite. Which ever occurs depends on the metal chemistry and various other process variables. Graphite is the more stable of the two and is the true equilibrium structure. Its formation is promoted by slow cooling, high carbon and silicon contents, heavy or thick section sizes, **inoculation** practices, and the presence of sulfur, phosphorus, aluminum, magnesium, antimony, tin, copper, nickel, and cobalt. Cementite (Fe_3C) formation is favored by fast cooling, low carbon and silicon levels, thin sections, and alloy additions of titanium, vanadium, zirconium, chromium, manganese, and molybdenum.

Cast iron is really a generic term that is applied to a variety of metal alloys. Depending on which type of high-carbon phase is present and the form or nature of that phase, the cast iron can be classified as gray, white, malleable, ductile, or nodular or compacted graphite. The ferrous metals, including steels, stainless steels, tool steels, and cast irons, will be presented in more detail in Chapter 6.

Review Questions

1. What kind of questions can be answered by equilibrium phase diagrams?

2. What are some features that are useful in defining a phase?

3. Supplement the examples provided in the text with another example of a single phase that is each of the following: continuous, discontinuous, gaseous, and a liquid solution.

4. What is an equilibrium phase diagram?

5. What three primary variables are generally considered in equilibrium phase diagrams?

6. Use the pressure–temperature diagram for water to describe the freeze-drying process.

7. Why is a pressure–temperature phase diagram not that useful for most engineering applications?

8. What form of equilibrium phase diagram is most commonly used?

9. What is a cooling curve?

10. What features in a cooling curve indicate some form of change in a material's structure? What causes a constant-temperature hold? A slope change?

11. What is a liquidus line?

12. What is a solubility limit, and how might it be determined?

13. Describe the conditions of complete solubility, partial solubility, and insolubility.

14. What types of changes occur upon cooling through a liquidus line? A solidus line? A solvus line?

15. What three pieces of information can be obtained for each point in an equilibrium phase diagram?

16. What is a tie-line? For what types of phase diagram regions would it be useful?

17. What points on a tie-line are used to determine the chemistry (or composition) of the component phases?

18. What tool can be used to compute the relative amounts of the component phases in a two-phase mixture? How does this tool work?

19. What is a cored structure? Under what conditions is it produced?

20. What is the difference between a cored structure and macrosegregation?

21. What features in a phase diagram can be used to identify three-phase reactions?

22. What is the general form of a eutectic reaction?

23. What is the general form of the eutectic structure?

24. Why are alloys of eutectic composition attractive for casting and as filler metals in soldering and brazing?

25. For the various three-phase reactions, what does the suffix "-*ic*" denote? The suffix "-*oid*?"

26. What is a stoichiometric intermetallic compound, and how would it appear in a temperature–composition phase diagram? How would a nonstoichiometric intermetallic compound appear?

27. What type of mechanical properties would be expected for intermetallic compounds?

28. In what form(s) might intermetallic compounds be undesirable in an engineering material? In what form(s) might they be attractive?

29. What are the four single phases in the iron–iron carbide diagram? Provide both the phase diagram notation and the assigned name.

30. What features of austenite make it attractive for forming operations? What features make it attractive as a starting structure for many heat treatments?

31. What feature in the iron–carbon diagram is used to distinguish between cast irons and steels?

32. Which of the three-phase reactions in the iron–carbon diagram is most important in understanding the behavior of steels? Write this reaction in terms of the interacting phases and their composition.

33. Describe the relative ability of iron to dissolve carbon in solution when in the form of austenite (the elevated temperature phase) and when in the form of ferrite at room temperature.

34. What is pearlite? Describe its structure.

35. What is a hypoeutectoid steel, and what structure will it assume upon slow cooling? What is a hypereutectoid steel, and how will its structure differ from that of a hypoeutectoid?

36. What are some attractive properties of cast irons?

37. In addition to iron and carbon, what other element is present in rather large amounts in cast iron?

38. What is carbon equivalent, and how is it computed?

39. What are the two possible high-carbon phases in cast irons? What features tend to favor the formation of each?

40. What are some of the families or types of cast iron?

Problems

1. Obtain a binary (two-component) phase diagram for a system not discussed in this chapter. Identify each:

a. Single phase

b. Three-phase reaction

c. Intermetallic compound

2. Copper and aluminum are both extremely ductile materials, as evidenced by the manufacture of fine copper wire and aluminum foil. Equal weights of copper and aluminum are melted together to produce an alloy and solidified in a pencil-shaped mold to produce short-length rods approximately ¼ in. or 6.5 mm in diameter. These rods appear extremely bright and shiny, almost as if they had been chrome plated. When dropped on a concrete floor from about waist height, however, the rods shatter into a multitude of pieces, a behavior similar to that observed with glass.

a. How might you explain this result? [HINT: Use the aluminum–copper phase diagram provided in Figure 5.3 of this text to determine the structure of the 50–50 weight percent alloy.]

b. Several of the high-strength aerospace aluminum alloys are aluminum–copper alloys. Explain how the observations above might be useful in providing the desired properties in these alloys.

3. Steel can be galvanized (zinc coated) in various ways. Because zinc is molten at temperatures well below the melting temperature of steel, steel can be immersed in molten zinc in a process known as hot-dip galvanizing. Another process is zinc electroplating, in which the zinc is deposited onto the steel at room temperature. Obtain a binary (two-component) iron–zinc phase diagram and consider the two processes.

a. Electroplating creates a distinct interface across which the steel transitions to the deposited zinc. During hot-dip immersion, however, the elevated temperature of the molten zinc drives interdiffusion, creating a region where the zinc content gradually transitions from 0 to 100%. Using the phase diagram, identify the various intermetallic compounds that might form at the temperature of molten zinc.

b. If you were given a piece of zinc-coated steel, how might you use metallographic microscopy to determine how the zinc coating had been applied?

CHAPTER 5

Heat Treatment

5.1 Introduction

In the previous chapters, you were introduced to the interrelationships among the structure, properties, processing, and performance of engineering materials. Chapters 3 and 4 considered the aspects of structure, whereas Chapter 2 focused on the properties. In this chapter, we begin to incorporate processing as a means of manipulating and controlling the structure and the companion properties of materials.

Many engineering materials can be characterized not by a single set of properties but by an entire spectrum of possibilities that can be selected and varied at will. **Heat treatment** is the term used to describe *the controlled heating and cooling of materials for the purpose of altering their structures and properties.* The same material can be made weak and ductile for ease in manufacture and then retreated to provide high strength and good fracture resistance for use and application. Because both physical and mechanical properties (such as strength, toughness, machinability, wear resistance, and corrosion resistance) can be altered by heat treatment, and these changes can be induced with no concurrent change in product shape, heat treatment is one of the most important and widely used manufacturing processes.

Technically, the term *heat treatment* applies only to the processes in which the heating and cooling are performed for the specific purpose of altering properties, but heating and cooling often occur as incidental phases of other manufacturing processes, such as hot forming or welding. The structure and properties of the material will be altered, however, just as though an intentional heat treatment had been performed, and the results can be either beneficial or harmful. For this reason, both the individual who selects material and the person who specifies its processing must be fully aware of the possible changes that can occur during heating or cooling activities. Heat treatment should be fully integrated with other manufacturing processes if effective results are to be obtained. To provide a basic understanding, this chapter will present both the theory of heat treatment and a survey of the more common heat treatment processes. Because more than 90% of all heat treatment is performed on steel and other ferrous metals, these materials will receive the bulk of our attention.

5.2 Processing Heat Treatments

The term *heat treatment* is often associated with those thermal processes that increase the strength and wear resistance of a material, but the broader definition permits inclusion of another set of processes that we will call **processing heat treatments**. These are often performed as a means of preparing the material for subsequent fabrication. Specific objectives might be the improvement of machining characteristics, the reduction of forming forces, or the restoration of ductility to enable further processing. Other uses include the relief of residual stresses from previous operations, such as casting, forming or welding, the removal of dissolved gases, and homogeneously distributing alloy elements. For the latter objectives, the processing heat treatments might be one of the final manufacturing operations.

Equilibrium Diagrams as Aids

Most of the processing heat treatments involve rather slow cooling or extended times at elevated temperatures. These conditions tend to approximate equilibrium, and the resulting structures, therefore, can be reasonably predicted through the use of an **equilibrium phase diagram** (presented in Chapter 4). These diagrams can be used to determine the temperatures that must be attained to produce a desired starting structure and to describe the changes that will then occur on subsequent cooling. It should be noted, however, that these diagrams are for true equilibrium conditions, and any departure from equilibrium might lead to substantially different results.

Processing Heat Treatments for Steel

Because many of the processing heat treatments are applied to plain-carbon and low-alloy steels, they will be presented here with the simplified iron–carbon equilibrium diagram of Figure 4-10 serving as a reference guide. **Figure 5.1** shows this diagram with the key transition lines labeled in standard

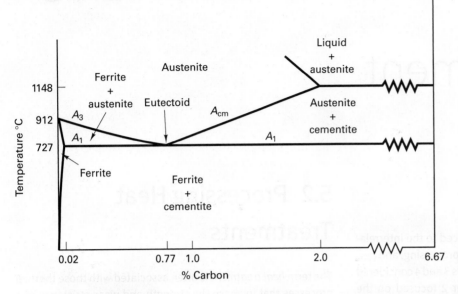

FIGURE 5.1 Simplified iron–carbon phase diagram for steels with transition lines labeled in standard notation as A_1, A_3, and A_{cm}.

History: Iron-Carbon Phase Diagram Nomenclature

Historically, an A_2 line once appeared between the A_1 and A_3. This line designated the magnetic property change known as the Curie point. Because this transition was later shown to be an atomic change and *not a change in phase, the line was deleted from the equilibrium phase diagram without a companion relabeling.*

notation. The eutectoid line is designated by the symbol **A_1**, and **A_3** designates the boundary between austenite and ferrite + austenite. The transition from austenite to austenite + cementite is designated as the **A_{cm}** line.

A number of process heat-treating operations have been classified under the general term of **annealing**. These might be employed to reduce strength or hardness, remove residual stresses, improve machinability, improve toughness, restore ductility, refine grain size, reduce segregation, stabilize dimensions, or alter the electrical or magnetic properties of the material. By producing a certain desired structure, characteristics can be imparted that will be favorable to subsequent operations (such as machining or forming) or applications. Because of the variety of anneals, it is important to designate the specific treatment, which is usually indicated by a preceding adjective. The specific temperatures, cooling rate, and details of the process will depend on the material being treated and the objectives of the treatment.

In the **full annealing** process, hypoeutectoid steels (less than 0.77% carbon) are heated to 30° to 60°C (50° to 100°F) above the A_3 temperature, held for sufficient time to convert the structure to homogeneous single-phase austenite of uniform composition and temperature, and then slowly cooled at a controlled rate through the A_1 temperature. Cooling is usually done in the furnace by decreasing the temperature by 10° to 30°C (20° to 50°F) per hour to at least 30°C (50°F) below the A_1 temperature. At this point all structural changes are complete, and the metal can be removed from the furnace and air

cooled to room temperature. The resulting structure is one of coarse pearlite (widely spaced layers or lamellae) with excess ferrite in amounts predicted by the equilibrium phase diagram. In this condition, the steel is quite soft and ductile.

The procedure to full-anneal a hypereutectoid alloy (greater than 0.77% carbon) is basically the same, except that the original heating is only into the austenite plus cementite region (30° to 60°C above the A_1). If the material were to be slow cooled from the all-austenite region, a continuous network of cementite might form on the grain boundaries and make the entire material brittle. When properly annealed, a hypereutectoid steel will have a structure of coarse pearlite with excess cementite in dispersed spheroidal form.

Although full anneals produce the softest and weakest properties, they are quite time consuming, and considerable amounts of energy must be spent to maintain the elevated temperatures required during soaking and furnace cooling. When maximum softness and ductility are not required and cost savings are desirable, **normalizing** might be specified. In this process, the steel is heated to 60°C (100°F) above the A_3 (hypoeutectoid) or A_{cm} (hypereutectoid) temperature, held at this temperature to produce uniform austenite, and then removed from the furnace and allowed to cool in still air.[1] The

---

[1] When successive batches of material are to be processed, it might be significant to note that the furnace operates at a steady temperature during normalizing, whereas the full annealing operation ends at a temperature below the A_1. In a successive full anneal treatment, the entire furnace must be reheated to the starting temperature along with the material being treated.

resultant structures and properties will depend on the subsequent cooling rate. Wide variations are possible, depending on the size and geometry of the product, but fine pearlite with excess ferrite or cementite is generally produced.

One should note a key difference between full annealing and normalizing. In the full anneal, the furnace imposes identical cooling conditions at all locations within the metal, which results in identical structures and properties. With normalizing, the cooling will be different at different locations. Properties will vary between surface and interior, and different thickness regions will also have different properties. When subsequent processing involves a substantial amount of machining that can be automated, the added cost of a full anneal might be justified because it produces a product with uniform machining characteristics at all locations.

If cold working has severely strain hardened a metal, it is often desirable to restore the ductility, either for service or to permit further processing without danger of fracture. This is often achieved through the **recrystallization** process described in Chapter 3. When the material is a low-carbon steel (<0.25% carbon), the specific procedure is known as a **process anneal**. The steel is heated to a temperature slightly below the A_1, held long enough to induce recrystallization of the dominant ferrite phase, and then cooled at a desired rate (usually in still air). Because the entire process is performed at temperatures within the same phase region, the process simply induces a change in phase morphology (size, shape, and distribution). The material is not heated to as high a temperature as in the full anneal or normalizing process, so a process anneal is somewhat cheaper and tends to produce less scaling.

A **stress-relief anneal** might be employed to reduce the **residual stresses** in large steel castings, welded assemblies, and cold-formed products. Parts are heated to temperatures below the A_1 (between 550° and 650°C or 1000° and 1200°F), held for a period of time, and then slow cooled to prevent the creation of additional stresses. Times and temperatures vary with the condition of the component, but the basic microstructure and associated mechanical properties generally remain unchanged.

When high-carbon steels (>0.60% carbon) are to undergo extensive machining or cold-forming, a heat treatment known as **spheroidization** is often employed. Here the objective is to produce a structure in which all of the cementite is in the form of small spheroids or globules dispersed throughout a ferrite matrix. This can be accomplished by a variety of techniques, including (1) prolonged heating at a temperature just below the A_1 followed by relatively slow cooling, (2) prolonged cycling between temperatures slightly above and slightly below the A_1, or (3) in the case of tool or high-alloy steels, heating to 750° to 800°C (1400° to 1500°F) or higher and holding at this temperature for several hours, followed by slow cooling.

Although the selection of a processing heat treatment often depends on the desired objectives, steel composition strongly influences the choice. Process anneals are restricted to low-carbon steels, and spheroidization is a treatment for high-carbon material. Normalizing and full annealing can be applied to all carbon contents, but even here, preferences are noted. Because different cooling rates do not produce a wide variation of properties in low-carbon steels, the air cool of a normalizing treatment often produces acceptable uniformity. For higher carbon contents, different cooling rates can produce wider property variations, and the uniform furnace cooling of a full anneal is often preferred. For plain-carbon steels the recommended treatments are: 0 to 0.4% carbon—normalize; 0.4% to 0.6% carbon—full anneal; and above 0.6% carbon—spheroidize. Because of the effects of hardenability (to be discussed later in this chapter), the transition between normalizing and full annealing for alloy steels is generally lowered to 0.2% carbon.

Figure 5.2 provides a graphical summary of the process heat treatments.

Heat Treatments for Nonferrous Metals

Most of the nonferrous metals do not have the significant phase transitions observed in the iron–carbon system, and for them, the process heat treatments do not play such a significant role. Aside from the strengthening treatment of precipitation hardening, which is discussed later, the nonferrous metals are usually heat-treated for three purposes: (1) to produce a uniform, homogeneous structure, (2) to provide stress relief, or (3) to bring about recrystallization. Castings that have been cooled too rapidly can possess a segregated solidification structure known as coring (discussed more fully in Chapter 4). **Homogenization** can be achieved by heating to moderate temperatures and then holding for a sufficient time to allow thorough diffusion to take place. Similarly, heating for several hours at relatively low temperatures can reduce the internal stresses that are often produced by forming, welding, or brazing.

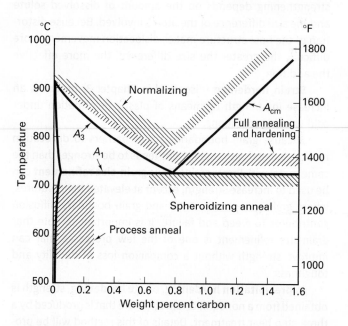

FIGURE 5.2 Graphical summary of the process heat treatments for steels on an equilibrium diagram.

Recrystallization (discussed in Chapter 3) is a function of the particular metal, the amount of prior deformation, and the desired recrystallization time. In general, the more a metal has been strained, the lower the recrystallization temperature or the shorter the time. Without prior straining, however, recrystallization will not occur, and heating will only produce undesirable grain growth.

5.3 Heat Treatments Used to Increase Strength

Six major mechanisms are available to increase the strength of metals:

1. Solid-solution strengthening
2. Strain hardening
3. Grain-size refinement
4. Precipitation hardening
5. Dispersion hardening
6. Phase transformations

Although all of these might not be applicable to a given metal or alloy, these heat treatments can often play a significant role in inducing or altering the final properties of a product.

In **solid-solution strengthening**, a base metal dissolves other atoms, either as **substitutional solutions**, where the new atoms occupy sites in the host crystal lattice, or as **interstitial solutions**, where the new atoms squeeze into "holes" between the atoms of the base lattice. The amount of strengthening depends on the amount of dissolved solute and the size difference of the atoms involved. Because distortion of the host structure makes dislocation movement more difficult, the greater the size difference, the more effective the addition.

Strain hardening (discussed in Chapter 3) produces an increase in strength by means of plastic deformation under cold-working conditions.

Because grain boundaries act as barriers to dislocation motion, a metal with small grains tends to be stronger than the same metal with larger grains. Thus **grain-size refinement** can be used to increase strength, except at elevated temperatures, where grain growth can occur and grain boundary diffusion contributes to creep and failure. It is important to note that grain-size refinement is one of the few processes that can improve strength without a companion loss of ductility and toughness.

In **precipitation hardening**, or **age hardening**, strength is obtained from a nonequilibrium structure that is produced by a three-step heat treatment. Details of this method will be provided in the following section.

Strength obtained by dispersing second-phase particles throughout a base material is known as **dispersion hardening**. To be effective, the dispersed particles should be stronger than the matrix, adding strength through both their reinforcing action and the additional interfacial surfaces that present barriers to dislocation movement.

Phase transformation strengthening involves those alloys that can be heated to form a single phase at elevated temperature and subsequently transform to one or more low temperature phases on cooling. When this feature is used to increase strength, the cooling is usually rapid, and the phases that are produced are usually of a nonequilibrium nature.

5.4 Strengthening Heat Treatments for Nonferrous Metals

All six of the mechanisms just described have been used to increase the strength of nonferrous metals. Solid-solution strengthening can impart strength to single-phase materials. Strain hardening can be quite useful if sufficient ductility is present to permit deformation. Alloys containing eutectic structure exhibit considerable dispersion hardening. Among all of the possibilities, however, the most effective strengthening mechanism for the nonferrous metals tends to be precipitation hardening.

Precipitation or Age Hardening

To be a candidate for precipitation hardening, an alloy system must exhibit solubility that decreases with decreasing temperature, such as the aluminum-rich portion of the aluminum-copper system whose equilibrium phase diagram is shown in **Figure 5.3** and enlarged in **Figure 5.4**. Consider the alloy with 4% copper. Liquid metal cools through the alpha plus liquid region and solidifies into a single-phase solid (α phase). At 1000°F, the full 4% of copper would be dissolved and distributed throughout the alpha phase crystals. As the temperature drops, however, the maximum solubility of copper in aluminum decreases from 5.65% at 1018°F to less than 0.2% at room temperature. Upon cooling through the solvus (or solubility limit) line at 930°F, the 4% copper alloy enters a two-phase region, and copper-rich theta-phase precipitates begin to form and grow within the alpha-phase crystalline solid. (*Note:* Theta-phase is actually a hard, brittle intermetallic compound with the chemical formula of $CuAl_2$.) The equilibrium structure at room temperature, therefore, would be an aluminum-rich alpha-phase structure with coarse theta-phase precipitates,

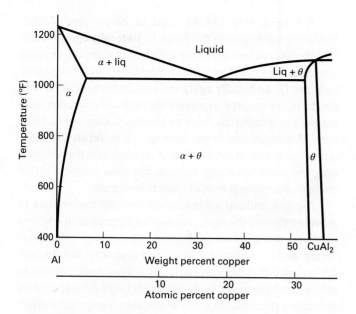

FIGURE 5.3 High-aluminum section of the aluminum–copper equilibrium phase diagram.

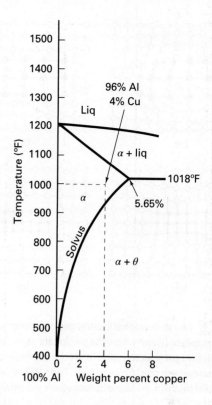

FIGURE 5.4 Enlargement of the solvus-line region of the aluminum–copper equilibrium diagram of Figure 5.3.

generally lying along alpha-phase grain boundaries where the nucleation of second-phase particles can benefit from the existing interfacial surface.

Whenever two or more phases are present, the material exhibits dispersion strengthening. Dislocations are confined to their own crystal and cannot cross interfacial boundaries. Therefore, each interface between the alpha-phase and theta-phase precipitate is a strengthening boundary. Now take a particle of theta precipitate, cut it into two halves, and separate the segments. Forming the two half-size precipitates has just added two additional interfaces, corresponding to both sides of the cut. If the particle were to be further cut into quarters, eighths, and sixteenths, we would expect strength to increase as we continually add interfacial surface. Ideally, we would like to have millions of ultra-small particles dispersed throughout the alpha-phase structure. When we try to create this more-desirable nonequilibrium configuration (nonequilibrium because energy is added each time new interfacial surface is created), we gain an unexpected benefit that adds significant strength. This new nonequilibrium treatment is known as *age hardening* or *precipitation hardening*.

The process of precipitation hardening is actually a three-step sequence. The first step, known as **solution treatment**, erases the room-temperature structure and redissolves any existing precipitate. The metal is heated to a temperature above the solvus and held in the single-phase region for sufficient time to redissolve the second phase and uniformly distribute the solute atoms (in this case, copper).

If the alloy were slow cooled, the second-phase precipitate would nucleate and the material would revert back to a structure similar to equilibrium. To prevent this from happening, age hardening alloys are **quenched** from their solution treatment temperature. The rapid-cool quenching,

usually in water, suppresses diffusion, trapping the dissolved atoms in place. The result is a room temperature *supersaturated* solid solution. For the alloy being considered, the alpha phase would now be holding 4% copper in solution at room temperature—far in excess of its equilibrium maximum of < 0.2%. In this nonequilibrium quenched condition, the material is often soft and can be easily straightened, formed, or machined.

If the supersaturated material is then reheated to a temperature where atom movement (diffusion) can occur but still within the two-phase region, the alloy will attempt to form its equilibrium two-phase structure as the excess solute atoms precipitate out of the supersaturated matrix. This stage of the process, known as **aging**, is actually a progressive transition. Solute atoms begin to cluster at locations within the parent crystal, still occupying atom sites within the original lattice. Various intermediates might then occur as the clusters grow, leading ultimately to the formation of distinct second phase particles with their own characteristic chemistry and crystal structure.

A key concept in the aging sequence is that of **coherency** or crystalline continuity. If the clustered solute atoms continue to occupy lattice sites within the parent structure, the crystal planes remain continuous in all directions, and the clusters of solute atoms (which are of different size and possibly different valence from the host material) tend to distort or strain the adjacent lattice for a sizable distance in all directions, as illustrated two-dimensionally in **Figures 5.5a** and **5.5c**. For this

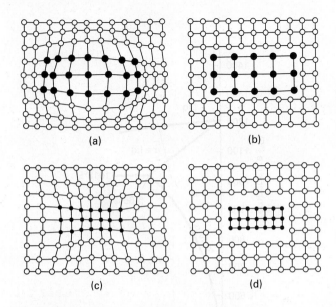

(a)   (b)

(c)   (d)

FIGURE 5.5 Two-dimensional illustrations depicting: (a) a coherent precipitate cluster where the precipitate atoms are larger than those in the host structure, and (b) its companion overaged or discrete second-phase precipitate particle. Parts (c) and (d) show equivalent sketches where the precipitate atoms are smaller than the host.

reason, each small cluster appears to be much larger with respect to its ability to interfere with dislocation motion (i.e., impart strength). When the clusters reach a certain size, however, the associated strain becomes so great that the clusters can lower their energy by breaking free from the parent structure to form distinct second phase particles with their own crystal structure and well-defined interphase boundaries, as shown in **Figures 5.5b** and **5.5d**. Coherency is lost, and the strengthening reverts to dispersion hardening, where dislocation interference is limited to the actual size of the particle. Strength and hardness decrease, and the material is said to be **overaged**.

Aging can be performed at any temperature within the two-phase region where diffusion is sufficiently rapid to form the desired precipitate in a reasonable time. **Figure 5.6** presents a family of aging curves for the 4% copper–96% aluminum alloy. For higher aging temperatures, the peak properties are achieved in a shorter time, but the peak hardness (or strength) is not as great as can be achieved at lower aging temperatures. The higher peak strength at lower aging temperature can be attributed to the combined effect of the increased amount of precipitate that forms at lower temperature (use a lever-law calculation as described in Chapter 4) and the fineness of the precipitate that forms when diffusion is more limited. Actual selection of the aging conditions (temperature and time) is a decision that is made on the basis of desired strength, available equipment, and production constraints.

The aging step can be used to divide precipitation-hardening materials into two types: (1) **naturally aging** materials, where room temperature is sufficient to move the unstable supersaturated solution toward the stable two-phase structure, and (2) **artificially aging** materials, where elevated temperatures are required to provide the necessary diffusion. With natural aging materials, such as aluminum alloy rivets, some form of refrigeration might be required to retain the after-quench condition of softness. Upon removal from the refrigeration, the rivets are easily headed, but then progress to full strength after several days at room temperature.

Because artificial aging requires elevated temperature to provide diffusion, the aging process can be stopped at any time by simply dropping the temperature (quenching). Diffusion is halted, and the current structure and properties are "locked-in," provided that the material is not subsequently exposed to elevated temperatures that would reactivate diffusion. When diffusion is possible, the material will always attempt to revert to its equilibrium structure! According to Figure 5.6, if the 4% copper alloy were aged for one day at 375°F and then quenched to prevent overaging, the metal would attain a hardness of 94 Vickers (and the associated strength) and retain these properties throughout its useful lifetime provided subsequent diffusion did not occur. If a higher strength is required, a lower aging temperature and longer time could be selected.

Precipitation hardening is an extremely effective strengthening mechanism and is responsible for the attractive engineering properties of many aluminum, copper, magnesium, and titanium alloys, as well as nickel-based superalloys. In most cases, strength can more than double that observed upon conventional cooling, and a 10-fold or greater increase is possible with some alloys. Although other strengthening methods are traditionally used with steels and cast irons, age hardening

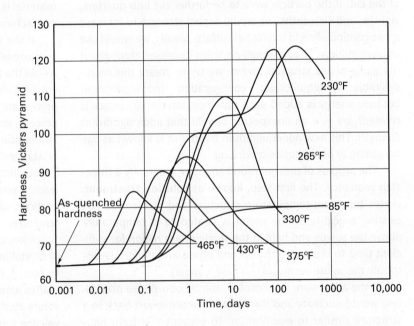

FIGURE 5.6 Aging curves for the Al–4% Cu alloy at various temperatures showing peak strengths and times of attainment. (Adapted from *Journal of the Institute for Metals*, Vol. 79, p. 321, 1951.)

has been combined with those methods to produce some of the highest strength ferrous alloys, such as the maraging steels and precipitation hardenable stainless.

5.5 Strengthening Heat Treatments for Steel

Iron-based metals have been heat treated for centuries, and the striking changes that resulted from plunging red-hot metal into cold water or some other quenching medium were awe-inspiring to the ancients. Those who performed these acts in the making of swords and armor were looked on as possessing unusual powers, and much superstition arose regarding the process. Because quality was directly related to the act of quenching, great importance was placed on the quenching medium. Urine was found to be a superior quenching medium, and that from a red-haired boy was deemed particularly effective, as was that from a 3-year-old goat fed only ferns.

Although the overwhelming majority of heat treatments are performed on steel, it has only been within the last century that the art of heat treating has become a science. One of the major barriers to understanding was the fact that the strengthening treatments were nonequilibrium. Minor variations in cooling often produced major variations in structure and properties.

Isothermal Transformation (I-T) or Time-Temperature-Transformation (T-T-T) Diagrams

A useful aid to understanding nonequilibrium heat treatment processes is the **isothermal transformation (I-T)** or **time-temperature-transformation (T-T-T) diagram**. The information in this diagram is obtained by heating thin specimens of a particular steel to produce elevated-temperature uniform-chemistry **austenite**, "instantaneously" quenching to a temperature where austenite is no longer the stable phase, holding for variable periods of time at this new temperature, and observing the resultant structures via metallographic photomicrographs (i.e., optical microscope examination).

For simplicity, consider a carbon steel of eutectoid composition (0.77% carbon) and its T-T-T diagram shown as **Figure 5.7**. Above the A_1 temperature of 1341°F (727°C), austenite is the stable phase and will persist regardless of the time. Below this temperature, the face-centered austenite would like to transform to body-centered ferrite and carbon-rich cementite. Two factors control the rate of transition: (1) the motivation or driving force for the change, and (2) the ability to form the desired products (i.e., the ability to redistribute the atoms through atomic diffusion). The region below 1341°F in Figure 5.7 can be

interpreted as follows. Zero time corresponds to a sample "instantaneously" quenched to its new lower temperature. The structure is usually unstable austenite. As time passes (moving horizontally across the diagram), a line is encountered representing the start of transformation and a second line indicating completion of the phase change. At elevated temperatures (just below 1341°F), atom movement within the solid (diffusion) is rapid, but the rather sluggish driving force dominates the kinetics. At a low temperature, the driving force is high but diffusion is quite limited. The kinetics of phase transformation are most rapid at a compromise intermediate temperature, resulting in the characteristic C-curve shape. The portion of the C that extends farthest to the left is known as the *nose* of the T-T-T or I-T diagram.

If the transformation occurs between the A_1 temperature and the nose of the curve, the departure from equilibrium is not very great. The austenite transforms into alternating layers of ferrite and cementite, producing the **pearlite** structure that was introduced with the equilibrium phase diagram description in Chapter 4. Because the diffusion rate is greater at higher temperatures, pearlite produced under those conditions has a larger lamellar spacing (separation distance between similar layers). The pearlite formed near the A_1 temperature is known as **coarse pearlite**, whereas the closer-spaced structures formed near the nose are called **fine pearlite**. Because the resulting structures and properties are similar to those of the near-equilibrium process heat treatments, the constant-temperature transformation procedure just described is called an **isothermal anneal**.

If the austenite is quenched to a temperature between the nose and the temperature designated as M_s, a different structure is produced. These transformation conditions are a significant departure from equilibrium, and the amount of diffusion required to form the continuous layers within pearlite is no longer available. The metal still has the goal of changing crystal structure from face-centered austenite to body-centered ferrite, with the excess carbon being accommodated in the form of cementite. The resulting structure, however, does not contain cementite layers but rather a dispersion of discrete cementite particles dispersed throughout a matrix of ferrite. Electron microscopy might be required to resolve the carbides in this structure, which is known as **bainite**. Because of the fine dispersion of carbide, it is stronger than fine pearlite, and ductility is retained because the soft ferrite is the continuous matrix.

If austenite is quenched to a temperature below the M_s line, a different type of transformation occurs. The steel still wants to change its crystal structure from face-centered-cubic to body-centered cubic, but it can no longer expel the amount of carbon necessary to form ferrite. Responding to the severe nonequilibrium conditions, it simply undergoes an abrupt change in crystal structure with no significant movement of carbon. The excess carbon becomes trapped, distorting the structure into a body-centered tetragonal crystal lattice (distorted body-centered cubic), with the amount of distortion being proportional to the amount of excess carbon. The new

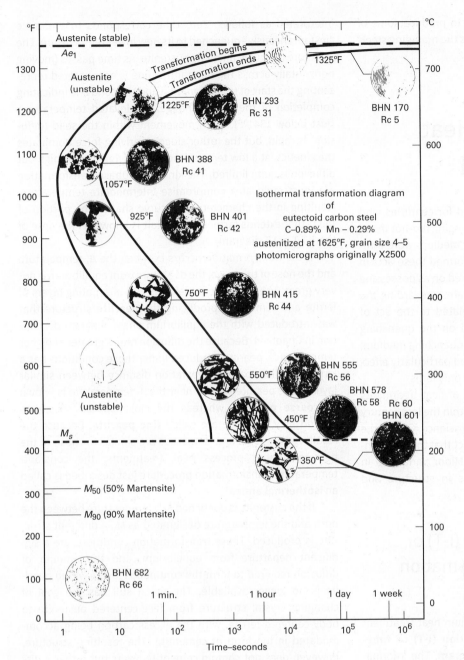

FIGURE 5.7 Isothermal transformation diagram (T-T-T diagram) for eutectoid composition steel. Structures resulting from transformation at various temperatures are shown as insets. (Courtesy of United States Steel Corporation)

structure, shown in **Figure 5.8**, is known as **martensite**, and, with sufficient carbon, it is exceptionally strong, hard, and brittle. The highly distorted lattice effectively blocks the dislocation motion that is necessary for metal deformation. It is significant to note that with each successive departure from equilibrium (coarse pearlite → fine pearlite → bainite → martensite) strength and hardness increases, whereas ductility and fracture resistance decline.

If we allow the carbon content to vary, **Figure 5.9** shows that the hardness and strength of steel with the martensitic structure are strong functions of the carbon content. Below 0.10% carbon, martensite is not very strong. Because no diffusion occurs during the transformation, higher-carbon steels form higher-carbon martensite, with an increase in strength and hardness and a concurrent decrease in toughness and ductility. From 0.3% to 0.7% carbon, strength and hardness

increase rapidly. Above 0.7% carbon, however, the rise is far less dramatic and might actually be a decline, a feature related to the presence of retained austenite (to be described later).

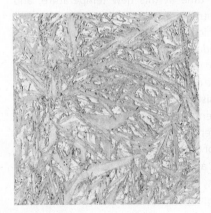

FIGURE 5.8

Photomicrograph of martensite; 1000X. (Courtesy of United States Steel Corporation)

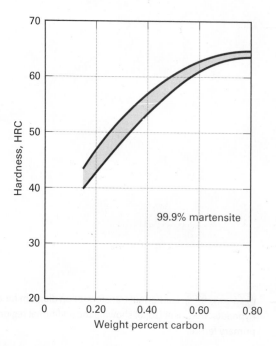

FIGURE 5.9 Effect of carbon on the hardness of martensite.

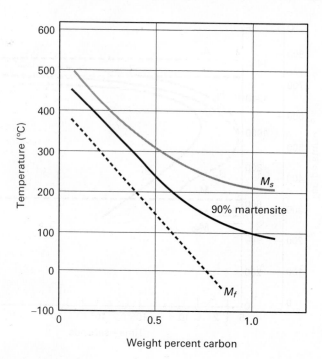

FIGURE 5.11 Variation of M_s and M_f temperatures with carbon content. Note that for high carbon steels, completion of the martensite transformation requires cooling to below room temperature.

Unlike the other structure transformations, the *amount* of martensite that forms is not a function of time, but depends only on the lowest temperature that is encountered during the quench. This feature is shown in **Figure 5.10**, where the amount of martensite is recorded as a function of temperature. Returning to the C curve of Figure 5.7, there is a temperature designated as M_{50}, where the structure is 50% martensite and 50% untransformed austenite. At the lower M_{90} temperature, the structure has become 90% martensite. If no further cooling were to occur, the untransformed austenite could remain within the structure. This **retained austenite** can cause loss of strength or hardness, dimensional instability, and cracking or brittleness. Because many quenches are to room temperature, retained austenite becomes a significant problem when the martensite finish, or 100% martensite, temperature lies below room temperature. **Figure 5.11** presents the martensite start

and martensite finish temperatures for a range of carbon contents. Higher carbon contents, as well as most alloy additions, decrease all martensite-related temperatures, and materials with these chemistries might require refrigeration or a quench in dry ice or liquid nitrogen to produce full hardness.[2]

It is important to note that all of the transformations that occur below the A_1 temperature are one-way transitions (austenite to something). The steel is simply seeking to change its crystal structure, and the various products are the result of this change. It is impossible, therefore, to convert one transformation product to another without first reheating to above the A_1 temperature to again form the face-centered-cubic austenite.

T-T-T diagrams can be quite useful in determining the kinetics of transformation and the nature of the products. The left-hand curve shows the elapsed time (at constant temperature) before the transformation begins, and the right-hand curve shows the time required to complete the transformation. If hypo- or hypereutectoid steels were considered (carbon contents below or above 0.77%), additional regions would have to be added to the diagram to incorporate the primary equilibrium phases that form below the A_3 or A_{cm} temperatures. These regions would not extend below the nose, however, because the nonequilibrium bainite and martensite structures can exist with variable amounts of carbon, unlike the near-equilibrium pearlite. **Figure 5.12** shows the T-T-T curve for a 0.5%-carbon hypoeutectoid steel, showing the additional region for the primary ferrite beginning with the ferrite start, or F_s line.

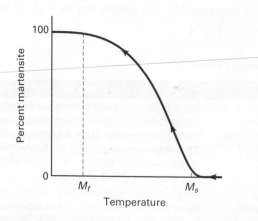

FIGURE 5.10 Schematic representation depicting the amount of martensite formed upon quenching to various temperatures from M_s through M_f.

[2] With a quench to room temperature, carbon steels with less than 0.5% carbon generally have less than 2% retained austenite. As the amount of carbon increases, so does the amount of retained austenite, rising to about 6% at 0.8% carbon and to more than 30% when the carbon content is 1.25%.

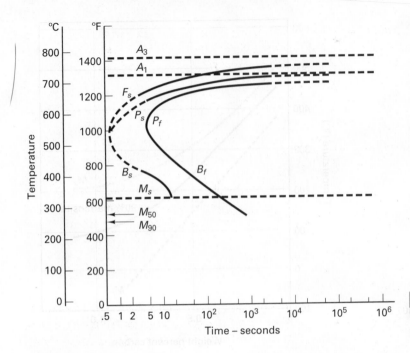

FIGURE 5.12 Isothermal transformation diagram for a hypoeutectoid steel (1050) showing the additional region for primary ferrite.

Tempering of Martensite

Despite its great strength, medium- or high-carbon martensite in its as-quenched form lacks sufficient toughness and ductility to be a useful engineering structure. A subsequent heating, known as **tempering**, is usually required to impart the necessary ductility and fracture resistance and relax undesirable residual stresses. As with most property-changing processes, however, there is a concurrent drop in other features, most notably strength and hardness.

Martensite is a supersaturated solid solution of carbon in alpha ferrite and, therefore, is a metastable structure. When heated into the range of 100° to 700°C (200° to 1300°F), the excess carbon atoms are rejected from solution, and the structure moves *toward* a mixture of the stable phases of ferrite and cementite. This decomposition of martensite into ferrite and cementite is a time- and temperature-dependent, diffusion-controlled phenomenon with a continuous spectrum of intermediate and transitory structures.

Table 5.1 presents a chart-type comparison of the previously discussed precipitation hardening process and the austenitize-quench-and-temper sequence. Both are nonequilibrium heat treatments that involve three distinct stages. In both, the first step is an elevated temperature soaking designed to erase the prior structure, redissolving material to produce a uniform-chemistry, single-phase starting condition. Important features include heating rate, soaking temperature, and soaking time. Both treatments follow this soak with a rapid-cool quench. In precipitation hardening, the purpose of the quench is to prevent nucleation of the second phase, thereby producing a supersaturated solid solution. This material is usually soft, weak, and ductile, with good toughness. Subsequent aging (reheating within the temperatures of the stable two-phase region) allows the material to move toward the formation of the stable two-phase structure and sacrifices toughness and ductility for an increase in strength. When the proper balance is achieved, the temperature is dropped, diffusion ceases, and the current structure

TABLE 5.1 Comparison of Age Hardening with the Quench-and-Temper Process

Heat Treatment	Step 1	Step 2	Step 3
Age hardening	*Solution treatment.* Heat into the stable single-phase region (above the solvus) and hold to form a uniform-chemistry single-phase solid solution.	*Quench.* Rapid cool to form a nonequilibrium supersaturated single-phase solid solution (crystal structure remains unchanged, material is soft and ductile).	*Age.* A controlled reheat in the stable two-phase region (below the solvus). The material moves toward the formation of the stable two-phase structure, becoming stronger and harder. The properties can be "frozen in" by dropping the temperature to stop further diffusion.
Quench and temper for steel	*Austenitize.* Heat into the stable single-phase region (above the A_3 or A_{cm}) and hold to form a uniform-chemistry single-phase solid solution (austenite).	*Quench.* Rapid cool to form a nonequilibrium supersaturated single-phase solid solution (crystal structure changes to body-centered martensite, which is hard but brittle).	*Temper.* A controlled reheat in the stable two-phase region (below the A_1). The material moves toward the formation of the stable two-phase structure, becoming weaker but tougher. The properties can be "frozen in" by dropping the temperature to stop further diffusion.

All of the furnaces that have been described can heat in air, but most commercial furnaces can also employ **artificial gas atmospheres.** These are often selected to prevent scaling or tarnishing, to prevent decarburization, or even to provide car-bon or nitrogen for surface modification. Many of the artificial atmospheres are generated from either the combustion or decomposition of natural gas, but nitrogen-based atmospheres frequently offer reduced cost, energy savings, increased safety, and environmental attractiveness. Other common atmos-pheres include argon, dissociated ammonia, dry hydrogen, helium, steam, and vacuum.

The heating rates of gas atmosphere furnaces can be sig-nificantly increased by incorporating the **fluidized-bed** con-cept. These furnaces consist of a bed of dry, inert particles, such as aluminum oxide (a ceramic), which are heated and flu-idized (suspended) in a stream of upward-flowing gas. Products introduced into the bed become engulfed in the particles, which then radiate uniform heat. Temperature and atmosphere can be altered quickly, and high heat-transfer rates, high ther-mal efficiency, good temperature uniformity, and low fuel con-sumption have been observed. Because atmosphere changes can be performed in minutes, a single furnace can be used for nitriding, stress relieving, carburizing, carbonitriding, anneal-ing, and hardening.

When a liquid heating medium is preferred, **salt bath fur-naces** or **pot furnaces** are a popular choice. Electrically con-ductive salt can be heated by passing a current between two electrodes suspended in the bath, which also causes the bath to circulate and maintain uniform temperature. Nonconductive salts can be heated by some form of immersion heater, or the containment vessels can be externally fired. In these furnaces, the molten salt not only serves as a uniform source of heat but can also be selected to prevent decarburization. The immer-sion isolates the hot metal from air and prevents oxidation and scaling. Parts can be partially immersed for localized or selec-tive heating. A *lead pot* is a similar device, where molten lead replaces salt as the heat transfer medium.

Electrical induction is another popular means of heating electrically conductive materials, such as metal. Small parts can be through-heated and hardened. Long products can be heated and quenched in a continuous manner by passing them through a stationary heating coil or by having a moving coil traverse a stationary part. Localized or selective heating can also be performed at rapid production rates. Flexibility is another attractive feature because a standard induction unit can be adapted to a wide variety of products simply by chang-ing the induction coil and adjusting the equipment settings. If the power settings are adjusted, the same unit that austeni-tizes can also be used to temper.

High-temperature **microwave processing** of materials is quickly becoming an industrial reality. Because microwave fur-naces generally heat only the objects to be processed, and not the furnace walls and atmospheres, they are extremely energy efficient. Moreover, conventional heating transfers heat through the outer surface of a material to the interior, while microwave heating, like induction, places the energy directly into the volume of the material. Processing times can be reduced up to 90% with a corresponding decrease of up to 80% in energy consumption, and the environmental impact is also reduced. Originally applied to the sintering of ceramics, micro-wave processing has recently expanded to include powder metal binder removal and sintering, melting of metal, brazing, and surface treating.

Furnace Control

All heat treatment operations should be conducted with rigid control if the desired results are to be obtained in a consistent fashion. Most furnaces are equipped with one or more temper-ature sensors, which can be coupled to a controller or com-puter to regulate the temperature and the rate of heating or cooling. It should be remembered, however, that it is the tem-perature of the workpiece, and not the temperature of the fur-nace, that controls the result, and it is this temperature that should be monitored.

5.8 Heat Treatment and Energy

Because of the elevated temperatures and the time required at those temperatures, heat treatments can consume considera-ble amounts of energy. However, if one considers the broader picture, heat treatment might actually prove to be an energy conservation measure. The manufacture of higher-quality, more-durable products can often eliminate the need for fre-quent replacements. Higher strengths can also permit the use of less material in the manufacture of a product, thereby saving additional energy.

Further savings can often be obtained by integrating the manufacturing operations. For example, a direct quench and temper from hot forging can be used to replace the conven-tional sequence of forge, air cool, reheat, quench, and temper. One should note, however, that the integrated procedure quenches from the conditions of forging, which generally have greater variability in temperature, temperature uniformity, and austenite grain size. If these variations are too great, the addi-tional energy for the reheat and soak might be well justified.

Heat treatment, a $15 billion–$20 billion a year business in the United States, impacts nearly every industrial market sec-tor. It is both capital intensive (specialized and dedicated equipment) and energy intensive. Industry goals currently include reducing energy consumption, reducing processing times, reducing emissions, increasing furnace life, improving heat transfer during heating and cooling, reducing distortion, and improving uniformity of structure and properties, both within a given part and throughout an entire produc-tion quantity.

Review Questions

1. What is heat treatment?

2. What types of properties can be altered through heat treatment?

3. Why should people performing hot forming or welding be aware of the effects of heat treatment?

4. What is the broad goal of the processing heat treatments? Cite some of the specific objectives that might be sought.

5. Why might equilibrium phase diagrams be useful aids in designing and understanding the processing heat treatments?

6. What are the A_1, A_3, and A_{cm} lines?

7. What are some possible objectives of annealing operations?

8. Why might it be important to include a preceding adjective when specifying an annealing operation?

9. Describe the cooling conditions of a full anneal.

10. Why are the hypereutectoid steels not furnace-cooled from the austenite region?

11. Although full anneals often produce the softest and most ductile structures, what might be some of the objections or undesirable features of these treatments?

12. What is the major process difference between full annealing and normalizing?

13. Although normalizing is less expensive than a full anneal, some manufacturers cite cost saving through the use of a full anneal. How is this achieved?

14. What are some of the process heat treatments that can be performed without reaustenitizing the material (heating above the A_1 temperature)?

15. What types of steel would be candidates for a process anneal? Spheroidization?

16. How might steel composition influence the selection of a processing heat treatment?

17. Other than increasing strength, for what three purposes are nonferrous metals often heat treated?

18. What are the six major mechanisms that can be used to increase the strength of a metal?

19. What is the most effective strengthening mechanism for the nonferrous metals?

20. What is required for a metal to be a candidate for precipitation hardening?

21. What are the three steps in an age-hardening treatment? Describe the material structure at the end of each of the stages.

22. What is the difference between a coherent precipitate and a distinct second-phase particle? Why does coherency offer significant strengthening?

23. What is overaging? Why does strength decrease?

24. Describe the various aging responses (maximum attainable strength and time for attaining that strength) that can occur over the range of possible aging temperatures.

25. What is the difference between natural and artificial aging? Which offers more flexibility? Over which does the engineer have more control?

26. Why might naturally aging aluminum rivets be preferred for the assembly of large products like a passenger airplane?

27. Why is it important not to expose precipitation hardened material to subsequent elevated temperatures?

28. Why is it more difficult to understand the nonequilibrium strengthening treatments?

29. What types of heating and cooling conditions are imposed in an I-T or T-T-T diagram? Are they realistic for the processing of commercial items?

30. What are the stable equilibrium phases for steels at temperatures below the A_1 temperature?

31. What are some nonequilibrium structures that appear in the T-T-T diagram for a eutectoid composition steel?

32. Which steel structure is produced by a diffusionless phase change?

33. What is the major factor that influences the strength and hardness of martensite?

34. For a given steel, describe the relative strengths of pearlite, bainite, and martensite structures.

35. Most structure changes proceed to completion over time. The martensite transformation is different. What must be done to produce more martensite in a partially transformed structure?

36. What is retained austenite, and why is it an undesirable structure in heat-treated steels?

37. What types of steels are more prone to retained austenite?

38. Why are martensitic structures usually tempered before being put into use? What properties increase during tempering? Which ones decrease?

39. Why does tempering offer a spectrum of possible structures and companion properties?

40. In what ways is the quench-and-temper heat treatment similar to age hardening? How are the property changes different in the two processes?

41. What is a C-C-T diagram? Why is it more useful than a T-T-T diagram?

42. What is the "critical cooling rate," and how is it determined?

43. What two features combine to determine the structure and properties of a heat-treated steel?

44. What conditions are used to standardize the quench in the Jominy test?

45. How do the various locations of a Jominy test specimen correlate with cooling rate?

46. How does the data collected from a Jominy test correlate with strength?

47. What is the assumption that allows the data from a Jominy test to be used to predict the properties of various locations on a manufactured product?

48. What is hardenability? How is it different from hardness?

49. What capabilities are provided by high-hardenability materials?

50. When selecting a steel for an application, what features influence the selection of carbon content? Alloy content?

51. What are the three stages of liquid quenching?

52. What are some of the major advantages and disadvantages of a water quench?

53. Why does brine provide faster cooling than water?

54. Why is an oil quench less likely to produce quench cracks than water or brine?

55. What are some of the attractive qualities of a polymer or synthetic quench?

56. What are some undesirable design features that might be present in parts that are to be heat treated?

57. What is the cause of thermally induced residual stresses?

58. Why would the residual stresses in steel be different from the residual stresses in an identically processed aluminum part?

59. What causes quench cracking to occur when steel is rapidly cooled?

60. How might the thermally induced residual stresses be reduced or minimized?

61. What are some of the potentially undesirable effects of residual stresses?

62. How does press quenching suppress distortion?

63. Describe several techniques that suppress cracking and distortion by inducing simultaneous transformation of the material.

64. What is thermomechanical processing?

65. What are some of the benefits of ausforming?

66. What are some possible mechanisms for the improvements that have been noted on materials that have been cryogenically processed?

67. What are some of the methods that can be used to selectively alter the surface properties of metal parts?

68. What types of steel can benefit from surface hardening by selective surface heating? What determines maximum surface hardness? The depth of the hardened layer?

69. How can the depth of heating be controlled during induction hardening of surfaces?

70. What are some of the attractive features of surface hardening with a laser beam?

71. How can a laser beam be manipulated and focused? How are these operations performed with an electron beam?

72. What is the major limitation of electron-beam hardening?

73. What is carburizing?

74. Why does a carburized part have to be further heat-treated after the carbon is diffused into the surface? What are the various options?

75. How do carburizing and nitriding differ with regard to subsequent thermal treatment?

76. In what ways might ionitriding be more attractive than conventional nitriding or carburizing?

77. For what type of products or product mixes might a batch furnace be preferred to a continuous furnace?

78. Describe the distinguishing features of a box furnace, car-bottom furnace, bell furnace, elevator furnace, vertical pit furnace, and a rotary-hearth furnace.

79. What are some possible functions of artificial atmospheres in a heat-treating furnace?

80. How are parts heated in a fluidized-bed furnace? What are some of the attractive features?

81. What are some of the potential benefits of microwave processing or heating?

82. Heat treatments consume energy. In what ways might the heat treatment of metals actually be an energy conservation measure?

83. What are some current goals of the heat treatment industry?

Problems

1. This chapter presented four processing-type heat treatments whose primary objective is to soften, weaken, enhance ductility, or promote machinability. Consider each of the following processes as they are applied to steels:

a. Full annealing

b. Normalizing

c. Process annealing

d. Spheriodizing

Provide information relating to

(1) A basic description of how the process works and what its primary objectives are

(2) Typical materials on which the process is performed

(3) Type of equipment used

(4) Typical times, temperatures, and atmospheres required

(5) Recommended rates of heating and cooling

(6) Typical properties achieved

2. A number of different quenchants were discussed in the chapter, including brine, water, oil, synthetic polymer mixes, and even high-pressure gas flow. Select two of these, and investigate the environmental concerns that might accompany their use.

3. It has been noted that *hot oil* is often a more effective quench than *cold oil*. Can you explain this apparent contradiction?

4. Traditional manufacturing generally separates mechanical processing (such as forging, extrusion, presswork, or machining) and thermal processing (heat treatment) and applies them as sequential operations. Ausforming was presented as an example of a thermomechanical process where the mechanical and thermal processes are performed concurrently. When this is done, the resulting structures and properties are often quite different from the traditional. Identify another thermomechanical process and discuss its use and attributes.

5. A number of heat treatments have been devised to harden the surfaces of steel and other engineering metals. Consider the following processes:

a. Flame hardening

b. Induction hardening

c. Laser-beam hardening

d. Carburizing

e. Nitriding

f. Ionitriding

For each of these processes, provide information relating to:

(1) A basic description of how the process works

(2) Typical materials on which the process is performed

(3) Type of equipment required

(4) Typical times, temperatures, and atmospheres required

(5) Typical depth of hardening and reasonable limits

(6) Hardness achievable

(7) Subsequent treatments or processes that might be required

(8) Information relating to distortion and/or stresses

(9) Ability to use the process to harden selective areas

6. Select one of the lesser-known surface modification techniques, such as ion implantation, boriding, chromizing, or other similar technique, and investigate the nine numbered areas of Problem 5.

CHAPTER 6

Ferrous Metals and Alloys

6.1 Introduction to History-Dependent Materials

Engineering materials are available with a wide range of useful properties and characteristics. Some of these are inherent to the particular material, but many others can be varied by controlling the manner of production and the details of processing. Metals are classic examples of such "history-dependent" materials. The final properties are clearly affected by their past processing history. The particular details of the smelting and refining process control the resulting purity and the type and nature of any influential contaminants. The solidification process imparts structural features that might be transmitted to the final product. Preliminary operations such as the rolling of sheet or plate often impart directional variations to properties, and their impact should be considered during subsequent processing and use. Thus, although it is easy to take the attitude that "metals come from warehouses," it is important to recognize that aspects of their prior processing can significantly influence further operations as well as the final properties of a product. The breadth of this book

does not permit full coverage of the processes and methods involved in the production of engineering metals, but certain aspects will be presented because of their role in affecting subsequent performance.

6.2 Ferrous Metals

In this chapter, we will introduce the major **ferrous** (iron-based) **metals** and **alloys**, summarized in **Figure 6.1**. These materials made possible the Industrial Revolution 150 years ago, and they continue to be the backbone of modern civilization. We see them everywhere in our lives—in the buildings where we work, the cars we drive, the homes in which we live, the cans we open, and the appliances that enhance our standard of living. Numerous varieties have been developed over the years to meet the specific needs of various industries. These developments and improvements have continued, with recent decades seeing the introduction of a number of new varieties and even classes of ferrous metals. According to the American Iron and Steel Institute, the number of available grades of steel has

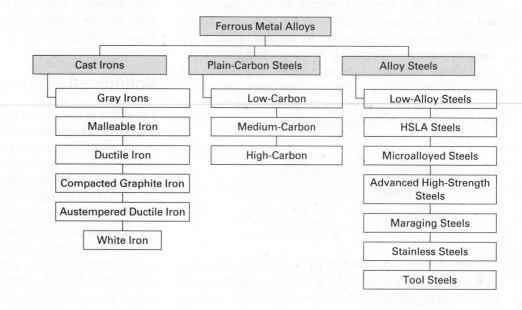

FIGURE 6.1 Classification of common ferrous metals and alloys.

doubled since 2000. The newer steels are stronger than ever, rolled thinner, easier to shape, and more corrosion resistant. As a result, steel still accounts for more than half of the metal used in an average vehicle in North America.

In addition, all steel is recyclable, and this recycling does not involve any loss in material quality. In fact, more steel is recycled each year than all other materials combined, including aluminum, glass, plastic, and paper. Because steel is magnetic, it is easily separated and recovered from demolished buildings, junked automobiles, and discarded appliances. The overall recycling rate for steel is approximately 88%–92.5% for automobiles, 90% for appliances, and 72% for steel packaging. Structural beams and plates are the most recycled products at 97.5%. Two-thirds of all new steel is produced from old steel, and each ton of recycled steel saves more than 4,000 pounds of raw materials (including 1,400 pounds of coal) and 74% of the energy required to make new steel.

6.3 Iron

For centuries, **iron** has been the most important of the engineering metals. Although iron is the fourth most plentiful element in the earth's crust, it is rarely found in the metallic state. Instead, it occurs in a variety of mineral compounds, known as ores, the most attractive of which are iron oxides coupled with companion impurities. To produce metallic iron, the ores are processed in a manner that breaks the iron–oxygen bonds (chemical reducing reactions). Ore, limestone, coke (carbon), and air are continuously introduced into specifically designed furnaces, and molten metal is periodically withdrawn.

Within the furnace, other oxides (that were impurities in the original ore) will also be reduced. All of the phosphorus and most of the manganese will enter the molten iron. Oxides of silicon and sulfur compounds are partially reduced, and these elements also become part of the resulting metal. Other contaminant elements, such as calcium, magnesium, and aluminum, are collected in the limestone-based slag and are largely removed from the system. The resulting **pig iron** tends to have roughly the following composition:

Carbon	3.0%–4.5%
Manganese	0.15%–2.5%
Phosphorus	0.1%–2.0%
Silicon	1.0%–3.0%
Sulfur	0.05%–0.1%

Although the bulk of molten pig iron is further processed into steel, a small portion is cast directly into final shape and is classified as cast iron. Most commercial cast irons, however, are produced by recycling scrap iron and steel, with the possible addition of some newly produced pig iron. Cast iron was introduced in Chapter 4, and the various types of cast iron will

be discussed later in this chapter. The conversion of this material to cast products will be developed when we present the casting processes in Chapters 13 through 15.

6.4 Steel

Steel is an extremely useful engineering material, offering strength, rigidity, and durability. From a manufacturing perspective, its formability, joinability, and paintability, as well as repairability, are all attractive characteristics. For the past 20 years, steel has accounted for about 55% of the weight of a typical passenger car. Although the automotive and construction industries are indeed the major consumers of steel, the material is also used extensively in containers, appliances, and machinery and provides much of the infrastructure of such industries as oil and gas.

The manufacture of steel is essentially an oxidation process that decreases the amount of carbon, silicon, manganese, phosphorus, and sulfur in a molten mixture of pig iron and/or steel scrap. In 1856, the Kelly–Bessemer process opened up the industry by enabling the manufacture of commercial quantities of steel. The open-hearth process surpassed the Bessemer process in tonnage produced in 1908 and was producing more than 90% of all steel in 1960. Most of our commercial steels are currently produced by a variety of oxygen and electric arc furnaces.

In many of the steelmaking processes, air or oxygen passes over or through the molten metal to drive a variety of exothermic refining reactions. Carbon oxidizes to form gaseous CO or CO_2, which then exits the melt. Other elements, such as silicon and phosphorus, are similarly oxidized and, being lighter than the metal, rise to be collected in a removable slag. At the same time, however, oxygen and other elements from the reaction gases dissolve in the molten metal and might later become a cause for concern.

Solidification Concerns

Regardless of the method by which the steel is made, it must undergo a change from liquid to solid before it can become a usable product. Prior to solidification, however, we want to remove as much contamination as possible. The molten metal is first poured from the steelmaking furnaces into containment vessels, known as **ladles**. Historically, the ladles simply served as transfer and pouring containers, but they now provide a site for additional processing. **Ladle metallurgy** refers to a variety of processes designed to provide final purification and to fine-tune both the chemistry and temperature of the melt. Alloy additions can be made, carbon can be further reduced, and dissolved gases can be reduced or removed. Stirring, degassing, and reheating can all be performed.

The liquid can then be poured into molds to produce finish-shape steel castings, or solidified into a form suitable for

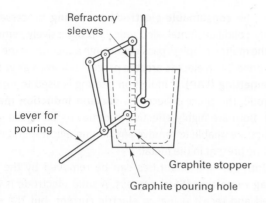

Diagram of a bottom-pouring ladle.

further processing. In most cases, some form of **continuous casting** produces the feedstock material for subsequent forging or rolling operations. The molten metal is usually introduced

into the continuous caster through a bottom-pouring process such as the one shown schematically in **Figure 6.2**. By extracting the metal from the bottom of the ladle, slag and floating matter are not transferred, and a cleaner product results. **Figure 6.3a** illustrates a typical continuous caster, in which molten metal flows from a ladle, through a tundish, into a bottomless, water-cooled mold, usually made of copper. Cooling is controlled to ensure that the outside has solidified before the metal exits the mold. Closely spaced rolls support the emerging product, and direct water sprays complete the solidification. The newly solidified metal can then be cut to desired length, or, because the cast solid is still hot, it can be bent and fed horizontally through a short reheat furnace or directly to a rolling operation. By varying the size and shape of the mold, products can be cast with a variety of cross sections with names such as slab, bloom, billet, and strand. **Figure 6.3b** depicts the simultaneous casting of multiple strands. Compared to the casting of discrete ingots, continuous casting offers significant reduction

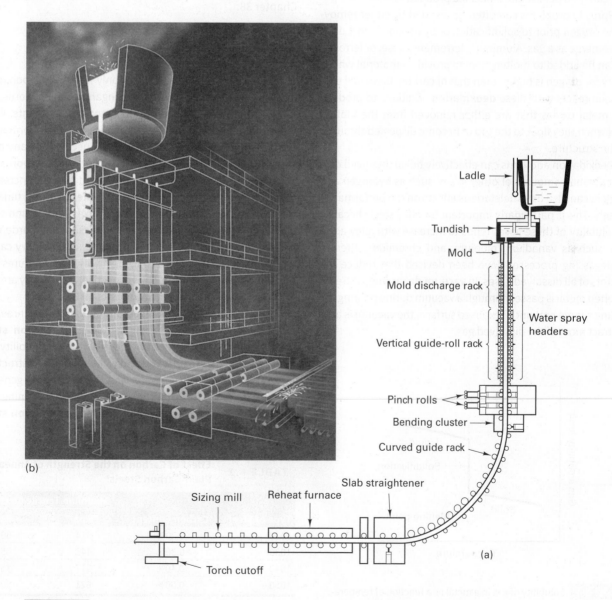

FIGURE 6.3 (a) Schematic representation of the continuous casting process for producing billets, slabs, and bars. (b) Simultaneous continuous casting of multiple strands. (Reproduced with permission from Penton Media)

in cost, energy, and scrap. In addition, the products have improved surfaces, more uniform chemical composition, and fewer oxide inclusions.

Deoxidation and Degasification

During the steelmaking process, large amounts of oxygen can become dissolved in the molten metal. During the subsequent cooling and solidification, the solubility levels decrease significantly, as shown in **Figure 6.4**. The excess oxygen can no longer be held within the material and frequently links with carbon to produce carbon monoxide gas. This gas might then escape through the liquid or might become trapped to produce pores within the solid, ranging from small, dispersed voids to large blowholes. Although these pores can often be welded shut during subsequent hot forming, some might not be fully closed, and others might not weld on closure. Cracks and internal voids can then persist into a finished product.

Porosity problems can often be avoided by either removing the oxygen prior to solidification or by making sure it does not reemerge as a gas. Aluminum, ferromanganese, or ferrosilicon can be added to molten steel to provide a material whose affinity for oxygen is higher than that of carbon. Dissolved oxygen then reacts with these **deoxidation** additions to produce solid metal oxides that are either removed from the molten metal when they float to the top or become dispersed throughout the structure.

Deoxidation additions can effectively tie up dissolved oxygen, but small amounts of other gases, such as hydrogen and nitrogen, can also have deleterious effects on the performance of steels. This is particularly important for alloy steels because the solubility of these gases tends to increase with alloy additions, such as vanadium, niobium, and chromium. Alternative degassing processes have been devised that reduce the amounts of all dissolved gases. In **vacuum degassing**, a stream of molten metal is passed through a vacuum during pouring. By creating a large amount of exposed surface, the vacuum is able to extract most of the dissolved gas.

In the **consumable-electrode remelting** processes, an already solidified metal electrode is progressively remelted, with the molten droplets passing through a vacuum. If the melting is done by an electric arc, the process is known as **vacuum arc remelting (VAR)**. If induction heating is used to melt the electrode, the process becomes **vacuum induction melting (VIM)**. Both are highly effective in removing dissolved gases, but they are unable to remove any nonmetallic impurities that might be present in the metal.

Both gas and impurities can be removed by the **electroslag remelting (ESR)** process. A solid electrode is again melted and recast using an electric current, but the entire remelting is conducted under a blanket of molten flux. Nonmetallic impurities float and are collected in the flux, and the progressive freezing permits easy escape for the rejected gas. The result is a newly solidified metal structure with much-improved quality. This process is simply a large-scale version of the electroslag welding process that will be discussed in Chapter 38.

Plain-Carbon Steel

Although theoretically an alloy of only iron and carbon, commercial steel actually contains manganese, phosphorus, sulfur, and silicon in significant and detectable amounts. When these four additional elements are present in their normal percentages and no minimum amount is specified for any other constituent, the product is referred to as **plain-carbon steel**. Strength is primarily a function of carbon content, increasing with increasing carbon, as shown in **Table 6.1**. Unfortunately, the ductility, toughness, and weldability of plain-carbon steels decrease as the carbon content is increased, and hardenability is quite low. In addition, the properties of ordinary carbon steels are diminished by both high and low temperatures (loss of strength and embrittlement, respectively), and they are subject to corrosion in most environments.

Plain-carbon steels are generally classified into three subgroups based on their carbon content. **Low-carbon steels** have less than 0.20% carbon and possess good formability (can be strengthened by cold work) and weldability. Their structures are usually ferrite and pearlite, and the material is generally used as it comes from the hot-forming or cold-forming processes, or in the as-welded condition. **Medium-carbon steels**

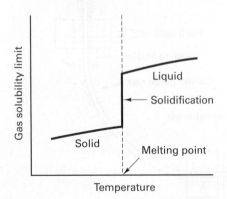

FIGURE 6.4 Solubility of gas in a metal as a function of temperature showing significant decrease upon solidification.

TABLE 6.1	Effect of Carbon on the Strength of Annealed Plain-Carbon Steels[a]		
		Minimum Tensile Strength	
Type of Steel	Carbon Content	Mpa	ksi
1020	0.20%	414	60
1030	0.30%	448	65
1040	0.40%	517	75
1050	0.50%	621	90

[a]Data are from ASTM Specification A732.

have between 0.20% and 0.50% carbon, and they can be quenched to form martensite or bainite if the section size is small and a severe water or brine quench is used. The best balance of properties is obtained at these carbon levels, where the high toughness and ductility of the low-carbon material is in good compromise with the strength and hardness that come with higher carbon contents. These steels are extremely popular and find numerous mechanical applications. **High-carbon steels** have more than 0.50% carbon. Toughness and formability are low, but hardness and wear resistance are high. Severe quenches can form martensite, but hardenability is still poor. Quench cracking is often a problem when the material is pushed to its limit. **Figure 6.5** depicts the characteristic properties of low-, medium-, and high-carbon steels using a balance of properties that shows the offsetting characteristics of "strength and hardness" and "ductility and toughness."

Compared to other engineering materials, the plain carbon steels offer high strength and high stiffness, coupled with reasonable toughness. Unfortunately, they also rust easily and generally require some form of surface protection, such as paint, galvanizing, or other coating. Because the plain carbon steels are generally the lowest-cost steel material, they are often given first consideration for many applications. Their limitations, however, might become restrictive. When improved performance is required, these steels can often be upgraded by the addition of one or more alloying elements.

Alloy Steels

The differentiation between plain carbon and alloy steel is often somewhat arbitrary. Both contain carbon, manganese, and usually silicon. Copper and boron are possible additions to both classes. Steels containing more than 1.65% manganese, 0.60% silicon, or 0.60% copper are usually designated as **alloy steels**. Also, a steel is considered to be an alloy steel if a definite or minimum amount of other alloying element is specified. The most common alloy elements are chromium, nickel, molybdenum, vanadium, tungsten, cobalt, boron, and copper, as well as manganese, silicon, phosphorus, and sulfur in amounts greater than are normally present. If the steel contains less than 8% of total alloy addition, it is considered to be a **low-alloy steel**. Steels with more than 8% alloying elements are **high-alloy steels**.

In general, alloying elements are added to steels in small percentages (usually less than 5%) to improve strength or hardenability, or in much larger amounts (often up to 20%) to produce special properties such as corrosion resistance or stability at high or low temperatures. Additions of manganese,

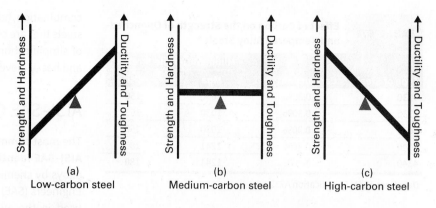

FIGURE 6.5 A comparison of low-carbon, medium-carbon, and high-carbon steels in terms of their relative balance of properties. (a) Low-carbon has excellent ductility and fracture resistance, but lower strength. (b) Medium-carbon has balanced properties. (c) High-carbon has high strength and hardness at the expense of ductility and fracture resistance.

silicon, or aluminum can be made during the steelmaking process to remove dissolved oxygen from the melt. Manganese, silicon, nickel, and copper add strength by forming solid solutions in ferrite. Chromium, vanadium, molybdenum, tungsten, and other elements increase strength by forming dispersed second-phase carbides. Nickel and copper can be added in small amounts to improve corrosion resistance. Nickel has been shown to impart increased toughness and impact resistance, and molybdenum helps resist embrittlement. Zirconium, cerium, and calcium can also promote increased toughness by controlling the shape of inclusions. Machinability can be enhanced through the formation of manganese sulfides or by additions of lead, bismuth, selenium, or tellurium. Still other additions can be used to provide ferrite or austenite grain-size control. Phosphorus and sulfur, as well as hydrogen, oxygen, and nitrogen are considered to be undesirable elements.

Compared to carbon steels, the alloy steels can be considerably stronger. For a specified level of strength, the ductility and toughness tend to be higher. For a specified level of ductility or toughness, the strength is higher. In essence, there is an improvement in the overall combination of properties. Mechanical properties at both low and high temperatures can be increased, but all these improvements come with an increase in cost. Weldability, machinability, and formability generally decline.

Selection of an alloy steel still begins with identifying the proper carbon content. **Table 6.2** shows the effect of carbon on the strength of quenched-and-tempered alloy steels. The strength values are significantly higher than those of Table 6.1, reflecting the difference between the annealed and quenched-and-tempered microstructures. The 4130 steel has about 1.2% total alloying elements, 4330 has 3.0%, and 8630 has about 1.3%, yet all have the same quenched-and-tempered tensile strength. Strength and hardness depend primarily on carbon content. The primary role of an alloy addition, therefore, is usually to increase **hardenability**, but other effects such as

TABLE 6.2	Effect of Carbon on the Strength of Quenched-and-Tempered Alloy Steels [a]		
		Minimum Tensile Strength	
Type of Steel	Carbon Content	Mpa	ksi
4130	0.30%	1030	150
4330	0.30%	1030	150
8630	0.30%	1030	150
4140	0.40%	1241	180
4340	0.40%	1241	180

[a]Data from ASTM Specification A732.

modified toughness or machinability are also possible. The most common hardenability-enhancing elements (in order of decreasing effectiveness) are manganese, molybdenum, chromium, silicon, and nickel. Boron is an extremely powerful hardenability agent. Only a few thousandths of a percent are sufficient to produce a significant effect in low-carbon steels, but the results diminish rapidly with increasing carbon content. Because no carbide formation or ferrite strengthening accompanies the addition, improved machinability and cold-forming characteristics might favor the use of boron in place of other hardenability additions. Small amounts of vanadium can also be quite effective, but the response drops off as the quantity is increased.

Table 6.3 summarizes the primary effects of the common alloying elements in steel. A working knowledge of this information might be useful in selecting an alloy steel to meet a given set of requirements. Alloying elements are often used in combination, however, resulting in the immense variety of alloy steels that are commercially available. To provide some degree of simplification, a classification system has been developed and has achieved general acceptance in a variety of industries.

AISI–SAE Classification System

The most common classification scheme for alloy steels is the **AISI–SAE identification system**. This system, which classifies alloys by chemistry, was started by the Society of Automotive Engineers (SAE) to provide some standardization for the steels used in the automotive industry. It was later adopted and expanded by the American Iron and Steel Institute (AISI) and has been incorporated into the Universal Numbering System that was developed to include all engineering metals. Both plain-carbon and low-alloy steels are identified by a four-digit number, where the first number indicates the major alloying elements and the second number designates a subgrouping within the major alloy system. These first two digits can be interpreted by consulting a list, such as the one presented in **Table 6.4**. The last two digits (three if the number contains five digits) indicate the approximate amount of carbon, expressed as "points," where one point is equal to 0.01%. Thus, a 1080 steel would be a plain carbon steel with 0.80% carbon. Similarly, a 4340 steel would be a Mo–Cr–Ni alloy with 0.40% carbon. Because of the meanings associated with the numbers, the designation is not read as a series of single digits, such as four-three-four-zero, but as a pair of double-digit groupings, such as ten-eighty or forty-three forty.

TABLE 6.3	Principal Effects of Alloying Elements in Steel	
Element	**Percentage**	**Primary Function**
Aluminum	0.95–1.30	Alloying element in nitriding steels
Bismuth	—	Improves machinability
Boron	0.001–0.003	Powerful hardenability agent
Chromium	0.5–2	Increase of hardenability
	4–18	Corrosion resistance
Copper	0.1–0.4	Corrosion resistance
Lead	—	Improved machinability
Manganese	0.25–0.40	Combines with sulfur to prevent brittleness
	>1	Increases hardenability by lowering transformation points and causing transformations to be sluggish
Molybdenum	0.2–5	Stable carbides; inhibits grain growth
Nickel	2–5	Toughener
	12–20	Corrosion resistance
Silicon	0.2–0.7	Increases strength and hardenability
	2	Spring steels
	Higher percentages	Improves magnetic properties
Sulfur	0.08–0.15	Free-machining properties
Titanium	—	Fixes carbon in inert particles
		Reduces martensitic hardness in chromium steels
Tungsten	—	Hardness at high temperatures
Vanadium	0.15	Stable carbides; increases strength while retaining ductility, Promotes fine grain structure

TABLE 6.4   AISI– SAE Standard Steel Designations and Associated Chemistries

AISI Number	Type	Alloying Elements (%)					
		Mn	Ni	Cr	Mo	V	Other
1xxx	Carbon steels						
10xx	Plain carbon						
11xx	Free cutting (S)						
12xx	Free cutting (S) and (P)						
13xx	High manganese	1.60–1.90					
15xx	High manganese						
2xxx	Nickel steels		3.5–5.0				
3xxx	Nickel–chromium		1.0–3.5	0.5–1.75			
4xxx	Molybdenum						
40xx	Mo				0.15–0.30		
41xx	Mo, Cr			0.40–1.10	0.08–0.35		
43xx	Mo, Cr, Ni		1.65–2.00	0.40–0.90	0.20–0.30		
44xx	Mo				0.35–0.60		
46xx	Mo, Ni (low)		0.70–2.00		0.15–0.30		
47xx	Mo, Cr, Ni		0.90–1.20	0.35–0.55	0.15–0.40		
48xx	Mo, Ni (high)		3.25–3.75		0.20–0.30		
5xxx	Chromium						
50xx				0.20–0.60			
51xx				0.70–1.15			
6xxx	Chromum–vanadium						
61xx				0.50–1.10		0.10–0.15	
8xxx	Ni, Cr, Mo						
81xx			0.20–0.40	0.30–0.55	0.08–0.15		
86xx			0.40–0.70	0.40–0.60	0.15–0.25		
87xx			0.40–0.70	0.40–0.60	0.20–0.30		
88xx			0.40–0.70	0.40–0.60	0.30–0.40		
9xxx	Other						
92xx	High silicon						1.20–2.20Si
93xx	Ni, Cr, Mo		3.00–3.50	1.00–1.40	0.08–0.15		
94xx	Ni, Cr, Mo		0.30–0.60	0.30–0.50	0.08–0.15		

AISI-SAE Steel Identification

If an individual informs you that he is working with a four-one-four-five steel, what information has he provided you?

a. *The steel is an alloy steel.*

b. *The steel contains approximately 0.45% carbon.*

c. *The steel is a molybdenum-containing steel with additional chromium.*

d. *He is totally unfamiliar with alloy steels and the identification system.*

Correct answer – all of the above! It is a forty-one forty-five steel.

Letters can also be incorporated into the designation. The letter *B* between the second and third digits indicates that the base metal has been supplemented by the addition of boron. Similarly, an *L* in this position indicates a lead addition for enhanced machinability. A letter prefix might also be employed to designate the process used to produce the steel, such as *E* for electric furnace.

When hardenability is a major requirement, one might consider the H grades of AISI steels, designated by an *H* suffix attached to the standard designation. The chemistry specifications are somewhat less stringent, but the steel must now meet a hardenability standard. The hardness values obtained for each location on a Jominy test specimen (see Chapter 5) must fall within a predetermined band for that particular type of steel. When the AISI designation is followed by an RH suffix (restricted-hardenability), an even narrower range of hardness values is imposed.

Other designation organizations, such as the American Society for Testing and Materials (ASTM) and the U.S. government (MIL and federal), have specification systems based more on specific products and applications. Acceptance into a given classification is generally determined by physical or mechanical properties rather than the chemistry of the metal. ASTM designations are often used when specifying structural steels.

Selecting Alloy Steels

From the previous discussion it is apparent that two or more alloying elements can often produce similar effects. When properly heat-treated, steels with substantially different chemical compositions can possess almost identical mechanical properties. **Figure 6.6** clearly demonstrates this fact, which becomes particularly important when one realizes that some alloying elements can be very costly and others might be in short supply or difficult to obtain. Overspecification has often been employed to guarantee success despite sloppy manufacturing and heat-treatment practice. The correct steel, however, is usually the least expensive one that can be consistently processed to achieve the desired properties. This usually involves taking advantage of the effects provided by all of the alloy elements.

When selecting alloy steels, it is also important to consider both use and fabrication. For one product, it might be permissible to increase the carbon content to obtain greater strength.

For another application, such as the one involving assembly by welding, it might be best to keep the carbon content low and use a balanced amount of alloy elements, obtaining the desired strength while minimizing the risk of weld cracking. Steel selection involves defining the required properties, determining the best microstructure to provide those properties (strength can be achieved through alloying, cold work, and heat treatment, as well as combinations thereof), determining the method of part or product manufacture (casting, machining, metal forming, etc.), and then selecting the steel with the best carbon content and hardenability characteristics to facilitate those processes and achieve the desired goals.

High-Strength Steels (HSLA, Microalloyed, and Bake-Hardenable)

Among the general categories of alloy steels are (1) the **constructional alloys**, where the desired properties are typically developed by a separate thermal treatment and the specific alloy elements tend to be selected for their effect on hardenability; (2) **conventional high-strength steels**, which rely largely on chemical composition to develop the desired properties in a single-phase ferritic microstructure, usually in the as-rolled or normalized condition; and (3) **advanced high-strength steels (AHSS)**, which are primarily multiphase steels (ferrite, martensite, bainite, and/or retained austenite) that provide high strength with unique mechanical properties. The constructional alloys are usually purchased by AISI–SAE

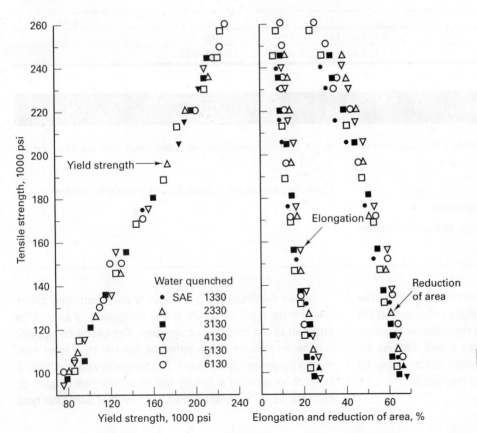

FIGURE 6.6 Relationships between the mechanical properties of a variety of properly heat-treated AISI–SAE 0.3% carbon alloy steels. (Courtesy of ASM International, Materials Park, OH)

identification, which effectively specifies chemistry. The high-strength designations generally focus on product (size and shape) and desired properties. When steels are specified by mechanical properties, the supplier or producer is free to adjust the chemistry (within limits), and substantial cost savings can result. To ensure success, however, it is important that all of the necessary properties be specified.

The conventional high-strength steels provide increased strength-to-weight, good weldability, and acceptable corrosion resistance for only a modest increase in cost compared to the low-carbon, plain carbon steels, often referred to as **mild steels**. Because of their higher yield strength (up to 690–760 MPa or 100–110 ksi), weight savings of 20% to 30% can often be achieved with no sacrifice to strength or safety. They are available in a variety of forms, including sheet, strip, plate, structural shapes, and bars. Ductility and hardenability might be somewhat limited, however. The increase in strength, and the resistance to martensite formation in a weld zone, is obtained by controlling the amounts of carbon (0.05% to 0.25%), manganese (up to 2.0%), and silicon, with the addition of small amounts of niobium, vanadium, titanium, or other alloys. About 0.2% copper can be added to improve corrosion resistance.

One of the largest groups of conventional high-strength steels, the **high-strength low-alloy (HSLA)**, or **microalloyed steels**, occupy a position between carbon steels and the quenched-and-tempered alloy grades in terms of both cost and performance and are being used increasingly as substitutes for heat-treated steels in the manufacture of small- to medium-sized discrete parts. These low- and medium-carbon steels contain solid-solution-strengthening alloys (such as manganese and silicon), along with small amounts (0.05% to 0.15%) of alloying elements, such as niobium, vanadium, titanium, molybdenum, zirconium, boron, rare earth elements, or combinations thereof. Many of these additions form alloy carbides, nitrides, or carbonitrides, whose primary effect is to provide grain refinement and/or precipitation strengthening. Yield strengths between 500 and 750 MPa (70 and 110 ksi) can be obtained without heat treatment. Weldability can be retained or even improved if the carbon content is simultaneously decreased. In essence, these steels offer maximum strength with minimum carbon, while simultaneously preserving weldability, machinability, and formability. Compared to a quenched-and-tempered alternative, however, ductility and toughness are generally somewhat inferior.

Cold-formed HSLA or microalloyed steels require less cold work to achieve a desired level of strength, so they tend to have greater residual ductility. Hot-formed products, such as forgings, can often be used in the air-cooled condition. By means of accurate temperature control and controlled-rate cooling directly from the forming operation, mechanical properties can be produced that approximate those of quenched-and-tempered material. Machinability, fatigue life, and wear resistance can all be superior to those of the heat-treated counterparts. In applications where the properties are adequate, microalloyed steels can often provide attractive cost savings. Energy savings can be substantial, straightening or stress relieving after heat treatment is no longer necessary, and quench cracking is not a problem. Because of the increase in material strength, the size and weight of finished products can often be reduced. As a result, the cost of a finished product can be reduced by 5% to 25%.

Bake-hardenable steels are low-carbon steels that are processed in such a way that they are resistant to aging during normal storage but begin to age during sheet-metal forming. A subsequent exposure to heat during the paint-baking operation (a finishing operation in automotive manufacture) completes the aging process and adds an additional 35 to 70 MPa (5 to 10 ksi), raising the final yield strength to approximately 275 MPa (40 ksi). Because the increase in strength occurs after the forming operation, the material offers good formability coupled with improved dent resistance in the final product. In addition, it allows weight savings to be achieved without compromising the attractive features of steel sheet, which include spot weldability, good crash energy absorption, low cost, and full recyclability.

Advanced High-Strength Steels (AHSS)

Traditional methods of producing high-strength steel have included adding carbon and/or alloy elements followed by heat treatment or cold-working to a high level followed by a partial anneal to restore some ductility. As strength increased, however, ductility and toughness decreased and often became a limiting feature. The high-strength low-alloy (HSLA) and microalloyed steels, introduced about 40 years ago, used thermomechanical processing to further increase strength, but were accompanied by an even further decline in ductility. Beginning in the mid-1990s, enhanced thermomechanical processing capabilities and controls have led to the development of a variety of new high-strength steels that go collectively by the name **advanced high-strength steels (AHSS)**. Many were developed for weight savings in automotive applications (higher strength enabling reduced size or thickness) while preserving or enhancing energy absorption. As a result, large amounts of low-carbon and HSLA steels are being replaced by the advanced high-strength steels. One group (including the dual-phase [DP] and transformation-induced plasticity [TRIP] steels) provides greater formability for already-existing levels of strength. Another group (including complex-phase and martensitic varieties) provides higher levels of strength while retaining current levels of ductility. Because of the improved formability, the AHSS materials can often be stamped or formed into more complex parts. Parts can often be integrated into single pieces, eliminating the cost and time associated with assembly, and the higher strength can provide weight reduction accompanied by improved fatigue and crash performance.

The various types of advanced high-strength steels are primarily ferrite-phase, soft steels with varying amounts of martensite, bainite, or retained austenite that offer high strength with enhanced ductility. Each of the types will be described briefly next.

Dual-phase steels form when we cool material to a temperature that is above the A_1 but below the A_3 to form a structure that consists of ferrite and high-carbon austenite, and then follow with a rapid-cool quench. During the quench, the ferrite remains unaffected, while the high-carbon austenite transforms to high-carbon martensite. A low- or medium-carbon steel now has a mixed microstructure of a continuous, weak, ductile ferrite matrix combined with islands of high-strength, high-hardness, high-carbon martensite. The dual-phase structure offers strengths that are comparable to the conventional high-strength materials, coupled with improved forming characteristics and no loss in weldability. The high work-hardening rates and excellent elongation lead to a high ultimate tensile strength (590 to 1400 MPa or 85 to 200 ksi) coupled with low initial yield strength. The high strain-rate sensitivity means that the faster the steel is crushed, the more energy it absorbs—a feature that further enhances the crash resistance of automotive structures. Dual-phase steels also exhibit the bake-hardenable effect, a precipitation-induced increase in yield strength when stamping or forming is followed by the elevated temperature of a paint-bake oven.

Whereas the dual-phase steels have structures of ferrite and martensite, **transformation-induced plasticity (TRIP) steels** contain a matrix of ferrite combined with hard martensite or bainite and at least five volume percent of retained austenite. Because of the hard phases dispersed in the soft ferrite, deformation behavior begins much like the dual-phase steels. At higher strains, however, the retained austenite transforms progressively to martensite, enabling the high work-hardening to persist to greater levels of deformation. At lower levels of carbon, the austenite transformation begins at lower levels of strain, and the extended ductility of the lower-carbon TRIP steels offers significant advantages in operations such as stretch-forming and deep-drawing. At higher carbon levels, the retained austenite is more stable and requires greater strains to induce transformation. If the retained austenite can be carried into a finished part, subsequent deformation (such as a crash) can induce transformation. The conversion of retained austenite to martensite, and the companion high rate of work-hardening, can then be used to provide excellent energy absorption. Tensile strengths range from 590 to 1180 MPa (85 to 170 ksi).

Complex-phase (CP) steels and **martensitic (Mart) steels** offer even higher strengths with useful capacity for deformation and energy absorption. The CP steels have a microstructure of ferrite and bainite, combined with small amounts of martensite, retained austenite, and pearlite and are strengthened further by grain refinement created by a fine precipitate of niobium, titanium, or vanadium carbides or nitrides. Strength ranges from 800 to 1180 MPa (115 to 170 ksi). The Mart steels are almost entirely martensite and can have tensile strengths up to 1700 MPa (245 ksi) depending on carbon content, but elongation and formability are lower than the other types.

Still other types are making the transition from research into production. These include the *Ferritic-Bainitic Steels (FB)*—also known as Stretch-Flangeable (SF) and High Hole Expansion (HHE) because of the improved stretch formability of sheared edges—*Twinning-Induced Plasticity Steels (TWIP), Nano steels,* and others. The FB steels have a microstructure of soft ferrite and hard bainite, coupled with grain refinement. Formability is excellent, and weldability and fatigue properties are both good. TWIP steels contain between 17% and 24% manganese, making the steel fully austenitic at room temperature. Deformation occurs by twinning inside the grains, with the newly created twin boundaries providing increased strength and a high rate of strain hardening. The result is a high strength (more than 1000 MPa or 145 ksi) combined with extremely high ductility (as high as 70% elongation). Nano steels replace the hard phases that are present in the DP and TRIP steels with an array of ultra-fine nano-sized precipitates (diameters less than 10 nm). During sheet forming operations, fractures often initiate at the interface between the very soft and very hard phases. Because the interfaces are so small in the nano steels, these fractures are avoided, and formability is increased.

Hot-forming steel (HF) offers tensile strengths in excess of 1600 MPA (230 ksi) with elongation of 4%–8%. A manganese-boron steel is heated to above 900°C (1650°F) and then rapidly transferred to a press, where it is shaped while the structure is austenite, the strength is less than 100 MPa (15 ksi), and ductility is excellent. The part then remains in the water-cooled die (with a cooling rate in excess of 50°C/sec), where the structure transforms to martensite in a process known as **press hardening**. Complex shapes can be produced with little or no springback and final strength of 1500 to 1800 MPa (215 to 260 ksi).

Figure 6.7 shows the relative strengths and formability (elongation) of mild steel, conventional high-strength steels, and the newer AHSS materials. Note how the newer AHSS steels increase strength or formability or both. Also included is

Nomenclature of Advanced High-Strength Steels (AHSS)

More than 30 different AHSS steels are projected to be commercially available between 2015 and 2020 offering enhancements in strength and formability. The specific grades are identified by the type-designating letters [Dual phase = DP; Transformation-Induced Plasticity = TRIP; Complex Phase = CP; Martensite = MS; Hot Formed = HF;

and Twinning-Induced Plasticity = TWIP] followed by the minimum yield strength and minimum tensile strength in megapascals, where 6.9 MPa is equivalent to 1 ksi. As an example, DP 350/600 is a dual-phase steel with a 350 MPa (50.7 ksi) minimum yield strength, and a 600 MPa (87 ksi) minimum tensile strength.

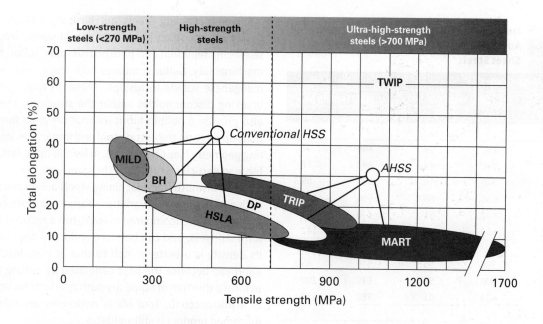

FIGURE 6.7 Relative strength and formability (elongation) of conventional, high-strength low-alloy, and advanced high-strength steels. BH = bake hardenable; DP = dual phase; and Mart = martensitic. Also included is the newer TWIP steels.

TABLE 6.5 Material Content of a North American Light Vehicle in 1975 and 2007 with a Projection for 2015

	Material Content in Pounds			
	1975	**2007**	**2015**	**Change from 1975 to 2015**
Mild Steel	2,180	1,748	1,314	Down 866 lbs.
High Strength Steel	140	334	315	Up 175 lbs.
Advanced HSS	—	149	403	Up 403 lbs.
Other Steels	65	76	77	Up 12 lbs.
Iron	585	284	244	Down 341 lbs.
Aluminum	84	327	374	Up 290 lbs.
Magnesium	—	9	22	Up 22 lbs.
Other Metals	120	149	145	Up 25 lbs.
Plastic/Composites	180	340	364	Up 184 lbs.
Other Materials	546	634	650	Up 104 lbs.
Total Pounds	3,900	4,050	3,908	Up 8 lbs.

Note: Data from Drucker Worldwide presentation at the AISI "Great Designs in Steel" Conference, March, 2007.

the TWIP steels, which occupy a region of very high formability that lies outside of the overlapping band. Some useful distinctions between low-strength steel (UTS below 270 MPa or 40 ksi), high-strength steel, and ultra-high-strength steel (UTS above 700 MPa or 100 ksi) have also been included in this figure.

Table 6.5 shows the material content of a North American light vehicle for 1975 and 2007 with an estimate for 2015. **Table 6.6** looks only at the metal content of the vehicle body and enclosure. Note the increased role of the advanced high-strength steels! Automotive forecasts in the 1960s, 1980s, and into the 1990s all predicted an increasing role for the lightweight alternative materials, such as aluminum, magnesium, and fiber-reinforced polymers and a declining role for steel. Recent studies, however, have shown that the percentage of steel has actually risen over the years. This has largely been because of the development of new types of

steels that are stronger, more energy absorbent, and easy to fabricate into thinner-gage, reduced-weight structures. Alloy development continues along with research into improved

TABLE 6.6 Metallic Material Content of the Body and Enclosure of a North American Light Vehicle

Material	2007 Percentage	Projected 2015 Percentage
Mild Steel	54.6%	29.0%
Bake-Hardenable and Medium HSS	22.4%	23.5%
Conventional HSS	12.7%	10.2%
Advanced HSS	9.5%	34.8%
Aluminum and Magnesium	0.8%	2.5%

Note: Data from Drucker Worldwide presentation at the AISI "Great Designs in Steel" Conference, March, 2007.

TABLE 6.7	Mechanical Properties of Various Grades of Advanced High-Strength Automotive Sheet Steels [a]			

Cold-Rolled AHSS Grades

Type	Min. Yield Str		Min. Tensile Str.	
	MPa	ksi	MPa	ksi
Dual-phase				
	250	36	440	64
	290	42	490	71
	340	49	590	85.5
	550	80	690	100
	420	61	780	113
	550	80	980	142
TRIP Steels				
	380	55	590	85.5
	400	58	690	100
	420	61	780	113
Martensitic Steels				
	700	101.5	900	130.5
	860	125	1100	1
	1030	149.5	1300	188.5
	1200	174	1500	217.5

Hot-Rolled AHSS Grades

Type	Min. Yield Str		Min. Tensile Str.	
	MPa	ksi	MPa	ksi
Dual-phase				
	300	43.5	590	85.5
	380	55	780	113
TRIP Steels				
	400	58	590	85.5
	450	65	780	113

[a]Data from society of Automotive Engineers document SAE J2745, July, 2007.

processing methods including induction heating and innovative cooling systems. **Table 6.7** provides the mechanical properties of some grades of advanced high-strength automotive sheet steels.

Free-Machining Steels

The increased use of high-speed, automated machining has spurred the use and development of several varieties of **free-machining steels**. These steels machine readily and form small chips when cut. The smaller chips reduce the length of contact between the chip and cutting tool, thereby reducing the associated friction and heat, as well as required power and wear on the cutting tool. The formation of small chips also reduces the likelihood of chip entanglement in the machine and makes chip removal much easier. On the negative side, free-machining steels often carry a cost premium of 15% to 20% over conventional alloys, but this increase can be easily recovered through higher machining speeds, larger depths of cut, and extended tool life.

Free-machining steels are basically carbon steels that have been modified by an addition of sulfur, lead, bismuth, selenium, tellurium, or phosphorus plus sulfur to enhance machinability. Sulfur combines with manganese to form soft manganese sulfide inclusions. These, in turn, serve as chip-breaking discontinuities within the structure. The inclusions also provide a built-in lubricant that prevents formation of a built-up edge on the cutting tool and imparts an improved cutting geometry (see Chapter 21). In leaded materials, the insoluble lead particles work much the same way.

The bismuth free-machining steels are an attractive alternative to the previous varieties. Bismuth is more environmentally acceptable (compared to lead), has a reduced tendency to form stringers, and can be more uniformly dispersed because its density is a better match to that of iron. Machinability is improved because the heat generated by cutting is sufficient to form a thin film of liquid bismuth that lasts for only fractions of a microsecond. Tool life is noticeably extended, and the machined product is still weldable.

The use of free-machining steels is not without compromise, however. Ductility and impact properties are somewhat reduced compared to the unmodified steels. Weldability can be limited if the free-machining component wets weld metal grain boundaries. Copper-based braze joints tend to embrittle when used to join bismuth free-machining steels, and the machining additions reduce the strength of shrink-fit assemblies. If these compromises are objectionable, other methods might be used to enhance machinability. For example, the machinability of steels can be improved by cold working the metal. As the strength and hardness of the metal increase, the metal loses ductility, and subsequent machining produces chips that tear away more readily and fracture into smaller segments.

Precoated Steel Sheet

Traditional sheet metal fabrication involves the shaping of components from bare steel, followed by the finishing (or coating) of these products on a piece-by-piece basis. In this sequence, it is not uncommon for the finishing processes to be the most expensive and time-consuming stages of manufacture because they involve handling, manipulation, and possible curing or drying, as well as adherence to the various EPA (environmental) and OSHA (safety and health) requirements.

An alternative to this procedure is to purchase **precoated steel sheet**, where the steel supplier applies the coating when the material is still in the form of a long, continuous strip. Cleaning, pretreatment, coating, and curing can all be performed in a continuous manner, producing a coating that is uniform in thickness and offers improved adhesion. Numerous coatings can be specified, including the entire spectrum of dipped and plated metals (including aluminum, zinc, and chromium), vinyls, paints, primers, and other polymers or organics. Many of these coatings are specially formulated to endure the rigors of subsequent forming and bending. The continuous sheets can also be printed, striped, or embossed to provide a number of visual effects. Extra caution must be exercised during handling and fabrication to prevent damage to the coating, but the additional effort and expense are often less than the cost of finishing individual pieces.

Steels for Electrical and Magnetic Applications

Soft magnetic materials can be magnetized by relatively low-strength magnetic fields, but lose almost all of their magnetism when the applied field is removed. They are widely used in products such as solenoids, transformers, motors, and generators. The most common soft magnetic materials are high-purity iron, low-carbon steels, iron-silicon electrical steels, amorphous ferromagnetic alloys, iron-nickel alloys, and soft ferrites (ceramic material).

In recent years, the **amorphous metals** have shown attractive electrical and magnetic properties. Because the material has no crystal structure, grains or grain boundaries: (1) the magnetic domains can move freely in response to magnetic fields, (2) the properties are the same in all directions, and (3) corrosion resistance is improved. The high magnetic strength and low hysteresis losses offer the possibility of smaller, lighter-weight magnets. When used to replace silicon steel in power transformer cores, this material has the potential of reducing core losses by as much as 50%.

To exhibit permanent magnetism, materials must remain magnetized when removed from the applied field. Although most permanent magnets are ceramic materials or complex metal alloys, cobalt alloy steels (containing up to 36% cobalt) might be specified for electrical equipment where high magnetic densities are required.

Maraging Steels

When super-high strength is required, the **maraging** grades become a very attractive option. These alloys contain between 15% and 25% nickel, plus significant amounts of cobalt, molybdenum, and titanium—all added to a very-low-carbon steel. They can be hot worked at elevated temperatures and machined or cold worked in the air-cooled condition. Yield strengths in excess of 1725 MPa (250 ksi) coupled with good residual elongation are the result of a low-carbon, tough and ductile, iron-nickel martensite, which is further strengthened by the precipitation of intermetallic compounds during a subsequent age hardening process.

Maraging alloys are very useful in applications where ultra-high strength and good toughness are important. (The fracture toughness of maraging steel is considerably higher than that of the conventional high-strength steels.) They can be welded, provided the weldment is followed by the full solution and aging treatment. As might be expected from the large amount of alloy additions (more than 30%) and the multistep thermal processing, maraging steels are quite expensive and should be specified only when their outstanding properties are absolutely required.

Steels for High-Temperature Service

As a general rule of thumb, plain-carbon steels should not be used at temperatures in excess of about 250°C (500°F). Conventional alloy steels extend this upper limit to around 350°C (650°F). Numerous application areas, however, require materials with good strength characteristics, corrosion resistance, and creep resistance at operating temperatures in excess these values.

The high-temperature ferrous alloys tend to be low-carbon materials with less than 0.1% carbon. At their peak operating temperatures, 1000-hour rupture stresses tend to be quite low, however, often in the neighborhood of 50 MPa (7 ksi). Iron can be a major component of other high-temperature alloys, but when the amount is less than 50% the metal is not classified as a ferrous material. High strength at high temperature usually requires the more expensive nonferrous materials that will be discussed in Chapter 7.

6.5 Stainless Steels

Low-carbon steel with the addition of 4% to 6% chromium exhibits good resistance to many of the corrosive media encountered in the chemical industry. This behavior is attributed to the formation of a strongly adherent iron chromium oxide on the surface. If more improved corrosion resistance and outstanding appearance are required, materials should be specified that use a superior oxide that forms when the amount of chromium in solution (excluding chromium carbides and other forms where the chromium is no longer available to react with oxygen) exceeds 12%. When damaged, this tough, adherent, transparent, corrosion-resistant oxide (which is only 1–2 nanometers thick) actually heals itself, provided oxygen is present, even in very small amounts. Materials that form this superior protective oxide are known as **stainless steels**, or more specifically, **true stainless steels**.

Several classification schemes have been devised to categorize these alloys. The American Iron and Steel Institute (AISI) groups the metals by chemistry and assigns a three-digit number that identifies the basic family and the particular alloy within that family. In this text, however, we group these alloys into microstructural families because it is the basic structure that controls the engineering properties of the metal. **Table 6.8** presents the AISI designation scheme for stainless steels and correlates it with the microstructural families.

Because chromium has a body-centered cubic structure, it tends to stabilize the body-centered ferrite structure in a steel, increasing the temperature range over which ferrite is the stable structure. With sufficient chromium and a low level of

TABLE 6.8 AISI Designation Scheme for Stainless Steels

Series	Alloys	Structure
200	Chromium, nickel, manganese, or nitrogen	Austenitic
300	Chromium and nickel	Austenitic
400	Chromium and possibly carbon	Ferritic or martensitic
500	Low chromium (<12%) and possibly carbon	Martensitic

carbon, a corrosion-resistant iron alloy can be produced that is ferrite at all temperatures below solidification. These alloys are known as the **ferritic stainless steels**. They possess rather limited ductility and poor toughness, but are readily weldable. No martensite can form in the welds because there is no possibility of forming the FCC austenite structure that can then transform during cooling. These alloys cannot be heat-treated, and poor ductility limits the amount of strengthening by cold work. The primary source of strength is the body-centered-cubic (BCC) crystal structure combined with the effects of solid solution strengthening. Characteristic of the BCC metals, the ferritic stainless steels exhibit a ductile-to-brittle transition as the temperature is reduced. The ferritic alloys are the cheapest type of stainless steel, however, and as such, they should be given first consideration when a stainless alloy is required.

If increased strength is needed, the **martensitic stainless steels** should be considered. For these alloys, carbon is added and the chromium content is reduced to a level where the material can be austenite (FCC) at high temperature and ferrite (BCC) at low. Upon heating, the carbon will dissolve in the face-centered-cubic austenite, which can then be quenched to trap it in the body-centered martensitic structure. The carbon contents can be varied up to 1.2% to provide a wide range of strengths and hardnesses. Caution should be taken, however, to ensure more than 12% chromium remains in solution. Slow cools can allow the carbon and chromium to react and form chromium carbides. When this occurs, the chromium is no longer available to react with oxygen and form the protective oxide. As a result, the martensitic stainless steels can only exhibit good corrosion resistance when in the martensitic condition (when the chromium is trapped in atomic solution) and might be susceptible to red rust when annealed or normalized for ease of machining or fabrication. The martensitic stainless steels cost about 1½ times as much as the ferritic alloys, with part of the increase being because of the additional heat treatment, which generally consists of an austenitization, quench, stress relief, and temper. They are less corrosion resistant than the other varieties, are the least weldable of the stainless steels, and have a ductile-to-brittle transition at low temperatures. The martensitic stainlesses tend to be used in applications such as cutlery, where strength and hardness are the dominant requirements.

Nickel, being face-centered cubic, is an austenite stabilizer, and with sufficient amounts of both chromium and nickel (and

low carbon), it is possible to produce a stainless steel in which austenite is the stable structure from elevated to cryogenic temperatures. Known as **austenitic stainless steels**, these alloys can cost two to three times as much as the ferritic variety, but here the added expense is attributed to the cost of the nickel and chromium alloys. (The most widely used austenitic stainless steel, Type 304, is also known as 18-8 because it contains 18% chromium and 8% nickel.) Manganese and nitrogen are also austenite stabilizers and can be substituted for some of the nickel to produce a lower-cost, somewhat lower-quality, austenitic stainless steel (the AISI 200-series).

Austenitic stainless steels are easily identified by their nonmagnetic characteristic (the ferritic and martensitic stainlesses are attracted to a magnet). They are highly resistant to corrosion in almost all media (except hydrochloric acid and other halide acids and salts) and can be polished to a mirror finish, thereby combining attractive appearance and corrosion resistance. Toughness is excellent, particularly at low temperatures. Formability is outstanding (because of the low yield strength and high elongation that is characteristic of the FCC crystal structure), and these steels strengthen significantly when cold worked. The following table shows the response of the popular 304 alloy to a small amount of cold work:

	Water Quench	Cold Rolled 15%
Yield strength [MPa (ksi)]	260 (38)	805 (117)
Tensile strength [MPA (ksi)]	620 (90)	965 (140)
Elongation in 2 in. (%)	68	11

Among the stainless varieties, the austenitic stainless steels offer the best combination of corrosion resistance and toughness, are easily welded, and do not embrittle at low temperatures. Because they are also some of the costliest, they should not be specified where the less-expensive ferritic or martensitic alloys would be adequate or where a true stainless steel is not required. **Figure 6.8** lists some of the popular alloys from each of the three major structural classifications and links them to some associated properties. **Table 6.9** summarizes the basic types and the primary mechanism of strengthening.

A fourth and special class of stainless steel is the **precipitation-hardening stainless steel**. These alloys are basically martensitic or austenitic types, modified by the addition of alloying elements such as copper, aluminum, and titanium that permit the precipitation of hard intermetallic compounds at the temperatures used to temper martensite. With the addition of age hardening, these materials are capable of attaining high-strength properties such as a 1790 MPa (260 ksi) yield strength, 1825 MPa (265 ksi) tensile strength, and a 2% elongation. Because the additional alloys and extra processing make the precipitation-hardening alloys some of the most expensive stainless steels, they should be used only when their high-strength feature is absolutely required.

Although the four structures described earlier constitute the bulk of stainless steels, there are also some additional

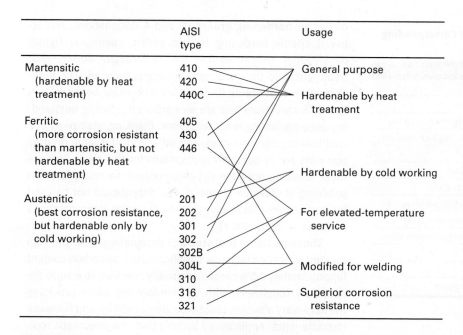

	AISI type	Usage
Martensitic (hardenable by heat treatment)	410 420 440C	General purpose
		Hardenable by heat treatment
Ferritic (more corrosion resistant than martensitic, but not hardenable by heat treatment)	405 430 446	
		Hardenable by cold working
Austenitic (best corrosion resistance, but hardenable only by cold working)	201 202 301 302 302B 304L 310 316 321	For elevated-temperature service
		Modified for welding
		Superior corrosion resistance

FIGURE 6.8 Different types of stainless steels, along with popular alloys within each type and key properties.

variants. **Duplex stainless steels** contain between 18% and 25% chromium, 4% and 7% nickel, and up to 4% molybdenum and can be water quenched from a hot-working temperature to produce a microstructure that is approximately half ferrite and half austenite. This mixed structure offers good toughness and a high yield strength, coupled with greater resistance to both stress corrosion cracking and pitting corrosion than either the full-austenitic or full-ferritic grades.

Stainless steels are difficult to machine because of their work-hardening properties and their tendency to seize during cutting. To overcome these limitations, special **free-machining** alloys have been produced within each family. Additions of sulfur, phosphorus, or selenium can raise machinability to approximately that of a medium-carbon steel. The free-machining grades are designated by the letters F or Se following the three-digit alloy code.

The preceding discussion has focused on the wrought stainless alloys. **Cast stainless steels** have structures and properties that are similar to the wrought grades. Most are specified by properties using standards of ASTM or ISO (International Organization for Standardization), while chemistries have been classified by the High Alloy Product Group of the Steel Founders Society of America. The C-series are used primarily to impart corrosion resistance and are used in valves,

pumps, and fittings. The H-grades (heat-resistant) have been designed to provide useful properties at elevated temperature and are used for furnace parts and turbine components.

Several potential problems are unique to the family of stainless steels. Because the protective oxide provides the excellent corrosion resistance, this feature can be lost whenever the amount of chromium in solution drops below 12%. A localized depletion of chromium can occur when elevated temperatures allow chromium carbides to form along grain boundaries (**sensitization**). To prevent their formation, one can keep the carbon content of stainless steels as low as possible, usually below 0.10%. (Special low-carbon grades are designated by the letter L after their three-digit number.) Another method is to tie up existing carbon with small amounts of stabilizing elements, such as titanium or niobium, that have a stronger affinity for carbon than does chromium. A letter suffix again designates the modification. Rapidly cooling these metals through the carbide-forming range of 480° to 900°C (900° to 1650°F) also works to prevent carbide formation.

Another problem with high-chromium stainless steels is an embrittlement that can occur after long times at elevated temperatures. This is attributed to the formation of a brittle compound that forms at elevated temperature and coats grain boundaries. Known as **sigma phase**, this material then provides a brittle crack path through the metal. Stainless steels used in high-temperature service should be checked periodically to detect and monitor sigma-phase formation.

6.6 Tool Steels

Tool steels are high-carbon, high-strength, ferrous alloys that have been modified by alloy additions to provide a desired balance of strength (resistance to deformation), toughness (ability to absorb shock or impact), and wear resistance (ability to resist erosion between the tool steel and contact material) when properly heat-treated. Several classification systems have been developed, some using chemistry as a basis, whereas others employ a hardening method or major mechanical property. The AISI system uses a letter designation to identify basic features such as quenching method, primary application, special alloy, or dominant characteristic. **Table 6.10** lists seven basic families of tool steels, the corresponding AISI letter and the associated feature or characteristic. Individual alloys within the letter grades are then listed numerically to produce a letter–number identification system, such as W1, D2, or H13.

TABLE 6.9 Primary Strengthening Mechanism for the Various Types of Stainless Steel

Type of Stainless Steel	Primary Strengthening Mechanism
Ferritic	Solid-solution strengthening
Martensitic	Phase transformation strengthening (martensite)
Austenitic	Cold work (deformation strengthening)

<table>
<tr><td colspan="3">**TABLE 6.10**   Basic Types of Tool Steel and Corresponding AISI Grades</td></tr>
</table>

Type	AISI Grade	Significant Characteristic
1. Water-hardening	W	
2. Cold-work	O	Oil-hardening
	A	Air-hardening medium alloy
	D	High-carbon–high-chromium
3. Shock-resisting	S	
4. High-speed	T	Tungsten alloy
	M	Molybdenum alloy
5. Hot-work	H	H1–H19: chromium alloy
		H20–H39: tungsten alloy
		H40–H59: molybdenum alloy
6. Plastic-mold	P	
7. Special-purpose	L	Low alloy
	F	Carbon–tungsten

Water-hardening tool steels (W-grade) are essentially high-carbon plain-carbon steels. They are the least expensive variety and are used for a wide range of parts that are usually quite small and not subject to severe usage or elevated temperature. Because strength and hardness are functions of the carbon content, a wide range of properties can be achieved through composition variation. Hardenability is low, so these steels must be quenched in water to attain high hardness. They can be used only for relatively thin sections if the full depth of hardness is desired. They are also rather brittle, particularly at higher hardness.

Typical uses of the various plain-carbon steels are as follows:

0.60%–0.75% carbon: machine parts, chisels, setscrews, and similar products where medium hardness is required coupled with good toughness and shock resistance

0.75%–0.90% carbon: forging dies, hammers, and sledges

0.90%–1.10% carbon: general-purpose tooling applications that require good balance of wear resistance and toughness, such as drills, cutters, shear blades, and other heavy-duty cutting edges

1.10%–1.30% carbon: small drills, lathe tools, razor blades, and other light-duty applications in which extreme hardness is required without great toughness

In applications where improved toughness is required, small amounts of manganese, silicon, and molybdenum are often added. Vanadium additions of about 0.20% are used to form strong, stable carbides that retain fine grain size during heat treatment. One of the main weaknesses of the plain-carbon tool steels is their loss of hardness at elevated temperature, which can occur with prolonged exposure to temperatures higher than 150°C (300°F).

When larger parts must be hardened or distortion must be minimized, the **cold-work tool steels** are usually recommended. The alloy additions and higher hardenability of the

oil- or **air-hardening grades** (O and A designations, respectively) enable hardening by less severe quenches. Tighter dimensional tolerances can be maintained during heat treatment, and the cracking tendency is reduced. The **high-chromium tool steels** (D designation) contain between 10% and 18% chromium and are air-hardened, offering outstanding deep-hardening wear resistance. Blanking, stamping, and cold forming dies, punches, and other tools for large production runs are all common applications for this class. Because these steels do not have the alloy content necessary to resist softening at elevated temperatures, they should not be used for applications that involve prolonged service at temperatures in excess of 250°C (500°F).

Shock-resisting tool steels (S designation) offer the high toughness needed for impact applications. Low carbon content (approximately 0.5% carbon) is usually specified to ensure the necessary toughness, with carbide-forming alloys providing the necessary abrasion resistance, hardenability, and hot-work characteristics. Applications include parts for pneumatic tooling, chisels, punches, and shear blades.

High-speed tool steels are used for cutting tools and other applications where strength and hardness must be retained at temperatures up to or exceeding red-heat (about 760°C or 1400°F). One popular member of the *tungsten high-speed tool steels* (T designation) is the T1 alloy, which contains 0.7% carbon, 18% tungsten, 4% chromium, and 1% vanadium. It offers a balanced combination of shock resistance and abrasion resistance and is used for a wide variety of cutting applications. The *molybdenum high-speed steels* (M designation) were developed to reduce the amount of tungsten and chromium required to produce the high-speed properties, and M1 has become quite popular for drill bits.

Hot-work tool steels (H designation) were developed to provide strength and hardness during prolonged exposure to elevated temperature. All employ substantial additions of carbide-forming alloys. H1 to H19 are chromium-based alloys with about 5.0% chromium; H20 to H39 are tungsten-based types with 9% to 18% tungsten coupled with 3% to 4% chromium; and H40 to H59 are molybdenum-based. The chromium types tend to be less expensive than the tungsten or molybdenum alloys.

Other types of tool steels include (1) the *plastic mold steels* (P designation), designed to meet the requirements of zinc die casting and plastic injection molding dies; (2) the *low-alloy special-purpose tool steels* (L designation), such as the L6 extreme toughness variety; and (3) the *carbon-tungsten type* of special-purpose tool steels (F designation), which are water hardening but substantially more wear resistant than the plain-carbon tool steels.

Most tool steels are wrought materials, but some are designed specifically for fabrication by casting. Powder metallurgy processing has also been used to produce special compositions that are difficult or impossible to produce by wrought or cast methods, or to provide key structural enhancements. By subjecting water atomized powders to hot-isostatic pressing (HIP), 100% dense billets can be produced with fine grain size and small, uniformly distributed carbide particles. These materials offer superior wear resistance (approaching that of the ceramic carbides), combined with useful levels of toughness.

6.7 Cast Irons

The term **cast iron** applies to an entire family of metals that are alloys of iron, carbon (in excess of 2.0%), and silicon (0.5% to 3.0%) and offer a wide variety of properties. Various types of cast iron can be produced, depending on the chemical composition, cooling rate, and the type and amount of inoculants that are used. (Inoculants and inoculation practice will be discussed shortly.) Each of the types, however, can be described as having a structure consisting of an iron-based metal matrix (such as ferrite, pearlite, bainite, or martensite) and a high-carbon second phase, that is either graphite (pure carbon) or iron carbide (Fe_3C). The basic types of cast iron are gray, white, malleable, ductile or nodular, austempered ductile, compacted graphite, and high-alloy.

Types of Cast Iron

Gray cast iron is the least expensive and most common variety and can be characterized by those features that promote the formation of graphite (discussed at the end of Chapter 4). Typical compositions range from 2.5% to 4.0% carbon, 1.0% to 3.0% silicon, and 0.4% to 1.0% manganese. The microstructure consists of three-dimensional, interconnected, graphite flakes (which form during the eutectic reaction) dispersed in a matrix of ferrite, pearlite, or other iron-based structure that forms from the cooling of austenite. **Figure 6.9** presents a typical section through gray cast iron, showing the graphite flakes dispersed throughout the metal matrix. Because the graphite

flakes have no appreciable strength, they act essentially as voids in the structure. The pointed edges of the flakes act as preexisting notches or crack initiation sites, giving the material its characteristic brittle nature. Because a large portion of any fracture follows the graphite flakes, the freshly exposed fracture surfaces have a characteristic gray appearance (hence the name—gray iron), and a graphite smudge can usually be obtained if one rubs a finger across the fracture. On a more positive note, the formation of the lower-density graphite reduces the amount of shrinkage that occurs when the liquid goes to solid, making possible the production of more complex iron castings.

The size, shape, and distribution of the graphite flakes have a considerable effect on the overall properties of gray cast iron. When maximum strength is desired, small, uniformly distributed flakes with a minimum amount of intersection are preferred. A more effective means of controlling strength, however, is through control of the metal matrix structure, which is in turn controlled by the carbon and silicon contents and the cooling rate of the casting. Gray cast iron is normally sold by **class**, with the class number corresponding to the minimum tensile strength in thousands of pounds per square inch.[1] Class 20 iron (minimum tensile strength of 20,000 psi) consists of high-carbon, high-silicon metal with a ferrite matrix. Higher strengths, up to class 40, can be obtained with lower carbon and silicon and a pearlite matrix. To go above class 40, alloying is required to provide solid solution strengthening, and heat-treatment practices must be performed to modify the matrix. Gray cast irons can be obtained up through class 80, but regardless of strength the presence of the graphite flakes results in extremely low ductility.

Gray cast irons offer excellent compressive strength (compressive forces do not promote crack propagation, so compressive strength is typically 2.5–4 times tensile strength), excellent machinability (the graphite flakes act to break up the chips and lubricate contact surfaces), good resistance to adhesive wear and galling (graphite flakes self-lubricate), and outstanding sound and vibration damping characteristics (graphite flakes absorb transmitted energy). **Table 6.11** compares the relative

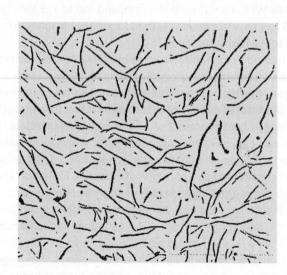

FIGURE 6.9 Photomicrograph showing the distribution of graphite flakes in gray cast iron; unetched, 100X. (Courtesy Ronald Kohser)

TABLE 6.11 Relative Damping Capacity of Various Metals

Material	Damping Capacity [a]
Gray iron (high carbon equivalent)	100–500
Gray iron (low carbon equivalent)	20–100
Ductile iron	5–20
Malleable iron	8–15
White iron	2–4
Steel	4
Aluminum	0.4

[a]Natural log of the ratio of successive amplitudes.

---

[1] The ASTM class uses numbers reflecting thousands of pounds per square inch, while the International Organization for Standardization (ISO) assigns class numbers using megapascals.

damping capacities of various engineering metals and clearly shows the unique characteristic of the high-carbon-equivalent (high-carbon and high-silicon) gray cast irons. This material is 20–25 times better than steel and 250 times better than aluminum! High silicon contents promote good corrosion resistance and provide the enhanced fluidity desired for casting operations. For these reasons, coupled with low cost, high thermal conductivity, low rate of thermal expansion, good stiffness, resistance to thermal fatigue, and 100% recyclability, gray cast iron is specified for a number of applications, including automotive engine blocks, heads, and cylinder liners; transmission housings; machine tool bases; and large equipment parts that are subjected to compressive loads and vibrations. Weldability, however, is poor, and the material cannot be shaped by deformation.

White cast iron has all of its excess carbon in the form of iron carbide and receives its name from the white surface that appears when the material is fractured. Features promoting its formation are those that favor cementite over graphite: a low carbon equivalent (1.8% to 3.6% carbon, 0.5% to 1.9% silicon, and 0.25% to 0.8% manganese) and rapid cooling.

Because the large amount of iron carbide (as much as 50%) dominates the microstructure, white cast iron is very hard and brittle and finds applications where high abrasion resistance is the overwhelming requirement. For these uses it is also common to pursue the hard, wear-resistant martensite structure (described in Chapter 5) as the metal matrix. In this way, both the metal matrix and the high-carbon second phase contribute to the wear-resistant characteristics of the material.

White cast iron surfaces can also be applied over a base of another material. For example, mill rolls that require extreme wear resistance can have a white cast iron surface on top of a steel interior. By accelerating the cooling rate and controlling chemistry, white iron surfaces or regions can be produced in gray iron castings. Tapered sections or metal chill bars placed in the molding sand provide the accelerated cooling. When regions of white and gray cast iron occur in the same component, there is generally a transition zone containing regions of both white and gray irons, known as the **mottled zone**.

If white cast iron is exposed to an extended heat treatment at temperatures in the range of 900°C (1650°F), the cementite will dissociate into its component elements, and some or all of the carbon will be converted into irregularly shaped clusters of graphite (also referred to as clump or popcorn graphite). The product, known as **malleable cast iron**, has significantly improved ductility compared to gray cast iron because the more favorable graphite shape no longer resembles an internal crack or notch. The rapid cooling required to produce the starting white iron structure, however, restricts the size and thickness of malleable iron products such that most weigh less than 5 kilograms (10 pounds).

Various types of malleable iron can be produced, depending on the type of heat treatment that is employed. If the white iron is heated and held for a prolonged time just below the melting point, the carbon in the cementite converts to graphite (first-stage graphitization). Subsequent slow cooling

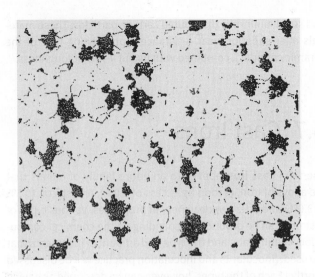

FIGURE 6.10 Photomicrograph of malleable iron showing the irregular graphite clusters, etched to reveal the ferrite matrix, 100×. (Courtesy Ronald Kohser)

through the eutectoid reaction causes the carbon-containing austenite to transform to ferrite and more graphite (second-stage graphitization), and the resulting product, known as *ferritic malleable cast iron*, has a structure of irregular particles of graphite dispersed in a ferrite matrix (**Figure 6.10**). Typical properties of this material are: 10% elongation, 240 MPa (35 ksi) yield strength, 345 MPa (50 ksi) tensile strength, and excellent impact strength, corrosion resistance, and machinability. The heat-treatment times, however, are quite lengthy, often involving more than 100 hours at elevated temperature.

If the material is cooled more rapidly through the eutectoid transformation, the carbon in the austenite does not form additional graphite but is retained in a pearlite or martensite matrix. The resulting *pearlitic malleable cast iron* is characterized by higher strength and lower ductility than its ferritic counterpart. Properties range from 1% to 4% elongation, 310 to 590 MPa (45 to 85 ksi) yield strength, and 450 to 725 MPa (65 to 105 ksi) tensile strength, with reduced machinability compared to the ferritic material.

The modified graphite structure of malleable iron provided quite an improvement in properties compared to gray cast iron. However, it would be even more attractive if a similar structure could be obtained directly on solidification rather than through a prolonged heat treatment at high elevated temperature. If a high-carbon-equivalent cast iron is sufficiently low in sulfur (either by original chemistry or by desulfurization), the addition of certain materials can promote graphite formation and change the morphology (shape) of the graphite product. If ferrosilicon is injected into the melt (**inoculation**), it will promote the formation of graphite. If magnesium (in the form of MgFeSi or MgNi alloy) is also added just prior to solidification, the graphite will form as smooth-surface spheres. The latter addition is known as a **nodulizer**, and the product becomes **ductile** or **nodular cast iron**. One should note that the magnesium nodulizer volatilizes easily. Its effectiveness will diminish

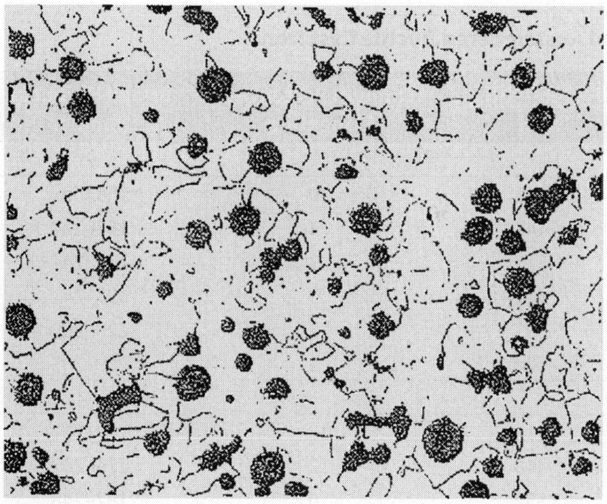

FIGURE 6.11 Ductile cast iron with a ferrite matrix. Note the spheroidal shape of the graphite. 100×. (Courtesy Ronald Kohser)

or be lost with time, and the graphite structure will transition to the flake form of gray cast iron. This phenomenon is known as **fading**.

Subsequent control of cooling can produce a variety of matrix structures, with as-cast ferrite and/or pearlite being the most common (**Figure 6.11**) and alloying and/or heat treatment extending these to include martensite, bainite, or austenite. By controlling the matrix structure, properties can be produced that span a wide range from 2% to 18% elongation, 275 to 620 MPa (40 to 90 ksi) yield strength, and 415 to 825 MPa (60 to 120 ksi) tensile strength. The combination of good ductility, high strength, toughness, wear resistance, machinability, low-melting-point castability, and up to a 10% weight reduction compared to steel makes ductile iron an attractive engineering material. High silicon-molybdenum ductile irons offer excellent high temperature strength and good corrosion resistance. Unfortunately, the costs of a nodulizer, higher-grade melting stock, better furnaces, and the improved process control required for its manufacture combine to place it among the more expensive of the cast irons.

Austempered ductile iron (ADI), ductile iron that has undergone a special austempering heat treatment to modify and enhance its properties,[2] has emerged as a significant engineering material during the past 30 years. It combines the ability to cast intricate shapes with strength, fatigue-resistance, and wear-resistance properties that are similar to those of heat-treated steel. Compared to conventional as-cast ductile iron, it offers nearly double the strength at the same level of ductility. Compared to steel, it offers reduced cost, an 8% to 10% reduction in density (so strength-to-weight is excellent)

and enhanced damping capability, both because of the graphite nodules. Machinability is generally poorer, thermal conductivity is lower, and there is about a 20% drop in elastic modulus. ADI is approximately three times stronger than aluminum, more than two times stiffer, with better fatigue strength and wear resistance. **Table 6.12** compares the typical mechanical properties of some malleable cast irons, the five grades of ductile cast iron specified in ASTM Standard A536, and the six grades of austempered ductile cast iron that are specified in ASTM Standard A897.

Compacted graphite cast iron (CGI) is also attracting considerable attention. Produced by a method similar to that used to make ductile iron (a Mg–Ce–Ti addition is made), compacted graphite iron is characterized by a graphite structure that is intermediate to the flake graphite of gray iron and the nodular graphite of ductile iron, and it tends to possess some of the desirable properties and characteristics of each. **Table 6.13** shows how the properties of compacted graphite iron bridge the gap between gray and ductile. Strength, stiffness, and ductility are greater than those of gray iron, whereas castability, machinability, thermal conductivity, and damping capacity all exceed those of ductile. Impact and fatigue properties are good.

ASTM Specification A842 identifies five grades of compacted graphite cast iron—250, 300, 350, 400, and 450—where the numbers correspond to tensile strength in megapascals. Areas of application tend to be those where the mechanical properties of gray iron are insufficient and those of ductile iron, along with its higher cost, are considered to be overkill. More specific, compacted graphite iron is attractive when the desired properties include high strength, castability, machinability, thermal conductivity, and thermal shock resistance.

The Role of Alloys in Cast Irons

Because the effects of alloying elements are often the same regardless of the process used to produce the final shape, much of what was presented earlier for wrought steel also applies to cast irons. When the desired shape is to be made by casting, however, some alloys can be used to enhance process-specific features, such as fluidity and as-solidified properties.

Many cast iron products are used in the as-cast condition, with the only heat treatment being a stress relief or annealing. For these applications, the alloy elements are selected for their ability to alter properties by (1) affecting the formation of graphite or cementite, (2) modifying the morphology of the carbon-rich phase, (3) strengthening the matrix material, or (4) enhancing wear resistance through the formation of alloy carbides. Nickel, for example, promotes graphite formation and tends to promote finer graphite structures. Chromium retards graphite formation and stabilizes cementite. These alloys are frequently used together in a ratio of two or three parts of nickel to one part of chromium. Between 0.5% and 1.0% molybdenum is often added to gray cast iron to impart additional strength, form alloy carbides, and help to control the size of the graphite flakes.

---

[2] The austempering process begins by heating the metal to a temperature between 1500° and 1750 °F (815° and 955 °C) and holding for sufficient time to saturate the austenite with carbon. The metal is then rapidly cooled to an austempering temperature between 450° and 750 °F (230° and 400 °C), where it is held until all crystal structure changes have completed, and then cooled to room temperature. High austempering temperatures give good toughness and fatigue properties, whereas lower austempering temperatures give better strength and wear resistance.

TABLE 6.12 Typical Mechanical Properties of Malleable, Ductile, and Austempered Ductile Cast Irons

Class or Grade	Minimum Yield Strength		Minimum Tensile Strength		Minimum Percentage Elongation	Brinell Hardness Number
	ksi	MPa	ksi	MPa		
Malleable Iron[a]						
M3210	32	224	50	345	10	156 max
M4504	45	310	65	448	4	163–217
M5003	50	345	75	517	3	187–241
M5503	55	379	75	517	3	187–241
M7002	70	483	90	621	2	229–269
M8501	85	586	105	724	1	269–302
Ductile Iron[b]						
60–40–18	40	276	60	414	18	149–187
65–45–12	45	310	65	448	12	170–207
80–55–06	55	379	80	552	6	187–248
100–70–03	70	483	100	689	3	217–269
120–90–02	90	621	120	827	2	240–300
Austempered Ductile Iron[c]						
1	70	500	110	750	11	241–302
2	90	550	130	900	9	269–341
3	110	750	150	1050	7	302–375
4	125	850	175	1200	4	341–444
5	155	1100	200	1400	2	388–477
6	185	1300	230	1600	1	402–512

[a]ASTM Specification A602 (Also SAE J 158).
[b]ASTM Specification A536.
[c]ASTM Specification A897.

TABLE 6.13 Typical Properties of Pearlitic Gray, Compacted Graphite, and Ductile Cast Irons

Property	Gray	CGI	Ductile
Tensile strength (MPa)	250	450	750
Elastic modulus (Gpa)	105	145	160
Elongation (%)	0	1.5	5
Thermal conductivity (w/mk)	48	37	28
Relative damping capacity (Gray = 1)	1	0.35	0.22

High-alloy cast irons have been designed to provide enhanced corrosion resistance and/or good elevated temperature service. Within this family, the austenitic gray cast irons, which contain about 14% nickel, 5% copper, and 2.5% chromium, offer good corrosion resistance to many acids and alkalis at temperatures up to about 800°C (1500°F). The addition of up to 1% molybdenum and 5% silicon to ductile iron greatly enhances high temperature tensile strength and creep strength. They are attractive for applications involving temperatures between 650°C and 875°C (1200°F and 1600°F).

6.8 Cast Steels

If a ferrous casting alloy contains less than about 2.0% carbon, it is considered to be a **cast steel**. (Alloys with more than 2% carbon are *cast irons*.) Cast steels are generally used whenever a cast iron is not adequate for the application. Compared to cast irons, the cast steels offer enhanced stiffness, toughness, and ductility over a wide range of operating temperatures and can be readily welded. They are usually heat-treated to produce a final quenched-and-tempered structure, and the alloy additions are selected to provide the desired hardenability and balance of properties. The enhanced properties come with a price, however, because the cast steels have a higher melting point (more energy to melt and higher cost refractories are necessary), lower fluidity (leading to increased probability of incomplete die or mold filling), and increased shrinkage (because graphite is not formed during solidification). The diverse applications take advantage of the material's structural strength and its ability to contain pressure, resist impacts, withstand elevated temperatures, and resist wear.

6.9 The Role of Processing on Cast Properties

Although typical properties have been presented for the various types of cast materials, it should be noted that the properties of all metals are influenced by how they are processed. For cast materials, properties will often vary with the manner of solidification and cooling. Because cast components often have complex geometries, the cooling rate can vary from location to location, with companion variation in properties. To ensure compliance with industry specifications, standard geometry test bars are often cast along with manufactured products so the material can be evaluated and quality can be assured independent of product geometry.

Alloy cast irons and cast steels are usually specified by their ASTM designation numbers, which relate the materials to their mechanical properties and intended service applications. The Society of Automotive Engineers (SAE) also has specifications for cast steels used in the automotive industry.

Review Questions

1. Why might it be important to know the prior processing history of an engineering material?

2. What is a ferrous material?

3. How does the amount of steel that is recycled compare to aluminum and other materials?

4. Why is the recycling of steel so attractive compared to the manufacture of new steel from virgin ore?

5. When iron ore is reduced to metallic iron, what other elements are generally present in the metal?

6. What properties or characteristics have made steel such an attractive engineering material?

7. What is involved in the conversion of pig iron into steel?

8. What are some of the modification processes that can be performed on steel during ladle metallurgy operations?

9. What is the advantage of pouring molten metal from the bottom of a ladle?

10. What are some of the attractive economic and processing advantages of continuous casting?

11. What problems are associated with oxygen or other gases that are dissolved in molten steel?

12. What are some of the techniques used to reduce the amount of dissolved oxygen in molten steel?

13. How might other gases, such as nitrogen and hydrogen, be reduced?

14. What are some of the attractive features of electroslag remelting?

15. What is plain-carbon steel?

16. What is considered a low-carbon steel? Medium-carbon? High-carbon?

17. What properties account for the high-volume use of medium-carbon steels?

18. Why should plain-carbon steels be given first consideration for applications requiring steel?

19. What are some of the common alloy elements added to steel?

20. For what different reasons might alloying elements be added to steel?

21. What are some of the alloy elements that tend to form stable carbides within a steel?

22. While strength and hardness are dependent on the level of carbon, what is primary role of the alloy additions in a heat-treated steel?

23. What alloys are particularly effective in increasing the hardenability of steel?

24. What is the basis of the AISI–SAE classification system (i.e., strength, structure, chemistry)?

25. What is the significance of the last two digits in a typical four-digit AISI–SAE steel designation?

26. How are letters incorporated into the AISI–SAE designation system for steel, and what do some of the more common ones mean?

27. What is an H-grade steel, and when should it be considered in a material specification?

28. Why should the proposed fabrication processes enter into the considerations when selecting a steel?

29. How are the final properties usually obtained in the constructional alloy steels? In the conventional high-strength steels?

30. How are the high-strength steels typically specified?

31. What are microalloyed steels?

32. What are some of the potential benefits that might be obtained through the use of microalloyed steels?

33. What is the primary attraction of the bake-hardenable steels?

34. What are advanced high-strength steels (AHSS)?

35. What are the two phases that are present in dual-phase steels?

36. What is the "transformation" that occurs during the deformation of the "transformation-induced plasticity" (TRIP) steels?

37. What are some other types of advanced high-strength steels (AHHS)?

38. What is press hardening?

39. What nomenclature has been applied to identify the advanced-high-strength steels?

40. Describe the role of steel in the automotive industry during the past 50 years. What is the projection for the near future?

41. What is the role of chip-breakers in free-machining steels?

42. What are some of the various alloy additions that have been used to improve the machinability of steels?

43. What are some of the compromises associated with the use of free-machining steels?

44. What factors might be used to justify the added expense of pre-coated steel sheet?

45. What are some of the coating materials that have been applied to precoated steel sheet?

46. What are soft magnetic materials?

47. Why have the amorphous metals attracted attention as potential materials for magnetic applications?

48. What are maraging steels, and for what conditions might they be required?

49. What are the typical elevated temperature limits of plain carbon and alloy steels?

50. What feature is responsible for the observed corrosion resistance of stainless steels?

51. What are the three primary families of stainless steels based on microstructures?

52. Why should ferritic stainless steels be given first consideration when selecting a stainless steel?

53. Which of the major types of stainless steel is likely to contain significant amounts of carbon? Why?

54. Under what conditions might a martensitic stainless steel "rust" when exposed to a hostile environment?

55. What are some of the unique properties of austenitic stainless steels?

56. How can an austenitic stainless steel be easily identified?

57. What is the dominant mechanism providing strength for each of the three primary types of stainless steels?

58. What two microstructures are present in a duplex stainless steel?

59. What is sensitization of a stainless steel, and what are the several methods to minimize or prevent it?

60. What problem is associated with the presence of sigma phase in a stainless steel?

61. What is a tool steel?

62. How does the AISI–SAE designation system for tool steels differ from that of plain-carbon and alloy steels?

63. What is the least expensive variety of tool steel?

64. For what types of applications might an oil- or air-hardenable tool steel be attractive?

65. What alloying elements are used to produce the high-speed and hot-work tool steels?

66. What assets can be provided by the hot-isostatic-pressed powder metallurgy tool steels?

67. What are the three primary chemical elements in a cast iron?

68. What is a generic description of the structure of cast irons?

69. Describe the microstructure of gray cast iron.

70. Which of the structural units is generally altered to increase the strength of a gray cast iron?

71. What are some of the attractive engineering properties of gray cast iron?

72. What are some of the key limitations to the engineering use of gray cast iron?

73. What is the dominant mechanical property of white cast iron?

74. How is malleable cast iron produced?

75. What structural feature is responsible for the increased ductility and fracture resistance of malleable cast iron?

76. What is unique about the shape or form of graphite in ductile cast iron, and what is required to produce it?

77. What is the purpose of inoculation when making ductile cast iron? Of nodulizing?

78. What is fading? Why should ductile iron be solidified immediately following nodulizing?

79. What requirements of ductile iron manufacture are responsible for its increased cost over materials such as gray cast iron?

80. What are some of the attractive features of austempered ductile cast iron?

81. Compacted graphite iron has a structure and properties intermediate to what two other types of cast irons?

82. What are some of the reasons that alloy additions are made to cast irons that will be used in their as-cast condition?

83. What properties are enhanced in the high-alloy cast irons?

84. When should cast steel be used instead of a cast iron?

85. In what ways might a cast steel be more difficult to process than a cast iron?

86. Why are standard geometry test bars often cast along with manufactured products?

Problems

1. Investigate the various ways in which the steel industry has sought to combat the shift to lighter-weight materials as a means of improving fuel economy. What problems needed to be overcome? What, if any, are current challenges?

2. Select from among the common hand tools in the following list, and try to identify the type of steel that would be most commonly used in their production. Consider: low-, medium-, and high-carbon contents; plain-carbon or alloy; standard or stainless; etc. Coatings are often required for corrosion resistance. What type of coating would you recommend for your selected tool? How might your answers be different if the target market was "lowest cost" or "top-of-the line?"

    a. Adjustable crescent wrench

    b. Wood chisel

    c. Phillips-head screwdriver

    d. Wire cutters

    e. Pliers

    f. Spark-plug socket

    g. $1/4$-in. drill bit (for drilling holes in wood)

3. Find an example where an advanced high-strength steel is being used in an automotive component. What specific type was selected, and why was it chosen over the other varieties?

4. Identify a particular product that has been manufactured using precoated steel sheet.

5. What type of stainless steel would you recommend for the following products?

    a. Inexpensive cafeteria flatware (forks and spoons)

    b. A set of high-quality flatware

    c. A set of steak knives

    d. A top-quality pocket knife

6. Select among the components in the following list, and recommend an appropriate tooling material, justifying your selection.

    a. Blanking dies to shear the "click-it" hardware of automotive seat belts from 2-mm thick 1050 steel sheet

    b. The cutting blade segments of a pair of industrial-quality tin snips

    c. Drop forging dies to hot forge open-end wrenches from 4147 steel bar stock using a power hammer

    d. Deep-drawing tooling to shape the outer shell of an automotive oil filter from discs of 1008 steel sheet

    e. Die casting dies to produce the right and left handle segments of a common utility knife from ZA-8 zinc-alloy

    f. Blanking dies to shear commercial-grade key blanks (building and office keys) from 2-mm thick free-machining brass

    g. Embossing/coining dies to press a decorative pattern onto the handles of a set of flatware made from 316 austenitic stainless steel

    h. Extrusion dies to form the various moldings used in aluminum window frames

    i. Chisel tips for a pneumatic concrete-breaking jack hammer

7. Identify at least one easily identified product or component that is currently being produced from each of the following types of cast irons:

    a. Gray cast iron

    b. White cast iron

    c. Malleable cast iron

    d. Ductile cast iron

    e. Compacted graphite cast iron

8. Find an example where one of the types of cast iron has been used in place of a previous material. What feature or features might have prompted the substitution?

CHAPTER 7

Nonferrous Metals and Alloys

7.1 Introduction

Nonferrous metals and alloys have assumed increasingly important roles in modern technology. Because of their number and the fact that their properties vary widely, they provide an almost limitless range of properties for the design engineer. Although they tend to be costlier than iron or steel, these metals often possess certain properties or combinations of properties that are not available in the ferrous metals, such as:

1. Resistance to corrosion
2. Ease of fabrication
3. High electrical and thermal conductivity
4. Light weight
5. Strength at elevated temperatures
6. Color

Nearly all of the nonferrous alloys possess at least two of these listed qualities, and some possess nearly all. **Figure 7.1** groups some of the nonferrous metals by attractive engineering property.

As a whole, the strength of the nonferrous alloys is generally inferior to that of steel, and the modulus of elasticity is usually lower, a fact that places them at a distinct disadvantage when stiffness is required. Ease of fabrication, however, is often

attractive. Those alloys with low melting points are easy to cast in sand molds, permanent molds, or dies. Many alloys have high ductility coupled with low yield points, the ideal combination for cold working. Good machinability is also characteristic of many nonferrous alloys. The savings obtained through ease of fabrication can often overcome the higher cost of the nonferrous material and justify its use in place of steel. Weldability is the one fabrication area where the nonferrous alloys tend to be somewhat inferior to steel. With modern joining techniques, however, it is generally possible to produce satisfactory weldments in all of the nonferrous metals.

7.2 Copper and Copper Alloys

General Properties and Characteristics

Copper has been an important engineering metal for more than 6,000 years. As a pure metal, it has been the backbone of the electrical industry. It is also the base metal of a number of alloys, generically known as brasses and bronzes. Compared to other engineering materials, copper and copper alloys offer three important properties: (1) *high electrical and thermal conductivity*, (2) *useful strength with high ductility*, and (3) *corrosion*

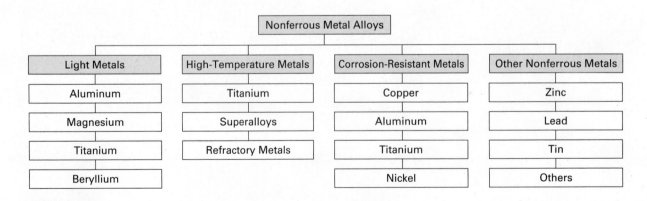

FIGURE 7.1 Some common nonferrous metals and alloys, classified by attractive engineering property.

resistance to a wide range of media. Because of its excellent conductivity, nearly two-thirds of all copper produced is used in some form of electrical application, including numerous types of wiring and products like those shown in **Figure 7.2**. Other large areas of use include plumbing (also illustrated in Figure 7.2), heating, air conditioning, and industrial valves and fittings. Copper is recyclable without loss of quality and is the third-most-recycled metal after iron and aluminum.

Pure copper in its annealed state, has a tensile strength of only about 200 MPa (30 ksi), with an elongation of nearly 60%. Through cold working, the tensile strength can be more than doubled to greater than 450 MPa (65 ksi), with a decrease in elongation to about 5%. Because of its relatively low strength and high ductility, copper is a very desirable metal for applications where extensive forming is required. Because the recrystallization temperature for copper is less than 260°C (500°F), the hardening effects of cold working can be easily removed to establish desired final properties or permit further deformation. Copper and copper alloys also lend themselves nicely to the whole spectrum of fabrication processes, including casting, machining, joining, and surface finishing by either plating or polishing.

Unfortunately, copper is *heavier than iron*. Although strength can be quite high, the strength/weight ratio for copper alloys is usually less than that for the weaker aluminum and magnesium materials. In addition, problems can occur when copper is used at an elevated temperature. Copper alloys tend to soften when heated above 220°C (400°F), and if copper is stressed for a long period of time at high temperature, intercrystalline failure can occur at about half of its normal room temperature strength. While offering low friction and good resistance to adhesive wear, copper and copper alloys have poor abrasive wear characteristics.

The low-temperature properties of copper are quite attractive, however. Strength tends to increase as temperatures drop, and the material does not embrittle, retaining attractive ductility even under cryogenic conditions. Conductivity also tends to increase with a drop in temperature.

Copper and copper alloys respond well to strengthening methods, with the strongest alloy being 15 to 20 times stronger than the weakest. Because of the wide range of properties, the material can often be tailored to the specific needs of a design. Elastic stiffness is between 50% and 60% of steel. In addition, copper alloys are nonmagnetic, nonpyrophoric or nonsparking (slivers or particles do not burn in air), nonbiofouling (inhibits marine organism growth—a property unique to copper), and available in a wide spectrum of available colors (including yellow, red, brown, and silver).

The U.S. Environmental Protection Agency (EPA) has confirmed the antimicrobial properties of more than 350 copper alloys (all with at least 60% copper), killing more than 99.9% of bacteria (including *Staphylococcus* and *E. coli*) within two hours of contamination and delivering continuous antibacterial action even after repeated wear and recontamination. Normal tarnishing does not impair antimicrobial effectiveness, but the metals cannot have any applied waxes or coatings. Because of this property, copper alloys might well replace stainless steel in applications involving surfaces that are repeatedly touched by multiple individuals, such as door knobs and sink faucet handles. Copper components can also reduce airborne pathogens in heating, ventilation and air-conditioning systems. Use is expected to expand in hospitals, schools, public buildings, exercise facilities, nursing homes, shopping malls, mass transit facilities, airports, interiors of airplanes and cruise ships, and even private homes.

(a) (b)

FIGURE 7.2 Copper and copper alloys are used for a variety of electrical and plumbing applications, as shown in these illustrations. (FactoryTh/iStockphoto; Jeff J Daly/Alamy)

Another expanding market is the hybrid or electric car, with each using up to 100 pounds of copper in the various motors and wiring.

Commercially Pure Copper

Being second only to silver in conductivity, commercially pure copper is used primarily for electrical applications. Refined copper containing between 0.02% and 0.05% oxygen is called **electrolytic tough-pitch (ETP) copper**. It is often used as a base for copper alloys and can be used for electrical applications such as wire and cable when the highest conductivity is not required. For superior conductivity, additional refining can reduce the oxygen content and produce **oxygen-free high-conductivity (OFHC) copper**. The better grades of conductor copper now have a conductivity rating of about 102% IACS, reflecting metallurgical improvements made since 1913, when the International Annealed Copper Standard (IACS) was established and the conductivity of pure copper was set at 100% IACS.

Copper-Based Alloys

As a pure metal, copper is not used extensively in manufactured products, except in electrical applications, and even here alloy additions of silver, arsenic, cadmium, and zirconium are used to enhance various properties without significantly impairing conductivity. More often, copper is the base metal for an alloy, where the copper imparts its good ductility, corrosion resistance, and electrical and thermal conductivity. A full spectrum of mechanical properties is available, ranging from pure copper, which is soft and ductile, through alloys, such as copper–beryllium and manganese bronze, whose properties can rival those of quenched and tempered steel.

Copper-based alloys are commonly designated using a system of numbers standardized by the Copper Development Association (CDA). **Table 7.1** presents a breakdown of this system, which has been further adopted by the American Society for Testing and Materials (ASTM), the Society of Automotive Engineers (SAE), and the U.S. government. Alloys numbered from 100 to 199 are coppers with less than 2% alloy addition. Numbers 200 to 799 are **wrought alloys**, and the 800 and 900 series are **cast alloys**. When converted to the Unified Numbering System for metals and alloys, the three-digit numbers are converted to five digits by placing two zeros at the end and the letter *C* as a prefix to denote the copper base.

Copper–Zinc Alloys

Zinc is by far the most popular alloying addition, and the resulting alloys are generally known as some form of **brass**. If the zinc content is less than 36%, the brass is a single-phase solid solution. Because this structure is identified as the alpha phase, these alloys are often called *alpha brasses*. They are quite ductile and formable, with both strength and ductility increasing with the amount of zinc throughout the single-phase region. The alpha brasses can be strengthened significantly by cold working and are commercially available in various degrees of cold-worked strength and hardness. *Cartridge brass*, the 70% copper–30% zinc alloy, offers the best overall combination of strength and ductility. As its name implies, it has become a popular material for sheet-forming operations like deep drawing.

Wrought Alloys vs. Cast Alloys

The term wrought *means "shaped or fabricated in the solid state." Key properties for wrought material generally relate to ductility. Cast* alloys are shaped as a liquid, where the attractive features include low melting point, high fluidity, and good as-solidified strength.

TABLE 7.1 **Designation System for Copper and Copper Alloys (Copper Development Association System)**

Wrought Alloys		Cast Alloys	
100–155	Commercial coppers	833–838	Red brasses and leaded red brasses
162–199	High-copper alloys	842–848	Semired brasses and leaded semired brasses
200–299	Copper–zinc alloys (brasses)	852–858	Yellow brasses and leaded yellow brasses
300–399	Copper–zinc–lead alloys (leaded brasses)	861–868	Manganese and leaded manganese bronzes
400–499	Copper–zinc–tin alloys (tin brasses)	872–879	Silicon bronzes and silicon brasses
500–529	Copper–tin alloys (phosphor bronzes)	902–917	Tin bronzes
532–548	Copper–tin–lead alloys (leaded phosphor bronzes)	922–929	Leaded tin bronzes
600–642	Copper–aluminum alloys (aluminum bronzes)	932–945	High-leaded tin bronzes
647–661	Copper–silicon alloys (silicon bronzes)	947–949	Nickel–tin bronzes
667–699	Miscellaneous copper–zinc alloys	952–958	Aluminum bronzes
700–725	Copper–nickel alloys	962–966	Copper nickels
732–799	Copper–nickel–zinc alloys (nickel silvers)	973–978	Leaded nickel bronzes

Many applications of these alloys result from the high electrical and thermal conductivity coupled with useful engineering strength. The wide range of colors (red, orange, yellow, silver, and white), enhanced by further variations that can be produced through the addition of a third alloy element, account for a number of decorative uses. Because the plating characteristics are excellent, the material is also a frequently used base for decorative chrome or similar coatings. Another attractive property of alpha brass is its ability to have rubber vulcanized to it without any special treatment except thorough cleaning. As a result, brass is widely used in mechanical rubber goods.

With more than 36% zinc, the copper–zinc alloys enter a two-phase region involving a brittle, zinc-rich phase, and ductility drops markedly. Although cold-working properties are rather poor for these high-zinc brasses, deformation can be performed easily at elevated temperature.

Most brasses have good corrosion resistance. In the range of 0 to 40% zinc, the addition of a small amount of tin imparts improved resistance to seawater corrosion. Cartridge brass with tin becomes admiralty brass, and the 40% zinc Muntz metal with a tin addition is called *naval brass*. Brasses with 20% to 36% zinc, however, are subject to a selective corrosion, known as **dezincification**, when exposed to acidic or salt solutions. Brasses with more than 15% zinc often experience **season cracking** or **stress-corrosion cracking**. Both stress and exposure to corrosive media are required for this failure to occur (but residual stresses and atmospheric moisture might be sufficient!). As a result, cold-worked brass is usually stress relieved (to remove the residual stresses) before being placed in service.

When high machinability is required, as with automatic screw machine stock, 2% to 3% lead can be added to the brass to ensure the formation of free-breaking chips. There are both wrought and cast brasses, and the casting alloys are quite popular for use in plumbing fixtures and fittings, low-pressure valves, and a variety of decorative hardware. They have good fluidity during pouring and attractive low melting points. An alloy containing between 50% and 55% copper and the remainder zinc is often used as a filler metal in brazing. It is an effective material for joining steel, cast iron, brasses, and copper, producing joints that are nearly as strong as those obtained by welding. Copper–zinc alloys with an addition of manganese are called manganese bronze.

Table 7.2 lists some of the more common copper–zinc alloys and their composition, properties, and typical uses.

Copper–Tin Alloys

Because **tin** is costlier than zinc, alloys of copper and tin, commonly called **tin bronzes** (or **phosphor bronzes** because of the small addition of phosphorus), are usually specified when they offer some form of enhanced property or characteristic. The term **bronze** is often confusing, however, because it can be used to designate any copper alloy where the major alloy addition is not zinc or nickel. To provide clarification, the major alloy addition is usually included in the designation name.

The tin bronzes usually contain less than 12% tin, often with a small addition of phosphorus for deoxidation. Tin bronzes offer good strength, toughness, wear resistance, and

TABLE 7.2   **Composition, Properties, and Uses of Some Common Copper–Zinc Alloys**

CDA Number	Common Name	Composition(%)					Condition	Tensile Strength		Elongation in 2 in. (%)	Typical Uses
		Cu	Zn	Sn	Pb	Mn		ksi	MPa		
220	Commercial bronze	90	10				Soft sheet Hard sheet	38 64	262 441	45 4	Screen wire, hardware, screws. jewelry
240	Low brass	80	20				Spring Annealed sheet Hard	73 47 75	503 324 517	3 47 7	Drawing, architectural work, ornamental
260	Cartridge brass	70	30				Spring Annealed sheet Hard	91 53 76	627 365 524	3 54 7	Munitions, hardware, musical instruments, tubing
270	Yellow brass	65	35				Spring Annealed sheet Hard	92 46 76	634 317 524	3 64 7	Cold forming, radiator cores, springs, screws
280	Muntz metal	60	40				Hot-rolled Cold-rolled	54 80	372 551	45 5	Architectural work; condenser tube
443–445	Admiralty metal	71	28	1			Soft Hard	45 95	310 655	60 5	Condenser tube (salt water), heat exchangers
360	Free-cutting brass	61.5	35.3		3		Soft Hard	47 62	324 427	60 20	Screw-machine parts
675	Manganese bronze	58.5	39	1		0.1	Soft Bars, half hard	65 84	448 579	33 19	Clutch disks, pump rods, valve stems, highstrength propellers

corrosion resistance (including resistance to seawater). They offer a low coefficient of friction and resist thermal softening. Because of these properties, they are often used for parts that are subjected to heavy compressive loads, such as bearings, bushings, gears, pump parts, springs, and fittings. When made by mixing powders of copper and tin, followed by low-density powder metallurgy processing (described in Chapter 16), the porous products that result can be used as a filter for high-temperature or corrosive media, or can be infiltrated with oil to produce self-lubricating bearings. When the copper–tin alloys are used for bearing applications, up to 10% lead can be added.

Copper–Nickel Alloys

Copper and nickel exhibit complete solubility (as shown previously in Figure 4.6), and a wide range of useful alloys has been developed. Key features include high thermal conductivity, high-temperature strength, and corrosion resistance to a range of materials, including seawater. These properties, coupled with a high resistance to stress-corrosion cracking, make the copper–nickel alloys a good choice for heat exchangers, cookware, desalination apparatus, and a wide variety of coinage.

Cupronickels contain 2% to 30% nickel. **Nickel silvers** contain no silver, but 10% to 30% nickel and at least 5% zinc. The bright silvery luster makes them attractive for ornamental applications, and they are also used for musical instruments. An alloy with 45% nickel is known as **constantan**, and the 67% nickel material is called *Monel*. Monel will be discussed later in the chapter as a nickel alloy.

Other Copper-Based Alloys

The copper alloys discussed previously acquire their strength primarily through solid solution strengthening and cold work. Within the copper-alloy family, alloys containing aluminum, silicon, or beryllium can be strengthened by precipitation hardening.

Aluminum bronze alloys are best known for their combination of high strength and excellent corrosion resistance and are often considered as cost-effective alternatives to stainless steel and nickel-based alloys. The wrought alloys can be strengthened by solid-solution strengthening, cold work, and the precipitation of iron- or nickel-rich phases.

With less than 8% aluminum, the alloys are very ductile. When aluminum exceeds 9%, however, the ductility drops, and the hardness approaches that of steel. Still higher aluminum contents result in brittle, but wear-resistant, materials. By varying the aluminum content and heat treatment, the tensile strength can range from about 415 to 1,000 MPa (60 to 145 ksi). Typical applications include marine hardware, power shafts, sleeve bearings, and pump and valve components for handling seawater, sour mine water, and various industrial fluids. Cast alloys are available for applications where casting is the preferred means of manufacture, such as boat propellers and pump impellers. Because aluminum bronze exhibits large amounts of solidification shrinkage, castings made of this material should be designed with specific attention to shrinkage.

Silicon bronzes contain up to 4% silicon and 1.5% zinc. Strength, formability, machinability, and corrosion resistance are all quite good. Low melting points and high fluidity are attractive for permanent mold and die-casting. Tensile strengths range from a soft condition of about 380 MPa (55 ksi) through a maximum that approaches 900 MPa (130 ksi). Uses include boiler, tank, and stove applications, which require a combination of weldability, high strength, and corrosion resistance.

Copper–beryllium alloys, which contain up to 2.8% beryllium, can be age hardened to produce the highest strengths of the copper-based metals but are quite expensive to use. When annealed, the material has a yield strength of 170 MPa (25 ksi), tensile strength of 480 MPa (70 ksi), and an elongation of 50%. After heat treatment, these properties can rise to 1,100 MPa (160 ksi), 1,250 MPa (180 ksi), and 5%, respectively. Cold work coupled with age hardening can produce even stronger material. The modulus of elasticity is about 125,000 MPa (8×10^6 psi) and the endurance limit is around 275 MPa (40 ksi). These properties make the material an excellent choice for electrical contact springs, but cost limits application to small components requiring long life and high reliability. Other applications, such as spark-resistant safety tools and spot welding electrodes, utilize the unique combination of properties: (1) the material has the strength of heat-treated steel, but is also (2) nonsparking, nonmagnetic, and electrically and thermally conductive. Concerns about the toxicity of beryllium have created a demand for substitute alloys with similar properties, but no clear alternative has emerged.

Safety Tools

Natural gas or propane is leaking into an area, and a valve must be shut off, requiring the use of a wrench. Any spark in the area could ignite an explosion, and it is possible that the tool could be dropped on the concrete floor or strike another object. It is imperative, therefore, that the wrench be made of a nonpyrophoric material (i.e., nonsparking, chips or slivers do not burn in air). A standard material for

safety tools is the age-hardened, 2% beryllium, copper–beryllium alloy. It can be heat-treated to achieve strength levels equivalent to medium- or even high-carbon steels. Copper beryllium is considerably more expensive than traditional tool materials, and it is heavier than steel. However, the unique combination of strength and nonsparking make the tools well worth the expense and inconvenience.

Lead Additions and Lead-Free Casting Alloys

For many years, **lead** has been a common alloy additive to copper alloys. Insoluble lead particles act as a lubricant for cutting tool edges and promote the formation of small, discontinuous chips, enhancing both machinability and machined surface finish. In castings, lead helps to fill and seal the microporosity that forms during solidification, thereby providing the pressure tightness required for use with pressurized gases and fluids. For parts like bearings and bushings, lead provides surface lubricity.

Because of the attractive properties of the lead addition (machinability and pressure-tightness), many plumbing components have been made from leaded red and semired brass casting alloys. However, with increased concern for lead in drinking water and the introduction of environmental regulations, efforts have been made to develop lead-free copper-based casting alloys. Among the most common are the EnviroBrass alloys, which use **bismuth** and **selenium** as substitutes for lead. Bismuth is not known to be toxic for humans and has been used in a popular remedy for upset stomach. Selenium is an essential nutrient for humans. Although somewhat lower in ductility and impact strength, the new alloys have been shown to have melting and casting characteristics, mechanical properties, machinability, and platability that are quite similar to the traditional leaded materials.

7.3 Aluminum and Aluminum Alloys

General Properties and Characteristics

Although **aluminum** has only been a commercial metal for about 125 years, it now ranks second to steel in both worldwide quantity and expenditure and is clearly the most important of the nonferrous metals. It has achieved importance in virtually all segments of the economy, with principal uses in transportation, containers and packaging, building construction, electrical applications, consumer durables, and mechanical equipment. We are all familiar with uses such as aluminum cookware, window frames, aluminum siding, and the ever-present aluminum beverage can.

A number of unique and attractive properties account for the engineering significance of aluminum. These include its light weight; corrosion resistance; good electrical and thermal conductivity; optical reflectivity; ability to be formed, cast, or machined with relative ease; and a nearly limitless array of available finishes. It is nonmagnetic and nonsparking. In contrast to steel, titanium, and other materials that become brittle at very low (cryogenic) temperatures, aluminum remains ductile and even gains strength as temperature drops.

Aluminum has a specific gravity of 2.7 compared to 7.85 for steel, making aluminum about one-third the weight of steel for an equivalent volume.

Aluminum can be recycled repeatedly with no loss in quality, and recycling saves 95% of the energy required to produce aluminum from ore. Since the 1980s, the overall reclamation rate for aluminum has been more than 50%. The aluminum can is the most recycled beverage container in North America, and nearly 90% of all aluminum used in cars is recovered at the end of their useful life.

A serious weakness of aluminum from an engineering viewpoint is its relatively low modulus of elasticity, which is also about one-third that of steel. Under identical loadings, an aluminum component will deflect three times as much as a steel component of the same design. Because the modulus of elasticity cannot be significantly altered by alloying or heat treatment, it is usually necessary to provide stiffness and buckling resistance through design features such as ribs or corrugations. These can be incorporated with relative ease, however, because aluminum adapts easily to the full spectrum of fabrication processes.

Commercially Pure Aluminum

In its pure state, aluminum is soft, ductile, and not very strong. In the annealed condition, pure aluminum has only about one-fifth the strength of hot-rolled structural steel and is used primarily for its physical rather than its mechanical properties.

Electrical-conductor-grade aluminum is used in large quantities and has replaced copper in many applications, such as electrical transmission lines. Commonly designated by the letters EC, this grade contains a minimum of 99.45% aluminum and has an electrical conductivity that is 62% that of copper for the same size wire and 200% that of copper on an equal-weight basis.

Cost Comparison—Steel vs. Aluminum—Cost per unit weight versus cost per unit volume

Because metals are frequently purchased on the basis of weight, cost comparisons are often made on that basis, where aluminum is at a distinct disadvantage—generally appearing as four to five times more expensive than carbon steel. When producing parts with a fixed size and shape, however, a more appropriate comparison would be based on cost per unit volume. A pound of aluminum would produce three times as many same-size parts as a pound of steel. On a per unit volume basis, the cost difference becomes markedly less.

Aluminums for Mechanical Applications

For nonelectrical applications, most aluminum is used in the form of alloys. These have much greater strength than pure aluminum, yet retain the advantages of light weight, good conductivity, and corrosion resistance. Although usually weaker than steel, some alloys are now available that have tensile properties (except for ductility) that are comparable to those of the HSLA structural steel grades. Because alloys can be as much as 30 times stronger than pure aluminum, designers can frequently optimize their design and then tailor the material to their specific requirements. Some alloys are specifically designed for casting, whereas others are intended for the manufacture of wrought products.

On a strength-to-weight basis, most of the aluminum alloys are superior to steel and other structural metals, but wear, creep, and fatigue properties are generally rather poor. Aluminum alloys have a finite fatigue life at all reasonable values of applied stress. In addition, aluminum alloys rapidly lose their strength and dimensions change by creep when temperature is increased. As a result, most aluminum alloys should not be considered for applications involving service temperatures much above 150°C (300°F). At subzero temperatures, however, aluminum is actually stronger than at room temperature with no loss in ductility. Both the adhesive and the abrasive varieties of wear can be extremely damaging to aluminum alloys.

The selection between steel and aluminum for any given component is often a matter of cost, but considerations of light weight, corrosion resistance, low maintenance expense, and high thermal or electrical conductivity might be sufficient to justify the added cost of aluminum. At the close of World War II, the average American car contained just 5.5 kg (12 lb) of aluminum, but his number has risen every year to 156 kg (343 lb)

in 2012 and is projected to reach 247 kg (550 lb) by 2025. Because a 6%–8% fuel savings is obtained for each 10% reduction in vehicle weight, aluminum alloys now find themselves being used for engines, transmissions, wheels, heat exchangers (radiators), hoods, trunks, doors, roofs, and bumpers. Nearly 70% of all automotive engine blocks are now cast in aluminum. An aluminum space frame, such as the one shown in **Figure 7.3** for the 2012 Audi S8 sedan, can reduce the overall weight of the structure, enhance recyclability, and reduce the number of parts required for the primary body structure.

Although composites are making inroads in aerospace, most current commercial aircraft are nearly 80% aluminum by weight. Compared to composites, aluminum continues to offer reduced cost to manufacture, operate, and repair. Sporting goods is another popular market area, with usage in bicycles, boats, and hiking equipment. Consumer items include walkers, wheelchairs, and strollers.

Corrosion Resistance of Aluminum and its Alloys

Pure aluminum is very reactive and forms a tight, adherent oxide coating on the surface as soon as it is exposed to air. This oxide is resistant to many corrosive media and serves as a corrosion-resistant barrier to protect the underlying metal. Like stainless steels, the corrosion resistance of aluminum is actually a property of the oxide, not the metal itself. Because the oxide formation is somewhat retarded when alloys are added, aluminum alloys do not have quite the corrosion resistance of pure aluminum.

The oxide coating also causes difficulty when welding. To produce consistent quality resistance welds, it is usually

Audi S8
ASF - Audi Space Frame aus Aluminium
ASF - Audi Space Frame in Aluminum
10/11

Audi

CARICOS.COM

FIGURE 7.3   The all-aluminum space frame of the 2012 Audi S8 sedan. (Courtesy of Audi)

necessary to remove the tenacious oxide immediately before welding. For fusion welding, special fluxes or protective inert gas atmospheres must be used to prevent material oxidation. Although welding aluminum might be more difficult than steel, suitable techniques have been developed to permit the production of high-quality, cost-effective welds with most of the welding processes.

Classification System

Aluminum alloys can be divided into two major groups based on the method of fabrication. *Wrought alloys* are those that are shaped as solids and are therefore designed to have attractive forming characteristics, such as low yield strength, high ductility, good fracture resistance, and good strain hardening. *Casting alloys* achieve their shape as they solidify in molds or dies. Attractive features for the casting alloys include low melting point, high fluidity, resistance to hot-cracking during and after solidification, good as-cast surface finish, and attractive as-solidified structures and properties. Clearly, these properties are distinctly different, and the alloys that have been designed to meet them are also different. As a result, separate classification systems exist for the wrought and cast aluminum alloys.

Wrought Aluminum Alloys

The wrought aluminum alloys are generally identified using the standard four-digit designation system for aluminums. The first digit indicates the major alloy element or elements according to the following list:

Major Alloying Element	
Aluminum, 99.00% and greater	1xxx
Copper	2xxx
Manganese	3xxx
Silicon	4xxx
Magnesium	5xxx
Magnesium and silicon	6xxx
Zinc	7xxx
Other element	8xxx

For the 1xxx series, the remaining digits indicate the level of purity. For all remaining series, the second digit is usually zero, with nonzero numbers being used to indicate some form of modification or improvement to the original alloy. The last two digits simply indicate the particular alloy within the family, specifying the chemical composition and allowable chemistry ranges. For example, 2024 simply means alloy number 24 within the 2xxx, or aluminum–copper, system.

The four digits of a wrought aluminum designation identify the chemistry of the alloy. Additional information about the processing history of the alloy (i.e., its condition) is then provided through a **temper designation**, in the form of a letter or letter–number suffix using the following system:

- F: as fabricated
- H: strain-hardened (cold worked)
  - H1: strain-hardened by working to desired dimensions; a second digit, 1 through 9, indicates the degree of hardening, 8 being commercially full-hard and 9 extra-hard
  - H2: strain-hardened by cold working, followed by partial annealing
  - H3: strain-hardened and stabilized
- O: annealed
- T: thermally treated (heat treated)
  - T1: cooled from hot working and naturally aged
  - T2: cooled from hot working, cold-worked, and naturally aged
  - T3: solution-heat-treated, cold-worked, and naturally aged
  - T4: solution-heat-treated and naturally aged
  - T5: cooled from hot working and artificially aged
  - T6: solution-heat-treated and artificially aged
  - T7: solution-heat-treated and stabilized
  - T8: solution-heat-treated, cold-worked, and artificially aged
  - T9: solution-heat-treated, artificially aged, and cold-worked
  - T10: cooled from hot working, cold-worked, and artificially aged
- W: solution-heat-treated only

These treatments or tempers can be used to: increase strength, hardness, wear resistance, or machinability; increase chemical homogeneity; stabilize mechanical or physical properties; alter electrical characteristics; reduce residual stresses; or improve corrosion resistance. By identifying both the chemistry and the condition, one can reasonably estimate the mechanical and physical properties.

Wrought alloys can be further divided into two basic types: those that achieve strength by solid-solution strengthening and cold working, and those that can be strengthened by an age-hardening heat treatment. **Table 7.3** lists some of the common wrought aluminum alloys in each of these families. It can be noted that the work-hardenable alloys (those that cannot be age-hardened) are primarily those in the 1xxx (pure aluminum), 3xxx (aluminum–manganese), and 5xxx (aluminum–magnesium) series. A comparison of the annealed (O-suffix) and cold-worked (H-suffix) conditions reveals the amount of strengthening achievable through strain hardening.

The precipitation-hardenable alloys are found primarily in the 2xxx, 6xxx, and 7xxx series. By comparing the properties in the heat-treated condition to those of the strain-hardened alloys, we see that heat-treatment offers significantly higher strength. Alloy 2017, the original **duralumin**, is probably the oldest age-hardenable aluminum alloy. The 2024 alloy is

TABLE 7.3 Composition and Properties of Some Wrought Aluminum Alloys in Various Conditions

Designation[a]	Composition (%) Aluminum = Balance					Form Tested	Tensile Strength		Yield Strength[b]		Elongation in 2 in. (%)	Brinell Hardness	Uses and Characteristics
	Cu	Si	Mn	Mg	Others		ksi	MPa	ksi	MPa			
Work-Hardening Alloys—Not Heat-Treatable													
1100-O	0.12				99 Al	$\frac{1}{16}$-in. sheet	13	90	5	34	35	23	Commercial Al: good forming properties
1100-H14						$\frac{1}{16}$-in. sheet	16	110	14	97	9	32	Good corrosion resistance, low yield strength
110-H18						$\frac{1}{16}$-in. sheet	24	165	21	145	5	44	Cooking utensils; sheet and tubing
3003-0	0.12		1.2			$\frac{1}{16}$-in. sheet	16	110	6	41	30	28	Similar to 1100
3003-H14						$\frac{1}{16}$-in. sheet	22	152	21	145	8	40	Slightly stronger and less ductile
3003-H18						$\frac{1}{16}$-in. sheet	29	200	27	186	4	55	Cooking utensils; sheet-metal work
5052-0				2.5	0.25 Cr	$\frac{1}{16}$-in. sheet	28	193	13	90	25	45	Strongest work-hardening alloy
5052-H32						$\frac{1}{16}$-in. sheet	33	228	28	193	12	60	Highly yield strength and fatigue limit
5052-H36						$\frac{1}{16}$-in. sheet	40	276	35	241	8	73	Highly stressed sheet-metal products
Precipitation-Hardening Alloys—Heat-Treatable													
2017-0	4.0	0.5	0.7	0.6		$\frac{1}{16}$-in. sheet	26	179	10	69	20	45	Duralumin, original strong alloy
2017-T4						$\frac{1}{16}$-in. sheet	62	428	40	276	20	105	Hardened by quenching and aging
2024-0	4.4		0.6	1.5		$\frac{1}{16}$-in. sheet	27	186	11	76	20	42	Stronger than 2017
2024-T4						$\frac{1}{16}$-in. sheet	64	441	45	290	19	120	Used widely in aircraft construction
2014-0	4.4	0.8	0.8	0.5		$\frac{1}{2}$-in. extruded shapes	27	186	14	97	12	45	Strong alloy for extruded shapes
2014-T6						Forgings	65	448	55	379	10	125	Strong forging alloy
2014-T6						$\frac{1}{16}$-in. sheet	70	483	60	413	8		Higher yield strength than Alclad 2024
Alclad 2014-T6	4.5	1.0	0.8	0.4		$\frac{1}{16}$-in. sheet	63	434	56	386	7		Clad with heat-treatable alloy[c]
7075-0	1.6		0.2	2.5	0.3 Cr 5.6 Zn	$\frac{1}{16}$-in. sheet	33	228	15	103	17	60	Alloy of highest strength
7075-T6						$\frac{1}{16}$-in. sheet	76	524	67	462	11	150	Lower ductility than 2024
Alclad 7075-T6						$\frac{1}{16}$-in. sheet	76	524	67	462	11		Strongest Alclad product
7075-T6						$\frac{1}{2}$-in. extruded shapes	80	552	70	483	6		Strongest alloy for extrusions
6061-T6	0.28	0.6		1.0	0.20 Cr	$\frac{1}{2}$-in. extruded shapes	42	290	40	276	12	95	Strong, corrosion resistant
6063-T6		0.4		0.7		$\frac{1}{2}$-in. rod extruded	35	241	31	214	12	80	Good forming properties and corrosion resistance
6151-T6		0.9		0.6	0.20 Cr	Forgings	48	331	43	297	17	90	For intricate forgings
2025-T6	4.5	0.8	0.8			Forgings	55	379	30	207	18	100	Good forgeability, lower cost
2018-T6	4			0.7	2 Ni	Forgings	55	379	40	276	10	100	Strong at elevated temperatures; forged pistons
4032-T6	0.9	12.2		1.1	0.9 Ni	Forgings	55	379	46	317	9	115	Forged aircraft pistons
2011-T3	5.5	(0.5 Bi)			0.5 Pb	$\frac{1}{2}$-in. rod	55	379	43	297	15	95	Free cutting, screw-machine products

[a]O, annealed; T, quenched and aged; H, cold-rolled to hard temper.
[b]Yield strength taken at 0.2% permanent set.
[c]Cladding alloy; 1.0 Mg, 0.7 Si, 0.5 Mn.

stronger and has seen considerable use in aircraft applications. An attractive feature of the 2xxx series is the fact that ductility does not significantly decrease during the strengthening heat treatment. Within the 7xxx series are some newer alloys with strengths that approach or exceed those of the high-strength structural steels. Ductility, however, is generally low, and fabrication is more difficult than for the 2xxx-type alloys. Nevertheless, the 7xxx series alloys have also found wide use in aircraft applications. To maintain properties, age-hardened alloys should not be used at temperatures greater than 175°C (350°F). Welding should be performed with considerable caution because the exposure to elevated temperature will significantly diminish the strengthening achieved through either cold working or age hardening.

Because of their two-phase structure, the heat-treatable alloys tend to have poorer corrosion resistance than either pure aluminum or the single-phase work-hardenable alloys. When both high strength and superior corrosion resistance are desired, wrought aluminum is often produced as **Alclad** material. A thin layer of corrosion-resistant aluminum is bonded to one or both surfaces of a high-strength alloy during rolling, and the material is further processed as a composite.

Because only moderate temperatures are required to lower the strength of aluminum alloys, extrusions and forgings are relatively easy to produce and are manufactured in large quantities. Deep drawing and other sheet-metal-forming operations can also be carried out quite easily. In general, the high ductility and low yield strength of the aluminum alloys make them appropriate for almost all forming operations. Good dimensional tolerances and fairly intricate shapes can be produced with relative ease.

The machinability of aluminum-based alloys, however, can vary greatly, and special tools and techniques might be desirable if large amounts of machining are required. Free-machining alloys, such as 2011, have been developed for screw-machine work. These special alloys can be machined at very high speeds and have replaced brass screw machine stock in many applications.

Color anodizing offers an inexpensive and attractive means of surface finishing. A thick aluminum oxide is produced on the surface. Colored dye then penetrates the porous surface and is sealed by immersion into hot water. The result is the colored metallic finish commonly observed on products such as bicycle frames and softball bats.

Aluminum Casting Alloys

Although its low melting temperature tends to make it suitable for casting, pure aluminum is seldom cast. Its high shrinkage on solidification (about 7%) and susceptibility to hot cracking cause considerable difficulty, and scrap is high. By adding small amounts of alloying elements, however, very suitable casting characteristics can be obtained and strength can be increased. Aluminum alloys are cast in considerable quantity by a variety of processes. Many of the most popular alloys contain enough silicon to produce the eutectic reaction, which is characterized by a low melting point and high as-cast strength. Silicon also improves the fluidity of the metal, making it easier to produce complex shapes or thin sections, but high silicon also produces an abrasive, difficult-to-cut material. Copper, zinc, and magnesium are other popular alloy additions that permit the formation of age-hardening precipitates.

Table 7.4 lists some of the commercial aluminum casting alloys and uses the three-digit designation system of the Aluminum Association to designate alloy chemistry. The first digit indicates the alloy group as follows:

Major Alloying Element	
Aluminum, 99.00% and greater	1xx.x
Copper	2xx.x
Silicon with Cu and/or Mg	3xx.x
Silicon	4xx.x
Magnesium	5xx.x
Zinc	7xx.x
Tin	8xx.x
Other elements	9xx.x

The second and third digits identify the particular alloy or aluminum purity, and the last digit, separated by a decimal point, indicates the product form (e.g., finished casting or ingot to be used in casting manufacture). A letter before the numerical designation indicates a modification of the original alloy, such as a small variation in the amount of an alloying element or impurity.

Aluminum casting alloys are selected for both properties and process. When the strength requirements are low, as-cast properties are usually adequate. High-strength castings usually require the use of alloys that can subsequently be heat-treated. In addition, not all alloys can be used with all casting methods. Sand casting has the fewest process restrictions and the widest range of aluminum casting alloys, with alloy 356 being the most common. The aluminum alloys used for permanent mold casting are designed to have lower coefficients of thermal expansion (or contraction) because the molds offer restraint to the dimensional changes that occur on cooling. Die-casting alloys require high degrees of fluidity because they are often cast in thin sections. Most of the die-casting alloys are also designed to produce high "as-cast" strength without heat treatment, using the rapid cooling conditions of the die-casting process to promote a fine grain size and fine eutectic structure. Alloy 380 comprises about 85% of aluminum die casting production. Tensile strengths of the aluminum permanent-mold and die-casting alloys can be in excess of 275 MPa (40 ksi).

Aluminum–Lithium Alloys

Lithium is the lightest of all metallic elements, and in the search for aluminum alloys with higher strength, greater stiffness, and lighter weight, aluminum–lithium alloys have emerged. Each percent of lithium reduces the overall weight by 3% and increases stiffness by 5%. Although the initial alloys offered lower density, greater stiffness, strengths comparable to those

TABLE 7.4   Composition, Properties, and Characteristics of Some Aluminum Casting Alloys

Alloy Designation[a]	Process[b]	Composition (%) (Major Alloys > 1%)						Temper	Tensile Strength		Elongation in 2 in. (%)	Uses and Characteristics
		Cu	Si	Mg	Zn	Fe	Other		ksi[c]	MPa		
201	S, P	4.6					0.7Ag	T6	65	448	8.0	Auto engine components
206	S	4.5						T7	63	434	12.0	Strength at elevated temps.
208	S	4.0	3.0		1.0	1.2		F	19	131	1.5	General-purposes and castings, can be heat treated
242	S, P	4.0		1.6		1.0	2.0 Ni	T61	40	276	—	Withstands elevated temperatures
295	S	4.5	1.0			1.0		T6	32	221	3.0	Structural castings, heat-treatable
296	P	4.5	2.5		1.0	1.2		T6	35	241	2.0	Permanent-mold version of 295
308	P	4.5	5.5		1.0	1.0		F	24	166	—	General-purpose permanent mold
319	S, P	3.5	6.0		1.0	1.0		T6	31	214	1.5	Superior casting characteristics
355	S, P	1.3	5.0					T6	32	221	2.0	High strength and pressure tightness
C355	S, P	1.3	5.0					T61	40	276	3.0	Stronger and more ductile than 355
356	S, P		7.0					T6	30	207	3.0	Excellent castability and impact strength
A356	S, P		7.0					T61	37	255	5.0	Stronger and more ductile than 356
357	S, P		7.0					T6	45	310	3.0	High strength-to-weight castings
359	S, P		9.0					—	—	—	—	High-strength aircraft usage
360	D		9.5			2.0		F	44[d]	303	2.5[d]	Good corrosion resistance and strength
A360	D		9.5			2.0		F	46[d]	317	3.5[d]	Similar to 360
380	D	3.5	8.5		3.0	2.0		F	46[d]	317	2.5[d]	High strength and hardness
A380	D	3.5	8.5		3.0	1.3		F	47[d]	324	3.5[d]	Similar to 380
384	D	3.75	11.3		1.0	1.3		F	48[d]	331	2.5	High strength and hardness
390	S, P	4.5	17.0	0.6				T6	45	310	—	High wear resistance
413	D	1.0	12.0			2.0		F	43[d]	297	2.5[d]	General-purpose, good castability
A413	D	1.0	12.0			1.3		F	42[d]	290	3.5[d]	Similar to 413
443	D		5.25			2.0		F	33[d]	228	9.0[d]	General-purpose, good castability
B443	S		5.25			2.0		F	17	117	3.0	General-purpose casting alloy
514	S			4.0				F	22	152	6.0	High corrosion resistance
518	D			8.0		1.8		F	45[d]	310	5.0[d]	Good corrosion resistance, strength, and toughness
520	S			10.0				T4	42	290	12.0	High strength with good ductility
535	S			6.9				F	35	241	9.0	Good corrosion resistance and machinability
712	S				5.8			F	34	234	4.0	Good properties without heat treatment
713	S, P				7.5	1.1		F	32	221	3.0	Similar to 712
850	S, P	1.0					6.3 Sn, 1.0 Ni	T5	16	110	5.0	Bearing alloy

(Note the compatibility with specific casting processes presented in the second column.)

[a] Aluminum Association.

[b] Sand-cast; P, permanent-mold-cast; D, die cast.

[c] Minimum figures unless noted.

[d] Typical values.

of existing alloys, and good resistance to fatigue crack propagation, fracture toughness, ductility, and stress-corrosion resistance were poorer than for conventional alloys. Successive iterations have emerged, and the current third-generation alloys offer properties that are equivalent or superior to carbon-fiber composite and can be fabricated by conventional processes.

Because aluminum alloys can comprise as much as 80% of the weight of commercial aircraft, even small percentage reductions can be significant. Improved strength and stiffness can further facilitate weight reduction. Fuel savings over the life of the airplane can more than compensate for any additional manufacturing expense.

Aluminum Foam

A material known as **stabilized aluminum foam** can be made by mixing ceramic particles with molten aluminum and blowing gas into the mixture. The bubbles remain through solidification, yielding a structure that resembles metallic Styrofoam. Large panels can be cast with densities ranging from 2.5% to 20% that of solid aluminum. Originally developed for automotive, aerospace, and military applications, the material has found additional uses in architecture and design. Strength-to-weight is outstanding, and the material offers excellent energy absorption from impacts, crashes, and explosive blasts. The fuel cells of race cars have been shrouded with aluminum foam, and foam fill has been inserted between the front of cars and the driver compartment. Tubular structures can be filled with foam to increase strength, absorb energy, and provide resistance to crushing. Still other applications capitalize on the excellent thermal insulation, vibration damping, and sound absorption that result from the numerous trapped air pockets. The metal foams are easily machined and can be joined by adhesive bonding and brazing, as well as laser and gas-tungsten-arc welding.

7.4 Magnesium and Magnesium Alloys

General Properties and Characteristics

Magnesium is the third-most-commonly-used structural metal, following iron and aluminum, and is the lightest of the commercially important metals, having a specific gravity of about 1.74 (two-thirds that of aluminum, one-quarter that of steel, and only slightly higher than fiber-reinforced plastics). Like aluminum, magnesium is relatively weak in the pure state and for engineering purposes is almost always used as an alloy. Even in alloy form, however, the metal is characterized by poor wear, creep, and fatigue properties. It has the highest thermal expansion of all engineering metals. Strength drops rapidly when the temperature exceeds 100°C (200°F), so magnesium

should not be considered for elevated-temperature service. Its modulus of elasticity is even less than that of aluminum, being between one-fourth and one-fifth that of steel. Thick sections are required to provide adequate stiffness, but the alloy is so light that it is often possible to use thicker sections for the required rigidity and still have a lighter structure than can be obtained with any other metal. Cost per unit volume is low, so the use of thick sections is generally not prohibitive. Moreover, because a large portion of magnesium components are cast, the thicker sections actually become a desirable feature. Ductility is frequently low, a characteristic of the hexagonal-close-packed (HCP) crystal structure, but some alloys have values exceeding 10%.

On the more positive side, magnesium alloys have a relatively high strength-to-weight ratio, with some commercial alloys attaining strengths as high as 380 MPa (55 ksi). High energy absorption provides good damping of noise and vibration, as well as impact and dent resistance, and electrical and thermal conductivity are relatively good. Although many magnesium alloys require enamel or lacquer finishes to impart adequate corrosion resistance, this property has been improved markedly with the development of high-purity alloys.[1] In the absence of unfavorable galvanic couples, these materials have excellent corrosion resistance. Although aluminum alloys are often used for the load-bearing members of mechanical structures, magnesium alloys are best suited for those applications where light weight is the primary consideration and strength is a secondary requirement. Compared to reinforced plastics, the strengths and densities are quite comparable, but magnesium is three to six times more rigid and can be cast with thinner walls. As a result, magnesium alloys are finding applications in a wide range of markets, including automotive, aerospace, power tools, sporting goods, luggage, and electronic products (where it offers the combination of electromagnetic shielding, light weight, and durability).

Magnesium Alloys and Their Fabrication

Like aluminum, magnesium alloys are classified as either cast or wrought, and some are heat-treatable by age hardening. A designation system for magnesium alloys has been developed by ASTM, identifying both chemical composition and temper, and is presented in specification B93. Two prefix letters designate the two largest alloying metals in order of decreasing amount, using the following format:

A aluminum	F iron	M manganese	R chromium
B bismuth	H thorium	N nickel	S silicon
C copper	K zirconium	P lead	T tin
D cadmium	L beryllium	Q silver	Z zinc
E rare earth			

[1] Iron, nickel, copper, and cobalt have low solubility limits in magnesium and tend to precipitate out as intermetallic compounds that are cathodic to the magnesium. Reducing the amount of these metals improves corrosion resistance, but increases cost.

TABLE 7.5   **Composition, Properties, and Characteristics of Common Magnesium Alloys**

Alloy	Temper	Composition (%)						Tensile Strength[a]		Yield Strength[a]		Elongation in 2 in. (%)	Uses and Characteristics
		Al	Rare Earths	Mn	Th	Zn	Zr	ksi	MPa	Ksi	MPa		
AM60A	F	6.0		0.13				30	207	17	117	6	Die castings
AM100A	T4	10.0		0.1				34	234	10	69	6	Sand and permanent-mold castings
AZ31B	F	3.0				1.0		32	221	15	103	6	Sheet, plate, extrusions, forgings
AZ61A	F	6.5				1.0		36	248	16	110	7	Sheet, plate, extrusions, forgings
AZ63A	T5	6.0				3.0		34	234	11	76	7	Sand and permanent-mold castings
AZ80A	T5	8.5				0.5		34	234	22	152	2	High-strength forgings, extrusions
AZ81A	T4	7.6				0.7		34	234	11	76	7	Sand and permanent-mold castings
AZ91A	F	9.0				0.7		34	234	23	159	3	Die castings
AZ92A	T4	9.0				2.0		34	234	11	76	6	High-strength sand and permanent-mold castings
EZ33A	T5		3.2			2.6	0.7	20	138	14	97	2	Sand and permanent-mold castings
HK31A	H24				3.2		0.7	33	228	24	166	4	Sheet and plates; castings in T6 temper
HM21A	T5			0.8	2.0			33	228	25	172	3	High-temperature (800°F) sheets, plates, forgings
HZ32A	T5				3.2	2.1		27	186	13	90	4	Sand and permanent-mold castings
ZH62A	T5				1.8	5.7	0.7	35	241	22	152	5	Sand and permanent-mold castings
ZK51A	T5					4.6	0.7	34	234	20	138	5	Sand and permanent-mold castings
ZK60A	T5					5.5	0.45	38	262	20	138	7	Extrusions, forgings

[a]Properties are minimums for the designated temper.

Aluminum is the most common alloying element and, along with zinc, zirconium, and thorium, promotes precipitation hardening. Manganese improves corrosion resistance, and tin improves castability. The two letters are then followed by two or three numbers and a possible suffix letter. The numbers correspond to the rounded-off whole-number percentages of the two main alloy elements and are arranged in the same order as the letters. Thus the AZ91 alloy would contain approximately 9% aluminum and 1% zinc. A suffix letter is used to denote variations of the same base alloy, such as AZ91A. The temper designation is quite similar to that used with the aluminum alloys. **Table 7.5** lists some of the more common magnesium alloys together with their properties and uses.

Sand, permanent mold, die, semisolid, and investment casting are all well developed for magnesium alloys and take advantage of the low melting points and high fluidity. Die casting is clearly the most popular manufacturing process, accounting for 70% of all magnesium castings. Although the magnesium alloys typically cost about twice as much as aluminum, the hot-chamber die-casting process used with magnesium is easier, more economical, and 40 to 50% faster than the cold-chamber process generally required for aluminum. Wall thickness, draft angle, and dimensional tolerances are all lower than for both aluminum die castings and thermoplastic moldings. Die life is significantly greater than that observed with aluminum. As a result, magnesium die castings compete well with aluminum[2] and often replace plastic injection-molded components when improved stiffness or dimensional stability, or the benefits of electrical or thermal conductivity, are required.

Forming behavior is poor at room temperature, but most conventional processes can be performed when the material is heated to temperatures between 250° and 500°C (480° and 775°F). Because these temperatures are easily attained and generally do not require a protective atmosphere, many formed and drawn magnesium products are manufactured. Magnesium extrusions and sheet metal products have properties similar to the more common wrought aluminum alloys. Although slightly heavier than plastics, they offer an order of magnitude or greater improvement in stiffness or rigidity.

---

[2] The most common magnesium die casting alloy, AZ91, has the same yield strength and ductility as the most common die cast aluminum, alloy 380.

The machinability of magnesium alloys is the best of any commercial metal, and in many applications, the savings in machining costs, achieved through deeper cuts, higher cutting speeds, and longer tool life, more than compensate for the increased cost of the material. It is necessary, however, to keep the tools sharp and provide adequate cooling for the chips.

Magnesium alloys can be spot welded almost as easily as aluminum, but scratch brushing or chemical cleaning is necessary before forming the weld. Fusion welding is best performed with processes using an inert shielding atmosphere of argon or helium gas.

Although heat treatments can be used to increase strength, the added increment achieved by age hardening is far less than observed with aluminum. In fact, the strongest magnesium alloy is only about three times stronger than the weakest. Because of this, designs must be made to accommodate the material, rather than the material being tailored to the design.

Considerable misinformation exists regarding the fire hazards when processing or using magnesium alloys. It is true that magnesium alloys are highly combustible when in a finely divided form, such as powder or fine chips, and this hazard should never be ignored. In the form of sheet, bar, extruded product, or finished castings, however, magnesium alloys rarely present a fire hazard. When the metal is heated above 500°C (950°F), a noncombustible, oxygen-free atmosphere is recommended to suppress burning, which will initiate around 600°C (1100°F). Casting operations often require additional precautions because of the reactivity of magnesium with certain mold materials, such as those containing sand and water.

7.5 Zinc and Zinc Alloys

More than 50% of all metallic **zinc** is used in the **galvanizing** of iron and steel. In this process the iron-based material is coated with a layer of zinc by one of a variety of processes that include direct immersion in a bath of molten metal (hot dipping) and electrolytic plating. The resultant coating provides excellent corrosion resistance, even when the surface is badly scratched or marred, and this corrosion resistance will persist until all of the sacrificial zinc has been depleted.

Zinc is also used as the base metal for a variety of die-casting alloys. For this purpose, zinc offers low cost, a low melting point (only 380°C or 715°F), and the attractive property of not adversely affecting steel dies when in molten metal contact. Unfortunately, pure zinc is almost as heavy as steel and is also rather weak and brittle. Therefore, when alloys are designed for die casting, the alloy elements are usually selected for their ability to increase strength and toughness in the as-cast condition while retaining the low melting point.

The composition and properties of common zinc die-casting alloys are presented in **Table 7.6**. Alloy AG40A (also known as alloy 903 or Zamak 3) is widely used because of its excellent dimensional stability, and alloy AC41A (also known as alloy 925 or Zamak 5) offers higher strength and better corrosion resistance. As a whole, the zinc die-casting alloys offer a reasonably high strength and impact resistance, along with the ability to produce complex shapes, with high dimensional precision, fine surface finish, and extremely thin sections (as thin as 0.5 mm or 0.02 in). The dimensions are quite stable, and the products

TABLE 7.6 **Composition and Properties of Some Zinc Die-Casting Alloys**

Alloy	#3 SAE 903 ASTM AG40A	#5 SAE 925 ASTM AC41A	#7 ASTM AG408	ZA-8 S[a]	ZA-8 P	ZA-8 D	ZA-12 S	ZA-12 P	ZA-12 D	ZA-27 S	ZA-27 P	ZA-27 D
Composition[b]												
Aluminum	3.5–4.3	3.5–4.3	3.5–4.3	8.0–8.8			10.5–11.5			25.0–28.0		
Copper	0.25 max	0.75–1.25	0.25 max	0.8–1.3			0.75–1.2			2.0–2.5		
Zinc	balance	balance	balance	balance			balance			balance		
Properties												
Density (g/cc)	6.6	6.6	6.6	6.3			6.0			5.0		
Yield strength (MPa)	221	228	221	200	206	290	214	269	317	372		379
(ksi)	32	33	32	29	30	42	31	39	46	54		55
Tensile strength (MPa)	283	328	283	263	255	374	317	345	400	441		421
(ksi)	41	48	41	38	37	54	46	50	58	64		61
Elongation (% in 2 in.)	10	7	13	2	2	10	3	3	7	6		3
Impact strength (J)	58	65	58	20		42	25		29	47		5
Modulus of elasticity (GPa)	85.5	85.5	85.5	85.5			82.7			77.9		
Machinability[c]	E	E	E	E			VG			G		

[a]S, sand-cast; P, permanent-mold cast; D, die-cast.
[b]Also contains small amounts of Fe, Pb, Cd, Sn, and Ni.
[c]E, excellent; VG, very good; G, good.

can be finish machined at a minimum of cost. Resistance to surface corrosion is adequate for a number of applications, and the material can be surface finished by a variety of means that include polishing, plating, painting, anodizing, or a chromate conversion coating. Energy costs are low (low melting temperature), tool life is excellent, and the zinc alloys can be efficiently recycled. Although the rigidity is low compared to other metals, it is far superior to engineering plastics, and thin-wall zinc die castings often compete with plastic injection moldings.

The attractiveness of zinc die casting has been further enhanced by the zinc–aluminum casting alloys (ZA-8, ZA-12, and ZA-27, with 8, 12, and 27% aluminum, respectively). Initially developed for sand, permanent mold, and graphite mold casting, these alloys can also be die-cast to achieve higher strength (up to 415 MPa or 60 ksi), higher hardness (up to 120 BHN), improved creep resistance and wear resistance, and lighter weight than is possible with any of the conventional alloys. Because of their lower melting and casting costs, these materials are becoming attractive alternatives to the conventional aluminum, brass, and bronze casting alloys, as well as cast iron.

7.6 Titanium and Titanium Alloys

Titanium is a strong, lightweight, corrosion-resistant metal that has been of commercial importance since about 1950. Because its properties are generally between those of steel and aluminum, its importance has been increasing rapidly. The yield strength of commercially pure titanium is about 210 MPa (30 ksi), but this can be raised to 1300 MPa (190 ksi) or higher through alloying and heat treatment, a strength comparable to that of many heat-treated alloy steels. Density, on the other hand, is approximately 60% that of steel (making strength-to-weight quite attractive), and the modulus of elasticity is a little more than one-half that of steel. Titanium alloys are often considered to be high-temperature engineering materials because the good mechanical properties are retained up to temperatures between 425° and 595°C (800° to 1100°F). The coefficient of thermal expansion is lower than that of steel and less than half that of aluminum. On the negative side, titanium and its alloys suffer from high cost, fabrication difficulties, a high energy content (they require about 10 times as much energy to produce as steel), and a high reactivity at elevated temperatures (above 535°C).

Titanium alloys are designated by major alloy and amount (see ASTM specification B-265) and are generally grouped into three classes based on their microstructural features. These classes are known as *alpha-*, *beta-*, and *alpha-beta-titanium alloys*, where the terms denote the stable phase or phases at room temperature. Alloying elements can be used to stabilize the room-temperature hexagonal-close-packed alpha phase

or the elevated-temperature body-centered-cubic beta phase, and heat treatments can be applied to manipulate structure and improve properties. Fabrication can be by casting (generally investment or graphite-mold), forging, rolling, extrusion, or welding, provided that special process modifications and controls are implemented. Advanced processing methods include powder metallurgy, mechanical alloying, rapid-solidification processing (RSP), superplastic forming, diffusion bonding, and hot-isostatic pressing (HIP).

Although titanium is an abundant metal, it is difficult to extract from ore, difficult to process, and difficult to fabricate. These difficulties make it significantly more expensive than either steel or aluminum, so its uses relate primarily to its light weight, high strength-to-weight ratio, good stiffness, good fatigue strength and fracture toughness, good thermal conductivity, excellent corrosion resistance (the result of a thin, tenacious oxide coating), and the retention of mechanical properties at elevated temperatures, as shown for several alloys in **Figure 7.4**. Titanium is also nontoxic and biocompatible with human tissues and bones. Aluminum, magnesium, and beryllium are the only base metals that are lighter than titanium, and none of these come close in either mechanical performance or elevated temperature properties. Aerospace applications tend to dominate, with titanium comprising up to 40% of the structural weight of high-performance military fighters. Titanium and titanium alloys are also used in such diverse areas as chemical- and electrochemical-processing equipment, food-processing equipment, heat exchangers, marine implements, medical implants, high-performance bicycle and automotive components, and sporting goods. They are

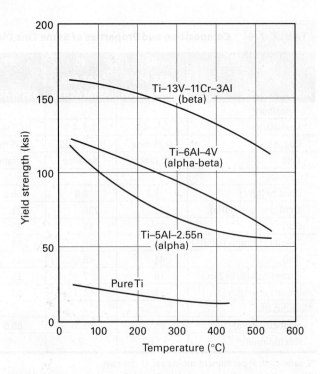

FIGURE 7.4 Strength retention at elevated temperature for various titanium alloys.

often used in place of steel where weight savings are desired, and to replace aluminums where high-temperature performance is necessary. Some bonding applications utilize the unique property that titanium wets glass and some ceramics. The titanium–6% aluminum–4% vanadium alloy (Ti-6-4) is the most popular titanium alloy, accounting for nearly 50% of all titanium usage worldwide.

7.7 Nickel-Based Alloys

Nickel-based alloys are most noted for their outstanding strength and corrosion resistance, particularly at high temperatures, and are available in a wide range of wrought and cast grades. Wrought alloys are generally known by trade names, such as Monel, Hastelloy, Inconel, Incoloy, and others. Cast alloys are generally identified by Alloy Casting Institute or ASTM designations. General characteristics include good formability (FCC crystal structure), good creep resistance, and the retention of strength and ductility at cold or even cryogenic temperatures.

Monel metal, an alloy containing about 67% nickel and 30% copper, has been used for years in the chemical and food-processing industries because of its outstanding corrosion characteristics. In fact, Monel probably has better corrosion resistance to more media than any other commercial alloy. It is particularly resistant to saltwater, sulfuric acid, and even high-velocity, high-temperature steam. For the latter reason, Monel has been used for steam turbine blades. It can be polished to have an excellent appearance, similar to that of stainless steel, and is often used in ornamental trim and household ware. In its most common form, Monel has a tensile strength ranging from 500 to 1200 MPa (70 to 170 ksi), with a companion elongation between 2 and 50%.

Nickel-based alloys have also been used for electrical resistors and heating elements. These materials are primarily nickel–chromium alloys and are known as **nichromes**. They have excellent resistance to oxidation while retaining useful strength at red heats. **Invar**, an alloy of nickel and 36% iron, has a near zero thermal expansion and is used where dimensions cannot change with a change in temperature. **Nitinol**, a nickel–titanium alloy, possesses the unique properties of shape memory or superelasticity, depending on its heat treatment. In the superelastic (or hardened) condition, its superb corrosion resistance, wear and erosion resistance, and ability to withstand compressive loading in excess of 2400 MPa (350 ksi) without permanent deformation make it an attractive material for bearing applications.

Other nickel-based alloys have been designed to provide good mechanical properties at extremely high temperatures and are generally classified as *superalloys*. These alloys will be discussed along with other similar materials in the following section.

7.8 Superalloys, Refractory Metals, and Other Materials Designed for High-Temperature Service

Titanium and titanium alloys have already been cited as being useful in providing strength at elevated temperatures, but the maximum temperature for these materials is approximately 535°C (1000°F). Jet engine, gas turbine, rocket, and nuclear applications often require materials that possess high strength, creep resistance, oxidation and corrosion resistance, and fatigue resistance at temperatures up to and in excess of 1100°C (2000°F). Other high-temperature applications include heat exchangers, chemical reaction vessels, and furnace components.

One group of materials offering these properties is the **superalloys**, first developed in the 1940s for use in the elevated-temperature areas of turbojet aircraft. These alloys are based on **nickel**, **iron and nickel**, or **cobalt** and have the ability to retain most of their strength even after long exposures to extremely high temperatures. Strength comes from solid-solution strengthening, precipitation hardening and dispersed alloy carbides or oxides. The nickel-based alloys tend to have higher strengths at room temperature, with yield strengths up to 1200 MPa (175 ksi) and ultimate tensile strengths as high as 1450 MPa (210 ksi). The 1,000-hour rupture strengths of the nickel-based alloys at 815°C (1500°F) are also higher than those of the cobalt-based material. Nickel alloys currently comprise approximately 50% of the weight of a jet engine. The cobalt-based alloys have greater ductility and good thermal shock, corrosion resistance, and wear resistance. One of the more common uses of cobalt is as the binder in cemented carbide cutting tools, providing good strength into and above the red-heat range of temperatures. Unfortunately, the density of all the superalloy metals is significantly greater than iron, so their use is often at the expense of additional weight.

Most of the superalloys are difficult to form or machine, so methods such as electrodischarge, electrochemical, or ultrasonic machining are often used, or the products are made to final shape as investment castings. Powder metallurgy techniques are also used extensively. Because of their ingredients, all of the alloys are quite expensive, and this limits their use to small or critical parts where cost is not the determining factor and component failure is not an option.

Still other engineering applications require materials whose temperature limits exceed those of the superalloys. **Figure 7.5** shows the high temperature exhaust of a jet engine. One reference estimates that the exhaust of future jet engines will reach temperatures in excess of 1425°C (2600°F). Rocket nozzles go well beyond this point. Materials such as TD-nickel

Example: History of Aerospace Turbine Blade Materials and Manufacturing Processes

Thermodynamics dictates that the efficiency of an airplane engine increases with increased operating temperatures. The turbine blades in the exhaust region of an engine see the continuous flow of hot combustion gases, along with direct contact with all of the various combustion gases and products. The following figure shows the

70-year history of material and process advancements as reflected in operating temperature capability. Note the material improvements that occur within a process, as well as the significant improvements that can occur with new processes. The interrelation of materials and processes is the primary subject of this text.

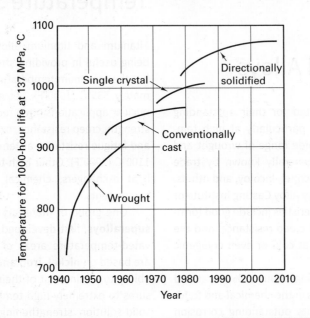

(a powder metallurgy nickel alloy containing 2% dispersed thorium oxide) can operate at service temperatures somewhat above 1100°C (2000°F). Going to higher temperatures, we look to the **refractory metals**, which include **niobium**, **molybdenum**, **tantalum**, **rhenium**, and **tungsten**. All have

FIGURE 7.5 Superalloys and refractory metals are needed to withstand the high temperatures of jet engine exhaust as seen when this F/A 18 Hornet launches from an aircraft carrier. (U.S. Navy Photo by Mass Communications Specialist 2nd Class Michael D. Cole)

melting points near or in excess of 2500°C (4500°F) and low thermal expansion. They retain a significant fraction of their strength at elevated temperature, resist creep, and can be used at temperatures as high as 1650°C (3000°F) provided that protective ceramic coatings effectively isolate them from gases in their operating environment. Coating technology is quite challenging, however, because the ceramic coatings must (1) have a high melting point, (2) not react with the metal it is protecting, (3) provide a diffusion barrier to oxygen and other gases, and (4) have thermal expansion characteristics that match the underlying metal. Although the refractory metals could be used at higher temperatures, the uppermost temperature is currently being set by limitations and restrictions imposed by the coating. In addition to its high-temperature properties, tantalum is resistant to chemical attack from virtually all environments at temperatures below 150°C (300°F) and is frequently used as a corrosion-resistant material.

Table 7.7 presents key properties for several refractory metals. Unfortunately, all are heavier than steel, and several are significantly heavier. In fact, tungsten, with a density about 1.7 times that of lead, is often used in counterbalances, compact flywheels, and weights, with applications as diverse as military projectiles, gyratory compasses, and golf clubs.

Other materials and technologies that offer promise for high-temperature service include intermetallic compounds, engineered ceramics and ceramic composites, graded materials,

TABLE 7.7   Properties of Some Refractory Metals

Metal	Melting Temperature [°F(°C)]	Room Temperature				Elevated Temperature [1832°F (1000°C)]	
		Density (g/cm³)	Yield Strength (ksi)	Tensil Strength (ksi)	Elongation (%)	Yield Strength (ksi)	Tensile Strength (ksi)
Molybdenum	4730 (2610)	10.22	80	120	10	30	50
Niobium	4480 (2470)	8.57	20	45	25	8	17
Tantalum	5430 (3000)	16.6	35	50	35	24	27
Tungsten	6170 (3410)	19.25	220	300	3	15	66

and advanced coating systems. The **intermetallic compounds** provide properties that are between metals and ceramics and are excellent candidates for high-temperature applications. They are hard, stiff, creep resistant, and oxidation resistant, with good high-temperature strength and wear resistance that often increases with temperature. The titanium and nickel aluminides offer the additional benefit of being significantly lighter than the superalloys. Unfortunately, the intermetallics are also characterized by poor ductility, poor fracture toughness, and poor fatigue resistance. They are difficult to fabricate using traditional techniques, such as forming and welding. On a positive note, research and development efforts have begun to overcome these limitations, and the intermetallics have been appearing in commercial products. Titanium aluminide has been used in the compressor section of aircraft engines that operate at temperatures above 600°C (1100°F) and is being considered as a replacement for nickel-based alloys in the turbine section. The high-temperature ceramics will be discussed in a future chapter.

Figure 7.6 provides a temperature scale depicting the upper limit for useful mechanical properties for a variety of engineering metals, ranging from aluminum through the refractory metals. **Figure 7.7** graphically presents the densities of the various metals. Note that the superalloys (cobalt and nickel) and the refractory metals (niobium, molybdenum, tantalum, tungsten, and rhenium) are all heavier than steel. Titanium is the only elevated temperature metal that is also lightweight, and its temperature range is limited.

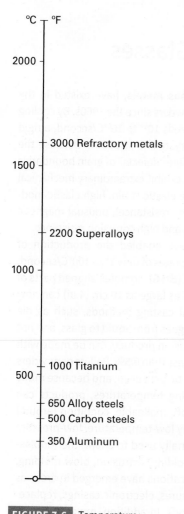

FIGURE 7.6 Temperature scale indicating the upper limit to useful mechanical properties for various engineering metals.

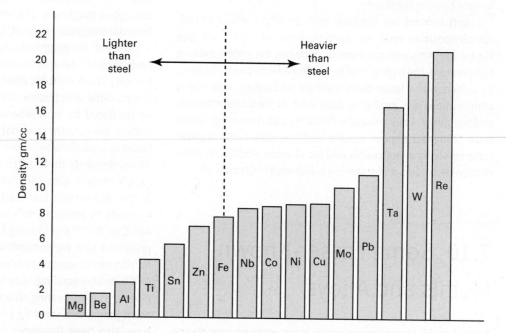

FIGURE 7.7 Densities of the various engineering metals. The elevated-temperature superalloys and refractory metals are all heavier than steel.

7.9 Lead and Tin and Their Alloys

The dominant properties of **lead** and lead alloys are high density coupled with strength and stiffness values that are among the lowest of the engineering metals. The principal uses of lead as a pure metal include storage batteries, cable cladding, and radiation-absorbing or sound- and vibration-dampening shields. Lead-acid batteries are clearly the dominant product, and more than 60% of U.S. lead consumption is generated from battery recycling. Other applications utilize the properties of good corrosion resistance, low melting point, and the ease of casting or forming. Structural applications are severely limited because lead is susceptible to creep under low loads, even at room temperature. As a pure metal, **tin** is used primarily as a corrosion resistant coating on steel.

In the form of alloys, lead and tin are almost always used together. Bearing material and **solder** are the two most important uses. One of the oldest and best bearing materials is an alloy of 84% tin, 8% copper, and 8% antimony, known as genuine or tin **babbitt**. Because of the high cost of tin, however, lead babbitt, composed of 85% lead, 5% tin, 10% antimony, and 0.5% copper, is a more widely used bearing material. The tin and antimony combine to form hard particles within the softer lead matrix. The shaft rides on the harder particles with low friction. The softer matrix acts as a cushion that can distort sufficiently to compensate for misalignment and ensure a proper fit between the two surfaces. For slow speeds and moderate loads, the lead-based babbitts have proven to be quite adequate. Babbitt coatings have also been applied to brass- or bronze-bearing substrates.

Soft solders are basically lead–tin alloys with a chemical composition near the eutectic value of 61.9% tin (see Figure 4.5). Although the eutectic alloy has the lowest melting temperature, the high cost of tin has forced many users to specify solders with a lower-than-optimum tin content. A variety of compositions are available, each with its own characteristic melting range. Environmental concerns and recent legislation have prompted a move toward lead-free solders for applications involving water supply and distribution. Additional information on solders and soldering is provided in Chapter 39.

7.10 Some Lesser-Known Metals and Alloys

Several of the lesser-known metals have achieved importance as a result of their somewhat unique physical and mechanical properties. **Bismuth** has an extremely low melting temperature (271°C or 520°F). **Beryllium** combines a density less than aluminum with a stiffness greater than steel and is transparent to X-rays. **Hafnium**, **thorium**, and *beryllium* are used in nuclear reactors because of their low neutron-absorption characteristics. Depleted **uranium**, because of its very high density (19.1 g/cm^3), is useful in special applications where maximum weight must be put into a limited space, such as counterweights or flywheels. **Zirconium** is used for its outstanding corrosion resistance to most acids, chlorides, and organic acids. It offers high strength, good weldability and fatigue resistance, and attractive neutron absorption characteristics. **Rare-earth metals** have been incorporated into magnets that offer increased strength compared to the standard ferrite variety. Neodymium–iron–boron and samarium–cobalt are two common varieties.

Although the **precious metals** (*gold*, *silver*, and the platinum group metals—*platinum, palladium, rhodium, ruthenium, iridium, and osmium*) might seem unlikely as engineering materials, they offer outstanding corrosion resistance and electrical conductivity, often under extreme conditions of temperature and environment

7.11 Metallic Glasses

Metallic glasses, or **amorphous metals**, have existed in the form of thin ribbon and fine powders since the 1960s. By cooling liquid metal at a rate that exceeds 10^5° to 10^6°C/second, a rigid solid is produced that lacks crystalline structure. Because the structure also lacks the crystalline "defects" of grain boundaries and dislocations, the materials exhibit extraordinary mechanical properties (high strength, large elastic strain, high elastic modulus, good toughness, and wear resistance), unusual magnetic behavior (magnetically "soft"), and high corrosion resistance.

Recent developments have enabled the production of amorphous metal with cooling rates of only 1° to 100°C/second. Known as **bulk metallic glass (BMG)**, complex-shaped parts of this material with thicknesses as large as 10 cm (4 in) can now be produced by conventional casting methods, such as die casting. Because the material goes from liquid to glass, and not liquid to crystalline solid, precision products can be made with a total shrinkage that is often less than 0.5%. Pellets or powders of bulk metallic glass can also be produced, and because many of the alloys have low melting temperatures, products can be made by reheating to a soft, malleable, condition (around 400°C or 750°F) and forming by low-temperature, low-pressure processes that are conventionally used to shape thermoplastic polymers (compression molding, extrusion, blow molding, and injection molding). Applications have emerged in areas as diverse as load-bearing structures, electronic casings, replacement joints, and sporting goods. In addition, metallic glasses have also been developed that retain their glassy structure at temperatures as high as 870°C (1600°F), but most soften or weaken when heated above 260°C (500°F).

7.12 Graphite

Although technically not a metal, **graphite** is an engineering material with considerable potential. It offers properties of both a metal and nonmetal, including good thermal and electrical conductivity, inertness, the ability to withstand high temperature, and lubricity. In addition, it possesses the unique property of increasing in strength as the temperature is elevated. Polycrystralline graphites can have mechanical strengths up to 70 MPa (10 ksi) at room temperature, which double when the temperature reaches 2500°C (4500°F). The material is stable in air at temperatures up to 500°C (930°F) and in vacuum or inert atmospheres up to 3000°C (5430°F).

Large quantities of graphite are used as electrodes in arc furnaces, but other uses are developing rapidly. The addition of small amounts of borides, carbides, nitrides, and silicides greatly lowers the oxidation rate at elevated temperatures and improves the mechanical strength. This makes the material highly suitable for use as rocket-nozzle inserts and as permanent molds for casting various metals, where it costs less than tool steel, requires no heat treating, and has a lower coefficient of thermal expansion. It can be machined quite readily to excellent surface finishes. Graphite fibers have also found extensive use in composite materials. This application will be discussed in Chapter 8.

Carbon nanotubes and **graphene** (mono-atomic layer carbon sheets with atoms in a hexagonal arrangement) have emerged as revolutionary new materials and have begun to make the transition from research to application. The nanotubes are approximately one billionth of a meter in diameter (50,000 times thinner than a human hair) and are 16 times stronger than steel. They are stable to 2750°C (5000°F) in a vacuum and 760°C (1400°F) in air, conduct electricity as well as silicon, and have good thermal conductivity. Current use includes both structural and electromagnetic applications.

Graphene, first manufactured in 2003, has many extreme properties: ultra-light weight (its density of 0.16 kg/m^2 is less than helium), ultra-high tensile strength (130 GPa—about 500 times stronger than low-carbon steel), and stiffness (1TPa or 150 × 10^{6}psi); can conduct electricity 100 times faster than silicon; is possibly the most effective conductor of heat; and is more transparent than glass. On the negative side, graphene is brittle, like a ceramic (fracture toughness is in the range of 15 to 50 MPa/m) and is the most reactive form of carbon, burning in air at a temperature of 350°C or 662°F. As of 2014, graphene was not used in any commercial application, but many have been proposed or are in development. Solar collectors could be 50 to 100 times more efficient, semiconductors 50 to 100 times faster, aircraft and automotive bodies 70% lighter, batteries charge 10 times faster and hold 10 times more energy, and electronic displays could be "foldable." Volume production and a significant reduction in cost will do much to make graphene a significant material in numerous applications.

7.13 Materials for Specific Applications

Materials for **outer space applications** must withstand ionizing radiation, wide extremes of temperature (both positive and negative), and possible impacts from micrometeorites. In addition, light weight is a definite asset.

Smart materials can be used to control automated processes and equipment. **Sensors** are materials for which some form of input triggers a response in the form of a signal. **Actuators** respond to input, which might be a sensor signal, by some form of property change. Just as the human body has a wide array of sensors and actuators, there are a number of smart materials with parallel functions.

With the expanded development of hybrid vehicles and a significant increase in the number of computer-activated motors in automobiles, there is enhanced demand for strong **permanent magnets**. The three most common families are aluminum–nickel–cobalt (AlNiCo), samarium cobalt, and neodymium iron boron (the most powerful).

A number of materials have been used or adapted for **medical applications**, including joint replacements, medical implants, and medical devices. Titanium continues to be a dominant material because of its strength, light weight, and corrosion resistance. Tantalum and niobium have been used for similar reasons. Nitinol (a nickel–titanium alloy with superelastic shape memory properties) has been used where its change of shape is useful, as in artery-expanding stents. Copper wires are useful for electrical transmission and the sending and receiving of signals. Cobalt-based alloys, such as cobalt–chrome, are used in various joint replacements. Polymers, such as polyetheretherketone (PEEK) offer light weight, chemical resistance, and biocompatibility for non-load-bearing applications like facial implants. Various ceramic materials offer excellent wear resistance, chemical resistance, and electrical resistivity. Others closely mimic bone.

7.14 High Entropy Alloys

High entropy alloys are a new class of multicomponent alloys composed of five or more constituent elements, each with a concentration between 5 and 35 atomic percent. Although still in the research and development stage, current studies have produced materials that are considerably lighter than conventional alloys, with high fracture resistance, tensile strength, hardness, corrosion resistance, and oxidation resistance, along with structural stability and strength retention at extreme elevated temperatures—ideal properties to replace and extend the role of the superalloys.

Other varieties become tougher, stronger, and more ductile as temperatures drop, offering good fracture toughness at cryogenic temperatures.

Although most of the studied alloys are single-phase, solid solutions with simple structures such as face-centered cubic and body-centered cubic, other structures have been produced, including multiple phase, nanocrystalline, and amorphous. Various methods of production have been employed,

including melt casting, mechanical alloying of powders followed by spark plasma sintering and the sputter deposition of thin films.

With the almost limitless number of combinations of five or more elements, the potential for significant advances is immense. However, research is still needed to identify specific compositions and develop the production processes that will enable manufacture on a commercial scale.

Review Questions

1. What types of properties do nonferrous metals possess that might not be available in the ferrous metals?

2. For what respects are the nonferrous metals generally inferior to steel?

3. In what ways might the nonferrous metals offer attractive ease of fabrication?

4. What are the three properties of copper and copper alloys that account for many of their uses and applications?

5. What properties make copper attractive for cold-working processes?

6. What are some of the limiting properties of copper that might restrict its area of application?

7. What properties of copper make it attractive for low-temperature applications?

8. What are some potential applications that would make use of copper's antimicrobial properties?

9. What is the primary use of commercially pure copper?

10. Why does the copper designation system separate wrought and cast alloys? What properties are attractive for each group?

11. What are some of the attractive engineering properties that account for the wide use of the copper–zinc alpha brasses?

12. Why might cold-worked brass require a stress relief prior to being placed in service?

13. Why might the term *bronze* be potentially confusing when used in reference to a copper-based alloy?

14. What are some attractive engineering properties of copper–nickel alloys?

15. Which of the copper alloys can be strengthened by precipitation hardening?

16. Describe the somewhat unique properties available with heat-treated copper–beryllium alloys. What has limited their use in recent years?

17. What alloys have been used to replace lead in copper casting alloys being targeted to drinking water applications?

18. What are some of the attractive engineering properties of aluminum and aluminum alloys?

19. How does aluminum compare to steel in terms of weight? Discuss the merits of comparing cost per unit weight versus cost per unit volume.

20. What is the primary benefit of aluminum recycling compared to making new aluminum from ore?

21. When designing with aluminum, why might there be concern regarding rigidity or stiffness?

22. How does aluminum compare to copper in terms of electrical conductivity?

23. What features might limit the mechanical uses and applications of aluminum and aluminum alloys?

24. What features make aluminum attractive for transportation applications?

25. How is the corrosion-resistance mechanism observed in aluminum and aluminum alloys similar to that observed in stainless steels?

26. Why do aluminum and aluminum alloys present special problems when welding?

27. How are the wrought alloys distinguished from the cast alloys in the aluminum designation system? Why would these two groups of metals have distinctly different properties?

28. What feature in the wrought aluminum designation scheme is used to describe the condition or structure of a given alloy?

29. What is the primary strengthening mechanism in the high-strength "aircraft-quality" aluminum alloys?

30. What unique combination of properties is offered by the composite Alclad materials?

31. What surface finishing technique is used in the production of numerous metallic-colored aluminum products?

32. Why are aluminum–silicon alloys popular for casting operations?

33. What specific material properties might make an aluminum casting alloy attractive for permanent mold casting? For die casting?

34. What are the attractive features of aluminum–lithium alloys?

35. What are some possible applications of aluminum foam?

36. What are some attractive and restrictive properties of magnesium and magnesium alloys?

37. Describe the designation system applied to magnesium alloys.

38. What is the most popular fabrication process applied to magnesium alloys?

39. In what way can ductility be imparted to magnesium alloys so that they can be formed by conventional deformation processes?

40. How does the ability to strengthen magnesium compare with the ability to strengthen aluminum?

41. Under what conditions should magnesium be considered to be a flammable or explosive material?

42. What is the primary application of pure zinc? Of the zinc-based engineering alloys?

43. What are some of the attractive characteristics of the zinc die-casting alloys?

44. What are some of the attractive features of the zinc–aluminum casting alloys?

45. What are some of the attractive engineering properties of titanium and titanium alloys? Limiting characteristics?

46. What feature is used in the designation of titanium alloys, and how are they grouped metallurgically?

47. What are some of the primary application areas for titanium and titanium alloys?

48. Under what conditions might titanium replace steel? Replace aluminum?

49. What conditions favor the selection and use of nickel-base alloys?

50. What property of Monel dominates most of its applications?

51. What are the distinct characteristics of nichromes? Invar? Nitinol?

52. What metals or combinations of metals form the bases of the superalloys?

53. What class of metals or alloys must be used when the operating temperatures exceed the limits of the superalloys?

54. Which metals are classified as refractory metals?

55. What are some of the requirements of the protective coating applied to the refractory metals?

56. What are some general characteristics of intermetallic compounds?

57. What temperature is generally considered to be the upper limit for which titanium alloys retain their useful engineering properties? The superalloys? Refractory metals?

58. Why is weight often a problem when the elevated-temperature metals are being considered?

59. What is the dominant product for which lead is used?

60. What features make beryllium a unique, lightweight metal?

61. What are some applications of depleted uranium? Rare-earth metals?

62. What are some of the attractive properties of metallic glasses?

63. What unique property of graphite makes it attractive for elevated-temperature applications?

64. What are some of the unique conditions encountered in outer space?

65. What are some of the medical applications of engineering materials?

66. What are some of the attractive properties of the high entropy alloys currently undergoing research and development?

Problems

1. Aluminum, magnesium, and titanium are each being considered for the frame of a tennis racquet. What are the pros and cons of each of these materials? What would you recommend for an inexpensive beginner's model? Would your answer change if you wanted to increase the quality?

2. Identify or select a product in which a magnesium alloy would be an appropriate material. What features or requirements would favor the use of magnesium?

3. Your company is considering the production of small replica automobiles to compete with the traditional Matchbox and HotWheels lines. Which of the metals presented in this chapter do you think would be most appropriate for such an application? Why?

4. The purchase price of an electrical-resistance-heated, laboratory size, furnace increases significantly when the maximum operating temperature exceeds 1100°–1200°C (2000°–2200°F). Using the information in this chapter, determine what materials might be used for heating elements both below and above this range. Would the required change in heating element material account for the change in cost?

5. Your company is considering an expansion of its line of conventional hand tools to include safety tools, capable of being used in areas such as gas leaks where the potential of explosion or fire exists. Conventional irons and steels are pyrophoric (i.e., small slivers or fragments can burn in air, forming sparks if dropped or impacted on a hard surface).

You are asked to evaluate potential materials and processes that might be used to manufacture a nonsparking pipe wrench. This product is to be produced in the same shape and range of sizes as conventional pipe wrenches and possess all the same characteristic properties (strength in the handle, hardness in the teeth, fracture resistance, corrosion resistance, etc.). In addition, the new safety wrench must be nonsparking (or nonpyrophoric).

Your initial review of the nonferrous metals reveals that aluminum is nonpyrophoric, but lacks the strength and wear resistance needed in the teeth and jaw region of the wrench. Copper is also nonpyrphoric, but is heavier than steel, and this might be unattractive for the larger wrenches. Copper–2% beryllium can be age hardened to provide the strength and hardness properties equivalent to the steel that is currently being used for the jaws of the wrench, but the cost of this material is also quite high. Titanium is difficult to fabricate and might not possess the needed hardness and wear resistance. Mixed materials might create an unattractive galvanic corrosion cell. Both forging and casting appear to be viable means of forming the desired shape. You want to produce a quality product but also wish to make the wrench in the most economical manner possible so that the new line of safety tools is attractive to potential customers.

Suggest some alternative manufacturing systems (materials coupled with companion methods of fabrication) that could be used to produce the desired wrench. What might be the advantages and disadvantages of each? Which of your alternatives would you recommend to your supervisor?

CHAPTER 8

Nonmetallic Materials: Plastics,* Elastomers,** Ceramics, and Composites

8.1 Introduction

Because of their wide range of attractive properties, the **nonmetallic materials** have always played a significant role in manufacturing. Wood has been a key engineering material down through the centuries, and artisans have learned to select and use the various types and grades to manufacture a broad spectrum of quality products. Stone and rock continue to be key construction materials, and clay products can be traced to antiquity. Even leather has been a construction material and was used for fenders in early automobiles.

More recently, however, the family of nonmetallic materials has expanded from the natural materials just described and now includes an extensive list of plastics (polymers), elastomers, ceramics, and composites. Most of these are manufactured materials, so a wide variety of properties and characteristics can be obtained. New variations are being created on a continuous basis. Many observers now refer to a materials revolution as these new materials compete with and complement steel, aluminum, and the other more traditional engineering metals. New products have emerged, utilizing the new properties, and existing products are continually being reevaluated for the possibility of material substitution. As the design requirements of products continue to push the limits of traditional materials, the role of the manufactured nonmetallic materials will no doubt continue to expand.

Because of the breadth and number of nonmetallic materials, we will not attempt to provide information about all of them. Instead, the emphasis will be on the basic nature and properties of the various families so that the reader will be able to determine if they might be reasonable candidates for specific products and applications. For detailed information about specific materials within these families, more extensive and dedicated texts, handbooks, and compilations should be consulted.

8.2 Plastics

It is difficult to provide a precise definition of the term **plastics**. From a technical viewpoint, the term refers to engineered materials characterized by large molecules that are built up by the joining of smaller molecules. They are natural or synthetic resins, or their compounds, that can be molded, extruded, cast, or used as thin films or coatings. They offer low density, low tooling costs, good resistance to corrosion and chemicals, cost reduction, and design versatility. From a chemical viewpoint, most are organic substances containing hydrogen, oxygen, carbon, and nitrogen.

In less than a century, we have gone from a world without plastic to a world where its uses and applications are limitless. The United States currently produces more plastic than steel, aluminum, and copper combined. Plastics are used to save lives in applications such as artificial organs, shatter-proof glass, and bullet-proof vests. They reduce the weight of cars, provide thermal insulation to our homes, and encapsulate our medicines. They form the base material in products as diverse as shower curtains, contact lenses, and clothing, and compose some of the primary components in televisions, computers, cell

---

*The technically correct term here would be **polymers** because not all polymers exhibit the property of plasticity—the ability to deform without rupture. The term **plastics**, however, has been used to describe all of the various organic compounds produced by polymerization, regardless of their plasticity, and has been accepted in this context, as evidenced by the industry-wide Society of the Plastics Industry and the Society of Plastics Engineers. In this text, the words *polymer* and *plastic* will be used interchangeably.

**Elastomers are a special group of polymers whose elastic properties are similar to those of natural rubber. Elastomers are generally treated separately from the nonelastomeric "plastics."

phones, and furniture. Even the Statue of Liberty has a plastic coating to protect it from corrosion.

Molecular Structure of Plastics

To understand the properties of plastics, it is important to first understand their molecular structure. For simplicity, let's begin with the paraffin-type hydrocarbons, in which carbon and hydrogen combine in the relationship C_nH_{2n+2}. Theoretically, the atoms can link indefinitely to form very large molecules, extending the series depicted in **Figure 8.1**. The bonds between the various atoms are all pairs of shared electrons (covalent bonds). Bonding within the molecule, therefore, is quite strong, but the attractive forces between adjacent molecules are much weaker. Because there is no provision for additional atoms to be added to the chain, these molecules are said to be **saturated monomers**.

Carbon and hydrogen can also form molecules where the carbon atoms are held together by double or triple covalent bonds. Ethylene and acetylene are common examples (**Figure 8.2**). Because these molecules do not have the maximum number of hydrogen atoms, they are said to be **unsaturated monomers** and are important in the polymerization process, where small molecules link to form large ones with the same constituent atoms.

In all the described molecules, four electron pairs surround each carbon atom, and one electron pair is shared with each hydrogen atom. Other atoms or structures can be substituted for carbon and hydrogen. Chlorine, fluorine, or even a benzene ring can take the place of hydrogen. Oxygen, silicon, sulfur, or nitrogen can take the place of carbon. Because of these substitutions, a wide range of organic compounds can be created.

Isomers

The same kind and number of atoms can also unite in different structural arrangements, known as **isomers**, and these ultimately behave as different compounds with different properties. **Figure 8.3** shows an example of this feature, involving propyl and isopropyl alcohol. Isomers can be considered

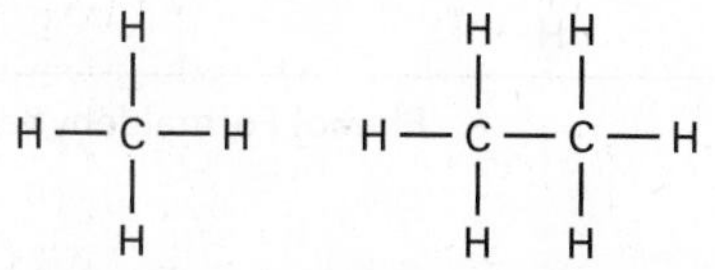

FIGURE 8.3 Linking of eight hydrogen, one oxygen, and three carbon atoms to form two isomers: propyl alcohol and isopropyl alcohol. Note the different locations of the –OH attachment.

analogous to allotropism or polymorphism in crystalline materials, where the same material possesses different properties because of different crystal structures.

Forming Molecules by Polymerization

The **polymerization** process, or linking of molecules, occurs by either an **addition** or **condensation** mechanism. **Figure 8.4** illustrates polymerization by addition, where a number of basic units (**monomers**) link to form a large molecule (**polymer**) in which there is a repeated unit (**mer**). Activators or catalysts, such as benzoyl peroxide, initiate and terminate the chain as each molecule goes through the stages of initiation, propagation or growth, and termination. The amount of activator relative to the amount of monomer determines the average molecular weight (or average length) of the polymer chain. The average number of mers in the polymer, known as the **degree of polymerization**, ranges from 75 to 750 for most commercial plastics. Chain length controls many of the properties of a plastic. Increasing the chain length tends to increase toughness, creep resistance, melting temperature, melt viscosity, and difficulty in processing.

Copolymers are a special category of polymer where two different types of mers are combined into the same addition chain. The formation of copolymers (**Figure 8.5**), analogous to alloying in metals, greatly expands the possibilities of creating new types of plastics with improved physical and mechanical properties. **Terpolymers** further extend the possibilities by combining three different monomers.

In contrast to polymerization by addition, where all the original atoms appear in the product molecule, *condensation polymerization* occurs when reactive molecules combine with one another to produce a polymer plus small, by-product

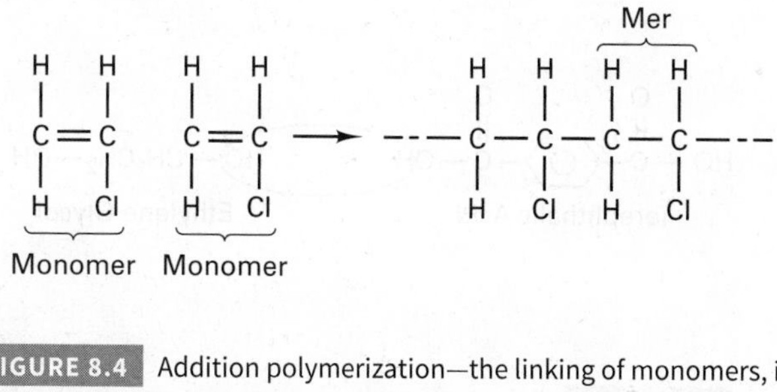

FIGURE 8.1 The linking of carbon and hydrogen to form methane and ethane molecules. Each dash represents a shared electron pair or covalent bond.

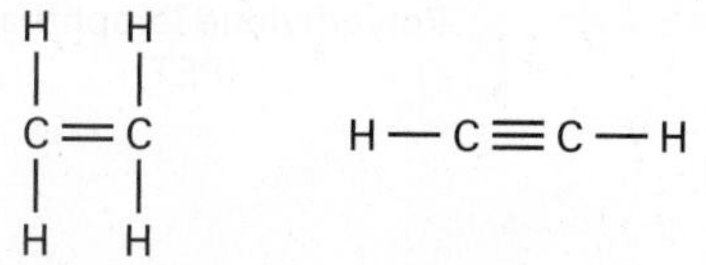

FIGURE 8.2 Double and triple covalent bonds exist between the carbon atoms in unsaturated ethylene and acetylene molecules.

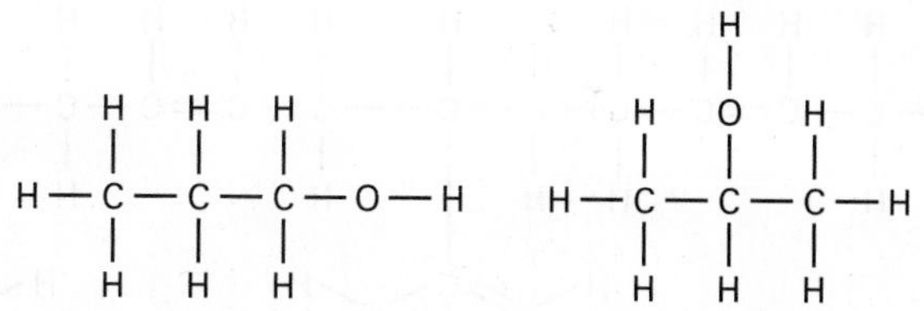

FIGURE 8.4 Addition polymerization—the linking of monomers, in this case, identical ethylene molecules.

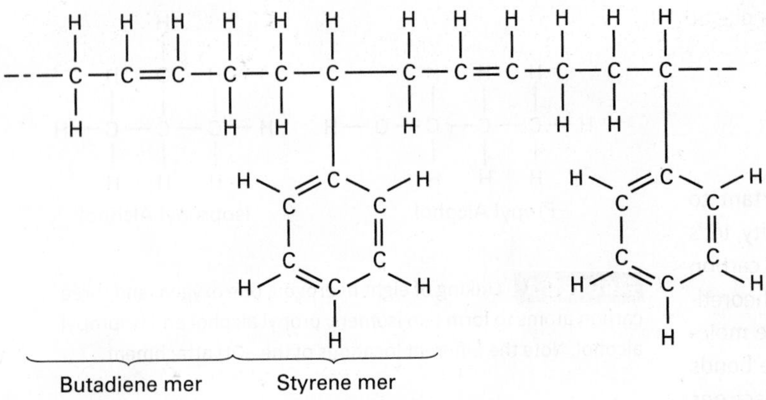

Butadiene mer   Styrene mer

FIGURE 8.5 Addition polymerization with two kinds of mers—here, the copolymerization of butadiene and styrene.

molecules, such as water or alcohol. Heat, pressure, and catalysts are often required to drive the reaction. **Figure 8.6a** illustrates the reaction between phenol and formaldehyde to form phenol-formaldehyde, otherwise known as Bakelite, first performed in 1910. **Figure 8.6b** shows the condensation reaction to produce polyethylene terephthalate (PET). The structure of condensation polymers can be either linear chains or a three-dimensional framework in which all atoms are linked by strong, primary bonds.

Thermosetting and Thermoplastic Materials

The terms **thermosetting** and **thermoplastic** refer to the material's response to elevated temperature. Addition polymers (or linear condensation polymers) can be viewed as long chains of bonded carbon atoms with attached pendants of hydrogen, fluorine, chlorine, or benzene rings. All of the bonds within the molecules are strong covalent bonds. The attraction between neighboring molecules is through the much weaker van der Waals forces. For these materials, the intermolecular forces strongly influence the mechanical and physical properties. In general, the linear polymers tend to be flexible and tough. Because the intermolecular bonds are weakened by elevated temperature, plastics of this type soften with increasing temperature, and the individual molecules can slide over each other in a molding process. When the material is cooled, it becomes harder and stronger. The softening and hardening of these thermoplastic or heat-softening materials can be repeated as often as desired, and no chemical change is involved.

Because *thermoplastic materials* contain molecules of different lengths, they do not have a definite melting temperature, but instead soften over a range of temperatures. Above the temperature required for melting, the material can be poured and cast as a liquid or formed by injection molding. When cooled to a temperature where it is fully solid, the material can retain its amorphous structure, but with companion properties that are somewhat rubbery. The application of a force produces both elastic and plastic deformation. Large amounts of permanent deformation are possible and make this range attractive for molding and extrusion. At still lower temperatures, the bonds become stronger and the polymer is stiffer and somewhat leathery. Many commercial polymers, such as polyethylene, have useful strength in this condition. When further cooled below the glass transition temperature, however, the linear polymer retains its amorphous structure but becomes hard, brittle, and glasslike.

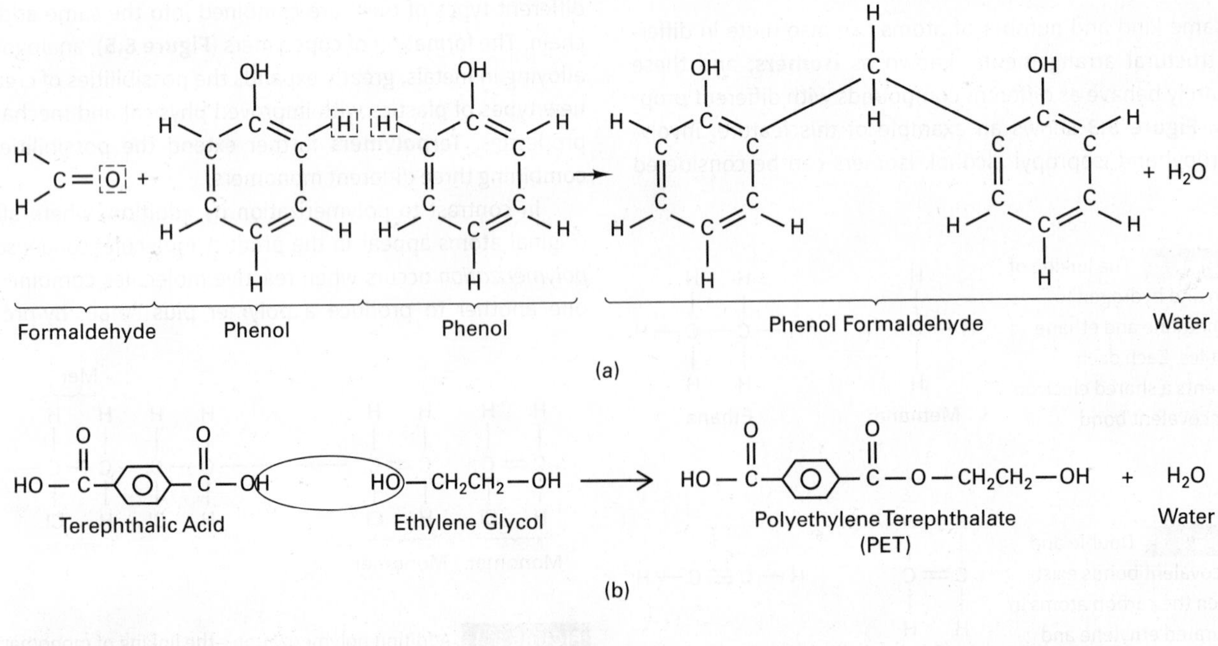

(a)

Formaldehyde   Phenol   Phenol   →   Phenol Formaldehyde   Water

(b)

Terephthalic Acid   Ethylene Glycol   →   Polyethylene Terephthalate (PET)   Water

FIGURE 8.6 The formation of (a) phenol–formaldehyde (Bakelite) and (b) polyethylene terephthalate (PET) by condensation polymerization. Note the H_2O or water by-product.

Thermoplastic vs. Thermoset

As a helpful analogy, thermoplastic polymers are a lot like candle wax. They can be softened or melted by heat and then cooled to assume a solid shape. Thermosets are more like egg whites or bread dough. Heating changes their structure and properties in an irreversible fashion.

Many thermoplastics can partially **crystallize**[1] when cooled below the melting temperature. This should not be confused with the crystal structures discussed previously in this text. When polymers "crystallize," the chains closely align through intramolecular folding or the stacking of adjacent chains, producing small three-dimensional regions of order within an otherwise amorphous structure. Density increases, and the polymer becomes stiffer, harder, less ductile, and more resistant to solvents and heat. The ability of a polymer to crystallize depends on the complexity of its molecules, the degree of polymerization (length of the chains), the cooling rate, and the amount of deformation during cooling.

The mechanical behavior of an amorphous (noncrystallized) thermoplastic polymer can be modeled by a common cotton ball. The individual molecules are bonded within by strong covalent bonds and are analogous to the individual fibers of cotton. The bonding forces between molecules are much weaker and are similar to the friction forces between the strands of cotton. When pulled or stretched, plastic deformation occurs by slippage between adjacent fibers or molecular chains. Methods to increase the strength of thermoplastics, therefore, focus on restricting intermolecular slippage. Longer chains have less freedom of movement and are therefore stronger. Connecting adjacent chains to one another with primary bond cross-links, as with the sulfur links when vulcanizing rubber, can also impede deformation. Because the strength of the secondary bonds is inversely related to the separation distance between the molecules, processes such as deformation or crystallization can be used to produce a tight parallel alignment of adjacent molecules and a concurrent increase in strength, stiffness, and density. Polymers with larger side structures, such as chlorine atoms or benzene rings, might be stronger or weaker than those with just hydrogen, depending on whether the dominant effect is the impediment to slippage or the increased separation distance. Branched polymers, where the chains divide in a Y with primary bonds linking all segments, are often weaker because branching reduces the density and close packing of the chains. Physical, mechanical, and electrical properties all vary with the aforementioned changes in structure.

The four most common thermoplastic polymers are polyethylene (PE), polypropylene (PP), polystyrene (PS), and polyvinylchloride (PVC). Other thermoplastics include polycarbonate (PC), polyethylene terephthalate (PET), polymethylmethacrylate (PMMA), and acrylonitrile-butadiene-styrene (ABS).

In contrast to the thermoplastic polymers, *thermosetting plastics* usually have a highly cross-linked or three-dimensional framework structure in which all atoms are connected by strong, covalent bonds. These materials are generally produced by condensation polymerization where elevated temperature promotes an irreversible reaction, hence the term *thermosetting*. Once set, subsequent heating will not produce the softening observed with the thermoplastics. Instead, thermosetting materials maintain their mechanical properties up to the temperature at which they char or burn. Because deformation requires the breaking of primary bonds, the thermosetting polymers are significantly stronger and more rigid than the thermoplastics. They can resist higher temperatures and have greater dimensional stability, but they also have lower ductility and poorer impact properties. Some common thermosetting polymers include polyurethane (PUR), phenol-formaldehyde or Bakelite (PF), and various epoxies.

Although classification of a polymer as thermosetting or thermoplastic provides insight as to properties and performance, it also has a strong effect on fabrication. For example, thermoplastics can be easily molded. After the hot, soft material has been formed to the desired shape, however, the mold must be cooled so that the plastic will harden and be able to retain its shape on removal. The repetitive heating and cooling cycles affect mold life, and the time required for the thermal cycles influences productivity. When a part is produced from thermosetting materials, the mold can remain at a constant temperature throughout the entire process, but the setting or curing of the resins now determines the time in the mold. Because the material hardens as a result of the reaction and has strength and rigidity even when hot, product removal can be performed without cooling the mold.

Properties and Applications

Because there are so many varieties of plastics, and new ones are being developed almost continuously, it is helpful to have knowledge of both the general properties of plastics as well as the unique or specific properties of the various families. General properties of plastics include the following:

1. *Light weight.* Most plastics have specific gravities between 1.1 and 1.6, compared with about 1.75 for magnesium (the lightest engineering metal).

---

[1]It should be noted that the term *crystallize*, when applied to polymers, has a different meaning than when applied to metals and ceramics. Metals and ceramics are crystalline materials, meaning that the atoms occupy sites in a regular, periodic array, known as a lattice. In polymers, it is not the atoms that become aligned, but the molecules. Because van der Waals bonding has a bond strength that is inversely related to the separation distance, the parallel alignment of the crystallized state is a lower energy configuration and is promoted by slow cooling and equilibrium-type processing conditions.

2. *Corrosion resistance.* Many plastics perform well in hostile, corrosive, or chemical environments. Some are notably resistant to acid corrosion.

3. *Electrical resistance.* Plastics are widely used as insulating materials.

4. *Low thermal conductivity.* Plastics are relatively good thermal insulators.

5. *Variety of optical properties.* Through the incorporation of pigments and dyes, many plastics have an almost unlimited color range, and the color goes throughout, not just on the surface. Both transparent and opaque materials are available.

6. *Formability or ease of fabrication.* Objects can frequently be produced from plastics in a single operation. Raw material can be converted to final shape through such processes as casting, extrusion, and molding. Relatively low temperatures are used in the forming of plastics.

7. *Surface finish.* The same processes that produce the shape also produce excellent surface finish. Additional surface finishing might not be required.

8. *Comparatively low cost.* The low cost of plastics generally applies to both the material itself and the manufacturing process. Plastics frequently offer reduced tool costs and high rates of production.

9. *Low energy content.*

Whereas the attractive features of plastics tend to be in the area of physical properties, the inferior features generally relate to mechanical strength. Plastics can be flexible or rigid, but none of the plastics possess strength properties that approach those of the engineering metals unless they are reinforced in the form of a composite. Their low density allows them to compete effectively on a strength-to-weight (or specific strength) basis, however. Many have low-impact strength, although several (such as ABS, high-density polyethylene, and polycarbonate) are exceptions to this rule. Aluminum is nearly 10 times more rigid than a high-rigidity plastic, and steel is 30 times more rigid.

The dimensional stability of plastics tends to be greatly inferior to that of metals, and the coefficient of thermal expansion is rather high. Thermoplastics are quite sensitive to heat, and their strength often drops rapidly as temperatures increase above normal environmental conditions. Thermosetting materials offer good strength retention at elevated temperature but have an upper limit of about 250°C (500°F). Low-temperature properties are generally inferior to those of other materials. Although the corrosion resistance of plastics is generally good, they often absorb moisture, and this, in turn, decreases strength. Some thermoplastics can exhibit a 50% drop in tensile strength as the humidity increases from 0 to 100%. Radiation, both ultraviolet and particulate, can markedly alter the properties. Many plastics used in an outdoor environment have ultimately failed because of the cumulative effect of ultraviolet radiation. Plastics are also difficult to repair if broken.

Table 8.1 summarizes the properties of a number of common plastics. By considering the information in this table along with the preceding discussion of general properties, it becomes apparent that plastics are best used in applications that require materials with low to moderate strength, light weight, low electrical and/or thermal conductivity, a wide range of available colors, and ease of fabrication into finished products. No other family of materials can offer this combination of properties. Because of their light weight, attractive appearance, and ease of fabrication, plastics have been selected for many packaging and container applications. This classification includes such items as household appliance housings, clock cases, and exteriors of electronic products, where the primary role is to contain the interior mechanisms. Applications such as insulation on electrical wires and handles for hot articles capitalize on the low electrical and thermal conductivities. Soft, pliable, **foamed plastics** are used extensively as cushioning material. Rigid foams are used inside sheet metal structures to provide compressive strength. Nylon has been used for gears, acrylic for lenses, and polycarbonate for safety helmets and unbreakable windows.

There are many applications where only one or two of the properties of plastics are sufficient to justify their use. When special characteristics are desired that are not normally found in the commercial plastics, composite materials can often be designed that use a polymeric matrix. For example, high directional strength can be achieved by incorporating a fabric or fiber reinforcement within a plastic resin. These materials will be discussed in some detail later in this chapter.

Common Types or Families of Plastics

The following is a brief descriptive summary of some of the types of plastics listed in Table 8.1.

Thermoplastics
ABS (acrylonitrile butadiene styrene): Low weight; good strength; rigid, hard, and very tough, even at low temperatures; opaque; resists heat, weather, and chemicals quite well; dimensionally stable but flammable.

Acrylics: Hard, brittle (at room temperature); high impact, flexural, tensile, and dielectric strengths; transparent or easily colored; resist weathering and UV light; resistant to chemicals and fuels; biocompatible; the most common example is *PMMA (polymethyl methacrylate):* highest optical clarity, transmitting more than 90% of light; shatterproof; common trade names include Lucite and Plexiglas; used in automotive taillight lenses, aircraft windows, and other optical applications; scratches easily, however.

Cellulosics: Commercially important examples include: *cellulose acetate:* wide range of colors; good insulating qualities; easily molded; high moisture absorption in most grades; and *cellulose acetate butyrate:* higher impact strength and moisture resistance than cellulose acetate; will withstand rougher usage.

TABLE 8.1 Properties and Major Characteristics of Some Common Types of Plastics

Material	Specific Gravity	Tensile Strength (1000 lb/in.²)	Impact Strength Izod (ft-lb/in. of Notch)	Top Working Temperature [°F(°C)]	Dielectric Strength[b] (V/mil)	24-Hour Water Absorption (%)	Weatherability	Colorability	Optical Clarity	Chemical Resistance	Injection Molding	Extrusions	Formable Sheet	Film	Fiber	Compression or Transfer Moldings	Castings	Reinforced Plastics	Moldings	Industrial Thermosetting Laminates	Foam
Thermoplastics																					
ABS material	1.02–1.06	4–8	1.3–10.0	300–400		0.2–0.3	0	x		0	•	•	•								
Acetal	1.4	10	1.5	250(121)	1200	0.22		x		0	•	•									
Acrylics	1.12–1.19	5.5–10	0.2–2.3	200(93)	400–530	0.2–0.4	x	x	x	0	•	•	•								
Cellulose acetate	1.25–1.50	3–8	0.75–4.0	260(127)	300–600	2.0–6.0	x	x	x		•	•	•	•							
Cellulose acetate butyrate	1.18–1.24	2–6	0.6–3.2	130(54)	250–350	1.8–2.1	x	x	x		•		•								
Cellulose propionate	1.19–1.24	1–5	0.8–9	140(60)	300	1.8–2.1		x	x		•	•	•	•							
Chlorinated polyether	1.4	6	3.3	300(149)	400	0.01		x		x	•	•									
Ethyl cellulose	1.16	3–6	1.8–4.0	150(66)	350	1.6–2.2		x			•	•	•								
TFE-fluorocarbon	2.1–2.3	1.5–3	2.5–4.0	500(260)	450	0	x			x		•		•							
CFE-fluorocarbon	2.1–2.15	4.5–6	3.5–3.6	390(199)	550	0	x			x		•		•		•					
Nylon	1.1–1.2	8–10	2	250(121)	385–470	0.4–5.5		0	0	0	•	•	•		•						
Polycarbonate	1.2	9.5	14	250(121)	400	0.15		0	0	x	•	•	•								
Polyethylene	0.96	4	10	200(93)	440	0.003	x	0	0	x	•	•	•	•							•
Polypropylene	0.9–1.27	3.4–5.3	1.02	230(110)	520–800	0.03		0	x	x	•	•	•	•	•						
Polystyrene	1.05–1.15	5–9	0.3–0.6	190(88)	400–600	<0.2		x	x	0	•	•	•	•							
Modified polystyrene	1.0–1.1	2.5–6	0.25–11.0	212(100)	300–600	0.03–02		x	x	x	•	•	•	•							
Vinyl	1.16–1.55	1–5.9	0.25–2.0	220(104)	25–500	0.2–1		x	x	x	•	•	•	•	•						•
Thermosetting plastics																					
Epoxy	1.1–1.7	4–13	0.4–1.5	325(163)	500	0.1–0.5	x	x		x						•	•	•		•	•
Melamine	1.76–1.98	5–8	0.25–5	350(177)	460	0.1		x		0						•		•		•	•
Phenolic	1.2–1.45	5–9	0.25–5	300(149)	100–500	0.2–0.6				0		•				•	•	•		•	•
Polyester (other than molding compounds)	1.06–1.46	4–10	0.18–0.4	300(149)	340–570	0.5		x	x	0				•	•			•	•		
Polyester (alkyd, DAP)	1.6–1.75	3.2–8	3.6–8			0.16–0.67										•	•	•			
Silicone	2.0	3–5	0.2–3.0	550(288)	250–350	0.4–0.5										•	•	•		•	•
Urea	1.41–1.80	4–8.5	0.2–0.5	185(85)	300–600	1–3	x	x	0	0						•				•	•

[a] X denotes a principal reason for its use; 0 indicates a secondary reason.
[b] Short-time ASTM test.

Fluorocarbons: Inert to most chemicals (solvents, acids, and bases); high temperature resistance; high strength; low moisture absorption; good weathering; very low coefficients of friction (polyfluoroethylene or Teflon accounts for about 85% of this family), used for nonlubricated bearings and nonstick coatings for cooking utensils and electrical irons.

Polyamides: Nylons are the most important member; good strength, toughness, stiffness, abrasion resistance, and chemical resistance; low coefficient of friction; excellent dimensional stability; good heat resistance; used for small gears and bearings, zip fasteners, and as monofilaments for textiles, carpets, fishing line, and ropes. *Aromatic polyamides* or *aramids* form another important subgroup, with the most important member being *Kevlar.* As a reinforcing fiber, it offers the strength of steel at $^1/_5$ the weight.

Polycarbonates: High strength and outstanding toughness; excellent heat resistance; good dimensional stability; transparent or easily colored; easy to process and readily recyclable; most recognizable as the base material for CDs and DVD discs, but also used for office machine housings, safety helmets, and pump impellers.

Polyesters: Can be either thermoplastic or thermosetting depending on the presence or absence of cross-linking; key example is *polyethylene terephthalate (PET)*; easily processed by injection- or blow-molding or extrusion; commonly used for food packaging, such as soft-drink bottles; polyester fibers are common in wearing apparel.

Polyethylenes: The most common polymer (30% of global plastics consumption); inexpensive, tough, good chemical resistance to acids, bases, and salts; high electrical resistance; low strength; easy to shape and join; easily recycled; reasonably clear in thin-film form; subject to weathering via ultraviolet light; limited resistance to elevated temperatures; flammable; used for grocery bags, milk jugs and other food containers, tubes, pipes, sheeting, and electrical wire insulation. Variations include low-density polyethylene (LDPE—floats in water), high-density polyethylene (HDPE—stronger than low-density), ultra-high molecular weight polyethylene (UHMW).

Polypropylene: Lightest of the plastics; inexpensive; stronger, stiffer, and better heat resistance than polyethylene; transparent; reasonable toughness; used for beverage containers, luggage, pipes, and ropes.

Polystyrenes: High dimensional stability and stiffness with low water absorption; best all-around dielectric; excellent thermal insulator; clear, hard, and brittle at room temperature; often used for rigid packaging such as CD cases; can be foamed to produce expanded polystyrene (trade name of Styrofoam); burns readily; softens at about 95°C; easily recycled; *high-impact polystyrenes* contain additions of rubber to improve toughness.

Polyurethane: Can be thermoplastic, thermosetting, or elastomeric; the thermoplastic polyurethanes bridge the gap between flexible rubbers and rigid thermoplastics; easy to process; strong, tough, and durable; flexible at low temperatures; cut and tear resistant; resistant to oil, grease, fuels, solvents, and other chemicals; common applications include flexible foam (seat cushions and carpet underlays), coatings, sealants, and adhesives.

Polyvinylchloride (PVC): Extensively used general-purpose thermoplastic; strong, light, and durable; good resistance to ultraviolet light (good for outdoor applications); easily molded or extruded; easily recycled; always used with fillers, plasticizers, and pigments; used largely in building construction (gas and water pipes, window frames, flooring).

Vinyls: Wide range of types, from thin, rubbery films to rigid forms; tear resistant; good aging properties; good dimensional stability and water resistance in rigid forms; used for floor and wall covering, upholstery fabrics, and lightweight water hose; common trade names include Saran and Tygon.

Thermosets
Amino Resins: Two primary groups are melamines and urea-formaldehydes. *Melamines*: excellent resistance to heat, water, and many chemicals; full range of translucent and opaque colors; excellent electric-arc resistance; tableware and counter tops (trade name Formica); used in treating paper and cloth to impart water-repellent properties. *Urea–formaldehyde*: properties similar to those of phenolics but available in lighter colors; useful in containers and housings, but not outdoors; used in lighting fixtures because of translucence in thin sections; as a foam, can be used as household insulation.

Epoxies: Good strength, toughness, elasticity, chemical resistance, moisture resistance, and dimensional stability; easily compounded to cure at room temperature; used as adhesives, bonding agents, coatings, and as a matrix in fiber composites.

Phenolics: Oldest of the plastics but still widely used; hard, strong, low cost, and easily molded, but rather brittle; resistant to heat and moisture; dimensionally stable; opaque, but with a range of dark colors; wide variety of forms: sheet, rod, tube, and laminate; uses include molded products, printed circuit boards, insulation of electrical components, thermal insulation foams, countertops, and as a bonding material in grinding wheels; phenol-formaldehyde goes by the trade name of Bakelite.

Polyesters (can be thermoplastic or thermoset): Thermoset polyesters are strong with good resistance to environmental influences; uses include fiber-reinforced polymer-matrix composites (boat and car bodies, pipes, tanks, and construction panels), vents and ducts, textiles, adhesives, coatings, and laminates.

Polyimides: Good chemical resistance; high strength and stiffness; good elevated temperature stability (considered to be a high-temperature polymer).

Polyurethanes: The thermosetting polyurethanes are often used as rigid foams where they provide support, rigidity, and thermal insulation.

Silicones: Heat and weather resistant; low moisture absorption; chemically inert; high dielectric properties; excellent sealant.

Additive Agents in Plastics

For most uses, additional materials are incorporated into plastics to (1) impart or improve properties, (2) reduce cost, (3) improve moldability, and/or (4) impart color. These **additive agents** are usually classified as *fillers and reinforcements, plasticizers, lubricants or release agents, coloring agents, stabilizers, antioxidants, flame retardants, conductive compounds,* and *foaming agents.* The amount of additives ranges from 0 to 50%, with an average content of additives of 20% by weight.

Fillers often comprise a large percentage of the total volume of a molded plastic product. Their primary roles are to improve strength, stiffness, or toughness; reduce shrinkage; reduce weight; or simply serve as an extender, providing cost-saving bulk (often at the expense of reduced moldability). To a large degree, they determine the general properties of a molded plastic. Selection tends to favor materials that are relatively inert and much less expensive than the plastic resin. Some of the most common fillers and their properties are

1. *Wood flour* (fine sawdust): a general-purpose filler; low cost with fair strength; good moldability
2. *Cloth fibers*: improved impact strength; fair moldability
3. *Macerated cloth*: high impact strength; limited moldability
4. *Glass fibers*: high strength; dimensional stability; translucence
5. *Mica*: excellent electrical properties and low moisture absorption
6. *Calcium carbonate, silica, talc, and clay*: serve primarily as extenders

When fillers are used with a plastic resin, the resin acts as a binder, surrounding the filler material and holding the mass together. The surface of a molded part, therefore, will be almost pure resin with no exposed filler. Cutting or scratching through the shiny surface, however, will expose the less-attractive filler.

Plasticizers can be added in small amounts to reduce viscosity and improve the flow of the plastic during molding or to increase the flexibility of thermoplastic products by reducing the intermolecular contact and strength of the secondary bonds between the polymer chains. When used for molding purposes, the amount of plasticizer is governed by the intricacy of the mold. In general, it should be kept to a minimum because it is likely to affect the stability of the finished product through a gradual aging loss. When used for flexibility, plasticizers should be selected with minimum volatility to impart the desired property for as long as possible. Examples of products where plasticizers improve flexibility include wire and cable coatings, tubes and hoses, a wide variety of sheets and films from construction membranes to food wraps, coatings, adhesives, and sealants.

Lubricants and **mold release agents**, such as waxes, silicones, stearates, and soaps can be used to improve the moldability of plastics and to facilitate removal of parts from the mold. They are also used to keep thin polymer sheets from sticking to each other when stacked or rolled. When applied directly onto mold surfaces between cycles, these materials are called external mold releases. In processes such as injection molding, it might be more economical to incorporate the release as a resin additive (internal mold release) than to periodically interrupt the process to spray the mold surfaces. When used in this manner, however, only a minimum amount should be used because the lubricants adversely affect most engineering properties.

Coloring agents can be used to provide almost any color and frequently eliminates the need for secondary coating operations. They can be either **dyes**, which are soluble in the resins, or **pigments**, which are insoluble and impart color simply by their presence as they are dispersed in the polymer matrix. In general, dyes are used for transparent plastics and pigments for the opaque or translucent ones. Optical brighteners can also be used to enhance appearance. Carbon black can provide both a black color and electrical conductivity. Titanium dioxide provides both a white color and UV stabilization.

Heat, light (especially ultraviolet), and oxidation tend to degrade polymers, but **stabilizers** and **antioxidants** can be added to retard these effects. Heat stabilizing additives seek to prevent changes in the chemical structure caused by elevated temperatures, both during processing and in the final products. UV stabilizers absorb ultraviolet radiation and convert it to heat, preventing photolytic decomposition of the polymer. Polymers can undergo chemical reactions, both during processing and during use, that lead to degradation and loss of favorable properties. The most common of these degradation reactions are based on oxidative free radicals. Antioxidant additives or stabilizers can intercept these radicals and significantly slow or halt the degradation.

Flame retardants can be added when nonflammability is important. **Electrically conductive additives**, such as carbon-based powder and fibers, metal powder and fibers, metal coated carbon or glass fibers, and carbon nanotubes, allow for the migration of electrical charge and combat problems associated with the generation and accumulation of static electricity, electrostatic discharges, and electromagnetic or radio-frequency interference. Conductive polymers are being used for a number of applications, such as electronics packaging. The antistatic agents can also reduce the attraction of dust and contaminants, making the materials more suitable for use in food or drug applications. Metal flakes, fibers, or powders can also modify magnetic properties. **Antimicrobial additives** can provide long-term protection from both fungus (such as mildew) and bacteria. **Table 8.2** summarizes the purposes of the various additives.

Oriented Plastics

Because the intermolecular bond strength increases with reduced separation distance, any processing that aligns the molecules parallel to the applied load can be used to give the

TABLE 8.2	Additive Agents in Plastics and Their Purpose
Type	**Purpose**
Fillers	Enhance mechanical properties, reduce shrinkage, reduce weight, or provide bulk
Plasticizer	Increase flexibility, improve flow during molding, reduce elastic modulus
Lubricant	Improve moldability and extraction from molds
Coloring agents (dyes and pigments)	Impart color
Stabilizers	Retard degradation due to heat or light
Antioxidants	Retard degradation due to oxidation
Flame retardants	Reduce flammability
Conductive/ antistatic additives	Impart various degrees of electrical conductivity

long-chain thermoplastics high strength in a given direction.[2] This orientation can be accomplished by forming processes, such as stretching, rolling, or extrusion. The material is often heated prior to the orienting process to aid in overcoming the intermolecular forces and is cooled immediately afterward to "freeze" the molecules in the desired orientation.

Orienting can increase the tensile strength by more than 50%, but a 25% increase is more typical. In addition, the elongation can be increased by several hundred percent. If the **oriented plastics** are reheated, however, they tend to deform back toward their original shape, a phenomenon known as **viscoelastic memory**. The various shrink-wrap materials are examples of this effect.

Engineering Plastics

The standard polymers tend to be lightweight, corrosion-resistant materials with low strength and low stiffness. They are relatively inexpensive and are readily formed into a wide range of useful shapes, but are not suitable for use at elevated temperatures.

In contrast, a group of plastics has been developed with improved thermal properties (up to 350°C, or 650°F), enhanced impact and stress resistance, high rigidity, superior electrical characteristics, excellent processing properties, and little dimensional change with varying temperature and humidity. These true engineering plastics include the polyamides, polyacetals, polyacrylates, polycarbonates, modified polyphenylene oxides, polybutylene terepthalates, polyketones, polysulfones, polyetherimides, and liquid crystal polymers. Although stabilizers, fibrous reinforcements, and particulate fillers can upgrade the conventional plastics, there is usually an accompanying reduction in other properties. The engineering plastics offer a more balanced set of properties. They are usually produced in small quantities, however, and are often quite expensive.

Materials producers have also developed electroconductive polymers with tailored electrical and electronic properties and high-crystalline polymers with properties comparable to some metals.

Plastics as Adhesives

Polymeric adhesives are used in many industrial applications. They are quite attractive for the bonding of dissimilar materials, such as metals to nonmetals, and have even been used to replace welding or riveting. A wide range of mechanical properties is available through variations in composition and additives, and a variety of curing mechanisms can be used. Examples can be found from the thermoplastics (hot melt glues), thermosets (two-part epoxies), and even elastomers (silicone adhesives). The seven most common structural adhesives are epoxies, urethanes, cyanoacrylates, acrylics, anaerobics, hot melts, and silicones. Selection usually involves consideration of the manufacturing conditions, the substrates to be bonded, the end-use environment, and cost. The various features of adhesive bonding are discussed in greater detail in Chapter 40.

Plastics for Tooling

Polymers can also provide inexpensive tooling for applications where pressures, temperatures, and wear requirements are not extreme. Because of their wide range of properties, their ease of conversion into desired shapes, and their excellent properties when loaded in compression, plastics have been widely used in applications such as jigs, fixtures, and a wide variety of forming-die components. Both thermoplastic and thermoset polymers (particularly the cold-setting types) have been used. By using plastics in these applications, costs can be reduced and smaller quantities of products can be economically justified. In addition, the tooling can often be produced in a much shorter time, enabling quicker production.

Foamed Plastics

A number of polymeric materials can be produced in the form of foams that incorporate arrays of voids throughout their structure. Chemical **foaming agents** are added to the resins, often in the form of powders or pellets. When activated by temperature, they generate gas by a decomposition reaction. This gas dissolves in the polymer melt, and when the pressure is dropped, it emerges as gas pockets or bubbles.

The resulting foamed materials are extremely versatile, with properties ranging from soft and flexible to hard and rigid. The softer foams are generally used for cushioning in upholstery and automobile seats and in various applications as vibration absorbers. Semirigid foams find use in floatation devices, refrigerator insulation, disposable food trays and containers,

---

[2]The effects of orienting can be observed in the common disposable flexible-walled plastic drinking cup. Start at the top lip. Place a sharp bend in the lip and then tear down the side wall. The material tears easily. Move around the lip about ½-in. and make another side-wall tear—also easy. Now try to tear across the strip that you have created. This tear is much more difficult because you are tearing across molecules that have been oriented vertically along the cup walls by the cup-forming operation.

building insulation panels, and sound attenuation. Rigid foams have been used as construction materials for boats, airplane components, electronic encapsulation, and furniture.

Foamed products can be made by a wide variety of processes and can be made as either discrete products or used as a "foamed-in-place" material. In addition to the sound and vibration attenuation and thermal insulation properties mentioned earlier, foams offer light weight (often as much as 95% air) and the possibility of improved stiffness and reduced cost (less material to make the part).

Polymer Coatings

Polymer coatings are used extensively to enhance appearance, but they have also assumed a significant role in providing corrosion protection. The tough, thick coatings must adhere to the substrate; not chip or peel; and resist exposure to heat, moisture, salt, and chemicals. Polymer coatings have been replacing chrome and cadmium because of environmental concerns relating to the heavy metals. In addition, polymers provide better resistance to the effects of acid rain.

Plastics vs. Other Materials

Polymeric materials have successfully competed with traditional materials in a number of areas. Plastics have replaced glass in containers and other transparent products. PVC pipe and fittings compete with copper and brass in many plumbing applications. Plastics have even replaced ceramics in areas as diverse as sewer pipe and lavatory facilities.

Although plastics and metals are often viewed as competing materials, their engineering properties are really quite different. Many of the attractive features of plastics have already been discussed. In addition to these, we can add (1) the ability to be fabricated with lower tooling costs; (2) the ability to be molded at the same rate as product assembly, thereby reducing inventory; (3) a possible reduction in assembly operations and easier assembly through snap fits, friction welds, or the use of self-tapping fasteners; (4) the ability to reuse manufacturing scrap; and (5) reduced finishing costs.

Metals, on the other hand, are often cheaper and offer faster fabrication speeds and greater impact resistance. They are considerably stronger and more rigid and can withstand traditional paint cure temperatures. In addition, resistance to flames, acids, and various solvents is significantly better. **Table 8.3** compares the mechanical properties of selected polymers to annealed commercially pure aluminum and annealed 1040 steel. Note the mechanical superiority of the metals, even though they are being presented in their weakest condition. **Table 8.4** compares the cost per pound and elastic modulus of several engineering plastics with values for steel and aluminum. When the size of the part is fixed, cost per cubic inch becomes a more valid comparison, and the figures show plastics to be quite competitive because of their low density.

TABLE 8.3	Property Comparison of Metals and Polymers			
Material	Condition	TS (ksi)	E (10^6 psi)	Elongation
Polyethylene	Branched	2	0.025	90–650
Polyethylene	Crystallized	4	0.100	50–800
Polyvinylchloride	Cl-sides	8	0.375	2–40
Polystyrene	Benzene-sides	7	0.500	1–3
Bakelite	Framework	7	1.0	1
Aluminum	Annealed	13	10.0	15–30
1040 steel	Annealed	75	30.0	30

TABLE 8.4	Comparison of Various Materials (Modulus and Cost)*		
Material	Modulus (x 10^6 psi)	$/pound	$/in³
Aluminum	10.0	0.92	0.093
Steel	30.0	0.30–0.50	0.075–0.125
Nylon	0.1	1.80	0.129
ABS	0.3	1.20	0.040
High-density polyethylene	0.1	1.10	0.041
Polycarbonate	0.35	2.30	0.100
Polypropylene	0.2	1.10	0.034
Polystyrene	0.3	1.40	0.043
Epoxy (bisphenol)	0.45	1.05	0.033

*Cost figures are fall 2014 values and are clearly subject to change. Polymer prices from www.plasticsnews.com.

The automotive industry is a good indication of the expanding use of plastics. Polymeric materials now account for more than 350 pounds of a typical vehicle, and this figure is projected to rise to 400 pounds by 2025. Compare these figures to those of the past—25 pounds in 1960, 105 in 1970, 195 in 1980, and 229 in 1990. In addition to the traditional application areas of dashboards, interiors, body panels, and trim, plastics are now being used for bumpers, intake manifolds, valve covers, fuel tanks, and fuel lines and fittings. If we include clips and fasteners, there are now more than 1,000 plastic parts in a typical automobile.

Recycling of Plastics

More than 200 million tons of plastic are manufactured annually around the world, much of it for disposable, low-value items such as food-wrap and product packaging. Unfortunately, the features that make plastics so durable also make them resistant to environmental and biological degradation. Burning them often releases toxic chemicals, and recycling is difficult because there are so many different kinds, and each has to be recycled by a different process. Moreover, many conventional plastics are often commingled with organic wastes, such as food scraps, wet paper, and liquids. Expensive cleaning and sanitizing procedures would be required prior to recycling.

Because of the wide variety of types and compositions, all with similar physical properties, the recycling of mixed plastics is far more difficult than the recycling of mixed metals. These materials must be sorted not only on the basis of resin type, but also by type of filler, color, and other additive features, and this sorting is usually done by hand. If the various types of resins can be identified and kept separate, many of the thermoplastic materials can be readily recycled into useful products. Packaging is the largest single market for plastics, and there is currently a well-established network to collect and recycle PET (the polyester used in soft-drink bottles) and high-density polyethylene (the plastic used in milk, juice, and water jugs). Fortunately, these two materials represent nearly 96% of the plastic bottles used in the United States. The properties generally deteriorate with recycling, however, so applications must often be downgraded with reuse. PET is being recycled into new bottles, fiber-fill insulation, and carpeting. Recycled polyethylene is used for new containers, plastic bags, and recycling bins. Polystyrene has been recycled into cafeteria trays and the cases for CDs and DVDs. Plastic "lumber," made from low-density polyethylene, offers weather and insect resistance and a reduction in required maintenance (but at higher cost than traditional wood). Polyvinyl chloride (PVC) and polypropylene are also readily recycled. **Table 8.5** presents the standard recycling symbols often incorporated onto plastic products to identify the material, along with a list of some typical recycled uses.

Because of the cross-linking or network bonding, thermosets and elastomers cannot be recycled by simple remelting. Thermoset materials can be ground into particulates and used as fillers in other plastic parts. Rubber tires can be ground into chunks or nuggets and used for purposes such as landscape mulch or playground surfaces.

When thermoplastics and thermosets are mixed or separation is difficult, the material is often regarded more as an alternative fuel (competing with coal and oil) than as a resource for recycling into quality products. On an equivalent-weight basis, polystyrene and polyethylene have heat contents greater than fuel oil and far in excess of paper and wood. In another alternative, decomposition processes can be used to break polymers down into useful building blocks. Hydrolysis (exposure to high-pressure steam) and pyrolysis (heating in the absence of oxygen) methods can be used to convert plastics into simple petrochemical materials. These processes, however, require good control of the input material. Because of all of the recycling difficulties, only about 5% of all plastic (and less than one-third of plastic containers) now finds a second life.

Biodegradable Plastics

Whereas additives like UV- and heat-stabilizers are often added to plastics to increase durability, biodegradable plastics are designed to break down more quickly in either natural aerobic

TABLE 8.5	Symbols for Recyclable Plastics and Some Common Uses		
Symbol	Abbreviation[a]	Polymer Name	Recycled Uses
♳ 1	PETE or PET	Polyethylene terephthalate	Polyester fibres, thermoformed sheet, strapping, and soft drink bottles
♴ 2	HDPE	High density polyethylene	Bottles, grocery bags, milk jugs, recycling bins, agricultural pipe, base cups, car stops, playground equipment, and plastic lumber
♵ 3	PVC or V	Polyvinyl chloride	Pipe, fencing, and non-food bottles
♶ 4	LDPE	Low density polyethylene	Plastic bags, 6 pack rings, various containers, dispensing bottles, wash bottles, tubing, and various molded laboratory equipment
♷ 5	PP	Polypropylene	Auto parts, industrial fibers, food containers, and dishware
♸ 6	PS	Polystyrene	Desk accessories, cafeteria trays, plastic utensils, toys, video cassettes and cases, and insulation board and other expanded polystyrene products (e.g., Styrofoam)
♹ 7	OTHER or O	Other plastics, including acrylic, acrylonitrile butadiene styrene, fiberglass, nylon, polycarbonate, and polylactic acid	Bottles, plastic lumber applications

[a]Developed by the Society of the Plastics Industry in 1988.
[b]Often found below symbol.

(composting) or anaerobic (landfill) environments. **Bioplastics** are made from renewable natural materials, such as cornstarch. Lactic acid, obtained from corn, can be polymerized to form *polylactide* or *polylactic acid (PLA)*, a thermoplastic that looks and behaves like polyethylene and polypropylene. With a natural filler, such as cellulose, this material has been used in the manufacture of a variety of quickly discarded food containers and packaging items. When discarded, these products absorb water and swell up, breaking into small fragments that are easily processed by microorganisms. Chicken feathers, composed of the same keratin found in the hair, hooves, and horns of other animals, have also been used as a base for bioplastics.

Biodegradable plastics are made using conventional petrochemical polymers and a natural (biodegradable) filler that breaks down on exposure to sunlight, water or dampness, bacteria, enzymes, or even insect attack. The degradation of the filler converts the material to a sponge-like consistency that might be more easily broken down over time. These materials might also contain bioactive additives and swelling agents that further accelerate the decay.

There is some debate regarding the benefit of these materials, however. All require a specific environment for biodegradation. In the absence of such an environment (such as natural-material biodegradable plastics placed in an anaerobic landfill), these materials remain extremely durable. When they do degrade, some release potentially harmful greenhouse gases (methane or carbon dioxide) or toxic residues. The *bioplastics* (all natural materials) and *biodegradable plastics* (petrochemical based) cannot be easily recycled. When PLA plastics are mixed with PET (polyethylene terephthalate), the entire mix becomes impossible to recycle—thereby impairing existing recycling efforts.

8.3 Elastomers

The term **elastomer**, a contraction of the words *elastic polymer*, refers to a special class of linear polymers that display an exceptionally large amount of elastic deformation when a force is applied. Many can be stretched to several times their original length. On release of the stretching force, the deformation is completely recovered as the material quickly returns to its original shape. In addition, the cycle can be repeated numerous times with identical results, as with the stretching of a rubber band.

The elastic properties of most engineering materials are the result of a change in the distance between adjacent atoms (i.e., bond length) when loads are applied. Hooke's Law is commonly obeyed, where twice the force produces twice the stretch. When the applied load is removed, the interatomic forces return all the atoms to their original position, and the elastic deformation is completely recovered.

In the elastomeric polymers, the linear chain-type molecules are twisted or curled, much like a coil spring. When a force is applied, the polymer stretches by uncoiling. When the load is removed, the molecules recoil as the bond angles return to their original, unloaded values, and the material returns to its original size and shape. The relationship between force and stretch, however, does not follow Hooke's Law.

In reality, the behavior of elastomers is a bit more complex. Although the chains indeed uncoil when placed under load, they can also slide with respect to one another to produce a small degree of viscous deformation. When the load is removed, the molecules return to their coiled shape, but the viscous deformation is not recovered, and there is some permanent change in shape.

By linking the coiled molecules to one another by strong covalent bonds, a process known as **cross-linking**, it is possible to restrict the viscous deformation while retaining the large elastic response. The elasticity or rigidity of the product can be determined by controlling the number of cross-links. Small amounts of cross-linking leave the elastomer soft and flexible, as in a rubber band. Additional cross-linking further restricts the uncoiling, and the material becomes harder, stiffer, and more brittle, like the rubber used in bowling balls. Because the cross-linked bonds can only be destroyed by extremely high temperatures, the engineering elastomers can be tailored to possess a wide range of stable properties and stress-strain characteristics.

If placed under constant strain, however, even highly cross-linked material will exhibit some viscous flow over time. Consider a rubber band stretched between two nails. Although the dimensions remain fixed, the force or stress being applied to the nails will continually decrease. This phenomenon is known as **stress relaxation**. The rate of this relaxation depends on the material, the force, and the temperature.

Natural Rubber

Natural **rubber**, the oldest commercial elastomer, is made from latex, a secretion from the inner bark of a tropical tree. In its crude form it is an excellent adhesive, and many cements can be made by dissolving it in suitable solvents. Its use as an engineering material dates from 1839, when Charles Goodyear discovered that it could be vulcanized (cross-linked) by the addition of about 30% sulfur followed by heating to a suitable temperature. The cross-linking restricted the movement of the molecular chains and imparted strength. Subsequent research found that the properties could be further improved by various additives (such as carbon black), which act as stiffeners, tougheners, and antioxidants. Accelerators have been found that speed up the **vulcanization** process. These have enabled a reduction in the amount of sulfur, such that most rubber compounds now contain less than 3% sulfur. Softeners can be added to facilitate processing, and fillers can be used to add bulk.

Rubber (polyisoprene) can now be compounded to provide a wide variety of characteristics, ranging from soft and gummy to extremely hard. When additional strength is required, textile cords or fabrics can be coated with rubber. The fibers carry the load, and the rubber serves as a matrix to join the cords while isolating them from one another to prevent chafing. For severe service, steel wires can be used as the load-bearing medium. Vehicle tires and heavy-duty conveyor belts are examples of this technology.

Natural rubber compounds are outstanding for their high tensile and tear strength; good resilience (ability to recover shape after deformation); good electrical insulation; low internal friction; and resistance to most inorganic acids, salts, and alkalis. However, they have poor resistance to petroleum products, such as oil, gasoline, and naphtha. In addition, they lose their strength at elevated temperatures, so it is advisable that they not be used at temperatures above 80°C (175°F). Unless they are specially compounded, they also deteriorate fairly rapidly in direct sunlight.

Artificial Elastomers

To overcome some of these limitations, as well as the uncertainty in the supply and price of natural rubber, a number of synthetic or artificial elastomers have been developed and have come to assume great commercial importance. Although some are a bit inferior to natural rubber, others offer distinctly different and, frequently, superior properties. Polyisoprene can also be synthesized chemically and offers properties closest to natural rubber. Styrene-butadiene or polybutadiene is today's largest-tonnage elastomer. It is an oil-derivative, high-volume substitute for natural rubber that has become the standard material for passenger-car tires. For this material, some form of reinforcement is generally required to provide the desired tensile strength, tear resistance and durability. Polychloroprene or Neoprenes (trade name) have properties similar to natural rubber, with improved stiffness and better resistance to oils, ozone, oxidation, and flame. They are used for a wide range of applications, including fuel hoses, automotive hoses and belts, footwear, wet suits, and seals.

Polyurethane can be a thermoplastic, thermoset, or elastomer, depending on the degree of cross-linking. With a small amount of cross-linking, the material has elastomeric properties and is frequently used in the form of foams for cushioning in furniture and automobile seats. In its unfoamed condition, it can be molded into products like car bumpers.

Silicone rubbers look and feel like organic rubber but are based on a linear chain of silicon and oxygen atoms (not carbon). Various mixes and blends offer retention of physical properties at elevated temperatures [as hot as 230°C (450°F)]; flexibility at low temperatures [as low as −100°C (−150°F)]; resistance to acids, bases, and other aqueous and organic fluids; resistance to flex fatigue; ability to absorb energy and provide damping; good weatherability; ozone resistance; and availability in a variety of different hardnesses. A number of other artificial elastomers are available and are identified by both chemical and commercial trade names.

Elastomers are often classified as thermosetting elastomers and thermoplastic elastomers. The thermoset materials are formed during the irreversible vulcanization (cross-linking) process, which might be somewhat time consuming. Thermoplastic elastomers do not employ cross-links, but derive their properties from a complex combination of soft and hard component phases and can be processed into products by all of the conventional thermoplastic polymer processes (injection molding, extrusion, blow molding, thermoforming, and others). They soften at elevated temperatures, which the thermosets easily withstand, but offer good low-temperature flexibility, scrap recyclability, availability in a variety of colors, and high gripping friction. Unfortunately, many are costlier than the conventional rubber materials.

Selection of an Elastomer

Elastomeric materials can be used for a wide range of engineering applications where they impart properties that include shock absorption, noise and vibration control, sealing, corrosion protection, abrasion protection, friction modification, electrical and thermal insulation, waterproofing, and load bearing. Selection of an elastomer for a specific application requires consideration of many factors, including the mechanical and physical service requirements, the operating environment (including temperature), the desired lifetime, the ability to manufacture the product, and cost. There are a number of families, and within each family, there exists a wide range of available properties. Moreover, almost any physical or mechanical property can be altered through additives, which can also be used to enhance processing or reduce cost, and modifications of the processing parameters.

Table 8.6 lists some of the more common artificial elastomers, along with natural rubber for comparison, and gives their properties and some typical uses.

Elastomers for Tooling Applications

When an elastomer is confined, it acts like a fluid, transmitting force uniformly in all directions. For this reason, elastomers can be substituted for one-half of a die set in sheet-metal-forming operations. Elastomers are also used to perform bulging and to form reentrant sections that would be impossible to form with rigid dies except through the use of costly multipiece tooling. The engineering elastomers have become increasingly popular as tool materials because they can be compounded to range from very soft to very hard; hold up well under compressive loading; are impervious to oils, solvents, and other similar fluids; and can be made into a desired shape quickly and economically. In addition, the elastomeric tooling will not mark or damage highly polished or prepainted surfaces. The urethanes are currently the most popular elastomer for tooling applications.

TABLE 8.6 Properties and Uses of Some Common Elastomers

Elastomer	Specific Gravity	Durometer Hardness	Tensile Strength (psi) Pure Gum	Tensile Strength (psi) Black	Elongation (%) Pure Gum	Elongation (%) Black	Service Temperature F(C) Min.	Service Temperature F(C) Max.	Resistance to:[a] Oil	Resistance to:[a] Water Swell	Resistance to:[a] Tear	Typical Application
Natural rubber	0.93	20–100	2500	4000	75	650	−65(54)	180(82)	P	G	G	Tires, gaskets, hose
Polyacrylate	1.10	40–100	350	2500	600	400	0(−18)	300(149)	G	P	F	Oil hose, O-rings
EDPM (ethylene propylene)	0.85	30–100	1	3		500	−40(−40)	300(149)	P	G	G	Electric insulation, footwear, hose, belts
Chlorosulfonated polyethylene	1.10	50–90	4	2		400	−65(−54)	250(121)	G	E	G	Tank lining, chemical hose; shoes, soles, and heels
Polychloroprene (neoprene)	1.23	20–90	3500	4000	800	550	−50(−46)	225(107)	G	G	G	Wire insulation, belts, hose, gaskets, seals, linings
Polybutadiene	1.93	30–100	1000	3000	800	550	−80(−62)	212(100)	P	P	G	Tires, soles and heels, gaskets, seals
Polyisoprene	0.94	20–100	3000	4000		600	−65(−54)	180(82)	P	G	G	Same as natural rubber
Polysulfide	1.34	20–80	350	00600	600	400	−65(−54)	180(82)	E	G	G	Seals, gaskets, diaphragms, valve disks
SBR (styrene-butadiene)	0.94	40–100	2			1200	−65(−54)	225(107)	P	G	G	Molded mechanical goods, disposable pharmaceutical items
Silicone	1.1	25–90		1200		450	−120(−84)	450(232)	F	E	P	Electric insulation, seals, gaskets, O-rings
Epichlorohydrin	1.27	40–90	2			325	−50(−46)	250(121)	G	G	G	Diaphragms, seals, molded goods, low-temperature parts
Urethane	0.85	62–95	5000		700		−54(−65)	212(100)	E	F	E	Caster wheels, heels, foam padding
Fluoroelastomers	1.65	60–90	1	3		400	−40(−40)	450(232)	E	E	F	O-rings, seals, gaskets, roll coverings

[a]P, poor; F, fair; G, good; E, excellent.

8.4 Ceramics

The first materials used by humans were natural materials such as wood and stone. The discovery that certain clays could be mixed, shaped, and hardened by firing led to what was probably the first man-made material. Many people today, when they hear the word **ceramics**, think of a coffee mug or some other form of pottery, or dinnerware, or even decorative tiles. Although traditional ceramic products, such as cement, bricks, tile, and pottery, have continued to be key materials throughout history, ceramic materials have also assumed important roles in a number of engineering applications. They are used in aircraft engines and catalytic converters; to make lightweight body armor; as the basis of artificial bones and bioimplants; as the fiber in fiberoptic communications; as filters for molten materials; as wear-resistant cutting tools, bearings, and seals; in a variety of sensor and control devices; and in a multitude of electronic applications that make possible many of the devices in our everyday world. Many of these applications utilize their outstanding physical properties, including the ability to withstand high temperatures, resist chemical attack, provide a wide variety of electrical and magnetic properties, and resist wear. In general, ceramics are hard, brittle, high-melting-point materials with low electrical and thermal conductivity, low thermal expansion, good chemical and thermal stability, good creep resistance, high elastic modulus, and high compressive strengths that are retained at elevated temperature. A family of "structural ceramics" has also emerged, and these materials now provide enhanced mechanical properties that make them attractive for many load-bearing applications.

Glass and glass products now account for about half of the ceramic materials market. Advanced ceramic materials (including the structural ceramics, electrical and magnetic ceramics, and fiberoptic material) compose another 20%. Whiteware and porcelain enameled products (such as household appliances) account for about 10% each. Refractories and structural clay products make up most of the difference.

Nature and Structure of Ceramics

Ceramic materials are compounds of metallic and nonmetallic elements (often in the form of oxides, carbides, and nitrides) and exist in a wide variety of compositions and forms. Most have crystalline structures, but unlike metals, the bonding electrons are generally captive in strong ionic or covalent bonds. The absence of free electrons makes the ceramic materials poor electrical conductors and results in many being transparent in thin sections. Because of the strength of the primary bonds, most ceramics have high melting temperatures, high rigidity, and high compressive strength.

The crystal structures of ceramic materials can be quite different from those observed in metals. In many ceramics, atoms of significantly different sizes must be accommodated within the same structure, and the interstitial sites, therefore, become extremely important. Charge neutrality must be maintained throughout ionic structures. Covalent materials must have structures with a limited number of nearest neighbors, set by the number of shared-electron bonds. These features often dictate a less-efficient packing, and hence lower densities, than those observed for metallic materials. As with metals, the same chemistry material can often exist in more than one structural arrangement (polymorphism), depending on the conditions of temperature and pressure. Silica (SiO_2) for example, can exist in three forms—quartz, tridymite, and crystobalite.

Ceramic materials can also exist in the form of chains, similar to the linear molecules in plastics. Like the polymeric materials having this structure, the bonds between the chains are not as strong as those within the chains. Consequently, when forces are applied, cleavage or shear can occur between the chains. In other ceramics, the atoms bond in the form of sheets, producing layered structures. Relatively weak bonds exist between the sheets, and these interfacial surfaces become the preferred sites for fracture. Mica is a good example of such a material. A noncrystalline structure is also possible in solid ceramics. This **amorphous** condition is referred to as the *glassy state*, and the materials are known as *glasses*.

Elevated temperatures can be used to decrease the viscosity of glass, allowing the atoms to move as groups and the material to be shaped and formed. When the temperature is dropped, the material again becomes hard and rigid. The crystalline ceramics do not soften, but they can creep at elevated temperature by means of grain boundary sliding. Therefore, when ceramic materials are produced for elevated temperature service, large grain size is generally desired.

Ceramics Are Brittle, but Can Be Tough

Both crystalline and noncrystalline ceramics tend to be brittle. The glass materials have a three-dimensional network of strong primary bonds that impart brittleness. The crystalline materials do contain dislocations, but for ceramic materials, brittle fracture tends to occur at stresses lower than those required to induce plastic deformation.

There is little that can be done to alter the brittle nature of ceramic materials. However, the energy required to induce brittle fracture (the material toughness) can often be increased. **Tempered glass** uses rapid cooling of the surfaces to induce residual surface compression. The surfaces cool, contract, and harden. As the center then cools and tries to contract, it compresses or squeezes the surface. Because fractures initiate on the surface, the applied stresses must first cancel the residual compression before they can become tensile. Tempered glass is four to five times stronger than annealed glass. **Cermet** materials surround particles of brittle ceramic with a continuous matrix of tough, fracture-resistant metal. **Ceramic–ceramic composites** use weak interfaces that separate or delaminate to become crack arrestors or crack diverters, allowing the remaining structure to continue carrying the load.

Example: Corning's "Gorilla Glass"

An ion-exchange process is used to toughen and enhance the scratch resistance of the glass used in the touch screens of cell phones and other portable electronic devices. Some of the sodium ions in the *sodium aluminosilicate glass are swapped for larger potassium ions. The difference in ion size creates the compressive stresses that toughen the surface, much like those in tempered glass.*

Stabilization involves compounding or alloying to eliminate crystal structure changes that might occur over a desired temperature range, thereby eliminating the dimensional expansions or contractions that would accompany them. Non-uniform heating or cooling can now occur without the stresses that induce fracture. **Transformation toughening** stops the progress of a crack by a crystal structure change that occurs whenever volume expansion is permitted. Fine grain size, high purity, and high density can be promoted by enhanced processing, and these all act to improve toughness.

Clay and Whiteware Products

Many ceramic products are still based on **clay**, to which various amounts of quartz, feldspar, and other materials are added. Selected proportions are mixed with water, shaped, dried, and fired to produce the structural clay products of brick, roof and structural tiles, drainage pipe, and sewer pipe, as well as the **whiteware** products of sanitary ware (toilets, sinks, and bathtubs), dinnerware, decorative floor and wall tile, pottery, and other artware. The terms *earthenware*, *stoneware*, *china,* and *porcelain* have been applied to products with increasing firing temperatures and decreasing amounts of residual porosity.

Refractory Materials

Refractory materials are ceramics that have been designed to provide acceptable mechanical or chemical properties at high operating temperatures, generally in excess of 550°C or 1000°F. Most are based on stable oxide compounds, where the coarse oxide particles are bonded by finer refractory material. Various carbides, nitrides, and borides can also be used in refractory applications. Desirable properties include high melting temperature, excellent hot strength, resistance to elevated-temperature chemical attack, low thermal conductivity, low thermal expansion, resistance to creep, and resistance to thermal shock.

Refractory ceramics fall into three distinct chemical classes: *acidic, basic,* and *neutral.* Common acidic refractories are based on silica (SiO_2) and alumina (Al_2O_3) and can be compounded to provide high-temperature resistance along with high hardness and good mechanical properties. Magnesium oxide (MgO) is the core material for most basic refractories. These are generally more expensive than the acidic materials but provide better chemical resistance and are often required in metal-processing applications to provide compatibility with the metal. Neutral refractories, containing chromite (Cr_2O_3), are often used to separate the acidic and basic materials because they tend to attack one another. The combination is often attractive when a basic refractory is necessary on the surface for chemical reasons, and the cheaper, acidic material is used beneath to provide strength and insulation. Refractories that exhibit extraordinary properties for special applications can be based on silicon carbide (abrasion resistance), zircon or zirconia, fused silica (temperatures up to 1450°C or 3000°F), and carbon or graphite (not wetted by most molten metals or slags).

Refractory ceramics are the principal materials in the construction of furnaces, in the lining of ladles and molten material containment vessels, in metal casting molds (like the one shown in **Figure 8.7**), and in the flues and stacks through which hot gases are conducted. They can take the form of bricks and shaped products, bulk materials (often used as coatings), and insulating ceramic fibers. Insulating bricks, designed with a high degree of porosity, are often used behind surface refractories to provide enhanced thermal insulation. Bulk or monolithic (continuous structure) refractories that do not require firing for their manufacture are extremely attractive. There is

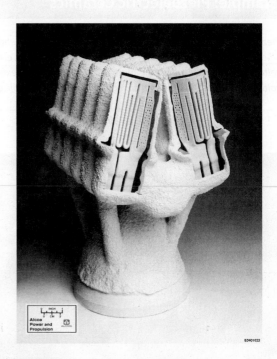

FIGURE 8.7 Ceramic investment-casting mold for casting gas turbine engine rotor blades. The positioned cores create intricate internal cooling passages. (Photo courtesy of Alcoa Howmet)

a significant energy saving by eliminating the firing operation, the coatings are joint-free, and repairs are easily made.

Figure 8.8 shows a variety of high-strength alumina components.

Abrasives

Because of their high hardness, ceramic materials, such as silicon carbide and aluminum oxide (alumina), are often used as the basis for **abrasives** in the following forms: *bonded* (such as grinding wheels), *coated* (like sandpaper), and *loose* (like grit blasting and polishing compounds). Tungsten carbide is often used in wear applications, and manufactured diamond and cubic boron nitride have such phenomenal properties that they are often termed **superabrasives**. **Table 8.7** presents hardness values for various ceramics. Materials used for abrasive applications are discussed in greater detail in Chapter 28.

Ceramics for Electrical and Magnetic Applications

Ceramic materials also offer a variety of useful electrical and magnetic properties. Some ceramics, such as silicon carbide, are used as resistors and high-temperature heating elements for electric furnaces. Others have semiconducting properties and are used for thermistors and rectifiers. Dielectric, piezoelectric, and ferroelectric behavior can also be utilized in many applications. Barium titanate, for example, is used in capacitors and transducers. High-density clay-based ceramics and aluminum oxide make excellent high-voltage insulators, such as those found on automotive spark plugs. The magnetic ferrites have been used in a number of magnetic applications. Considerable attention has also been directed toward the "high-temperature" ceramic superconductors.

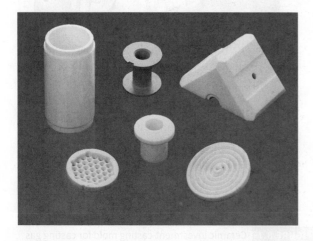

FIGURE 8.8 A variety of high-strength alumina components. (Supplied by Morgan Technical Ceramics)

TABLE 8.7 **Hardness Values of Selected Ceramics**

Material	Knoop Hardness (kg/mm²)
Diamond	7000–9500
Cubic boron nitride (CBN)	3500–4750
Boron carbide (B₄C)	3200
Titanium carbide (TiC)	2800
Silicon carbide (SiC)	2300–2900
Aluminum oxide (Al₂O₃)	2000
Tungsten carbide (WC)	1880
Tungsten carbide cermet (94% WC – 6% cobalt)	1500
Zirconium oxide (ZrO₂)	1100–1300
Silicon dioxide (SiO₂)	550–750
Window glass	300–500
Hardened steel (R_c60.5)	740

Glasses

When some molten ceramics are cooled at a rate that exceeds a critical value, the material solidifies into a hard, rigid, non-crystalline (i.e., amorphous) solid, known as a **glass**. Most commercial glasses are based on silica (SiO_2), lime ($CaCO_3$), and sodium carbonate ($NaCO_3$), with additives to alter the structure or reduce the melting point. Various chemistries can be used to optimize or adjust: fluidity of the molten glass, optical properties including color and index of refraction, thermal stability, and resistance to thermal shock.

Glass is soft and moldable when hot, making shaping rather straightforward. When cool and solid, glass is strong in compression, but brittle and weak in tension. In addition, most glasses exhibit excellent resistance to weathering and attack by most chemicals. Traditional applications include automotive and window glass, bottles and other containers, light bulbs, mirrors, lenses, and fiberglass insulation. There is also a wide variety of specialty applications, including glass fiber for fiber-optic communications and glass fiber to reinforce composites, glass cookware, and a variety of glass uses in electronics, medical, and biological products. Glass and other ceramic fibers have also been used for filtration where they provide a chemical inertness and the possibility to withstand elevated temperature. Formulated glass coatings can block or channel certain wavelengths of light, such as ultraviolet or infrared.

Because the material never loses its quality, purity, or clarity when remelted and resolidified, glass is recyclable, and manufacturers use as much recycled glass or cullet as possible. In addition, the use of recycled glass enables lower furnace temperatures to be used, reducing the amount of energy needed to produce a product.

Glass Ceramics

These materials are first shaped as a glass and then heat-treated to promote partial devitrification or crystallization of the material, resulting in a structure that contains large amounts of fine-grain-size crystalline material within an amorphous base. Because they were initially formed as a glass, **glass ceramics** do not have the strength-limiting or fracture-inducing porosity that is characteristic of the conventional sintered ceramics. This feature, coupled with the fine grain size, provides strength that is considerably greater than that of the traditional glasses. In addition, the crystalline phase makes the material opaque

and helps to retard creep at high temperatures. Because the thermal expansion coefficient is near zero, the material has good resistance to thermal shock. The white Pyroceram (trade name) material commonly found in Corningware is a common example of a glass ceramic.

Cermets

Cermets are combinations of metals and ceramics (usually oxides, carbides, nitrides, or carbonitrides), united into a single product by the procedures of powder metallurgy. This usually involves pressing mixed powders at pressures ranging from 70 to 280 MPa (10 to 40 ksi) followed by sintering in a controlled-atmosphere furnace at about 1650°C (3000°F). Cermets combine the high hardness and refractory characteristics of ceramics with the toughness and thermal shock resistance of metals. They are used as crucibles, jet engine nozzles, and aircraft brakes, and in other applications requiring hardness, strength, and toughness at elevated temperature. Cemented tungsten carbide (tungsten carbide particles cemented in a cobalt binder) has been used in dies and cutting tools for quite some time. The more advanced cermets now enable higher cutting speeds than those achievable with high-speed tool steel, tungsten carbide, or the coated carbides. See Chapter 22.

Cements

Various ceramic materials can harden by chemical reaction, enabling their use as a binder that does not require firing or sintering. Sodium silicate hardens in the presence of carbon dioxide and is used to produce sand cores in metal casting. Plaster of Paris and Portland cement both harden by hydration reactions.

Ceramic Coatings

A wide spectrum of enamels, glazes, and other ceramic coatings has been developed to decorate, seal, and protect substrate materials. Porcelain enamel can be applied to carbon steel in the perforated tubs of washing machines where the material must withstand the scratching of zippers, buttons, and snaps along with the full spectrum of laundry products. Chemical reaction vessels are often glass lined.

Example: Titanium Nitride

Titanium nitride is an extremely hard ceramic material, often used as a coating where it imparts wear resistance, corrosion resistance, and low sliding friction, or enhanced appearance because of its gold coloration. It is commonly found on gold-colored drill bits and milling cutters. With minor additives, it can closely resemble 18-carat gold at a cost that is 20 to 25 times lower and with a hardness or scratch resistance that is 5 to 10 times greater—hence its use in both costume and higher-end jewelry. It can be deposited by several methods, most commonly physical vapor deposition (PVD) and chemical vapor deposition (CVD).

Ceramics for Mechanical Applications: The Structural and Advanced Ceramics

Because of the strong ionic or covalent bonding and high shear resistance, ceramic materials tend to have low ductility and high compressive strength. Theoretically, ceramics could also have high tensile strengths. However, because of their high melting points and lack of ductility, most ceramics are processed in the solid state, where products are made from powdered material. After various means of compaction, voids remain between the powder particles, and a portion of these persists through the sintering process (described in Chapter 16). Contamination can also occur on particle surfaces and then become part of the internal structure of the product. As a result, full theoretical density is extremely difficult to achieve, and small cracks, pores, and impurity inclusions tend to be an integral part of most ceramic materials. These act as mechanical stress concentrators. As loads are applied, the effect of these flaws cannot be reduced through plastic flow, and the result is generally a brittle fracture. Applying the principles of fracture mechanics,[3] we find that ceramics are sensitive to very small flaws. Tensile failures typically occur at stress values between 20 and 210 MPa (3 and 30 ksi), more than an order of magnitude less than the corresponding strength in compression.

Because the number, size, shape, and location of the flaws are likely to differ from part to part, ceramic parts produced from identical material by identical methods often fail at very different applied loads. As a result, the mechanical properties of ceramic products tend to follow a statistical spread that is much less predictable than for metals. This feature tends to limit the use of ceramics in critical high-strength applications.

If the various flaws and defects could be eliminated or reduced to very small size, high and consistent tensile strengths could be obtained. Hardness, wear resistance, and strength at elevated temperatures would be attractive properties, along with light weight (specific gravities of 2.3 to 3.85), high stiffness, dimensional stability, low thermal conductivity, corrosion resistance, and chemical inertness. Failure would still occur by brittle fracture, however. Because of the poor thermal conductivity, thermal shock might also be a problem. The cost of these "flaw-free" or "restricted-flaw" materials would be rather high. Joining to other engineering materials and machining would be extremely difficult, so products would have to be fabricated to final shape through the use of net-shape processing.

---

[3]According to the principles of fracture mechanics, fracture will occur in a brittle material when the fracture toughness, K, is equal to a product involving a dimensionless geometric factor (α), the applied stress, (σ), and the square root of the number π (3.14) times the size of the most critical flaw, (a).

$$K = \alpha \sigma (\pi a)^{1/2}$$

When the right-hand side is less than the value of K, the material bears the load without breaking. Fracture occurs when the combination of applied stress and flaw size equals the critical value, K. Because K is a material property, any attempt to increase the load or stress a material can withstand must be achieved by a companion reduction in flaw size.

Advanced (also known as *structural or engineering*) **ceramics** is an emerging technology with a broad base of current and potential applications. The base materials currently include silicon nitride, silicon carbide, partially stabilized zirconia, transformation toughened zirconia, alumina, sialons, boron carbide, boron nitride, titanium diboride, and ceramic composites (such as ceramic fibers in a glass, glass–ceramic, or ceramic matrix). The materials and products are characterized by high strength and hardness, high fracture toughness, resistance to heat and chemicals, fine grain size, and little or no porosity. Applications include a wide variety of wear-resistant parts including cutting tools, punches, dies, and engine components (such as bearings, seals, and valves), as well as use in heat exchangers, gas turbines, and furnaces. Porous products have been used as substrate material for catalytic converters and as filters for streams of molten metal.

Alumina (or aluminum oxide) ceramics are the most common for industrial applications. They are relatively inexpensive and offer high hardness and abrasion resistance, low density, and high electrical resistivity. Alumina is strong in compression and retains useful properties at temperatures as high as 1900°C (3500°F), but it is limited by low toughness, low tensile strength, and susceptibility to thermal shock and attack by highly corrosive media. Because of its high melting point, it is generally processed in a powder form. Alumina can be joined to metals or other ceramics through metalizing and brazing techniques.

Silicon carbide and silicon nitride offer excellent strength, extreme hardness, adequate toughness, high thermal conductivity, low thermal expansion, and good thermal shock resistance, coupled with light weight and corrosion resistance. They work well in high-stress, high-temperature applications, such as turbine blades, and might well replace nickel- or cobalt-based superalloys. **Figure 8.9** shows gas-turbine rotors made from injection-molded silicon nitride. They are designed to operate at 1250°C (2300°F), where the material retains over half of its room-temperature strength and does not require external cooling. **Figure 8.10** shows some additional silicon nitride products.

Boron carbide offers a hardness that is only less than diamond, cubic boron nitride, and boron carbide. Its high elastic modulus, high compressive strength, and light weight make it ideal for lightweight body armor.

Sialon (a *si*licon–*al*uminum–*o*xygen–*n*itrogen structural ceramic) is really a solid solution of alumina and silicon nitride, and it bridges the gap between them. More aluminum oxide enhances hardness, whereas more silicon nitride improves toughness. The resulting material is stronger than steel, extremely hard, and as light as aluminum. It has good resistance to corrosion, wear, and thermal shock; is an electrical insulator; and retains good tensile and compressive strength up to 1400°C (2550°F). It has excellent dimensional stability, with a coefficient of thermal expansion that is only one-third that of steel and one-tenth that of plastic. When overloaded, however, it exhibits the ceramic property of failure by brittle fracture.

Zirconia is inert to most metals, and retains strength to temperatures higher than 2200°C (4000°F). Partially stabilized

toughness brought about by doping the material with oxides of calcium, yttrium, or magnesium. Transformation-toughened zirconia has even greater toughness as a result of dispersed second phases throughout the ceramic matrix. When a crack approaches the metastable phase, it transforms to a more stable structure, increasing in volume to compress and stop the crack.

The high cost of the structural ceramics continues to be a barrier to their widespread acceptance. High-grade ceramics are currently several times more expensive than their metal counterparts. Even factoring in enhanced lifetime and improved performance, there is still a need to reduce cost. Work continues, however, toward the development of a low-cost, high-strength, high-toughness ceramic with a useful temperature range. Parallel efforts are under way to ensure flaw detection in the range of 10 to 50 mm. With success in these efforts, ceramics could compete where tool steels, powdered metals, coated materials, and tungsten carbide are now being used. Potential applications include engines, turbochargers, gas turbines, bearings, pump and valve seals, and other products that operate under high-temperature, high-stress environments.

In addition to the attractive properties described earlier, the advanced ceramics are also biocompatible (chemically inert, promote tissue and bone growth, and are not susceptible to attack by the body's immune system), which make them ideal for medical implant applications. Ceramic-on-ceramic artificial joints now have a predicted life in excess of 20 years. Ceramic materials have also emerged as the material of choice in numerous dental procedures.

Table 8.8 provides the mechanical properties of some of today's advanced or structural ceramics.

Advanced Ceramics as Cutting Tools

Because of their high hardness, retention of hardness at elevated temperature, and low reactivity with metals, ceramic materials are attractive for a variety of cutting applications. Silicon carbide is a common abrasive in many grinding wheels. Cobalt-bonded tungsten carbide has been a popular alternative to high-speed tool steels for many tool and die applications. Tool steel and carbide tools are often enhanced by a variety of vapor-deposited ceramic coatings. Thin layers of titanium carbide, titanium nitride, and aluminum oxide can inhibit reactions between the metal being cut and the tool steel or binder phase of the carbide. This results in a significant reduction in friction and wear and enables faster rates of cutting. Silicon

FIGURE 8.9 Gas-turbine rotors made from advanced ceramic silicon nitride. The lightweight material (one-half the weight of stainless steel) offers strength at elevated temperature as well as excellent resistance to corrosion and thermal shock. (Kyocera Industrial Ceramics Corp.)

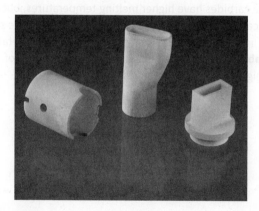

FIGURE 8.10 A variety of components manufactured from silicon nitride. (Supplied by Morgan Technical Ceramics)

zirconia combines the zirconia characteristics of resistance to thermal shock, wear, and corrosion; low thermal conductivity; and low friction coefficient, with the enhanced strength and

Example: The Ceramic Automobile Engine

A ceramic automobile engine has been discussed for a number of years. By allowing higher operating temperatures, engine efficiency could be increased. Sliding friction would be reduced, and there would be no need for cooling. The radiator, water pump, coolant, fan belt, and water lines could all be eliminated. The net result would be up to a 30% reduction in fuel consumption. Unfortunately, this is still a dream because of the inability to produce large, complex-shaped products with few, small-sized flaws.

TABLE 8.8 **Properties of Some Advanced or Structural Ceramics**

Material	Density (g/cm³)	Tensile Strength (ksi)	Compressive Strength (ksi)	Modulus of Elasticity (10⁶ psi)	Fracture Toughness (ksi√in.)
Al_2O_3	3.98	30	400	56	5
Sialon	3.25	60	500	45	9
SiC	3.1	25	560	60	4
ZrO_2 (partially stabilized)	5.8	65	270	30	10
ZrO_2 (transformation toughened)	5.8	50	250	29	11
Si_3N_4 (hot pressed)	3.2	80	500	45	5

nitride, boron carbide, cubic boron nitride, and polycrystalline diamond cutting tools offer even greater tool life, higher cutting speeds, and reduced machine downtime. Cutting tool materials are developed in greater detail in Chapter 22.

With advanced tool materials, cutting speeds can be increased to as high as 1500 m/min (5000 ft/min). The use of these ultra-high-speed materials, however, requires companion developments in the machine tools themselves. High-speed spindles must be perfectly balanced, and workholding devices must withstand high centrifugal forces. Chip-removal methods must be able to remove the chips as fast as they are formed.

As environmental regulations become more stringent, dry machining might be pursued as a means of reducing or eliminating coolant- and lubricant-disposal problems. Ceramic materials are currently the best materials for dry operations. Ceramic tools have also been used in the direct machining of materials that once required grinding, a process sometimes called **hard machining**. **Figure 8.11** shows the combination of toughness and hardness for a variety of cutting-tool materials.

Ultra-High-Temperature Ceramics

Ultra-high-temperature ceramics can be defined as compounds with melting temperatures in excess of 3000°C (5400°F), most of which are the borides, carbides, or nitrides of the early transition metals (niobium, zirconium, tantalum, and hafnium). Because of their chemical and structural stability, they are materials of interest in applications like the handling of high-temperature molten metals, as electrodes in electric arcs, and in aerospace applications such as hypersonic flight, scramjet propulsion, rocket propulsion, and atmospheric re-entry. The boride compounds offer the ceramic properties of high melting point, high elastic modulus, and high hardness, coupled with metal-like values of high electrical and thermal conductivity. The carbides have higher melting temperatures, but lower electrical and thermal conductivities. Both families are significantly lighter than the high-temperature refractory metals.

Table 8.9 presents the melting temperatures of some ultra-high-temperature ceramics.

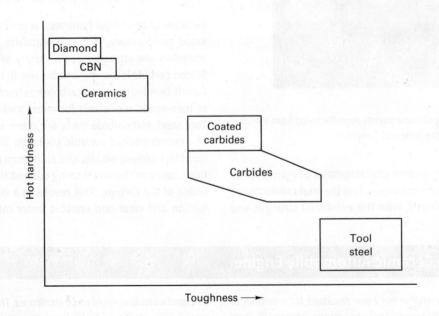

FIGURE 8.11 Graphical mapping of the combined toughness and hardness for a variety of cutting-tool materials. Note the superior hardness of the ceramic materials. [CBN = cubic boron nitride].

TABLE 8.9 Melting Temperatures of Ultra-High-Temperature Ceramics*

Material	Melting Temperature (°C /°F)
Titanium diboride (TiB$_2$)	3225 / 5837
Zirconium diboride (ZrB$_2$)	3247 / 5877
Niobium diboride (NbB$_2$)	3036 / 5497
Hafnium diboride (HfB$_2$)	3380 / 6116
Tantalum diboride (TaB$_2$)	3037 / 5499
Titanium carbide (TiC)	3067 / 5553
Zirconium carbide (ZrC)	3445 / 6233
Niobium carbide (NbC)	3610 / 6530
Hafnium carbide (HfC)	3928 / 7102
Tantalum carbide (TaC)	3997 / 7227

*Data collected from *Phase Diagrams for Ceramists*, Vol. X, ed. by A. E. McHale, American Ceramic Society, Westerville, OH (1994) and Phase Diagrams for Ceramists, Volume 1, ed. by E. M. Levins, C. R. Robbins, and H. F. McMurdie, The American Ceramic Society, Columbus, OH (1964).

8.5 Composite Materials

A **composite material** is a nonuniform solid consisting of two or more different materials that are mechanically or metallurgically bonded together. Each of the various components retains its identity in the composite and maintains its characteristic structure and properties. There are recognizable interfaces between the component materials. The composite, however, generally possesses characteristic properties (or combinations of properties), such as stiffness, strength, weight, high-temperature performance, corrosion resistance, hardness, and conductivity, which are not possible with the individual components by themselves. Analysis of these properties shows that they depend on (1) the properties of the individual components; (2) the relative amounts of the components; (3) the size, shape, and distribution of the discontinuous components; (4) the orientation of the various components; and (5) the degree of bonding between the components. The materials involved can be natural materials, man-made organics, metals, or ceramics. Hence a wide range of freedom exists, and composite materials can often be designed to meet a desired set of engineering properties and characteristics.

There are many types of composite materials and several methods of classifying them. One method is based on geometry and consists of three distinct families: laminar or layered composites, particulate composites, and fiber-reinforced composites.

Laminar or Layered Composites

Laminar composites are those having distinct layers of material bonded together in some manner and include thin coatings, thicker protective surfaces, claddings, bimetallics, laminates, sandwiches, and others. They are used to impart properties such as reduced cost, enhanced corrosion resistance or wear resistance, electrical insulation or conductivity, unique expansion characteristics, lighter weight, improved strength, or altered appearance.

Plywood is probably the most common engineering material in this category and is an example of a laminate material. Layers of wood veneer are adhesively bonded with their grain orientations at various angles to one another. Strength and fracture resistance are improved, properties are somewhat uniform within the plane of the sheet, swelling and shrinkage tendencies are minimized, and large pieces are available at reasonable cost. Automotive tires have a series of bonded belts or plies. Safety glass is another laminate in which a layer of polymeric adhesive is placed between two pieces of glass and serves to retain the fragments when the glass is broken. *Aramid-Aluminum-Laminates* (Arall) consist of thin sheets of aluminum bonded with woven adhesive-impregnated aramid fibers. The combination offers light weight coupled with high fracture, impact, and fatigue resistance.

Laminated plastics are made from layers of reinforcing material that have been impregnated with thermosetting resins, bonded together, and cured under heat and pressure. They can be produced as sheets or rolled around a mandrel to produce a tube or rolled tightly to form a rod. Various resins have been used with reinforcements of paper, cotton or nylon fabric, asbestos, or glass fiber (usually in woven form). Common applications include a variety of decorative items, such as Formica countertops, imitation hardwood flooring, furniture, and sporting items (such as snow skis). When combined with a metal layer on one or both surfaces, the material is used for printed circuit boards.

Bimetallic strip is a laminate of two metals with significantly different coefficients of thermal expansion. As **Figure 8.12** illustrates, changes in temperature now produce flexing or curvature in the product. This unique property of shape varying with temperature is often employed in thermostat and other heat-sensing or temperature-control applications. Still other laminar composites are designed to provide enhanced surface characteristics while retaining a low-cost, high-strength, or lightweight core. Many **clad materials** fit

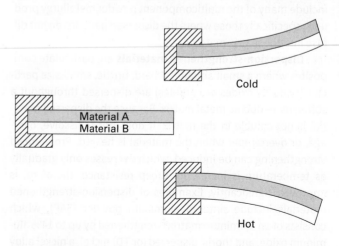

FIGURE 8.12 Schematic of a bimetallic strip where material A has the greater coefficient of thermal expansion. Note the response to cold and hot temperatures.

Example: The Current U.S. Quarter

In 1964, the value of the silver in a U.S. quarter exceeded 25 cents, and a less-expensive replacement was needed. The replacement needed to be accepted in coin-operated vending machines, however, and these machines assessed both size and weight. The replacement quarter, therefore, had to have the same overall density as the original silver coin. The solution involved creating a laminar composite of metal alloys that are heavier than silver (copper) and lighter than silver (nickel). By placing the high-nickel alloy on the surface, the brightness and appearance of the previous coin could be retained.

this description. Alclad metal, for example, consists of high-strength, age-hardenable aluminum with an exterior cladding of one of the weaker, but more corrosion-resistant, single-phase, non-heat-treatable aluminum alloys. Stainless steel has been applied to cheaper, less-corrosion-resistant, substrates. Other laminates have surface layers that have been selected primarily for enhanced wear resistance, improved appearance, or electrical conductivity.

Sandwich material is a laminar structure composed of a thick, low-density core placed between thin, high-density surfaces. Corrugated cardboard is an example of a sandwich structure. Other engineering sandwiches incorporate cores of a polymer foam or honeycomb structure to produce a lightweight, high-strength, high-rigidity composite (see Figure 20.19).

It should be noted that the properties of laminar composites are always **anisotropic**—that is they are not the same in all directions. Because of the variation in structure, properties will always be different in the direction perpendicular to the layers.

Particulate Composites

Particulate composites consist of discrete particles of one material surrounded by a matrix of another material. Concrete is a classic example, consisting of sand and gravel particles surrounded by hydrated cement. Asphalt consists of mineral aggregate in a matrix of bitumen, a thermoplastic polymer. In both of these examples, the particles are rather coarse. Other particulate composites involve extremely fine particles and include many of the multicomponent powder metallurgy products, specifically those where the dispersed particles do not diffuse into the matrix material.

Dispersion-strengthened materials are particulate composites where a small amount of hard, brittle, small-size particles (typically oxides or carbides) are dispersed throughout a softer, more ductile metal matrix. Because the dispersed material is not soluble in the matrix, it does not redissolve, over-age, or overtemper when the material is heated. Pronounced strengthening can be induced, which decreases only gradually as temperature is increased. Creep resistance, therefore, is improved significantly. Examples of dispersion-strengthened materials include sintered aluminum powder (SAP), which consists of an aluminum matrix strengthened by up to 14% aluminum oxide, and thoria-dispersed (or TD) nickel, a nickel alloy containing 1 to 2 wt% thoria (ThO_2). Silicon carbide dispersed in aluminum has been shown to be a low-cost, easily fabricated material with isotropic properties and has been used in

automotive connecting rods. Because of the metal-ceramic mix and the desire to distribute materials of differing density, the dispersion-strengthened composites are often produced by powder metallurgy techniques.

Other types of particulate composites are known as **true particulate composites** and contain large amounts of coarse particles. They are usually designed to produce some desired combination of properties rather than increased strength. Cemented carbides, for example, consist of hard ceramic particles, such as tungsten carbide, tantalum carbide, or titanium carbide, embedded in a metal matrix, which is usually cobalt or nickel. Although the hard, stiff carbide could withstand the high temperatures and pressure of cutting, it is extremely brittle. Toughness is imparted by combining the carbide particles with metal powder, pressing the material into the desired shape, heating to melt the metal, and then resolidifying the compacted material. Varying levels of toughness can be imparted by varying the amount of metallic material in the composite.

Grinding and cutting wheels are often formed by bonding abrasives, such as alumina (Al_2O_3), silicon carbide (SiC), cubic boron nitride (CBN), or diamond, in a matrix of glass or polymeric material. As the hard particles wear, they fracture or pull out of the matrix, exposing fresh, new cutting edges. By combining tungsten powder and powdered silver or copper, electrical contacts can be produced that offer both high conductivity and resistance to wear and arc erosion. Foundry molds and cores are often made from sand (particles) and an organic or inorganic binder (matrix).

Metal-matrix composites of the particulate type have been made by introducing a variety of ceramic or glass particles into aluminum or magnesium matrices. Particulate-toughened ceramics using zirconia and alumina matrices are being used as bearings, bushings, valve seats, die inserts, and cutting tool inserts. Many plastics could be considered to be particulate composites because the additive fillers and extenders are actually dispersed particles. Designation as a particulate composite, however, is usually reserved for polymers where the particles are added for the primary purpose of property modification. One such example is the combination of granite particles in an epoxy matrix that is currently being used in some machine tool bases. This unique material offers high strength and a vibration damping capacity that exceeds that of gray cast iron.

Because of their unique geometry, the properties of particulate composites are usually **isotropic**, that is, uniform in all directions. This might be particularly important in engineering applications.

Fiber-Reinforced Composites

The most popular type of composite material is the **fiber-reinforced composite** geometry, where continuous or discontinuous thin fibers of one material are embedded in a matrix of another. The objective is usually to enhance strength, stiffness, fatigue resistance, or strength-to-weight ratio by incorporating strong, stiff, but possibly brittle fibers in a softer, more ductile, matrix. The matrix supports and transmits forces to the fibers, protects them from environments and handling, provides ductility and toughness as well as the necessary corrosion resistance, and sets the material's processability. The fibers carry most of the load and impart enhanced stiffness. Tensile strength depends primarily on the fibers, whereas compressive strength depends more on the matrix. Shear strength is strongly dependent on the bonding between the fibers and the matrix, and coupling agents are often incorporated to enhance bond strength.

Wood and bamboo are two naturally occurring fiber composites, consisting of cellulose fibers in a lignin matrix. Bricks of straw and mud might well have been the first human-made material of this variety, dating back to near 800 B.C. Automobile tires now use fibers of nylon, rayon, aramid (Kevlar), or steel in various numbers and orientations to reinforce the rubber and provide added strength and durability. Steel-reinforced concrete is actually a double composite, consisting of a particulate matrix reinforced with steel fibers.

Glass-fiber-reinforced resins, the first of the modern fibrous composites, were developed shortly after World War II in an attempt to produce lightweight materials with high strength and high stiffness. Glass fibers about 10 μm in diameter are bonded in a variety of polymers, generally epoxy or polyester resins. Between 30 and 60% by volume is made up of fibers of either E-type borosilicate glass (tensile strength of 500 ksi and elastic modulus of 10.5×10^6 psi) or the stronger, stiffer, high-performance S-type magnesia–alumina–silicate glass (with tensile strength of 670 ksi and elastic modulus of 12.4×10^6 psi).[4]

Glass fibers are still the most widely used reinforcement, primarily because of their lower cost and adequate properties for many applications. Current uses of glass-fiber-reinforced plastics include sporting goods, boat hulls, and bathtubs. Limitations of the glass-fiber material are generally related to strength and stiffness. Alternative fibers have been developed for applications requiring enhanced properties. Boron–tungsten fibers (boron deposited on a tungsten core) offer an elastic modulus of 55×10^6 psi with tensile strengths in excess of 400 ksi. Silicon carbide filaments (SiC on tungsten) have an even higher modulus of elasticity.

Graphite or carbon and aramid (DuPont tradename is Kevlar) are other popular reinforcing fibers. Carbon fibers can be either the PAN type, produced by the controlled thermal pyrolysis of synthetic organic fibers, primarily polyacrylonitrile, or pitch type, made from petroleum pitch, coal tar, or asphalt. They have low density, a range of high tensile strengths (600 to 750 ksi) and high elastic moduli (40 to 65×10^6 psi), and good electrical and thermal conductivity. They do not melt or soften with heat, in fact, strength actually increases with increased temperature. Graphite's negative thermal-expansion coefficient can also be used to offset the positive values of most matrix materials, leading to composites with low or zero thermal expansion. Kevlar is an organic aramid fiber with a tensile strength up to 650 ksi, elastic modulus of 27×10^6 psi, a density approximately one-half that of aluminum, and good toughness. In addition, it is flame retardant, cut and tear resistant, and transparent to radio signals, making it attractive for a number of military and aerospace applications where the service temperature is not excessive.

Ceramic fibers, metal wires, and specially grown whiskers have also been used as reinforcing fibers for high-strength, high-temperature applications. Metal fibers can also be used to provide electrical conductivity or shielding from electromagnetic interference to a lightweight polymeric matrix. With the demand for less expensive, lightweight, environmentally friendly (renewable and compostable) materials, the natural fibers have also assumed an engineering material role. Cotton, hemp, flax, jute, coir (coconut husk), and sisal have found use in various composites. Thermoplastic fibers, such as nylon and polyester, have been used to enhance the toughness and impact strength of the brittle thermoset resins.

Table 8.10 lists some of the key engineering properties for several of the common reinforcing fibers. Because the objectives are often high strength coupled with light weight, or high stiffness coupled with light weight, properties are often reported as **specific strength** and **specific stiffness**, where the strength or stiffness values are divided by density.

The orientation of the fibers within the composite is often key to properties and performance. Sheet-molding compound (SMC), bulk-molding compound (BMC), and fiberglass generally contain short, randomly oriented fibers. Long, unidirectional fibers, either monofilament or multifilament bundles, can be used to produce highly directional properties, with the fiber directions being tailored to the direction of loading. Woven fabrics or tapes can be produced and then layered in various orientations to produce a laminar or plywood-like product. The layered materials can then be stitched together (knitted) to add a third dimension to the weave, and complex three-dimensional shapes can be woven from fibers and later injected with a matrix material.

The properties of fiber-reinforced composites depend strongly on several characteristics: (1) the properties of the fiber material; (2) the volume fraction of fibers; (3) the **aspect ratio** of the fibers, that is, the length-to-diameter ratio; (4) the orientation of the fibers; (5) the degree of bonding between the fiber and the matrix; and (6) the properties of the matrix. Although more fibers tend to provide greater strength and stiffness, the volume fraction of fibers generally cannot exceed 80% to allow

---

[4] It is important to note that a fiber of material tends to be stronger than the same material in bulk form. The size of any flaw is limited to the diameter of the fiber, and the complete failure of a given fiber does not propagate through the assembly, as would occur in an identical bulk material.

TABLE 8.10 **Properties and Characteristics of Some Common Reinforcing Fibers**

Fiber Material	Specific Strength[a] (10^6 in.)	Specific Stiffness[b] (10^6 in.)	Density (lb/in.^{3})	Melting Temperature[c] (°F)
Al_2O_3 whiskers	21.0	434	0.142	3600
Boron	4.7	647	0.085	3690
Ceramic fiber (mullite)	1.1	200	0.110	5430
E-type glass	5.6	114	0.092	<3140
High-strength graphite	7.4	742	0.054	6690
High-modulus graphite	5.0	1430	0.054	6690
Kevlar	10.1	347	0.052	—
SiC whiskers	26.2	608	0.114	4890

[a]Strength divided by density.
[b]Elastic modulus divided by density.
[c]Or maximum temperature of use.

Of Interest: Use of Renewable Natural Materials

Ford Motor Company used a composite with 20% wheat straw in a polypropylene matrix to produce two internal storage bins used in a recent crossover vehicle and is considering use in center console bins and trays, door trim panels, and other interior components. The wheat straw acts as a weight-reducing filler and simultaneously increases stiffness. Rice hulls replaced a talc-based reinforcement in a polypropylene composite electrical harness used on the 2014 Ford F-150 pickup. Processed corn cobs and sunflower hull fibers have also been employed as polymer reinforcements.

for a continuous matrix. Long, thin fibers (higher aspect ratio) provide greater strength, and a strong bond is usually desired between the fiber and the matrix.

Polymer-Matrix Fiber-Reinforced Composites

Polymer-matrix composites comprise more than 90% of all composite materials. The matrix materials should be strong, tough, and ductile so that they can transmit the loads to the fibers and prevent cracks from propagating through the composite. In addition, the matrix material is often responsible for providing the electrical properties, chemical behavior, and elevated-temperature stability. For polymer-matrix composites, both thermosetting and thermoplastic resins have been used. The thermosets provide high strength and high stiffness, and the low-viscosity, uncured liquid resins readily impregnate the fibers. Popular thermosets include polyesters and vinyls, epoxies, bismaleimides, and polyimides. Although epoxies are the most common thermosets for high-strength use, they are often limited to temperatures of 120°C (250°F). Bismaleimides extend this range to 170°–230°C (350°–400°F), and the polyimides can be used up to 260°–320°C (500°–600°F).

Bulk-molding compound (BMC) is a viscous, putty-like material, consisting of thermosetting resin, 5%–50% discontinuous fiber reinforcement consisting of chopped strand (usually glass) 0.75 to 12.5 mm (⅟₃₂ to ½ in.) in length, and various additives. It can be shaped by various thermoset molding operations to produce parts with good dimensional precision, good mechanical properties, high strength-to-weight ratio (up to

one-third lighter than metal), integral color, and surfaces receptive to powder coating, painting, or other coating techniques.

Sheet-molding compound (SMC) is the sheet equivalent of BMC with 10%–65% reinforcement fibers that range from 12.5 to 50 mm (½ to 2 in.) in length. Parts ranging from 30 to 150 cm (12 to 60 in.) in width can be formed from the various thermoset sheet-forming techniques.

From a manufacturing viewpoint, it might be easier and faster to heat and cool a thermoplastic than to cure a thermoset. Moreover, the thermoplastics are tougher, more tolerant to damage, and often recyclable or remoldable. Polyethylene, polystyrene, and nylon are traditional thermoplastic matrix materials. Improved high-temperature and chemical-resistant properties can be achieved with the thermoplastic polyimides, polyphenylene sulfide (PPS), polyether ether ketone (PEEK), and the liquid-crystal polymers. When reinforced with high-strength, high-modulus fibers, these materials can show dramatic improvements in strength, stiffness, toughness, and dimensional stability compared to traditional thermoplastics.

Advanced Fiber-Reinforced Composites

Advanced composites are materials that have been developed for applications requiring exceptional combinations of strength, stiffness, and light weight. Fiber content generally exceeds 50% (by weight), and the modulus of elasticity is typically greater than 16×10^6 psi. Superior creep and fatigue resistance, low thermal expansion, low friction and wear, vibration-damping characteristics, and environmental stability are other properties that might also be required in these materials.

There are four basic types of advanced composites where the matrix material is matched to the fiber and the conditions of application:

1. The advanced **polymer-matrix composites** frequently use high-strength, high-modulus fibers of graphite (carbon), aramid (Kevlar), or boron, with epoxy, bismaleimide, or polyimide matrix to create a high-strength, lightweight, fatigue-resistant material. Properties can be put in desired locations or orientations at about one-half the weight of aluminum or one-fifth that of steel. Thermal expansion can be designed to be low or even negative, and the fibers frequently provide or enhance electrical and thermal conductivity. Unfortunately, these materials have a maximum service temperature of about 315°C (600°F) because the polymer matrix loses strength when heated. **Table 8.11** compares the properties of some of the common resin-matrix composites with those of several of the lightweight or low-thermal-expansion metals. Typical applications include sporting equipment (tennis rackets, skis, golf clubs, and fishing poles), lightweight armor plate, and myriad low-temperature automotive and aerospace components.

2. **Metal-matrix composites (MMCs)** can be used for operating temperatures up to 1250°C (2300°F), where the conditions require high strength, high stiffness, good electrical and/or thermal conductivity, exceptional wear resistance, and good ductility and toughness. The ductile matrix material can be aluminum, copper, magnesium, titanium, nickel, superalloy, or even intermetallic compound, and the reinforcing fibers can be graphite, boron carbide, alumina, or silicon carbide. Fine whiskers (tiny needlelike single crystals of 1 to 10 μm in diameter) of sapphire, silicon carbide, and silicon nitride have also been used as the reinforcement, as well as wires of titanium, tungsten, molybdenum, beryllium, and stainless steel. The reinforcing fibers can be either continuous or discontinuous and typically comprise between 10% and 60% of the composite by volume. Reinforcement

can also be added to only select regions of a product, optimizing material usage and reducing cost.

Compared to the engineering metals, these composites offer higher stiffness and strength (especially at elevated temperatures); a lower coefficient of thermal expansion; better elevated temperature properties; and enhanced resistance to fatigue, abrasion, and wear. Compared to the polymer-matrix composites, they offer higher heat resistance, as well as improved electrical and thermal conductivity. They are nonflammable, do not absorb water or gases, and are corrosion resistant to fuels and solvents. Unfortunately, these materials are quite expensive and are often difficult to machine. The vastly different thermal expansions of the components can lead to debonding, and the assemblies might be prone to degradation through interdiffusion or galvanic corrosion.

Graphite-reinforced aluminum has been shown to be twice as stiff as steel at one-third to one-fourth the weight, with near-zero thermal expansion in the fiber direction. Aluminum oxide-reinforced aluminum has been used in automotive connecting rods to provide stiffness and fatigue resistance with lighter weight. Aluminum reinforced with silicon carbide exhibits increased strength and stiffness at both room and elevated temperatures and has been fabricated into automotive drive shafts, cylinder liners, and brake drums, as well as aircraft wing panels, all offering significant weight savings. Thermal expansion is reduced, but so is ductility and both thermal and electrical conductivity. Fiber-reinforced superalloys might well become a preferred material for applications such as turbine blades.

Table 8.12 presents some of the properties of fiber-reinforced metal-matrix composites. Note the significant difference between the properties measured along the fiber direction and those measured across the fibers.

3. **Ceramic-matrix composites (CMCs)** offer light weight, high-temperature strength and stiffness, wear resistance, and good dimensional and environmental stability. The matrix

TABLE 8.11	Properties of Several Polymer-Matrix Fiber-Reinforced Composites (in the Fiber Direction) Compared to Lightweight or Low-Thermal-Expansion Metals				
Material	Specific Strength[a] (10^6 in.)	Specific Stiffness[b] (10^6 in.)	Density (lb/in.^{3})	Thermal Expansion Coefficient [in./(in.−°F)]	Thermal Conductivity [Btu/(hr−ft−°F)]
Boron–epoxy	3.3	457	0.07	2.2	1.1
Glass–epoxy (woven cloth)	0.7	45	0.065	6	0.1
Graphite–epoxy: high modulus (unidirectional)	2.1	700	0.063	−0.5	75
Graphite–epoxy: high strength (unidirectional)	5.4	400	0.056	−0.3	3
Kevlar–epoxy (woven cloth)	1	80	0.5	1	0.5
Aluminum	0.7	100	0.10	13	100
Beryllium	1.1	700	0.07	7.5	120
Invar[c]	0.2	70	0.29	1	6
Titanium	0.8	100	0.16	5	4

[a]Strength divided by density.
[b]Elastic modulus divided by density.
[c]A low-expansion metal containing 36% Ni and 64% Fe.

TABLE 8.12   **Properties of Some Fiber-Reinforced Metal-Matrix Composites***

Material	Orientation	Specific Str (10^6 in.)	Specific Stiffness (10^6 in.)	Density (lb/in.^{3})
Aluminum with 50 vol % boron fiber	Along fiber	2.26	312	0.096
	Across fiber	0.21	229	
Aluminum with 50 vol % silicon carbide fiber	Along fiber	0.35	437	0.103
	Across fiber	0.15	—	
Titanium-6Al-4V with 35 vol % silicon carbide fiber	Along fiber	1.83	316	0.139
	Across fiber	0.42	—	
Aluminum-Lithium with 60 vol % Al_2O_3 fiber	Along fiber	0.80	304	0.125
	Across fiber	0.22	176	
Magnesium alloy with 38 vol % graphite fiber	Along fiber	1.14	—	0.065
Aluminum with 30 vol % graphite fiber	Along fiber	1.12	258	0.089

*Note the variation of properties with fiber orientation.

provides high-temperature resistance. Glass matrices can operate at temperatures as high as 1500°C (2700°F). The crystalline ceramics, usually based on alumina, silicon carbide, silicon nitride, boron nitride, titanium diboride, or zirconia, can be used at even higher temperatures. The fibers add directional strength, increase fracture toughness, improve thermal shock resistance, and can be incorporated in unwoven, woven, knitted, and braided form. Typical reinforcements include carbon fiber, glass fiber, fibers of the various matrix materials, and ceramic whiskers. Composites with discontinuous fibers tend to be used primarily for wear applications, such as cutting tools, forming dies, and automotive parts such as valve guides and high-performance brake rotors. Other applications include lightweight armor plate and radomes. Continuous-fiber ceramic composites are used for applications involving the combination of high temperatures and high stresses and have been shown to fail in a noncatastrophic manner. Application examples include gas-turbine components, high-pressure heat exchangers, and high-temperature filters. Unfortunately, the cost of ceramic–ceramic composites ranges from high to extremely high, so applications are restricted to those where the benefits are quite attractive.

4. **Carbon–carbon composites** (graphite fibers in a graphite or carbon matrix) offer the possibility of a heat-resistant material that could operate at temperatures above 2000°C (3600°F), along with a strength that is 20 times that of conventional graphite, a density that is 30% lighter (1.38 g/cm³) and a low coefficient of thermal expansion. Not only does this material withstand high temperatures, but it also actually gets stronger when heated. Companion properties include good toughness, good thermal and electrical conductivity, and resistance to corrosion and abrasion. For temperatures higher than 540°C (1000°F), however, the composite requires some form of coating to protect it from oxidizing. Various coatings can be used for different temperature ranges. It was used on the nose cone and leading edges of the U.S. space shuttle. Current applications include aircraft and racing car disc brakes, automotive clutches, aerospace turbines and jet engine components, rocket nozzles, and surgical implants.

Hybrid Composites

Hybrid composites involve two or more different types of fibers in a common matrix. The particular combination of fibers is usually selected to balance strength and stiffness, provide dimensional stability, reduce cost, reduce weight, or improve fatigue and fracture resistance. Types of hybrid composites include (1) interply (alternating layers of fibers), (2) intraply (mixed strands in the same layer), (3) interply–intraply, (4) selected placement (where the costlier material is used only where needed), and (5) interply knitting (where plies of one fiber are stitched together with fibers of another type).

Design and Fabrication

The design of composite materials involves the selection of the component materials; the determination of the relative amounts of each component; the determination of size, shape, distribution, and orientation of the components; and the selection of an appropriate fabrication method. Many of the possible fabrication methods have been specifically developed for use with composite materials. For example, fibrous composites can be manufactured into useful shapes through compression molding, filament winding, pultrusion (where bundles of coated fibers are drawn through a heated die), cloth lamination, and autoclave curing (where pressure and elevated temperature are applied simultaneously). A variety of fiber-containing thermoset resins premixed with fillers and additives (bulk-molding compounds) can be shaped and cured by compression, transfer, and injection molding to produce three-dimensional fiber-reinforced products for numerous applications. Sheets of glass-fiber-reinforced thermoset resin, again with fillers and additives, (sheet-molding compound) can be press formed to provide lightweight, corrosion-resistant products that are similar to those made from sheet metal. The reinforcing fibers can be short and random, directionally oriented, or fully continuous in a specified direction. With a wide spectrum of materials, geometries, and processes, it is now possible to tailor a composite material product for a specific application. As one example, consider the cargo beds for pickup trucks, where composite products offer reduced weight coupled with resistance to dents, scratches, and corrosion.

A significant portion of Chapter 20 is devoted to a more complete description of the fabrication methods that have been developed for composite materials.

Assets and Limitations

Figure 8.13 graphically presents the strength-to-weight ratios of various aerospace materials as a function of temperature. The superiority of the various advanced composites over the conventional aerospace metals is clearly evident. The weight of a graphite–epoxy composite I-beam is less than one-fifth that of steel, one-third that of titanium, and one-half that of aluminum. Its ultimate tensile strength equals or exceeds that of the other three materials, and it possesses an almost infinite fatigue life. The greatest limitations of this and other composites are their relative brittleness and the high cost of both materials and fabrication.

Although there has been considerable advancement in the field, composite materials remain relatively expensive, and manufacturing with composites can still be quite labor intensive and involve long cycle times. It is often difficult to predict the interfacial bond strength, the strength of the composite and its response to impacts, and the probable modes of failure. Reliable methods of quality control and inspection are not well established. Defects can involve delaminations, voids, missing layers, contamination, fiber breakage, and (hard-to-detect) improperly cured resin. There is often concern about heat resistance. Many composites with polymeric matrices are sensitive to moisture, acids, chlorides, organic solvents, oils, and ultraviolet radiation and tend to cure forever, causing continually changing properties. In addition, most composites have limited ability to be repaired if damaged, preventive maintenance procedures are not well established, and recycling is often extremely difficult. Assembly operations with composites generally require the use of industrial adhesives.

On the positive side, the availability of a corrosion-resistant material with strength and stiffness greater than those of steel at only one-fifth the weight might be sufficient to justify some engineering compromises. Reinforcement fibers can be oriented in the direction of maximum stiffness and strength. In addition, products can often be designed to significantly reduce the number of parts (part consolidation), number of fasteners, assembly time, and both tooling cost and overall cost. As an example, Adam Aircraft manufactures two styles of high-performance aircraft using carbon-fiber composites. Their planes have 240 structural parts in comparison to more than 5,000 for traditional aluminum aircraft.

Areas of Application

Many composite materials are stronger than steel, lighter than aluminum, and stiffer than titanium. They can also possess tailored electrical and thermal conductivity, good heat resistance, good fatigue life, low corrosion rates, and adequate wear resistance. For these reasons they have become well established in several areas, often ones involving relatively low production rates or small production quantities.

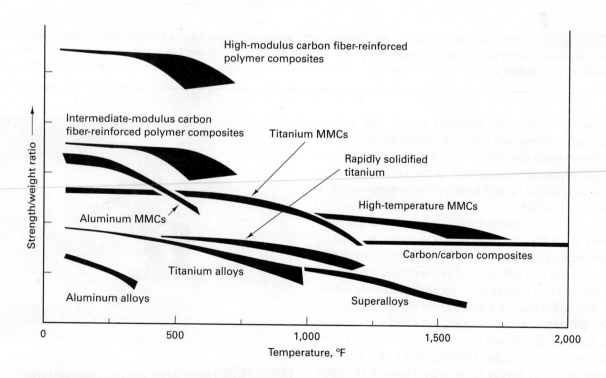

FIGURE 8.13 The strength/weight ratio of various aerospace materials as a function of temperature. Note the superiority of the various fiber-reinforced composites. (Adapted with permission of DuPont Company, Wilmington, DE)

Example: Composite Usage in Aerospace

Figure 8.14 shows a schematic of the F-22 Raptor fighter airplane. Traditional materials, such as aluminum and steel, make up only about 20% of the F-22 structure by weight. Its higher speed, longer range, greater agility, and reduced detectability are made possible through the use of 42% titanium and 24% composite material. In 1977, only 12% of the weight of the Boeing 777 was composites (50% was aluminum). Eighteen years later, Boeing's 787 Dreamliner, a 200-seat intercontinental commercial airliner, reduced the use of aluminum to 20% and increased the amount of composites to 50%. The majority of its primary structure, including wings and fuselage, is made of polymer-matrix (carbon/epoxy) composites, and a titanium/graphite composite is used in the wings. As a result, the aircraft will require 15%–20% less fuel than previous wide-body planes. The Airbus Industries' A380 wide-body plane marked the introduction of a new composite material, known as glass-reinforced aluminum or GLARE, a laminate composite of alternating layers of aluminum and glass prepreg. This material, which enabled a 25% reduction in the weight of fuselage skin, is more fatigue-resistant than aluminum and less expensive that a full composite. Composites are also being used in the unmanned air vehicles (drones) and single-engine piston aircraft.

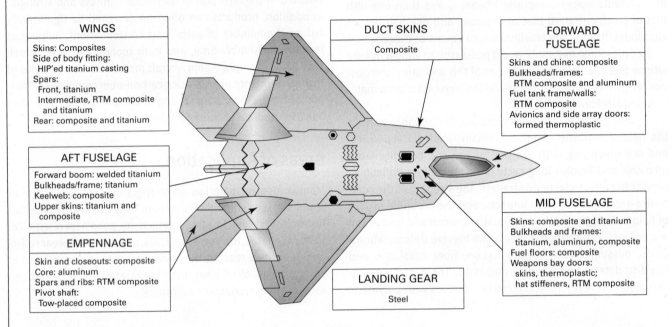

WINGS

Skins: Composites
Side of body fitting:
  HIP'ed titanium casting
Spars:
  Front, titanium
  Intermediate, RTM composite
    and titanium
Rear: composite and titanium

AFT FUSELAGE

Forward boom: welded titanium
Bulkheads/frame: titanium
Keelweb: composite
Upper skins: titanium and
  composite

EMPENNAGE

Skin and closeouts: composite
Core: aluminum
Spars and ribs: RTM composite
Pivot shaft:
  Tow-placed composite

DUCT SKINS

Composite

LANDING GEAR

Steel

FORWARD FUSELAGE

Skins and chine: composite
Bulkheads/frames:
  RTM composite and aluminum
Fuel tank frame/walls:
  RTM composite
Avionics and side array doors:
  formed thermoplastic

MID FUSELAGE

Skins: composite and titanium
Bulkheads and frames:
  titanium, aluminum, composite
Fuel floors: composite
Weapons bay doors:
  skins, thermoplastic;
  hat stiffeners, RTM composite

FIGURE 8.14 Schematic diagram showing the materials used in the various sections of the F-22 Raptor fighter airplane. Traditional materials, such as aluminum and steel, comprise only 20 wt%. Titanium accounts for 42%, and 24% is composite material. The plane is capable of flying at Mach 2. (*Note:* RTM is resin-transfer molding.) (Reprinted with permission of ASM International, Materials Park, OH.)

Aerospace applications frequently require light weight, high strength, stiffness, and fatigue resistance. For the past 40 years, almost every new aircraft design has introduced more composite structural materials than its predecessor. Composites currently account for 30% to 60% of the airframe weight of new fixed-wing aircraft.

Sports are highly competitive, and fractions of a second or tenths of a millimeter often decide victories. As a result, both professionals and amateurs are willing to invest in athletic equipment that will improve performance. The materials of choice have evolved from naturally occurring wood, twine, gut, and rubber to a wide variety of high-technology metals, polymers, ceramics, and composites. Golf club shafts, baseball bats, fishing rods, archery bows, tennis rackets, bicycle frames, skis, and snowboards are now available in a wide variety of fibrous composites. **Figure 8.15** shows an example.

In addition to body panels, automotive uses of composite materials include drive shafts, springs, and bumpers. Weight

FIGURE 8.15 Composite materials are often used in sporting goods, like this snowboard, to improve performance through light weight, high stiffness, and high strength, and also to provide attractive styling. (VP Photo Studio/Shutterstock)

savings compared to existing parts is generally 20% to 25%. Truck manufacturers now use fiber-reinforced composites for cab shells and bodies, oil pans, fan shrouds, instrument panels, and engine covers. Some pickup trucks now have cargo boxes formed from sheet-molding compound.

Other applications of composites include such diverse products as boat hulls, bathroom shower and tub structures, chairs, architectural panels, agricultural tanks and containers, pipes and vessels for the chemical industry, and external housings for a variety of consumer and industrial products.

Review Questions

1. Discuss the terms *plastics* and *polymers*. In what sense are they interchangeable, and in what sense are they not?

2. What are some naturally occurring nonmetallic materials that have been used for engineering applications?

3. What are some material families that would be classified under the general term *nonmetallic engineering materials*?

4. How might plastics be defined from the viewpoints of chemistry, structure, fabrication, and processing?

5. What is the primary type of atomic bonding within polymers?

6. What is the difference between a saturated and an unsaturated molecule?

7. What is an isomer?

8. Describe and differentiate the two means of forming polymers: addition polymerization and condensation polymerization.

9. What is a mer?

10. What is degree of polymerization?

11. How does increasing the chain length tend to affect the properties of a polymer?

12. In what way are copolymers similar to alloys in metals?

13. Describe and differentiate thermoplastic and thermosetting plastics.

14. Describe the mechanism by which thermoplastic polymers soften under heat and deform under pressure.

15. What does it mean when a polymer "crystallizes"? How is this different from the crystal structures observed in metals and ceramics?

16. What are some of the ways that a thermoplastic polymer can be made stronger?

17. What are the four most common thermoplastic polymers?

18. Why are thermosetting polymers characteristically brittle?

19. How do thermosetting polymers respond to subsequent heating?

20. Describe how thermoplastic or thermosetting characteristics affect productivity during the fabrication of a molded part.

21. What are some attractive engineering properties of polymeric materials?

22. What are some limiting properties of plastics, and in what general area do they fall?

23. What are some environmental conditions that might adversely affect the engineering properties plastics?

24. What is the most common polymer (accounting for 30% of global consumption)?

25. What are some reasons that additive agents are incorporated into plastics?

26. What are some functions of a filler material in a polymer?

27. What are some of the more common filler materials used in plastics?

28. What is the function of a plasticizer?

29. What is the difference between a dye and a pigment?

30. What is the role of a stabilizer or antioxidant?

31. How might electrical conductivity be imparted to a plastic?

32. What is an oriented plastic, and what is the primary engineering benefit?

33. What are some properties and characteristics of the "engineering plastics"?

34. Describe the use of plastic materials as adhesives. Give an example of a thermoplastic, thermoset, and elastomeric adhesive.

35. What are some potential benefits of using plastics in tooling applications?

36. Describe some of the applications for foamed plastics.

37. Provide some examples where plastics have competed with or replaced other materials.

38. What are some features of plastics that make them attractive from a manufacturing perspective?

39. In a cost comparison, why might cost per unit volume be a more valid figure than cost per unit weight?

40. How has the use of plastics grown in the automotive industry?

41. What kinds of plastics are most easily recycled?

42. Why is the recycling of mixed plastics more difficult than the recycling of mixed metals?

43. What are some recycling alternatives for thermosetting polymers?

44. What are some of the natural materials used to produce bioplastics?

45. What are some of the approaches to producing a biodegradable plastic?

46. What is the unique mechanical property of elastomeric materials, and what structural feature is responsible for it?

47. How can cross-linking be used to control the engineering properties of elastomers?

48. What is the cause of stress relaxation in elastomers?

49. What are some of the materials that can be added to natural rubber, and for what purpose?

50. What are some of the limitations of natural rubber?

51. What are some of the common artificial elastomers?

52. What are some of the attractive properties of the silicone rubbers?

53. What factors should be considered when selecting an elastomer for a specific application?

54. What are some outstanding physical properties of ceramic materials?

55. Which class of ceramic material accounts for nearly half of the ceramic materials market?

56. Why are the crystal structures of ceramics frequently more complex than those observed for metals?

57. What is the common name given for ceramic material in the noncrystalline, or amorphous, state?

58. What are some of the ways that toughness can be imparted to ceramic materials?

59. How does one produce tempered glass?

60. What kinds of ceramic products are classified as whiteware?

61. What is the dominant characteristic of refractory ceramics?

62. What are some common applications of refractory ceramics?

63. What is the dominant property of ceramic abrasives?

64. What is the unique feature of piezoelectric ceramics?

65. How are glass products formed or shaped?

66. What are some of the specialty applications of glass?

67. How is a glass ceramic different from a glass?

68. What are cermets, and what properties or combination of properties do they offer?

69. What are some of the attractive features of a titanium nitride coating?

70. Why do most ceramic materials fail to possess their theoretically high tensile strength?

71. Why do the mechanical properties of ceramics generally show a wider statistical spread than the same properties of metals?

72. If all significant flaws or defects could be eliminated from the structural ceramics, what properties might be present and what features might still limit their possible applications?

73. What properties are characteristic of the advanced ceramics?

74. What are some attractive and limiting properties of sialon (one of the advanced ceramics)?

75. What are some potential applications of the advanced ceramics? What is limiting their use?

76. What are some ceramic materials that are currently being used for cutting-tool applications, and what features or properties make them attractive?

77. What properties are desired in ultra-high-temperature ceramics, and what are some possible applications?

78. What is a composite material?

79. What are the basic features of a composite material that influence and determine its properties?

80. What are the three primary geometries of composite materials?

81. Give several examples of laminar composites.

82. What feature in a bimetallic strip makes its shape sensitive to temperature?

83. What are some reasons for creating clad composites?

84. What is the attractive aspect of the strength that is induced by the particles in a dispersion-strengthened particulate composite material?

85. Provide some examples of true particulate composites.

86. Which of the three primary composite geometries is most likely to possess isotropic properties?

87. What are the primary roles of the matrix in a fiber-reinforced composite? Of the fibers?

88. What are some of the more popular fiber materials used in fiber-reinforced composite materials?

89. What is specific strength? Specific stiffness?

90. What are some possible fiber orientations or arrangements in a fiber-reinforced composite material?

91. What are some features that influence the properties of fiber-reinforced composites?

92. What is the most common matrix material used in fiber-reinforced composites?

93. What is bulk-molding compound? Sheet-molding compound?

94. What are the attractive features of a thermosetting polymer matrix in a polymer-matrix fiber-reinforced composite? A thermoplastic matrix?

95. What are "advanced composites"?

96. What feature often limits the use of polymer-matrix composites?

97. In what ways are metal-matrix composites superior to straight engineering metals? To polymer-matrix composites?

98. Consider the data in Table 8.12. How significant are the property variations between directions along the fiber and across the fiber?

99. What features might be imparted by the fibers in a ceramic-matrix composite?

100. What are the primary application conditions for the carbon–carbon composites?

101. What are hybrid composites?

102. Compare the weight of a graphite-epoxy composite to steel, titanium, and aluminum.

103. What are some of the limitations that might restrict the use of composite materials in engineering applications?

104. The use of composite materials is generally favored by what level of production rate and production quantity?

105. Give some examples of the use of composite materials in sporting equipment.

Problems

1. a. One of Leonardo da Vinci's sketchbooks contains a crude sketch of an underwater boat (or submarine). Leonardo did not attempt to develop or refine this sketch further, possibly because he recognized that the engineering materials of his day (wood, stone, and leather) were inadequate for the task. What properties would be required for the body of a submersible vehicle? What materials might you consider?

b. Another of Leonardo's sketches bears a crude resemblance to a helicopter—a flying machine. What properties would be desirable in a material that would be used for this type of application?

c. Try to identify a possible engineering product that would require a material with properties that do not exist among today's engineering materials. For your application, what are the demanding features or requirements? If a material were to be developed for this application, from what family or group do you think it would emerge? Why?

2. Select a product (or component of a product) that can reasonably be made from materials from two or more of the basic materials families (metals, polymers, ceramics, and composites).

a. Describe briefly the function of the product or component.

b. What properties would be required for this product or component to perform its function?

c. What two materials groups might provide reasonable candidates for your product?

d. Select a candidate material from the first of your two families and describe its characteristics. In what ways does it meet your requirements? How might it fall short of the needs?

e. Repeat part (d) for a candidate material from the second material family.

f. Compare the two materials to one another. Which of the two would you prefer? Why?

3. Many of the materials presented in this chapter are lighter than steel and become prime candidates for weight reduction in automobiles. Select several automotive components where these materials can be used, describe the function of these components, and identify key properties (in addition to light weight) that must be present for these components to function.

4. Coatings have been applied to cutting tool materials since the early 1970s. Desirable properties of these coatings include high-temperature stability, chemical stability, low coefficient of friction, high hardness for edge retention, and good resistance to abrasive wear. Consider the ceramic material coatings of titanium carbide (TiC), titanium nitride (TiN), and aluminum oxide (Al_2O_3), and compare them with respect to the conditions required for deposition and the performance of the resulting coatings.

5. Ceramic engines continue to constitute an area of considerable interest and are frequently discussed in the popular literature. If perfected, they would allow higher operating temperatures with a companion increase in engine efficiency. In addition, they would lower sliding friction and permit the elimination of radiators, fan belts, cooling system pumps, coolant lines, and coolant. The net result would be reduced weight and a more compact design. Estimated fuel savings could amount to 30% or more.

a. What are the primary limitations to the successful manufacture of such a product?

b. What types of ceramic materials would you consider to be appropriate?

c. What methods of fabrication could produce a product of the required size and shape?

d. What types of special material properties or special processing might be required?

6. Material recyclability has become an important requirement in many manufactured products.

a. Consider each of the four major materials groups (metals, polymers, ceramics, and composites) and evaluate each for recyclability. What properties or characteristics tend to limit or restrict recyclability?

b. Which materials within each group are currently being recycled in large or reasonable quantities?

c. Europe has recently legislated extensive recycling of automobiles and electronic products. How might this legislation change the material make-up of these products?

d. Consider a typical family automobile and discuss how the factors of (1) recyclability, (2) fuel economy, and (3) energy required to produce materials and convert them to products might favor or oppose various candidate engineering materials.

7. Biodegradable plastics are attractive from an environmental perspective. Identify one or more applications where the biodegradable material is currently being used. Are there additional advantages to the use of this material? Are there any cons or compromises that must be made?

8. Locate a current or recent article describing an advance in polymeric, ceramic, or composite materials. Describe that advance, and briefly discuss possible applications.

CHAPTER 9

Material Selection

9.1 Introduction

The objective of manufacturing operations is to make products or components that adequately perform their intended task. Meeting this objective implies the manufacture of components from selected engineering materials, with the required geometrical shape and precision, and with companion material structures and properties that are optimized for the intended service environment. The ideal product is one that will just meet all requirements. Anything better will usually incur added cost through excess or higher-grade materials, enhanced processing, or improved properties that may not be necessary. Anything worse will likely cause product failure, dissatisfied customers, and the possibility of unemployment.

It was not that long ago that each of the materials groups had its own well-defined uses and markets. Metals were specified when strength, toughness, and durability were the primary requirements. Ceramics were generally limited to low-value applications where heat or chemical resistance was required and any loadings were compressive. Glass was used for its optical transparency, and plastics were relegated to low-value applications where low cost and light weight were attractive features, and performance properties were secondary.

Such clear delineations no longer exist. Many of the metal alloys in use today did not exist as little as 30 years ago, and the common alloys that have been in use for a century or more have been much improved due to advances in metallurgy and production processes. New on the scene are amorphous metals, dispersion-strengthened alloys produced by powder metallurgy, mechanical alloyed products, and directionally solidified materials. Ceramics, polymers, and composites are now available with specific properties that often transcend the traditional limits and boundaries. Advanced structural materials offer higher strength and stiffness, strength at elevated temperature, light weight, and resistance to corrosion, creep, and fatigue. Still other materials offer enhanced thermal, electrical, optical, magnetic, and chemical properties.

To the inexperienced individual, "wood is wood," but to the carpenter or craftsman, oak is best for one application, whereas maple excels for another, and yellow pine is preferred

for a third. The ninth edition of *Woldman's Engineering Alloys*[1] includes over 56,000 metal alloys, and that doesn't consider polymers, ceramics, or composites. Even if we eliminate the obsolete and obscure, we are still left with tens of thousands of options from which to select the "right" or "best" material for the task at hand.

Unfortunately, the availability of so many alternatives has often led to poor materials selection. Money can be wasted in the unnecessary specification of an expensive alloy or one that is difficult to fabricate. At other times, these materials may be absolutely necessary, and selection of a cheaper alloy would mean certain failure. It is the responsibility of the design and manufacturing engineer, therefore, to be knowledgeable in the area of engineering materials and to be able to make the best selection among the numerous alternatives.

In addition, it is also important that the **material selection** process must be one of constant reevaluation. New materials are continually being developed, others may no longer be available, and prices are always subject to change. Concerns regarding environmental pollution, recycling, and worker health and safety may impose new constraints. Desires for weight reduction, energy savings, or improved corrosion resistance may well motivate a change in engineering material. Pressures from domestic and foreign competition, increased demand for quality and serviceability, or negative customer feedback can all prompt a reevaluation. Finally, the proliferation of manufacturer recalls and product liability actions, many of which are the result of improper material use, has further emphasized the need for constant reevaluation of the engineering materials in a product.

The automotive industry alone consumes approximately 60 million metric tons of engineering materials worldwide every year—including substantial amounts of steel, aluminum, cast iron, copper, glass, lead, polymers, rubber, and zinc. In recent years, the shift toward lighter, more fuel-efficient, or alternative-powered vehicles has led to an increase in the use of the lightweight metals, high-strength steels, plastics, and composites.

A million metric tons of engineering materials go into aerospace applications every year. The principal materials tend to be aluminum, magnesium, titanium, superalloys,

---

[1] *Woldman's Engineering Alloys*, 9th ed., edited by J. Frick, ASM International, Metals Park, Ohio, 2000.

polymers, rubber, steel, metal-matrix composites, and polymer-matrix composites. Competition is intense, and materials substitutions are frequent. The use of advanced composite materials in aircraft construction has risen from less than 2% in 1970 to the point where they now account for one-quarter of the weight of the U.S. Air Force's Advanced Tactical Fighter and are already being used in the main fuselage of commercial planes. Titanium is used extensively for applications that include the exterior skins surrounding the engines, as well as the engine frames. The cutaway section of the Rolls Royce Trent 900 jet engine in **Figure 9.1a** reveals numerous components—each with their own characteristic shape, precision, stresses, and operating temperatures—that require a variety of engineering materials. **Figure 9.1b** shows an actual engine in a manner that reveals both its size and complexity. The intake fan diameter is nearly 3 meters in diameter (9 ft. 8 in.).

The earliest two-wheeled bicycle frames were constructed of wood, with various methods and materials employed at the joints. Then, for nearly a century, the requirements of yield strength, stiffness, and acceptable weight were met by steel tubing, either low-carbon, plain-carbon, or thinner-walled, higher-strength chrome-moly steel tubing (such as 4130), with either a brazed or welded assembly. In the 1970s a full-circle

occurred. Where a pair of bicycle builders (the Wright brothers) pioneered aerospace, the aerospace industry returned to revolutionize bicycles. Lightweight frames were constructed from the aerospace materials of high-strength aluminum, titanium, graphite-reinforced polymer, and even beryllium. Wall thickness and cross-section profiles were often modified to provide strength and rigidity. Materials paralleled function as bicycles specialized into road bikes, high-durability mountain bikes, and ultralight racing bikes. Further building on the aerospace experience, the century-old tubular frame has recently been surpassed by one-piece monocoque frames of either die-cast magnesium or continually wound carbon-fiber epoxy tapes with or without selective metal reinforcements. One top-of-the-line carbon fiber frame now weighs only 2.5 pounds! **Figure 9.2** compares a traditional tubular frame with one of the newer designs.

Window frames were once made almost exclusively from wood. Although wood remains a competitive material, a trip to any building supply will reveal a selection that includes anodized aluminum in a range of colors, as well as frames made from colored vinyl and other polymers. Each has its companion advantages and limitations. Auto bodies have been traditionally fabricated from steel sheet and assembled by resistance

(a)

(b)

FIGURE 9.1 (a) Cutaway drawing showing the internal design of a Rolls-Royce jet engine. Notice the number and intricacy of the components. Ambient air enters the front, and hot exhaust exits the back. A wide range of materials and processes will be used in its construction. (b) A full-size Rolls-Royce engine showing the size and complexity of the product. (Courtesy of Rolls-Royce Corporation, London, England)

(a)

(b)

FIGURE 9.2 (a) A traditional two-wheel bicycle frame (1970s vintage) made from joined segments of metal tubing (Courtesy Ronald Kohser). (b) A top-of-the-line (Tour de France or triathlon-type) bicycle with one-piece frame made from fiber-reinforced polymer-matrix composite. (Courtesy of Trek Bicycle Corporation, Waterloo, WI)

spot welding. Designers now select from steel, aluminum, and polymeric sheet molding compounds and may use adhesive bonding to produce the joints.

The listing of available engineering materials now includes metals and alloys, ceramics, plastics, elastomers, glasses, concrete, composite materials, and others. It is not surprising, therefore, that a single person might have difficulty making the necessary decisions concerning the materials in even a simple manufactured product. More frequently, the design engineer or design team will work in conjunction with various materials specialists to select the materials that will be needed to convert conceptual designs into tomorrow's reality.

9.2 Material Selection and Manufacturing Processes

The interdependence between materials and their processing must also be recognized. New processes frequently accompany new materials, and their implementation can often cut production costs and improve product quality. A change in material may well require a change in the manufacturing process. Conversely, improvements in processes may enable a reevaluation of the materials being processed. Improper processing of a well-chosen material can definitely result in a defective product. If satisfactory products are to be made, considerable care must be exercised in selecting *both* the **engineering materials** and the **manufacturing processes** used to produce the product.

Most textbooks on materials and manufacturing processes spend considerable time discussing the interrelationships between the structure and properties of engineering materials,

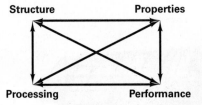

FIGURE 9.3 Schematic showing the interrelation between material, properties, processing, and performance.

the processes used to produce a product, and the subsequent performance. As **Figure 9.3** attempts to depict, each of these aspects is directly related to all the others. An engineering material may possess different properties depending on its structure. Processing of that material can alter the structure, which in turn will alter the properties. Altered properties certainly alter performance. The objective of manufacturing, therefore, is to devise an optimized system of material and processes to produce the desired product.

9.3 The Design Process

The first step in the manufacturing process is **design**—defining in considerable detail the objective or the target, that is, what it is that we want to produce. Design usually takes place in several distinct stages: (1) conceptual, (2) functional, and (3) production. During the **conceptual design** stage, the designer is primarily concerned with the functions that the product is to fulfill. Several concepts are often considered, and a determination is made that the concept is either not practical, or is sound and should be developed further. Here the only concern about materials is that materials exist that could provide the desired properties. If such materials are not available, consideration is given to whether there is a reasonable prospect that new ones could be developed within the limitations of cost and time.

At the **functional design**, or engineering, stage, a workable design is developed, including a detailed plan for manufacturing. Geometric features are determined, and dimensions are specified, along with allowable tolerances. In addition to the various functional or performance factors, consideration is given to appearance (styling and marketability), producibility, reliability and serviceability, possibly recyclability, and definitely cost. It is important to have a complete understanding of the functions and performance requirements of each component and to perform a thorough materials analysis, selection, and specification. If these decisions are deferred, they may end up being made by individuals who are less knowledgeable about all the functional aspects of the product.

Often, a **prototype** or working model is constructed to permit a full evaluation of the product. It is possible that the prototype evaluation will show that some changes have to be made in either the design or material before the product can be advanced to production. This should not be taken, however, as an excuse for not doing a thorough job. It is strongly recommended that all prototypes must be built with the same materials that will be used in production and, where possible, with the same manufacturing techniques. It is of little value to have a perfectly functioning prototype that cannot be manufactured economically in the desired volume, or one that is substantially different from the future production units.[2]

In the **production design** stage, we look to full production and determine if the proposed solution is compatible with production speeds and quantities. Can the parts be processed economically, and will they be of the desired quality? As actual manufacturing begins, changes in both the materials and processes may be suggested. Changes made after the tooling and machinery have been placed in production; however, they tend to be quite costly. Good upfront material selection and thorough product evaluation can do much to eliminate the need for change.

As production continues, the availability of new materials and new processes may present possibilities for cost reduction or improved performance. Before adopting new materials, however, the candidates should be evaluated very carefully to ensure that all their characteristics related to both processing and performance are well established. It is indeed rare that as much is known about the properties and reliability of a new material as an established one. Numerous product failures and product liability cases have resulted from new materials being utilized before their long-term properties were fully known.

[2]Because of the prohibitive cost of a dedicated die or pattern, as might be required for forging or casting, one-of-a-kind or limited quantity prototype parts are often made by machining or one of the newer direct-digital manufacturing processes. If the objective is simply to verify dimensional fit and interaction, the prototype material may be selected for compatibility with the prototype process. If performance is to be verified, however, it is best to use the proper material and process. Machining, for example, simply cuts through the material structure imparted in the manufacture of the starting bar or plate. Casting erases all prior structure during melting and establishes a new structure during solidification. Metal forming processes reorient the starting structure by plastic flow. The altered features caused by these processes may lead to altered performance.

9.4 Approaches to Material Selection

The selection of an appropriate material and its subsequent conversion into a useful product with desired shape and properties can be a rather complex process. As depicted in **Figure 9.4**, nearly every engineered item goes through a sequence of activities:

Design → Material selection → Process selection → Manufacture or fabrication → Evaluation → and Feedback

Manufacture involves both the production of the desired shape and the establishment of the required properties. Feedback can involve redesign or modification at one or more of the preceding steps. Numerous engineering decisions must be made at each of the steps, and several iterations may be necessary.

Several methods have been developed for approaching a design and selection problem. The **case-history method** is one of the simplest. Begin by evaluating what has been done in the past in terms of engineering material and method of manufacture, or what a competitor is currently doing. This can yield important information that will serve as a starting base. Then, either duplicate or modify the details of that solution. The basic assumption of this approach is that similar requirements can be met with similar solutions.

The case-history approach is quite useful, and many manufacturers continually examine and evaluate their competitors' products just for this purpose. The real issue here, however, is "how similar is similar." A minor variation in service requirement, such as a different operating temperature or a different corrosive environment, may be sufficient to justify a totally

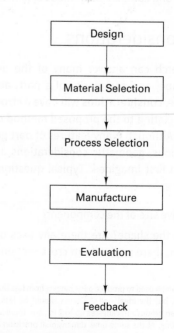

FIGURE 9.4 Sequential flowchart showing activities leading to the production of a part or product.

different material. Different production quantities may justify a different manufacturing method. In addition, this approach tends to preclude the use of new materials, new technology, and any manufacturing advances that may have occurred since the formulation of the original solution. It is equally unwise, however, to totally ignore the benefits and insights that can be gained through past experience.

Other design and selection activities may be related to the *modification of an existing product*, generally in an effort to reduce cost, improve quality, or overcome a problem or defect that has been encountered. A customer may have requested a product like the current one, but capable of operating at higher temperatures, or in an acidic environment, or at higher loads or pressures. Efforts here generally begin with an evaluation of the current product and its present method of manufacture. The most frequent pitfall, however, is to overlook one of the original design requirements and recommend a change that in some way compromises the total performance of the product. Examples of such oversights are provided in Section 9.8, where material substitutions have been recommended to meet specific objectives.

The safest and most comprehensive approach to part manufacture is to follow the full sequence of design, material selection, and process selection, considering all aspects and all alternatives. This is the approach one would take in the *development of an entirely new product*.

Before any decisions are made, take the time to fully define the needs of the product. What exactly is the "target" that we wish to hit? Develop a clear picture of all the characteristics necessary for this part to adequately perform its intended function, and do so with no prior biases about material or method of fabrication. These requirements will fall into three major areas: (1) shape or geometry considerations, (2) property requirements, and (3) manufacturing concerns. By first formulating these requirements, we will be in a better position to evaluate the candidate materials and companion methods of fabrication.

Geometric Considerations

A dimensioned sketch can answer many of the questions about the size, shape, and complexity of a part, and these **geometric**, or shape, **considerations** will have a strong influence on decisions relating to the proposed method or methods of fabrication.[3] Although many features of part geometry are somewhat obvious, geometric considerations are often more complex than first imagined. Typical questions might include the following:

1. What is the relative size of the component?
2. How complex is the shape? Are there any axes or planes of symmetry? Are there any uniform cross sections? Could the component be divided into several simpler shapes that might be easier to manufacture?
3. How many dimensions must be specified?
4. How precise must these dimensions be? Are all precise? How many are restrictive, and which ones?
5. How does this component interact geometrically with other components? Are there any restrictions imposed by the interaction?
6. What are the surface finish requirements? Must all surfaces be finished? Which ones do not?
7. How much can each dimension change by wear, corrosion, thermal expansion, or thermal contraction and the part still function adequately?
8. Could a minor change in part geometry increase the ease of manufacture or improve the performance (fracture resistance, fatigue resistance, etc.) of the part?

Producing the right shape is only part of the desired objective. If the part is to perform adequately, it must also possess the necessary **mechanical** and **physical properties**, as well as the ability to endure anticipated environments for a specified period of time. **Environmental considerations** should include all aspects of shipping, storage, and use! Some key questions include the following:

Mechanical Properties

1. How will the loads be applied, and what will be the magnitudes? Tension? Compression? Torsion? Shear?
2. How much static strength is required?
3. If the part is accidentally overloaded, is it permissible to have a sudden brittle fracture, or is plastic deformation and distortion a desirable or mandatory precursor to failure?
4. How much can the material bend, stretch, twist, or compress under load and still function properly?
5. Are any impact loadings anticipated? If so, of what type, magnitude, and velocity?
6. Can you envision vibrations or cyclic loadings? If so, of what type, magnitude, and frequency?
7. Is wear resistance desired? Where? How much? How deep?
8. Will all of the aforementioned requirements be needed over the entire range of operating temperature? If not, which properties are needed at the lowest extreme? At the highest extreme?

Physical Properties (Electrical, Magnetic, Thermal, and Optical)

1. Are there any electrical requirements? Conductor? Insulator?
2. Are any magnetic properties desired?
3. Are thermal properties significant? Thermal conductivity? Changes in dimension with change in temperature?

---

[3] Die casting, for example, can be used to produce parts ranging from less than an ounce to more than 100 pounds, but the ideal wall thickness should be less than $\frac{5}{16}$ in. Permanent mold casting can produce thicknesses up to 2 in., and there is no limit to the thickness for sand casting. At the same time, dimensional precision and surface finish become progressively worse as we move from die casting, to permanent mold, to sand. Extrusion and rolling can be used to produce long parts with constant cross section. Powder metallurgy parts must be able to be ejected from a compacting die.

Accuracy versus Precision in Processes

It is vitally important that the difference between accuracy and precision be understood. **Accuracy** refers to the ability to hit what is aimed at (the bull's-eye of the target). **Precision** refers to the repeatability of the process. Suppose that five sets of five shots are fired at a target from the same gun. **Figure 10.4** shows some of the possible outcomes. In **Figure 10.4a**, inspection of the target shows that this is a good process—accurate and precise. **Figure 10.4b** shows precision (repeatability) but poor accuracy. The agreement with a standard is not good. In **Figure 10.4c**, the process is on the average quite accurate, as the X (average) is right in the middle of the bull's-eye, but the process has too much scatter or variability; it does not repeat. Finally, in **Figure 10.4d**, a failure to repeat accuracy between samples with respect to time is observed; the process is not stable. These four outcomes are typical but not all-inclusive of what may be observed.

In Chapter 12 more discussion on accuracy and precision is presented as they relate to **process capability**. This term is used to describe how well a manufacturing process performs in its part making.

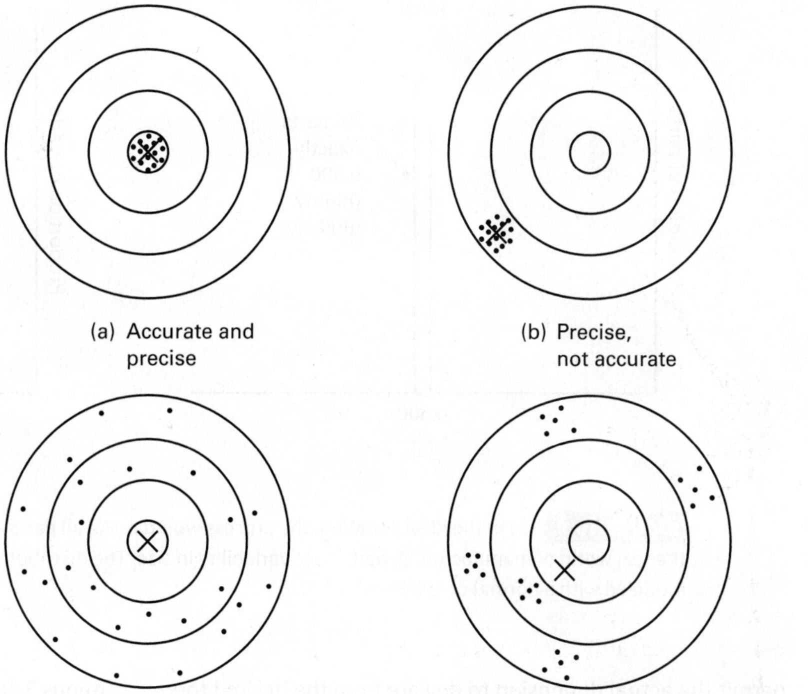

In measuring instruments used in the factory, precision (or **repeatability**) is critical because the devices must be very repeatable as well as accurate. For manually operated instruments, the skill of the operator must also be considered—this is called **reproducibility**.

So accuracy, repeatability, and reproducibility are characteristics of what is called **gage capability**, also discussed in Chapter 12.

(a) Accurate and precise

(b) Precise, not accurate

(c) Accurate, not precise

(d) Precise within sample
Not precise between samples
Not accurate overall or within sample

FIGURE 10.4 Accuracy versus precision. Dots in targets represent location of shots. Cross (✗) represents the location of the average position of all shots. (a) Accurate and precise; (b) precise, not accurate; (c) accurate, not precise; (d) precise within sample, not precise between samples, not accurate overall or within sample.

interference exists between the mating parts. In the case of a shaft and mating hole, it is the difference in diameters of the largest shaft and the smallest hole. With **clearance fits**, the largest shaft is smaller than the smallest hole, whereas with **interference fits**, the hole is smaller than the shaft.

Tolerance is an undesirable but permissible deviation from a desired dimension. There is variation in all processes, and no part can be made *exactly* to a specified dimension. Furthermore, high precision or repeatability in a process is neither necessary nor economical. Consequently, it is necessary to

10.3 Allowance and Tolerance

Most assembly operations depend on individual components interacting (mating) with one another. If the desired **fit** between mating parts is to be obtained, the designer must specify two factors, allowance and tolerance. **Allowance** is the intentional, desired difference between the dimensions of two mating parts. It is the difference between the dimension of the largest interior-fitting part (shaft) and that of the smallest exterior-fitting part (hole). **Figure 10.5** shows shaft A designed to fit into the hole in block B. This difference (0.5035–0.5025) thus determines the condition of *tightest* fit between mating parts. Allowance may be specified so that either *clearance* or

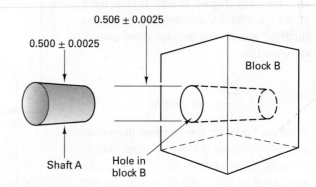

0.500 ± 0.0025

0.506 ± 0.0025

Block B

Shaft A

Hole in block B

FIGURE 10.5 When mating parts are designed, each shaft must be smaller than each hole for a clearance fit.

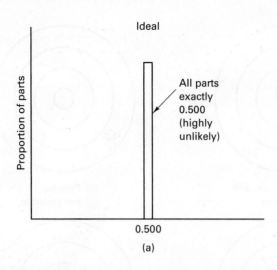

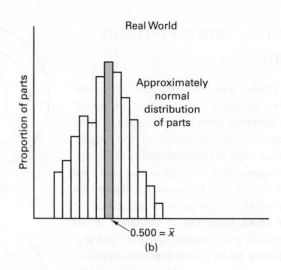

FIGURE 10.6 (a) In the ideal situation, the process would make all parts exactly the same size. (b) In the real world of manufacturing, parts have variability in size. The distribution of sizes can often be modeled with a normal distribution.

permit the actual dimension to deviate from the desired theoretical dimension (called the **nominal**) and to control the degree of deviation so that satisfactory functioning of the mating parts within the permissible deviation range will still be ensured.

Now we can see that the objective of inspection, by means of measurement techniques, is to provide feedback information on the actual size of the parts as manufactured compared to the size specified by the designer on the part drawing.

The manufacturing processes that make the shaft are different from those that make the hole, but both the hole and the shaft are subject to deviations in size because of variability in the processes and the materials. Thus, although the designer wishes ideally that all the shafts would be exactly 0.500 (**Figure 10.6a**) and all the holes 0.506, the reality of processing is that there will be deviations in size around these nominal or ideal sizes.

Most manufacturing processes result in products whose measurements of the geometrical features and sizes are distributed normally (**Figure 10.6b**). That is, most of the (0.5035–0.5025) measurements are clustered around the average dimension, $\bar{X}$ calculated as

$$\bar{X} = \frac{\sum_{n}^{n} X_i}{n} \quad \text{for } n \text{ items} \quad (10\text{-}1)$$

$\bar{X}$ will be equal to the nominal dimension only if the process is 100% accurate, that is, perfectly centered. More likely, parts will be distributed on either side of the average, and the process might be described (modeled) with a normal distribution. In normal distributions, as shown in **Figure 10.7**, 99.73% of the measurements (X_i) will fall within plus or minus 3 standard deviations ($\pm 3\sigma$) of the mean, 95.46% will be within $\pm 2\sigma$ and 68.26% will be within $\pm 1\sigma$ where σ is calculated by:

$$\sigma = \sqrt{\frac{\sum_{i=1}^{n}\left(X_i - \bar{X}\right)^2}{n}} \quad (10\text{-}2)$$

In summary, the designer applies nominal values to the mating parts according to the desired fit between the parts. Tolerances are added to those nominal values in recognition of the fact that all processes have some natural amount of variability.

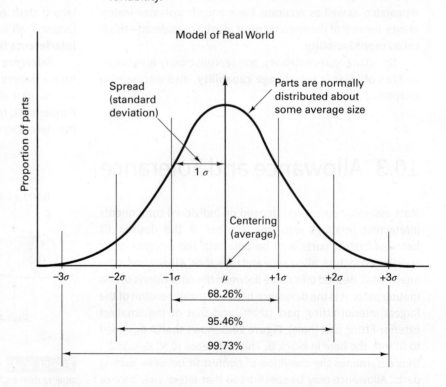

FIGURE 10.7 The normal distributional can be used to model many processes.

Assume that the data for both the hole and the shaft are normally distributed. The ($\pm 3\sigma$) added to the mean (μ) gives the **upper** and **lower natural tolerance limits**, defined as

$$\mu + 3\sigma = UNTL$$
$$\mu - 3\sigma = LNTL$$

As shown in **Figure 10.8a**, the average fit of two mating parts is equal to the difference between the mean of the shaft distribution and the mean of the hole distribution. The **range of fit** would be the difference between the minimum diameter shaft and the maximum diameter hole. The minimum clearance would be the difference between the smallest hole and the largest shaft. The tighter the distributions (i.e., the more precise the process), the better the fit between the parts.

During machining processes, the cutting tools wear and change size. If tool wear is considered, the diameter of the shaft will tend to get larger as the tool wears. However, the diameter of the hole in block B of Figure 10.5 decreases in size with tool wear. If no corrective action is taken, the means of the distributions will drift toward each other, and the fit will become increasingly tight (the clearance will be decreased). If the hole and the shaft distributions overlap by 2 standard deviations, as shown in **Figure 10.8b**, 6 parts out of every 10,000 could not be assembled. As the shaft and the hole distributions move closer together, more interference between parts will occur, and the fit will become tighter, eventually becoming an interference fit.

The designer must specify the tolerances according to the function of the mating parts. Suppose that the mating parts are the cap and the body of an ink pen. The cap must fit snugly but must be able to be easily removed by hand. A snug fit would be too tight for a dead bolt in a door lock. A snug fit is also too tight for a high-speed bearing for rotational parts but is not tight enough for permanently mounting a wheel on an axle. In the next section, the manner in which tolerances and allowances are specified will be introduced, but design engineers are expected to have a deeper understanding of this topic.

Specifying Tolerance And Allowances

Tolerance can be specified in four ways: bilateral, unilateral, limits, and geometrically. **Bilateral tolerance** is specified as a plus or minus deviation from the nominal size, such as

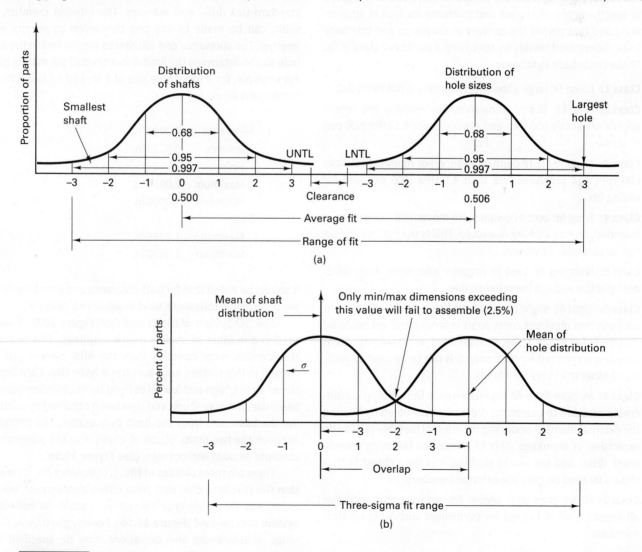

FIGURE 10.8 (a) The manner in which the distributions of the two mating parts interact determines the fit. UNTL (upper natural tolerance limit) = $\mu + 3\sigma$; LNTL (lower natural tolerance limit) = $\mu - 3\sigma$. (b) Shifting the means of the distributions toward each other results in some interface fits.

2.000 ± 0.002 in. Modern practice uses the **unilateral tolerance** system, where the deviation is in one direction from the basic size, such as

$$2.000 \text{ in.} \quad \begin{array}{l} +.004 \text{ in.} \\ -.000 \text{ in.} \end{array}$$

In the first case, that of bilateral tolerance, the dimension of the part could vary between 1.998 and 2.002 in., a total tolerance of 0.004 in. For the unilateral tolerance, the dimension could vary between 2.000 and 2.004 in., again a tolerance of 0.004 in. Obviously, to obtain the same maximum and minimum dimensions with the two systems, different basic sizes must be used. The maximum and minimum dimensions that result from application of the designated tolerance are called **limit dimensions**, or *limits*. (*Geometric* tolerances are discussed in the next section.)

There can be no rigid rules for the amount of clearance that should be provided between mating parts; the decision must be made by the designer, who considers how the parts are to function. The American National Standards Institute (ANSI) has established eight classes of fits that serve as a useful guide in specifying the allowance and tolerance for typical applications and that permit the amount of allowance and tolerance to be determined merely by specifying a particular class of fit. These classes are as follows:

Class 1: *Loose fit:* large allowance. Accuracy is not essential.

Class 2: *Free fit:* liberal allowance. For running fits where speeds are above 600 rpm and pressures are 4.1 MPa (600 psi) or above.

Class 3: *Medium fit:* medium allowance. For running fits under 600 rpm and pressures less than 4.1 MPa (600 psi) and for sliding fits.

Class 4: *Snug fit:* zero allowance. No movement under load is intended, and no shaking is wanted. This is the tightest fit that can be assembled by hand.

Class 5: *Wringing fit:* zero to negative allowance. Assemblies are selective and not interchangeable.

Class 6: *Tight fit:* slight negative allowance. An interference fit for parts that must not come apart in service and are not to be disassembled or are to be disassembled only seldomly. Light pressure is required for assembly. It is not to be used to withstand other than very light loads.

Class 7: *Medium-force fit:* an interference fit requiring considerable pressure to assemble; ordinarily assembled by heating the external member or cooling the internal member to provide expansion or shrinkage. This fit is used for fastening wheels, crank disks, and the like to shafting. It is the tightest fit that should be used on cast iron external members.

Class 8: *Heavy-force and shrink fits:* considerable negative allowance. This fit is used for permanent shrink fits on steel members.

The allowances and tolerances that are associated with the ANSI classes of fits are determined according to the theoretical relationship shown in **Table 10.3**. The actual resulting

TABLE 10.3 ANSI-recommended Allowances and Tolerances

Class of Fit	Allowance	Average Interference	Hole Tolerance	Shaft Tolerance
1	$0.0025\sqrt[3]{d^2}$	–	$0.0025\sqrt[3]{d}$	$0.0025\sqrt[3]{d}$
2	$0.0014\sqrt[3]{d^2}$	–	$0.0013\sqrt[3]{d}$	$0.0013\sqrt[3]{d}$
3	$0.0009\sqrt[3]{d^2}$	–	$0.0008\sqrt[3]{d}$	$0.0008\sqrt[3]{d}$
4	0	–	$0.0006\sqrt[3]{d}$	$0.0004\sqrt[3]{d}$
5	–	0	$0.0006\sqrt[3]{d}$	$0.0004\sqrt[3]{d}$
6	–	$0.0025d$	$0.0006\sqrt[3]{d}$	$0.0006\sqrt[3]{d}$
7	–	$0.0005d$	$0.0006\sqrt[3]{d}$	$0.0006\sqrt[3]{d}$
8	–	$0.001d$	$0.0006\sqrt[3]{d}$	$0.0006\sqrt[3]{d}$

dimensional values for a wide range of basic sizes can be found in tabulations in drafting and machine design books.

In the ANSI system, the hole size is always considered basic because the majority of holes is produced through the use of standard-size drills and reamers. The internal member, the shaft, can be made to any one dimension as readily as to another. The allowance and tolerances are applied to the basic hole size to determine the limit dimensions of the mating parts. For example, for a basic hole size of 2 in. and a Class 3 fit, the dimensions would be:

Allowance	0.0014 in.
Tolerance	0.0010 in.
Hole	
Maximum	2.0010 in.
Minimum	2.0000 in.
Shaft	
Maximum	1.9986 in.
Minimum	1.9976 in.

It should be noted that for both clearance and interference fits, the permissible tolerances tend to result in a looser fit.

The *ISO System of Limits and Fits* (**Figure 10.9**) is widely used in a number of leading metric countries. This system is considerably more complex than the ANSI system just discussed. In this system, each part has a basic size. Each limit of size of a part, high and low, is defined by its *deviation* from the basic size, the magnitude and sign being obtained by subtracting the basic size from the limit in question. The difference between the two limits of size of a part is called *tolerance*, an absolute amount without sign. (See **Figure 10.9a**).

There are three **classes of fits**: (1) *clearance fits*, (2) **transition fits** (the assembly may have either clearance or interference), and (3) *interference fits*. Either a **shaft-** or **hole-basis system** may be used (**Figure 10.9b**). For any given basic size, a range of tolerances and deviations may be specified with respect to the line of zero deviation, called the **zero line**. The tolerance is a function of the basic size and is designated by a number symbol, called the **grade** (e.g., the **tolerance grade**).

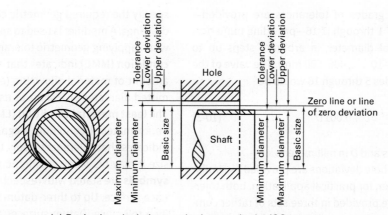

(a) Basic size, deviation, and tolerance in the ISO system.

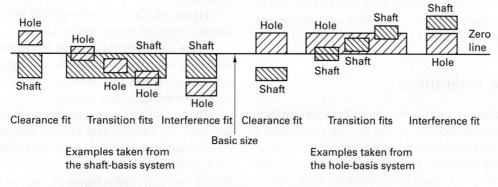

Clearance fit   Transition fits   Interference fit   Clearance fit   Transition fits   Interference fit

Examples taken from
the shaft-basis system

Examples taken from
the hole-basis system

(b) Shaft-basis and hole-basis system for specifying fits in the ISO system.

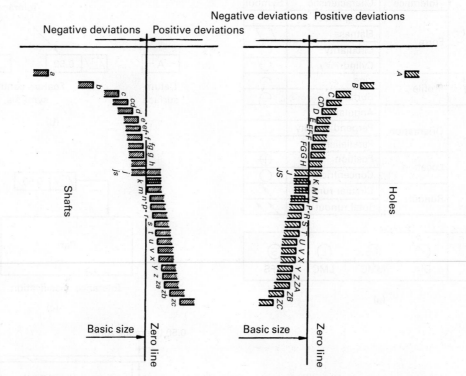

(c) Position of the various tolerance zones for a given diameter in the ISO system.

FIGURE 10.9   The ISO System of Limits and Fits. (By permission from Recommendations R286-1962, *System of Limits and Fits*, copyright 1962, American Standards Institute, NJ)

The **position** of the tolerance with respect to the zero line, also a function of the basic size, is indicated by a letter symbol (or two letters)—a capital letter for holes and a lowercase letter for shafts—as illustrated in **Figure 10.9c**. Thus, the specification for a hole and a shaft having a basic size of 45 mm might be 45 H8/g7.

Eighteen standard grades of tolerances are provided—called IT 01, IT 0, and IT 1 through IT 16—providing numerical values for each nominal diameter, in arbitrary steps up to 500 mm (i.e., 0–3, 3–6, 6–10, ..., 400–500 mm). The valve of the tolerance unit, i, for grades 5 through 16 would be

$$i = 0.45\sqrt{D} + 0.001\,D \qquad (10\text{-}3)$$

where i is in micrometers and D in millimeters.

Standard shaft and hole deviations are provided by similar sets of formulas. However, for practical application, both tolerances and deviations are provided in three sets of rather complex tables. Additional tables give the values for basic sizes above 500 mm and for "commonly used shafts and holes" in two categories: "general purpose" and "fine mechanisms and horology" (horology is the art of making timepieces).

Geometric Tolerances

Geometric tolerances state the maximum allowable deviation of a form or a position from the perfect geometry implied by a drawing. These tolerances specify the diameter or the width of a tolerance zone necessary for a part to meet its required accuracy. **Figure 10.10** shows the various symbols used to specify the required geometric characteristics of dimensioned drawings. A modifier is used to specify the limits of size of a part when applying geometric tolerances. The **maximum material condition (MMC)** indicates that a part is made with the largest amount of material allowable (e.g., a hole at its smallest permitted diameter or a shaft at its largest permitted diameter). The **least material condition (LMC)** is the converse of the maximum material condition. **Regardless of feature size (RFS)** indicates that tolerances apply to a geometric feature for any size it may be. Many geometric tolerances or **feature control symbols** are stated with respect to a particular datum or reference surface. Up to three datum surfaces can be given to specify a tolerance. Datum surfaces are generally designated by a letter symbol.

Figure 10.10b gives examples of the symbols used for datum planes and feature control symbols.

There are four tolerances that specify the permitted variability of forms: *flatnesss, straightness, roundness*, and *cylindricity*. Form tolerances describe how an actual feature may vary from a geometrically ideal feature.

The surface of a part is ideally flat if all its elements are coplanar. The flatness specification describes the tolerance zone formed by two parallel planes that bound all the elements on a surface. A 0.5-mm tolerance zone is described by the feature control symbol in **Figure 10.10c**. The distance between

	Tolerance	Characteristic	Symbol
Individual features	Form	Straightness	—
		Flatness	▱
		Circularity	◯
		Cylindricity	⌀
Individual or related features	Profile	Line	⌒
		Surface	⌓
Related features	Orientation	Angularity	∠
		Perpendicularity	⊥
		Parallelism	//
	Location	Position	⊕
		Concentricity	◎
	Runout	Circular runout	↗
		Total runout	⟋⟋

Notes	⌀	Ⓜ	Ⓛ	Ⓢ
	DIA	MMC	LMC	RFS

(a)

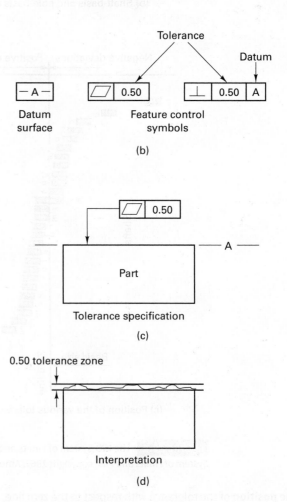

(b)

Tolerance specification

(c)

Interpretation

(d)

FIGURE 10.10 (a) Geometric tolerancing symbols; (b) feature control symbols for part drawings; (c) how a geometric tolerance for flatness is specified; (d) what the specification means.

the highest point on the surface to the lowest point on the surface may not be greater than 0.5 mm, as shown in **Figure 10.10d**.

Coordinate measuring machines discussed later provide additional examples of geometric tolerances.

10.4 Inspection Methods for Measurement

The field of **metrology**, even when limited to geometrical or dimensional measurements, is far too large to cover here. This chapter concentrates on basic linear measurements and the measurement devices most commonly found in a company's metrology or quality control facility. At a minimum, such labs would typically contain optical flats: one or two granite measuring tables; an assortment of indicators, calipers, micrometers, and height gages; an optical comparator; a set or two of Grade 1 gage blocks; a coordinate measuring machine; a laser scanning device; a laser interferometer; a toolmaker's microscope; and pieces of equipment specially designed to inspect and/or test the company's products.

The discrete digital readout on a clear liquid-crystal display (LCD) eliminates reading interpretations associated with analog scales and can be entered directly into dedicated microprocessors or computers for permanent recording and analysis. The added speed and ease of use for this type of equipment have allowed it to be routinely used on the plant floor instead of in the metrology lab. In summary, the trend toward tighter tolerances (greater precision) and accuracy associated with the need for superior quality and reliability has greatly enhanced the need for improved measurement methods.

Factors in Selecting Inspection Equipment

Many inspection devices use electronic outputs to communicate directly to microprocessors. Inspection devices are being built into the processes themselves and are often computer aided. In-process inspection generates feedback sensory data from the process or its output to the computer control of the machine, which is the first step in making the processes responsive to changes (adaptive control). In addition to in-process inspection, many other quality checks and measurements of parts and assemblies are needed. In general, six factors should be considered when selecting equipment for an inspection job by measurement techniques:

1. *Gage capability.* The measurement device (or working gage) should be 10 times more precise than the tolerance to be measured. This is known on the factory floor as the **rule of 10**. The rule actually applies to all stages in the inspection sequence, as shown in **Figure 10.11**. The master gage should be 10 times more precise than that of the inspection device. The reference standard used to check the master gage should be 10 times more precise than the master gage. The application of the rule greatly reduces the probability of rejecting good parts or accepting bad components and performing additional work on them. Additional discussion of gage capability is found in Chapter 12.

2. *Linearity.* The **linearity** factor refers to the calibration accuracy of the device over its full working range. Is it linear? What is its degree of nonlinearity? Where does it become nonlinear, and what, therefore, is its real linear working region?

3. *Repeat accuracy.* How repeatable is the device in taking the same reading over and over on a given standard?

4. *Stability.* How well does this device retain its calibration over a period of time. **Stability** is also called **drift**. As devices become more accurate, they often lose stability and become more sensitive to small changes in temperature and humidity.

5. *Magnification.* This refers to the amplification of the output portion of the device over the actual input dimension. The more accurate the device, the greater must be its **magnification** factor, so that the required measurement can be read out (or observed) and compared with the desired standard. Magnification is often confused with resolution, but they are not the same.

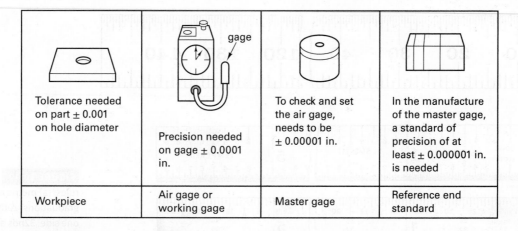

Tolerance needed on part ± 0.001 on hole diameter	Precision needed on gage ± 0.0001 in.	To check and set the air gage, needs to be ± 0.00001 in.	In the manufacture of the master gage, a standard of precision of at least ± 0.000001 in. is needed
Workpiece	Air gage or working gage	Master gage	Reference end standard

FIGURE 10.11 The **rule of 10** states that for reliable measurements each successive step in the inspection sequence should have 10 times the *precision* of the preceding step.

6. *Resolution.* **Resolution** is sometimes called **sensitivity** and refers to the smallest unit of scale or dimensional input that the device can detect or distinguish. The greater the resolution of the device, the smaller will be the features it can detect or identify (resolve) and the greater will be the magnification required to expand these measurements up to the point where they can be observed by the naked eye.

Some other factors of importance in selecting inspection devices include the type of measurement information desired; the range or the span of sizes the device can handle versus the size and geometry of the workpieces; the environment; the cost of the device; and the cost of installing, training, and using the device. The last factor depends on the speed of measurement, the degree to which the system can be automated, and the functional life of the device in service.

10.5 Measuring Instruments

Because of the great importance of measuring in manufacturing, a great variety of instruments are available that permit measurements to be made routinely, ranging in accuracy from 1/64 to 0.00001 in. and from 0.5 to 0.0003 mm. Machine-mounted measuring devices (probes and lasers) for automatically inspecting the workpiece during manufacturing are beginning to compete with postprocess gaging and inspection, in which the part is inspected, automatically or manually, after it has come off the machine. In-process inspection for automatic size control has been used for some years in grinding to compensate for the relatively rapid wear of the grinding wheel. Touch trigger probes, with built-in automatic measuring systems, are being used on CNC machine tools to determine cutting tool offsets and compensations for tool wear.

For manually operated analog instruments, the ease of use, precision, and accuracy of measurements can be affected by (1) the least count of the subdivisions on the instrument, (2) line matching, and (3) the parallax in reading the instrument. Elastic deformation of the instrument and workpiece and temperature effects must be considered. Some instruments are more subject to these factors than others. In addition, the skill of the person making the measurements is very important. Digital readout devices in measuring instruments lessen or eliminate the effect of most of these factors, simplify many measuring problems, and lessen the chance of making a math error.

Linear Measuring Instruments

Linear measuring instruments are of two types: direct reading and indirect reading. **Direct-reading instruments** contain a line-graduated scale so that the size of the object being measured can be read directly on this scale. **Indirect-reading instruments** do not contain line graduations and are used to transfer the size of the dimension being measured to a direct-reading scale, thus obtaining the desired size information indirectly.

The simplest and most common direct-reading linear measuring instrument is the **machinist's rule**, shown in **Figure 10.12**. Metric rules usually have two sets of line graduations on each side, with divisions of ½ and 1 mm; English rules have four sets, with divisions of 1/16, 1/32, 1/64, and 1/100 in. Other combinations can be obtained in each type.

The machinist's rule is an end- or line-matching device. For the desired reading to be obtained, an end and a line, or two lines, must be aligned with the extremities of the object or the distance being measured. Thus, the accuracy of the resulting reading is a function of the alignment and the magnitude of the smallest scale division. Such scales are not ordinarily used for accuracies greater than 1/64 in. (0.01 in.), or about ½ mm.

Several attachments can be added to a machinist's rule to extend its usefulness. The **square head** (**Figure 10.13**) can be used as a miter or tri-square or to hold the rule in an upright position on a flat surface for making height measurements. It also contains a small bubble-type level so that it can be used by itself as a level. The **bevel protractor** permits the measurement or layout of angles. The **center head** permits the center of cylindrical work to be determined.

The **vernier caliper** (**Figure 10.14**) is an end-measuring instrument, available in various sizes, that can be used to make both outside and inside measurements to theoretical

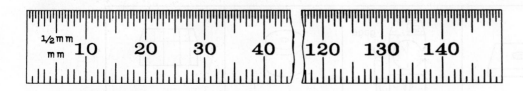

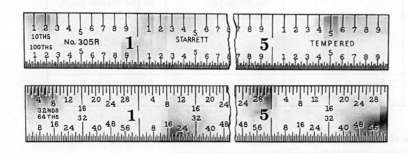

FIGURE 10.12 Machinist's rules: (a) metric and (b) inch graduations; 10ths and 100ths on one side, 32nds and 64ths on the opposite side. (Images Courtesy of The L. S. Starrett, Co.)

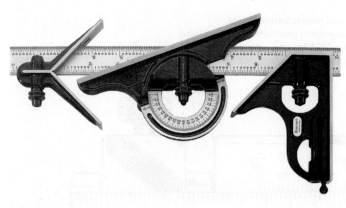

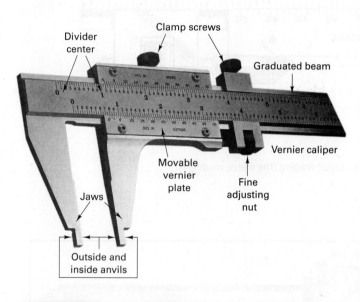

FIGURE 10.14 This vernier caliper can make measurements using both inside (for holes) and outside (shafts) anvils. (Courtesy J T. Black)

accuracies of 0.01 mm or 0.001 in. End-measuring instruments are more accurate and somewhat easier to use than line-matching types because their jaws are placed against either end of the object being measured, so any difficulty in aligning edges or lines is avoided. However, the difficulty remains in obtaining uniform contact pressure, or "feel," between the legs of the instrument and the object being measured.

A major feature of the vernier caliper is the auxiliary scale (**Figure 10.15**). The caliper shown has a graduated beam with a metric scale on the top, a metric vernier plate, and an English scale on the bottom with an English vernier. The manner in which readings are made is made as follows. The vernier plate zero line is 1 in. (1.000 in.) plus ½₀ (0.50 in.) beyond the zero line on the bar, or 1.050 in. The 29th graduation on the vernier plate coincides with a line on the bar (as indicated by stars); 29 × 0.001 (0.029 in.) is therefore added to the 1.050-in. bar reading, and the total is 1.079 in.

Figure 10.16 shows a vernier height gage for measuring the length of shoulders on parts. **Figure 10.17** shows a caliper that has a digital readout or a dial indicator that replaces the

vernier. The latter two calipers are capable of making inside and outside measurements as well as depth measurements.

The **micrometer caliper**, more commonly called a **micrometer**, is one of the most widely used measuring devices. Until recently, the type shown in **Figure 10.18** was virtually standard. It consists of a fixed anvil and a movable spindle. When the thimble is rotated on the end of the caliper, the spindle is moved away from the anvil by means of an accurate screw thread. On English types, this thread has a lead of 0.025 in., and one revolution of the thimble moves the spindle this distance. The barrel, or sleeve, is calibrated in 0.025-in. divisions, with each ¹⁄₁₀ of an inch being numbered. The circumference at the edge of the thimble is graduated into 25 divisions, each representing 0.001 in.

A major difficulty with this type of micrometer is making the reading of the dimension shown on the instrument. To read the instrument, the division on the thimble that coincides with the longitudinal line on the barrel is added to the largest reading exposed on the barrel.

Micrometers graduated in ten-thousandths of an inch are the same as those graduated in thousandths, except that an additional vernier scale is placed on the sleeve so that a reading of ten-thousandths is obtained and added to the thousandths reading. The vernier consists of 10 divisions on the sleeve, shown in inset B, which occupy the same space as 9 divisions on the thimble. Therefore, the difference between the width of one of the 10 spaces on the vernier and one of the 9 spaces on the thimble is one-tenth of a division on the thimble, or one-tenth of one-thousandth, which is one ten-thousandth. To read a ten-thousandths micrometer, first obtain the thousandths reading, then see which of the lines on the vernier coincides with a line on the thimble. If it is the line marked "1," add one ten-thousandth; if it is the line marked "2," add two ten-thousandths; and so on.

EXAMPLE: REFER TO THE INSETS IN FIGURE 10.18.

The "2" line on sleeve is visible, representing	0.200 in.
Two additional lines, each representing 0.025"	2 × 0.025" = 0.050 in.
Line "O" on the thimble coincides with the reading line on the sleeve, representing	0.000 in.
The "O" lines on the vernier coincide with lines on the thimble, representing	0.000 in.
The micrometer reading is	0.2500 in.

Now you try to read inset C.

The "2" line on sleeve is visible, representing	0.200 in.
Two additional lines, each representing 0.025"	2 × 0.025" = 0.050 in.
The reading line on the sleeve lies between the "O" and "1" on the thimble, so ten-thousandths of an inch is to be added as read from the vernier.	
The "7" line on the vernier coincides with a line on the thimble, representing	7 × 0.00010" = 0.007 in.
The micrometer reading is	0.2507 in.

Refer to the upper bar graduations and metric vernier plate.
Each bar graduation is 1.00 mm. Every tenth graduation is numbered
in sequence—10 mm, 20 mm, 30 mm, 40 mm, etc.—over the full range
of the bar. This provides for direct reading in millimeters.

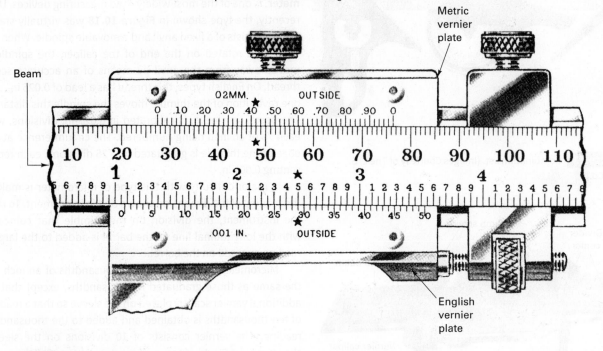

FIGURE 10.15 Vernier caliper graduated for English and metric (direct) reading. The metric reading is 27 + 0.42 = 27.42 mm. (Courtesy J T. Black)

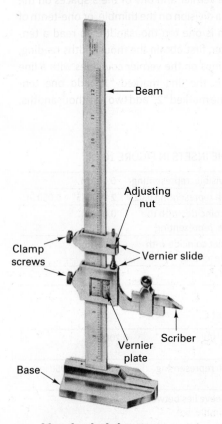

Vernier height gage

FIGURE 10.16 Variations in the vernier caliper design result in other basic gages, like the vernier height gage. (Courtesy J T. Black)

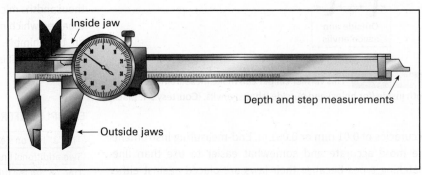

(a) Dial caliper with 0.001-in. accuracy

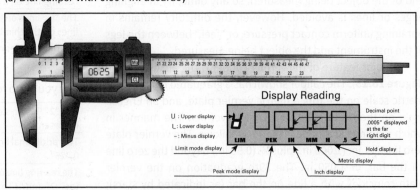

(b) Digital electronic caliper with 0.001-in. (0.03-mm) accuracy and 0.0001-in. resolution with inch/metric conversion.

FIGURE 10.17 Two styles of calipers in common use today: (a) dial caliper with 0.001-in. accuracy; (b) digital electronic caliper with 0.001-in. (0.03-mm) accuracy and 0.0001-in. resolution with inch/metric conversion.

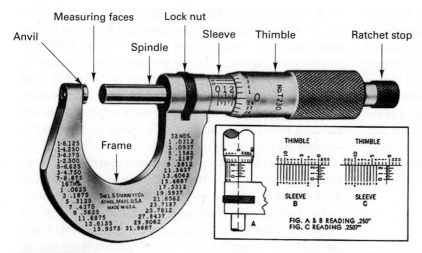

Anvil · Measuring faces · Spindle · Lock nut · Sleeve · Thimble · Ratchet stop · Frame

THIMBLE · THIMBLE
SLEEVE B · SLEEVE C
FIG. A & B READING .250"
FIG. C READING .2507"

FIGURE 10.18 Micrometer caliper graduated in ten-thousandths of an inch with insets A, B, and C showing two example readings. (Images Courtesy of The L. S. Starrett, Co.)

However, owing to the lack of pressure control, micrometers can seldom be relied on for accuracy beyond 0.0005 in., and such vernier scales are not used extensively. On metric micrometers, the graduations on the sleeve and thimble are usually 0.5 mm and 0.01 mm, respectively (see the Problems at the end of the chapter).

Many errors have resulted from the ordinary micrometer being misread, the error being 60.025 or 60.5 mm. Consequently, direct-reading micrometers have been developed. **Figure 10.19** shows a digital outside micrometer that reads to 0.001 in. on the digit counter and 0.0001 in. on the vernier on the sleeve. The range of a micrometer is limited to 1 in. Thus, a number of micrometers of various sizes are required to cover a wide range of dimensions.

To control the pressure the anvil, the spindle, and the piece being measured, most micrometers are equipped with a ratchet or a friction device, as shown in Figures 10.18 and 10.19.

Calipers that do not have this device may be overtightened and sprung by several thousandths by applying excess torque to the thimble. Micrometer calipers should not usually be relied on for measurements of greater accuracy than 0.01 mm or 0.001 in., unless they are digital.

Micrometer calipers are available with a variety of specially shaped anvils and/or spindles, such as point, balls, and disks, for measuring special shapes, including screw threads. Micrometers are also available for inside measurements, and the micrometer principle is also incorporated into a **micrometer depth gage**.

Bench micrometers with direct readout to data processors are becoming standard inspection devices on the plant floor (see **Figure 10.20**). The data processor provides a record of measurement, as well as control charts and histograms. Direct-gaging height gages, calipers, indicators, and micrometers are also available with statistical analysis capability. See Chapter 12 for more discussion on control charts and process capability.

Larger versions of micrometers, called **supermicrometers**, are capable of measuring 0.0001 in. when equipped with an indicator that shows that a selected pressure between the anvils has been obtained. The addition of a digital readout permits the device to measure to ±0.00005 in. (0.001 mm) directly when it is used in a controlled-temperature environment.

The **toolmaker's microscope** is a versatile instrument that measures by optical means; no pressure is involved. Thus, it is very useful for making accurate measurements on small or delicate parts. The base, on which the special microscope is mounted, has a table that can be moved in two mutually perpendicular, horizontal directions (X and Y) by means of accurate micrometer screws that can be read to 0.0001 in. or, if so equipped, by means of the digital readout. Parts to be measured are mounted on the table, the microscope is focused, and one end of the desired part feature is aligned with the cross-line in the microscope. The reading is then noted, and the table is moved until the other extremity of the part coincides with the cross-line. From the final reading, the desired measurement can be determined. In addition to a wide variety of linear measurements, accurate angular measurements can also be made by means of a special protractor eye-piece. These microscopes are available with digital readouts and image analyses software.

The **optical projector** or **comparator** (**Figure 10.21**) is a large optical device on which both linear and angular measurements can be made. As with the toolmaker's microscope, the part to be measured is mounted on a table that can be moved in X and Y directions by accurate micrometer screws. The optical system projects the image of the part on a screen, magnifying it

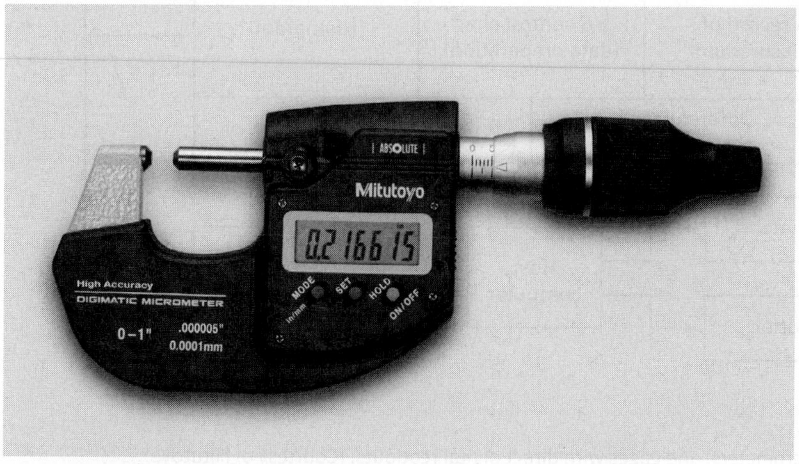

FIGURE 10.19 Digital micrometer for measurements from 0 to 1 in., in 0.0001-in. graduations. (Courtesy of Mitutoyo America Corporation, Aurora, IL)

from 5 to more than 100 times. Measurements can be made directly by means of the micrometer dials, the digital readouts, or the dial indicators, or on the magnified image on the screen by means of an accurate rule. A very common use for this type of instrument is the checking of parts, such as dies and screws. A template is drawn to an enlarged scale and is placed on the screen. The projected contour of the part is compared to the desired contour on the screen. Digital touch screens and software that allow for more sophisticated 2D analyses are available. Some projectors also function as low-power microscopes by providing surface illumination.

Measuring with Lasers

One of the earliest and most common metrological uses of low-power lasers has been in **interferometry**. The interferometer uses light interference bands to determine distance and thickness of objects (**Figure 10.22**). First, a beam splitter divides a beam of light into a measurement beam and a reference beam. The measurement beam travels to a reflector (optical glass plate A), resting on the part whose distance is to be measured, while the reference beam is directed at fixed reflector B. Both beams are reflected back through the beam splitter, where they are recombined into a single beam before traveling to the observer. This recombined beam produces interference fringes, depending on whether the waves of the two returning beams are in phase (called **constructive interference**) or out of phase (termed **destructive interference**). In-phase waves produce a series of bright bands, and out-of-phase waves produce dark bands. The number of fringes can be related to the size of the object, measured in terms of light waves of a given frequency. The following example will explain the basics of the method.

To determine the height of object U in **Figure 10.23**, a calibrated reference standard S, an optical flat, and a toolmaker's flat are needed, along with a monochromatic light source **Optical flats** are quartz or special glass disks, from 2 to 10 in. (50 to 250 mm) in diameter and about ½ to 1 in. (12 to 25 mm) thick, whose surfaces are very nearly true planes and nearly parallel. Flats can be obtained with the surfaces within

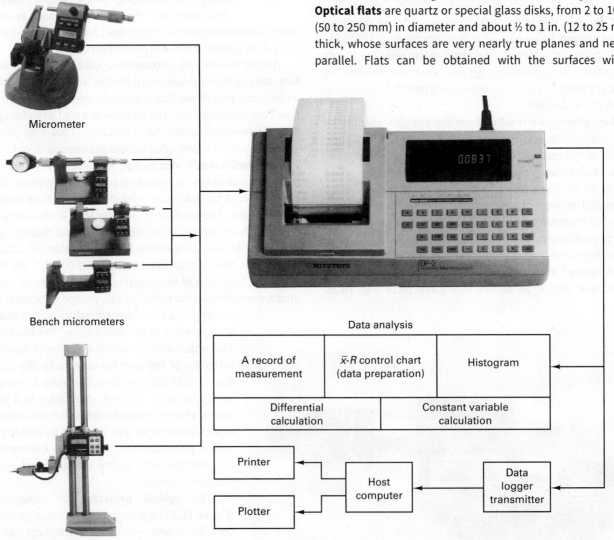

Micrometer

Bench micrometers

Height gage

Data analysis

A record of measurement	$\bar{x}$-R control chart (data preparation)	Histogram
Differential calculation	Constant variable calculation	

Printer

Plotter

Host computer

Data logger transmitter

FIGURE 10.20 Toolmaker's micrometer, bench micrometers, and gages with direct digital readouts. (Courtesy of Mitutoyo America Corporation, Aurora, IL)

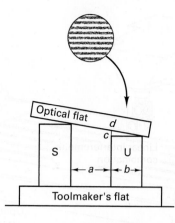

FIGURE 10.23 Method of measuring object U using calibrating gage block S and light-wave interference.

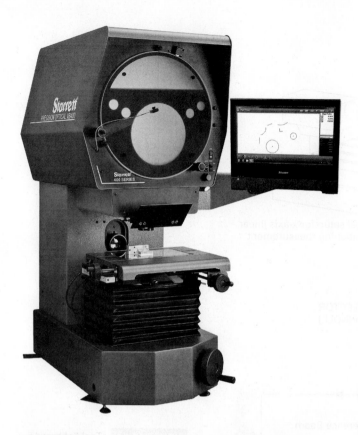

FIGURE 10.21 The optical comparator uses light to show the magnified contour of a workpiece, projecting the image on a screen for measurement. (Courtesy of The L.S. Starrett Company)

0.000001 in. (0.00003 mm) of true flatness. It is not essential that both surfaces be accurate or that they be exactly parallel, but one must be certain that only the accurate surface is used in making measurements. A **toolmaker's flat** is similar to an optical flat but is made of steel and usually has only one surface that is accurate. A **monochromatic light source**, light of a single wavelength, must be used. Selenium, helium, or cadmium light sources are commonly used along with helium-neon lasers.

The block to be measured is U and the calibrated block is S. Distances *a* and *b* must be known but do not have to be measured with great accuracy. By counting the number of **interference bands** shown on the surface of block U, the distance *c–d* can be determined. Because the difference in the distances between the optical flat and the surface of U is one-half wavelength, each dark band indicates a change of one-half wavelength in the elevation. If a monochromatic light source having a wavelength of 23.2 μin. (0.589 mm) is used, each interference band represents 11.6 μin. (0.295 μm). Then, by simple geometry, the difference in the heights of the two blocks can be computed. The same method is applicable for making precise measurements of other objects by comparing them with a known gage block.

Accurate measurement of distances greater than a few inches was very difficult until the development of laser interferometry, which permits accuracies of 60.5 ppm over a distance of 6.1 m with 0.01 μm resolution. Such equipment is particularly useful in checking the movement of machine tool tables, aligning and checking large assembly jigs, and making measurements of intricate parts such as tire-tread molds.

The **laser interferometer** in **Figure 10.24** uses a helium–neon laser beam split into two beams, each of different

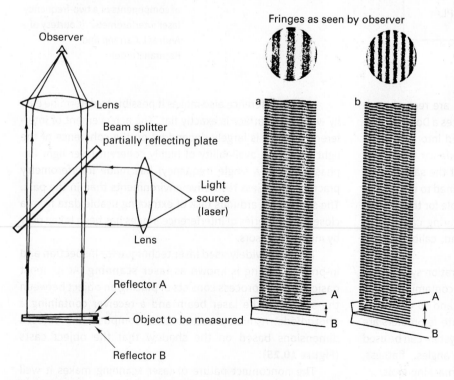

Fringes as seen by observer

FIGURE 10.22 Interference bands can be used to measure the size of objects to great accuracy. (Based on the Michelson interferometer, invented in 1882)

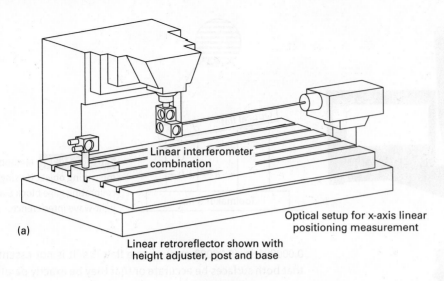

(a)

Linear interferometer combination

Optical setup for x-axis linear positioning measurement

Linear retroreflector shown with height adjuster, post and base

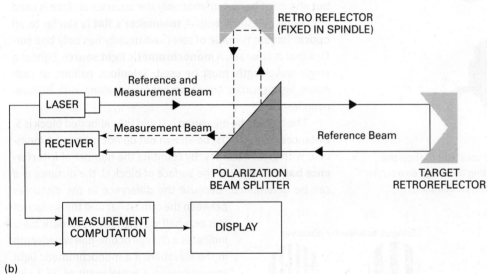

(b)

RETRO REFLECTOR (FIXED IN SPINDLE)

Reference and Measurement Beam

LASER

Measurement Beam

RECEIVER

Reference Beam

POLARIZATION BEAM SPLITTER

TARGET RETROREFLECTOR

MEASUREMENT COMPUTATION

DISPLAY

FIGURE 10.24 (Top) Calibrating the x-axis linear table displacement of a vertical-spindle milling machine; (middle) schematic of optical setup; (bottom) schematic of components of a two-frequency laser interferometer. (Courtesy of Andres L.Carrano and Kamran Kardel)

frequency and polarized. When the beams are recombined, any relative motion between the optics creates a Doppler shift in the frequency. This shift is then converted into a distance measurement. The laser light has less tendency to diverge (spread out) and is also monochromatic (of the same wavelength). A process that has been largely confined to the optical industry and the metrology lab is now suitable for the factory, where its extremely precise distance-measuring capabilities have been applied to the alignment and calibration of machine tools.

The two-frequency interferometer calibration system was introduced in 1970 to overcome workplace contamination by thermal gradients, air turbulence, oil mist, and so on, which affect the intensity of light. Doppler laser interferometers are relatively insensitive to such problems. The system can be used to measure linear distances, velocities, angles, flatness, straightness, squareness, and parallelism in machine tools.

Lasers provide for accurate machine tool alignment. Large, modern machine tools can move out of alignment in a matter of months, causing production problems often attributed to the cutting tools, the workholders, the machining conditions, or the numerical control part program.

Light interference also makes it possible to determine easily whether a surface is exactly flat. The achievement of interference fringes is largely dependent on the coherence of the light used. The availability of highly coherent laser light (in-phase light of a single frequency) has made interferometry practical in far less restrictive environments than in the past. The sometimes arduous task of extracting usable data from a close-packed series of interference fringes has been taken over by microprocessors.

The most widely used laser technique for inspection and in-process gaging is known as **laser scanning**. At its most basic level, the process consists of placing an object between the source of the laser beam and a receiver containing a photodiode. A microprocessor then computes the object's dimensions based on the shadow that the object casts (**Figure 10.25**).

The noncontact nature of laser scanning makes it well suited to in-process measurement, including such difficult tasks as the inspection of hot-rolled or extruded material, and its comparative simplicity has led to the development of highly portable systems. The bench gage versions can measure to resolutions of 0.0001 mm.

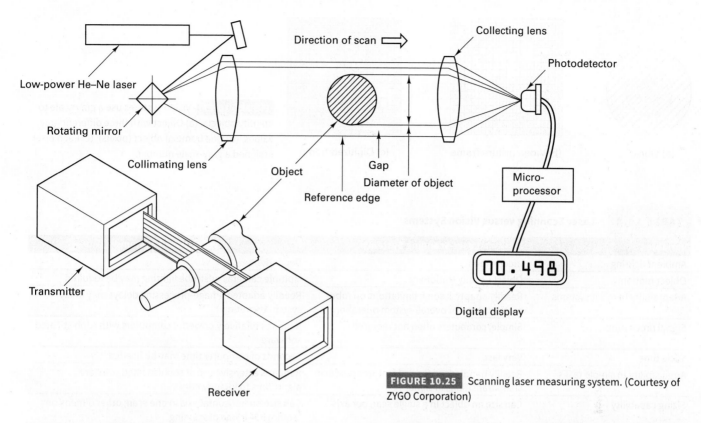

FIGURE 10.25 Scanning laser measuring system. (Courtesy of ZYGO Corporation)

10.6 Vision Systems

If a picture is worth a thousand words, then **vision systems** are the tome of inspection methods (see **Figure 10.26**). Machine vision is used for **visual inspection**, for guidance and control, or for both. Normal TV image formation on photosensitive surfaces or arrays is used, and the video signals are analyzed to obtain information about the object. Each picture frame represents the object at some brief interval of time. Each frame must be dissected into picture elements (called **pixels**). Each pixel is **digitized** (has binary numbers assigned to it) by fixing the brightness or gray level of each pixel to produce a **bit-map** of the object (**Figure 10.27**). That is, each pixel is assigned a numerical value based on its shade of gray. Image preprocessing improves the quality of the image data by removing unwanted detail. The bit-map is stored in a buffer memory. By analyzing and processing the digitized and stored bit-map, the patterns are extracted, edges are located, and dimensions are determined.

Sophisticated computer algorithms using artificial intelligence have greatly reduced the computer operations needed to achieve a result, but even the most powerful video-based systems currently require 1 to 2 s to achieve a measurement. This may be too long a time for many online production applications. **Table 10.4** provides a comparison of vision systems to laser scanning.

With the recent emphasis on quality and 100% inspection, applications for inspection by machine vision have increased markedly. Vision systems can check hundreds of parts per hour for multiple dimensions. Resolutions of ±0.01 in. have been demonstrated, but 0.02 in. is more typical for part location. Machine vision is useful for robot guidance in material handling, welding, and assembly, but nonrobotic inspection and part location applications are still more typical. The use of vision systems in inspection, quality control, sorting, and machining tool monitoring will continue to expand. Systems can cost $10,000 or more to install and must be justified on the basis of improved quality rather than labor replacement.

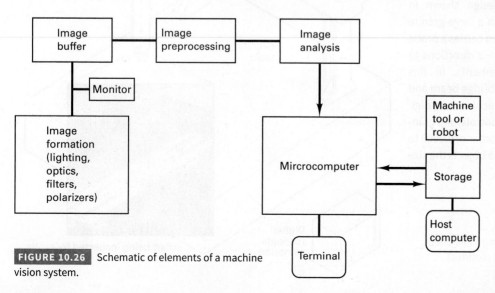

FIGURE 10.26 Schematic of elements of a machine vision system.

(a) Object

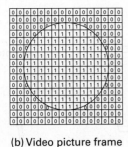

(b) Video picture frame

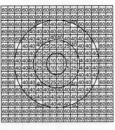

(c) Digitized frame

FIGURE 10.27 Vision systems use a gray scale to identify objects. (a) Object with three different gray values. (b) One frame of object (pixels). (c) Each pixel assigned a gray-scale number.

TABLE 10.4 Laser Scanning versus Vision Systems

Variable	Laser-Scanning Systems	Video-based Systems
Ambient lighting	Independent	Dependent
Object motion	Object usually stationary	Multiple cameras or strobe lighting may be required
Adaptability to robot systems	Readily adapted; some limitations on robot motion speed or overall system operation	Readily adapted; image-processing delays may delay system operation
Signal processing	Simple; computers often not required	Requires relatively powerful computers with sophisticated software
Cycle time	Very fast	Seconds of computer time may be needed
Applicability to simple tasks	Readily handled; edges and features produce sharp transitions in signal	Requires extensive use of sophisticated software algorithms to identify edges
Sizing capability	Can size an object in a single scan per axis	Can size on horizontal axis in one scan; other dimensions require full-frame processing
Three-dimensional capability	Limited three dimensionality; needs ranging capability	Uses two views of two cameras with sophisticated software or structured light
Accuracy and precision	Submicrometer 0.001 to 0.0001 in. or better accuracy; highly repeatable	Depends on resolution of cameras and distance between camera and object; systems with 0.004-in. precision and 0.006-in. accuracy are typical

10.7 Coordinate Measuring Machines

Precision measurements in three-dimensional Cartesian coordinate space can be made with **coordinate measuring machines (CMMs)** of the design shown in **Figure 10.28**. The parts are placed on a large granite flat, a table or the floor. The vertical arm carries a probe that can be precisely moved in $x - y - z$ directions to produce three-dimensional measurements. In this design, the vertical column rides on a bridge beam and carries a touch-trigger probe. Such machines use digital readouts, air bearings, computer controls, and granite tables to achieve accuracies of the order of 0.0002 to 0.0004 in. over spans of 30 in. to 30 ft so objects the size

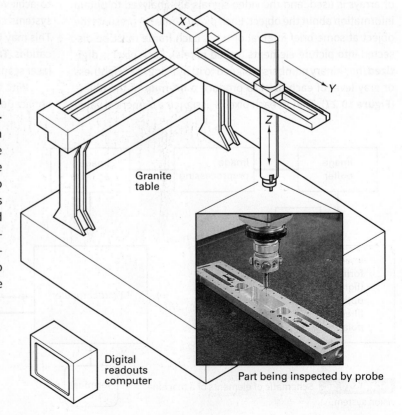

Granite table

Digital readouts computer

Part being inspected by probe

FIGURE 10.28 Examples of geometric form tolerances developed by probing surface with a CMM. (Courtesy J T. Black)

of a car can be inspected. CMMs, although expensive, have the following benefits:

- One-time setup with direct computer control.
- Reduced human error and variances.
- Flexibility eliminates special gaging devices and fixtures for a particular part.
- Electronic digital readouts with statistical calculations.

- Geometric calculations to establish feature measurements and form tolerances.

CMM systems typically have computer routines that give the best fit to feature measurements and that provide the means of establishing geometric tolerances, discussed earlier in this chapter. **Figure 10.29** gives a partial listing of the results that can be achieved with these machines.

Straightness 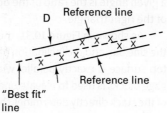	**Straightness** Measured or previously calculated points may be used to determine a "best fit" line. The form routine establishes two reference lines that are parallel to the "best fit" line and that just contain all of the measured or calculated points. Straightness is defined as the distance D between these two reference lines.
Flatness 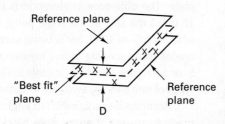	**Flatness** Measured or previously calculated points may be used to determine the "best fit" plane. The form routine establishes two reference planes that are parallel to the "best fit" plane and that just contain all of the measured or calculated points. Flatness is defined as the distance D between these two reference planes.
Roundness 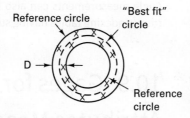	**Roundness** Measured or previously calculated points may be used to determine the "best fit" circle. The form routine establishes two reference circles that are concentric with the "best fit" circle and that just contain all of the measured or calculated points. Roundness is defined as the difference D in radius of these two reference circles.
Cylindricity 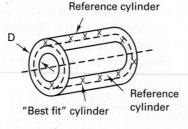	**Cylindricity** Measured or previously calculated points may be used to determine the "best fit" cylinder. The form routine establishes two reference cylinders that are co-axial to the "best fit" cylinder and that just contain all of the measured or calculated points. Cylindricity is the difference D in radius of these two reference cylinders. Also applicable to stepped cylinders.
Conicity 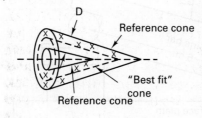	**Conicity** Measured or previously calculated points may be used to determine the "best fit" cone. The form routine establishes two reference cones that are co-axial with and similar to the "best fit" cone and that just contain all of the measured or calculated points. Conicity is defined as the distance D between the side of these two reference cones.

FIGURE 10.29 Coordinate measuring machine with inset showing probe and a part being measured.

10.8 Angle-measuring Instruments

Accurate angle measurements are usually more difficult to make than linear measurements. Angles are measured in degrees (a degree is 1/360 of a circle) and decimal subdivisions of a degree (or in minutes and seconds of arc). The SI system calls for measurements of plane angles in radians, but degrees are permissible. The use of degrees will continue in manufacturing, but with minutes and seconds of arc possibly being replaced by decimal portions of a degree.

The bevel protractor (**Figure 10.30**) is the most general angle-measuring instrument. The two movable blades are brought into contact with the sides of the angular part, and the angle can be read on the vernier scale to 5 min of arc. A clamping device is provided to lock the blades in any desired position so that the instrument can be used for both direct measurement and layout work. As indicated previously, an angle attachment on the combination set can also be used to measure angles, similar to the way a bevel protractor is used but usually with somewhat less accuracy.

The toolmaker's microscope is very satisfactory for making angle measurements, but its use is restricted to small parts. The accuracy obtainable is 5 min of arc. Similarly, angles can be measured on the optical contour projector. Angular measurements can also be made by means of an angular interferometer with the laser system.

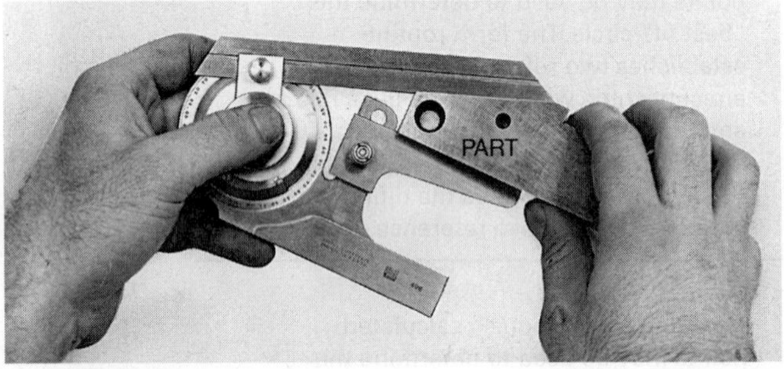

A **sine bar** may be used to obtain accurate angle measurements if the physical conditions permit. This device (**Figure 10.31**) consists of an accurately ground bar on which two accurately ground pins of the same diameter are mounted an exact distance apart. The distances used are usually either 5 or 10 in., and the resulting instrument is called a 5- or 10-in. sine bar. Sine bars are also available with millimeter dimensions. Measurements are made by using the principle that the sine of a given angle is the ratio of the opposite side to the hypotenuse of the right triangle.

As shown in Figure 10.31, the part being measured is set on the sine bar, and the inclination of the assembly is raised until the top surface is exactly parallel with the surface plate. A stack of gage blocks is used to elevate one end of the sine bar. The height of the stack directly determines the difference in height of the two pins. A dial indicator gage or any other type of gage can be used to determine when there is no deviation (parallel) from the surface plate. The difference in elevation is then equal to either 5 or 10 times the sine of the angle being measured, depending on whether a 5- or 10-in. bar is being used. Tabulated values of the angles corresponding to any measured elevation difference for 5- or 10-in. sine bars are available in various handbooks. Several types of sine bars are available to suit various requirements.

Accurate measurements of angles to 1 s of arc can also be made by means of **angle gage blocks**. These come in sets of 16 blocks that can be assembled in desired combinations. Angle measurements can also be made to 60.001 degrees on rotary indexing tables having suitable numerical control.

10.9 Gages for Attributes Measuring

In manufacturing, particularly in lean manufacturing systems, (see Chapters 42-43), it may not be necessary to know the exact dimensions of a part, only that it is within previously established limits. Limits can often be determined more easily than

FIGURE 10.30 Measuring an angle on a part with a bevel protractor. (Image Courtesy of Hexagon Metrology, Inc.)

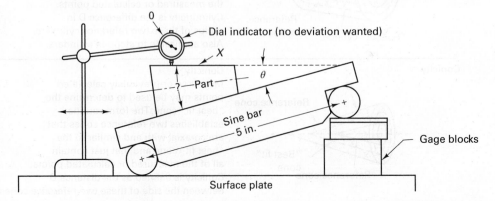

FIGURE 10.31 Setup to measure an angle on a part using a sine bar. The dial indicator is used to determine when the part surface *X* is parallel to the surface plate. (Image Courtesy of Hexagon Metrology, Inc.)

specific dimensions by the use of attribute-type instruments called **gages**. They may be of either fixed type or deviation type, may be used for both linear and angular dimensions, and may be used manually or mechanically (automatically).

FIXED-TYPE GAGES

Fixed-type gages are designed to gage only one dimension and to indicate whether it is larger or smaller than the previously established standard.

They do not determine how much larger or smaller the measured dimension is than the standard. Because such gages fulfill a simple and limited function, they are relatively inexpensive and usually quick and easy to use.

Gages of this type are ordinarily made of hardened steel of proper composition and are heat-treated to produce dimensional stability. Hardness is essential to minimize wear and maintain accuracy. Because steels of high hardness can become dimensionally unstable, some fixed gages are made of softer steel, then given a hard chrome plating to provide surface hardness. Chrome plating can also be used for reclaiming some worn gages. Where gages are to be subjected to extensive use, they may be made of tungsten carbide at the wear points. Gages of this type are also used in lean manufacturing cells as part of the poka-yoke methodology, where the operator inspects parts to prevent defects from occurring.

One of the most common fixed gages is the **plug gage**. As shown in **Figure 10.32**, plug gages are accurately ground cylinders used to gage internal dimensions, such as holes. The gaging element of a *plain plug gage* has a single diameter. To control the minimum and maximum limits of a given hole, two plug gages are required. The smaller, or **go gage**, controls the minimum because it must go (slide) into any hole that is larger than the required minimum. The larger, or **no-go gage**, controls the maximum dimension because it will not go into any hole unless that hole is over the maximum permissible size. The go and no-go plugs are often designed with two gages on a single handle for convenience in use. The no-go plug is usually much shorter than the go plug; it is subjected to little wear because it seldom slides into any holes. **Figure 10.33** shows a **step-type go/no-go gage** that has the go and no-go diameters on the same end of a single plug, the go portion being the outer end. The user knows that the part is good if the go gage goes into the hole but the no-go gage does not go. Such gages require careful use and should never be forced into (or onto) the part. Obviously, these plug gages were specially designated and made for checking a specific hole on a part.

FIGURE 10.33 Step-type plug gage with go and no-go elements on the same end. (Image Courtesy of Hexagon Metrology, Inc.)

In designing plug and snap ring gages, the key principle is this: *it is better to reject a good part than declare a bad part to be within specifications*. All gage design decisions are made with this principle in mind. Gages must have tolerances like any manufactured components. All gages are made with gage and wear tolerances. Gage tolerance allows for the permissible variation in the manufacture of the gage. It is typically 5 to 20% (depending on the industry) of the tolerance on the dimension being gaged. Wear tolerances compensate for the wear of the gage surface as a result of repeated use. Wear tolerance is applied only to the go side of the gage because the no-go side should seldom see contact with a part surface. It is typically 5 to 20% of the dimensional tolerance.

Plug-type gages are also made for gaging shapes other than cylindrical holes. Three common types are **taper plug gages, thread plug gages**, and **spline gages**. Taper plug gages gage both the angle of the taper and its size. Any deviation from the correct angle is indicated by looseness between the plug and the tapered hole. The size is indicated by the depth to which the plug fits into the hole, the correct depth being denoted by a mark on the plug. Thread plug gages come in go and no-go types. The go gage must screw into the threaded holes, and the no-go gage must not enter.

Ring gages are used to check shafts or other external round members. These are also made in go and no-go types, as shown in **Figure 10.34**. "Go" ring gages have plain knurled exteriors, whereas no-go ring gages have a circumferential groove in the knurling, so that they can easily be distinguished. **Ring thread gages** are made to be slightly adjustable because it is almost impossible to make them exactly to the desired size.

FIGURE 10.32 Plain plug gage having the go member on the left end (1.1250-in. diameter) and no-go member on the right end. (Image Courtesy of Hexagon Metrology, Inc.)

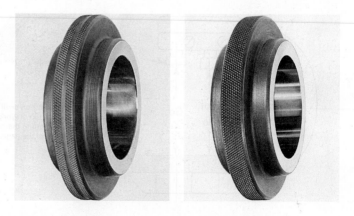

FIGURE 10.34 Go and no-go (on right) ring gages for checking a shaft. (Image Courtesy of Hexagon Metrology, Inc.)

Thus, they are adjusted to exact, final size after the final grinding and polishing have been completed.

Snap gages are the most common type of fixed gage for measuring external dimensions. As shown in **Figure 10.35**, they have a rigid, U-shaped frame on which are two or three gaging surfaces, usually made of hardened steel or tungsten carbide. In the adjustable type shown, one gaging surface is fixed, and the other(s) may be adjusted over a small range and locked at the desired position(s). Because in most cases one wishes to control both the maximum and the minimum dimensions, the **progressive** or **step-type snap gage** is used most frequently. These gages have one fixed anvil and two adjustable surfaces to form the outer go and the inner no-go openings, thus eliminating the use of separate go and no-go gages. Snap gages are available in several types and a wide range of sizes. The gaging surfaces may be round or rectangular. They are set to the desired dimensions with the aid of gage blocks.

Several types of **form gages** are available for use in checking the **profile** of various objects. Two of the most common types are **radius gages** (**Figure 10.36**) and **screw-thread pitch gages** (**Figure 10.37**). Many types of special gages are available or can be constructed for special applications. The **flush-pin gage** (**Figure 10.38**) is an example for gaging the depth of a shoulder. The main section is placed on the higher of the two surfaces, with the movable step pin resting on the lower surface. If the depth between the two surfaces is sufficient but not too great, the top of the pin, but not the lower step, will be slightly above the top surface of the gage body. If the depth is too great, the top of the pin will be below the surface. Similarly, if the depth is not great enough, the lower step on the top of the pin will be above the surface of the gage body. When a finger or finger-nail is run across the top of the pin, the pin's position with respect to the surface of the gage body can readily be determined.

DEVIATION-TYPE GAGES

A large amount of gaging, and some measurement, is done through the use of **deviation-type gages**, which determine the amount by which a measured part deviates, plus or minus, from a standard dimension to which the instrument has been

FIGURE 10.36 Set of radius gages, showing how they are used. (Images Courtesy of The L.S. Starrett, Co.)

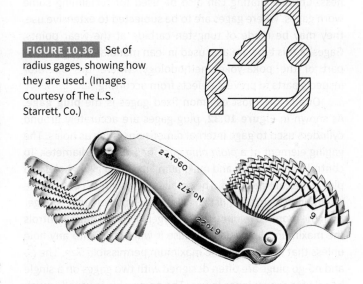

FIGURE 10.37 Thread pitch gages. (Image Courtesy of The L.S. Starrett, Co.)

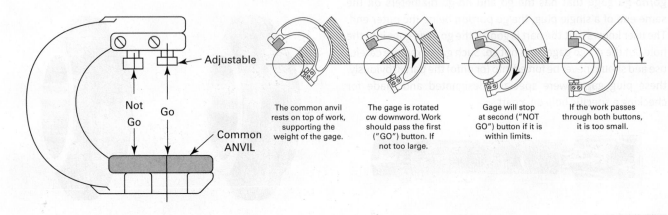

Adjustable

Not Go

Go

Common ANVIL

The common anvil rests on top of work, supporting the weight of the gage.

The gage is rotated cw downword. Work should pass the first ("GO") button. If not too large.

Gage will stop at second ("NOT GO") button if it is within limits.

If the work passes through both buttons, it is too small.

FIGURE 10.35 Adjustable go-not go limit snap gages come in sizes up-to 12 in. (Image Courtesy of Hexagon Metrology, Inc.)

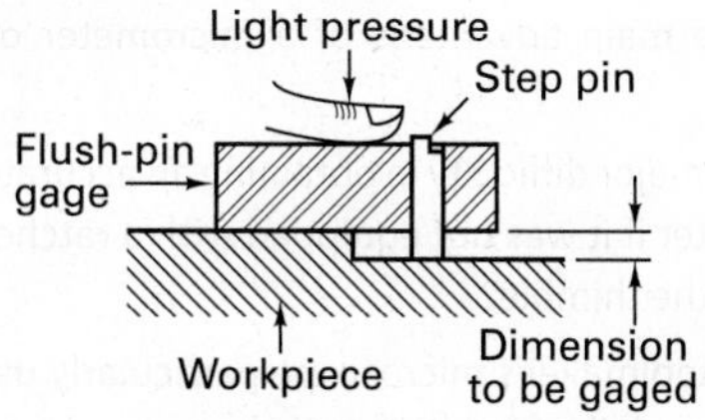

Light pressure
Step pin
Flush-pin
gage
Workpiece
Dimension
to be gaged

FIGURE 10.38 Flush-pin gage being used to check height of step.

FIGURE 10.39 A digital dial indicator with 1-in. range and 0.0001-in. accuracy is an example of a deviation type gage. (Image Courtesy of The L.S. Starrett, Co.)

set. In most cases, the deviation is indicated directly in units of measurement, but in some cases, the gage shows only whether the deviation is within a permissible range. A good example of a deviation-type gage is a flashlight battery checker, which shows whether the battery is good (green), bad (red), or borderline (yellow), but not how much voltage or current is generated. Such gages use mechanical, electrical, or fluidic amplification techniques so that very small linear deviations can be detected. Most are quite rugged, and they are available in a variety of designs, amplifications, and sizes.

Dial indicators, shown in Figure 10.2, are a widely used form of deviation-type gage. Movement of the gaging spindle is amplified mechanically through a rack and pinion and a gear train and is indicated by a pointer on a graduated dial. Most dial indicators have a spindle travel equal to about 2.5 revolutions of the indicating pointer and are read in either 0.001 or 0.0001 in. (or 0.02 or 0.002 mm).

The dial can be rotated by means of the knurled bezel ring to align the zero point with any position of the pointer. The indicator is often mounted on an adjustable arm to permit its being brought into proper relationship with the work. It is important that the axis of the spindle be aligned exactly with the dimension being gaged if accuracy is to be achieved. Digital dial indicators are also readily available (See **Figure 10.39**).

Dial indicators should be checked occasionally to determine if their gage capability has been lost through wear in the gear train. Also, it should be remembered that the pressure of the spindle on the work varies because of spring pressure as the spindle moves into the gage. This spring pressure normally causes no difficulty unless the spindles are used on soft or flexible parts.

Linear variable-differential transformers (LVDT) are used as sensory elements in many electronic gages, usually with a solid-state diode display or in automatic inspection set-ups. These devices can frequently be combined into multiple units for the simultaneous gaging of several dimensions. Ranges and resolutions down to 0.0005 and 0.00001 in. (0.013 and 0.00025 mm, respectively) are available.

Air gages have special characteristics that make them especially suitable for gaging holes or the internal dimensions of various shapes. A typical gage of this type, shown earlier, indicates the clearance between the gaging head and the hole by measuring either the volume of air that escapes or the pressure drop resulting from the airflow. The gage is calibrated directly in 0.0001-in. (0.02-mm) divisions. Air gages have an advantage over mechanical or electronic gages for this purpose in that they detect not only linear size deviations but also out-of-round conditions. Also, they are subject to very little wear because the gaging member is always slightly smaller than the hole and the airflow minimizes rubbing. Special types of air gages can be used for external gaging.

Review Questions

1. What are some of the advantages to the consumer of standardization and of interchangeable parts?

2. *DFM* stands for "design for manufacturing." Why is it important for designers to interface with manufacturing as early as possible with the design phase?

3. Explain the difference between attributes and variables inspection.

4. Why have so many variable-type devices in autos been replaced with attribute-type devices?

5. What are the four basic measures upon which all others depend?

6. What are gage blocks?

7. Why do gage blocks come in sets?

8. When gage blocks are "wrung together," what keeps them together?

9. What is the difference between accuracy and precision?

10. What is the difference between tolerance and allowance?

11. What type of fit would describe the following situations.

   a. The cap of a ball-point pen

   b. The lead in a mechanical lead pencil, at the tip

   c. The bullet in a barrel of a gun

12. What does the word *shrink* imply in a shrink fit?

13. Why might you use a shrink fit to join the wheels of trains to the axle rather than welding them?

14. Explain the difference between repeatability and reproducibility.

15. When measuring time, is it more important to be accurate or precise? Why?

16. Which kind of inspection is interferometry?

17. What factors should be considered in selecting measurement equipment?

18. Explain what is meant by the statement that usable magnification is limited by the resolution of the device.

19. What is parallax? (Why do linesmen in tennis sit looking down the line?)

20. Explain the rule of 10 in terms of tolerances.

21. How does the vernier caliper work to make measurements?

22. What are the two most likely sources of error in using a micrometer caliper?

23. What is the major disadvantage of a micrometer caliper as compared with a vernier caliper?

24. What is the main advantage of a micrometer over the vernier caliper?

25. What is the major difficulty in obtaining an accurate measurement with a micrometer if it was not equipped with a ratchet or friction device for turning the thimble?

26. Why is the toolmaker's microscope particularly useful for making measurements on delicate parts?

27. What are the ways that linear measurements can be made using an optical projector?

28. What type of instrument would you select for checking the accuracy of the linear movement of a machine tool table through a distance of 50 in.?

29. What are the chief disadvantages of using a vision system for measurement compared to laser scanning?

30. What is a CMM (coordinate measuring machine)?

31. What is the principle of a sine bar?

32. How can the no-go member of a plug gage be easily distinguished from the go member?

33. What is the primary precaution that should be observed in using a dial gage?

34. What tolerances are added to gages when they are being designed?

35. Explain how a go/no-go ring gage works to check a shaft.

36. Why are air gages particularly well suited for gaging the diameter of a hole?

37. Explain the principle of measurement by light-wave interference.

38. How does a toolmaker's flat differ from an optical flat?

Problems

1. Read the 25-division vernier graduated in English (Figure 10.A).

2. Read the 25-division vernier graduated in metric (direct reading) (Figure 10.B).

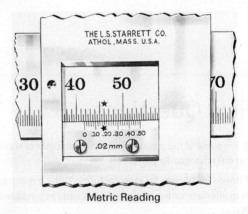

English Reading

Metric Reading

3. In Figure 10.C, the sleeve-thimble region of three micrometers graduated in thousandths of an inch is shown. What are the readings for these three micrometers? (*Hint:* Think of the various units as if you were making change from a $10 bill. Count the figures on the sleeve as dollars, the vertical lines on the sleeve as quarters, and the divisions on the thimble as cents. Add up your change, and put a decimal point instead of a dollar sign in front of the figures.)

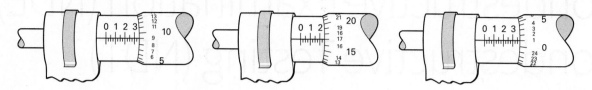

FIGURE 10.C

4. Suppose that in Figure 10.31 the height of the gage blocks is 3.2500 in. What is the angle θ, assuming that the dial indicator is reading zero?

5. What is the estimated error in this measurement, given that Grade 3 working gage blocks are being used?

6. Figure 10.D shows the sleeve-thimble region of two micrometers graduated in thousandths of an inch with a vernier for an additional ten-thousandths. What are the readings?

7. In Figure 10.E, two examples of a metric vernier micrometer are shown. The micrometer is graduated in hundredths of a millimeter (0.01 mm), and an additional reading in two-thousandths of a millimeter (0.002 mm) is obtained from vernier on the sleeve. What are the readings?

8. Suppose you had a 2-ft steel bar in your supermicrometer. Could you detect a length change if the temperature of the bar changed by 20 °F?

9. Figure 10.F shows a section of a vernier caliper. What is the reading for the outside caliper?

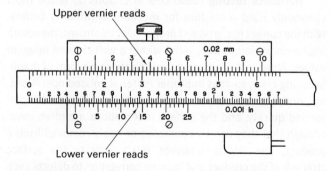

FIGURE 10.F

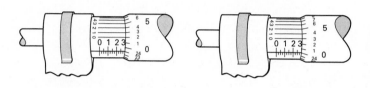

FIGURE 10.D

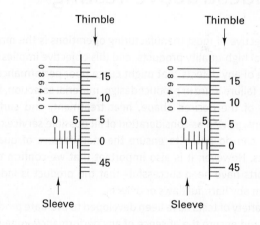

FIGURE 10.E

10. Here is a table that provides a description of fits from clearance to interference. Try to think of an example of each of these fits.

	ISO Symbol		Example
	Hole Basis	Shaft Basis	
Clearance Fits	H11/c11	C11/h11	*Loose-running fit:* for wide commercial tolerances or allowances on external members
	H9/d9	D9/h9	*Free-running fit:* not for use where accuracy is essential, but good for large temperature variations
	H8/f7	F8/h7	*Close-running fit:* for running on accurate machines and for accurate location at moderate speeds and journal pressures
	H7/g6	G7/h6	*Sliding fit:* not intended to run freely, but to move and turn freely and locate accurately
	H7/h6	H7/h6	*Locational-clearance fit:* provides snug fit for locating stationary parts, but can be freely assembled and disassembled
Transition Fits	H7/k6	K7/h6	*Locational-transition fit:* for accurate location; a compromise between clearance and interference
	H7/n6	N7/h6	*Locational-transition fit:* for more accurate location where greater interference is permissible
	H7/p6	P7/h6	*Locational-interference fit:* for parts requiring rigidity and alignment with prime accuracy of location, but without special bore pressure requirements
Interference Fits	H7/s6	S7/h6	*Medium-drive fit:* for ordinary steel parts or shrink fits on light sections; the tightest fit usable with cast iron
	H7/u6	U7/h6	*Force fit:* for highly stressed parts or for shrink fits where the heavy pressing forces required are impractical

CHAPTER 11

Nondestructive Examination (NDE) / Nondestructive Testing (NDT)

11.1 Destructive vs. Nondestructive Testing

The objective of most manufacturing operations is the manufacture of high-quality products, and this objective implies the absence of any defects that might cause poor performance or product failure. Care in product design, material selection, fabrication of the desired shape, heat treatment, and surface treatment, as well as consideration of all possible service conditions, can do much to ensure the manufacture of quality products. However, it is also important that we confirm that our efforts have been successful—that the product is indeed free from any harmful flaws or defects.

A variety of tests have been developed to evaluate product quality and ensure the absence of any performance-impairing flaws. **Destructive testing** provides one such means of product assessment. Components or assemblies are selected and then subjected to conditions that induce failure. Determining the specific conditions where failure occurs can provide insight into the performance characteristics and quality of the remaining products. Statistical methods are used to determine the probability that the remaining products would exhibit similar behavior. For example, assume that 100 parts are produced and a randomly selected one is tested to failure with satisfactory results. Is it safe to assume that the remaining 99 would also be acceptable? A satisfactory test of another randomly selected part (or better yet, the first and last of the 100 parts) would further increase our confidence in the remaining 98. Additional tests would enhance this confidence, but the cost of destroying each of the tested (i.e., destroyed) products must be borne by the remaining quantity. Regardless of the amount of testing, there will still be some degree of uncertainty because none of the remaining products has actually been subjected to any form of property assessment.

Proof testing is another means of ensuring product quality. Here a product is subjected to a load or pressure of some determined magnitude (generally equal to or greater than the designed capacity or the condition expected during operation). If the part remains intact, there is reason to believe that it will subsequently perform in an adequate fashion, provided it is not subjected to abuse or service conditions that exceed its rated level. Proof tests can be conducted under laboratory conditions or at the site of installation or assembly, as with large manufactured assemblies such as pressure vessels.

Hardness testing (described in Chapter 2) is the most commonly used procedure for mechanical property testing. With the correct material and proper heat treatment, the resulting hardness values should fall within a well-defined range of values. Abnormal results usually indicate some form of manufacturing error, such as improper material, missed operations, or poorly controlled processes. Hardness tests can be performed quickly, and the surface indentations are often small enough that they can be concealed or easily removed from a product. The results, however, relate only to the surface strength of the product and bear no correlation to defects such as cracks or voids.

Table 11.1 provides a summary of the advantages and limitations of destructive testing and compares that approach with **nondestructive testing**. In nondestructive testing, the product is examined in a manner that retains its usefulness for future service. Tests can be performed to determine: (1) if the starting material is acceptable prior to fabrication; (2) if a part is acceptable after each fabrication step, where it can also provide process control feedback; (3) if a final product is acceptable after fabrication; and (4) if a part that is already in service is acceptable for continued use. An entire production lot can be inspected, or representative samples can be taken. Different tests can be applied to the same item, either simultaneously or sequentially, and the same test can be repeated on the same specimen for additional verification. Little or no specimen

Example: Aging Infrastructure

In the United States and many other countries, an aging infrastructure has become a major concern. Because of the cost and magnitude of the problem, replacement is less of an option, and long-term maintenance has become the trend for systems that are currently operating beyond their designed lifetime. Examples include buildings, roads and highways, bridges, airports, airplanes, trains and their tracks, and electrical power generation systems (including nuclear reactors). Can they be repaired or maintained, or should they be replaced? On what basis should the decision be made?

TABLE 11.1	Advantages and Limitations of Destructive and Nondestructive Testing

DESTRUCTIVE TESTING

Advantages

1. Provides a direct and reliable measurement of how a material or component will respond to service conditions.
2. Provides quantitative results, useful for design.
3. Does not require interpretation of results by skilled operators.
4. Usually find agreement as to meaning and significance of test results.

Disadvantages

1. Applied only to a sample; must show that the sample is representative of the group.
2. Tested parts are destroyed during testing.
3. Usually cannot repeat a test on the same item or use the same specimen for multiple tests.
4. May be restricted for costly or few-in-number parts.
5. Hard to predict cumulative effect of service usage.
6. Difficult to apply to parts in use; if done, testing terminates their useful life.
7. Extensive machining or preparation of test specimens is often required.
8. Capital equipment and labor costs are often high.

NONDESTRUCTIVE TESTING

Advantages

1. Can be performed directly on production items without regard to cost or quantity available.
2. Can be performed on 100% of production lot (when high variability is observed) or a representative sample (if sufficient similarity is noted).
3. Different tests can be applied to the same item, and a test can be repeated on the same specimen.
4. Can be performed on parts that are in service; the cumulative effects of service life can be monitored on a single part.
5. Little or no specimen preparation is required.
6. The test equipment is often portable.
7. Labor costs are usually low.

Disadvantages

1. Results often require interpretation by skilled operators.
2. Different observers may interpret the test results differently.
3. Properties are measured indirectly, and results are often qualitative or comparative.
4. Some test equipment requires a large capital investment.

preparation is required, and the equipment is often portable, permitting on-site testing in most locations.

Nondestructive tests can detect internal or surface flaws or discontinuities, measure a product's dimensions, determine a material's structure or chemistry, or evaluate a material's physical or mechanical properties. In general, nondestructive tests incorporate the following aspects: (1) some means of probing a material or product; (2) a means by which a flaw, defect, material property, or specimen feature interacts with or modifies whatever is probing; (3) a sensor to detect the response; (4) a device to indicate or record the response; and (5) a way to interpret and evaluate quality.

How you choose to look at a material or product generally depends on what you are looking at, what you wish to see, and in how much detail you wish to examine it. Each of the various inspection processes has characteristic advantages and limitations. Some can be performed only on certain types of materials (such as electrical conductors or ferromagnetic materials). Many are limited in the type, size, and orientation of flaws that they can detect. There may be geometric restrictions relating to part size, part complexity, or the accessibility of critical surfaces or locations. The availability of required equipment, the cost of the operation, the need for a skilled operator or technician, and the possibility of producing a permanent test record are additional considerations when selecting a test procedure.

Regardless of the selected method, nondestructive testing can be a vital element in good manufacturing practice. Its potential value has been widely recognized as productivity and production rates increase, consumers demand higher-quality products, and product liability continues to be a concern. Rather than being an added manufacturing cost, nondestructive testing can actually expand profit by ensuring product reliability and customer satisfaction. In addition to its role in quality control, nondestructive testing can also be used as an assessment aid in product design. Periodic testing can provide a means of controlling a manufacturing process, reducing overall manufacturing costs by preventing the continued manufacture of out-of-specification, defective, or poor-quality parts.

The remainder of this chapter consists of an overview of the various nondestructive test methods. Each is presented along with a discussion of its underlying principle, associated advantages and limitations, compatible materials, and typical applications.

Example: Nondestructive Examination in Everyday Life

People use NDE every day and never recognize it. Consider the purchase of an item of produce like peaches, or an apple, or a melon. You probably look at it rather closely (visual examination). You may tap the melon to determine ripeness (acoustic examination). You may use a variety of other techniques – squeeze the peach slightly to determine ripeness or smell the melon. Then, on the basis of your examinations, you decide to purchase or reject the item. If you observe others, you may notice that people possess different standards of choice and different ability to test and examine – a potential problem when applied to manufacturing because we would like a procedure that is independent of inspector judgment and would always produce the same results.

11.2 Visual Inspection

Probably the simplest and most widely used nondestructive testing method is **visual inspection**, summarized in **Table 11.2**, and it is normally the first step in the examination process. The human eye is a very discerning instrument, and with training, the brain can readily interpret the signals. Optical aids such as mirrors, magnifiers, and microscopes can expand the capabilities of this system. Cameras can provide a permanent record. Video cameras and computer systems, such as digital image analyzers, can be used to automate the inspection and perform quantitative geometrical evaluations. Borescopes, fiberoptics, and similar tools can provide accessibility to otherwise inaccessible locations. Only the surfaces of a product can be examined, but that is often sufficient to reveal dimensional irregularities, corrosion, contamination, surface finish flaws, wear or discoloration, and a wide variety of surface discontinuities.

11.3 Liquid Penetrant Inspection

Liquid penetrant testing, also called **dye penetrant inspection**, is an effective method of detecting surface defects in metals and other nonporous materials and is illustrated schematically in **Figure 11.1**. The piece to be tested is first subjected to a thorough cleaning and is dried prior to the test. Then a **penetrant**, a liquid material capable of wetting the entire surface and being drawn into fine openings, is applied to the surface of the workpiece by dipping, spraying, or brushing. Sufficient time is given for capillary action to draw the penetrant into any surface discontinuities, including ones so small that they could not be detected by the unaided eye, and the excess penetrant liquid is then removed by wiping, water wash, or solvent. The surface is then coated with a thin film of **developer**, an absorbent material capable of drawing traces of penetrant from the defects back onto the surface. Brightly colored dyes or fluorescent materials that glow under ultraviolet light are generally added to the penetrant to make these traces more visible, and the developer is often selected to provide a contrasting background, such as red dye with white developer. Radioactive tracers can also be added and used in conjunction with photographic paper

TABLE 11.2	Visual Inspection

Principle: Illuminate the test specimen and observe the surface. Can reveal a wide spectrum of surface flaws and geometric discontinuities. Use of optical aids or assists (such as magnifying glass, microscopes, illuminators, and mirrors) is permitted. Although most inspection is by human eye, video cameras and computer-vision systems can be employed.

Advantages: Simple, easy to use, relatively inexpensive, fast, minimal surface preparation.

Limitations: Depend on skill and knowledge of inspector. Limited to detection of surface flaws.

Material limitations: None.

Geometrical limitations: Any size or shape providing viewing accessibility of surfaces to be inspected.

Permanent record: Photographs or videotapes are possible. Inspectors' reports also provide valuable records.

Remarks: Should always be the initial and primary means of inspection and is the responsibility of everyone associated with parts manufacture.

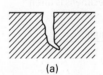
(a)

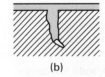
(b)

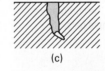
(c)

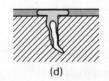
(d)

FIGURE 11.1 Liquid-penetrant testing: (a) initial surface with open crack; (b) penetrant is applied and is pulled into the crack by capillary action; (c) excess penetrant is removed; and (d) developer is applied, some penetrant is extracted, and the product inspected.

to produce a permanent image of the defects. Cracks, laps, seams, lack of bonding, pinholes, gouges, and tool marks can all be detected. After inspection, the developer and residual penetrant are removed by a second cleaning operation.

TABLE 11.3	Liquid Penetrant Inspection

Principle: A liquid penetrant containing fluorescent material or dye is drawn into surface flaws by capillary action and subsequently revealed by developer material in conjunction with visual inspection. Can detect cracks as small as 5 μm deep and 10 μm wide

Advantages: Simple, inexpensive, versatile, portable, easily interpreted, and applicable to complex shapes.

Limitations: Can only detect flaws that are open to the surface; surfaces must be cleaned before and after inspection; deformed surfaces and surface coatings may prevent detection; and the penetrant may be wiped or washed out of large defects. Cannot be used on hot products.

Material limitations: Applicable to all materials with a nonporous surface.

Geometrical limitations: Any size or shape permitting accessibility of surfaces to be inspected.

Permanent record: Photographs, videotapes, and inspectors' reports provide the most common records.

TABLE 11.4	Magnetic Particle Inspection

Principle: When magnetized, ferromagnetic materials will have a distorted magnetic field in the vicinity of flaws and defects. Magnetic particles will be strongly attracted to regions where the magnetic flux breaks the surface.

Advantages: Relatively simple, fast, easy-to-interpret; portable units exist; can reveal both surface and subsurface flaws and inclusions (as much as or 6 mm deep) and small, tight cracks.

Limitations: Parts must be relatively clean; alignment of the flaw and the field affects the sensitivity so that multiple inspections with different magnetizations may be required; can only detect defects at or near surfaces; must demagnetize part after test; high current source is required; some surface processes can mask defects; postcleaning may be required.

Material limitations: Must be ferromagnetic; nonferrous metals such as aluminum, magnesium, copper, lead, tin, and titanium and the ferrous (but not ferromagnetic) austenitic stainless steels cannot be inspected.

Geometrical limitations: Size and shape are almost unlimited, most restrictions relate to the ability to induce uniform magnetic fields within the piece. Hard to use on rough surfaces.

Permanent record: Photographs, videotapes, and inspectors' reports are most common. In addition, the defect pattern can be preserved on the specimen by an application of transparent lacquer, or transferred to a piece of transparent tape that has been applied to the specimen and peeled off.

To be successful, the inspection for surface defects must be correlated with the manufacturing operations. If previous processing involved techniques that might have induced the flow of surface material, such as shot peening, honing, burnishing, machining, or various forms of cold working, a chemical etching may be required to remove material that might be covering critical flaws. An alternative procedure is to penetrant test before any surface-finishing operations, when significant defects will still be open and available for detection. Penetrant inspection systems can range from aerosol spray cans of cleaner, penetrant, and developer (for portable applications), to automated, mass-production equipment using sophisticated computer vision systems. **Table 11.3** shows a summary of the process and its advantages and limitations.

11.4 Magnetic Particle Inspection

Magnetic particle inspection, summarized in **Table 11.4**, is based on the principle that ferromagnetic materials (such as the alloys of iron, nickel, and cobalt), when magnetized, will have distorted magnetic fields in the vicinity of material defects. As shown in **Figure 11.2**, surface and subsurface flaws, such as cracks and inclusions, will produce magnetic anomalies that can be mapped with the aid of magnetic particles on the specimen surface.

As with the previous method, the specimen must be cleaned prior to inspection. A suitable magnetic field is then established in the part. As shown in **Figure 11.3**, orientation can be quite important. For a flaw to be detected, it must produce a significant disturbance of the magnetic field at or near the surface. If a bar of steel is placed within an energized coil, a magnetic field will be produced whose lines of flux travel along the axis of the bar. Any defect perpendicular to this axis will significantly alter the field. If the perturbation is sufficiently large and close enough to the surface, the flaw can be detected. However, if the flaw is in the form of a crack aligned with the specimen axis, there will be little perturbation of the lines of flux, and the flaw is likely to go undetected.

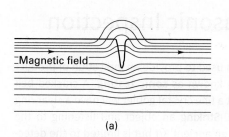
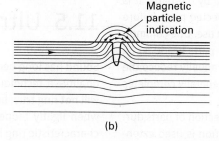
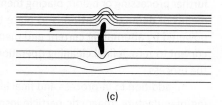

FIGURE 11.2 (a) Magnetic field showing disruption by a surface crack; (b) magnetic particles are applied and are preferentially attracted to field leakage; (c) subsurface defects can also produce surface-detectable disruptions if they are sufficiently close to the surface.

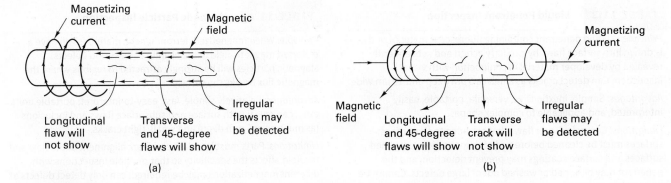

Magnetizing current

Magnetic field

Longitudinal flaw will not show

Transverse and 45-degree flaws will show

Irregular flaws may be detected

(a)

Magnetizing current

Magnetic field

Longitudinal and 45-degree flaws will show

Transverse crack will not show

Irregular flaws may be detected

(b)

FIGURE 11.3 (a) A bar placed within a magnetizing coil will have an axial magnetic field. Defects parallel to this field may go unnoticed, while those that disrupt the field and are sufficiently close to a surface are likely to be detected. (b) When magnetized by a current passing through it, the bar has a circumferential magnetic field, and the geometries of detectable flaws are reversed.

If the cylindrical specimen is then magnetized by passing a current through it, a circumferential magnetic field will be produced. Any axial defect now becomes a significant perturbation, and a defect perpendicular to the axis will likely go unnoticed. To fully inspect a product, therefore, a series of inspections may be required using various forms of magnetization. Passing a current between various points of contact is a popular means of inducing the desired fields. Electromagnetic coils of various shapes and sizes are also used. Alternating-current methods are most sensitive to surface flaws, whereas direct-current inspections are better for detecting subsurface defects, such as nonmetallic inclusions.

After the specimen has been subjected to a magnetic field, fine ferromagnetic particles are applied to the surface in the form of either a dry powder or a suspension in a liquid carrier. These particles are preferentially attracted to places where the lines of magnetic flux break the surface, revealing anomalies that can then be interpreted. To better reveal the orientation of the lines of flux, the particles are often made in an elongated form. They can also be treated with a fluorescent material to enhance observation under ultraviolet light or coated with a lubricant to prevent oxidation and enhance their mobility. **Figure 11.4** shows a component of a truck front-axle assembly: as manufactured, under straight magnetic particle inspection, and under ultraviolet light with fluorescent particles.

After magnetic particle inspection, it is not uncommon for some residual magnetization to be retained by the parts. It is usually necessary to demagnetize the inspected parts before further processing or before placing them in use. One common means of **demagnetization** is to place the parts inside a coil powered by alternating current and then gradually reduce the current to zero. A final cleaning operation generally completes the process.

In addition to in-process and final inspection of parts during manufacture, magnetic particle inspection is used extensively during the maintenance and overhaul of equipment and machinery. The testing equipment ranges from small, portable units to complex automated systems.

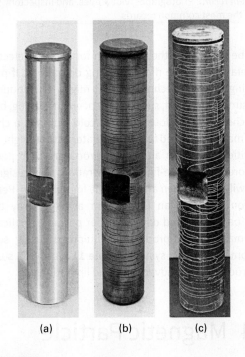

(a) (b) (c)

FIGURE 11.4 Front-axle king pin for a truck: (a) As manufactured and apparently sound; (b) inspected under conventional magnetic particle inspection to reveal numerous grinding-induced cracks; (c) fluorescent particles and ultraviolet light make the cracks even more visible. (Courtesy of Magnaflux, a division of Illinois Tool Works, Inc.)

11.5 Ultrasonic Inspection

Sound has long been used to provide an indication of product quality—a technique known as **sonic testing**. A cracked bell will not ring true, but a fine crystal goblet will have a clear ring when lightly tapped. Striking an object and listening to the characteristic ring is an ancient art but is limited to the detection of large defects because the wavelength of audible sound is rather large compared to the size of most defects. By reducing the wavelength of the signal to the ultrasonic range, typically

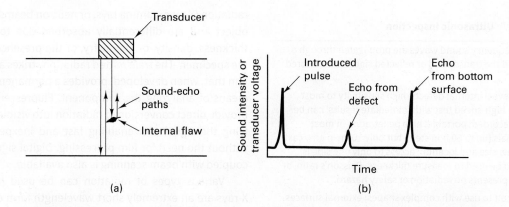

FIGURE 11.5 (a) Ultrasonic inspection of a flat plate with a single transducer; (b) plot of sound intensity or transducer voltage versus time showing the initial pulse and echoes from the bottom surface and intervening defect.

between 100 kHz and 25 MHz, ultrasonic inspection can be used to detect rather small defects and flaws.

Ultrasonic inspection involves sending high-frequency waves through a material and observing the response. [Note: The material must be capable of transmitting sound. Metals, composite laminates, ceramics, and polymers can all be tested.] Within the specimen, sound waves can be affected by voids, impurities, changes in density, delaminations, interfaces with materials having a different speed of sound, and other imperfections. At any interface, part of the ultrasonic wave will be reflected and part will be transmitted. If the incident beam is at an angle to an interface where materials change, the transmitted portion of the beam will be bent to a new angle by the phenomenon of refraction. By receiving and interpreting either transmitted or reflected signals, ultrasonic inspection can be used to detect flaws within the material, measure thickness from only one side, or characterize metallurgical structure. Flaws can be detected with a surface area as small as 1.3 mm² (0.002 in²).

An ultrasonic inspection system begins with a pulsed oscillator and **transducer**, a device that transforms electrical energy into mechanical vibrations. The pulsed oscillator generates a burst of alternating voltage with a characteristic principal frequency, duration, profile, and repetition rate. This burst is then applied to a sending transducer, which uses a piezoelectric crystal to convert the electrical oscillations into mechanical vibrations. Because air is a poor transmitter of ultrasonic waves, an acoustic **coupling medium**—such as oil, water, glycerin, or grease—is required to link the transducer to the piece to be inspected and transmit the vibrations into the part. The pulsed vibrations then propagate through the part with a velocity that depends on the density and elasticity of the test material. A receiving transducer is then used to convert the transmitted or reflected vibrations back into electrical signals. The receiving transducer is often identical to the sending unit, and the same transducer can actually perform both functions. A receiving unit then amplifies, filters,

and processes the signal for display, possible recording, and final interpretation. An electronic clock is generally integrated into the system to time the responses and provide reference signals for comparison purposes.

Depending on the test objectives and part geometry, several different inspection methods can be employed:

1. In the **pulse-echo technique**, an ultrasonic pulse is introduced into the piece to be inspected, and the echoes from opposing surfaces and any intervening flaws are detected by the receiver. The time interval between the initial emitted pulse and the various echoes can be displayed on the horizontal axis of a display screen. The position and amplitude of the various echoes provides significant information. By correlating the position of a defect echo with those from known surfaces, the location of the flaw or defect can be determined. The relative amplitude of reflected pulses is proportional to the size of the reflecting surface. **Figure 11.5** shows a schematic of a single-transducer pulse-echo inspection and the companion signal as it would appear on a display. **Figure 11.6a** depicts a dual-transducer pulse-echo examination. Both cases require access to only one side of the specimen.

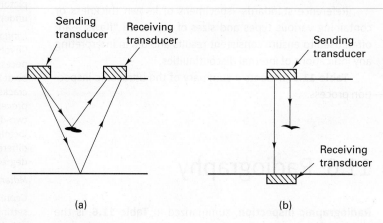

FIGURE 11.6 (a) Dual-transducer ultrasonic inspection in the pulse-echo mode; (b) dual transducers in through-transmission configuration.

TABLE 11.5	Ultrasonic Inspection

Principle: High-frequency sound waves are propagated through a test specimen and the transmitted or reflected signal is monitored and interpreted.

Advantage: Can reveal internal defects; high sensitivity to most cracks and flaws; high-speed test with immediate results; can be automated and recorded; portable; high penetration in most important materials (up to 60 ft in steel) but only 10cm in gray cast iron; indicates flaw size and location; access to only one side is required; can also be used to measure thickness, Poisson's ratio, or elastic modulus; presents no radiation or safety hazard.

Limitations: Difficult to use with complex shapes; external surfaces and defect orientation can affect the test (may need dual transducer or multiple inspections); a couplant is required; the area of coverage is small (inspection of large areas requires scanning); trained, experienced, and motivated technicians may be required; requires smooth surfaces.

Material limitations: Few—can be used on metals, plastics, ceramics, glass, rubber, graphite, and concrete, as well as joints and interfaces between materials. Material must conduct sound.

Geometric limitations: Small, thin, or complex-shaped parts or parts with rough surfaces and nonhomogeneous structure pose the greatest difficulty.

Permanent record: Ultrasonic signals can be recorded for subsequent playback and analysis. Strip charts can also be used.

2. The **through-transmission technique** requires separate sending and receiving transducers. As shown in **Figure 11.6b**, a pulse is emitted by the sending transducer and detected by a receiver on the opposite surface. Flaws in the material decrease the amplitude of the transmitted signal because of back-reflection and scattering.

3. **Resonance testing** can be used to determine the thickness of a plate or sheet from one side of the material. Input pulses of varying frequency are fed into the material. When resonance is detected by an increase in energy at the transducer, the thickness can be calculated from the speed of sound in the material and the time of traverse. Ultrasonic thickness gages can be calibrated to provide direct digital readout of the thickness of a material.

Reference standards—specimens of known thickness or containing various types and sizes of machined "flaws"—are often used to ensure consistent results and aid in interpreting any indications of internal discontinuities.

Table 11.5 presents a summary of the ultrasonic inspection process.

11.6 Radiography

Radiographic inspection, summarized in **Table 11.6**, is the second most used method of nondestructive inspection and employs the same principles and techniques as those of medical X-rays. A shadow pattern is created when certain types of radiation (X-rays, gamma rays, or neutron beams) penetrate an object and are differentially absorbed due to variations in thickness, density, or chemistry, or the presence of defects in the specimen. The transmitted radiation strikes a photographic film that, when developed, provides a permanent record and a means of analyzing the component. Fluorescent screens can provide direct conversion of radiation into visible light, intensifying the signal and enabling fast and inexpensive viewing without the need for film processing. Digital signal processing coupled with beam scanning is also available.

Various types of radiation can be used for inspection. X-rays are an extremely short wavelength form of electromagnetic radiation that are capable of penetrating many materials that reflect or absorb visible light. They are generated by high-voltage electrical apparatus—the higher the voltage, the shorter the X-ray wavelength and the greater the energy and penetrating power of the beam. Gamma rays, another useful form of electromagnetic radiation, are emitted during the disintegration of radioactive nuclei. Various radioactive isotopes can be selected as the radiation source. Neutron beams for radiography can be obtained from nuclear reactors, nuclear accelerators, or radioisotopes. For most applications it is necessary to moderate the energy and collimate the beam before use.

The absorption of X-rays and gamma rays depends on the thickness, density, and composition of the material being inspected. For X-ray radiation, the higher the atomic number, the greater the attenuation of the beam. In contrast to X-ray absorption, neutron absorption varies widely from atom to atom, with no pattern in terms of atomic number. Unusual

TABLE 11.6	Radiography

Principle: Some form of radiation (X-ray, gamma ray, or neutron beam) is passed through the sample and is differentially absorbed depending on the thickness, type of material, and the presence of internal flaws or defects.

Advantages: Probes the internal regions of a material; provides a permanent record of the inspection; can be used to determine the thickness of a material; very sensitive to density changes; the pictorial image matches the geometry of the part (easily understood).

Limitations: Most costly of the nondestructive testing methods (involves expensive equipment); radiation precautions are necessary (potentially dangerous to human health); the defect must be at least 2% of the total section thickness to be detected (thin cracks can be missed if oriented perpendicular to the beam); film processing requires time, facilities, and care; the image is a two-dimensional projection of a three-dimensional object, so the location of an internal defect requires a second inspection at a different angle; complex shapes can present problems; a high degree of operator training is required; two-side access is necessary.

Material limitations: Applicable to most engineering materials.

Geometric limitations: Complex shapes can present problems in setting exposure conditions and obtaining proper orientation of source, specimen, and film. Two-side accessibility is required.

Permanent record: A photographic image is part of the standard test procedure.

Example: Radiography of the Liberty Bell

Figure 11.7 shows a radiograph of the historic Liberty Bell. The famous crack is clearly visible, along with the internal spider (installed to support the clapper in 1915) and the steel beam and bolts installed in the wooden yoke in 1929. Other radiographs disclosed previously unknown shrinkage separations and additional cracks in the bell, as well as a crack in the bell's clapper.

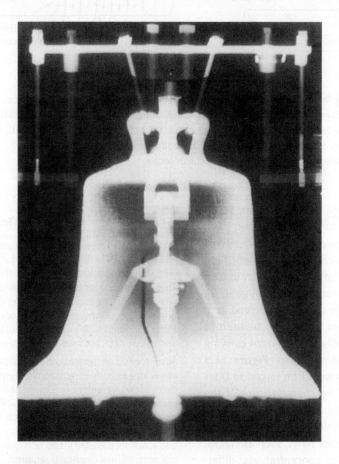

FIGURE 11.7 Radiograph of the Liberty Bell. The photo reveals the famous crack, as well as the iron spider installed in 1915 to support the clapper and the steel beam and supports, which were set into the yoke in 1929. (Used with Permission of Kodak)

contrasts can be obtained that would be impossible with other inspection methods. For example, hydrogen has a high neutron absorption. The presence of water in a product can be easily detected by neutron radiography. X-rays, on the other hand, are readily transmitted through water, and its presence could be missed. Nitrogen has a similar response, and nitrides can be readily detected. Detection by differential absorption generally requires that the size of flaws or defects be greater than 1 to 2% of the total thickness, so features like narrow or hairline cracks can easily be missed.

A standard test piece, or **penetrameter**, is often included in a radiographic exposure. Penetrameters are made of the same or similar material as the specimen and contain features with known dimensions, such as incremental thickness steps or flaws of known dimension. The image of the penetrameter is compared to the image of the product being inspected. Regions of similar intensity are considered to be of similar thickness.

Radiography is not inexpensive, however. Many users, therefore, recommend extensive use only during the development of a new product or process, followed by spot checks and statistical methods during subsequent production. Other motivations may be the presence of a permanent record that can be stored or viewed by multiple personnel, or the availability of an inspection image that corresponds to the geometry of the product.

11.7 Eddy-Current Testing

When an electrically conductive material is exposed to an alternating magnetic field such as that generated by a coil of wire carrying an alternating current, small electric currents are

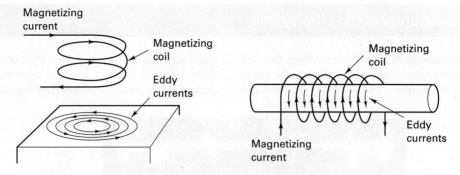

FIGURE 11.8 Relation of the magnetizing coil, magnetizing current, and induced eddy currents. The magnetizing current is actually an alternating current, producing a magnetic field that forms, collapses, and reforms in the opposite direction. This dynamic magnetic field induces the eddy currents, and the changes in the eddy currents produce a secondary magnetic field that interacts with the sensor coil or probe.

induced on or near the surface of the material (**Figure 11.8**). These induced **eddy currents** in turn generate their own opposing magnetic field, which then reduces the strength of the original field from the coil. This change in magnetic field causes a change in the **impedance** of the coil, which in turn changes the magnitude of the current flowing through it. By monitoring the impedance or voltage of the exciting coil, or a separate indicating coil, eddy-current testing can be used to detect any condition that would affect the current-carrying ability (or conductivity) of the test specimen. **Figure 11.9** shows how the eddy-current paths would be forced to alter around a crack, thereby changing the characteristics of the induced magnetic field in that vicinity.

Eddy-current testing, summarized in **Table 11.7**, can be used to detect surface and near-surface flaws, such as cracks, voids, inclusions, and seams. Stress concentrations, differences in metal chemistry, or variations in heat treatment (i.e., microstructure, grain size, and hardness) will all affect the magnetic permeability and conductivity of a metal and therefore alter the eddy-current characteristics. Material mix-ups and

TABLE 11.7	Eddy-Current Testing

Principle: When an electrically conductive material is brought near an alternating-current coil that produces an alternating magnetic field, surface currents (eddy currents) are generated in the material. These surface currents generate their own magnetic field, which interacts with the original, modifying the impedance of the originating coil. Various material properties and/or defects can affect the magnitude and direction of the induced eddy currents and can be detected by the electronics.

Advantages: Can detect both surface and near-surface irregularities; applicable to both ferrous and nonferrous metals; versatile—can detect flaws, variations in alloy or heat treatment, variations in plating or coating thickness, wall thickness, and crack depth; intimate contact with the specimen is not required; can be automated; electrical circuitry can be adjusted to select sensitivity and function; pass/fail inspection is easily conducted; high speed; low cost; no final cleanup is required.

Limitations: Response is sensitive to a number of variables, so interpretation may be difficult; sensitivity varies with depth, and depth of inspection depends on the test frequency; reference standards are needed for comparison; trained operators are generally required.

Material limitations: Only applicable to conductive materials; some difficulties may be encountered with ferromagnetic materials.

Geometric limitations: Depth of penetration is limited; must have accessibility of coil or probe; constant separation distance between coils and specimen is required for good results.

Permanent record: Electronic signals can be recorded using devices such as strip-chart recorders.

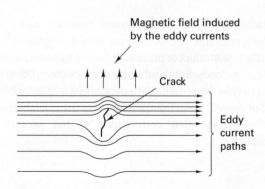

FIGURE 11.9 Eddy currents are constrained to travel within the conductive material, but the magnitude and path of the currents will be affected by defects and changes in material properties. By focusing on the magnitude of the eddy currents, features such as differences in heat treatment can be detected.

processing errors can therefore be detected. Specimens can be sorted by hardness, case depth, residual stresses, or any other structure-related property. Thickness (or variation in thickness) of platings, coatings, or even corrosion can be detected and measured.

Eddy-current test equipment can range from simple, portable units with handheld probes to fully automated systems with computer control and analysis. Each system, however, includes the following:

1. *A source of changing magnetic field capable of inducing eddy currents in the part being tested.* This source generally takes the form of a coil (or coil-containing probe) carrying alternating current of a specific frequency and amplitude. (*Note:* High frequency gives shallow penetration and a higher eddy-current density on the surface. Low frequency gives deeper penetration. Eddy-current frequencies vary from 10 Hz to the MHz range.) Various coil geometries are used for different-shaped specimens.

2. *A means of sensing the field changes caused by the interaction of the eddy currents with the original magnetic field.* Either the exciting coil itself or a secondary sensing coil can be used to detect the impedance changes. Differential testing can be performed using two oppositely wound coils wired in series. In this method, differences in the signals between the two coils are detected as one or both coils are scanned over the specimen.

3. *A means of measuring and interpreting the resulting impedance changes.* The simplest method is to measure the induced voltage of the sensing coil, a reading that evaluates the cumulative effect of all variables affecting the eddy-current field. Phase analysis can be used to determine the magnitude and direction of the induced eddy-current field. Familiarity with characteristic impedance responses or comparison with signals from other parts or other locations can be used to identify or interpret the desired features in the specimen.

When comparing alternative techniques, eddy current is usually not as sensitive as penetrant testing in detecting small, open flaws, but it requires none of the cleanup operations and is noticeably faster. In a similar manner, it is not as sensitive as magnetic particle inspection to small subsurface flaws but it can be applied to all metals (ferromagnetic and nonferromagnetic alike). In addition, eddy-current testing offers capabilities that cannot be duplicated by the other methods, such as the ability to differentiate between various chemistries and heat treatments.

11.8 Acoustic Emission Monitoring

Materials experiencing the dynamic events of deformation, fracture, crack growth, or phase transformation emit stress waves in frequencies between 20 and 1200 kHz. Although these sounds are inaudible to the human ear, they are detectable through the use of sophisticated electronics. Sensitive piezoelectric transducers, amplifiers, filters, counters, and computers can be used to detect, isolate and analyze the sonic emissions of a cracking or deforming material. Much like the warning sound of ice cracking underneath boots or skates, the acoustic emissions of materials can be used to provide a warning of impending danger. They can detect deformations as small as 10^{-12} cm/cm or in/in (that occur in short intervals of time), initiation or propagation of cracks (including stress-corrosion cracking), delamination of layered materials, and fiber failure in composites. By using multiple sensors, it is possible to accurately pinpoint the source of these sounds by a time-of-arrival triangulation method similar to that used to locate seismic sources (earthquakes) in the earth.

Acoustic emission monitoring, summarized in **Table 11.8**, involves listening for indications of failure. Temporary monitoring can be used to detect the formation of cracks in materials during selected production operations, such as welding and subsequent cooling of the weld region. Monitoring can also be employed to ensure the absence of plastic deformation during preservice proof testing. Continuous surveillance may be used when the product or component is particularly critical, like nuclear reactor pressure vessels and fatigue-prone structures, such as bridges. The sensing electronics can be coupled to an alarm and safety system to protect and maintain the integrity of the structure. Other uses include the detection of tool contact or monitoring tool wear in automated machining and detection of increased wear or loss of lubrication in operating equipment.

In contrast to the previous inspection methods, acoustic emission cannot detect an existing defect in a static product. Instead, it is a monitoring technique designed to detect a dynamic change in the material, such as the formation or growth of a crack or defect, or the onset of plastic deformation.

TABLE 11.8   Acoustic Emission Monitoring

Principle: Almost all materials will emit high-frequency sound (acoustic emissions) when stressed, deformed, or undergoing structural changes, such as the formation or growth of a crack or defect. These emissions can now be detected and provide an indication of dynamic change within the material.

Advantages: The entire structure can be monitored with near-instantaneous detection and response; continuous surveillance is possible; defects inaccessible to other methods can be detected; inspection can be in harsh environments; and the location of the emission source can be determined; can accurately determine the stress at which damage occurs.

Limitations: Only growing or "active" flaws can be detected (the mere presence of defects is not detectable); background signals may cause difficulty; there is no indication of the size or shape of the flaw; expensive equipment is required, and experience is required to interpret the signals.

Material limitations: Virtually unlimited, provided that they are capable of transmitting sound.

Geometric limitations: Requires continuous sound-transmitting path between the source and the detector. Size and shape of the component affect the strength of the emission signals that reach the detector.

11.9 Other Methods of Nondestructive Testing and Inspection

Leak Testing

Leak testing is a form of nondestructive testing designed to determine the existence or absence of leak sites and the rate of material loss through the leaks. Various testing methods have been developed, ranging from the rather crude bubble-emission test (pressurize, immerse, and look for bubbles), through simple pressure drop tests with either air or liquid as the pressurized media, to advanced techniques involving tracers, detectors, and sophisticated apparatus. Each has its characteristic advantages, limitations, and sensitivity. Selection should be on the basis of cost, sensitivity, reliability, and compatibility with the specific product to be tested.

Thermography and Other Thermal Methods

Thermography and other thermal methods can also be used to evaluate the soundness of engineering materials and components. Contact and noncontact temperature sensing devices include: thermometers, thermocouples, pyrometers, temperature-sensitive paints and coatings, liquid crystals, infrared scanners, infrared film, and others. Parts can be heated and then inspected during cooldown to reveal abnormal temperature distributions that are the result of faults or flaws. The identification of **hot spots** on an operating component is often an indication of a flaw or defect and may provide advanced warning of impending failure. For example, faulty electrical components tend to be hotter than defect-free devices. Composite materials (difficult to inspect by many standard techniques) can be subjected to brief pulses of intense heat and then inspected to reveal the temperature pattern produced by subsequent thermal conductivity. Thermal anomalies tend to appear in areas where the bonding between the components is poor or incomplete. In another technique, ultrasonic waves are used to produce heat at internal defects, which are then detected by infrared examination.

Microwaves

Microwaves, with wavelengths of 1 mm or greater, penetrate deeply into materials and are reflected from internal boundaries. **Microwave inspection** has been used to detect voids, delaminations, macroporosity, and inclusions, and microwaves can also measure the thickness of products or coatings. They are particularly useful with plastics and ceramics, where other probing techniques are less effective.

Strain Sensing

Although used primarily during product development, **strain-sensing techniques** can also be used to provide valuable insight into the stresses and stress distribution within a part. Brittle coatings, photoelastic coatings, or electrical resistance strain gages can be applied to the external surfaces of a part, which are then subjected to an applied stress. The extent and nature of cracking, the photoelastic pattern produced, or the electrical resistance changes then provide insight into the strain at various locations. X-ray diffraction methods and extensometers have also been used.

Advanced Optical Methods

Although visual inspection is often the simplest and least expensive of the nondestructive inspection methods, there are also several advanced optical methods. **Monochromatic laser light** can be used to detect differences in the backscattered pattern from a part and a master. The presence or absence of geometrical features such as holes or gear teeth is readily detected. **Holograms** can provide three-dimensional images of an object, and **holographic interferometry** can detect minute changes in the shape of an object under stress.

Resistivity Methods

The **electrical resistivity** of a conductive material is a function of its chemistry, processing history, and structural soundness. Measurement of resistivity can therefore be used for alloy identification, flaw detection, or the assurance of proper processing. Tests can be developed to evaluate the effects of heat treatment, the amount of cold work, the integrity of welds, or the depth of case hardening. The development of sensitive microohm-meters has greatly expanded the possibilities in this area.

Computed Tomography

Whereas X-ray radiography provides a single image of the X-ray intensity being transmitted through an object, X-ray **computed tomography (CT)** is an inspection technique that provides a cross-sectional view of the interior of an object along a plane parallel to the X-ray beam. This is the same technology that has revolutionized medical diagnostic imaging (CAT scans), with the process parameters (such as the energy of the X-ray source) being adapted to permit the nondestructive probing of industrial products. Basic systems include an X-ray source, an array of detectors, a mechanical system to move and rotate the test object, and a dedicated computer system. The intensity of the received signal is recorded at each of the numerous detectors with the part in a variety of orientations. Complex numerical algorithms are then used to construct an image of the interior of the component. Internal boundaries and surfaces can be determined clearly, enabling inspection and dimensional analysis of a product's interior. The presence of cracks, voids, or inclusions can also be detected, and their precise location can be determined.

CT inspections are slow and costly, so they are currently used only when the component is critical and the more standard inspection methods prove to be inadequate due to features such as shape complexity, thick walls, or poor resolution of detail. The video images of the CT technique also permit easy visualization and interpretation.

Acoustic holography is another computer reconstruction technique, this time based on ultrasound reflections from within the part.

Chemical Analysis and Surface Topography

Whereas nondestructive inspection is usually associated with the detection of flaws and defects, various nondestructive techniques can also be employed to determine the chemical and elemental analysis of surface and near-surface material. These techniques include Auger electron spectroscopy (AES), energy-dispersive X-ray analysis (EDX), electron spectroscopy for chemical analysis (ESCA), and various forms of secondary-ion mass spectroscopy (SIMS). Because of its large depth of focus, the scanning electron microscope has become an extremely useful tool for observing the surfaces of materials. More recently, the atomic-force microscope and scanning tunneling microscope have extended this capability and can now provide information about surface topography with resolution to the atomic scale.

11.10 Dormant vs. Critical Flaws

There was a time when the detection of a flaw was considered to be sufficient cause for rejecting a material or component (Detection = Rejection), and material specifications often contained the term **flaw-free**. Such a criterion, however, is no longer practical, because the sensitivity of detection methods has increased dramatically. If materials were rejected upon detection of a flaw, we would find ourselves rejecting nearly all commercial engineering materials and products. If a defect is sufficiently small, it is possible for it to remain dormant throughout the useful lifetime of a product, never changing in size or shape. Such a defect is clearly allowable. Larger defects, or defects of a more undesirable geometry, may grow or propagate under the same conditions of loading, often causing sudden or catastrophic failure. These flaws would be clearly unacceptable. The objective (or challenge), therefore, is to identify the conditions below which a flaw remains **dormant** and above which it becomes **critical** and a cause for rejection. This issue is addressed in the section on "Fracture Toughness and the Fracture Mechanics Approach" in Chapter 2.

11.11 Current and Future Trends

NDT technology has advanced rapidly in the last several years. Vastly increased computing capability has expanded the ability to gather, store, and manipulate images and data. Many techniques can be partially or fully automated. Embedded sensors can provide continuous monitoring as a supplement to or replacement of the traditional periodic inspections. Computer modeling can better define the point at which a flaw becomes critical.

Review Questions

1. What is the purpose of nondestructive examination or testing?

2. Why must destructive testing be performed on a statistical basis?

3. What is a proof test, and what assurance does it provide?

4. What quality-related features can a hardness test reasonably ensure?

5. What exactly is nondestructive testing, and what are some attractive features of the approach?

6. What are some possible objectives of nondestructive testing?

7. What are some factors that should be considered when selecting a nondestructive testing method?

8. How might the costs of nondestructive testing actually be considered as an asset rather than a liability?

9. Why should visual inspection be considered as the initial and primary means of inspection?

10. What types of visual aids or equipment might be used in visual inspection?

11. What is the primary limitation of a visual inspection?

12. Describe the sequence of activity in a liquid penetrant inspection.

13. What types of defects can be detected in a liquid penetrant test?

14. What is the basic principle of magnetic particle inspection?

15. Magnetic particle inspection is limited to the inspection of what materials?

16. Describe how the orientation of a flaw with respect to a magnetic field can affect its detectability.

17. What is the major limitation of sonic testing, where one listens to the characteristic ring of a product in an attempt to detect defects?

18. What is the basic principle of ultrasonic inspection?

19. What is the role of a coupling medium in ultrasonic inspection?

20. What are three types of ultrasonic inspection methods?

21. What are some of the different techniques for capturing the image in radiographic inspection?

22. What types of radiation can be used in radiographic inspection of manufactured products?

23. What is a reasonable detection limit for radiographic inspection?

24. What are penetrameters, and how are they used in radiographic inspection?

25. Although radiographs offer a graphic image that looks like the part being examined, the technique has some significant limitations. What are some of these limitations?

26. Describe the basic principle of eddy-current testing.

27. Why would we not expect eddy-current examination to be useful with ceramics or polymeric materials?

28. What types of detection capabilities are offered by eddy-current inspection that cannot be duplicated by the other methods?

29. How does eddy-current compare with competing techniques, such as penetrant testing or magnetic particle inspection?

30. Why can't acoustic emission methods be used to detect the presence of an existing but static defect?

31. How can acoustic emission be used to determine the location of a flaw or defect?

32. How can temperature be used to reveal defects?

33. Under what conditions might microwave inspection be attractive?

34. What are some of the ways to evaluate strains within a stressed component?

35. What kinds of product features can be evaluated by electrical resistivity methods?

36. What type of information can be obtained through computed tomography?

37. What are some of the techniques that can be used to determine the chemical composition of surface and near-surface material?

38. Why is it necessary to determine the distinction between allowable and critical flaws, as opposed to rejecting all materials that contain detectable flaws?

Problems

1. A manufacturing company routinely specifies X-ray radiography to ensure the absence of cracks in its cast metal products. The primary reason for selecting radiography is the availability of a hard-copy record of each inspection for use in any possible liability litigation. Discuss the pros and cons of their selection. What other processes might you want to consider? If some form of permanent record is desirable, discuss how these might be obtained for the various alternative processes.

2. For each of the inspection methods listed, cite one major limitation to its use.

   a. Visual inspection

   b. Liquid penetrant inspection

   c. Magnetic particle inspection

   d. Ultrasonic inspection

   e. Radiography

   f. Eddy-current testing

   g. Acoustic emission monitoring

3. Which of the major nondestructive inspection methods might you want to consider if you want to detect surface flaws and internal flaws in products made from each of the following materials?

   a. Ceramics

   b. Polymers

   c. Fiber-reinforced composites with (i) polymer matrix and (ii) metal matrix (Consider various fiber materials.)

4. Discuss the application of nondestructive inspection methods to powder metallurgy (metallic) products with low, average, and high density.

5. Provide an example of how nondestructive inspection methods could be used to separate mixed materials (i.e., similar, but different materials that have gotten into the same warehouse bin or location).

6. It has been said, "Total Quality Control is superior to 100% inspection because inspection, like any other process, has its own defect rate. Some substandard product, therefore, is always likely to slip through." Do you agree or disagree? Discuss.

7. The pulse-echo ultrasonic technique can be used to determine the thickness of a part or structure. By accurately measuring the time it takes for a short ultrasonic pulse to travel through the thickness of a material, reflect from the back or inside surface, and return to the transducer, the distance can be calculated by:

$$d = Vt / 2$$

Where:

d is the thickness of the test piece.

V is the speed of sound in the material being tested.

t is the measured round-trip time.

What are some alternative means of measuring thickness? Briefly discuss their relative pros and cons. Consider such features as geometric and material constraints and the ability to measure from one side only.

8. If V for a particular metal is 5000 m/sec and a part made of that material is 3 mm thick, what is the transit time for the pulse to cross the material and reflect back to the source/receptor? (*Note:* The transit time is a consideration in evaluating equipment capabilities and may well influence cost!)

9. With nondestructive inspection methods using wave phenomena, the detection limit or resolution is dependent on the wavelength being used. Compare the wavelengths of:

a. Ultrasonic waves of frequency 500,000 Hz

b. X-rays

c. Acoustic emission stress waves of frequency 1 MHz

Note: Velocity = Frequency × Wavelength

CHAPTER 12

Process Capability and Quality Control

12.1 Introduction

All manufacturing processes display some level of variation. No two items coming from the process will be exactly the same. The primary objective of quality engineering is the systematic reduction of variability, or precision *Variability* can be measured by a statistic called sigma, σ, the **standard deviation**, which decreases with the reduction in variability. Statistical process control (SPC) efforts based on sampling techniques gave way to companywide quality control (CWQC) at Toyota and total quality control (TQC) programs in the United States that were part of single-piece flow manufacturing and subassembly techniques in U-shaped manufacturing cells. The drive toward zero **defects** has been led by Toyota, which achieved exceptional levels of quality by redesigning the manufacturing system so that each step in the making of the car and all its components are checked before the part moves to the next step or stage in its manufacturing sequence. This system redesign is called **lean manufacturing**. See Chapter 43. Variation can be further reduced by the application of advanced statistical techniques, like multiple variable analysis, designed experiments, and Taguchi methods—techniques that are part of Six Sigma efforts that many companies are implementing.

In a manufacturing process, the variation may be due to **chance causes** that produce random variations—these causes are said to be inherent and represent a stable source of variation. In addition, there are **assignable causes** of variation that can be detected and eliminated to help improve the process. For example, suppose that we view "shooting at a metal target" as a "process" for putting holes in a piece of metal. I hand you the gun and tell you to take nine shots at the bull's-eye. **Figure 12.1** shows some possible results.

You are the operator of the process. To measure the **process capability (PC)**—that is, your ability to consistently hit the bull's-eye you are aiming at—the target is inspected after you have finished shooting. So, the capability of manufacturing processes is determined by measuring the output of the process. In **quality control (QC)**, the product is examined to determine whether or not the processing accomplished was what was specified by the designer in the design (usually the nominal size and the tolerance). Of course, there are many other

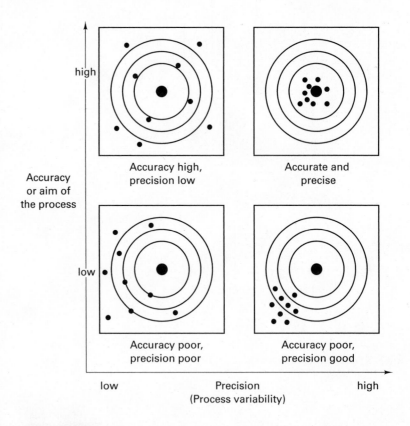

FIGURE 12.1 The concepts of accuracy (aim) and precision (repeatability) are shown in the four target outcomes. Accuracy refers to the ability of the process to hit the true value (nominal) on the average, while precision is a measure of the inherent variability of the process.

aspects of quality that quality engineers must address, such as performance, reliability, durability, aesthetics, and more. In this chapter, we will cover PC studies that are directed at the machine tools used in the processing, rather than the quality of the output or products from the processes. The quality tools used in lean manufacturing and the Six Sigma methodology will be introduced without covering any of the advanced statistical tools. Looking at our PC example, a study would quantify the inherent accuracy and precision in the shooting process. Accuracy is reflected in your aim (the average of all your shots), whereas precision reflects the repeatability of the process. The objective is to root out problems that can cause defective products during production. Traditionally, the objective has been to find defects in the process. The more progressive point of view is to design the process to prevent the problems that can cause defective products from occurring during production.

12.2 Determining Process Capability

The *nature of the process* refers to both the **variability** (or inherent uniformity) and the **accuracy** or the *aim* of the process. Thus, in the target-shooting example, a perfect process would be capable of placing nine shots right in the middle of the bull's-eye, one right on top of the other. The process would display no variability with perfect accuracy. Such performance would be very unusual in a real industrial process. The variability may have assignable causes and may be correctable if the cause can be found and eliminated. That variability to which no cause can be assigned and that cannot be eliminated is said to be inherent in the process and is, therefore, its nature.

Some examples of assignable causes of variation in processes include multiple machines for the same components, operator blunders, defective materials, or progressive wear in the tools during machining. Sources of inherent variability in the process include variation in material properties, operator variability, vibrations and chatter, and the wear of the sliding components in the machine, perhaps resulting in poorer operation of the machine. These kinds of variations, which occur naturally in processes, usually display a random nature and often cannot be eliminated. In quality control terms, these are referred to as chance causes. Sometimes, the causes of assignable variation cannot be eliminated because of cost. In general, it is easier to correct the aim of the process than it is to reduce the variability. Almost every process has multiple causes of variability occurring simultaneously, so it is extremely difficult to separate the effects of the different sources of variability during the analysis.

Making Process Capability Studies by the Traditional Method

The object of the PC study is to determine the inherent nature of the process as compared to the desired specifications. The output of the process must be examined under normal conditions, or what is typically called **hands-off conditions**. The inputs (e.g., materials, setups, cycle times, temperature, pressure, and operator) are fixed or standardized. The process is allowed to run without tinkering or adjusting, while the output (i.e., the product or units or components) is documented with respect to (1) time, (2) source, and (3) order of production. A sufficient amount of data needs to be taken to ensure confidence in the statistical analysis of the data. The capability of the gage (its precision) used to measure the products must exceed the expected tolerance on the part by one order of magnitude. (See the discussion of the rule of 10 in Chapter 10.)

Prior to any data collection, these steps must be taken:

1. Design the PC experiment (standard method): use normal or hands-off process conditions; specify machine settings for speed, feed, volume, pressure, material, temperature, operator, and so on.
2. Define the inspection method and the inspection means (the procedure and the instrumentation). In selecting the gage, consider these aspects:
   a. Features that the gage will be checking.
   b. Speed or rate of operation.
   c. Level of accuracy and precision.
   d. Skill of the operator.
   e. Portability of gages or part, or both.
   f. Environment (clean and stable, cutting fluids).
   g. Workpiece (clean, lubricants present).
   h. Cost (initial, maintenance, daily).
3. Decide how many items (measurements) will be needed to perform the statistical analysis.
4. For a standard PC study, use homogeneous input material, and try to contrast it with normal (more variable) input material.
5. Data sheets must be designed to record date, time, source, order of production, and all the process parameters being used (or measured) while the data are being gathered.
6. Assuming that the standard PC study approach is being used, the process is run, and the parts are made and measured.

Now follow the steps outlined in **Figure 12.2**. Assume that the designer specified the part to be 1.000 ± 0.005 in. After manufacturing engineering has developed a process plan, some units are manufactured according to the process plan without

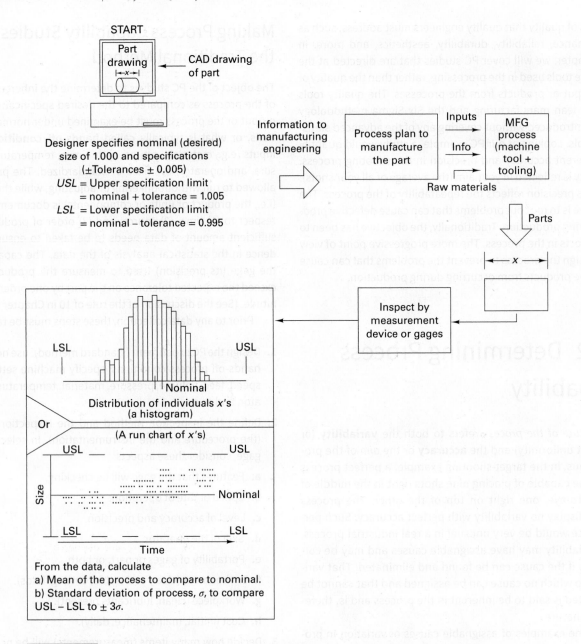

FIGURE 12.2 The process capability study compares the part as made by the manufacturing process to the specifications called out by the designer. Measurements from the parts are collected into run charts and for histograms for analysis.

any adjustment of the process. Each unit is measured, and the data are recorded on the data sheet. A frequency distribution, in the form of a **histogram**, or a **run chart**, is developed. This histogram shows the raw data and the desired **nominal value**, along with the upper and lower **specification limits**, LSL represents the lower specification limit and USL the upper specification limit. The statistical data are used to estimate the mean and the standard deviation of this distribution. The run chart shows the same data, but here the data are plotted against time.

The mechanics of this statistical analysis are outlined in **Figure 12.3**. The true mean of the **distribution**, designated μ (mu), is to be compared with the nominal value specified by the designer. The estimate of the true standard deviation, designated σ (sigma), is used to determine how the process compares

with the desired tolerance. *The purpose of the analysis is to obtain estimates of μ and σ values, the true process parameters, because they are not known.* σ is also designated as σ' (sigma prime).

The process capability is defined by $\pm 3\sigma$ or 6σ. Thus, $\mu \pm 3\sigma$ defines the natural capability limits of the process, assuming the process is approximately normally distributed. Note that a distinction is made between a sample and a population. A **sample** is of a specified, limited size and is drawn from the population. The **population** is the large source of items, which can include all the items the process will ever produce under the specified conditions. Our calculations assume that this distribution was normal or bell-shaped. **Figure 12.4** shows a typical normal curve and the areas under the curve as defined by the

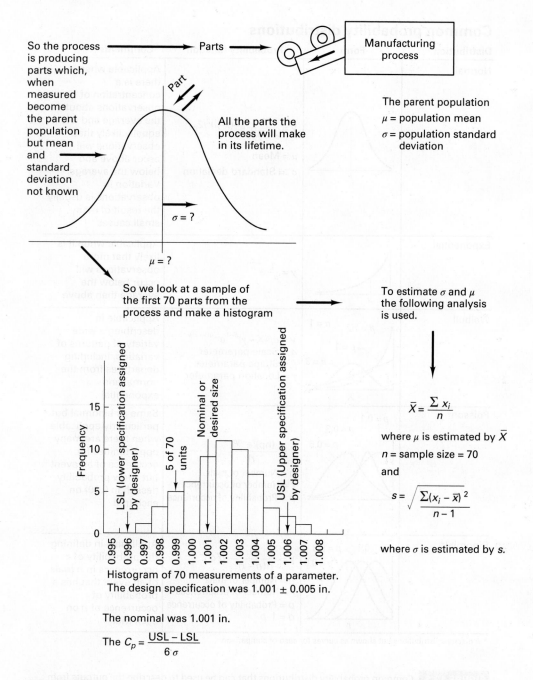

So the process is producing parts which, when measured become the parent population but mean and standard deviation not known

$\sigma = ?$

$\mu = ?$

Parts

All the parts the process will make in its lifetime.

Manufacturing process

The parent population

μ = population mean

σ = population standard deviation

So we look at a sample of the first 70 parts from the process and make a histogram

To estimate σ and μ the following analysis is used.

$$\bar{X} = \frac{\sum x_i}{n}$$

where μ is estimated by $\bar{X}$

n = sample size = 70

and

$$s = \sqrt{\frac{\sum(x_i - \bar{x})^2}{n - 1}}$$

where σ is estimated by s.

LSL (lower specification assigned by designer)

5 of 70 units

Nominal or desired size

USL (Upper specification assigned by designer)

Histogram of 70 measurements of a parameter. The design specification was 1.001 ± 0.005 in.

The nominal was 1.001 in.

The $C_p = \dfrac{USL - LSL}{6\,\sigma}$

FIGURE 12.3 Examples of calculations to obtain estimates of the mean (μ) and standard deviation (σ) of a process.

standard deviation. Other distributions shown in **Figure 12.5** are possible but, in this case, the histogram clearly suggested that this process can best be described by a normal probability distribution. Now it remains for the process engineer and the operator to combine their knowledge of the process with the results from the analysis to draw conclusions about the ability of this process to meet specifications.

FIGURE 12.4 The normal or bell-shaped curve with the areas within ±1σ, ±2σ, and ±3σ for a normal distribution; 68.26% of the observations will fall within ±1 σ from the mean, and 99.73% will fall within ±3σ from the mean.

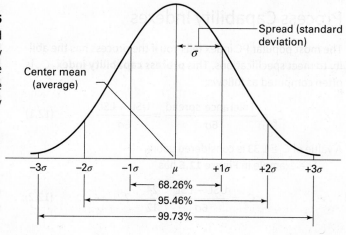

Spread (standard deviation)

Center mean (average)

σ

-3σ $\quad -2\sigma$ $\quad -1\sigma$ $\quad \mu$ $\quad +1\sigma$ $\quad +2\sigma$ $\quad +3\sigma$

68.26%

95.46%

99.73%

Common probability distributions

Distribution	Form	Probability function	Comments
Normal	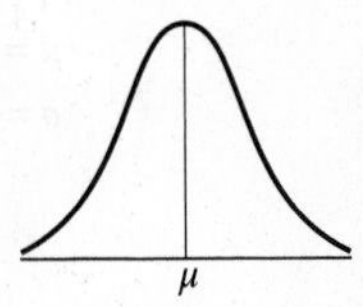	$y = \dfrac{1}{\sigma'\sqrt{2\pi}} e^{-(x-\mu)^2/2\sigma^2}$ μ = Mean σ' = Standard deviation	Applicable when there is a concentration of observations about the average and it is equally likely that observations will occur above and below the average. Variation in observations is usually the result of many small causes
Exponential	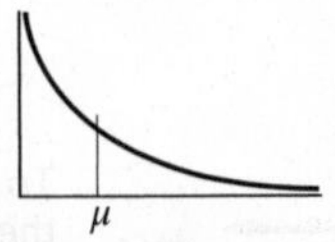	$y = \dfrac{1}{\mu} e^{-\frac{x}{\mu}}$	Applicable when it is likely that more observations will occur below the average than above
Weibull	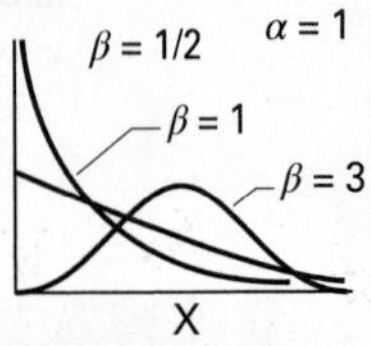	$y = \alpha\beta(X - \gamma)^{\beta-1} e^{-\alpha(x-\gamma)\beta}$ α = Scale parameter β = Shape parameter γ = Location parameter	Applicable in describing a wide variety of patterns of variation, including departures from the normal and exponential
Poisson*	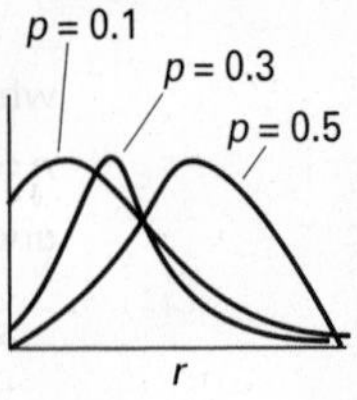	$y = \dfrac{(np)^r e^{-np}}{r!}$ n = Number of trials r = Number of occurrences p = Probability of occurrence	Same as binomial but particularly applicable when there are many opportunities for occurence of an event but a low probability (less than 0.10) on each trial
Binomial*	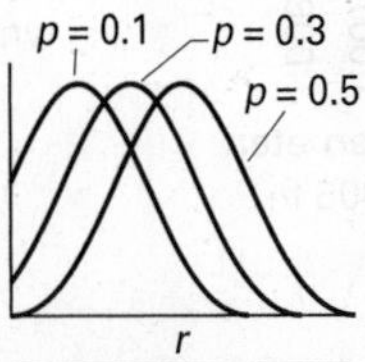	$y = \dfrac{n!}{r!(n-r)!} p^r q^{n-r}$ n = Number of trials r = Number of occurrences p = Probability of occurrence $q = 1-p$	Applicable in defining the probability of r occurrences in n trials of an event that has a probability of occurrence of p on each trial

* = discrete distributions but shown as curves for ease of comparison

FIGURE 12.5 Common probability distributions that can be used to describe the outputs from manufacturing processes. (Source: *Quality Control Handbook*, 3rd ed.)

Process Capability Indexes

The most popular PC index tells you if the process has the ability to meet specifications. This **process capability index,** C_p, is often computed as follows;

$$C_p = \frac{\text{Tolerance spread}}{6\sigma} = \frac{\text{USL} - \text{LSL}}{6\sigma} \qquad (12.1)$$

A value of $C_p \geq 1.33$ is considered good.
The example in **Figure 12.6** has

$$C_p = \frac{\text{USL} - \text{LSL}}{6\sigma} = \frac{24}{12} = 2 \qquad (12.2)$$

The process capability ratio, C_p, does not, however, take into account the location of the process mean, μ, with respect to the nominal or the specifications. C_p only looks at the variability or spread of process (compared to specifications) in terms of sigmas. Thus, another process capability ratio has been developed for off-center processes. This ratio is called C_{pk}, where

$$C_{pk} = \min C_{pu}, C_{pl}$$
$$= \min\left(C_{pu} = \frac{\text{USL} - \mu}{3\sigma}, \ C_{pl} = \frac{\mu - \text{LSL}}{3\sigma} \right) \qquad (12.3)$$

C_{pk} is simply a one-sided ratio for the specification nearest to the process average, μ. To compare the two, look at **Figure 12.6**.

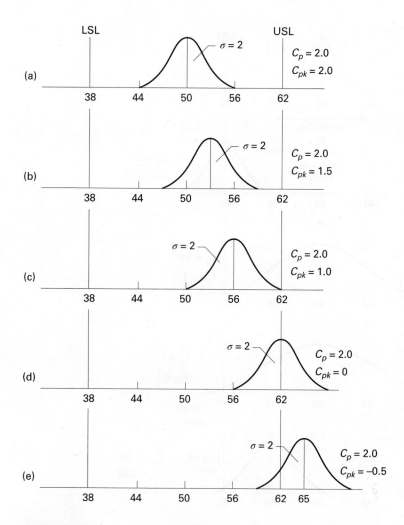

FIGURE 12.6 The output from the process is shifting toward the USL, which changes the C_{pk} ratio but not C_p ratio.

All the histograms have the same standard deviation ($\sigma = 2$) and the same USL – LSL specifications (62 – 38). For Figure 12.6a, the two indexes are the same. For Figure 12.6b, the mean has shifted to $\mu = 54$, so $C_{pk} = (62 - 54)/3(2) = 1.5$.

Can you calculate C_{pk} for (c), (d), and (e)? Answers are on Figure 12.6.

The capability indexes can tell you about the variance, where the width of the histogram is compared with the specifications (**Figure 12.7**). The natural spread of the process, 6σ, is computed and is then compared with the upper and lower tolerance limits. These situations can exist:

1. $6\sigma < \text{USL} - \text{LSL}$, or $C_p > 1$, or process variability less than tolerance spread (see Figure 12.7a).

2. $6\sigma < \text{USL} - \text{LSL}$, but process has shifted (see Figure 12.7b).

3. $6\sigma < \text{USL} - \text{LSL}$, or $C_p = 1$, or process variability is just equal to tolerance spread (see Figure 12.7c).

4. $6\sigma < \text{USL} - \text{LSL}$, or $C_p < 1$, or process variability greater than tolerance spread (see Figure 12.7d).

5. The process mean and variability have both changed (see Figure 12.7e).

Discussion of Process Capability Scenarios

In the situation shown in Figure 12.7a, the process is capable of meeting the tolerances applied by the designer. Generally speaking, if process capability is on the order of two-thirds to three-fourths of the design tolerance, there is a high probability that the process will produce all good parts over a long period of time. If the PC is on the order of one-half or less of the design tolerance, it may be that the selected process is too good; that is, the company may be producing ball bearings when what is called for is marbles. In this case, it may be possible to trade off some precision in this process for looser specifications elsewhere, resulting in an overall economic gain. Quality in well-behaved processes can be maintained by checking the first, middle, and last part of a lot or production run. If these parts are good, then the lot is certain to be good. This is called $N = 3$. Naturally, if the lot size is 3 or less, this is 100% inspection. Sampling and control charts are also used under these conditions to maintain the process aim and variability.

When the process is not capable of meeting the design specifications, there are a variety of alternatives, including:

1. Shifting this job to another machine with greater process capability.

2. Getting a review of the specifications to see if they may be relaxed.

3. Sorting the product, to separate the good from the bad. This entails 100% inspection of the product, which may not be a feasible economic alternative unless it can be done automatically. Automatic sorting of the product on a 100% basis can ensure near-perfect quality of all the accepted parts. The automated station shown in **Figure 12.8** checks parts for the proper diameter with the aid of a **linear variable-differential transformer (LVDT)**. As a part approaches the inspection station on a motor-driven conveyor system, a computer-based controller activates a clamping device. Embedded in the clamp is an LVDT position sensor with which the control computer can measure the diameter of the part. Once the measurement has been made, the computer releases the clamp, allowing the part to be carried away. If the diameter of the part is within a given tolerance, a solenoid-actuated gate operated by the computer lets the part pass. Otherwise, the part is ejected into a bin. With the fast-responding LVDT, 100% of manufactured parts can be automatically sorted quickly and economically.

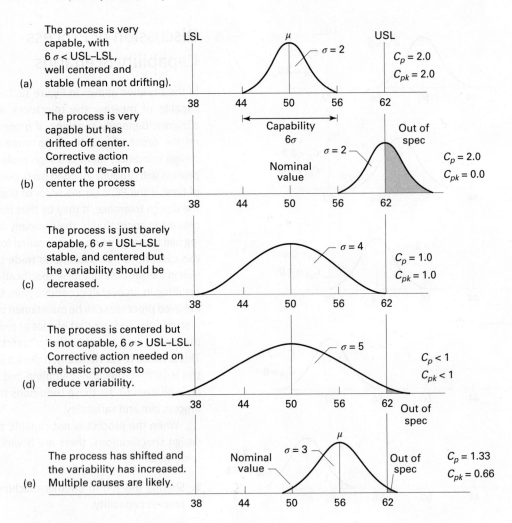

(a) The process is very capable, with $6\sigma < USL–LSL$, well centered and stable (mean not drifting).

$C_p = 2.0$
$C_{pk} = 2.0$

(b) The process is very capable but has drifted off center. Corrective action needed to re-aim or center the process

$C_p = 2.0$
$C_{pk} = 0.0$

(c) The process is just barely capable, $6\sigma = USL–LSL$ stable, and centered but the variability should be decreased.

$C_p = 1.0$
$C_{pk} = 1.0$

(d) The process is centered but is not capable, $6\sigma > USL–LSL$. Corrective action needed on the basic process to reduce variability.

$C_p < 1$
$C_{pk} < 1$

(e) The process has shifted and the variability has increased. Multiple causes are likely.

$C_p = 1.33$
$C_{pk} = 0.66$

FIGURE 12.7 Five different scenarios for a process output versus the designer's specifications for the nominal (50) and upper and lower specifications of 65 and 38, respectively.

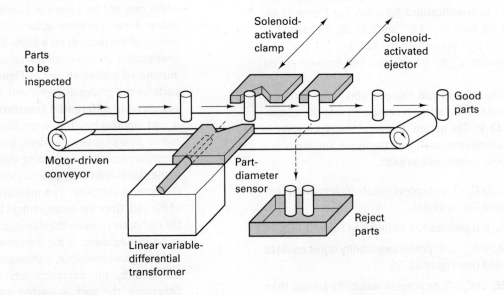

FIGURE 12.8 A linear variable-differential transformer (LVDT) is a key element in an inspection station checking part diameters. Momentarily clamped into the sensor fixture, a part pushed the LVDT armature into the device winding. The LVDT output is proportional to the displacement of the armature. The transformer makes highly accurate measurements over a small displacement range.

Sorting to find defects by automatic inspection is bad because you already paid to produce the defects. Also automated sorting does not determine what caused the defects, so this example is an "automated defect finder." How would one change this inspection system to make it "inspect to prevent" the defect from occurring?

4. Determining whether the **precision** (or **repeatability**) of the process can be improved by:

   a. Switching cutting tools, workholding devices, or materials.

   b. Overhauling the existing process and/or developing a preventive maintenance program.

   c. Finding and eliminating the causes of variability, using cause-and-effect diagrams.

   d. Combinations of (1), (2), and (3).

   e. Using designed experiments and Taguchi methods to reduce the variability of the process.

In Figure 12.7c, the process capability is almost exactly equal to the assigned tolerance spread, so if the process is not perfectly centered, defective products will always result. Thus, this situation should be treated like situation Figure 12.7d unless the process can be perfectly centered and maintained. Tool wear, which causes the distribution to shift, must be negligible.

PC studies can evaluate the ability of the process to maintain centering so that the average of the distribution comes as close as possible to the desired nominal value (see Figure 12.7b, where the process needs to be recentered so that the mean of the process distribution is at or near the nominal value). Most processes can be re-aimed. Poor accuracy is often due to assignable causes, which can be eliminated.

In addition to direct information about the accuracy and the precision of the process, PC studies can also tell the manufacturing engineer how **pilot processes** compare with **production processes**, and vice versa. If the source and time of the manufacture of each product are carefully recorded, information about the instantaneous reproducibility can be found and compared with the repeatability of the process with respect to time (time-to-time variability). More important, because almost all processes are duplicated, PC studies generate information about machine-to-machine variability.

Going back to our target-shooting example, suppose that nine different guns were used, all of the same make and type. The results would have been different, just as having nine marksmen use the same gun would have resulted in yet another outcome. Thus, PC studies generate information about the homogeneity and the differences in multiple machines and operators.

It is quite often the case in such studies that one variable dominates the process. Target shooting viewed as a process is "operator dominated" in that the outcome is highly dependent on the skill (the capability) of the "worker." Processes that are not well engineered nor highly automated, or in which the worker is viewed as "highly skilled," are usually operator-dominated. Processes that change or shift uniformly with time

but that have good repeatability in the short run are often machine dominated. For example, the mean of a process (μ) will usually shift after a tool change, but the variability may decrease or remain unchanged. Machines tend to become more precise (to have less variability within a sample) after they have been "broken in" (i.e., the rough contact surfaces have smoothed out because of wear) but will later become less precise (will have less repeatability) due to poor fits between moving elements (called **backlash**) of the machine under varying loads. Other variables that can dominate processes are setup, input parameters, and even information.

In many machining processes in use today, the task of tool setting has been replaced by an automatic tool positioning capability, which means that one source of variability in the process has been eliminated, thus making the process more repeatable. In the same light, it will be very important in the future for manufacturing engineers to know the process capability of robots they want to use in the workplace.

The discussion to this point has assumed that the **parent population** is normally distributed, that is, has the classic bell-shaped distribution in which the percentages (shown in Figure 12.4) are dictated by the number of standard deviations from the central value or mean. The shape of the histogram may reveal the nature of the process to be skewed to the left or the right (unsymmetrical), often indicating some natural limit in the process. Drilled holes exhibit such a trend as the drill tends to make the hole oversize. Figure 12.5 shows some of the other common probability distributions that can capture the output from different manufacturing processes.

12.3 Introduction to Statistical Quality Control

In virtually all manufacturing, it is extremely important that the dimensions and quality of individual parts be known and maintained. This is of particular importance where large quantities of parts, often made in widely separated plants, must be capable of interchangeable assembly. Otherwise, difficulty may be experienced in subsequent assembly or in service, and costly delays and failures may result. In recent years, defective products resulting in death or injury to the user have resulted in expensive litigation and damage awards against manufacturers. **Inspection** is the function that controls the quality (e.g., the dimensions, the performance, and the color) manually, by using operators or inspectors, or automatically, with machines, as discussed previously.

The economics-based question, "How much should be inspected?" has three possible answers:

1. *Inspect every item being made.* 100% checking with prompt execution of feedback and immediate corrective action can ensure perfect quality.

2. *Sample.* Inspect some of the product by sampling, and make decisions about the quality of the process based on the sample.

3. *None.* Assume that everything made is acceptable or that the product is inspected by the consumer, who will exchange it if it is defective. (This is not a recommended procedure).

The reasons for not inspecting all of the product (i.e., for sampling) include

1. Everything has not yet been manufactured—the process is continuing to make the item—so we have to look at some before we are done with all.

2. The test is destructive.

3. There is too much product for all of it to be inspected.

4. The testing takes too much time or is too complex or too expensive.

5. It is not economically feasible to inspect everything, even though the test is simple, cheap, and quick.

Some characteristics are nondissectible; that is, they cannot be measured during the manufacturing process because they do not exist until after a whole series of operations has taken place. The final edge geometry of a razor blade is a good example, as is the yield strength of a rolled bar of steel.

Sampling (looking at some percentage of the whole) requires the use of statistical techniques that permit decisions about the acceptability of the whole based on the quality found in the sample. This is known as statistical process control (SPC).

Statistical Process Control

Looking at some (sampling) and deciding about the behavior of the whole (the parent population) is common in industrial inspection operations. The most widely used basic **statistical process control (SPC)** technique is the control charts.

Control charts for variables are used to monitor the output of a process by sampling (looking at some), by measuring selected quality characteristics, by plotting the sample data on the chart, and then by making decisions about the performance of the process.

Figure 12.9 shows an industrial example of two charts commonly used for variable types of measurements. The $\bar{X}$ chart tracks the aim (accuracy) of the process. This chart plots sample averages against time. The sample size is 5, and 25 samples have been taken. The R chart (or σ chart) tracks the precision or variability of the process. This chart plots the range statistic versus time. Usually only the $\bar{X}$ chart and the R chart are used unless the sample size is large, and then σ charts are used in place of R charts. The calculations for these two charts are outlined in **Figure 12.10**.

After k samples of size n have been drawn, calculate $\bar{X}$ and R for each sample.

$$\bar{X}(\text{mean}) = \frac{\sum x_i}{n} \quad \text{Where } n = \text{sample size.} \quad (12.4)$$

$$R(\text{range}) = x_{\text{HIGH}} - x_{\text{LOW}} \quad (12.5)$$

Sometimes the standard deviation is calculated for each sample:

$$\sigma(\text{sigma}) = \sqrt{\frac{\sum(x_i - \bar{x})^2}{n-1}} \quad (12.6)$$

$$n = \text{sample size}$$

The samples are drawn over time, in this case, six days.

Because some sample statistics tend to be normally distributed about their own mean, $\bar{X}$ values are normally distributed about $\bar{\bar{X}}$, R values are normally distributed about $\bar{R}$, and σ values are normally distributed about $\bar{\sigma}$.

Quality control charts are widely used as aids in maintaining quality and in achieving the objective of detecting trends in quality variation before defective parts are actually produced. These charts are based on the previously discussed concept that if only chance causes of variation are present, the deviation from the specified dimension or attribute will fall within predetermined limits.

When sampling inspection is used, the typical sample sizes are from 3 to about 12 units. The $\bar{X}$ chart tracks the sample averages ($\bar{X}$ values). The R chart plots the range values (R values). Figure 12.9 shows one example of $\bar{X}$ and R charts for measuring a dimension of a gap on a part called the retainers. Twenty-five samples of size 5 were taken over six days, and these sample data will be used to prepare the control charts.

The centerline of the $\bar{X}$ chart was computed prior to actual usage of the charts in control work,

$$\bar{\bar{X}} = \frac{\sum_{i=1}^{k} \bar{X}}{k} \quad (12.7)$$

Where $\bar{X}$ is a sample average and k is the number of sample averages. The horizontal axis for the charts is time, thus indicating *when* the sample was taken; $\bar{\bar{X}}$ serves as an estimate for μ, the true center of the process distribution; $\bar{\bar{X}}$ is also the centerline of the $\bar{X}$ chart. The upper and lower **control limits** are commonly based on three standard error units, $3\sigma_{\bar{x}}$ ($\sigma_{\bar{x}}$ is the standard deviations for the distribution of $\bar{X}$'s about $\bar{\bar{X}}$).

Thus,

$$\text{UCL}_{\bar{x}} = \text{Upper control limit on } \bar{X} \text{ chart}$$
$$= \mu + 3\sigma_{\bar{x}} \text{ or} \quad (12.8)$$
$$= \mu' + A_2\bar{R}$$
$$\text{LCL}_{\bar{x}} = \text{Lower control limit, } \bar{X}\text{ chart} = \mu - 3\sigma_{\bar{x}}$$
$$= \mu' - A_2R$$

(See Figure 12.10 for A_2 values, and a note how A_2 is calculated.)

The upper and lower control limits are entered as dashed lines on the chart. The $\bar{X}$ chart is used to track the central tendency (aim) of the process. In this example, the samples were being taken four times a day. The R chart is used to track the variability or dispersion of the process. A σ chart could also be

VARIABLES CONTROL CHART X̄&R
Averages & Ranges

Part/Asm. Name *Retainer*	Operation *Bend Clip*	Specification .50 – .90 mm	Nominal Size .70 mm
Part No. 1234567	Department 105	Gage *Depth Gage Micrometer*	
Parameter *Gap. Dim. "A"*	Machine 030	Sample Size/Frequency 5/2 *Hours*	

Date		6/8				6/9				6/10				6/11				6/12				6/15				6/16
Time of day		8	10	12	2	8	10	12	2	8	10	12	2	8	10	12	2	8	10	12	2	8	10	12	2	8
Operator																										
Sample Measurements — Value of X	1	.65	.75	.75	.60	.70	.60	.75	.60	.65	.60	.80	.85	.70	.65	.90	.75	.75	.75	.65	.60	.50	.60	.80	.65	.65
	2	.70	.85	.80	.70	.75	.75	.80	.70	.80	.70	.75	.75	.70	.70	.80	.80	.70	.70	.65	.60	.55	.80	.65	.60	.70
	3	.65	.75	.80	.70	.65	.75	.65	.80	.85	.60	.90	.85	.75	.85	.80	.75	.75	.60	.85	.65	.65	.65	.75	.65	.70
	4	.65	.85	.70	.75	.85	.85	.75	.75	.85	.80	.60	.65	.75	.75	.75	.80	.70	.70	.65	.60	.80	.65	.65	.60	.60
	5	.85	.65	.75	.65	.80	.70	.70	.75	.75	.65	.80	.70	.70	.60	.85	.65	.80	.60	.70	.65	.80	.75	.65	.70	.65
Sum		3.50	3.85	3.80	3.40	3.75	3.65	3.65	3.60	3.90	3.35	3.75	3.80	3.60	3.55	4.10	3.75	3.80	3.35	3.50	3.10	3.30	3.45	3.50	3.20	3.30
Average X̄		.70	.77	.76	.68	.75	.73	.73	.72	.78	.67	.75	.76	.72	.71	.82	.75	.76	.67	.70	.62	.66	.69	.70	.64	.66
Range R		.20	.20	.10	.15	.20	.25	.15	.20	.20	.20	.40	.20	.05	.25	.15	.15	.15	.15	.20	.05	.30	.20	.15	.10	.10

$\overline{\overline{X}} = (.70 + .77 + \ldots + .64 + .66)/25 = 17.90/25 = .716$ $UCL_{\overline{X}} = .716 + (.58 \times .178) = .819$

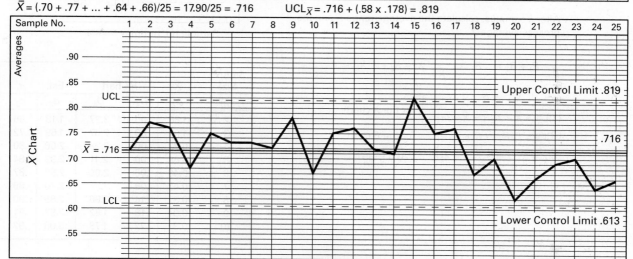

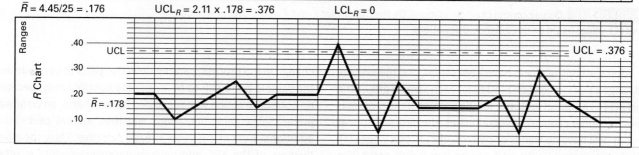

$\overline{R} = 4.45/25 = .176$ $UCL_R = 2.11 \times .178 = .376$ $LCL_R = 0$

FIGURE 12.9 Example of $\overline{X}$ and R charts and the data set of 25 samples [$k = 25$ of size 5 ($n = 5$)]. (Source: *Continuing Process Critical and Process Capability Improvement*, Statistical Methods Office, Ford Motor Co., 1985)

used. R is computed for each sample ($x_{HIGH} - x_{LOW}$). The value of $\overline{R}$ is calculated as:

$$\overline{R} = \frac{\sum\limits_{i=1}^{k} R}{k} \qquad (12.9)$$

where $\overline{R}$ represents the average range of k range values. The range values are normally distributed about $\overline{R}$, with standard deviation σ_R. To determine the upper and lower control limits for the charts, the following relationships are used.

UCL_R = Upper control limit,

R chart = $\overline{R} + 3\sigma_R = D_4\overline{R} = 2.11\overline{R}$ for $n = 5$

LCL_R = Lower control limit = $R = 3\sigma_R = D_3\overline{R} = 0$

where D_4 and D_3 are constants and are given in Figure 12.10. For small values of n, the distance between centerline $\overline{R}$ and LCL_R is more than $3\sigma_R$, but LCL_R cannot be negative because negative range values are not allowed, by definition. Hence, $D_3 = 0$ for values of n up to 6.

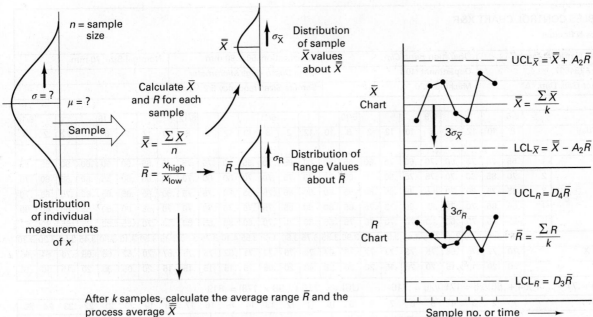

FIGURE 12.10   Quality control chart calculations. On the charts, plot $\bar{X}$ and R values over time. UCL and LCL values based on three standard deviations.

After the control charts have been set up, and the average and range values have been plotted for each sample group, the chart acts as a control indicator for the process. If the process is operating under chance cause conditions, the data will appear random (will have no trends or pattern). If $\bar{X}$, R, or σ values fall outside the control limits, or if nonrandom trends occur (like seven points on one side of the central line or six successive increasing or decreasing points appear), an assignable cause or change may have occurred, and some action should be taken to correct the problem.

Trends in the control charts often indicate the existence of an assignable cause factor before the process actually produces a point outside the control limit. In grinding operations, wheel wear (wheel undersize) results in the parts becoming oversized, and corrective action should be taken. (Redress and reset wheel, or replace with new wheel.) Note that defective parts can be produced even if the points on the charts indicate the process is in control. That is, it is possible for something to

change in the process, causing defective parts to be made and the sample point still to be within the control limits. Because no corrective action was suggested by the charts, an error was made. Subsequent operations will then involve performing additional work on products already defective. Thus, the effectiveness of the SPC approach in improving quality is often deterred by the lag in time between the discovery of an abnormality and the corrective action.

With regard to control charts in general, it should be kept in mind that the charts are only capable of indicating that something has happened, not what happened, and that a certain amount of detective work will be necessary to find out what has occurred to cause a break from the random, normal pattern of sample points on the charts. Keeping careful track of when and where the sample was taken will be very helpful in such investigations, but the best procedure is to have the operator take the data and run the chart. In this way, quality feedback is very rapid and the causes of defects is readily found.

Process Capability Determination from Control Chart Data

After the process is determined to be "under control," the data can be used to estimate the process capability parameters.

For example, examine the data in Figure 12.9. These are measurements of the gaps in 125 retainers. They are supposed to have a nominal size of 0.70 mm. The population is assumed to be normal, and only chance variations are occurring. (Could you make a histogram of these 125 measurements?) The mean for the all data can be obtained by

$$\overline{X} = \frac{\sum\limits_{i=1}^{125} x_i}{n}$$

where x_i is an individual measurement and the number of items is 125. The **mean** is a measure of the central tendency around which the individual measurements tend to group. The variability of the individual measurements about the average may be indicated by the standard deviation, σ, where

$$\sigma = \frac{\sqrt{\sum\limits_{i=1}^{n}\left(x_i - X^2\right)}}{n} \quad \text{Where } n = 125$$

But more commonly, the sample data are used to make determinations of the process capability.

A sample size of five was used in this example, so $n = 5$. Twenty-five groups or samples were drawn from the process, so $k = 25$. For each sample, the **sample mean, $\overline{X}$**, and the **sample range, R**, are computed. [For large samples ($n > 12$), the standard deviation of each sample should be computed rather than the range.] Next, the average of the sample averages, $\overline{\overline{X}}$, is computed as shown in Figure 12.10. This is sometimes called the **grand average**, which is used to estimate the mean of the process, μ. The standard deviation of the process, which is a measure of the spread or variability of the process, is estimated from either the average of the sample ranges, $\overline{R}$, or the average of the sample standard deviations, $\overline{\sigma}$, using either $\frac{\overline{R}}{d_2}$ or $\frac{\overline{\sigma}}{c_2}$. The factors d_2 and c_2 depend on the **sample size (n)**, and are given in the table in Figure 12.10. Thus, μ is estimated by $\overline{\overline{X}}$, and σ is estimated by $\frac{\overline{R}}{d_2}$ or $\frac{\overline{\sigma}}{c_2}$.

And remember, the standard deviation of the distribution of the $\overline{X}$ values about $\overline{\overline{X}}$, $\sigma_{\overline{x}}$, is related to the standard deviation of the parent population, σ by

$$\sigma_{\overline{x}} = \frac{\sigma}{\sqrt{n}} \tag{12.10}$$

Now these estimates can be used to determine the process capability of the process in the same way the histogram was used.

12.4 Sampling Errors

It is important to understand that in sampling, two kinds of decision errors are always possible (see **Figure 12.11**). Suppose that the process is running perfectly, but the sample data indicate that something is wrong. You, the quality engineer, decide to stop the process to make adjustments. This is called a type I error. Alternately, suppose that the process was not running perfectly and was making defective products. However, the sample data did not indicate that anything was wrong. You (the quality engineer) decided to *not stop* the process and set it right. This is called a type II error. Both types of errors are possible in sampling. For a given sample size, reducing the chance of one type of error will increase the chance of the other. Increasing the sample size or the frequency of sampling reduces the probability of errors but increases the cost of the inspection. It is common practice in control chart work to set the upper and lower control limits at three standard deviations. This makes the probability of an alpha error very small and the probability of a beta error quite large. Many companies determine the size of the errors they are willing to accept according to the overall cost of making the errors plus the cost of inspection. If, for example, a type II error is very expensive in terms of product recalls or legal suits, the company may be willing to make more type I errors, to sample more, or even to go to 100% inspection on very critical items to ensure that the company is not accepting defective materials and passing them on to the customer. The inspection should take place immediately after the processing.

As mentioned earlier, in any continuing manufacturing process, variations from established standards are of two types: **assignable cause variations**, such as those due to malfunctioning equipment or personnel, or to defective material, or to a worn or broken tool; **normal chance variations**, resulting from the inherent nonuniformities that exist in materials

	The process is running. It really has:	
	Changed (making defects)	Not changed
Based on the sample, you decide the process has changed	No error	Type I α error
Based on the sample, you decide the process has not changed	Type II β error	No error

Type I (α error) Saying the process has changed when it has not changed
Type II (β error) Saying the process has not changed when it has changed

FIGURE 12.11 When you look at some of the output from a process and decide about the whole (i.e., the quality of the process), you can make two kinds of errors.

and in machine motions and operations. Deviations due to assignable causes may vary greatly. Their magnitude and occurrence are unpredictable and thus should be prevented. However, if the assignable causes of variation are removed from a given operation, the magnitude and frequency of the chance variations can be predicted with great accuracy. Thus, if one can be assured that only chance variations will occur, the quality of the product will be better known, and manufacturing can proceed with assurance about the results. By using SPC procedures, one may detect the presence of an assignable cause variation and, after investigation to find the cause, remove the cause before it causes quality to become unacceptable. Also, the astute application of statistical experimental design methods (Taguchi experiments) can help identify some assignable causes.

To sum up, PC analysis and process improvements help get the process "under control." Control charts help keep the process on center ($\bar{X}$ chart) with no increases in variability (R chart or σ chart).

12.5 Gage Capability

The instrument (gage) used to measure the process will also have some inherent precision and accuracy, often referred to as **gage capability**. In other words, the observed variation in the component part being measured is really composed of the actual process variation plus the variation in the measured system; see **Figure 12.12**. The measuring system will display:

- *Bias:* poor accuracy or aim.
- *Linearity:* accuracy changes over the span of measurements.
- *Stability:* accuracy changes over time.
- *Repeatability:* loss of precision in the gage (variability).
- *Reproducibility:* variation due to different operators.

Measurements made by different operators will have different means and different variation about the mean when performing a measurement. Determining the capability of the gage is called an **R and R study**. In selecting a gage, the engineer tries to get the variation in the gage (as measured by the standard deviation) to be less than 10% of the total tolerance spread (USL – LSL). This is called the **10% rule**, and the R and R study can determine the magnitude of the gage variability.

Space does not permit a full discussion here of R and R studies, but detailed descriptions are found in sources cited in the reference section of the book. In particular, students involved in ISO 9000 studies should examine the *Measurement System Analysis Reference Manual*, published by the Automotive Industry Action Group. See Chapter 10 for further discussion.

12.6 Just in Time/Total Quality Control

The next level of quality is called JIT/TQC (just in time/total quality control). The phrase **total quality control (TQC)** was first used by A.V. Feigenbaum in *Industrial Quality Control* in May 1957. TQC means that all departments of a company must participate in quality control (**Table 12.1**), a concept implemented by Toyota as part of lean production and their companywide quality control (CWQC) methods. Quality control is the responsibility of workers at every level in every department, all of whom have had quality control training in the seven basic tools of quality control. It begins at the product design stages and carries through the manufacturing system, where the emphasis is on making it right the first time.

Note, changes is the manufacturing system design and company culture are needed to give the responsibility for quality to the worker, along with the authority to stop the process when something goes wrong. An attitude of defect prevention and a habit of constant improvement of quality are fundamental to lean production, a system developed by Toyota. Companies such as Toyota have accomplished TQC by extensive education of the workers, giving them the analysis tools they need (control charts with cause-and-effect diagrams) to find and expose the problems. Workers are encouraged to correct their own errors, and 100% inspection (often done automatically) is the rule. Passing defective products on to the next process is not allowed. The goal is perfection.

The Seven Tools for Lean Production

In an earlier section, control charts were discussed along with histograms. In lean manufacturing these two basic quality control tools are placed in the hands of operators. A histogram is a representation of a frequency distribution that uses rectangles

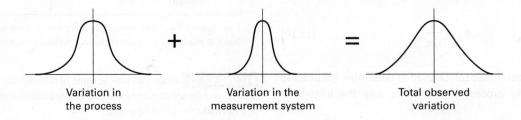

FIGURE 12.12 Gage capability (variation) contributes to the total observed variation in the measurement of a part.

TABLE 12.1	JIT/Total Quality Control: Concepts and Categories
TQC Category	**TQC Concept**
1. Organization	• Manufacturing engineering has responsibility for quality—quality circles
2. Goals	• Habit of improvement for everyone everywhere in the manufacturing system (companywide quality control)
	• Perfection—zero defects—not a program, a goal
3. Basic principles	• Process control—defect prevention, not detection
	• Easy-to-see quality—quality on display so customers can see and inspect processes—easy to understand quality
	• Insist on compliance with maintenance
	• Line stop when something goes wrong
	• Correct your own errors
	• 100% check in manufacturing and subassembly cells, MO-CO-MOO
4. Facilitating concepts	• QC department acts as facilitator
	Audit suppliers
	Help in quality improvement projects
	Training workers, supervisors, suppliers
	• Small lot sizes through rapid changeover (SMED)
	• Housekeeping (5 S methods)
	• Less-than-full-capacity scheduling (70%)
	• Total preventive maintenance (TPM)
	• 8-4-8-4 two-shift scheduling
5. Techniques and aids	• Remove some inventory, expose problems, solve problems
	• Defect prevention, poka-yokes for checking 100% of parts
	• $N = 2$, for checking first and last item in lot (or $N = 3$, for large lots)
6. The 7 analysis tools for lean production	1. Check sheets
	2. Cause-and-effect diagrams
	3. Histograms
	4. Control charts ($\bar{X}, R, \sigma$) Run charts
	5. Scatter diagrams
	6. Pareto charts
	7. Process flow charts and Value Stream Mapping
7. Advanced quality control	• Multivariant analysis, and other 6 sigma tools Taguchi methods

Source: Schoenberger, 1983, updated by Black

whose widths represent class intervals and whose heights are proportional to the corresponding frequencies. The *frequency histogram* is a type of diagram in which data are grouped into cells (or intervals) and the frequency of observations falling into each interval can be noted. All the observations within a cell are considered to have the same value, which is the midpoint of the cell. Thus, a histogram is a picture that describes the variation in a process. It is good to have this visual impression of the distribution of values, along with the mean and standard deviation. Histograms are used in many ways in QC, for example,

• To determine the process capability (central tendency and dispersion).

• To compare the process with the specifications.

• To suggest the shape of the population (e.g., normality).

• To indicate discrepancies in data, such as gaps.

There are several types of histograms. A histogram shows either absolute frequency (actual occurrence) or relative frequency (percentage). *Cumulative histograms* show cumulative frequency and reliability cumulative frequency. Each type has its own advantages and is used in different situations. **Figure 12.13** shows frequency versus location for 150 measurements. The aim (accuracy) of the process (μ) is a bit low compared to the nominal, but all the data are well within the tolerances. The disadvantage of the histogram is that it does not show trends and does not take time into account. We can take the data from the histogram and spread them out over time to create a run chart or run diagram.

Run Chart or Diagram

A run diagram is a plot of individual data—that is, a quality characteristic as a function of time. It provides some idea of general trends and degree of variability. **Run charts** reveal information that histograms

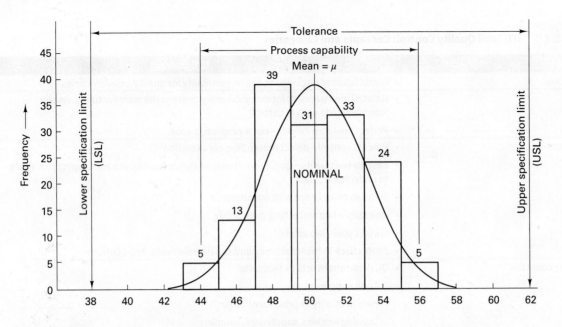

FIGURE 12.13 Histogram shows the output mean, μ, from the process versus nominal and the tolerance specified by the designer versus the spread as measured by the standard deviation, σ. Nominal 49.2 USL = 60 LSL = 38 μ = 50.2 σ = 2.

cannot, such as certain trends over time or at certain times of day. Individual measurements (not samples) are taken at regular time intervals, and the points are plotted on a connected line graph as a function of time. The graph can be used to find obvious trends in the process as shown in **Figure 12.14**, a run diagram with measurements made every hour over four shifts.

Run diagrams are very important at startup to identify the basic nature of a process. Without this information, one may use an inappropriate tool in analyzing the data. For example, a control chart or histogram might hide tool wear if frequent tool changes and adjustments are made between groups of observations. As a result, run diagrams (with 100% inspection where feasible) should always precede the use of control charts for averages and ranges.

Check Sheet

Suppose you had a manufacturing cell as used in lean production (see Chapters 42–43) and you were using a **check sheet** to gather data on one of the operations. The time to perform an assembly task is being recorded and the data are bimodal (see **Figure 12.15**). Upon investigation, you learn that this operation was performed by two operators. The data would suggest that they do not perform the manual assembly process the same way because there appears to be a 5- or 6-s difference in their average time to complete this task. Clearly, further study is needed to determine what the real problem is. For the next check sheet, use a different symbol for each worker.

The check sheet is an excellent way to view data while it is being collected. It can be constructed using predetermined parameters based on experience with the cell or system. The appropriate interval is checked as the data are being collected.

The lengths of manufactured components are measured. A run diagram is constructed to determine how the process is behaving. During 34 hours, measurements are made every hour (60 minutes) and plotted in the order that rack bars are produced.

First shift	35	40	27	30	30	34	26	31
Second shift	24	23	28	15	23	17	16	21
Third shift	15	13	28	8	20	9	5	11
First shift	16	5	9	13	16	10	9	10

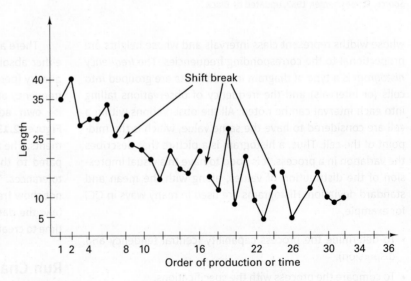

FIGURE 12.14 An example of a run chart or graph, which can reveal trends in the process behavior not shown by the histogram.

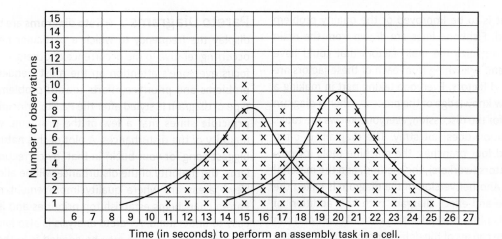

FIGURE 12.15 Example of a check sheet for gathering data on a process.

This often allows the central tendency and the spread of the data to be seen. The check sheet can provide basically the same information as a histogram, but it is easier to build (once the check sheet is formatted). The possibilities are endless and require a careful recording of all the sources of the data to track down the factors that result in loss of precision and accuracy in the process. Rapid feedback on quality is perhaps the most important factor, so these data-gathering tasks are done right on the factory floor, by the operators.

Cause-and-Effect Diagram

The best way to quickly isolate quality problems is to make everyone an inspector. This means every worker, foreman, supervisor, engineer, manager, and so forth, is responsible for making it right the first time and every time. One very helpful tool in this effort is the **cause-and-effect**, or **fishbone, diagram**. As shown in **Figure 12.16**, the cause-and-effect diagram can be used in conjunction with the run chart to root out the causes of problems. The problem can have multiple causes, but in general, the cause will lie in the process, operators, materials, or method (i.e., the four main branches on the chart as discussed in the following paragraph). Every time a quality problem is caused by one of these events, it is noted by the observer, and corrective action is taken. As before, experimental design procedures to be discussed later can help identify causes that affect performance.

Cause-and-effect diagrams are also known as fishbone diagrams because of their structure. Initially developed by Kaoru Ishikawa in 1943, this diagram organizes theories about the probable cause of a problem. On the main line is a quality

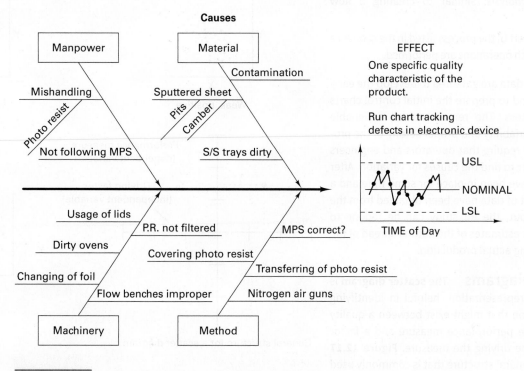

FIGURE 12.16 Example of a cause-and-effect diagram using a RUN chart to track effects.

characteristic that is to be improved or the quality problem being investigated. Fishbone lines are drawn from the main line. These lines organize the main factors that could have caused the problem. Branching from each of these factors are even more detailed factors. Everyone taking part in making a diagram gains new knowledge of the process. When a diagram serves as a focus for the discussion, everyone knows the topic, and the conversation does not stray. The diagram is often structured around four branches: the machine tools (or processes), the operators (workers), the method, and the material being processed. Another version of the diagram is called the *CEDAC*: the cause-and-effect diagram with the addition of cards. The effect is often tracked with a control chart or a run chart. The possible causes of the defect or problem are written on cards and inserted in slots in the charts.

There are three main applications of cause-and-effect diagrams:

1. *Cause enumeration*: listing of every possible cause and subcause.
   a. *Visual presentation:* one of the most widely used graphical techniques for QC.
   b. Better understanding of the relationships within the process yielding a better understanding of the process as a whole.
2. *Dispersion analysis*: causes grouped under similar headings; the 4 Ms are men, machines, materials, and methods.
   a. Each *major* cause is thoroughly analyzed.
   b. Possibility of not identifying root cause (may not fall into main categories).
3. *Process analysis*: similar to creating a flow diagram.
   a. Each part of the process listed in the sequence in which operations are performed.

In summary, data are gathered to develop the early PC studies and to prepare the initial control charts for the process. The removal of all assignable causes for variability and proper setting of the process average require that operators and engineers work together to find the causes for variation. After the charts have been in place for some time and a large amount of data have been obtained from the process output, the PC study can be redone to obtain better estimates of the natural spread of the process during actual production.

Scatter Diagrams
The **scatter diagram** is a graphical representation, helpful in identifying the correlation that might exist between a quality or productive performance measure and a factor that might be driving the measure. **Figure 12.17** shows the general structure that is commonly used for the scatter diagram and some typical patterns.

Pareto Diagrams
Pareto diagrams are bar charts that display the frequency to which a particular phenomenon is occurring relative to the occurrence of others. As such, it helps focus everyone's attention on the most frequently occurring problems and prioritize efforts toward problem solving. This type of diagram is styled after the Pareto principle. The **Pareto principle** states that a few of the problems will cause the majority of the disturbances. A plot of this nature shows "the biggest bang for your buck" in that, by correcting a few problems, the majority of the disturbance will be alleviated. These "few" areas are where quality improvement must focus. A Pareto chart helps establish top priorities and is visually very easy to understand. A Pareto analysis is also fairly easy to do. The only expertise that may be needed is in the area of data gathering. **Figure 12.18** shows a Pareto analysis of the usage rate of various TQC methods reported by attendees at a QC conference.

Process Flow Chart
Perhaps the first task in system analysis should be charting the process. A **process flow chart (PFC)** is a pictorial representation of a process as well as decisions and the interconnectivity between the steps and the

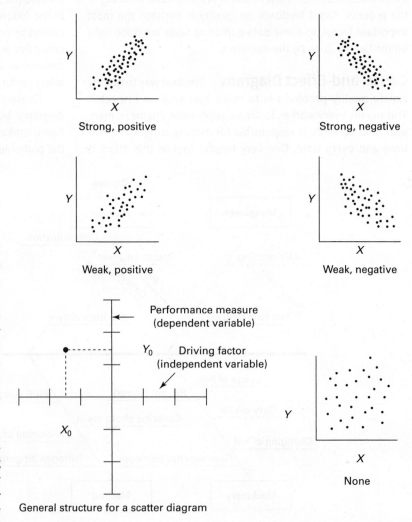

Strong, positive

Strong, negative

Weak, positive

Weak, negative

Performance measure (dependent variable)

Y_0 Driving factor (independent variable)

X_0

None

General structure for a scatter diagram

FIGURE 12.17 Typical patterns of correlation for scatter diagrams.

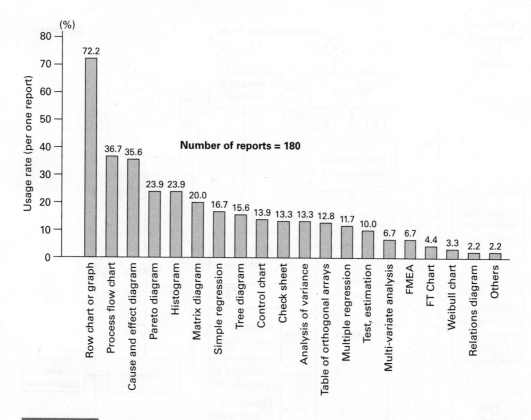

FIGURE 12.18 Usage rate of TQC methods and techniques reported at a QC conference.

decisions. These charts provide excellent documentation of a process at a particular time and can be used to show the relationships between the process steps.

Process flow charts are constructed using easily recognized symbols to represent the type of processing performed. A flattened oval can represents the start or stopping points of process, while diamonds can represent decisions and rectangles can represent process steps. Lines connecting each of the process steps end with arrows showing the flow of parts or information.

Making a PFC will often lead to the discovery of omissions in a process, identification of steps that everyone thought were taking place but are not, identification of a sequence of operations that does not make sense when analyzed or the discovery of steps that are no longer being used.

An example of a flow chart is shown in **Figure 12.19**. This example is complex but it deals with the steps involved in the manufacture of a refrigerator. This example is for manufacturing but flow charts can be used at any level within an organization, from the highest-level management process, to the process used to ship an order to a customer.

There are a few simple rules to follow when constructing a flow chart:

1. List all known steps in the process.
2. Use the simplest symbols possible.
3. Make sure every feedback loop has an escape.
4. Every process should have only one arrow proceeding outward.

In recent years, PFCs have been upgraded to value-stream maps. **(VSM) Value-stream mapping** documents a process or the movement of a product from raw material to finished part. A visual diagram, called a map is drawn, using VSM symbols that represent every step in the process, showing both the material and the information flows (see **Figure 12.20**). The bar across the bottom records the time for production (adding value) versus total lead time. In this example, 58 seconds adding values (processing) for 46.5 days of produce from through put time.

After analysis and process improvements, a new process flow is developed, and a future state of the value stream is drawn to create the desired flow, often reducing the lead time. Value stream mapping will be discussed in more detail in Chapter 43.

Line Stop in Lean Production

A pair of yellow and red lights hanging above the workers on the assembly line can be used to alert everyone in the area to the status of the processes. Many companies use Andon boards, which hang above the aisles. The number on the board reflects stations on the line. A worker can turn on a yellow light when assistance is needed, and nearby workers will move to assist the worker having a problem. The line keeps moving, however, until the product reaches the end of the station. Only then is a red light turned on if the problem cannot be solved quickly and the line needs to be stopped. When the problem is solved and everyone is ready to go again, the red light goes off and everyone starts back to work, all in synch.

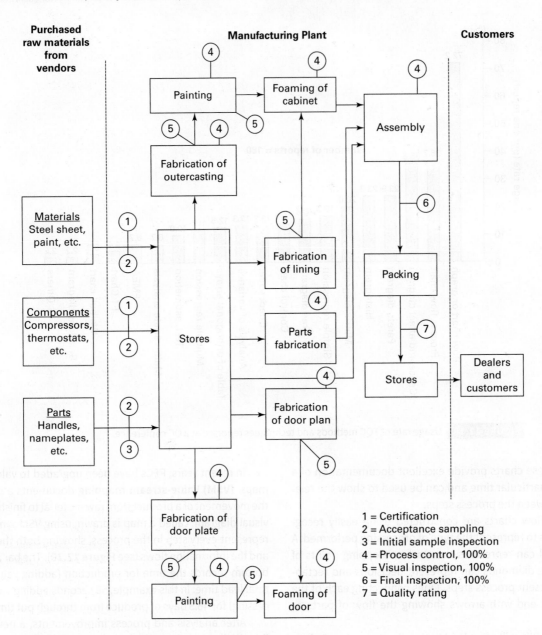

FIGURE 12.19 Process flow diagram for refrigerator manufacturing showing monitoring points.

1 = Certification
2 = Acceptance sampling
3 = Initial sample inspection
4 = Process control, 100%
5 = Visual inspection, 100%
6 = Final inspection, 100%
7 = Quality rating

Every worker should be given the authority to stop the production line to correct quality problems. In systems using *poka-yoke* or *autonomation*, devices may stop the line automatically preventing a defective step. The assembly line or manufacturing cell should be stopped immediately and started again only when the necessary corrections have been made. Although stopping the line takes time and money, it is advantageous in the long run. Problems can be found immediately, and the workers have more incentive to be attentive because they do not want to be responsible for stopping the line.

Implementing Quality Companywide

In lean production, the basic idea of **integration** is to shift functions that were formerly done in the staff organization (called the production system) into the manufacturing system. What

happens to the quality control department? The department serves as the facilitator and therefore acts to promote quality concepts throughout the plant. In addition, its staff educates and trains the operators in statistical and process control techniques and provides engineering assistance on visual and automatic inspection installations. Its most important function will be training the entire company in quality control.

Another important function of the QC department will be to work with and audit the vendors. The vendor's quality must be raised to the level at which the buyer does not need to inspect incoming material, parts, or subassemblies. The vendor simply becomes an extension of the buyer's plant. Ultimately, each vendor will deliver to the plant perfect materials that need no incoming inspection. Note that this means the **acceptable quality level (AQL)** of incoming material is 0%. Perfection is the goal. For many years, this country lived with the unwritten rule that 2 or 3% defective level was about as

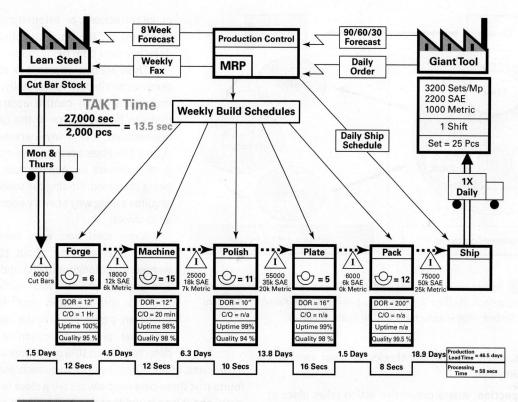

FIGURE 12.20 Example of a current-value-stream map.

good as you could get: better quality just cost too much. For the mass production systems, this was true. To achieve the kinds of quality that Toyota and many others have achieved, a company has to implement lean production by restructuring the production system, integrating the quality function directly into the linked-cell manufacturing system (L-CMS).

The quality control department also performs complex or technical inspections, total performance checks (often called **end-item inspection**), chemical analysis, nondestructive testing, X-ray analysis, destructive tests, or tests of long duration discussed in Chapter 11.

Making Quality Visible

Visual displays on quality should be placed throughout manufacturing facilities to make quality evident. These displays tell workers, managers, customers, and outside visitors what quality factors are being measured, what the current quality improvement projects are, and who has won awards for quality. Examples of visible quality are signs showing quality improvements, framed quality awards presented to or by the company, and displays of high-precision measuring equipment.

These displays have several benefits. When customers visit your plant to inspect your processes, they want to see measurable standards of quality. Highly visible indicators of quality such as control charts and displays should be posted in every department. Everyone is informed on current quality goals and the progress being made. Displays and quality awards are also an effective way to show the workforce that the company is serious about quality.

Source, Self-, and Successive Checks and Poka-Yokes

Many companies have developed an extensive QC program based on having many inspections. However, inspections can only *find* defects, not *prevent* them. Adding more inspectors and inspections merely uncovers more defects, but does little to prevent them. Clearly, the least costly system is one that produces no defects. But is this possible? Yes, it can be accomplished through methods like self- and source inspection, where quality control is in the hands of the operators.

Many people do not believe that the goal of zero defects is possible to reach, but many companies have achieved this goal or have reduced their defect level to virtually zero using techniques like poka-yokes and source inspection. When you are inspecting to find defects, the components are compared with standards, and defective items are removed. **Sampling inspection** is used when 100% inspection is not an option, but this assumes that defects are inevitable and that more rigorous inspection can reduce defects.

The truth is, to reduce the defects within a process, it must be recognized that defects are generated by the process itself and that most inspection techniques merely discover defects that already exist.

To achieve zero defects, working within a lean manufacturing concept where all operators are responsible for quality, you can perform

- **Successive checks**, where the next operator checks the work of the previous worker.

- **Self-checks**, where the operator checks his or her own work before passing it on.
- **Source inspection**, where preventive action takes place at the error stage, to prevent errors from turning into defects.

Source inspection (see **Figure 12.21**) involves rethinking the inspection part of the manufacturing process. First off, although it is necessary to have efficient inspection operations, they add little value in the product. Even the most efficient inspection operations are merely efficient forms of non-value-adding activity. Inspection plays a passive role in manufacturing and cannot by itself reduce defects.

Defects and *errors* are not the same thing in manufacturing. **Errors** can cause **defects**. For example, not setting the oven temperature correctly can burn the roast (too high) or have it undercooked (too low). Incorrectly loading the original into the fax machine (the error) results in sending blank pages (the defect). When you discover the effect, you make corrections but you have already spent the money to produce the defect. **Table 12.2** outlines some common errors associated with manufacturing.

Thus, source inspection looks for errors before they become defects. These techniques either stop the system or

TABLE 12.2	**Common Errors That Can Produce Defects But Are Preventable**

- Omissions in processing steps
- Errors in setting up the job in the machine
- Omission in the assembly process (missing parts)
- Inclusion of incorrect part
- Size errors due to wrong measurement
- Errors due to adjustments
- Errors in cutting tool geometry or cutting tool setting
- Errors in processing components (heat Treating)

make corrections, or automatically compensate or correct for the error condition to prevent a defective item from being made.

There are two ways to look at source inspections: vertically and horizontally. *Vertical-source inspections* try to control upstream processes that can be the source or the cause of defects downstream. It is always necessary to examine source processes because they may have a much greater impact on quality than the processes being examined. Finding the source of a problem requires asking why at every opportunity. Here is an example.

Some steel bars were being cylindrically ground. After grinding, about 10% of the bars warped (bent longitudinally) and were rejected. The grinding process was studied extensively, and no sure solution was found. Looking upstream, a problem with the heat-treating process that preceded cylindrical grinding was detected. About 10% of the bars were not getting a complete, uniform heating prior to quench. Asking why, it was found that these bars were always lying close to the door of the oven, and it was found that the door was not properly sealed, which resulted in a temperature gradient inside. Quenching of the bars induced a residual stress that was released by the grinding and caused the warping. Asking why five times uncovered the source of the problem.

Horizontal-source inspections detect defect sources within the processes and then introduce corrections to keep from turning errors into defects. In metal cutting and forming, this is commonly called adaptive control (for preventing defects) or in-process quality control. One of the best ways to prevent errors from occurring is through the use of poka-yokes.

Poka-yoke is a Japanese word for defect prevention. Poka-yoke devices and procedures are often devised mainly for preserving the safety of operations. The idea is to develop a method, mechanism, or device that will prevent the defect from occurring rather than to find the defect after it has occurred. Poka-yokes can be attached to machines to automatically check the products or parts in a process. Poka-yokes are source inspections that are usually attributes inspections. The production of a bad part is prevented by the device. Some devices may automatically shut down a machine if a defect is produced, preventing the production of an additional defective part. The poka-yoke system uses 100% inspection to guard against unavoidable human error.

Modern cars are equipped with many poka-yoke devices: you can't turn off the motor unless it is in Park; you can't open a door while the car is moving; the headlights come on with the windshield wipers. These are all devices to prevent you from making a mistake.

Such devices work very well when physical detection is needed, but many items can be checked only by sensory detection methods, such as the surface finish on a bearing race or the flatness of a glass plate. Variations in nonvisible conditions (air pressure, fluid velocity, temperature, electrical voltage, etc.) require detection devices where critical conditions are readily visible. For such problems, a system of self-checks and successive checks also can be used.

Teams (aka Quality Circles)

Several popular programs are built on the concept of participative management, such as **quality circles**, improvement teams, and task groups. These programs have been very successful in many companies, but have failed miserably in others. The difference is often due to the way the program was implemented by management. Programs must be integrated and managed within the context of a lean manufacturing system design strategy. For example, asking an employee for a suggestion that management does not use (or cannot explain why it does not use) defeats a suggestion system. Management must learn to trust the employees' ideas and decisions and move the decision making to the factory floor. (See *Ideas are Free* by Alan Robinson.)

A quality circle is usually a group of employees within the same department or factory floor area. Meetings are held to work on problems. An organization structure is usually composed of members, a team leader, a facilitator, a manufacturing engineer, and a steering committee.

Quality circles usually have the following main objectives: provide all workers with a chance to demonstrate their ideas; raise employee morale; and encourage and develop workers' knowledge, quality control techniques, and problem-solving methods. They also unify companywide QC activities, clarify managerial policies, and develop leadership and supervisory capabilities.

Quality circles have been implemented in U.S. companies with limited success when they are not part of a lean manufacturing strategy. It is possible for quality circles to work in the United States, but they must be encouraged and supported by management. Everyone must be taught the importance and benefits of integrated quality control.

Superior Quality in Manufacturing/ Assembly Cells

In lean manufacturing, subassembly and manufacturing cells are designed for a **ma+ke one–check one–move one on (MO-CO-MOO)** strategy. The part receives successive checks after each processing or assembly step. For successive checks to be successful, several rules should be followed. All the possible variables and attributes are not measured, because this would eventually lead to errors and confusion in the inspection process. The part should be analyzed so that only one or two points are inspected after each step in the process. This is the heart of MO-CO-MOO. Only the most important elements just produced are inspected.

Another important rule is that the immediate feedback of a defect leads to immediate action. Because the parts are produced in an integrated manufacturing system, this will be very effective in preventing the production of more defective parts. Suppose the cell has only one or two workers and they are not in a position to directly check each other's work after each step. Here is where the decouplers can play a role by providing automatic successive checking of the parts' critical features before proceeding to the next step. Only perfect parts are pulled from one process to the next through the decoupler.

In assembly lines, a worker may inspect each part immediately after producing it. This is called self-checking. There is an immediate feedback to the worker on quality. However, it would be difficult for many workers not to allow a certain degree of bias to creep into their inspection, whether they were aware of it or not, because they are inspecting their own work. Within cells operated by multiple workers, the operator of the downstream station or process can inspect the parts produced by the upstream operator. If there is a problem with the parts, the defective item is immediately passed back to the worker at the previous station. There, the defect is verified and the problem corrected. Action is immediately taken to prevent any more defective parts. While this is going on, the line is shut down.

12.7 Six Sigma

To meet the quality challenge of the Japanese, an American company, Motorola, developed the **Six Sigma** concept. The concept is shown in **Figure 12.22** in terms of four-sigma and six-sigma capability. Most people do not know how sigma is determined, (σ is estimated from $\bar{R}/d_2$ or σ/C_2 sample data), that a sigma is a standard deviation, or what sigma measures (sigma measures the repeatability or variability or lack of precision in a process.)

In essence, the Six Sigma concept calls for the process to be improved to the place where there are 12 standard deviations between USL and LSL. So the concept should be called 12 sigma, so a true six-sigma process, by definition, is obtained when $\mu \pm 6\sigma$ equals the product specification tolerance interval and when the process mean is centered between the upper and lower product specifications. Assuming a normal distribution for the process what C_p would be associated with six-sigma production and if the process is centered on its target, what C_{pk} should be realized?

Given that the process is a six-sigma process, then the USL − LSL = 6σ + 6σ = 12σ. Hence,

$$C_p = \frac{USL - LSL}{6\sigma} = \frac{12\sigma}{6\sigma} = 2.0$$

$$C_{pk} = \text{minimum} \left\{ \frac{\mu - LSL}{3\sigma}, \frac{USL - \mu}{3\sigma} \right\}$$

$$= \text{minimum} \left\{ \frac{6\sigma}{3\sigma}, \frac{6\sigma}{3\sigma} \right\} = 2.0$$

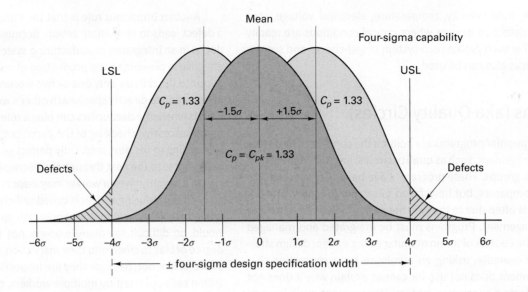

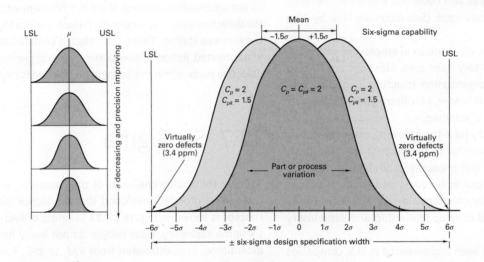

FIGURE 12.22 To move to six-sigma capability from four-sigma capability requires that the process capability (variability) be greatly improved (σ reduced). The curves in these figures represent histograms or curves fitted to histograms.

As the variability of a process changes, so does sigma. A reduction in sigma (a reduction in spread) reflects an improvement in process or an improvement in precision (better repeatability). As the process is improved, sigma decreases. So the question is, "How do I improve the process?" This is the essence of process capability work outlined in this chapter.

Here is an example. A foundry was having a problem with cores breaking in the molds during pouring. A PC study determined that the core strength was widely variable and that it was the low-strength cores that broke when the molten metal hit them during filling. Increasing the resin content in the cores and changing the gating in the mold did not eliminate the problem, so a Taguchi study was run that revealed that core strength was highly dependent on the grain size of the sand, which was also highly variable. The sand preparation process was revised to yield more uniform-sized grains, which, when used in the core-packing process, reduced the variability in the strength characteristic and eliminated the core breakage problem.

Design of Experiments and Taguchi Methods

Foreign competition has forced American manufacturers to take a second look at quality, as evidenced by the major emphasis (reemphasis) on statistical process control (SPC) in American industry. This drive toward superior quality has led to the introduction of **Taguchi methods** for improvement in products, product design, and processes. Basically, SPC looks at processes and control, the latter loosely implying "improvement." **Design of experiments (DOE)** and Taguchi methods, however, span a much wider scope of functions and include the design aspects of products and processes—areas that were seldom, if ever, formally treated from the quality standpoint. Another threshold has been reached in quality control, witnessed by an expanding role of quality in the production of goods and services. The consumer is the central focus of attention on quality, and the methods of quality design and control have been incorporated into all phases of production.

The Taguchi methods incorporate the following general features:

1. Quality is defined in relation to the total loss to the consumer (or society) from the less-than-perfect quality of the product. The methods include placing a monetary value on quality loss. Anything less than perfect is waste.

2. In a competitive society, continuous quality improvement and cost reduction are necessary for staying in business.

3. Continuous quality improvement requires continuous reduction in the variability of product performance characteristics with respect to their target values.

4. The quality and cost of a manufactured product are determined by the engineering designs of the product and its manufacturing system.

5. The variability in product and process performance characteristics can be reduced by exploiting the nonlinear (interactive) effects of the process or product parameters on the performance characteristics.

6. Statistically planned (Taguchi) experiments can be used to determine settings for processes and parameters that reduce the performance variation.

7. Design and improvement of products and processes can make them *robust,* or less sensitive to uncontrollable or difficult-to-control variations, called *noise* by Taguchi.

In the Taguchi approach, specified combinations of all of the input parameters, at various levels, that are believed to influence the quality characteristics being measured are used. These combinations should be run with the objective of selecting the best level for individual factors. For example, speed levels may be high, normal, and low, and operators may be fast or slow. For a Taguchi approach, material is often an input variable specified at different levels: normal, homogeneous, and highly variable. If a material is not controllable, it is considered a noise factor.

Designed experiments and Taguchi methods can be used as alternative approaches to making a PC study. The Taguchi approach uses a truncated experimental design (called an **orthogonal array**) to determine which process inputs have the greatest effect on process variability (i.e., precision) and which have the least. Those inputs that have the greatest influence are set at levels that minimize their effect on process variability. As shown in **Figure 12.23**, factors A, B, C, and D all have an effect on process variability Y. By selecting a high level of A and low levels of B and C, the inherent variability of the process can be reduced. Those factors that have little effect on process variability, like factor D, are used to adjust or recenter the

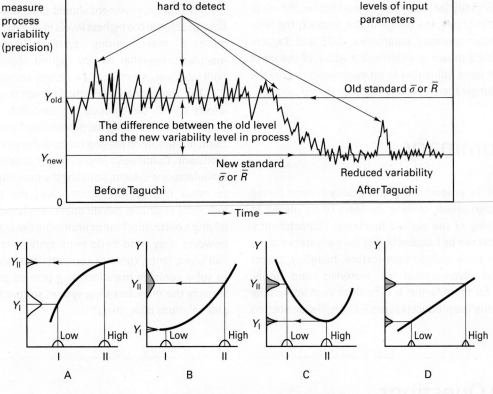

Selecting high and low values of input parameters changes the process output.

FIGURE 12.23 The use of Taguchi methods can reduce the inherent process variability as shown in the upper figure. Factors A, B, C, and D versus process variable Y shown in lower figure.

process aim. In other words, Taguchi methods seek to minimize or dampen the effect of the causes of variability and thus to reduce the total process variability. This is also the goal of the Six Sigma method, which was discussed earlier.

The methods are, however, more than just mechanical procedures. They infuse an overriding new philosophy into manufacturing management that basically makes quality the primary issue in manufacturing. The manufacturing world is rapidly becoming aware that the consumer is the ultimate judge of quality. Continuous quality improvement toward perfect quality is the ultimate goal. Finally, it is recognized that the ultimate quality and lowest cost of a manufactured product are determined to a large extent by the engineered designs of

1. The product.
2. The manufacturing process technology and the sequences of processes.
3. The manufacturing system (integration of the product and the process).

So, a new understanding of quality has emerged. Process variability is not fixed. It can be improved! The noise level of a process can be reduced by exploring the nonlinear effects of the products (or process) parameters on the performance characteristics.

Process capability must be also addressed in the context of machining centers, (programmable NC) machines. In machine tools, accuracy and precision in processing are affected by machine alignment, the setup of the workholder, the design and rigidity (accuracy) of the workholder, the accuracy of the cutting tools, the design of the product, the temperature, and the operating parameters. DOE and Taguchi methods provide a means of determining which of the input parameters are most influential in product quality when the operator is no longer there to compensate for the variability.

12.8 Summary

The designer of the product must have quality in mind during the quality design phase, seeking the least costly means to ensure the quality of the desired functional characteristics. Major factors that can be handled during the early stages of the product design cycle include temperature, humidity, power variations, and deterioration of materials and tools. Compensation for these factors is difficult or even impossible to implement after the product is in production. The distinction

between superior- and poor-quality products can be seen in their variability in the face of internal and external causes. This is where Taguchi parameter design methods can be important.

The secret to successful process control is putting the control of quality in the hands of the workers. Many companies in this country are currently engaged in SQC (statistical quality control), but they are still inspecting to *find* defects. The number of defects will not be reduced merely by making the inspection stage better or faster or automated. You are simply more efficient at discovering defects. The trick is to *inspect* to *prevent* defects. How can this be done? Here are the basic ideas:

- Use source inspection techniques that control quality at the stage where defects originate.
- Use 100% inspection by the operators with immediate feedback rather than sampling.
- Make every worker an inspector, trained in the seven quality tools.
- Minimize the time it takes to carry out corrective action.
- Remember that people are human and not infallible.
- Devise methods and devices to prevent them from making errors. (Can you think of such a device? Does your car have a procedure that prevents you from locking the ignition keys inside the car? That is a poka-yoke.)
- Do not simply rely on inspection to control quality. Sorting to find defects by inspection is bad because you already paid to produce those defects.

Process improvement should drive toward defect prevention. To achieve the highest levels in quality, you have to implement a manufacturing system design—that is, lean manufacturing—that has the highest objective (zero defects) built into (integrated into) the system design.

The use of analysis methods made popular by Six Sigma efforts can further reduce the variability in the processes. Concentrate on making the process less variable and more efficient, not simply on making the operators and operations more proficient. Continuous improvement requires redesign of the manufacturing system, reducing the time required for products to move through the system (i.e., the throughput time). Industrial engineers can do operations improvement work like buying a better machine or improving the ergonomics of a task. However, they need to do more systems improvement work. Too often, fancy, complex computerized solutions are devised to solve complex manufacturing process problems. Why not simplify the manufacturing system so that the need for complex solutions disappears?

Review Questions

1. Define *accuracy* or *precision* in terms of a process capability study.

2. What does the *nature of the process* refer to?

3. Suppose you have a "pistol-shooting" process that is accurate and precise. What might the target look like if, occasionally while shooting, a sharp gust of wind blew left to right?

4. Review the steps required to make a PC study of a process.

5. Why don't standard tables exist detailing the natural variability of a given process, like rolling, extruding, or turning?

6. Here are some common, everyday processes with which you are familiar. What variable or aspect to the process might dominate the process, in terms of quality, not output?

  a. Baking a cake (from scratch) drilling a steak

  b. Mowing the lawn

  c. Washing dishes in a dishwasher

7. Why might the diameter measurements for holes produced by the process of drilling have a skewed rather than a normal distribution?

8. What are common manufactured items that may

  a. Receive 100% inspection?

  b. Receive no final inspection?

  c. Receive some final inspection, that is, sampling?

9. What are common reasons for sampling inspection rather than 100% inspection?

10. Fill in this table with one of the four following statements: no error—the process is good; no error—the process is bad; type I or alpha error; type II or beta error.

		In reality, if we looked at everything the process made, we would know that it had:	
		Changed	Not Changed
The sample suggested that the process had:	Changed		
	Not Changed		

11. Now explain why when we sample, we cannot avoid making type I and type II errors.

12. Which error can lead to legal action from the consumer for a defective product that caused bodily injury?

13. Define and explain the difference between

  a. σ and $\sigma_{\bar{x}}$

  b. σ and $\bar{\sigma}$

  c. $\sigma_{\bar{x}}$, σ_R, and σ_σ

14. What is C_p, and why is a value of 0.80 not good? How about a value of 1.00? 1.3?

15. The designer of a component usually sets the nominal and tolerance values when designing the part. How do these decisions affect the decisions of the manufacturing engineer?

16. What are some of the alternatives available to you when you have the situation where $6\sigma >$ USL – LSL?

17. C_{pk} is another process capability index. How does it differ from C_p?

18. In a sigma chart, are σ values for the samples normally distributed about $\bar{\sigma}$? Why or why not?

19. What is an assignable cause, and how is it different from a chance cause?

20. Why is the range used to measure variability when the standard deviation is really a better statistic?

21. How is the standard deviation of a distribution of sample means related to the standard deviation of the distribution from which the samples were drawn?

22. In the last two decades, the quality in automobiles has significantly improved. What do *you* think is the main cause for this marked quality improvement?

23. Figure 12.15 shows a bimodal check sheet indicating that the two operators performing an assembly task (in a cell) do the jobs at different rates. What would you recommend here?

24. Control charts use upper and lower control limits. How is a UCL different from a USL?

25. In Figure 12.10, what are the USL and the LSL, and why are they *not* shown on the charts?

26. What are four major branches (fishbones) on a cause-and-effect diagram?

27. How does variation in the measuring device (instrument) effect the measurements obtained on a component?

28. What are Taguchi or factorial experiments, and how might they be used to perform a process capability study?

29. How does the Taguchi approach differ from the standard experimental method outlined in this chapter?

30. Why are Taguchi experiments so important compared to classical DOE type experiments?

31. Explain what happed to improve the process in Figure 12.23.

32. In Table 12.1, explain

  a. MO-CO-MOO.

  b. 8-4-8-4 scheduling.

  c. $N = 2$ inspection.

  d. Pareto chart.

33. What is a quality circle, and how might you apply this concept to your college life?

34. Which of the seven QC tools is being used in Figure 12.16 to measure the effect?

Problems

1. Go to a subway restaurant and develop a process flow diagram for making sandwich.

2. For the items listed in the following chart, obtain a quantity of 48. Measure the indicated characteristics, and determine the process mean and standard deviation. Use a sample size of 4, so that 12 samples are produced.

Item	Characteristic(s) You Can Measure
Flat washer	Weight, width, diameter of hole, outside diameter
Paper clip	Length, diameter of wire
Coin (penny, dime)	Diameter, thickness at point, weight
Your choice	Your choice

3. Perform a process capability study to determine the PC of the process that makes M&M candies. You will need to decide what characteristics you want to measure (weight, diameter, thickness, etc.), how you will measure it (use rule of 10), and what kind of M&Ms you want to inspect (how many bags of M&Ms you wish to sample). Take sample size of 4 ($n = 4$). Make a histogram of the individual data, and estimate $\overline{X}'$ and σ as outlined in the chapter. If you decide to measure the weight characteristics, you can check your estimate of $\overline{X}'$ by weighing all the M&Ms together and dividing by the total number of M&Ms.

4. For the data given in Figure 12.3, compute the mean and standard deviation for the histogram and then C_p and C_{pk} values making any assumptions needed to perform the calculations.

5. For the data given in Figure 12.6, compute C_p, and C_{pk} for (a) through (e).

6. Calculate $\overline{\overline{X}}$ and $\overline{R}$ and the control limits for the $\overline{X}$ and R control charts shown in **Figure 12.A**. The sample mean, $\overline{X}$ and range, R, for the first few subgroups and the data for each sample are given. There are 25 samples of size 4. Therefore $k = 25$, and $n = 4$. Complete the bottom part of the table and then compute the control limits for both charts. Construct the charts, plotting $\overline{\overline{X}}$ and $\overline{R}$ bar as solid lines and control limits as dashed lines as shown in Figure 12.6. Plot the data on the charts and comment on your findings. (Use Figure 12.6 to check your results.)

7. For the data given in Figure 12.A, estimate the mean and standard deviation for the process from which these samples were drawn (i.e., the parent population) and discuss the process capability in terms of C_p and C_{pk}. The USL and LSL for this dimension are 0.9 and 0.5, respectively, and the nominal is 0.7, as shown in Figure 12.6.

8. **Figure 12.B** contains data from a machining process that produces holes (drilling) with limits of 6.00 to 6.70 mm.

a. Construct the control charts for $\overline{X}$ and R using $n = 5$ and $k = 25$ mean values and the range values for the 25 samples, and then check the calculations for $\overline{\overline{X}}$ and $\overline{R}$ and the control limits for the charts.

b. Insert the center lines for $\overline{\overline{X}}$ and $\overline{R}$ on the charts.

c. Plot the points on the charts.

d. Discuss the charts.

e. Using the data, develop the process capability indexes, and discuss the capability of this process.

f. Using the data $n = 5$ and $k = 25$, develop the σ control chart and use $\overline{\sigma}$ to estimate σ' for the process capability indexes and C_p and C_{pk}.

g. Develop $\overline{X}$ and R charts for samples sizes of 4 (or 3) by ignoring X_5 (X_3 and X_5) or any combination of individual values. Use the charts to perform a process capability study. Did the findings change?

Date		6/8				6/9				6/10				6/11				6/12				6/15				6/16
	Time	8				8	10	12	2	8	10	12	2	8	10	12	2	8	10	12	2	8	10	12	2	8
R e a d i n g s	1	.65	.75	.75	.60	.70	.60	.75	.60	.65	.60	.80	.85	.70	.65	.90	.75	.75	.75	.65	.60	.50	.60	.80	.65	.65
	2	.70	.85	.80	.70	.75	.75	.80	.70	.80	.70	.75	.75	.70	.70	.80	.80	.70	.70	.65	.60	.55	.80	.65	.60	.70
	3	.65	.75	.80	.70	.65	.75	.65	.80	.85	.60	.90	.85	.75	.85	.80	.75	.85	.60	.85	.65	.65	.65	.75	.65	.70
	4	.65	.85	.70	.75	.85	.85	.75	.75	.85	.80	.50	.65	.75	.75	.75	.80	.70	.70	.65	.60	.80	.65	.65	.60	.60
	Sum	2.65	3.20	3.05	2.75																					
$\overline{X} = \frac{\text{Sum}}{\text{No. of readings}}$		.66	.80	.76	.69																					
$R = \frac{\text{Highest−}}{\text{Lowest}}$		.05	.10	.10	.15	.20	.25	.15	.20																	

* For sample sizes of less than seven, there is no lower control limit for ranges.

For first subgroup:
Sum = .65 + .70 + .65 + .65 = 2.65
$\overline{X}$ = 2.65 ÷ 4 = .6625
R = .05

FIGURE 12.A

Product name	Cylinder		Sample size		3, 4, or 5		1996, 10.15	$\overline{X}$ Control chart	R Chart
Quality characteristic	Hole diameter	Samples	Timing of taking samples		Daily	Period	1975, 10.30	Center line $\overline{\overline{X}}$ = 6.299	Center line $\overline{R}$ = 0.274
Limits of allowable range	Max. 6.70 mm	Section		00 00	Person in charge	C. Black	UCL$_{\overline{x}}$ = $\overline{\overline{X}}$ + $A_2\overline{R}$	UCL 2.11 × $\overline{R}$ = 0.578	
	Min. 6.00 mm	Measuring instrument serial number		103037	Person in charge of inspection	Pogi Bear	LCL$_{\overline{x}}$ = $\overline{\overline{X}}$ − $A_2\overline{R}$	LCL = 0	

	Lot No.	1	2	3	4	5	6	7	8	9	10	11	12	13	14	15	16	17	18	19	20	21	22	23	24	25	Total	Mean
Measured values	X_1	47	19	19	29	28	40	15	35	27	23	28	31	22	37	25	7	38	35	31	12	52	20	29	28	42		
	X_2	32	37	27	29	12	35	30	44	37	45	44	25	37	32	40	31	0	12	20	27	42	31	47	27	34		
	X_3	44	31	21	42	45	11	12	32	26	26	40	24	19	12	24	23	41	29	35	38	52	15	41	22	15		
	X_4	35	25	15	59	36	38	33	11	20	37	31	32	47	38	50	18	40	48	24	40	24	3	32	32	29		
	X_5	20	34	19	38	25	33	26	38	35	32	18	22	14	30	19	32	37	20	47	31	25	28	22	54	21		
	Total	178	146	101	197	146	157	116	160	145	163	161	134	139	149	158	111	156	144	157	148	195	97	171	163	141		
	Mean $\overline{x}$	35.6	29.2	20.2	39.4	29.2	31.4	23.2	32.0	29.0	32.6	32.2	26.8	27.8	29.8	31.6	22.2	31.2	28.8	31.4	29.6	39.0	19.4	34.2	32.6	28.2	746.6	29.86
	Range R	27	18	33	30	33	29	21	33	17	22	26	10	33	26	31	25	41	36	27	28	28	28	25	32	27	686	27.44

FIGURE 12.B

CHAPTER 13

Fundamentals of Casting

13.1 Introduction to Materials Processing

Almost every manufactured product (or component of a product) goes through a series of activities that include (1) design, defining what we want to produce; (2) material selection; (3) process selection; (4) manufacture; (5) inspection and evaluation; and (6) feedback. Previous chapters have presented the fundamentals of *materials engineering*, the study of the structure, properties, processing, and performance of engineering materials and the systems interactions between these aspects. Other chapters have addressed the use of heat treatment to achieve desired properties and various aspects of quality assurance. In this chapter, we begin a focus on **materials processing**, the science and technology through which a material is converted into a useful shape with structure and properties that are optimized for the proposed service environment. A less-technical definition of materials processing might be "whatever must be done to convert stuff into things."

A primary objective of materials processing is the production of a desired shape in the desired quantity. Shape-producing processes can be grouped into five basic "families," as indicated in **Figure 13.1**. **Casting** processes exploit the properties of a liquid as it flows into and assumes the shape of a prepared container and then solidifies on cooling. The **material removal** processes remove selected segments from an initially oversized piece. Traditionally, these processes have often been referred to as **machining**, a term used to describe the mechanical cutting of materials. The more general term, *material removal*, includes a wide variety of techniques, including those based on chemical, thermal, and physical processes. **Deformation processes** exploit the ductility or plasticity of certain materials, mostly metals, to produce the desired shape by mechanically moving or rearranging the solid. **Consolidation processes** build a desired shape by putting smaller pieces together.

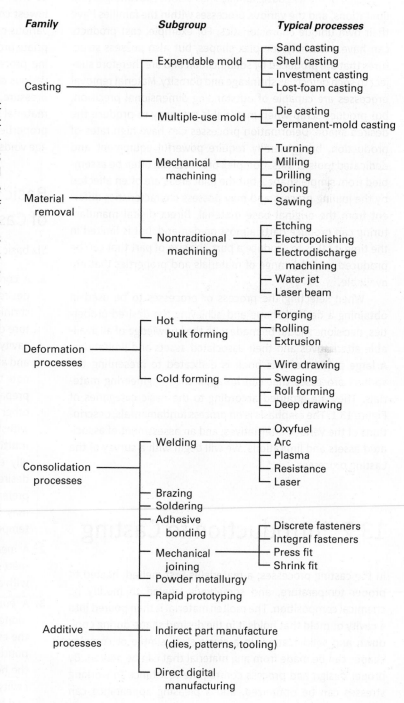

Family	Subgroup	Typical processes
Casting	Expendable mold	Sand casting / Shell casting / Investment casting / Lost-foam casting
	Multiple-use mold	Die casting / Permanent-mold casting
Material removal	Mechanical machining	Turning / Milling / Drilling / Boring / Sawing
	Nontraditional machining	Etching / Electropolishing / Electrodischarge machining / Water jet / Laser beam
Deformation processes	Hot bulk forming	Forging / Rolling / Extrusion
	Cold forming	Wire drawing / Swaging / Roll forming / Deep drawing
Consolidation processes	Welding	Oxyfuel / Arc / Plasma / Resistance / Laser
	Brazing / Soldering / Adhesive bonding / Mechanical joining / Powder metallurgy	Discrete fasteners / Integral fasteners / Press fit / Shrink fit
Additive processes	Rapid prototyping / Indirect part manufacture (dies, patterns, tooling) / Direct digital manufacturing	

FIGURE 13.1 The five materials processing families with subgroups and typical processes.

Included here are welding, brazing, soldering, adhesive bonding, and mechanical fasteners. **Powder metallurgy** is the manufacture of a desired shape from particulate material, a type of consolidation that can also involve aspects of casting and forming. The newest grouping or classification is **additive manufacturing** or **direct digital manufacturing**, which includes a variety of processes developed to directly convert a computer-file "drawing" to a finished product by a layer-by-layer deposition of material. Products can be made without any intervening patterns, dies, or other tooling.

Each of the five basic families has distinct advantages and limitations, and the various processes within the families have their own unique characteristics. For example, cast products can have extremely complex shapes, but also possess structures that are produced by solidification and are therefore subject to such defects as shrinkage and porosity. Material removal processes are capable of outstanding dimensional precision, but produce scrap when material is cut away to produce the desired shape. Deformation processes can have high rates of production, but generally require powerful equipment and dedicated tools or dies. Complex products can often be assembled from simple shapes, but the joint areas are often affected by the joining process and may possess characteristics different from the original base material. Direct digital manufacturing can produce parts almost on-demand, but is limited in the time it takes to produce a part, the size of part that can be produced, and the range of materials and properties that are available.

When selecting the process or processes to be used in obtaining a desired shape and achieving the desired properties, decisions should be made with the knowledge of all available alternatives and their associated assets and limitations. A large portion of this book is dedicated to presenting the various processes that can be applied to engineering materials. They are grouped according to the basic categories of Figure 13.1. The emphasis is on process fundamentals, descriptions of the various alternatives, and an assessment of associated assets and limitations. We will begin with a survey of the casting processes.

13.2 Introduction to Casting

In the casting processes, a material is first melted, heated to proper temperature, and sometimes treated to modify its chemical composition. The molten material is then poured into a cavity or mold that holds it in the desired shape during cool-down and solidification. In a single step, simple or complex shapes can be made from any material that can be melted. By proper design and process control, the resistance to working stresses can be optimized, and a pleasing appearance can be produced.

An extremely high percentage of manufactured goods contain at least one metal casting. Cast parts range in size from a fraction of a centimeter and a fraction of a gram (such as the individual teeth on a zipper), to over 10 meters and many tons (as in the huge propellers and stern frames of ocean liners). Moreover, the casting processes have distinct advantages when the production involves complex shapes, parts having hollow sections or internal cavities, parts that contain irregular curved surfaces (except those that can be made from thin sheet metal), very large parts, or parts made from metals that are difficult to machine.

It is almost impossible to design a part that cannot be cast by one or more of the commercial casting processes. However, as with all manufacturing techniques, the best results and lowest cost are only achieved if the designer understands the various options and tailors the design to use the most appropriate process in the most efficient manner. The variety of casting processes uses different mold materials (sand, metal, or various ceramics) and pouring methods (gravity, vacuum, low pressure, or high pressure). All share the requirement that the material should solidify in a manner that would maximize the properties and avoid the formation of defects, such as shrinkage voids, gas porosity, and trapped inclusions.

Basic Requirements of Casting Processes

Six basic steps are present in most casting processes:

1. A container must be produced with a *cavity* having the desired shape and size (*mold cavity*), with due allowance for shrinkage of the solidifying material. Any geometrical feature desired in the finished casting must be present in the cavity. The mold material must provide the desired detail and also withstand the high temperatures and not contaminate the molten material. In some processes, a new mold is prepared for each casting (single-use molds), whereas in other processes, the mold is made from a material that can withstand repeated use, such as metal or graphite. The **multiple-use molds** tend to be quite costly and are generally employed with products where large quantities are desired. The more economical **single-use molds** are usually preferred for the production of smaller quantities, but may also be required when casting the higher melting-temperature materials.

2. A *melting process* must be capable of providing molten material at the proper temperature, in the desired quantity, with acceptable quality, and at a reasonable cost.

3. A *pouring technique* must be devised to introduce the molten metal into the mold. Provision should be made for the escape of all air or gases present in the cavity prior to pouring, as well as those generated by the introduction of the hot metal. The molten material must be free to fill the cavity, producing a high-quality casting that is fully dense and free of defects.

4. The *solidification process* should be properly designed and controlled. Castings should be designed so that solidification and solidification shrinkage can occur without

producing internal porosity or voids. In addition, the molds should not provide excessive restraint to the shrinkage that accompanies cooling, a feature that may cause the casting to crack when it is still hot and its strength is low.

5. It must be possible to remove the casting from the mold. With single-use molds that are broken apart and destroyed after each casting, mold removal presents no serious difficulty. With multiple-use molds, however, the removal of a complex-shaped casting may be a major design problem.

6. Various *cleaning, finishing, and inspection* operations may be required after the casting is removed from the mold. Extraneous material is usually attached where the metal entered the cavity; excess material may be present along mold parting lines (segment separation interfaces); and mold material may adhere to the casting surface. All of these must be removed from the finished casting.

Each of these six steps will be considered in more detail as we move through the chapter. The fundamentals of solidification, pattern design, gating, and risering will all be developed. Various defects will also be considered, together with their causes and cures.

13.3 Casting Terminology

Before we proceed to the process fundamentals, it is helpful to first become familiar with a bit of casting vocabulary. **Figure 13.2** shows a two-part mold, its cross section, and a variety of features or components that are present in a typical casting process. To produce a casting, we begin by constructing a **pattern**—an approximate duplicate of the final casting.

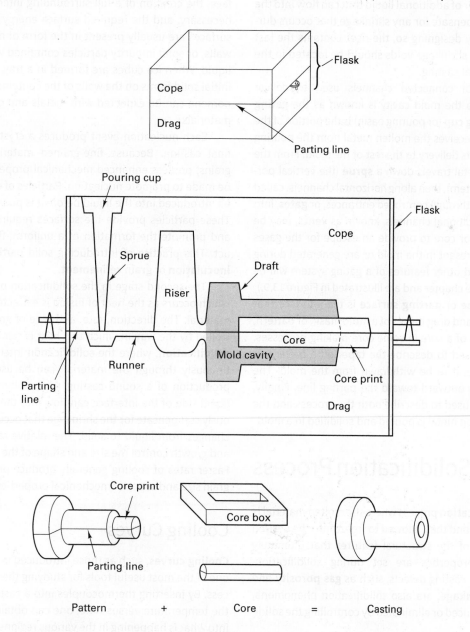

FIGURE 13.2 Cross section of a typical two-part sand mold indicating various mold components and terminology.

Molding material will then be packed around the pattern, and the pattern is removed to create all or part of the mold cavity. The rigid metal or wood frame that holds the molding aggregate is called a **flask**. In a horizontally parted two-part mold, the top half of the pattern, flask, mold, or core is called the **cope**. The bottom half of any of these features is called the **drag**. A **core** is a sand (or metal) shape that is inserted into a mold to produce the internal features of a casting, such as holes or passages. Cores are produced in wood, metal, or plastic tooling, known as **core boxes**. A **core print** is a feature that is added to a pattern, core, or mold and is used to locate and support a core within the mold. The mold material and the cores then combine to produce a completed **mold cavity**, a shaped hole into which the molten metal is poured and solidified to produce the desired casting. A **riser** is an additional void in the mold that also fills with molten metal. Its purpose is to provide a reservoir of additional liquid that can flow into the mold cavity to compensate for any shrinkage that occurs during solidification. By designing so, the riser contains the last material to solidify, shrinkage voids should be located in the riser and not the final casting.

The network of connected channels used to deliver the molten metal to the mold cavity is known as the **gating system**. The **pouring cup** (or pouring basin) is the portion of the gating system that receives the molten metal from the pouring vessel and controls its delivery to the rest of the mold. From the pouring cup, the metal travels down a **sprue** (the vertical portion of the gating system), then along horizontal channels, called **runners**, and finally through controlled entrances, or **gates**, into the mold cavity. Additional channels, known as **vents**, may be included in a mold or core to provide an escape for the gases that are originally present in the mold or are generated during the pour. (These and other features of a gating system will be discussed later in the chapter and are illustrated in Figure 13.9.)

The **parting line** or **parting surface** is the interface that separates the cope and drag halves of a mold, flask, or pattern, and also the halves of a core in some core-making processes. **Draft** is the term used to describe the taper on a pattern or casting that permits it to be withdrawn from the mold. The draft usually tapers outward toward the parting line. Finally, the term **casting** is used to describe both the process and the product when molten metal is poured and solidified in a mold.

13.4 The Solidification Process

Casting is a **solidification** process where the molten material is poured into a mold and then allowed to freeze into the desired final shape. Many of the structural features that ultimately control product properties are set during solidification. Furthermore, many casting defects, such as **gas porosity** and **solidification shrinkage**, are also solidification phenomena, and they can be reduced or eliminated by controlling the solidification process.

Solidification is a two-stage, nucleation and growth, process, and it is important to control both of these stages. **Nucleation** occurs when stable particles of solid form from within the molten liquid. When a material is at a temperature below its melting point, the solid state has a lower energy than the liquid. As solidification occurs, there is a release of energy. At the same time, however, interface surfaces are created between the new solid and the parent liquid. Formation of these surfaces requires energy. For nucleation to proceed, there must be a net reduction or release of energy. As a result, nucleation generally begins at a temperature somewhat below the equilibrium melting point (the temperature where the internal energies of the liquid and solid are equal). The difference between the melting point and the actual temperature of nucleation is known as the amount of **undercooling**.

If nucleation can occur on some form of existing surface, the creation of a full surrounding interface is no longer necessary, and the required surface energy is reduced. Such surfaces are usually present in the form of mold or container walls, or solid impurity particles contained within the molten liquid. When ice cubes are formed in a tray, for example, the initial solid forms on the walls of the container. The same phenomena can be expected with metals and other engineering materials.

Each nucleation event produces a crystal or grain in the final casting. Because fine-grained materials (many small grains) possess enhanced mechanical properties, efforts may be made to promote nucleation. Particles of existing solid may be introduced into the liquid before it is poured into the mold. These particles provide the surfaces required for nucleation and promote the formation of a uniform, fine-grained product. This practice of introducing solid particles is known as **inoculation** or **grain refinement**.

The second stage in the solidification process is **growth**, which occurs as the heat of fusion is extracted from the liquid material. The direction, rate, and type of growth can be controlled by the way in which this heat is removed. **Directional solidification**, where the solidification interface sweeps continuously through the material, can be used to ensure the production of a sound casting. The molten material on the liquid side of the interface can flow into the mold to continuously compensate for the shrinkage that occurs as the material changes from liquid to solid. The relative rates of nucleation and growth control the size and shape of the resulting crystals. Faster rates of cooling generally produce products with finer grain size and superior mechanical properties.

Cooling Curves

Cooling curves, such as those introduced in Chapter 4, can be one of the most useful tools for studying the solidification process. By inserting thermocouples into a casting and recording the temperature versus time, one can obtain valuable insight into what is happening in the various regions.

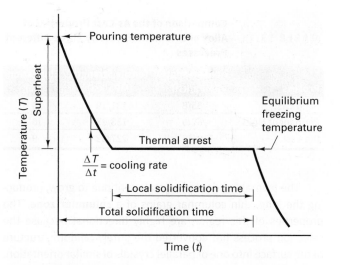

FIGURE 13.3 Cooling curve for a pure metal or eutectic-composition alloy (metals with a distinct freezing point) indicating major features related to solidification.

Figure 13.3 shows a typical cooling curve for a pure or eutectic-composition material (one with a distinct melting point) and is useful for depicting many of the features and terms related to solidification. The **pouring temperature** is the temperature of the liquid metal when it first enters the mold. **Superheat** is the difference between the pouring temperature and the freezing temperature of the material. Most metals are poured at temperatures of 100–200°C (200–400°F) above which solid begins to form. The higher the superheat, the more time is given for the material to flow into the intricate details of the mold cavity before it begins to freeze. The **cooling rate** is the rate at which the liquid or solid is cooled and can be viewed as the slope of the cooling curve at any given point. The **thermal arrest** is the plateau in the cooling curve that occurs during the solidification of a material with fixed melting point.

At this temperature, the energy or heat being removed from the mold comes from the latent heat of fusion that is being released during the solidification process. The time from the start of pouring to the end of solidification is known as the **total solidification time**. The time from the start of solidification to the end of solidification is the **local solidification time**.

If the metal or alloy does not have a distinct melting point, such as the copper–nickel alloy shown in **Figure 13.4**, solidification will occur over a range of temperatures. The **liquidus** temperature is the lowest temperature where the material is all liquid, and the **solidus** temperature is the highest temperature where it is all solid. The region between the liquidus and solidus temperatures is known as the **freezing range**. The onset and termination of solidification appear as slope changes in the cooling curve.

The actual form of a cooling curve will depend on the type of material being poured, the nature of the nucleation process, and the rate and means of heat removal from the mold. By analyzing experimental cooling curves, we can gain valuable insight into both the casting process and the cast product. Fast cooling rates and short solidification times generally lead to finer structures and improved mechanical properties.

Prediction of Solidification Time: Chvorinov's Rule

The amount of heat that must be removed from a casting to cause it to solidify depends on both the amount of superheating and the volume of metal in the casting. Conversely, the ability to remove heat from a casting is directly related to the amount of exposed surface area through which the heat can be extracted and the environment surrounding the molten material (i.e., the mold and mold surroundings). These observations

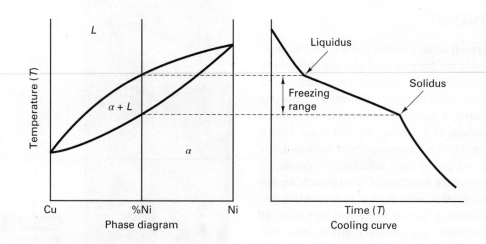

FIGURE 13.4 Phase diagram and companion cooling curve for an alloy with a freezing range. The slope changes indicate the onset and termination of solidification.

are reflected in **Chvorinov's Rule,**[1] which states that the total solidification time, t_s, can be computed by:

$$t_s = B\left(V/A\right)^n \quad \text{where } n = 1.5 \text{ to } 2.0 \quad (13\text{-}1)$$

The total solidification time, t_s, is the time from pouring to the completion of solidification; V is the volume of the casting; A is the surface area through which heat is extracted; and B is the **mold constant**. The mold constant, B, incorporates the characteristics of the metal being cast (heat capacity and heat of fusion), the mold material (heat capacity and thermal conductivity), the mold thickness, initial mold temperature, and the amount of superheat.

Test specimens can be cast to determine the value of B for a given mold material, casting material, and condition of casting. This value can then be used to compute the solidification times for other castings made under the same conditions. Because a riser and casting both lie within the same mold and fill with the same metal under the same conditions, Chvorinov's rule can be used to compare the solidification times of each and thereby ensure that the riser will solidify after the casting. This condition is absolutely essential if the liquid within the riser is to effectively feed the casting and compensate for solidification shrinkage. Aspects of riser design, including the use of Chvorinov's rule, will be developed later in this chapter.

Different cooling rates and solidification times can produce substantial variation in the structure and properties of the resulting casting. Die casting, for example, uses water-cooled metal molds, and the faster cooling produces higher-strength products than sand casting, where the mold material is more thermally insulating. Even variations in the type and condition of sand can produce different cooling rates. Sands with high moisture contents extract heat faster than ones with low moisture. **Table 13.1** presents a comparison of the properties of aluminum alloy 443 cast by the three different processes: sand casting (slow cool), permanent mold casting (intermediate cooling rate), and die casting (fast cool).

The Cast Structure

The products that result when a molten metal is poured into a mold and permitted to solidify may have as many as three distinct regions or zones. The rapid nucleation that occurs when molten metal contacts the cold mold walls results in the production of a **chill zone**, a narrow band of randomly oriented crystals on the surface of a casting. As additional heat is removed, the grains of the chill zone begin to grow inward, and the rate of heat extraction and solidification decreases. Because most crystals have directions of rapid growth, a selection process begins. Crystals with rapid-growth direction perpendicular to the casting surface grow fast and shut off adjacent grains whose rapid-growth direction is at some intersecting angle.

[1] N. Chvorinov, "Theory of Casting Solidification," *Giesserei*, Vol. 27, 1940, pp. 177–180, 201–208, 222–225.

	Comparison of the As-Cast Properties of Alloy 443 Aluminum Cast by Three Different Processes		
TABLE 13.1			
Process	**Yield Strength (MPa/ksi)**	**Tensile Strength (MPa/ksi)**	**Elongation (%)**
Sand Cast	55/8	131/19	8
Permanent Mold	62/9	158/23	10
Die Cast	110/16	227/33	9

The favorably oriented crystals continue to grow, producing the long, thin columnar grains of a **columnar zone**. The properties of this region are highly directional because the selection process has converted the purely random structure of the surface into one of parallel crystals of similar orientation. **Figure 13.5** shows a cast structure containing both chill and columnar zones.

In many materials, new crystals then nucleate in the interior of the casting and grow to produce another region of spherical, randomly oriented crystals, known as the **equiaxed zone**. Low pouring temperatures, addition of alloy, and addition of inoculants can be used to promote the formation of this region, whose isotropic properties (uniform in all directions) are far more desirable than those of columnar grains.

Molten Metal Problems

Castings begin with molten metal, and there are a number of chemical reactions that can occur between molten metal and its surroundings. These reactions and their products can often

FIGURE 13.5 Internal structure of a cast metal bar showing the chill zone at the periphery, columnar grains growing toward the center, and a central shrinkage cavity. (Courtesy Ronald Kohser)

lead to defects in the final casting. For example, oxygen and molten metal can react to produce metal oxides (a nonmetallic or ceramic material), which can then be carried away with the molten metal during the pouring and filling of the mold. **Dross** or **slag** can become trapped in the casting and impair surface finish, machinability, and mechanical properties. Material eroded from the linings of furnaces and pouring ladles, and loose sand particles from the mold surfaces, can also contribute nonmetallic components to the casting.

Dross and slag can be controlled by using special precautions during melting and pouring and by good mold design. Lower pouring temperatures or superheat slows the rate of dross-forming reactions. Fluxes can be used to cover and protect molten metal during melting, or the melting and pouring can be performed under a vacuum or protective atmosphere. Measures can be taken to agglomerate the dross and cause it to float to the surface of the metal, where it can be skimmed off prior to pouring. Special ladles can be used that extract metal from beneath the surface, such as those depicted in **Figure 13.6**. Gating systems can be designed to trap any dross, sand, or eroded mold material and keep it from flowing into the mold cavity. In addition, ceramic **filters** can be inserted into the feeder channels of the mold. These filters are available in a variety of shapes, sizes, and materials and will be discussed later in this chapter.

Liquid metals can also contain significant amounts of dissolved gas. When these materials solidify, the solid structure cannot accommodate as much gas, and the rejected atoms form bubbles or **gas porosity** within the casting. **Figure 13.7** shows the maximum solubility of hydrogen in aluminum as a function of temperature. Note the substantial decrease that occurs as the material goes from liquid to solid. **Figure 13.8** shows a small demonstration casting that has been made from aluminum where the liquid had been saturated with dissolved hydrogen

Several techniques can be used to prevent or minimize the formation of gas porosity. One approach is to prevent the gas from initially dissolving in the molten metal. Melting can be performed under vacuum, in an environment of low-solubility gases, or under a protective flux that excludes contact with the air. Superheat temperatures can be kept low to minimize gas

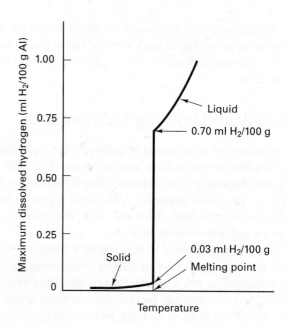

The maximum solubility of hydrogen in aluminum as a function of temperature.

solubility. In addition, careful handling and pouring can do much to control the flow of molten metal and minimize the turbulence that brings air and molten metal into contact.

Another approach is to remove the gas from the molten metal before it is poured into castings. **Vacuum degassing** sprays the molten metal through a low-pressure environment. Spraying creates a large amount of surface area, and the amount of dissolved gas is reduced as the material seeks to establish equilibrium with its new surroundings. (see a discussion of Sievert's law in any basic chemistry text.) Passing small bubbles of inert or reactive gas through the melt, known as **gas flushing**, can also be effective. In seeking equilibrium, the dissolved gases enter the flushing gas and are carried away. Bubbles of nitrogen or chlorine, for example, are particularly effective in removing hydrogen from molten aluminum. Ultrasonic vibrations, alone or with an assist gas, have also been shown to be quite effective in degassing aluminum alloys.

The dissolved gas can also be reacted with something to produce a low-density compound, which then floats to the surface and can be removed with the dross or slag. Oxygen can be removed from copper by the addition of phosphorus. Steels can be deoxidized with addition of aluminum or silicon. The

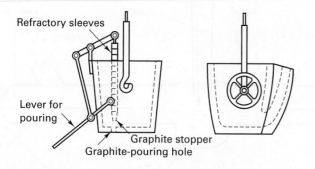

Bottom – pour ladle   **Tea pot ladle**

FIGURE 13.6 Two types of ladles are used to pour castings. Note how each extracts molten material from the bottom, avoiding transfer of the impure material from the top of the molten pool.

FIGURE 13.8 Demonstration casting made from aluminum that has been saturated in dissolved hydrogen. Note the extensive gas porosity. (Courtesy Ronald Kohser)

resulting phosphorus, aluminum, or silicon oxides are then removed by skimming, or are left on the top of the container as the remaining high-quality metal is extracted from beneath the surface.

Fluidity and Pouring Temperature

When molten metal is poured to produce a casting, it should first *flow* into all regions of the mold cavity and then *freeze* into this new shape. It is vitally important that these two functions occur in the proper sequence. If the metal begins to freeze before it has completely filled the mold, defects known as **misruns** and **cold shuts** are produced.

The ability of a metal to flow and fill a mold, the "runniness" of the liquid, is known as **fluidity**, and casting alloys are often selected for this property. Fluidity affects the minimum section thickness that can be cast, the maximum length of a thin section, the fineness of detail, and the ability to fill mold extremities. Although no single method has been accepted to measure fluidity, various "standard molds" have been developed where the results are sensitive to metal flow. One popular approach, illustrated in **Figure 13.9**, produces castings in the form of a long, thin spiral that progresses outward from a central sprue. The length of the resulting casting will increase with increased fluidity.

Fluidity is dependent on the composition, freezing temperature, and freezing range of the metal or alloy, as well as the surface tension of oxide films. The most important controlling factor, however, is usually the pouring temperature or the amount of superheat. The higher the pouring temperature, the higher the fluidity. Excessive temperatures should be avoided, however. At high pouring temperatures, chemical reactions between the metal and the mold, and the metal and its pouring atmosphere, are all accelerated. Dross formation is promoted, and larger amounts of gas can be dissolved.

If the metal is too runny, however, it may also flow into the small voids between the particles that compose a sand mold. The result of this intrusion is a casting surface that contains small particles of embedded sand, a defect known as **penetration**.

The Role of the Gating System

When molten metal is poured into a mold, the gating system conveys the material and delivers it to all sections of the mold cavity. The speed or rate of metal movement is important, as well as the amount of cooling that occurs while it is flowing. Slow filling and high loss of heat can result in misruns and cold shuts. Rapid rates of filling, on the other hand, can produce erosion of the gating system and mold cavity and might result in the entrapment of mold material in the final casting. It is imperative that the cross-sectional areas of

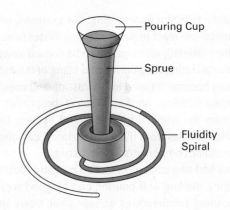

FIGURE 13.9 Fluidity test using a spiral mold. The distance that the liquid travels prior to solidification is taken as a measure of metal fluidity.

the various channels be selected to regulate flow. The shape and length of the channels affect the amount of temperature loss. When heat loss is to be minimized, short channels with round or square cross sections (minimum surface area) are the most desirable. The gates are usually attached to the thickest or heaviest sections of a casting to control shrinkage, and to the bottom of the casting to minimize turbulence and splashing. For large castings, multiple gates and runners may be used to introduce metal to more than one point of the mold cavity.

Gating systems should be designed to minimize **turbulent flow**, which tends to promote absorption of gases, oxidation of the metal, and erosion of the mold. **Figure 13.10** shows a typical gating system for a mold with a horizontal parting line and can be used to identify some of the key components that can be optimized to promote the smooth flow of molten metal. Short sprues are desirable because they minimize the distance that the metal must fall when entering the mold and the kinetic

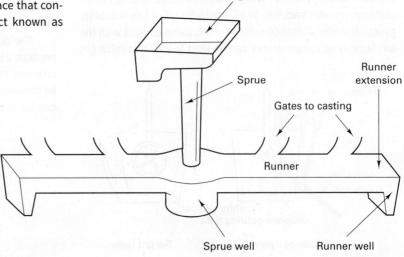

FIGURE 13.10 Typical gating system for a horizontal-parting-plane mold, showing key components involved in controlling the flow of metal into the mold cavity.

energy that the metal acquires during that fall. Rectangular pouring cups prevent the formation of a vortex or spiraling funnel, which tends to suck gas and oxides into the sprue. Tapered sprues also prevent vortex formation. A large **sprue well** can be used to dissipate the kinetic energy of the falling stream and prevent splashing and turbulence as the metal makes the turn into the runner.

The smallest cross-sectional area in the gating system, known as the **choke**, serves to control the rate of metal flow. If the choke is located near the base of the sprue, flow through the runners and gates is slowed, and flow is rather smooth. If the choke is moved to the gates, the metal might enter the mold cavity with a fountain effect, an extremely turbulent mode of flow, but the small connecting area would enable easier separation of the casting and gating system.

Gating systems can also be designed to trap dross and sand particles and keep them from entering the mold cavity. Given sufficient time, the lower-density contaminants will rise to the top of the molten metal. Long, flat runners can be beneficial (but these promote cooling of the metal), as well as gates that exit from the lower portion of the runners. Because the first metal to enter the mold is most likely to contain the foreign matter (dross from the top of the pouring ladle and loose particles washed from the walls of the gating system), **runner extensions** and **runner wells** (see Figure 13.10) can be used to catch and trap this first metal and keep it from entering the mold cavity. These features are particularly effective with aluminum castings because aluminum oxide has approximately the same density as molten aluminum.

Screens or ceramic **filters** of various shapes, sizes, and materials can also be inserted into the gating system to trap foreign material. Wire mesh can often be used with the nonferrous metals, but ceramic materials are generally required for irons and steel. **Figure 13.11** shows several ceramic filters and depicts the two basic types—extruded and foam. The pores on the extruded ceramics are uniform in size and shape and provide parallel channels. The foams contain interconnected pores of various size and orientation, forcing the material to change direction as it negotiates its passage through the filter. Contaminant removal can be by either particle entrapment or by a wetting action, whereby the nonmetallic contaminant adheres to the filter surface as the metal, which does not wet, flows freely through.

Because these devices can also restrict the fluid velocity, streamline the fluid flow, or reduce turbulence, proper placement is an important consideration. To ensure removal of both dross and eroded sand, the filter should be as close to the mold cavity as possible, but because a filter can also act as the choke, it may be positioned at other locations, such as the base of the pouring cup, base of the sprue, or in one or more of the runners.

The specific details of a gating system often vary with the metal being cast. Turbulent-sensitive metals (such as aluminum and magnesium) and alloys with low melting points generally employ gating systems that concentrate on eliminating turbulence and trapping dross. Turbulent-insensitive alloys (such as steel, cast iron, and most copper alloys) and alloys

FIGURE 13.11 **FIGURE 13.11** Various types of ceramic filters that may be inserted into the gating systems of metal castings. (Courtesy Ronald Kohser)

with a high melting point generally use short, open gating systems that provide for quick filling of the mold cavity.

Solidification Shrinkage

Most metals and alloys undergo a noticeable volumetric contraction once they enter the mold cavity and begin to cool. **Figure 13.12** shows the typical changes experienced by a metal column as the material goes from superheated liquid to room-temperature solid. There are three principal stages of **shrinkage**: (1) *shrinkage of the liquid* as it cools to the temperature

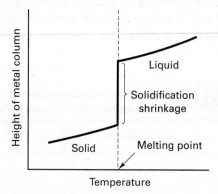

FIGURE 13.12 Dimensional changes experienced by a metal column as the material cools from a superheated liquid to a room-temperature solid. Note the significant shrinkage that occurs on solidification.

where solidification begins, (2) *solidification shrinkage* as the liquid turns into solid, and (3) *solid metal contraction* as the solidified material cools to room temperature.

The amount of liquid metal contraction depends on the coefficient of thermal contraction (a property of the metal being cast) and the amount of superheat. Liquid contraction is rarely a problem, however, because the metal in the gating system continues to flow into the mold cavity as the liquid already in the cavity cools and contracts.

As the metal changes state from liquid to crystalline solid, the new atomic arrangement is usually more efficient, and significant amounts of shrinkage can occur. The actual amount of shrinkage varies from metal to metal or alloy to alloy, as provided in **Table 13.2**. As indicated in that table, not all metals contract upon solidification. Some actually expand, such as gray cast iron, where low-density graphite flakes form as part of the solid structure.

When solidification shrinkage does occur, however, it is important to control the form and location of the resulting void. Metals and alloys with short freezing ranges, such as pure metals and eutectic alloys, tend to form large cavities or pipes. These can be avoided by designing the casting to have directional solidification where freezing begins farthest away from the feed gate or riser and moves progressively toward it. As the metal solidifies and shrinks, the mold cavity is continually being filled with additional liquid metal. When the flow of additional liquid is exhausted and solidification is complete, we hope that the final shrinkage void is located external to the desired casting in either the riser or the gating system.

Alloys with large freezing ranges have a period of time when the material is in a slushy (liquid plus solid) condition. As the material cools between the liquidus and solidus, the relative amount of solid increases and tends to trap small, isolated pockets of liquid. It is almost impossible for additional liquid to feed into these locations, and the resultant casting tends to contain small but numerous shrinkage pores dispersed throughout. This type of shrinkage is far more difficult to prevent by means of gating and risering, and a porous product may be inevitable. If a gas- or liquid-tight product is required, these castings may need to be impregnated (the pores filled with a resinous material or lower-melting-temperature metal) in a subsequent operation. Castings with dispersed porosity tend to have poor ductility, toughness, and fatigue life.

After solidification is complete, the casting will contract further as it cools to room temperature. This solid metal contraction is often called *patternmaker's contraction* because compensation for these dimensional changes should be made when the mold cavity or pattern is designed. Examples of these compensations will be provided later in this chapter. Concern arises, however, when the casting is produced in a rigid mold, such as the metal molds used in die casting. If the mold provides constraint during the time of contraction, tensile forces can be generated within the hot, weak casting, and cracking can occur (**hot tears**). It is often desirable, therefore, to eject the hot castings as soon as solidification is complete.

Risers and Riser Design

Risers are added reservoirs designed to fill with liquid metal, which is then fed to the casting as a means of compensating for solidification shrinkage. To effectively perform this function, the risers must solidify after the casting. If the reverse were true, liquid metal would flow from the casting toward the solidifying riser, and the casting shrinkage would be even greater. Hence, castings should be designed to produce directional solidification that sweeps from the extremities of the mold cavity toward the riser. In this way, the riser can continuously feed molten metal and will compensate for the solidification shrinkage of the entire mold cavity. **Figure 13.13** shows

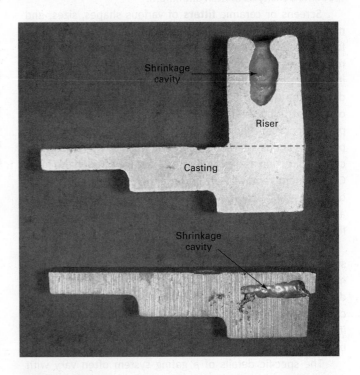

FIGURE 13.13 A three-tier step-block aluminum casting made with (top) and without (bottom) a riser. Note how the riser has moved the shrinkage void external to the desired casting. (Courtesy Ronald Kohser)

TABLE 13.2	Solidification Shrinkage of Some Common Engineering Metals (Expressed in Percent)
Aluminum	6.6
Copper	4.9
Magnesium	4.0
Zinc	3.7
Low-carbon steel	2.5–3.0
High-carbon steel	4.0
White cast iron	4.0–5.5
Gray cast iron	−1.9

a three-level step block cast in aluminum with and without a riser. Note that the riser is positioned, therefore directional solidification moves from thin to thick, and the shrinkage void is moved from the casting to the riser. If a single directional solidification is not possible, multiple risers may be required, with various sections of the casting each solidifying toward their respective riser.

The risers should also be designed to conserve metal. If we define the **yield** of a casting as the casting weight divided by the total weight of metal poured (complete gating system, risers, and casting), it is clear that there is a motivation to make the risers as small as possible, yet still able to perform their task. This is usually done through proper consideration of riser size, shape, and location, as well as the type of connection between the riser and casting.

A good shape for a riser would be the one that has a long freezing time. According to Chvorinov's rule, this would favor a shape with small surface area per unit volume. Although a sphere would make the most efficient riser, this shape presents considerable difficulty to both patternmaker and moldmaker. The most popular shape for a riser, therefore, is a cylinder, where the height-to-diameter ratio varies depending on the nature of the alloy being cast, the location of the riser, the size of the flask, and other variables. A one-to-one height-to-diameter ratio is generally considered to be ideal.

Risers should be located so that directional solidification occurs from the extremities of the mold cavity back toward the riser. Because the thickest regions of a casting will be the last to freeze, risers should feed directly into these locations. Various types of risers are possible. A **top riser** is one that sits on top of a casting. Because of their location, top risers have shorter feeding distances and occupy less space within the flask. They give the designer more freedom for the layout of the pattern and gating system. **Side risers** are located adjacent to the mold cavity and displaced horizontally along the parting line. **Figure 13.14** depicts both a top and a side riser. If the riser is contained entirely within the mold, it is known as a **blind riser**. If it is open to the atmosphere, it is called an **open riser**. Blind risers are usually larger than open risers because of the additional heat loss that occurs where the top of the riser is in contact with mold material.

Live risers (also known as *hot risers*) receive the last hot metal that enters the mold and generally do so at a time when the metal in the mold cavity has already begun to cool and solidify. Thus, they can be smaller than **dead** *(or cold)* **risers**, which fill with metal that has already flowed through the mold cavity. As shown in Figure 13.14, top risers are almost always dead risers. Risers that are part of the gating system are generally live risers.

The minimum size of a riser can be calculated from Chvorinov's rule by setting the total solidification time for the riser to be greater than the total solidification time for the casting. Because both cavities receive the same metal and are in the same mold, the mold constant, B, will be the same for both regions. Assuming that $n = 2$, and a safe difference in solidification time is 25% (the riser takes 25% longer to solidify than the casting), we can write this condition as

$$t_{riser} = 1.25\, t_{casting} \tag{13-2}$$

or

$$\left(V/A\right)^2_{riser} = 1.25 \left(V/A\right)^2_{casting} \tag{13-3}$$

Calculation of the riser size then requires selection of riser geometry, which is generally cylindrical. For a cylinder of diameter D and height H, the volume and surface area can be written as:

$$V = \pi D^2 H/4$$
$$A = \pi D H + 2(\pi D^2/4)$$

Selecting a specific height-to-diameter ratio for the riser then enables equation 13-3 to be written as a simple expression with one unknown, D. The volume-to-area ratio for the casting is computed for its particular geometry, and equation 13-3 can then be solved to provide the size of the required riser. One should note that if the riser and casting share a surface, as with a blind top riser, the area of the common surface should be subtracted from both components because it will not be a

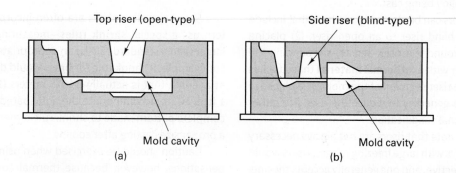

Top riser (open-type)   Side riser (blind-type)

Mold cavity   Mold cavity
(a)   (b)

FIGURE 13.14 Schematic of a sand casting mold showing (a) an open-type top riser and (b) a blind-type side riser (right). The side riser is a live riser receiving the last hot metal to enter the mold. The top riser is a dead riser receiving metal that has flowed through the mold cavity.

surface of heat loss to either. It should also be noted that there are a number of methods to calculate riser size. The Chvorinov's rule method will be the only one presented here.

A final aspect of riser design is the connection between the riser and the casting. Because the riser must ultimately be separated from the casting, it is desirable that the connection area be as small as possible. On the other hand, the connection area must be sufficiently large so that the link does not freeze before solidification of the casting is complete. If the risers are placed close to the casting with relatively short connections, the mold material surrounding the link will receive heat from both the casting and the riser. It should heat rapidly and remain hot throughout the cast, thereby preventing solidification of the metal in the channel.

Risering Aids

Various methods have been developed to assist the risers in performing their job. Some are intended to promote directional solidification, whereas others seek to reduce the number and size of the risers, thereby increasing the yield of a casting. These techniques generally work by either speeding the solidification of the casting (**chills**) or retarding the solidification of the riser (**sleeves** or toppings).

External chills are masses of high-heat-capacity and high-thermal-conductivity materials (such as steel, iron, graphite, or copper) that are placed in the mold adjacent to the casting to absorb heat and accelerate the cooling of various regions. Chills can promote directional solidification or increase the effective feeding distance of a riser. They can also be used to reduce the number of risers required for a casting. External chills are frequently covered with a protective wash, silica flour, or other refractory material to prevent bonding with the casting.

Internal chills are pieces of metal that are placed within the mold cavity to absorb heat and promote more rapid solidification. When the molten metal of the pour surrounds the chill, it absorbs heat as it seeks to come to equilibrium with its surroundings. Internal chills ultimately become part of the final casting, so they must be made from an alloy that is the same or compatible with the alloy being cast.

The cooling of risers can be slowed by methods that include (1) switching from a blind riser to an open riser, (2) placing **insulating sleeves** around the riser, and (3) surrounding the sides or top of the riser with **exothermic material** that supplies added heat just to the riser segment of the mold. The objective of these techniques is generally to reduce the riser size rather than promote directional solidification.

It is important to note that risers are not always necessary or functional. For alloys with large freezing ranges, risers would not be particularly effective, and one generally accepts the fine, dispersed porosity that results. For processes such as die casting, low-pressure permanent molding, and centrifugal casting, the positive pressures associated with the process provide the feeding action that is required to compensate for solidification shrinkage.

13.5 Patterns

Casting processes can be divided into two basic categories: (1) those for which a new mold must be created for each casting (the **expendable-mold processes**), and (2) those that employ a permanent, **reusable mold**. Most of the expendable mold processes begin with some form of reusable **pattern**—a physical representation of the object to be cast, modified dimensionally to reflect both the casting process and the material being cast. Patterns can be made from a variety of materials. In order of increasing longevity, these include Styrofoam or wax (single-use), soft wood (100 molds), hard wood (500 molds), epoxies and urethanes (750 to 1000 molds), aluminum (2000 molds), and iron (5000 molds). Urethane is currently the material of choice for nearly half of all casting patterns.

The dimensional modifications that are incorporated into a pattern are called **allowances**, and the most important of these is the **shrinkage allowance**. Following solidification, a casting continues to contract as it cools to room temperature, the amount of this contraction being as much as 2% or 20 mm/m (¼ in/ft). To produce the desired final dimensions, the pattern (which sets the dimensions on solidification) must be slightly larger than the room-temperature casting. The exact amount of this shrinkage compensation depends on the metal that is being cast and can be estimated by the equation:

$$\Delta \text{ length} = \text{length } \alpha \text{ } \Delta T \qquad (13\text{-}4)$$

where α is the coefficient of thermal expansion and ΔT is the difference between the freezing temperature and room temperature. Typical allowances for some common engineering metals are

Cast iron	0.8–1.0%
Steel	1.5–2.0%
Aluminum	1.0–1.3%
Magnesium	1.0–1.3%
Brass	1.5%

Shrinkage allowances are often incorporated into a pattern using special **shrink rules**—measuring devices that are larger than a standard rule by an appropriate shrink allowance. For example, a shrink rule for brass would designate 0.1 meter at a length that is actually 0.1015 meters (1 ft would become 1 ft 3⁄16 in), to accommodate the anticipated 1.5% shrinkage. A complete pattern made to shrink rule dimensions will produce a proper size casting after cooling.

Caution should be exercised when using shrink rule compensations, however, because thermal contraction may not be the only factor affecting the final dimensions. The various phase transformations discussed in Chapter 4 are often accompanied by significant dimensional expansions or contractions. Examples include eutectoid reactions, martensitic reactions, and graphitization.

In many casting processes, mold material is formed around the pattern, and the pattern is then extracted to create the mold cavity. To facilitate pattern removal, molds are often made in two or more sections that separate along mating surfaces of the parting line or parting plane. A flat parting line is usually preferred, but the casting design or molding practice may dictate the use of irregular or multiple parting surfaces. In general, the best parting line will be a flat plane that allows for proper metal flow into the mold cavity, requires the fewest cores and molding steps, and provides adequate core support and venting.

If the pattern contains surfaces that are perpendicular to the parting line (parallel to the direction of pattern withdrawal), friction between the pattern and the mold material as well as any horizontal movement of the pattern during extraction could induce damage to the mold. This damage could be particularly severe at the corners where the mold cavity intersects the parting surface. Such extraction damage can be minimized by incorporating a slight taper, or **draft**, on all pattern surfaces that are parallel to the direction of withdrawal. A slight withdrawal of the pattern will free it from the mold material on all surfaces, and it can then be further removed without damage to the mold. **Figure 13.15** illustrates the use of draft to facilitate pattern removal.

The size and shape of the pattern, the depth of the mold cavity, the method used to withdraw the pattern, the pattern material, the mold material, and the molding procedure all influence the amount of draft required. Draft is seldom less

allowance depends to a great extent on the casting process and the mold material. Ordinary sand castings have rougher surfaces than those of shell-mold castings. Die castings have smooth surfaces that may require little or no metal removal, and the surfaces of investment castings are even smoother. It is also important to consider the location of the desired machining and the presence of other allowances because the draft allowance may provide part or all of the extra metal needed for machining.

Some casting shapes require yet an additional allowance for **distortion**. Consider a U-shaped section where the arms are restrained by the mold at a time when the base of the U is shrinking. The result will be a final casting with outwardly sloping arms. If the design is modified to have the arms originally slope inward, the subsequent distortion will produce the desired final shape. Distortion depends greatly on the particular configuration of the casting, and casting designers must use experience and judgment to provide an appropriate distortion allowance.

If a casting is to be made in a multiuse metal mold, all of the "pattern allowances" should be incorporated into the machined cavity. The dimensions of this cavity will further change, however, as sequential casts raise the mold temperature to a steady-state level. An additional correction should be added to compensate for the thermal effect.

Similar allowances should be applied to the cores that create the holes and interior passages of a casting.

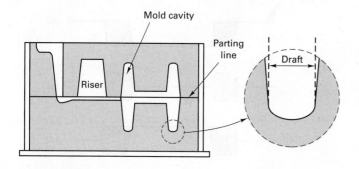

FIGURE 13.15 Two-part mold showing the parting line and the incorporation of a draft allowance on vertical surfaces.

13.6 Design Considerations in Castings

To produce the best-quality product at the lowest possible cost, it is important that the designers of castings give careful attention to several process requirements. It is not uncommon for minor and readily permissible changes in design to greatly facilitate and simplify the casting of a component and also reduce the number and severity of defects.

One of the first features that must be considered by a designer is the *location* and *orientation of the parting plane*, an important part of all processes that use segmented or separable molds. The location of the parting plane can affect (1) the number of cores, (2) the method of supporting the cores, (3) the use of effective and economical gating, (4) the weight of the final casting, (5) the final dimensional accuracy, and (6) the ease of molding.

In general, it is desirable to minimize the use of cores. A change in the location or orientation of the parting plane can often assist in this objective. The change illustrated in **Figure 13.16** not only eliminates the need for a core but also reduces the weight of the casting by eliminating the need for draft. **Figure 13.17** shows another example of how a core can

than 1° or 10 mm/m (⅛ in/ft), with a minimum taper of about 1.6 mm (1/16-in) over the length of any surface. Because draft allowances increase the size of a pattern (and thus the size and weight of a casting), it is generally desirable to keep them to the minimum that will permit satisfactory pattern removal. Molding procedures that produce higher strength molds and the use of mechanical pattern withdrawal can often enable reductions in draft allowances. By reducing the taper, casting weight and the amount of subsequent machining can both be reduced.

When smooth machined surfaces are required, it may be necessary to add an additional **machining allowance**, or **finish allowance**, to the pattern. The amount of this

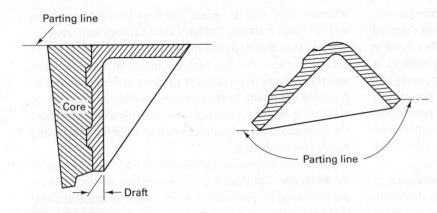

FIGURE 13.16 Elimination of a core by changing the location or orientation of the parting plane.

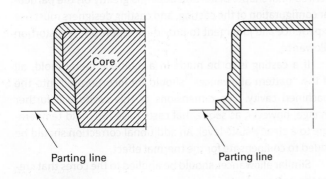

FIGURE 13.17 Elimination of a dry-sand core by a change in part design.

be eliminated by a simple design change. **Figure 13.18** shows six different parting line arrangements for the casting of a simple ring. Arrangements (a) through (e) provide flat and parallel side faces with draft tapers on the inner and outer diameters. The (f) alternative requires the use of a core, but might be preferred if draft cannot be tolerated on the inner and outer diameter surfaces. This figure also shows that simply noting the desired shape and the need to provide sufficient draft can provide considerable design freedom. Because mold closure may not always be consistent, consideration should also be given to the fact that dimensions across the parting plane are subject to greater variation than those that lie entirely within a given segment of the mold.

Control of the solidification process is also related to design. Those portions of a casting that have a high ratio of surface area to volume will experience more rapid cooling and will be stronger and harder than the other regions. Thicker or heavier sections will cool more slowly and may contain shrinkage cavities and porosity, or have weaker, large-grain-size structures.

An ideal casting would have uniform thickness at all locations. Ribs or other geometric features can often be used in place of thicker sections to impart additional strength while maintaining uniform wall thickness. When the section thickness must change, it is best if these changes are gradual,

as indicated in the recommendations of **Figure 13.19**.

When sections of castings intersect, as in **Figure 13.20a**, two problems can arise. The first of these is **stress concentration**. Generous **fillets** (inside radii) at all interior corners can better distribute stresses and help to minimize potential problems. If the fillets are excessive, however, the additional material can augment the second problem, known as **hot spots**. Thick sections, like

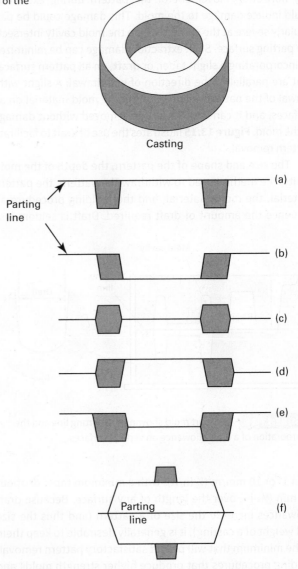

FIGURE 13.18 Multiple options to cast a simple ring with draft to the parting line. Evaluate the six options with respect to the following possible concerns: (1) flat and parallel side surfaces, (2) flat and parallel inner and outer diameters, (3) amount of material that must be removed if no tapers are allowed on any surfaces, and (4) possibility for nonuniform wall thickness if one of the mold segments is shifted with respect to the other.

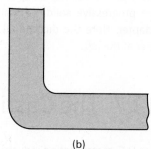

FIGURE 13.19 Typical guidelines for section change transitions in castings.

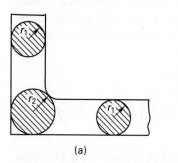

FIGURE 13.20 (a) The "hot spot" at section r_2 is caused by intersecting sections. (b) An interior fillet and exterior radius leads to more uniform thickness and more uniform cooling.

those at the intersection in Figure 13.20a and those illustrated in **Figure 13.21**, cool more slowly than other locations and tend to be the sites of localized shrinkage. Where thick sections must exist, an adjacent riser is often used to feed the section during solidification and ensure that the shrinkage cavity will form external to the actual casting. Sharp exterior corners tend to cool faster than the other sections of a casting. By providing an exterior radius, the surface area can be reduced and cooling slowed to be more consistent with the surrounding material. **Figure 13.20b** shows a recommended modification to Figure 13.20a.

If sections intersect to form continuous ribs, like those in **Figure 13.22**, contraction occurs in opposite directions as each of the arms cool and shrink, and cracking may occur at the intersections, which are also local hot spots. By staggering the ribs, as shown in the lower portion of the left-side figure, the negative effects of thermal contraction and hot spots can be minimized. The bad design in the right-hand segment of Figure 13.22 can be improved by placing a cored

hole through the center of the intersection, creating uniform thickness walls throughout. An even better design might be to use an oval-shaped core with the major dimension in the direction of the longest arms to provide the greatest reduction in shrinkage stresses. The best design, shown at the bottom of the right-hand segment, would be to use honeycomb ribs with thickness about 80% of the exterior walls.

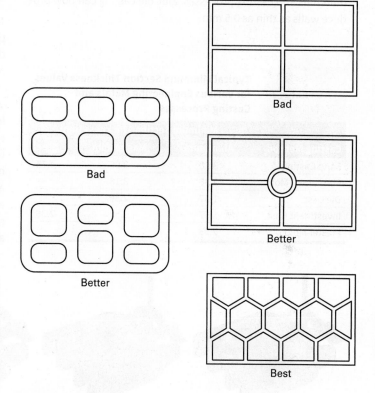

FIGURE 13.22 Design modifications to reduce cracking and hot spot shrinkage in ribbed castings.

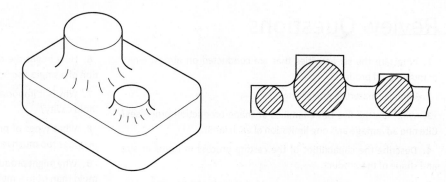

FIGURE 13.21 Hot spots often result from intersecting sections of various thickness.

The location of the parting line may also be an appearance consideration. A small amount of fin, or **flash**, is often present at the parting line, and when the flash is removed (or left in place if it is small enough), a line of surface imperfection results. If the location is in the middle of a flat surface, the imperfection will be clearly visible on the product. If the parting line can be moved to coincide with a corner, however, the associated "defect" will go largely unnoticed.

Thin-walled castings are often desired because of their reduced weight, but thin walls can often present manufacturing problems related to mold filling (premature freezing before complete fill). Minimum section thickness should always be considered when designing castings. Specific values are rarely given, however, because they tend to vary with the shape and size of the casting, the type of metal being cast, the method of casting, and the practice of an individual foundry. **Table 13.3** presents typical minimum thickness values for several cast materials and casting processes. Zinc die casting can now produce walls as thin as 0.5 mm.

TABLE 13.3	Typical Minimum Section Thickness Values for Various Engineering Metals and Casting Processes		
	Minimum Section Thickness (mm.)		
Casting Method	**Aluminum**	**Magnesium**	**Steel**
Sand Casting	3.18	3.96	4.75
Permanent Mold	2.36	3.18	—
Die Cast	1.57	2.36	—
Investment Cast	1.57	1.57	2.36
Plaster Mold	2.03	—	—

Casting design can often be aided by **computer simulation**. The mathematics of fluid flow can be applied to **mold filling**, and the principles of heat transfer can be used for **solidification modeling**. The mathematical tools of finite element or finite difference calculations can be coupled with the use of high-speed computers to enable beneficial design changes before the manufacture of patterns or molds. Resulting mechanical properties, thermally induced casting stresses, and the amount and type of distortion can all be quantitatively predicted. The computer model in **Figure 13.23** shows the progressive solidification of a cast steel mining shovel adapter. Note the directional solidification back toward the riser at the left.

13.7 The Casting Industry

The U.S. metal casting industry ships approximately 14 million pounds of castings every year, valued at over $18 billion, with ductile iron, gray iron, aluminum alloys, steel, and copper-base metals comprising the major portion. Metal castings form primary components in agricultural implements, construction equipment, mining equipment, valves and fittings, metalworking machinery, power tools, pumps and compressors, railroad equipment, power transmission equipment, heating, refrigeration, and air conditioning equipment. Ductile iron pipe is a mainstay for conveying pressurized fluids, and household appliances and electronics all utilize metal castings. The average new vehicle contains more than 270 kg (600 lb) of castings, ranging from engine blocks and wheels to seat-belt retractors and air-bag frames.

FIGURE 13.23 Computer model showing the progressive solidification of a cast steel mining shovel adapter. As time passes (left-to-right) the material directionally solidifies back toward the riser at the left side of the casting. Light material is liquid; dark material is solid. (Copyright 2009, Giesseri-Verlag GmbH, Dusseldorf, Germany)

Review Questions

1. What are the six activities that are conducted on almost every manufactured product?

2. What is "materials processing"?

3. What are the five basic families of shape-production processes? Cite one advantage and one limitation of each family.

4. Describe the capabilities of the casting process in terms of size and shape of the product.

5. What are some of the various mold materials and pouring methods used in casting?

6. How might the desired production quantity influence the selection of a single-use or multiple-use molding process?

7. Why is it important to provide a means of venting gases from the mold cavity?

8. What types of problem or defect can occur if the mold material provides too much restraint to the solidifying and cooling metal?

9. Why might product removal be less of a problem from a single-use mold than from a multiple-use mold?

10. What is a casting pattern? Flask? Core? Mold cavity? Riser?

11. In a horizontally parted two-part mold, what is the cope? The drag?

12. What are some of the components that combine to make up the gating system of a mold?

13. What is a parting line or parting surface?

14. What is draft, and why is it used?

15. Why is it important to control the solidification process in metal casting?

16. What are the two stages of solidification, and what occurs during each?

17. Why is it that most solidification does not begin until the temperature falls somewhat below the equilibrium melting temperature (i.e., undercooling is required)?

18. Why might it be desirable to promote nucleation in a casting through inoculation or grain refinement processes?

19. Nucleation generally begins at preferred sites within a mold. What are some probable sites for nucleation?

20. Why might directional solidification be desirable in the production of a cast product?

21. Describe some of the key features observed in the cooling curve of a pure metal.

22. What is superheat?

23. What causes the thermal arrest in the cooling curve of a pure metal?

24. What is a liquidus temperature? A solidus temperature?

25. What is the freezing range for a metal or alloy?

26. Discuss the roles of casting volume and surface area as they relate to the total solidification time and Chvorinov's rule.

27. What characteristics of a specific casting process are incorporated into the mold constant, B, of Chvorinov's rule?

28. What is the correlation between cooling rate and final properties of a casting?

29. What is the chill zone of a casting, and why does it form?

30. Which of the three regions of a cast structure is least desirable? Why are its properties highly directional?

31. What is dross or slag, and how can it be prevented from becoming part of a finished casting?

32. What are some of the possible approaches that can be taken to prevent the formation of gas porosity in a metal casting?

33. What is a misrun or cold shut, and what causes them to form?

34. What is fluidity, and how can it be measured?

35. What is the most important factor controlling the fluidity of a casting alloy?

36. What defect can form in sand castings if the pouring temperature is too high and fluidity is too great?

37. Why is it important to design the geometry of the gating system to control the rate of metal flow as it travels from the pouring cup into the mold cavity? What are the pros and cons of slow fill? Fast fill?

38. Why might it be preferable to attach gates to the thickest or heaviest segments of a casting? To the bottom of a casting?

39. What are some of the undesirable consequences that could result from turbulence of the metal in the gating system and mold cavity?

40. What are some desirable features in the sprue region of a gating system?

41. What is a choke, and how does its placement affect metal flow?

42. What features can be incorporated into the gating system to aid in trapping dross and loose mold material that is flowing with the molten metal?

43. What are some of the materials and designs of liquid metal filters?

44. What factors might influence the positioning of filters within a gating system?

45. What features of the metal being cast tend to influence whether the gating system is designed to minimize turbulence and reduce dross, or promote rapid filling to minimize temperature loss?

46. What are the three stages of contraction or shrinkage as a liquid is converted into a finished casting?

47. Why is it more difficult to prevent shrinkage voids from forming in metals or alloys with large freezing ranges?

48. What steps can be taken to compensate for the various types of shrinkage?

49. During what stage of shrinkage might hot tears form?

50. What is the role of a riser?

51. Why is it desirable to design a casting to have directional solidification sweeping from the extremities of a mold toward a riser?

52. What is "yield," and how does it relate to the number and size of risers?

53. Based on Chvorinov's rule, what would be an ideal shape for a casting riser? A desirable shape from a practical perspective?

54. Define the following riser-related terms: top riser, side riser, open riser, blind riser, live riser, and dead riser.

55. What assumptions were made when using Chvorinov's rule to calculate the size of a riser in the manner presented in the text? Why is the mold constant, B, not involved in the calculations?

56. Discuss aspects relating to the connection between a riser and the casting.

57. What is the purpose of a chill? Of an insulating sleeve? Of exothermic material?

58. What are some materials that are commonly used to produce casting patterns?

59. What types of modifications or allowances are generally incorporated into a casting pattern?

60. What is a shrink rule, and how does it work?

61. What is the purpose of a draft or taper on pattern surfaces?

62. Why is it desirable to make the pattern allowances as small as possible?

63. What additional adjustment or correction must be incorporated into the dimensions of a multiuse metal mold?

64. What are some of the features of the casting process that are directly related to the location and orientation of the parting plane?

65. What are "hot spots," and what sort of design features cause them to form?

66. What are some design recommendations for inside corners? Exterior corners?

67. What are some appearance considerations in parting line location?

68. What determines the minimum section or wall thickness for various casting materials and processes?

69. How can computer simulation be used in the design of successful castings?

Problems

1. Using Chvorinov's rule as presented in the text with $n = 2$, calculate the dimensions of an effective riser for a casting that is a 2 in.by 4 in. by 6 in. rectangular plate. Assume that the casting and riser are not connected, except through a gate and runner, and that the riser is a cylinder of height/diameter ratio $H/D = 1.5$. The finished casting is what fraction of the combined weight of the riser and casting?

2. Reposition the riser in Problem 1 so that it sits directly on top of the flat rectangle, with its bottom circular surface being part of the surface of the casting, and recompute the size and yield fraction. Which approach is more efficient?

3. A rectangular casting having the dimensions 3 in. by 5 in. by 10 in. solidifies completely in 13.5 minutes. Using $n = 2$ in Chvorinov's rule, calculate the mold constant B. Then compute the solidification time of a 0.5 in. by 8 in. by 8 in. casting poured under the same conditions.

4. A cylinder with a diameter of 2.5 in. and a height of 2.5 in. takes three minutes to solidify in a green sand mold. What would the solidification time be if the diameter would be doubled? If the height were doubled? If both were doubled?

5. Figure 13.A shows the wall profile of a cast iron coupling and the shrinkage porosity that was observed in the cast product. If the interior profile must be maintained, suggest possible changes to the design that would enable the casting of a defect-free product.

6. Investigate various experimental techniques to evaluate molten metal fluidity.

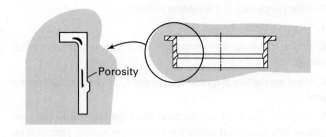

FIGURE 13.A

7. Porosity within a casting can be either shrinkage-induced or gas-induced. What features could be used to determine the actual cause?

8. The chapter text describes various materials that have been used to produce casting patterns (i.e., the tooling that is used to produce molds), including Styrofoam, soft woods, hard woods, epoxy/urethane polymers, aluminum, and iron. For each of these materials, briefly discuss the pros and cons, considering such factors as: number of castings to be produced, the size and shape of the casting, the desired precision of the cast product, pattern cost, dimensional stability (both wear and environmental factors such as temperature and humidity), susceptibility to damage, ability to be repaired or refurbished, process limitations, and storage concerns.

CHAPTER 14

Expendable-Mold Casting Processes

14.1 Introduction

The versatility of metal casting is made possible by a number of distinctly different processes, each with its own set of characteristic advantages and benefits. Selection of the best process requires a familiarization with the various options and capabilities as well as an understanding of the needs of the specific product. Some factors to be considered include the size of the product, the desired dimensional precision and surface quality, the number of castings to be produced, the type of pattern and core box that will be needed, the cost of making the required mold or die, and restrictions imposed by the selected material.

As we begin to survey the various casting processes, it is helpful to have some form of process classification. One approach focuses on the molds and patterns and utilizes the following three categories:

1. Single-use molds with multiple-use patterns
2. Single-use molds with single-use patterns
3. Multiple-use molds

Categories 1 and 2 are often combined under the more general heading of **expendable-mold casting processes** and will be presented in this chapter. Sand, plaster, ceramics, or other refractory materials are combined with binders to form the mold. Those processes where a mold can be used multiple times (Category 3) will be presented in Chapter 15. The multiple-use molds are usually made from metal.

Because the casting processes are primarily used to produce metal products, the emphasis of the casting chapters will be on metal casting. The metals most frequently cast are iron, steel, stainless steel, aluminum alloys, brass, bronze and other copper alloys, magnesium alloys, certain zinc alloys, and nickel-based superalloys. Among these, cast iron and aluminum are the most common, primarily because of their low cost, good fluidity, adaptability to a variety of processes, and the wide range of product properties that are available. The processes used to fabricate products from polymers, ceramics (including glass), and composites, including casting processes, will be discussed in Chapter 20.

14.2 Sand Casting

Sand casting is by far the most common and possibly the most versatile of the casting processes, accounting for over 70% of all metal castings. It is the most widely used method for the production of gray iron, ductile iron, and steel castings and is used for about 12% of aluminum castings, where die casting is the dominant process.

Granular refractory material (such as silica, zircon, olivine, or chromite sand) is mixed with small amounts of other materials, such as clay and water, and is then packed around a pattern that has the shape of the desired casting. Because the grains can pack into thin sections and can be economically used in large quantities, products spanning a wide range of sizes and detail can be made by this method. If the pattern is to be removed before pouring, the mold is usually made in two or more segments. An opening called a **sprue** is cut from the top of the mold through the sand and connected to a system of horizontal channels called **runners**. The molten metal is poured down the sprue, flows through the runners, and enters the mold cavity through one or more openings, called **gates**. Gravity flow is the most common means of introducing the metal into the mold. The metal is allowed to solidify, and the mold is then broken to permit removal of the finished casting. Because the mold is destroyed in product removal, a new mold must be made for each casting. **Figure 14.1** shows the essential steps and basic components of a sand casting process. A two-part cope-and-drag mold is illustrated, and the casting incorporates both a core and a riser (discussed in Chapter 13).

Patterns and Pattern Materials

The first step in making a sand casting is the design and construction of a **pattern**. This is a duplicate of the part to be cast, modified in accordance with the requirements of the casting process, the metal being cast, and the particular molding technique that is being used. Selection of the pattern material is determined by the number of castings to be made, the size and shape of the casting, the desired dimensional precision, and

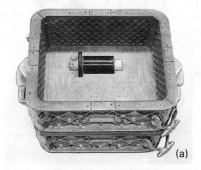
(a)

(b)

Riser pin
Sprue pin
(c)

(d)

(e)

(e')

(f)

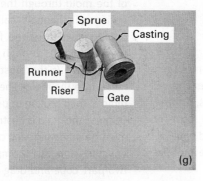
Sprue
Casting
Runner
Riser
Gate
(g)

FIGURE 14.1 Sequential steps in making a sand casting. (a) A pattern board is placed between the bottom (drag) and top (cope) halves of a flask, with the bottom side up. (b) Sand is then packed into the bottom or drag half of the mold. (c) A bottom board is positioned on top of the packed sand, and the mold is turned over, showing the top (cope) half of pattern with sprue and riser pins in place. (d) The upper or cope half of the mold is then packed with sand. (e) The mold is opened, the pattern board is drawn (removed), and the runner and gate are cut into the bottom parting surface of the sand. (e') The parting surface of the upper or cope half of the mold is also shown with the pattern and pins removed. (f) The core is positioned, the pattern board is removed, the mold is reassembled, and molten metal is poured through the sprue. (g) The contents are shaken from the flask, and the metal segment is separated from the sand, ready for further processing. (E. Paul DeGarmo)

the molding process. Wood patterns are relatively easy to make and are frequently used when small quantities of castings are required. Wood, however, is not very dimensionally stable. It may warp or swell with changes in humidity, and it tends to wear with repeated use. Metal patterns are more expensive but are more stable and durable. Hard plastics, such as urethanes, offer another alternative and are often preferred with processes that use strong, organically bonded sands that tend to stick to other pattern materials. In the full-mold and lost-foam processes, expanded polystyrene (EPS) is used, and investment casting uses patterns made from wax. In the latter processes, both the pattern and the mold are single-use, each being destroyed when a casting is produced.

Types of Patterns

Many types of patterns are used in the foundry industry, with selection being based on the number of duplicate castings required and the complexity of the part.

One-piece or *solid* **patterns**, such as the one shown in **Figure 14.2**, are the simplest and often the least expensive type. They are essentially a duplicate of the part to be cast, modified only by the various allowances discussed in Chapter 13 and by the possible addition of core prints. One-piece patterns are relatively cheap to construct, but the subsequent molding process is usually slow. As a result, they are generally used when the shape is relatively simple and the number of duplicate castings is rather small.

If the one-piece pattern is simple in shape and contains a flat surface, it can be placed directly on a **follow board**. The entire mold cavity will be created in one segment of the mold, with the follow board forming the parting surface. If the parting plane is to be more centrally located,

FIGURE 14.2 Single-piece pattern for a pinion gear. (E. Paul DeGarmo)

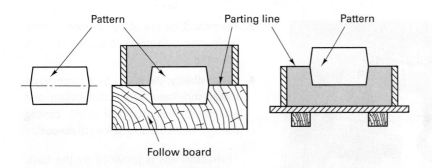

FIGURE 14.3 Method of using a follow board to position a single-piece pattern and locate a parting surface. The final figure shows the flask of the previous operation (the drag segment) inverted in preparation for construction of the upper portion of the mold (cope segment).

FIGURE 14.4 Split pattern, showing the two sections together and separated. The light-colored portions are core prints. (E. Paul DeGarmo)

special follow boards are produced with inset cavities that position the one-piece pattern at the correct depth for the parting line. **Figure 14.3** illustrates this technique, where the follow board again forms the parting surface.

Split patterns are used when moderate quantities of a casting are desired. The pattern is divided into two segments along what will become the parting plane of the mold. The bottom segment of the pattern is positioned in the **drag** portion of a **flask**, and the bottom segment of the mold is produced. This portion of the flask is then inverted, and the upper segment of the pattern and flask are attached. Tapered pins in the cope half of the pattern align with holes in the drag segment to ensure proper positioning. Mold material is then packed around the full pattern to form the upper segment (**cope**) of the mold. The two segments of the flask are separated, and the pattern pieces are removed to produce the mold cavity. Sprues and runners are cut, and the mold is then reassembled, ready for pour. **Figure 14.4** shows a split pattern that also contains several core prints (lighter color).

Match-plate patterns, like the one shown in **Figure 14.5**, further simplify the process and can be coupled with modern molding machines to produce large quantities of duplicate molds. The cope and drag segments of a split pattern are permanently fastened to opposite sides of a wood or metal **match plate**. The match plate is positioned between the upper and lower segments of a flask using holes that align with pins on one of the flask segments. Mold material is then packed on both sides of the match plate to form the cope and drag segments of a two-part mold. The mold sections are then separated, and the match-plate pattern is removed. The segments are then reassembled with the pins and guide holes ensuring that the cavities in the cope and drag are in proper alignment. The necessary gates, runners, and risers are usually incorporated on the match plate, as well. This guarantees that these features will be uniform and of the proper size in each mold, thereby reducing

the possibility of defects. The sprue is cut and the mold is ready for pouring. Figure 14.1 is an example of match-plate molding, and Figure 14.5 further illustrates a common practice of including multiple patterns in a single mold.

When large quantities of identical parts are to be produced, or when the casting is quite large, it may be desirable to have the cope and drag halves of split patterns attached to separate pattern boards. These **cope-and-drag patterns** enable independent molding of the cope and drag segments of a mold. Large molds can be handled more easily in separate segments, and small molds can be made at a faster rate if a machine is only producing one segment. **Figure 14.6** shows the mating pieces of a typical cope-and-drag pattern.

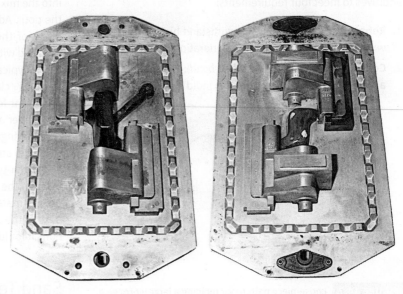

FIGURE 14.5 Match-plate pattern used to produce two identical parts in a single flask. (Left) Cope side; (right) drag side. (*Note:* The views are opposite sides of a single pattern board.) (E. Paul DeGarmo)

FIGURE 14.6 Cope-and-drag pattern for producing two heavy parts. (Left) Cope section; (right) drag section. (*Note:* These are two separate pattern boards.) (E. Paul DeGarmo)

When the geometry of the product is such that a one-piece or two-piece pattern could not be removed from the molding sand, a **loose-piece pattern** can sometimes be developed. Separate pieces are joined to a primary pattern segment by beveled grooves or pins (**Figure 14.7**). After molding, the primary segment of the pattern is withdrawn. The hole that is created then permits the remaining segments to be sequentially extracted. Loose-piece patterns are expensive. They require careful maintenance, slow the molding process, and increase molding costs. They do, however, enable the sand casting of complex shapes that would otherwise require the full-mold, lost-foam, or investment processes.

Sands and Sand Conditioning

The sand used to make molds must be carefully prepared if it is to provide satisfactory and uniform results. Ordinary silica (SiO_2), zircon, olivine, or chromite sands are compounded with additives to meet four requirements:

1. **Refractoriness**: the ability to withstand high temperatures without melting, fracture, or deterioration

2. **Cohesiveness** (also referred to as *bond*): the ability to retain a given shape when packed into a mold

FIGURE 14.7 Loose-piece pattern for molding a large worm gear. After sufficient sand has been packed around the pattern to hold the pieces in position, the wooden pins are withdrawn. The mold is then completed, after which the pieces of the pattern can be removed in a designated sequence. (E. Paul DeGarmo)

3. **Permeability**: the ability of mold cavity, mold, and core gases to escape through the sand

4. **Collapsibility**: the ability to accommodate metal shrinkage after solidification and provide for easy removal of the casting through mold disintegration (**shakeout**).

Refractoriness is provided by the basic nature of the sand. Cohesiveness, bond, or strength is obtained by coating the sand grains with clays, such as bentonite, kaolinite, or illite, that become cohesive when moistened. Permeability is a function of the size of the sand particles, the amount and type of clay or bonding agent, the moisture content, and the compacting pressure. Collapsibility is sometimes enhanced by adding cereals or other organic materials, such as cellulose, that burn out when they come in contact with the hot metal. The combustion of these materials reduces both the volume and strength of the restraining sand.

Good molding sand always represents a compromise between competing factors. The size of the sand particles, the amount of bonding agent (such as clay), the moisture content, and the organic additives are all selected to obtain an acceptable compromise of the four basic requirements. Additional factors that affect mold material selection include chemical inertness with respect to the molten metal, flowability when introduced into the mold, availability, and cost. Particle size and shape will have a strong influence on the surface finish of the resultant casting.

Because molding material is often reclaimed and recycled, the temperature of the mold during pouring and solidification is also important. If organic materials have been incorporated into the mix to provide collapsibility, a portion will burn during the pour. Adjustments will be necessary, and ultimately some or all of the mold material may have to be discarded and replaced with new.

A typical **green-sand** mixture contains about 88% silica sand, 9% clay, and 3% water. To achieve good molding, it is important for each grain of sand to be coated uniformly with the proper amount of additive agents. This is achieved by putting the ingredients through a **muller**, a device that kneads, rolls, and stirs the sand. **Figure 14.8** shows both a continuous and batch-type muller, with each producing the desired mixing through the use of rotating blades that lift, fluff, and redistribute the material and wheels that compress and squeeze. After mixing, the sand is often discharged through an aerator, which fluffs it for further handling.

Sand Testing

If a foundry is to produce high-quality products, it is important that it maintain a consistent quality in its molding sand. The sand itself can be characterized by grain size, grain shape,

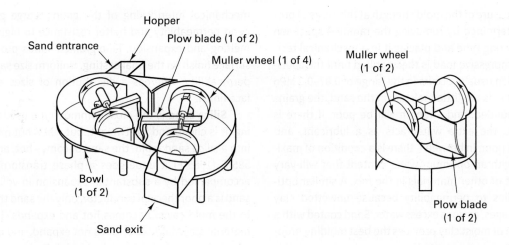

FIGURE 14.8 Schematic diagram of a continuous (left) and batch-type (right) sand muller. Plow blades move and loosen the sand, and the muller wheels compress and mix the components. (Courtesy of ASM International, Materials Park, OH)

surface smoothness, density, and contaminants. After blending, important features of the molding sand are moisture content, clay content, and *compactability*. Key properties of compacted sand or finished molds include **mold hardness**, *permeability*, and *strength*. Standard tests and procedures have been developed to evaluate many of these properties.

Grain size can be determined by shaking a known amount of clean, dry sand downward through a set of 11 standard screens or sieves of decreasing mesh size. After shaking for 15 minutes, the amount of material remaining on each sieve is weighed, and these weights are used to compute an AFS (American Foundry Society) grain fineness number.

Moisture content can be determined by a special device that measures the electrical conductivity of a small sample of compressed sand. A more direct method is to measure the weight lost by a 50-gram sample after it has been subjected to a temperature of about 110°C (230°F) for sufficient time to drive off all the water.

Clay content is determined by washing the clay from a 50-gram sample of molding sand, using water that contains sufficient sodium hydroxide to make it alkaline. Several cycles of agitation and washing may be required to fully remove the clay. The remaining sand is then dried and weighed to determine the amount of clay removed from the original sample.

Permeability and strength tests are conducted on compacted sands, using a **standard rammed specimen**. Sand is first placed into a 2-inch-diameter steel tube. A 14-pound weight is then dropped on it three times from a height of 2 in., and the height of the resulting specimen must be within $\frac{1}{32}$-in. of a targeted 2-in. height.

Permeability is a measure of how easily gases can pass through the narrow voids between the sand grains. Air in the mold before pouring, plus the steam that is produced when the hot metal contacts the moisture in the sand, along with various combustion gases, must all be allowed to escape, rather than prevent mold filling or be trapped in the casting as porosity or blow holes. During the permeability test, shown schematically

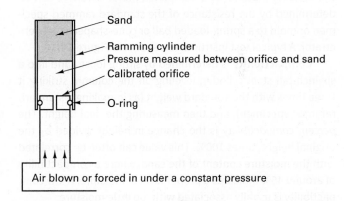

FIGURE 14.9 Schematic of a permeability tester in operation. A standard sample in a metal sleeve is sealed by an O-ring onto the top of the unit while air is passed through the sand. (Courtesy of Dietert Foundry Testing Equipment, Inc., Detroit, MI)

in **Figure 14.9**, a sample tube containing the standard rammed specimen is subjected to an air pressure of 10 g/cm². By means of either a flow rate determination or a measurement of the steady-state pressure between the orifice and the sand specimen, an **AFS permeability number**[1] can be computed or directly read from the instrument.

The molding material must have sufficient strength to retain the integrity of the mold cavity while the mold is being handled between molding and pouring. It must also withstand the erosion of the liquid metal as it flows into the mold and the pressures induced by a column of molten metal. The **compressive strength** of the sand (also referred to as **green compressive**

---

[1] The AFS permeability number is defined as:

$$AFS\ Number = (V \times H) / (P \times A \times T)$$

where V is the volume of air (2000 cm³), H is the height of the specimen (5.08 cm), P is the pressure (10 g/cm²), A is the cross-section area of the specimen (20.268 cm²), and T is the time in seconds to pass a flow of 2000 cm³ of air through the specimen. Substituting each of the preceding constants, the permeability number becomes equal to $3000.2/T$.

strength) is a measure of the mold strength at this stage of processing. It is determined by removing the rammed specimen from the compacting tube and placing it in a mechanical testing device. A compressive load is then applied until the specimen breaks, which usually occurs in the range of 0.07–0.2 MPa (10–30 psi). If there is too little moisture in the sand, the grains will be poorly bonded, and strength will be poor. If there is excess moisture, the extra water acts as a lubricant, and strength is again poor. In between, there is a condition of maximum strength with an optimum water content that will vary with the content of other materials in the mix. A similar optimum also applies to permeability because unwetted clay blocks vent passages, as does excess water. Sand coated with a uniform thin film of moist clay provides the best molding properties. A ratio of one part water to three parts clay (by weight) is often a good starting point.

The **hardness** of compacted sand can give additional insight into the strength and permeability characteristics of the molding material or a completed mold. Hardness can be determined by the resistance of the standard rammed specimen or mold to a spring-loaded ball or cone-shaped steel penetrator. A typical test instrument is shown in **Figure 14.10**.

Compactibility is determined by sifting loose sand into a six-inch-tall steel cylinder, leveling off the column, striking it three times with the standard weight (as in making a standard rammed specimen), and then measuring the final height. The *percent compactibility* is the change in height divided by the original height, times 100%. This value can often be correlated with the moisture content of the sand, where a compactibility of around 45% indicates a proper level of moisture. A low compactibility is usually associated with too little moisture.

Sand Properties and Sand-Related Defects

The characteristics of the sand granules themselves can be very influential in determining the properties of foundry molding material. Round grains give good permeability and minimize the amount of clay required because of their low surface area. Angular sands give better green strength because of the mechanical interlocking of the grains. Large grains provide good permeability and better resistance to high-temperature melting and expansion. Fine-grained sands produce a better surface finish on the final casting. Uniform size sands give good permeability, whereas a distribution of sizes enhances surface finish.

Silica sand is the most common type used in metal casting. It is cheap and lightweight, but when hot metal is poured into a silica sand mold, the sand becomes hot, and at or about 585°C (1085°F) it undergoes a phase transformation that is accompanied by a substantial expansion in volume. Because sand is a poor thermal conductor, only the sand that is adjacent to the mold cavity becomes hot and expands. The remaining material stays fairly cool, does not expand, and often provides a high degree of mechanical restraint. Because of this uneven heating, the sand at the surface of the mold cavity may buckle or fold. Castings with large, flat surfaces are more prone to **sand expansion defects** because a considerable amount of expansion must occur in a single direction.

Sand expansion defects can be minimized in a number of ways. Certain particle geometries permit the sand grains to slide over one another, thereby relieving the expansion stresses. Excess clay can be added to absorb the expansion, or volatile additives, such as cellulose, can be added to the mix. When the casting is poured, the cellulose burns, creating voids that can accommodate the sand expansion. Another alternative is the use of olivine or zircon sand in place of silica. Because these sands do not undergo phase transformations on heating, their expansion is only about one-half that of silica sand. Unfortunately, these sands are much more expensive and heavier in weight than the more commonly used silica.

Trapped or evolved gas can create gas-related voids or **blows** in finished castings. The most common causes are low sand permeability (often associated with angular, fine, or wide-size distribution sands; fine sand additives; and over-compaction) and large amounts of evolved gas due to high mold-material moisture or excessive amounts of volatiles. If adjustments to the mold composition are not sufficient to eliminate the voids, **vent passages** may have to be cut into the mold, a procedure that may add significantly to the mold-making cost.

Molten metal can also penetrate between the sand grains, causing the mold material to become embedded in the surface of the casting. This defect, known as **penetration** can be the result of high pouring temperatures (excess fluidity), high metal pressure (possibly due to excessive cope height or pouring from too high an elevation above the mold), or the use of high permeability sands with coarse, uniform particles. Fine-grained materials, such as silica flour (fine powdered sand), can be blended in to fill the voids, but this reduces permeability and increases the likelihood of both gas and expansion defects.

Hot tears or **cracks** can form in castings made from metals or alloys with large amounts of solidification shrinkage and a high degree of thermal contraction as they cool to room temperature. As the metal contracts, it may find itself restrained by a strong mold or core. Tensile stresses can develop while the

FIGURE 14.10   Sand mold hardness tester. (Courtesy of Dietert Foundry Testing Equipment, Inc., Detroit, MI)

metal is still partially liquid, or fully solidified, but still hot and weak. If these stresses become great enough, the casting can crack or tear. Hot tears are often attributed to poor mold collapsibility. Additives, such as cellulose, can be used to improve the collapsibility of sand molds.

Table 14.1 describes some components of sand-based molding materials, and **Table 14.2** summarizes many of its desirable properties.

The Making of Sand Molds

Mold making usually begins with a pattern, like the match-plate pattern discussed earlier, and a flask. The flasks may be straight-walled containers with guide pins or removable jackets, and they are generally constructed of lightweight aluminum or magnesium. **Figure 14.11** shows a **snap flask**, so named because it is designed to snap open for easy removal after the mold material has been packed in place.

When only a few castings are to be made, **hand ramming** is often the preferred method of packing sand. Hand ramming, however, is slow, is labor intensive, and usually results in

TABLE 14.1	Components of Sand-Based Molding Material

1. Types of base sands

   Silica sand—Most common, low cost, light weight, but high thermal expansion and low thermal conductivity, can cause silicosis

   Olivine, chromite, and zircon sands—Low thermal expansion, high thermal conductivity, high melting point, but more expensive and heavier than silica. Used with higher melting point metals and for cores

2. Binders

   Clay and water—Bentonite and kaolinite clays are most common

   Oils—Including linseed oil and vegetable oils, costly and requires careful baking

   Resins—Natural and synthetic, both heat- and catalyst-curing. Offer good collapsibility and good surface finish

   Sodium silicate—Room temperature cures with exposure to CO_2, poor collapsibility and shakeout

3. Additives— Improve surface finish, dry strength, and refractoriness or provide "cushioning"

   Reducing agents—Such as powdered coal that create gases at the surface of the mold cavity preventing the metal from adhering to the mold

   Cushioning materials—Including wood flour, sawdust, straw, and others that burn off during the pour, creating voids in the mold that allow for sand expansion and promote good collapsibility and shakeout

   Cereal binders—Such as starch and molasses that improve mold strength after curing and improve surface finish

   Refractoriness enhancers—Generally fine ceramic materials

4. Parting compounds— These materials are applied to the mold cavity surface to assist in removal of the casting. Dry materials include talc, graphite, and dry silica. Liquids include mineral oil and water-based silicon solutions.

TABLE 14.2	Desirable Properties in Sand-Based Molding Materials

1. Is inexpensive in bulk quantities
2. Retains properties through transportation and storage
3. Uniformly fills a flask or container
4. Can be compacted or set by simple methods.
5. Has sufficient elasticity to remain undamaged during pattern withdrawal
6. Can withstand high temperatures and maintains its dimensions until the metal has solidified
7. Is sufficiently permeable to allow the escape of gases
8. Is sufficiently dense to prevent metal penetration
9. Is sufficiently cohesive to prevent wash-out of mold material into the pour stream
10. Is chemically inert to the metal being cast
11. Can yield to solidification and thermal shrinkage, thereby preventing hot tears and cracks
12. Has good collapsibility to permit easy removal and separation of the casting
13. Can be recycled

nonuniform compaction. For normal production, sand molds are generally made using specially designed molding machines. The various methods differ in the type of flask, the way the sand is packed within the flask, whether mechanical assistance is provided to turn or handle the mold, and whether a flask is even required. In all cases, however, the molding machines greatly reduce the labor and required skill and lead to castings with good dimensional accuracy and consistency.

Several techniques have been developed to pack the mixed sand (mold material) into the flask. A **sand slinger** uses a rotating impeller to fling or throw sand against the pattern, progressively depositing compacted sand into the mold. Sand slinging is a common method of achieving uniform sand compaction when making large molds and large castings.

In a method known as **jolting**, a flask is positioned over a pattern and filled with sand. The pattern, flask, and sand are then lifted and dropped several times, as shown in **Figure 14.12**. The weight and kinetic energy of the sand produces optimum packing at the bottom of the mass, directly around the pattern.

FIGURE 14.11 Bottom and top halves of a snap flask. *(Left)* drag segment in closed position, *(right)* cope segment with latches opened for easy removal from the mold. (E. Paul DeGarmo)

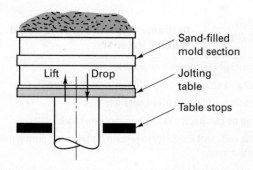

FIGURE 14.12 Jolting a mold section. (*Note:* The pattern is on the bottom where the greatest packing is expected.)

Jolting machines can be used on the first half of a match-plate pattern or on both halves of a cope-and-drag operation.

Squeezing machines use an air-operated squeeze head, a flexible diaphragm, or small individually activated squeeze heads to compact the sand. The squeezing motion provides firm packing adjacent to the squeeze head, with density diminishing as you move farther into the mold. **Figure 14.13** illustrates the squeezing process, and **Figure 14.14** compares the density achieved by squeezing with a flat plate and squeezing with a flexible diaphragm.

In **match-plate molding**, a combination of jolting and squeezing is often used to produce a more uniform density throughout the mold. The match-plate pattern is positioned between the cope and drag sections of a flask, and the assembly is placed drag side up on the molding machine. A parting compound is sprinkled on the pattern, and the drag section of the flask is filled with mixed sand. The entire assembly is then jolted a specified number of times to pack the sand around the drag side of the pattern. A squeeze head is then swung into place, and pressure is applied to complete the drag portion of the mold. The entire flask is then inverted and a squeezing operation is performed to compact loose sand in the cope segment. (*Note:* Jolting is not performed here because it might cause the already-compacted sand to break free of the inverted drag section of the pattern!) Because the drag segment sees both jolting and squeezing, while the cope is only squeezed, the pattern side with the greatest detail is generally placed in the drag. If the cope and drag segments of a mold are made on

separate machines (using separate cope and drag patterns), the combination of jolting and squeezing can be performed on each segment of the mold.

The sprue hole is most often cut by hand, with this operation being performed before removal of the pattern to prevent loose sand from falling into the mold cavity. The pouring basin may also be hand cut, or it may be shaped by a protruding segment on the squeeze board. The gates and runners are usually included on the pattern.

The pattern board is removed, and the segments of the mold are reassembled ready for pour. Heavy metal weights are often placed on top of the molds to prevent the cope section from rising and "floating" when the hydrostatic pressure of the molten metal presses upward. The weights are left in place until solidification is complete, and they are then moved to other molds.

For mass-production molding, a number of automatic mold-making methods have been developed. These include **automatic match-plate molding**, automatic cope-and-drag molding, and methods that produce some form of stacked segments. **Figure 14.15** shows the production sequence for one of the variations of automatic match-plate molding where the sand is introduced into the cope and drag mold segments from the side and then vertically compressed. The two-part cope-and-drag mold is produced in one station, with a single pattern, and one machine squeeze cycle. The compressed blocks are extracted from the molding machine and are poured in a flaskless condition.

Figure 14.16 depicts the **vertically parted flaskless molding** process, where the pattern has been rotated into a vertical position, and the cope-and-drag impressions are now incorporated into rams on opposing sides of a compaction machine. Molding sand is deposited between the patterns and squeezed with a horizontal motion. The patterns are withdrawn, cores are set, and the mold block is then pressed against those that were previously molded. Because each block contains both a right-hand cavity and a left-hand cavity, an entire mold is made with each cycle of the machine. A vertical gating system is usually included on one side of the pattern, and the assembled molds are usually poured individually. Vertically

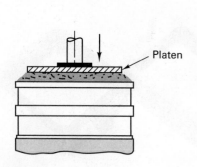

FIGURE 14.13 Squeezing a sand-filled mold section. While the pattern is on the bottom, the highest packing will be directly under the squeeze head.

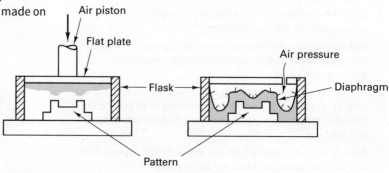

FIGURE 14.14 Schematic diagram showing relative sand densities obtained by flat-plate squeezing where all areas get vertically compressed by the same amount of movement (left) and by flexible-diaphragm squeezing where all areas flow to the same resisting pressure (right).

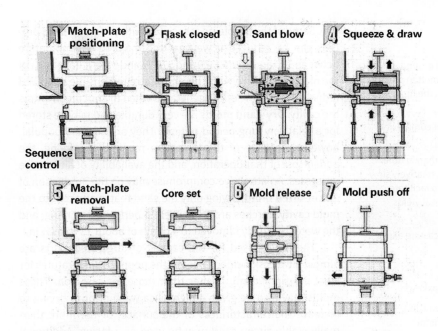

flasks can be placed directly on the foundry floor. Various types of mechanical aids, such as a sand slinger, can then be used to add and pack the sand. Pneumatic rammers can provide additional tamping. Even larger molds can be constructed in sunken pits. Because of the size, complexity, and need for strength, pit molds are often assembled by joining smaller sections of baked or dried sand. Added binders may be required to provide the strength required for these large molds.

Green-Sand, Dry-Sand, and Skin-Dried Molds

Green-sand casting (where the term "green" implies that the mold material has not been fired or cured) is the most widely used process for casting both ferrous and nonferrous metals. The mold material is composed of sand blended with clay, water, and additives. Tooling costs are low, and the entire process is one of the least expensive of the casting methods. Almost any metal can be cast (except titanium), and there are few limits on the size, shape, weight, and complexity of the products. Over the years, green-sand casting has evolved from a manually intensive operation to a mechanized and automated system capable of producing over 300 molds per hour. As a result, it can be economically applied to both small (manual molding) and large (automated system) production runs.

Design limitations are usually related to the rough surface finish and poor dimensional accuracy—and the resulting need for finish machining. Still other problems can be attributed to the low strength of the mold material and the moisture that is present in the clay-and-water binder. **Table 14.3** provides a process summary for green-sand casting, and **Figure 14.17** shows a variety of parts that have been produced in aluminum.

FIGURE 14.15 Activity sequence for automatic match-plate molding. Green sand is blown from the side and compressed vertically. The final mold is ejected from the flask and poured in a flaskless condition. (Copyright 2000, American Foundry Society, Schaumberg, IL)

parted molding machines can produce as many as 550 sand molds per hour.

If a common horizontal runner is used to connect multiple vertically parted mold segments, the method is known as the **H-process**. Because metal cools as it travels through long runners, the individual cavities of the H-process often fill with different temperature metal. To ensure product uniformity, most producers reject the H-process and prefer to pour their vertically parted molds individually.

In **stack molding**, sections containing a cope impression on the bottom and a drag impression on the top are stacked vertically on top of one another. Metal is poured down a common vertical sprue, which is connected to horizontal gating systems at each of the parting plane interfaces.

For molds that are too large to be made by either hand ramming or by one of the previous molding processes, large

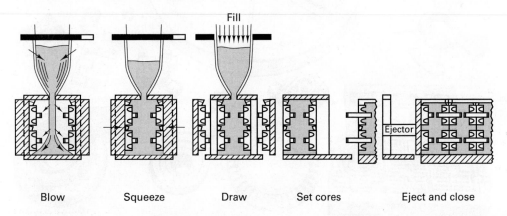

FIGURE 14.16 Vertically parted flaskless molding with inset cores. Note how one mold block now contains both the cope and drag impressions.

TABLE 14.3	Green-Sand Casting

Process: Sand, bonded with clay and water, is packed around a wood or metal pattern. The pattern is removed, and molten metal is poured into the cavity. When the metal has solidified, the mold is broken and the casting is removed.

Advantages: Almost no limit on size, shape, weight, or complexity; low cost; almost any metal can be cast.

Limitations: Tolerances and surface finish are poorer than in other casting processes; some machining is often required; relatively slow production rate; a parting line and draft are needed to facilitate pattern removal; due to sprues, gates, and risers, typical yields range from 50% to 85%.

Common metals: Cast iron, steel, stainless steel, and casting alloys of aluminum, copper, magnesium, and nickel.

Size limits: 30 g to 3000 kg (1 oz to 7000 lb).

Thickness limits: As thin as 0.25 cm ($\frac{3}{32}$ in.), with no maximum.

Typical tolerances: 0.8 mm for first 15 cm ($\frac{1}{32}$ in. for first 6 in.), 0.003 cm for each additional cm; additional increment for dimensions across the parting line

Draft allowances: 1–3°.

Surface finish: 2.5–25 microns (100–1000 μin.) rms.

Some of the problems associated with the green-sand process can be reduced if we heat the mold to a temperature between 150 and 300°C (300 and 575°F) and bake it until most of the moisture is driven off. This drying strengthens the mold and reduces the volume of gas generated when the hot metal enters the cavity. **Dry-sand molds** are very durable and may be stored for a relatively long period of time. They are not very popular, however, because of the long time required for drying, the added cost of that operation, and the availability of alternative processes. An attractive compromise may be the production of a **skin-dried mold**, drying only the sand that is adjacent to the mold cavity. Torches are often used to perform the drying, and the water is usually removed to a depth of about 13 mm (½ in.).

The molds used for the casting of large steel parts are almost always skin dried because the pouring temperatures for steel are significantly higher than those for cast iron. These molds may also be given a high-silica wash prior to drying to increase the refractoriness of the surface, or the more thermally stable zircon sand may be used as a facing. Additional binders, such as molasses, linseed oil, or corn flour, can be added to the facing sand to enhance the strength of the skin-dried segment.

FIGURE 14.17 A variety of sand cast aluminum parts. (Courtesy of Bodine Aluminum Inc., St. Louis, MO)

Sodium Silicate–CO₂ Molding

Molds (and cores) can also be made from sand that receives its strength from the addition of 3–6% **sodium silicate**, a clear inorganic liquid binder, commonly known as **water glass**. The sand can be mixed with the liquid sodium silicate in a standard muller and can be packed into flasks by any of the methods discussed previously in this chapter. It remains soft and moldable until it is exposed to a flow of CO_2 gas, when it hardens in a matter of seconds by the reaction:

$$Na_2SiO_3 + CO_2 \rightarrow Na_2CO_3 + SiO_2\,(colloidal)$$

The CO_2 gas is nontoxic, nonflammable, and odorless, and no heating is required to initiate or drive the reaction. The sands achieve a tensile strength of about 0.3 MPa (40 psi) after 5 seconds of CO_2 gassing, with strength increasing to 0.7–1.4 MPa (100–200 psi) after 24 hours of aging. Unlike most other sands, however, the heating that occurs as a result of the metal pour further increases the strength of the material (a phenomenon similar to the firing of a ceramic material). As a result, the sodium silicate sands have extremely poor collapsibility, restraining the casting during postsolidification shrinkage and making shakeout and core removal quite difficult. Additives that will burn out during the pour are frequently used to enhance the collapsibility of sodium silicate molds. Care must also be taken to prevent the carbon dioxide in the air from hardening the premixed sand before the mold-making process is complete. As a result, the **bench life** of sodium silicate sands (available time between mixing and the completion of molding) is quite short.

A modification of this process can be used when certain portions of a mold require better accuracy, thinner sections, or deeper draws than can be achieved with ordinary molding sand. Sand mixed with sodium silicate is packed around a special metal pattern to a thickness of about 2½ cm. (1 in.), followed by regular molding sand as a backing material. After the sand is fully compacted, CO_2 is introduced through vents in the metal pattern. The adjacent sand is further hardened, and the pattern can be withdrawn with less possibility of damage to the mold.

No-Bake, Air-Set, or Chemically Bonded Sands

An alternative to the sodium silicate-CO_2 process involves room-temperature chemical reactions that can occur between organic or inorganic resin binders and liquid curing agents or catalysts. The various components are mixed with sand just prior to the molding operation, and the curing reactions begin immediately. The molds (or cores) are then made in a reasonably rapid fashion because the mix remains workable for only a short period of time. After a few minutes to a few hours at room temperature (depending on the specific binder and curing agent), the sands harden sufficiently to permit removal from the pattern without concern for distortion. After time for additional curing and the possible application of a refractory coating, the molds are then ready for pour or storage where they can be stored almost indefinitely.

No-bake molding can be used with virtually all engineering metals over a wide range of product sizes and weights. Because the time for mold curing slows production, no-bake molding is generally limited to low-to-medium production quantities (less than 5000 castings per year). The cost of no-bake molding is about 20–30% greater than green sand, so no-bake is generally used where offsetting savings, such as reduced machining, can be achieved. Products can be designed with thinner sections, deeper draws, and smaller draft, and the rigid, brick-like molds enable high dimensional precision, along with good surface finish. Because no-bake sand can be compacted by only light vibrations, patterns can often be made from wood, plastic, fiberglass, or even Styrofoam, thereby reducing pattern cost.

Because of its higher strength, no-bake molding is commonly used in the production of cores, cast components with higher complexities, and large (even multiton) castings. A wide variety of systems are available, with selection being based on the metal being poured, the cure time desired, the complexity and thickness of the casting, and possible desire for sand reclamation. Like the molds produced by the sodium silicate process, no-bake offers good hot strength and high resistance to mold-related casting defects. In contrast to the sodium silicate material, however, the no-bake molds decompose readily after the metal has been poured, providing excellent shakeout characteristics. Permeability must be good because the heat causes the resins to decompose to hydrogen, water vapor, carbon oxides, and various hydrocarbons—all gases that must be vented.

Air-set molding is another term used to describe the no-bake process. **Chemically bonded sands** is a more generic classification that includes no-bake, as well as the variety of gas-hardened and heat-activated processes. **Table 14.4** summarizes the features of no-bake casting.

Shell Molding

A popular heat-activated sand casting process is **shell molding**, the basic steps of which are described next and illustrated in **Figure 14.18**.

1. The individual grains of fine silica sand are first precoated with a thin layer of thermosetting resin and heat-sensitive liquid catalyst. This material is then dumped, blown, or shot onto a metal pattern (usually some form of cast iron) that has been preheated to a temperature between 230 and 340°C (450 and 650°F). During a period of sustained contact, heat from the pattern partially cures (polymerizes and crosslinks) a layer of material. This forms a strong, solid-bonded region adjacent to the pattern. The actual thickness of cured material depends on the pattern temperature and the time of contact, but typically ranges between 10 and 20 mm (0.4 and 0.8 in.).

TABLE 14.4   No-Bake Casting

Process: Sand is mixed with a chemical binder and catalyst and then molded around a pattern. After a specified period of time, the sand mixture hardens, the pattern is withdrawn, and the mold is assembled and poured. Also used in the production of cores.

Advantages: High precision, thin walls, and reduced draft can significantly reduce machining cost; economical for production runs between 1 and 5,000 parts/yr; compatible with most pattern materials, including wood, plastic, metal, fiberglass, and Styrofoam because strong force compaction is not required; minimum pattern wear.

Limitations: Mold curing slows production; cost is 20–30% higher than green sand.

Common metals: Virtually all castable metals.

Size limits: Less than 0.5 kg to multiple tonnes, 0.2–70 kg (0.5–150 lb) most common.

Thickness limits: Less than 0.25 cm (0.1 in).

Typical tolerances: 0.13–0.4 mm (0.005–0.015 in).

Draft allowances: 1–2°.

Surface finish: 4–15 microns (150–600 μin) rms.

2. The pattern and sand mixture are then inverted, allowing the excess (uncured) sand to drop free. Only the layer of partially cured material remains adhered to the pattern.

3. The pattern with adhering shell is then placed in an oven where additional heating completes the curing process.

4. The hardened shell, with tensile strength between 2.4 and 3.1 MPa (350 and 450 psi), is then stripped from the pattern.

5. Two or more shells are then clamped or glued together with a thermoset adhesive to produce a mold, which may be poured immediately or stored almost indefinitely.

6. To provide extra support during the pour, shell molds are often placed in a pouring jacket and surrounded with metal shot, sand, or gravel.

Because the shell is formed and partially cured around a metal pattern, the process offers excellent dimensional accuracy. Tolerances of 0.08–0.13 mm (0.003–0.005 in.) are quite common. Shell-mold sand is typically finer than ordinary foundry sand and, in combination with the plastic resin, enables fine detail and a very smooth casting surface. Cleaning, machining, and other finishing costs can be significantly reduced, and the mold process offers an excellent level of product consistency.

Figure 14.19 shows a set of metal patterns, the two shells before clamping, and the resulting shell-mold casting. Machines for making shell molds vary from simple ones for small operations, to large, completely automated devices for mass production. The cost of a metal pattern is often rather high, and its design must include the gate and runner system because these cannot be cut after molding. Large amounts of expensive binder are required, but the amount of material actually used to form a thin shell is not that great. High productivity, low labor costs, smooth surfaces, and a level of precision that reduces the amount of subsequent machining all combine to make the process economical for even moderate quantities. The thin shell provides for the easy escape of gases that evolve during the pour, and the volume of evolved gas is rather low because of the absence of moisture in the mold material. When the shell becomes hot, some of the resin binder burns out, providing excellent collapsibility and shakeout characteristics. In addition, both the molding sand and completed shells can be stored for indefinite periods of time. **Table 14.5** summarizes the features of shell molding.

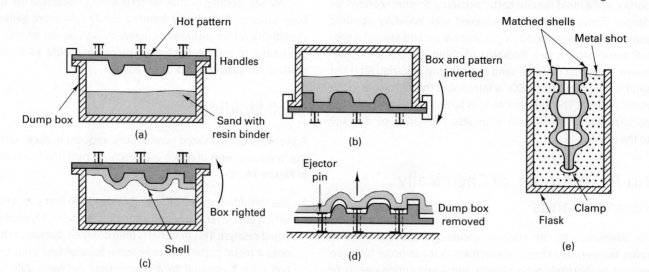

<u>FIGURE 14.18</u> Schematic of the dump-box version of shell molding. (a) A heated pattern is placed over a dump box containing granules of resin-coated sand. (b) The box is inverted and the heat forms a partially cured shell around the pattern. (c) The box is righted, the top is removed, and the pattern and partially cured sand is placed in an oven to further cure the shell. (d) The shell is stripped from the pattern. (e) Matched shells are then joined and supported in a flask ready for pouring.

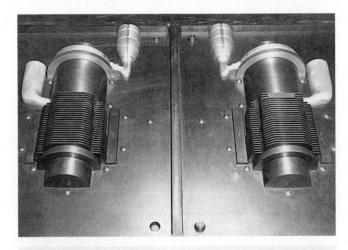

FIGURE 14.19 (Top) Two halves of a shell-mold pattern. (Bottom) The two shells before clamping, and the final shell-mold casting with attached pouring basin, runner, and riser. (Courtesy Roberts Sinto Corp./Shalco Systems, Lansing, MI)

TABLE 14.5	Shell-Mold Casting

Process: Sand coated with a thermosetting plastic resin is dropped onto a heated metal pattern, which cures the resin. The shell segments are stripped from the pattern and assembled. When the poured metal solidifies, the shell is broken away from the finished casting.

Advantages: Faster production rate than sand molding, high dimensional accuracy with smooth surfaces. Thin shells and hollow cores enable use of more expensive molding materials.

Limitations: Requires expensive metal patterns. Plastic resin adds to cost; part size is limited.

Common metals: Cast irons and casting alloys of aluminum and copper.

Size limits: 30 g (1 oz) minimum; usually less than 10 kg (25 lb); mold area usually less than 0.3 m² (500 in²).

Thickness limits: Minimums range from 0.15 to 0.6 cm $\left(\frac{1}{16} \text{ to } \frac{1}{16} \text{ in.}\right)$, depending on material.

Typical tolerances: Approximately 0.005 cm/cm or in/in.

Draft allowance: $\frac{1}{4}$ or $\frac{1}{2}$ degree.

Surface finish: $\frac{1}{3}$ – 4.0 microns (50–150 μin.) rms.

Other Sand-Based Molding Methods

Over the years, a variety of processes have been proposed to overcome some of the limitations or difficulties of the more traditional sand-based molding methods. Although few have become commercially significant, several are included here to illustrate the nature of these efforts.

In the **V-process or vacuum molding**, a vacuum performs the role of the sand binder. **Figure 14.20** depicts the production sequence, which begins by draping a thin sheet of heat-softened plastic (similar to Saran Wrap) over a special vented pattern, which is often made of urethane or other plastic. A vacuum is applied within the pattern, drawing the sheet tight to its surface. A special vacuum flask is then placed over the pattern; the flask is filled with fine, dry unbonded sand; and vibrated to fill pattern details and produce maximum bulk density. A sprue and pouring cup are then formed, and a second sheet of heated plastic is placed over the mold. A vacuum is then drawn on the flask itself, compressing the sand to provide the necessary strength and hardness. The pattern vacuum is released, and the pattern is then withdrawn. The other segment of the two-part cope-and-drag mold is made in a similar fashion, cores are set in place, and the mold halves are assembled to produce a plastic-lined cavity. The mold is then poured with a vacuum of 300–600 mm Hg (40–80 kPa) being maintained in both the cope and drag segments of the flask. During the pour, the thin plastic film melts and vaporizes and is replaced immediately by metal, allowing the vacuum to continue holding the sand in shape until the casting has cooled and solidified. When the vacuum is released, the sand reverts to its loose, unbonded state and falls away from the casting.

Products can be made with zero draft, and walls can be as thin as 3 mm (0.125 in.) over large areas. There is no pattern wear because the sand never touches the pattern. With the vacuum serving as the binder, there is a total absence of moisture-related defects; binder cost is eliminated; and the loose, dry sand is completely and directly reusable. With no clay, water, or other binder to impair permeability, finer sands can be used, resulting in better surface finish in the resulting castings. With no burning binders (only the thin plastic sheets are burned), few fumes are generated during the pouring operation. Shakeout characteristics are exceptional because the mold collapses when the vacuum is released. Unfortunately, the process is relatively slow because of the additional steps and the time required to produce a sufficient vacuum. The V-process is used primarily for the production of prototype, frequently modified, or low- to medium-volume parts (more than 10 but less than 15,000).

In the **Eff-set process**, wet sand with just enough clay to prevent mold collapse is packed around a pattern. The pattern is removed, and the surface of the mold is sprayed with liquid nitrogen. The ice that forms serves as the binder, and the molten metal is poured into the mold while the surface is in its frozen condition. This process offers low binder cost and excellent shakeout, but is not being used in a commercial operation.

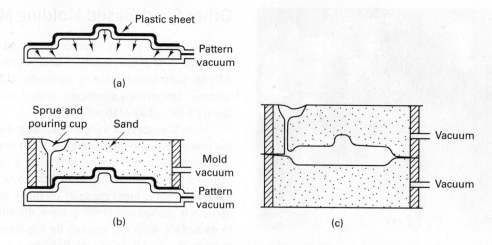

Schematic of the V-process or vacuum molding. (a) A vacuum is pulled on a pattern, drawing a heated shrink-wrap plastic sheet tightly against it. (b) A vacuum flask is placed over the pattern and filled with dry unbonded sand; a pouring basin and sprue are formed; the remaining sand is leveled; a second heated plastic sheet is placed on top; and a mold vacuum is drawn to compact the sand and hold the shape. (c) With the mold vacuum being maintained, the pattern vacuum is then broken and the pattern is withdrawn. The cope and drag segments are assembled, and the molten metal is poured.

14.3 Cores and Core Making

Casting processes are unique in their ability to easily incorporate internal cavities, often intricate in shape. To produce these features, however, it is often necessary to use **cores**—preformed masses of bonded sand that keep the metal from filling the entire mold cavity. Internal cores can also be used to create weight-reducing cavities, while maintaining a desired external profile and full functionality of a casting. External cores can enable the withdrawal of patterns for products with reentrant angles.

Cores can often be used to improve casting design and permit the optimization of processes. Although they constitute an added cost, cores significantly expand the capabilities of the process. The belt pulley shown schematically in **Figure 14.21** is a simple example. Various methods of fabrication are suggested in the four sketches, beginning with the casting of a solid form and the subsequent machining of the through-hole for the drive shaft. A large volume of metal would have to be

removed by a secondary machining process. A more economical approach would be to make the pulley with a cast-in hole. In Figure 14.21b each half of the pattern includes a tapered hole, which fills with the same green sand being used for the remainder of mold. These protruding sections are an integral part of the mold, but they are also known as **green-sand cores**. Green-sand cores have a relatively low strength. If the protrusions are long or narrow, it might be difficult to withdraw the pattern without breaking them, or they may not have enough strength to even support their own weight. For long cores, a considerable amount of machining may still be required to remove the draft that must be provided on the pattern. In addition, green-sand cores are not an option for more complex shapes where it is often impossible to withdraw the pattern.

Dry-sand cores can overcome some of the cited difficulties. These cores are produced separate from the remainder of the mold and are then inserted into core prints that hold them in position. The sketches in Figure 14.21c and Figure 14.21d show dry-sand cores in the vertical and horizontal positions. Dry-sand cores can be made in a number of ways. In each, the sand, mixed with some form of binder, is packed into a wood or metal core box that contains a cavity of the desired shape. A **dump core box** such as the one shown in **Figure 14.22** offers the simplest approach. Sand is packed into the cavity and scraped level with the top surface (which acts like the parting line in a traditional mold). A wood or metal plate is then placed over the top of the box, and the box is inverted and lifted, leaving the molded sand segment resting on the plate. After baking or hardening, the various core segments are assembled with hot-melt glue or some other bonding agent. Rough spots along the parting line are removed with files or sanding belts, and the final core may be given a thin coating to provide a smoother

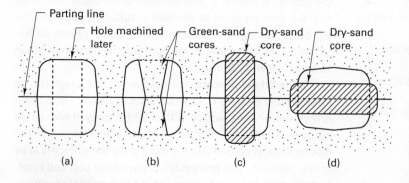

Four methods of making a hole in a cast pulley. Three involve the use of a core.

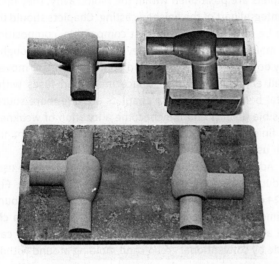

FIGURE 14.22 (Upper right) A dump-type core box; (bottom) Two core halves ready for baking; and (upper left) a completed core made by gluing two opposing halves together. (E. Paul DeGarmo)

surface or greater resistance to heat. Graphite, silica, or mica can be sprayed or brushed onto the surface.

Single-piece cores can be made in a **split core box**. Two halves of a core box are clamped together, with an opening in one or both ends through which sand is introduced and rammed. After the sand is compacted, the halves of the box are separated to permit removal of the core. Cores with a uniform cross section can be formed by a core-extruding machine and cut to the desired length as the product emerges. The individual cores are then placed in core supports for subsequent hardening. More complex cores can be made in core-blowing machines that use separating dies and receive the sand in a manner similar to injection molding or die casting.

Cores are frequently the most fragile part of a mold assembly. To provide the necessary strength, the various core-making processes utilize a number of special binders. In the **core-oil process**, sand is blended with about 1% vegetable or synthetic oil, along with 2–4% water and about 1% cereal or clay to help develop green strength (i.e., to help retain the shape prior to curing). The wet sand is blown or rammed into a relatively simple corebox at room temperature. The fragile uncured cores are then gently transferred to flat plates or special supports and placed in convection ovens at 200–260°C (400–500°F) for curing. The heat causes the oil to cross-link or polymerize, producing a strong organic bond between the grains of sand. Although the process is simple and the materials are inexpensive, the dimensional accuracy of the resultant cores is often difficult to maintain.

In the **hot-box method**, sand blended with a liquid thermosetting binder and catalyst is packed into a core box that has been heated to around 230°C (450°F). When the sand is heated, the initial stages of curing begin within 10–30 seconds. After this brief period, the core can be removed from the pattern and will hold its shape during subsequent handling. For some materials, the cure completes through an exothermic curing reaction. For others, further baking is required to complete the process.

In the aforementioned methods, cores must be handled in an uncured or partially cured state, and breakage or distortion is not uncommon. Processes that produce finished cores while still in the corebox, and do not require heating operations, would appear to offer distinct advantages.

In the **cold-box or gas-hardened processes**, resin-coated sand is first blown into a room temperature core box, which can be made from wood, metal, or even plastic. The box is sealed, and a gas or vaporized catalyst is then passed through the permeable sand to polymerize the resin binder. In a variation of the process, hollow cores are produced by introducing small amounts of curing gas through holes in the corebox pattern, with the uncured sand in the center being dumped and reused. Unfortunately, the required gases tend to be either toxic (an amine gas with phenolic urethane resin) or odorous (SO_2 with furan or acrylic), making special handling of both incoming and exhaust gas a process requirement. The cold-box process is most attractive for large numbers of small-sized products. Curing speed is faster than most of the alternative core-making methods, and cold-box cores can be stored for lengthy periods of time prior to use (**shelf life**). The sodium silicate-CO_2 process is also a gas-hardened system, but the poor collapsibility limits its use to easily extracted cores.

Room-temperature cores can also be made with the air-set or no-bake sands. These systems eliminate the gassing operation of the cold-box process through the use of a reactive organic resin and a curing catalyst. Because of the short bench life, there is only a brief period of time to form the core once the components have been mixed. Shell molding is another core-making alternative, producing hollow cores with excellent strength and permeability.

Selecting the actual method of core production is usually based on a number of considerations, including production quantity, production rate, required precision, required surface finish, and the metal being poured. Certain metals may be sensitive to gases that are emitted from the cores when they come into contact with the hot metal. Other materials with low pouring temperatures may not break down the binder sufficiently to provide collapsibility and easy removal from the final casting.

To function properly, casting cores must have the following characteristics:

1. Sufficient strength before hardening if they will be handled in the "green" condition.

2. Sufficient hardness and strength after hardening to withstand handling and the forces of the casting process. As metal fills the mold, most cores want to "float." The cores must be strong enough to resist the induced stresses, and the supports must be sufficient to hold them in place. Flowing metal can also cause surface erosion. Compressive strength should be between 0.7 and 3.5 MPa (100 and 500 psi).

3. A smooth surface.

4. Minimum generation of gases when heated by the pour.

5. Adequate permeability to permit the escape of gas. Because cores are largely surrounded by molten metal, the gases must escape through the core.

6. Adequate refractoriness. Being surrounded by hot metal, cores can become quite a bit hotter than the adjacent mold material. They should not melt or adhere to the casting.

7. Collapsibility. After pouring, the cores must be weak enough to permit the casting to shrink as it cools, thereby preventing cracking. In addition, the cores must be easily removed from the interior of the finished product via shakeout.

Various techniques have been developed to enhance the natural properties of cores and core materials. Strength can be increased by the addition of internal wires or rods. Collapsibility can be enhanced by producing hollow cores, or by placing a material such as straw in the center. Hollow cores may also be used to provide for the escape of trapped or evolved gases. Other techniques to enhance venting include pushing small wires into the core and removing them to create vent channels or placing coke or cinders in the center of large cores.

To permit the expulsion of vent gases and the ultimate removal of the core material to create the desired internal feature, the cores must be connected to the outer surfaces of the mold cavity. Recesses at these connection points, known as **core prints**, are used to support the cores and hold them in proper position during mold filling. The dry-sand cores in Figure 14.21c and 14.21d are supported by core prints.

If the cores do not pass completely through the casting where they can be supported on both ends, a single core print may not be able to provide adequate support. Additional measures may also be necessary to support the weight of large cores or keep lighter ones from becoming buoyant as the molten metal fills the cavity. Small metal supports, called **chaplets**, can be placed between cores and the surfaces of a mold cavity, as illustrated in **Figure 14.23**. Because the

chaplets are positioned within the mold cavity, they become an integral part of the finished casting. Chaplets should therefore be of the same, or at least comparable, composition as the material being poured. They should be large enough that they do not completely melt and permit the core to move, but small enough that their surface melts and fuses with the metal being cast. Because chaplets are one more source of possible defects and may become a location of weakness in the finished casting, efforts are generally made to minimize their use.

Additional sections of mold material can also be used to produce castings with reentrant angles or features. **Figure 14.24** depicts a round pulley with a recessed groove around its perimeter. By using a third segment of flask, called a **cheek** and adding a second parting plane, the entire mold can be made by conventional green-sand molding around withdrawable patterns. Although additional molding operations are required, this may be an attractive approach for small production runs.

If we want to produce a large number of identical pulleys, rapid machine molding of a simple green-sand mold might be preferred. As shown in **Figure 14.25**, the pattern would be modified to include a seat for an inserted ring-shaped core. Molding time is reduced at the expense of a core box and a separate core-making operation.

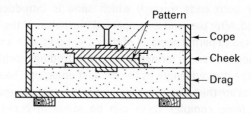

FIGURE 14.24 Method of making a reentrant angle or inset section by using a three-piece flask.

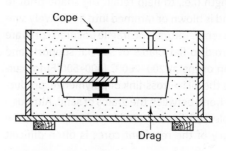

FIGURE 14.23 (Left) Typical chaplets. (Right) Method of supporting a core by use of chaplets (relative size of the chaplets is exaggerated). (E. Paul DeGarmo)

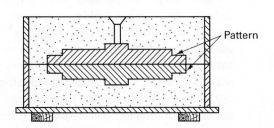
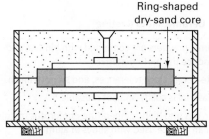

Ring-shaped
dry-sand core

Pattern

FIGURE 14.25 Molding an inset section using a dry-sand core.

14.4 Other Expendable-Mold Processes with Multiple-Use Patterns

Plaster Mold Casting

In **plaster molding** the mold material is plaster of Paris (also known as calcium sulfate or gypsum), combined with various additives to improve green strength, dry strength, permeability, and castability. Talc or magnesium oxide can be added to prevent cracking and reduce the setting time. Lime or cement helps to reduce expansion during baking. Glass fibers can be added to improve strength, and sand can be used as a filler.

The mold material is first mixed with water, and the creamy slurry is then poured over a metal or plastic pattern (wood patterns tend to warp or swell) and allowed to set. Hydration of the plaster produces a hard mold that can be easily stripped from the pattern. (*Note:* Flexible rubber patterns can be used when complex angular surfaces or reentrant angles are required. The plaster is strong enough to retain its shape during pattern removal.) The plaster mold is then baked to remove excess water, assembled, and poured. When baking, just the right amount of water should be left in the mold material. Too much can cause casting defects; too little sacrifices strength.

With metal patterns and plaster mold material, surface finish and dimensional accuracy are both excellent. Cooling is slow because the plaster has low heat capacity and low thermal conductivity. The poured metal stays hot and can flow into thin sections and replicate fine detail, which can often reduce machining cost. Unfortunately, plaster casting is limited to the lower-melting-temperature nonferrous alloys (such as aluminum, copper, magnesium, and zinc). At the high temperatures of ferrous metal casting, the plaster would first undergo a phase transformation and then melt, and the water of hydration can cause the mold to explode. **Table 14.6** summarizes the features of plaster mold casting.

One of the biggest problems with plaster casting is the lack of permeability. The **Antioch process** is a variation of plaster

TABLE 14.6 **Plaster Casting**

Process: A slurry of plaster, water, and various additives is poured over a pattern and allowed to set. The pattern is removed, and the mold is baked to remove excess water. After pouring and solidification, the mold is broken and the casting is removed.

Advantages: High dimensional accuracy and smooth surface finish; can reproduce thin sections and intricate detail to make net- or near-net-shaped parts.

Limitations: Lower-temperature nonferrous metals only; long molding time restricts production volume or requires multiple patterns; mold material is not reusable; maximum size is limited; permeability is poor.

Common metals: Primarily aluminum and copper.

Size limits: As small as 30 g (1 oz) but usually less than 7 kg (15 lb).

Thickness limits: Section thickness as small as 0.06 cm (0.025 in.).

Typical tolerances: 0.01 cm on first 5 cm (0.005 in. on first 2 in.), 0.002 cm per additional cm (0.002 in. per additional in.)

Draft allowance: ½–1 degree.

Surface finish: 0.8–4 microns (30–125 μin.) rms.

mold casting where the mold material is comprised of 50% plaster and 50% sand, mixed with water. Autoclaving in steam reduces solidification time and produces molds with improved permeability. Aeration or the addition of a foaming agent to the plaster–water mix can also improve permeability by adding fine air bubbles that increase the material volume by 50–100%.

Ceramic Mold Casting

Ceramic mold casting (summarized in **Table 14.7**) is similar to plaster mold casting, except that the mold is now made from a ceramic material that can withstand the higher-melting-temperature metals (ferrous metals including stainless steels and tool steels and high-temperature nonferrous), and a firing operation is required to harden the mold. Much like the plaster process, ceramic molding can produce thin sections, fine detail, and smooth surfaces, thereby eliminating a considerable amount of finish machining. These advantages, however, must be weighed against the greater cost of the mold material and long mold preparation time, which limit the process to the

TABLE 14.7	Ceramic Mold Casting

Process: Stable ceramic powders are combined with binders and gelling agents to produce the mold material.

Advantages: Intricate detail, close tolerances, and smooth finish.

Limitations: Mold material is costly and not reusable.

Common metals: Ferrous and high-temperature nonferrous metals are most common; can also be used with alloys of aluminum, copper, magnesium, titanium, and zinc.

Size limits: 100 grams to several thousand kilograms (several ounces to several tons).

Thickness limits: As thin as 0.06 cm (0.025 in.); no maximum.

Typical tolerances: 0.01 cm on the first 2.5 cm (0.005 in. on the first in.), 0.003 cm per each additional cm (0.003 in. per each additional in.).

Draft allowances: 1° preferred.

Surface finish: 2–4 microns (75–150 μin.) rms.

production of small-size castings in small to medium quantities. For large molds, the ceramic can be used to produce a facing around the pattern, which is then backed up by a less-expensive material such as reusable fireclay.

One of the most popular of the ceramic molding techniques is the **Shaw process**. A reusable pattern is positioned inside a slightly tapered flask, and a slurry-like mixture of refractory aggregate, hydrolyzed ethyl silicate, alcohol, and a gelling agent is poured on top. This mixture sets to a rubbery state that permits removal of both the pattern and the flask. The mold surface is then ignited with a torch. Most of the volatiles are consumed during the "burn-off," and a three-dimensional network of microscopic cracks (microcrazing) forms in the ceramic. The gaps are small enough to prevent metal penetration but large enough to provide venting of air and gas (permeability), accommodate both the thermal expansion of the ceramic particles during the pour and the subsequent shrinkage of the solidified metal, and aid in providing collapsibility. A baking operation then removes all of the remaining volatiles, making the mold hard and rigid. Ceramic molds are often preheated prior to pouring to ensure proper filling and to control the solidification characteristics of the metal.

Expendable Graphite Molds

For metals such as titanium, which tend to react with many of the more common mold materials, powdered **graphite** can be combined with additives, such as cement, starch, and water, and compacted around a pattern. After "setting," the pattern is removed and the mold is fired at 1000°C (1800°F) to consolidate the graphite. The casting is poured, and the mold is broken to remove the product.

Rubber-Mold Casting

Artificial elastomers can also be compounded in liquid form and poured over a pattern to produce a semirigid mold. These molds are sufficiently flexible to permit stripping from an intricate shape or patterns with reverse-taper surfaces. Unfortunately, **rubber molds** are generally limited to small castings and low melting-point materials. The wax patterns used in investment casting are often made by rubber-mold casting, as are small quantities of finished parts made from plastics or metals, which can be poured at temperatures below 250°C (500°F). Molds for plaster casting can also be made by pouring the plaster into reverse-pattern rubber molds.

14.5 Expendable-Mold Processes Using Single-Use Patterns

Investment Casting

Investment casting is actually a very old process—used in ancient China and Egypt and more recently performed by dentists and jewelers for a number of years. It was not until the end of World War II, however, that it attained a significant degree of industrial importance. Products such as rocket components and jet engine turbine blades required the fabrication of high-precision complex shapes from high-melting-point metals that are not easily machined. Investment casting offers almost unlimited freedom in both the complexity of shapes and the types of materials that can be cast, and millions of investment castings are now produced each year.

Investment casting uses the same type of molding aggregate as the ceramic molding process and typically involves the following sequential steps:

1. *Produce a master pattern—* a modified replica of the desired product made from metal, wood, plastic, or some other easily worked material.

2. *From the master pattern, produce a master die.* This can be made from low-melting-point metal, steel, or possibly even wood. If a low-melting-point metal is used, the die might be cast directly from the master pattern. Rubber molds can also be made directly from the master pattern. Steel dies are often machined directly, eliminating the need for step 1.

3. *Produce wax patterns.* Patterns are then made by pouring molten wax into the master die, or injecting it under pressure (injection molding), and allowing it to harden. Release agents, such as silicone sprays are used to assist in pattern removal. Polystyrene plastic is an alternate pattern material and may be preferred for producing thin and complex surfaces where its higher strength and greater durability are desired. If cores are required, they can generally be made from soluble wax or ceramic. The soluble wax cores are dissolved out of the patterns prior to further processing, while the ceramic cores remain and are not removed until after solidification of the metal casting.

4. *Assemble the wax patterns onto a common wax sprue.* Using heated tools and melted wax, a number of wax patterns can be attached to a central sprue and runner system to create a pattern cluster, or a **tree**. If the product is sufficiently complex that its pattern could not be withdrawn from a single master die, the pattern may be made in pieces and assembled prior to attachment.

5. *Coat the cluster or tree with a thin layer of investment material.* This step is usually accomplished by dipping into a watery slurry of finely ground refractory material. A thin but very smooth layer of investment material is deposited onto the wax pattern, ensuring a smooth surface and good detail in the final product.

6. *Form additional investment around the coated cluster.* After the initial layer has dried, the cluster can be redipped, but this time sand or other refractory aggregate is rained over the wet surface, a process called stuccoing. After drying, this process is repeated until the investment coating has the desired thickness (typically 5–15 mm or 3/16–5/8 in. with up to eight layers). As an alternative, the single-dipped cluster can be placed upside down in a flask and liquid investment material poured around it. The flask is then vibrated to remove entrapped air and ensure that the investment material now surrounds all surfaces of the cluster.

7. *Allow the investment to fully harden.*

8. *Remove the wax pattern from the mold by melting or dissolving.* Molds or trees are generally placed upside down in an oven, where the wax can melt and run out, and any subsequent residue vaporizes. This step is the most distinctive feature of the process because it enables a complex pattern to be removed from a single-piece mold. Extremely complex shapes can be readily cast. (*Note:* In the early years of the process, only small parts were cast, and when the molds were placed in the oven, the molten wax was absorbed into the porous investment. Because the wax "disappeared," the process was called the **lost-wax process**, and the name is still used.) Figure 8.7, in the chapter on ceramic materials, shows an investment casting mold at this stage of the process with part of the wall cut away to reveal the inner cavities.

9. *Heat the mold in preparation for pouring.* Heating to 550–1100°C (1000–2000°F) ensures complete removal of the mold wax, cures the mold to give added strength, and allows the molten metal to retain its heat and flow more readily into all of the thin sections and details. Mold heating also gives better dimensional control because the mold and the metal can shrink together during cooling.

10. *Pour the molten metal.* While gravity pouring is the simplest, other methods may be used to ensure complete filling of the mold. When complex, thin sections are involved, mold filling may be assisted by positive air pressure, evacuation of the air from the mold, or some form of centrifugal process.

11. *Remove the solidified casting from the mold.* After solidification, techniques such as mechanical chipping or vibration, high-pressure water jet, or sand blasting are used to break the mold and remove the mold material from the metal casting. Cutoff and finishing completes the process.

Figure 14.26 depicts the investment procedure where the investment material fills the entire flask, and **Figure 14.27** shows the shell-investment method. **Table 14.8** summarizes the features of investment casting.

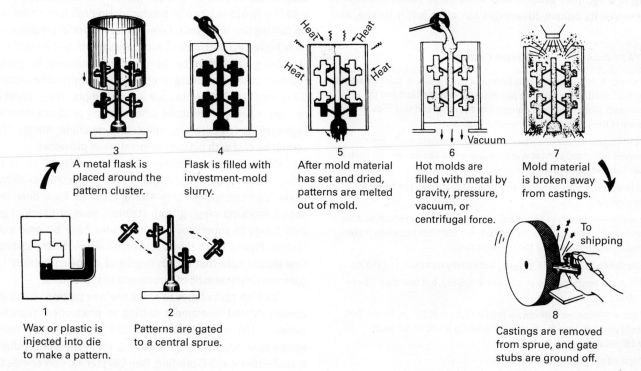

3 A metal flask is placed around the pattern cluster.

4 Flask is filled with investment-mold slurry.

5 After mold material has set and dried, patterns are melted out of mold.

6 Hot molds are filled with metal by gravity, pressure, vacuum, or centrifugal force.

7 Mold material is broken away from castings.

1 Wax or plastic is injected into die to make a pattern.

2 Patterns are gated to a central sprue.

8 Castings are removed from sprue, and gate stubs are ground off.

FIGURE 14.26 Investment casting steps for the flask-cast method. (Courtesy of Investment Casting Institute, Dallas, TX)

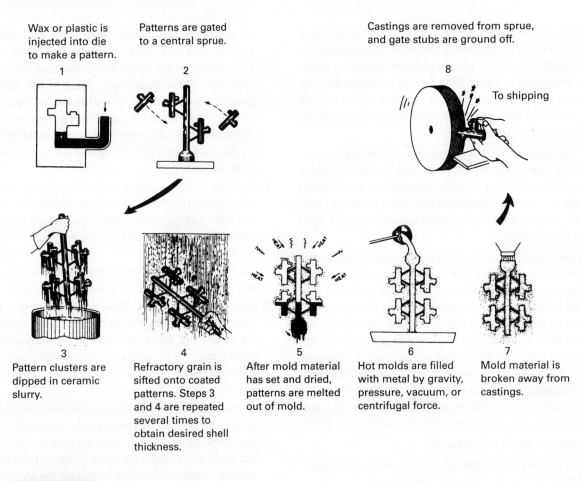

1 — Wax or plastic is injected into die to make a pattern.

2 — Patterns are gated to a central sprue.

3 — Pattern clusters are dipped in ceramic slurry.

4 — Refractory grain is sifted onto coated patterns. Steps 3 and 4 are repeated several times to obtain desired shell thickness.

5 — After mold material has set and dried, patterns are melted out of mold.

6 — Hot molds are filled with metal by gravity, pressure, vacuum, or centrifugal force.

7 — Mold material is broken away from castings.

8 — Castings are removed from sprue, and gate stubs are ground off.

To shipping

FIGURE 14.27   Investment casting steps for the shell-casting procedure. (Courtesy of Investment Casting Institute, Dallas, TX)

Compared to other methods of casting, investment casting is a complex process and tends to be rather expensive. However, its unique advantages can often justify its use, and

TABLE 14.8   Investment Casting

Process: A refractory slurry is formed around a wax or plastic pattern and allowed to harden. The pattern is then melted out and the mold is baked. Molten metal is poured into the mold and solidifies. The mold is then broken away from the casting.

Advantages: Excellent surface finish; high dimensional accuracy; almost unlimited intricacy; almost any metal can be cast; no flash or parting line concerns.

Limitations: Costly patterns and molds; labor costs can be high; limited size.

Common metals: Just about any castable metal. Aluminum, copper, and steel dominate; also performed with stainless superalloys steel, nickel, magnesium, and the precious metals.

Size limits: As small as 3 g $\left(\frac{1}{10}\text{oz}\right)$ but usually less than 5 kg (10 lb).

Thickness limits: As thin as 0.06 cm (0.025 in.), but less than 7.5 cm (3.0 in.).

Typical tolerances: 0.01 cm for the first 2.5 cm (0.005 in. for the first inch) and 0.002 cm for each additional cm (0.002 in. for each additional in.).

Draft allowances: None required.

Surface finish: 1.3–4 microns (50–125 μin.) rms.

many of the steps can be easily automated. Extremely complex shapes can be cast as a single piece. Thin sections, down to 0.40 mm (0.015 in.), can be produced. No draft is required, and no parting line is present. Excellent dimensional precision can be achieved in combination with very smooth as-cast surfaces. Machining can often be completely eliminated or greatly reduced. When machining is required, allowances of as little as 0.4–1 mm (0.015–0.040 in.) are usually ample. These capabilities are especially attractive when making products from the high-melting-temperature, difficult-to-machine metals that cannot be cast with plaster- or metal-mold processes.

Although most investment castings are less than 10 cm (4 in.) in size and weigh less than ½ kg (1 lb), complex aircraft engine parts weighing up to 450 kg (1000 lb) have been produced. Products ranging from stainless steel or titanium golf club heads to superalloy turbine blades have become quite routine. **Figure 14.28** shows some typical investment castings. One should note that a high degree of shape complexity is a common characteristic of investment cast products.

The high cost of dies to make the wax patterns has traditionally limited investment casting to production quantities between 100 and 10,000 pieces per year. However, recent advances in additive manufacturing, also called direct-digital manufacturing or 3-D printing, (see Chapter 33) now enable the production of wax-like patterns directly from computer design

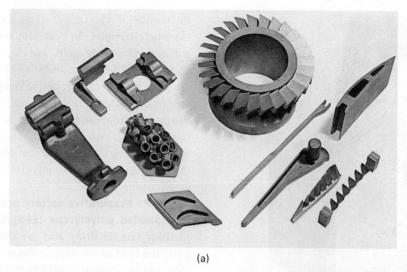

(a)

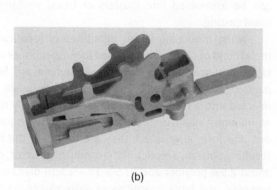

(b)

(c)

FIGURE 14.28 Parts produced by investment casting. (a) Nickel and cobalt superalloys, (b) copper alloy, and (c) Aluminum fuel metering unit for an aircraft engine. [Reproduced by permission of (a) Haynes International, Inc. (b) O'Fallon Casting, and (c) Arconic Power and Propulsion]

data. The absence of part-specific tooling now enables the economical casting of one-of-a-kind or small-quantity products using the investment methods. Another attractive feature of direct-digital pattern making is the opportunity to produce wax patterns that have solid exteriors but honeycomb-like interiors with a high percentage of open volume. There is less material to melt out and considerably less volume expansion during the wax-removal operation. A common problem during wax removal is expansion-induced **shell cracking**, which destroys the shell and results in loss of the time and expense of reaching that stage of manufacture. A polyurethane foam has also been developed for investment cast patterns, with claims of enhanced dimensional stability, simpler pattern removal, and significantly reduced shell cracking.

Counter-Gravity Investment Casting

Counter-gravity investment casting turns the pouring process upside down. In one variation of the process, the ceramic shell mold is encapsulated within a chamber, and the open end of the sprue, which protrudes out the bottom, is lowered into a pool of molten metal. A vacuum is then induced within the chamber, and as the air is withdrawn, the vacuum draws metal up through the central sprue and into the mold. The castings are allowed to solidify, the vacuum is released, and any unsolidified metal in the sprue and gating flows back into the melt. In another variation, a low-pressure inert gas is used to push the molten metal upward into the mold. **Figure 14.29** illustrates the basic counter-gravity investment

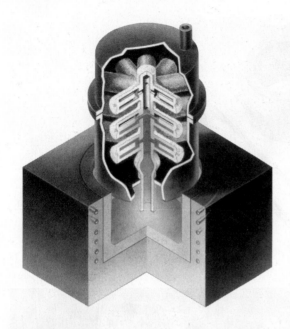

FIGURE 14.29 Illustration of the basic counter-gravity investment casting process. (Courtesy of O'Fallon Castings, O'Fallon, MO)

casting process. The vacuum and low-pressure approaches are also discussed in more detail and further illustrated in the section on low-pressure permanent-mold casting in Chapter 15.

The counter-gravity processes have a number of distinct advantages. Because the molten metal is withdrawn from below the surface of its ladle, it is generally free of slag and dross and has a very low level of inclusions. The vacuum or low-pressure filling allows the metal to flow with little turbulence, further enhancing metal quality. The reduction in metallic inclusions improves machinability and enables mechanical properties to approach or equal those of wrought material. Because the molten metal is taken directly from the furnace or larger holding vessel, the temperature of the metal can be more consistently maintained from mold-to-mold.

Because the gating system does not need to control turbulence, simpler gating systems can be used, reducing the amount of metal that does not become product. In the counter-gravity process, between 60 and 95% of the withdrawn metal becomes cast product, compared to a 15–50% level for gravity-poured castings. The pressure differential enables metal to flow into thinner sections, and lower "pouring" temperatures can be used, resulting in improved grain structure and better surface finish.

Evaporative Pattern (Full-Mold and Lost-Foam) Casting

Several limitations are common to most of the casting processes that have been presented. Some form of pattern is usually required, and this pattern may be costly to design and fabricate. Pattern costs may be hard to justify, especially when

the number of identical castings is rather small or the part is extremely complex. In addition, reusable patterns must be withdrawn from the mold, and this withdrawal often requires some form of design modification or compromise, division into multiple pieces, or special molding procedures. Investment casting overcomes the withdrawal limitations through single-use patterns that can be removed by melting and vaporization. Unfortunately, investment casting has its own set of limitations, including the large number of individual operations and the need to remove the investment material from the finished casting.

In the **evaporative pattern process**, the pattern is made of **expanded polystyrene (EPS)**, or expanded polymethyl-methacrylate (EPMMA), and remains in the mold. During the pour, the heat of the molten metal melts and burns the polymer, and the molten metal fills the space that was previously occupied by the pattern.

When small quantities are required, patterns can be cut by hand or machined from pieces of foamed polystyrene (a material similar to that used in Styrofoam drinking cups). This material is extremely light in weight and can be cut by a number of methods, using tools as simple as an electrically heated wire. Preformed material in the form of a pouring basin, sprue, runner segments, and risers can be attached with hot melt glue to form a complete gating and pattern assembly. Small products can be assembled into clusters or trees, similar to investment casting.

When producing larger quantities of identical parts, a metal mold or die is generally used to mass-produce the evaporative patterns. Hard beads of polystyrene are first preexpanded and stabilized. The preexpanded beads are then injected into a heated metal die or mold, usually made from aluminum. A steam cycle causes them to further expand, fill the die, and fuse, after which they are cooled in the mold. The resulting pattern, a replica of the product to be cast, consists of about 2.5% polymer and 97.5% air. Pattern dies can be quite complex, and large quantities of patterns can be accurately and rapidly produced. When size or complexity is great, or geometry prevents easy removal, the pattern can be divided into multiple segments, or slices, which are then assembled by hot-melt gluing. The ideal glue should be strong, be fast setting, and produce a minimum amount of gas when it decomposes or combusts.

After a polystyrene gating system is attached to the polystyrene pattern, there are several options for the completion of the mold. In the **full mold process**, shown schematically in **Figure 14.30**, green sand or some type of chemically bonded (no-bake) sand is compacted around the pattern and gating system, taking care not to crush or distort it. The mold is then poured like a conventional sand-mold casting.

In the **lost-foam casting** process, depicted schematically in **Figure 14.31**, the polystyrene assembly is first dipped into a water-based ceramic that wets both external and internal surfaces and dries to form a thin refractory coating. The coating must be thin enough and sufficiently permeable to permit the escape of the molten and gaseous pattern material while preventing metal penetration. At the same time, it must be thick enough and rigid enough to support the foam pattern and prevent mold collapse during pouring. The pattern assembly is then suspended in a one-piece flask and surrounded by fine, unbonded, low-thermal-expansion sand. Vibration ensures that the sand compacts around the pattern and fills all cavities and passages. During the pour, the molten metal melts, vaporizes, and ultimately replaces the expanded polystyrene, while the coating isolates the metal from the loose, unbonded sand and provides a smooth surface finish. After the casting has cooled and solidified, the loose sand is then dumped from the flask, freeing the casting and attached gating system. The backup sand can then be reused, provided the coating residue is removed and the organic condensates are periodically burned off. **Figure 14.32** shows the Styrofoam pattern and finished casting of a five-cylinder engine block.

FIGURE 14.32 The Styrofoam pattern and the finished casting of an automotive engine block produced by lost foam casting. (Courtesy of General Motors Corporation, Detroit, MI)

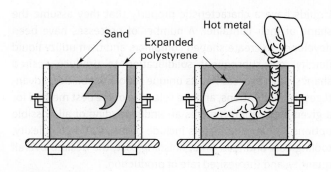

FIGURE 14.30 Schematic of the full-mold process. (Left) An uncoated expanded polystyrene pattern is surrounded by bonded sand to produce a mold. (Right) Hot metal progressively vaporizes the expanded polystyrene pattern and fills the resulting cavity.

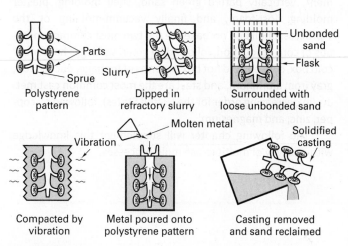

FIGURE 14.31 Schematic of the lost-foam casting process. In this process, the polystyrene pattern is dipped in a ceramic slurry, and the coated pattern is then surrounded with loose, unbonded sand.

The full-mold and lost-foam processes can produce castings of any size in both ferrous and nonferrous metals. Because the pattern need not be withdrawn, no draft is required in the design. Complex patterns can be produced to make shapes that would ordinarily require multiple cores, loose-piece patterns, or extensive finish machining. Component integration is a major attraction—multicomponent assemblies can often be replaced by a single casting. Because of the high precision and smooth surface finish, finishing operations can often be reduced or totally eliminated. Fragile or complex-geometry cores are no longer required, and the absence of parting lines eliminates the need to remove associated lines or fins on the metal casting.

Unlike gravity-pour casting, where the mold cavity fills from the bottom, the evaporative-pattern molds fill by the progressive melting and volatilization of the foam. As the molten metal progresses through the pattern, it loses heat. As a result, the material farthest from the gate is the coolest, and solidification tends to proceed in a directional manner back toward the gate. For many castings, risers are not required. Metal yield (product weight versus the weight of poured metal) tends to be rather high. For these and other reasons, evaporative-pattern casting has grown rapidly in popularity and use. **Table 14.9** summarizes the process and its capabilities.

In **pressurized lost foam casting**, the mold is placed in a pressure vessel, which closes within 5 seconds of pouring, and a pressure of approximately 10 atmospheres is applied during solidification. The applied pressure serves to reduce porosity of aluminum castings by more than an order of magnitude over that experienced in normal sand casting. Accompanying the decrease in porosity is a significant increase in high-cycle fatigue strength.

TABLE 14.9	Lost-Foam Casting

Process: A pattern containing a sprue, runners, and risers is made from single or multiple pieces of foamed plastic, such as polystyrene. It is dipped in a ceramic material, dried, and positioned in a flask, where it is surrounded by loose sand. Molten metal is poured directly onto the pattern, which vaporizes and is vented through the sand.

Advantages: Almost no limits on shape and size; most metals can be cast; no draft is required and no flash is present (no parting lines).

Limitations: Pattern cost can be high for small quantities; patterns are easily damaged or distorted because of their low strength.

Common metals: Aluminum, iron, steel, and nickel alloys; also performed with copper and stainless steel.

Size limits: 0.5 kg to several thousand kg (1 lb to several tons).

Thickness limits: As small as 2.5 mm (0.1 in.) with no upper limit.

Typical tolerances: 0.003 cm/cm (0.003 in./in.) or less.

Draft allowance: None required.

Surface finish: 2.5–25 microns (100–1000 μin.) rms.

14.6 Shakeout, Cleaning, and Finishing

In each of the casting processes presented in this chapter, the final step involves removing the castings from the molds and mold material. **Shakeout** operations are designed to separate the molds and sand from the flasks (i.e., containers), separate the castings from the molding sand, and remove the cores from within the castings. Punchout machines can be used to force the entire contents of a flask (both molding sand and casting) from the container. Vibratory machines, which can operate on either the entire flasks or the extracted contents, are available in a range of styles, sizes, and vibratory frequencies. Rotary separators remove the sand from castings by placing the mold contents inside a slow-turning, large-diameter, rotating drum. The tumbling action breaks the gates and runners from the castings, crushes lumps of sand, and extracts the cores. Because of possible damage to lightweight or thin-sectioned castings, rotary tumbling is usually restricted to cast iron, steel, and brass castings of reasonable thickness.

Processes such as blast cleaning can be used to remove adhering sand, oxide scale, and parting line burrs. Compressed air or centrifugal force is used to propel abrasive particles against the surfaces of the casting. The propelled media can be metal shot (usually iron or steel), fine aluminum oxide, glass beads, or naturally occurring quartz or silica. The blasting action may be combined with some form of tumbling or robotic manipulation to expose the various surfaces. Additional finishing operations may include grinding, trimming, or various forms of machining.

In an emerging process, known as **ablation**, the sand and binder of a sand mold is rapidly removed to facilitate rapid cooling of the casting. Shortly after the metal is poured, a liquid or liquid–gas mixture is directed onto the mold, where it dissociates the binder and allows the cooling liquid to come into direct contact with the metal. The high cooling rates produce a fine metal structure that enhances mechanical properties, and progressive movement of the ablation stream can create directional solidification.

14.7 Summary

Liquids have a characteristic property that they assume the shape of their container. A number of processes have been developed to create shaped containers and then utilize liquid fluidity and subsequent solidification to produce desired shapes. Each process has its unique set of capabilities, advantages, and limitations, and the selection of the best method for a given application requires an understanding of all possible options. Influencing factors include part size and complexity, tooling cost and lead time, finishing cost, the desired overall quantity, and the desired rate of production.

This chapter has presented processes that produce castings with a single-use (expendable) mold. Within those processes, horizontally parted green-sand molding is the most common, with no-bake molding a close second. Following in order of decreasing use are gas-hardened/cold-box, investment, vertically parted green sand, shell molding, plaster molding, lost-foam, and finally vacuum-molding or the V-process. Shell and no-bake are the two most common methods of core production, followed by green sand, gas hardened/cold box, core oil, and hot box. In terms of tonnage, ductile iron, gray iron, aluminum, and steel are the most common cast metals (in decreasing order for the United States), followed by copper, zinc, and magnesium.

The following chapter will supplement this knowledge with a survey of multiple-use mold processes.

Review Questions

1. What are some of the factors that influence the selection of a specific casting process as a means of making a product?

2. What are the three basic categories of casting processes when classified by molds and patterns?

3. What metals are frequently cast into products?

4. What features combine to make cast iron and aluminum the most common cast metals?

5. Which type of casting is the most common and most versatile?

6. What is a casting pattern?

7. What are some of the materials used in making casting patterns? What features should be considered when selecting a pattern material?

8. What is the simplest and least expensive type of casting pattern?

9. What is a match plate, and how does it aid molding?

10. How is a cope-and-drag pattern different from a match-plate pattern? When might this be attractive?

11. What is the benefit of incorporating gates, runners, and risers onto a match plate?

12. When might separate cope and drag patterns be used instead of a match plate?

13. For what types of products might a loose-piece pattern be required?

14. What are the four primary requirements of molding sand?

15. In what ways might molding sand be a compromise material?

16. What are the components of a typical green-sand mixture?

17. What is a muller, and what function does it perform?

18. What are some of the properties or characteristics of foundry sands that can be evaluated by standard tests?

19. What is a standard rammed specimen for evaluating foundry sands, and how is it produced?

20. What is permeability, and why is it important in molding sands?

21. How does the ratio of water to clay affect the compressive strength of green sand? What is a good starting point ratio?

22. How is the hardness of molding sand determined?

23. How might compactability correlate with moisture content?

24. How does the size and shape of the sand grains relate to molding sand properties?

25. What are the attractive features of silica sand? A negative characteristic?

26. What is a sand expansion defect, and what is its cause?

27. How can sand expansion defects be minimized?

28. What are the pros and cons of olivine and zircon sands?

29. What causes "blows" to form in a casting, and what can be done to minimize their occurrence?

30. What features can cause the penetration of molten metal between the grains of the molding sand?

31. What are hot tears, and what can cause them to form?

32. When might hand ramming be the preferred method of packing sand?

33. Describe the distribution of sand density after compaction by sand slinging, jolting, squeezing, and a jolt–squeeze combination.

34. Describe the sequence of activity in match-plate molding.

35. How does vertically parted flaskless molding reduce the number of mold sections required to produce a series of castings?

36. What concern has limited the acceptance of the H-process?

37. What is stack molding?

38. How might extremely large molds be made?

39. What are the components of green sand?

40. What are some of the limitations or problems associated with green sand as a mold material?

41. What restricts the use of dry-sand molding?

42. What is a skin-dried mold?

43. What are some of the advantages and limitations of the sodium silicate–CO_2 process?

44. What provides the strength of no-bake sands?

45. What features or characteristics might justify the higher cost of no-bake sands (compared to green sand)? Make it more attractive than sodium silicate–CO_2?

46. What are some other terms used to describe no-bake molding material?

47. What is bench life? How does it relate to the sodium silicate–CO_2 process? No-bake? What may determine the bench life of green sand?

48. What material serves as the binder in the shell-molding process, and how is it cured?

49. Describe the steps of the shell-molding process.

50. Why do shell molds have excellent permeability and collapsibility?

51. What is the sand binder in the V-process? The Eff-set process?

52. What are a few of the attractive features of the V-process?

53. What types of geometric features might require the use of cores?

54. What is the primary limitation of green-sand cores?

55. How might core segments be joined prior to insertion into the mold?

56. What is the sand binder in the core-oil process, and how is it cured?

57. What is the binder in the hot-box core-making process?

58. What is the primary attraction of the cold-box core-making process? The primary negative feature?

59. What is shelf life? How is it different from bench life?

60. What is an attractive feature of air-set or no-bake core making? Of shell-molded cores?

61. Why is it common for greater permeability, collapsibility, and refractoriness to be required of cores than for the base molding sand?

62. How can the properties of cores be enhanced beyond those offered by the various core materials?

63. What is the function of a core print?

64. What is the role of chaplets, and why is it important that they not completely melt during the pouring and solidification of a casting?

65. Describe the steps of plaster molding.

66. Why are plaster molds only suitable for the lower-melting-temperature nonferrous metals and alloys?

67. How does the Antioch process provide permeability to plaster molds?

68. What is the primary performance difference between plaster and ceramic molds?

69. How is permeability provided to ceramic molds through the Shaw process?

70. For what materials might a graphite mold be required?

71. What materials are used to produce the expendable patterns for investment casting?

72. Describe the progressive construction of an investment casting mold.

73. Why are investment casting molds generally preheated prior to pouring?

74. Why are investment castings sometimes called "lost-wax" castings?

75. What are some of the attractive features of investment casting?

76. What recent development has made one-of-a-kind or small quantities of investment castings an economic feasibility?

77. How can hollow-like investment casting patterns reduce the possibility of costly shell cracking?

78. What are some of the advantages of counter-gravity investment casting over the conventional gravity pour approach?

79. What are some of the benefits of not having to remove the pattern from the mold (as in investment casting, full-mold casting, and lost-foam casting)?

80. What are some of the ways by which expanded polystyrene patterns can be made?

81. Because both use expanded polystyrene as a pattern, what is the primary difference between full-mold and lost-foam casting?

82. What are some of the attractive features of the evaporative pattern processes?

83. What is the objective of pressurized lost foam casting?

84. What are some of the objectives of a shakeout operation?

85. How might castings be cleaned after shakeout?

86. What is the objective of ablation? The benefits?

87. What are the most common single-use mold processes?

88. What are the most common methods of core production?

Problems

1. Although cores increase the cost of castings, they also provide a number of distinct advantages. The most significant is the ability to produce complex internal passages. They can also enable the production of difficult external features, such as undercuts, or allow the production of zero draft walls. Cores can reduce or eliminate additional machining, reduce the weight of a casting, and reduce or eliminate the need for multipiece assembly. Answer the following questions about cores.

    a. The cores themselves must be produced, and generally have to be removed from coreboxes or molds. What geometric limitations might this impose? How might these limitations be overcome?

    b. Cores must be positioned and supported within a mold. Discuss some of the limitations associated with core positioning and orientation. Consider the weight of a core, prevention of core fracture, minimization of core deflection, and possible buoyancy.

    c. Because cores are internal to the casting, adequate venting is necessary to eliminate or minimize porosity problems. Discuss possible features to aid in venting.

    d. How might core behavior vary with different materials being cast—steel versus aluminum, for example?

    e. Discuss several of the reasons why cores may be made from a different material than the molding material used in the primary mold.

    f. Core removal is another design concern. Discuss how several different core-making processes might perform in the area of removal. What are some ways to assist or facilitate core removal?

2. Several of the additive manufacturing processes can selectively deposit binder material between the grains of sand in a layer-by-layer approach. Molds or cores can be produced directly from computer files, eliminating the need for any patterns or tooling.

    a. For what type or types of products might this approach be attractive?

    b. What concerns might you have regarding permeability, collapsibility, or any of the other mold or mold material characteristics?

3. Additive manufacturing processes can also build shapes from various polymers and metals directly from computer files. These products could be used as patterns for the various sand casting and core-making processes.

    a. Briefly discuss the advantages of this approach.

    b. For the various additive materials and the intended use as a sand pattern, discuss features such as cost, wear resistance and mold life, and other consideration factors.

    c. Wax or selected polymers could be used to produce the expendable patterns for investment casting. How might this expand the use of the investment casting process?

CHAPTER 15

Multiple-Use-Mold Casting Processes

15.1 Introduction

In each of the expendable-mold casting processes discussed in Chapter 14, a separate mold had to be created for each pour. Variations in mold consistency, mold strength, moisture content, pattern removal, and other factors contribute to dimensional and property variation from casting to casting. In addition, the need to create and then destroy a separate mold for each pour results in rather low production rates.

The multiple-use-mold casting processes overcome many of these limitations, but they, in turn, have their own assets and liabilities. Because the molds are generally made from metal, many of the processes are restricted to casting the lower-melting-point nonferrous metals and alloys. Part size is often limited, and the dies or molds can be rather costly.

15.2 Permanent-Mold Casting

In the **permanent-mold casting** process, also called *gravity die casting*, a reusable mold is machined from gray cast iron, alloy cast iron, steel, bronze, graphite, or other material. The molds are usually made in segments, which are often hinged to permit rapid and accurate opening and closing. After preheating, a refractory or other material coating is applied to the preheated mold, and the mold is clamped shut. Molten metal is then poured into the pouring basin and flows through the feeding system into the mold cavity by simple gravity flow. After solidification, the mold is opened, and the product is removed. Because the heat from the previous cast is usually sufficient to maintain mold temperature, the process can be immediately repeated, with a single refractory coating serving for several pouring cycles. Aluminum-, magnesium-, zinc-, lead-, and copper-based alloys are the metals most frequently cast. If graphite is used as the mold material, iron and steel castings can also be produced.

Numerous advantages can be cited for the permanent-mold process. Near-net shapes can be produced that require little finish machining. The mold is reusable, and a good surface finish is obtained if the mold is in good condition. Dimensions are consistent from part to part, and dimensional accuracy can often be held to within 0.25 mm (0.010 in.). Directional solidification can be achieved through good design and can be promoted by selectively heating or chilling various portions of the mold or by varying the thickness of the mold wall. The result is usually a sound, defect-free casting with good mechanical properties. The faster cooling rates of the metal mold produce a finer grain structure and higher-strength products than would result from a sand casting process. Because permanent mold products have lower levels of porosity compared to die castings, they can often acquire enhanced properties through heat treatment, and machining operations do not expose porosity. Cores of expendable sand, plaster, or retractable metal can be used to increase the complexity of the casting, and multiple cavities can often be included in a single mold. When sand cores are used, the process is often called **semipermanent-mold casting**.

On the negative side, the process is generally limited to the lower-melting-point alloys, and high mold costs can make low production runs prohibitively expensive. The useful life of a mold is generally set by molten metal erosion or thermal fatigue. When making products of steel or cast iron, mold life can be extremely short. For the lower-temperature metals, one can usually expect somewhere between 10,000 and 120,000 cycles. The actual mold life will depend on the following:

1. *Alloy being cast.* The higher the melting point, the shorter the mold life.

2. *Mold material.* Gray cast iron is used most frequently for permanent molds because it has about the best resistance to thermal fatigue and machines easily. Other materials include steel, bronze, and graphite.

3. *Pouring temperature.* Higher pouring temperatures reduce mold life, increase shrinkage problems, and induce longer cycle times.

4. *Mold temperature.* If the temperature is too low, one can expect misruns and large temperature differences in the

mold. If the temperature is too high, excessive cycle times result, and mold erosion is aggravated

5. *Mold configuration.* Differences in section sizes of either the mold or the casting can produce temperature differences within the mold and reduce its life.

The permanent molds contain the mold cavity, pouring basin, sprue, runners, risers, gates, possible core supports, alignment pins, and some form of ejection system. The molds are usually heated at the beginning of a run, and continuous operation then maintains the mold at a fairly uniform elevated temperature. This minimizes the degree of thermal fatigue, facilitates metal flow, and controls the cooling rate of the metal being cast. Because the mold temperature rises when a casting is produced, it may be necessary to provide a mold-cooling delay before the cycle is repeated. Refractory washes or graphite coatings can be applied to the mold walls to protect the die surface, control or direct the cooling, prevent the casting from sticking, improve surface finish, and prolong the mold life by minimizing thermal shock and fatigue. When pouring cast iron, an acetylene torch is often used to apply a coating of carbon black to the mold.

Because the molds are not permeable, special provision must be made for **venting**. This is usually accomplished through the slight cracks between mold halves or by very small vent holes that permit the escape of trapped air but not the passage of molten metal. Although turbulence-related defects are also more common with the gravity pour, these can often be minimized by reducing the free-fall height at the beginning of the pouring process and using ceramic-foam filters to smooth the flow of the molten metal. Because gravity is the only means of inducing metal flow, risers must still be employed to compensate for solidification shrinkage, and with the necessary sprues and runners, yields are generally less than 60%.

Mold complexity is often restricted because the rigid cavity offers no collapsibility to compensate for the solid-state shrinkage of the casting. As a best alternative, it is common practice to open the mold and remove the casting immediately after solidification. This prevents the formation of hot tears that may form if the product is restrained during the cooldown to room temperature.

For permanent-mold casting, high-volume production is usually required to justify the high cost of the metal molds. Automated machines can be used to coat the mold, pour the metal, and remove the casting. **Figure 15.1** shows several automotive pistons that are mass-produced by the permanent-mold process, which is summarized in **Table 15.1**.

Tilt-Pour, Low-Pressure, and Vacuum Permanent-Mold Casting

Gravity pouring is the oldest, simplest, and most traditional form of permanent-mold casting. In a variation known as **tilt-pour permanent-mold casting**, the mold is rotated into a horizontal position, and a cup attached to the mold is filled

FIGURE 15.1 An example of automotive pistons that have been mass-produced by the millions using permanent-mold casting. (Courtesy of United Engine & Machine Co., Carson City, NV)

TABLE 15.1	Permanent-Mold Casting

Process: Mold cavities are machined into mating metal die blocks, which are then preheated and clamped together. Molten metal is then poured into the mold and enters the cavity by gravity flow. After solidification, the mold is opened and the casting is removed.

Advantages: Good surface finish and dimensional accuracy; metal mold gives rapid cooling and fine-grain structure; multiple-use molds (up to 120, 000 uses); metal cores or collapsible sand cores can be used.

Limitations: High initial mold cost; shape, size, and complexity are limited; yield rate rarely exceeds 60%, but runners and risers can be directly recycled; mold life is very limited with high-melting-point metals such as steel.

Common metals: Alloys of aluminum, magnesium, and copper are most frequently cast; irons and steels can be cast into graphite molds; alloys of lead, tin, and zinc are also cast.

Size limits: 100 grams to 75 kilograms (several ounces to 150 pounds).

Thickness limits: Minimum depends on material but generally greater than 3 mm ($\frac{1}{8}$ in.); maximum thickness about 50 mm (2.0 in.).

Geometric limits: The need to extract the part from a rigid mold may limit certain geometric features. Uniform section thickness is desirable.

Typical tolerances: 0.4. mm for the first 2.5. cm (0.015 in. for the first inch) and 0.02 mm for each additional centimeter (0.002 in. for each additional inch); 0.25mm (0.01 in.) added if the dimension crosses a parting line.

Draft allowance: 2°–3°.

Surface finish: 4.0 to 12.0 μm (150–400 μin.) rms.

with molten metal. The mold is then tilted upward, enabling the molten metal to flow through the gating system and into the mold cavity with reduced free-fall and far less turbulence than when the metal is poured vertically down a sprue. **Figure 15.2** shows a tilt-pour permanent mold in its vertical position where the pouring cup extends to become the sprue and riser.

In **low-pressure permanent-mold (LPPM) casting** and **vacuum permanent-mold casting**, the mold is turned upside down and positioned above a sealed airtight chamber that contains a crucible of molten metal. In the low-pressure process,

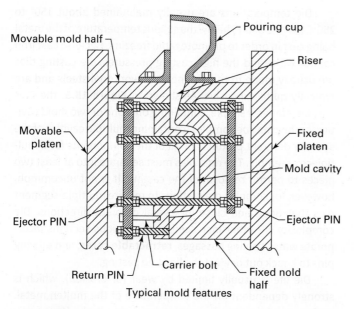

FIGURE 15.2 Schematic of the tilt-pour permanent-mold process, showing the mold assembly in its tilted or vertical position. (Courtesy of C.M.H. Manufacturing, Lubbock, TX)

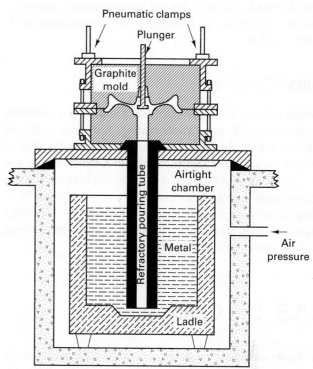

FIGURE 15.3 Schematic of the low-pressure permanent-mold process. (Courtesy of Amsted Industries, Chicago, IL)

illustrated in **Figure 15.3**, a low-pressure gas (150 to 775 torr or 3 to 15 psi) is introduced into the sealed chamber, causing the molten metal to flow upward through a refractory fill tube and into the gating system or cavity of the metal mold. This metal is exceptionally clean because it flows from the center of the melt and is fed directly into the mold (a distance of about 10 cm, or 3 to 4 in.), never passing through the atmosphere. Product quality is further enhanced by the nonturbulent mold filling that helps to minimize gas porosity and dross formation.

Through design and cooling, the products directionally solidify from the top down. The molten metal in the pressurized fill tube acts as a riser to continually feed the casting during solidification. When solidification is complete, the pressure is released, and the unused metal in the feed tube simply drops back into the crucible. The reuse of this metal, coupled with the absence of additional risers, leads to yields that are often greater than 85%.

Nearly all low-pressure permanent-mold castings are made from aluminum or magnesium, but some copper-base alloys can also be used. Mechanical properties are typically about 5% better than those of conventional permanent mold-castings. Cycle times are somewhat longer, however, than those of conventional permanent molding.

Figure 15.4 depicts a similar variation of permanent-mold casting where a vacuum is drawn on the die assembly, and atmospheric pressure in the chamber forces the metal upward. All of the benefits and features of the low-pressure process are retained, including the subsurface extraction of molten metal from the melt, the bottom feed

to the mold, the low metal turbulence during filling, the downward directional solidification, and the self-risering action. Thin-walled castings can be produced with high metal yield and excellent surface quality. Because of the vacuum, the cleanliness of the metal and the dissolved gas content are superior to that of the low-pressure process. Final castings typically range

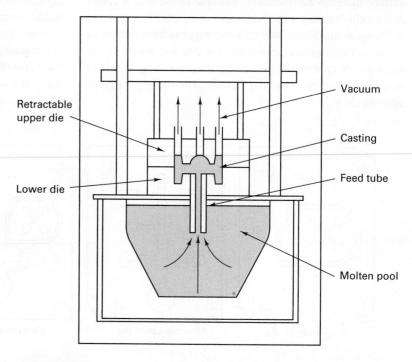

FIGURE 15.4 Schematic illustration of vacuum permanent-mold casting. Note the similarities to the low-pressure process.

from 0.2 to 5 kg (0.4 to 10 lb) and have mechanical properties that are even better than those of the low-pressure permanent-mold products.

Slush Casting

Hollow castings can be produced by a variant of permanent-mold casting known as **slush casting**. Hot metal is poured into the metal mold and is allowed to cool until a shell of desired thickness has formed. The mold is then inverted, and the remaining liquid is poured out. The resulting casting is a hollow shape with good surface detail but variable wall thickness. Common applications include the casting of ornamental objects such as candlesticks, lamp bases, and statuary from the low-melting-temperature metals.

15.3 Die Casting

In the **die-casting** process—or more specifically **pressure die casting**—molten metal is forced into preheated metal molds under pressures of tens of MPa (several thousand pounds per square inch) and held under high pressure during solidification. Most die castings are made from nonferrous metals and alloys, with special zinc-, copper-, magnesium-, and aluminum-based alloys having been designed to produce excellent properties when die cast. Ferrous-metal die castings are possible, but are generally considered to be uncommon. Production rates are high, the products exhibit good strength, shapes can be quite intricate, and dimensional precision and surface qualities are excellent. Because of the thin sections and excellent detail, there is almost a complete elimination of subsequent machining. Most die castings can be classified as small- to medium-sized parts, but the size and weight of die castings are continually increasing. Parts can now be made with weights up to 50 kg (100 lb) with dimensions as large as 600 mm (24 in.).

Die temperatures are usually maintained about 150° to 250°C (300° to 500°F) below the solidus temperature of the metal being cast in order to promote rapid freezing. Because cast iron cannot withstand the high casting pressures, die-casting dies are usually made from hardened hot-work tool steels and are typically quite expensive. As shown in **Figure 15.5**, the dies may be relatively simple, containing only one or two mold cavities, or they may be complex, containing multiple cavities of the same or different products or all of the components for a multipiece assembly. The rigid dies must separate into at least two pieces to permit removal of the casting. It is not uncommon, however, for complex die castings to require multiple-segment dies that open and close in several different directions. Die complexity is further increased as the various sections incorporate water-cooling passages, **retractable cores**, and moving pins to knock out or eject the finished casting.

Die life is usually limited by wear (or erosion), which is strongly dependent on the temperature of the molten metal. Surface cracking can also occur in response to the large number of heating and cooling cycles that are experienced by the die surfaces. If the rate of temperature change is the dominant feature, the problem is called **heat checking**. If the number of cycles is the primary cause, the problem is called **thermal fatigue**. Typical die life is 100,000 shots, and production runs of 10,000 pieces or more are generally required to justify die setup.

In the basic die-casting process, the mold cavity is first sprayed with a lubricant that helps control the temperature of the die and assists in the removal of the casting. The water-cooled dies are then clamped tightly together, and a premeasured amount of molten metal is injected under high pressure. Because high injection pressures cause turbulence and air entrapment, the specified values of pressure and the time and duration of application vary considerably. The pressure need not be constant, and there has been a trend toward the use of larger gates and lower injection pressures, followed by the application of higher pressure after the mold has completely filled and the metal has started to solidify. By reducing turbulence and solidifying under high pressure, both the porosity and inclusion content of the finished casting are reduced. After

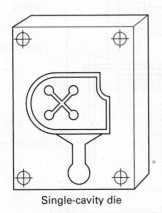 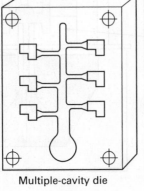 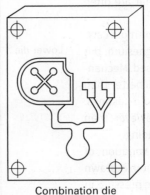 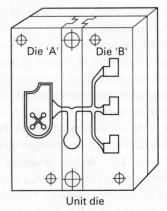

| Single-cavity die | Multiple-cavity die | Combination die | Unit die |

FIGURE 15.5 Various types of die-casting dies. (Courtesy of American Die Casting Institute, Inc., Des Plaines, IL)

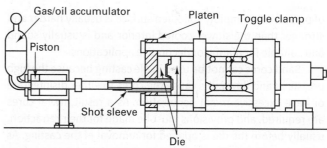

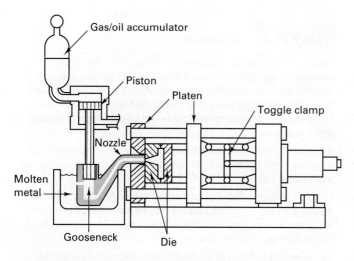

FIGURE 15.6 Principal components of a hot-chamber die-casting machine. (Adapted from *Metals Handbook,* 9th Ed., Vol. 15, p. 287, ASM International, Materials Park, OH)

FIGURE 15.7 Principal components of a cold-chamber die-casting machine. (Adapted from *Metals Handbook,* 9th Ed., Vol. 15, p. 287, ASM International, Materials Park, OH)

solidification is complete, the pressure is released, the dies separate, and ejector pins extract the finished casting along with its attached runners and sprues.

There are two basic types of die-casting machines. **Figure 15.6** schematically illustrates the **hot-chamber**, or **gooseneck**, variety. A gooseneck chamber is partially submerged in a reservoir of molten metal. With the plunger raised, molten metal flows through an open port and fills the chamber. The mechanical plunger then applies a pressure of about 20 MPa (3000 psi), which forces the metal up through the gooseneck, through the runners and gates, and into the die, where it rapidly solidifies. Retraction of the plunger allows the gooseneck to refill as the casting is being ejected, and the cycle repeats at speeds ranging from 100 shots per hour for large castings to over 1000 shots per hour for small products.

Hot-chamber die-casting machines offer fast cycling times (set by the ability of the water-cooled dies to cool and solidify the metal) and the added advantage that the molten metal is injected from the same chamber in which it is melted (i.e., there is no handling or transfer of molten metal and minimal heat loss). Unfortunately, the hot-chamber design cannot be used for the higher-melting-point metals, and it is unattractive for aluminum because the molten aluminum tends to pick up some iron during the extended time of contact with the casting equipment. Hot-chamber machines, therefore, see primary use with zinc-, tin-, and lead-based alloys.

Zinc die castings can also be made by a process known as **heated-manifold direct-injection die casting** (also known as direct-injection die casting or runnerless die casting). The molten zinc is forced through a heated manifold and then through heated mini-nozzles directly into the die cavity. This approach totally eliminates the need for sprues, gates, and runners. Scrap is reduced, energy is conserved (less molten metal per shot and no need to provide excess heat to compensate for cooling in the gating system), and product quality is increased. Existing die-casting machines can be converted

through the addition of a heated manifold and modification of the various dies.

Cold-chamber machines are usually employed for the die casting of materials that are not suitable for the hot-chamber design. These include alloys of aluminum, magnesium, and copper, high-aluminum zinc, and even some ferrous alloys. As illustrated in **Figure 15.7**, metal that has been melted in a separate furnace is transported to the die-casting machine, where a measured quantity is transferred to an unheated shot chamber (or injection cylinder) and subsequently driven into the die by a hydraulic or mechanical plunger under pressures of around 70 MPa (10,000 psi). The pressure is then maintained or increased until solidification is complete. Because molten metal must be transferred to the chamber for each shot, the cold-chamber process has a longer operating cycle compared to hot-chamber machines. Nevertheless, productivity is still high.

In all variations of the process, die-casting dies fill with metal so fast that there is little time for the air in the runner system and mold cavity to escape, and the metal molds offer no permeability. The air can become trapped and cause a variety of defects, including blow holes, porosity, and misruns. To minimize these defects, it is crucial that the dies be properly vented, usually by wide, thin (0.13 mm or 0.005 in.) vents positioned along the parting line. Proper positioning is a must because all of the air must escape before the molten metal contacts the vents. The long thin slots allow the escape of gas, but promote rapid freezing of the metal and a plugging of the hole. After the casting has been ejected, special trimming dies remove the metal that solidified in the vents along with the sprues and runners. The use of a vacuum during die filling (vacuum die casting), larger ingates and lower injection velocities (squeeze casting), or semisolid casting (processes to be discussed) are other methods to reduce entrapped gas and produce parts that can be heat treated or welded.

Risers are not used in the die-casting process because the high injection pressures ensure the continuous feed of molten metal from the gating system into the casting. The porosity that is often found in die castings is not shrinkage porosity, but is more likely to be the result of either entrapped air or the turbulent mode of die filling. This porosity tends to be confined to the interior of castings, and its formation can often be minimized by smooth metal flow, good venting, and proper application of pressure. The rapidly solidified surface is usually harder and stronger than the slower-cooled interior and is usually sound and suitable for plating or decorative applications.

Sand cores cannot be used in die casting because the high pressures and flow rates cause the cores to either disintegrate or have excessive metal penetration. As a result, metal cores are required, and provisions must be made for their retraction, usually before the die is opened for removal of the casting. As with all mating segments and moving components, a close fit must be maintained to prevent the pressurized metal from flowing into the gap. Loose core pieces (also metal) can also be positioned into the die at the beginning of each cycle and then removed from the casting after its ejection. This procedure permits more complex shapes to be cast, such as holes with internal threads, but production rate is slowed, and costs increase.

Cast-in **inserts** can also be incorporated in the die-casting process. Examples include prethreaded bosses, electrical heating elements, threaded studs, and high-strength bearing surfaces. These high-temperature components are positioned in the die before the lower-melting-temperature metal is injected. Suitable recesses must be provided in the die for positioning and support, and the casting cycle tends to be slowed by the additional operations.

Die casting produces more products than any other casting process. **Table 15.2** summarizes the key features. Attractive aspects include smooth surfaces and excellent dimensional accuracy. For aluminum-, magnesium-, zinc-, and copper-based alloys, linear tolerances of 3 mm/m (0.003 in./in.) are not uncommon. Thinner sections can be cast than with either sand or permanent mold casting. The minimum section thickness and draft vary with the type of metal, with typical values as follows:

Metal	Minimum Section	Minimum Draft
Aluminum alloys	0.89 mm (0.035 in.)	1:100 (0.010 in./in.)
Brass and bronze	1.27 mm (0.050 in.)	1: 80 (0.015 in./in.)
Magnesium alloys	1.27 mm (0.050 in.)	1:100 (0.010 in./in.)
Zinc alloys	0.63 mm (0.025 in.)	1:200 (0.005 in./in.)

Because of the precision and finish, most die castings require no finish machining except for the removal of excess metal fin, or flash, around the parting line and the possible drilling or tapping of holes. Production rates are high, and a set of dies can produce many thousands of castings without significant change in dimensions. Although die casting is most economical for large production volumes, quantities as low as 2000 can be justified if extensive secondary machining or surface finishing can be eliminated.

Thin-wall zinc die castings are now considered to be significant competitors to plastic injection moldings. The die castings are stronger, stiffer, more dimensionally stable, and more heat resistant. In addition, the metal parts are more resistant to ultraviolet radiation, weathering, and stress cracking when exposed to various reagents. The metal casting also provides an excellent base for a wide variety of decorative platings and coatings.

TABLE 15.2	**Die Casting**

Process: Molten metal is injected into closed metal dies under pressures ranging from 10 to 175 MPa (1500–25,000 psi). Pressure is maintained during solidification, after which the dies separate and the casting is ejected along with its attached sprues and runners. Cores must be simple and retractable and take the form of moving metal segments.

Advantages: Extremely smooth surfaces and excellent dimensional accuracy; rapid production rate; product tensile strengths as high as 415 Mpa (60 ksi).

Limitations: High initial die cost; limited to high-fluidity nonferrous metals; part size is limited; porosity may be a problem; some scrap in sprues, runners, and flash, but this can be directly recycled; large production quantity required; cores must pull straight out.

Common metals: Alloys of aluminum, zinc, magnesium, and lead; also possible with alloys of copper and tin.

Size limits: Less than 30 grams (1 oz) through 7 kg (15 lb) most common. Over 50 kg (100 lb) possible.

Thickness limits: As thin as 0.75 mm (0.03 in.), but generally less than 13 mm ($\frac{1}{2}$in.).

Typical tolerances: Varies with metal being cast; typically 0.1mm for the first 2.5 cm (0.005 in. for the first inch) and 0.02 mm for each additional centimeter (0.002 in. for each additional inch).

Draft allowances: 1°–3°.

Surface finish: 1–5.0 μm(40–200 μin.) rms.

Figure 15.8 presents a variety of aluminum and zinc die castings, and **Figure 15.9** shows a large high-pressure aluminum die casting of an automotive transmission. **Table 15.3** compares the key features of the four dominant families of die casting alloys, and **Table 15.4** compares the mechanical properties of various die cast alloys with the properties of other engineering materials.

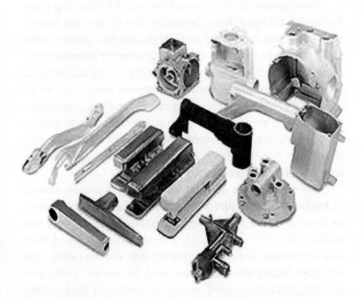

FIGURE 15.8 A variety of die cast products: zinc door handles and executive desk staplers; magnesium housings for pneumatic staplers and nailers; and aluminum aircraft food tray arms. (Courtesy of Chicago White Metal Casting, Inc., Bensenville, IL)

FIGURE 15.9 High-pressure aluminum die casting of an automotive transmission housing. Dimensions are 15 x 13 x 14.25 in. (Reproduced from *Metal Casting Design & Purchasing* by permission of the American Foundry Society, Schaumburg, IL)

TABLE 15.3	Key Properties of the Four Major Families of Die-Cast Metal
Metal	**Key Properties**
Aluminum	Lowest cost per unit volume; second lightest to magnesium; highest rigidity; good machinability, electrical conductivity, and heat-transfer characteristics; corrosion resistant.
Magnesium	Lowest density, faster production than aluminum since hot-chamber cast, highest strength-to-weight ratio, good vibration damping, best machinability, can provide electromagnetic shielding.
Zinc	Attractive for small parts; tooling lasts 3–5 times longer than for aluminum; heaviest of the die-castable metals but can be cast with thin walls for possible weight savings; good impact strength, machinability, electrical conductivity, and thermal conductivity; easily plated.
Zinc–Aluminum	Highest yield and tensile strength, lighter than conventional zinc alloys, good machinability.

15.4 Squeeze Casting and Semisolid Casting

Squeeze casting and **semisolid casting** are methods that enable the production of high-quality, near-net-shape, thin-walled parts with good surface finish, dimensional precision, and properties that approach those of forgings. Both processes can be viewed as derivatives of conventional high-pressure die casting because they employ tool steel dies and apply high pressure during solidification. Although the majority of applications involve alloys of aluminum, each of the processes has been successfully applied to magnesium, zinc, copper, and a limited number of ferrous alloys.

In the squeeze casting[1] process, molten metal is introduced into the die cavity of a metal mold, using large gate areas and slow metal velocities to avoid turbulence. When the cavity has filled, high pressure (20 to 175 MPa, or 3000 to 25,000 psi) is then applied and maintained during the subsequent solidification. Parts must be designed to directionally solidify toward the gates, and the gates must be sufficiently large that they freeze after solidification in the cavity, thereby allowing the pressurized runner to feed additional metal to compensate for shrinkage.

Intricate shapes can be produced at lower pressures than would normally be required for hot or cold forging. Both retractable and disposable cores can be used to create holes and internal passages. Gas and shrinkage porosity are substantially reduced, and mechanical properties are enhanced. In addition to the production of aluminum and magnesium castings, squeeze casting has also been adapted to the production of metal-matrix composites, where the pressurized metal is forced around or through foamed or fiber reinforcements that have been positioned in the mold.

For most alloy compositions, there is a range of temperatures where liquid and solid coexist, and several techniques have been developed to produce shapes from this semisolid material. In the **rheocasting** process, molten metal is cooled to the semisolid state with constant stirring. The stirring or shearing action breaks up the dendrites, producing a slurry of rounded particles of solid in a liquid melt. This slurry, with about a 30% solid content, can be readily shaped by high-pressure injection into metal dies. Because the slurry contains no superheat and is already partially solidified, it freezes quickly.

In the **thixocasting** variation, there is no handling of molten metal. The material is first subjected to special processing (stirring during solidification as in rheocasting) to produce solid blocks or bars with a nondendritic structure. The solid material is then cut to a prescribed length and reheated to a semisolid state where the material is about 40% liquid and 60% solid. In this condition, the **thixotropic material**[2] can be handled like a solid, but flows like a liquid when agitated or squeezed. It is then mechanically transferred to the shot chamber of a cold-chamber die-casting machine, and injected under pressure. In a variation of the process, known as **thixomolding**, solid metal granules or pellets are fed into a barrel chamber, where a rotating screw shears and advances the material through heating zones that raise the temperature to the semisolid region. When a sufficient volume of thixotropic material has

---

[1]The term *squeeze casting* has also been applied to another process that should, more appropriately, go by its other name, *liquid-metal forging*. Molten metal is poured into the bottom half of a preheated die set. As the metal starts to solidify, the upper die closes and applies pressure during the further solidification. The pressure is considerably less than that required for conventional forging, and parts of great detail can be produced. Cores can be used to produce holes and recesses. The process has been applied to both ferrous and nonferrous metals, but aluminum and magnesium are the most common materials.

[2]Thixotropic material flows or shears when under pressure, but thickens when stationary.

TABLE 15.4 Comparison of Properties (Die Cast Metals vs. Other Engineering Materials)

Material	Yield Strength		Tensile Strength		Elastic Modulus	
	MPa	ksi	MPa	ksi	GPa	10⁶ psi
Die-cast alloys						
360 aluminum	170	25	300	44	71	10.3
380 aluminum	160	23	320	46	71	10.3
AZ91D magnesium	160	23	230	34	45	6.5
Zamak 3 zinc (AG4OA)	221	32	283	41	—	—
Zamak 5 zinc (AC41A)	269	39	328	48	—	—
ZA-8 (zinc–aluminum)	283–296	41–43	365–386	53–56	85	12.4
ZA-27 (zinc–aluminum)	359–379	52.–55	407–441	59–64	78	11.3
Other metals						
Steel sheet	172–241	25–35	276	40	203	29.5
HSLA steel sheet	414	60	414	60	203	29.5
Powdered iron	483	70	—	—	120–134	17.5–19.5
Plastics						
ABS	—	—	55	8	7	1.0
Polycarbonate	—	—	62	9	7	1.0
Nylon 6[a]	—	—	152	22	10	1.5
PET[a]	—	—	145	21	14	2.0

[a]30% glass reinforced.

accumulated at the end of the barrel, a shot system drives it into the die or mold at velocities of 1 to 2.5 m/sec (40–100 in./sec). The injection system of this process is a combination of the screw feed used in plastic injection molding and the plunger used in conventional die casting.

In all the semisolid casting processes, the absence of turbulent flow during the casting operation minimizes gas pickup and entrapment. Because the material is already partially solid, the lower injection temperatures and reduced solidification time act to extend tool life. The prior solidification coupled with further solidification under pressure results in a significant reduction in solidification shrinkage and related porosity. The minimization of porosity enables the use of high-temperature heat treatments, such as the T6 solution treatment and artificial aging of aluminum, to further enhance strength, producing mechanical properties that are superior to those of other casting processes. Because the thixocasting process does not use molten metal, both wrought and cast alloys have been successfully shaped. Walls have been produced with thickness as low as 0.2 mm (0.01 in.).

15.5 Centrifugal Casting

The inertial forces of rotation or spinning are used to distribute the molten metal into the mold cavity or cavities in the **centrifugal casting** processes, a category that includes true centrifugal casting, semicentrifugal casting, and centrifuging. In **true centrifugal casting**, a dry-sand, graphite, or metal mold is rotated about either a horizontal or vertical axis at speeds of 300 to 3000 rpm. As the molten metal is introduced, it is flung to the surface of the mold, where it solidifies into some form of hollow product. The exterior profile is usually round (as with pipes, tubes, and gun barrels), but hexagons and other symmetrical shapes are also possible.

No core or mold surface is needed to shape the interior, which will always have a round profile because the molten metal is uniformly distributed by the centrifugal forces. When rotation is about the horizontal axis, as illustrated in **Figure 15.10**, the inner surface is always cylindrical. If the mold is oriented vertically, as in **Figure 15.11**, gravitational forces cause the

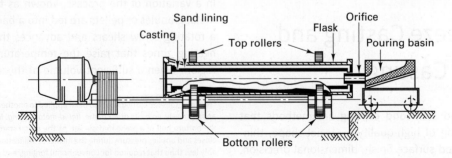

FIGURE 15.10 Schematic representation of a horizontal centrifugal casting machine. (Courtesy of American Cast Iron Pipe Company, Birmingham, AL)

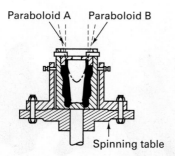

Paraboloid A   Paraboloid B

Spinning table

FIGURE 15.11 Vertical centrifugal casting, showing the effect of rotational speed on the shape of the inner surface. Paraboloid A results from fast spinning, whereas slower spinning will produce paraboloid B.

inner surface to become parabolic, with the exact shape being a function of the speed of rotation. Horizontal machines are preferred for long, thin cylinders; vertical machines, for ring-shaped products. Wall thickness can be controlled by varying the amount of metal that is introduced into the mold.

During the rotation, the metal is forced against the outer walls of the mold with considerable force, and solidification begins at the outer surface. Centrifugal force continues to feed molten metal as solidification progresses inward. Because the process compensates for shrinkage, no risers are required. The final product has a strong, fine-grained exterior with all of the lighter impurities (including dross and pieces of the refractory mold coating) collecting on the inner surface of the casting. This surface is often left in the final casting, but for some products it may be removed by a light boring operation.

Products can have outside diameters ranging from 7.5 cm. to 1.4 meters (3 to 55 in.) and wall thickness up to 25 cm (10 in.). Pipe (up to 12 m, or 40 ft., in length), pressure vessels, cylinder liners, brake drums, the starting material for bearing rings, and all of the parts illustrated in **Figure 15.12** can be manufactured by centrifugal casting. The equipment is rather specialized and can be quite expensive for large castings. The permanent molds can also be expensive, but they offer a long service life, especially when coated with some form of refractory dust or wash. Because no sprues, gates, or risers are required, yields can be greater than 90%. Composite products can also be made by the centrifugal casting of a second material on the inside surface of an already cast product giving different properties on the outer and inner diameters. **Table 15.5** summarizes the features of the centrifugal casting process.

In **semicentrifugal casting** (**Figure 15.13**) the centrifugal force assists the flow of metal from a central reservoir to the extremities of a rotating, vertical axis, symmetrical mold. The rotational speeds are usually lower than for true centrifugal casting, and the molds may be either expendable or multiple-use. Several molds may also be stacked on top of one another, so they can be fed by a common pouring basin and sprue. In general, the mold shape is more complex than for true centrifugal casting, and cores can be placed in the mold to further increase the complexity of the product.

FIGURE 15.12 Brass and bronze bushings that have been produced by centrifugal casting. (Courtesy Machine Rebuilding Services, Milford, CT)

TABLE 15.5	Centrifugal Casting

Process: Molten metal is introduced into a rotating sand, metal, or graphite mold and held against the mold wall by centrifugal force until it is solidified.

Advantages: Can produce a wide range of cylindrical parts, including ones of large size; good dimensional accuracy, soundness, and cleanliness.

Limitations: Shape is limited; spinning equipment can be expensive.

Common metals: Iron; steel; stainless steel; and alloys of aluminum, copper, and nickel.

Size limits: Up to 3 m (10 ft) in diameter and 15 m (50 ft) in length.

Thickness limits: Wall thickness 2.5 to 125 mm (0.1–5 in.).

Typical tolerances: O.D. to within 2.5 mm (0.1 in.); I.D. to about 4 mm (0.15 in.).

Draft allowance: 10 mm/m ($\frac{1}{8}$ in./ft); 0°–1°

Surface finish: 2.5–12.5 μm (100–500 μin.) rms.

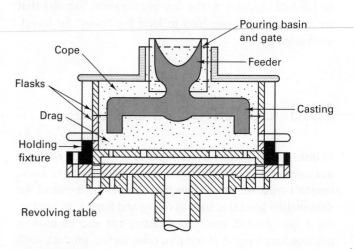

FIGURE 15.13 Schematic of a semicentrifugal casting process.

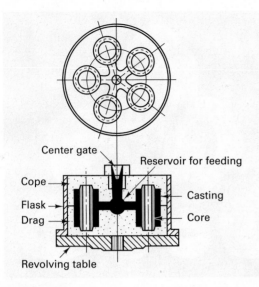

FIGURE 15.14 Schematic of a centrifuging process. Metal is poured into the central pouring sprue and spun into the various mold cavities. (Courtesy of American Cast Iron Pipe Company, Birmingham, AL)

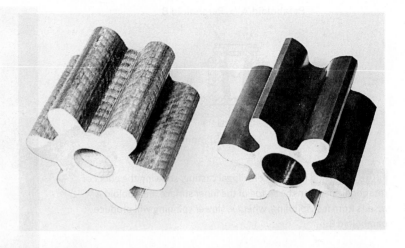

FIGURE 15.15 Gear produced by continuous casting. (Left) As-cast material; (right) after machining. (Courtesy of ASARCO, Tucson, AZ)

The central reservoir acts as a riser and must be large enough to ensure that it will be the last material to freeze. Because the lighter impurities concentrate in the center, the process is best used for castings where the central region will ultimately be hollow. Common products include gear blanks, pulley sheaves, wheels, impellers, and electric motor rotors.

Centrifuging, or *centrifuge centrifugal casting* (**Figure 15.14**) uses centrifugal action to force metal from a central pouring reservoir or sprue, through spoke-type runners, into separate mold cavities that are offset from the axis of rotation. Relatively low rotational speeds are required to produce sound castings with thin walls and intricate shapes. Centrifuging is often used to assist in the pouring of multiple-product investment casting trees.

Centrifuging can also be used to drive pewter, zinc, or wax into spinning rubber molds to produce products with close tolerances, smooth surfaces, and excellent detail. These can be finished products or the low-melting-point patterns that are subsequently assembled to form the "trees" for investment casting.

15.6 Continuous Casting

As discussed in Chapter 6 and depicted in Figure 6.3, **continuous casting** is usually employed in the solidification of basic, constant-cross-section, shapes that become the feedstock for deformation processes such as rolling and forging. By producing a special mold, continuous casting can also be used to produce long lengths of complex cross-section product, such as the one depicted in **Figure 15.15**. Because each product is simply a cutoff section of the continuous strand, a single mold is all that is required to produce a large number of pieces.

Near-net-shape continuous casting can significantly reduce the number of intermediate steps in the manufacture of a product, simultaneously reducing material losses, energy consumption and required manpower. Quality is high as well because the metal can be protected from contamination during melting and pouring, and only a minimum of handling is required.

Other variations of the continuous casting process include beam-blank casting (where the cast product has the cross-section profile of a structural beam), thin slab and strip casting (currently enabling the casting of material with thickness as small as 2 mm or 0.08 in.), and wire and rod casting. Hot-rolling operations can be reduced or eliminated; low volumes can be economically produced; and products can be made from materials that are not easily hot-rolled.

15.7 Melting

All casting processes begin with molten metal, which ideally should be available in an adequate amount, at the desired temperature, with the desired chemistry and minimum contamination. The melting furnace should be capable of holding material for an extended period of time without deterioration of quality, be economical to operate, and be capable of being operated without contributing to the pollution of the environment. Except for experimental or very small operations, virtually all foundries use cupolas, air furnaces (also known as direct fuel-fired furnaces), electric-arc furnaces, electric resistance furnaces, or electric-induction furnaces. In locations such as primary steel mills, molten metal may be taken directly from a steelmaking furnace and poured into casting molds. This practice is usually reserved for exceptionally large castings. For small operations, gas-fired crucible furnaces are also common, but these have rather limited capacities. Furnace operations generally include melting the charge, refining the melt (removing deleterious gases and elements from the molten metal), adjusting the melt chemistry, and tapping to a transport vessel.

Selection of the most appropriate melting method depends on such factors as (1) the temperature needed to melt and superheat the metal, (2) the alloy being melted and the form of available charge material, (3) the desired melting rate or the desired quantity of molten metal, (4) the desired quality of the metal, (5) the availability and cost of various fuels, (6) the variety of metals or alloys to be melted, (7) whether melting is to be batch or continuous, (8) the required level of emission control, and (9) the various capital and operating costs.

The feedstock entering the melting furnace may take several forms. Although prealloyed ingot may be purchased for remelt, it is not uncommon for the starting material to be a mix of commercially pure primary metal and commercial scrap, along with recycled gates, runners, sprues, and risers, as well as defective castings. The chemistry can be adjusted through alloy additions in the form of either pure materials or master alloys that are high in a particular element but are designed to have a lower melting point than the pure material and a density that allows for good mixing. Preheating the metal being charged is another common practice, and it can increase the melting rate of a furnace by as much as 30%.

Cupolas

A significant amount of gray, nodular, and white cast iron is still melted in **cupolas**, although many foundries have converted to electric induction furnaces. A cupola is a refractory-lined, vertical steel shell into which alternating layers of coke (carbon), iron (pig iron and/or scrap), limestone or other flux, and possible alloy additions are charged and melted under forced air draft. The operation is similar to that of a blast furnace, with the molten metal collecting at the bottom of the cupola to be tapped off either continuously or at periodic intervals.

Cupolas are simple and economical, can be obtained in a wide range of capacities, and can produce cast iron of excellent quality if the proper raw materials are used and good control is practiced. Control of temperature and chemistry can be somewhat difficult, however. The nature of the charged materials and the reactions that occur within the cupola can all affect the product chemistry. Moreover, by the time the final chemistry is determined through analysis of the tapped product, a substantial charge of material is already working its way through the furnace. Final chemistry adjustments, therefore, are often performed in the ladle, using the various techniques of ladle metallurgy discussed in Chapter 6.

Various methods can be used to increase the melting rate and improve the economy of a cupola operation. In a hot-blast cupola, the stack gases are put through a heat exchanger, to preheat the incoming air. Oxygen-enriched blasts can also be used to increase the temperature and accelerate the rate of melting. Plasma torches can be employed to melt the iron scrap. With typical enhancements, the melting rate of a continuously operating cupola can be quite high, such that production of 120 tons of hot metal per hour is not uncommon.

Indirect Fuel-Fired Furnaces (or Crucible Furnaces)

Small batches of nonferrous metal are often melted in **indirect fuel-fired furnaces** that are essentially crucibles or holding pots whose outer surface is heated by an external flame. The containment crucibles are generally made from clay and graphite, silicon carbide, cast iron, or steel. Stirring action, temperature control, and chemistry control are often poor, and furnace size and melting rate are limited. Nevertheless, these furnaces do offer low capital and operating cost. Electrical resistance heating can improve the control of both temperature and chemistry.

Direct Fuel-Fired Furnaces or Reverberatory Furnaces

In **direct fuel-fired furnaces**, also known as **reverberatory furnaces** and illustrated in **Figure 15.16**, combustion gases from a fuel-fired flame pass directly over the pool of molten metal, with heat being transferred to the metal through both radiant heating from the refractory roof and walls and convective heating from the hot gases. Capacity is significantly greater than the crucible furnace, but the operation is still limited to the batch melting of nonferrous metals and the holding of cast iron that has been previously melted in a cupola. The rate of heating and melting and the temperature and composition of the molten metal are all easily controlled.

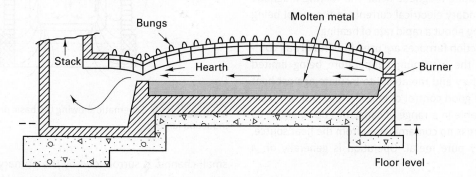

FIGURE 15.16 Cross section of a direct fuel-fired furnace. Hot combustion gases pass across the surface of a molten metal pool.

Arc Furnaces

Arc furnaces are the preferred method of melting in many foundries because of the (1) rapid melting rates, (2) ability to hold the molten metal for any desired period of time, and (3) greater ease of incorporating pollution control equipment. The basic features and operating cycle of a *direct-arc furnace* can be described with the aid of **Figure 15.17**. The top of the wide, shallow unit is first lifted or swung aside to permit the introduction of charge material. The top is then repositioned, and the electrodes are lowered to create an arc between the electrodes and the metal charge. The path of the heating current is usually through one electrode, across an arc to the metal charge, through the metal charge, and back through another arc to another electrode.

Fluxing materials are usually added to create a protective slag over the pool of molten metal. Reactions between the slag and the metal serve to further remove impurities, and are efficient because of the large interface area and the fact that the slag is as hot as the metal. Because the metal is covered and can be maintained at a given temperature for long periods of time, arc furnaces can be used to produce high-quality metal of almost any desired composition. They are available in sizes up to about 200 tons (but capacities of 25 tons or less are most common), and up to 50 tons per hour can be melted conveniently in batch operations. Arc furnaces are generally used with ferrous alloys, especially steel, and provide good mixing and homogeneity to the molten bath. Unfortunately, the noise and level of particle emissions can be rather high, and the consumption of electrodes, refractories, and power results in high operating costs. **Figure 15.18** shows the pouring of an electric arc furnace. Note the still-glowing electrodes at the top of the furnace.

Induction Furnaces

Because of their very rapid melting rates and the relative ease of controlling pollution, electric **induction furnaces** have become another popular means of melting metal. There are two basic types of induction furnaces. The *high-frequency*, or *coreless* units, shown schematically in **Figure 15.19**, consist of a crucible surrounded by a water-cooled coil of copper tubing. A high-frequency electrical current passes through the coil, creating an alternating magnetic field. The varying magnetic field induces secondary electrical currents in the metal being melted, which bring about a rapid rate of heating.

Coreless induction furnaces are used for virtually all common alloys, with the maximum temperature being limited only by the refractory and the ability to insulate against heat loss. They provide good control of temperature and composition and are available in a range of capacities up to about 65 tons. Because there is no contamination from the heat source, they produce very pure metal. Operation is generally on a batch basis.

Low-frequency or *channel-type* induction furnaces are also seeing increased use. As shown in **Figure 15.20**, only a

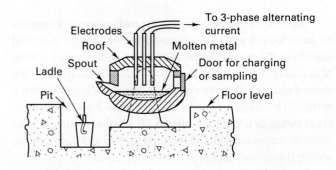

FIGURE 15.17 Schematic diagram of a three-phase electric-arc furnace.

FIGURE 15.18 Electric-arc furnace, tilted for pouring. (Courtesy of Lectromelt Corporation, Pittsburgh, PA)

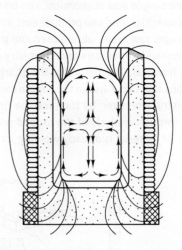

FIGURE 15.19 Schematic showing the basic principle of a coreless induction furnace.

small channel is surrounded by the primary (current-carrying or heating) coil. A secondary coil is formed by a loop, or channel, of molten metal, and all the liquid metal is free to circulate

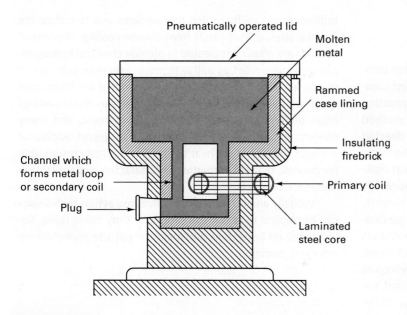

FIGURE 15.20 Cross section showing the principle of the low-frequency or channel-type induction furnace.

through the loop and gain heat. To start, enough molten metal must be placed into the furnace to fill the secondary coil, with the remainder of the charge taking a variety of forms. The heating rate is high and the temperature can be accurately controlled. As a result, channel-type furnaces are often preferred as holding furnaces, where the molten metal is maintained at a constant temperature for an extended period of time. Capacities can be quite large, up to about 250 tons.

15.8 Pouring Practice

Some type of pouring device, or ladle, is usually required to transfer the metal from the melting furnace to the molds. The primary considerations for this operation are (1) to maintain the metal at the proper temperature for pouring, and (2) to ensure that only high-quality metal is introduced into the molds. The specific type of **pouring ladle** is determined largely by the size and number of castings to be poured. In small foundries, a handheld, shank-type ladle is used for manual pouring. In larger foundries, either bottom-pour or teapot-type ladles are used, like the ones illustrated in Figure 13.6. These are often used in conjunction with a conveyor line that moves the molds past the pouring station. Because metal is extracted from beneath the surface, slag and other impurities that float on top of the melt are not permitted to enter the mold. To prevent the formation of new oxide inclusions, some form of protective shroud is often used to isolate the molten metal as it transfers from the ladle to the mold.

High-volume, mass-production foundries often use automatic pouring systems, like the one shown in **Figure 15.21**. Molten metal is transferred from a main melting furnace to a holding furnace. A programmed amount of molten metal is further transferred into individual pouring ladles and is then poured into the corresponding molds as they traverse by the pouring station. Laser-based control units position the pouring ladle over the sprue and control the flow rate into the pouring cup and fill height in the mold. Load cells can monitor the weight poured per mold.

15.9 Cleaning, Finishing, Heat Treating, and Inspection

Cleaning and Finishing

After solidification and removal from the mold, most castings require some additional cleaning and finishing. Specific operations may include all or several of the following:

1. Removing cores
2. Removing gates and risers
3. Removing fins, flash, and rough spots from the surface
4. Cleaning the surface

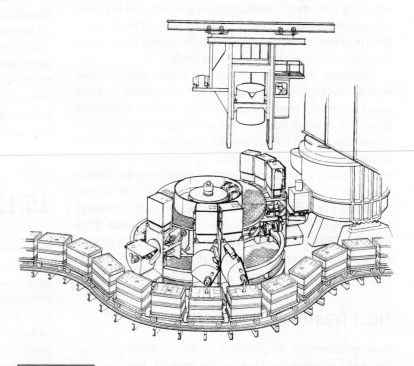

FIGURE 15.21 Automatic pouring of molds on a conveyor line. (Courtesy of Roberts Sinto Corporation, Lansing, MI)

5. Machining operations

6. Repairing any defects

Because cleaning and finishing operations can often comprise a considerable portion of the final cost of a cast component, consideration should be given to their minimization when designing the product and selecting the specific method of casting. In addition, consideration should also be directed toward the possibility of automating the cleaning and finishing.

Sand cores can usually be removed by mechanical shaking. At times, however, they must be removed by chemically dissolving the core binder. On small castings, sprues, runners, gates, and risers can sometimes be knocked off. For larger castings, a cutting operation is usually required. Most nonferrous metals and cast irons can be cut with an abrasive cutoff wheel, power hacksaw, or band saw. Steel castings frequently require an oxyacetylene torch. Plasma arc cutting can also be used. For high-volume castings, a trim press can be used to shear off the gates, riser connections, and parting-line material. After the initial removal, the remaining segments of gates and excess material along parting lines are generally removed by abrasive grinding.

The specific method of cleaning often depends on the size and complexity of the casting. After the gates and risers have been removed, small castings are often tumbled in barrels to remove fins, flash, and sand that may have adhered to the surface. Tumbling may also be used to remove cores and, in some cases, gates and risers. Metal shot or abrasive material is often added to the barrel to aid in the cleaning. Conveyors can be used to pass larger castings through special cleaning chambers, where they are subjected to blasts of abrasive or cleaning material either propelled by compressed air or hurled by a shot wheel. Extremely large castings usually require manual finishing, using pneumatic chisels, portable grinders, and manually directed blast hoses. Grinding, sanding or machining operations may then be required to achieve the desired dimensional accuracies, physical shape, or surface finish.

Although defect-free castings are always desired, flaws such as cracks, voids, and laps are not uncommon. In some cases, especially when the part is large and the production quantity is small, it may be more attractive to repair the part rather than change the pattern, die, or process. If the material is weldable, repairs are often made by removing the defective region (usually by chipping or grinding) and filling the created void with deposited weld metal. Porosity that is at or connected to free surfaces can be filled with resinous material, such as polyester, by a process known as **impregnation**. If the pores are filled with a lower-melting-point metal, the process becomes **infiltration**. (See Chapter 16 for a further discussion of these processes.)

Heat Treatment and Inspection

Heat treatment is an attractive means of altering properties while retaining the shape of the product. Steel castings are frequently given a full anneal to reduce the hardness and brittleness of rapidly cooled thin sections and to reduce the internal stresses that result from uneven cooling. Nonferrous castings are often heat-treated to provide chemical homogenization or stress relief, as well as to prepare them for subsequent machining. For final properties, virtually all of the treatments discussed in Chapter 5 can be applied. Ferrous-metal castings often undergo a quench-and-temper treatment, and many nonferrous castings are age hardened to impart additional strength. The variety of heat treatments is largely responsible for the wide range of properties and characteristics available in cast metal products.

Virtually all of the nondestructive **inspection techniques** can be applied to cast metal products. X-ray radiography, liquid penetrant inspection, and magnetic particle inspection are extremely common.

15.10 Automation in Foundry Operations

Many of the operations that are performed in a foundry are ideally suited for robotic automation because they tend to be dirty, dangerous, or dull. Robots can dry molds, coat cores, vent molds, and clean or lubricate dies. They can tend stationary, cyclic equipment, such as die-casting machines, and if the machines are properly grouped, one robot can often service two or three machines. In the investment casting process, robots can be used to dip the wax patterns into refractory slurry and produce the desired molds. In a similar manner, robots have been used in the full mold and lost foam processes to dip the Styrofoam patterns in their refractory coating and hang them on conveyors to dry. In a fully automated lost-foam operation, robots could be used to position the pattern, fill the flask with sand, pour the metal, and use a torch to remove the sprue. In the finishing room, robots can be equipped with plasma cutters or torches to remove sprues, gates, and runners. They can perform grinding and blasting operations, as well as various functions involved in the heat treatment of castings.

15.11 Process Selection

As shown in the individual process summaries that have been included throughout Chapters 14 and 15, each of the casting processes has a characteristic set of capabilities, assets, and limitations. The requirements of a particular product (such as size, complexity, required dimensional precision, desired surface finish, total quantity to be made, and desired rate of production) often limit the number of processes that should be considered as production candidates. Further selection is usually based on cost.

Some aspects of product cost, such as the cost of the material and the energy required to melt it, are somewhat independent of the specific process. The cost of other features, such as patterns, molds, dies, melting and pouring equipment, scrap material, cleaning, inspection, and all related labor, can vary markedly and be quite dependent on the process. For example, pattern and mold costs for sand casting are quite a bit less than the cost of die-casting dies. Die casting, on the other hand, offers high production rates and a high degree of automation. When a small quantity of parts is desired, the cost of the die or tooling must be distributed over the total number of parts, and unit cost (or cost per casting) is high. When the total quantity is large, the tooling cost is distributed over many parts, and the cost per piece decreases.

Figure 15.22 shows the relationship between unit cost and production quantity for a product that can be made by both sand and die casting. Sand casting is an expendable mold process. Because an individual mold is required for each pour, increasing quantity does not lead to a significant drop in unit cost. Die casting involves a multiple-use mold, and the cost of the die can be distributed over the total number of parts. As shown in the figure, sand casting is often less expensive for small production runs, and processes such as die casting are preferred for large quantities. One should note that although the die-casting curve in Figure 15.22 is a smooth line, it is not uncommon for an actual curve to contain abrupt discontinuities. If the lifetime of a set of tooling is 50,000 casts, the cost *per part* for 45,000 pieces, using one set of tooling, would actually be less than for 60,000 pieces because the latter would require a second set of dies.

In most cases, multiple processes are reasonable candidates for production, and the information for all the options should be considered. The final selection is often based on a combination of economic, technical, and management considerations.

The summary section at the close of Chapter 14 listed the various expendable-mold casting processes in order of decreasing usage. Among the multiple-use-mold processes, pressure die casting is clearly dominant, followed in decreasing order by: permanent mold casting (both gravity and tilt-pour), low-pressure and vacuum permanent mold, centrifugal, squeeze and semisolid, and continuous.

Table 15.6 presents a comparison of the more dominant casting processes, including green-sand casting, chemically bonded sand molds (shell, sodium silicate, and air-set), ceramic mold and investment casting, permanent mold casting, and die casting. The processes are compared on the basis of cost for both small and large quantities, thinnest section, dimensional precision, surface finish, ease of casting a complex shape, ease of changing the design while in production, and range of castable materials.

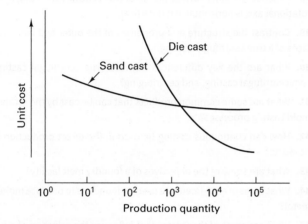

FIGURE 15.22 Typical unit cost of castings comparing sand casting and die casting. Note how the large cost of a die-casting die diminishes as it is spread over a larger quantity of parts.

TABLE 15.6 **Comparison of Casting Processes**

Property or Characteristic	Green-Sand Casting	Chemically Bonded Sand (Shell, Sodium Silicate, Air-Set)	Ceramic Mold and Investment Casting	Permanent-Mold Casting	Die Casting
Relative cost for small quantity	Lowest	Medium high	Medium	High	Highest
Relative cost for large quantity	Low	Medium high	Highest	Low	Lowest
Thinnest section (mm)	2.5	2.5	1.5	3.2	0.08
Dimensional precision (+/− in mm)	0.25–0.75	0.15–0.40	0.25–0.50	0.25–1.25	0.025–0.40
Relative surface finish	Fair to good	Good	Very good	Good	Best
Ease of casting complex shape	Fair to good	Good	Best	Fair	Good
Ease of changing design while in production	Best	Fair	Fair	Poor	Poorest
Castable metals	Unlimited	Unlimited	Unlimited	Low-melting-point metals	Low-melting-point metals

Review Questions

1. What are some of the major disadvantages of the expendable-mold casting processes?

2. What are some possible limitations of multiple-use molds?

3. What are some common mold materials for permanent-mold casting? What are some of the metals more commonly cast?

4. Describe some of the process advantages of permanent-mold casting.

5. Why do permanent-mold castings generally have higher strength than sand castings made from the same material?

6. What types of cores can be used in permanent mold casting?

7. What is semipermanent-mold casting?

8. Why might low production runs be unattractive for permanent-mold casting?

9. What features affect the life of a permanent mold?

10. In addition to the mold cavity, what additional features are typically contained within permanent molds?

11. How is venting provided in the permanent-mold process?

12. Why are permanent-mold castings generally removed from the mold immediately after solidification has been completed?

13. What is the benefit of the tilt-pour version of permanent-mold casting?

14. How does low-pressure permanent-mold casting differ from the traditional gravity-pour process?

15. What are some of the attractive features of the low-pressure permanent-mold process?

16. What are some additional advantages of vacuum permanent-mold casting over the low-pressure process?

17. What types of products would be possible candidates for manufacture by slush casting?

18. Contrast the feeding pressures on the molten metal in low-pressure permanent molding and die casting.

19. What are the most common die-cast materials?

20. Contrast the materials used to make dies for gravity-pour permanent-mold casting and die casting. Why is there a notable difference?

21. By what mechanisms do die-casting dies typically fail?

22. Why might it be advantageous to vary the pressure on the molten metal during the die-casting cycle?

23. Describe the casting cycle for hot-chamber die casting.

24. For what types of materials would a hot-chamber die-casting machine be appropriate?

25. What is the benefit of heated-manifold direct-injection die casting?

26. What metals are routinely cast with cold-chamber die-casting machines?

27. How does the air in the mold cavity escape in the die-casting process?

28. Are risers employed in die casting? Can sand cores be used?

29. What is the most likely source of the porosity observed in die castings?

30. Give some examples of cast-in inserts.

31. What are some of the attractive features of die casting compared to alternative casting methods?

32. When might low quantities be justified for the die-casting process?

33. In what ways might a thin-walled zinc die casting be more attractive than a plastic injection molding?

34. Describe the squeeze-casting process.

35. Describe rheocasting.

36. What is a thixotropic material? How does it provide an attractive alternative to squeeze casting or rheocasting?

37. What are some of the attractive features of semisolid casting?

38. Describe the inner surface profile of true centrifugal casting if the rotational axis is horizontal. If it is vertical.

39. Contrast the structure and properties of the outer and inner surfaces of a true centrifugal casting.

40. What are the key differences between true centrifugal casting, semicentrifugal casting, and centrifuging?

41. What are some common products that can be cast by the rotating mold casting processes?

42. How can continuous casting be used in the direct production of products?

43. What are some of the objectives of a foundry melt facility?

44. What types of equipment are used by foundries to produce molten metal?

45. What are some of the factors that influence the selection of a furnace type or melt procedure in a casting operation?

46. What are some of the possible feedstock materials that may be put in foundry melt furnaces?

47. What types of metals are commonly melted in cupolas?

48. What are some of the ways that the melting rate of a cupola can be increased?

49. What are some of the pros and cons of indirect fuel-fired furnaces?

50. What are some of the attractive features of arc furnaces in foundry applications?

51. What are some of the pros and cons of arc furnaces in a foundry operation?

52. Why are channel induction furnaces attractive for metal-holding applications where molten metal must be held at a specified temperature for long periods of time?

53. What are the primary functions of a pouring operation?

54. Why is it desirable to extract metal from the bottom of a pouring ladle, and not simply pour out of the top?

55. What are some of the typical cleaning and finishing operations that are performed on castings?

56. What are some common ways to remove cores from castings? To remove sprues, runners, gates, and risers?

57. What are some of the methods used to clean and finish castings?

58. How might defective castings be repaired to permit successful use in their intended applications?

59. What is the difference between infiltration and impregnation?

60. What are some of the heat treatments that are applied to metal castings?

61. What are some of the ways that industrial robots can be employed in metal-casting operations?

62. Describe some of the features that affect the cost of a cast product. Why might the cost vary significantly with the quantity to be produced?

63. What are some of the key factors that should be considered when selecting a casting process?

Problems

1. Attractive properties for casting alloys include low melting points, high fluidity (or runniness), and high as-solidified strength. Wrought alloys (those fabricated as solids by processes such as rolling, forging and extrusion) are best if they possess low yield strength, high ductility, and good strain-hardening characteristics. Reflecting these differences, the Aluminum Association uses distinctly different designation systems—a four-digit number for wrought alloys and a three-digit number for cast. Aluminum casting alloys are often further classified as "recommended for sand casting," "recommended for permanent mold casting," and "recommended for die casting." Why might different material properties be required for casting into the three different mold materials—sand, cast iron, and water-cooled tool steel? What features or characteristics would be desired for each? Why?

2. Small, intricate-shaped products can often be made by a variety of processes, including die casting, automated machining (screw machines), powder metallurgy, injection molding, and direct-digital or additive manufacturing. What features would favor manufacture by die casting over the alternative or competing processes? What features might favor each of the cited alternatives?

3. Identify a small component that you feel could be made as an injection-molded polymer, conventional zinc die casting, or thin-wall zinc die casting. For the part that you identified, discuss the pros and cons of the various options.

4. Using the Internet or other resource, identify a recent technical advance in one of the processes presented in this chapter. In what area or areas might this advance expand the capabilities of manufacturing?

5. Select two of the methods used to produce and hold molten metal, and compare their relative efficiencies. Under what conditions might the lower efficiency method actually be preferred?

CHAPTER 16

Powder Metallurgy (Particulate Processing)

16.1 Introduction

Powder metallurgy is the name given to the process by which fine particulate materials are blended, pressed into a desired shape (compacted), and then heated (sintered) in a controlled atmosphere to bond the contacting surfaces of the particles and establish desired properties. The process, commonly designated as P/M, readily lends itself to the mass production of small, intricate parts of high precision, often eliminating the need for additional machining or finishing. There is little material waste, unusual materials or mixtures can be utilized, and controlled degrees of porosity or permeability can be produced. Major areas of application tend to be those for which the P/M process has strong economical advantage (compared to machined components, castings or forgings) or where the desired properties and characteristics would be difficult to obtain by any other method (such as products made from tungsten, molybdenum, or tungsten carbide; porous bearings; filters; and various magnetic components). Because of its level of manufacturing maturity, powder metallurgy should actually be considered as a possible means of manufacture for any part where the geometry and production quantity are appropriate.

Mass manufacturing of P/M products did not begin until the mid- or late-nineteenth century, with early products including copper coins and medallions, platinum ingots, lead printing type, and tungsten wires, the primary material for lightbulb filaments. The tips of tungsten carbide cutting tools and nonferrous bushings were being produced by the 1920s. Self-lubricating bearings and metallic filters were other early products. A period of rapid technological development occurred after World War II, based primarily on automotive applications, and iron and steel replaced copper as the dominant P/M material. Aerospace and nuclear developments created demand for refractory and reactive metal components. Full-density products emerged in the 1960s, and high-performance superalloy components, such as aircraft turbine engine parts, were a highlight of the 1970s. Commercialization of rapidly solidified and amorphous powders, powder forging, warm compacting, and P/M injection molding followed in the 1980s

and 1990s. More recent technologies include submicron and nanophase powders; direct powder rolling; high-velocity and ultra-high-pressure compaction; and high-temperature, plasma- and microwave-sintering. With products now incorporating ceramics, ceramic fibers, and intermetallic compounds, some being entirely of nonmetallic material, the term *powder metallurgy* may need to be replaced by the broader term, **particulate processing**.

Most P/M parts weigh less than 2.25 kg (5 pounds), and many are still less than 50 mm (2 in.) in size. Some, however, have been produced with weights up to 45 kg (100 lb) with linear dimensions up to 500 mm (20 in.). Automotive applications now account for over 70% of the powder metallurgy market. Other areas where powder metallurgy products are used extensively include household appliances, recreational equipment, hand and power tools, hardware items, office equipment, industrial motors, and hydraulics. Areas of rapid growth include aerospace applications, advanced composites, electronic components, magnetic materials, metalworking tools, and a variety of biomedical and dental applications. Iron and low-alloy steels now account for 85% of all P/M usage, with copper and copper-based powders comprising about 7%. Stainless steel, high-strength and high-alloy steels, and aluminum and aluminum alloys are other high-volume materials. Titanium, magnesium, refractory metals, particulate composites, and intermetallics are seeing increased use.

16.2 The Basic Process

The powder metallurgy process generally consists of four basic steps: (1) powder manufacture, (2) mixing or blending, (3) compacting, and (4) sintering. Compaction is generally performed at room temperature, and the elevated temperature process of sintering is usually conducted at atmospheric pressure. Optional secondary processing often follows to obtain special properties or enhanced precision. **Figure 16.1** presents a simplified flowchart of the conventional press-and-sinter P/M process.

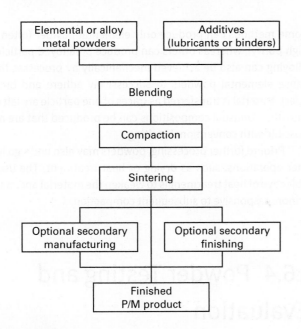

FIGURE 16.1 Simplified flowchart of the basic powder metallurgy process.

16.3 Powder Manufacture

The properties of powder metallurgy products are highly dependent on the characteristics of the starting powders. Some important properties and characteristics include **chemistry** and **purity**, **particle size**, **size distribution**, **particle shape**, and the **surface texture** of the particles. Several processes can be used to produce powdered material, with each imparting distinct properties and characteristics to the powder and hence to the final product.

More than 80% of all commercial powder is produced by some form of melt **atomization**, where liquid material is fragmented into small droplets that cool and solidify before they come into contact with each other or with a solid surface.

Various methods have been used to form the droplets, several of which are illustrated in **Figure 16.2**. Part (a) illustrates **gas atomization**, where jets of high pressure gas (usually nitrogen, argon, or helium) strike a stream of liquid metal as it emerges from an orifice. Pressurized liquid (usually water) can replace the pressurized gas, converting the process to **liquid atomization** or **water atomization**. In part (b), an electric arc impinges on a rapidly rotating electrode. Centrifugal force causes the molten droplets to fly from the surface of the electrode and freeze in flight. Particle size is very uniform and can be varied by changing the speed of rotation.

Regardless of the specific process, atomization is an extremely useful means of producing **prealloyed powders**. By starting with an alloyed melt or prealloyed electrode, each powder particle will have the desired alloy composition. Powders have been commercially produced from alloys of aluminum, copper, nickel, titanium, cobalt, zinc and tin, stainless steel and various low-alloy steels, and nonmetallic materials that can be melted to form the starting liquid. The size, shape, and surface texture of the powder particles varies, depending on such process features as the velocity and media of the atomizing jets or the speed of electrode rotation, the starting temperature of the liquid (which affects the time that surface tension can act on the individual droplets prior to solidification), and the environment provided for cooling. When cooling is slow (such as in gas atomization) and surface tension is high, smooth-surface spheres can form before solidification as shown in **Figure 16.3a**. With the more rapid cooling of water atomization, irregular shapes tend to be produced, as illustrated in **Figure 16.3b**. Still faster cooling can produce particles with ultrafine or microcrystalline grain size, or a noncrystalline (**amorphous**) structure. Both possibilities exhibit unusual and often very desirable properties.

Other methods of powder manufacture include the following:

1. *Chemical reduction of particulate compounds* (generally crushed oxides or ores). The powders that result from these

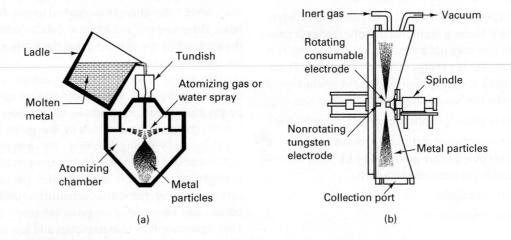

FIGURE 16.2 Two methods for producing metal powders: (a) melt atomization; (b) atomization from a rotating consumable electrode.

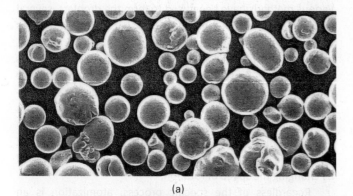

(a)

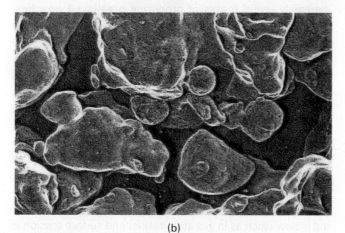

(b)

FIGURE 16.3 (a) Scanning electron microscope image of gas-atomized nickel powder showing the smooth surfaces that form by surface tension prior to solidification; (b) similar image of water-atomized iron powder showing the irregular shapes created by the rapid freezing of the shapes created by the water stream. (Courtesy Ronald Kohser)

solid-state reactions are usually soft, irregular in shape, and spongy in texture. Powder purity depends on the purity of the starting materials. A large amount of iron powder is produced by reducing iron ore or rolling mill scale.

2. *Electrolytic deposition* from solutions or fused salts with process conditions favoring the production of a spongy or powdery deposit that does not adhere to the cathode. Purity is generally high, but the energy required is also high. Therefore, electrolysis is usually restricted to the production of high-value powders, such as high-conductivity copper.

3. *Pulverization* or *grinding* of brittle materials (comminution).

4. *Thermal decomposition of particulate hydrides or carbonyls.* Iron and nickel powders are produced by carbonyl decomposition, resulting in extremely high purity.

5. *Precipitation from solution.*

6. *Condensation of metal vapors.*

Almost any metal, metal alloy, or nonmetal (ceramic, polymer, or wax or graphite lubricant) can be converted into powder form by one or more of the powder production methods.

Some methods can produce only elemental powder (often of high purity), whereas others can produce prealloyed particles. Alloying can also be achieved mechanically by processes that cause elemental powders to successively adhere and break apart. Material is transferred as traces of one particle are left on the other. Unusual compositions can be produced that are not possible with conventional melting.

Prior to further processing, powders may also undergo further operations, such as drying or heat treatment. The usual objective of heat treatment is to weaken the material and make it more responsive to subsequent compaction.

16.4 Powder Testing and Evaluation

Key properties of powdered material include bulk chemistry, surface chemistry, particle size and size distribution, particle shape and shape distribution, surface texture, and internal structure. In addition, powders should also be evaluated for their suitability for further processing. **Flow rate** measures the ease by which powder can be fed and distributed into a die. Poor flow characteristics can result in nonuniform die filling, as well as nonuniform density and nonuniform properties in a final product.

Associated with the flow characteristics is the **apparent density**, a measure of a powder's ability to fill available space without the application of external pressure. A low apparent density means that there is a large fraction of unfilled space in the loose-fill powder. **Compressibility** tests evaluate the effectiveness of applied pressure in raising the density of the powder, and **green strength** is used to describe the strength of the pressed powder immediately after compacting. It is well established that higher product density correlates with superior mechanical properties, such as strength and fracture resistance. Good green strength is required to maintain smooth surfaces, sharp corners, and intricate details during ejection from the compacting die or tooling and the subsequent transfer to the sintering operation.

The overall objective is often to achieve a useful balance of the key properties. The smooth-surface spheres produced by gas atomization, for example, tend to pour and flow well, but the compacts have extremely low green strength, disintegrating easily during handling. The irregular particles of water-atomized powder have better compressibility and green strength, but poorer flow characteristics. The sponge iron powders produced by chemical reduction of iron oxide are extremely porous and have highly irregular, extremely rough surfaces. They have poor flow characteristics and low compacted density, but green strength is quite high. The same material, therefore, can have widely different performance characteristics, depending on the specifics of powder manufacture.

16.5 Powder Mixing and Blending

It is rare that a single powder will possess all the characteristics desired in a given process and product. Most likely, the starting material will be a mixture of various grades or sizes of powder, or powders of different compositions, along with additions of **lubricants** or **binders**.

In powder products, the final chemistry is often obtained by mixing pure metal or nonmetal powders, rather than starting with a prealloyed material. To produce a uniform chemistry and structure in a product of mixed material, sufficient diffusion must occur during the sintering operation. Compositions can be created that cannot be made in any other way. Unique **composites** can be produced, such as the distribution of an immiscible reinforcement material in a matrix, or the blended combination of metals and nonmetals in a single product such as a tungsten carbide–cobalt matrix cutting tool used for high-temperature service.

Some powders, such as graphite, can even play a dual role, serving as a lubricant during compaction and a source of carbon as it alloys with iron during sintering to produce steel. Lubricants such as graphite or stearic acid improve the flow characteristics and compressibility of the powder and assist in the ejection from the compaction tooling, but also reduce green strength. Binders produce the reverse effect. Because most lubricants or binders are not wanted in the final product, they are removed (volatilized or burned off) in the early stages of sintering, leaving holes that are reduced in size or closed during subsequent heating.

Blending or **mixing** operations can be performed either dry or wet, where water or other solvent is used to enhance particle mobility, reduce dust formation, and lessen explosion hazards. Large lots of powder can be homogenized with respect to both chemistry and distribution of components, sizes, and shapes. Quantities up to 16,000 kg (35,000 lb) have been blended in single lots to ensure uniform behavior during processing and large-run production of a consistent product.

Most compacting is done with mechanical presses and rigid tools, but hydraulic and hybrid (combinations of mechanical, hydraulic, and pneumatic) presses can also be used. **Figure 16.4** shows a typical mechanical press for compacting powders and a removable set of compaction tooling. The removable die sets allow the time-consuming tool alignment and synchronization of tool movements to be performed while the press is producing parts with another die set. Compacting pressures generally range between 40 and 1650 MPa (3 and 120 tons/in.²) depending on material and application (see **Table 16.1**), with the range of 140 to 690 MPa (10 to 50 tons/in.²) being the most common. Because most P/M presses have total capacities of less than 100 tons, powder metallurgy products are often limited to pressing areas of less than 65 cm² (10 in.²).

FIGURE 16.4 An 880-ton compacting press, capable of compacting multilevel powder metallurgy products. A second die set can be seen at the lower right side of the press, enabling quick change of the compaction tooling. (Courtesy Cincinnati Incorporated, Cincinnati, OH)

16.6 Compacting

One of the most critical steps in the P/M process is **compaction**. Loose powder is compressed and densified into a shape known as a green compact, usually at room temperature. High product density and the uniformity of that density throughout the compact are generally desired characteristics. In addition, the mechanical interlocking and cold welding of the particles should provide sufficient green strength for in-process handling and transport to the sintering furnace.

TABLE 16.1	Typical Compaction Pressures for Various Applications	
	Compaction Pressures	
Application	tons/in.²	Mpa
Porous metals and filters	3–5	40–70
Refractory metals and carbides	5–15	70–200
Porous bearings	10–25	146–350
Machine parts (medium-density iron & steel)	20–50	275–690
High-density copper and aluminum part	18–20	250–275
High-density iron and steel parts	50–120	690–1650

Some P/M presses, however, now have capacities up to 3000 tons and are capable of pressing areas up to 650 cm² (100 in.²). When even larger products are desired, compaction can be performed by dynamic methods, such as use of an explosively induced shock wave. Metal forming processes, such as rolling, forging, extrusion, and swaging, have also been adapted for powder compaction.

Figure 16.5 shows the typical compaction sequence for a mechanical press. With the bottom punch in its fully raised position, a feed shoe moves into position over the die. The feed shoe is an inverted container filled with powder, connected to a much larger powder container by a flexible feed tube. With the feed shoe in position, the bottom punch descends to a pre-set fill depth, and the shoe retracts, with its edges leveling the powder. The upper punch then descends and compacts the powder as it penetrates the die. The upper punch retracts, and the bottom punch then rises to eject the green compact. As the die shoe advances for the next cycle, its forward edge clears the compacted product from the press, and the cycle repeats.

During uniaxial or one-direction compaction, the powder particles move primarily in the direction of the applied force. Because the loose-fill dimensions are two to two-and-a-half times the pressed dimensions, the amount of particle travel in the pressing direction can be substantial. The amount of lateral flow, however, is quite limited. In fact, it is rare to find a particle in the compacted product that has moved more than three particle diameters off its original axis of pressing. Thus, the powder does not flow like a liquid but simply compresses until an equal and opposing force is created. This opposing force is probably a combination of (1) resistance by the bottom punch, and (2) friction between the particles and the die sidewall surfaces. Densification occurs by particle movement, as well as plastic deformation of the individual particles.

As illustrated in **Figure 16.6**, when the pressure is applied by only one moving punch, maximum density occurs below the punch and decreases as one moves down the column. It is very difficult to transmit uniform pressure and produce uniform density throughout a compact, especially when the thickness

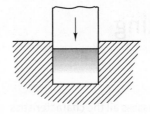

FIGURE 16.6 Compaction with a single moving punch, showing the resultant nonuniform density (*shaded*), highest where particle movement is the greatest.

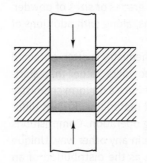

FIGURE 16.7 Density distribution obtained with a double-acting press and two moving punches. Note the increased uniformity compared to Figure 16.5. Thicker parts can be effectively compacted.

is large. By use of a double-action press, where pressing movements occur from both top and bottom (**Figure 16.7**), thicker products can be compacted to a more uniform density. Because sidewall friction is a key factor in compaction, the resulting density shows a strong dependence on both the thickness and width of the part being pressed. For uniform compaction, the ratio of thickness/width should be kept below 2.0 whenever possible. When the ratio exceeds 2.0, the products tend to exhibit considerable variation in density.

As shown in **Figure 16.8**, the average density of the compact depends on the amount of pressure that is applied, with the specific response being strongly dependent on the characteristics of the powder being compressed (its size, shape, surface texture, mechanical properties, etc.). The final density may be reported as either an absolute density in units such as grams per cubic centimeter, or as a percentage of the pore-free or theoretical density. The difference between this percentage and 100% corresponds to the amount of void space remaining within the compact.

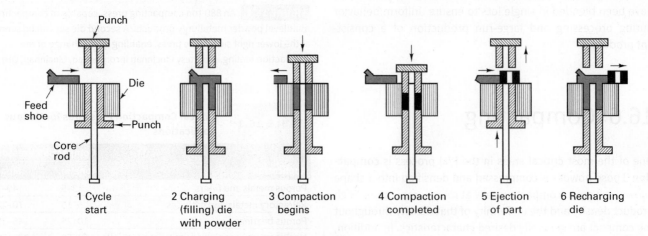

FIGURE 16.5 Typical compaction sequence for a single-level part, showing the functions of the feed shoe, die, core rod, and upper and lower punches. Loose powder is shaded; compacted powder is solid black.

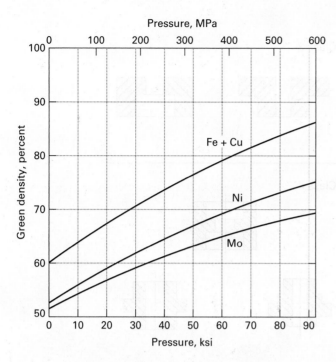

FIGURE 16.8 Effect of compacting pressure on green density (the density after compaction but before sintering). Separate curves are for several commercial powders.

Figure 16.9 shows that a single displacement will produce different degrees of compaction in different thicknesses of powder. It is impossible, therefore, for single punch movement to produce uniform density in a multithickness part. When more than one thickness is required, more complicated presses or compaction methods must be employed. **Figure 16.10** illustrates two methods of compacting a dual-thickness part. By providing different amounts of motion to the various punches and synchronizing these movements to provide simultaneous compaction, a uniform-density product can be produced.

Because the complexity of the part dictates the complexity of equipment, powder metallurgy components have been grouped into classes. Class 1 components are the simplest and

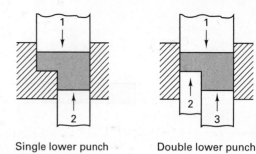

Single lower punch   Double lower punch

FIGURE 16.10 Two methods of compacting a double-thickness part to near-uniform density. Both involve the controlled movement of two or more punches.

easiest to compact. They are thin, single-level parts that can be pressed with a force from one direction. The thickness is generally less than 6.35 mm (¼ in.). Class 2 parts are single-level parts of any thickness that require pressing from two directions. These are usually thicker parts. Class 3 parts are double-level parts that require pressing from two directions. The most complex of the parts produced by rigid die compaction are Class 4 parts. They are multilevel and require two or more pressing motions. These four classes are summarized in both **Table 16.2** and **Figure 16.11**.

If a large part with complex shape is desired, the powder is generally encapsulated in a flexible mold, usually made of rubber or some elastomeric material, which is then immersed in liquid and pressurized up to 400 Mpa (60 ksi). This procedure is known as **isostatic** (*uniform pressure*) **compaction**. Because

TABLE 16.2	Features that Define the Various Classes of Press-and-Sinter P/M Parts	
Class	**Levels**	**Press Actions**
1	1	Single
2	1	Double
3	2	Double
4	More than 2	Double or multiple

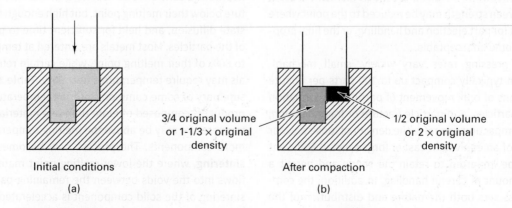

Initial conditions
(a)

3/4 original volume or 1-1/3 × original density

After compaction
(b)

1/2 original volume or 2 × original density

FIGURE 16.9 Compaction of a two-thickness part with only one moving punch. (*a*) Initial conditions; (*b*) after compaction by the upper punch. Note the drastic difference in compacted density.

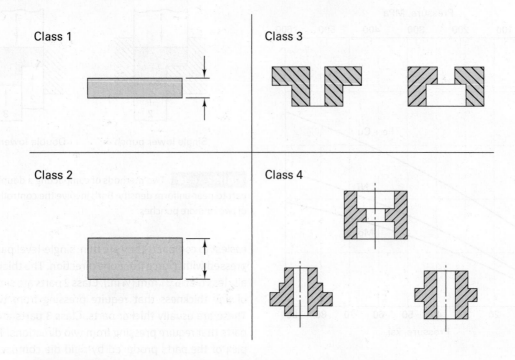

FIGURE 16.11 Sample geometries of the four basic classes of press-and-sinter powder metallurgy parts shown in side view. Note the increased pressing complexity that would be required as class increases.

the pressure is applied in all directions, compaction is uniform throughout the mass, and lower pressures produce densities higher than conventional punch-and-die compaction. Production rates are extremely low, but parts with weights up to several hundred pounds have been effectively compacted.

Warm compaction emerged as a common practice in the 1990s. By preheating the powder prior to pressing, the metal is softened and responds better to the applied pressures. The better compaction results in improved properties, both in the as-compacted state and after final processing.

Compaction can also be enhanced by increasing the amount of lubricant in the powder. This reduces the friction between the powder and the die wall, improves the transmission of pressure through the powder, and makes it easier to eject the compact from the die. If too much lubricant is used, however, the green strength may be reduced to the point where it is insufficient for part ejection and handling, or the final properties may become unacceptable.

Although pressing rates vary widely, small mechanical presses can typically compact up to 100 parts per minute (ppm). By means of bulk movement of particles, deformation of individual particles, and particle fracture or fragmentation, mechanical compaction can raise the density of loose powder to about 80% of an equivalent cast or forged metal. Sufficient strength can be imparted to retain the shape and permit a reasonable amount of careful handling. In addition, the compaction process sets both the nature and distribution of the porosity remaining in the product.

Because powder particles tend to be somewhat abrasive and high pressures are involved during compaction, wear of the tool components is a major concern. Consequently, compaction tools are usually made of hardened tool steel. For particularly abrasive powders, or for high-volume production, cemented carbides may be employed. Die surfaces should be highly polished, and the dies should be heavy enough to withstand the high pressing pressures. Lubricants are also used to reduce die wear.

16.7 Sintering

In the **sintering** operation, the pressed-powder compacts are heated in a controlled atmosphere environment to a temperature below their melting point, but high enough to permit solid-state diffusion, and held for sufficient time to permit bonding of the particles. Most metals are sintered at temperatures of 70 to 80% of their melting point, while certain refractory materials may require temperatures near 90%. **Table 16.3** presents a summary of some common sintering temperatures. When the product is composed of more than one material, the sintering temperature may be above the melting temperature of one or more components. The process now becomes **liquid-phase sintering**, where the lower-melting-point material melts and flows into the voids between the remaining particles and the sintering of the solid component is accelerated through solution and reprecipitation. In another modification known as **activated sintering**, a small amount of selected additive is used to increase the rate of diffusion.

TABLE 16.3	Typical Sintering Temperatures for Some Common Metals and Materials	
	Sintering Temperature	
Metal	°C	°F
Aluminum alloys	590–620	1095–1150
Brass	850–950	1550–1750
Copper	750–1000	1400–1850
Iron/steel	1100–1200	2000–2200
Stainless steel	1200–1280	2200–2350
Cemented carbides	1350–1450	2450–2650
Molybdenum	1600–1700	2900–3100
Tungsten	2200–2300	4000–4200
Various ceramics	1400–2100	2550–3800

Most sintering operations involve *three stages*, and many sintering furnaces employ three corresponding zones. The *first operation*, the *preheat, burn-off* or *purge*, is designed to combust any air, volatilize and remove lubricants or binders that would interfere with good bonding, and slowly raise the temperature of the compacts in a controlled manner. Rapid heating would produce high internal pressure from air entrapped in closed pores or volatilizing lubricants and would result in swelling or fracture of the compacts. When the compacts contain appreciable quantities of volatile materials, their removal creates additional *porosity* and *permeability* within the pressed shape. The manufacture of products such as metal filters is designed to take advantage of this feature. When the products are load-bearing components, however, high amounts of porosity are undesirable, and the amount of volatilizing lubricant is kept to an optimized minimum. The *second, or high-temperature, stage* is where the desired solid-state diffusion and bonding between the powder particles take place. As the material seeks to lower its surface energy, atoms move toward the points of contact between the particles. The areas of contact become larger, and the part ultimately becomes a solid mass with small isolated pores of various sizes and shapes. The mechanical bonds that formed during compaction become true metallurgical bonds. The time in this stage must be sufficient to produce the desired density and final properties, usually varying from 10 minutes to several hours. Finally, a *cooling period* is required to lower the temperature of the products while retaining them in a controlled atmosphere. This feature serves to prevent oxidation that would occur on direct discharge into air, as well as possible thermal shock from rapid cooling. Both batch and continuous furnaces are used for sintering.

All three stages of sintering must be conducted in the oxygen-free conditions of a *vacuum* or **protective atmosphere**. This is critical because the compacted shapes typically have 10 to 25% residual porosity, and some of the internal voids are connected to exposed surfaces. At elevated temperatures, rapid oxidation would occur and significantly impair the quality of interparticle bonding. **Reducing atmospheres**, commonly based on hydrogen, dissociated ammonia, or cracked hydrocarbons, are preferred because they can reduce any oxide

already present on the particle surfaces and combust harmful gases that are liberated during the sintering. **Inert gases** cannot reduce existing oxides but will prevent the formation of any additional contaminants. **Vacuum sintering** is frequently employed with stainless steel, titanium, and the refractory metals. **Nitrogen atmospheres** are also common.

During the sintering operation, a number of changes occur in the compact. Metallurgical bonds form between the powder particles as a result of solid-state atomic diffusion, and strength, ductility, toughness, and electrical and thermal conductivities all increase. If different chemistry powders were blended, interdiffusion promotes the formation of alloys or intermetallic phases. As the lubricant is removed and the pores reduce in size, there will be a concurrent increase in density and contraction in product dimensions. To meet final tolerances, the dimensional shrinkage will have to be compensated through the design of oversized compaction dies. During sintering, not all of the porosity is removed, however. Conventional pressed-and-sintered P/M products generally contain between 5 and 25% residual porosity.

Sinter brazing is a process in which two or more pieces are joined while they are being sintered. The individual pieces are compacted separately and are assembled with braze metal positioned so it will flow into the joint. When the assembly is heated for sintering, the braze metal melts, and capillary action pulls it into the joint. As sintering continues, much of the braze metal diffuses into the surrounding metal, producing a final bond that is often stronger than the materials being joined.

16.8 Advances in Sintering (Shorter Time, Higher Density, Stronger Products)

Because product properties improve with increases in density, various techniques have been developed to produce higher-density components. One way of achieving this while using the conventional "press-and-sinter" approach is to increase the temperature of sintering. Although **high-temperature sintering** may seem easily attainable, for iron and steel components, the increased temperatures generally require significant changes in furnace design and materials. The higher product densities, however, can often enable the use of less costly materials, such as chromium- or silicon-alloyed steels in place of the nickel- or molybdenum-steels.

Sinter hardening integrates a strengthening heat-treatment directly into the sintering operation. At the temperatures of sintering, iron and steel parts are austenite, and a rapid cool from this condition can produce the stronger, nonequilibrium microstructures. (See Chapter 5.) In place of the usual slow cool under protective atmosphere, parts are exposed to rapid convection cooling. Liquid quenches are generally avoided due to product porosity, but oil quenches are sometimes used. Energy is saved by eliminating a heat-treatment reheat, and distortion is reduced by minimizing elevated temperature handling.

Microwave sintering has recently moved to full-scale production. Unlike convection heating, where heat is transmitted through external surfaces, microwaves permeate through the entire volume of material, uniformly heating the whole part, thereby reducing both processing time and energy consumption to as little as 20% of that required for traditional processing. Microwave sintering has been used with both ceramic and metal powders. With bulk metals, microwaves induce surface eddy currents and skin heating, but are otherwise reflected. Powdered metals, however, heat up well due to the large amounts of surface area and poor electrical connectivity. Surface oxides, moisture, and other surface contaminants aid in the initial heating. Once the interparticle connectivity improves and conductivity increases, microwave heating becomes less effective.

Microwave heating can also be used to reduce the time required for removal of binders and lubricants. In traditional processing, heat is conducted from the surface of the part into the interior. If the part is heated too quickly, volatilization can create gas pockets that cause volume expansion or even cracking. If the temperature is too high, the surface can densify, trapping binder or lubricant in the interior. Optimal removal can often be achieved by combining conventional heating with a microwave assist that heats the lubricant or binder from the inside and drives it to the surface.

Compaction and sintering are combined in **spark-plasma sintering**, where axial force compaction is coupled with high-frequency, high-amperage, low-voltage pulses of direct current that are applied through the punches. Spark discharges occur in the gaps between particles, while electrical resistance heating occurs at points of particle contact. Some surface melting is observed, and the heating of the particles combined with the axial pressure causes the particles to deform, further aiding densification. No binders are required, and full density can be achieved with both metal and ceramic powders. Sintering can be achieved at lower overall temperatures, and processing time can be greatly reduced.

16.9 Hot-Isostatic Pressing

In conventional press-and-sinter powder metallurgy, the pressing or compaction is usually performed at room temperature, and the sintering, at atmospheric pressure. **Hot-isostatic pressing (HIP)** combines powder compaction and sintering into a single operation by using gas-pressure squeezing at elevated temperature. Although this may seem to be an improvement over the two-step approach, it should be noted that heated powders often need to be "protected" or isolated from harmful

environments, and the pressurizing media must be prevented from entering the voids between the particles. One approach to hot-isostatic pressing begins by sealing the powder in a flexible, airtight, evacuated container, which is then subjected to a high-temperature, high-pressure environment. Conditions for processing irons and steels involve pressures around 70 to 100 MPa (10,000 to 15,000 psi) coupled with temperatures in the neighborhood of 1250°C (2300°F). For the nickel-based superalloys, refractory metals, and ceramic powders, the equipment must be capable of 310 MPa (45,000 psi) and 1500°C (2750°F). Multiple pieces, totaling up to several tons, can be processed in a single cycle that typically lasts several hours.

After processing, the products emerge at full density with uniform, isotropic properties that are often superior to those of other processes. Near-net shapes are possible, thereby reducing material waste and costly machining operations. Because the powder is totally isolated and compaction and sintering occur simultaneously, the process is attractive for reactive or brittle materials, such as beryllium, uranium, zirconium, and titanium. Difficult-to-compact materials, such as superalloys, tool steels, and stainless steels, can be readily processed. Because die compaction is not required, large parts are now possible, and shapes can be produced that would be impossible to eject from rigid compaction dies. Hot-isostatic pressing has also been employed to densify existing parts (such as those that have been conventionally pressed and sintered), heal internal porosity in castings, and seal internal cracks in a variety of products. The elimination or reduction of defects yields startling improvements in strength, toughness, fatigue resistance, and creep life.

Several aspects of the HIP process make it expensive and unattractive for high-volume production. The first is the high cost of **canning** the powder in a flexible isolating medium that can resist the subsequent temperatures and pressures, and then later removing this material from the product (**decanning**). Sheet metal containers are most common, but glass and even ceramic molds have been used. The second problem involves the relatively long time for the HIP cycle. Although the development of advanced cooling methods has reduced cycle times from 24 hours to 6 to 8 hours, production is still limited to several loads a day, and the number of parts per load is limited by the ability to produce and maintain uniform temperature throughout the pressure chamber.

The **sinter-HIP** process and **pressure-assisted sintering** are techniques that have been developed to produce full-density powder products without the expense of canning and decanning. Conventionally compacted P/M parts are placed in a pressurizable chamber and sintered (heated) under vacuum for a time that is sufficient to seal the surface and isolate all internal porosity. (*Note:* This generally requires achieving a density greater than 92–95%.) While maintaining the elevated temperature, the vacuum is broken and high pressure is then applied for the remainder of the process. The sealed surface produced during the vacuum sintering acts as an isolating can during the high-pressure stage. Because these processes start with as-compacted powder parts, they eliminate the additional heating and cooling cycle that would be required if parts were first sintered in the conventional manner and then subjected to the HIP process for further densification.

16.10 Other Techniques to Produce High-Density P/M Products

Metal deformation processes can also be used to produce high-density P/M parts. Sheets of sintered powder (produced by roll compaction and sintering) can be reduced in thickness and further densified by hot rolling in the process depicted in **Figure 16.12**. Rods, wires, and small billets can be produced by the hot extrusion of encapsulated powder or pressed-and-sintered slugs. Forging can be applied to form complex shapes from canned powder or simple-shaped sintered preforms. By using powdered material, these processes offer the combined benefits of powder metallurgy and the respective forming process, such as the production of fabricated shapes with uniform fine grain size, uniform chemistry, or unusual alloy composition.

The **Ceracon process** is another method of raising the density of conventional pressed-and-sintered P/M products without requiring encapsulation or canning. A heated preform is surrounded by hot granular material, usually smooth-surface ceramic particles. When the assembly is then compacted in a conventional hydraulic press, the granular material transmits a somewhat uniform pressure. Encapsulation is not required because the pressurizing medium is not capable of entering pores in the material. When the pressure cycle is complete, the part and the pressurizing medium separate freely, and the pressure-transmitting granules are reheated and reused.

Yet another means of producing a high-density shape from fine particles is **in situ compaction** or **spray forming** (also known as the **Osprey process**). Consider an atomizer similar to that of Figure 16.2a, in which jets of inert or harmless gas

(nitrogen or carbon dioxide) propel molten droplets down into a collecting container. If the droplets solidify before impact, the container fills with loose powder. If the droplets remain liquid during their flight, the container fills with molten metal, which then solidifies into a conventional casting. However, if the cooling of the droplets is controlled so that they are semisolid (and computers can provide the necessary process control), they act as "slush balls" and flatten on impact. The remaining freezing occurs quickly, and the resultant product is a uniform chemistry, fine-grain-size, high-density (in excess of 98%) solid. Depending on the shape of the collecting container, the spray-formed product can be a finished part, a strip or plate, a deposited coating, or a preform for subsequent operations, such as forging. Both ferrous and nonferrous products can be produced with deposition rates as high as 200 kg/min (400 lb/min). Unique composites can be produced by the simultaneous deposition of two or more materials, injecting secondary particles into the stream, or promoting in-stream reactions.

16.11 Metal Injection Molding (MIM)

For many years, injection molding has been used to produce small, complex-shaped components from plastic. A thermoplastic resin is heated to impart the necessary degree of fluidity and is then pressure injected into a die, where it cools and hardens. Die casting is a similar process for metals but is restricted to alloys with relatively low melting temperature, such as lead-, zinc-, aluminum-, and copper-based materials. Small, complex-shaped products of the higher-melting-point metals are generally made by more costly processes, which include investment casting, machining directly from metal stock, or conventional powder metallurgy. **Metal injection molding (MIM)** is an extension of conventional powder metallurgy that combines the shape-forming capability of plastics, the precision of die casting, and the materials flexibility of powder metallurgy.

Because powdered material does not flow like a fluid, complex shapes are produced by first combining ultra-fine (usually in the range of 3–20 μm) spherical-shaped metal, ceramic, or carbide powder with a low-molecular-weight thermoplastic or wax material in a mix that is typically 60% powder by volume. This mixture is frequently produced in the form of pellets or granules, which become the feedstock for the injection process. After heating to a pastelike consistency (about 260°C, or 500°F), the material is injected into the heated mold cavity of a conventional injection molding machine under sufficient pressure (about 70 MPa or 10 ksi) to ensure die filling.[1] After cooling and ejection, the binder material is removed by one of a variety

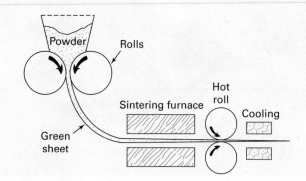

FIGURE 16.12 One method of producing continuous sheet products from powdered feedstock.

---

[1] Metals are good thermal conductors, so the metal-containing MIM feedstock freezes more rapidly than a comparable plastic injection molding. Strategic gate placement, larger gates, or multiple gates may be required to ensure adequate mold filling.

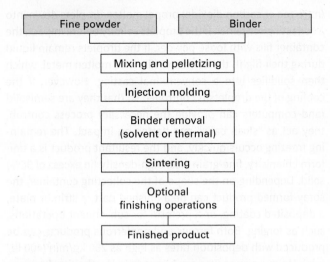

FIGURE 16.13 Flowchart of the metal injection molding process (MIM) used to produce small, intricate-shaped parts from metal powder.

of processes that include solvent extraction, controlled heating to above the volatilization temperature, or heating in the presence of a catalyst that breaks the binder down into products that volatilize easily. Removing the binder is currently the most expensive and time-consuming part of the process. Heating rates, temperatures, and debinding times must be carefully controlled and adjusted for part thickness. The parts then undergo conventional sintering, where any remaining binder is first removed, and the diffusion processes then set the final properties of the product. During sintering, MIM parts typically shrink 15 to 25%, and the density increases from about 60% up to as much as 99% of ideal. (*Note:* Because MIM parts are molded without density variations, the subsequent shrinkage tends to be both uniform and repeatable.) Secondary processes may take the form of surface cleaning or finishing, plating, machining, or heat-treating. The high final density enables the secondary processes to be conducted in the same manner as for wrought materials. **Figure 16.13** summarizes the full sequence of activities, and **Table 16.4** provides a summary comparison of conventional powder metallurgy and MIM.

Although the size of conventional P/M products is generally limited by press capacity, the size of P/M injection moldings is more limited by economics (cost of the fine powders) and binder removal.

TABLE 16.4 Comparison of Conventional Powder Metallurgy and Metal Injection Molding

Feature	P/M	MIM
Particle size	20–250 μm	<20 μm
Particle response	Deform plastically	Undeformed
Porosity (% nonmetal)	10–20%	30–40%
Amount of binder/lubricant	0.5–2%	30–40%
Homogeneity of green part	Nonhomogeneous	Homogeneous
Final sintered density	<92%	>96%

The best candidates for P/M injection molding are complex-shaped parts (generally too complex to compact by conventional powder metallurgy) with thicknesses of less than 0.6 cm (¼ in), weights under 20 g (2 oz), and made from a metal that cannot be economically die cast. MIM parts compete with and frequently replace machined components and investment castings. Section thicknesses as small as 0.25 mm (0.010 in) are possible because of the fineness of the powder. As a general rule, the smaller the part and the greater the complexity, the more likely MIM will be an attractive alternative to machining, casting, stamping, cold forming, or traditional powder metallurgy. **Figure 16.14** shows a variety of MIM products.

Medium-to-large production volumes (more than 2000 to 5000 identical parts) are generally required to justify the cost of die design and manufacture, but advances in die

FIGURE 16.14 Metal injection molding (MIM) is ideal for producing small, complex parts. (Courtesy of Megamet Solid Metals, Inc., St Louis, MO)

manufacture have recently made small lots (as small as 25 parts) economically viable. The relatively high final density (95 to 99% compared to 75 to 90% for conventional P/M parts), the uniformity of that density, the close tolerances (0.3 to 0.5%), and excellent surface finish (about 125 μ-in.) all combine to make the process attractive for many applications. Parts can be made from a wide selection of metal alloys, including steel, stainless steel, tool steel, brass, copper, titanium, tungsten, nickel-based superalloys, ceramics, and many specialty materials. The final properties are superior to those of conventional powder metallurgy and are generally close to those of wrought or cast equivalents. Major areas of application include firearms components, medical and dental tools, and electronic and computer parts.

16.12 Secondary Operations

Powder metallurgy products are often ready to use when they emerge from the sintering furnace. Many P/M products, however, utilize one or more secondary operations to provide enhanced precision, improved properties, or special characteristics.

During sintering, product dimensions shrink due to densification. In addition, warping or distortion may occur due to nonuniform cooldown from elevated temperatures. As a result, a second, room-temperature, pressing operation, known as **repressing**, **coining**, or **sizing**, may be required to restore or improve dimensional precision. The part is placed in a die and subjected to pressures equal to or greater than the initial pressing pressure. A small amount of plastic flow takes place, resulting in high dimensional accuracy, sharp detail, and improved surface finish. The associated cold working and increase in part density may combine to increase part strength by 25 to 50%. (*Note:* Because of the shrinkage that occurs during sintering, repressing cannot be performed with the same set of tooling that was used for the original powder compaction.)

If substantial metal deformation takes place in the second pressing, the operation is known as **P/M forging** or **powder forging**. Conventional press-and-sinter powder metallurgy is used to produce a preform, which is one forging operation removed from the finished shape. The normal forging sequence of billet or bloom production, shearing, reheating, and sequential deformation is replaced by the manufacture of a comparatively simple-shaped powder metallurgy preform followed by a single closed-die hot-forging operation. The forging stage produces a more-complex shape, adds precision, provides the benefits of metal flow, and increases the density (often up to 99%). The increase in density is accompanied by a significant improvement in mechanical properties, which are often equivalent or superior to those of wrought materials. Although protective atmospheres or coatings may be required to prevent oxidation of the powder preform during heating and hot forging, the added cost and operations can often be offset by a significant reduction in scrap or waste. (By controlling preform weight to within 0.5%, flash-free forging can often be performed.) Forged products can benefit from the improved properties of powder metallurgy, such as the absence of segregation, the uniform fine grain size, and the use of novel alloys or unique composites. The conventional powder metallurgy process can be expanded to larger size products with increased complexity. Size and shape constraints are set by the limitations of tooling and press tonnage. The tolerance requirements of cams, splines, and gears can often be met without subsequent machining. **Figure 16.15** illustrates the reduction in scrap by comparing the same part made by conventional forging and the P/M forge approach. Annual production volumes in excess of 25,000 are typically required, and all current commercial P/M forgings are of steel. P/M forged connecting rods, like those shown in **Figure 16.16**, have been produced by the millions.

Impregnation and infiltration are secondary processes that utilize the interconnected porosity or permeability of low-density P/M products. **Impregnation** refers to the forcing of oil or other liquid, such as a polymeric resin, into the porous network. This can be done by immersing the part in a bath and applying pressure or by a combination vacuum-pressure process. The most common application is that of oil-impregnated bearings. After impregnation, the bearing material will contain

FIGURE 16.15 Comparison of conventional forging and the forging of a powder metallurgy preform to produce a gear blank (or gear). Moving left to right, the top sequence shows the sheared stock, upset section, forged blank, and exterior and interior scrap associated with conventional forging. The finished gear is generally machined from the blank with the generation of additional scrap. The bottom pieces are the powder metallurgy preform and forged gear produced entirely without scrap by P/M forging. (Courtesy of GKN Sinter Metals, Auburn Hills, MI)

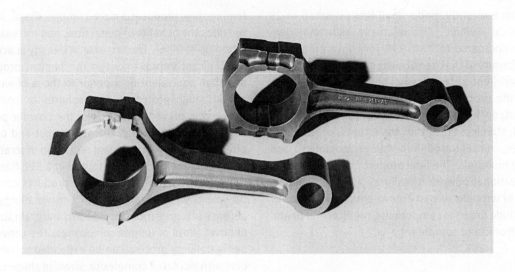

FIGURE 16.16 P/M forged connecting rods have been produced by the millions. (Courtesy of Metal Powder Industries Federation, Princeton, NJ)

from 10 to 40% oil by volume, which will provide lubrication over an extended lifetime of operation. In a similar manner, P/M parts can be impregnated with fluorocarbon resin (such as Teflon) to produce products offering a combination of high strength and low friction.

When the presence of pores is undesirable, P/M products may be subjected to metal **infiltration**. In this process a molten metal or alloy with a melting point lower than the P/M constituent flows into the interconnected pores of the product under pressure or by capillary action. Steel parts are often infiltrated with copper, for example. After infiltration, the engineering properties such as strength and toughness are improved to a level where they are generally comparable to those of solid metal products. Infiltration can also be used to seal pores prior to plating, improve machinability or corrosion resistance, or make the components gas- or liquid-tight. Additional heating after infiltration can cause interdiffusion between the infiltrant and base metal, further enhancing mechanical properties.

Powder metallurgy products can also be subjected to the conventional finishing operations of *heat treatment, machining,* and *surface treatment.* If the part is of high density (<10% porosity) or has been metal impregnated, conventional processing can often be employed. Special precautions must be taken, however, when processing low-density P/M products. During heat treatment, protective atmospheres must again be used, and certain liquid quenchants (such as water or brine) should be avoided. Speeds and feeds must be adjusted when machining, and care should be taken to avoid pickup of lubricant or coolant. In general, P/M products should be machined using sharp tools, light cuts, and high feed rates. When a large amount of machining is required, special machinability-enhancing additions may be incorporated into the initial powder blend. Nearly all common methods of surface finishing can be applied to P/M products, including platings and coatings, diffusion treatments, surface hardening, and steam treatment (which is used to produce a hard, corrosion-resistant oxide on

ferrous parts). As with the other secondary processes, some process modifications may be required if the part has a reasonable amount of porosity or permeability. Because most parts are small and are produced in large quantity, barrel tumbling is another common means of cleaning, deburring, and surface modification.

16.13 Properties of P/M Products

Because the properties of powder metallurgy products depend on so many variables—type and size of powder, amount and type of lubricant, pressing pressure, sintering temperature and time, finishing treatments, and so on—it is difficult to provide generalized information. Products can range all the way from low-density, highly porous parts with tensile strengths as low as 70 MPa (10 ksi) up to high-density pieces with tensile strengths of 1250 MPa (180 ksi) or greater.

Many of the properties of P/M parts are closely related to **final density**, reported as either straight mass per unit volume or as relative density (the ratio of the P/M part density to that of a pore-free equivalent). P/M parts with less than 75% relative density are considered to be low density. Parts with more than 90% relative density are high-density products. **Porosity** is the percentage of void volume. A medium density P/M product, therefore, will have between 10 and 25% porosity. This can be a network of interconnected voids extending from the surface, or a multitude of isolated holes. **Permeability** is the ability of a part to pass fluids or gas, a measure of the interconnectedness of the porosity. Permeable products can filter materials, diffuse the flow of liquids or gases, regulate flow or pressure drops in lines, or act as flame arrestors by cooling gases below combustion temperatures.

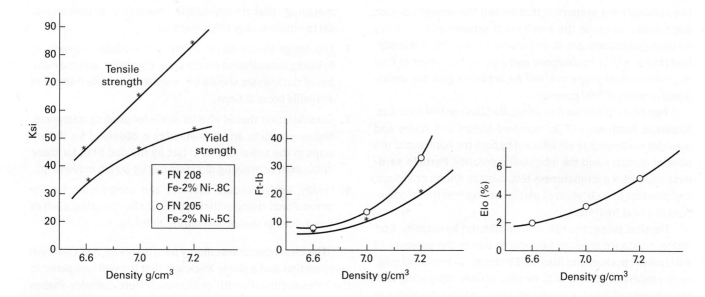

FIGURE 16.17 Mechanical properties versus as-sintered density for two iron-based powders. Properties depicted include yield strength, tensile strength, Charpy impact energy (shown in ft-lbs), and percent elongation in a 1-in. gage length.

As shown in **Figure 16.17**, most mechanical properties of P/M products exhibit a strong dependence on product density, with the fracture-limited properties of toughness, ductility, and fatigue life being more sensitive than strength and hardness. The voids in the P/M part act as stress concentrators and assist in starting and propagating fractures. The yield strength of P/M products made from the weaker metals is often equivalent to the same material in wrought form. When higher-strength materials are used or the fracture-related

tensile strength is specified, the properties of P/M product tend to fall below those of wrought equivalents by varying but usually substantial amounts. **Table 16.5** shows the properties of a few powder metallurgy materials compared with those of wrought material of similar composition. When larger presses or processes such as P/M forging or hot isostatic pressing are used to produce higher density, the strength of P/M products approaches that of the wrought material. If the processing results in full density with fine grain size, P/M parts

TABLE 16.5   **Comparison of Properties of Powder Metallurgy Materials and Equivalent Wrought Metals**

Material [a]	Form and Composition	Condition [b]	Percent of Theoretical Density	Tensile Strength 10³ psi	Mpa	Elongation in 2 in. (%)
Iron	Wrought	HR	—	48	331	30
	P/M—49% Fe min	As sintered	89	30	207	9
	P/M—99% Fe min	As sintered	94	40	276	15
Steel	Wrought AISI 1025	HR	—	85	586	25
	P/M—0.25% C.	As sintered	84	34	234	2
	99.75% Fe					
Stainless steel	Wrought type 303	Annealed	—	90	621	50
	P/M type 303	As sintered	82	52	358	2
Aluminum	Wrought 2014	T6	—	70	483	20
	P/M 201 AB	T6	94	48	331	2
	Wrought 6061	T6	—	45	310	15
	P/M 601 AB	T6	94	36.5	252	2
Copper	Wrought OFHC	Annealed	—	34	234	50
	P/M copper	As sintered	89	23	159	8
		Repressed	96	35	241	18
Brass	Wrought 260	Annealed	—	44	303	65
	P/M 70% Cu-30% Zn	As sintered	89	37	255	26

[a] Equivalent wrought metal shown for comparison.
[b] HR. hot rolled: 16 age hardened.
Note how porosity diminishes mechanical performance.

can actually have properties that exceed the wrought or cast equivalents. Because the mechanical properties of powder metallurgy products are so dependent on density, *it is important that products be designed and materials selected so that the desired final properties will be achieved with the anticipated amount of final porosity.*

Two types of hardness readings are taken on P/M products. **Apparent hardness** utilizes standard testers and scales and provides readings that are affected by *both* the hardness of the powder particles and the intervening porosity. **Particle hardness** requires a microhardness test, such as Knoop or Vickers, and provides an indication of particle strength or the effectiveness of a heat treatment.

Physical properties can also be affected by porosity. Corrosion resistance tends to be reduced due to the presence of entrapment pockets and fissures. Electrical, thermal, and magnetic properties all vary with density, usually decreasing with the presence of pores. Porosity actually increases the ability to damp both sound and vibration, however, and many P/M parts have been designed to take advantage of this feature.

16.14 Design of Powder Metallurgy Parts

Powder metallurgy is a manufacturing system whose ultimate objective is to economically produce products for specific applications. Success begins with good design and follows with good material and proper processing. In designing parts that are to be made by powder metallurgy, it must be remembered that P/M is a special manufacturing process and provision should be made for its unique factors. Products that are converted from other manufacturing processes without modification in design rarely perform as well as parts designed specifically for manufacture by powder metallurgy. Some basic rules for the design of P/M parts are as follows:

1. The shape of the part must permit ejection from the die. Sidewall surfaces should be parallel to the direction of pressing. Holes or recesses should have uniform cross section with axes and sidewalls parallel to the direction of punch travel.

2. The shape of the part should be such that powder is not required to flow into small cavities such as thin walls, narrow splines, or sharp corners.

3. The shape of the part should permit the construction of strong tooling.

4. The thickness of the part should be within the range for which P/M parts can be adequately compacted.

5. The part should be designed with as few changes in section thickness as possible.

6. Parts can be designed to take advantage of the fact that certain forms and properties can be produced by powder metallurgy that are impossible, impractical, or uneconomical to obtain by any other method.

7. The design should be consistent with available equipment. Pressing areas should match press capability, and the number of thicknesses should be consistent with the number of available press actions.

8. Consideration should also be made for product tolerances. Higher precision and repeatability is observed for dimensions in the radial direction (set by the die) than for those in the axial or pressing direction (set by punch movement).

9. Finally, design should consider and compensate for the dimensional changes that will occur after pressing, such as the shrinkage that occurs during sintering.

The ideal powder metallurgy part, therefore, has a uniform cross section and a single thickness that is small compared to the cross-sectional width or diameter. More complex shapes are indeed possible, but it should be remembered that uniform strength and properties require uniform density. Holes that are parallel to the direction of pressing are easily accommodated. Holes at angles to this direction, however, must be made by secondary processing. Multiple-stepped diameters, reentrant holes, grooves, and undercuts should be eliminated whenever possible. Abrupt changes in section, narrow deep flutes, and internal angles without generous fillets should also be avoided. Straight serrations can be readily molded, but diamond knurls cannot. Punches should be designed to eliminate sharp points or thin sections that could easily wear or fracture. **Figure 16.18** illustrates some of these design recommendations and restrictions.

16.15 Powder Metallurgy Products

The products that are commonly produced by powder metallurgy can generally be classified into six groups.

1. *Porous or permeable products, such as bearings, filters, and pressure or flow regulators.* Oil-impregnated bearings, made from either iron or copper alloys, constitute a large volume of P/M products. They are widely used in home appliance and automotive applications because they require no lubrication or maintenance during their service life. P/M filters can be made with pores of almost any size, some as small as 0.0025 mm (0.0001 in.). Unlike many alternative filters, powder metallurgy filters can withstand the conditions of elevated temperature, high applied stresses, and corrosive environments.

2. *Products of complex shapes that would require considerable machining when made by other processes.* Because of the dimensional accuracy and fine surface finish that are characteristic of the P/M process, many parts require no further

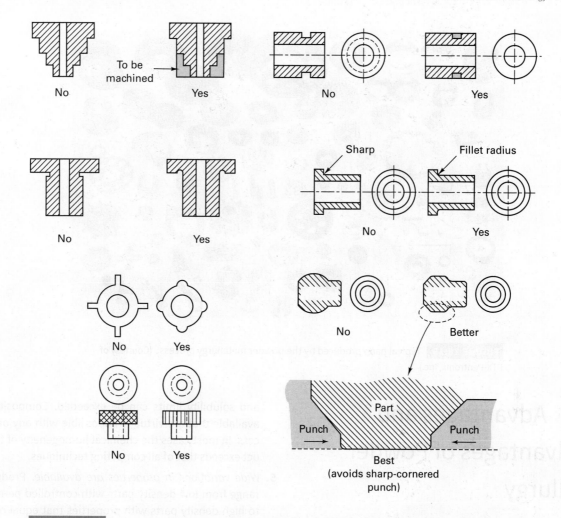

FIGURE 16.18 Examples of poor and good design features for powder metallurgy products. Recommendations are based on ease of pressing, design of tooling, uniformity of properties, and ultimate performance.

processing, and others require only a small amount of finish machining. Tolerances can generally be held to within 0.1 mm (0.005 in.). Large numbers of small gears are currently being made by the powder metallurgy process. Other complex shapes, such as pawls, cams, and small activating levers, can be made quite economically.

3. *Products made from materials that are difficult to machine or materials with high melting points.* The production of tungsten lamp filaments and tungsten carbide cutting tools were among the first modern uses of powder metallurgy.

4. *Products where the combined properties of two or more metals (or metals and nonmetals) are desired.* The unique capability of mixing any materials in any amounts has been applied to a number of products. In the electrical industry, copper and graphite are frequently combined in applications like motor or generator brushes where copper provides the current-carrying capacity and graphite provides lubrication. Bearings have been made of iron or copper combined with graphite, or from mixtures of two metals, such as tin and copper, where the harder material provides wear resistance and the softer material deforms in a way that better distributes the load. Electrical contacts often combine copper or silver with tungsten, nickel,

or molybdenum. Here, the copper or silver provides high conductivity, and the high melting temperature material provides resistance to fusion when the contacts experience arcing during closure.

5. *Products where the powder metallurgy process produces clearly superior properties.* The development of processes that produce full density has resulted in P/M products that are clearly superior to those produced by competing techniques. In areas of critical importance such as aerospace applications, the additional cost of the processing may be justified by the enhancement of properties. Some of the premium grades of tool steel are full-density P/M products. In the production of P/M magnets, a magnetic field can be used to align particles prior to sintering, resulting in a product with extremely high flux density.

6. *Products where the powder metallurgy process offers definite economic advantage.* The process advantages described in the next section often make powder metallurgy the most economical among competing ways to produce a part.

Figure 16.19 shows an array of typical powder metallurgy products.

FIGURE 16.19 Typical parts produced by the powder metallurgy process. (Courtesy of PTX-Pentronix, Inc.)

16.16 Advantages and Disadvantages of Powder Metallurgy

Like all other manufacturing processes, powder metallurgy has distinct advantages and disadvantages that should be considered if the technique is to be employed economically and successfully. Among the important advantages are these:

1. *Elimination or reduction of machining.* The **dimensional accuracy** and **surface finish** of P/M products are such that subsequent machining operations can be eliminated for many applications. If unusual dimensional accuracy is required, simple coining or sizing operations can often produce accuracies equivalent to those of most production machining. Reduced machining is especially attractive for difficult-to-machine materials.

2. *High production rates.* All steps in the P/M process are simple and readily automated. Labor requirements are low, and product uniformity and reproducibility are among the highest in manufacturing.

3. *Complex shapes can be produced.* Subject to the limitations discussed previously, complex shapes can be produced, such as combination gears, cams, and internal keys. It is often possible to produce parts by powder metallurgy that cannot be economically machined or cast.

4. *Wide variations in compositions are possible.* Parts of very high purity can be produced. Metals and ceramics can be intimately mixed. Immiscible materials can be combined,

and solubility limits can be exceeded. Compositions are available that are virtually impossible with any other process. In most cases the chemical homogeneity of the product exceeds that of all competing techniques.

5. *Wide variations in properties are available.* Products can range from low-density parts with controlled permeability to high-density parts with properties that equal or exceed those of equivalent wrought counterparts. Damping of noise and vibration can be tailored into a P/M product. Magnetic properties, wear properties, friction characteristics, and others can all be designed to match the needs of a specific application.

6. *Scrap is eliminated or reduced.* Whereas scrap can often exceed 50% of the starting material in casting, machining, and press forming, more than 97% of the starting material typically appears in finished P/M parts. This is particularly important when expensive materials are involved and may make it possible to use more costly materials without increasing the overall cost of the product. The production of rare-earth magnets is an example of such a product.

The major disadvantages of the powder metallurgy process are these:

1. *Inferior strength properties.* Because of the residual porosity, powder metallurgy parts generally have mechanical properties that are inferior to wrought or cast products of the same size, shape, and material. Their use may be limited when high stresses are involved. The required strength and fracture resistance, however, can often be obtained by using different materials or by employing alternate or secondary processing techniques that are unique to powder metallurgy.

2. *Relatively high tooling cost.* Because of the high pressures and severe abrasion involved in the process, the P/M dies must be made of expensive materials and be relatively massive. Because of the need for part-specific tooling, production quantities of less than 10,000 identical parts are normally not practical.

3. *High material cost.* On a unit weight basis, powdered metals are considerably more expensive than wrought or cast stock. However, the absence of scrap and the elimination of machining can often offset the higher cost of the starting material. In addition, powder metallurgy is usually employed for rather small parts where the material cost per part is not very great.

4. *Size and shape limitations.* The powder metallurgy process is simply not feasible for many shapes. Parts must be able to be ejected from the die. The thickness/diameter (or thickness/width) ratio is limited. Thin vertical sections are difficult, and the overall size must be within the capacity of available presses. Few parts exceed 150 cm^2 (25 in.^{2}) in pressing area.

5. *Dimensions change during sintering.* Although the actual amount depends on a variety of factors, including as-pressed density, sintering temperature, and sintering time, it can often be predicted and controlled.

6. *Density variations produce property variations.* Any nonuniform product density that is produced during compacting generally results in property variations throughout the part. For some products, these variations may be unacceptable.

7. *Health and safety hazards.* Many metals, such as aluminum, titanium, magnesium, and iron, are pyrophoric—they can ignite or explode when in particle form with large surface/volume ratios. Fine particles can also remain airborne for long times and can be inhaled by workers. To minimize the health and safety hazards, the handling of powdered material frequently requires the use of inert atmospheres, dry boxes, and hoods, as well as special cleanliness of the working environment.

16.17 Process Summary

For many years, powder metallurgy products carried the stigma of "low strength" or "inferior mechanical properties." This label was largely the result of comparisons where "identical" parts were made of the same material, but by various methods of manufacture. In essence, the size, shape, *and material* were all specified. In such a comparison, any product with 10 to 25% residual porosity would naturally be inferior to a fully dense product made by casting, forming, or machining processes. Unfortunately, it is this type of comparison that is frequently made when one considers converting an existing design or existing part to P/M manufacture.

A far more valid comparison would be obtained by specifying size, shape, *and desired mechanical properties.* Each process could be optimized by the selection of *both* material and process conditions. Casting would use a castable alloy; forming would use wrought material; and machining might utilize a material with free-machining additives. Powder metallurgy could then use its unique materials, such as iron-copper blends for which there are no cast or wrought equivalents, and the products would be designed to provide the targeted properties while containing the typical amounts of residual porosity. Because the various alternatives would all possess the targeted mechanical properties, process comparison could then be based on economic factors, such as total production cost. On this basis, powder metallurgy has emerged as a significant manufacturing process, and its products no longer carry the stigma of "inferior mechanical properties."

Table 16.6 summarizes some of the important manufacturing features of four powder processing methods. Note the variations in product size, production rate, production quantity, mechanical properties, and cost.

TABLE 16.6 Comparison of Four Powder Processing Methods

Characteristic	Conventional Press and Sinter	Metal Injection Molding (MIM)	Hol-Isostatic Pressing (HIP)	P/M Forging
Size of workpiece	Intermediate	Smallest	Largest	Intermediate
	<5 pounds	<1/4 pounds	1–1000 pounds	<5 pounds
Shape complexity	Good	Excellent	Very good	Good
Production rate	Excellent	Good	Pour	Excellent
Production quantity	>5000	>5000	1–1000	>10,000
Dimensional	Excellent	Good	Poor	Very good
precision	±0.001 in./in.	±0.003 in./in.	0.020 in./in.	±0.0015 in./in.
Density	Fair	Very good	Excellent	Excellent
Mechanical	80–90% of	90–95% of	Greater than	Equal to
properties	wrought	wrought	wrought	wrought
Cost	Low	Intermediate	High	Somewhat low
	$0.50–5.00/lb	$1.00–10.00/lb	>$100.00/lb	$1.00–5.00/lb

Review Questions

1. What type of product would be considered to be a prospect for powder metallurgy manufacture?

2. What were some of the earliest powder metallurgy products?

3. What are some of the newest technologies in powder metallurgy?

4. Why might the term "particulate processing" be a more accurate description than "powder metallurgy"?

5. What are some of the primary market areas for P/M products?

6. Which metal family currently dominates the powder metallurgy market?

7. What are the four basic steps that are usually involved in making products by powder metallurgy?

8. What are some of the important properties and characteristics of metal powders to be used in powder metallurgy?

9. What is the most common method of producing metal powders?

10. What are some of the other techniques that can be employed to produce particulate material?

11. Which of the powder manufacturing processes are likely to be restricted to the production of elemental (unalloyed) metal particles?

12. Why might the powdered material be heat-treated prior to compaction?

13. Why is flow rate an important powder characterization property?

14. What is apparent density, and how is it related to the final density of a P/M product?

15. What is green strength, and why is it important to the manufacture of high-quality P/M products?

16. How do the various powder properties relate to the method of powder manufacture?

17. What are some of the objectives of powder mixing or blending?

18. How does the addition of a lubricant affect compressibility? Green strength?

19. How might the use of a graphite lubricant be fundamentally different from the use of wax or stearates?

20. What types of composite materials can be produced through powder metallurgy?

21. What are some of the objectives of the compaction operation?

22. What is the benefit of a removable die set in a P/M compaction press?

23. What limits the cross-sectional area of most P/M parts to several square inches or less?

24. Describe the movement of powder particles during uniaxial compaction.

25. For what conditions might a double-action pressing be more attractive than compaction with a single moving punch?

26. Why is it more difficult to compact a multiple-thickness part?

27. Describe the four classes of conventional powder metallurgy products.

28. What is isostatic compaction? For what product shapes might it be preferred?

29. What is the benefit of warm compaction?

30. What is a typical as-compacted density? How much residual porosity is still present?

31. How do the common sintering temperatures compare to material melting points?

32. What is liquid-phase sintering? Activated sintering?

33. What are the three stages associated with most P/M sintering operations?

34. Why is it necessary to raise the temperature of P/M compacts slowly to the temperature of sintering?

35. Why is a protective atmosphere required during sintering? During the cool-down period?

36. What types of atmospheres are used during sintering?

37. What are some of the changes that occur to the compact during sintering?

38. What is the purpose of the sinter brazing process?

39. What are some benefits of high-temperature sintering? Sinter hardening? Microwave sintering?

40. How can microwave heating help in the removal of binders and lubricants?

41. Describe the process of spark-plasma sintering.

42. The combined heating and pressing of powder would seem to be an improvement over separate operations. What features act as deterrents to this approach?

43. What are some of the attractive properties of hot isostatic pressed products?

44. What is canning and decanning, and how do these operations relate to the HIP process?

45. What types of materials have been used as isolation barriers in hot-isostatic pressing?

46. What is the attractive feature of the sinter-HIP and pressure-assisted sintering processes?

47. What are some of the other methods that can produce high-density P/M products?

48. Describe the spray-forming process and the unique feature that enables production of high-density, fine-grain-size products.

49. How is the injection molding of powdered material similar to the injection molding of plastic or polymeric products?

50. In the MIM process, what is done to enable metal powder to flow like a fluid under pressure?

51. How is the metal powder used in metal injection molding (MIM) different from the metal powder used in a conventional press-and-sinter production?

52. What are some of the ways that the binder can be removed from metal injection molded parts?

53. Why are MIM products injection molded to sizes that are considerably larger than the desired product?

54. For what types of parts is P/M injection molding an attractive manufacturing process?

55. How does the final density of a MIM product compare to a press-and-sinter P/M part?

56. What is the purpose of repressing, coining, or sizing operations?

57. Why can we not use the original compaction tooling to perform repressing?

58. What is the major difference between repressing and P/M forging?

59. What is the difference between impregnation and infiltration? How are they similar?

60. Why might different conditions be required for the heat treatment, machining, or surface treatment of a powder metallurgy product?

61. What is the difference between "porosity" and "permeability"?

62. The properties of P/M products are strongly tied to density. Which properties show the strongest dependence?

63. What is apparent hardness? How is it different from particle hardness?

64. How do the physical properties of P/M products vary with density?

65. What advice would you want to give to a person who is planning to convert the manufacture of a component from die casting to powder metallurgy?

66. What is the shape of an "ideal" powder metallurgy product?

67. What are some P/M products that utilize the porosity or permeability features of the P/M process?

68. Give an example of a product where two or more materials are mixed to produce a composite P/M product with a unique set of properties.

69. What are the primary assets or advantages of the powder metallurgy method of parts manufacture?

70. Why is finish machining such an expensive component in parts manufacture?

71. Describe some of the materials that can be made into P/M parts that could not be used for processes such as casting and forming.

72. Why is P/M not attractive for parts with low production quantities?

73. What features of the P/M process often compensate for the higher cost of the starting material?

74. Why is it so important to achieve uniform density in a P/M product?

75. How might you respond to the criticism that P/M parts have inferior properties compared to wrought (machined, cast, or formed) products of the same material?

76. When comparing different manufacturing alternatives, why might the chosen material be different, even though the size, shape, and desired properties are the same?

Problems

1. When specifying the starting material for casting processes, the primary variables are chemistry and purity. Any structural features of the starting material will be erased by the melting. For forming processes, the material remains in the solid state, so the principal concerns relating to the starting material are chemistry and purity, ductility, yield strength, strain-hardening characteristics, grain size, and so on. What are some of the characteristics that should be specified for the starting powder to ensure the success of a powder metallurgy process? In what ways are these similar or different from those mentioned for casting and forming processes?

2. Various techniques are used to "measure" the size of small particles. One method evaluates the volume of fluid that a particle displaces and computes a *volume diameter*—the diameter of a sphere having the same volume as the particle. Sedimentation methods measure the terminal settling velocity and compute a *"Stokes' diameter"*—the diameter of a sphere having the same density that would have the same settling velocity in a fluid. Surface area methods determine the total surface for a given mass of powder. The *surface diameter* is the size of spheres that would have the same surface area as that measured. Direct observation methods can be used to determine linear dimensions, projected areas, and projected particle perimeters, all of which can be adjusted to the diameter of a spherical particle that would produce the same result (*projected*

area diameter and *perimeter diameter*). *Sieve diameter* is the minimum width of a square aperture that would permit passage of the particle.

a. Select a simple geometrical shape (such as a cube or rectangular solid), and compute several of the diameters cited earlier. Note the variation in results for the different methods.

b. Which of the techniques could be used to determine the relative distribution of sizes (size distribution) within a mass of powder?

3. In conventional powder metallurgy manufacture, the material is compacted with applied pressure at room temperature and then sintered by elevated temperature at atmospheric pressure. With P/M hot pressing, the loose powder is subjected to pressure while it is also at elevated temperature. It would appear, therefore, that hot pressing could produce a finished part in a single operation and would be a more economical and attractive manufacturing process. What features have been overlooked in this argument that would tend to favor the press-and-sinter sequence for conventional manufacture?

4. Investigate the method(s) used to produce tungsten incandescent lamp filaments. How does the method used today compare to the method developed by Coolidge in the late 1800s?

5. Select a product or component that could be manufactured by powder metallurgy and one or more other manufacturing processes.

    a. Identify the pros and cons of each of the manufacturing alternatives.

    b. If the desired quantity were small (25 or less), which of your alternatives would you favor?

    c. If the desired quantity were large (10,000 or more), which would you prefer?

6. Particulate materials can be made by a variety of methods, each producing characteristic size, shape, and surface texture.

    a. For smooth-surface spherical particles, which P/M properties would you expect to be "good"? Which would you expect to be poor? Identify a product where this type of powder might be attractive.

    b. For a powder with highly irregular shapes and rough surface texture, answer the questions posed in part (a).

7. Consider the various tooling components in Figure 16.5—the die, punches, and core rod. In selecting tool materials, key properties include strength, toughness, and wear resistance, along with other features. For each of these components, describe their function in terms of forces, motions, and the flow of powder over contact surfaces. What are the most important properties for each of the three components of compaction tooling? Would the same tool material be best for all components?

CHAPTER 17

Fundamentals of Metal Forming

17.1 Introduction

Chapters 13 through 16 have already presented a variety of methods for producing a desired shape from an engineering material. Each of those methods had its characteristic set of capabilities, advantages, and limitations. If we are to select the best method to make a given product, however, we must have a reasonable understanding of the entire spectrum of available techniques for shape production and their related features.

The next several chapters will expand the study of shape production methods by considering the family of **deformation processes**. These processes have been designed to exploit a remarkable property of some engineering materials (most notably metals) known as **plasticity**, the ability to flow as solids without deterioration of their properties. Because all processing is done in the solid state, there is no need to handle molten material or deal with the complexities of solidification. Because the material is simply moved (or rearranged) to produce the shape, as opposed to the cutting away of unwanted regions, the amount of waste can be substantially reduced. Unfortunately, the required forces are often high. Machinery and tooling can be quite expensive, and large production quantities may be necessary to justify the approach.

The overall usefulness of metals is largely due to the ease of fabrication into useful shapes. A majority of metal products undergo deformation at some stage of their manufacture. By rolling, cast ingots, strands, and slabs are reduced in size and converted into basic forms such as sheets, rods, plates, or structural shapes, such as I-beams. These forms may then undergo further deformation to produce wire, or the numerous finished products formed by processes such as forging, extrusion, sheet metal forming, and others. The deformation may be **bulk flow** in three dimensions, simple **shearing**, simple or compound **bending**, or complex combinations of these. The stresses producing these deformations can be tension, compression, shear, or any of the other varieties included in **Table 17.1**. **Table 17.2** depicts a wide variety of specific processes, and identifies the primary state of stress responsible for the deformation. For most of these processes, a wide range of speeds, temperatures, tolerances, surface finishes, and amounts of deformation are possible.

17.2 Forming Processes: Independent Variables

Forming processes tend to be complex systems consisting of independent variables, dependent variables, and independvent–dependent interrelations. **Independent variables** are those aspects of a process over which the engineer or operator has direct control, and they are generally selected or specified

TABLE 17.1	Classification of States of Stress

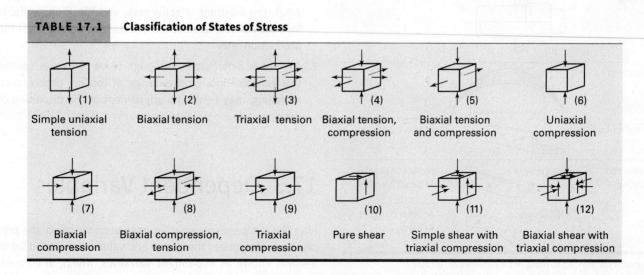

(1)	(2)	(3)	(4)	(5)	(6)
Simple uniaxial tension	Biaxial tension	Triaxial tension	Biaxial tension, compression	Biaxial tension and compression	Uniaxial compression
(7)	(8)	(9)	(10)	(11)	(12)
Biaxial compression	Biaxial compression, tension	Triaxial compression	Pure shear	Simple shear with triaxial compression	Biaxial shear with triaxial compression

TABLE 17.2	Common Forming Operations and Their Stress States	
Process	**Schematic Diagram**	**State of Stress in Main Part During Forming[a]**
Rolling		7
Forging		9
Extrusion		9
Shear spinning		12
Tube spinning		9
Swaging or kneading		7
Deep drawing		In flange of blank, 5 In wall of cup, 1
Wire and tube drawing	(a) (b)	8
Stretching		2
Straight bending		At bend, 2 and 7
Contoured flanging	(a) Convex	At outer flange, 6 At bend, 2 and 7
	(b) Concave	At outer flange, 1 At bend, 2 and 7

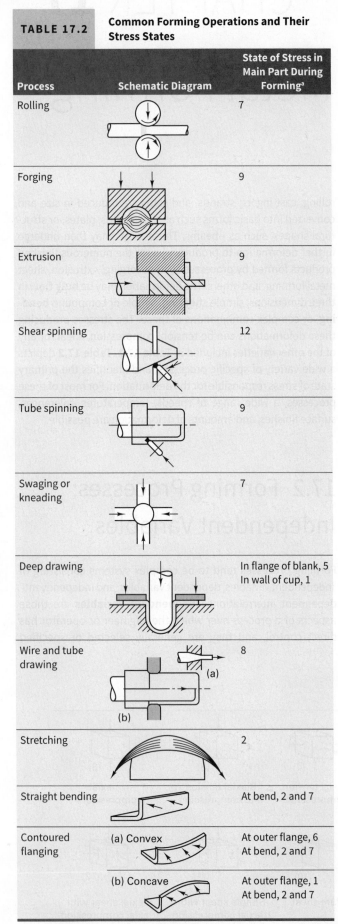

[a] Numbers correspond to those in parentheses in Table 17.1.

during setup. Consider some of the independent variables in a typical forming process:

1. *Starting material.* When specifying the starting material, we may define not only the chemistry of that material, but also its condition (as-cast, annealed, as-previously hot rolled, etc.). In so doing, we define the initial properties and characteristics. These may be chosen entirely for ease of fabrication, or they may be restricted by the desire to achieve the required final properties upon completion of the deformation process.

2. *Starting geometry of the workpiece.* The starting geometry may be dictated by previous processing, or it may be selected from a variety of available shapes. Economic considerations often influence this decision.

3. *Tool or die geometry.* This is an area of major significance and has many aspects, such as the diameter and profile of a rolling mill roll, the bend radius in a sheet-forming operation, the die angle in wire drawing or extrusion, and the cavity details when forging. Because the **tooling** will induce and control the metal flow as the material goes from starting shape to finished product, success or failure of a process often depends on tool geometry.

4. *Lubrication.* It is not uncommon for friction between the tool and the workpiece to account for more than 50% of the power supplied to a deformation process. In addition to reducing friction, lubricants can also act as coolants, thermal barriers, corrosion inhibitors, and parting compounds. Hence, their selection is an important aspect in the success of a forming operation. Specification includes type of lubricant, amount to be applied, and the method of application.

5. *Starting temperature.* Because material properties can vary greatly with temperature, temperature selection and control are often key to the success or failure of a metal forming operation. Specification of starting temperatures may include the temperatures of both the workpiece and the tooling.

6. *Speed of operation.* Most deformation equipment can be operated over a range of speeds. Because speed can directly influence the forces required for deformation (see Figure 2-32), the lubricant effectiveness, and the time available for heat transfer, its selection affects far more than just the production rate.

7. *Amount of deformation.* Although some processes control this variable through the design of tooling, others, such as rolling, may permit its adjustment at the discretion of the operator.

17.3 Dependent Variables

After the independent variables have been specified, the process then determines the nature and values of a second set of features known as **dependent variables**, which, in essence, are the consequences of the independent variable selection. Examples of dependent variables include the following:

1. *Force or power requirements.* A certain amount of force or power is required to convert a selected material from a starting shape to a final shape, with a specified lubricant, tooling geometry, speed, and starting temperature. A change in any of the independent variables will result in a change in the required force or power, but the effect is indirect. We cannot directly specify the force or power; we can only specify the independent variables and then experience the consequences of that selection.

   It is extremely important, however, that we be able to predict the forces or powers that will be required for any forming operation. Without a reasonable estimate, we would be unable to specify the equipment for the process, select appropriate tool or die materials, compare various die designs or deformation methods, or ultimately optimize the process.

2. *Material properties of the product.* Although we can easily specify the properties of the starting material, the combined effects of deformation and the temperatures experienced during forming will certainly change them. The starting properties of the material may be of interest to the manufacturer, but the customer is far more concerned with receiving the desired final shape with the desired final properties. It is important to know, therefore, how the initial properties will be altered by the shape-producing process.

3. *Exit (or final) temperature.* Deformation generates heat within the material. Hot workpieces cool when in contact with colder tooling. Lubricants can change viscosity, decompose when overheated, or react with the workpiece. The properties of an engineering material will be altered by both the mechanical and thermal aspects of a deformation process. Therefore, if we are to produce quality products, it is important to control the temperature of the material throughout the deformation. (*Note:* The fact that temperature may vary from location to location within the product further adds to the complexity of this variable.)

4. *Surface finish and precision.* The **surface finish** and **dimensional precision** of the resultant product will depend on the specific details of the forming process.

5. *Nature of the material flow.* In deformation processes, dies or tooling control the movement of the external surfaces of the workpiece. Although the objective of an operation is the production of a desired shape, the internal flow of material may actually be of equal importance. The production of a defect-free product with the desired shape and required properties requires attention and control of all aspects of material flow.

6. *Anisotropy and residual stresses.* Because of the flow of material, deformation products generally have anisotropy—directional variation in both structure and properties. In addition, variation in deformation and thermal history tend to create complex residual stresses. Directional anisotropy and residual stresses can be beneficial or detrimental and should be considered in all deformation products.

17.4 Independent–Dependent Relationships

Figure 17.1 serves to illustrate the major problem facing metal-forming personnel. On the left side are the *independent variables*—those aspects of the process for which control is direct and immediate. On the right side are the *dependent variables*—those aspects for which control is entirely indirect. Unfortunately, although it is the dependent variables that we want to control, their values are determined by the process, as complex consequences of the independent variable selection. If we want to change a dependent variable, we must determine which independent variable (or combination of independent variables) is to be changed, in what manner, and by how much. To make appropriate decisions, therefore, it is important for us to develop an understanding of the *independent variable–dependent variable interrelations*.

Understanding the links between independent and dependent variables is truly the most important area of knowledge for a person in metal forming. Unfortunately, this knowledge is often difficult to obtain. Metal forming processes are complex systems composed of the material being deformed, the tooling performing the deformation, lubrication at surfaces and interfaces, and various other process parameters such as temperature and speed. The number of different forming processes (and variations thereof) is quite large. In addition, different materials often behave differently in the same process, and there are multitudes of available lubricants. Some processes are sufficiently complex that they may have 15 or more interacting independent variables.

We can gain information on the independent–dependent relationships in three distinct ways:

1. *Experience.* Unfortunately, this generally requires longtime exposure to a process and is often limited to the specific materials, equipment, and products encountered during past contact. Younger employees may not have the experience

Independent variables	Links	Dependent variables
Starting material		Force or power requirements
Starting geometry	-Experience-	Product properties
Tool geometry		Exit temperature
Lubrication	-Experiment-	Surface finish
Starting temperature		Dimensional precision
Speed of deformation	-Modeling-	Material flow details
Amount of deformation		Anisotropy and residual stresses

FIGURE 17.1 Schematic representation of a metal forming system showing independent variables, dependent variables, and the various means of linking the two.

necessary to solve production problems. Moreover, a single change in an area such as material, temperature, speed, or lubricant may make the bulk of past experience irrelevant.

2. *Experiment.* Although possibly the least likely to be in error, direct experiment can be both time consuming and costly. Size and speed of deformation are often reduced when conducting laboratory studies. Unfortunately, lubricant performance and heat transfer behave differently at different speeds and sizes, and their effects are generally altered. The most valid experiment, therefore, is one conducted under full-size and full-speed production conditions—generally too costly to consider to any great degree. Although laboratory experiments can provide valuable insight, caution should be exercised when extrapolating lab-scale results to more realistic, often orders of magnitude different, production conditions.

3. *Process modeling.* Here, the process is approached through high-speed computing and one or more mathematical models. Numerical values are selected for the various independent variables, and the models are used to compute predictions for the dependent outcomes. Most techniques rely on the applied theory of plasticity with various simplifying assumptions. Alternatives vary from crude, first-order approximations to sophisticated, computer-based methods, such as finite element analysis. Various models may incorporate strain hardening, thermal softening, heat transfer, and other phenomena. Solutions may be algebraic relations that describe the process and reveal trends and relations between the variables, or simply numerical values based on the specific input features.

17.5 Process Modeling

Metal forming simulations using the finite element modeling method became common in the 1980s but generally required high-power minicomputers or engineering workstations. By the mid-1990s, the rapid increase in computing power made it possible to model complex processes on desktop personal computers. With the continued expansion of computing power and speed, process simulations are now quick, inexpensive, and quite accurate. As a result, modeling is being used in all areas of manufacturing, including part design, manufacturing process design, heat-treatment and surface-treatment optimization, and others. Models can predict how a material will respond to a rolling process, fill a forging die, flow through an extrusion die, or solidify in a casting. Loads on equipment and tooling can be predicted, and defects can be prevented. Hot forming models can predict the effects of temperature and the variation in temperature that would occur due to heat transfer to the environment and tooling. Cold forming models incorporate the effects of strain hardening. Entire heat treatments can be simulated, including cooling rates in various quenchants. Models can even predict the strain distribution, residual stresses, microstructure, and final properties at all locations within a product.

Advanced simulation techniques can provide a clear and thorough understanding of a process, eliminating costly trial-and-error development cycles. Product design and manufacturing methods can be optimized for quality and reliability, while reducing production costs and minimizing lead times. When coupled with appropriate sensors, the same models can be used to determine the type of adjustments needed to provide real-time process control. Process models can also serve as laboratory tools to explore new ideas or new products. New employees can become familiar with what works and what doesn't in a quick and inexpensive manner.

It is important to note, however, that the accuracy of any model can be no better than that of the input variables. For example, when modeling a metal forming operation, the mechanical properties of the deforming material (i.e., yield strength, ductility) must be known for the specific conditions of temperature, strain (amount of prior deformation), and strain rate (speed of deformation) being considered. The mathematical descriptions of material behavior as a function of the process conditions are known as **constitutive relations**. The development of such relationships is not an easy task, however, because the same material may respond differently to the same conditions if its microstructure is different. A medium-carbon steel that has been annealed (ferrite and pearlite) will not have the same properties as a quenched and tempered (tempered martensite) steel of the same chemistry. Microstructure and its effects on properties are difficult to describe in quantitative terms that can be input to a model.

Another rather elusive variable is the **friction** between the tool and the workpiece. Studies have shown that friction depends on contact pressure, contact area, lubricant, speed, and the surface finish and mechanical properties of the two contacting materials. We know that these parameters often vary from location to location and also change during a process, but many models tend to describe friction with a single variable of constant magnitude. Any variations with time and location are simply ignored in favor of mathematical simplicity, or because of a lack of any better information.

At first glance, problems such as those just discussed appear to be a significant barrier to the use of mathematical models. It should be noted, however, that the same difficulties apply to the person trying to document, characterize, and extrapolate the conditions of experience or experiments. **Process modeling** often reveals features that might otherwise go unnoticed and can be quite useful when attempting to prevent or eliminate defects, optimize performance, or extend a process into a previously unknown area.

17.6 General Parameters

Although much metal forming knowledge is specific to a given process, there are certain features that are common to all processes.

It is extremely important to characterize the *material being deformed*. What is its strength or resistance to deformation at the relevant conditions of temperature, speed of deformation, and amount of prior straining? What are the formability limits and conditions of anticipated fracture? What is the effect of temperature or variations in temperature? To what extent will the material strain-harden? What are the recrystallization kinetics? Will the material react with various environments or lubricants? These and many other questions must be answered to assess the suitability of a material to a given deformation process. Because the properties of engineering materials vary widely, the details will not be presented at this time. The reader is referred to the various chapters on engineering materials, as well as the more in-depth references cited in Chapter 9 and the reference appendix.

Another general parameter is the *speed of deformation* and the various related effects. Some rate-sensitive materials may shatter or crack if impacted, but will deform plastically when subjected to slow-speed loadings. Other materials appear to be stronger when deformed at higher speeds. For these *speed-sensitive materials*, more energy is needed to produce the same result if we wish to do it faster, and stronger tools may be required. Mechanical data obtained at slow strain rates (as in tensile tests) may be totally useless if the deformation process operates at a significantly greater rate of deformation. **Speed sensitivity** is also greatest when the material is at elevated temperature, a condition that is frequently encountered in metal forming operations. The selection of hammer or press for the hot forging of a small product may well depend on the speed sensitivity of the material being forged.

In addition to the changes in mechanical properties, faster deformation speeds tend to promote improved lubricant efficiency. Faster speeds also reduce the time for heat transfer and cooling. During hot working, workpieces stay hotter, and less heat is transferred to the tools.

Other general parameters include *friction, lubrication* and *temperature*. Both of these are of sufficient importance that they will be discussed in some detail.

Production rates, tool design, tool wear, and process optimization all depend on the ability to determine and control friction between the tool and workpiece.

In most cases, we would like to economically reduce the effects of friction. However, some deformation processes, such as rolling, can only operate when sufficient friction is present. Regardless of the process, friction effects are hard to measure. As previously noted, the specific friction conditions depend on a number of variables, including contact area, contact pressure, surface finish, speed, type and amount of lubricant, and temperature. Because of the many variables, the effects of friction are extremely difficult to scale down for laboratory testing, or extrapolate from laboratory tests to production conditions.

It should be noted that friction under metalworking conditions is significantly different from the friction encountered in most mechanical devices. The friction conditions of gears, bearings, journals, and similar components generally involve (1) two surfaces of similar material and similar strength, (2) experiencing elastic loads such that neither body undergoes permanent change in shape, (3) with wear-in cycles that produce surface compatibility, and (4) operating temperatures in the low-to-moderate range. In contrast, metal forming operations involve a hard, nondeforming tool interacting with a soft workpiece at pressures sufficient to cause plastic flow in the weaker material. Only a single pass is involved as the tool and workpiece interact, the workpiece is often at elevated temperature, and the contact area is frequently changing as the workpiece deforms.

Figure 17.2 shows the typical relationship between frictional resistance and contact pressure. For light, elastic loads, friction is directly proportional to the applied pressure, with the proportionality constant, μ, being known as the **coefficient of friction** or, more specifically, the *Coulomb coefficient of friction*. At high pressures, friction becomes independent of contact pressure and is more closely related to the strength of the weaker material.

An understanding of these results can be obtained from modern friction theory, whose primary premise is that "flat

17.7 Friction, Lubrication, and Wear under Metalworking Conditions

In metal forming, high forces and pressures are applied through tools to induce deformation, and there is relative motion between the workpiece and the tooling. An important consideration, therefore, is the friction that exists at the tool–workpiece interface. For some processes, more than 50% of the input energy is spent in overcoming friction. Changes in lubrication can alter the mode of material flow during forming, create or eliminate defects, alter the surface finish and dimensional precision of the product, and modify product properties.

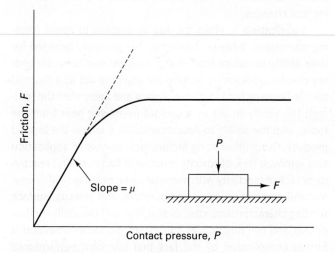

FIGURE 17.2 The effect of contact pressure on the frictional resistance between two surfaces.

surfaces are not flat" but have some degree of roughness. When two irregular surfaces interact, sufficient contact is established to support the applied load. At the lightest of loads, only three points of contact may be necessary to support a plane. As the load is increased, the contacting points deform and the contact area increases, initially in a linear fashion. As the load continues to increase, more area comes into contact. Finally, at some high value of load, there is full contact between the surfaces. Additional loads can no longer bring additional area into contact, and friction can now be described by a constant, independent of pressure.

Friction is the resistance to sliding along an interface. From a mechanistic viewpoint, this resistance can be attributed to (1) **abrasion**, the force necessary to plow the peaks of a harder material through a softer one, and/or (2) **adhesion**, the force necessary to rip apart microscopic weldments that form between the two materials. Because the weldment tears generally occur in the weaker of the two materials, it is reasonable to assume that the resistance attributed to both features would be proportional to the strength of the weaker material and also to the actual area of metal-to-metal contact. Thus, the curve depicted in Figure 17.2 could also be viewed as a plot of actual contact area at the interface versus contact pressure. Unfortunately, that figure and the associated theory applies only to unlubricated metal-to-metal contact. The addition of a lubricant, as well as any variation in its type or amount, can significantly alter the frictional response.

Surface deterioration or **wear** is another phenomenon that is directly related to friction. Because the workpiece only interacts with the tooling during its own forming operation, any wear experienced by the workpiece is usually not objectionable. In fact, the shiny, fresh-metal surface produced by wear is often viewed as desirable. Wear on the tooling, however, is quite the reverse. Tooling is expensive and is expected to shape many workpieces while maintaining consistent control of product dimensions. Tool wear changes those dimensions, and at some point the tools will have to be replaced. Other consequences of tool wear include increased frictional resistance (increased required power and decreased process efficiency), poor surface finish on the product, and loss of production during tool changes.

Lubrication is often the key to success in metal forming operations. Whereas lubricants are generally selected for their ability to reduce friction and suppress tool wear, secondary considerations may include the ability to act as a thermal barrier (keeping heat in the workpiece and away from the tooling), the ability to act as a coolant (removing heat from the tools), and the ability to retard corrosion if left on the formed product. Other influencing factors include ease of application and removal; lack of toxicity, odor, and flammability; reactivity or lack of reactivity with material surfaces; adaptability over a useful range of pressure, temperature, and velocity; surface wetting characteristics; cost; availability; and the ability to flow or thin and still function as a lubricant. Lubricant selection is further complicated by the fact that lubricant performance may change with any change in the interface conditions. The exact response is often dependent on such factors as the finish of both surfaces, the area of contact, the applied load, the speed, the temperature, and the amount of lubricant.

The ability to select an appropriate lubricant can be a critical factor in determining whether a process is successful or unsuccessful, efficient or inefficient. Considerable effort, therefore, has been directed to the study of friction and lubrication, a subject known as **tribology**, as it applies to both general metalworking conditions and specific metal forming processes. There are thousands of lubricant chemistries available, and most can be further modified by various additives. All of them, however, can be grouped into several general types. *Straight oils* are petroleum-based compounds that are easy to apply and offer good corrosion protection, but can be difficult to clean. *Water-soluble oils* (emulsions) are easier to clean, but compromise on corrosion and long-term product stability. *Synthetic lubricants* are water based and free of petroleum. *Semi-synthetics* are a blend of water, petroleum oils, and emulsifiers. *Dry-film lubricants* are often composed of soaps or polymers that are applied in aqueous form and dried before the forming operation. *Chemically bonded agents* include plated coatings or products that form by chemical reaction with the workpiece.

Extreme pressure additives form a chemical film that bonds to both the workpiece and the tooling, enhancing lubrication under high pressure and high temperature. Boundary additives (including fats and solids such as graphite) attach themselves to the metal surfaces and provide cushioning or separation under high pressure. Other additives can help to promote **hydrodynamic lubrication**, where the combination of lubricant and the speed of relative motion is sufficient to create a lubricant layer thick enough to prevent mechanical contact between the tool and the workpiece. The forces and power required for the operation may decrease by as much as 30 to 40%, and tool wear becomes almost nonexistent.

17.8 Temperature Concerns

In metalworking operations, workpiece temperature can be one of the most important process variables. The role of temperature in altering the properties of a material has been discussed in Chapter 2. In general, an increase in temperature brings about a decrease in strength, an increase in ductility, and a decrease in the rate of strain hardening—all effects that would tend to promote ease of deformation.

Forming processes tend to be classified as hot working, cold working, or warm working based on both the temperature and the material being formed. In hot working, the deformation is performed under conditions of temperature and strain rate where recrystallization occurs simultaneously with the deformation. To achieve this, the temperature of deformation is usually in excess of 0.6 times the melting point of the material on an absolute temperature scale (Kelvin or Rankine). Cold working is deformation under conditions where the recovery processes are not active. Here the working temperatures are usually less than 0.3 times the workpiece melting temperature. Warm working is deformation under the conditions of transition (i.e., a working temperature between 0.3 and 0.6 times the melting point).

Hot Working

Hot working is defined as the plastic deformation of metals at a temperature above the recrystallization temperature. It is important to note, however, that the recrystallization temperature varies greatly with different materials. Tin is near hot-working conditions at room temperature; steels require temperatures in excess of 1100°C (2000°F); and tungsten does not enter the hot-working regime until about 2200°C (4000°F). Thus, the term *hot working* does not necessarily correlate with high or elevated temperature, although such is usually the case.

As shown in Figures 2.30 and 2.31, elevated temperatures bring about a decrease in the yield strength of a metal and an increase in ductility. At the temperatures of hot working, recrystallization eliminates the effects of strain hardening, so there is no significant increase in yield strength or hardness, or corresponding decrease in ductility. The true stress–true strain curve is essentially flat once we exceed the yield point, and deformation can be used to drastically alter the shape of a metal without fear of fracture, and without the requirement of excessively high forces. In addition, the elevated temperatures promote diffusion that can remove or reduce chemical inhomogeneities, pores can be welded shut or reduced in size during the deformation; and the metallurgical structure is often altered through recrystallization to improve the final properties. An added benefit is observed for steels, where hot working involves the deformation of the weak, ductile, face-centered-cubic austenite structure, which then cools and transforms to the stronger body-centered-cubic ferrite, or much stronger nonequilibrium structures, such as martensite.

From a negative perspective, the high temperatures of hot working may promote undesirable reactions between the metal and its surroundings. Tolerances are poorer due to thermal contraction, and warping or distortion can occur due to nonuniform cooling. The metallurgical structure may also be nonuniform because the final grain size depends on the amount of deformation, the temperature of the last deformation and recrystallization, the cooling history after the deformation, and other factors, all of which may vary throughout a workpiece.

Recrystallization sets the minimum temperature for hot working, and the upper limit for hot working is usually determined by factors such as excess oxidation, grain growth, or undesirable phase transformations. To keep the forming forces as low as possible and enable hot deformation to be performed for a reasonable amount of time, the starting temperature of the workpiece is usually set at or near the highest temperature for hot working.

Structure and Property Modification by Hot Working.

When metals solidify into the large sections that are typical of ingots or continuously cast slabs or strands, coarse structures tend to form with a certain amount of chemical segregation. The size of the grains is usually not uniform, and undesirable grain shapes can be quite common, such as the columnar grains that have been revealed in **Figure 17.3**. Small gas cavities or shrinkage porosity can also form during solidification.

If a cast metal is reheated without prior deformation, it will simply experience grain growth and the accompanying deterioration in engineering properties. However, if the metal first experiences a sufficient amount of deformation, the distorted structure will be rapidly replaced by new strain-free grains. This **recrystallization** is then followed by (1) grain growth, (2) additional deformation and recrystallization, or (3) a drop in temperature that will terminate diffusion and "freeze in" the recrystallized structure. The structure in the final product is that formed by the last recrystallization and the thermal history that followed. By replacing the initial structure with a new one consisting of fine, spherical-shaped grains, it is possible to produce an increase not only in strength but also in ductility and toughness—a somewhat universal enhancement of properties.

Engineering properties can also be improved through the reorientation of inclusions or impurity particles that are present within the metal. With normal melting and cooling, many impurities tend to locate along grain boundary interfaces. If these are unfavorably oriented or intersect surfaces, they can initiate a crack or assist its propagation. When a metal is plastically deformed, the impurities tend to flow along with the base metal or fracture into rows of fragments (**stringers**) that are aligned in the direction of working. These nonmetallic impurities do not recrystallize with the base metal but retain their distorted shape and orientation. The product exhibits an **oriented** or **flow structure**, like the one shown in **Figure 17.4**, and final properties tend to exhibit directional variation. Through proper design of the deformation, impurities can often be reoriented into a "crack-arrestor" configuration where they are parallel

FIGURE 17.3 Cross section of a 4-inch-diameter cast copper bar polished and etched to show the as-cast grain structure. (Courtesy Ronald Kohser)

FIGURE 17.4 Etched surface of a forged component showing the grain flow pattern that gives forged parts enhanced strength, fracture resistance and fatigue life. (Courtesy of HHI Forging, Royal Oak, MI)

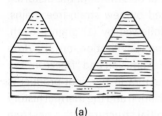

(a)

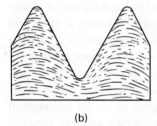

(b)

FIGURE 17.5 Schematic comparison of the grain flow in a machined thread (a) and a rolled thread (b). The rolling operation further deforms the axial structure produced by the previous wire- or rod-forming operations, while machining simply cuts through it.

to the exterior surface (perpendicular to the direction of crack propagation). This is especially important in regions of high stress, such as interior corners. The impurities would have crack initiator or crack propagator orientations only where they intersect perpendicular to free surfaces, which hopefully are low-stress or noncritical locations.

An example of property improvement through reorientation of defects can be seen in **Figure 17.5** that schematically compares a machined thread and a rolled thread in a threaded fastener. By reorienting the axial defects in the starting wire or rod to be parallel to the thread profile, the rolled thread will offer improved strength and fracture resistance. (Note: Threads are usually rolled under cold conditions, but is presented here as a very common example of defect reorientation.)

Temperature Variations.

The success or failure of a hot deformation process often depends on the ability to control the temperatures within the workpiece. More than 90% of the energy imparted to a deforming workpiece will be converted into heat. If the deformation process is sufficiently rapid, the temperature of the workpiece may actually increase. More common, however, is the cooling of the workpiece in its lower-temperature environment. Heat is lost through the workpiece surfaces, with the majority of the loss occurring where the workpiece is in direct contact with lower-temperature tooling. Nonuniform temperatures are produced,

and flow of the hotter, weaker, interior may well result in cracking of the colder, less ductile, surfaces. Thin sections cool faster than thick sections, and this may further complicate the flow behavior.

To minimize problems, it is desirable to keep the workpiece temperatures as uniform as possible. Heated dies can reduce the rate of heat transfer, but die life tends to be compromised. For example, dies are frequently heated to 325 to 450°C (600 to 850°F) when used in the hot forming of steel. Tolerances could be improved and contact times could be increased if the tool temperatures could be raised to 550 to 650°C (1000 to 1200°F), but tool life drops so rapidly that these conditions become quite unattractive.

A final concern is the cooldown from the temperatures of hot working. Nonuniform cooling can introduce significant amounts of **residual stress** in hot-worked products. Associated with these stresses may be warping or distortion and possible cracking.

Cold Working

The plastic deformation of metals below the recrystallization temperature is known as **cold working**. Here, the deformation is usually performed at room temperature, but mildly elevated temperatures may be used to provide increased ductility and reduced strength. From a manufacturing viewpoint, cold working has a number of distinct advantages, and the various cold-working processes have become quite prominent. Recent advances have expanded the capabilities, and a trend toward increased cold working appears likely to continue.

When compared to hot working, the advantages of cold working include the following:

1. No heating is required.
2. Better surface finish is obtained.
3. Superior dimensional control is achieved because the tooling sets dimensions at room temperature. As a result, little, if any, secondary machining is required.
4. Products possess better reproducibility and interchangeability.
5. Strength, fatigue, and wear properties are all improved through strain hardening.
6. Directional properties can be imparted.
7. Contamination problems are minimized.

Some disadvantages associated with cold-working processes include the following:

1. Higher forces are required to initiate and complete the deformation.
2. Heavier and more powerful equipment and stronger tooling are required.
3. Less ductility is available.
4. Metal surfaces must be clean and scale-free.

5. Intermediate anneals may be required to compensate for the loss of ductility that accompanies strain hardening.

6. The imparted directional properties may be detrimental.

7. Undesirable residual stresses may be produced.

The strength levels induced by **strain hardening** are often comparable to those produced by the strengthening heat treatments. Even when the precision and surface finish of cold working are not required, it may be cheaper to produce a product by cold working a less-expensive material (achieving the strength by strain hardening) than by heat-treating parts that have been hot formed from a heat-treatable alloy. As an added benefit, most cold-working processes eliminate or minimize the production of waste material and the need for subsequent machining—a significant feature with today's emphasis on conservation and materials recycling.

Because the cold-forming processes require powerful equipment and product-specific tools or dies, they are best suited for large-volume production of precision parts where the quantity of products can justify the cost of the equipment and tooling. Considerable effort has been devoted to developing and improving cold-forming machinery along with methods to enable these processes to be economically attractive for modest production quantities. By grouping products made from the same starting material and using quick-change tooling, cold-forming processes can often be adapted to small-quantity or just-in-time manufacture.

Metal Properties and Cold Working.

The suitability of a metal for cold working is determined primarily by its mechanical properties, and these are a direct consequence of its metallurgical structure. Cold working then alters that structure, thereby altering the mechanical properties of the resulting product. It is important for both the incoming and outgoing properties to be considered when selecting metals that are to be processed by cold working.

Figure 17.6 presents the true stress–true strain curves for both a low- and high-carbon steel. Focusing on the low-carbon material, we note that plastic deformation cannot occur until the strain exceeds the strain associated with the elastic limit, point a on the stress–strain curve. Plastic deformation then continues until the strain reaches the value x_4, where the metal ruptures. From the viewpoint of cold working, two features are significant: (1) the magnitude of the yield-point stress, which determines the force required to initiate permanent deformation, and (2) the extent of the strain region from x_1 to x_4, which indicates the amount of plastic deformation (or ductility) that can be achieved without fracture. If a considerable amount of deformation is desired, a material like the low-carbon steel is more desirable than the high-carbon variety. Greater ductility would be available, and less force would be required to initiate and continue the deformation. The curve on the right, however, has a higher strain-hardening coefficient (see Chapter 2 for discussion). If strain hardening is being used to impart strength, this material would have a greater increase in strength for the same amount of cold work. In addition, the material on the right would be more attractive for shearing operations and may be easier to machine (see Chapter 21).

Springback is another cold-working phenomenon that can be explained with the aid of a stress–strain diagram. When a metal is deformed by the application of a load, part of the resulting deformation is elastic. For example, if a metal is stretched to point x_1 on either of the curves of Figure 17.6 and the load is removed, it will return to its original size and shape because all the deformation is elastic. If, on the other hand, the metal is stretched by an amount x_3, corresponding to point b on the stress–strain curve, the total strain is made up of two parts: a portion that is elastic and another that is plastic. When the deforming load is removed, the stress relaxation will follow line bx_2, and the final strain will only be x_2. The decrease in strain, $x_3 - x_2$, is known as **elastic springback**.

In cold-working processes, springback can be extremely important. If a desired size is to be achieved, the deformation must be extended beyond that point by an amount equal to the springback. Because different materials have different elastic moduli, the amount of springback from a given load will change from one material to another. A substitution in material, therefore, may well require adjustments in the forming process. Fortunately, springback is a predictable phenomenon, and most difficulties can be prevented by proper design procedures.

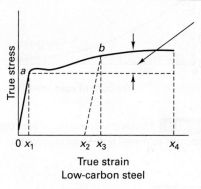

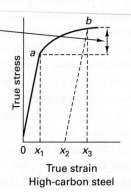

Increase in tensile strength due to work hardening. The high-carbon steel will also have more springback.

True stress / True strain / Low-carbon steel

True stress / True strain / High-carbon steel

FIGURE 17.6 Use of true stress–true strain diagrams to assess the suitability of two metals for cold working. (Courtesy Ronald Kohser)

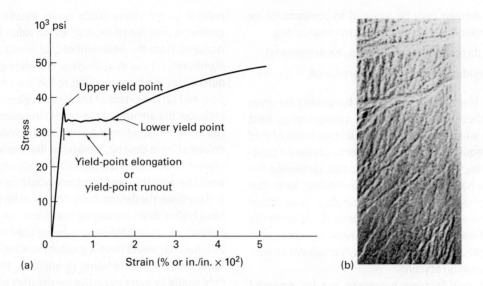

(a)   (b)

FIGURE 17.7 (*Left*) Stress–strain curve for a low-carbon steel showing the commonly observed yield-point runout; (*Right*) Luders bands or stretcher strains that form when this material is stretched to an amount less than the yield-point runout. (Courtesy Ronald Kohser)

Initial and Final Properties in a Cold Working Process.

The quality of the starting material is often key to the success or failure of a cold-working operation. To obtain a good surface finish and maintain dimensional precision, the starting material must be clean and free of oxide or scale that might cause abrasion and damage to the dies or rolls. Scale can be removed by **pickling**, a process in which the metal is dipped in acid and then washed. In addition, sheet metal and plate are sometimes given a light cold rolling prior to the major deformation. The rolling operation not only ensures uniform starting thickness but also produces a smooth starting surface.

The light cold-rolling pass can also serve to remove the **yield-point** phenomenon and the associated problems of nonuniform deformation and surface irregularities in the product. **Figure 17.7** presents an expansion of the left-hand region of Figure 2-6 or the low-carbon portion of Figure 17.6. After loading to the upper yield point, the material exhibits a **yield-point runout** wherein the material can strain up to several percent with no additional force being required. Consider a piece of sheet metal that is to be formed into an automotive body panel. If a segment of that panel were to receive a total stretch less than the magnitude of the yield-point runout, it would be induced by a stress equal to the yield-point stress. Because the stress is constant in the runout region, segments of the material are free to not deform at all, to deform the entire amount of the yield-point runout, or to select some point in between. It is not uncommon for some regions to deform the entire amount and thin correspondingly, while adjacent regions resist deformation and retain the original thickness. The resulting ridges and valleys, shown in Figure 17.7, are referred to as **Luders bands** or **stretcher strains** and are very difficult to remove or conceal. By first cold rolling the material to a strain near or past the yield-point runout, all subsequent forming occurs in a region where a well-defined strain corresponds to each value of stress. If the body panel were shaped from prerolled material,

the deformation and thinning would be uniform throughout the piece.

Figure 17.8 shows how the mechanical properties of pure copper are affected by cold working. Individual tensile tests were conducted on specimens that had experienced progressively greater amounts of cold work.[1] As the graph shows, yield strength and tensile strength increase significantly with

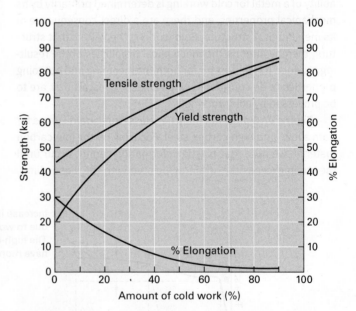

FIGURE 17.8 Mechanical properties of pure copper as a function of the amount of cold work (expressed in percent).

---

[1] A block of copper was rolled in increments to progressively reduced thickness, the decrease in height being compensated by an increase in length with width remaining constant. The amount of cold work can then be computed as: Percent cold work = $[(A_o - A_f) / A_o]$ × 100%, where A_o is the original cross-sectional area and A_f is the cross-sectional area at each of the computed increments.

increased deformation. Hardness is not presented on the graph, but generally follows tensile strength. Elongation, along with reduction in area, electrical conductivity, and corrosion resistance will all decline. Because the ductility decreases, the amount of cold working is generally limited by the onset of fracture.

To maximize the starting ductility, an **annealing** heat treatment is often applied to a metal prior to cold working. If the required amount of deformation exceeds the fracture limit, one or more **intermediate anneals** may be necessary to restore ductility. If the desired final properties require a specific amount of cold work, as in Problem 1 at the end of this chapter, the last anneal can be judiciously positioned in the deformation cycle. In this way, the desired shape can be produced along with the mechanical properties that accompany the amount of cold work imparted after the last anneal. In all annealing operations, care should be exercised to control the grain size of the resulting material. Grain sizes that are too large or too small can both be detrimental.

Like hot working, cold working also produces an anisotropic structure—one whose properties vary with direction. Here, the **anisotropy** is related to the distorted crystal structure and is not simply a function of the nonmetallic inclusions. Also associated with cold working is the generation of **residual stresses**. Although anisotropy and residual stresses can be beneficial, they can also be quite harmful. Because they occur as a consequence of cold working, their effect on performance should always be considered.

Warm Working

Deformation performed at temperatures intermediate to hot and cold forming is known as **warm working**. Compared to cold working, warm working offers the advantages of reduced loads on the tooling and equipment, increased material ductility, and a possible reduction in the number of anneals due to a reduction in the amount of strain hardening. The use of higher forming temperatures can often expand the range of materials and geometries that can be formed by a given process or piece of equipment. High-carbon steels may be formed without a spheroidization treatment.

Compared to hot working, the lower temperatures of warm working produce less scaling and decarburization and enable production of products with better dimensional precision and smoother surfaces. Finish machining is reduced, and less material is converted into scrap. Because of the finer structures and the presence of some strain hardening, the as-formed properties may be adequate for many applications, enabling the elimination of final heat-treatment operations. The warm regime generally requires less energy than hot working due to the decreased energy in heating the workpiece (lower temperature), energy saved through higher precision (less material being heated), and the possible elimination of postforming heat treatments. Although the tools must exert 25 to 60% higher forces, they last longer because there is less thermal shock and thermal fatigue.

When energy was cheap, metal forming was usually conducted in either the hot- or cold-working regimes, and warm working was largely ignored. Even today, material behavior is less well characterized for the warm-working temperatures (the warm-working temperatures for steel are between 550 and 800°C or 1000 and 1500°F). Lubricants have not been as fully developed for the warm-working temperatures and pressures, and die design technology is not as well established. Nevertheless, the pressures of energy and material conservation, coupled with the other cited benefits, strongly favor the continued development of warm working. Cold forming is still the preferred method for fabricating small components, but warm forming is considered to be attractive for larger parts (up to about 5 kg or 10 lb) and steels with more than 0.35% carbon and/or high alloy content.

Hot working and warm working are usually applied to bulk forming processes, like forging and extrusion. For sheet material, the surface-to-volume ratio is sufficiently large that the workpiece can rapidly lose its heat. As auto manufacturers seek to increase fuel efficiency, there has been significant interest in aluminum sheet as a replacement for steel. Unfortunately, the formability of high-strength aluminum is much lower than similar strength low-carbon steels. If the steel is simply replaced with aluminum, and the design and tooling remain unchanged, fracture often occurs in the more heavily worked regions. If the material, die, and blank holder are all heated to 200 to 300°C (400 to 575°F), however, the aluminum sheet shows a significant increase in formability and a decrease in elastic springback. Satisfactory parts can generally be produced. **Figure 17.9** shows how the properties of aluminum change with temperature in the warm-working regime.

Isothermal Forming

Figure 17.10 shows the relationship between yield strength (or forging pressure) and temperature for several engineering metals. The 1020 and 4340 steels show a moderate increase in strength with decreasing temperature. In contrast, the strength of the titanium alloy (open circles) and the A-286 nickel-based superalloy (solid circles) shows a much stronger variation. During hot working operations, surfaces cool when in contact with tooling; and small changes in temperature can produce significant increases in strength. Variations in temperature can lead to nonuniform deformation and cracking of the less ductile surface.

To successfully deform temperature-sensitive materials, deformation may have to be performed under **isothermal forming** (constant temperature) conditions. The dies or tooling must be heated to the same temperature as the workpiece, sacrificing die life for product quality. Deformation speeds must be slowed so that any heat generated by deformation can be removed in a manner that would maintain a uniform and constant temperature. Inert atmospheres may be required because of the long times at elevated temperature. Although such methods are indeed costly, they are often the only means of producing satisfactory products from certain materials.

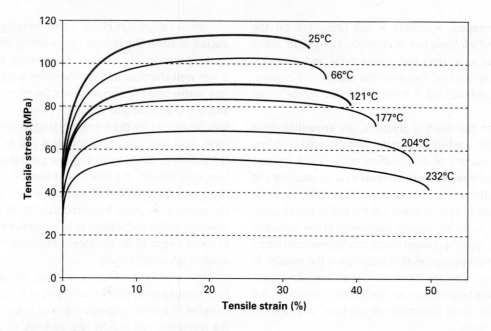

FIGURE 17.9 Increasing the temperature increases the formability of aluminum, decreasing the strength and increasing the ductility. (Adapted from data provided by Interlaken Technology Corporation, Chaska, MN)

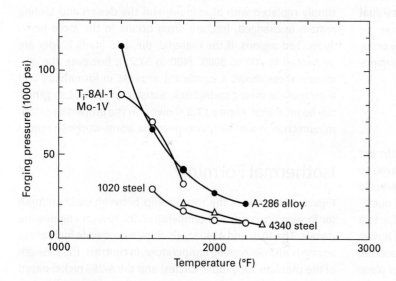

FIGURE 17.10 Yield strength of various materials (indicated by pressure required to forge a standard specimen) as a function of temperature. Materials with steep curves require isothermal forming. (From "A Study of Forging Variables," ML-TDR-64-95, March 1964; courtesy of Battelle Columbus Laboratories, Columbus, OH)

Because of the uniform temperatures and slow deformation speeds, isothermally formed components generally exhibit close tolerances, low residual stresses, and fairly uniform metal flow.

17.9 Formability

As discussed in Section 2-5, the word **formability**, although often used to describe the ability of a material to be deformed, is really an extremely complex and rather nebulous concept. There is no universal formability scale. A material that responds well to one forming process and one set of process conditions may fail miserably when used in another process or under a different set of process conditions. All of the independent variables—part design, die design, lubrication, speed of deformation, material temperature, die or tooling temperature, and even the specific operator—can have a significant effect on the observed formability.

Review Questions

1. What is plasticity?

2. What are some of the general assets of the metal deformation processes? Some general liabilities?

3. Why might large production quantities be necessary to justify metal deformation as a means of manufacture?

4. What types of deformation may occur in forming processes?

5. What is an independent variable in a metal forming process?

6. What are some considerations regarding selection of the starting material for a forming process?

7. What is the significance of tool and die geometry in designing a successful metal forming process?

8. Why is lubrication often a major concern in metal forming?

9. What are some of the secondary effects that may occur when the speed of a metal forming process is varied?

10. What is a dependent variable in a metal forming process?

11. Why is it important to be able to predict the forces or powers required to perform specific forming processes?

12. Why is it important to control the final properties of a deformation product?

13. Why is it important to know and control the thermal history of a metal as it undergoes deformation?

14. Why is it important to control not only the external shape of a formed product, but also its internal flow?

15. Why is it often difficult to determine the specific relationships between independent and dependent variables?

16. What are the three distinct ways of determining the interrelation of independent and dependent variables?

17. What features limit the value of laboratory experiments in modeling metal forming processes?

18. What are some of the features that may be incorporated into metal forming process models?

19. What features have contributed to the expanded use of process modeling?

20. What are some of the uses or applications of process models?

21. What features may limit the accuracy of a mathematical model?

22. What is a constitutive relation for an engineering material?

23. What simplifying assumptions are often made regarding friction between the tool and workpiece?

24. What type of information about the material being deformed may be particularly significant to a metal forming engineer?

25. How might a material's performance vary with changes in the speed of deformation?

26. Why is friction such an important parameter in metalworking operations?

27. Why are friction effects in metalworking difficult to scale down for laboratory testing, or scale-up from laboratory conditions to production conditions?

28. What are several ways in which the friction conditions during metalworking differ from the friction conditions found in most mechanical equipment?

29. How might friction be mathematically described under conditions of light, elastic loading? Under conditions of heavy loads, sufficient to induce plastic deformation?

30. How can actual contact area be used to explain the observed friction versus contact pressure curve?

31. According to modern friction theory, frictional resistance can be attributed to what two physical phenomena?

32. Discuss the significance of wear in metal forming: (a) wear on the workpiece, and (b) wear on the tooling.

33. Lubricants are often selected for properties in addition to their ability to reduce friction. What are some of these additional properties?

34. What is tribology?

35. What are some of the common types of metal forming lubricants?

36. What is hydrodynamic lubrication? What are some of its benefits?

37. If the temperature of a material is increased, what changes in properties might occur that would promote the ease of deformation?

38. Define the various regimes of cold working, warm working, and hot working in terms of the melting point of the material being formed.

39. What is an acceptable definition of hot working? Is a specific temperature involved?

40. What are some of the attractive manufacturing and metallurgical features of hot-working processes?

41. What are some of the negative aspects of hot working?

42. What factors generally set the upper temperature limit for hot working?

43. How can hot working be used to improve the grain structure of a metal?

44. If the deformed grains recrystallize during hot working, how can hot deformation impart an oriented or flow structure (and directionally dependent properties)?

45. Why might a rolled thread offer improved strength and fracture resistance compared to a machined thread?

46. How might the temperature of a deforming workpiece actually increase, even though it is being deformed in a colder environment?

47. Why are heated dies or tools often employed in hot-working processes?

48. What generally restricts the upper temperature to which dies or tooling is heated?

49. What is the primary cause of residual stresses in hot-worked products?

50. What is cold working?

51. Compared to hot working, what are some of the advantages of cold-working processes?

52. What are some of the disadvantages of cold-forming processes?

53. How could cold working be used to reduce the cost of a moderate-to-high-strength product?

54. Why are cold forming processes best suited for large-volume production of precision parts?

55. How can the tensile test properties of a metal be used to assess its suitability for cold forming?

56. Why is elastic springback an important consideration in cold-forming processes?

57. What is pickling, and how does it remove surface oxide or scale?

58. What are Luders bands or stretcher strains, and what causes them to form? How can they be eliminated?

59. What engineering properties are likely to decline during the cold working of a metal?

60. How can the selective placement of the final intermediate anneal be used to establish desired final properties in a cold formed product?

61. Is the anisotropy induced by cold working an asset or a liability? What about the residual stresses?

62. What is warm working?

63. What are some of the advantages of warm forming compared to cold forming? Compared to hot forming?

64. Why is warm forming attractive for the shaping of aluminum products?

65. What material feature is considered to be the driving force for isothermal forming?

66. Why is isothermal forming considerably more expensive than conventional hot forming?

67. Why is there no standard test to assess or quantify formability?

Problems

1. Select a metal forming process and an important desired outcome (i.e., dependent variable). For that desired outcome, discuss the various independent variables that would influence the result.

2. Copper is being reduced from a hot-rolled ⅜-in. diameter rod to a final diameter of 0.100 in. by wire drawing through a series of dies. The final wire should have a yield strength in excess of 50,000 psi and an elongation greater than 10%. Use Figure 17.8 to determine a desirable amount of final cold work. Compute the placement of the last intermediate anneal so that the final product has both the desired size and the desired properties.

3. a. List and discuss the various economic factors that should be considered when evaluating a possible switch from cold forming to warm forming.

  b. Repeat part (a) for a possible conversion from hot forming to warm forming.

4. An advertisement for automobile spark plugs has cited the superiority of rolled threads over machined threads. Figure 17.5 shows such a comparison for hot forming, where the deformation process reorients flaws and defects without significantly changing the structure and properties of the metal. The spark plug threads, however, were formed by cold rolling. Do the same benefits apply? Discuss the assets and liabilities of the cold rolling of threads compared to thread formation by conventional machining.

5. Computer modeling of metal deformation processes is a powerful and extremely useful tool. At the same time, there are several areas of limitation that can significantly compromise or even invalidate the final results. Consider each of the following areas of limitation,

investigating what is currently being used or what current options are available:

  a. *A mathematical description of material behavior (a constitutive equation):* In almost all cases, some simplification of actual flow behavior is assumed. For accurate modeling, flow behavior should be known and mathematically characterized as a function of strain, strain rate, and temperature.

  b. *Interfacial friction between the tooling and the workpiece:* How is this being modeled? Does it consider the effects of surface finish, sliding velocity, interface temperature, and numerous other factors? As the process commences, lubricants may thin or be wiped from surfaces, forces and pressures change, temperatures change, and surface roughness or texture is modified. Does the model reflect any of these changes? Some models assign a single value to friction over the entire contact surface. This value may also remain constant throughout the entire operation.

  c. *Assignment of boundary conditions:* Often the mathematical solutions must conform to assigned features, such as defined motions or stresses at specified surfaces. The boundary conditions have a profound effect on the results that are calculated. Poor choices or choices made to facilitate easy analysis can often produce misleading or erroneous results.

6. Computer control of processes has enabled a relatively new family known as thermomechanical processes, in which heating or cooling is synchronized with deformation to produce new or improved structures and properties, eliminate the need for subsequent heat treatment, and reduce overall cost. Identify a thermomechanical process and describe the control required and the obtained benefits.

CHAPTER 18

Bulk Forming Processes

18.1 Introduction

The shaping of metal by deformation is as old as recorded history. The Bible, in the fourth chapter of Genesis, introduces Tubal-cain and cites his ability as a worker of metal. Although we have no description of his equipment, it is well established that metal forging was practiced long before written records. Processes such as rolling and wire drawing were common in the Middle Ages and probably date back much further. In North America, by 1680 the Saugus Iron Works near Boston had an operating drop forge, rolling mill, and slitting mill.

Although the basic concepts of many forming processes have remained largely unchanged throughout history, the details and equipment have evolved considerably. Manual processes were converted to machine processes during the Industrial Revolution. The machinery then became bigger, faster, and more powerful. Water wheel power was replaced by steam and then by electricity. More recently, computer-controlled, automated operations have become the norm with many support operations being performed by industrial robots.

18.2 Classification of Deformation Processes

A wide variety of processes have been developed to mechanically shape material, and a number of classification methods have been proposed. One approach divides the processes into primary and secondary. *Primary processes* reduce a cast material into intermediate shapes, such as slabs, plates, or billets. *Secondary processes* further convert these shapes into finished or semifinished products. Unfortunately, some processes clearly fit both categories, depending on the particular product being made.

In Chapter 17, we discussed the temperature of deformation and presented the various regimes based on the temperature of the workpiece. These included cold forming, warm forming, hot forming, and isothermal forming. This classification has also become somewhat blurred, especially with the increased emphasis on energy conservation. Processes that were traditionally performed hot are now being performed

cold, and cold-forming processes can often be enhanced by some degree of heating. Warm working has experienced considerable growth.

Chapters 18 and 19 will utilize a division that focuses on the size and shape of the workpiece and how that size and shape is changed. **Bulk deformation processes** are those in which the thicknesses or cross sections are reduced or shapes are significantly changed. Because the volume of the material remains constant, changes in one dimension require proportionate changes in others. Thus, the enveloping surface area changes significantly, usually increasing as the product lengthens or the shape becomes more complex. The bulk forming operations can be performed in all of the temperature regimes. Common processes include rolling; forging; extrusion; cold forming; and wire, rod, and tube drawing.

In contrast, **sheet-forming** operations involve the deformation of a material where the thickness and surface area remain relatively constant. Common processes include shearing or blanking, bending, and deep drawing. Because of the large surface-to-volume ratio, sheet material tends to lose heat rapidly, and most sheet-forming operations are performed cold.

Even this division is not without confusion, however. Coining, for example, begins with sheet material but alters the thickness in a complex manner that is essentially bulk deformation. The bulk deformation processes will be presented here in Chapter 18. Sheet forming processes can be found in Chapter 19.

18.3 Bulk Deformation Processes

The bulk deformation processes that will be presented in this chapter include the following:

1. Rolling
2. Forging
3. Extrusion
4. Wire, rod, and tube drawing
5. Cold forming, cold forging, and impact extrusion
6. Piercing
7. Other squeezing processes

These processes can be further divided in several ways. One grouping separates the processes by focusing on the size and shape of the deforming region. In continuous flow processes, such as forging, the size and shape are continually changing, and process analysis must reflect this change. In processes such as rolling or wire drawing, material moves through the deforming region, but the size and shape of that region remains unchanged. Some form of steady-state analysis can often be applied.

In all of the bulk forming processes, the primary deformation stress is compression. This may be applied directly by tools or dies that squeeze the workpiece, or indirectly as in wire drawing, where the workpiece is pulled in tension but the resisting die generates compression in the region undergoing deformation.

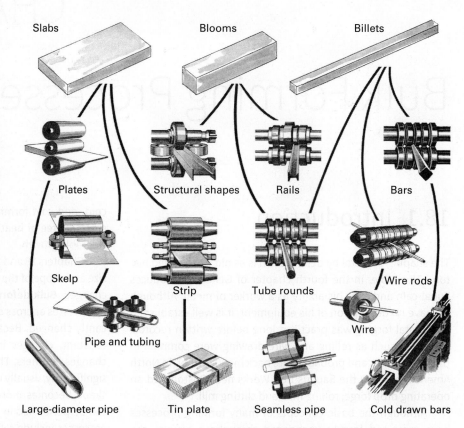

FIGURE 18.1 Flowchart for the production of various finished and semifinished steel shapes. Note the abundance of rolling operations. (Courtesy of American Iron and Steel Institute, Washington, DC)

18.4 Rolling

Rolling operations reduce the thickness or change the cross section of a material through compressive forces exerted by rolls. As shown in **Figure 18.1**, rolling is often the first process that is used to convert material into a finished wrought product. Thick starting stock can be rolled into blooms, billets, or slabs, or these shapes can be obtained directly from continuous casting. A **bloom** has a square or rectangular cross section, with a thickness greater than 15 cm (6 in.) and a width no greater than twice the thickness. A **billet** is usually smaller than a bloom and has a square or circular cross section. Billets are usually produced by some form of deformation process, such as rolling or extrusion. A **slab** is a rectangular solid where the width is greater than twice the thickness. Slabs can be further rolled to produce **plate**, **sheet**, and **strip**. Plates have thickness greater than 6 mm (¼ in.), whereas sheet and strip ranges from 6 mm to 0.1mm (¼ in. to 0.004 in.).

These hot-rolled products often form the starting material for subsequent processes, such as cold forming or machining. Sheet and strip can be fabricated into products or further cold rolled into thinner, stronger material, or even into **foil** (thicknesses less than 0.1 mm). Blooms and billets can be further rolled into finished products, such as **structural shapes** or railroad rail, or they can be processed into semifinished shapes, such as **bar**, **rod**, **tube**, or **pipe**.

From a tonnage viewpoint, rolling is clearly predominant among all manufacturing processes, with approximately 90% of all metal products experiencing at least one rolling

operation. Rolling equipment and rolling practices are sufficiently advanced that standardized, uniform-quality products can be produced at relatively low cost. Because shaped rolls are both massive and costly, shaped products are only available in standard forms and sizes where there is sufficient demand to permit economical production.

Basic Rolling Process

In the basic rolling process, shown in **Figure 18.2**, metal is passed between two rolls that rotate in opposite directions, the gap between the rolls being somewhat less than the thickness of the entering metal. Because the rolls rotate with a surface velocity that exceeds the speed of the incoming metal, friction along the contact interface acts to propel the metal forward. The metal is then squeezed and elongates to compensate for the decrease in thickness or cross-sectional area. The amount of deformation that can be achieved in a single pass between a given pair of rolls depends on the friction conditions along the interface. If too much is demanded, the rolls cannot advance the material and simply skid over its surface. If too little deformation is taken, the operation will be successful, but the additional passes required to produce a given part will increase the cost of production.

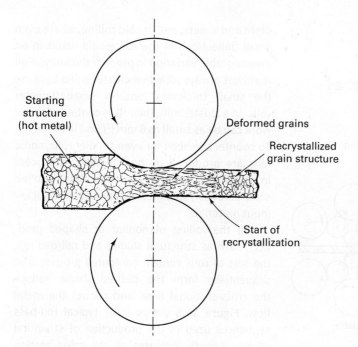

Starting structure (hot metal)

Deformed grains

Recrystallized grain structure

Start of recrystallization

FIGURE 18.2 Schematic representation of the hot-rolling process, showing the deformation and recrystallization of the metal being rolled.

Hot Rolling and Cold Rolling

In **hot rolling**, as with all hot-working processes, temperature control is required for success. The starting material should be heated to a uniform elevated temperature. If the temperature is not uniform, the subsequent deformation will not be uniform. Consider a piece being reheated for rolling. If the soaking time is insufficient, the hotter exterior will flow in preference to the cooler, stronger interior. Conversely, if a uniform-temperature material is allowed to cool prior to working or has cooled during previous working operations, the cooler surfaces will tend to resist deformation. Cracking and tearing of the surface may result as the hotter, weaker interior tries to deform.

It is not uncommon for high-volume producers to begin with continuous-cast feedstock. The cooling from solidification is controlled to enable direct insertion into a hot-rolling operation without additional handling or reheating. For smaller operations or secondary processing, the starting material is often a room-temperature solid, such as an ingot, slab, or bloom. This material must first be brought to the desired rolling temperature, usually in gas- or oil-fired soaking pits or furnaces. For plain-carbon and low-alloy steels, the soaking temperature is usually about 1200°C (2200°F). For smaller cross sections, induction coils may be used to heat the material prior to rolling.

Hot-rolling operations are usually terminated when the temperature falls to about 50° to 100°C (100°° to 200°F) above the recrystallization temperature of the material being rolled. Such a **finishing temperature** ensures the production of a uniform fine grain size and prevents the possibility of unwanted

strain hardening. If additional deformation is required, a period of reheating will be necessary to reestablish desirable hot-working conditions.

Cold rolling can be used to produce sheet, strip, bar, and rod products with extremely smooth surfaces and accurate dimensions. Cold-rolled sheet and strip can be obtained in various conditions, including *skin-rolled*, *quarter-hard*, *half-hard*, and *full-hard*. Skin-rolled metal is subjected to only a 0.5 to 1% reduction to produce a smooth surface and uniform thickness, and to remove or reduce the yield-point phenomenon (i.e., prevent formation of Luders bands on further forming). This material is well suited for subsequent cold-working operations where good ductility is required. Quarter-hard, half-hard, and full-hard sheet and strip experience greater amounts of cold reduction, up to 50%. Their yield points are higher, properties have become directional, and ductility has decreased. Quarter-hard steel can be bent back on itself across the grain without breaking. Half-hard and full-hard can be bent back 90° and 45°, respectively, about a radius equal to the material thickness.

For products with a uniform cross section and cross-sectional dimensions less than about 5 cm (2 in.), cold rolling of rod or bar may be an attractive alternative to extrusion or machining. Strain hardening can provide up to 20% additional strength to the material, and the process offers the smooth surfaces and high dimensional precision of cold working. Like the rolling of structural shapes, however, the process generally requires a series of shaping operations. Separate roll passes (and roll grooves) may be required for sizing, breakdown, roughing, semiroughing, semifinishing, and finishing. Although the various grooves may be in a single set of rolls, a minimum order of several tons may be required to justify the cost of tooling.

Rolling Mill Configurations

As illustrated in **Figure 18.3**, rolling mill stands are available in a variety of roll configurations. Early reductions, often called primary, roughing, or breakdown passes, usually employ a two- or three-high configuration with rolls 60 to 140 cm (24 to 55 in.) in diameter. The *two-high nonreversing mill* is the simplest design, but the material can only pass through the mill in one direction. A *two-high reversing mill* permits back-and-forth rolling, but the rolls must be stopped, reversed, and brought back to rolling speed between each pass. A *three-high mill* eliminates the need for roll reversal but requires some form of elevator on each side of the mill to raise or lower the material and possibly some form of mechanical manipulators to turn or shift the product between passes.

As shown in **Figure 18.4**, smaller-diameter rolls produce less length of contact for a given reduction and therefore require lower force and less energy to produce a given change in shape. The smaller cross section, however, provides reduced stiffness, and the rolls are prone to flex elastically because they are supported on the ends and pressed

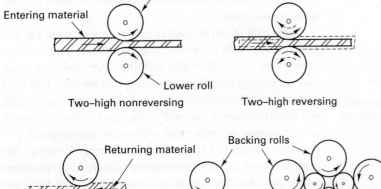

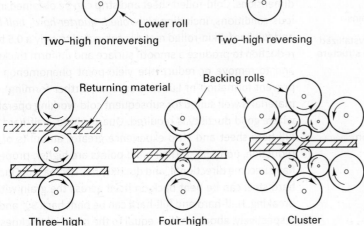

FIGURE 18.3 Various roll configurations used in rolling operations.

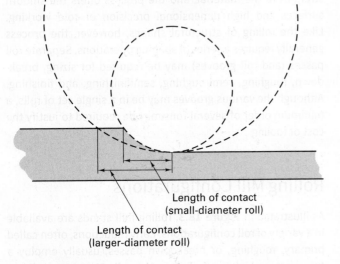

FIGURE 18.4 The effect of roll diameter on the length of contact for a given reduction.

apart by the metal passing through the middle (a condition known as three-point bending). *Four-high* and *cluster* arrangements use backup rolls to support the smaller work rolls. These configurations are used in the hot rolling of wide

plate and sheets, and in cold rolling, where even small deflections in the roll would result in an unacceptable variation in product thickness. Foil is almost always rolled on **cluster mills** because the small thickness requires small-diameter rolls. In a cluster mill, the roll in contact with the work can be as small as 6 mm ($^1/_4$ in.) in diameter. To counter the need for even smaller rolls, some foils are produced by **pack rolling**, a process in which two or more layers of metal are rolled simultaneously as a means of providing a thicker input material.

In the rolling of nonflat or shaped products, such as structural shapes and railroad rail, the sets of rolls contain contoured grooves that sequentially form the desired shape, reduce the cross-sectional area, and control the metal flow. **Figure 18.5** shows some typical roll-pass sequences used in the production of structural shapes. Length increases as the cross section is reduced.

Continuous (or Tandem) Rolling Mills

When the volume of a product justifies the investment, rolling may be performed on a **continuous** or **tandem rolling mill**. Billets, blooms, or slabs are heated and fed through an integrated series of nonreversing rolling mill stands, where each stand consists of a frame and set of rolls that perform a reduction on

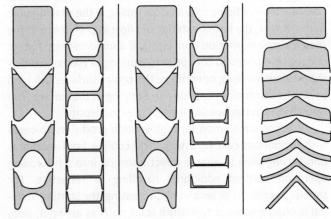

FIGURE 18.5 Typical roll-pass sequences used in producing structural shapes.

Example: Household Aluminum Foil

One side of household aluminum foil is shiny, and the opposite side is dull. This is the result of a pack rolling operation where two sheets are simultaneously passed through the rolling operation to further reduce their thickness. The shiny side is in contact with the rubbing roll, and the dull side is the surface in contact with the other aluminum sheet. A lubricant prevents the two sheets from welding to one another.

the material. Continuous mills for the hot rolling of steel strip, for example, often consist of a roughing train of approximately four four-high mill stands and a finishing train of six or seven additional four-high stands. In a continuous structural mill, the rolls in each stand contain only one set of shaped grooves, in contrast to the multigrooved rolls used when the product is produced by back-and-forth passes through a single stand.

If a single piece of material is in multiple rolling stations at the same time, it is imperative that the same volume pass through each stand in the same amount of time. If the cross section is reduced, speed must be increased proportionately. Therefore, as a material is reduced in size, the rolls of each successive stand must turn faster than those of the preceding one. If a subsequent stand is running too slow, material will accumulate between stands. If the demand for incoming material exceeds the output of the previous stand, the material is placed in tension and may tear or rupture.

The synchronization of six or seven mill stands is not an easy task, especially when key variables such as temperature and lubrication may vary during a single run, and the product may be exiting the final stand at speeds in excess of 110 km/hr (70 mph). Computer control is basic to successful rolling, and modern mills are equipped with numerous sensors to provide the needed information. When continuous casting units feed directly into continuous rolling mills, the time lapse from final solidification to finished rolled product is often a matter of a few minutes.

Ring Rolling

Ring rolling is a special rolling process where one roll is placed through the hole of a thick-walled ring, and a second roll presses in from the outside (**Figure 18.6**). As the rolls squeeze and rotate, the wall thickness is reduced and the diameter of the ring increases. Shaped rolls can be used to produce a wide variety of cross-section profiles. The resulting seamless rings have a circumferential grain orientation and find application in products such as rockets, turbines, airplanes, pipelines, and pressure vessels. Diameters can be as large as 8 m (25 ft) with face heights as great as 2 m (80 in.).

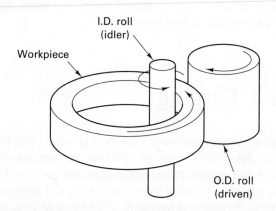

FIGURE 18.6 Schematic of a horizontal ring rolling operation. As the thickness of the ring is reduced, its diameter will increase.

Thread Rolling

Thread rolling is a deformation alternative to the cutting of threads, and is illustrated in Figure 17-5 and discussed in Chapter 30.

Characteristics, Quality, and Precision of Rolled Products

Because hot-rolled products are formed and finished above their recrystallization temperature, they have little directionality in their properties and are relatively free of deformation-induced residual stresses. These characteristics may vary, however, depending on the thickness of the product and the presence of complex sections. Nonmetallic inclusions do not recrystallize, so they may impart some degree of directionality. In addition, residual stresses can be induced by nonuniform cooling from the temperatures of hot working. Thin sheets often show directional characteristics, whereas thicker plate (above 20 mm or 0.8 in.) will usually have very little. Because of high residual stresses in the rapidly cooled edges, a complex shape, such as an I- or H-beam, may warp in a noticeable fashion if a portion of one flange is cut away.

As a result of the hot deformation and the good control that is maintained during processing, hot-rolled products are normally of uniform and dependable quality. It is quite unusual to find any voids, seams, or laminations when produced by reliable manufacturers. The surfaces of hot-rolled products are usually a bit rough, however, and are originally covered with a tenacious high-temperature oxide, known as **mill scale**. This can be removed by an acid pickling operation, resulting in a surprisingly smooth surface finish. The dimensional tolerances of hot-rolled products vary with the kind of metal and the size of the product. For most products produced in reasonably large tonnages, the tolerances are within 2 to 5% of the specified dimension (either height or width).

Cold rolled products exhibit superior surface finish and dimensional precision and can offer the enhanced strength obtained through strain hardening.

Flatness Control and Rolling Defects

To roll a flat product with uniform thickness, the gap between the rolls must be a uniform one. Attaining such an objective, however, may be difficult. Consider the upper roll in a set that is rolling sheet or plate. As shown in **Figure 18.7**, the material presses upward in the middle of the roll, while the roll is held in place by bearings that are mounted on either end and are supported in the mill frame. The roll, therefore, is loaded in three-point bending and tends to flex in a manner that produces a thicker center and thinner edge. Because the thicker center will not lengthen as much as the thinner edge, the result is often a product with either a wavy edge or a fractured center.

If the rolls are always used to reduce the same material at the same temperature by the same amount, the forces and

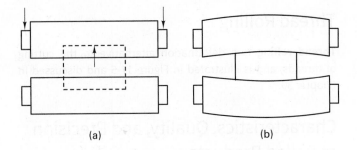

(a) Loading on a rolling mill roll. The top roll is pressed upward in the center while being supported on the ends. (b) The elastic response to the three-point bending.

deflections can be predicted, and the roll can be designed to have a specified profile. If a **crowned**, or barrel-shaped, **roll** is subjected to the designed load, it will deflect into flatness, as illustrated in **Figure 18.8**. If the applied load is not of the designed magnitude, however, the resulting profile will not be flat, and defects may result. If the correction is insufficient, for example, wavy edges or center fractures may still occur. If the correction is excessive, the center becomes thinner and longer, and the result may be a wavy center or cracking of the edges.

Because roll deflections are proportional to the forces applied to the rolls, product flatness can also be improved by measures that reduce these forces. If possible, friction could be reduced, smaller-diameter rolls could be used (reduced area of contact), and smaller reductions could be employed. Heating the workpiece generally makes it weaker, so increased workpiece temperature will also reduce the force on the rolls. Horizontal tensions can be applied to the piece as it is being rolled (strip tension in sheet metal rolling). Because these tensions combine with the vertical compression to deform the piece (stretching while squeezing), the roll forces and associated deflections are less. Other techniques to improve flatness include an increase in the elastic modulus of the rolls themselves through material selection or providing some form of backup support to oppose deflection, as with the four-high and cluster mill configurations.

Successful rolling requires the balancing of many factors relating to the material being rolled, the variables of the rolling process, and lubrication between the workpiece and the rolls. Common defects include the nonuniform thickness previously

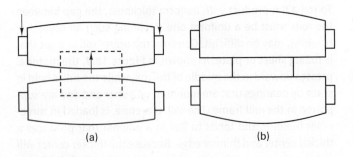

Use of a "crowned" roll to compensate for roll flexure. When the roll flexes in three-point bending, the crowned roll flexes into flatness.

discussed, dimensional variations caused by changes in workpiece temperature, surface flaws (such as rolled-in scale and roll marks), laps, seams, and various types of distortions.

Thermomechanical Processing and Controlled Rolling

As with most deformation processes, rolling is generally considered to be a way of changing the shape of a material. Heat may be used to reduce forces and promote plasticity, however, the thermal processes that produce or control product properties (heat treatments) are usually performed as subsequent operations. **Thermomechanical processing**, of which **controlled rolling** is an example, consists of integrating deformation and thermal processing into a single process that will produce not only the desired shape, but also the desired properties, such as strength and toughness. The heat for the property modification is the same heat used in the rolling operation, and subsequent heat treatment becomes unnecessary.

A successful thermomechanical operation begins with process design. The starting material must be specified and the composition closely maintained. Then a time–temperature–deformation system must be developed to achieve the desired objective. Possible goals include production of a uniform fine grain size; controlling the nature, size, and distribution of the various transformation products (such as ferrite, pearlite, bainite, and martensite in steels); controlling the reactions that produce solid solution strengthening or precipitation hardening; and producing a desired level of toughness. Starting structure (controlled by composition and prior thermal treatments), deformation details, temperature during the various stages of deformation, and the conditions of cooldown from the working temperature must all be specified and controlled. Moreover, the attainment of uniform properties requires uniform temperatures and deformations throughout the product. Computer-controlled facilities are an absolute necessity if thermomechanical processing is to be successfully performed.

Possible benefits of thermomechanical processing include improved product properties; substantial energy savings (by eliminating subsequent heat treatment); and the possible substitution of a cheaper, less-alloyed metal for a highly alloyed one that responds to heat treatment.

18.5 Forging

Forging is a term applied to a family of processes that induce plastic deformation through localized compressive forces applied through dies. The equipment can take the form of hammers, presses, or special forging machines. The deformation can be performed in all temperature regimes (hot, cold, warm, or isothermal), but most forging is done with workpieces above the recrystallization temperature.

Forging is clearly the oldest known metalworking process. From the days when prehistoric peoples discovered that they could heat sponge iron and beat it into a useful implement by hammering with a stone, forging has been an effective method of producing many useful shapes. Modern forging is simply an extension of the ancient art practiced by the armor makers and immortalized by the village blacksmith. High-powered hammers and mechanical presses have replaced the strong arm and the hammer, and tool steel dies have replaced the anvil. Metallurgical knowledge has supplemented the art and skill of the craftsman, as we seek to control the heating and handling of the metal. Parts can range in size from ones whose largest dimension is less than 2 cm (1 in.) to others weighing more than 170 metric tons (450,000 lb).

The variety of forging processes offers a wide range of capabilities. A single piece can be economically fashioned by some methods, whereas others can mass-produce thousands of identical parts. The metal may be (1) *drawn out* to increase its length and decrease its cross section, (2) *upset* to decrease the length and increase the cross section, or (3) *squeezed in closed impression dies* to produce multidirectional flow. As indicated in Table 15-2, the state of stress in the work is primarily uniaxial or multiaxial compression.

Common forging processes include

1. Open-die drop-hammer forging
2. Impression-die drop-hammer forging
3. Press forging
4. Orbital forging
5. Upset forging
6. Automatic hot forging
7. Roll forging
8. Swaging
9. Net-shape and near-net-shape

Open-Die Hammer Forging

In concept, **open-die hammer forging** is the same type of forging done by the blacksmith of old, but massive mechanical equipment is now used to impart the repeated blows. The metal is first heated to the proper temperature using a furnace or electrical induction heating. An impact is then delivered by some type of mechanical hammer. The simplest industrial hammer is a **gravity drop hammer**, where a free-falling ram strikes the workpiece, and the energy of the

blow is varied by adjusting the height of the drop. Most forging hammers now employ some form of energy augmentation, however, where pressurized air, steam, or hydraulic fluids are used to raise and propel the hammer. Higher striking velocities are achieved, with more control of striking force, easier automation, and the ability to shape pieces up to several tons. **Computer-controlled hammers** can provide blows of differing impact speed (energy) for different products or for each of the various stages of a given operation. Their use can greatly increase the efficiency of the process and also minimize the amount of noise and vibration, which are the most common outlets for the excess energy not absorbed in the deformation of the workpiece. **Figure 18.9** shows a large double-frame hammer along with a labeled schematic.

Open-die forging does not fully control the flow of metal. To obtain the desired shape, the operator must orient and position the workpiece between blows. The hammer may contact the workpiece directly, or specially shaped tools can be inserted to assist in making concave or convex surfaces, forming holes, or performing a cutoff operation. Manipulators may be used to position larger workpieces, which may weigh several tons. Although some finished parts can be made by this technique, open-die forging is usually employed to pre-shape metal in preparation for further operations. For example, consider parts like turbine rotors and generator shafts with dimensions up to 20 m (70 ft) in length and up to 1 m (3 ft) in diameter. Open-die forging induces oriented plastic flow and minimizes the amount of subsequent machining.

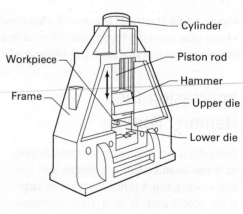

FIGURE 18.9 (Left) Double-frame drop hammer. (Right) Schematic diagram of a forging hammer. (Courtesy of Erie Press Systems, Erie, PA)

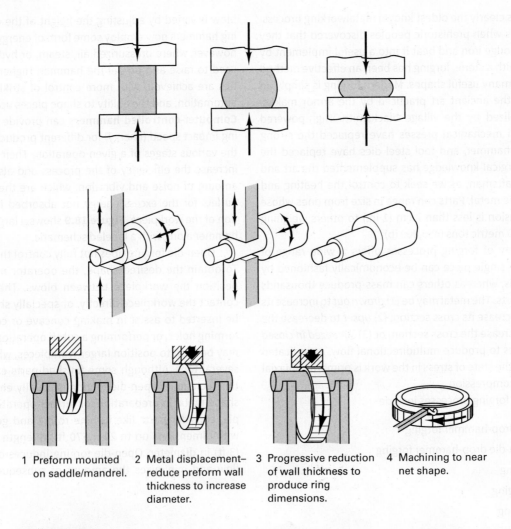

1 Preform mounted on saddle/mandrel.

2 Metal displacement– reduce preform wall thickness to increase diameter.

3 Progressive reduction of wall thickness to produce ring dimensions.

4 Machining to near net shape.

FIGURE 18.10 *(Top)* Illustration of the unrestrained flow of material in open-die forging. Note the barrel shape that forms due to friction between the die and material. *(Middle)* Open-die forging of a multidiameter shaft. *(Bottom)* Forging of a seamless ring by the open-die method. (Courtesy of Forging Industry Association, Cleveland, OH)

Figure 18.10 illustrates the unrestricted flow of material and shows how open-die forging can be used to shape a multidiameter cylindrical shaft and a seamless metal ring.

the hammer and the lower piece to the anvil. Heated metal is positioned in the lower cavity and struck one or more blows by the upper die. The hammering causes the metal to flow

Impression-Die Hammer Forging

Open-die hammer forging (or *smith forging*, as it has been called) is a simple and flexible process, but it is not practical for large-scale production. It is a slow operation, and the shape and dimensional precision of the resulting workpiece are dependent on the skill of the operator. As shown in **Figure 18.11**, **impression-die** or **closed-die forging** overcomes these difficulties by using shaped dies to control the flow of metal. **Figure 18.12** shows a typical set of multicavity dies. The upper piece attaches to

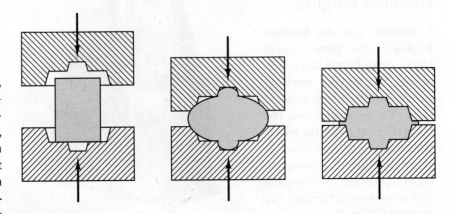

FIGURE 18.11 Schematic of the impression-die forging process showing partial die filling and the beginning of flash formation in the center sketch, and the final shape with flash in the right-hand sketch.

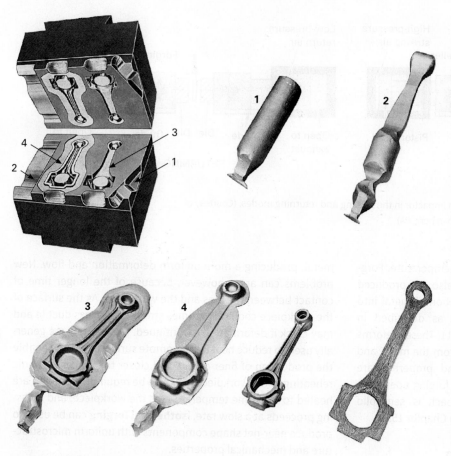

away. Intermediate impressions are for **blocking** the metal to approximately its final shape, with generous corner and fillet radii. For small production lots, the cost of further cavities may not be justified, and the blocker-type forgings are simply finished by machining. More often, the final shape and size are imparted by an additional forging operation in a final or **finisher impression**, after which the flash is trimmed from the part. Figure 18.12 shows an example of these steps and the shape of the part at the conclusion of each. Because every part is shaped in the same die cavities, each mass-produced part is a close duplicate of all the others.

Conventional closed-die forging begins with a simple hot-rolled shape and utilizes reheating and working to progressively convert it into a more complex geometry. The shape of the various cavities controls the flow of material, and the flow, in turn, imparts the oriented structure discussed in Chapter 17. Grain flow that follows the external contour of the component is in the crack arrestor orientation, improving strength, ductility, and resistance to impact and fatigue. By controlling the size and shape of various cross sections, the metal can be distributed as needed to resist the applied loads. Because of these factors, coupled with a fine recrystallized grain structure (hot working) and the absence of voids (compressive forming stresses),

FIGURE 18.12 Impression drop-forging dies and the product resulting from each impression. The flash is trimmed from the finished connecting rod in a separate trimming die. The sectional view shows the grain flow resulting from the forging process. (Courtesy of Forging Industry Association, Cleveland, OH)

and completely fill the die cavity. Excess metal is squeezed out along the parting line to form a **flash** around the periphery of the cavity. This material cools rapidly, increases in strength, and by resisting deformation, effectively blocks the formation of additional flash. By trapping material within the die, the flash then ensures the filling of all of the cavity details. The flash is ultimately trimmed from the part in a final forging operation.

In **flashless forging**, also known as *true closed-die forging*, the metal is deformed in a cavity that provides total confinement. Accurate workpiece sizing is required because complete filling of the cavity must be ensured with no excess material. Accurate workpiece positioning is also necessary, along with good die design and control of lubrication. The major advantage of this approach is the elimination of the scrap generated during flash formation, an amount that is often between 20 and 45% of the starting material.

Most conventional forgings are impression-die with flash. They are produced in dies with a series of cavities where one or more blows of the hammer are used for each step in the sequence. The first impression is often an *edging*, *fullering*, or *bending* impression to distribute the metal roughly in accordance with the requirements of the later cavities. Edging gathers material into a region, whereas fullering moves material

forgings often have about 20% higher strength-to-weight ratios compared with cast or machined parts of the same material.

Board hammers, steam hammers, and air hammers have all been used in impression die forging. An alternative to the hammer and anvil arrangement is the **counterblow machine**, or **impactor**, illustrated in **Figure 18.13**. These machines have two horizontal hammers that simultaneously impact a workpiece that is positioned between them. Excess energy simply becomes recoil, in contrast to the hammer and anvil arrangement, where energy is lost to the machine foundation, and a heavy machine base is required. Impactors also operate with less noise and less vibration and produce distinctly different flow of material, as illustrated in **Figure 18.14**.

Direct quenching or controlled cooling of the hot parts as they emerge from the forging operation can significantly reduce or eliminate heat treatment costs. Energy conservation can also be achieved through several processes that have been designed to produce a product that is somewhere between a conventional forging and a conventional casting. In one approach, a forging preform is cast from liquid metal, removed from the mold while still hot, and then finish-forged in a single-cavity die. The flash is then trimmed and the part is

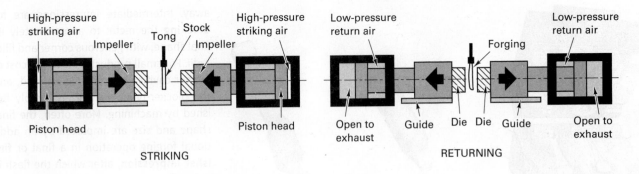

FIGURE 18.13 Schematic diagram of an impactor in the striking and returning modes. (Courtesy of Chambersburg Engineering Company, Chambersburg, PA)

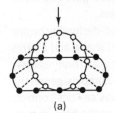

(a)

Conventional forged disk with paths of flow

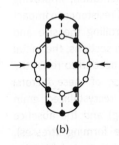

(b)

Disk formed by impactor with paths of flow

FIGURE 18.14 A comparison of metal flow in conventional forging and impacting.

quenched to room temperature. Forging preforms can also be produced by the spray deposition of metal into shaped containers, as described in Chapter 16, Section 11. These preforms are then removed from the mold, and the final shape and properties are imparted by a final forging operation. Still another approach is semisolid forging, discussed in Chapter 15.

Press Forging

In hammer or impact forging, the metal flows to dissipate the energy imparted in the hammer–workpiece collision. Speeds are high, so the forming time is short. Contact times under load are on the order of milliseconds. There is little time for heat transfer and cooling of the workpiece, and the adiabatic heating that occurs during deformation helps to minimize chilling. It is possible, however, for all of the energy to be dissipated by deformation of just the surface region of the metal (coupled with additional absorption by the anvil and foundation) and the interior of the workpiece to remain essentially undeformed.

If large pieces or thick products are to be formed, **press forging** may be required. The deformation is now analyzed in terms of forces or pressures (rather than energy), and the slower squeezing action penetrates completely through the

metal, producing a more uniform deformation and flow. New problems can arise, however, because of the longer time of contact between the dies and the workpiece. As the surface of the workpiece cools, it becomes stronger and less ductile and may crack if deformation is continued. Heated dies are generally used to reduce heat loss, promote surface flow, and enable the production of finer details and closer tolerances. Periodic reheating of the workpiece may also be required. If the dies are heated to the same temperature as the workpiece, and pressing proceeds at a slow rate, **isothermal forging** can be used to produce near-net shape components with uniform microstructure and mechanical properties.

Forging presses are of two basic types, mechanical and hydraulic, and are usually quite massive. **Mechanical presses** use cams, cranks, or toggles to produce a preset and reproducible stroke. Because of their mechanical drives, different forces are available at the various stroke positions. Production presses are quite fast, capable of up to 50 strokes per minute, and are available in capacities ranging from 300 to 18,000 tons (3 to 160 MN). **Hydraulic presses** move in response to fluid pressure in a piston and are generally slower, more massive, and more costly to operate. On the positive side, hydraulic presses are much more flexible and can have greater capacity. Because motion is in response to the flow of pressurized drive fluids, hydraulic presses can be programmed to have different strokes for different operations and even different speeds within a stroke. Presses can be used to perform all types of forging, including open-die and impression-die. Impression-die press forgings usually require less draft than hammer forgings and have higher dimensional accuracy. In addition, press forgings can often be completed in a single closing of the dies as opposed to the multiple blows of a hammer. Machines with capacities up to 50,000 tons (445 MN) are currently in operation in the United States.

Example: Wood-Splitting Wedge

Consider the deformation of a metal wood-splitting wedge after it has been struck repeatedly by a sledge hammer. The top is usually "mushroomed," *while the remainder retains the original geometry and taper.*

A third type of press is the **screw press**, which in many ways acts like a hammer. A large flywheel stores a predetermined amount of energy. This energy is then transmitted to a vertical screw, which drives a descending ram. Downward motion stops when all of the energy from the flywheel has been dissipated.

Additional information about the various types of presses and drive mechanisms can be found in the closing section of Chapter 19.

Design of Impression-Die Forgings and Associated Tooling

The geometrical possibilities for impression-die forging are quite numerous, with complex shapes like connecting rods, crankshafts, wrenches, and gears being commonly produced. **Figure 18.15** shows a forged-and-machined steel automotive crankshaft that significantly outperformed similar components made of austempered ductile cast iron (higher strength, higher impact toughness, better fatigue resistance, higher ductility, and greater modulus of elasticity). Parts typically range from under 1.5 kg (3 lb) up to about 350 kg (750 lb), with the major dimension being between 20 and 50 cm (7 and 20 in.). Steels, stainless steels, and alloys of aluminum, copper, and nickel can all be forged with fair to excellent results.

The forging dies are usually made of high-alloy or tool steel and can be expensive to design and construct. Impact resistance, wear resistance, strength at elevated temperature, and the ability to withstand cycles of rapid heating and cooling must all be outstanding. In addition, considerable care is required to produce and maintain a smooth and accurate cavity and parting plane. Better and more economical results will be obtained if the following rules are observed:

1. The dies should part along a single, flat plane if at all possible. If not, the parting plane should follow the contour of the part.
2. The parting surface should be a plane through the center of the forging and not near an upper or lower edge.
3. Adequate draft should be provided—at least 3° for aluminum and 5 to 7° for steel.
4. Generous fillets and radii should be provided.
5. Ribs should be low and wide.
6. The various sections should be balanced to avoid extreme differences in metal flow.
7. Full advantage should be taken of fiber flow lines.
8. Dimensional tolerances should not be closer than necessary

The various design details, such as the number of intermediate steps, the shape of each, the amount of excess metal required to ensure die filling, and the dimensions of the flash at each step, are often a matter of experience. Each component is a new design entity and brings its own unique challenges. Computer-aided design has made notable advances, however, and the development and accessibility of high-speed, large-memory computers have enabled the accurate modeling of many complex shapes.

Good dimensional accuracy is a characteristic of impression-die forging. With reasonable care, the dimensions for steel products can be maintained within the tolerances of 0.50 to 0.75 mm (0.02 to 0.03 in.). It should be noted, however, that the dimensions across the parting plane are affected by closure of the dies and are therefore dependent on die wear and the thickness of the final flash. Dimensions contained entirely within a single die segment can be maintained at a significantly greater level of accuracy. Surface finish values range from 80 to 300 μin. Draft angles can sometimes be reduced, occasionally approaching zero, but this is not recommended for general practice.

Selection of a lubricant is also critical to successful forging. The lubricant not only affects the friction and wear and associated metal flow, but it may also be expected to act as a thermal barrier (restricting heat flow from the workpiece to the dies) and a parting compound (preventing the part from sticking in the cavities).

Orbital Forging

Uniform-radius disk-shaped or conical products (such as gear blanks or bevel gears) can be formed by a process known as **orbital forging**, depicted in **Figure 18.16**. A conical-faced upper die is tilted or inclined so a portion of the cone is in contact with the workpiece. The upper die then orbits about a central axis, sweeping the contact area around the workpiece circumference. As the upper die is orbiting, it then descends or the lower die is raised to

FIGURE 18.15 A forged-and-machined automobile engine crankshaft. Forged steel crankshafts provide superior performance, compared to those of ductile cast iron. (© Sergiy Goruppa/Stockphoto)

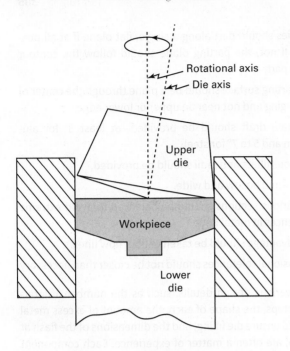

FIGURE 18.16 Schematic depiction of orbital forging.

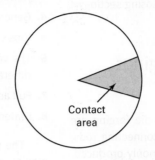

FIGURE 18.17 Set of upset forging dies and punches. The product resulting from each of the four positions is shown along the bottom. (Courtesy of Ajax Manufacturing Company, Euclid, OH)

produce a continuous deformation that is similar to a combined rolling and forging—forming the material much like a rolling pin moving across a lump of dough. Parts are generally formed with 10 to 20 orbits of the upper die.

The forming force is reduced due to the small area of contact between the tool and the workpiece and is further reduced due to the improved lubrication that results from the higher speed of movement between the orbiting tool and the workpiece. Significantly smaller presses can be used (6 to 10 times smaller than conventional), reducing cost and space, noise level is low, and tool life is extended. Strength, hardness, and surface finish can all be improved by cold working.

Upset Forging

Upset forging increases the diameter of a material by compressing its length. Because of its use with myriad fasteners, it is the most widely used of all forging processes when evaluated in terms of the number of pieces produced. Parts can be upset forged both hot and cold, with the operation generally being performed on special high-speed machines. The forging motion is usually horizontal, and the workpiece is rapidly moved from station to station. Although most operations start with wire or rod, some machines can upset bars up to 25 cm (10 in.) in diameter.

Upset forging generally employs split dies that contain multiple positions or cavities, as seen in the typical die set of **Figure 18.17**. The dies separate enough for the bar or rod to advance between them and move into position. They are then clamped together, and a heading tool or ram moves longitudinally against the material, upsetting it into the cavity. Separation of the dies then permits transfer to the next position or removal of the product. If a new piece is started with each die

separation, and an operation is performed in each cavity simultaneously, a finished product can be made with each cycle of the machine. By including a shearing operation as the initial piece moves into position, the process can operate with continuous coil or long-length rod as its incoming feedstock.

Upset-forging machines are often used to form heads on bolts and other fasteners and to shape valves, couplings, and many other small components. The upset region can be on the end or central portion of the workpiece, and the final diameter may be up to three times the original. The following three rules, illustrated in **Figure 18.18**, should be followed when designing parts that are to be upset forged:

1. The length of unsupported metal that can be gathered or upset in one blow without injurious buckling should be limited to three times the diameter of the bar.

2. Lengths of stock greater than three times the diameter may be upset successfully provided that the increase in diameter is not more than 1 times the diameter of the bar.

3. In an upset requiring stock length greater than three times the diameter of the bar, and where the increase in diameter is not more than 1 times the diameter of the bar (the conditions of rule 2), the length of unsupported metal beyond the face of the die must not exceed the diameter of the bar.

Automatic Hot Forging

Several equipment manufacturers now offer highly automated upset equipment in which mill-length steel bars (typically 7 m or 24 ft long) are fed into one end at room temperature, and hot-forged products emerge from the other end at rates of up to 180 parts per minute (86,400 parts per 8-hour shift). These parts can be solid or hollow, round or symmetrical, up to 6 kg (12 lb) in weight, and up to 18 cm (7 in.) in diameter.

The process begins with the lowest-cost steel bar stock—hot-rolled and air-cooled carbon or alloy steel. The bar is first

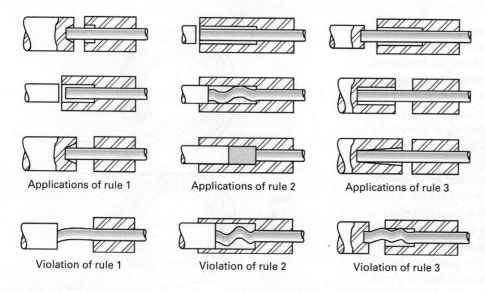

Applications of rule 1   Applications of rule 2   Applications of rule 3

Violation of rule 1   Violation of rule 2   Violation of rule 3

FIGURE 18.18 Schematics illustrating the rules governing upset forging. (Courtesy of National Machinery Company, Tiffin, OH)

heated to 1200 to 1300°C (2200 to 2350°F) in under 60 seconds as it passes through high-power induction coils. It is then descaled by rolls, sheared into individual blanks, and transferred through several successive forming stages, during which it is upset, preformed, final forged, and pierced (if necessary). Small parts can be produced at up to 180 ppm, and larger parts at rates on the order of 90 ppm. **Figure 18.19** shows a typical deformation sequence and a variety of hot-forged ferrous products.

Automatic hot forging has a number of attractive features. Low-cost input material and high production speeds have already been cited. Minimum labor is required, and because no flash is produced, material usage can be as much as

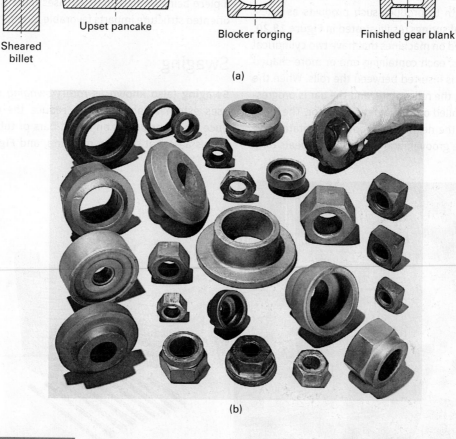

Sheared billet   Upset pancake   Blocker forging   Finished gear blank

(a)

(b)

FIGURE 18.19 (a) Typical four-step sequence to produce a spur-gear forging by automatic hot forging. The sheared billet is progressively shaped into an upset pancake, blocker forging, and finished gear blank. (b) Samples of ferrous parts produced by automatic hot forging at rates between 90 and 180 parts per minute. (Courtesy of National Machinery Company, Tiffin, OH)

20 to 30% greater than with conventional forging. With a consistent finishing temperature near 1050°C (1900°F), an air cool can often produce a structure suitable for machining, eliminating the need for an additional anneal or normalizing treatment. Tolerances are generally within 0.3 mm (0.012 in.), surfaces are clean, and draft angles need only be 0.5 to 1° (as opposed to the conventional 3 to 5°). Tool life is nearly double that of conventional forging because the contact times are only on the order of 0.06 second.

Automatic hot formers can also be coupled with high-rate, cold-forming operations. Preform shapes that are hot formed at rates that approach 180 ppm can then be cold formed to final shape on machines that operate at speeds near 90 ppm. The benefits of the combined operations include high-volume production at low cost, coupled with the precision, surface finish, and strain hardening that are characteristic of a cold-finished material.

To justify an automatic hot forging operation, however, large quantities of a given product must be required. A single production line may well require an initial investment in excess of $10 million.

Roll Forging

In **roll forging**, round or flat bar stock is reduced in thickness and increased in length to produce such products as axles, tapered levers, and leaf springs. As illustrated in **Figure 18.20**, roll forging is performed on machines that have two cylindrical or semicylindrical rolls, each containing one or more shaped grooves. A heated bar is inserted between the rolls. When the bar encounters a stop, the rolls rotate, and the bar is progressively shaped as it is rolled out toward the operator. The piece is then transferred to the next set of grooves (or rotated and reinserted in the same groove), and the process repeats until

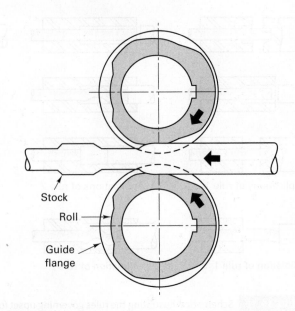

FIGURE 18.21 Schematic of the roll-forging process showing the two shaped rolls and the stock being formed. (Courtesy of Forging Industry Association, Cleveland, OH)

the desired size and shape is produced. Figure 18.20 also shows a set of rolls and the product formed by each set of grooves. **Figure 18.21** shows the cross section of one set of grooves and a piece being formed. In most cases there is no flash, and the oriented structure imparts favorable forging-type properties.

Swaging

Swaging (also known as *rotary swaging* or *radial forging*) uses external hammering to reduce the diameter or produce tapers or points on round bars or tubes. **Figure 18.22** shows a typical swaging machine, and **Figure 18.23** shows

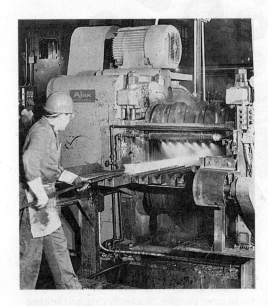

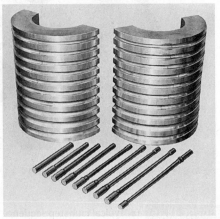

FIGURE 18.20 (Left) Roll-forging machine in operation. (Right) Rolls from a roll-forging machine and the various stages in roll forging a part. (Courtesy of Ajax Manufacturing Company, Euclid, OH)

FIGURE 18.22 Tube being reduced in a rotary swaging machine. (Courtesy of the Timkin Company, Canton, OH)

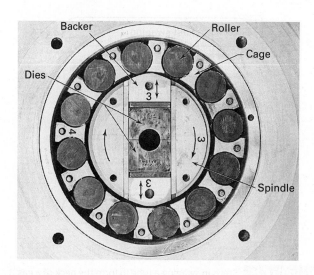

FIGURE 18.23 Basic components and motions of a rotary swaging machine. (*Note:* The cover plate has been removed to reveal the interior workings.) (Courtesy of the Timkin Company, Canton, OH)

a schematic of its internal components. The split dies, located in the center of the apparatus, consist of two blocks of hardened tool steel. They combine to form a central hole that generally has a conical input transitioning to a cylinder. An external motor drives a large, massive flywheel, which is connected to the central spindle of the machine. High-speed rotation of the central unit generates centrifugal force, which causes the matching die segments and backing blocks to separate. As the spindle rotates, the backing blocks are driven into opposing rollers that have been mounted in a massive machine housing. To pass beneath the rollers, the backer blocks must squeeze the dies tightly together. Once the assembly clears the rollers, the dies once again separate and the cycle repeats, generating between 1000 and 3000 blows per minute.

With the machine in motion, the operator simply inserts a rod or tube between the dies and advances it during the periods of die separation. Because the dies rotate, the repeated blows are delivered from various angles, reducing the diameter and increasing the length. Because the rotating spindle is usually hollow, the workpiece can be passed completely through the machine, or withdrawn after a preset length has been reduced.

Swaging operations can also be used to form tubular products with internal cavities of constant cross section. A shaped mandrel is inserted into a thick-walled tube (or hollow-end workpiece), and the metal is collapsed around it to simultaneously shape and size both the interior and exterior of the product. Swaging over a mandrel can be used to form parts with internal gears, splines, recesses, or sockets. **Figure 18.24**

shows a variety of swaged products, many of which contain shaped holes.

The term *swaging* has also been applied to a process where material is forced into a confining die to reduce its diameter. This process is usually performed on heated material. **Figure 18.25** shows a hot swaging sequence being used to form the end of a pressurized gas cylinder.

Net-Shape and Near-Net-Shape Forging

As much as 80% of the cost of a forged gear is incurred during the machining operations that follow forging, and a finished aerospace wing spar may contain as little as 4% of the original billet (the remaining 96% being lost as scrap in the

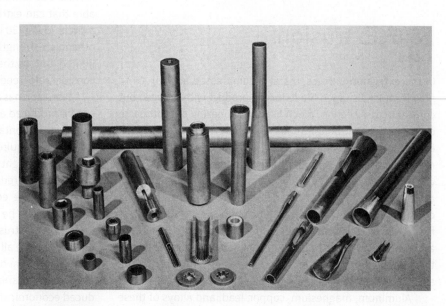

FIGURE 18.24 A variety of swaged parts, some with internal details. (Courtesy of Cincinnati Milacron, Inc., Cincinnati, OH)

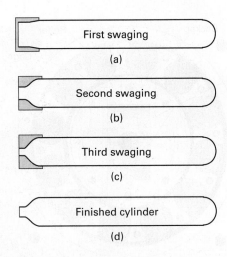

FIGURE 18.25 Steps in swaging a tube to form the neck of a gas cylinder. (Courtesy of United States Steel Corporation, Pittsburgh, PA)

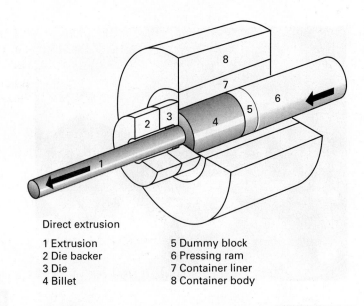

Direct extrusion

1 Extrusion   5 Dummy block
2 Die backer   6 Pressing ram
3 Die   7 Container liner
4 Billet   8 Container body

FIGURE 18.26 Direct extrusion schematic showing the various equipment components. (Courtesy of Danieli Wean United, Cranberry Township, PA)

forging and subsequent machining operations). To minimize both the expense and waste, considerable effort has been made to develop forging processes that can form complex shape parts with sufficient dimensional precision that little or no final machining is required. These are known as **net-shape**, or **near-net-shape forging** and may also be referred to as **precision forging**. Draft angles can be reduced to less than 1° (or even zero). Cost savings result from the reduction or elimination of secondary machining (and the associated handling, positioning, and fixturing), the companion reduction in scrap, and an overall decrease in the amount of energy required to produce the product. Because the design and implementation of net-shape processing can be rather expensive, however, application is usually reserved for parts where a significant cost reduction is possible.

18.6 Extrusion

In the **extrusion** process, metal is compressed and forced to flow through a shaped die to form a product with reduced but constant cross section, much like the squeezing of toothpaste out of a tube. Although metal extrusion may be performed either hot or cold, hot extrusion is commonly employed for many metals to reduce the forces required, eliminate cold-working effects, and reduce directional properties. A common arrangement is to have a heated billet placed inside a confining chamber. A ram advances from one end, causing the billet to first upset and conform to the chamber. As the ram continues to advance, the pressure builds until the material flows plastically through the die and *extrudes*, as depicted in **Figure 18.26**. The stress state within the material is one of triaxial compression.

Aluminum, magnesium, copper, lead, and alloys of these metals are commonly extruded, taking advantage of the relatively low yield strengths and low hot-working temperatures.

Steels, stainless steels, nickel-based alloys, and titanium are far more difficult to extrude. Their yield strengths are high, and the metals tend to weld to the walls of the die and confining chamber under the required conditions of temperature and pressure. With the development and use of phosphate-based and molten glass lubricants, however, hot extrusions can be routinely produced from the high-strength, high-temperature metals. The lubricants are able to withstand the required temperatures and adhere to the billet, flowing and thinning in a way that prevents metal-to-metal contact throughout the process.

As shown in the left-hand segment of **Figure 18.27**, almost any cross-sectional shape can be extruded from the nonferrous metals. Size limitations are few because presses are now available that can extrude any shape that can be enclosed within a circle 75 cm (30 in.) in diameter. In the case of steels and the other high-strength metals, the shapes and sizes are a bit more limited, but, as the right-hand segment of Figure 18.27 shows, considerable freedom still exists.

Extrusion has a number of attractive features. Many shapes can be produced as extrusions that are not possible by rolling, such as ones containing reentrant angles or longitudinal holes. No draft is required, so extrusions can offer savings in both metal and weight. Because the deformation is compressive, the amount of reduction in a single step is limited only by the capacity of the equipment. Billet-to-product cross-sectional area ratios can be in excess of 100-to-1 for the weaker metals. In addition, extrusion dies can be relatively inexpensive, and one die may be all that is required to produce a given product. Conversion from one product to another requires only a single die change, so small quantities of a desired shape can be produced economically. The major limitation of the process is the requirement that the cross section be uniform for the entire length of the product.

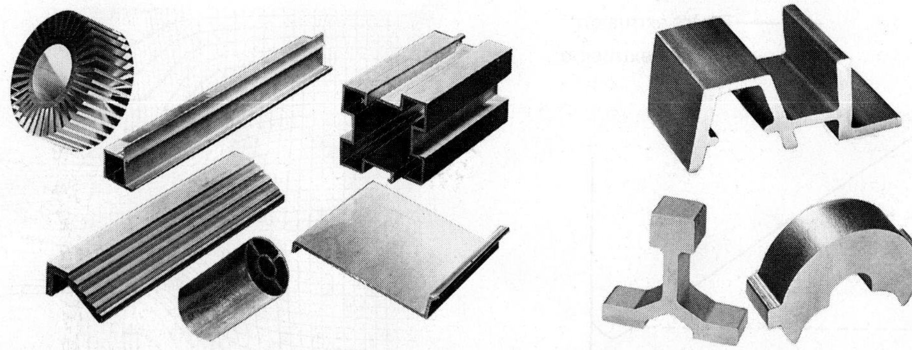

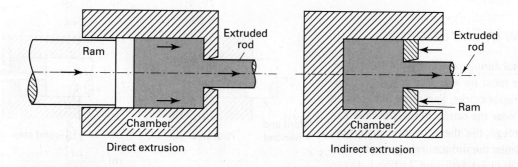

FIGURE 18.27 Typical shapes produced by extrusion. (Left) Aluminum products. (Right) Steel products. (Left) Alcoa Fastening Systems; (Right) Courtesy of Allegheny Ludlum Steel Corporation, Pittsburgh, PA)

Extruded products have good surface finish and dimensional precision. For most shapes, tolerances of 0.003 cm/cm or in./in. with a minimum of 0.075 mm (0.003 in.) are easily attainable. Grain structure is typical of other hot-worked metals, but strong directional properties (longitudinal versus transverse) are usually observed. Standard product lengths are about 6 to 7 m (20 to 24 ft), but lengths in excess of 12 m (40 ft) have been produced. Because little scrap is generated, billet-to-product yields are rather high.

Extrusion Methods

Extrusions can be produced by various techniques and equipment configurations. Hot extrusion is usually done by either the direct or indirect method, both of which are illustrated in **Figure 18.28**. In **direct extrusion**, a solid ram drives the entire billet to and through a stationary die and must provide additional power to overcome the frictional resistance between the surface of the moving billet and the confining chamber. With **indirect extrusion**, also called *reverse*, *backward*, or *inverted extrusion*, a hollow ram pushes the die back through a stationary, confined billet. Because there is no relative motion, friction between the billet and the chamber is eliminated. The required force is lower, and longer billets can be used with no penalty in power or efficiency.

Figure 18.29 shows the ram force versus ram position curves for both direct and indirect extrusion. The areas below the lines have units of Newton-meters or foot-pounds and are therefore proportional to the work required to produce the product. The area between the two curves is the work required to overcome the billet–chamber friction during direct extrusion, an amount that can be saved by converting to indirect extrusion. Unfortunately, the added complexity of the indirect process (applying force through a hollow ram, extracting the product through the hollow, and removing residual billet material at the end of the stroke) serves to increase the purchase price and maintenance cost of the required equipment.

With either process, the speed of hot extrusion is usually rather fast to minimize the cooling of the billet within the chamber. Extruded products can emerge at rates up to 300 m/min (1000 ft/min). The extrusion speed may be restricted, however, by the large amount of heat that is generated by the massive deformation and the associated rise in temperature. Sensors are often used to monitor the temperature of the emerging product and feed this information back to a control system. For materials whose properties are not sensitive to strain rate, ram speed may be maintained at the highest level that will keep the product temperature below some predetermined value.

Lubrication is another important area of concern. If the reduction ratio (cross-section area of billet to cross-section area

FIGURE 18.28 Direct and indirect extrusion. In direct extrusion, the ram and billet both move, and friction between the billet and the chamber opposes forward motion. For indirect extrusion, the billet is stationary. There is no billet–chamber friction because there is no relative motion.

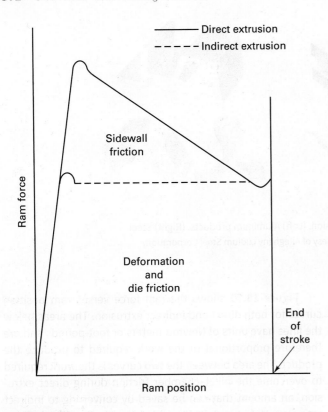

FIGURE 18.29 Diagram of the ram force versus ram position for both direct and indirect extrusion of the same product. The area under the curve corresponds to the amount of work (force x distance) performed. The difference between the two curves is attributed to billet-chamber friction.

of product) is 100, the product will be 100 times longer than the starting billet. If the product has a complex cross section, its perimeter can be significantly greater than a circle of equivalent area. Because the surface area of the product is the length times the perimeter, this value can be more than an order of magnitude greater than the surface area of the original billet. A lubricant that is applied to the starting piece must thin considerably as the material passes through the die and is converted to product. An acceptable lubricant is expected to reduce friction and act as a barrier to heat transfer at all stages of the process.

Metal Flow in Extrusion

The flow of metal during extrusion is often complex, and some care must be exercised to prevent surface cracks, interior cracks, and other flow-related defects. Metal near the center of the chamber can often pass through the die with little distortion, whereas metal near the surface undergoes considerable shearing. In direct extrusion, friction between the forward-moving billet and both the stationary chamber and die serves to further impede surface flow. The result is often a deformation pattern similar to the one shown in **Figure 18.30**. If the surface

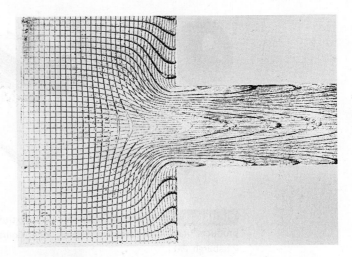

FIGURE 18.30 Grid pattern showing the metal flow in a direct extrusion. The billet was sectioned and the grid pattern was engraved prior to extrusion. (Courtesy Ronald Kohser)

regions of the billet undergo excessive cooling, surface deformation is further impeded, often leading to the formation of surface cracks. If quality is to be maintained, process control must be exercised in the areas of design, lubrication, extrusion speed, and temperature.

Extrusion of Hollow Shapes

Hollow shapes, and shapes with multiple longitudinal cavities (like several shown in the left-hand segment of Figure 18.27), can be extruded by several methods. For tubular products, the stationary or moving **mandrel** processes of **Figure 18.31** are

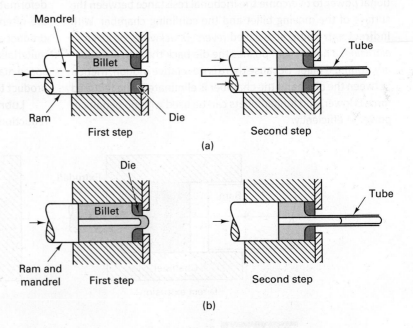

FIGURE 18.31 Two methods of extruding hollow shapes using internal mandrels. (a) The mandrel and ram have independent motions; (b) they move as a single unit.

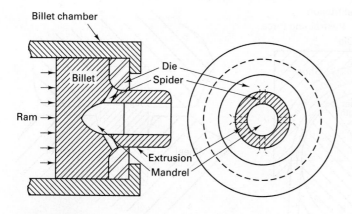

FIGURE 18.32 Hot extrusion of a hollow shape using a spider-mandrel die. Note the four arms connecting the external die and the central mandrel.

quite common. The die forms the outer profile, while the mandrel shapes and sizes the interior.

For products with multiple or more complex cavities, a **spider-mandrel die** (also known as a *porthole*, *bridge*, or *torpedo die*) may be required. As illustrated in **Figure 18.32**, metal flows around the arms of a "spider," and a further reduction then forces the material back together. Because the metal is never exposed to contamination, perfect welds result. Unfortunately, lubricants cannot be used because they will contaminate the surfaces to be welded. The process is therefore limited to materials that can be extruded without lubrication and can also be easily pressure welded.

Because additional tooling is required, hollow extrusions will obviously cost more than solid ones, but a wide variety of continuous cross-section shapes can be produced that cannot be made economically by any other process.

Hydrostatic Extrusion

Another type of extrusion, known as **hydrostatic extrusion**, is illustrated schematically in **Figure 18.33**. Here high-pressure fluid surrounds the workpiece and applies the force necessary to extrude it through the die. The product emerges into either atmospheric pressure or a lower-pressure fluid-filled chamber. The process resembles direct extrusion, but the pressurized fluid surrounding the billet prevents any upsetting. Because the billet does not come into contact with the surrounding chamber, billet-chamber friction is eliminated. In addition, the pressurized fluid can also emerge between the billet and the die, acting in the form of a lubricant.

Although the efficiency can be significantly greater than most other extrusion processes, there are problems related to the fluid and the associated high pressures (which typically range between 900 and 1700 MPa, or 125 and 250 ksi). Temperatures are limited because the fluid acts as a heat sink, and many of the pressurizing fluids (typically light hydrocarbons and oils) burn or decompose at moderately low temperatures. Seals must be designed to contain the pressurized fluid without leaking, and measures must be taken to prevent the complete ejection of the product, often referred to as **blowout**. Because of these features, hydrostatic extrusion is usually employed only where the process offers unique advantages that cannot be duplicated by the more conventional methods.

Pressure-to-pressure extrusion is one of those unique capabilities. In this variant, the product emerges from an extremely high-pressure chamber into a second pressurized chamber. In effect, the metal deformation is performed in a highly compressed environment. Crack formation begins with void formation, void growth, and void coalescence.

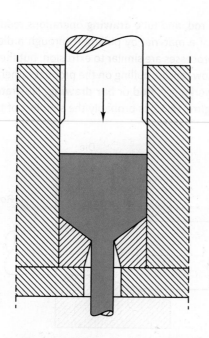

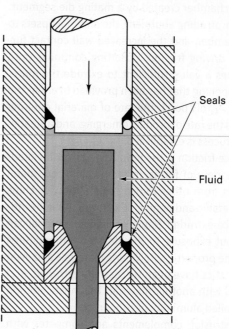

FIGURE 18.33 Comparison of conventional (left) and hydrostatic (right) extrusion. Note the addition of the pressurizing fluid and the O-ring and miter-ring seals on both the die and ram.

Because voids are suppressed in a compressed environment, the result is a phenomenon known as **pressure-induced ductility**. Relatively brittle materials such as molybdenum, beryllium, tungsten, and various intermetallic compounds can be plastically deformed without fracture, and materials with limited ductility become highly formable. Products can be made that could not be otherwise produced, and materials can be considered that would have been rejected because of their limited ductility at room temperature and atmospheric pressure.

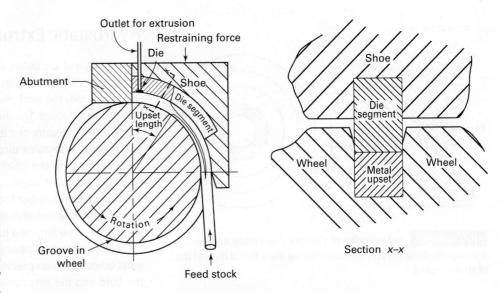

FIGURE 18.34 Cross-sectional schematic of the Conform continuous extrusion process. The material upsets at the abutment and extrudes. Section x–x shows the material in the shoe.

Continuous Extrusion

Conventional extrusion is a discontinuous process, converting finite-length billets into finite-length products. If the pushing force could be applied to the periphery of the feedstock, rather than the back, continuous feedstock could be converted into continuous product, and the process could become one of **continuous extrusion**. Since the first continuous extrusion of solid metal feedstock in 1970, a number of techniques have been proposed with varying degrees of success. In terms of commercial application, the most significant is probably the **Conform process**, illustrated schematically in **Figure 18.34**. Continuous feedstock is inserted into a grooved wheel and is driven by surface friction into a chamber created by a mating die segment. Upon impacting a protruding abutment, the material upsets to conform to the chamber, and the increased wall contact further increases the driving friction. Upsetting continues until the pressure reaches a value sufficient to extrude the material through a die opening that has been provided in either the shoe or abutment. At this point, the rate of material entering the machine equals the rate of product emerging, and a steady-state continuous process is established.

Because surface friction is the propulsion force, the feedstock can take a variety of forms, including solid rod, metal powder, punchouts from other forming operations, or chips from machining. Metallic and nonmetallic powders can be intimately mixed and co-extruded. Rapidly solidified material can be extruded without exposure to the elevated temperatures that would harm the properties. Polymeric materials and even fiber-reinforced plastics have been successfully extruded. Rod or wire can be clad with another material. The most common feed, however, is coiled aluminum or copper rod.

Continuous extrusion complements and competes with wire drawing and shape rolling as a means of producing nonferrous products with small, but uniform, cross sections. It is particularly attractive for complex profiles and cross sections that

contain one or more holes. Because extrusion operations can perform massive reductions through a single die, one Conform operation can produce an amount of deformation equivalent to 10 conventional drawing or cold-rolling passes. In addition, sufficient heat can be generated by the deformation that the product will emerge in an annealed condition, ready for further processing without intermediate heat treatment.

18.7 Wire, Rod, and Tube Drawing

Wire, rod, and tube **drawing** operations reduce the cross section of a material by pulling it through a die. In many ways, the processes are similar to extrusion, but the applied stresses are now tensile, pulling on the product rather than pushing on the workpiece. **Rod** or **bar drawing**, illustrated schematically in **Figure 18.35**, is probably the simplest of these operations.

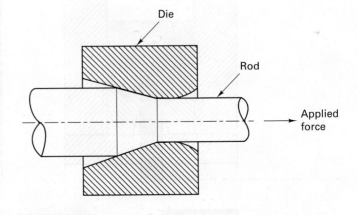

FIGURE 18.35 Schematic diagram of the rod- or bar-drawing process.

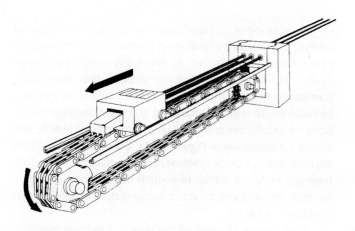

FIGURE 18.36 Diagram of a chain-driven, multiple-die, draw bench used to produce finite lengths of straight rod or tube. (Courtesy of Danieli Wean United, Inc., Cranberry Township, PA)

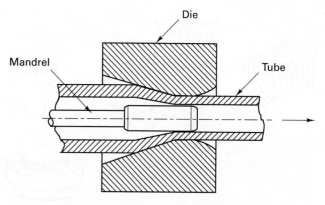

FIGURE 18.37 Cold-drawing smaller tubing from larger tubing. The die sets the outer dimension while the stationary mandrel sizes the inner diameter.

One end of a rod is reduced or pointed, so that it can pass through a die of somewhat smaller cross section. The protruding material is then placed in grips and pulled in tension, drawing the remainder of the rod through the die. The rods reduce in section, elongate, and become stronger (strain harden). Because the product cannot be readily bent or coiled, straight-pull **draw benches** are generally employed with finite-length feedstock. Hydraulic cylinders can be used to provide the pull for short-length products, while chain drives, as depicted in **Figure 18.36**, can be used to draw products up to 30 m (100 ft) in length.

The reduction in area is usually restricted to between 20 and 50%, because higher values would require higher pulling forces that may exceed the tensile strength of the reduced product. (The die would act as a second "grip," and tensile failure of the emerged product would result.) To produce a desired size or shape, multiple draws may be required through a series of progressively smaller dies. Intermediate anneals may also be required to restore ductility and enable further deformation.

Starting with tubular shapes that have been made by welding the edges of roll-shaped strip, extrusion or piercing, **tube drawing** can be used to produce high-quality tubing where the product requires the smooth surfaces, thin walls, accurate dimensions, and added strength (from the strain hardening) that are characteristic of cold forming. Internal mandrels are often used to control the inside diameter of tubes, which range from about 12 to 250 mm (0.5 to 10 in.) in diameter. As shown in **Figure 18.37**, these mandrels are inserted through the incoming stock and are held in place during the drawing operation. Products are generally limited to lengths of 30 meters (100 feet) or less.

Thick-walled tubes and those less than 12 mm (0.5 in.) in diameter are often drawn without a mandrel in a process known as **tube sinking**. Precise control of the inner diameter is sacrificed in exchange for process simplicity and the ability to draw long lengths of product. The wall thickness can increase, decrease, or remain the same, depending on the die angle and

other process variables. Low die angles tend to favor wall thickening, whereas larger angles promote thinning.

If a controlled internal diameter must be produced in a long-length product, it is possible to utilize a **floating plug**, like the one shown in **Figure 18.38**. This plug must be designed for the specific conditions of material, reduction, and friction. If the friction on the plug surface is too great, the flowing tube will pull it too far forward, pinching off or fracturing the tube wall. If the amount of friction is insufficient, the plug will chatter or vibrate within the tube. When properly designed, the floating plug will assume a stable position within the die and size the internal diameter, while the external die shapes and sizes the outside of the tube.

The drawing of bar stock can also be used to make products with shaped cross sections. By using cold drawing instead of hot extrusion, the material emerges with precise dimensions and excellent surface finish. Inexpensive materials strengthened by strain hardening can replace stronger alloys or ones that would require additional heat treatment. Small parts with complex but constant cross sections can often be economically produced by sectioning long lengths of cold-drawn shaped bars. Steels, copper alloys, and aluminum alloys have all been cold drawn into shaped bars. For steel, the cross-sectional area is usually less than 4 cm^2 (0.6 in.^{2}), and the largest cross-sectional dimension is generally less than 3 cm (1.25 in.).

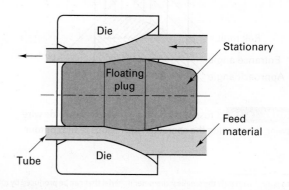

FIGURE 18.38 Tube drawing with a floating plug.

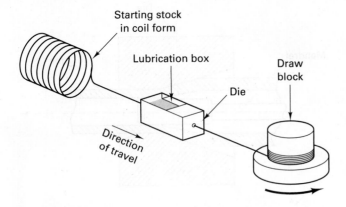

FIGURE 18.39 Schematic of wire drawing with a rotating draw block. The rotating motor on the draw block provides a continuous pull on the incoming wire.

Wire drawing is essentially the same process as bar drawing except that it involves smaller-diameter material. Because the material can now be coiled, the process can be conducted in a more continuous manner on rotating draw blocks, like the one illustrated schematically in **Figure 18.39**. Wire drawing usually begins with large coils of hot-rolled rod stock approximately 9 mm (⅜ in.) in diameter.[1] After descaling or other forms of surface preparation, one end of the coil is pointed and fed through a die where it is gripped and pulled.

Wire dies generally have a configuration similar to the one shown in **Figure 18.40**. The contact regions are usually made of wear-resistant tungsten carbide or polycrystalline, manufactured diamond. Single-crystal diamonds can be used for the drawing of very fine wire, and wear-resistant and low-friction coatings can be applied to the various die material substrates. Lubrication boxes often precede the individual dies to help reduce friction drag and wear of the dies.

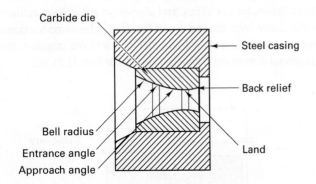

FIGURE 18.40 Cross section through a typical carbide wire drawing die showing the characteristic regions of the contour.

---

[1] This is approximately the smallest diameter material that can be produced by rolling with grooved rolls. With smaller diameter grooves, the concentrated stresses are likely to crack the rolls.

Because the tensile load is applied to the already reduced product, the amount of reduction is severely limited. Multiple draws are usually required to affect any significant change in size. To convert hot-rolled rod stock to fine wire, reduction passes through as many as 20 or 30 individual dies may be required. To minimize handling and labor, multiple reductions are usually performed on tandem machines, like the one shown schematically in **Figure 18.41**. Between 3 and 12 dies are mounted in a single machine, and the material moves continuously from one station to another in a synchronized manner that prevents any localized accumulation or tension that might induce fracture.

After passing through all the dies in a tandem machine, the material usually requires an intermediate anneal before it can be subjected to further deformation. By controlling the placement of the last anneal so the final product has a selected amount of cold work, wires can be made with a wide range of strengths (or tempers) for applications as diverse as electrical conductor wire and high-strength cable. (see Problem 1 at the end of Chapter 17.) When maximum ductility and conductivity is desired, the wire should be annealed after the final draw.

18.8 Cold Forming, Cold Forging, and Impact Extrusion

Large quantities of products are now being made by **cold forming**, a family of processes in which slugs of material are squeezed into or extruded from shaped die cavities to produce finished parts of precise shape and size. Workpiece temperature varies from room temperature to more than 100°C.

Cold heading is a form of the previously discussed upset forging. As illustrated in **Figure 18.42**, it is used for making enlarged sections on the ends of rod or wire, such as the heads of nails, bolts, rivets, or other fasteners. Two variations of the process are common. In the first, rod or wire is sheared to a preset length and then transferred to a holder-ejector assembly. Heading punches then strike one or more blows on the exposed end to perform the upsetting. If intermediate shapes are required, the piece is transferred from station to station, or the various heading punches sequentially rotate into position. When the heading is completed, the ejector stop advances and expels the product. In the second variation, a continuous rod (or wire) is fed forward to produce a preset extension, clamped, and the head is formed. The rod is then advanced to a second preset length and sheared, and the cycle repeats. This procedure is particularly attractive for producing nails because the point can be formed in the shearing or cutoff operation. Enlarged sections can also be produced at locations other than the ends of a rod or wire, in the manner illustrated in **Figure 18.43**.

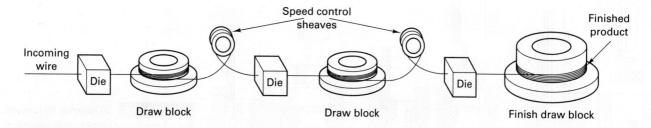

FIGURE 18.41 Schematic of a multistation synchronized wire drawing machine. To prevent accumulation or breakage, it is necessary to ensure that the same volume of material passes through each station in a given time. The loops around the sheaves between the stations use wire tensions and feedback electronics to provide the necessary speed control.

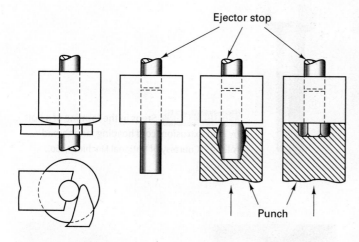

FIGURE 18.42 Typical steps in a shearing and cold-heading operation.

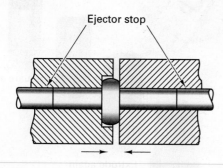

FIGURE 18.43 Method of upsetting the center portion of a rod. The stock is supported in both dies during upsetting.

Cold heading generally produces symmetrical parts, however, the expanded regions can be round, square, hexagonal, or even offset. Production speeds tend to vary with the diameter of the incoming material. When the blanks are less than 6 mm (0.25 in.) in diameter, speeds of 400 to 600 pieces per minute are typical. For larger diameters, the speeds may reduce to 40 to 100 pieces per minute. Alloys of aluminum and copper have excellent formability, whereas mild steel and stainless steel are rated fair to good. Alloy steels are a bit more difficult because of their higher strength and lower ductility.

A variety of extrusion operations, commonly called **impact extrusion**, can also be incorporated into cold forming. In these processes, a metal slug of predetermined size is positioned in a die cavity, where it is struck a single blow by a rapidly moving punch. The metal may flow forward through the die, backward around the punch, or in a combination mode. **Figure 18.44** illustrates the forward and backward variations, using both open and closed dies. In **forward extrusion**, the diameter is decreased while the length increases. **Backward extrusion** shapes hollow parts with a solid bottom. The punch controls the inside shape, while the die shapes the exterior. The wall thickness is determined by the clearance between the punch and die, and the bottom thickness is set by the stop position of the punch. **Figure 18.45** provides additional schematics of forward, backward, and combination impacting. Typical production speeds range from 20 to 60 strokes per minute.

The impact extrusion processes were first used to shape low-strength metals such as lead, tin, zinc, and aluminum into products such as collapsible tubes for toothpaste, medications, and other creams; small "cans" for shielding electronic components; zinc cases for flashlight batteries; and larger cans for food and beverages. In recent years, impact extrusion has

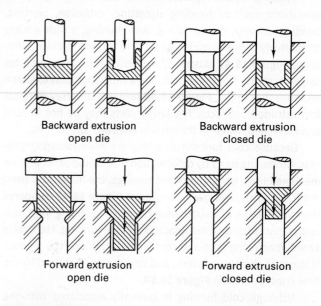

Backward extrusion open die

Backward extrusion closed die

Forward extrusion open die

Forward extrusion closed die

FIGURE 18.44 Backward and forward extrusion with open and closed dies.

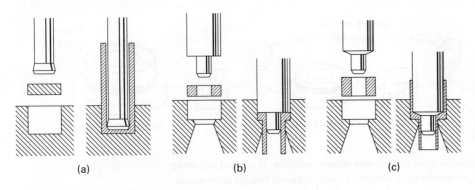

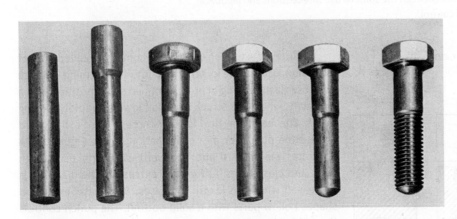

(a)   (b)   (c)

FIGURE 18.45 (a) Reverse, (b) forward, and (c) combined forms of cold extrusion. (Courtesy of The Aluminum Association, Arlington, VA)

FIGURE 18.46 Steps in the forming of a bolt by cold extrusion, cold heading, and thread rolling. (Courtesy of National Machinery Co., Tiffin, OH)

expanded to the forming of mild steel parts, where it is often used in combination with cold heading, as in the example of **Figure 18.46**. When heading alone is used, there is a definite limit to the ratio of the head and stock diameters (as presented in Figure 18.18 and related discussion). The combination of forward extrusion and cold heading overcomes this limitation by using an intermediate starting diameter. The shank portion is reduced by forward extrusion while upsetting is used to increase the diameter of the head.

By using various types of dies and combining high-speed operations such as heading, upsetting, extrusion, piercing, bending, coining, thread rolling, and knurling, a wide variety of relatively complex parts can be cold formed to close tolerances. **Figure 18.47** illustrates an operation that incorporates two extrusions, a central upset, and a final operation to shape and trim that upset. **Figure 18.48** presents an array of upset and extruded products. The larger parts may be hot formed and machined, whereas the smaller ones are cold formed.

Because cold forming is a chipless manufacturing process, producing parts by deformation that would otherwise be machined from bar stock or hot forgings, the material is used more efficiently, and waste is reduced. **Figure 18.49** compares the manufacture of a spark-plug body by machining from hexagonal bar stock with manufacture by cold forming. Material is saved, machining time and cost are reduced, and the product is stronger, due to cold work, and tougher, as illustrated by the flow lines revealed in **Figure 18.50**.

Although cold forming is generally associated with the manufacture of small parts from the weaker nonferrous metals, the process has seen extensive use on steel and stainless steel, and parts have been produced up to 45 kg (100 lb) in

Cutoff

Square ends

Extrude

Extrude again

Upset

Trim hex

Trim washer

FIGURE 18.47 Cold-forming sequence involving cutoff, squaring, two extrusions, an upset, and a trimming operation. Also shown are the finished part and the trimmed scrap. (Courtesy of National Machinery Co., Tiffin, OH)

FIGURE 18.48 Typical parts made by upsetting and related operations. (Courtesy of National Machinery Co., Tiffin, OH)

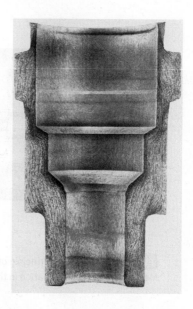

FIGURE 18.50 Section of the cold-formed spark-plug body of Figure 18.49, etched to reveal the flow lines. The cold-formed structure produces an 18% increase in strength over the machined product. (Courtesy of National Machinery Co., Tiffin, OH)

weight and 18 cm (7 in.) in diameter. At the small end of the scale, microformers are now cold forming extremely small electronic components with dimensional accuracies within 0.005 mm (0.0002 in.). Shapes are usually axisymmetric or those with relatively small departures from symmetry. Production rates are high, dimensional tolerances and surface finish are excellent, and there are no draft angles, parting lines, or flash to trim off. There is almost no material waste, and a considerable amount of machining can often be eliminated when used in place of alternate processes. Strain hardening can provide additional strength (up to 70% stronger than machined parts), and favorable grain flow can enhance toughness and fatigue life. As a result, parts can often be made smaller or thinner, or from lower-cost materials. Unfortunately, the cost of the required tooling, coupled with the high production speed, generally requires large-volume production, typically in excess of 50,000 parts per year.

18.9 Piercing

Thick-walled **seamless tubing** can be made by **rotary piercing**, a process illustrated in **Figure 18.51**. A heated billet is fed longitudinally into the gap between two large, convex-tapered rolls that are rotated in the same direction, but with their axes offset from the axis of the billet by about 6°, one to the right and the other to the left. The clearance between the rolls is set at a value less than the diameter of the incoming billet. As the billet is caught by the rolls, it is simultaneously rotated and driven forward. The reduced clearance between the rolls forces the billet to deform into a rotating ellipse. As shown in the right-hand segment of Figure 18.51, rotation of the elliptical section causes the metal to shear about the major axis. A crack tends to form down the center axis of the billet, and the cracked material is then forced over a pointed mandrel that enlarges and shapes the opening to create a seamless tube. The result is a short length of thick-walled seamless tubing, which can then be passed through sizing rolls to reduce the diameter and/or wall thickness. Seamless tubes can also be expanded in diameter by passing them over an enlarging mandrel. As the diameter and circumference increase, the walls correspondingly thin.

The **Mannesmann mills** commonly used in hot piercing can be used to produce tubing up to 300 mm (12 in.) in diameter. Larger-diameter tubes can be produced on **Stiefel mills**, which use the same principle but replace the convex rolls of the Mannesmann mill with larger-diameter conical disks.

Cutting (74% waste)   Cold forming (6% waste)

FIGURE 18.49 Manufacture of a spark plug body: (left) by machining from hexagonal bar stock; (right) by cold forming. Note the reduction in waste. (Courtesy of National Machinery Co., Tiffin, OH)

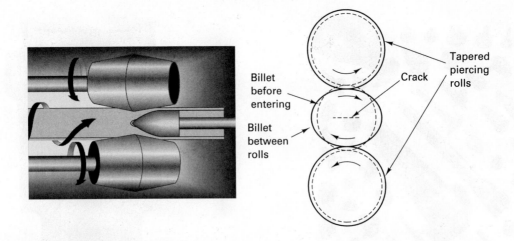

18.10 Other Squeezing Processes

Roll Extrusion or Flow Forming

Thin-walled cylinders or cups can be produced from thicker-wall material by **roll-extrusion** or **flow forming**. In the variant depicted in **Figure 18.52a**, internal rollers expand the internal diameter as they squeeze the rotating material against an external confining ring. The tube elongates as the wall thickness is reduced. In **Figure 18.52b**, the internal diameter is maintained as external rollers squeeze the material against a rotating mandrel. Precision is improved by using multiple rollers in arrangements such as two rollers 180° apart, three rollers at 120°, or four rollers at 90°. By moving the rollers in and out, different wall thicknesses can be produced in different areas. Although the process has been used to produce cylinders from 2 cm to 4 m (0.75 to 156 in.) in diameter, most products have diameters between 7.5 and 50 cm (3 and 20 in.).

Sizing

Sizing involves squeezing all or selected regions of forgings, ductile castings, or powder metallurgy products to achieve a prescribed thickness or enhanced dimensional precision. By incorporating sizing, designers can make the initial tolerances of a part more liberal, enabling the use of less-costly production methods. Those dimensions that must be precise are then set by one or more sizing operations that are usually performed on simple, mechanically driven presses.

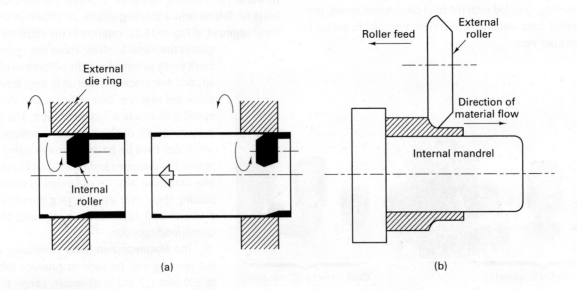

FIGURE 18.52 The roll extrusion process: (a) with internal rollers expanding the inner diameter; (b) with external rollers reducing the outer diameter.

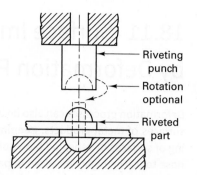

FIGURE 18.53 Joining components by riveting.

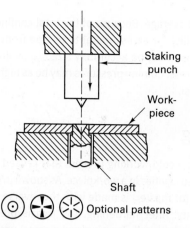

FIGURE 18.55 Permanently attaching a shaft to a plate by staking.

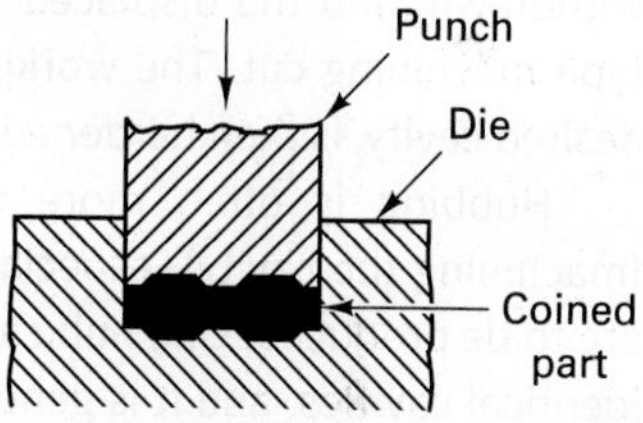

Riveting

In **riveting**, an expanded head is formed on the shank end of a fastener to permanently join sheets or plates of material. Although riveting is usually done hot in structural applications, it is almost always done cold in manufacturing. Where there is access to both sides of the work, the method illustrated in **Figure 18.53** is commonly used. The shaped punch may be driven by a press or contained in a special, hand-held riveting hammer. When a press is used, the rivet is usually headed in a single squeezing action, although the heading punch may also rotate to shape the head in a progressive manner, an approach known as **orbital forming**. Special riveting machines, like those used in aircraft assembly, can punch the hole, place the rivet in position, and perform the heading operation—all in about 1 second.

It is often desirable to use riveting in situations where there is access to only one side of the assembly. **Figure 18.54** shows two types of special rivets that can be used for one-side-access applications. The shank on the "blind" or inaccessible side of an **explosive rivet** expands on detonation to form a retaining head when a heated tool is touched against the accessible segment. In the pull type, or **pop-rivet**, a pull-up pin is used to expand a tubular shank. After performing its function, the pull pin breaks or is cut off flush with the accessible head.

Staking

Staking is a method of permanently joining parts together when a segment of one part protrudes through a hole in the other. As shown in **Figure 18.55**, a shaped punch is driven into

FIGURE 18.56 The coining process.

the exposed end of the protruding piece. The deformation causes radial expansion, mechanically locking the two pieces together. Because the tooling is simple and the operation can be completed with a single stroke of a press, staking is a convenient and economical method of fastening when permanence is desired and the appearance of the punch mark is not objectionable. Figure 18.55 includes some of the decorative punch designs that are commonly used.

Coining

The term **coining** refers to the cold squeezing of metal while all of the surfaces are confined within a set of closed dies. The process, illustrated schematically in **Figure 18.56**, is used to produce coins, medals, and other products where exact size and fine detail are required and where thickness varies about a

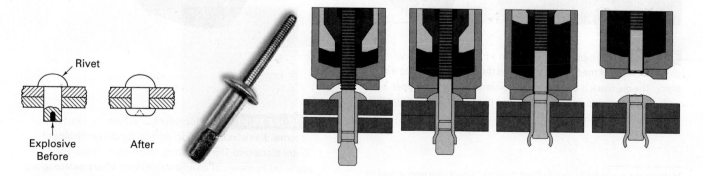

FIGURE 18.54 Rivets for use in "blind" riveting: (left) explosive type;(center) shank-type pull-up; (right) installation sequence for the shank-type pull-up rivet (Alcoa Fastening Systems, Magna-Lok)

well-defined average. Because of the total confinement (there is no possibility for excess metal to escape from the die), the input material must be accurately sized to avoid breakage of the dies or press. Coining pressures may be as high as 1400 MPa or 200,000 psi.

Hubbing

Hubbing[2] is a cold-working process that is used to plastically form recessed cavities in a workpiece. As shown in **Figure 18.57**, a male hub (or master) is made with the reverse profile of the desired cavity. After hardening, the hub is pressed into an annealed block (usually by a hydraulic press) until the desired impression is produced. (*Note*: Production of the cavity is often aided by machining away some of the metal in regions where large amounts of material would be displaced.) The hub is withdrawn, and the displaced metal is removed by a facing-type machining cut. The workpiece, which now contains the desired cavity, is then hardened by heat treatment.

Hubbing is often more economical than die sinking (machining the cavity), especially when multiple impressions are to be produced. One hub can be used to form a number of identical cavities, and it is generally easier to machine a male profile (with exposed surfaces) than a female cavity (where you are cutting in a hole).

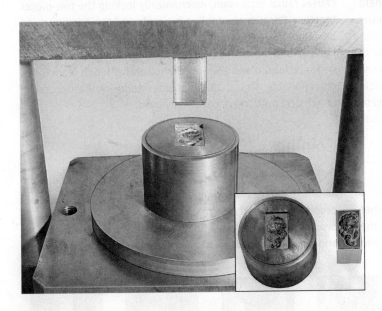

FIGURE 18.57 Hubbing a die block in a hydraulic press. Inset shows close-up of the hardened hub and the impression in the die block. The die block is contained in a reinforcing ring. The upper surface of the die block is then machined flat to remove the bulged metal. (E. Paul DeGarmo)

[2] This process should not be confused with "hobbing," a machining process used for cutting gears.

18.11 Surface Improvement by Deformation Processing

Deformation processes can also be used to improve or alter the surfaces of metal products. **Peening** is the mechanical working of surfaces by repeated blows of impelled shot or a round-nose tool. The highly localized impacts attempt to flatten and broaden the metal surface, but the underlying material restricts spread, resulting in a surface with residual compression.

Because the net loading on a material surface is the applied load minus the residual compression, peening tends to enhance the fracture resistance and fatigue life of tensile-loaded components. For this reason, shot impellers are frequently used to peen shafting, crankshafts, connecting rods, gear teeth, and other cyclic-loaded components.

FIGURE 18.58 Tools for roller burnishing. (a) Tool for internal diameter burnishing; (b) Tool for the burnishing of outer diameters. The burnishing rollers move inward or outward by means of an adjustable taper. (Courtesy Monaghan & Associates Inc., Dayton, OH)

Manual or pneumatic hammers are frequently used to peen the surfaces of metal weldments. Solidification shrinkage and thermal contraction produce surfaces with residual tension. Peening can reduce or cancel this effect, thereby reducing associated distortion and preventing cracking.

Burnishing involves rubbing a smooth, hard object (under considerable pressure) over the minute surface irregularities that are produced during machining or shearing. The edges of sheet metal stampings can be burnished by pushing the stamped parts through a slightly tapered die having its entrance end a little larger than the workpiece and

its exit slightly smaller. As the part rubs along the sides of the die, the pressure is sufficient to smooth the slightly rough edges that are characteristic of a blanking operation (see Figure 19-2).

Roller burnishing, illustrated in **Figure 18.58**, can be used to improve the size and finish of internal and external cylindrical and conical surfaces. The hardened rolls of a burnishing tool press against the surface and deform the protrusions to a more-nearly-flat geometry. The resulting surfaces possess improved wear and fatigue resistance, because they have been cold worked and are now in residual compression.

Review Questions

1. Briefly describe the evolution of forming equipment from ancient to modern.

2. What are some of the possible means of classifying metal-deformation processes?

3. How are bulk deformation processes different from sheet-forming operations? Describe each.

4. What are some common bulk deformation processes? Sheet-forming operations?

5. Why might the method of analysis be different for a process like forging and a process like rolling?

6. What are some of the common terms applied to the various sizes and shapes of rolled products?

7. Why are hot-rolled shaped products generally limited to standard forms and sizes?

8. Why is it undesirable to minimize friction between the workpiece and tooling in a rolling operation?

9. Why is it desirable to have a uniform temperature when hot rolling a material?

10. Why is it important to control the finishing temperature of a hot-rolling operation?

11. What are some of the attractive attributes of cold rolling?

12. What are some of the possible conditions to which cold-rolled material can be purchased?

13. Discuss the relative advantages and typical uses of two-high rolling mills with large-diameter rolls, three-high mills, and four-high mills.

14. Roll flexing can result in nonuniform thickness in the rolled product. How do four-high and cluster mills minimize this flexing?

15. Why is foil almost always rolled on a cluster mill?

16. What is pack rolling? For what types of products would it be used?

17. Describe a tandem rolling mill.

18. Why is speed synchronization of the various rolls so vitally important in a continuous or multistand rolling mill?

19. What types of products are produced by ring rolling?

20. Explain how hot-rolled products can have directional properties and residual stresses.

21. What is mill scale, and how can it be removed?

22. Discuss the problems in producing uniform thickness in a rolled product and some of the associated defects.

23. Why is a "crowned" roll always designed for a specific operation on a specific material?

24. How might the addition of horizontal tensions act to improve the thickness uniformity of rolled products?

25. What are some other techniques to reduce roll flexure?

26. What is thermomechanical processing, and what are some of its possible advantages?

27. Provide a concise description of the forging process.

28. What are some of the types of flow that can occur in forging operations?

29. Why are steam, air, or hydraulic hammers more attractive than gravity drop hammers for hammer forging?

30. What are some of the attractive features of computer-controlled forging hammers?

31. What is the difference between open-die and impression-die forging?

32. Why is open-die forging not a practical technique for large-scale production of identical products?

33. How might flash formation be an asset to impression-die forging?

34. What additional controls must be exercised to perform flashless forging?

35. What is a blocker impression in a forging sequence?

36. What attractive features are offered by counterblow forging equipment, or impactors?

37. Describe several ways to save energy or reduce costs by eliminating traditional forging or heat treatment steps.

38. For what types of forging products or conditions might a press be preferred over a hammer?

39. Why are heated dies generally employed in hot-press forging operations?

40. Describe some of the primary differences among hammers, mechanical presses, hydraulic presses, and screw presses.

41. What are some common examples of impression-die forgings?

42. What are some of the significant requirements of forging die materials?

43. Why are different tolerances usually applied to dimensions contained within a single die cavity and dimensions across the parting plane?

44. What are some of the roles played by lubricants in forging operations?

45. What are some of the attractive features of orbital forging?

46. What types of product geometry can be produced by orbital forging?

47. What is upset forging?

48. What are some of the typical products produced by upset-forging operations?

49. What types of products can be produced by automatic hot forging?

50. What are some of the attractive features of automatic hot forging? What is a major limitation?

51. How does roll forging differ from a conventional rolling operation?

52. Describe the swaging process.

53. What kind of products are produced by swaging?

54. How can the swaging process impart different sizes and shapes to an interior cavity and the exterior of a product?

55. What are some possible objectives of near-net-shape forging?

56. Provide a concise definition of extrusion.

57. What metals can be shaped by extrusion?

58. What are some of the attractive features of the extrusion process?

59. What is the primary shape limitation of the extrusion process?

60. What is the primary benefit of indirect extrusion?

61. What are some temperature considerations in hot extrusion?

62. Why might lubricant selection be more critical in an extrusion operation than for other metal forming processes?

63. What are some possible causes of surface cracks in extruded products?

64. How might tubular products be made by extrusion?

65. What types of products are made using a spider-mandrel die?

66. Why can lubricants not be used in spider-mandrel extrusion?

67. What are some of the attractive features of hydrostatic extrusion?

68. What are some unique concerns and limitations of hydrostatic extrusion?

69. What is the unique capability provided by pressure-to-pressure hydrostatic extrusion?

70. How is the feedstock pushed through the die in continuous extrusion processes?

71. Describe the Conform process of continuous extrusion.

72. What types of feedstock can be used in continuous extrusion other than the conventional solid rod?

73. How is wire, rod, and tube drawing different from extrusion?

74. Why are rods generally drawn on draw benches, while wire is drawn on draw block machines?

75. Why is the reduction in area significantly restricted during wire, rod, and tube drawing?

76. What is the difference between tube drawing and tube sinking?

77. For what types of products might a floating plug be employed?

78. What are some of the benefits of cold drawing of bar stock?

79. What types of materials are used for wire-drawing dies?

80. What is the benefit of a tandem wire drawing machine?

81. What is cold forming?

82. What types of products are produced by cold heading?

83. What is impact extrusion and what variations exist? Describe each.

84. If a product contains a large-diameter head and a small-diameter shank, how can the processes of cold extrusion and cold heading be combined to save metal?

85. What are some of the attractive properties or characteristics of cold-forming operations?

86. What process can be used to produce seamless pipe or tubing?

87. What type of products can be made by the roll-extrusion process?

88. What types of rivets can be used when there is access to only one side of a joint?

89. How is coining different from a process known as embossing?

90. Why might hubbing be an attractive way to produce a number of identical recessed cavities?

91. How might a peening operation increase the fracture resistance and fatigue life of a product?

92. What is burnishing?

Problems

1. Some snack foods, such as rectangular corn chips, are often formed by a rolling-type operation and are subject to the same types of defects common to rolled sheet and strip. Obtain a bag of such a snack and examine the chips to identify examples of rolling-related defects such as those discussed in Section 18.4 and shown in **Figure 18.A.**

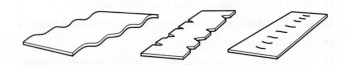

FIGURE 18.A Some typical defects that occur during rolling: wavy edges, edge cracking, and center cracking.

2. Consider the extrusion of a cylindrical billet, and compute the following.

a. Assume the starting billet to have a length of 0.3 m and a diameter of 15 cm. This is extruded into a cylindrical product that is 3 cm in diameter and 7.5 m long (a reduction ratio of 25). Neglecting the areas on the two ends, compute the ratio between the product surface area (wraparound cylinder) and the surface area of the starting billet.

b. How would this ratio change if the product were a square with the same cross-sectional area as that of the 3-cm-diameter circle?

c. Consider a cylinder-to-cylinder extrusion with a reduction ratio of R. Derive a general expression of the relative surface areas of product to billet as a function of R. (*Hint:* Start with a cylinder with length and diameter both equal to 1 unit. Because the final area will be $1/R$ times the original, the final length will be R units, and the final diameter will be proportional to $1/R$).

d. If the final product had a more complex cross-sectional shape than a cylinder. Would the final area be greater than or less than that computed in part c?

e. Relate your preceding answers to a consideration of lubrication during large-reduction extrusion operations.

3. The force required to compress a cylindrical solid between flat parallel dies (see Figure 18.10) has been estimated (by a theory of plasticity analysis) to be

$$\text{Force} = \pi R^2 \sigma_o \frac{1 + 2\,mR}{3\sqrt{3}T}$$

where:

R = radius of the cylinder

T = thickness of the cylinder

σ_o = yield strength of the material

m = friction factor (between 0 and 1 where 0 is frictionless and 1 is complete sticking)

An engineering student is attempting to impress his date by demonstrating some of the neat aspects of metal forming. He places a shiny penny between the platens of a 60,000-lb capacity press and proceeds to apply pressure. Assume that the coin has a ¾-in. diameter and is ¹⁄₁₆-in. thick. The yield strength is estimated as 50,000 psi, and because no lubricant is applied, friction is that of complete sticking, or $m = 1.0$.

a. Compute the force required to induce plastic deformation.

b. If this force is greater than the capacity of the press (60,000 lb), compute the pressure when the full-capacity force of 60,000 lb is applied.

c. If the press surfaces are made from thick plates of steel with a yield strength of 120,000 psi, describe the results of the demonstration.

d. A simple model of forging force uses the equation:

$$\text{Force} = K\sigma_o A$$

where:

K = a dimensionless multiplying factor

σ_o = yield strength of the material

A = projected area of the forging

K is assigned a value of 3–5 for simple shapes without flash, 5–8 for simple shapes with flash, and 8–12 for complex shapes with flash. Consider the two equations for forging force, and discuss their similarities and differences.

4. Mathematical analysis of the rolling of flat strip reveals that the roll-separation force (the squeezing force required to deform the strip) is directly proportional to the term:

$$1 + \frac{K_1 mL}{t_{av}}$$

where:

K_1 = geometric constant

m = friction factor

L = length of contact

t_{av} = average thickness of the strip in the roll bite

Because L is proportional to the roll radius, R, $K_1 L$ can be replaced by $K_2 R$ so the force becomes proportional to the term:

$$1 + \frac{K_2 mR}{t_{av}}$$

If K_2 and m are both positive numbers, how will the roll-separation force change as the strip becomes thinner? How can this effect be minimized? Relate your observations to the types of rolling mills used for various thicknesses of product.

5. In Figure 18.29, the vertical axis is force and the horizontal axis is position. If force is measured in pounds and position in feet, the area under the curves has units of foot-pounds and is a measure of the work performed. If the area under the indirect extrusion curve is proportional to the work required to extrude a product without billet-chamber frictional resistance, how could the relative regions of the direct extrusion curve be used to determine a crude measure of the mechanical "efficiency" of direct extrusion?

6. Compare the forming processes of wire drawing, conventional extrusion, and continuous extrusion with respect to continuity, reduction in area possible in a single operation, possible materials, speeds, typical temperatures, and other important processing variables.

7. Figure 18.B shows the rolling of a wide, thin strip where the width remains constant as thickness is reduced. Material enters the mill at a rate equal to $t_o w_o v_o$ and exits at a rate of $t_f w_o v_f$. Because material cannot be created or destroyed, these rates must be equal, and the w_o terms will cancel. As a result, v_f is equal to $(t_o/t_f) v_o$. The material enters at velocity v_o and accelerates to velocity v_f as the material passes through the mill. For stable rolling, the velocity of the roll surface, v_r, which is a constant, must be a value between v_o and v_f. For these conditions, describe the relative sliding between the strip and the rolls as the strip moves through the region of contact.

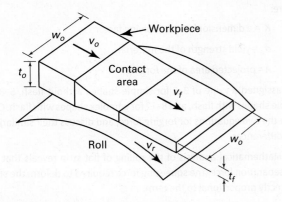

FIGURE 18.B Strip rolling where the width of the strip remains unchanged. The lines across the workpiece identify the area of contact with the rolls. The top roll has been removed for ease of visualization.

8. Various claims have been made for the application of ultrasonic vibrations to the tooling components of rod, tube, and wire drawing, including reduction in required draw force, ability to significantly increase draw speed, improved product surface finish, and elimination of sticking-related chatter—all tied to modification of lubricant effectiveness and other aspects of the tool–workpiece interface.

   a. Do you feel that these claims have validity? Why or why not?

   b. What might be the benefit to vibrating the die? The mandrel or plug in tube drawing?

   c. Possible vibration modes for the die include radial, torsional, and longitudinal. What might be the pros and cons of each option? Which do you think would be most attractive and offer the greatest benefit?

   d. If the claims are indeed valid, what other metal-forming processes might benefit from the application of ultrasonic vibrations?

Sheet-Forming Processes

19.1 Introduction

The various classification schemes for metal deformation processes have been presented at the beginning of Chapter 18, with the indication that our text will be grouping by **bulk processes** (Chapter 18) and **sheet processes** (Chapter 19). Bulk forming uses heavy machinery to apply three-dimensional stresses, and most of the processes are considered to be primary operations. Sheet metal processes, on the other hand, generally involve plane stress loadings and lower forces than bulk forming. Almost all sheet metal forming is considered to be secondary processing.

The classification into bulk and sheet is far from distinct, however. Some processes can be considered as either, depending on the size, shape, or thickness of the workpiece. The bending of rod or bar is often considered to be bulk forming, whereas the bending of sheet metal is sheet forming. Tube bending can be either, depending on the wall thickness or diameter of the tube. Similar areas of confusion can be found in deep drawing, roll forming, and other processes. The squeezing processes were described in Chapter 18. Presented here will be the processes that involve *shearing*, *bending*, and *drawing*. **Table 19.1** lists some of the processes that fit these categories.

19.2 Shearing Operations

Shearing is the mechanical cutting of materials without the formation of chips or the use of burning or melting. It is often used to prepare materials between 0.025 and 20 mm (0.001 and 0.8 in.) in thickness for subsequent operations, and its success helps to ensure the accuracy and precision of the finished product. When the two cutting blades are straight, the process is called *shearing*. When the blades are curved, the processes have special names, such as *blanking*, *piercing*, *notching*, and *trimming*. In terms of tool design and material behavior, however, all are shearing-type operations.

A simple type of shearing operation is illustrated in **Figure 19.1**. As the punch (or upper blade) pushes on the workpiece, the metal responds by flowing plastically into the die (or over the lower blade). Because the clearance between the two tools is small, usually between 5% and 25% of the thickness of the metal being cut (the ideal value depending on workpiece material and thickness), the deformation occurs as highly localized shear. As the punch pushes downward on the metal, the material flows into the die, with the opposite surface bulging slightly. An instability arises when the penetration is between 15% and 60% of the metal thickness, the actual amount

TABLE 19.1	Classification of the Nonsqueezing Metal-Forming Operations	
Shearing	**Bending**	**Drawing & Stretching**
1. Simple Shearing	1. Angle Bending	1. Spinning
2. Slitting	2. Roll Bending	2. Shear Forming or Flow Turning
3. Piercing	3. Draw Bending	3. Stretch Forming
4. Blanking	4. Compression Bending	4. Deep Drawing and Shallow Drawing
5. Fineblanking	5. Press Bending	5. Rubber-tool Forming
6. Lancing	6. Tube Bending	6. Sheet Hydroforming
7. Notching	7. Roll Forming	7. Tube Hydroforming
8. Nibbling	8. Seaming	8. Hot Drawing
9. Shaving	9. Flanging	9. High-energy-rate Forming
10. Trimming	10. Straightening	10. Ironing
11. Cut off		11. Embossing
12. Dinking		12. Superplastic Sheet Forming

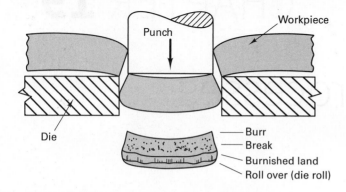

FIGURE 19.1 Simple blanking with a punch and die.

FIGURE 19.2 (Top) Conventionally sheared surface showing the distinct regions of deformation and fracture. (Bottom) Magnified view of the sheared edge. (Courtesy of Feintool Equipment Corp., Cincinnati, OH)

depending on the strength and ductility of the material. The applied stress exceeds the shear strength of the remaining material, and the metal tears or ruptures through the rest of its thickness, linking the cutting edges of the punch and die to produce an inwardly inclined fracture and a ragged edge or burr. As shown in **Figure 19.2**, the two distinct stages of the shearing process—deformation (producing the smooth burnished band) and fracture—are often visible on the edges of sheared parts.

Because of the normal inhomogeneities in a metal and the possibility of nonuniform clearance between the shear blades, the final shearing does not occur in a uniform manner. Fracture and tearing begin at the weakest point and proceed progressively or intermittently to the next-weakest location. This usually results in a rough and ragged edge that, combined with possible microcracks and work hardening of the sheared edge, can adversely affect subsequent forming processes.

Changing the clearance between the punch and the die can greatly change the condition of the cut edge. If the punch and die (or upper and lower shearing blades) have proper alignment and clearance and are maintained in good condition, sheared edges may be produced that have sufficient smoothness to permit use without further finishing. The quality of the sheared edge can often be improved by clamping the starting stock firmly against the die (from above) and restraining the movement of the sheared piece by a plunger or rubber die cushion that applies opposing pressure from below. Each of these measures causes the shearing to take place more uniformly around the perimeter of the cut.

If the entire shearing operation is performed in a compressive environment, fracture is suppressed and the relative fraction of smooth edge (produced by deformation) is increased. Above a certain pressure, no fracture occurs, and the entire edge is smooth, deformed metal. **Figure 19.3** shows one method of producing a compressive environment. In the **fineblanking** process, a V-shaped protrusion is incorporated into the hold-down or pressure plate at a location slightly external to the contour of the cut. As pressure is applied to the hold-down or pressure plate, the protrusion is driven into the material, preventing lateral movement of the workpiece and compressing the region to be cut. Matching upper and lower punches then squeeze the material from above and below and descend in unison, completely extracting the desired segment. With punch-die clearances of about ¹⁄₁₀ those of conventional blanking or about 0.5% of material thickness, the sheared edges are now both smooth and square (taper is less than half a degree), as shown in **Figure 19.4**.

Fineblanked parts are usually less than 6 mm (¼ in.) in thickness, but can be as thick as 19 mm (¾ in.), and they typically have complex-shaped perimeters. Dimensional accuracy is often within 0.025 mm (0.001 in.), and holes, slots, bends, and semipierced projections can be incorporated as part of the fineblanking operation. Holes and web sections can be smaller than the material thickness. Because the parts are pressed

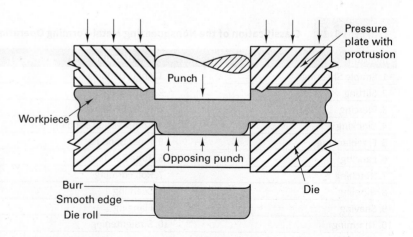

FIGURE 19.3 Fineblanking method of obtaining a smooth edge in shearing by using a shaped pressure plate to put the metal into localized compression and a punch and opposing punch descending in unison. The resulting product shows die roll and burr formation on opposite sides and a smooth edge surface.

FIGURE 19.4 Fine-blanked surface of the same component shown in Figure 19.2 . (Courtesy of Feintool Equipment Corp., Cincinnati, OH)

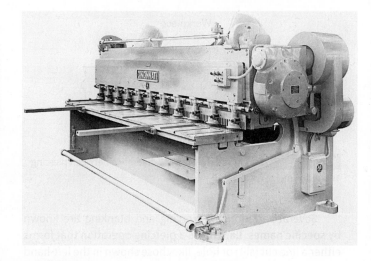

FIGURE 19.6 A 3-m (10 ft) power shear for 6.5-mm (1/4-in.) steel. (Courtesy of Cincinnati Incorporated, Cincinnati, OH)

from both sides during the shearing, excellent flatness is maintained. Secondary edge finishing can often be eliminated, and the work hardening that occurs during the shearing process enhances wear resistance. Production rates vary between 10 and 60 parts per minute.

In fineblanking, however, a triple-action mechanical or hydraulic press is generally required. The fineblanking force is about 40% greater than conventional blanking of the same contour, and the extra material required for the impinging protrusion often forces a greater separation between nested parts. The added costs must be offset by a reduction in secondary operations, such as drilling, reaming, grinding, or other machining.

Figure 19.5 illustrates a technique for shearing bar stock under compression. Bar stock is pressed against the closed end of a feed hole, placing the stock in a state of compression. A transverse punch then shears the material into smooth-surface, burr-free slugs, ready for further processing.

Simple Shearing

When sheets of metal are to be sheared along a straight line, **squaring shears**, like the one shown in **Figure 19.6**, are frequently used. As the upper ram descends, a clamping bar or set of clamping fingers presses the sheet of metal against the machine table to hold it firmly in position. A moving blade then comes down across a fixed blade and shears the metal. On larger shears, the moving blade is often set at an angle or "rocks" as it descends, so the cut is made in a progressive fashion from one side of the material to the other, much like a

pair of household scissors. This action significantly reduces the amount of cutting force required, replacing a high force—short stroke operation with one of low force and longer stroke.

The upper blade may also be inclined about 0.5° to 2.5° with respect to the lower blade and set to descend along this line of inclination. Although squareness and edge quality may be compromised, this action helps to ensure that the sheared material does not become wedged between the blades.

Slitting

Slitting is the lengthwise shearing process used to cut coils of sheet metal into several rolls of narrower width. Here, the shearing blades take the form of cylindrical rolls with circumferential mating grooves. The raised ribs of one roll match the recessed grooves on the other. The process is now continuous and can be performed rapidly and economically. Moreover, because the distance between adjacent shearing edges is fixed, the resultant **strips** have accurate and constant width, more consistent than that obtained from alternative cutting processes.

Piercing and Blanking

Piercing and **blanking** are shearing operations where a part is removed from sheet material by forcing a shaped punch through the sheet and into a shaped die. Any two-dimensional shape can be produced, with one surface having a slightly rounded edge and the other containing a slight burr. Because both processes involve the same basic cutting action, the primary difference is one of definition. **Figure 19.7** shows that in blanking, the piece being punched out becomes the workpiece. In piercing, the punchout is the scrap and the remaining strip is the workpiece. The term *piercing* can also be used to describe the formation of a hole by a pointed punch where no metal is removed from the sheet (like driving a nail through sheet metal). Piercing and blanking are usually done on some form of mechanical press.

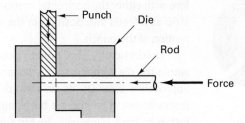

FIGURE 19.5 Method of smooth shearing a rod by putting it into compression during shearing.

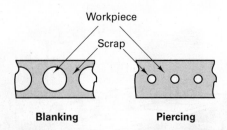

Blanking   **Piercing**

FIGURE 19.7 Schematic showing the difference between piercing and blanking.

Several variations of piercing and blanking are known by specific names. **Lancing** is a piercing operation that forms either a line cut (slit) or hole, like those shown in the left-hand portion of **Figure 19.8**. Lancing can be combined with bending to form tabs or openings like those found in vents or louvers. Lancing is also used to permit the adjacent metal to flow more readily in subsequent forming operations. In the case illustrated in Figure 19.8, the lancing makes it easier to shape the recessed grooves, which were formed before the product was blanked from the strip stock and shallow drawn. **Perforating** consists of producing a large number of closely spaced holes where a slug is removed from each hole. **Notching** is used to remove segments from along the edge of an existing product.

In **nibbling**, a contour is progressively cut by producing a series of overlapping slits or notches, as shown in **Figure 19.9**. In this manner, simple tools can be used to cut a complex shape from sheets of metal up to 6 mm (¼ in.) thick. The process is widely used when the quantities are insufficient to justify the expense of a dedicated blanking die. Edge smoothness is determined by the shape of the tooling and the degree of overlap in successive cuts.

Shaving is a finishing operation in which a small amount of metal is sheared away from the edge of an already blanked part. Its primary use is to obtain greater dimensional accuracy, but it may also be employed to produce a squared or smoother edge. Because only a small amount of metal is removed, the punches and dies must be made with very little clearance. Blanked parts, such as small gears, can be shaved to produce dimensional accuracies within 0.025 mm (0.001 in.).

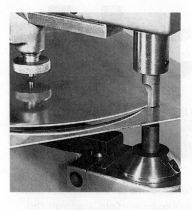

FIGURE 19.9 Shearing operation being performed on a nibbling machine. (Courtesy of Pacific Press Technologies, Mt. Carmel, IL)

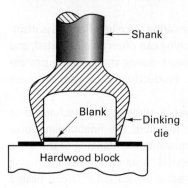

FIGURE 19.10 The dinking process.

In a **cutoff** operation, a punch and die are used to separate a stamping or other product from a strip of stock. The contour of the cutoff frequently completes the periphery of the workpiece. Cutoff operations are quite common in progressive die sequences, like several to be presented shortly.

Dinking is a modified shearing operation that is used to blank shapes from low strength materials, such as rubber, fiber, or cloth. As illustrated in **Figure 19.10**, the shank of a die is either struck with a hammer or mallet or the entire die is driven downward by some form of mechanical press.

Tools and Dies for Piercing and Blanking

As shown in **Figure 19.11**, the basic components of a piercing and blanking die set are a **punch**, a **die**, and a **stripper plate**, which is attached above the die to keep the strip material from ascending with the retracting punch. The position of the stripper plate and the size of its hole should be such that it does not interfere with either the horizontal motion of the strip as it feeds into position or the vertical motion of the punch.

Theoretically, the punch should fit within the die with a uniform clearance that approaches zero. On its downward stroke, it should not enter the die but should stop just as its base aligns with the top surface of the die. In general practice, the clearance is between 5% to 7% of the stock thickness, and the punch enters slightly into the die cavity.

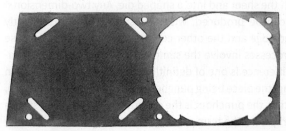

FIGURE 19.8 (Left to right) Piercing, lancing, and blanking precede the forming of the final disc-shaped piece. The small round holes assist positioning and alignment. (E. Paul DeGarmo)

If the face of the punch is normal to the axis of motion, the entire perimeter is cut simultaneously. The maximum cutting force can then be computed as:

Maximum cutting force = Perimeter length × Thickness × Material shear stress[1]

By tilting the punch face on angle, a feature known as **shear** or **rake angle**, the cutting force can be reduced substantially. As shown in **Figure 19.12**, the periphery is now cut in a progressive fashion, similar to the action of a pair of scissors or the opening of a "pop-top" beverage can. Variation in the shear angle controls the amount of cut that is made at any given time and the total stroke that is necessary to complete the operation. Adding shear increases the stroke, but reduces the force, providing an attractive way to cut thicker or stronger material on an existing piece of equipment.

Punches and dies should also be in proper alignment so that a uniform clearance is maintained around the entire periphery. The die is usually attached to the bolster plate of the press, which, in turn, is attached to the main press frame. The punch is attached to the movable ram, enabling motion in and out of the die with each stroke of the press. Punches and dies can also be mounted on a separate **punch holder** and **die shoe**, like the one shown in **Figure 19.13**, to create an **independent die set**. The holder and shoe are permanently aligned and guided by two or more guide pins. By pre-aligning the punch and die and fastening them to the die set, an entire unit can be inserted directly into a press. This can significantly reduce the amount of production time lost during tool change.

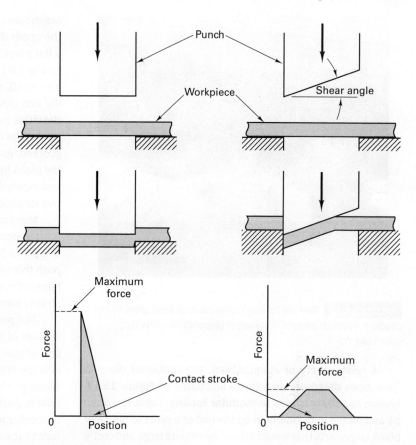

FIGURE 19.12 Blanking with a square-faced punch (left) and one containing angular shear (right). Note the difference in maximum force and contact stroke. The total work (the area under the curve) is the same for both processes.

Moreover, when a given punch and die are no longer needed, they can be removed and new tools attached to the shoe and holder assembly.

In most cases, the punch holder attaches directly to the ram of the press, and ram motion acts to both raise and lower the punch. On smaller die sets, springs can be incorporated to provide the upward motion. The ram simply pushes on the top of the punch holder, forcing it downward. When the ram retracts, the springs cause the punch to return to its starting position. This form of construction makes the die set fully self-contained. It is simply positioned in the press (not attached) and can be easily removed, thereby reducing setup time.

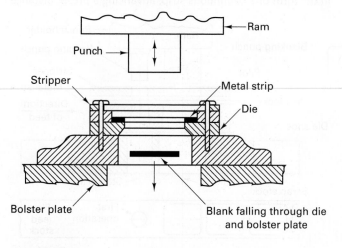

FIGURE 19.11 The basic components of piercing and blanking dies.

FIGURE 19.13 Typical die set having two alignment guideposts. (Courtesy of Danly IEM, Ithaca, MI)

---

[1] Tensile strength is often substituted for shear strength to provide a margin of safety when designing tooling or specifying machinery.

FIGURE 19.14 Modular tooling (subpress dies) assembled to produce a desired pattern. (Courtesy of Unipunch Products, Inc., Clear Lake, WI)

A wide variety of standardized, self-contained die sets have been developed like those illustrated in **Figure 19.14**. Known as **subpress dies** or **modular tooling**, these can often be assembled and combined on the bed of a press to pierce or blank large parts that would otherwise require large and costly complex die sets. A single downward motion of the press activates each of the subpress dies, producing a variety of holes and slots, each in proper relation to one another.

Punches and dies are usually made from low-distortion or air-hardenable tool steel so they can be hardened after machining with minimal warpage. The die profile is maintained for a depth of about 3 mm (⅛ in.) from the upper face, beyond which an angular clearance or back relief is generally provided (see Figure 19.11) to reduce friction between the part and the die and to permit the part to fall freely from the die after being sheared. The 3-mm depth provides adequate strength and sufficient metal so that the shearing edge can be resharpened by grinding a few thousandths of an inch from the face of the die.

Dies can be made as a single piece, or they can be made in component sections that are assembled on the punch holder and die shoe. The component approach simplifies production and enables the replacement of single sections in the event of wear or fracture. Substantial savings can often be achieved by modifying the design of parts to enable the use of standard die components. A further advantage of this approach is that when the die set is no longer needed, the components can be removed and used to construct tooling for another product.

When the cut periphery is composed of simple lines, and the material being cut is either soft metal or other soft material (such as plastics, wood, cork, felt, fabrics, and cardboard), **steel-rule**, or cookie-cutter, **dies** can often be used. The cutting die is fashioned from hardened steel strips, known as **steel rule**, that

are mounted on edge in grooves that have been machined in the upper die block. The mating piece of tooling can be either a flat piece of hardwood or steel, a male shape that conforms to the part profile (such that the protruding strips descend around it), or a set of matching grooves into which the upper die can descend. Rubber pads are usually inserted between the strips to replace the stripper plate. During the compression stroke, the rubber compresses and allows the cutting action to proceed. As the ram ascends, the rubber then expands to push the blank free of the steel-rule cavity. Steel-rule dies are usually less expensive to construct than solid dies and are quite attractive for producing small quantities of parts.

Many parts require multiple cutting-type operations, and it is often desirable to produce a completed part with each cycle of a press. Several types of dies have been designed to accomplish this task. For simplicity, their operations are discussed in terms of manufacturing simple, flat washers from a continuous strip of metal.

The **progressive die** set, depicted in **Figure 19.15**, is the simpler of the two types. Basically, it consists of two or more sets of punches and dies mounted in tandem. Strip stock is fed into the first die, where a hole is pierced as the ram descends. When the ram raises, the stock advances, and the pierced hole is positioned under the blanking punch. On the second descent, a pilot on the bottom of the blanking punch enters the hole that was pierced on the previous stroke to ensure accurate alignment. Further descent of the punch blanks the completed washer from the strip, and at the same time, the first punch pierces the hole for the next washer. As the process continues, a finished part is completed with each stroke of the press.

Progressive dies can be used for many combinations of piercing, blanking, forming, lancing, and drawing, as shown by the examples in **Figures 19.8** and **19.16**. They are relatively simple to construct and are economical to maintain and repair because a defective punch or die does not require replacement of the entire die set. The material moves through the operations in the form of a continuous strip, advancing a preset distance

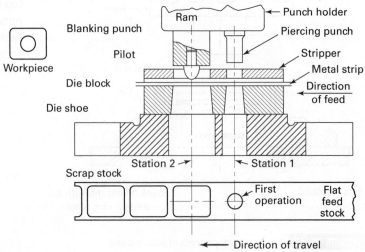

FIGURE 19.15 Progressive piercing and blanking die for making a square washer. Note that the punches are of different lengths.

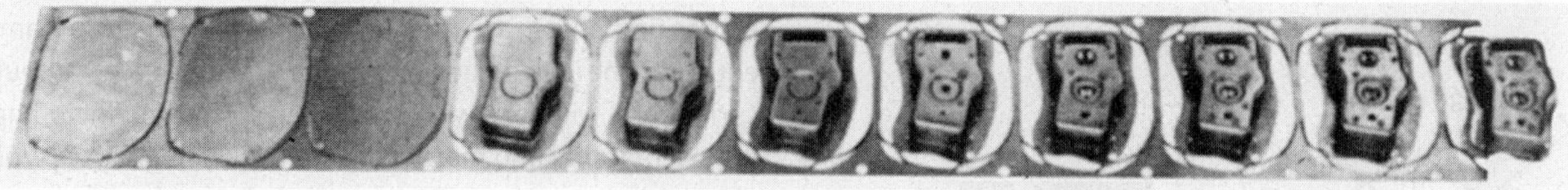

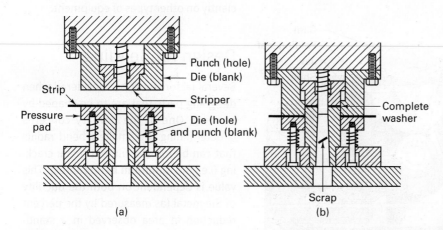

FIGURE 19.16 The various stages of an 11-station progressive die. (Courtesy of the Minster Machine Company, Minster, OH)

with each cycle of the press. As the products are shaped, they remain attached to the strip or carrier until a final cutoff operation. Although the attachment may restrict some of the forming operations and prevents part reorientation between steps, it also enables the quick and accurate positioning of material in each of the die segments.

If individual parts are mechanically moved from operation to operation within a single press, the dies are known as **transfer dies**. Part handling must operate in harmony with the press motions to move, orient, and position the pieces as they travel through the die.

In **compound dies** like the one shown schematically in **Figure 19.17**, piercing and blanking, or other combinations of operations, occur sequentially during a single stroke of the ram. Dies of this type are usually more expensive to construct and are more susceptible to breakage, but they generally offer more precise alignment of the sequential operations.

If many holes of varying sizes and shapes are to be placed in sheet components, **turret-type punch presses** may be specified. In these machines, as many as 60 separate punches and dies are contained within a turret that can be quickly rotated to provide the specific tooling required for an operation. Between operations, the workpiece is repositioned through numerically controlled movements of the worktable. This type of machine is particularly attractive when a variety of materials and thicknesses (0.4 to 8.0 mm or 0.015 to 0.3 in.) are being processed. Even greater flexibility can be achieved by single machines that combine mechanical punch pressing with laser, plasma, or waterjet cutting.

Design for Piercing and Blanking

The construction, operation, and maintenance of piercing and blanking dies can be greatly facilitated if designers of the parts to be fabricated keep a few simple rules in mind:

1. Diameters of pierced holes should not be less than the thickness of the metal, with a minimum of 0.3 mm (0.025 in.). Smaller holes can be made, but with difficulty.

2. The minimum distance between holes, or between a hole and the edge of the stock, should be at least equal to the metal thickness.

3. The width of any projection or slot should be at least 1 time the metal thickness and never less than 2.5 mm (3/32 in.).

4. Keep tolerances as large as possible. Tolerances below about 0.075 mm (0.003 in.) will require shaving.

5. Arrange the pattern of parts on the strip to minimize scrap.

19.3 Bending

Bending is the plastic deformation of metals about a linear axis with little or no change in the surface area. Multiple bends can be made simultaneously, but to be classified as true bending, and treatable by simple bending theory, each axis must be linear and independent of the others. If multiple bends are made with a single die, the process is often called **forming**. When the axes of deformation are not linear or are not independent, the processes are known as **drawing** and/or **stretching**, and these operations will be treated later in the chapter.

As shown in **Figure 19.18**, simple bending causes the metal on the outside to be stretched while that on the inside is compressed. The location that is neither stretched nor compressed is known as the **neutral axis** of the bend. Because the yield strength of metals in compression is somewhat higher than the yield strength in tension, the metal on the outer side yields first, and the neutral axis is displaced from the midpoint of the material. The neutral axis is generally located between one-third and one-half of the way from the

Punch (hole)
Die (blank)
Strip
Stripper
Pressure pad
Die (hole) and punch (blank)
Complete washer
Scrap
(a)
(b)

FIGURE 19.17 Method for making a simple washer in a compound piercing and blanking die. Part is (a) blanked and (b) subsequently pierced in the same stroke. The blanking punch contains the die for piercing.

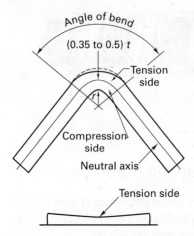

FIGURE 19.18 (Top) Nature of a bend in sheet metal showing tension on the outside and compression on the inside. (Bottom) The upper portion of the bend region, viewed from the side, shows how the center portion will thin more than the edges.

inner surface, depending on the bend radius and the material being bent. Because of this lack of symmetry and the dominance of tensile deformation, the metal is generally thinned at the bend. In a linear bend, thinning is greatest in the center of the sheet and less near the free edges where inward movement can provide some compensation.

On the inner side of a bend, the compressive stresses can induce upsetting and a companion thickening of material. Although this thickening somewhat offsets the thinning of the outer section, the upsetting can also produce an outward movement of the free edges. This contraction of the tensile segment and expansion of the compression segment can produce significant distortion of the edge surfaces that terminate a linear bend. This distortion is particularly pronounced when bends are produced across the width of thick but narrow plates.

Still another consequence of the combined tension and compression is the elastic recovery that occurs when the bending load is removed. The stretched region contracts and the compressed region expands, resulting in a small amount of "unbending," known as **springback**. To produce a product with

a specified angle, the metal must be overbent by an amount equal to the subsequent springback. The actual amount of springback will vary with a number of factors, including the type of material and material thickness. Springback is typically about 0.5 degree for softer metals, 1 degree for steels, and as much as 3 degrees for stainless steel. Because the amount of springback increases with an increase in material strength, the newer advanced high-strength steels have been shown to have springback amounts as great as eight times that observed with annealed low-carbon steel.

Angle Bending (Bar Folder and Press Brake)

Machines like the **bar folder**, shown in **Figure 19.19**, can be used to make angle bends up to 150° in sheet metal under 1.5 mm (1/16 in.) thick. The workpiece is inserted under the folding leaf and aligned in the proper position. Raising the handle then actuates a cam, causing the leaf to clamp the sheet. Further movement of the handle bends the metal to the desired angle. These manually operated machines can be used to produce linear bends up to about 3.5 m (12 ft) in length.

Bends in heavier sheet or more complex bends in thin material are generally made on **press brakes**, like the one shown in **Figure 19.20**. These are mechanical or hydraulic presses with a long, narrow bed and short strokes. The metal is bent between interchangeable dies that are attached to both the bed and the ram. As illustrated in **Figures 19.20** and **19.21**, different dies can be used to produce many types of bends. The metal can be repositioned between strokes to produce complex contours or repeated bends, such as corrugations. **Figure 19.22** shows how a roll bead can be formed with repeated strokes, repositioning, and multiple sets of tooling. Seaming, embossing, punching, and other operations can also be performed with press brakes, but these operations can usually be done more efficiently on other types of equipment.

Design for Bending

Several factors must be considered when designing parts that are to be shaped by bending. One of the primary concerns is determining the smallest bend radius that can be formed without metal cracking (i.e., the **minimum bend radius**). This value is dependent on both the ductility of the metal (as measured by the percent reduction in area observed in a standard tensile test) and the thickness of the material being bent. **Figure 19.23** shows how the ratio of the minimum bend radius R to the thickness of the material t varies

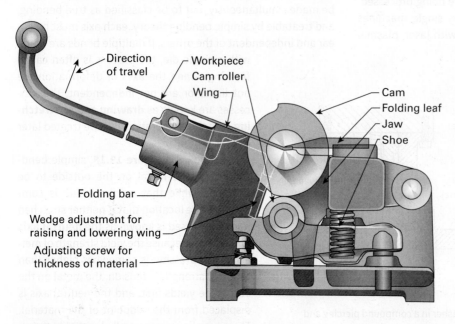

FIGURE 19.19 Phantom section of a bar folder, showing position and operation of internal components. (Courtesy of Niagara Machine and Tool Works, Buffalo, NY)

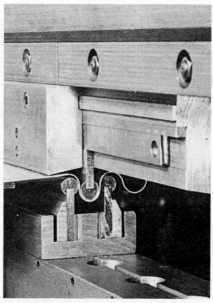

FIGURE 19.20 (Left) A commercial press brake. (Right) Close-up view of press brake dies forming corrugations. [(Left). Courtesy of Accurpress, Rapid City, SD; (Right) Courtesy of Cincinnati Incorporated, Cincinnati, OH]

     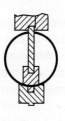

FIGURE 19.21 Press brake dies can form a variety of angles and contours. (Courtesy of Cincinnati Incorporated, Cincinnati, OH)

 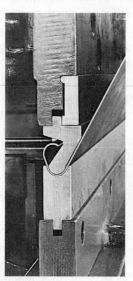 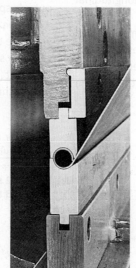

FIGURE 19.22 Dies and operations used in the press brake forming of a roll bead. (Courtesy of Cincinnati Incorporated, Cincinnati, OH)

FIGURE 19.23 Relationship between the minimum bend radius, (relative to thickness) and the ductility of the metal being bent (as measured by the reduction in area in a uniaxial tensile test).

with material ductility. As this plot reveals, an extremely ductile material is required if we wish to produce a bend with radius less than the thickness of the metal. If possible, bends should be designed with large bend radii. This permits easier forming and allows the designer to select from a wider variety of engineering materials.

If the punch radius is large and the bend angle is shallow, large amounts of springback are often encountered. The sharper the bend, the more likely the surfaces will be stressed beyond the yield point. Less severe bends have large amounts of elastically stressed material, and large amounts of springback. In general, when the bend radius is greater than four times the material thickness, the tooling or process must provide springback compensation.

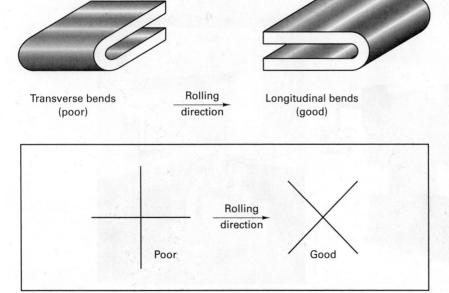

FIGURE 19.24 Bends should be made with the bend axis perpendicular to the rolling direction. When intersecting bends are made, both should be at angle to the rolling direction, as shown.

If the metal has experienced previous cold work or has marked directional properties, these features should be considered when designing the bending operation. Whenever possible, it is best to make the bend axis perpendicular to the direction of previous working, as shown in the upper portion of **Figure 19.24**. The explanation for this recommendation has little to do with the grain structure of the metal, but is more closely related to the mechanical loading applied to the weak, oriented inclusions. Cracks can easily start along tensile-loaded inclusions and propagate to full cracking of the bend. If intersecting or perpendicular bends are required, it is often best to place each at an angle to the rolling direction, as shown in the lower portion of Figure 19.24, rather than have one longitudinal and one transverse.

Another design concern is determining the dimensions of a flat blank that will produce a bent part of the desired precision. As discussed earlier in the chapter, metal tends to thin and lengthen when it is bent. The amount of lengthening is a function of both the material thickness and the bend radius. **Figure 19.25** illustrates one method that has been found to give satisfactory results for determining the blank length for bent products. In addition, the minimum length of any protruding leg should be at least equal to the bend radius plus 1.5 times the thickness of the metal.

Whenever possible, the tolerance on bent parts should not be less than 0.8 mm ($^1/_{32}$ in.). Bends of 90° or greater should not be specified without first determining whether the material and bending method will permit them. Parts with multiple bends should be designed with most (or preferably all) of them to be of the same bend radius. This will reduce setup time and tooling costs. Consideration should also be given to providing regions for adequate clamping or support during manufacture. Bending near the edge of a material will distort the edge. If an undistorted edge is required, additional material must be included and a trimming operation performed after bending.

Air-Bend, Bottoming, and Coining Dies

Yet another design decision is the use of air-bend, bottoming, or coining dies. As shown in **Figure 19.26**, **bottoming dies** contact and compress the full area within the tooling. The angle of the resulting bend is set by the geometry of the tooling, adjusted for subsequent springback, and the inside bend radius is that machined on the nose of the punch. Bottoming dies are designed for a specific material and material thickness, and they form bends of a single configuration. If the results are outside specifications, or the material is changed and produces a different amount of springback, the geometry of the tooling will have to be modified. Once the geometry of the tool is successfully set, however, reproducibility of the bend geometry is excellent (usually within 0.25 degree), provided there is consistency within the size and properties of the material being bent.

In contrast, **air-bend dies** produce the desired geometry by simple three-point bending. Because the resulting angle is controlled by the bottoming position of the upper die, a single set of tooling can produce a range of bend geometries from 180° through the included angle of the die. Air bending can also accommodate a variety of materials in a range of material gages and requires the least force of the three options (bottoming, air-bending, and coining). Product reproducibility depends on the ability to control the

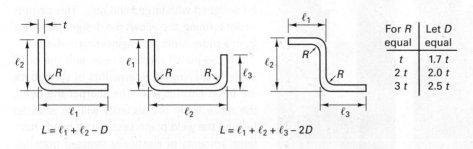

For R equal	Let D equal
t	1.7 t
2 t	2.0 t
3 t	2.5 t

$$L = \ell_1 + \ell_2 - D$$

$$L = \ell_1 + \ell_2 + \ell_3 - 2D$$

FIGURE 19.25 One method of determining the starting blank size (L) for several bending operations. Due to thinning, the product will lengthen during forming. ℓ_1, ℓ_2, and ℓ_3 and are the desired product dimensions. See table to determine D based on size of radius R where t is the stock thickness.

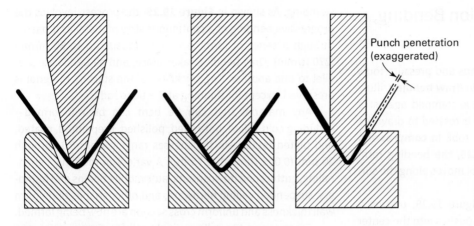

Punch penetration (exaggerated)

FIGURE 19.26 Comparison of air-bend (left), bottoming (center), and coining (right) press brake dies. With the air-bend die, the amount of bend is controlled by the bottoming position of the upper die.

stroke of the press. Adaptive control and on-the-fly corrections are frequently used with air-bend tooling and can generally produce consistent bends within 0.5-degree accuracy.

If bottoming dies continue to close beyond the full-contact position, the thickness of the bent material is reduced. Because of the extensive plastic deformation, the operation becomes one of **coining** (or bottoming with penetration). Springback can be significantly reduced, and more consistent results can be achieved with materials having variation in structure and thickness. Unfortunately, the loading on both the press and the tools is typically 5 to 10 times greater than with bottoming or air-bending.

Reproducible-stroke mechanical presses are generally used for bottom bending and coining, whereas adjustable-stroke hydraulic presses are preferred for air bending.

Yet another approach is to replace the bottom die with a urethane (rubber) pad, which is usually mounted in a steel retainer. The urethane acts like a bladder of hydraulic fluid, maintaining full contact with the workpiece as the bend forms from the centerline outward. The pad, and sheet material,

follows the punch geometry. Because both the sheet material and the urethane are deforming, bending with a urethane tool requires more force than air bending, but less than bottoming with tool-steel tooling. Reinforcing bars and relief holes can be designed into the urethane tool to control its deformation. Prepainted materials, perforated sheets, and textured materials can all be bent with minimum marring and scratching. Depending on the range of materials and bend profiles, a single urethane pad may be able to act as a universal bottom die. The use of urethane tooling in blanking, drawing, and bulging will be discussed later in this chapter.

Roll Bending

Roll bending is a continuous form of three-point bending where plates, sheets, beams, pipe, and even rolled shapes and extrusions are bent to a desired curvature using forming rolls. As shown in **Figure 19.27**, roll-bending machines usually have three rolls in the form of a triangle. The two lower rolls are driven and the position of the upper roll is adjusted to control the degree of curvature in the product. The rolls on the machine pictured in the right-hand segment of Figure 19.27 are supported on only one end. When wider material is being formed, the longer rolls often require support on both ends. The support frame on one end may be swung clear, however, to permit the removal of closed circular shapes or partially rolled product. Because of the variety of applications, roll-bending machines are available in a wide range of sizes, some being capable of bending plate up to 25 cm (10 in.) thick.

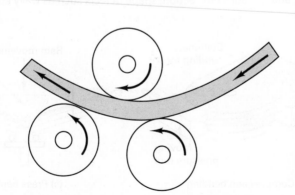 

FIGURE 19.27 (Left) Schematic of the roll-bending process. (Right) Roll bending of an I-beam section. Note how the material is continuously subjected to three-point bending. (Courtesy of Buffalo Machines, Inc., Lockport, NY)

Draw Bending, Compression Bending, and Press Bending

Bending machines can also utilize clamps and pressure tools to bend material against a **form block**. In **draw bending**, illustrated in **Figure 19.28**, the workpiece is clamped against a bending form, and the entire assembly is rotated to draw the workpiece along a stationary pressure tool. In **compression bending**, also illustrated in Figure 19.28, the bending form remains stationary, and the pressure tool moves along the surface of the workpiece.

Press bending, also shown in Figure 19.28, utilizes a downward descending bend die, which pushes into the center of material that is supported on either side by wing dies. As the ram descends, the wing dies pivot up, bending the material around the form on the ram. The flexibility of each of the preceding processes is somewhat limited because a certain length of the product must be used for clamping or support.

Tube Bending

Quite often the material being bent is a tube or pipe (**tube bending**), and this geometry presents additional problems. Key parameters are the outer diameter of the tube, the wall thickness, and the radius of the bend. Small diameter, thick-wall tubes usually present little difficulty. As the outer diameter increases, the wall thickness decreases, or the bend radius becomes smaller, the outside of the tube tends to pull to the center, flattening the tube, and the inside surface may wrinkle. For many years, a common method of overcoming these problems was to pack the tube with wet sand, produce the bend, and then remove the sand from the interior. Flexible mandrels have now replaced the sand and are currently available in a wide variety of styles and sizes.

Roll Forming

The continuous **roll forming** of flat strip into complex sections has become a highly developed forming technique that competes directly with press brake forming, extrusion, and stamping. As shown in **Figure 19.29**, the process involves the progressive bending of a continuous strip of metal as it passes through a series of forming rolls at speeds up to 80 m/min (270 ft/min). Only bending takes place, and all bends are parallel to one another. The thickness of the starting material is preserved, except for thinning at the bend radii.

Any material that can be bent can be roll formed—including cold-rolled, hot-rolled, polished, prepainted, coated, and plated metals—in thicknesses ranging from 0.1 through 20 mm (0.005 through 0.75 in.). A variety of moldings, channeling, gutters and downspouts, automobile beams and bumpers, suspended ceiling hardware, and other shapes of uniform wall thickness and uniform cross section are now being formed.

By changing the rolls, a single roll-forming machine can produce a wide variety of different shapes. However, changeover, setup, and adjustment may take several hours, so a production run of at least 3000 m (10,000 ft) is usually required for any given product. To produce pipe or tubular products, a resistance welding unit or seaming operation is often integrated with the roll forming.

Seaming and Flanging

Seaming is a bending operation that can be used to join the ends of sheet metal in some form of mechanical interlock. **Figure 19.30** shows several of the more common seam designs that can be formed by a series of small rollers. Seaming machines range from small hand-operated types to large automatic units capable of producing hundreds of seams per minute. Common products include cans, pails, drums, and other similar containers.

Flanges can be rolled on sheet metal in essentially the same manner as seams. In many cases, however, **flanging** and seaming are drawing operations because the bending often occurs along a curved axis.

Straightening, Flattening, or Leveling

As described in the rolling segment of Chapter 18, flatness control is a significant issue. If one section is rolled longer than an adjacent section, shape defects such as wavy edge and center

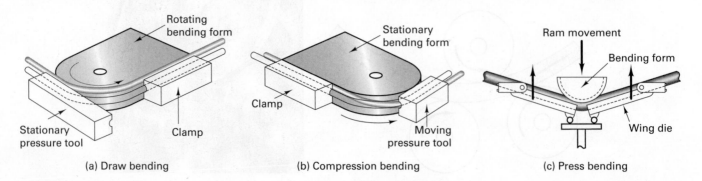

(a) Draw bending   (b) Compression bending   (c) Press bending

FIGURE 19.28 (a) Draw bending, in which the form block rotates; (b) compression bending, in which a moving tool compresses the workpiece against a stationary form; and (c) press bending, where the press ram moves the bending form.

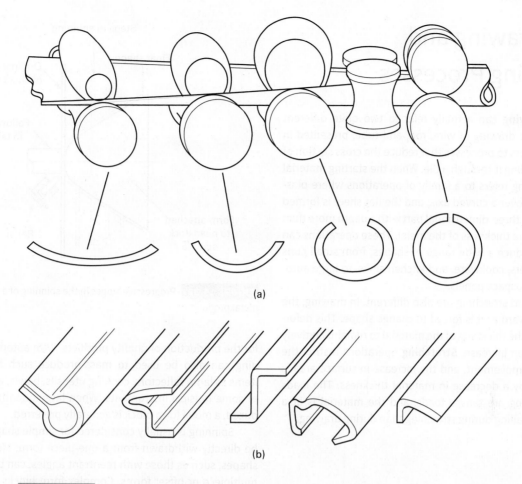

(a)

(b)

FIGURE 19.29 (a) Schematic representation of the cold roll-forming process being used to convert sheet or plate into tube. (b) Some typical shapes produced by roll forming.

FIGURE 19.30 Various types of seams used on sheet metal.

buckles result. The objective of the shape correction methods of **straightening**, **flattening**, or **leveling** is the opposite of bending, and these operations are often performed before subsequent forming to ensure the use of flat or straight material that is reasonably free of residual stresses. **Roll straightening** or **roller leveling**, illustrated in **Figure 19.31**, subjects the material to a series of reverse bends. The sheet (can also be rod or wire) is passed through a series of alternating upper and lower rolls with progressively decreased offsets. This configuration produces deep bends at the entry decreasing to little or no bend at the exit. As the material is bent, first up and then down, the outside surfaces are placed in tension and are stressed beyond their elastic limit, replacing any permanent set with a flat or straight profile. Additional tension can be applied along the length of the product to help induce the required deformation, a modification known as **tension leveling**.

FIGURE 19.31 Method of straightening rod or sheet by passing it through a set of straightening rolls. For rods, another set of rolls is used to provide straightening in the transverse direction.

Sheet material can also be straightened by a process called **stretcher leveling**. Finite-length sheets are gripped mechanically and linearly stretched beyond the elastic limit to produce the desired flatness.

19.4 Drawing and Stretching Processes

The term **drawing** can actually refer to two quite different operations. The drawing of wire, rod, and tube, presented in Chapter 18, refers to processes that reduce the cross section of material by pulling it through a die. When the starting material is sheet, drawing refers to a family of operations where plastic flow occurs over a curved axis, and the flat sheet is formed into a recessed, three-dimensional part with a depth more than several times the thickness of the metal. These operations can be used to produce a wide range of shapes, from small cups through oil filters, cookware, and kitchen sinks, to large automobile and aerospace panels.

Drawing and stretching are also different. In drawing, the metal flows inward as it is forced to change shape. This deformation allows the thickness of the material to remain relatively constant or even increase. **Stretching** operations restrict the compensating movement, and the increase in surface area is accompanied by a decrease in material thickness. The loads during stretching are tensile loads, and the material is at a greater risk of failing during stretching than during drawing.

Spinning

Spinning is a cold-forming operation where a rotating disk of sheet metal is progressively shaped over a male form, or mandrel, to produce rotationally symmetrical shapes, such as cones, hemispheres, cylinders, bells, and parabolas. A form block possessing the shape of the desired part is attached to a rotating spindle, such as the drive section of a simple lathe. A disk of metal (or preformed shape) is centered on the small end of the form and held in place by a pressure pad. As the disk and form rotate, localized pressure is applied through a round-ended wooden or metal tool or small roller that traverses the entire surface of the part, causing it to flow progressively against the form. **Figure 19.32** depicts the progressive operation, and **Figure 19.33** shows an array of parts that were produced by spinning. Because the final diameter of the part is less than that of the starting disk, the circumferential length decreases. This decrease must be compensated by either an increase in thickness, a radial elongation, or circumferential buckling. Control of the process is often dependent on the skill of the operator, and multiple passes may be required to complete the shape.

During spinning, the form block sees only localized compression, and the metal does not move across it under pressure. As a result, form blocks can often be made of hardwood or even plastic. The primary requirement is simply to replicate the shape with a smooth surface. As a result, tooling cost can be extremely low, making spinning an attractive process for producing small quantities of a single part. Skilled labor is required, however, to make the minute adjustments necessary

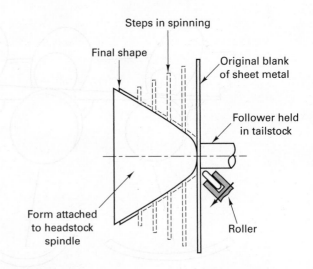

FIGURE 19.32 Progressive stages in the spinning of a sheet metal product.

for the production of quality products. With automation, spinning can also be used to mass-produce such high-volume items as lamp reflectors, cooking utensils, bowls, and the bells of some musical instruments. When large quantities are produced, a metal form block is generally preferred.

Spinning is usually considered for simple shapes that can be directly withdrawn from a one-piece form. More complex shapes, such as those with reentrant angles, can be spun over multipiece or offset forms. Complex form blocks can also be made from frozen water, which is melted out of the product after spinning. Any ductile material with more than 2% elongation can be formed by metal spinning, and workpiece sizes can be as large as 6 m (20 ft) in diameter.

A recent modification of metal spinning is **laser-assisted metal spinning**. A laser beam is focused directly ahead of the

FIGURE 19.33 An array of metal parts produced by spinning. (Photo courtesy of ACME Metal Spinning, Minneapolis, MN)

roller to heat and soften the material. Materials like titanium, nickel-based alloys, and austenitic stainless steels can be spun with less difficulty, and the operation can be performed with less force and power.

Shear Forming or Flow Turning

Cones, hemispheres, and similar shapes are often formed by **shear forming** or **flow turning**, a modification of the spinning process in which each element of the blank maintains its distance from the axis of rotation and the spinning flange remains vertical throughout the process. The roller provides highly localized pressure that plastically deforms and cold-works the metal. Because there is no circumferential shrinkage, the metal flow is entirely by shear, and no compensating stretch has to occur. As shown in **Figure 19.34**, the wall thickness of the product, t_c, will vary with the angle of the particular region according to the relationship: $t_c = t_b \sin \alpha$, where t_b is the thickness of the starting blank. If α is less than 30°, it may be necessary to complete the forming in two stages with an intermediate anneal in between. Reductions in wall thickness as high as 8:1 are possible, but the limit is usually set at about 5:1 or 80%.

Conical shapes are usually shear formed by the direct process depicted in Figure 19.34. The bottom of the product is held against the face of the form block or mandrel, while the material being formed moves in the same direction as the roller. Products can also be formed by a reverse process, like that illustrated in **Figure 19.35**. By controlling the position and feed of the forming rollers, the reverse process can be used to shape concave, convex, or conical parts without a matching form block or mandrel. Cylinders can be shear formed by both the direct and reverse processes. As shown in **Figure 19.36**, the direct process restricts the length of the product to the length of the mandrel. No schematic is provided for the reverse process because it is essentially the same as the roll extrusion process, depicted in Figure 18.52.

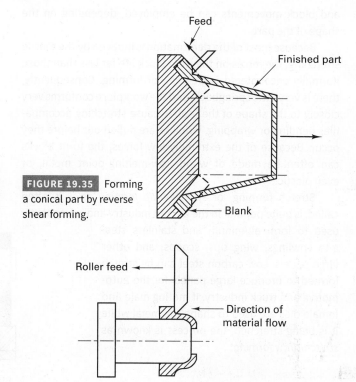

FIGURE 19.35 Forming a conical part by reverse shear forming.

FIGURE 19.36 Shear forming a cylinder by the direct process.

Stretch Forming

Stretch forming, illustrated in principle in **Figure 19.37**, is an attractive means of producing large sheet metal parts in low or limited quantities. A sheet of metal is gripped along its edges by two or more sets of jaws that stretch it and wrap it around a single form block. Various combinations of stretching, wrapping,

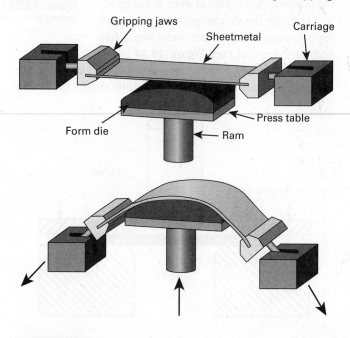

FIGURE 19.37 Schematic of a stretch-forming operation. (Courtesy of *MetalForming Magazine*, a publication of the Precision Metalforming Association, Independence, OH)

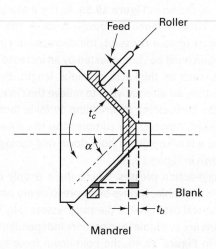

FIGURE 19.34 Schematic representation of the basic shear-forming process.

and block movements can be employed, depending on the shape of the part.

Because most of the deformation is induced by the tensile stretching, the forces on the form block are far less than those normally encountered in bending or forming. Consequently, there is very little springback, and the workpiece conforms very closely to the shape of the tool. Because stretching accompanies bending or wrapping, wrinkles are pulled out before they occur. Because of the extremely low forces, the form blocks can often be made of wood, low-melting-point metal, or even plastic.

Stretch forming, or *stretch-wrap forming* as it is often called, is quite popular in the aircraft industry and is frequently used to form aluminum and stainless steel into cowlings, wing tips, scoops, and other large panels. Low-carbon steel can be stretch formed to produce large panels for the automotive and truck industry. If mating male and female dies are used to shape the metal while it is being stretched, the process is known as *stretch-draw forming.*

Deep Drawing and Shallow Drawing

The forming of solid-bottom cylindrical or rectangular containers from metal sheet is one of the most widely used manufacturing processes. When the depth of the product is less than its diameter (or the smallest dimension of its opening), the process is considered to be **shallow drawing**. If the depth is greater than the diameter, it is known as **deep drawing**.

Consider the simple operation of converting a circular disk of sheet metal into a flat-bottom cylindrical cup. **Figure 19.38** shows the blank positioned over a die opening and a circular punch descending to pull or draw the material into the die cavity. The

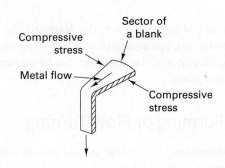

FIGURE 19.39   Flow of material during deep drawing. Note the circumferential compression as the radius is pulled inward.

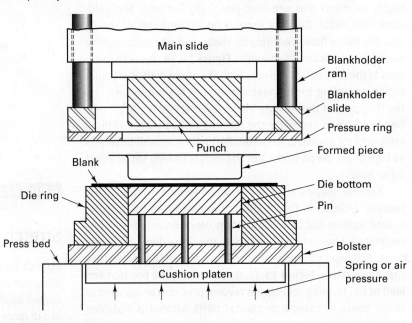

FIGURE 19.40   Drawing on a double-action press, where the blankholder uses the second press action.

material beneath the punch remains largely unaffected and simply becomes the bottom of the cup. The cup wall is formed by pulling the remainder of the disk inward and over the radius of the die, as shown in **Figure 19.39**. As the material is pulled inward, its circumference decreases. Because the volume of material must remain constant, the decrease in circumferential dimension must be compensated by an increase in another dimension, such as thickness or radial length. Because the material is thin, an alternative is to relieve the circumferential compression by buckling or wrinkling. Wrinkle formation can be suppressed, however, by compressing the sheet between the die and a blankholder or hold-down ring during the forming (as shown in Figure 19.38).

In single-action presses, where there is only one movement that is available, springs or air pressure are often used to clamp the metal between the die and pressure ring. When multiple-actions are available (two or more independent motions), as shown in **Figure 19.40**, the hold-down force can now be applied in a manner that is independent of the punch position. This restraining force can also be varied during the drawing

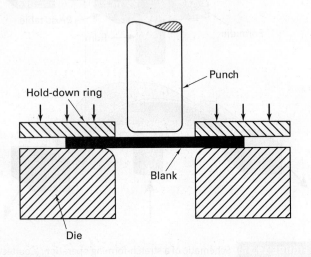

FIGURE 19.38   Schematic of the deep-drawing process.

operation. For this reason, multiple-action presses are usually specified for the drawing of more complex parts, whereas single-action presses can be used for the simpler operations.

Key variables in the deep-drawing process include the blank diameter and the punch diameter (which combine to determine the **draw ratio** and the height of the side walls), the die radius, the punch radius, the clearance between the punch and the die, the thickness of the blank, blankholder geometry, the surface finish of the die face and blankholder, lubrication, and the hold-down pressure. The punch-nose radius cannot be too small or the punch will cut or pierce the blank instead of pulling it into the die. If the punch radius is too large, the material will stretch and thin as it passes over the punch as opposed to pulling material in from the flange. If the die radius is too small, tearing may occur as the material bends over the tight radius. Excessive die radius, however, may lead to wrinkling in the unsupported region between the punch face and die face.

Once a process has been designed and the tooling manufactured, the primary variable for process adjustment is the **hold-down pressure** or **blankholder force**. If the force or pressure is too low, wrinkling may occur at the start of the stroke. If it is too high, there is too much restraint, and stretching and thinning of the material will occur. At the extreme, the descending punch will simply tear the disk or some portion of the already-formed cup wall.

When drawing a shallow cup, there is little change in circumference, and a small area is being confined by the blankholder. As a result, the tendency to wrinkle or tear is low. As cup depth increases, there is an increased tendency for forming both of the defects. In a similar manner, thin material is more likely to wrinkle or tear than thick material. **Figure 19.41** summarizes the effects of cup depth, blank thickness, and blankholder force or pressure. Note that for thin materials

and deep draws, a defect-free product may not be possible in a single operation, as the defects simply transition from wrinkling, to wrinkle plus tear, to just tearing as blankholder force is increased. To produce a defect-free product, the draw ratio (blank diameter/cup diameter) is often limited to values less than about 2.2.

The limitations of wrinkling and tearing (and the limiting draw ratio) can often be overcome by using multiple operations that include **redrawing**. **Figure 19.42** shows two alternatives for converting drawn parts into deeper cups. In the **forward redraw** option, the material undergoes reverse bending as it flows into the die. In **reverse redrawing**, the starting cup is placed over a tubular die, and the punch acts to turn it inside out. Because all bending is in one direction, reverse redraws can produce greater changes in diameter. Because there is bending and straightening of material that has been previously worked, redraws are always performed with successively lower diameter reductions than initial draws.

When the part geometry becomes more complex, as with rectangular or asymmetric parts, it is best if the surface area and thickness of the material can remain relatively constant. Different regions may need to be differentially constrained. One technique that can produce variable constraint is the use of **draw beads**—vertical projections and matching grooves in the die and blankholder. The added force of bending and unbending as the material flows over the draw bead restricts the flow of material. The degree of constraint can be varied by adjusting the height, shape, and size of the bead and bead cavity. An alternative approach is to apply different blankholder pressures at different locations.

Because of prior rolling and other metallurgical and process features, the flow of sheet metal is generally not uniform, even in the simplest drawing operation. Excess material may be required to ensure final dimensions, and **trimming** may be required to establish both the size and uniformity of the final part. **Figure 19.43** shows a shallow-drawn part before and after trimming. Trimming obviously adds to the production cost because it not only converts some of the starting material to scrap but also adds another operation to the manufacturing process—one that must be performed on an already-produced shape.

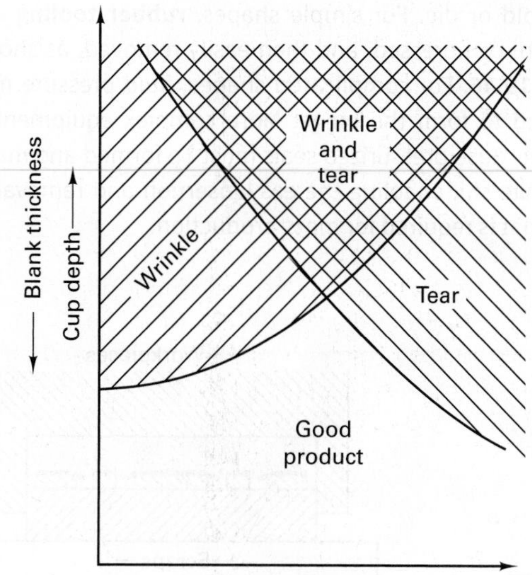

FIGURE 19.41 Defect formation in deep drawing as a function of blankholder force, blank thickness, and cup depth.

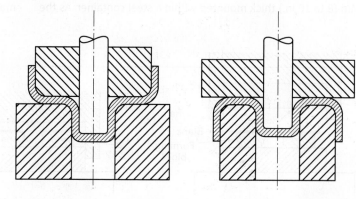

FIGURE 19.42 Cup redrawing to further reduce diameter and increase wall height. (Left) Forward redraw; (right) reverse redraw.

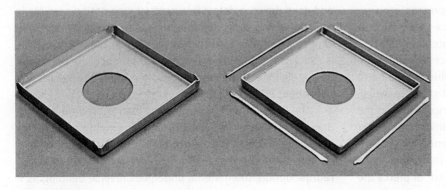

FIGURE 19.43 Pierced, blanked, and drawn part before and after trimming.
(E. Paul DeGarmo)

Forming with Rubber Tooling or Fluid Pressure

Blanking and drawing operations usually require mating male and female, or upper and lower die sets, and process setup requires that the various components be properly positioned and aligned. When the amount of deformation is great, multiple operations may be required, each with its own set of dedicated tooling, and intermediate anneals may also be necessary. Numerous processes have been developed that seek to: (1) reduce tooling cost, (2) decrease setup time and expense, or (3) extend the amount of deformation that can be performed with a single set of tools. Although most of these methods have distinct limitations, such as complexity of shape or types of metal that can be formed, they also have definite areas of application.

Several forming methods replace either the male or female member of the die set with rubber or fluid pressure. The **Guerin process** (also known as *rubber-die forming*) is depicted in **Figure 19.44**. It is based on the phenomenon that rubber of the proper consistency, *when totally confined*, acts as a fluid and transmits pressure uniformly in all directions. Blanks of sheet metal are placed on top of form blocks, which can be made of wood, Bakelite, polyurethane, epoxy, or low-melting-point metal. The upper ram contains a pad of rubber 20 to 25 cm (8 to 10 in.) thick mounted within a steel container. As the ram descends, the rubber pad becomes confined and transmits force to the metal, causing it to bend to the desired shape. Because no female die is used and inexpensive form blocks replace the male die, the total tooling cost is quite low. There are no mating tools to align, process flexibility is quite high (different shapes can even be formed at the same time), wear on the material and tooling is low, and the surface quality of the workpiece is easily maintained, a feature that makes the process attractive for forming prepainted or specially coated sheet. When reentrant sections are produced (as with product b in Figure 19.44), it must be possible to slide the parts lengthwise from the form blocks or to disassemble a multipiece form from within the product.

The Guerin process was developed by the aircraft industry, where the production of small numbers of duplicate parts clearly favors the low cost of tooling. It can be used on aluminum sheet up to 3 mm (⅛ in.) thick and on stainless steel up to 1.5 mm (1/16 in.). Magnesium sheet can also be formed if it is heated and shaped over heated form blocks.

Most of the forming done with the Guerin process is multiple-axis bending, but some shallow drawing can also be performed. The process can also pierce or blank thin gages of aluminum, as illustrated in **Figure 19.45**. For this application, the blanking blocks are shaped the same as the desired workpiece, with a sharp face, or edge, of hardened steel. Round-edge supporting blocks are positioned a short distance from the blanking blocks to support the scrap skeleton and permit the metal to bend away from the sheared edges.

In **bulging**, fluid or rubber transmits the pressure required to expand a metal blank or tube outward against a split female mold or die. For simple shapes, **rubber tooling** can be inserted, compressed, and then easily removed, as shown in **Figure 19.46**. For complicated shapes, fluid pressure may be required to form the bulge. More complex equipment is required because pressurized seals must be formed and maintained, while still enabling the easy insertion and removal of material that is required for mass production.

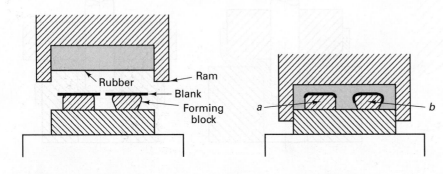

FIGURE 19.44 The Guerin process for forming sheet metal products.

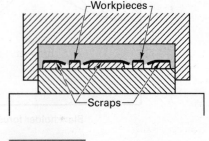

FIGURE 19.45 Method of blanking sheet metal using the Guerin process.

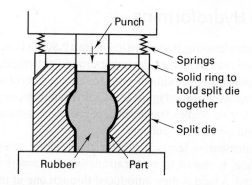

FIGURE 19.46 Method of bulging tubes with rubber tooling.

Sheet Hydroforming

Sheet hydroforming is really a family of processes in which a rubber bladder backed by fluid pressure, or pockets of pressurized liquid, becomes a universal die half, replacing either the solid punch or female die of the traditional tool set. In a variant known as *high-pressure flexible-die forming*, or **flexforming**, the rubber pad of the Guerin process is replaced by a flexible rubber diaphragm backed by controlled hydraulic pressure at values between 140 and 200 MPa (20,000 and 30,000 psi). As illustrated in **Figure 19.47**, the solid punch drives the sheet into

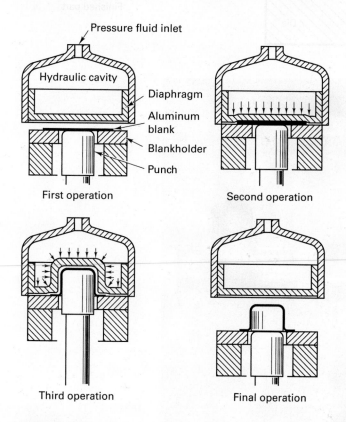

FIGURE 19.47 High-pressure flexible-die forming, showing (1) the blank in place with no pressure in the cavity; (2) press closed and cavity pressurized; (3) ram advanced with cavity maintaining fluid pressure; and (4) pressure released and ram retracted. (Courtesy of Aluminum Association, Arlington, VA)

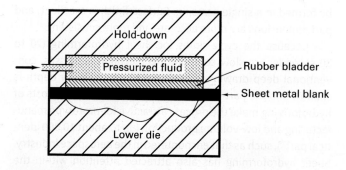

FIGURE 19.48 One form of sheet hydroforming.

the resisting bladder, whose pressure is adjusted throughout the stroke.

In a variation of this process that shares similarity to stretch forming, the sheet is first clamped against the opening and the punch is retracted downward. The fluid is then pressurized, causing the workpiece to balloon downward toward the retracted punch. Because the pressure is uniformly distributed over the workpiece, the sheet is uniformly stretched and thinned. The punch then moves upward, causing the prestretched metal to conform to the punch profile.

The flexible membrane can also be used to replace the hardened male punch, as shown in **Figure 19.48**. The ballooning action now causes the material to descend and conform fully to the female die, which may be made of epoxy or other low-cost material. **Parallel-plate hydroforming**, or **pillow forming**, extends the process to the simultaneous production of upper and lower contours. As shown in **Figure 19.49**, two sheet metal blanks are laser welded around their periphery or are firmly clamped between upper and lower dies. Pressurized fluid is then injected between the sheets, simultaneously forming both upper and lower profiles. This may be a more attractive means of producing complex sheet metal containers because the manufacturer no longer has to cope with the problems of aligning and welding separately formed upper and lower pieces.

Although the most attractive feature of sheet hydroforming is probably the reduced cost of tooling, there are other positive attributes. Materials exhibit greater formability because a more uniform distribution of strain is produced. Deeper parts can be formed without fracture (drawing limits are generally about 1.5 times those of conventional deep drawing), and complex shapes that would require multiple operations can often

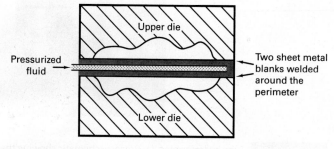

FIGURE 19.49 Two-sheet hydroforming or pillow forming.

be formed in a single pressing. Surface finish is excellent, and part dimensions are more accurate and more consistent.

Because the cycle times for sheet hydroforming (20 to 30 seconds) are slow compared to mechanical presses, conventional deep drawing is preferred when the draw depth is shallow and the part is not complex. The reduced tool costs of hydroforming make the process attractive for prototype manufacturing and low-volume production (up to about 10,000 identical parts), such as that encountered in the aerospace industry. Sheet hydroforming has also attracted attention within the automotive community because of its ability to not only produce low-volume parts in an economical manner, but also successfully shape lower-formability materials, such as alloyed aluminum sheet and high-strength steels. Warm hydroforming has further expanded this capability.

Tube Hydroforming

Tube hydroforming, illustrated in **Figure 19.50**, has become a significant process for manufacturing strong, lightweight, tubular components, that frequently replace an assembly of welded stampings. As shown in **Figure 19.51**, current automotive parts include engine cradles, frame rails, roof headers, radiator supports, and exhaust components.

In elementary terms, a tubular blank, either straight or preshaped, is placed in an encapsulating die, and the ends are sealed. A fluid is then introduced through one of the end plugs, achieving sufficient pressure to expand the material to the shape of the die. At the same time, the end closures may move inward to help drive radial expansions and compensate or overcome the wall thinning that would otherwise occur. In

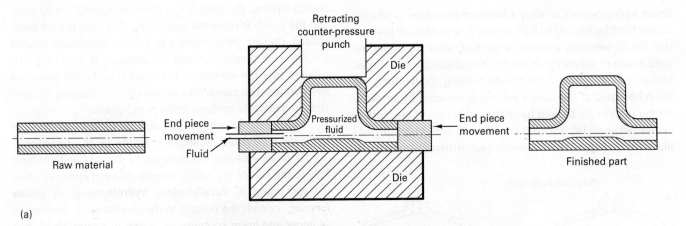

(a)

(b)

FIGURE 19.50 Tube hydroforming. (a) Process schematic showing the independent controls of axial feeding, internal pressure, and counterpressure; (b) actual copper product. Note the inward movement of the tube ends and the nonuniform wall thickness of the nonsymmetric product. (Courtesy Ronald Kohser)

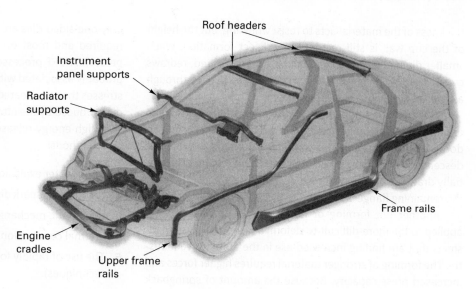

Roof headers

Instrument
panel supports

Radiator
supports

Frame rails

Engine
cradles

Upper frame
rails

FIGURE 19.51 Use of hydroformed tubes in automotive applications. (Courtesy of *MetalForming Magazine*, a publication of PMA Services, Inc., for the Precision Metalforming Association, Independence, OH; www.metalformingmagazine.com)

actual operation, the process may use combinations or even sequences of internal pressure, axial motions, and even external counter-pressure applied to bulging regions, to control the flow and final thickness of the material. Product length is currently limited by the ability of end movements to create axial displacements within the die.

In *low-pressure tube hydroforming* (pressures up to 35 MPa or 5000 psi), the tube is first filled with fluid and then the dies are closed around the tube. The primary purpose of the fluid is to act as a liquid mandrel that prevents collapsing as the tube is bent to the contour of the die. Although the cross-section shape can be changed, the shapes must be simple, corner radii must be large, and the perimeter of the tube must remain essentially constant. In *high-pressure tube hydroforming*, an internal pressure between 100 and 700 MPa (15,000 and 100,000 psi) is used to expand the diameter of the tube, forming tight corner radii and significantly altered cross sections. *Pressure-sequence hydroforming* begins by applying low internal pressure as the die is closing. This supports the inside wall of the tube and allows it to conform to the cavity. When the die is fully closed, high pressure is then applied to complete the forming of the tube walls.

Another modification of tube hydroforming uses *pulsing or vibrating pressure* (also called hammering action) to reduce the frictional drag as the tube moves across the dies. By oscillating the pressure between two preset values, the tube can move more freely, increasing the amount of allowable deformation and reducing the amount of thinning in regions of surface expansion. The pulsation frequency is generally between 1 and 3 cycles per second (Hz).

Attractive features of tube hydroforming include the ability to use lightweight, high-strength materials; the increase in strength that results from strain hardening; and the ability to utilize optimized designs with varying thickness or cross sections (including circular, rectangular, triangular, oblong, or any other shape) that can vary continuously throughout the length of the part. Stamped-and-welded assemblies can often be replaced by one-piece components with improved structural strength and rigidity, and secondary operations can often be reduced. Parts can be produced from materials as diverse as brass, stainless steel, carbon steel, aluminum, and even high-strength low-alloy and dual-phase steel. Disadvantages include the long cycle time (low production rate) and relatively high cost of tooling and process setup.

An emerging alternative to tube hydroforming is **high-temperature-metal-gas-forming**. In this process, a straight or preformed tube is heated to forming temperature (near but below its melting point) by induction heating, placed in heated tooling, and then pressurized internally with inert gas to drive the material outward to the enclosed die cavity. By using high temperatures, the parts can be elongated or stretched to greater degree than with the cold- or warm-forming fluid methods. The forming pressures are typically an order of magnitude lower than tube hydroforming, and the formed parts can be quenched directly from the high-temperature forming operation. Expectations for the process include finer part detail, faster speed, lower cost, and greater flexibility compared to tube hydroforming. Lower formability metals, such as ferritic stainless steels, may be used in place of the higher formability, but often more expensive types, like the austenitic grades.

Hot Forming and Hot Stamping

Because sheet material has a large surface area and small thickness, it cools rapidly in a lower-temperature environment. For this reason, most sheet forming is performed at room or mildly elevated temperature. Cold drawing uses relatively thin metal, changes the thickness very little or not at all, and produces parts in a wide variety of shapes. In contrast, hot drawing may be used for forming relatively thick-walled parts of simple geometries, and the material thickness may change significantly during the operation.

Hot drawing operations are extremely similar to previously discussed processes. A simple disk-to-cup drawing operation can be performed without a hold-down. The increased thickness of the material acts to resist wrinkling, but the height of the cup wall is still restricted by defect formation. When smaller diameter and higher wall height are desired, redraws can be used. As an alternative, the cup can be pushed through a series of dies with a single punch.

If the drawn products are designed to utilize part of the original disk as a flange around the top of the cup, the punch does not push the material completely through the die, but descends to a predetermined depth and then retracts. The partially drawn cup is then ejected upward, and the perimeter of the remaining flange is trimmed to the desired size and shape.

Another hot forming or hot stamping process is being applied to the more-difficult-to-deform advanced high-strength steels that are finding increased use in the automotive industry. The forming of stronger material requires higher forces and increased press capacity. Because the amount of springback is proportional to the yield strength after forming (doubling strength doubles the amount of springback), both springback and residual stresses are also significant problems.

Hot stamping, also called **press hardening**, can overcome some of these constraints. The material blank (often a boron-based steel) is first heated to between 850° and 950° C (1550° and 1750° F) to create an austenite microstructure with a tensile strength of about 100 MPa (15 ksi) and elongation of 50% to 60%. It is then transferred to a water-cooled (chilled) die, where it is immediately formed while in a soft, weak, and very ductile condition. The formed part is then held in the die, where it rapidly cools, forming a martensitic structure with tensile strength in the range of 1025 to 1600 MPa (150 to 230 ksi) and elongation of about 5% to 6%. The entire forming process takes approximately 20 seconds and produces high-strength, complex-shaped stampings in a single step with virtually no springback because the part is stress-relieved as it forms. Secondary operations cannot be performed, and the emerging part must be trimmed with a laser or hot trimmed in the same stroke as the forming operation because the resulting material is too hard for traditional methods.

Because of their high strength, hot stamped parts can weigh 25% to 30% less than traditional metal parts of similar strength. Where the design requires it, variation in properties within a single piece can be achieved by either differential cooling within the forming die or differential tempering of the formed product.

High-Energy-Rate Forming

A number of methods have been developed to form metals through the application of large amounts of energy in a very short time (high strain rate). These are known as **high-energy-rate forming** processes and often go by the abbreviation **HERF**. Many metals tend to deform more readily under the ultrarapid load application rates used in these processes (deformation velocities can exceed 100 m/sec). As a consequence, HERF makes it possible to form large workpieces and difficult-to-form metals with less-expensive equipment and cheaper tooling than would otherwise be required. In many cases,

only one-sided dies are required. Lubricants are generally not required and most operations are performed at room temperature. HERF processes also produce less springback. This is probably associated with two factors: (1) the high compressive stresses that are created, and (2) the elastic deformation of the die produced by the ultrahigh pressure.

High-energy-release rates can be obtained by five distinct methods:

1. Underwater explosions (**explosive forming**).
2. Underwater spark discharge (electro-hydraulic techniques).
3. Pneumatic-mechanical means.
4. Internal combustion of gaseous mixtures.
5. The use of rapidly formed magnetic fields (electromagnetic techniques).

Specific processes were developed around each of these approaches and attracted considerable attention during the 1960s and 1970s when they were used to produce one-of-a-kind and small quantities of parts for the space program.

Ironing

Deep-drawn parts generally do not have uniform wall thickness and vertical walls. In deep drawing, the gap between the punch and die is greater than the thickness of the material. The resulting walls have an outward taper as they progress upward from the punch radius to the die radius. As the blank is pulled inward, its circumference decreases. This decrease is often offset by an increase in thickness. The cup wall, therefore, tends to be thinnest at the base where it stretches around the punch face and increases in thickness as we move up from the base. If uniform wall thickness and straight, nontapered walls are required, an ironing process may be necessary. **Ironing** thins the walls of a drawn cylinder by passing it between a punch and die where the gap is less than the incoming wall thickness. As shown in **Figure 19.52**, the walls reduce to a uniform thickness and lengthen, while the thickness of the base remains unchanged. The most common example of an ironed product is the thin-walled beverage can.

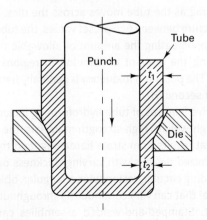

FIGURE 19.52 The ironing process.

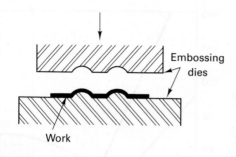

Embossing

Embossing, shown in **Figure 19.53**, is a pressworking process in which raised lettering or other designs are impressed in sheet material while the material thickness remains largely unchanged. Basically, it is a very shallow drawing operation where the depth of the draw is limited to one to three times the thickness of the metal. A common example of an embossed product is the patterned or textured industrial stair tread.

Superplastic Sheet Forming

Conventional metals and alloys typically exhibit tensile elongations in the range of 10% to 30%. By producing sheet materials with ultrafine grain size and performing the deformation at low strain rates and elevated temperatures, elongations in some materials can exceed 100% and may be as high as 2000 to 3000%. This **superplastic forming** behavior can be used to form material into large, complex-shaped products with compound curves. Deep or complex shapes can be made as single-piece, single-operation pressings rather than multistep conventional pressings or multipiece assemblies.

At the elevated temperatures required to promote superplasticity (about 900°C or 1650°F for titanium, between 450° and 520°C or 850° and 970°F for aluminum, or generally above half of the melting point of the material on an absolute scale), the strength of material is sufficiently low that many of the superplastic forming techniques are adaptations of processes used to form thermoplastics (discussed in Chapter 20). The tooling doesn't have to be exceptionally strong, so form blocks can often be used in place of die sets. In thermoforming, a vacuum or pneumatic pressure causes the sheet to conform to a heated male or female die. Blow forming, vacuum forming, deep drawing, and combined superplastic forming and diffusion bonding are other possibilities. Precision is excellent, and fine details or surface textures can be reproduced accurately. Springback and residual stresses are almost nonexistent, and the products have a fine, uniform grain size.

The major limitation to superplastic forming is the low forming rate that is required to maintain superplastic behavior. Cycle times may range from 2 minutes to as much as 2 hours per part, compared to the several seconds that is typical of conventional presswork. As a result, applications tend to be limited to low-volume products such as those common to the aerospace industry. By making the products larger and eliminating assembly operations, the weight of products can often be reduced, there are fewer fastener holes to initiate fatigue cracks, tooling and fabrication costs are reduced, and there is a shorter production lead time.

Free-Form Fabrication

Small quantities and one-of-a-kind sheet metal parts can be formed by a technique that completely eliminates part-specific dies or tooling. A piece of sheet metal is clamped around its edges, and a pair of stylus-type tools mounted in robot arms (one above and one below the sheet) move in unison following computer-generated tool paths to press on the metal and form the three-dimensional shape. Replacement parts could be made on demand, and prototypes can be manufactured in a few days.

Properties of Sheet Material

A wide variety of materials have been used in sheet-forming operations, including hot- and cold-rolled steel, stainless steel, copper alloys, magnesium alloys, aluminum alloys, and even some types of plastics. These materials are available in a range of tempers (amounts of prior cold working) and surface finishes. Sheet material can also have a variety of platings or coatings, or even be a clad or laminated composite. With the desire to reduce the weight of automotive vehicles, sheet formers are now working with thinner-gauge, higher-strength steels, some of which undergo significant metallurgical and property changes during deformation, and more-difficult-to-deform aluminum alloys.

The success or failure of many sheet-forming operations is strongly dependent on the properties of the starting material. In acquiring the sheet geometry, the material has already undergone a number of processes, such as casting, hot rolling, and cold rolling, and has acquired distinct properties and characteristics. A simple uniaxial tensile test of the sheet material can be quite useful by providing values for the yield strength, tensile strength, elongation, and strain-hardening exponent. The amount of elongation prior to necking, or the uniform elongation, is probably a more useful elongation number because localized thinning is actually a form of sheet metal failure. A low yield, high tensile, and high uniform elongation all combine to indicate a large amount of useful plasticity. A high value of the **strain-hardening exponent, n** [obtained by fitting the true stress and true strain to the equation: stress = K (strain)^n] indicates that the material will have both greater allowable and more uniform stretch.

In a uniaxial tensile test, the sheet is stretched in one direction and is permitted to contract and compensate in both width and thickness. Many sheet-forming operations subject the material to stretching in more than one direction, however. When the material is in biaxial tension, as in deep drawing or the various stretch-forming operations, all of the elongation must be compensated by a decrease in thickness. The strain to fracture is typically about one-third of the value observed in a uniaxial test. As a result, some form of biaxial stretching test may be preferred to assess formability, such as a dome-height or hydraulic bulge test.[2]

Sheet metal is often quite aniso-tropic, with properties varying with direction or orientation. A useful assessment of this variation can be obtained through the **plastic strain ratio, R**, defined in ASTM specifica-tion E517. During a uniaxial tensile test, as the length increases, the width and thickness both reduce. The R-value is simply the ratio of the width strain to the thickness strain. Materials with values greater than one tend to compensate for stretching by flow within the sheet, and resist the thinning that leads to fracture. Hence materials with high R-values have good formability.

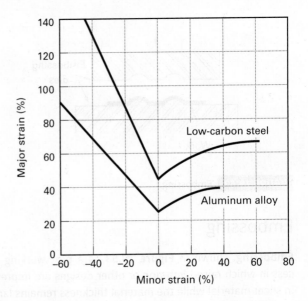

FIGURE 19.54 (Left) Typical pattern for sheet metal deformation analysis; (Right) forming limit diagram used to determine whether a metal can be shaped without risk of fracture. Fracture is expected when strains fall above the lines.

Sheet materials can also have directional variations within the plane of the sheet. These are best evaluated by performing four uniaxial tensile tests, one cut longitudinally (generally along the direction of prior rolling), one transverse (perpendicular to the first), and one along each of the 45° axes. Values are then computed for the average **normal anisotropy** (the sum of the four values of R divided by four) and the **planar anisotropy, ΔR** [computed as $(R_{longitudinal} + R_{transverse} - R_{45\,right} - R_{45\,left})/2$]. Because a ΔR of zero means that the material is uniform in all directions within the plane of the sheet, an ideal drawing material would have a high value of R and a low value of ΔR. Unfortunately, the two values tend to be coupled, such that high formability is often accompanied by directional variations within the plane of the sheet. Due to these variations, discs being drawn into cylindrical cups will have variations in the final wall height.

Sheet-forming properties have also been shown to change significantly with variations in both temperature and speed of deformation, or strain rate. Low-carbon steels generally become stronger when deformed at higher speeds, whereas many aluminum alloys weaken and become more prone to fail-ure during the operation.

Design Aids for Sheet Metal Forming

The majority of sheet metal failures occur due to thinning or fracture, and both are the result of excessive deformation in a given region. A quick and economical means of evaluating the severity of deformation in a formed part is to use **strain analy-sis** coupled with a **forming limit diagram**. A pattern or grid, such as the one on the left side of **Figure 19.54**, is placed on the surface of a sheet by scribing, printing, or etching. The cir-cles generally have diameters between 2.5 and 5 mm (0.1 and 0.2 in.) to enable detection of point-to-point variations in strain distribution. During deformation, the circles convert into ellip-ses, and the distorted pattern can be measured and evaluated. Regions where the enclosed area has expanded are locations of sheet thinning and possible failure. Regions where the enclosed area has decreased are locations of sheet thickening and may be sites of possible buckling or wrinkles.

Using the ellipses on the deformed grid, the major strains (strain in the direction of the largest radius or diameter) and the associated minor strains (strain 90° from the major) can be determined for a variety of locations. These values can then be plotted on a forming limit diagram such as the one shown on the right side of Figure 19.54. If both major and minor strains are positive (right-hand side of the diagram), the deforma-tion is known as *stretching*, and the sheet metal will definitely decrease in thickness. If the minor strain is negative, the con-traction may partially or wholly compensate for any positive stretching in the major direction. This combination of tension and compression is known as *drawing*, and the thickness of the material may decrease, increase, or stay the same, depend-ing on the relative magnitudes of the two strains. Regions where both strains are negative do not appear on the diagram because its purpose is to reveal locations of possible fracture, and fractures only occur in a tensile environment.

Those strains that fall above the forming limit line indicate regions of probable fracture. Possible corrective actions include modification of the lubricant, change in the die design, or vari-ation in the clamping or hold-down pressure. Strain analysis can also be used to determine the best orientation of blanks

---

[2] Examples of formability assessment tests include Olsen stretch, limiting dome height, 4-in.-diameter hemispherical dome, Swift cup, Fukui conical, hole expansion, hydraulic bulge, and Erichsen stretch, along with numerous bending tests. As with all small-size simulation tests, there is concern regarding any significant variation in process parameters (such as size, temperature, lubrication, interface pressure, form-ing speed, and others) between the simulation test and actual production conditions.

relative to the rolling direction, assist in the design of dies for complex-shaped products, or compare the effectiveness of various lubricants. Because of the large surface-to-volume ratio in sheet material, lubrication can play a key role in process success or failure.

Dies can now be designed using solid modeling, providing a three-dimensional visualization that often eliminates confusion and interpretation errors. Forming simulation software often eliminates the need for prototype tooling and testing. Fractures, splits, wrinkles, and excessive thinning appear in the simulated part just as they would if an actual die had been built. Features such as blank size, hold-down pressures, and draw bead details can all be determined prior to die construction. Simulation software can also calculate the approximate springback that might be expected, permitting consideration of design changes prior to actual fabrication.

19.5 Alternative Methods of Producing Sheet-Type Products

Electroforming

Several manufacturing processes have been developed to produce sheet-type products by directly depositing metal onto preshaped forms or mandrels. In a process known as **electroforming**, the metal is deposited by plating. Nickel, iron, copper, or silver can be deposited in thicknesses up to 16 mm (⅝ in.). When the desired thickness has been attained, plating is stopped, and the product is stripped from the mandrel.

A wide variety of sizes and shapes can be made by electroforming, and the fabrication of a product requires only a single pattern or mandrel. Low production quantities can be made in an economical fashion, with the principal limitation being the need to strip the product from the mandrel. Replication of the contact surface and profile are extremely good, but the uniformity of thickness and external profile may present problems. For applications like the production of multiple molds from a single master pattern, the interior surface is the critical one, and the wall thickness serves only to provide the necessary strength. Exterior irregularities are not critical, and various types of backup material can be employed to provide additional support. For applications where the exterior dimensions are also important, uniform deposition is required.

Spray Forming

Similar parts can also be formed by **spray deposition**, described in more detail in Chapter 38 . One approach is to inject powdered material into a plasma torch (a stream of hot

ionized gas with temperatures up to 11,000°C or 20,000°F). The particles melt and are propelled onto a shaped form or mandrel. On impact, the droplets flatten and undergo rapid solidification to produce a dense, fine-grained product. Multiple layers can be deposited to build up a desired size, shape, and thickness. Because of the high adhesion, the mandrel or form is often removed by machining or chemical etching. Most applications of plasma spray forming involve the fabrication of specialized products from difficult-to-form or ultrahigh-melting-point materials.

The **Osprey process**, described in Chapter 16 (Section 16.10), can also be adapted to produce thin, spray-formed products. Here, molten metal flows through a nozzle, where it is atomized and carried by high-velocity nitrogen jets. The semisolid particles are propelled toward a target, where they impact and complete their solidification. Tubes, plates, and simple forms can be produced from a variety of materials. Layered structures can also be produced by sequenced deposition.

19.6 Seamed Pipe Manufacture

The manufacture of seamless pipe has been presented in Chapter 18 under the heading of piercing. Large quantities of seamed pipe are made by two processes that use the hot forming of steel strip coupled with welding of the free edges. Both of these processes, **butt welding** of pipe and **lap welding** of pipe, begin with steel in the form of **skelp**—long strips with specified width, thickness, and edge configuration.

Butt-Welded Pipe

In the manufacture of butt-welded pipe, steel skelp is heated to a specified hot-working temperature by passing it through a tunnel-type furnace. On exiting the furnace, the skelp is pulled through forming rolls that shape it into a cylinder and bring the free ends into contact. The pressure exerted between the opposite edges of the skelp may be sufficient to upset the metal and produce a welded seam, or resistance or arc welding may be employed. Additional sets of rollers then size and shape the pipe, and it is cut to standard, preset lengths. Product diameters range from 3 mm (⅛ in.) to 75 mm (3 in.), and speeds can approach 150 m/min (500 ft/min).

Lap-Welded Pipe

The lap-welding process for making pipe differs from the butt-welding technique in that the skelp now has beveled edges, and the rolls form the weld by forcing the overlapped edges down against a supporting mandrel. This process is used primarily for larger sizes of pipe, with diameters from about

50 mm (2 in.) to 400 mm (14 in.). Because the product is driven over an internal mandrel, product length is limited to about 6 to 7 m (20 to 25 ft).

19.7 Presses

Classification of Presses

The primary tool for performing most of the sheet-forming operations discussed in this chapter, and many of the bulk forming operations presented in Chapter 18, is some form of press, and successful manufacture often depends on using the right kind of equipment. When selecting a press for a given application, consideration should be given to the capacity required, the type of power (manual, mechanical, hydraulic, or servo), the number of slides or drives, the type of drive, the stroke length for each drive, the type of frame or construction, and the speed of operation. **Table 19.2** lists some of the major types of presses and groups them by the type of drive.

Manually operated presses such as foot-operated or **kick presses** are generally used for very light work such as shearing small sheets and thin material.

Moving toward larger equipment, we find that **mechanical presses** tend to provide fast motion and excellent accuracy and repeatability with respect to displacement. Once built, however, the flexibility of a mechanical press is limited because the length of the stroke is set by the design of the drive. The available force usually varies with position, so mechanical presses are preferred for operations that require the maximum pressure near the bottom of the stroke, such as cutting, shallow forming, drawing (up to about 10 cm), and progressive and transfer die operations. Typical capacities range up to about 9000 metric tons.

Figure 19.55 depicts some of the basic types of mechanical press drive mechanisms that have been designed to provide high forces, specific ram velocities, and desired dwell time at

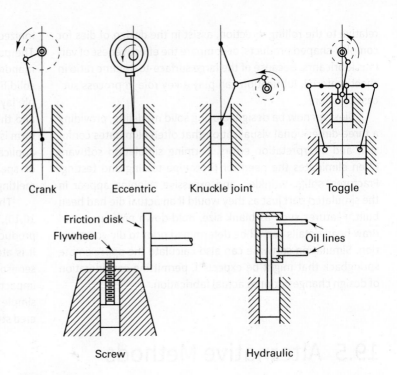

FIGURE 19.55 Schematic representation of the various types of press drive mechanisms.

the bottom of the stroke. *Crank-driven* presses are the most common type because of their simplicity in converting the rotary motion of a heavy, spinning flywheel to linear up-and-down movement. They are used for most piercing and blanking operations and for simple drawing. Double-crank presses offer a means of actuating blank holders or operating multiple-action dies. *Eccentric or cam drives* are used where the ram stroke is rather short. Cam action can also provide a dwell at the bottom of the stroke and is often the preferred method of actuating the blank holder in deep-drawing processes. *Knuckle-joint drives* provide a very high mechanical advantage along with fast action. They are often preferred for coining, sizing, and Guerin forming. *Toggle mechanisms* are used principally in drawing presses to actuate the blank holder, and *screw-type drives* offer great mechanical advantage coupled with an action that resembles a drop hammer (but slower and with less impact). For this reason, screw presses have become quite popular in the forging industry.

In contrast to the mechanical presses, **hydraulic presses** produce motion through movement of a piston. The stroke can be programmed to any length within the limits of the cylinder (which may be as long as 250 cm or 100 in.) reducing or eliminating excessive ram movement. Forces and pressures are more accurately controlled, and full pressure is available throughout the entire stroke. A built-in pressure relief valve provides overload protection to both the press and any inserted tooling. Moving parts are few in number, and most remain fully lubricated, being immersed in the pressurizing oil. Speeds can be varied within the stroke or remain constant during an operation, and the return stroke can be programmed for fast reset. Because position is varied through fluid displacement,

TABLE 19.2	Classification of the Drive Mechanisms of Commercial Presses		
Manual	**Mechanical**	**Hydraulic**	**Servo**
Kick presses	Crank	Single slide	
	Single	Multiple slide	
	Double		
	Eccentric		
	Cam		
	Knuckle joint		
	Toggle		
	Screw		
	Rack and pinion		

the reproducibility of position will have greater variation than a mechanical press, but the noise level will be considerably less.

Hydraulic presses are available in capacities exceeding 50,000 metric tons and are preferred for operations requiring a steady pressure throughout a substantial stroke (such as deep drawing), operations requiring wide variation in stroke length, and operations requiring high or widely variable forces. In general, hydraulic presses tend to be slower than the mechanical variety, but some are available that can provide up to 600 strokes per minute in a high-speed blanking operation. By using multiple hydraulic cylinders, programmed loads can be applied to the main ram. A separate force and timing are used on the blank holder.

Servo-drive presses use servo motors in place of the traditional flywheel and clutch to control the ram motion of the press, offering much of the versatility of a hydraulic press at production speeds that approach those of traditional mechanical presses. They are more expensive and more complex than mechanical presses and have a much more sophisticated electrical control system. Although servo presses are available in capacities only up to about 4000 tons, they offer variable stroke length and maximum torque anywhere in the stroke and at any speed. A single press, therefore, can be adapted to a variety of parts. With programmable motion, servo presses can change speeds or even stop within a stroke. Deep drawing is typically performed at slower speeds, whereas progressive dies use fast speeds and short strokes. While slower speeds are often used to form parts, fast speeds can be used for noncontact movements such as approach and retract, thereby increasing press productivity. Impact speed can be controlled, reducing shock on dies as well as noise. Stopping at the bottom of a stroke can be ideal for hot stamping operations, and several small reverse movement restrikes can reduce the amount of springback.

Types of Press Frame

As shown in **Table 19.3**, presses should also be selected with consideration for the type of frame. Frame design often imposes limitations on the size and type of work that can be accommodated, how that work is fed and unloaded, the overall stiffness of the machine, and the time required to change dies.

Presses that have their frames in the shape of an arch (arch-frame presses) are seldom used today, except with screw drives for coining operations. **Gap-frame presses**, where the frames have the shape of the letter *C*, are among the most versatile and commonly preferred presses. They provide unobstructed access to the dies from three directions and permit large workpieces to be fed into the press. Gap-frame presses are available in a wide range of sizes, from small bench types of about 1 metric ton up to 300 metric tons or more.

Popular design features include open back, inclinability, adjustable bed, and sliding bolster. **Open-back presses** allow the ejection of products or scrap through an opening in the back of the press frame. **Inclinable presses** can be tilted so that ejection can be assisted by gravity or compressed air jets. As a result of these features, open-back inclinable (OBI) presses are the most common form of gap-frame press. The addition of an **adjustable bed** allows the base of the machine to raise or lower to accommodate different workpieces. A **sliding bolster** permits a second die to be set up on the press while another is in operation. Die changeover then requires only a few minutes to unclamp the punch segment of the active die, move the second die set into position, clamp the new punch to the press ram, and resume operation. **Figure 19.56** shows an open-back, gap-frame press.

A **horn press** is a special type of gap-frame press where a heavy cylindrical shaft or "horn" appears in place of the usual bed. Curved or cylindrical workpieces can be placed over the horn for such operations as seaming, punching, and riveting. On some presses, both a horn and a bed are provided, with provision for swinging the horn aside when not needed.

Turret presses are especially useful in the production of sheet metal parts with numerous holes or slots that vary in size and shape. They usually employ a modified gap-frame structure and add upper and lower turrets that carry a number of punches and dies. The two turrets are geared together so that any desired tool set can be quickly rotated into position.

TABLE 19.3	Classification of Presses According to Type of Frame	
Arch	**Gap**	**Straight Sided**
Crank or eccentric	Foot	Many variations but all with straight-sided frames
Percussion	Bench	
	Vertical	
	Inclinable	
	Inclinable	
	Open back	
	Horn	
	Turret	

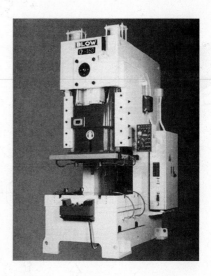

FIGURE 19.56 Gap-frame press. (Courtesy of Blow Press, Guelph, Ontario, Canada)

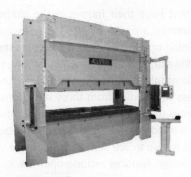

Straight-side presses have frames that consist of a crown, two uprights, a base or bed, and one or more moving slides. Accessibility is generally from the front and rear, but openings are often provided in the side uprights to permit feeding and unloading of workpieces. Straight-sided presses are available in a wide variety of sizes and designs and are the preferred design for most hydraulic, large-capacity, or specialized mechanical-drive presses. As an added benefit,

elastic deflections tend to be uniform across the working surface, as opposed to the angular deflections that are typical of the gap-frame design. **Figure 19.57** shows a typical straight-sided press.

Special Types of Presses

Presses have also been designed to perform specific types of operations. **Transfer presses** have a long moving slide or ram that enables multiple operations to be performed simultaneously in a single machine. Multiple die sets are mounted side by side along the slide. After the completion of each stroke, a continuous strip is advanced or individual workpieces are transferred to the next station by a mechanism like the one shown in **Figure 19.58**. Transfer presses can be used to perform blanking, piercing, forming, trimming, drawing, flanging, embossing, and coining. **Figure 19.59** illustrates the production of a part that incorporates a variety of these operations.

By using a single machine to perform multiple operations, transfer presses offer high production rates, high flexibility, and reduced costs (attributed to the reduced labor, floor space, energy, and maintenance). Because production is usually

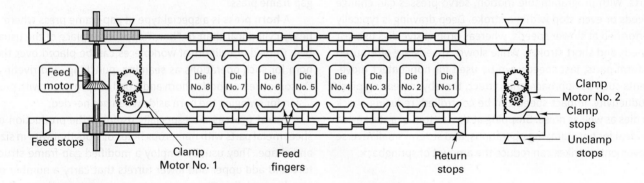

FIGURE 19.58 Schematic showing the arrangement of dies and the transfer mechanism used in transfer presses. (Courtesy of Verson Allsteel Press Company, Chicago, IL)

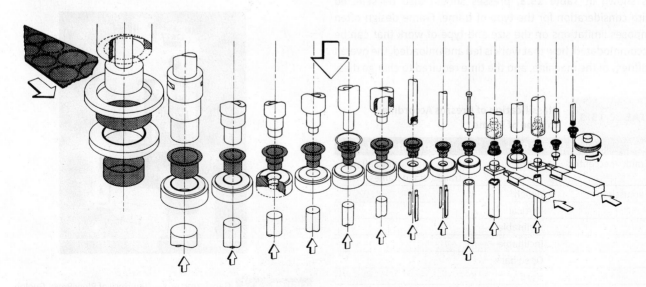

FIGURE 19.59 Various operations can be performed during the production of stamped and drawn parts on a transfer press. (Courtesy of U.S. Baird Corporation, Stratford, CT)

automotive and appliances, where large numbers of identical products are being produced.

Four-slide or **multislide machines** like the one shown in **Figure 19.60** are extremely versatile presses that are designed to produce small, intricately shaped parts from continuously fed wire or coil strip. The basic machine has four power-driven slides (or motions) set 90 degrees apart. The attached tooling is controlled by cams and is designed to operate in a progressive cycle. In the sheet metal variation, strip stock is fed into the machine, where it is straightened and progressively pierced, notched, bent, and cut off at the various slide stations. **Figure 19.61** presents the operating mechanism of one such machine. As the material moves from right to left, it undergoes a straightening, two successive pressing operations, various operations from all four directions, and a final cutoff. **Figure 19.62** shows the carrier strip and the successive operations as flat strip is pierced, blanked, and formed into a folded sheet metal product. The strip stock may be up to 75 mm (3 in.) wide and 2.5 mm (3/32 in.) thick. Wires up to about 6 mm (1/4 in.) in diameter are also commonly processed. Products such as hinges, links, clips, razor blades and those illustrated in **Figure 19.63** can be formed at very high rates approaching 15,000 pieces per hour. Setup times are long, so large production runs are preferred.

between 500 and 1500 parts per hour, these machines are usually restricted to operations where 4000 or more identical parts are required daily, each involving three or more separate operations. A total production run of 30,000 or more identical parts is generally desired between major changes in tooling. As a result, transfer presses are used primarily in industries such as

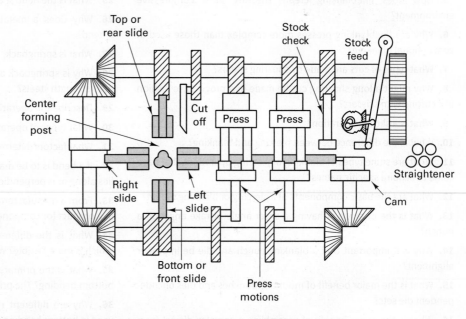

FIGURE 19.63 An array of typical multislide/four-slide products. (Ken-Tron Manufacturing, Inc., Owensboro, KY)

Press-Feeding Devices

Although hand feeding may still be used in some press operations, operator safety and the desire to increase productivity have motivated a strong shift to feeding by some form of mechanical device. When continuous strip is used it can be fed automatically by double-roll feeds mounted on the side of the press. Discrete products can be moved and positioned in a wide variety of ways. Dial-feed mechanisms enable an operator to insert workpieces into the front holes of a rotating dial, which then indexes with each stroke of the press to move the parts progressively into proper position between the punch and die. Lightweight parts can be fed by suction-cup mechanisms, vibratory-bed feeders, and similar devices. Robots are frequently used to place parts into presses and remove them after forming. See Chapter 34 for an expanded discussion of manufacturing automation.

Review Questions

1. What distinguishes sheet forming from bulk forming?

2. What is a definition of shearing?

3. Why are sheared or blanked edges generally not smooth? What are the various regions on a sheared edge?

4. What measures can be employed to improve the quality of a sheared edge?

5. How does fineblanking create shearing in a compressive environment?

6. Why are fineblanking presses more complex than those used in conventional blanking?

7. What types of cuts are made by squaring shears?

8. Why might a long shearing cut be made in a progressive fashion like cutting with scissors?

9. What is a slitting operation?

10. What is the difference between piercing and blanking?

11. What are some types of blanking or piercing operations that have come to acquire specific names?

12. What are the basic components of a piercing or blanking die set?

13. What is the purpose of having a shear angle or rake angle on a punch?

14. Why is it important that a blanking punch and die be in proper alignment?

15. What is the major benefit of mounting punches and dies on independent die sets?

16. What is the major benefit of assembling a complex die set from standard subpress dies?

17. What is the benefit of making dies as a multipiece assembly?

18. What is a "steel-rule die," and for what types of designs and materials is it used?

19. What is a progressive die set?

20. What is the difference between progressive dies and transfer dies?

21. How do compound dies differ from progressive dies?

22. What is the attractive feature of a turret-type punch press?

23. When making bends in sheet metal, what is the distinction between bending, forming, and drawing?

24. What are the stress states on the exterior surface and interior surface of a bend?

25. What is the neutral axis in bending?

26. Why does a metal usually become thinner in the region of a bend?

27. What is springback, and why is it a concern during bending?

28. Why is springback an increased concern for the newer advanced high-strength steels?

29. Describe the operation of a bar folder.

30. What types of operations can be performed on a press brake?

31. What factors determine the minimum bend radius for a material?

32. If a bend is to be made in a cold-rolled sheet, is it best if the bend lies along or is perpendicular to the direction of previous rolling?

33. From a manufacturing viewpoint, why is it desirable for all bends in a product (or component) to have the same radius?

34. What is the difference between air-bend and bottoming dies? Which is more flexible? Which produces more reproducible bends?

35. What is the primary benefit of incorporating a coining action in bottom bending? The primary negative feature?

36. Why are different types of presses (mechanical and hydraulic) used in bottom bending and air bending?

37. What is the benefit of using a urethane (rubber) bottom die in bending operations?

38. What is the objective of the roll bending process?

39. What is the role of the form block in draw bending and compression bending?

40. Describe the press bending process.

41. How can we prevent flattening or wrinkling when bending a tube?

42. What is the function of the rolls in a roll-forming operation?

43. What are some typical products that are shaped by roll forming?

44. Why is the roll-forming process not appropriate for making short lengths of specialized products?

45. What are some methods for straightening or flattening rod or sheet?

46. What two distinctly different metal-forming processes use the term *drawing*?

47. What is the difference between drawing and stretching?

48. What are some of the shapes that can be produced by a spinning operation?

49. Why is the tooling cost for a spinning operation relatively low?

50. What is the benefit of the laser assist in laser-assisted metal spinning?

51. How is shear forming different from spinning?

52. For what type of products is stretch forming an appropriate manufacturing technique?

53. Why is there very little springback in a stretch formed product?

54. What is the distinction between shallow drawing and deep drawing?

55. What is the function of the pressure ring or hold-down in a deep-drawing operation?

56. What are the key variables in a deep-drawing operation?

57. What type of defect can form if the hold-down pressure is too low? Too high?

58. Explain why thin material may be difficult to draw into a defect-free cup.

59. How can redraw operations be used to produce a taller, smaller diameter cup than can be produced in a single deep-drawing operation?

60. What are draw beads, and what function do they perform?

61. Why is a trimming operation often included in a deep-drawing manufacturing sequence?

62. How does the Guerin process reduce the cost of tooling in a drawing operation?

63. How can fluid pressure or rubber tooling be used to perform bulging?

64. What is sheet hydroforming?

65. What are some of the variations of sheet hydroforming?

66. What explanation can be given for the greater formability observed during sheet hydroforming?

67. What features have made sheet hydroforming attractive to the aerospace industry? The automotive industry?

68. Describe the sequence of activities in tube hydroforming.

69. What is the purpose of the inward movement of the end plugs during tube hydroforming?

70. What is the difference between low-pressure tube hydroforming and high-pressure tube hydroforming in terms of both pressures and the nature of the deformations produced?

71. What are some of the attractive features of tube hydroforming?

72. What is the advantage of hot-metal-gas-forming over conventional tube hydroforming?

73. Why is hot stamping attractive for the newer high-strength automotive steels?

74. How are forming and hardening being combined in a hot stamping/press hardening operation?

75. What are some benefits or attractions of high-energy-rate forming?

76. What are some of the basic methods that have been used to achieve the high energy-release rates needed in the HERF processes?

77. What are some well-known products that have been produced by processes that include ironing? By embossing?

78. What material and process conditions are associated with superplastic forming?

79. What are some of the different techniques used for superplastic forming?

80. What is the major limitation of the superplastic forming of sheet metal? What are some of the more attractive features?

81. For what production conditions might free-form fabrication be appropriate?

82. What properties from a uniaxial tensile test can be used to assess sheet metal formability?

83. How is the formability in biaxial tension different from that in uniaxial tension? Why?

84. What is "normal anisotropy, R" and "planar anisotropy, ΔR," and what conditions would be ideal?

85. How can strain analysis be used to determine locations of possible defects or failure in sheet metal components?

86. What is a forming limit diagram?

87. Explain the key difference between the right- and left-hand sections of a forming limit diagram. (These sections correspond to stretching and drawing)

88. Describe two alternative methods of producing complex-shaped thin products through deposition (i.e., do not require sheet metal deformation techniques).

89. What two hot-forming operations can be used to produce pipe from steel strip?

90. What features should be considered when selecting a press for a given application?

91. What are the primary assets and limitations of mechanical press drives?

92. What are some of the drive mechanisms employed on mechanical presses?

93. What are the some of the assets and limitations of hydraulic presses?

94. What are some of the attractive features of servo-drive presses?

95. What are some of the common types of press frames?

96. What are some features that may be included into a press design to add flexibility or increase productivity?

97. Describe how multiple operations are performed simultaneously in a transfer press.

98. For what type of production quantities might a transfer press be appropriate?

99. What types of products are produced on a four-slide or multislide machine?

100. What are some of the methods used in press feeding?

Problems

1. The maximum punch force in blanking can be estimated by the equation:

$$Force = StL$$

where:

  S = the material shear strength

  t = the sheet thickness, and

  L = the total length of sheared edge (circumference or perimeter)

  a. How would this number change if the rake angle is equal to a $1t$ change across the width or diameter of the part being sheared?

  b. How would this number change if the rake were increased to a $3t$ change across the width or diameter?

2. Consider the various means of producing tubular products, such as extrusion, seam welding, butt-welding during forming, piercing, and the various drawing operations. Describe the advantages, limitations, and typical applications of each.

3. Tube and sheet hydroforming have been undergoing rapid growth. Investigate current uses for these processes in automotive and other fields.

4. What are some of the techniques for minimizing the amount of springback in sheet-forming operations?

5. Select a forming process from either Chapter 18 or 19, and investigate the residual stresses that typically accompany or result from that process. If they are considered to be detrimental, how could these residual stresses be reduced or removed without damaging or deteriorating the product?

6. In most products, the painting or surface finishing operations are usually performed on individual pieces after the shape has been produced. An alternative approach is to paint, plate, or surface finish the sheet material while it is in the form of continuous coil (coil-coated sheet material), and then fabricate the shape in a manner so as to retain the integrity of the surface. Discuss some of the pros and cons of this approach, such as uncoated sheared edges or special handling or modified tooling.

CHAPTER 20

Fabrication of Plastics, Ceramics, and Composites

20.1 Introduction

In Chapters 6, 7, and 8, **plastics**, **ceramics**, and **composites** were shown to be substantially different from metals in both structure and properties. Although the specific material will still be selected for its ability to provide the required properties, and the fabrication processes for their ability to produce the desired shape in an economical and practical manner, it is reasonable to expect that there will be significant differences in product design, material selection, and fabrication.

Plastics, ceramics, and composites tend to be used closer to their design limits, and many of the fabrication processes convert the raw material into a finished product in a single operation. Large, complex shapes can often be formed as a single unit, eliminating the need for multipart assembly operations. Materials in these classes can often provide integral and variable color, and the processes used to manufacture the shape can frequently produce the desired finish and precision. Finishing operations are often unnecessary—an attractive feature because altering the final dimensions or surface would be both difficult and costly for some of these materials. The joining and fastening operations also tend to be different from those used with metals.

As with metals, the resulting properties are often affected by the processes used to produce the shape. The fabrication of an acceptable product, therefore, involves the selection of both (1) an appropriate material and (2) a companion method of processing, such that the resulting combination provides the desired shape, properties, precision, and finish.

20.2 Fabrication of Plastics

The plastics industry is a significant part of American economy. More than 85 billion pounds of polymer products are manufactured each year in North America. The manufacture of a successful plastic product requires satisfying the various mechanical and physical property requirements through the use of the most economical resin or compound that will perform satisfactorily, coupled with a manufacturing process that is compatible with both the part design and the selected material.

Chapter 8 presented material about the wide variety of plastics or polymers that are currently used as engineering materials. As we move our attention to the fabrication of parts and shapes, we find that there are also a number of processes from which to choose. Determination of the preferred method depends on the desired size, shape, and quantity, as well as whether the polymer is a *thermoplastic, thermoset*, or *elastomer*. **Thermoplastic polymers** can be heated to produce either a soft, formable solid or a liquid. The material can then be cast, injected into a mold, or forced into or through dies to produce a desired shape. **Thermosetting polymers** have far fewer options because once the polymerization has occurred, the framework structure is established, and no further deformation can occur. Thus, the polymerization reaction must take place during the shape-forming operation. **Elastomers** are sufficiently unique that they will be treated in a separate section of this chapter.

Casting, blow molding, compression molding, transfer molding, cold molding, injection molding, reaction injection molding, extrusion, thermoforming, rotational molding, and *foam molding* are all processes that are used to shape polymers. Each has its distinct set of advantages and limitations that relate to part design, compatible materials, and production cost. To make optimum selections, we must become familiar with the shape capabilities of a process as well as how the process affects the properties of the material.

Casting

Casting is the simplest of the shape forming processes because no fillers are used, and no pressure is required. Not all plastics can be cast, but there are castable members of both the thermoplastic and thermosetting families. Castable thermoplastics include acrylics, nylons, urethanes, and PVC plastisols. If the polymer is thermoplastic, it is simply melted, and the liquid is gravity poured into a container having the shape of the desired part where it cools and solidifies. Small products can be cast directly into shaped molds, but there are other variations of the process. Plate glass can be used as a mold to cast individual pieces of thick plastic sheet. Continuous sheets and films can be produced by injecting the liquid polymer between two moving belts of highly polished stainless steel, the width and thickness being set by resilient gasket strips on either end of the gap. Thin sheets can be made by ejecting molten liquid from a gap-slot die onto a temperature-controlled chill roll. Molten plastic can also be spun against a rotating mold wall (centrifugal casting) to produce hollow or tubular shapes.

Castable thermosets include phenolics, polyesters, epoxies, silicones, and urethanes, as well as any resin that will polymerize at low temperatures and atmospheric pressure. The liquid resins or ingredients are introduced into the mold, and catalysts or heat induce polymerization and cross-linking. The material must remain liquid during mold filling, and curing must be completed before removal from the mold.

Compared to other shaping processes, casting is often simpler and less expensive (molds see only atmospheric pressure) and is well suited to low production quantities. Because cast plastics contain no fillers, they have a distinctly lustrous appearance, and a wide range of transparent and translucent colors are available. Fiber or particulate reinforcement can be easily incorporated into the starting liquid. Typical products include sheets, plates, films, rods, and tubes, as well as small objects, such as jewelry, ornamental shapes, gears, and lenses. Although dimensional precision can be quite high, quality problems can occur because of inadequate mixing, air entrapment, gas evolution, and shrinkage. Hollow parts can be made by allowing the external surfaces to cool or cure, then pouring the remaining liquid back out of the mold, a variation of the slush casting process discussed in Chapter 15.

Blow Molding

A variety of **blow molding** processes have been developed, the most common being used to convert thermoplastic polyethylene, polyvinyl chloride (PVC), polypropylene, polyethylene terephthalate, and polyether ether ketone (PEEK) resins into bottles and other thin-wall, seamless, hollow-shape containers. A solid-bottom, hollow-tube preform, known as a **parison**, is made from heated plastic by either extrusion or injection molding. The heated preform is then positioned between the halves of a split mold, the mold closes, and the preform is expanded against the mold by air or gas pressure, usually in the range of 350 to 700 kPa (50 to 100 psi). The mold is then cooled, the halves are separated, and the product is removed. Any flash is then trimmed for direct recycling. **Figure 20.1** depicts a form of this process where the starting material is a simple tube and the solid bottom is created by the pinching action of die closure.

Although the primary product of blow molding is the disposable bottle for beverages and other liquids, the process has recently expanded to include the higher-strength engineering thermoplastics. It is now being used to produce products as diverse as large shipping drums and storage tanks, gasoline cans, and automotive parts, including fuel tanks, seat backs, ductwork, and bumper beams. Variations of blow molding have been designed to provide both axial and radial expansion of the plastic (for enhanced strength) as well as to produce multilayered products.

Because the thermoplastics must be cooled before removal from the mold, the molds for blow molding must contain the desired cavity as well as a cooling system, venting system, and other design features. The mold material must provide thermal conductivity and durability while being inexpensive and compatible with the resins being processed. Beryllium copper, aluminum, tool steels, and stainless steels are all popular mold materials.

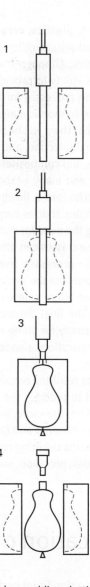

FIGURE 20.1 Steps in blow molding plastic parts: (1) a tube of heated plastic is placed in the open mold, (2) the mold closes over the tube, simultaneously sealing the bottom, (3) air expands the tube against the sides of the mold, and (4) after sufficient cooling, the mold opens to release the product.

Compression Molding or Hot-Compression Molding

Compression molding is one of the most widely used molding processes for thermosetting polymers. As illustrated schematically in **Figure 20.2**, a preshaped charge or premeasured amount of molding compound in the form of solid granules, powders, pellets, or liquid, often preheated to soften and reduce molding time, is first introduced into an open, heated cavity. A heated plunger then descends to close the cavity and apply pressure, typically in the range of 10 to 150 MPa (1500 to 20,000 psi). As the material melts and becomes fluid, it is driven into all portions of the cavity. The heat and pressure are maintained until the material has "set" (i.e., cured or polymerized). The mold is then opened, and the part is removed. A wide variety of heating systems and mold materials are used, and multiple cavities can be placed within a mold to produce more than one part in a single pressing. Although the process has been used primarily with the thermosetting polymers, recent developments permit the shaping of thermoplastics and various reinforced composites. Cycle times are set by the rate of heat transfer and the reaction or curing rate of the polymer. They typically range from under 1 minute to as much as 20 minutes or more.

The tool and machinery costs for compression molding are often lower than for competing processes, and the dimensional precision and surface finish are high, thereby reducing or eliminating secondary operations. Compression molding is most economical when it is applied to small production runs of parts requiring close tolerances, high impact strength, and low mold shrinkage. It is a poor choice when the part contains thick sections (the cure times become quite long), or when large quantities are desired. Most products have relatively simple shapes because the flow of material is rather limited. Typical compression-molded parts include gaskets, seals, exterior automotive panels, aircraft fairings, and a wide variety of interior panels.

More recently, compression molding has been used to form fiber-reinforced plastics, both thermoplastics and thermosets, into parts with properties that rival the engineering metals. In the thermoset family, polyesters, epoxies, and phenolics can be used as the base of fiber-containing sheet-molding compound, bulk-molding compound, or sprayed-up reinforcement mats. These are introduced into the mold and shaped and cured in the normal manner. Cycle times range from about 1 to 5 minutes per part, and typical products include wash basins, bathtubs, equipment housings, and various electrical components. If the starting material is a fiber-containing thermoplastic, precut blanks are first heated in an infrared oven to produce a soft, pliable material. The blanks are then transferred to the press, where they are shaped and cooled in specially designed dies. Compared to the thermosets, cycle times are reduced, and the scrap is often recyclable. In addition, the products can be joined or assembled using the thermal welding processes applied to plastics.

Compression molding equipment is usually rather simple, typically consisting of a hydraulic or pneumatic press with parallel platens that apply the heat and pressure. Pressing areas range from 15 cm² (6 in²) to as much as 2.5 m² (8 ft²), and the force capacities range from 6 to 9000 metric tons. The molds are usually made of tool steel and are polished or chrome plated to improve material flow and product quality. Mold temperatures typically run between 150 and 200°C (300 and 400°F) but can go as high as 650°C (1200°F). They are heated by a variety of means, including electric heaters, steam, oil, and gas.

Transfer Molding

Transfer molding is sometimes used to reduce the turbulence and uneven flow that can result from the high pressures of hot-compression molding. As shown in **Figure 20.3**, the *unpolymerized* raw material is now placed in a plunger cavity, where it is heated until molten. The plunger then descends, forcing the molten plastic through channels or runners into adjoining die cavities. Temperature and pressure are maintained until the thermosetting resin has completely cured. To shorten the cycle and extend the lifetime of the cavity, plunger, runner, and gates, the charge material may be preheated before being placed in the plunger cavity.

Because the material enters the die cavities as a liquid, there is little pressure until the cavity is completely filled. Thin

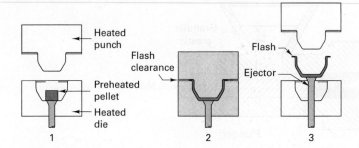

FIGURE 20.2 The hot-compression molding process: (1) solid granules or a preform pellet is placed in a heated die, (2) a heated punch descends and applies pressure, and (3) after curing (thermosets) or cooling (thermoplastics), the mold is opened and the part is removed.

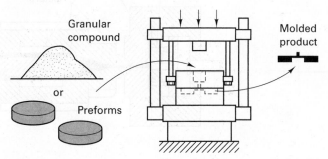

FIGURE 20.3 Diagram of the transfer molding process. Molten or softened material is first formed in the upper heated cavity. A plunger then drives the material into an adjacent die.

sections, excellent detail, and good tolerances and finish are all characteristics of the process. In addition, metal or ceramic inserts can be incorporated into the products of transfer molding. They are simply positioned within the cavity and maintained in place as the liquid resin easily flows around them. Composite products can be produced by infusing resin around or into reinforcing preforms.

Transfer molding is attractive for producing small-to-medium sized parts with relatively complex shapes and variable wall thickness. It combines elements of both compression molding and injection molding (to be discussed later) and enables some of the advantages of injection molding to be utilized with thermosetting polymers. The thermosetting resins can be reinforced with fillers—such as cellulose, glass, silica, alumina, or mica—to improve the mechanical or electrical properties and reduce shrinkage or warping. The main limitation is the loss of material because resin left in the pot or well, sprue, and runners also cures and must now be discarded. Common products include electrical switchgear and wiring devices, parts of household appliances that require heat resistance, structural parts that require hardness and rigidity under load, underhood automotive parts, and parts that require good resistance to chemical attack.

Cold Molding

In **cold molding**, uncured thermosetting material is pressed to shape while cold and is then removed from the mold and cured in a separate oven. Although the process is faster and more economical, the resulting products generally lack good surface finish and dimensional precision.

Injection Molding

Injection molding is the most widely used process for the high-volume production of relatively complex thermoplastic parts and is often considered to be the polymer equivalent to metal die casting. **Figure 20.4** illustrates one approach to the process where granules of raw material are fed by gravity from a hopper into a cavity that lies ahead of a moving plunger. As the plunger advances, the material is forced through a preheating chamber and on through a torpedo section, where it is mixed, melted, and superheated to a temperature between 150 and 260°C (300 and 500°F). The superheated material is then driven through a nozzle that seats against a mold. Other types of injection units control the flow of material and generate the injection pressure with screws that have either both rotational and axial movements, or combinations of screws and plungers. Alternative methods of heating the material include heated barrels and the shearing action as material moves through the screws.

Sprues and runners then channel the molten material into one or more closed-die cavities. Because the dies remain cool, the plastic solidifies almost as soon as the mold is filled. Premature solidification would cause defective parts, so the superheated material must be rapidly forced into the mold cavities by pressures in the range 35 to 200 MPa (5 to 30 ksi), which are maintained during solidification. The mold halves must clamp tightly together during molding and then be easily separated for part ejection. Various types of clamping designs have been developed, including toggle, hydraulic, and hydromechanical. As with the dies of die casting, the molds must provide for the escape of air from the mold cavity as it fills (venting) and incorporate a cooling system, usually in the form of internal water passages.

Control systems coordinate all the functions of the process, including the time required for cooling within the mold. By heating the material for the next part as the mold is separating for part ejection, a molding cycle can generally be completed in 10 to 30 seconds. The process is quite similar to the die casting of molten metal, and the result is usually a finished product needing no further work before assembly or use. Part size can be as small as 50 g (2 oz.) or a large as 25 kg (>50 lb.). Metal or ceramic inserts can be incorporated into the product. Different materials or materials with different colors can be coinjected or sequentially injected to produce single products with differing characteristics, like the multicolor light assemblies of current cars and trucks.

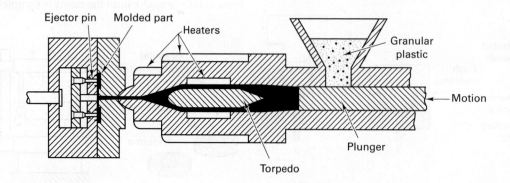

FIGURE 20.4 Schematic diagram of the injection molding process. A moving plunger advances material through a heating region (in this case, through a heated manifold and over a heated torpedo) and further through runners into a mold where the molten thermoplastic cools and solidifies.

Some injection molding machines incorporate a hot runner distribution system to transfer the material from the injection nozzle to the mold cavities. If the runners are cold, the material in the runner solidifies with each cycle and needs to be ejected and reprocessed or disposed. With hot runners, the thermoplastic material is maintained in a liquid state until it reaches the gate. Cycle times can be reduced because only the molded part must be cooled for ejection. The material in the runners can be incorporated into the subsequent shot. Quality is improved, because all material enters the mold at the same temperature, recycled sprues and runners are not incorporated into the charge, and there is less turbulence because pressurized material is not injected into empty runners. Hot runners do add an additional degree of complexity to the design, operation, and control of the system, so the additional cost must be weighed against the benefits cited.

Injection molding can also be applied to thermosetting and elastomeric materials, but the process must be modified to provide the temperature, pressure, and time required for curing and cross-linking. The injection chamber is now at a significantly lower temperature, and the mold is heated to 150 to 230°C (300 to 450°F). The time in the heated mold must be sufficient to complete the curing process, typically between 20 seconds and 2 minutes. The relatively long cycle times are the major deterrent to the injection molding of the thermosets. Speed-up alternatives include completing the cure outside the mold or having multiple dies that can be fed with one injector.

Reaction Injection Molding

Figure 20.5 depicts the **reaction injection molding** process, in which two or more liquid reactants (resin plus catalyst or mixing-activated ingredients) are metered into a unit where they are intimately mixed and directly injected into a mold at a relatively low pressure. An exothermic chemical reaction takes place between the two components, resulting in thermoset polymerization. Because no heating is required, the production rates are set primarily by the curing time of the polymer,

which may be less than 1 minute or up to 10 minutes. Molds are made from steel, aluminum, or nickel shell, with selection being made on the basis of the number of parts to be made and the desired quality. The molds are generally clamped in low-tonnage presses.

At present, the dominant materials for reaction injection molding are polyurethanes, polyamides, epoxies, and composites containing short fibers or flakes. Properties can span a wide range, depending on the combination and percentage of base chemicals and the additives that are used. Different formulations can result in elastomeric or flexible, structural foam (foam core with a hard, solid outer skin), solid (no foam core), or composite products. Part size can range from ½ to 50 kg (1 to 100 lb), shapes can be quite complex (with variable wall thickness), and surface finish is excellent. Automotive applications include steering wheels, airbag covers, instrument panels, door panels, armrests, headliners, and center consoles, as well as body panels, bumpers, spoilers, and wheel covers. Rigid polyurethanes are also used in such products as computer housings, household refrigerators, water skis, hot-water heaters, and picnic coolers.

From a manufacturing perspective, reaction injection molding has a number of attractive features. The low processing temperatures and low injection pressures make the process attractive for molding large parts, and the large size can often enable parts consolidation. Thermoset parts can generally be fabricated with less energy than the injection molding of thermoplastics, with similar cycle times and a similar degree of automation. The metering and mixing equipment and related controls tend to be quite sophisticated and costly, but the lower temperatures and pressures enable the use of cheaper molds, which can be quite large.

Extrusion

Long plastic products with uniform cross sections can be produced by the **extrusion** process depicted in **Figure 20.6**. Thermoplastic pellets or powders are fed through a hopper into the

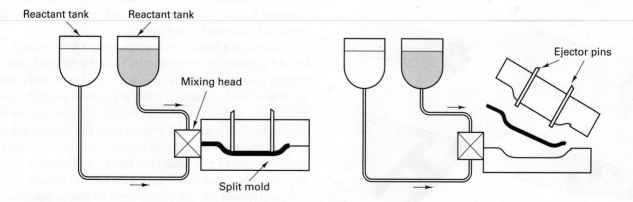

FIGURE 20.5 The reaction injection molding process. (*Left*) Measured amounts of reactants are combined in the mixing head and injected into the split mold. (*Right*) After sufficient curing, the mold is opened and the component is ejected.

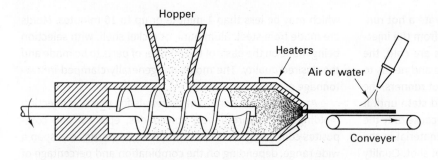

FIGURE 20.6 A screw extruder producing thermoplastic product. Some units may have a changeable die at the exit to permit production of different shaped parts.

barrel chamber of a screw extruder. A rotating screw propels the material through a preheating section, where it is heated, homogenized, and compressed, and then forces it through a heated die and onto a conveyor belt. To preserve its newly imparted shape, the material is cooled and hardened by jets of air, sprays of water, or immersion into a water bath. It continues to cool as it passes along the belt and is then either cut into lengths or coiled, depending on whether the material is rigid or flexible and the desires of the customer.

The process is continuous and provides a cheap and rapid method of molding. As shown in **Figure 20.7**, common production shapes include a wide variety of constant cross-section

profiles, including solid shapes (such as window and trim molding), hollow shapes (such as tubes and pipes), wires and cable coatings, sheets and films, and small-diameter fibers or filaments. Thermoplastic foam shapes can also be produced.

By using a narrow slit as the die opening, sheet (thicknesses between 0.5 and 12.5 mm or 0.02 to 0.5 in.) and film (thicknesses below 0.5 mm or 0.02 in.) can be produced in widths up to 3 m (10 feet). The extruded material is rapidly cooled by either wrapping around chilled rolls or direct immersion into a cooling bath. A combination of extrusion and blowing has been used in the manufacture of thin plastic bags, like those that are used as kitchen or bathroom trash can liners. This sequence begins with the extrusion of a thin-walled plastic tube through an open-ended metal die. Air flowing through the center of the die causes the diameter of the tube to expand substantially as it emerges from the die constraint. Air jets around the circumference of the expanded tube then cool the thin plastic material, after which it is passed through pinch rollers. The flattened tube is then periodically seam welded to form the bottom of the bag, perforated for tearing, and wound on a roll for easy dispensing. The product can also be cut on both edges to produce strips of thin film.

Thermoforming

In the **thermoforming** process, thermoplastic sheet material, as either discrete sheets or in a continuous roll, is heated to a working temperature and then formed into desirable shapes. If continuous material is used, it is usually heated to 200 to 425°C (400 to 800°F) by passing through an oven or other heating device. The material emerges over a single- or multi-impression, male or female, mold and is formed by the application of vacuum, pressure between 0.1 and 2 MPa (15 to 300 psi), or another mechanical tool. Cooling occurs on contact with the mold, which is usually at room temperature, and the product hardens in its new shape. After sufficient cooling, the part is removed from the mold and trimmed, and the unused strip material is diverted for recycling.

Figure 20.8 shows the process using a female mold cavity and a discrete sheet of material. Here the material is placed directly over the die or pattern and is heated in place, often by infrared radiant heaters. Either pressure or vacuum (sometimes both) is then applied, causing the material to draw into the cavity. The female die imparts both the dimensions and finish or texture to the exterior surface. The sheet material can also be stretched over male form blocks, and here the tooling controls dimensions and finish on the interior surface. Mating male and female dies can also be used. An entire cycle requires only a few minutes.

Although the starting material is a uniform thickness sheet, the thickness of the products will vary as the different regions

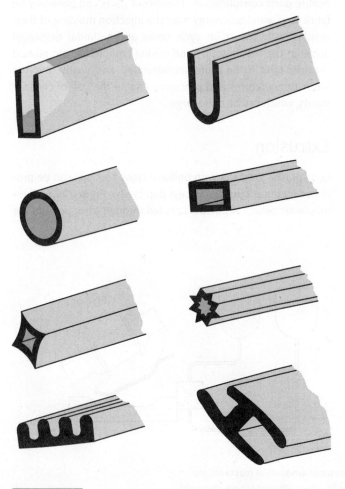

FIGURE 20.7 Typical shapes of polymer extrusions.

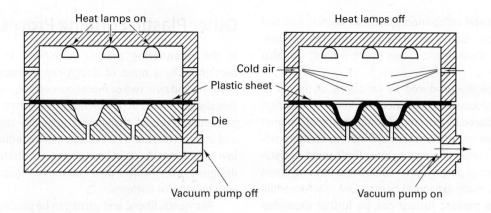

A type of thermoforming where thermoplastic sheets are shaped using a combination of heat and vacuum.

undergo stretching. The most common thin film products are small packaging containers for food products and commodities. Continuous lines may integrate sheet or film production, a thermoforming operation, and subsequent filling and sealing. Larger products, made from thicker sheet material, include plastic luggage, interior automotive door and body liners, panels for light fixtures, interior panels for refrigerators, shower enclosures, bathtubs, and boat hulls. Thermoforming of thin films has even been used to produce pages of Braille text for the blind.

Rotational Molding

Rotational molding or **rotomolding**, can be used to produce hollow, seamless products of a wide variety of sizes and shapes, including storage tanks, bins and refuse containers, small swimming pools, boat and canoe hulls, luggage, footballs, helmets, garbage cans, portable outhouses, septic tanks, and numerous automotive and truck panels and parts—generally parts with more complex external geometries, larger size, or lower production quantities than those produced by blow molding. Common thermoplastic materials include polyethylene, polypropylene, acrylonitrile butadiene styrene (ABS), and high-impact polystyrene. Some thermosets can be formed with process modifications.

The process begins with a split mold or cavity that has been filled with a premeasured amount of thermoplastic powder or liquid. The molds are either preheated or placed in a heated oven and are then rotated simultaneously about two perpendicular axes. Other designs rotate the mold about one axis while tilting or rocking about another. In either case, the

resin melts and is distributed, by gravity (not centrifugal force because the rotational speeds are slow) to produce a uniform-thickness coating over all of the surfaces of the mold. The mold is then transferred to a cooling chamber, where the motion is continued and air or water is used to slowly drop the temperature. After the material has solidified, the mold is opened and the hollow product is removed. All of the starting material is used in the product; no scrap is generated. The lightweight rotational molds are frequently made from cast aluminum, but sheet metal is often used for larger parts, and electroformed or vaporformed nickel is used when fine detail is to be reproduced. Production times are long compared to the other processes, as much as 10 minutes or more per cycle.

Foam Molding

Foamed plastic products have become an important and widely used form of polymer offering light weight, good thermal insulation, and good energy absorption. In **foam molding**, mechanical agitation introduces bubbles or a foaming agent is mixed with the plastic resin and releases gas or volatilizes when the material is heated during molding. The materials expand to 2 to 50 times their original size, resulting in products with densities ranging from 32 to 640 g/L (2 to 40 lb/ft³). **Open-cell foams** have interconnected pores that permit the permeability of gas or liquid. **Closed-cell foams** are gas- or liquid-tight.

Both rigid and flexible foams have been produced using thermoplastic and thermosetting materials. The rigid type is useful for structural applications, packaging, and shipping containers; as patterns for the full-mold and lost-foam casting processes (see Chapter 14); and for injection into the interiors

Why Can't Thermosets or Elastomers Be Thermoformed?

Thermoforming is a secondary shaping process—the starting sheet or film is produced in the primary process. If curing or cross-linking occurred during the forming of the starting sheet or film, subsequent

heating will not produce the softening required for shape modification forming. Only thermoplastic materials can be thermoformed.

of thin-skinned metal components, such as aircraft fins and stabilizers. Flexible foams are used primarily for cushioning and padding. Elastomeric foams (foam rubber) can also be produced.

Parts are typically produced by variations of the previously described processes. Large sheets of foamed insulation board can be produced by extruding foaming material. Thinner foamed sheets can be further thermoformed to create products such as the cushioning egg cartons. Prefoamed polystyrene beads (pellets of polystyrene containing a blowing agent that have been partially expanded by exposure to steam while being agitated to prevent fusion) can be further expanded and fused to create the various Styrofoam products, including drinking cups, disposable plates, and the expendable patterns for lost-foam casting. Reactive ingredients can be combined and projected through a spray gun, where polymerization and foaming occur simultaneously to create rigid insulating and sound damping foam for house and building construction. By introducing foaming material into the interior of a cold mold, parts can be produced with a solid outer skin and a rigid foam core. The rapid cooling against the cold mold surfaces produces a rigid skin that can be as thick as 2 mm (0.08 in.). Typical products include computer and business machine housings and home construction moldings that offer a wood look with significantly reduced expense. Dual structure products can also be produced by injecting the foaming polymer into a mold that already contains a hollow, partially formed product. **Figure 20.9** shows a light-weight garden planter with a foamed core and rigid skin.

FIGURE 20.9 This light-weight garden planter has a foamed core and a solid skin on both the interior and exterior surfaces. (Photo by Rachel Goldstein)

Other Plastic-Forming Processes

In the **calendering** process (described in more detail in Section 20.3), a mass of dough-like thermoplastic is forced between and over two or more counterrotating rolls to produce thin sheets or films of polymer, which are then cooled to induce hardening. Product thicknesses generally range between 0.3 and 1.0 mm (0.01 and 1/16 in.), but can be reduced further to as low as 0.05 mm (0.002 in.) by subsequent stretching. Embossed designs can be incorporated into the rolls to produce products with textures or patterns.

Filaments, fibers, and yarns can be produced by **spinning**, a modified form of extrusion in which molten thermoplastic polymer is forced through a die called a **spinneret** that contains many small holes. The emerging filaments are then drawn to further decrease their diameter and are air-cooled. Additional operations further elongate the fiber, aligning the structure and increasing the tensile strength. Most of the polymeric fibers, which include polyesters, nylons, rayons, and acrylics, are used in the carpet and textile industry. When multistrand yarns or cables are desired, the extrusion dies can rotate or spin to produce the necessary twists and wraps.

Traditional metal forming processes, such as rolling, forging, wire and deep drawing, and coining, can also be performed on room temperature and heated-but-still-solid thermoplastic polymers, provided they exhibit sufficient ductility. Conventional **drawing** can be used to produce fibers, and **rolling** can be performed to alter the shape of thermoplastic extrusions. In addition to changing the product dimensions, these processes can also serve to induce crystallization or produce a preferred orientation to the thermoplastic polymer chains.

Machining of Plastics

Plastics can be milled, sawed, drilled, and threaded much like metals, but their properties are so variable that it is impossible to give descriptions that would be correct for all types. Instead, let us first consider some of the general characteristics of plastics that affect their machinability. Because plastics tend to be poor thermal conductors, most of the heat generated during chip formation remains near the cut interface and is not conducted into the material or carried away in the chips. Thermoplastics tend to soften and swell, and they occasionally bind or clog the cutting tool. Considerable elastic flexing can also occur, and this couples with material softening to reduce the precision of final dimensions. Because the thermosetting polymers have higher rigidity and reduced softening, they generally

Filaments, Fibers, and Yarns

A fiber is a long, thin strand of material whose length is more than 100 times its diameter. A filament is a continuous fiber. Yarn is a continuous strand of multiple twisted filaments or fibers.

machine to greater dimensional precision provided that brittleness does not present problems.

The high temperatures that develop at the point of cutting also cause the tools to run very hot, and they may fail more rapidly than when cutting metal. Carbide tools may be preferred over high-speed tool steels if the cuts are of moderate duration or if high-speed cutting is performed. Coolants can often be used advantageously if they do not discolor the plastic or induce gumming. Water, soluble oil and water, and weak solutions of sodium silicate have been used effectively.

The tools that are used to machine plastics should also be kept sharp at all times. Drilling is best done by means of straight-flute drills or by "dubbing" the cutting edge of a regular twist drill to produce a zero rake angle. These configurations are shown in **Figure 20.10**. Drills, rotary files, saws, and milling cutters should be run at high speeds to improve cooling but with the feed and/or depth of cut carefully adjusted to avoid clogging the cutter.

Laser machining may be an attractive alternative to mechanical cutting. By vaporizing the material instead of forming chips, precise cuts can be achieved. Minute holes can be drilled, such as those in the nozzles of aerosol cans. Abrasive materials, such as filled and laminated plastics can be machined in a manner that also eliminates the fine machining dust that is often considered to be a health hazard.

Finishing and Assembly Operations

Polymeric materials frequently offer the possibility of integral color, and the as-formed surface is often adequate for final use. Some of the finishing processes that can be applied to plastics include printing, hot stamping, vacuum metallizing, electroplating, and painting. Chapter 32 presents the processes of surface finishing and surface engineering.

Thermoplastic polymers can often be joined by heating the relevant surfaces or regions. The joining heat can be provided by a stream of hot gases, applied through a tool like a soldering iron, or generated by ultrasonic vibrations. The welding techniques that are applied to plastics along with adhesive bonding (another popular means of joining plastic) are presented in Chapter 39. Because of the low modulus of elasticity, plastics can also be easily flexed, and **snap-fits** are another popular means of assembling plastic components. Because of the softness of some polymeric materials, self-tapping screws can also be used.

Designing for Fabrication

The primary objective of any manufacturing activity is the production of satisfactory components or products, in the necessary quantity, and at the desired rate of production. This activity begins with the selection of an appropriate material or materials. When polymers are selected as the material of construction, it is usually as a result of one or more of their somewhat unique properties, which include light weight, corrosion resistance, good thermal and electrical insulation, ease of fabrication, and the possibility of integral color. Although these properties are indeed attractive, one should also be aware of the more common limitations, such as strength and stiffness values that are lower than the engineering metals, high amounts of thermal expansion, creep under mechanical load, softening or burning at elevated temperatures, poor dimensional stability, and the deterioration of properties with age or exposure to sunlight or other forms of radiation.

The basic properties and characteristics of polymeric materials have been described in Chapter 8. One should note, however, that property evaluation tests are conducted under specific test conditions. Polymers are often speed-sensitive materials. Although a standard tensile test may show a polymer to have a moderately high strength value, a reduction in loading rate by two or three orders of magnitude may reduce this strength by as much as 80%. Conversely, an increase in loading rate may double or triple tensile strength. Polymers can also be extremely sensitive to changes in temperature. Strength values can vary by a factor of 10 over a temperature range of as little as 200°F (100°C). Polymeric materials should be selected, therefore, with full consideration to the specific conditions of temperature, loads and load rates, and operating environments that will be encountered.

The second area of manufacturing concern is selecting the process or processes to be used in producing the shape and establishing the desired properties. Each of the wide variety of fabrication processes has distinct advantages and limitations, as well as economical ranges of production quantity, and efforts should be made to utilize their unique features. Once a process has been selected, the production of quality products further requires an awareness of all the various aspects of that process. For example, consider a molding process in which a liquid or semifluid polymer is introduced into a mold cavity and allowed to harden. The proper amount of material must be introduced and caused to flow in such a way as to completely fill the cavity. Air that originally occupied the cavity needs to be vented and removed. Shrinkage will occur during solidification and/or cooling and may not occur in a uniform manner. Heat transfer must be provided to control the cooling and/or solidification. Finally, a means must be provided for part removal or ejection from the mold. Surface finish and appearance, the resultant engineering properties, and the ultimate cost of production are all

Straight-flute drill (left) and "dubbed" drill (right) used for drilling plastics. (E. Paul DeGarmo)

dependent on good design and proper execution of the molding process. Product properties can be significantly affected by such factors as: melt temperature, direction of flow, pressure during molding, thermal degradation, and cooling rate.

In all molded products, it is important to provide adequate fillets between adjacent sections to ensure smooth flow of the polymer into all sections of the mold and to eliminate stress concentrations at sharp interior corners. These fillets also make the mold less expensive to produce and reduce the danger of mold fracture during use. Even the exterior edges of the product should be rounded where possible. A radius of 0.25 to 0.40 mm (0.010 to 0.015 in.) is scarcely noticeable but will do much to prevent an edge from chipping. Sharp corners should also be avoided in products that will be used for electrical applications, because they tend to increase voltage gradients, which can lead to product failure.

Wall or section thickness is also very important because the hardening or curing time of a polymer is determined by the thickest section. If possible, sections should be kept nearly uniform in thickness because nonuniformity can lead to serious warpage and dimensional control problems. As a general rule, one should use the minimum thickness that will provide satisfactory end-use performance. The specific value will be determined primarily by the size of the part and, to some extent, the process and the type of polymer being used. Recommended minimum thicknesses for molded polymers are as follows:

Small parts	1.25 mm (0.050 in.)
Average-sized parts	2.15 mm (0.085 in.)
Large parts	3.20 mm (0.125 in.)

Thick regions and increased material at corners and joints should also be avoided because they can lead to locations of increased shrinkage, gas pockets, undercuring, or cracking. When extra strength is needed, it can often be provided by incorporating ribs into the design.

Economical production is also facilitated by appropriate dimensional tolerances. A minimum tolerance of 0.08 mm (0.003 in.) should be allowed in directions that are parallel to the parting line of a mold or contained within a mold segment. In directions that cross a parting surface, a minimum tolerance of 0.25 mm (0.010 in.) is desirable. In both cases, increasing these values by about 50% can simultaneously reduce both manufacturing difficulty as well as cost.

Because most molds are reusable, careful attention should also be given to the removal of the part. Rigid metal molds should be designed so that they can be easily opened and closed. A small amount of unidirectional taper should be provided to facilitate part withdrawal. Undercuts should be avoided whenever possible, because they will prevent part removal unless additional mold sections are used. These must move independent of the major segments of the mold, adding to the costs of mold production and maintenance and slowing the rate of production.

Inserts

Metal **inserts**, usually of brass or steel, are often incorporated into plastic products. Because molded threads are difficult to produce, machined threads require additional processing, and both types tend to chip or deform, threaded metal inserts are frequently used when assemblies require considerable strength or when frequent disassembly and reassembly is anticipated. **Figure 20.11** depicts one form of threaded insert, along with other types that provide pins or holes for alignment or mounting.

The successful use of inserts requires careful attention to design because they are generally held in place by only a mechanical bond that must resist both rotation and pullout. Knurling or grooving is often required to provide suitable sites for gripping. A medium or coarse knurl is usually adequate to resist torsional loads and moderate axial forces. Circumferential grooves are excellent for axial loads but offer little resistance to torsional rotation. Axial grooves resist rotation, but do little to prevent pullout. Other means of anchoring include bending, splitting, notching, and swaging. Headed parts with noncircular heads may be used as formed. Combinations of notches, grooves, and shoulders are also common. **Figure 20.12** depicts some of these common means of insert attachment.

If an insert is to act as a boss for mounting or serve as an electrical terminal, it should protrude slightly above the surface of the plastic. This permits a firm connection to be made without creating an axial load that would tend to pull the insert from its surroundings. If the insert serves to hold two mating parts together or align them, it should be flush with the surface. In this way, the parts can be held together snugly without danger of loosening the insert. In all cases, the wall thickness of the surrounding plastic must be sufficient to support any load that may be transmitted through the insert. For small inserts, the wall thickness should be at least half the diameter of the insert. For inserts larger than 13 mm (½ in.) in diameter, the wall thickness should be at least 6.5 mm (¼ in.).

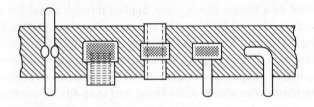

FIGURE 20.11 Typical metal inserts used to provide threaded cavities, holes, and alignment pins in plastic parts.

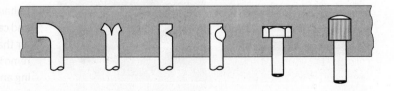

FIGURE 20.12 Various ways of anchoring metal inserts in plastic parts (left to right): bending, splitting, notching, swaging, noncircular head, and grooves and shoulders. Knurling is depicted in Figure 20.11.

Design Factors Related to Finishing

Because plastics are frequently used where consumer acceptance is of great importance, special attention should be given to finish and appearance. In many cases, plastic parts can be designed to require very little finishing or decorative treatment. For small parts, fins and rough spots can often be removed by a barrel tumbling with suitable abrasives or polishing agents. Smoothing and polishing occur in the same operation.

By etching the surfaces of a mold, decorations or letters can be produced that protrude approximately 0.01 mm (0.004 in.) above the surface of the plastic. When higher relief is required, the mold can be engraved, but this adds significantly to mold cost. Whenever possible, depressed letters or designs should be avoided. These features, when transferred to the mold, become raised above the surrounding surface. Mold making then requires a considerable amount of intricate machining, as the surrounding material must be cut away from the design or letters. When recessed features are absolutely required, manufacturing cost can be reduced if they can be incorporated into a small area that is originally raised above the primary surface.

When designing plastic parts, a prime objective is often the elimination of secondary machining, especially on surfaces that would be exposed to the customer. Even when fillers are used (as they are in most plastics), the surfaces of molded parts have a thin film of pure resin. This film provides the high luster that is characteristic of polymeric products. Machining cuts through the surface, exposing the underlying filler. The result is a poor appearance, as well as a site for the absorption of moisture.

One location that frequently requires machining is the parting line that is produced where the mold segments come together. Because perfect mating is difficult to achieve, a small fin, or **flash**, is usually produced around the part perimeter, as illustrated in **Figure 20.13**. When the flash is trimmed off, the resulting line of exposed filler may be objectionable. By locating the parting line along a sharp corner, it is easier to maintain satisfactory mating of the mold sections, and the exposed filler that is created by flash removal will be confined to a corner, where it is less noticeable.

Because plastics have a low modulus of elasticity, large flat areas are not rigid and should be avoided whenever possible.

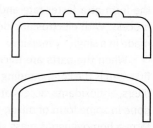

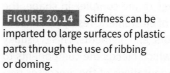

FIGURE 20.14 Stiffness can be imparted to large surfaces of plastic parts through the use of ribbing or doming.

Ribbing or doming, like that illustrated in **Figure 20.14**, can be used to provide the required stiffness. In addition, flat surfaces tend to reveal flow marks from the molding operation, as well as scratches that occur during handling or service. External ribbing then serves the dual function of increasing strength and rigidity while masking any surface flaws. Dimpled or textured surfaces can also be used to provide a pleasing appearance and conceal scratches.

Holes that are to be threaded or used to receive self-tapping screws should be countersunk. This not only assists in starting the tap or screw but also reduces chipping at the outer edge of the hole. If the threaded hole is less than 6.5 mm (¼ in.) in diameter, it is best to cut the threads after molding, using some form of thread tap. For diameters greater than 6.5 mm (¼ in.), the threads can be molded or an insert should be used. If the threads are molded, however, special provisions must be made to remove the part from the mold. Because the additional operations extend the molding time and reduce productivity, they are generally considered to be uneconomical.

20.3 Processing of Rubber and Elastomers

Rubber and elastomeric products can be produced by a variety of fabrication processes. Relatively thin parts with uniform wall thickness, such as boots and gloves, are often made by some form of **dipping**. A master form is first produced, usually from some type of metal. This form is then immersed into a liquid preparation or compound (usually based on natural rubber or latex, neoprene, or silicone), then removed and allowed to dry. With each dip, a certain amount of the liquid adheres to the surface, with repeated dips being used to produce a final desired thickness. After vulcanization, usually in steam, the products are stripped from the molds.

The dipping process can be accelerated by using electrostatic charges. A negative charge is introduced to the latex particles, and the form or mold receives a positive charge, either through an applied voltage or by a coagulant coating that releases positive ions when dipped into the solution. The attraction and neutralization of the opposite charges causes the elastomeric particles to be deposited on

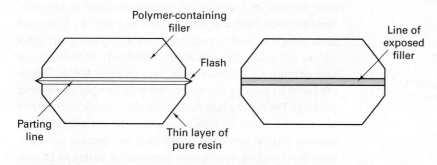

FIGURE 20.13 Trimming the flash from a plastic part ruptures the thin layer of pure resin along the parting line, and creates a line of exposed filler.

the form at a faster rate and in thicker layers than the basic process. With electrostatic deposition, many products can be made in a single immersion.

When the parts are thicker or are complex in shape, the first step is the compounding of elastomeric resin, vulcanizers, fillers, antioxidants, accelerators, and pigments. This is usually done in some form of mixer, which blends the components to form a homogeneous mass. Adaptations of the processes previously discussed for plastics are frequently used to produce the desired shapes, which include shoe soles and heels and various seals and gaskets. Injection, compression, and transfer molding are used, along with special techniques for foaming. Urethanes and silicones can also be directly cast to shape.

Rubber compounds can be made into sheets using calenders, like that shown in **Figure 20.15**. A warm mass of compound is fed into the gap between rotating rolls, and the emerging product is typically 0.3 to 1 mm (0.01 to 0.40 in.) in thickness. Three- or four-roll calenders can also be used to place a rubber or elastomer covering over cord or woven fabric. In the three-roll geometry, only one side of the fabric is coated in a single pass. The four-roll arrangement, shown in **Figure 20.16**, enables both sides to be coated simultaneously. Rubber-coated fabrics can also be produced by dipping the fabric into a rubber solution, spraying the solution onto the fabric, or skimming a thick solution of rubber compound and solvent onto the cloth and then driving off the solvent. The resultant material is used in products such as tires, conveyor belts, inflatable rafts, and raincoats.

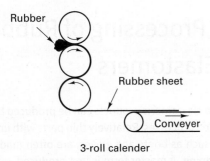

FIGURE 20.15 Schematic diagram showing the method of making sheets of rubber with a three-roll calender.

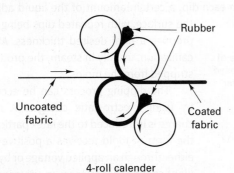

FIGURE 20.16 Arrangement of the rolls, fabric, and coating material for coating both sides of a fabric in a four-roll calender.

Rubber products such as inner tubes, garden hoses, tubing, and strip moldings can be produced by the extrusion process. The compounded elastomer is forced through a die by a screw device similar to that described for plastics.

Rubber or artificial elastomers can also be bonded to metal, such as brass or steel, using a variety of polymeric **adhesives**. Only moderate pressures and temperatures are required to obtain excellent adhesion.

Vulcanization or **cross-linking** is a critical step in all of the rubber processing sequences because it imparts the desired stiffness and strength while retaining the elastic behavior. Sulfur and other cross-linking materials are blended into the starting material. In molding operations, the mold is simply maintained at elevated temperature for sufficient time to impart curing. Other processes perform the vulcanization after the part has been shaped. In batch processing, the parts are heated in either an autoclave (a pressure-vessel containing hot steam) or an air-atmosphere furnace or oven. Continuous products can be vulcanized by passing them through hot-air tunnels, or over or through heated rolls.

20.4 Processing of Ceramics

The fabrication processes applied to ceramic materials generally fall into two distinct classes, based on the properties of the material. **Glasses** can be manufactured into useful articles by first heating the material to produce a molten or viscous state, shaping the material by means of **viscous flow**, and then cooling the material to produce a solid product. **Crystalline ceramics** have a characteristically brittle behavior and are normally manufactured into useful components by pressing moist aggregates or powder into a shape, followed by drying, and then bonding by one of a variety of mechanisms, which include chemical reaction, **vitrification** (cementing with a liquefied material), and **sintering** (solid-state diffusion).

Fabrication Techniques for Glasses

Glass is generally shaped at elevated temperatures where the viscosity can be controlled. A number of the processes begin with material in the liquid or molten condition at temperatures between 1000 and 1200°C (1850 and 2200°F). Sheet and plate glass is formed by processes such as extruding through a narrow slit, rolling through water-cooled rolls, or floating on a bath of molten tin. Glass shapes, such as larger lenses, can be produced by pouring the molten material directly into a mold (casting). The cooling rate during and after solidification is then controlled (usually as slow as possible) to minimize residual stresses and the tendency for cracking. In a process similar to centrifugal casting, rotationally symmetrical parts can be produced by introducing molten glass into a rotating mold. Constant cross-section shapes can be made by extrusion.

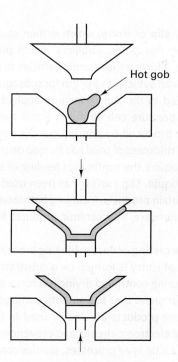

Hot gob

FIGURE 20.17 Viscous glass can be easily shaped by mating male and female die members.

Other glass-forming processes begin with viscous masses and use mating male and female die members to press the material into the desired shape, as illustrated in **Figure 20.17**. This **pressing** process is similar to the closed-die forging of metal. Typical products include dishes and bake ware. The millions of bottles, jars, and other thin-walled shapes are made by a process similar to the **blow molding** of plastics. Hollow gobs of viscous material are expanded against the outside of heated dies in a manner similar to that illustrated in **Figure 20.18**.

Glass fibers have been used for applications as diverse as thermal insulation, air and fluid filtration, sound damping, reinforcing polymers, weaving into cloth and fabric, and fiberoptic communication. When short length and random orientation are permissible, the fibers can be made by pouring molten glass into a rotating chamber with multiple small orifices around its exterior. Long, continuous fibers are generally made by extruding through a multiorifice die or pulling through a heated plate with multiple small orifices. Various coatings are often applied to the fibers to lubricate and protect the surface.

Special heat treatments are often applied to glass products before their manufacture is complete. The most common of these is **annealing**, performed to relieve the unfavorable residual stresses that form during shaping and subsequent cooling. The glass product is heated to an elevated temperature (typically around 500°C or 900°F), held for a period of time, and then slowly cooled to prevent the formation of new residual stresses.

Beneficial residual stresses can be imparted by a process known as **tempering**. Glass, generally in the form of precut sheets or plates, is heated to a temperature above that of annealing and held to create a uniform temperature through the thickness. The surfaces are then rapidly cooled, usually with jets of air. These regions contract on cooling, and the softer, still hot, interior flows to conform. Subsequently, the interior cools and attempts to contract, but is restrained by the already cold surface, placing the surface in compression. The resulting product, called **tempered glass**, is stronger and more fracture resistant because cracks tend to initiate on free surfaces, and these are now compressed. When tempered glass fails, the internal tension is released, causing the material to shatter into numerous small fragments that are less likely to injure an individual.

Similar results can be achieved by **chemical tempering**. The glass is heated in a bath of molten salt (potassium nitrate, sodium nitrate, or potassium sulfate), and the resulting diffusion and ion exchange (where larger atoms replace smaller surface atoms) creates the residual compression. More complex shapes can be tempered by the chemical process.

Glass-ceramics—materials that are part crystalline and part glass—are formed by another heat treatment process. Products are first fabricated into shape as a glass and are then subjected to a **devitrification** heat treatment that controls the nucleation and growth of the more-stable, lower-energy (i.e., equilibrium structure) crystalline component. Because of the dual structure, the final properties include good strength and toughness, along with low thermal expansion. Typical products include cookware (such as the white CorningWare products), dishes (Corelle), ceramic stove tops, and materials used in electrical and computer components.

Fabrication of Crystalline Ceramics

Crystalline ceramics are hard, brittle materials with high melting points. As a result, they cannot be formed by techniques requiring either plasticity (i.e., forming methods) or melting (i.e., casting methods). Instead, these materials are generally processed in the solid state by techniques that utilize particles or aggregates that have been produced by crushing or grinding, mixing with additives to impart certain characteristics (including binders, lubricants, plasticizers, and sintering assists), shaping by various techniques similar to

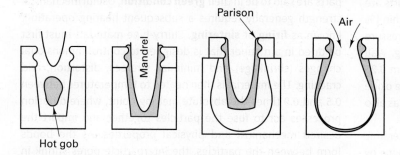

Parison

Mandrel

Air

Hot gob

FIGURE 20.18 Thin-walled glass shapes can be produced by a combination of pressing and blow molding.

those used in powder metallurgy, followed by drying, firing, and finishing.

Dry powders can be compacted and converted into useful shapes by pressing at either environmental or elevated temperatures. **Dry pressing** with rigid tooling, **hot pressing**, **isostatic pressing**, and **hot-isostatic pressing (HIP)** with flexible molds are common techniques and exhibit features and limitations similar to those discussed in Chapter 16. Binders are often added to impart strength and lubricants to prevent sticking during pressing and ejection. Density variation is common in dry pressing, as it is in powder metallurgy. Isostatic pressing produces a more uniform density. The hot techniques produce a denser product and may combine the operations of pressing and sintering. **Wet pressing**, accomplished by adding 10 to 15% moisture, can be used to produce more intricate shapes at lower pressures than dry pressing.

Clay products are based on special types of ceramics blended with water (typically 15 to 25%) and various additives to produce a material that can be shaped by most of the traditional forming methods. Plastic forming can also be applied to other ceramics if the ceramic particles are combined with additives that impart plasticity when subjected to pressure and heat. Extrusion can be used to produce products with constant cross sections.

Injection molding was discussed earlier in this chapter as a means of forming plastics, and metal injection molding (MIM) was presented in Chapter 16 as a way of producing small, complex-shaped metal parts. Injection molding can also be used to form complex, three-dimensional shapes from ceramic materials—**ceramic injection molding**. Ceramic powder is mixed with polymer material, wax, mold release, and other additives. After heating to 125 to 150°C (250 to 300°F), the mix is injected into a metal die—aluminum for a low-pressure variant and steel or carbide for high pressure molding where the injection pressure is on the order of 30 to100 MPa (5 to 15 ksi). The product cools in the mold (5 to 60 seconds) and becomes hard enough to permit ejection from the die. The additive materials are then removed by thermal, solvent, catalytic, or wicking methods, or combinations thereof, and the remaining ceramic is fused together by conventional sintering techniques. As with metal injection molding, the die shapes a part that is considerably oversized, and controlled shrinkage during additive removal and sintering produces the final dimensions. The major dimensions of most injection-molded ceramic parts are less than 10 cm (4 in.). Final dimensions are generally within 1% or 0.1 mm (0.005 in.), whichever is greater, for the low-pressure process, and 0.1% for high-pressure injection molding. Wall thickness in the high-pressure variation can range from 0.25 to 15 mm (0.01 to 0.6 in.). Most parts are made from the oxide ceramics, such as alumina or zirconia, but the process has also been used with silicon carbide and silicon nitride.

Several casting processes can be used to produce ceramic shapes beginning with a pourable slurry that strengthens by partial removal of the liquid or the gelation, polymerization, or crystallization of a matrix phase. In the **slip casting** process, ceramic powder is mixed with 25 to 40% liquid to form a low-viscosity **slip** or slurry, which is then cast into a mold containing very fine pores. Capillary action pulls the liquid from the slurry, allowing the ceramic particles to arrange into a "green" body with sufficient strength for subsequent handling. Pressure applied to the slurry, vacuum applied to the mold, or centrifugal pressure can all aid in liquid removal. Hollow shapes can be produced by pouring out the remaining slurry once a desired thickness of solid has formed on the mold walls. Solid shapes require the continuous feeding of slip to replace the absorbed liquid. Slip casting has been used to produce a variety of porcelain products, including bathroom fixtures, fine china and dinnerware, and ceramic products for the chemical industry.

In the **tape casting** or **doctor-blading** process, a controlled-thickness film of slurry is formed on a substrate. Evaporation of the liquid during controlled drying produces a thin, flexible, rubbery tape or sheet that has smooth surfaces and uniform thickness. These products are widely used in the multilayer construction of electronic circuits and capacitors.

In other casting-type processes, slurries containing bonding agents can be used to produce cast-in-place products, such as furnace linings or dental fillings. When mixed with a sticky binder, the material can be blown through a pipe to apply ceramic coatings or build up refractory linings.

The numerous variations of **sol-gel processing** can be used to produce ceramic films and coatings, fibers, and bulk shapes. These processes begin with a solution or colloidal dispersion (sol), which undergoes a molecular polymerization to produce a gel, which is then dried. This approach offers higher purity and homogeneity, lower firing temperatures, and finer grain size, at the expense of higher raw-material cost, large volume shrinkages during processing, and longer processing times.

Table 20.1 summarizes some of the primary processes used to fabricate shapes from crystalline ceramics.

Producing Strength in Particulate Ceramics

Each of the processes just described can be used to produce useful shapes from ceramic materials, but the products are largely just shapes of packed particles with strength levels similar to those of aspirin tablets. At this stage, the ceramic parts are said to be in their **green condition**. Useful mechanical strength generally requires a subsequent heating operation, known as **firing** or **sintering**. Slurry-type materials must first be dried in a manner that is designed to control dimensional changes (shrinkage) and minimize stresses, distortion, and cracking. The material is then heated to temperatures between 0.5 and 0.9 times the absolute melting point, where diffusion processes act to fuse the particles together and impart the desired mechanical and physical properties. As the bonds form between the particles, the interparticle pores shrink in size, density increases, and the overall part shrinks in size. The sintering temperature and sintering time are selected to control the resulting grain size, pore size, and pore shape. In some

TABLE 20.1 Processes Used to Form Products from Crystalline Ceramics

Process	Starting material	Advantages	Limitations
Dry axial pressing	Dry powder	Low cost; can be automated	Limited cross sections; density gradients
Isostatic processing	Dry powder	Uniform density; variable cross sections; can be automated	Long cycle times; small number of products per cycle
Slip casting	Slurry	Large sizes; complex shapes; low tooling cost	Long cycle times; labor-intensive
Injection molding	Ceramic–plastic blend	Complex cross sections; fast; can be automated; high volume	Binder must be removed; high tool cost
Forming processes (e.g., extrusion)	Ceramic–binder blend	Low cost; variable shapes (such as long lengths)	Binder must be removed; particles oriented by flow
Clay products	Clay, water, and additives	Easily shaped by forming methods; wide range of size and shape	Requires controlled drying

firing operations, surface melting (**liquid-phase sintering**) or component reactions (**reaction sintering**) can produce a substantial amount of liquid material (*vitrification*). The liquid then flows to produce a glassy bond between the ceramic particles and either solidifies as a glass or crystallizes.

Pressure and elevated temperature can be combined in the hot pressing and hot isostatic pressing operations. **Microwave sintering** and **spark-plasma sintering** are recent additions to the sintering options. These processes, along with more detail on the sintering process, will be presented in Sections 16-7 through 16-10 in the chapter on powder metallurgy.

Cementation is an alternative method of producing strength that does not require elevated temperature. A liquid binder material is used to coat the ceramic particles, and a subsequent chemical reaction converts the liquid to a solid, forming strong, rigid bonds.

Prototypes or small production quantities of ceramic products can be made by some of the **additive manufacturing/3-D printing** processes presented in Chapter 33, including the **laser sintering** process, where successive layers of heat-fusible powder are bonded. For ceramic parts, the powder particles are coated with a very thin thermoplastic polymer, and the laser acts on the polymer coating to produce the bond. After the shape is built, the parts undergo conventional debinding and sintering to about 55 to 65% of theoretical density. Isostatic pressing prior to sintering can raise the final density to 90 to 99% of ideal.

Machining of Ceramics

Most ceramic materials are brittle, and the techniques used to cut metals will generally produce uncontrolled or catastrophic cracks. In addition, ceramics are typically hard materials. Because ceramics are often used as abrasives or coatings on cutting tools, the tools needed to cut them have to be even harder.

Direct production to the desired final shape is clearly the most attractive alternative, but there are times when a material removal operation is necessary. This machining can be performed either before or after the final firing. Before firing, the material tends to be rather weak and fragile. Although fracture is always a concern, a more significant consideration might be the dimensional changes that will occur on subsequent firing. Shrinkage may be as much as 30%, so it may be difficult to achieve or maintain close final tolerances. For this reason, machining before firing, known as **green machining**, is usually rough machining designed to reduce the amount of finishing that will be required after firing.

When machining is performed after firing, the processes are generally ones that might be considered nonconventional. Grinding, lapping, and polishing with diamond abrasives; drilling with diamond-tipped tooling; cutting with diamond saws; ultrasonic machining; laser and electron-beam machining; water-jet machining; and chemical etching have all been used. When mechanical forces are applied, material support is quite critical (because ceramic materials are almost always brittle). Because of the hardness of the ceramic, the tools must be quite rigid. Selection and use of coolants are also important issues.

Materials producers have developed "machinable" ceramics that lend themselves to precision shaping by more traditional machining operations. It should be noted, however, that these are indeed special materials and not characteristic of ceramics as a whole.

Joining of Ceramics

When we consider joining operations, the unique properties of ceramics once again introduce limitations. Brittle ceramics cannot be joined by fusion welding or deformation bonding, and threaded assemblies should be avoided whenever possible. Therefore, most joining utilizes some form of adhesive bonding, brazing, diffusion bonding, or special cements. Even with these methods, the stresses that develop on the surfaces can lead to premature failure. As a result, most ceramic products are designed to be monolithic (single-piece) structures rather than multipart assemblies.

Design of Ceramic Components

Because ceramics are brittle materials, special care should be taken to minimize bending and tensile loading as well as design stress raisers. Sharp corners and edges should be avoided where possible. Outside corners should be chamfered to reduce the possibility of edge chips. Inside corners should have fillets of sufficient radius to minimize crack initiation. Undercuts are difficult to produce and should be avoided. Specifications should generally use the largest possible tolerances because these can often be met with products in the as-fired condition. Extremely precise dimensions usually require hand grinding, and costs can escalate significantly. In addition, consideration should be given to surface finish requirements because grinding, polishing, and lapping operations can increase production cost substantially.

20.5 Fabrication of Composite Materials

As shown in Chapter 8, composite materials can be designed to offer a number of attractive properties. In some market areas, such as aerospace and sporting goods, their acceptance and growth have been phenomenal. Use can only occur, however, if the material can be produced in useful shapes at an acceptable cost and rate of production. Many of the manufacturing processes designed for composites are slow, and some require considerable amounts of hand labor. There is often a degree of variability between nominally identical products, and inspection and quality control methods are not as well developed as for other materials. While these limitations may be acceptable for certain applications, they often restrict the use of composites for high-volume, mass-produced items. Faster production speeds, increased use of automation, reduced variability, and integrated quality control continue to be important issues in the expanded use of composite materials.

In Chapter 8, composite materials were classified by their basic geometry as particulate, laminar, and fiber-reinforced. Because the fabrication processes are often unique to a specific type of composite, they will also be grouped in the same manner.

Fabrication of Particulate Composites

Particulate composites usually consist of discrete particles dispersed in a ductile, fracture-resistant polymer or metal matrix. They offer isotropic properties and ease of fabrication relative to their fiber-reinforced counterparts. Their fabrication rarely requires processes unique to composite materials. Instead, the particles are simply dispersed in the matrix by introduction into a liquid melt or slurry, or by blending the various components as solids using powder metallurgy methods. Subsequent processing generally follows the conventional methods of casting or forming or utilizes the various techniques of powder metallurgy. These processes have been presented elsewhere in the text, and will not be repeated here.

Recent developments include the successful blending of reinforcement particles into the highly viscous slurries of rheocast material, the semisolid mixtures that are viscous when agitated but retain their shape when static. In the **sinter-forge** process, metal and ceramic powders (such as SiC and aluminum) are blended, compacted at room temperature, sintered, and then hot forged to produce near-net-shape parts at near-full density. Particle-reinforced composites have also been produced by spray forming multicomponent feeds. The various combinations of matrix and particulate have also expanded. Titanium has been strengthened by additions of TiC, TiB_2, and TiAl intermetallic particles.

Fabrication of Laminar Composites

Laminar composites include coatings and protective surfaces, claddings, bimetallics, laminates, and a host of other materials as previously presented in Chapter 8. Their production generally involves processes designed to form a high-quality bond between distinct layers of different materials. When the layers are metallic, as in claddings and bimetallics, the bonds are usually formed by one of the solid-state welding processes, such as roll bonding, ultrasonic welding, diffusion bonding, and explosive welding (all described in Chapter 37).

In the **roll bonding** process, sheets of the various materials are passed simultaneously through the rolls of a conventional rolling mill. If the amount of deformation is great enough, surface oxides and contaminants are broken up and dispersed as fresh metal surfaces are created, metal-to-metal contact is established, and the two surfaces become joined by a solid-state bond. U.S. coinage (dimes and quarters) is a common example of a roll-bonded material.

Explosive bonding is another means of joining layers of metal. A sheet of explosive material progressively detonates above the layers to be joined, causing a pressure wave to sweep across the interface. A small open angle is maintained between the two surfaces. As the pressure wave propagates, surface films are liquefied or scarfed off, and are jetted out the open interface. Clean metal surfaces are then forced together at high pressures, forming a solid-state bond with a characteristically wavy configuration at the interface. Large areas, wide plates (too wide to roll bond conveniently), and dissimilar materials with large differences in mechanical properties are attractive candidates for explosive bonding.

Diffusion bonding can be used to join a number of dissimilar metals and even metals to ceramics. A bond between any solid and any other solid can be achieved through **adhesive bonding**, described in Chapter 39. Plywood is an excellent example. By gluing the layers at various orientations, the directional effects of wood grain can be minimized within the plane of the sheet. Later in this chapter, we will discuss the lamination of fiber-reinforced polymer matrix composites where each ply is a fiber-containing or woven layer. Films of unpolymerized resin form between the layers. Pressing at elevated temperature cures the resin and completes the bond. In a manner

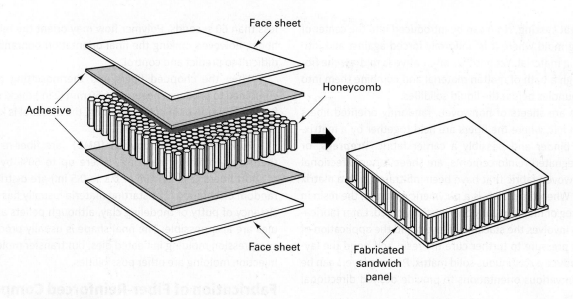

FIGURE 20.19 Fabrication of a honeycomb sandwich structure using adhesive bonding to join the facing sheets to the light-weight honeycomb filler. (Courtesy of ASM International, Materials Park, OH.)

similar to adhesive bonding, layers of metal can be joined by brazing to form composites that can withstand moderate elevated temperatures.

In **sandwich structures**, such as corrugated cardboard or the honeycomb shown in **Figure 20.19**, thin layers of facing material are bonded, usually by adhesive, to a light-weight filler material. Special fabrication methods may be employed to produce the foam, corrugated, or honeycomb filler.

Fabrication of Fiber-Reinforced Composites

In the fiber-reinforced composites, the matrix and fiber reinforcement combine to provide a system that offers properties not attainable by the individual components acting alone. The fiber reinforcement provides a significant increase in strength and stiffness, while the matrix functions as a binder, transfers the stresses, imparts toughness and fracture resistance, and withstands environmental attack and abrasion.

A number of processes have been developed to produce and shape the fiber-reinforced composites, with key differences relating to the orientation of the fibers, the length of continuous filaments, and the geometry of the final product. Each process seeks to embed the **fibers** in a selected **matrix** with the proper alignment and spacing necessary to produce the desired properties. Discontinuous fibers can be combined with a matrix to provide either a random or a preferred orientation. Continuous fibers are normally aligned in a unidirectional fashion in rods or tapes, woven into fabric layers, wound around a mandrel, or woven into a three-dimensional shape.

Some of the fiber-reinforced processes are identical to those previously-described for unreinforced plastics: compression, transfer and injection molding, extrusion, rotational molding, and thermoforming. Others are standard processes with simple modifications, such as reinforced reaction injection molding and resin transfer molding. Still others are specific to fiber-reinforced composites, such as hand lay-up, spray-up, vacuum-bag molding, pressure-bag molding, autoclave molding, filament winding, and pultrusion.

Production of Reinforcing Fibers.

A number of processes have been developed to produce the various types of reinforcement fibers described previously in Section 8-5. Metallic fibers, glass fibers, and many polymeric fibers (including the popular Kevlar) are produced by variations of conventional wire drawing and extrusion. Boron, carbon, and ceramic fibers such as silicon carbide are too brittle to be produced by the deformation methods. Boron fibers are produced by chemical vapor deposition around a tungsten filament. Carbon (graphite) fibers can be made by carbonizing (decomposing) an organic material that is more easily formed to the fiber shape. Natural fibers of flax, hemp and sisal are also being used.

The individual fine filaments are often bundled into **yarns** (twisted assemblies of filaments), **tows** (untwisted assemblies of fibers), and **rovings** (untwisted assemblies of yarns or tows), and these can be further woven into cloth. Fibers can also be chopped into short lengths, usually 12 mm (½ in.) or less, for incorporation into mats or the various sheet or bulk molding compounds where the fibers assume a random orientation.

Processes Designed to Combine Fibers and a Matrix.

A variety of processes have been developed to combine the fiber and the matrix into a unified material suitable for further processing. If the matrix material can be liquefied and the temperature is not harmful to the fibers, casting-type processes can be an attractive means of coating the reinforcement. The pouring of concrete around steel reinforcing rod is a macroscopic example of this approach. In the case of the fiber-reinforced plastics and metals, the liquid can be introduced between the fibers by means of **capillary action**, **vacuum infiltration**, or **pressure casting**. In a modification of

centrifugal casting, resin can be introduced into the center of a rotating mold where it is uniformly forced against and into reinforcing material. Yet another alternative is to draw the fibers through a bath of molten material and combine them into aligned bundles before the liquid solidifies.

Mats are sheets of nonwoven, randomly oriented fibers similar to felt, where the fibers are held together by a matrix-material binder and possibly a carrier fabric. **Prepregs**, or pre-impregnated reinforcements, are sheets of unidirectional fibers or woven fabric that have been infiltrated with a matrix material. When the matrix is a polymeric material, the resin in the prepreg or mat is usually only partially cured. Later fabrication then involves the stacking of layers and the application of heat and pressure to further cure the resin and bond the layers to produce a continuous solid matrix. Prepreg layers can be stacked in various orientations to provide desired directional properties.

Individual **filaments** can be coated with a matrix material by drawing through a molten bath, plasma spraying, vapor deposition, electrodeposition, or other techniques. The coated fibers can then be used, either individually or in various assemblies. They can also be wound around a mandrel with a specified spacing, and then cut to produce **tapes** that contain continuous, unidirectionally aligned filaments. These tapes are generally one fiber diameter in thickness and can be up to 1.2 m (48 in.) wide.

When the temperatures of the molten matrix become objectionable or potentially damaging to the fiber, bonding between the fiber and the matrix can often be achieved through lower-temperature diffusion or deformation processing (hot pressing or rolling). A common arrangement is to position aligned or woven fibers between sheets of matrix material in foil form. Loosely woven fibers can also be infiltrated with a particulate matrix, which is then compacted at high pressures and sintered to produce a continuous solid.

Sheet-molding compounds (SMC) are composed of chopped fibers (usually glass in lengths of 12.5 to 50 mm or ½ to 2 in.) and partially cured thermoset resin, along with fillers, pigments, catalysts, thickeners, and other additives, in sheets approximately 2.5 to 5 mm (0.1 to 0.2 in.) thick and 30 to 150 cm (12 to 60 in.) in width. With strengths in the range of 35 to 70 MPa (5 to 10 ksi) and the ability to be press-formed in heated dies, these compounds offer a feasible alternative to sheet metal in applications where light weight, corrosion resistance, and integral color are attractive features.

After initial compounding and a few days of curing, sheet-molding compounds generally take on the consistency of leather, making them easy to handle and mold. When they are placed in a heated mold, the viscosity is quickly reduced and the material flows easily under pressures of about 7 MPa (1000 psi). The elevated temperatures accelerate the chemical reactions, and final curing can often be completed in

less than 60 seconds. Polymer flow may orient the reinforcing fibers, however, making the final orientation nonrandom and difficult to predict and control.

When the chopped fibers and thermosetting resin are combined to produce sheets up to 50 mm (2 in.) thick, or billets for compression or injection molding, the material is known as **thick molding compound (TMC)**.

Bulk molding compounds (BMC) are fiber-reinforced, thermoset, molding materials, where up to 50% by volume of short fibers (2 to 12 mm or 0.1 to 0.5 in.) are distributed in random orientation. The starting material usually has the consistency of putty or modeling clay, although pellets and granules are also possible. The final shape is usually produced by compression molding in heated dies, but transfer molding and injection molding are other possibilities.

Fabrication of Fiber-Reinforced Composites into Final Shapes.

A number of processes have been developed for the production of finished products from fiber-reinforced material. Many are simply extensions or adaptations of processes that are used to shape the matrix materials (usually metals or polymers). Others are unique to the family of fiber-reinforced composites. The dominant techniques will be discussed individually in the sections that follow.

PULTRUSION Pultrusion is a continuous process that is used to produce long lengths of uniform cross-section, simple-complexity, shapes, such as round, rectangular, tubular, plate, sheet, and structural products. As shown in **Figure 20.20**, bundles of continuous fiber are first drawn through a bath of thermoset polymer resin, followed by dies to begin shaping the product. This pre-shaped material is then pulled through a long, heated die (1 to 1.5 m or 3 to 5 ft in length) which completes the shaping and cures the resin. When it emerges from the final die, the product is cooled by air or water and then cut to length. Some products, such as structural shapes, are complete at this stage, while others are further fabricated into products such as fishing poles, golf club shafts, and ski poles. Extremely high strengths and stiffnesses are possible because the reinforcement can be as much as 75% of the final structure. Tensile strengths of 210 MPa (30 ksi) and elastic modulus of 17 Gpa (2.5×10^6 psi) are coupled with densities about 20% that of steel or 60% that of aluminum. Cross sections can be as much as 1.5 m (60 in.) wide and 0.3 m (12 in.) thick.

The pultrusion process can be modified if the desired products have curvatures (such as leaf springs) or require variations

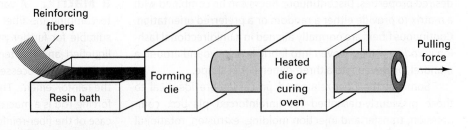

FIGURE 20.20 Schematic diagram of the pultrusion process. The heated dies cure the thermoset resin.

in cross-section (such as hammer handles). In the **pulforming** modification, product emerging from the shaping dies (but before curing) is fed into heated molds that complete the forming of the more-complex shape and cure the resin while it is held in that shape.

FILAMENT WINDING AND TAPE LAYING In the **filament winding** or **tape-laying** process, resin-coated (or resin-impregnated) continuous filaments, bundles, or tapes up to 200 mm or 8 in. in width, made from fibers of glass, graphite, boron, Kevlar (aramid), or similar materials, are used to produce cylinders, spheres, cones, and other container-type shapes with exceptional strength-to-weight ratios. The filaments or tapes are wound over a rotating form or mandrel, using longitudinal, circumferential, or helical patterns, or a combination of these, designed to take advantage of their highly directional strength properties. By adjusting the density of the filaments in various locations and selecting the orientation of the wraps, products can be designed to have strength where needed and lighter weight in less critical regions. After winding, the part and mandrel are placed in an oven for curing, after which the product is removed from the mandrel or form. [Note: When the part is too large for oven cure or oven cure is impractical, the resin-coated strands can be activated with an initiator or hardener prior to winding.] In some cases, part removal requires the use of an inflatable/deflatable mandrel, segmented mandrels, or mandrels made from soluble materials, such as salts or plaster. In the final product, the matrix, often an epoxy-type polymer, binds the structure together and transmits the stresses to the fibers.

Figure 20.21 shows a large tank being produced by filament winding. Products such as pressure tanks, airplane fuselage segments, and rocket motor casings can be made in virtually any size, some as large as 6 m (20 ft) in diameter and 20 m (65 ft) long. Smaller parts include helicopter rotor blades, baseball bats, and light poles. Moderate production quantities are feasible, and because the process can be highly mechanized, uniform quality can be maintained. A new form block is all that is required to produce a new size or design. Because the tooling is so inexpensive, the process offers tremendous potential for cost savings and flexibility. With advancements in computer software, and multiaxis computer-controlled equipment, parts no longer need to be axisymmetric. Filament-wound products can now be made with changing surfaces, nonsymmetric cross sections, and compound curvatures.

LAMINATION AND LAMINATION-TYPE PROCESSES In the **lamination** processes, prepregs, mats, or tapes are stacked, often in varying orientation, to produce a desired thickness and cured under pressure and heat. The resulting products possess unusually high strength as a result of the integral fiber reinforcement. Because the surface is a thin layer of pure resin, laminates usually possess a smooth, attractive appearance. If the resin is transparent, the fiber material is visible and can impart a variety of decorative effects. Other decorative laminates use a separate patterned face sheet that is bonded to the laminate structure.

Laminated materials can be produced as sheets, tubes, and rods. Flat sheets can be made using the method illustrated in **Figure 20.22**. Prepreg sheets or reinforcement sheets saturated in resin are stacked and then compressed under pressures on the order of 7 MPa (1000 psi). **Figure 20.23** depicts the technique used to produce rods or tubes. For tubing, the impregnated stock is wound around a mandrel of the desired

FIGURE 20.21 A large tank being made by filament winding. (Courtesy of Rohr Inc., Chula Vista, CA.)

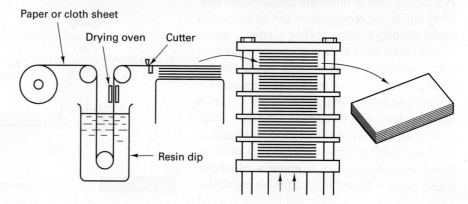

FIGURE 20.22 Method of producing multiple sheets of laminated plastic material.

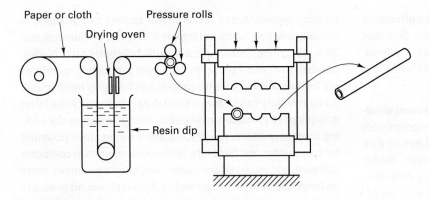

FIGURE 20.23 Method of producing laminated plastic tubing. In the final operation, the rolled tubes are cured by being held in heated tooling.

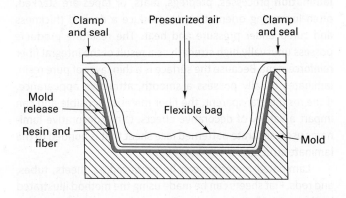

FIGURE 20.24 Schematic of the pressure-bag process

internal diameter. Solid rods are made by using a small-diameter mandrel, which is removed prior to curing, or by wrapping the material tightly about itself. Sheet laminating can also be a continuous process in which multiple reinforcement sheets are passed through a resin bath and then through squeeze rolls. In all of the above cases, the final operation is a curing, usually involving elevated temperature and possibly applied pressure. Because of their excellent strength properties, fiber-reinforced plastic laminates find a wide variety of uses. Some sheets can be easily blanked and punched. Gears machined from thick laminated sheets have unusually quiet operating characteristics when matched with metal gears.

Many laminated products are not flat but contain relatively simple curves and contours. Manufacturing processes that require zero to moderate pressures and relatively low curing temperatures can be performed where the only required tooling is often a female mold or male form block that can be made from metal, hardwood, or even particle board. The layers of prepreg or resin-dipped fabric are stacked in various orientations until the desired thickness is obtained. Care must be taken to avoid the entrapment of air bubbles and ensure that no impurities (such as oil, dirt, or other contaminants) are introduced between the layers. In **pressure-bag molding**, depicted in **Figure 20.24**, a flexible membrane

is positioned over the female mold cavity and is pressurized to force the individual plies together and drive out entrapped air and excess resin. Pressures usually range from 0.2 to 0.4MPa (30 to 50 psi) but can be as high as 2MPa (250 psi). This pressing is coupled with room- or low-temperature curing. Pressure-bag molding has been used to produce extremely large components, such as the skins of military aircraft and other aerospace panels, automobile and truck body panels, large air deflectors for tractor-trailers, boat bodies, and similar products. In the **vacuum-bag molding** process, the entire assembly (mold and material) is placed beneath or within a nonadhering flexible bag, as depicted in **Figure 20.25**. When the contained air is evacuated, pressure from the outside air forces the laminate against the mold while the resin is cured. While curing may occur at room temperature, moderately elevated temperatures may also be used.

Higher heats and pressures are used when parts are cured in an **autoclave**. The supporting molds and uncured layups are placed inside a heated pressure vessel where curing occurs under elevated temperatures and pressures in the range 0.4 to 0.7MPa (50 to 100 psi). Denser, void-free moldings are produced, and the properties can be further enhanced through the use of matrix resins that require higher-temperature cures. The size of the autoclave limits the size of the product.

When the quality demands are not as great, the reinforcement-to-resin ratio is not exceptionally high, and only one surface needs to be finished to high quality, the pressing operations can often be eliminated. In a process known as **hand lay-up** or **open mold processing**, depicted in **Figure 20.26**, successive layers of pliable resin-coated cloth are simply placed in an open mold or draped over a form. Squeegees or rollers are used to manually ensure good contact and remove any entrapped air, and the assembly is then allowed to cure, generally at room temperature. If prepreg layers are not used, a layer of mat, cloth, or woven roving can be put in place, and a layer of resin brushed, sprayed, or poured on. This process is then repeated to build the desired thickness.

While the hand lay-up process is slow and labor intensive, and has part-to-part and operator-to-operator variability, the tooling costs are sufficiently low that single items or small quantities become economically feasible. Molds or forms can

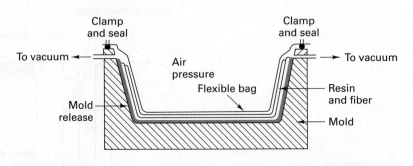

FIGURE 20.25 Schematic of the vacuum-bag process

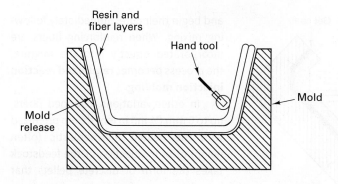

FIGURE 20.26 Schematic of the hand lay-up lamination process

be made from wood, plaster, plastics, aluminum, or steel, so design changes and the associated tool modifications are rather inexpensive, and manufacturing lead time can be quite short. In addition, large parts can be produced as single units, significantly reducing the amount of assembly, and various types of reinforcement can be incorporated into a single product, expanding design options. A high-quality surface can be produced by applying a pigmented gel coat to the mold before the lay-up.

The open mold lay-up process can be automated by using a programmed tape-laying machine that deposits strips of prepreg tape, typically in 75mm (3 in.) width. Unlike the filament-winding approach where the tape is deposited in a continuous strip, this adaptation uses a machine which deposits and cuts finite-length strips, each following a programmed three-dimensional contour, to build up the desired number of plies. While labor and time are reduced and quality and repeatability are improved, the desired quantity must be sufficient to justify the time and expense of programming the operation.

When production quantities are large, the quality needs to be high, part complexity is increased, and all surfaces need to have a smooth finish, matched metal dies mounted in a press can be used in place of the above techniques. This **closed-molding process** is essentially a modification of polymer **compression molding**. Sheet-molding compound, bulk-molding compound, or preformed mat is placed between the dies, and heat and pressure are applied. Temperatures typically range from 110 to 160°C (225 to 325°F), coupled with pressures from 1 to 7 MPa (250 to 1000 psi). With heated dies, the thermoset resin cures during the compression operation, with cycles repeating every 1 to 5 minutes. Because all surfaces now lie within the closed mold, precise tolerances can be maintained in any direction, including thickness. Part size is limited by the size of the molds.

Resin-transfer molding (RTM) is a low-pressure process that is intermediate to the slow, labor-intensive lay-up processes and the faster compression molding or injection

molding processes, which generally require more expensive tooling. Continuous fiber mat or woven material (usually employing glass fiber) is positioned dry in the bottom half of a matching mold, which is then closed and clamped. A low-viscosity catalyzed resin is then injected into the closed mold at pressures up to 2.1 MPa (300 psi), where it displaces the air, permeates the reinforcement, and subsequently cures when the mold is heated to low temperatures. Because of the low pressures employed in the process and the low curing temperatures, the mold tooling does not need to be steel, but can be electroformed nickel shells, epoxy composite, or aluminum. In addition, low-capacity presses can be used to clamp the mold segments, and inflatable bags can be used to produce simple holes or hollows, in much the same way that cores are used in conventional casting. By curing in the mold, the autoclave operation is eliminated, reducing both cost and production time. The resulting products can have excellent surfaces on both sides, because both mold surfaces can be precoated with a pigmented gel. Large parts can often be made as a single unit with a relatively low capital investment. Cycle times range from a few minutes to a few hours, depending on the part size and the resin system being used. The aerodynamic hood and fender assembly for Ford Motor Company's AeroMax heavy-duty truck (**Figure 20.27**) is an example of a large resin transfer molding.

In resin-transfer molding, the resin is pumped in under pressure. In a variant known as **vacuum-assisted resin-transfer molding (VARTM)**, a vacuum is drawn on the closed cavity, extracting the air and slowly infusing the dry fibers with inflowing resin. Large parts can be made by this process, originally used to produce large yacht hulls, but now being used extensively in the aerospace industry.

SPRAY MOLDING When continuous or woven fibers are not required to produce the desired properties, sheet-type parts

FIGURE 20.27 Aerodynamic styling and smooth surfaces characterize the hood and fender of Ford Motor Company's AeroMax truck. This one-piece panel was produced as a resin transfer molding by Rockwell International. (Courtesy of ASM International, Materials Park, OH.)

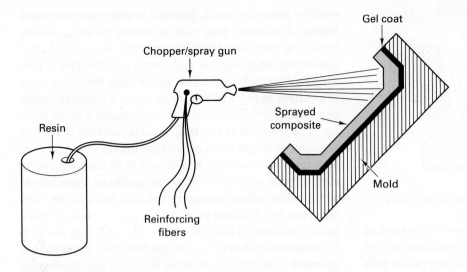

FIGURE 20.28   Schematic diagram of the spray forming of chopped-fiber-reinforced polymeric composite.

can be produced by mixing chopped fibers, fillers, and catalyzed liquid resin and spraying the combination into or onto a mold form in a process known as **spray molding**, depicted schematically in **Figure 20.28**. Rollers or squeegees are used to remove entrapped air and work the resin into the reinforcement. Room-temperature curing is usually preferred, but elevated temperatures are sometimes used to accelerate the cure. As with the hand lay-up process, an initial gel coat can be used to produce a smooth, pigmented surface.

To facilitate spray molding, the fiber content is limited to about 35%, significantly less than the laminated prepreg or mat approach, and the fibers assume a random orientation. Because the fibers are no longer long and oriented, the resultant mechanical properties are not as great as with the lamination processes. Some typical products include bathtubs, shower stalls, furniture, large architectural panels, and truck-bed liners and toppers.

SHEET STAMPING   Thermoplastic sheets that have been reinforced with nonwoven fiber can often be heated and press-formed in a manner similar to conventional sheet metal forming or polymeric thermoforming. Precut blanks are heated and placed between the halves of a matched metal mold that has been mounted in a vertical press. Ribs, bosses, and contours can be formed in parts with essentially uniform thickness. Because the parts must be cooled prior to removal from the mold, cycle times range from 25 to 50 seconds for most parts.

INJECTION MOLDING   The injection molding of fiber-reinforced plastics is a process that competes with metal die castings, and offers comparable properties at considerably reduced weight. In its simplest form, chopped or continuous fibers (or a fiber perform) are placed in a mold cavity that is then closed and injected with resin. If the resin is a thermoplastic, it is heated to a viscous condition and injected into a cold mold. Thermosetting resins, in contrast, are injected cold into a heated mold. In a variation known as **reaction injection molding**, reactive components are mixed just prior to injection

and begin their cure, immediately following mixing. When reinforcing fibers are incorporated directly into the mixture, the process becomes **reinforced reaction injection molding**.

In other variations, chopped fibers, up to 6 mm (¼ in.) in length, are premixed with the heated thermoplastic (often nylon) prior to injection or the feedstock takes the form of discrete pellets that have been manufactured by slicing continuous-fiber pultruded rods. The benefits of fiber reinforcement (compared to conventional plastic molding) include increased rigidity and impact strength, reduced possibility of brittle failure during impact, better dimensional stability at elevated temperatures and in humid environments, improved abrasion resistance, and better surface finish due to the reduced dimensional contraction and absence of related sink marks. The molding process is quite rapid, and the final parts can be both precise and complex. As with the conventional process, however, tooling cost is high, so large production volumes may be required.

BRAIDING, THREE-DIMENSIONAL KNITTING, AND THREE-DIMENSIONALWEAVING   The primary causes of failure in lamination-type composites are interlaminar cracking and delamination (layer-separation) upon impact. To overcome these problems, the high-strength reinforcing fibers can also be interwoven into three-dimensional preforms by processes that include weaving, braiding, and stitching through the thickness of stacked two-dimensional preforms. Resin is then injected into the assembly and the resultant product is cured for use. Complex shapes can be produced with the fiber orientations selected for optimum properties. Computers can be used to design and control the weaving, making the process less expensive than many of the more labor-intensive techniques.

Fabrication of Metal–Matrix Fiber-Reinforced Composites

Continuous-fiber **metal–matrix composites** can be produced by variations of filament winding, extrusion, and pultrusion. Fiber-reinforced sheets can be produced by electroplating, plasma spray deposition coating, or vapor deposition of metal onto a fabric or mesh. These sheets can then be shaped and bonded, often by some form of hot pressing. Diffusion bonding of foil–fabric sandwiches, roll bonding, and coextrusion are other means of producing fiber-reinforced metal products. Various casting processes have been adapted to place liquid metal around the fibers by means of capillary action, gravity, pressure (die casting and squeeze casting), or vacuum (countergravity casting). Products that incorporate discontinuous fibers can be produced by powder metallurgy or spray-forming techniques and further fabricated by hot pressing, superplastic forming, forging, or some types of casting. In general, efforts are made to reduce or eliminate the need for finish machining, which would often require the use of diamond or carbide tools, or methods such as EDM.

A critical concern with metal–matrix composites is the possibility of reactions between the reinforcement and the metal matrix during processing at the high temperatures required to melt and form metals or the temperatures encountered during subsequent service. These concerns often limit the kinds of materials that can be combined. Barrier coatings have been employed to isolate reactive fibers.

Fabrication of Ceramic–Matrix Fiber-Reinforced Composites

Unlike polymeric– or metal–matrix composites, where the objective is usually to increase strength and stiffness, the objective with **ceramic–matrix composites** is often to improve fracture resistance or toughness. Ceramic materials typically fail due to fractures induced by flaws in the matrix. If the reinforcement is bonded strongly to the matrix, a matrix crack would tend to propagate through the fibers and on through the remaining matrix. With a weak bond between the fiber and matrix, fractures are redirected along the fiber–matrix interface, preserving the integrity of the remaining matrix. [Note: This same technique is often applied to laminar ceramic composites where an interface delaminates to preserve the integrity of the remaining layers.]

The various matrix materials and reinforcement fibers for ceramic–ceramic composites have been discussed in Chapter 8, along with some of the unique property combinations that can be achieved. The fabrication techniques are often quite different from the other composite families. One approach is to pass the fibers or mats through a slurry mixture that contains the matrix material. The impregnated material is then dried, assembled, and fired. Other techniques include the chemical vapor deposition or chemical vapor infiltration of a coated fiber base, where the coating serves to weaken the otherwise strong bond. Silicon nitride matrices can be formed by reaction bonding. The reinforcing fibers are dispersed in silicon powder, which is then reacted with nitrogen. Hot-pressing techniques can also be used with the various ceramic matrices. When the matrix is a glass, the heated material behaves much like a polymer, and the processing methods are often similar to those used for polymer–matrix composites.

Secondary Processing and Finishing of Fiber-Reinforced Composites

The various fiber-reinforced composites can often be processed further with conventional equipment (sawed, drilled, routed, tapped, threaded, turned, milled, sanded, and sheared), but

special considerations should be exercised because composites are not uniform materials. Cutting some materials may be like cutting multilayer cloth, and precautions should be used to prevent the formation of splinters and cracks as well as frayed or delaminated edges. Sharp tools, high speeds, and low feeds are generally required. Cutting debris should be removed quickly to prevent the cutters from becoming clogged.

In addition, many of the reinforcing fibers are extremely abrasive and quickly dull most conventional cutting tools. Carbide, diamond-coated carbide, or polycrystalline diamond tooling is generally required to achieve realistic tool life. Abrasive slurrys can be used in conjunction with rigid tooling to ensure the production of smooth surfaces. Lasers and waterjets are alternative cutting tools. Lasers, however, can burn or carbonize the material or produce undesirable heat-affected zones. Waterjets can create moisture problems with some plastic resins, and pressurized water can cause delaminations, but the low heat and light cutting force are attractive characteristics. Elastic deflections are minimized during the cut. Parts can often be held in place by simple vacuum cups, and waterjets also minimize the generation of dust, which may be toxic.

When fiber-reinforced materials must be joined, the major concern is the lack of continuity of the fibers in the joint area. Thermoplastics can be softened and welded by applying pressure with heated tools, combining pressure and ultrasonic vibration, or using pressure and induction heating. Thermoset materials generally require the use of mechanical joints or adhesives, with each method having its characteristic advantages and limitations. Metal–matrix composites are often brazed.

Goals, Trends and Challenges for Fiber-Reinforced Composites

In recent years, there has been a dramatic shift away from the labor-intensive fabrication processes, such as hand lay-up, to the higher-output automated methods. Additional efforts to improve productivity have focused on reducing the shape-and-cure cycle time, with a goal of less than one minute per part. Both production time and energy consumption can be reduced by eliminating autoclaving by either out-of-autoclave curing or the use of resins that cure at room temperature. Recent research has shown that microwave curing, where only the product is heated and cooled and the oven chamber remains cool, has the potential for significant reduction in processing time (about 50%) and energy consumption (up to 90%). Material cost, particularly for components like carbon fiber, is significantly higher than the competing metals of steel and aluminum. While there have been reductions in the cost of both material and fabrication, a continuation of this trend is necessary for the growth of composites.

One of the attractive features of composites is the ability to consolidate components into a single-piece, but joining these products to materials such as steel or aluminum (as is often the case in the automotive and aerospace industries) can present difficulties, especially in the areas of galvanic corrosion and thermal expansion mismatch. Machining continues to present unique problems: fiber layers can delaminate and edges can fray; fibers and other hard reinforcements are abrasive and considerably reduce tool life; and the mix of materials and geometries complicates the selection of both tool materials and cutting parameters.

Because of the multicomponent nature of composites, the type of defects is often very different from those typically found

in metallic materials. Nondestructive examination techniques must be available for in-process, after-manufacture, and in-service inspection for a wide range of matrix materials, reinforcement materials, and inter-component geometries, as well as overall part sizes and shapes.

Recyclability is another significant problem, particularly when dictated by end-of-product-use legislation. Separation into component materials may not be physically or economically feasible, so reuse may be in areas such as reduction into small fragments and incorporation into alternate products.

Review Questions

1. Why are the fabrication processes applied to plastics, ceramics, and composites often different from those applied to metals? What are some of the key differences?

2. How does the fabrication of a shaped product from a thermoplastic polymer differ from the fabrication of a solid shape from a thermosetting polymer?

3. What are some of the methods used to shape polymers?

4. What are some of the ways that plastic sheet, plate, and tubing can be cast?

5. Why do cast plastic resins typically have a lustrous appearance?

6. What types of polymers are most commonly blow molded?

7. What are some common blow molded products, in addition to the common disposable beverage bottle?

8. Why do blow molding molds typically contain a cooling system?

9. Describe the compression molding process.

10. For what types of parts and production volumes would compression molding be an appropriate process?

11. What are typical mold temperatures for compression molding? What is the most common mold material?

12. What are some of the attractive features of the transfer molding process? Some process negatives?

13. Cold molding is faster and more economical than other types of molding. What limits its use?

14. What is the most widely used process for the fabrication of thermoplastic materials (in terms of number of parts produced)?

15. How is injection molding of plastic similar to the die casting of metal?

16. What are some of the benefits of a hot runner distribution system in plastic injection molding?

17. Why is the cycle time for the injection molding of thermosetting polymers significantly longer than that for the thermoplastics?

18. Since no heating is required, what drives the thermoset cure in reaction injection molding?

19. What are some of the attractive consequences of the low temperatures and low pressures of the reaction injection molding process?

20. What are some of the ways by which a polymer extrusion is rapidly cooled?

21. What are some of the typical production shapes that are produced by the extrusion of plastics?

22. How can the extrusion process be used to produce polymeric sheet and film? To produce thin plastic bags?

23. For what types of materials and products might thermoforming be considered to be attractive?

24. What types of products are produced by rotational molding?

25. Describe the rotational molding process.

26. What is the difference between open-cell and closed-cell foamed plastics?

27. What are some typical applications for rigid-type foamed plastics? For flexible foams?

28. What types of products are produced by calendaring?

29. What types of products are produced by the polymer spinning process?

30. What are some of the general properties of plastics that affect their machinability?

31. What are some of the attractive features of laser machining when applied to plastics?

32. What property of plastics is responsible for making snap-fit assembly a popular alternative for plastic products?

33. What are some of the attractive properties of plastics that favor their selection? What are some of the common limitations?

34. What are some cautions regarding the use of standard test data when designing with plastics?

35. What are some of the design concerns when specifying and setting up a plastic molding process?

36. When designing with plastics, why should adequate fillets be included between adjacent sections of a mold? What is a major benefit of rounding exterior corners?

37. Why is it most desirable to have uniform wall thickness in plastic products?

38. Why are product dimensions less precise when they cross a mold parting line?

39. Why might threaded inserts be preferred over other means of producing threaded holes in a plastic component?

40. What are some of the ways in which metal inserts are held in place in a plastic part?

41. When designing a decorative surface (design or lettering) on a plastic product, why is it desirable that the details be raised on the product rather than depressed?

42. Why does locating a parting line on a sharp corner make that feature less noticeable?

43. What is the benefit of countersinking holes that are to be threaded or used for self-tapping screws?

44. What types of products can be produced from elastomeric materials using the dipping process?

45. What process or equipment is used to form rubber compounds into sheets?

46. What property changes occur during vulcanization?

47. What are the two basic classes of ceramic materials, and how does their processing differ?

48. What are some of the most common processes used to shape glass?

49. How are glass fibers produced?

50. What are some of the special heat treatment operations performed on glass products?

51. How are residual compressive stresses imparted during chemical tempering of glass?

52. What are glass-ceramics? How are they produced?

53. What are some of the techniques for compacting crystalline ceramic powders?

54. Describe the differences between the injection molding of plastics and the injection molding of ceramics.

55. Describe the components of the starting slip in ceramic slip casting.

56. What is the difference between slip casting and tape casting?

57. What is the purpose of the firing or sintering operations in the processing of crystalline ceramic products?

58. What is the attractive feature of liquid-phase sintering or reaction sintering?

59. How does cementation differ from sintering?

60. What are the benefits and limitations of machining ceramic materials before firing versus after firing?

61. What are some of the nonconventional methods used to machine ceramics?

62. Why are joining operations usually avoided when fabricating products from ceramic materials?

63. If ceramic products must be joined, what are some of the usual methods?

64. Discuss some of the design guidelines that relate to the production of parts from ceramic material.

65. What are some concerns that may limit the expanded use of composite materials?

66. Why are the processes used to fabricate particulate composites essentially the same as those used for conventional material?

67. How are metals and ceramics combined in the sinter-forge process?

68. What are some of the processes that can be used to produce a high-quality bond between metallic layers in a laminar composite?

69. What conditions might be attractive for fabrication by explosive bonding?

70. What are some examples of a laminated sandwich structure?

71. List several fabrication processes for fiber reinforced products that are essentially the same as for unreinforced plastics. List several that are unique to reinforced materials.

72. What types of materials are used as reinforcing fibers in fiber-reinforced composites?

73. What are some of the forms in which reinforcement fibers appear in composite materials?

74. What are some of the ways that liquefied matrix material can be introduced between the fibers of a fiber-reinforced material?

75. What is a mat? A prepreg?

76. How are tapes produced from filaments?

77. What are sheet-molding compounds (SMC)? Bulk-molding compounds (BMC)? Thick-molding compound?

78. In what way is pultrusion similar to wire drawing?

79. How are the products of pulforming different from those of pultrusion?

80. What are some typical products that are made by filament winding?

81. What are some of the various molding processes that can be used to shape products from laminated sheets of woven fibers?

82. What are the benefits of using an autoclave instead of room-temperature and low-pressure curing?

83. For what conditions might hand lay-up be a preferred process?

84. How can an automated tape-layup machine produce a laminated-ply fiber-reinforced composite?

85. Describe the resin-transfer molding process.

86. What form of reinforcing fibers can be incorporated in the spray molding process?

87. What is the difference between reaction injection molding of fiber-reinforced plastics and reinforced reaction injection molding?

88. What is the major benefit of three-dimensional fiber reinforcement?

89. Describe some of the ways in which a metal matrix can be produced in a fiber-reinforced composite.

90. What is a common property objective in ceramic–matrix composites?

91. Why might it be desirable to have a weak bond between a reinforcing fiber and a ceramic–matrix material?

92. Discuss some of the concerns when cutting or machining fiber-reinforced composites.

93. What types of tools or techniques are commonly used when machining fiber-reinforced composites?

94. What is the major concern when considering the joining of fiber-reinforced composites?

95. What are some of the challenges that currently face composites manufacture?

Problems

1. Consider some of the more prominent sporting goods that are fabricated from composite materials, such as skis, snowboards, tennis rackets, golf club shafts, bicycle frames, and body panels for racing cars. For two specific products, identify composite materials that are currently being used and the companion shape-producing fabrication methods.

2. Figure 20.A depicts the handles of two large wrenches, ratchet wrench and a pipe wrench. These components are traditionally forged from a ferrous alloy, or made from a cast steel or cast iron. For various reasons, alternative materials may be desired. The ratchet wrench is quite long, and reduced weight may be a reasonable desire. Both of these tools could be used in areas, such as a gas leak, where a non-sparking safety tool would be required. Current specifications for the ratchet handle call for a yield strength in excess of 50 ksi and a minimum of 2% elongation in all directions to ensure prevention of brittle fracture. The pipe wrench most likely has similar requirements.

    a. Could a plastic or composite material be used to make a quality product with these additional properties? (NOTE: Metal jaw inserts can be used in the pipe wrench, enabling the other components to be considered as separate pieces.)

    b. If so, how would you propose to manufacture the new handles?

3. Tires are the dominant product of the rubber industry, accounting for nearly ¾ of all rubber material.

    a. What are some of the functions of a vehicle tire?

    b. Modern tires are built from multiple plies of material, where a ply is a layer of rubber-coated cord or fabric. What are some of the different constructions that can be made from these plies?

    c. Tire construction generally consists of three steps—manufacturing the individual components (plies and belts, filler strips, inner lining, tread strips, sidewall, and others), assembling these components in the desired sequence, and then molding and curing to produce an integral piece. At what stage is the tread design imparted to the tire?

    d. Tires are generally cured in the mold. Investigate both the methods, temperatures, and times of heating.

4. Rubber hoses often have to have different properties on the inside and outside surfaces, as well as some form of strengthening reinforcement. The inside surface comes in contact with the material flowing through the hose, while the outer surface must resist environmental conditions. The reinforcement is generally applied at the interface between the inner and outer materials. How might such a product be manufactured?

5. Lavatory wash basins (bathroom sinks) have been successfully made from a variety of engineering materials, including cast iron, steel, stainless steel, ceramics, and polymers (such as melamine). Your company, Diversified Household Products, Inc., is considering a possible entrance into this market and has assigned you the tasks of (1) assessing the competition and (2) recommending the "best" approach toward producing this product.

    a. For each of the materials (or families of materials), describe the material properties that are attractive for a washbasin application. What are the primary limitations or disadvantages?

    b. For each of the materials (or families of materials), describe possible means of fabricating lavatory wash basins. Consider

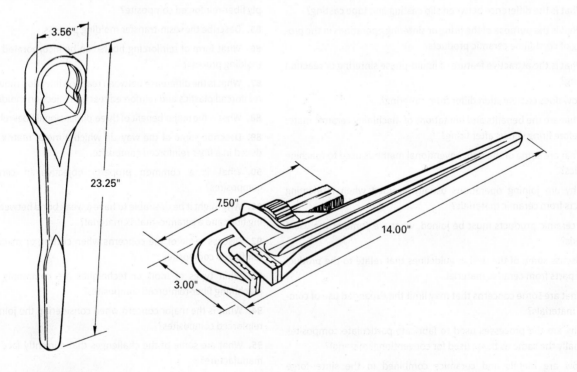

FIGURE 20.A Handle segment of a large ratchet wrench and components of a pipe wrench.

sheet metal forming, casting, molding, joining, and other types of fabrication processes. If multiple options exist, which one do you consider to be most attractive? Comment on the attractive features of the proposed system (materials and process) as well as the relative quality and cost.

c. Wash basins generally require a surface that is nonporous and stain resistant, scratch resistant, corrosion resistant, and attractive (and possibly available in a variety of colors). One approach to providing these properties on a steel or cast iron substrate is a coating of porcelain enamel. For each of the systems discussed in Question 2, discuss the need for additional surface treatment. What type of treatment would you recommend?

d. Most sinks contain an overflow feature that diverts excess water to the drain at a location beneath the stoppered basin. Discuss how this feature can be incorporated into each of your material-process manufacturing systems.

e. If your company were to consider producing lavatory wash basins on a competitive basis, which of the alternative manufacturing systems (material and manufacturing process) would you recommend? What features make it the most attractive?

6. How have the advances in additive manufacturing/3-D printing expanded the fabrication capabilities of the materials discussed in this chapter? Give specific examples of materials, additive process being used, and products being produced. Provide at least one example where two or more materials are being combined to additively produce a composite product.

7. Locate or identify a plastic, ceramic or composite material product where a joining operation has been performed. Briefly discuss the nature of the joint and the process or processes most likely used to produce it.

8. Research and report your findings relating to "machinable" or "weldable" ceramics.

CHAPTER 21

Fundamentals of Machining/ Orthogonal Machining

21.1 Introduction

Machining is the process of removing unwanted material from a workpiece in the form of chips. If the workpiece is metal, the process is often called **metal cutting** or *metal removal*. U.S. industries annually spend billions to perform metal removal operations because the vast majority of manufactured products require machining at some stage in their production, ranging from relatively rough or nonprecision work, such as cleanup of castings or forgings, to high-precision work involving tolerances of 0.0001 in. or less- and high-quality finishes. Thus, machining undoubtedly is the most important of the basic manufacturing processes.

Beginning with the work of F. W. Taylor at Midvale Steel in the 1880s, the process has been the object of considerable research and experimentation that have led to improved understanding of the nature of both the process itself and the surfaces produced by it. Although this research effort led to marked improvements in machining productivity, the complexity of the process has resulted in slow progress in obtaining a complete theory of chip formation. What makes this process so unique and difficult to analyze?

- Prior work-hardening of the material greatly affects the process.
- Different materials deform differently.
- The process is asymmetrical and unconstrained, bounded only by the cutting tool.
- The level of strain is very large.
- The strain rate is very high.
- The process is sensitive to variations in cutting tool geometry, cutting tool material, tool wear, temperature, environment (cutting fluids), and process dynamics (chatter and vibration).

The objective of this chapter is to put all this in perspective for the practicing engineer.

21.2 Fundamentals

The process of metal cutting is complex because it has such a wide variety of inputs (which are outlined in **Figure 21.1**):

- The machine tool selected to perform the process.
- The cutting tool selected (geometry and material).
- The properties and parameters of the workpiece.
- The cutting parameters selected (speed, feed, depth of cut).
- The workpiece holding devices or fixtures or jigs.

As we can see from Figure 21.1, the wide variety of inputs creates a host of outputs, most of which are critical to satisfactory performance of the component and product.

There are seven basic chip formation processes (see **Figure 21.2**): **turning**, **milling**, **drilling**, **sawing**, **broaching**, **shaping** *(planing)*, and **grinding** (also called *abrasive machining*), discussed in Chapters 23 through 28. Chapter 41 describes workholding devices that go into the machine tools and hold the work with respect to the cutting tools. Chapter 22 will provide additional insights into the selection of the cutting tools. The workpiece material is usually specified by the design engineer to meet the functional requirements of the part in service (See Chapter 9). The industrial, lean, or manufacturing engineer will be called on to do the process planning. Therefore the engineer will have to select the machine tool, the cutting tool materials and workholder parameters and then cutting parameters based on that work material decision. Let us begin with the assumption that the workpiece material has been selected as part of the design. To make the component, you decided to use a high-speed steel cutting tool for a turning operation (see **Figure 21.3**).

For all metal-cutting processes, it is necessary to determine the cutting parameters of speed, feed, and depth of cut. The turning process will be used to introduce these terms. In general, **speed** (*V*) is the primary cutting motion, which relates the velocity of the cutting tool relative to the **workpiece**. It is

INPUTS

Machine tool selection

- Lathe
- Milling machine
- Drill press
- Grinder
- Saw
- Broach
- Machining centers

Workpiece parameters

Predeformation (work hardening prior to machining)
Metal type
- BCC, FCC, HCP
- Stacking fault energy
- Purity

Cutting parameters

Depth of cut $\dfrac{D_1 - D_2}{2}$
Speed
Feed fr
Environment
- Oxygen
- Lubricant
- Temperature

Workholder

Fixtures
Jigs
Chucks
Collets

Cutting tool parameters

Tool design geometry
- Tool angles
- Nose radius
- Edge radius
- Material
- Hardness
- Finish
- Coating

Machining processes

I. Oblique (three-force) model

- Single-point cutting
- Multiple-edge tools

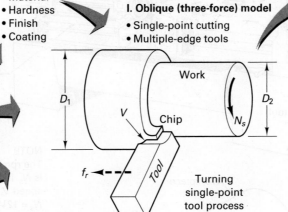

Turning single-point tool process

II. Orthogonal (two-force) model

- Macroindustrial studies performed on plates and tubes
- Microstudies carried out in microscopes using high-speed photography

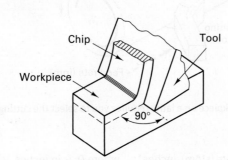

OUTPUTS

Measurements

Cutting forces
Chip dimensions
- Optical
- SEM
Shear process
Power
Surface finish
Tool wear, failures
Deflections
Temperatures
Vibrations
Part geometry

Determinations

Specific horsepower, HP_s
Flow stress, τ_s
Chip ratios, r_c
Shear front directions, ψ
Velocities (chip, shear)
Temperatures
Friction coefficients, μ
Strains, γ
Strain rates, $\dot{\gamma}$
Cutting stiffness, K_s
Heat in tool, chip, workpiece
Onset of shear direction ϕ

FIGURE 21.1 The fundamental inputs and outputs to machining processes.

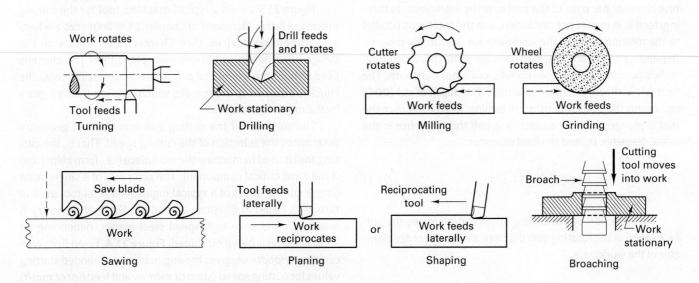

Turning

Drilling

Milling

Grinding

Sawing

Planing

Shaping

Broaching

FIGURE 21.2 The basic machining processes used in chip formation are widely varied.

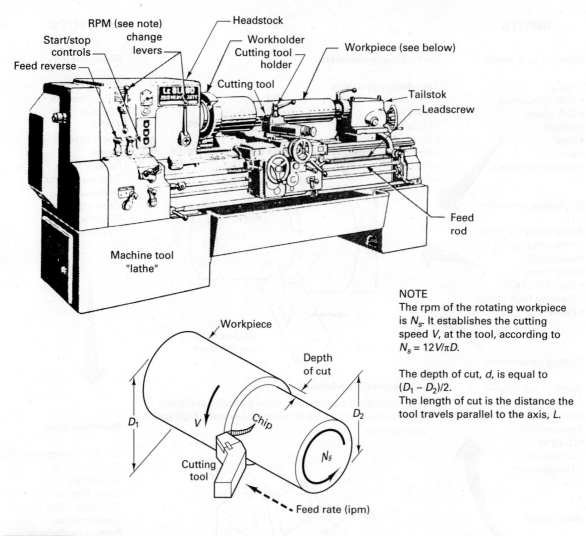

NOTE
The rpm of the rotating workpiece is N_s. It establishes the cutting speed V, at the tool, according to $N_s = 12V/\pi D$.

The depth of cut, d, is equal to $(D_1 - D_2)/2$.
The length of cut is the distance the tool travels parallel to the axis, L.

FIGURE 21.3 Turning a cylindrical workpiece on a lathe requires you to select the cutting speed, feed, and depth of cut.

generally given in units of surface feet per minute (sfpm), inches per minute (in./min), meters per minute (m/m), or meters per second (m/s). Speed or velocity (V) is shown with the heavy dark arrow. **Feed** (f_r) is the amount of material removed per revolution or per pass of the tool over the workpiece. In turning, feed is in inches per revolution, and the tool feeds parallel to the rotational axis of the workpiece for the turning process. Depending on the process, feed units are inches per revolution, inches per cycle, inches per minute, or inches per tooth. The feed rate is shown with dashed arrows. The **depth of cut (DOC)** represents the third dimension. In turning, it is the distance the tool is plunged into the surface. It is half the difference in the initial diameter, D_1, and the final diameter, D_2:

$$\text{DOC} = \frac{D_1 - D_2}{2} = d \qquad (21.1)$$

The selection of the cutting speed V determines the surface speed of the rotating part that is related to the outer diameter of the workpiece.

$$V = \frac{\pi D_1 N_s}{12} \qquad (21.2)$$

where D_1 is in inches, V is speed in surface feet per minute, and N_s is the revolutions per minute (rpm) of the workpiece. The operator inputs the desired revolutions per minute (rpm) of the spindle into the lathe to produce the desired cutting velocity.

Figure 21.3 shows a typical **machine tool** for the turning process, a lathe, discussed in Chapter 23. Workpieces are held in **workholding devices**. (See Chapter 41 for details on the design of workholders.) In this example, a three-jaw chuck is used to hold the workpiece and rotate it against the tool. The chuck is attached to the spindle, which is driven through gears by the motor in the lathe.

The selection of the cutting tool material and geometry determines the selection of the cutting speed. That is, the **cutting tool** is used to machine the workpiece (i.e., form chips) and is the most critical component. The geometry of a single point (single cutting edge) of a typical high-speed steel tool used in turning is found in Chapters 22 and 23. The tool geometry is usually ground onto high-speed steel blanks, depending on what material is being machined. **Figure 21.4**, taken from *Metcut's Data Handbook*, gives the engineer recommended starting values for cutting speed (sfpm or m/min) and feed (ipr or mm/r) for a given depth of cut, a given work material (hardness), and

Turning, Single Point and Box Tools

Material	Hard-ness Bhn	Condition	Depth of Cut* (in / mm)	HSS Speed (fpm m/min)	HSS Feed (ipr mm/r)	HSS Tool Material AISI ISO	Uncoated Speed Brazed (fpm m/min)	Uncoated Speed Indexable (fpm m/min)	Uncoated Feed (ipr mm/r)	Uncoated Tool Material Grade C ISO	Coated Speed (fpm m/min)	Coated Feed (ipr mm/r)	Coated Tool Material Grade C ISO
1. FREE MACHINING CARBON STEELS, WROUGHT (cont.) **Medium Carbon Leaded** **(cont.)** (materials listed on preceding page)	225 to 275	Hot Rolled, Normalized, Annealed, Cold Drawn or Quenched and Tempered	.040	160	.008	M2, M3	500	610	.007	C-7	925	.007	CC-7
			.150	125	.015	M2, M3	390	480	.020	C-6	600	.015	CC-6
			.300	100	.020	M2, M3	310	375	.030	C-6	500	.020	CC-6
			.625	80	.030	M2, M3	240	290	.040	C-6	—	—	—
			1	49	.20	S4, S5	150	185	.18	P10	280	.18	CP10
			4	38	.40	S4, S5	120	145	.50	P20	185	.40	CP20
			8	30	.50	S4, S5	95	115	.75	P30	150	.50	CP30
			16	24	.75	S4, S5	73	88	1.0	P40	—	—	—
	275 to 325	Hot Rolled, Normalized, Annealed or Quenched and Tempered	.040	135	.007	T15, M42†	460	545	.007	C-7	825	.007	CC-7
			.150	105	.015	T15, M42†	350	425	.020	C-6	525	.015	CC-6
			.300	85	.020	T15, M42†	275	380	.030	C-6	425	020	CC-6
			.625	—	—		—	—					
			1	41	.18	S9, S11†	140	165	.18	P10	250	.18	CP10
			4	32	.40	S9, S11†	105	130	.50	P20	160	.40	CP20
			8	26	.50	S9, S11†	84	100	.75	P30	130	.50	CP30
			16	—	—		—	—					
	325 to 375	Quenched and Tempered	.040	100	.007	T15, M42†	390	480	.007	C-7	725	.007	CC-7
			.150	80	.015	T15, M42†	300	375	.020	C-6	475	.015	CC-6
			.300	65	.020	T15, M42†	230	290	.030	C-6	375	.020	CC-6
			.625	—									
			1	30	.18	S9, S11†	120	145	.18	P10	220	.18	CP10
			4	24	.40	S9, S11†	90	115	.50	P20	145	.40	CP20
			8	20	.50	S9, S11†	70	88	.75	P30	115	.50	CP30
			16	—									
	375 to 425	Quenched and Tempered	.040	70	.007	T15, M42†	325	400	.007	C-7	600	.007	CC-7
			.150	55	.015	T15, M42†	250	310	.020	C-6	400	.015	CC-6
			.300	45	.020	T15, M42†	200	240	.030	C-6	325	.020	CC-6
			.625	—									
			1	21	.18	S9, S11†	100	120	.18	P10	185	.18	CP10
			4	17	.40	S9, S11†	76	95	.50	P20	120	.40	CP20
			8	14	.50	S9, S11†	60	73	.75	P30	100	.50	CP30
			16	—									
2. CARBON STEELS, WROUGHT **Low Carbon** 1005 1010 1020 1006 1012 1023 1008 1015 1025 1009 1017	85 to 125	Hot Rolled, Normalized, Annealed or Cold Drawn	.040	185	.007	M2, M3	535	700	.007	C-7	1050	.007	CC-7
			.150	145	.015	M2, M3	435	540	.020	C-6	700	.015	CC-6
			.300	115	.020	M2, M3	340	420	.030	C-6	550	.020	CC-6
			.625	90	.030	M2, M3	265	330	.040	C-6	—	—	—
			1	56	.18	S4, S5	165	215	.18	P10	320	.18	CP10
			4	44	.40	S4, S5	135	165	.50	P20	215	.40	CP20
			8	35	.50	S4, S5	105	130	.75	P30	170	.50	CP30
			16	27	.75	S4, S5	81	100	1.0	P40	—	—	—
	125 to 175	Hot Rolled, Normalized, Annealed or Cold Drawn	.040	150	.007	M2, M3	485	640	.007	C-7	950	.007	CC-7
			.150	125	.015	M2, M3	410	500	.020	C-6	625	.015	CC-6
			.300	100	.020	M2, M3	320	390	.030	C-6	500	.020	CC-6
			.625	80	.030	M2, M3	245	305	.040	C-6	—	—	—
			1	46	.18	S4, S5	150	195	.18	P10	290	.18	CP10
			4	38	.40	S4, S5	125	150	.50	P20	190	.40	CP20
			8	30	.50	S4, S5	100	120	.75	P30	150	.50	CP30
			16	24	.75	S4, S5	75	95	1.0	P40	—	—	—
	175 to 225	Hot Rolled, Normalized, Annealed or Cold Drawn	.040	145	.007	M2, M3	460	570	.007	C-7	850	.007	CC-7
			.150	115	.015	M2, M3	385	450	.020	C-6	550	.015	CC-6
			.300	95	.020	M2, M3	300	350	.030	C-6	450	.020	CC-6
			.625	75	.030	M2, M3	235	265	.040	C-6	—	—	—
			1	44	.18	S4, S5	140	175	.18	P10	260	.18	CP10
			4	35	.40	S4, S5	115	135	.50	P20	170	.40	CP20
			8	29	.50	S4, S5	90	105	.75	P30	135	.50	CP30
			16	23	.75	S4, S5	72	81	1.0	P40	—	—	—
	225 to 275	Annealed or Cold Drawn	.040	125	.007	M2, M3	410	510	.007	C-7	750	.007	CC-7
			.150	95	.015	M2, M3	360	400	.020	C-6	500	.015	CC-6
			.300	75	.020	M2, M3	285	315	.030	C-6	400	.020	CC-6
			.625	60	.030	M2, M3	220	240	.040	C-6	—	—	—
			1	38	.18	S4, S5	125	155	.18	P10	230	.18	CP10
			4	29	.40	S4, S5	110	120	.50	P20	150	.40	CP20
			8	23	.50	S4, S5	87	95	.75	P30	120	.50	CP30
			16	18	.75	S4, S5	67	73	1.0	P40	—	—	—

See section 15.1 for Tool Geometry.
*Caution: Check Horsepower requirements on heavier depths of cut.

See section 16 for Cutting Fluid Recommendations.
†Any premium HSS (T15,M33,M41–M47) or (S9,S10,S11,S12).

FIGURE 21.4 Examples of a table for selection of speed and feed for turning. (*Source:* Metcut's Machinability Data Handbook)

a given process (turning). Notice how speed decreases as DOC or feed increases. Cutting speeds can be increased with carbide and coated-carbide tool materials. Thus, we see that to process different metals, the input parameters to the machine tools must be determined. For the lathe, the input parameters are DOC, the feed rate, and the rpm value of the spindle. The rpm value depends on the selection of the cutting speed, V. Rewriting equation for N_s:

$$N_s = \frac{12V}{\pi D_1} \cong \frac{3.8V}{D_1} \qquad (21.3)$$

The selection of values of cutting speed, feed rate, and DOC depend on many factors, and a great deal of experience and experimentation are required to find the best combinations. Tables of recommended values, as shown in Figure 21.4 are good starting points. Tables like this one are arranged according to the process being used, the material being machined, the hardness, and the cutting tool material. The table given is only a sample, to be used only for solving turning problems in the book. For industrial calculations, standard references listed at the end of the book or cutting tool manufacturers should be consulted.

This table is for turning processes only. The amount of metal removed per pass is called the **metal removal rate**, and it depends on V, f_r, and DOC. In practice, roughing cuts

are heavier than finishing cuts in terms of DOC and feed and are run at a lower surface speed. Note that this table provides recommendations of V and f_r in both English and metric units based on the DOC needed to perform the job. Table values are usually conservative and should be considered starting points for determining the operational parameters for a process.

Once cutting speed (V) has been selected, equation 21.3 is used to determine the spindle rpm, N_s. The speed and feed can be used with the DOC to estimate the metal removal rate for the process, or MRR. For turning, the MRR is

$$MRR \cong 12Vf_r d \left(in.^3/min \right) \qquad (21.4)$$

This is an approximate equation for MRR. For turning, MRR values can range from 0.1 to 600 in.^{3}/min. The MRR can be used to estimate the horsepower needed to perform a cut, as will be shown later. For most processes, the MRR equation can be viewed as the volume of metal removed divided by the time needed to remove it:

$$MRR = \frac{Volume\ of\ cut}{T_m} \left(\frac{in.^3}{min} \right)$$

where T_m is the cutting time in minutes. For turning, the cutting time depends on the length of cut L divided by the rate of

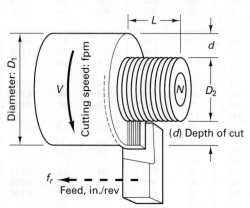

Turning

Speed, stated in surface feet per minute (sfpm), is the peripheral speed at the cutting edge. Feed per revolution in turning is a linear motion of the tool parallel to the rotating axis of the workpiece. The depth of cut reflects the third dimension.

L = length of cut
$$T_m = \frac{L + A}{f_r N_s}$$

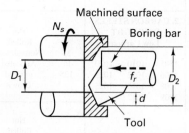

Boring

Enlarging hole of diameter D_1 to diameter D_2. Boring can be done with multiple point cutting tools. Feed in inches per revolution, f_r.

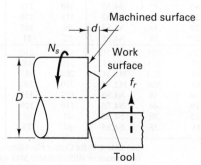

Facing

Tool feeds to center of workpiece so L = D/2. The cutting speed is decreasing as the tool approaches the center of the workpiece.

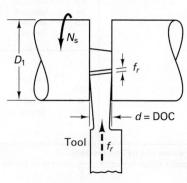

Grooving, parting, or cutoff

Tool feed perpendicular to the axis of rotation. The width of the tool produces the depth of cut (DOC).

FIGURE 21.5 Relationship of speed, feed, and depth of cut in turning, boring, facing, and cutoff operations typically done on a lathe.

traverse of the cutting tool past the rotating workpiece $f_r N_s$ in inches/minute, as shown in **Figure 21.5**. Therefore,

$$T_m = \frac{L + \text{Allowance}}{f_r N_s}(\text{min}) \qquad (21.5)$$

An allowance is usually added to the L term to allow for the tool to enter and exit the cut.

Basic Machining Processes

Turning is an example of a single-point tool process, as is shaping, milling, and drilling are examples of multiple-point tool processes. **Figures 21.5** through **21.8** show schematics of the basic processes. Speed (V) is shown in these figures with a dark heavy arrow. Feed in inches per revolution is the amount of material removed per pass of the tool over the workpiece and is shown as a dashed arrow.

Table 21.1 provides a summary of the equations for T_m and MRR for these basic processes. These equations are commonly referred to as **shop equations** and are as fundamental as the processes themselves, so the student should be as familiar with them as with the basic processes. If one keeps track of the units and visualizes the process, the equations are, for the most part, straightforward.

In addition to turning, other operations can be performed on the lathe. For example, as shown in Figure 21.5, a flat surface on the rotating part can be produced by facing or a cutoff operation. **Boring** can produce an enlarged hole after drilling and grooving puts a slot in the workpiece.

The process of milling requires two figures because it takes different forms depending on the selection of the machine tool and the cutting tool. Milling, a multiple-tooth process, has two feeds: the amount of metal an individual tooth removes, called the feed per tooth, f_t, and the rate at which the table translates past the rotating tool, called the table feed rate, f_m, in inches per minute. It is calculated from

$$f_m = f_t n N_s (\text{in.}/\text{min}) \qquad (21.6)$$

where n is the number of teeth in a cutter and N_s is the rpm value of the cutter. Just as was shown for turning, standard tables of speeds and feeds for milling provide values for the recommended cutting speeds and feeds per tooth, f_t.

Table 21.2 provides a summary on typical sizes (minimum–maximum), the production rates (part/hour), tolerances (precision or repeatability), and surface finish (roughness) for many of the machining processes. Milling has pretty much replaced shaping and planing, although gear shaping is still a viable process. Milling combined with other rotational multiple-edge tool processes (drilling or reaming) is often performed in machining centers rather than on milling machines. Turret lathe has been replaced by CNC turning centers with multiple turrets in many factories. See discussions in Chapters 23 and 26.

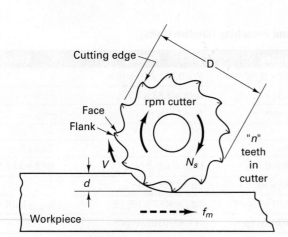

Slab milling—multiple tooth

Slab milling is usually performed on a horizontal milling machine. Equations for T_m and MRR derived in Chapter 24.

The tool rotates at rpm N_s. The workpiece translates past the cutter at feed rate f_m, in inches per minute, the table feed. The length of cut, L, is the length of workpiece plus allowance, L_A, where

FIGURE 21.6 Basics of milling processes for slab and face milling including equations for cutting time and metal removal rate (MRR).

$$L_A = \sqrt{\frac{D^2}{4} - \left(\frac{D}{2} - d\right)^2} = \sqrt{d(D - d)} \text{ inches}$$

$$T_m = (L + L_A)/f_m$$

The MRR = $W d f_m$ where W = width of the cut and d = depth of cut.

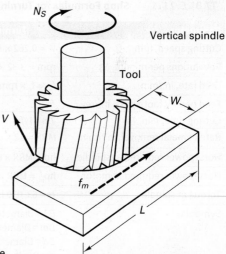

Face milling—Multiple-tooth cutting

Given a selected cutting speed V and a feed per tooth f_t, the rpm of the cutter is $N_s = 12V/\pi D$ for a cutter of diameter D. The table feed rate is $f_m = f_t n N_s$ for a cutter with n teeth.
The cutting time, $T_m = (L + L_A + L_o)/f_m$ where $L_o = L_A = \sqrt{W(D - W)}$ for $W < D/2$ or $L_o = L_A = D/2$ for $W \geqslant D/2$.
The MRR = $W d f_m$ where d = depth of cut.

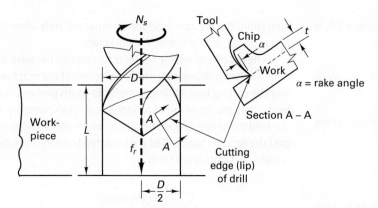

Select cutting speed V, fpm and feed, f_r, in./rev. Select drill, type and material. D = diameter of the drill, which rotates 2 cutting edges at rpm N_s. V = velocity of outer edge of the lip of the drill.
$N_s = 12V/\pi D$. T_m = cutting time = $(L + A)/f_rN_s$ where f_r is the feed rate in in. per rev. The allowance $A = D/2$.
The MRR = $(\pi D^2/4)f_rN_s$ in.^{3}/min which is approximately $3DVf_r$.

FIGURE 21.7 Basics of the drilling (hole-making) processes, including equations for cutting time and metal removal rate (MRR).

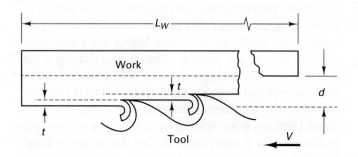

The T_m for broaching is $T_m = L/12V$. The MRR (per tooth) is $12tWV$ in.^{3}/min where V = cutting velocity in fpm, W is the width of cut, t = rise per tooth.

FIGURE 21.8 Process basics of broaching. Equations for cutting time and metal removal rate (MRR) are developed in Chapter 27.

TABLE 21.1	Shop Formulas for Turning, Milling, Drilling, and Broaching (English Units)			
Parameter	**Turning**	**Milling**	**Drilling**	**Broaching**
Cutting speed, fpm	$V = 0.262 \times D_1 \times$ rpm	$V = 0.262 \times D_m \times$ rpm	$V = 0.262 \times D_d \times$ rpm	V
Revolutions per minute, Ns	rpm $= 3.82 \times V_c / D_1$	rpm $= 3.82 \times V_c / D_m$	rpm $= 3.82 \times V_c / D_d$	—
Feed rate, in./min	$f_m = f_r \times$ rpm	$f_m = f_r \times$ rpm	$f_m = f_r \times$ rpm	—
Feed per rev tooth pass, in./rev	fr	ft	fr	—
Cutting time, min, Tm	$Tm = L/fm$	$Tm = L/fm$	$Tm = L/fm$	$Tm = L/12V$
Rate of metal removal, in.^{3}/min	MRR $= 12 \times d \times f_r \times V_c$	MRR $= w \times d \times f_m$	MRR $= \pi D^2 d / 4 \times f_m$	MRR $= 12 \times w \times d \times V$
Horsepower required at spindle	hp $=$ MRR $\times$ HP$_s$	hp $=$ MRR $\times$ HP$_s$	hp $=$ MRR $\times$ HP$_s$	—
Horsepower required at motor	hp$_m =$ MRR $\times$ HP$_s / E$	hp$_m =$ MRR $\times$ HP$_s / E$	hp$_m =$ MRR $\times$ HP$_s / E$	hp$_m =$ MRR $\times$ HP$_s / E$
Torque at spindle	ts $= 63,030$ hp/rpm	ts $= 63,030$ hp/rpm	ts $= 63,030$ hp/rpm	—
Symbols	$D1$ = Diameter of workpiece in turning, inches Dm = Diameter of milling cutter, inches Dd = Diameter of drill, inches d = Depth of cut, inches E = Efficiency of spindle drive fm = Feed rate, inches per minute fr = Feed, inches per revolution ft = Feed, inches per tooth hp$_m$ = Horsepower at motor MRR = Metal removal rate, in.^{3}/min		hp = horsepower at spindle L = Length of cut, inches n = Number of teeth in cutter HP$_s$ = Unit power, horsepower per cubic inch per minute, specific horsepower Ns = Revolution per minute of work or cutter ts = Torque at spindle, inch-pound Tm = Cutting time, minutes V = Cutting speed, feet per minute w = Width of cut, inches	

Values for specific horsepower (unit power) are given in Table 21.3.

TABLE 21.2 Basic Machining Process

Applicable Process	Raw Material Form	Size Maximum	Size Maximum	Typical Production Rate	Material Choice	Typical Tolerance	Typical Surface Roughness
Turning (engine lathes)	Cylinders preforms, castings forgings	78 in. dia. × 73 in. long	$\frac{1}{64}$ in. typical	1–10 parts/hour	All ferrous and nonferrous material considered machinable	±0.002 in. on dia. common; ±0.001 in. obtainable	125–250
Turning (CNC)	Bar, rod, tube preforms	36 in. dia. × 93 in. long	$\frac{1}{64}$ in. dia.	1–2 parts/minute to 1–4 parts/hour	Any material with good machinability rating	±0.001 in. on dia. where needed; ±0.0005 in. possible	63 or better
Turning (automatic screw machine)	Bar, rod	Generally 2 in. dia. × 6 in. long	$\frac{1}{16}$ in. dia. and less, weight less than 1 ounce	10–30 parts/minute	Any material with good machinability rating ±0.001 to ±0.003 in.	±0.0005 in. possible ±0.001 to ±0.003 in. common	63 average
Turning (Swiss automatic machining)	Rod	Collets adapt to $\frac{1}{2}$ in. dia.	Collets adapt to less than $\frac{1}{2}$ in.	12–30 parts/minute	Any material with good machinability rating	±0.0002 in. to ±0.001 in. common	63 and better
Boring (vertical)	Casting, preforms	98 in. × 72 in.	2 in. × 12 in.	2–20 hours/piece	All ferrous and nonferrous	±0.0005 in.	90–250
Milling	Bar, plate, rod, tube	4–6 ft long	Limited usually by ability to hold part	1–100 parts/hour	Any material with good machinability rating	±0.0005 in. possible; ±0.001 in. common	63–250
Hobbing (milling gears)	Blanks, preforms, rods	10-ft-dia. gears 14-in. face width	0.100 in. dia.	1 part/minute	Any material with good machinability rating	±0.001 in. or better	63
Drilling	Plate, bar, preforms	$3\frac{1}{2}$-in.-dia.drills (1-in.-dia. normal)	0.002-in. drill dia.	2–20 second/hole after setup	Any unhardened material; carbides needed for some case-hardened parts	±0.002–±0.010 in. common; ±0.001 in. possible	63–250
Sawing	Bar, plate, sheet	2-in. armor plate ($\frac{1}{2}$ in. is preferred)	0.010 in. thick	3–30 parts/hour	Any nonhardened material	±0.015 in. possible	250–1000
Broaching	Tube, rod, bar, plate	74 in. long	1 in.	300–400 parts/minute	Any material with good machinability rating	±0.0005–±0.001 in.	32–125
Grinding	Plate, rod, bars	36 in. wide × 7 in. dia.	0.020 in. dia.	1–1000 pieces/hour	Nearly all metallic materials plus many nonmetallic	0.0001 in. and less	16
Shaping	Bar, plate, casting	3 ft × 6 ft	Limited usually by ability to hold part	1–4 parts/hour	Low- to medium-carbon steels and nonferrous metals best; no hardened parts	±0.001–±0.002 in. (larger parts) ±0.0001–±0.0005 in. (small–medium parts)	63–250
Planing	Bar, plate, casting	42 ft wide × 18 ft high × 76 ft long	Parts too large for shaper work	1 part/hour	Low- to medium-carbon steels or nonferrous materials best	±0.001–±0.005 in.	63–125
Gear shaping	Blanks	120-in.-dia. gears 6-in. face width	1 in. dia.	1–60 parts/hour	Any material with good machinability rating	±0.001 in. or better at 200 D.P. to 0.0065 in. at 30 D.P.	63

21.3 Forces and Power in Machining

Most of the cutting operations process described to this point are examples of oblique or three-force cutting. The cutting force system in a conventional, oblique-chip formation process is shown schematically in **Figure 21.9**. Oblique cutting has three components:

1. F_c: Primary cutting force acting in the direction of the cutting velocity vector. This force is generally the largest force and accounts for 99% of the power required by the process.

2. F_f: Feed force acting in the direction of the tool feed. This force is usually about 50% of F_c but accounts for only a small percentage of the power required because feed rates are usually small compared to cutting speeds.

3. F_r: radial or thrust force acting perpendicular to the machined surface. This force is typically about 50% of F_r and contributes very little to power requirements because

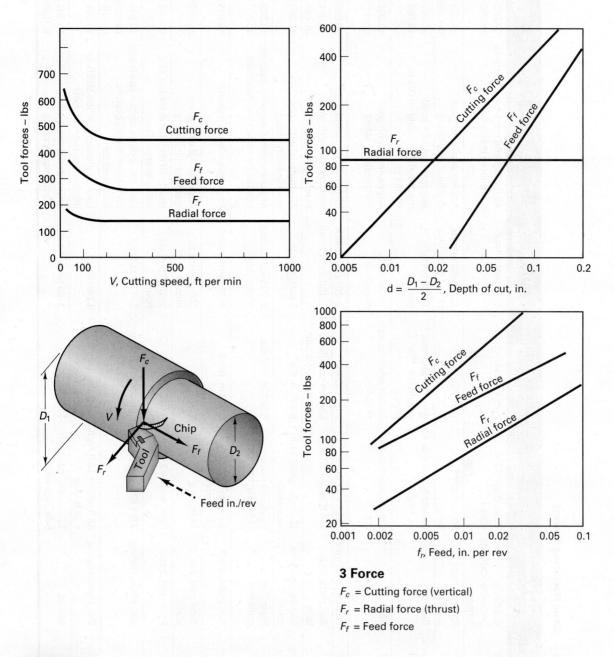

3 Force

F_c = Cutting force (vertical)

F_r = Radial force (thrust)

F_f = Feed force

FIGURE 21.9 Oblique (three-force) machining has three measurable components of forces acting on the tool. The forces vary with speed, depth of cut, and feed.

velocity in the radial direction is negligible. Figure 21.9 shows the general relationship between these forces and changes in speed, feed, and depth of cut. Note that these figures cannot be used to determine forces for a specific process.

The power required for cutting is

$$P = F_c V \left(\text{ft} - \text{lb}/\text{min} \right) \qquad (21.7)$$

The horsepower at the spindle of the machine is therefore

$$\text{hp} = \frac{F_c V}{33{,}000} \qquad (21.8)$$

In metal cutting, a very useful parameter is called the unit, or specific, horsepower, HP_s, which is defined as

$$HP_s = \frac{\text{hp}}{\text{MRR}} \left(\text{hp}/\text{in.}^3/\text{min} \right) \qquad (21.9)$$

In turning, for example, where $\text{MRR} \cong 12 V f_r d$, then

$$HP_s = \frac{F_c}{396{,}000 f_r d} \qquad (21.10)$$

Thus, this term represents the approximate power needed at the spindle to remove a cubic inch of metal per minute.

Values for **specific horsepower, HP_s**, which is also called unit power, are given in **Table 21.3**. These values are obtained through orthogonal metal-cutting experiments described later in this chapter.

Specific power can be used in a number of ways. First, it can be used to estimate the motor horsepower required to perform a machining operation for a given material.

HP_s values from the table are multiplied by the approximate MRR for the process. The motor horsepower HP_m, is then

$$HP_m = \frac{HP_s \times \text{MRR} \times CF}{E} \qquad (21.11)$$

where E is the efficiency of the machine. The E factor accounts for the power needed to overcome friction and inertia in the machine and drive moving parts. Usually, 80% is used. Usually the maximum MRR is used in this calculation. Correction factors (CFs) may also be used to account for variations in cutting speed, feed, and rake angle. There is usually a tool wear correction factor of 1.25, used to account for the fact that dull tools use more power than sharp tools.

The primary cutting force F_c can be roughly estimated according to

$$F_c \cong \frac{HP_s \times \text{MRR} \times 33{,}000}{V} \qquad (21.12)$$

This type of estimate of the major force F_c is useful in analysis of deflection and vibration problems in machining and in the proper design of workholding devices because these devices must be able to resist movement and deflection of the part during the process.

In general, increasing the speed, the feed, or the depth of cut will increase the power requirement. Doubling the speed doubles the horsepower directly. Doubling the feed or the depth of cut doubles the cutting force, F_c. In general, increasing the speed does not increase the cutting force, F_c, a surprising experimental result. However, speed has a strong effect on tool life because most of the input energy is converted into heat, which raises the temperature of the chip, the work, and the tool—to the latter's detriment. Tool life (or tool death) is discussed in Chapter 23.

Equation 21.13 can be used to estimate the maximum depth of cut, d, for a process as limited by the available power.

$$d_{max} = \frac{HP_m \times E}{12 HP_s V F_r \left(CF \right)} \qquad (21.13)$$

Another handbook value useful in chatter or vibration calculations is **cutting stiffness, K_s**. In this text, the term *specific energy, U,* will be used interchangeably with cutting stiffness, K_s, a term used in machining dynamics.

It is interesting to compute the total specific energy in the process and determine how it is distributed between the primary shear and the secondary shear that occurs at the interface between the chip and the tool. It is safe to assume that the majority of the input energy is consumed by these two regions.

Therefore,

$$U = U_s + U_f \qquad (21.14)$$

where specific energy (also called cutting stiffness) is

$$U = \frac{F_c V}{V f_r d} = \frac{F_c}{f_r d} = K_s \left(\text{turning} \right) \qquad (21.15)$$

Usually, 30 to 40% of the total energy goes into friction and 60 to 70% into the shear process.

Typical values for U are given in Table 21.3. This is experimental data developed by the orthogonal machining experiment described later.

In the metal-cutting process, the energy put into the process ($F_c V$) is largely converted into heat, elevating the temperatures of the chips, the tool, and the workpiece—all of which, along with the environment (the cutting fluids) act as heat sinks. In the next chapter, the effect of cutting speed on temperature will be discussed as a prime cause of tool wear and tool failure. The data that appear in tables like Table 21.1 for recommended speeds are usually based on the expected tool life of 30 to 60 min and are developed experimentally.

TABLE 21.3   Values for Unit Power and Specific Energy (Cutting Stiffness)

Material		Unit Power (hp-min. in.³) HP$_s$	Specific Energy (in.-lb/in.³) Ks or U	Hardness Brinell HB
Nonalloy carbon steel	C 0.15%	.58	268,000	125
	C 0.35%	.58	302,400	150
	C.0.60%	.75	324,800	200
Alloy steel	Annealed	.50	302,400	180
	Hardened and tempered	0.83	358,400	275
	Hardened and tempered	0.87	392,000	300
	Hardened and tempered	1.0	425,000	350
High-alloy steel	Annealed	0.83	369,000	200
	Hardened	1.2	560,000	325
Stainless steel, annealed	Martensitic/ferritic	0.75	324,800	200
Steel castings	Nonalloy	0.62	257,000	180
	Low-alloy	0.67	302,000	200
	High-alloy	0.80	336,000	225
Stainless steel, annealed	Austenitic	0.73	369,600	180
Heat-resistant alloys	Annealed	0.78		200
	Aged—Iron based	—		280
	Annealed—Nickel or cobalt	1.10		250
	Aged	1.20		350
Hard steel	Hardened steel	1.4	638,400	55HR$_c$
	Manganese steel 12%	1.0	515,200	250
Malleable iron	Ferritic	0.42	156,800	130
	Pearlitic	—	257,600	230
Cast iron, low tensile		0.62	156,800	180
Cast iron, high tensile		0.80	212,800	260
Nodular SG iron	Ferritic	0.55	156,800	160
	Pearlitic	0.76	257,600	250
Chilled cast iron		—	492,800	400
Aluminum alloys	Non-heat-treatable	.25	67,200	60
	Heat-treatable	.33	100,800	100
Aluminum alloys (cast)	Non-heat-treatable	.25	112,000	75
	Heat-treatable	.33	123,200	90
Bronze-brass alloys	Lead alloys, Pb>1%	.25	100,800	110
	Brass, cartridge brass	1.8–2.0	112,000	90
	Bronze and lead-free copper	0.33–0.83		
	Includes Electrolytic copper	0.90	246,400	100
Zinc alloy	Diecast	0.25	—	—
Titanium		.034	250,275	

Values assume normal feed ranges and sharp tools. Multiply values by 1.25 for a dull tool.

Calculation of unit power (HP$_s$)

    $HP = F_cV/33,000$

    $HP_s = HP/MRR$ Where

    $MRR = 12Vtw$ for tube turning

    $HP_s = F_cV/12Vtw \times 33,000 = F_c/tw \times 396,000$

Calculation of specific energy (U)

    $U = F_cV/Vtw = F_c/tw$ for tube turning

21.4 Orthogonal Machining (Two Forces)

Orthogonal machining (OM) is carried out mostly in research laboratories to better understand this complex process. In OM, the tool geometry is simplified from the three-dimensional (oblique) geometry, to a two-force geometry, as shown in Figure 21.1.

Using this simplified tool geometry, metals can be cut to test machining mechanics and theory and develop machining values for specific power and energy given in Table 21.2 and Table 21.3. There are basically three orthogonal machining setups.

1. Orthogonal plate machining (OPM), or machining a plate in a milling machine—low-speed cutting. See **Figure 21.10**.
2. Orthogonal tube turning (OTT), or end-cutting a tube wall in a turning setup—medium-speed ranges. See **Figure 21.11**.

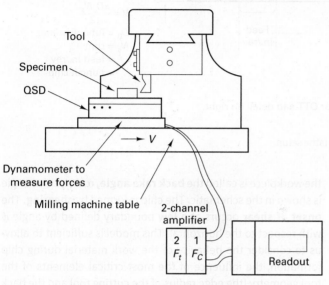

(a) OPM (Front view) of setup.

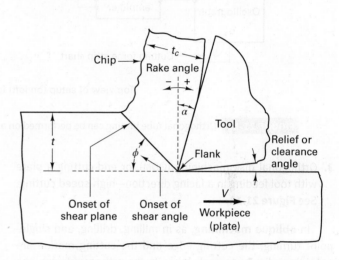

(b) Side view of orthogonal plate machining with fixed tool and moving plate. The input area is t, the uncut chip thickness and w, the plate thickness.

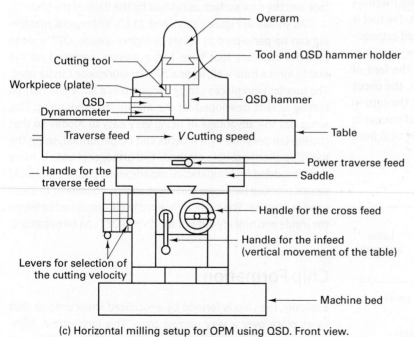

(c) Horizontal milling setup for OPM using QSD. Front view.

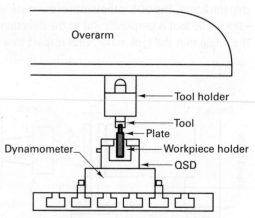

(d) End view of OPM setup on horizontal mill. Table feed is used for cutting speed.

FIGURE 21.10 Schematics of the orthogonal plate machining setup on a horizontal milling machine, using a quick-stop device (QSD) and dynamometer. The table feed mechanism is used to provide a low speed cut.

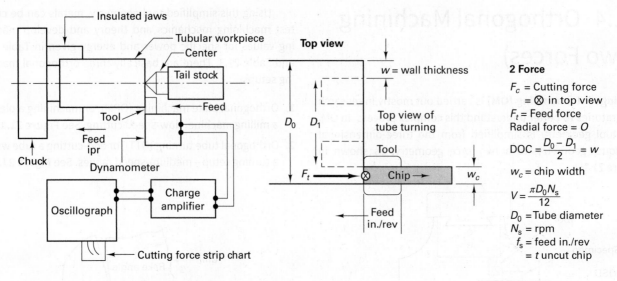

Top view of setup (on left) for OTT and detail on right.

FIGURE 21.11 Orthogonal tube turning can be performed on a lathe setup.

3. Orthogonal disk machining (ODM), or end-cutting a plate with tool feeding in a facing direction—high-speed cutting. See **Figure 21.12**.

In **oblique machining**, as in milling, drilling, and single-point turning, the cutting edge and the cutting motion are not perpendicular to each other. In the orthogonal case, the cutting velocity vector and the cutting edge are perpendicular. The OPM low-speed plate machining is shown in detail in Figure 21.10, using a modified horizontal milling machine where the table traverse provides the cutting speed. The tool is mounted in a tool holder in the overarm. Low-speed orthogonal plate machining uses a flat plate setup in a milling machine. The workpiece moves past the tool at velocity V. The feed of the work vertically up into the tool is now called t, the uncut chip thickness. The DOC is the width of the plate, w. The cutting edge of the tool is perpendicular to the direction of motion V. The angle that the tool makes with respect to a vertical from

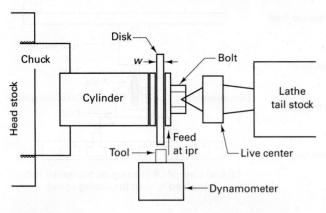

FIGURE 21.12 Orthogonal disk machining can be performed on a lathe setup and is used for high-speed machining experiments. The tool feeding in the facing direction provides the uncut chip thickness t.

the workpiece is called the **back rake angle, α**. A positive angle is shown in the schematic. The chip is formed by **shearing**. The **onset of shear** occurs at a low boundary defined by angle ϕ with respect to the horizontal. This model is sufficient to allow us to consider the behavior of the work material during chip formation, the influence of the most critical elements of the tool geometry (the edge radius of the cutting tool and the back rake angle, α), and the interactions that occur between the tool and the freshly generated surfaces of the chip against the rake face and the new surface as rubbed by the flank of the tool.

As shown in Figures 21.11 and 21.12, orthogonal machining can be performed in lathes at higher speeds. OTT is done on solid cylinders that have had a groove machined on the end to form a tube wall, w, or a tubular workpiece can be used. The tubular workpieces can be mounted in a lathe and normal cutting speeds developed for the machining experiment. This setup has the advantage of being very easy to modify so that cutting-temperature experiments can be performed, using the tool/chip thermocouple method. The orthogonal case is more easily modeled for temperature experiments. The OTT and ODM setups produce two-force cutting at speeds equivalent to these used in practice. The slight difference in cutting speed between the inside and outside edge of the chip in OTT can be neglected.

Chip Formation

Basically, the chip is formed by a localized shear process that takes place over a very narrow zone. This large-strain, high-strain-rate, plastic deformation evolves out of a radial compression zone that travels ahead of the tool as it passes over the workpiece. This radial compression zone has, like all plastic deformations, an elastic compression region that changes into a plastic compression region when the yield strength of the material is exceeded. The plastic compression generates

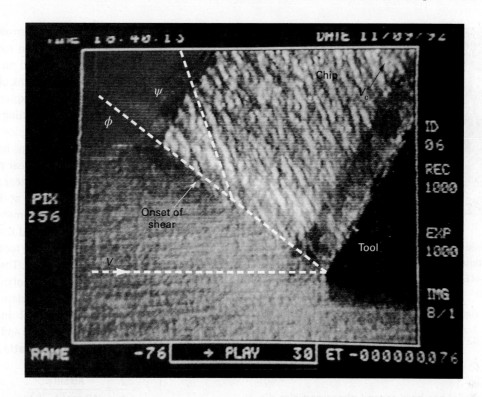

dislocation tangles and networks in annealed metals. The applied stress level increases as the material approaches the tool, where the material has no recourse but to yield in a shear process. The onset of the shear process takes place along the lower boundary of the shear zone defined by the shear angle, φ. The shear lamella (microscopic shear planes) lie at the angle ψ to the shear plane.

This can be seen in the videograph in **Figure 21.13** and the schematic made from the videograph (see **Figure 21.14**). The videograph was made by videotaping the orthogonal machining of an aluminum plate at more than 100 times magnification at 1000 frames per second using a high-speed videotaping camera. By machining at low speeds (*V* = 8.125 ipm), the behavior of the process was captured at high-frame rates and then observed at playback at very slow frame rates. The uncut chip thickness was *t* = 0.020. The termination of the shear process as defined by the upper boundary cannot be observed in the still videograph but can easily be seen in the videos. Videographic experiments showed how the onset of **shear angle**, ϕ, increases with metal hardness. This directly agrees with the material behavior observed in tensile/compression testing—that yield (and ultimate) strength increases with hardness. In steels the correlation is so good that hardness tests are used to estimate ultimate tensile strength. So in tensile testing, we observe that the onset of plastic deformation (yielding) is delayed by increased hardness or increased dislocation density. Correspondingly, in metal cutting, we observe that the onset of shear (to form the chip) is delayed by increased hardness (so ϕ increases directly with hardness). This is because as

the material being machined gets harder, dislocation motion becomes more difficult and plastic deformation (with continuous chips) gives way to fracture (discontinuous chips) just as it does in tensile testing. See **Figure 21.15** for examples of chips. At the same time, the angle *Φ* is increasing, the angle ψ is decreasing with increased hardness. Over the entire spectrum

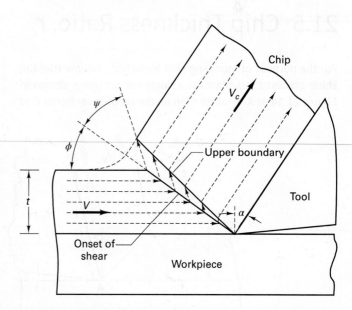

FIGURE 21.14 Schematic representation of the material flow, that is, the chip-forming shear process. Here, ϕ defines the onset of shear or lower boundary and ψ defines the direction of slip due to dislocation movement.

of hardness, from dead soft to full hard, Φ and ψ are related according to

$$\Phi + \psi = 45° + \frac{\alpha}{2}. \qquad (21.16)$$

If the work material has hard second-phase particles dispersed in it, they can act as barriers to the shear front dislocations, which cannot penetrate the particle. The dislocations create voids around the particles. If there are enough particles of the right size and shape, the chip will fracture through the shear zone, forming segmented chips. **Free-machining steels**, which have small percentages of hard second-phase particles added to them, use this metallurgical phenomenon to break up the chips for easier chip handling.

21.5 Chip Thickness Ratio, r_c

For the purpose of modeling chip formation, assume that the shear process takes place on a single narrow plane, shown in **Figure 21.16** as A–B rather than on the set of shear fronts that

actually comprise a narrow shear zone. Further, assume that the tool's cutting edge is perfectly sharp and no contact is being made between the flank of the tool and the new surface. The workpiece passes the tool with velocity V, the cutting speed. The uncut chip thickness is t. Ignoring the compression deformation, chips having thickness t_c are formed by the shear process. The chip has velocity V_c. The shear process then has velocity V_s and occurs at the onset of shear angle ϕ. The tool geometry is given by the back rake angle α and the clearance angle γ. The velocity triangle for V, V_c, and V_s is also shown in Figure 21.16. The chip makes contact with the rake face of the tool over length l_c. The workpiece is a plate thickness of w.

From orthogonal machining experiments, the chip thickness, t_c, is measured and used to compute the shear angle from the **chip thickness ratio, r_c**, defined as t/t_c:

$$r_c = \frac{t}{t_c} = \frac{AB \sin\phi}{AB \cos(\phi - \alpha)} \qquad (21.17)$$

where AB is the length of the shear plane from the tool tip to the free surface.

Equation 21.17 may be solved for the **shear angle, ϕ,** as a function of the measurable chip thickness ratio by expanding the cosine term and simplifying:

$$\tan\phi = \frac{r_c \cos\alpha}{1 - r_c \sin\alpha} \qquad (21.18)$$

There are numerous other ways to measure chip ratios and obtain shear angles both during (dynamically) and after (statically) the cutting process. For example, the ratio of the length of the chip L_c to the length of the cut L can be used to determine r_c. Many researchers use the chip compression ratio, which is the reciprocal of r_c, as a parameter. The shear angle can be measured statically by instantaneously interrupting

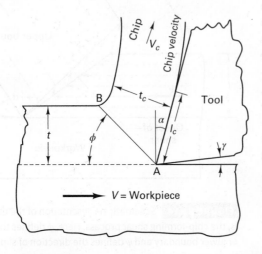

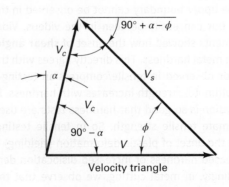

FIGURE 21.16 The orthogonal chip formation model is used to define the chip thickness ratio $r_c = t/t_c$.

the cut through the use of **quick-stop devices (QSDs)**. These devices disengage the cutting tool from the workpiece while cutting is in progress, leaving the chip attached to the workpiece. Optical and scanning electron microscopy is then used to observe the direction of shear. Figure 21.10 shows a QSD on a plate machining setup, and Figure 21.15 was made using a quick-stop device.

Dynamically, high-speed motion pictures and high-speed videographic systems have also been used to observe the process at frame rates as high as 30,000 frames per second. Figure 21.13 is a high-speed videograph.

Machining stages have been built that allow the process to be performed inside a scanning electron microscope and recorded on videotapes for high-resolution, high-magnification examination of the deformation process. Using sophisticated electronics and slow-motion playback, this technique can be used to measure the **shear velocity**. The vector sum of V_s and V_c equals V.

For consistency of volume, we observe that

$$r_c = \frac{t}{t_c} = \frac{\sin\phi}{\cos(\phi-\alpha)} = \frac{V_c}{V} \qquad (21.19)$$

indicating that the chip ratio (and therefore the onset of shear angle) can be determined dynamically if a reliable means to measure V_c can be found.

The ratio of V_s to V is

$$\frac{V_s}{V} = \frac{\cos\alpha}{\cos(\phi-\alpha)} \qquad (21.20)$$

These velocities are important in power calculations, heat and temperature calculations, and vibration analysis associated with chatter in chip formation.

21.6 Mechanics of Machining (Statics)

Orthogonal machining has been defined as a two-force system. Consider **Figure 21.17**, which shows a free-body diagram of a chip that has been separated at a shear plane. It is assumed that the resultant force, R, acting on the back of the chip is equal and opposite to the resultant force, R', acting on the shear plane. The resultant R is composed of the **friction force**, F, and the normal force N acting on the tool–chip interface contact area. The resultant force, R', is composed of a **shear force**, F_s, and normal force, F_n, acting on the shear plane area, A_s. Because neither of these two sets of forces can usually be measured, a third set is needed, which can be measured using a dynamometer (force transducer) mounted either in the workholder or the tool holder. Note that this set has resultant R, which is equal in magnitude to all the other resultant forces

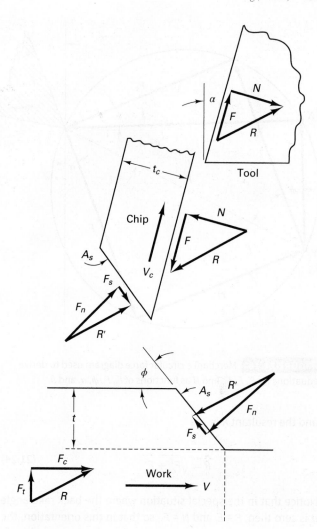

FIGURE 21.17 Free-body diagram of orthogonal chip formation process, showing equilibrium condition between resultant forces R and R'.

in the diagram. The resultant force, R, is composed of a **cutting force**, F_c, and a tangential (normal) force, F_t. Now it is necessary to express the desired forces (F_s, F_n, F, N) in terms of the measured dynamometer components, F_c and F_t, and appropriate angles. To do this, a circular force diagram is developed in which all six forces are collected in the same force circle (**Figure 21.18**). The only symbol in this figure as yet undefined is β, which is the angle between the normal force N and the resultant R. It is called friction angle β and is used to describe the friction coefficient μ on the tool–chip interface area, which is defined as F/N so that

$$\beta = \tan^{-1}\mu = \tan^{-1}\frac{F}{N} \qquad (21.21)$$

The friction force F and its normal N can be shown to be

$$F = F_c\sin\alpha + F_t\cos\alpha \qquad (21.22)$$

$$N = F_c\cos\alpha - F_t\sin\alpha \qquad (21.23)$$

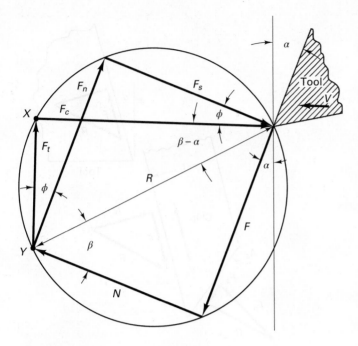

FIGURE 21.18 Merchant's circular force diagram used to derive equations for F_s, F_n, F_t, and N as functions of F_c, F_t, ϕ, α, and β.

and the resultant R is

$$R = \sqrt{F_c^2 + F_t^2} \tag{21.24}$$

Notice that in the special situation where the back rake angle α is zero then, $F = F_t$ and $N = F_c$, so that in this orientation, the friction force and its normal can be directly measured by the dynamometer.

The forces parallel and perpendicular to the shear plane can be shown from the circular force diagram (Figure 21.18) to be

$$F_s = F_c \cos \phi - F_t \sin \phi \tag{21.25}$$

$$F_n = F_c \sin \phi + F_t \cos \phi \tag{21.26}$$

F_s is of particular interest because it is used to compute the shear stress on the shear plane. This shear stress is defined as

$$\tau_s = \frac{F_s}{A_s} \tag{21.27}$$

where A_s is the area of the shear plane, where

$$A_s = \frac{tw}{\sin \phi} \tag{21.28}$$

recalling that t was the uncut chip thickness and w was the width of the workpiece. The **shear stress (flow stress)** is, therefore,

$$\tau_s = \frac{F_c \sin \phi \cos \phi - F_t \sin^2 \phi}{tw} \text{psi} \tag{21.29}$$

For a given polycrystalline metal, this shear stress has been shown to be not sensitive to variations in cutting parameters, tool material, or the cutting environment. **Figure 21.19** gives some typical values for the flow stress for a variety of metals, plotted against hardness.

Using orthogonal machining setups, values for HP_s as given in Table 21.3, can be computed. K_s can be computed from

$$K_s = F_c / tw \tag{21.30}$$

Specific horsepower is related to and correlates well with shear stress for a given metal. Unit power is sensitive to material properties (e.g., hardness), rake angle, depth of cut, and feed, whereas τ_s is sensitive to material properties only.

21.7 Shear Strain, γ, and Shear Front Angle, ϕ

Merchant's shear strain model[1] is cited in almost every textbook for manufacturing processes. It is widely used in theoretical studies and industrial practice. Merchant's model is based on the "stack of cards" model shown in **Figure 21.20a**. In this model, the material shears on a single plane defined by Φ, in the direction of the shear plane, so the model combines ψ and Φ into Φ, coupling them in any following analysis, Thus, Merchant's shear strain equation is:

$$\gamma = \cos \alpha / \sin \phi \cos(\phi - \alpha)$$

In the Black-Huang model,[2] the onset of shear begins at the lower boundary, defined by Φ with the material shearing in the direction of ψ relative to this lower boundary.

Using Merchant's chip formation bubble model shown in **Figure 21.20b** verified by videographic images like that in Figure 21.13 show that a new "stack-of-cards" model can be developed as shown in Figure 21.20c on the right. From the shear triangle, an equation for strain, γ, is

$$\gamma = \cos\alpha / \left[\sin(\phi + \varphi)\cos(\phi + \varphi - \alpha) \right] \tag{21.31}$$

where

ϕ = the angle of the onset of the shear plane
ψ = the shear front angle or direction of crystal elongation

Using the available machining data, for a given metal of some hardness, ψ is observed to decrease, reach a minimum, and rise again for all rake angles.

The minimum energy principle has been used in many scientific fields such as physics, metalforming processes, and

[1] M. E. Merchant, *Journal of Applied Physics*, 16, 267(a) and 318(b) (1945).
[2] J. T. Black and J. M. Huang, "Shear Strain Model in Machining," *Manufacturing Science and Energy*, MED-Vol 2-1 (1995), p. 81.

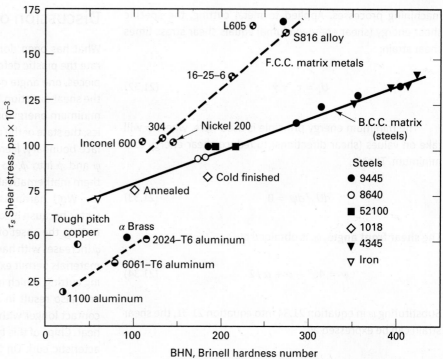

Shear stress τ_s variation with the Brinell hardness number for a group of steels and aerospace alloys. Data of some selected face-centered-cubic metals are also included. (Adapted with permission from S. Ramalingham and K. J. Trigger, *Advances in Machine Tool Design and Research*, 1971, Pergamon Press)

(a)

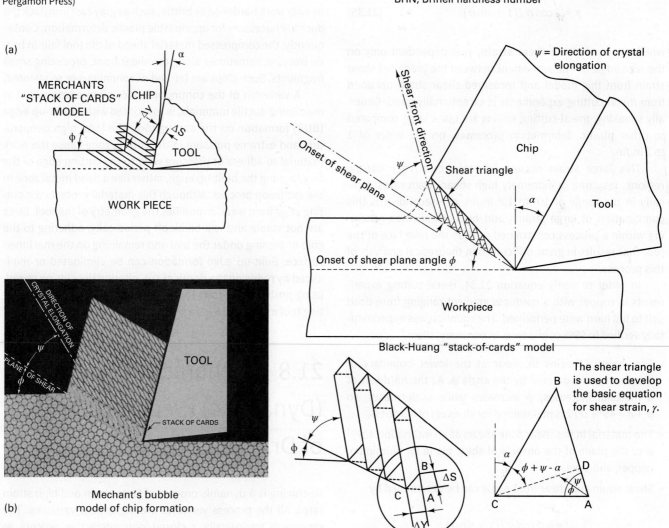

(b)

FIGURE 21.20 The Black-Huang "stack-of-cards" model for calculating shear strain in metal cutting is based on Merchant's bubble model for chip formation, shown on the left.

machining processes. Applied to metal cutting, the specific shear energy (shear energy/volume) equals shear stress times shear strain:

$$U_s = \tau_s \times \gamma \qquad (21.32)$$

The minimum energy principle is used here, where ψ will take on values (shear directions) to reduce shear energy to a minimum. That is,

$$dU_s / d\psi = 0 \qquad (21.33)$$

The **shear front angle,** ψ, is obtained by

$$\psi = 45° - \phi + \alpha / 2 \qquad (21.34)$$

Substituting ψ in equation 21.34 into equation 21.31, the shear strain can be expressed as

$$\gamma = 2 \cos \alpha / (1 + \sin \alpha) \qquad (21.35)$$

which shows that the **shear strain, γ,** is dependent only on the rake angle α. The agreement between the predicted shear strain from this model and measured shear strain obtained from metal-cutting experiments is exceptionally good. Generally speaking, metal-cutting strains are quite large compared to other plastic deformation processes, on the order of 1 to 2 in./in.

This large strain occurs, however, over very narrow regions, resulting in extremely high shear strain rates, typically in the range of 10^4 to 10^8 in./in. per second. It is this combination of large strains and high strain rates operating within a process constrained only by the rake face of the tool that results in great difficulties in theoretical analysis of this process.

In order to verify equation 21.34, metal-cutting experiments in copper with a hardness gradient ranging from dead soft to full hard were performed. The equation was experimentally verified to 99% confidence. In summary, then:

- The material begins to shear at the lower boundary of the shear zone, defined by the angle ϕ. As the hardness of a material increases, ϕ increases while ψ decreases, so $\psi + \phi = 45° + \alpha / 2$ is maintained for all levels of hardness.

- The material in the shear zone shears at an inclination angle ψ to the plane of the onset ϕ of shear plane for aluminum, copper, and steel.

- Shear strain and shear front angle can be determined by

$$\gamma = 2 \cos \alpha / (1 + \sin \alpha)$$
$$\psi + \phi = 45° + \alpha / 2$$

where ϕ and ψ vary with hardness.

Discussion of Black-Huang Equation

What has been done in the Black-Huang analysis is to separate the plastic deformation process (in chip formation) in two pieces, one angle defining the boundary and another defining the shear direction. The latter parameter, ψ, correctly obeys the minimum energy principle while ϕ is determined by mechanics, the state of the material (hardness), and the tool/chip interface boundary. Merchant's error was that his model combined ψ and ϕ into ϕ, so he coupled them and could not separate them mathematically.

Work hardness prior to machining is the most important factor because it controls the onset of shear. Just as in tensile testing, the onset of shear is delayed by increased hardness, so ϕ increases with hardness, as does τ_s, HP$_s$, and K$_s$. Highly ductile materials permit extensive plastic deformation of the chip during cutting, which increases heat generation and temperature, and also result in longer, "continuous" chips that remain in contact longer with the tool face, thus causing more frictional heat. Chips of this type are severely deformed and have a characteristic curl. On the other hand, materials that are already heavily work hardened or brittle, such as gray cast iron, lack the ductility necessary for appreciable plastic deformation. Consequently, the compressed material ahead of the tool fails in brittle fracture, sometimes along the shear front, producing small fragments. Such chips are termed *discontinuous* or *segmented*.

A variation of the continuous chip, often encountered in machining ductile materials, is associated with a **built-up edge (BUE)** formation on the cutting tool. The local high temperature and extreme pressure in the cutting zone cause the work material to adhere or pressure weld to the cutting edge of the tool forming the built-up edge, rather like a dead metal zone in the extrusion process. Although this material protects the cutting edge from wear, it modifies the geometry of the tool. BUEs are not stable and will break off periodically, adhering to the chip or passing under the tool and remaining on the machined surface. Built-up edge formation can be eliminated or minimized by reducing the depth of cut, altering the cutting speed, using positive rake tools, applying a coolant, or changing cutting tool materials.

21.8 Mechanics of Machining (Dynamics) (Section courtsey of Dr. Elliot Stern)

Machining is a dynamic process of large strain and high strain rates. All the process variables are dependent variables. The process is intrinsically a closed-loop interactive process as shown in **Figure 21.21**.

Starting at the top, inputs to the processes (speed, feed, depth of cut) determine the chip load on the tool. The chip load

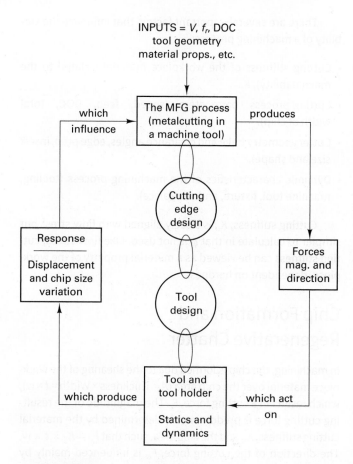

INPUTS = V, f, DOC
tool geometry
material props., etc.

The MFG process
(metalcutting in
a machine tool)

which
influence

produces

Cutting
edge
design

Response
____
Displacement
and chip size
variation

Forces
mag. and
direction

Tool
design

Tool and
tool holder
____
Statics and
dynamics

which produce

which act
on

FIGURE 21.21 Machining dynamics is a closed-loop interactive process that creates a force-displacement response.

determines the cutting forces (magnitude and direction; usually elastic), which alters the chip load on the tool. The altered chip load produces new forces. The cycle repeats, producing chatter and vibration.

Remember that plastic deformation is always preceded by elastic deformation. The elastic deflection behaves like a big spring. The mechanism by which a process dissipates energy is called **chatter** or **vibration**. In machining, it has long been observed in practice that rotational speed may greatly influence process stability and chatter. Experienced operators commonly listen to machining noise and interactively modify the speed when optimizing a specific application. In addition, experience demonstrates that the performance of a particular tool may vary significantly based on the machine tool employed and other characteristics such as the workpiece, fixture holder, and the like. Today more than ever, the manufacturing industry is more competitive and responsive, characterized by both high-volume and small-batch production, seeking economies of scale. High productivity is achieved by increased

machine and tooling capabilities along with the elimination of all non-value-added activities. Few companies can afford lengthy trial-and-error approaches to machining-process optimization or additional processes to treat the effect of chatter.

In metal cutting, chatter is a self-excited vibration that is caused by the closed-loop force-displacement response of the machining process. The process-induced variations in the cutting force may be caused by changes in the cutting velocity, chip cross section (area), tool–chip interface friction, built-up edge, workpiece variation, or, most commonly, process modulation resulting in regeneration of vibration. When more energy is input into the dynamic machining system than can be dissipated by mechanical work, damping, and friction, equilibrium (the state of minimum potential energy) is sought by the machining system through the generation of chatter vibration.

The proper classification of the type of vibration is the first step in identifying and solving the cause of unwanted vibration (see **Figure 21.22**):

1. **Free vibration** is the response to any initial condition or sudden change. The amplitude of the vibration decreases with time and occurs at the natural frequency of the system. Interrupted machining is an example that often appears as lines or shadows following a surface discontinuity.

2. **Forced vibration** is the response to a periodic (repeating with time) input. The response and input occur at the same frequency. The amplitude of the vibration remains constant for set input conditions and is linearly related to speed. Unbalance, misalignment, tooth impacts, and resonance of rotation systems are the most common examples.

• **Free Vibration** The response to an initial condition or sudden change. The amplitude of the vibration decreases with time and occurs at the natural frequency of the system often produced by interrupted machining. Often appears as lines or shadows following a surface discontinuity.

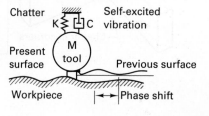

• **Forced Vibration** The response to a periodic (repeating with time) input. The response and input occur at the same frequency. The amplitude of the vibration remains constant for a set input condition and is nonlinearly related to speed. Unbalance, misalignment, tooth impacts, and resonance of rotating systems are the most common examples.

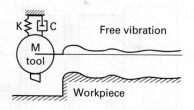

• **Self-Excited Vibration** The periodic response of the system to a constant input. The vibration may grow in amplitude (unstable) and occurs near the natural frequency of the system regardless of the input. Chatter due to the regeneration of surface waviness is the most common metal cutting example.

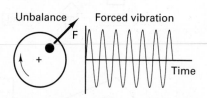

FIGURE 21.22 There are three types of vibration in machining.

3. **Self-excited vibration** is the periodic response of the system to a constant input. The vibration may grow in amplitude (become unstable) and occurs near the natural frequency of the system regardless of the input. Chatter due to the regeneration of waviness in the machined surface is the most common metal cutting example.

How do we know chatter exists? Listen and look! Chatter is characterized by the following:

1. There is a sudden onset of vibration (a screech or buzz or whine) that rapidly increases in amplitude until a maximum threshold (saturation) is reached.

2. The frequency of chatter remains very close to a natural frequency (critical frequency) of the machining system and changes little with variation of process parameters. The largest force-displacement response occurs at resonance and therefore the greatest energy dissipation.

3. Chatter often results in unacceptable surface finish, exhibited by a helical or angular pattern (pearled or fish scaled) superimposed over normal feed marks.

4. Visible surface undulations are found in the feed direction and corresponding wavy or serrated chips with variable thickness.

Figure 21.23 shows some typical examples of chatter visible in the surface finish marks.

There are several important factors that influence the stability of a machining process:

- Cutting stiffness of the workpiece material (related to the machinability), K_s.
- Cutting-process parameters (speed, feed, DOC, total width of chip).
- Cutter geometry (rake and clearance angles, edge prep, insert size and shape).
- Dynamic characteristics of the machining process (tooling, machine tool, fixture, and workpiece).

Cutting stiffness, K_s, is closely aligned with flow stress but simpler to calculate in that ϕ is not used. Like flow stress, cutting stiffness can be viewed as a material property of the workpiece, dependent on hardness.

Chip Formation and Regenerative Chatter

In machining, the chip is formed due to the shearing of the workpiece material over the chip area (A = Thickness × Width = $t \times w$), which results in a cutting force, F_c. The magnitude of the resulting cutting force is predominantly determined by the material cutting stiffness, K_s, and the chip area, such that $F_c = K_s \times t \times w$. The direction of the cutting force, F_c, is influenced mainly by

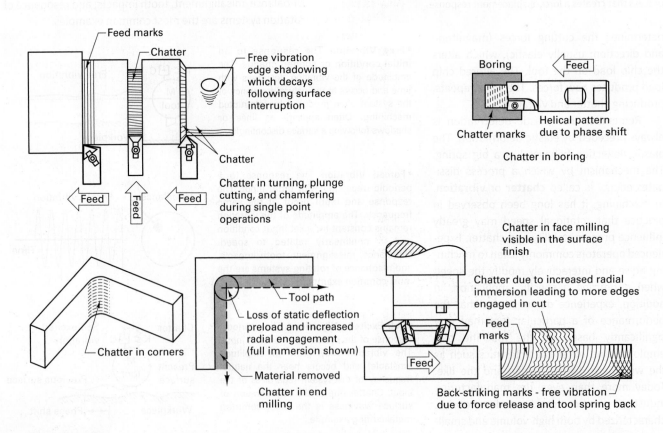

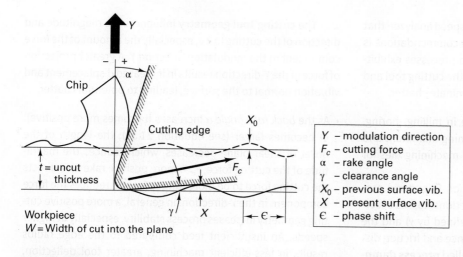

FIGURE 21.24 When the overlapping cuts get out of phase with each other, a variable chip thickness is produced, resulting in a change in F_c on the tool or workpiece.

the geometries of the rake and clearance angles as well as the edge prep.

Machining operations require an overlap of cutting paths that generate the machined surface (see **Figure 21.24**). In single-point operations, the overlap of cutting paths does not occur until one complete revolution. In milling or drilling, overlap occurs in a fraction of a revolution, depending on the number of cutting edges on the tool.

The cutting force causes a relative displacement, X, between the tool and workpiece, which affects the uncut chip thickness, t, and, in turn, the cutting force. This coupled relationship between displacement in the Y-direction (modulation direction) and the resulting cutting force forms a closed-loop response system. The modulation direction is normal to the surface defining the chip thickness.

A phase shift, ε, between subsequent overlapping surfaces results in a variable chip thickness and modulation of the displacement, causing chatter vibration. The phase shift between overlapping cutting paths is responsible for producing chatter.

However, there is a preferred speed that corresponds to a phase-locked condition ($\varepsilon = 0$) that results in a constant chip thickness, t. A constant chip thickness results in a steady cutting force and the elimination of the feedback mechanism responsible for **regenerative chatter**. This what the operators are trying to achieve when they vary cutting speeds (see **Figure 21.25**).

How Do the Important Factors Influence Chatter?

- *Cutting stiffness K_s.* This is a material property related to shear flow stress, hardness, and work hardening and is often described in a relative sense of the machinability of materials. Materials such as steel and titanium require much greater shear forces than aluminum or cast iron; therefore, the corresponding larger cutting forces lead to greater displacement in the Y-direction and less machining stability.

- *Speed.* The process parameters are the easiest factors to change chatter and its amplitude. The rotational speed of the tool affects the phase shift between overlapping surfaces and

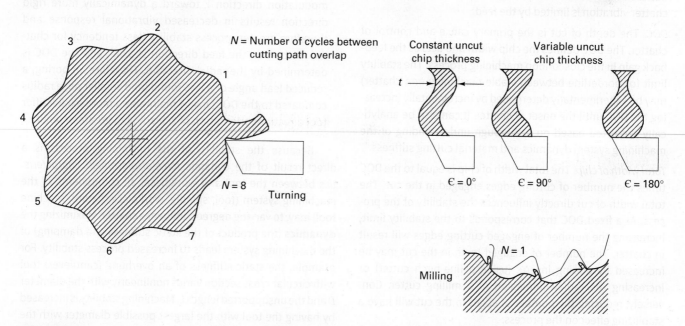

FIGURE 21.25 Regenerative chatter in turning and milling can be produced by variable uncut chip thickness.

the regeneration of vibration. A handheld speed analyzer[3] that produces dynamically preferred speed recommendations is commercially available. When applied to processes exhibiting a relative rotational motion between the cutting tool and workpiece, it recommends a speed to eliminate chatter.

The most successful applications are in milling, boring, and turning; in multipoint tools, in machining aluminum and cast iron; in high-speed machining and in machining dies for forming sheet metal.

At slow speeds (relative to the vibration frequency) process stability is mainly due to increased frictional losses occurring between the tool clearance angle (defined by γ) and the present surface vibration, X. This interference and friction dissipates energy in the form of heat and is called **process damping**. As machining speeds are increased, the wavelength of the surface vibration also increases, which reduces the slope of the surface and eliminates process-induced damping. Additionally, chatter becomes more significant as speeds increase because existing forces approach the natural frequencies of the machining system. The analyzer measures and identifies the vibrational frequencies of the chatter noise and determines which speeds will most closely result in $\varepsilon = 0$. A zero phase shift between overlapping surfaces eliminates the variation of the chip thickness and eliminates the modulation, resulting in chatter.

- *Feed.* The feed per tooth defines the average uncut chip thickness, t, and influences the magnitude of the cutting force. The feed does not greatly influence the stability of the machining process (i.e., whether chatter occurs) but does control the severity of the vibration. Because no cutting force exists if the vibration in the Y-direction results in the loss of contact between the tool and workpiece, the maximum amplitude of chatter vibration is limited by the feed.

- *DOC.* The depth of cut is the primary cause and control of chatter. The DOC defines the chip width and acts as the feedback gain in the closed-loop machining process. The stability limit (or borderline between stable machining and chatter) may be experimentally determined by incrementally increasing the DOC until the onset of chatter. It can also be analytically predicted based on a thorough understanding of the machining system dynamics and material cutting stiffness.

- *Total width of chip.* The total width of chip is equal to the DOC times the number of cutting edges engaged in the cut. The total width of cut directly influences the stability of the process. At a fixed DOC that corresponds to the stability limit, increasing the number of engaged cutting edges will result in chatter. The number of engaged teeth in the cut may be increased by adding inserts (using a fine-pitch cutter) or increasing the radial immersion of a milling cutter. Conversely, reducing the number of edges in the cut will have a stabilizing effect on the process.

The **cutting tool geometry** influences the magnitude and direction of the cutting force, especially the amount of the force component in the modulation direction Y. A greater projection of force in the Y-direction results in increased displacement and vibration normal to the surface, leading to potential chatter.

- As the *back rake angle* α increases (becomes more positive), Φ becomes larger (see equation 21.34); the length of the onset of shear plane decreases, which reduces the magnitude of the cutting force, F_c. A more positive rake also directs the cutting force to be more tangential and reduces the force component in the Y-direction. In general, a more positive cutting geometry increases process stability, especially at higher speeds. An insufficient feed compared to the edge radius results in less-efficient machining, greater tool deflection, and poorer machining stability.

- A *reduced clearance angle* γ, which increases the frictional contact between the tool and workpiece, may produce process damping. The stabilizing effect is due to energy dissipation in the form of heat, which potentially decreases tool life and may thermally distort the workpiece or increase the heat-affected zone in the workpiece. The initial wear of a new cutting edge may also have a stabilizing effect on chatter.

- The *size* (nose radius), *shape* (diamond, triangular, square, round), and *lead angle* of the insert all influence the chip area shape and the corresponding Y-direction (**Figure 21.26**). In milling, the feed direction is transverse to the tool axis (i.e., radial), and the DOC is defined by the axial immersion of the tool. In milling, as the lead angle of the cut increases or the shape of the insert becomes rounded (same effect as a large nose radius compared to a small DOC), the Y-orientation is directed away from the more flexible radial tool direction and toward the stiffer axial direction. The orientation of the modulation direction Y toward a dynamically more rigid direction results in decreased vibrational response and therefore greater process stability (less tendency for chatter). In boring, the feed direction is axial, and the DOC is determined by the radial direction. Therefore, in boring, a reduced lead angle or a less round (and smaller nose radius compared to the DOC) insert maintains a more axial (stiffer tool direction) orientation of Y, leading to greater stability.

Because the stability of the machining process is a direct result of the dynamic *force-displacement characteristics* between the tool and workpiece, all components of the machining system (tool, spindle, workpiece, fixture, machine tool) may, to varying degrees, influence chatter. Maximizing the **dynamics** (the product of the static stiffness and damping) of the machining system leads to increased process stability. For example, the static stiffness of an overhung (cantilever) tool with circular cross section varies nonlinearly with the diameter D and the unsupported length L. Machining stability is increased by having the tool with the largest possible diameter with the minimum overhang. The frequency of chatter occurs near the most flexible vibrational mode of the machining system.

---

[3] Best speed by Design Manufacturing Inc., Tampa, Florida.

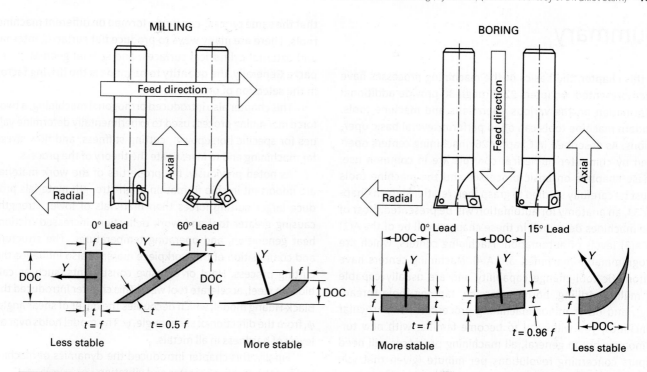

FIGURE 21.26 Milling and boring operations can be made more stable by correct selection of insert geometry.

Stability Lobe Diagram

A **stability lobe diagram** (**Figure 21.27**) relates the total width of cut that can be machined to the rotational speed of the tool with a specified number of cutting edges. If the total width of cut is maintained below a minimum level (although this may be of limited practical value for some machining systems), then the process stability exhibits speed independence or "unconditional" stability. At slow speeds, increased stability may be achieved within the process damping region. The "conditional" stability lobe regions allow increased total width of cut (DOC × Number of edges engaged in the cut) at dynamically preferred speeds at which the phase shift ε between overlapping or consecutive cutting paths approaches zero. The stability lobe number N indicates complete cycles of vibration that exist between overlapping surfaces. As can be seen by the diagram, the higher speeds correspond to lower lobe numbers and provide the greatest potential increase in the total width of cut and material removal rate (due to greater lobe height and width). If the total width of cut exceeds the borderline of stability, even if the process is operating at a preferred speed, chatter occurs. The greater the total width of cut above the stability limit, the more unstable and violent will be the chatter vibration.

When a chatter condition occurs, such as at point a on the stability lobe diagram, the rotational speed is adjusted to the first recommended speed (N = 1), which results in stable machining at point b on the diagram. The DOC may be incrementally increased until chatter again occurs as the stability border is crossed at point c. Using the analyzer again, chattering under the new operating conditions will result in a modified speed recommendation corresponding to point d. If desired, the DOC may again be incrementally increased (conservative steps promote safety) to point e. In general, do not attempt to maintain the DOC (and total width of cut) right up to the borderline of stability because workpiece variation affecting K_s, speed errors, or small changes in the dynamic characteristics of the machining system may result in crossing the stability limit into severe chatter. The amplitude of chatter vibration may be more safely limited by temporary reduction of the feed per tooth until a preferred speed and stable depth of cut have been established.

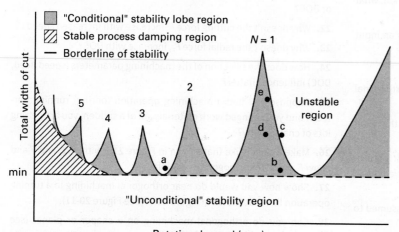

FIGURE 21.27 Dynamic analysis of the cutting process produces a stability lobe diagram, which defines speeds that produce stable and unstable cutting conditions.

Summary

In this chapter, the basics of the machining processes have been presented. Chapters 22 through 30 provide additional information on the various operations and machine tools. Modern machine tools can often perform several basic operations. As described in Chapter 26, machining centers operated by computer numerical controls are in common use. These chapters on basic processes and the machine tools must be carefully studied to grasp their relationship. In Chapter 34, an anatomy for automation will be presented. Most of the machines described in these chapters will be of the A(1) to A(3) levels of automation. Machining centers, which are programmable machines, are A(4). Machining centers have automatic tool-change capability and are usually capable of milling, drilling, boring, reaming, tapping (hole threading), and other minor machining processes. For particular machines, you will need to become familiar with new terminology, but in general, all machining processes will need inputs concerning revolutions per minute (given that you selected the cutting speed), feeds, and depths of cut. Note that the same process can be performed on different machine tools. There are many ways to produce flat surfaces, internal and external cylindrical surfaces, and special geometries in parts. Generally, the quantity to be made is the driving factor in the selection of processes.

This chapter also introduced orthogonal machining, a two-force machining process used to experimentally determine values for specific horsepower, cutting stiffness, and flow stress for machining and to investigate the theory of the process.

As noted previously, the properties of the work material are important in chip formation. High-strength materials produce larger cutting forces than materials of lower strength, causing greater tool and work deflection, increased friction, heat generation, and operating temperatures. The structure and composition of the workpiece material also influence the cutting process. Hard or abrasive constituents, such as carbides in steel, accelerate tool wear. This chapter introduced the Black-Huang model, which decouples the onset of shear angle, ϕ, from the direction of shear angle, ψ. The model holds over all levels of hardness in all metals.

Finally, this chapter introduced the dynamics of machining and the reality of chatter and vibration.

Review Questions

1. Why has the metal-cutting process resisted theoretical solution for so many years?

2. What variables must be considered in understanding a machining process?

3. Which of the seven basic chip formation processes are single point, and which are multiple point?

4. How is feed related to speed in the machining operations called turning?

5. Before you select speed and feed for a machining operation, what did you have to decide? (*Hint:* See Figure 21.4.)

6. Milling has two feeds. What are they, and which one is an input parameter to the machine tool?

7. What is the fundamental mechanism of chip formation?

8. What is the difference between oblique machining and orthogonal machining?

9. What are the implications of Figure 21.13, given that this videograph was made at a very low cutting speed?

10. Note that the units for the approximate equation for MRR for turning are not correct. When is the approximate equation not very good (yields a large error in MRR values)?

11. For orthogonal machining, the cutting edge radius is assumed to be small compared to the uncut chip thickness. Why?

12. How do the magnitude of the strain and strain rate values of metal cutting compare to those of tensile testing?

13. Why is titanium such a difficult metal to machine? (Note its high value of HP_s).

14. Explain why you get segmented or discontinuous chips when you machine cast iron.

15. Why is metal cutting shear stress such an important parameter?

16. Which of the three cutting forces in oblique cutting consumes most of the power?

17. How is the energy in a machining process typically consumed?

18. Where does the energy consumed in metal cutting ultimately go?

19. What are two ways of estimating the primary cutting force F_c.

20. What are the 3 different ways to perform orthogonal machining?

21. Why does the cutting force F_c increase with increased feed or DOC?

22. Why doesn't the cutting force F_c increase with increased speed V?

23. Why doesn't the radial force F_R, increase with DOC?

24. How does the selection of the machining parameters (speed, feed, DOC) influence chatter?

25. Suppose you had a machining operation (boring) running perfectly and you changed work materials. All of a sudden, you are getting lots of chatter. Why?

26. Make a sketch like that shown in Figure 21.1 with w, t, and V and show F_c and F_t for orthogonal machining.

27. Show how you would do near orthogonal machining in a turning operation like that shown in Figure 21.3. (See Figure 20-11).

28. Can you do orthogonal machining on a shaper or planer (see Chapter 27)?

29. What process and material combination would yield the largest cutting force, based on the data in Table 21.3.

30. What is meant by the statement that machining dynamics is a closed-loop interactive process?

31. In equation 21.8 what does the number 33,000 represent?

Problems

1. Figure 21.4 provides suggested cutting speeds and feeds for turning. You are cutting a wrought carbon steel (1012) at normalized to 100 Bhn and have have selected a high speed steel tool. For a DOC of 150 in., What speed and feed would you select?

2. For problem 1, suppose you selected a speed of 145 sfpm and a feed of 0.015 in. per revolution. The workpiece is 4 in. in diameter.

   a. What is the input rpm?

   b. What is the MRR?

   c. What is the cutting time for a 6-in. cut?

3. If the cutting forces is 1000 lb calculate the horsepower that a process operating at a speed of 80 fpm is going to use.

4. Explain how you would estimate the cutting force for a turning operation if you do not have a dynamometer?

5. For a turning operation, you have selected a high-speed steel (HSS) tool and turning a hot rolled free machining steel, BHN = 300. Your depth of cut will be 0.150 in. The diameter of the workpiece is 1.00 in.

   a. What speed and feed would you select for this job?

   b. Using a speed of 105 sfpm and a feed of 0.015, calculate the spindle rpm for this operation.

   c. Calculate the metal removal rate.

   d. Calculate the cutting time for the operation with a length of cut of 4 in. and 0.10-in. allowance.

6. For a slab milling operation using a 5-in.-diameter, 11-tooth cutter (see Figure 21.6), the feed per tooth is 0.005 in./tooth with a cutting speed of 100 sfpm (HSS steel). Calculate the rpm of the cutter and the feed rate (f_m) of the table, then calculate the metal removal rate, MRR, where the width of the block being machined is 2 in. and the depth of cut is 0.25 in. Calculate the time to machine (T_m) a 6-in.-long block of metal with this setup. Suppose you switched to a coated-carbide tool, so you increase the cutting speed to 400 sfpm. Now recalculate the machining time (T_m) with all the other parameters the same.

7. The power required to machine metal is related to the cutting force (F_c) and the cutting speed. For Problem 3, estimate cutting force F_c for this turning operation. (*Hint:* You have to estimate a value of HPs for this material.)

8. In order to drill a hole in the material described in Problem 3 using an HSS drill, you have to select a cutting speed and a feed rate. Using a speed of 105 sfpm for the HSS drill, calculate the rpm for a $\frac{3}{4}$-in.-diameter drill and the MRR if the feed rate is 0.008 in. per revolution.

9. Suppose you have the data in **Table 21.A** obtained from a metal-cutting experiment (orthogonal machining). Compute the shear angle, the shear stress, the specific energy, the shear strain, and the coefficient of friction at the tool–chip interface. How do your HP$_s$ and τ_s values compare with the values found in Table 21.3?

10. For the data in Problem 9, determine the specific shear energy and the specific friction energy.

11. Calculate the horsepower that a process is going to use if the cutting force is 563 lb and the speed is 90 fpm.

12. Explain how you would estimate the cutting force for this boring operation if you do not have a dynamometer.

13. Derive equations for F and N using the circular force diagram. (*Hint:* Make a copy of the diagram. Extend a line from point X intersecting force F perpendicularly. Extend a line from point Y intersecting the previous line perpendicularly. Find the angle α made by these constructions.)

14. Derive equations for F_s and F_n using the circular force diagram. (*Hint:* Construct a line through X parallel to vector F_n. Extend vector F_s to intersect this line. Construct a line from X perpendicular to F_n. Construct a line through point Y perpendicular to the line through X.)

15. For the data in Problem 9, calculate the shear strain and compare it to $1/r_c$. Comment on the comparison $1/r_c = t_c/t = L/L_c$ assuming that $W = W_c$.

16. A manufacturing engineer needs an estimate of the cutting force F_c to estimate the loss of accuracy of a machining process due to deflection. The material being machined is Inconel 600 with a BHN value of 100. The cutting speed was 250 fpm, the feed was 0.020 in./rev, and the depth of cut was 0.250 in. The chip from the process measured 0.080 in. thick. Estimate the cutting force F_c assuming that $F_t = F_c/2$.

17. Using Figure 21.4 for input data, determine the maximum and minimum MRR values for rough machining (turning) a 1020 carbon steel with a BHN value of 200. Repeat for finish machining assuming a DOC value equal to 10% of the roughing DOC.

18. Estimate the horsepower needed to remove metal at 550 in.³/min with a feed of 0.005 in./rev at a DOC value of 0.675 in. The cutting force F_c was measured at 10,000 lb. Comment on these values.

19. For a turning process, the horsepower required was 24 hp. The metal removal rate was 550 in.³/min. Estimate the specific horsepower and compare to published values for 1020 steel at 200 BHN.

TABLE 21.A

Run Number	F_c	F_t	Feed ipr ×1/1000	Chip Ratio r_c	ϕ	τ_s	U	HP$_s$	γ	μ
			Data							
1	330	295	4.89	0.331						
2	308	280	4.89	0.381						
3	410	330	7.35	0.426						
4	420	340	7.35	0.426						
5	510	350	9.81	0.458						
6	540	395	9.81	0.453						

Table 21.A has machining data for 1020 steel, as-received, in air with a K3H carbide tool, orthogonally (tube cutting on lathe) with tube OD = 2.875. The cutting speed was 530 fpm. The tube wall thickness was 0.200 in. The back rake angle was zero for all cuts. Only the feed rate was changed. Fill in the blanks.

CHAPTER 22

Cutting Tool Materials

Success in **metal cutting** depends on the selection of the proper cutting tool (material and geometry) for a given work material. A wide range of **cutting tool materials** is available with a variety of properties, performance capabilities, and costs. These include high-carbon steels and low-/medium-alloy steels, high-speed steels, cast cobalt alloys, cemented carbides, cast carbides, coated carbides, coated high-speed steels, ceramics, cermets, whisker-reinforced ceramics, sialons, sintered polycrystalline **cubic boron nitride (CBN)**, sintered polycrystalline diamond, and single-crystal natural diamond.

Improvements in cutting tool materials have led to significant increases in cutting speeds and productivity. As the speed (feed rate and DOC) increases, so does the metal removal rate. The time required to remove a given unit volume of material therefore decreases. Today, approximately 85% of carbide tools are coated, almost exclusively by the **chemical vapor deposition (CVD)** process. The cutting tool (material and geometry) is the most critical aspect of the machining process. Clearly, the cutting tool material, cutting parameters, and tool geometry selected directly influence the productivity of the machining operation. **Figure 22.1** outlines the input variables that influence the tool material selection decision. The elements that influence the decision are:

- Work material characteristics, hardness, chemical, and metallurgical state.
- Part characteristics (geometry, accuracy, finish, and surface-integrity requirements).
- Machine tool characteristics, including the workholders (adequate rigidity with high horsepower, and wide speed and feed ranges).
- Support systems (operator's ability, sensors, controls, method of lubrication, and chip removal).

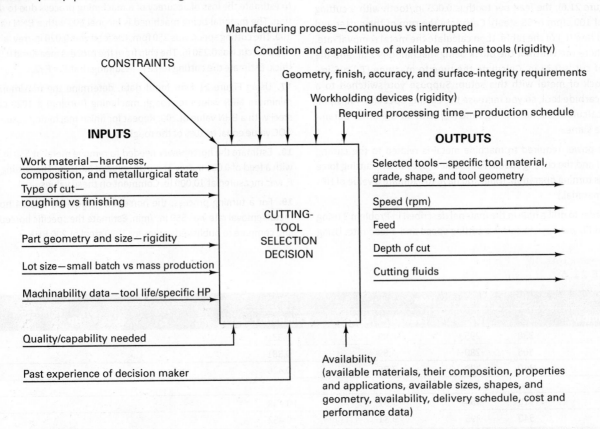

FIGURE 22.1 The selection of the cutting tool material and geometry followed by the selection of cutting conditions for a given application depends on many variables.

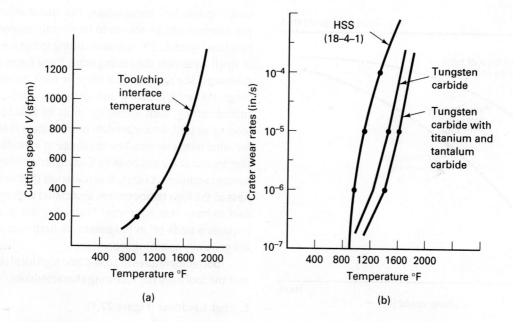

(a)

(b)

FIGURE 22.2 The typical relationship of temperature at the tool/chip interface to cutting speed shows a rapid increase. Correspondingly, the tool wears at the interface rapidly with increased temperature, often created by increased speed.

Tool material technology is continuosly improving, enabling many difficult to machine materials to be machined at higher removal rates and/or cutting speeds with greater performance reliability. Higher speed and/or removal rates usually improve productivity but also increase the temperature at the tool/chip interface as shown in **Figure 22.2**. The increase in temperature corresponds to increase in tool wear leading to tool failure. Predictable tool life is essential when machine tools are computer controlled with minimal operator interaction. Long tool life is desirable, especially when machines become automatic or are placed in manufacturing cells (see chapter 43).

There are three main sources of heat. Listed in order of their heat-generating capacity, they are shown in **Figure 22.3**:

1. The shear front itself, where plastic deformation results in the major heat source. Most of this heat stays in the chip.

2. The tool/chip interface contact region, where additional plastic deformation takes place in the chip and considerable heat is generated due to sliding friction. This heat goes into the chip and the tool.

3. The flank of the tool, where the freshly produced workpiece surface rubs the tool.

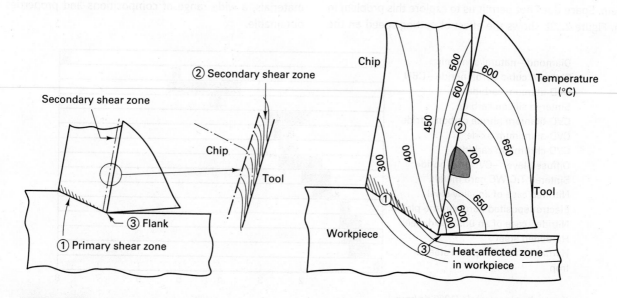

FIGURE 22.3 There are three main sources of heat in metal cutting: (1) primary shear zone; (2) secondary shear zone at the tool/chip interface; (3) tool flank. The peak temperature occurs at the center of the interface, in the shaded region, in the right schematic.

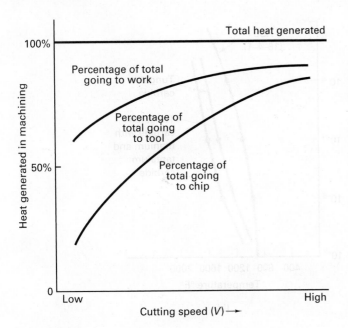

FIGURE 22.4 Distribution of heat generated in machining to the chip, tool, and workpiece. Heat going to the environment is not shown. Figure based on the work of A. O. Schmidt.

In metal cutting, the power put into the process is largely converted to heat, elevating the temperatures of the chip, the workpiece, and the tool. These three elements of the process, along with the environment (which includes the cutting fluid), act as the heat sinks. **Figure 22.4** shows the distribution of the heat to these three sinks as a function of cutting speed. As speed increases, a greater percentage of the heat ends up in the chip to the point where the chips can be cherry red or even burn at high cutting speeds.

There have been numerous experimental techniques developed to measure cutting temperatures and some excellent theoretical analyses of this "moving" multiple heat source problem. Space does not permit us to explore this problem in depth. Figure 22.2a shows the effect of cutting speed on the tool/chip interface temperature. The rate of wear of the tool at the interface can be shown to be directly related to temperature (see Figure 22.2b). Because cutting forces are concentrated in small areas near the cutting edge, these forces produce large pressures. The tool material must be hard (to resist wear) and tough (to resist cracking and chipping). Tools used in interrupted cutting, such as milling, must be able to resist impact loading as well. Tool materials must sustain their hardness at elevated temperatures. The challenge to manufacturers of cutting tools has always been to find materials that satisfy these severe conditions. Cutting tool materials that do not lose hardness at the high temperatures associated with high speeds are said to have "hot hardness." Obtaining this property usually requires a trade-off in toughness, as hardness and toughness are generally opposing properties.

Tool temperatures of 1000°C and high local stresses require that the tool have the following characteristics:

1. High **hardness** (**Figure 22.5**).
2. High hardness temperature, **hot hardness** (refer to **Figure 22.6**).
3. Resistance to abrasion, wear due to severe sliding friction.
4. Chipping of the cutting edges.
5. High toughness (impact strength) (refer to **Figure 22.7**).
6. Strength to resist bulk deformation.
7. Good chemical stability (inertness or negligible affinity with the work material).
8. Adequate thermal properties.
9. High elastic modulus (stiffness).
10. Correct geometry and surface finish.

Figure 22.8 compares these properties for various cutting tool materials. Overlapping characteristics exist in many cases. Exceptions to the rule are very common. In many classes of tool materials, a wide range of compositions and properties are obtainable.

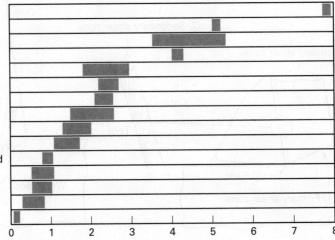

FIGURE 22.5 The hardness of various tool materials is compared, ranging from iron to diamond.

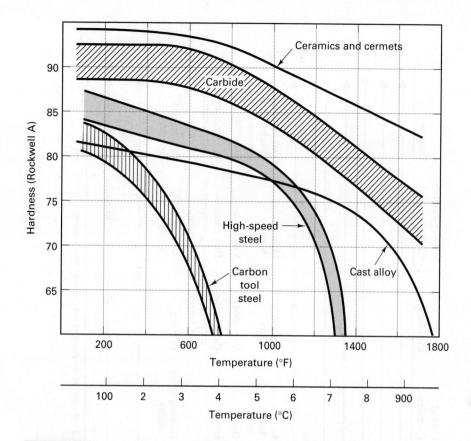

FIGURE 22.6 Decreasing hardness with increasing temperature is called hot hardness. Some materials display a more rapid drop in hardness above some temperatures. (From *Metal Cutting Principles*, 2nd ed. Courtesy of Ingersoll Cutting Tool Company)

Figure 22.5 compares various tool materials on the basis of hardness, the most critical characteristic. Hot hardness (hardness decreases slowly with temperature) is compared in Figure 22.6. Figure 22.7 compares hardness with **toughness,** or the ability to take impacts during interrupted cutting. Naturally, it would be wonderful if these materials were also easy to fabricate, readily available, and inexpensive, because cutting tools are routinely replaced, but this is not usually the case. Obviously, many of the requirements conflict and therefore, tool selection will always require trade-offs.

22.1 Cutting Tool Materials

In nearly all machining operations, cutting speed and feed are limited by the capability of the tool material. Speeds and feeds must be kept low enough to provide for an acceptable tool life. If not, the time lost changing tools may outweigh the productivity gains from increased cutting speed. Coated high-speed steel (HSS) and uncoated and coated carbides are currently the most extensively used tool materials.

Coated tools cost only about 15% to 20% more than uncoated tools, so a modest improvement in performance can justify the added cost. About 15% to 20% of all tool steels are coated, mostly by the **physical vapor deposition (PVD)** processes. Diamond and CBN are used for applications in which, despite higher cost, their use is justified. Cast cobalt alloys are being phased out because of the high raw-material cost and

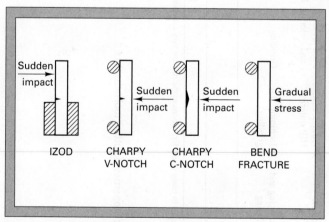

Methods of toughness testing

FIGURE 22.7 The most important properties for tool steels are hardness and roughness. Toughness (as considered for tooling materials) is the relative resistance of a material to breakage, chipping, or cracking under impact or stress. Toughness may be thought of as the opposite of brittleness. Toughness testing is not the same as standardized hardness testing. It may be difficult to correlate the results of different test methods. Common toughness tests are Izod Charpy impact tests and bend fracture tests.

	Carbon and Low-/Medium-Alloy Steels	High-Speed Steels	Sintered Cemented Carbides	Coated HSS	Coated Carbides	Ceramics	Polycrystalline CBN	Diamond
Toughness	← Decreasing							
Hot hardness	Increasing →							
Impact strength	← Decreasing							
Wear resistance	Increasing →							
Chipping resistance	← Decreasing							
Cutting speed	Increasing →							
Depth of cut	Light to medium	Light to heavy	Light to heavy	Light to heavy	Light to heavy	Light to heavy	Light to heavy	Very light for single-crystal diamond
Finish obtainable	Rough	Rough	Good	Good	Good	Very good	Very good	Excellent
Method of manufacture	Wrought	Wrought cast, HIP sintering	Cold pressing and sintering, PM	PVD[b] after forming	CVD[c]	Cold pressing and sintering or HIP sintering	High-pressure–high-temperature sintering	High-pressure–high-temperature sintering
Fabrication	Machining and grinding	Machining and grinding	Grinding	Machining and grinding, coating	Grinding before coating	Grinding	Grinding and polishing	Grinding and polishing
Thermal shock resistance	Increasing →							
Tool material cost	Increasing →							

[a] Overlapping characteristics exist in many cases. Exceptions to the rule are very common. In many classes of tool materials, a wide range of compositions and properties are obtainable.
[b] Physical vapor deposition.
[c] Chemical vapor deposition.

FIGURE 22.8 Salient properties of cutting tool materials.[a]

the increasing availability of alternate tool materials. New ceramic materials called *cermets* (ceramic material in a metal binder) are having a significant impact on future manufacturing productivity.

Tool requirements for other processes that use noncontacting tools, as in electrodischarge machining (EDM) and electrochemical machining (ECM), or *no tools at all* (as in laser machining), are discussed in Chapters 29 and 30. Grinding abrasives will be discussed in Chapter 28.

Tool Steels

Carbon steels and low-/medium-alloy steels, called **tool steels**, were once the most common cutting tool materials. Plain-carbon steels of 0.90% to 1.30% carbon when hardened and tempered have good hardness and strength and adequate toughness and can be given a keen cutting edge. However, tool steels lose hardness at temperatures above 400°F because of tempering and have largely been replaced by other materials for metal cutting.

The most important properties for tool steels are hardness, hot hardness, and toughness. Low-/medium-alloy steels have alloying elements such as molybdenum and chromium, which improve hardenability, and tungsten and molybdenum, which improve wear resistance. These tool materials also lose their hardness rapidly when heated to about their tempering temperature of 300° to 650°F, and they have limited abrasion resistance. Consequently, low-/medium-alloy steels are used in relatively inexpensive cutting tools (e.g., drills, taps, dies, reamers, broaches, and chasers) for certain low-speed cutting applications when the heat generated is not high enough to reduce their hardness significantly. High-speed steels, cemented carbides, and coated tools are also used extensively to make these kinds of cutting tools. Although more expensive, they have longer tool life and improved performance. These steels greatly benefit from **powder metallurgy (P/M)** manufacturing due to uniformly distributed carbides.

High-Speed Steels

First introduced in 1900 by F. W. Taylor and Mansel White, high-alloy steel is superior to tool steel in that it retains its cutting ability at temperatures up to 1100°F, exhibiting good "red hardness." Compared with tool steel, it can operate at about double or triple cutting speeds to about 100 sfpm with equal life, resulting in its name: **high-speed steel**, often abbreviated **HSS.**

Today's high-speed steels contain significant amounts of tungsten, molybdenum, cobalt, vanadium, and chromium besides iron and carbon. Tungsten, molybdenum, chromium, and cobalt in the ferrite (as a solid solution) provide strengthening of the matrix beyond the tempering temperature, thus increasing the hot hardness. Vanadium, along with tungsten, molybdenum, and chromium, improves hardness (R_c 65–70) and wear resistance. Extensive solid solutioning of the matrix also ensures good hardenability of these steels.

Although many formulations are used, a typical composition is that of the 18-4-1 type (tungsten 18%, chromium 4%, and vanadium 1 %), called T1. Comparable performance can also be obtained by the substitution of approximately 8% molybdenum for the tungsten, referred to as a tungsten equivalent. High-speed steel is still widely used for drills and many types of general-purpose milling cutters and in single-point tools used in general machining. For high-production machining, HSS has often been replaced by carbides, coated carbides, and coated HSS.

HSS main strengths are as follows:

- Great toughness—superior transverse rupture strength.
- Easily fabricated.
- Best for sever applications where complex tool geometry is needed (gear cutters, taps, drills, reamers, and dies).

High-speed steel tools are fabricated by three methods: cast, wrought, and sintered (using the powder metallurgy technique). Improper processing of cast and wrought products can result in carbide segregation, formation of large carbide particles and significant variation of carbide size, and nonuniform distribution of carbides in the matrix. The material will be difficult to grind to shape and will cause wide fluctuations of properties, inconsistent tool performance, distortion, and cracking.

To overcome some of these problems, a powder metallurgy technique has been developed that uses the hot-isostatic pressing (HIP) process on atomized, prealloyed tool steel mixtures. Because the various constituents of the P/M alloys are "locked" in place by the compacting procedure, the end product is a more homogeneous alloy. P/M high-speed steel cutting tools exhibit better grindability, greater toughness, better wear resistance, and higher red (or hot) hardness; they also perform more consistently. However they are about double the cost of regular HSS tools.

Titanium Nitride Coated High-Speed Steels

Coated high-speed steel provides significant improvements in cutting speeds, with increases of 10% to 20% being typical. First introduced in 1980 for gear cutters (hobs) and in 1981 for drills, TiN-coated HSS tools have demonstrated their ability to more than pay for the extra cost of the coating process.

In addition to hobs, gear-shaper cutters, and drills, HSS tooling coated by TiN now includes reamers, taps, chasers, spade-drill blades, broaches, bandsaw and circular saw blades, insert tooling, form tools, end mills, and an assortment of other milling cutters.

Physical vapor deposition has proven to be the most viable process for coating HSS, primarily because it is a relatively low-temperature process that does not exceed the tempering point of HSS. Therefore, no subsequent heat treatment of the cutting tool is required. Films 0.0001 to 0.0002 in. thick adhere well and withstand minor elastic, plastic, and thermal loads.

Thicker coatings tend to fracture under the typical thermomechanical stresses of machining.

There are many variations to the PVD process, as outlined in **Table 22.1**. The process usually depends on gas pressure and is performed in a vacuum chamber. PVD processes are carried out with the workpieces heated to temperatures in the range of 400° to 900°F. Substrate heating enhances coating adhesion and film structure.

Because surface pretreatment is critical in PVD processing, tools to be coated are subject to a vigorous cleaning process. Precleaning methods typically involve degreasing, ultrasonic cleaning, and freon drying. Deburring, honing, and more active cleaning methods are also used.

The main advantage of TiN-coated HSS tooling is reduced tool wear. Less tool wear results in less stock removal during tool regrinding, thus allowing individual tools to be reground more times. For example, a TiN hob can cut 300 gears per sharpening; the uncoated tool would cut only 75 parts per sharpening. Therefore, the cost per gear is reduced from 20¢ to 2¢. Naturally, reduced tool wear means longer tool life.

Higher hardness, with typical values for the thin coatings, "equivalent" to R_c 80–85, as compared to R_c 65–70 for hardened

TABLE 22.1	Surface Treatments for Cutting Tools			
Process	Method	Hardness[a] and Depth	Advantages	Limitations
Black oxide	HSS cutting tools are oxidized in a steam atmosphere at 1000°F	No change in prior steel hardness	Prevents built-up edge formations in machining of steel.	Strictly for HSS tools.
Nitriding case hardening	Steel surface is coated with nitride layer by use of cyanide salt at 900° to 1600°F, or ammonia gas or N_2 ions.	To 72 R_c; Case depth: 0.0001 to 0.100 in.	High production rates with bulk handling. High surface hardness. Diffuses into the steel surfaces. Simulates strain hardening.	Can only be applied to steel. Process has embrittling effect because of greater hardness. Post-heat treatment needed for some alloys.
Electrolytic electroplating	The part is the cathode in a chromic acid solution; anode is lead. Hard chronic plating is the most common process for wear resistance.	70-72 R_c 0.0002 to 0.100 in.	Low friction coefficient, antigalling Corrosion resistance. High hardness.	Moderate production: pieces must be fixtured. Part must be very clean. Coating does not diffuse into surface, which can affect impact properties.
Vapor deposition chemical vapor deposition (CVD)	Deposition of coating material by chemical reactions in the gaseous phase. Reactive gases replace a protective atmosphere in a vacuum chamber. At temperatures of 1800° to 1200°F, a thin diffusion zone is created between the base metal and the coating.	To 84 R_c; 0.0002 to 0.0004 in.	Large quantities per batch. Short reaction times reduce substrate stresses. Excellent adhesion. Recommended for forming tools. Multiple coatings can be applied (TiN-TiC. Al_2O_3), Line-of-sight not a problem.	High temperatures can affect substrate metallurgy, requiring post-heat treatment, which can cause dimensional distortion (except when coating sintered carbides). Necessary to reduce effects of hydrogen chloride on material properties, such as impact strength. Usually not diffused. Tolerances of + 0.001 in. required for HSS tools.
Physical vapor desposition (PVD sputtering)	Plasma is generated in a vacuum chamber by ion bombardment to dislodge particles from a target made of the coating material. Metal is evaporated and is condensed or attracted to substrate surfaces.	To 84 R_c; To 0.0002 in. thick	A useful experimental procedure for developing wear surfaces. Can coat substrates with metals, alloys, compounds, and refractories. Applicable for all tooling.	Not a high-production method. Requires care in cleaning. Usually not diffused.
PVD (electron beam)	A plasma is generated in vacuum by evaporation from a molten pool that is heated by an electron-beam gun.	To 84 R_c To 0.0002 in. thick	Can coat reasonable quantities per batch cycle. Coating materials are metals, compounds, alloys, and refractories. Substrate metallurgy is preserved. Very good adhesion. Fine particle deposition. Applicable for all tooling.	Parts require fixturing and orientation in line-of-sight process. Ultra-cleanliness required.
PVD/ARC	Titanium is evaporated in a vacuum and reacted with nitrogen gas. Resulting titanium nitride plasma is ionized and electrically attracted to the substrate surface. A high-energy process with multiple plasma guns.	To 85 R_c; To 0.0002 in. thick.	Process at 900°F preserves substrate metallurgy. Excellent coating adhesion. Controllable deposition of grain size and growth. Dimensions, surface finish, and sharp edges are preserved. Can coat all high-speed steels without distortion.	Parts must be fixtured for line-of-sight process. Parts must be very clean. No by-products formed in reaction. Usually only minor diffusion.

HSS, means reduced abrasion wear. Relative inertness (i.e., TiN does not react significantly with most workpiece materials) results in greater tool life through a reduction in adhesion. TiN coatings have a low coefficient of friction. This can produce an increase in the onset of shear angle, which in turn reduces the cutting forces, spindle power, and heat generated by the deformation processes. PVD coatings generally fail in high-stress applications such as cold extrusion, piercing, roughing, and high-speed machining.

Cast Cobalt Alloys

Cast cobalt alloys, popularly known as **stellite tools**, are cobalt-rich, chromium–tungsten–carbon cast alloys having properties and applications in the intermediate range between high-speed steel and cemented carbides. Although comparable in room temperature hardness to high-speed steel tools, cast cobalt alloy tools retain their hardness to a much higher temperature. Consequently, they can be used at higher cutting speeds (25% higher) than HSS tools. Cast cobalt alloys are hard as cast and cannot be softened or heat treated.

Cast cobalt alloys contain a primary phase of cobalt-rich solid solution strengthened by chromium and tungsten and dispersion hardened by complex hard, refractory carbides of tungsten and chromium. Other elements added include vanadium, boron, nickel, and tantalum. The casting provides a tough core and elongated grains normal to the surface. The structure is not, however, homogeneous.

Tools of cast cobalt alloys are generally cast to shape and finished to size by grinding. They are available only in simple shapes, such as single-point tools and saw blades, because of limitations in the casting process and the expense involved in the final shaping (grinding). The high cost of fabrication is primarily due to the high hardness of the material in the as-cast condition. Materials that can be machined with this tool material include plain-carbon steels, alloy steels, nonferrous alloys, and cast iron.

Cast cobalt alloys are currently being phased out for cutting-tool applications because of increasing costs, shortages of strategic raw materials (cobalt, tungsten, and chromium), and the development of other, superior tool materials of lower cost.

Carbide or Sintered Carbides

Carbide cutting tool inserts are traditionally divided into two primary groups:

1. Straight tungsten grades, which are used for machining cast irons, austenitic stainless steel, and nonferrous and nonmetallic materials.

2. Grades containing major amounts of titanium, tantalum, and/or columbium carbides, which are used for machining ferritic workpieces. There are also the titanium carbide

grades, which are used for finishing and semifinishing ferrous alloys.

The classification of carbide insert grades employs a C-classification system in the United States and ISO P and M classification system in Europe and Japan. These classifications are based on application, rather than composition or properties. Each cutting-tool vendor can provide proprietary grades and recommended applications.

Carbides, which are nonferrous alloys, are also called **sintered (or cemented) carbides** because they are manufactured by powder metallurgy techniques. The P/M process is discussed in Chapter 16. Here is a summary for carbide tools.

Tungsten is carburized in a high-temperature furnace, mixed with cobalt and blended in large ball mills. After ball milling, the powder is screened and dried. Paraffin is added *to* hold the mixture together for compacting. Carbide inserts are compacted using a pillpress. The compacted powder is sintered in a high-temperature vacuum furnace. The solid cobalt dissolves some tungsten carbide, then melts and fills the space between adjacent tungsten carbide grains. As the mixture is cooled, most of the dissolved tungsten carbide precipitates onto the surface of existing grains. After cooling, inserts are finish ground and honed or used in the pressed condition.

These materials became popular during World War II because they afforded a four- or fivefold increase in cutting speeds. The early versions had tungsten carbide as the major constituent, with a cobalt binder in amounts of 3% to 13%. Most carbide tools in use today are either straight WC or multicarbides of W–Ti or W–Ti–Ta, depending on the work material to be machined. Cobalt is the binder. These tool materials are much harder, are chemically more stable, have better hot hardness, have high stiffness, have lower friction, and operate at higher cutting speeds than HSS. They are more brittle and more expensive and use strategic metals (tungsten, tantalum, and cobalt) more extensively.

Cemented carbide tool materials based on TiC have been developed primarily for auto industry applications using predominantly nickel and molybdenum as a binder. These are used for higher-speed (>1000 ft/min) finish machining of steels and some malleable cast irons.

Cemented carbide tools are available in insert form in many different shapes: squares, triangles, diamonds, and rounds. They can be either brazed or mechanically clamped onto the tool shank. Mechanical clamping (**Figure 22.9**) is more popular because when one edge or corner becomes dull, the insert is rotated or turned over to expose a new cutting edge. Mechanical inserts can be purchased in the as-pressed state, or the insert can be ground to closer tolerances. Naturally, precision-ground inserts cost more. Any part tolerance less than ±0.003 normally cannot be manufactured without radial adjustment of the cutting tool, even with ground inserts. If no radial adjustment is performed, precision-ground inserts should be used only when the part tolerance is between ±0.006 and ±0.003.

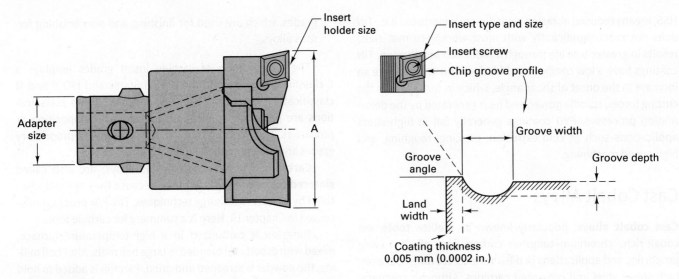

FIGURE 22.9 Boring head with two carbide inserts, with chip grooves.

Pressed inserts have an application advantage because the cutting edge is unground and thus does not leave grinding marks on the part after machining. Ground inserts can break under heavy cutting loads because the grinding marks on the insert produce stress concentrations that result in brittle fracture. Diamond grinding is used to finish carbide tools. Abusive grinding can lead to thermal cracks and premature (early) failure of the tool.

Brazed tools have the carbide insert brazed to the steel tool shank. These tools will have a more accurate geometry than the mechanical insert tools, but they are more expensive. Because cemented carbide tools are relatively brittle, a 90° corner angle at the cutting edge is desired. To strengthen the edge and prevent edge chipping, it is rounded off by honing, or an appropriate chamfer or a negative land (a T-land) on the rake face is provided. The preparation of the cutting edge can affect tool life. The sharper the edge (smaller edge radius), the more likely the edge is to chip or break. Increasing the edge radius will increase the cutting forces, so a trade-off is required. Typical edge radius values are 0.001 to 0.003 in.

A **chip groove** (see Figure 22.9) with a positive rake angle at the tool tip may also be used to reduce cutting forces without reducing the overall strength of the insert significantly. The groove also breaks up the chips by causing them to curl tightly, thus making disposal easier.

For very low-speed cutting operations, the chips tend to weld to the tool face and cause subsequent microchipping of the cutting edge. Cutting speeds for carbides are generally in the range of 150 to 600 ft/min. Higher speeds (>1000 ft/min) are recommended for certain less-difficult-to-machine materials (such as aluminum alloys) and much lower speeds (100 ft/min) for more difficult-to-machine materials (such as titanium alloys). In interrupted cutting applications, it is important to prevent edge chipping by choosing the appropriate cutter geometry and cutter position with respect to the workpiece. For interrupted cutting, finer grain size and

higher cobalt content improve toughness in straight WC–Co grades.

After use, carbide inserts (called disposable or throwaway inserts) are generally recycled in order to reclaim the tantalum, WC, and cobalt. This recycling not only conserves strategic materials but also reduces costs. A new trend is to regrind these tools for future use where the actual size of the insert is not of critical concern.

Coated-Carbide Tools

Beginning in 1969 with TiC-coated WC, **coated tools** became the norm in the metalworking industry because coating can consistently improve tool life 200% to 300% or more. In cutting tools, material requirements at the surface of the tool need to be abrasion resistant, hard, and chemically inert to prevent the tool and the work material from interacting chemically with each other during cutting. A thin, chemically stable, hard refractory coating accomplishes this objective. The bulk of the tool is a tough, shock-resistant carbide that can withstand high-temperature plastic deformation and resist breakage. The result is a composite tool as shown in **Figure 22.10**. The coatings must be fine grained, free of binders and porosity. Naturally, the coatings must be metallurgically bonded to the substrate. Interface coatings are graded to match the properties of the coating and the substrate. The coatings must be thick enough to prolong tool life but thin enough to prevent brittleness.

Coatings should have a low coefficient of friction so the chips do not adhere to the rake face. Coating materials include single coatings of TiC, TiN, Al_2O_3, HfN, or HfC. Multiple coatings are used, with each layer imparting its own characteristic to the tool. Successful coating combinations include TiN/TiC/TiCN/TiN and TiC/Al_2O_3/TiN.

Chemical vapor deposition is used to obtain coated carbides. The coatings are formed by chemical reactions that

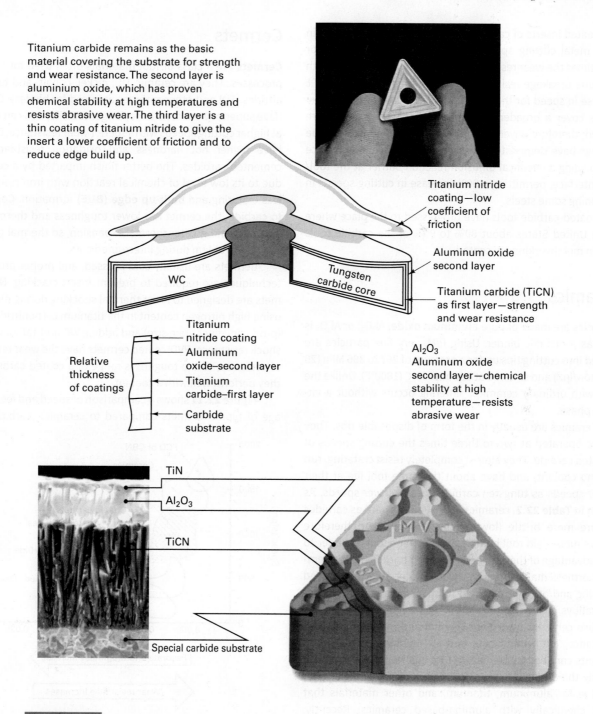

Titanium carbide remains as the basic material covering the substrate for strength and wear resistance. The second layer is aluminium oxide, which has proven chemical stability at high temperatures and resists abrasive wear. The third layer is a thin coating of titanium nitride to give the insert a lower coefficient of friction and to reduce edge build up.

Titanium nitride coating—low coefficient of friction

Aluminum oxide second layer

Titanium carbide (TiCN) as first layer—strength and wear resistance

WC

Tungsten carbide core

Relative thickness of coatings

Titanium nitride coating
Aluminum oxide—second layer
Titanium carbide—first layer
Carbide substrate

Al_2O_3 Aluminum oxide second layer—chemical stability at high temperature—resists abrasive wear

TiN
Al_2O_3
TiCN

Special carbide substrate

FIGURE 22.10 Triple-coated carbide tools provide resistance to wear and plastic deformation in machining of steel, abrasive wear in cast iron, and built-up edge formation. (Courtesy J T. Black)

take place only on or near the substrate. Like electroplating, CVD is a process in which the deposit is built up atom by atom. It is, therefore, capable of producing deposits of maximum density and of closely reproducing fine detail on the substrate surface.

Control of critical variables such as temperature, gas concentration, and flow pattern is required to ensure adhesion of the coating to the substrate. The coating-to-substrate adhesion must be better for cutting tool inserts than for most other coatings applications to survive the cutting pressure and temperature conditions without flaking off. Grain size and shape are controlled by varying temperature and/ or pressure.

The purpose of multiple coatings is to tailor the coating thickness for prolonged tool life. Multiple coatings allow a stronger metallurgical bond between the coating and the substrate and provide a variety of protection processes for machining different work materials, thus offering a more general-purpose tool material grade. A very thin final coat of TiN coating (in microns, or μm) can effectively reduce crater formation on the tool face by one to two orders of magnitude relative to uncoated tools.

Coated inserts of carbides are finding wide acceptance in many metal cutting applications. Coated tools have two or three times the wear resistance of the best uncoated tools with the same breakage resistance. This results in a 50% to 100% increase in speed for the same tool life. Because most coated inserts cover a broader application range, fewer grades are needed; therefore, inventory costs are lower. Aluminum oxide coatings have demonstrated excellent crater wear resistance by providing a chemical diffusion reaction barrier at the tool/chip interface, permitting a 90% increase in cutting speeds in machining some steels.

Coated-carbide tools have progressed to the place where in the United States about 80% to 90% of the carbide tools used in metalworking are coated.

Ceramics

Ceramics are made of pure **aluminum oxide**, Al_2O_3, or Al_2O_3 is used as a metallic binder. Using P/M, very fine particles are formed into cutting tips under a pressure of 267 to 386 MPa (20 to 28 ton/in.²) and sintered at about 1000°C (1800°F). Unlike the case with ordinary ceramics, sintering occurs without a vitreous phase.

Ceramics are usually in the form of disposable tips. They can be operated at two to three times the cutting speeds of tungsten carbide. They almost completely resist cratering, run with no coolant, and have about the same tool life at their higher speeds as tungsten carbide does at lower speeds. As shown in **Table 22.2**, ceramics are usually as hard as carbides but are more brittle (lower bend strength) and therefore require more rigid tool holders and machine tools in order to take advantage of their capabilities. Their hardness and chemical inertness make ceramics a good material for high-speed finishing and/or high-removal-rate machining applications of superalloys, hard-chill cast iron, and high-strength steels. Because ceramics have poor thermal and mechanical shock resistance, interrupted cuts and interrupted application of coolants can lead to premature tool failure. Edge chipping is usually the dominant mode of tool failure. Ceramics are not suitable for aluminum, titanium, and other materials that react chemically with alumina-based ceramics. Recently, whisker-reinforced ceramic materials that have greater transverse rupture strength have been developed. The whiskers are made from silicon carbide.

Cermets

Cermets are a class of tool materials often used for finishing processes. They are ceramic TiC, nickel, cobalt, and tantalum nitrides. TiN and other carbides are used for binders. Cermets have superior wear resistance, longer tool life, and can operate at higher cutting speeds with superior wear resistance. Cermets have higher hot hardness and oxidation resistance than cemented carbides. The better finish imparted by a cermet is due to its low level of chemical reaction with iron resulting in less cratering and **built-up edge (BUE)** formation. Compared to carbide, the cermet has lower toughness and thermal conductivity, and greater thermal expansion, so thermal cracking can be a problem during interrupted cuts.

Cermets are usually cold pressed, and proper processing techniques are required to prevent insert cracking. New cermets are designed to resist thermal shocking during milling by using high nitrogen content in the titanium carbonitride phase (producing finer grain size) and adding WC and TaC to improve shock resistance. PVD-coated cermets have the wear resistance of cermets and the toughness range of a coated carbide, and they perform well with a coolant.

Figure 22.11 shows a comparison of speed and feed coverage of typical cermets compared to ceramics, carbides, and

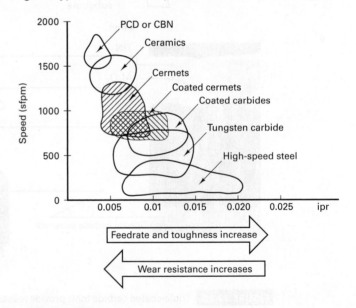

FIGURE 22.11 Cermets compared to other tool materials on the basis of speed and feed rate.

TABLE 22.2	Properties of Cutting Tool Materials Compared for Carbides, Ceramics, HSS, and Cast Cobalt[a]			
	Hardness Rockwell A or C	Transverse Rupture (bend) Strength (× 10³ psi)	Compressive Strength (× 10³ psi)	Modulus of Elasticity (e) (× 10⁶ psi)
Carbide C1–C4	90–95 R_A	250–320	750–860	89–93
Carbide C5–C8	91–93 R_A	100–250	710–840	66–81
High–speed steel	86 R_A	600	600–650	30
Ceramic (oxide)	92–94 R_A	100–125	400–650	50–60
Cast cobalt	46–62 R_C	80–120	220–335	40

[a]Exact properties depend on materials, grain size, bonder content, volume.

TABLE 22.3   **Comparison of cermets with various cutting tool materials.**

Tool Material Group	General Applications	Versus Cermet
PCD (polycrystal diamond)	High-speed machining of aluminum alloys, nonferrous metals, and nonmetals.	Cermets can machine same materials, but at lower speeds and significantly less cost per corner.
CBN (cubic boron nitride)	Hard workpieces and high-speed machining on cast irons.	Cermets cannot machine the harder workpieces that CBN can. Cermets cannot machine cast iron at the speeds CBN can. The cost per corner of cermets is significantly less.
Ceramics (cold press)	High-speed turning and grooving of steels and cast iron.	Cermets are more versatile and less expensive than cold press ceramics but cannot run at the higher speeds.
Ceramics (hot press)	Turning and grooving of hard workpieces; high-speed finish machining of steels and irons.	Cermets cannot machine the harder workpieces or run at the same speeds on steels and irons but are more versatile and less expensive.
Ceramics (silicon nitride)	Rough and semi rough machining of cast irons in turning and milling applications at high speeds and under unfavorable conditions.	Cermets cannot machine cast iron at the high speeds of silicon nitride ceramics, but in moderate-speed applications cermets may be more cost effective.
Coated carbide	General-purpose machining of steels, stainless steels, cast iron, etc.	Cermets can run at higher cutting speeds and provide better tool life at less cost for semiroughing to finishing applications.
Carbides	Tough material for lower-speed applications on various materials.	Cermets can run at higher speeds, provide better surface finishes and longer tool life for semiroughing to finishing applications.

coated carbides. See discussion in **Table 22.3**, showing that cermets can clearly cover a wide range of important metal-cutting applications.

Diamonds

Diamond is the hardest material known. Industrial diamonds are now available in the form of polycrystalline compacts, which are finding industrial application in the machining of aluminum, bronze, and plastics, at greatly reduced cutting forces compared to carbides. Diamond machining is done at high speeds, with fine feeds for finishing, and produces excellent finishes. Recently, *single-crystal* diamonds, with a cutting-edge radius of 100 Å or less, have been used for precision

machining of large mirrors. However, single-crystal diamonds have been used for years to machine brass watch faces, thus eliminating polishing. They have also been used to slice biological materials into thin films for viewing in transmission electron microscopes. This process, known as *ultramicrotomy,* is one of the few industrial versions of orthogonal machining in common practice.

The salient features of diamond tools include high hardness; good thermal conductivity; the ability to form a sharp edge of cleavage in single-crystal, natural diamond; very low friction; nonadherence to most materials; the ability to maintain a sharp edge for a long period of time, especially in machining soft materials such as copper and aluminum; and good wear resistance.

Diamonds do have some shortcomings, which include a tendency to interact chemically with elements of Group IVB to Group VIII of the periodic table. In addition, diamond wears rapidly when machining or grinding mild steel. It wears less rapidly with high-carbon alloy steels than with low-carbon steel and has occasionally machined gray cast iron (which has high carbon content) with long life. Diamond has a tendency to revert at high temperatures (700°C) to graphite and/or to oxidize in air. Diamond is very brittle and is difficult and costly to shape into cutting tools, the process for doing the latter being a tightly held industry practice.

Polycrystalline Diamonds

The limited supply of, increasing demand for, and high cost of natural diamonds led to the ultra-high-pressure (50 Kbar), high-temperature (1500°C) synthesis of diamond from graphite at the General Electric Company in the mid-1950s and the subsequent development of *polycrystalline* sintered diamond tools in the late 1960s.

Polycrystalline diamond (PCD) tools consist of a thin layer (0.5 to 1.5 mm) of fine grain size diamond particles sintered together and metallurgically bonded to a cemented carbide substrate. A high-temperature/high-pressure process, using conditions close to those used for the initial synthesis of diamond, is needed. Fine diamond powder (1 to 30 μm) is first packed on a support base of cemented carbide in the press. At the appropriate sintering conditions of pressure and temperature in the diamond stable region, complete consolidation and extensive diamond-to-diamond bonding take place. Laser cutting followed by grinding is used to shape, size, and accurately finish PCD tools, as outlined in **Figure 22.12**. The cemented carbide provides the necessary elastic support for the hard and brittle diamond layer above it. The main advantages of sintered polycrystalline tools over natural single-crystal tools are better quality, greater toughness, and improved wear resistance, resulting from the random orientation of the diamond grains and the lack of large cleavage planes. Diamond tools offer dramatic performance improvements over carbides. Tool life is often greatly improved, as is control over part size, finish, and surface integrity.

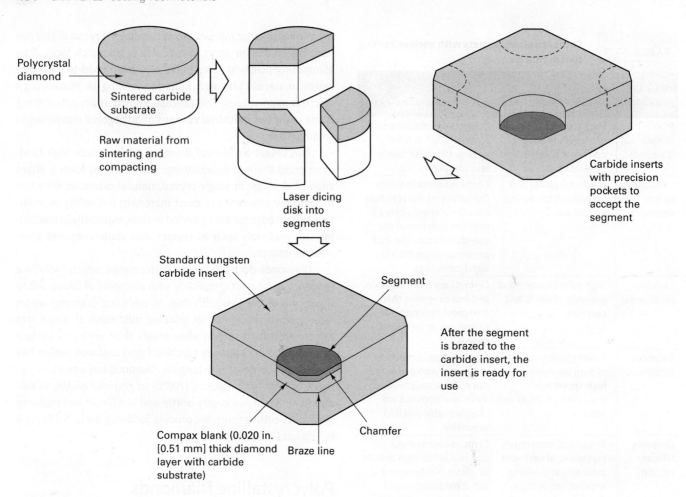

FIGURE 22.12 The process to make polycrystalline diamond insert tools is restricted to simple geometres

Positive rake tooling is recommended for the vast majority of diamond tooling applications. If BUE formation is a problem, increasing cutting speed and using more positive rake angles may eliminate it. If edge breakage and chipping are problems, the feed rate can be reduced. Coolants are not generally used in diamond machining unless, as in the machining of plastics, it is necessary to reduce airborne dust particles. Diamond tools can be reground.

There is much commercial interest in being able to coat HSS and carbides directly with diamond, but getting the diamond coating to adhere reliably has been difficult. Diamond-coated inserts would deliver roughly the same performance as PCD tooling when cutting nonferrous materials but could be given more complex geometries and chip breakers while reducing the cost per cutting edge.

Polycrystalline Cubic Boron Nitrides

Polycrystalline cubic boron nitride (PCBN) is a man-made tool material widely used in the automotive industry for machining hardened steels and superalloys. It is made in a compact form for tools by a process quite similar to that used for sintered polycrystalline diamonds. It retains its hardness at elevated temperatures (Knoop 4700 at 20°C, 4000 at 1000°C)

and has low chemical reactivity at the tool/chip interface. This material can be used to machine hard aerospace materials like Inconel 718 and René 95 as well as chilled cast iron.

Although not as hard as diamond, PCBN is less reactive with such materials as hardened steels, hard-chill cast iron, and nickel- and cobalt-based superalloys. PCBN can be used efficiently and economically to machine these difficult-to-machine materials at higher speeds (fivefold) and with a higher removal rate (fivefold) than cemented carbide, and with superior accuracy, finish, and surface integrity. PCBN tools are available in basically the same sizes and shapes as sintered diamond and are made by the same process. The cost of an insert is somewhat higher than either cemented carbide or ceramic tools, but the tool life may be five to seven times that of a ceramic tool. Therefore, to see the economy of using PCBN tools, it is necessary to consider all the factors.

The two predominant wear modes of PCBN tools are notching at the **depth-of-cut line (DCL)** and **microchipping.** In some cases, the tool will exhibit flank wear of the cutting edge. These tools have been used successfully for heavy interrupted cutting and for milling white cast iron and hardened steels using negative lands and honed cutting edges. See **Table 22.4** for suggested applications of CBN and diamonds along with carbides and ceramics.

TABLE 22.4	Suggested Application of Four (4) Cutting Tool Materials to Workpiece Materials				
	Applicable Tool Material				
Workplace Material	**Carbide-Coated Carbide**	**Ceramic, Cermet**	**(use for finishing cuts)**	**Cubic Boron Nitride**	**Diamond Compacts**
Cast irons or carbon steels	X	X			
Alloy steels or alloy cast iron	X	X	uninterrupted finishing cuts	X	
Aluminium, brass	X	X			X
High-silicon aluminium	X				X
Nickel-based	X	X		X	
Titanium	X				
Plastic composites	X			X	

Because diamond and PCBN are extremely hard but brittle materials, new demands are being placed on the machine tools and on machining practice in order to take full advantage of the potential of these tool materials. These demands include:

- Use of more rigid machine tools and machining practices involving gentle entry and exit of the cut in order to prevent microchipping.
- Use of high-precision machine tools, because these tools are capable of producing high finish and accuracy.
- Use of machine tools with higher power, because these tools are capable of higher metal removal rates and faster spindle speeds.

22.2 Tool Geometry

Selecting tool geometry is a critical part of selecting the cutting tool. Tool geometry can be very complex. **Figure 22.13** shows the cutting-tool geometry for a single-point HSS tool used in turning in oblique, threeforce machining. The **back rake angle** affects the ability of the tool to shear the work material and form the chip. It can be positive or negative. Positive rake angles reduce the cutting forces, resulting in smaller deflections of the workpiece, tool holder, and machine. In machining hard work materials, the back rake angle must be small, even negative for carbide and diamond tools. Generally speaking, the higher the hardness of the workpiece, the smaller the back rake angle. For high-speed steels, back rake angle is normally chosen in the positive range, depending on the type of tool (turning, planing, end milling, face milling, drilling, etc.) and the work material.

For carbide tools, inserts for different work materials and tool holders can be supplied with several standard values of back rake angle: −6 to +6°. The side rake angle and the back rake angle combine to form the *effective rake angle*. This is also called the *true* rake angle or *resultant* rake angle of the tool.

True rake inclination of a cutting tool has a major effect in determining the amount of chip compression and the onset of shear angle, ϕ. A small rake angle causes high compression, tool forces, and friction, resulting in a thick, highly deformed, hot chip. Increasing the back or side rake angles reduces the compression, the forces, and the friction, yielding a thinner, less deformed, cooler chip. In general, the power consumption is reduced by approximately 1% for each 1° change in the back rake angle. The end relief angle gamma (γ) is also called the clearnce angle.

Unfortunately, it is difficult to take much advantage of the desirable effects of larger positive rake angles because they are offset by the reduced strength of the cutting tool, due to the reduced tool section, which also greatly reduce the tool's capacity to conduct heat away from the cutting edge.

To provide greater strength at the cutting edge and better heat conductivity, zero or negative rake angles are commonly employed on carbide, ceramic, polydiamond, and PCBN cutting tools. These materials tend to be brittle, but their ability to hold their superior hardness at high temperatures results in their selection for high-speed and continuous machining operations. While the negative rake angle increases tool forces, it keeps the tool in compression and provides added support to the cutting edge. This is particularly important in making intermittent cuts, as in milling, and in absorbing the impact during the initial engagement of the tool and work.

The wedge angle, θ, determines the strength of the tool and its capacity to conduct heat and depends on the values of α and γ. The relief angles mainly affect the tool life and the surface quality of the workpiece. To reduce the deflections of the tool and the workpiece, and to provide good surface quality, larger relief values are required. For high-speed steel, relief angles in the range of 5 to 10° are normal, with smaller values being for the harder work materials. For carbides, the relief angles are lower to give added strength to the tool.

The side and end cutting-edge angles define the nose angle and characterize the tool design. The nose radius has a major influence on surface finish. Increasing the nose radius usually decreases tool wear and improves surface finish.

Tool nomenclature varies with different cutting tools, manufacturers, and users. Many terms are still not standard

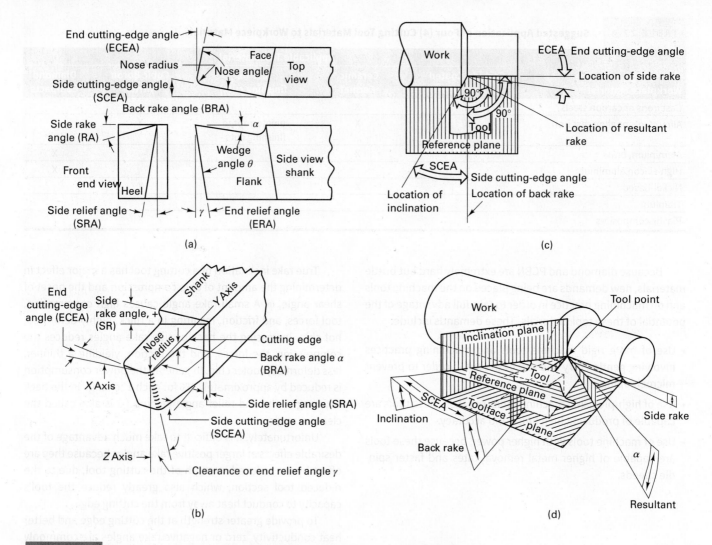

FIGURE 22.13 Standard terminology to describe the geometry of single-point tools: (a) three dimensional views of tool, (b) oblique view of tool from cutting edge, (c) top view of turning with single-point tool, (d) oblique view from shank end of single-point turning tool.

because of this variety. The most common tool terms will be used in later chapters to describe specific cutting tools.

The introduction of coated tools has spurred the development of improved tool geometries. Specifically, **low-force groove (LFG)** geometries have been developed that reduce the total energy consumed and break up the chips into shorter segments. These grooves effectively increase the rake angle, which increases the onset of shear angle and lowers the cutting force and power. This means that higher cutting speeds lower cutting temperatures and better tool lives are possible.

As a chip breaker, the groove deflects the chip at a sharp angle and causes it to break into short pieces that are easier to remove and are not as likely to become tangled in the machine and possibly causing injury to personnel. This is particularly important on high-speed and mass-production machines.

The shapes of cutting tools used for various operations and materials are compromises, resulting from experience and research so as to provide good overall performance. For coated tools, edge strength is an important consideration. A thin coat enables the edge to retain high strength, but a thicker coat

exhibits better wear resistance. Normally, tools for turning have a coating thickness of 6 to 12 μm. Edge strength is higher for multilayer coated tools. The radius of the edge should be 0.0005 to 0.005 in.

22.3 Tool-Coating Processes

The two most effective coating processes for improving the life and performance of tools are the chemical vapor deposition and physical vapor deposition of **titanium nitride (TiN)** and **titanium carbide (TiC).** The selection of the *cutting materials for cutting tools* depends on what property you are seeking. See **Table 22.5**.

The CVD process, used to deposit a protective coating onto carbide inserts, has been benefited the metal removal industry for many years and is now being applied with equal success to steel. The PVD processes have quickly become the preferred

TABLE 22.5 Cutting Tool material selection

For oxidation and corrosion resistance; high-temperature stability	select	Al_2O_3, TiN, TiC
For crater resistance	select	Al_2O_3, TiN, TiC
For hardness and edge retention	select	TiC, TiN, Al_2O_3
For abrasion resistance and flank wear	select	Al_2O_3, TiN, TiC
For low coefficient of friction and high lubricity	select	TiN, Al_2O_3, TiC
For fine grain size	select	TiN, TiC, Al_2O_3

TiN coating processes for high-speed steel and carbide-tipped cutting tools.

Chemical Vapor Deposition

Chemical vapor deposition is an atmosphere-controlled process carried out at temperatures in the range of 950° to 1050°C (1740° to 1920°F). **Figure 22.14** shows a schematic of the CVD process.

Cleaned tools ready to be coated are staged on precoated graphite work trays (shelves) and loaded onto a central gas distribution column (tree). The tree loaded with parts to be coated is placed inside the retort of the CVD reactor. The tools are heated under an inert atmosphere until the coating temperature is reached. The coating cycle is initiated by the introduction of titanium tetrachloride ($TiCL_4$), hydrogen, and methane (CH_4) into the reactor. $TiCL_4$ is a vapor and is transported into the reactor via a hydrogen carrier gas; CH_4 is introduced directly. The chemical reaction for the formation of TiC is:

$$TiCl_4 + CH_4 \rightarrow TiC + 4HCl \qquad (22.1)$$

To form titanium nitride, a nitrogen–hydrogen gas mixture is substituted for methane. The chemical reaction for TiN is:

$$2TiCl_4 + 2H_2 + N_2 \rightarrow 2TiN + 4HCl \qquad (22.2)$$

Physical Vapor Deposition

The simplest form of PVD is evaporation, where the substrate is coated by condensation of a metal vapor. The vapor is formed from a source material called the *charge,* which is heated to a temperature less than 1000°C. PVD methods currently being used include reactive sputtering, reactive ion plating, low-voltage electron-beam evaporation, triode high-voltage electron-beam evaporation, cathodic evaporation, and arc evaporation. In each of the methods, the TiN coating is formed by reacting free titanium ions with nitrogen away from the surface of the tool and relying on a physical means to transport the coating onto the tool surface.

All of these PVD processes share the following common features:

1. The coating takes place inside a vacuum chamber under a hard vacuum with the workpiece heated to 200° to 405°C (400° to 900°F).

2. Before coating, all parts are given a final cleaning inside the chamber to remove oxides and improve coating adhesion.

3. The coating temperature is relatively low for cutting and forming tools. It is typically about 450°C (842°F).

4. The metal source is vaporized in an inert gas atmosphere (usually argon), and the metal atoms react with gas to form the coating. Nitrogen is the reactive gas for nitrides. Methane or acetylene (along with nitrogen) is used for carbides.

5. These are ion-assisted deposition processes. The ion bombardment compresses the atoms on the growing film, yielding a dense, well-adhered coating.

A typical cycle time for the coating of functional tools, including heat-up and cool down, is about 6 hours.

In **Figure 22.15**, PVD arc evaporation is shown. The plasma sources are from several arc evaporators located on the sides and top of the vacuum chamber. Each evaporator generates plasma from multiple arc spots. In this way, a highly localized electrical arc discharge causes minute evaporation of the material of the cathode, and a self-sustaining

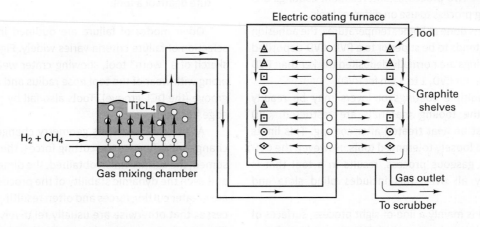

FIGURE 22.14 Chemical vapor deposition (CVD) is used to apply layers (TiC, TiN, etc.) to carbide cutting tools.

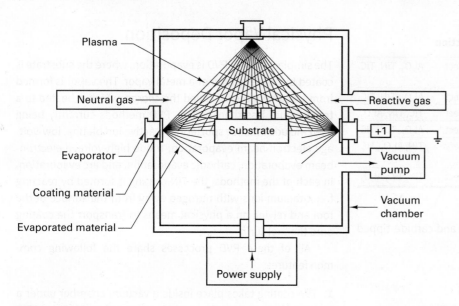

Plasma

Neutral gas

Evaporator

Coating material

Evaporated material

Reactive gas

Substrate

+1

Vacuum pump

Vacuum chamber

Power supply

FIGURE 22.15 Schematic of physical vapor deposition (PVC) arc evaporation process.

arc is produced that generates a high-energy and concentrated plasma.

The kinetic energy of deposition is much greater than that found in any other PVD method. During coating, this energy is of the order of 150 eV and more. Therefore, the plasma is highly reactive and the greater percentage of the vapor is atomic and ionized.

Coating temperatures can be selected and controlled so that metallurgy is preserved. This enables a coating of a wide variety of sintered carbide tools—for example, brazed tools and solid carbide tools such as drills, end mills, form tools, and inserts. The PVD arc evaporation process will preserve substrate metallurgy, surface finish, edge sharpness, geometrical straightness, and dimensions.

CVD and PVD—Complementary Processes

CVD and PVD are complementary coating processes. The differences between the two processes and resultant coatings dictate which coating process to use on different tools.

Because CVD is done at higher temperatures, the adhesion of these coatings tends to be superior to a PVD–CVD-deposited coating. CVD coatings are normally deposited thicker than PVD coatings (6 to 9 μm for CVD, 1 to 3 μm for PVD).

With CVD multiple coatings or layers may be readily deposited, but the tooling materials are restricted. CVD coated tools must be heat treated after coating. This limits the application to loosely toleranced tools. However, the CVD process, being a gaseous process, results in a tool that is coated uniformly all over; this includes blind slots and blind holes.

Because PVD is mainly a line-of-sight process, surfaces of the part not to be coated are masked. PVD also requires fixturing of each part in order to affect the substrate bias.

Applications

Applications for the two different processes are as follows:

CVD
- Loosely toleranced tooling.
- Piercing and blanking punches, trim dies, phillips punches, upsetting punches.
- AISI A, D, H, M, and air-hardening and tool steel parts.
- Solid carbide tooling.

PVD
- All HSS, solid carbide, and carbide-tipped cutting tools.
- Fine blanking punches, dies (0.001 in. tolerance or less).
- Non-composition-dependent process; virtually all tooling materials, including mold steels and bronze.

22.4 Tool Failure and Tool Life

In metal cutting, the failure of the cutting tool can be classified into two broad categories, according to the failure mechanisms that caused the tool to die (or fail):

1. *Physical failures* mainly include gradual tool wear on the flank(s) of the tool below the cutting edge (called **flank wear**) or wear on the rake face of the tool (called **crater wear**) or both. See **Figure 22.16**.
2. *Chemical failures*, which include wear on the rake face of the tool (crater wear) are rapid, usually unpredictable, and often catastrophic failures resulting from abrupt, premature death of a tool.

Other modes of failure are outlined in **Table 22.6**. The selection of failure criteria varies widely. Figure 22.16 shows a sketch of a "worn" tool, showing crater wear and flank wear, along with wear of the tool nose radius and an outer-diameter groove (the DCL groove). Tools also fail by edge chipping and edge fracture.

As the tool wears, its geometry changes. This geometry change will influence the cutting forces, the power being consumed, the surface finish obtained, the dimensional accuracy, and even the dynamic stability of the process. Worn tools create greater cutting forces and often resulting in chatter in processes that otherwise are usually relatively free of vibration. The actual wear mechanisms active in this high-temperature environment are abrasion, adhesion, diffusion, or chemical

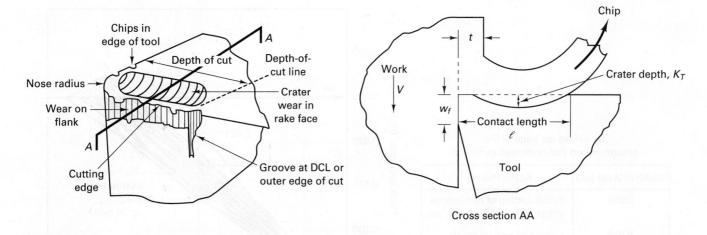

FIGURE 22.16 Tools can fail in many ways. Tool wear during oblique cutting can occur on the flank or the rake face; t = uncut chip thickness; k_t = crater depth; w_f = flank wear land length; DCL = depth-of-cut line.

TABLE 22.6	Modes of Tool Failure and Probable Causes		
No.	**Failure**		**Cause**
1–3	Flank wear		Due to the abrasive effect of hard grains contained in the work material
4–5	Groove		Due to wear at the DCL or outer edge of the cut
6	Chipping	Physical	Fine chips caused by high-pressure cutting, chatter, vibration, etc.
7	Partial fracture		Due to the mechanical impact when an excessive force is applied to the cutting edge
8	Crater wear		Carbide particles are removed due to degradation of tool performances and chemical reactions at high temperature
9	Deformation	Chemical	The cutting edge is deformed due to its softening at high temperature
10	Thermal crack		Thermal fatigue in the heating and cooling cycle with interrupted cutting
1	Built-up edge		A portion of the workpiece material adheres to the insert cutting edge

interactions. It appears that in metal cutting, any or all of these mechanisms may be operative at a given time in a given process.

Tool failure by plastic deformation, brittle fracture, fatigue fracture, or edge chipping can be unpredictable. Moreover, it is difficult to predict which mechanism will dominate and result in a tool failure in a particular situation. What can be said is that tools, like people, die (or fail) from a great variety of causes under widely varying conditions. Therefore, **tool life** should be treated as a random variable, or probabilistically, not as a deterministic quantity.

22.5 Taylor Tool Life

During machining, the tool is performing in a hostile environment in which high-contact stresses and high temperatures are commonplace; therefore, tool wear is always an unavoidable consequence. At lower speeds and temperatures, the tool most commonly wears on the flank. Suppose that the tool wear experiment were to be repeated 15 times without

changing any of the input parameters. The result would look like **Figure 22.17**, which depicts the variable nature of tool wear and shows why tool wear must be treated as a random variable. In Figure 22.17, the average time is denoted as μ_T and the standard deviation as σ_T, where the wear limit criterion was 0.025 in. At a given time during the test, 35 min, the tool displayed flank wear ranging from 0.013 to 0.021 in, with an average of μ_w = 0.0175 in. with standard deviation σ_w = 0.0013 in.

In **Figure 22.18**, four characteristic tool wear curves (average values) are shown for four different cutting speeds, V_1 through V_4; V_1 is the fastest cutting speed and therefore generates the fastest wear rates. Such curves often have three general regions, as shown in the figure. The central region is a steady-state region, or the region of secondary wear. This is the normal operating region for the tool. Such curves are typical for both flank wear and crater wear. When the amount of wear reaches the value w_f, the permissible tool wear on the flank, the tool is said to be "worn out." The value w_f is typically set at 0.025 to 0.030 in. for flank wear for high-speed steels and 0.008 to 0.050 in. for carbides, depending on the application. For crater wear, the depth of the crater, k_t, is used to determine tool failure.

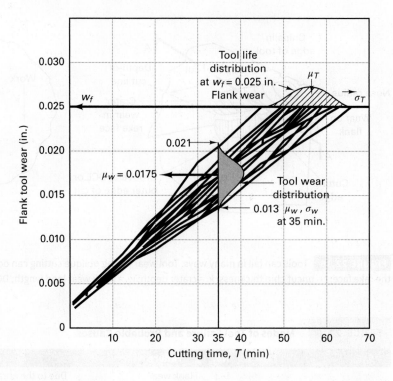

w_f values for general life determination (for cemented carbides)

Width of Wear (in.)	Applications
0.008	Finish cutting of nonferrous alloys, fine and light cut, etc.
0.016	Cutting of special steels
0.028	Normal cutting of cast irons, steels, etc.
0.040–0.050	Rough cutting of common cast irons

FIGURE 22.17 Tool wear on the flank displays a random nature, as does tool life. w_f = flank wear limit value.

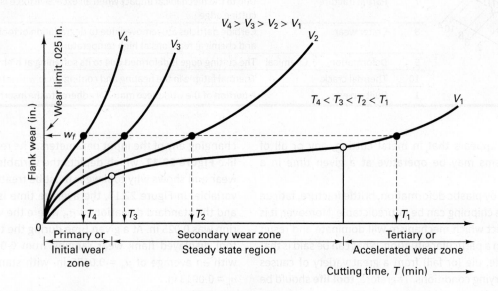

FIGURE 22.18 Typical tool wear curves for flank wear at different velocities. The initial wear is very fast, then it evens out to a more gradual pattern until the limit is reached; after that, the wear substantially increases.

Using the empirical tool wear data shown in Figure 22.18, which used the values of T (time in minutes) associated with V (cutting speed) for a given amount of tool wear, w_f (see the dashed-line construction), **Figure 22.19** was developed. When V and T are plotted on log-log scales, a linear relationship appears, described by the equation

$$VT^n = \text{Constant} = C \qquad (22.3)$$

This equation is called the Taylor tool life equation because in 1907, F. W Taylor published his now-famous paper, "On the

Art of Cutting Metals," in *ASME Transactions,* wherein tool life (T) was related to cutting speed (V) and feed (f). This equation had the form[1]

$$T = \frac{\text{Constant}}{f_x V_y} \qquad (22.4)$$

---

[1] Carl Barth, who was Taylor's mathematical genius, is generally thought to be the author of these formulations along with early versions of slide rules.

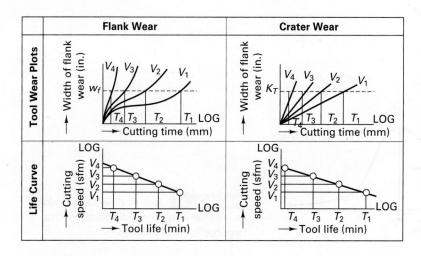

FIGURE 22.19 Construction of the Taylor tool life curve using data from deterministic tool wear plots like those of Figure 22.19. Curves like this can be developed for both flank and crater wear.

Over the years, the equation took the more widely published form of $VT^n = C$, where n is an exponent that depends mostly on tool material but is affected by work material, cutting conditions, and environment and C is a constant that depends on all the input parameters, including feed. Table in course material provides some data on Taylor tool life constants.

Figure 22.20 shows typical tool life curves for one tool material and three work materials. Notice that all three plots have about the same slope, n. Typical values for n are 0.14 to 0.16 for HSS, 0.21 to 0.25 for uncoated carbides, 0.30 for TiC inserts, 0.33 for poly-diamonds, 0.35 for TiN inserts, and 0.40 for ceramic-coated inserts.

It takes a great deal of experimental effort to obtain the constants for the Taylor equation because each combination of

tool and work material will have different constants. Note that for a tool life of 1 min, $C = V$, or the cutting speed that yields about 1 min of tool life for this tool.

A great deal of research has gone into developing more sophisticated versions of the Taylor equation, wherein constants for other input parameters (typically feed, depth of cut, and work material hardness) are experimentally determined. For example,

$$VT^n f^m d^p = K' \qquad (22.5)$$

where n, m, and p are exponents and K is a constant. Equations of this form are also deterministic and determined empirically.

The problem has been approached probabilistically in the following way. Because T depends on speed, feed, materials, and so on, one writes

$$T = \frac{K^{1/n}}{V^{1/n}} = \frac{K}{V^m} \qquad (22.6)$$

where K is now a random variable that represents the effects of all unmeasured factors and is an input variable.

The sources of tool life variability include factors such as:

1. Variation in work material hardness, from part to part and within a part.
2. Variability in cutting tool materials, geometry, and preparation.
3. Vibrations in machine tool, including rigidity of work and tool-holding devices.
4. Changing surface characteristics of workpieces.

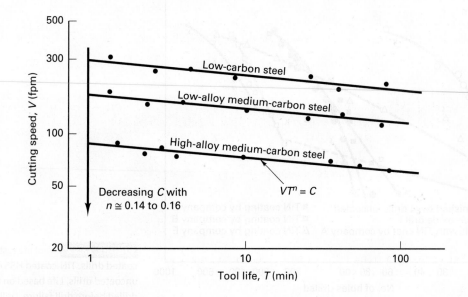

FIGURE 22.20 Log-log tool life plots for three steel work materials cut with HSS tool material.

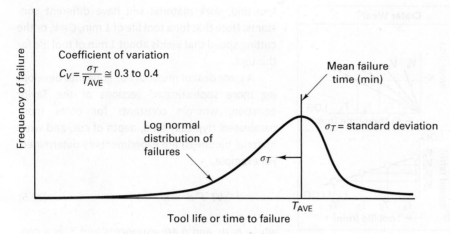

FIGURE 22.21 Tool life viewed as a random variable has a log normal distribution with a large coefficient of variation.

The examination of the data from a large number of tool life studies in which a variety of steels were machined shows that regardless of the tool material or process, tool life distributions are usually log normal and typically have a large standard deviation. As shown in **Figure 22.21**, tool life distributions have a large coefficient of variation, which means that tool life is not very predictable.

Other criteria can be used to define tool death in addition to wear limits:

• When surface finish deteriorates unacceptably.

• When workpiece dimension is out of tolerance.

• When power consumption or cutting forces increase to a specified limit.

• Sparking or chip discoloration and disfigurement.

• Cutting time or component quantity.

In automated processes, it is very beneficial to be able to monitor the tool wear online so that the tool can be replaced prior to failure, wherein defective products may also result. The feed force has been shown to be a good, indirect measure of tool wear. That is, as the tool wears and dulls, the feed force increases more than the cutting force increases.

Once criteria for failure have been established, tool life is that time elapsed between start and finish of the cut, in minutes. Other ways to express tool life, other than time, include:

1. Volume of metal removed between regrinds or replacement of tool.

2. Number of pieces machined per tool.

3. Number of holes drilled with a given tool (see **Figure 22.22**).

Drilling tool failure is also discussed more in Chapter 25 and is very complex because of the varied and complex geom-

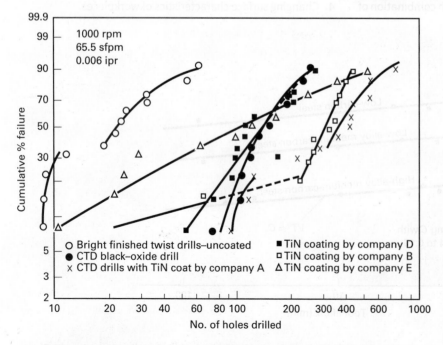

Drill performance based on the number of holes drilled with ¼-in.-diameter drills in T–1 structural steel.

FIGURE 22.22 Tool life test data for various coated drills. TiN-coated HSS drills outperform uncoated drills. Life based on the number of holes drilled before drill failure. Drill performance based on the number of holes drilled with ¼-in. diameter drills in T-1 structural steel.

etry of the tools and as shown here in Figure 22.22, the tool material.

Machinability

Machinability is a much-maligned term that has many different meanings but generally refers to the ease with which a metal can be machined to an acceptable surface finish. The principal definitions of the term are entirely different—the first based on material properties, the second based on tool life, and the third based on cutting speed.

1. Machinability is defined by the ease or difficulty with which the metal can be machined. In this light, specific energy, specific horsepower, and shear stress are used as measures, and, in general, the larger the shear stress or specific power values, the more difficult the material is to machine, requiring greater forces and lower speeds. In this definition, the material is the key.

2. Machinability is defined by the relative cutting speed for a given tool life while cutting some material, compared to a standard material cut with the same tool material. As shown in **Figure 22.23**, tool life curves are used to develop machinability ratings. For example, in steels, the material chosen for the standard material was B1112 steel, which has a tool life of 60 min at a cutting speed of 100 sfpm. Material X has a 70% rating, which implies that steel X has a cutting speed of 70% of B1112 for equal tool life. Note that this definition assumes that the tool fails when machining material X by whatever mechanism dominated the tool failure when machining the B1112. There is no guarantee that this will be the case. ISO standard 3685 has machinability index numbers based on 30 min of tool life with flank wear of 0.33 mm.

3. Cutting speed is measured by the maximum speed at which a tool can provide satisfactory performance for a specified time under specified conditions. (See ASTM standard E 618-81, "Evaluating Machining Performance of Ferrous Metals Using an Automatic Screw Bar Machine.")

4. Other definitions of machinability are based on the ease of removal of the chips (chip disposal), the quality of the surface finish of the part itself, the dimensional stability of the process, or the cost to remove a given volume of metal.

Further definitions are being developed based on the probabilistic nature of the tool failure, in which machinability is defined by a tool reliability index. Using such indexes, various tool replacement strategies can be examined and optimum cutting rates obtained. These approaches account for the tool life variability by developing coefficients of variation for common combinations of cutting tools and work materials.

The results to date are very promising. One thing is clear, however, from this sort of research: although many manufacturers of tools have worked at developing materials that have greater tool life at higher speeds, few have worked to develop tools that have less variability in tool life at all speeds. The reduction in variability is fundamental to achieving smaller coefficients of variation, which typically are of the order of 0.3 to 0.4. This means that a tool with a 100-min average tool life has a standard deviation of 30 to 40 min, so there is a good probability that the tool will fail early. In automated equipment, where early, unpredicted tool failures are extremely costly, reduction of the tool life variability will pay great benefits in improved productivity and reduced costs.

Reconditioning Cutting Tools

In the reconditioning of tools by sharpening and recoating, care must be taken in grinding the tool's surfaces. The following guidelines should be observed:

1. Resharpen to original tool geometry specifications. Restoring the original tool geometry will help the tool achieve consistent results on subsequent uses. Computer numerical control (CNC) grinding machines for tool resharpening have made it easier to restore a tool's original geometry.

2. Grind cutting edges and surfaces to a fine finish. Rough finishes left by poor and abusive regrinding hinder the performance of resharpened tools. For coated tools, tops of ridges left by rough grinding will break away in early tool use, leaving uncoated and unprotected surfaces that will cause premature tool failure.

3. Remove all burrs on resharpened cutting edges. If a tool with a burr is coated, premature failure can occur because the burr will break away in the first cut, leaving an uncoated surface exposed to wear.

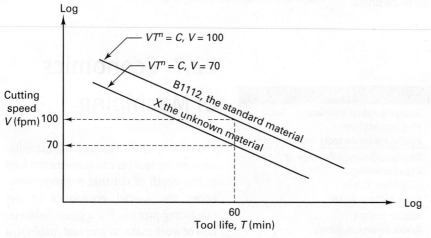

FIGURE 22.23 Machinability ratings defined by deterministic tool life curves.

4. Avoid resharpening practices that overheat and burn or melt (called *glazing over*) the tool surfaces, because this will cause problems in coating adhesion. Polishing or wire brushing of tools causes similar problems.

The cost of each recoating is about one-fifth the cost of purchasing a new tool. By recoating, the tooling cost per workpiece can be cut by between 20% and 30%, depending on the number of parts being machined.

22.6 Cutting Fluids

From the day that Frederick W. Taylor demonstrated that a heavy stream of water flowing directly on the cutting process allowed the cutting speeds to be doubled or tripled, **cutting fluids** have flourished in use and variety and have been employed in virtually every machining process. The cutting fluid acts primarily as a coolant and secondly as a lubricant, reducing the friction effects at the tool/chip interface and the work flank regions. The cutting fluids also carry away the chips and provide reductions in friction and forces in regions where the bodies of the tools rub against the workpiece. Thus, in processes such as drilling, sawing, tapping, and reaming, portions of the tool apart from the cutting edges come in contact with the work, and these sliding friction contacts greatly increase the power needed to perform the process, unless properly lubricated.

The reduction in temperature greatly aids in retaining the hardness of the tool, thereby extending the tool life or permitting increased cutting speed with equal tool life. In addition, the removal of heat from the cutting zone reduces thermal distortion of the work and permits better dimensional control. Coolant effectiveness is closely related to the thermal capacity and conductivity of the fluid used. Water is very effective in this respect but presents a rust hazard to both the work and tools and also is ineffective as a lubricant. Oils offer less effective coolant capacity but do not cause rust and have some lubricant value. In practice, straight cutting oils or emulsion combina-

tions of oil and water or wax and water are frequently used. Various chemicals can also be added to serve as wetting agents or detergents, rust inhibitors, or polarizing agents to promote formation of a protective oil film on the work. The extent to which the flow of a cutting fluid washes the very hot chips away from the cutting area is an important factor in heat removal. Thus, the application of a coolant should be copious and of some velocity.

The possibility of a cutting fluid providing lubrication between the chip and the tool face is an attractive one. An effective lubricant can modify the process, perhaps producing a cooler chip, discouraging the formation of a built-up edge on the tool, and promoting improved surface finish. However, the extreme pressure at the tool/chip interface and the rapid movement of the chip away from the cutting edge make it virtually impossible to maintain a conventional hydrodynamic lubricating film at the tool/chip interface. Consequently, any lubrication action is associated primarily with the formation of solid chemical compounds of low shear strength on the freshly cut chip face, thereby reducing tool/chip shear forces or friction. For example, carbon tetrachloride is very effective in reducing friction in machining several different metals and yet would hardly be classified as a good lubricant in the usual sense. Chemically active compounds, such as chlorinated or sulfurized oils, can be added to cutting fluids to achieve such a lubrication effect. Extreme-pressure lubricants are especially valuable in severe operations, such as internal threading (tapping), where the extensive tool-work contact results in high friction with limited access for a fluid.

In addition to functional effectiveness as a coolant and lubricant, cutting fluids should be stable in use and storage, noncorrosive to work and machines, and nontoxic to operating personnel. The cutting fluid should also be restorable by using a closed recycling system that will purify the used coolant and cutting oils. Cutting fluids become contaminated in three ways (**Table 22.7**). All these contaminants can be eliminated by filtering, hydrocycloning, pasteurizing, and centrifuging. Coolant restoration eliminates 99% of the cost of disposal and 80% or more of new fluid purchases. See course material for a schematic of a coolant recycling system.

22.7 Economics of Machining

The cutting speed has such a great influence on the tool life compared to the feed or the depth of cut that it greatly influences the overall economics of the machining process. For a given combination of work material and tool material, a 50% increase in speed results in a 90% decrease in tool life, while a 50% increase

TABLE 22.7	Cutting Fluid Contaminants	
Category	**Contaminants**	**Effects**
Solids	Metallic fines, chips	Scratch product's surface
	Grease and sludge	Plug coolant lines
	Debris and trash	Produce wear on tools and machines
Tramp fluids	Hydraulic oils (coolant)	Decrease cooling efficiency
	Water (oils)	Cause smoking
		Clog paper filters
		Grow bacteria faster
Biologicals (coolants)	Bacteria	Acidity coolant
	Fungi	**Break down** emulsions
	Mold	Cause rancidity, dermatitis
		Require toxic biocides

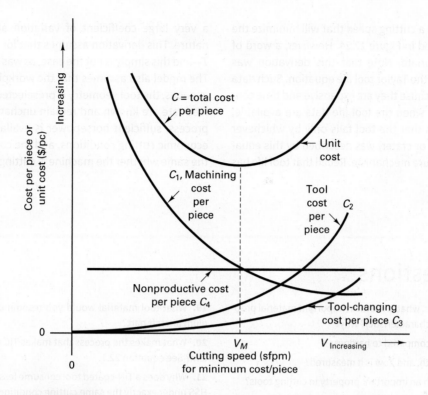

FIGURE 22.24 Cost per unit for a machining process vs cutting speed. Note that the "C" in this figure and related equations is not the same "C" used in the Taylor tool life (equation 22.3).

in feed results in a 60% decrease in tool life. A 50% increase in depth of cut produces only a 15% decrease in tool life. Therefore, in limited-horsepower situations, depth of cut and then feed should be maximized, while speed is held constant and horsepower consumed is maintained within limits. As cutting speed is increased, the machining time decreases, but the tools wear out faster and must be changed more often. In terms of costs, the situation is as shown in **Figure 22.24**, which shows the effect of cutting speed on the cost per piece.

The total cost per operation is composed of four individual costs: machining costs, tool costs, tool-changing costs, and handling costs. The machining cost is observed to decrease with increasing cutting speed because the cutting time decreases. Cutting time is proportional to the machining costs. Both the tool costs and the tool-changing costs increase with increases in cutting speeds. The handling costs are independent of cutting speed. Adding up each of the individual costs results in a total unit cost curve that is observed to go through a minimum point. For a turning operation, the total cost per piece C equals

$$C = C_1 + C_2 + C_3 + C_4$$
$$= \text{Machining cost} + \text{Tooling cost} \qquad (22.7)$$
$$+ \text{Tool-changing cost} + \text{Handling cost per piece}$$

Note: This "C" is not the same "C" used in the Taylor tool life equation. In this analysis, that "C" will be called "K."

Expressing each of these cost terms as a function of cutting velocity will permit the summation of all the costs.

$$C_1 = T_m \times C_o \qquad \text{where } C_o = \text{operating cost } (\$/\text{min})$$
$$T_m = \text{cutting time } (\text{min}/\text{piece})$$

$$C_2 = \left(\frac{T_m}{T}\right) C_t \qquad \text{where } T = \text{tool life } (\text{min}/\text{tool})$$
$$C_t = \text{initial cost of tool } (\$)$$

$$C_3 = t_c \times C_o \left(\frac{T_m}{T}\right) \qquad \text{where } t_c = \text{time to change tool } (\text{min}) \text{ and}$$
$$\frac{T_m}{T} = \text{number of tool changes per piece}$$

C_4 = Nonproductive costs, labor, overhead, and machine tool costs consumed while part is being loaded or unloaded, tools are being advanced, machine has broken down, and so on.

Because $T_m = L/Nf_r$ for turning $= \pi DL/12Vf_r$ and $T = (K/V)^{1/n}$, by rewriting equation **22.3**, and using "K" for the constant "C," the cost per unit, C, can be expressed in terms of V:

$$C = \frac{L\pi DC_o}{12Vf_r} + \frac{C_t V^{1/n}}{K^{1/n}} + \frac{t_c C_o V^{1/n}}{K^{1/n}} + C_4 \qquad (22.8)$$

To find the minimum, take $dc/dV = 0$ and solve for V:

$$V_m = K \left[\frac{n}{n-1} \cdot \frac{C_o}{C_o t_t + C_t} \right]^n \qquad (22.9)$$

Thus, V_m represents a cutting speed that will minimize the cost per unit, as depicted in Figure 22.24. However, a word of caution here is appropriate. Note that this derivation was totally dependent upon the Taylor tool life equation. Such data may not be available because they are expensive and time consuming to obtain. Even when the tool life data are available, this procedure assumes that the tool fails only by whichever wear mechanism (flank or crater) was described by this equation and by no other failure mechanism. Recall that tool life has a very large coefficient of variation and is probabilistic in nature. This derivation assumes that for a given V, there is one T—and this simply is not the case, as was shown in Figure 22.17. The model also assumes that the workpiece material is homogeneous, the tool geometry is preselected, the depth of cut and feed rate are known and remain unchanged during the entire process, sufficient horsepower is available for the cut at the economic cutting conditions, and the cost of operating time is the same whether the machine is cutting or not cutting.

Review Questions

1. For metal-cutting tools, what is the most important material property (i.e., the most critical characteristic)? Why?

2. What is hot hardness compared to hardness?

3. What is impact strength, and how is it measured?

4. Why is impact strength an important property in cutting tools?

5. Cemented carbide tools are made by a powder metallurgy method? What is powder matallurgy?

6. What are the primary considerations in tool selection?

7. What is the general strategy behind coated tools?

8. What is a cermet?

9. How is a CBN tool manufactured?

10. F. W. Taylor was one of the discoverers of high-speed steel. What else is he well known for?

11. What casting process do you think was used to fabricate cast cobalt alloys?

12. Discuss the constraints in the selection of a cutting tool.

13. What does *cemented* mean in the manufacture of carbides?

14. What advantage do ground carbide inserts have over pressed carbide inserts?

15. What is a chip groove?

16. What is the DCL?

17. Suppose you made four beams out of carbide, HSS, ceramic, and cobalt. The beams are identical in size and shape, differing only in material. Which beam would do each of the following?

   a. Deflect the most, assuming the same load.

   b. Resist penetration the most.

   c. Bend the farthest without breaking.

   d. Support the greatest compressive load.

18. Multiple coats or layers are put on the carbide base for what different purposes?

19. What tool material would you recommend for machining a titanium aircraft part?

20. What makes the process that makes TiC coatings for tools a problem? See equation 22.1.

21. Why does a TiN-coated tool consume less power than an uncoated HSS under exactly the same cutting conditions?

22. For what work material are CBN tools more commonly used, and why?

23. Why is CBN better for machining steel than diamond?

24. What is the typical coefficient of variation for tool life data, and why is this a problem?

25. What is meant by the statement "Tool life is a random variable"?

26. The typical value of a coefficient of variation in metal-cutting tool life distributions is 0.3. How could it be reduced?

27. Machinability is defined in many ways. Explain how a rating is obtained using tool life data.

28. What are the chief functions of cutting fluids?

29. How are CVD tools manufactured?

30. Why is the PVD process used to coat HSS tools?

31. Why is there no universal cutting tool material?

32. What is an 18-4-1 HSS composed of?

33. Over the years, tool materials have been developed that have allowed significant increases in MRR. Nevertheless, HSS is still widely used. Under what conditions might HSS be the material of choice?

34. Why is the rigidity of the machine tool an important consideration in the selection of the cutting tool material?

35. Explain how it can be that the tool wears when it may be four times as hard as the work material.

36. What is a honed edge on a cutting tool and why is it done?

Problems

1. Suppose you have a turning operation using a tool with a zero back rake and 5° end relief. The insert flank has a wear land on it of 0.020 in. How much has the diameter of the workpiece grown (increased) due to this flank wear, assuming the tool has not been reset to compensate for the flank wear?

2. A 2 in.-diameter bar of steel was turned at 284 rpm, and tool failure occurred in 10 min. The speed was changed to 132 rpm, and the tool failed in 30 min of cutting. Assume that a straight-line relationship exists. What cutting speed should be used to obtain a 60-min tool life of V60?

3. Tool cost is often used as the major criterion for justifying tool selection. Either silicon nitride or PCBN insert tips can be used to machine (bore) a cylinder block on a transfer line at a rate of 312, 000 part/yr (material: gray cast iron). The operation requires 12 inserts (2 per tool), as six bores are machined simultaneously. The machine was run at 2600 sfpm with a feed of 0.014 in. at 0.005 in. DOC for finishing. Here are some additional data:

	SiN	PCBN
Tips in use per part	12	12
Tool life (parts per tool)	200	4700
Cost per tip	$1.25	$28.50

a. Which tool material would you recommend?

b. On what basis?

4. The following data have been obtained for machining an Si-Al alloy:

5. In **Figure 22.A**, the insert is set with a 0° side cutting-edge angle. The insert at the right is set so that the edge contact length is increased from 0.250 in. to 0.289 in. The feed was 0.010 ipr.

a. Determine the side cutting-edge angle for the offset tool.

b. What is the uncut chip thickness in the offset position?

c. What effect will this have on the forces and the process?

6. The outside diameter of a roll for a steel (AISI 1015) rolling mill is to be turned. In the final pass, Starting diameter = 26.25 in. and Length = 48.0 in. The cutting conditions will be Feed = 0.0100 in./rev and Depth of cut = 0.125 in. A cemented carbide cutting tool is to be used, and the parameters of the Taylor tool life equation for this setup are $n = 0.25$ and $C = 1300$. It is desirable to operate at a cutting speed such that the tool will not need to be changed during the cut. Determine the cutting speed that will make the tool life equal to the time required to complete this turning operation. (Problem suggested by Groover, *Fundamentals of Modern Manufacturing: Materials, Processes, and Systems*, 2nd ed., John Wiley & Sons, 2002.)

7. Here is a single point tool. Identify angles A through G using tool nomenclature.

8. **Figure 22.B** gives data for cutting speed and tool life. Determine the constants for the Taylor tool life equation for these data. What do you think the tool material might have been?

Workpiece Material	Tool Material	Cutting Speed (m/min) for Tool Life (m/min) of		
		20 min	30 min	60 min
Sand casting	Diamond polycrystal	731	642	514
Permanent-mold casting	Diamond polycrystal	591	517	411
PMC with flood cooling	Diamond polycrystal	608	554	472
Sand casting	WC-K-20	175	161	139

Compute the *C* and *n* values for the Taylor tool life equation. How do these *n* values compare to the typical values?

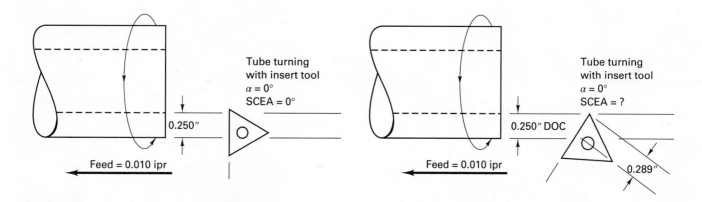

FIGURE 22.A

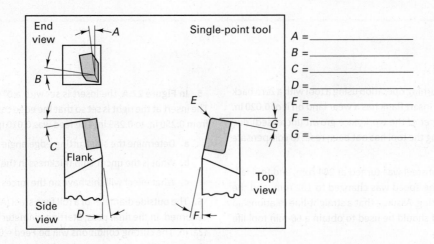

A = _____
B = _____
C = _____
D = _____
E = _____
F = _____
G = _____

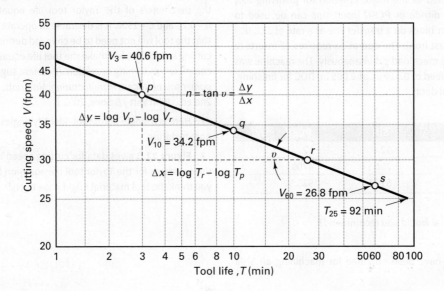

$$n = \tan v = \frac{\Delta y}{\Delta x}$$

$\Delta y = \log V_p - \log V_r$

$V_3 = 40.6$ fpm

$V_{10} = 34.2$ fpm

$\Delta x = \log T_r - \log T_p$

$V_{60} = 26.8$ fpm

$T_{25} = 92$ min

FIGURE 22.B

CHAPTER 23

Turning and Boring Processes

23.1 Introduction

Turning is the process of machining external cylindrical and conical surfaces (see **Figure 23.1**). It is usually performed on a machine tool called a lathe. An engine lathe is shown in **Figure 23.2**, using a cutting tool. The workpiece is held in a workholder as indicated in Figure 23.2. Relatively simple work and tool movements are involved in turning a cylindrical

surface. The workpiece is rotated and the single-point cutting tool is fed longitudinally into the workpiece and then travels parallel to the axis of workpiece rotation, reducing the diameter by the **depth of cut (DOC)**. The tool feeds at a rate, f_r, cutting at a speed, V, which is used to determine the revolutions per minute (rpm) of the workpiece, of diameter D_1, according to

$$N_s = \frac{12V}{\pi D_1} \tag{23.1}$$

If the tool is fed at an angle to the axis of rotation, an external conical surface results. This is called **taper turning** (see **Figure 23.3**). If the tool is fed to the axis of rotation, using a tool that is wider than the depth of the cut, the operation is called *facing,* and a flat surface is produced on the end of the cylinder.

By using a tool having a specific form or shape and feeding it radially or inward against the work, external cylindrical, conical, and irregular surfaces of limited length can also be turned. The shape of the resulting surface is determined by the shape and size of the cutting tool. Such machining is called **form turning** (see Figure 23.3). If the tool is fed all the way to the axis of

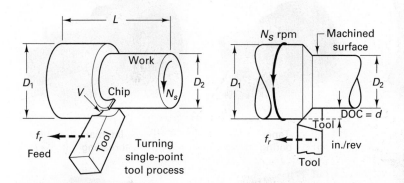

FIGURE 23.1 Basics of the turning process normally done on a lathe. The dashed arrows indicate the feed motion of the tool relative to the work.

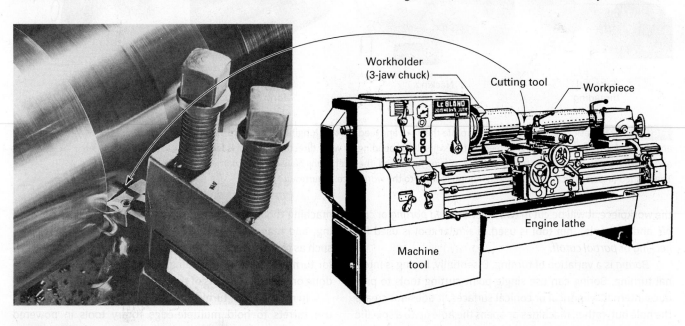

FIGURE 23.2 Schematic of a standard engine lathe performing a turning operation, with the cutting tool shown in inset. (Courtesy J T. Black)

Straight turning

Taper turning

Facing

Facing, grooving

Contour turning

Form turning

End facing

Parting or cutoff

Drilling

Boring

Taper boring

Internal threading

External threading

Necking

Internal forming

Knurling

FIGURE 23.3 Basic turning machines can rotate the work and feed the tool longitudinally for turning and can perform other operations by feeding transversely. Depending on what direction the tool is fed and on what portion of the rotating workpiece is being machined, the operations have different names. The dashed arrows indicate the tool feed motion relative to the workpiece. (Courtesy J T. Black)

the workpiece, it will be cut in two. This is called *parting* or *cut-off,* and a simple, thin tool is used. A similar tool is used for *necking* or *partial cutoff.*

Boring is a variation of turning. Essentially, boring is internal turning. Boring can use single-point cutting tools to produce internal cylindrical or conical surfaces. It does not create the hole but, rather, machines or opens the hole up to a specific size. Boring can be done on most machine tools that can do turning. However, boring can also be done using a rotating tool with the workpiece remaining stationary. Also, specialized

machine tools have been developed that will do boring, drilling, and reaming but will not do turning. Other operations, such as *threading* and *knurling,* can be done on machines used for turning. In addition, drilling, reaming, and tapping can be done on the rotation axis of the work.

In recent years, turning centers have been developed that use turrets to hold multiple-edge rotary tools in powered heads. Some new machine tools feature two opposing spindles with automatic transfer from one to the other and two turrets of tools (see **Figure 23.4**).

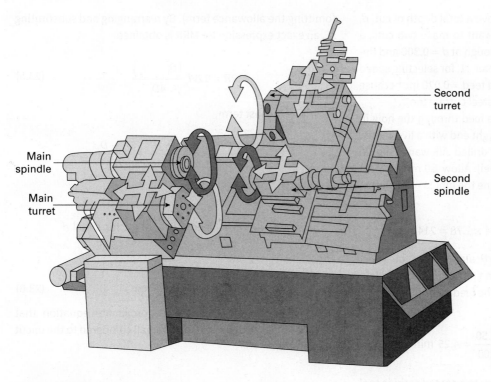

FIGURE 23.4 NC twin-spindle horizontal turning centers use co-axial spindles so work can be transferred from the main spindle to the second spindle.

Main spindle

Main turret

Second turret

Second spindle

23.2 Fundamentals of Turning, Boring, and Facing Turning

Turning constitutes the majority of lathe work. The cutting forces, resulting from feeding the tool from right to left, should be directed toward the headstock to force the workpiece against the workholder and thus provide better work support.

If good finish and accurate size are desired, one or more roughing cuts are usually followed by one or more finishing cuts. Roughing cuts may be as heavy as proper chip thickness, cutting dynamics, tool life, lathe horsepower, and the workpiece permit. Large depths of cut and smaller feeds are preferred to small DOCs and large feeds, because fewer cuts are required and less time is lost in reversing the carriage and resetting the tool for the following cut.

On workpieces that have a hard surface, such as castings or hot-rolled materials containing mill scale, the initial roughing cut should be deep enough to penetrate the hard materials. Otherwise, the entire cutting edge operates in hard, abrasive material throughout the cut, and the tool will dull rapidly. If the surface is unusually hard, the cutting speed on the first roughing cut should be reduced accordingly.

Finishing cuts are light, usually being less than 0.015 in. in depth, with the feed as fine as necessary to give the desired finish. Sometimes a special finishing tool is used, but often the same tool is used for both roughing and finishing cuts. In most cases, one finishing cut is all that is required. However, where exceptional accuracy is required, two finishing cuts may be made. If the diameter is controlled manually, a short finishing cut ($\frac{1}{4}$in. long) is made and the diameter checked before the cut is completed. Because the previous micrometer measurements were made on a rougher surface, it may be necessary to reset the tool in order to have the final measurement, made on a smoother surface, check exactly.

In turning, the primary cutting motion is rotational, with the tool feeding parallel to the axis of rotation. To determine the inputs to the machines, the depth of cut, cutting speed, and feed rate must be selected. The desired cutting speed establishes the necessary rpm (N_s) of the rotating workpiece. The feed f_r is given in inches per revolution (ipr).

The depth of cut is d, where

$$d = \text{DOC} = \frac{D_1 - D_2}{2} \text{ in.} \tag{23.2}$$

The length of cut is the distance traveled parallel to the axis L plus some allowance or overrun A to allow the tool to enter and/or exit the cut.

The inputs to the turning process are determined as follows. Based on the material being machined and the cutting tool material being used, the engineer selects the cutting speed, V, in feet per minute, the feed (f_r), and the depth of cut. The rpm value for the machine tool can be determined by calculating N_s (equation 23.1), (using the larger diameter), where the factor of 12 is used to convert feet to inches. The cutting time is

$$T_m = \frac{L + A}{f_r N_s} \tag{23.3}$$

where A is overrun allowance and f_r is the selected feed in inches per revolution.

Example of Turning

The 1.78-in.-diameter steel bar is to be turned down to a 1.10-in. diameter on a standard engine lathe. The overall length of the bar is 18.750 in., and the region to be turned is 16.50 in. The part is made from cold-drawn free-machining steel (this means the chips break up nicely) with a BHN of 250. Because you want to take the bar from 1.78 to 1.10, you have a total depth of cut, d, of 0.34 in. (0.68/2). You decide you want to make two cuts, a roughing pass and a finishing pass. Rough at $d = 0.300$ and finish with $d = 0.04$ in. Referring to Chapter 21, for selecting speed and feed, you select $V = 100$ fpm and feed = 0.020 ipr because you have decided to use high-speed steel cutting tools.

The bar is held in a chuck with a feed through the hole in the spindle and is supported on the right end with a live center. The ends of the bar have been center drilled. Allowance should be 0.50 in. for approach (no overtravel). Allow 1.0 min to reset the tool after the first cut. To determine the inputs to the lathe, calculate the spindle rpm:

$$N_s = 12V / \pi D_1 = 12 \times 100 / 3.14 \times 1.78 = 214 \text{ rpm}$$

But your lathe does not have this particular rpm, so you select the closest rpm, which is 200. You don't need any further calculations for lathe inputs as you input the feed in ipr directly.

The time to make the cut is

$$T_m = \frac{L + ALL}{f_r \times N_s} = \frac{16.50 + 0.50}{0.020 \times 200} = 4.25 \text{ min}$$

You could reduce this time by changing to a coated carbide tool that would allow you to increase the cutting speed to 925 sfpm. The time for the second cut will be different if you change the feed and/or the speed to improve the surface finish. Again from a table of speeds and feeds, you select a speed of 0.925 and a feed rate of 0.007, so

$$N_s = 12 \times 925 / 3.14 \times 1.10 = 3213 \text{ with 3200 rpm the closest}$$
value:

$$T_m = \frac{L + ALL}{0.007 \times 3200} = 0.75 \text{ min}$$

The **metal removal rate (MRR)** is

$$\text{MRR} = \frac{\text{Volume removed}}{\text{Time}} = \frac{\left(\pi D_1^2 - \pi D_2^2\right)L}{4L / f_r N_s} \quad (23.4)$$

(omitting the allowance term). By rearranging and substituting N_s, an exact expression for MRR is obtained:

$$\text{MRR} = 12 V f_r \frac{\left(D_1^2 - D_2^2\right)}{4D_1} \quad (23.5)$$

Rewriting the last term,

$$\frac{\left(D_1^2 - D_2^2\right)}{4D_1} = \frac{D_1 - D_2}{2} \times \frac{D_1 + D_2}{2D_1}$$

Therefore, because

$$d = \frac{\left(D_1 - D_2\right)}{2} \text{ and } \frac{D_1 + D_2}{2D_1} \cong 1 \text{ for small } d$$

then,

$$\text{MRR} \cong 12 \, V f_r d \text{ in.}^3 / \text{min} \quad (23.6)$$

Note that equation 23.6 is an approximate equation that assumes that the depth of cut d is small compared to the uncut diameter D_1.

Boring

Boring always involves the enlarging of an existing hole, which may have been made by drilling or may be the result of a core in a casting. An equally important and concurrent purpose of boring may be to make the hole concentric with the axis of rotation of the workpiece and thus correct any eccentricity that may have resulted from the drill starting or drifting off the centerline. Concentricity is an important attribute of bored holes.

When boring is done in a lathe, the work usually is held in a chuck or on a faceplate. Holes may be bored straight, tapered, with threads, or to irregular contours. **Figure 23.5a** shows the relationship of the tool and the workpiece for boring. Think of boring as internal turning while feeding the tool parallel to the rotation axis of the workpiece, with two important differences.

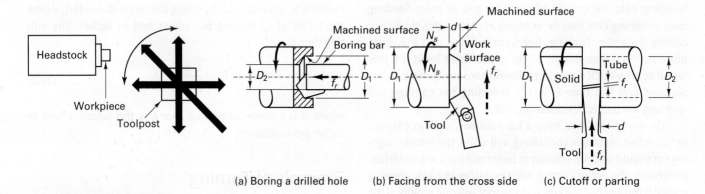

(a) Boring a drilled hole   (b) Facing from the cross side   (c) Cutoff or parting

FIGURE 23.5 Basic movement of boring, facing, and cutoff in a lathe, where cutting is performed by one-single-point cutting tool at a time and the tool can be fed in any direction.

First, the relief and clearance angles on the tool should be larger. Second, the tool overhang (length to diameter) must be considered with regard to stability and deflection problems.

As before V and f_r are selected. For a cut of length L, the cutting time is

$$T_m = \frac{L + A}{f_r N_s} \qquad (23.7)$$

where $N_s = 12V/\pi D_1$ for D_1, the diameter of bore, and A, the overrun allowance. The metal removal rate is

$$\text{MRR} = \frac{L\left(\pi D_1^2 - \pi D_2^2\right)/4}{L/f_r N_s}$$

where D_2 is the original hole diameter;

$$\text{MRR} \cong 12Vf_r d \qquad (23.8)$$

(omitting allowance term), where d is the depth of cut.

In most respects, the same principles are used for boring as for turning. Again, the tool should be set exactly at the same height as the axis of rotation. Larger end clearance angles help to prevent the heel of the tool from rubbing on the inner surface of the hole.

Second, because the tool overhang will be greater, smaller feeds and depths of cut may be selected to reduce the cutting forces that cause tool vibration and chatter. In some cases, the boring bar may be made of tungsten carbide because of this material's greater stiffness.

Facing

Facing is the producing of a flat surface as the result of the tool being fed across the end of the rotating workpiece, as shown in **Figure 23.5b**. Unless the work is held on a mandrel, if both ends of the work are to be faced, it must be turned end for end after the first end is completed and the facing operation is repeated.

The cutting speed should be determined from the largest diameter of the surface to be faced. Facing may be done either from the outside inward or from the center outward. In either case, the point of the tool must be set exactly at the height of the center of rotation. Because the cutting force tends to push the tool away from the work, it is usually desirable to clamp the carriage to the lathe bed during each facing cut to prevent it from moving slightly and thus producing a surface that is not flat.

In the facing of castings or other materials that have a hard surface, the depth of the first cut should be sufficient to penetrate the hard material to avoid excessive tool wear.

In facing, the tool feeds perpendicular to the axis of the rotating workpiece. Because the rpm is constant, the cutting speed is continually decreasing as the axis is approached. The length of cut L is $D_1/2$ or $(D_1 - D_2)/2$ for a tube.

$$T_m = \text{Cutting time} = \frac{L + A}{f_r N} \text{ min} = \frac{\dfrac{D_1}{2} + A}{f_r N_s}$$

$$\text{MRR} = \frac{\text{VOL}}{T_m} = \frac{\pi D_1^2 d f_r N_s}{4L} = 6Vf, d \text{ in.}^3/\text{min} \qquad (23.9)$$

where d is the depth of cut and $L = D_1/2$ is the length of cut.

Parting

Parting is the operation by which one section of a workpiece is severed from the remainder by means of a cutoff tool, as shown in **Figures 23.3** and **23.5c**. Because parting tools are quite thin and must have considerable overhang, this process is more difficult to perform accurately. The tool should be set exactly at the height of the axis of rotation, be kept sharp, have proper clearance angles, and be fed into the workpiece at a proper and uniform feed rate.

In parting or **cutoff** work, the tool is fed (plunged) perpendicular to the rotational axis, as it was in facing. The length of cut for solid bars is $D_1/2$. For tubes,

$$L = \frac{D_1 - D_2}{2}$$

In cutoff operations, the width of the tool is d in inches, the width of the cutoff operation. The equations for T_m and MRR are then basically the same as for facing.

Deflection

In boring, facing, and cutoff operations, the speeds, feeds, and depth of cut selected are generally less than those recommended for straight turning because of the large overhang of the tool often needed to complete the cuts. Recall the basic equation for deflection of a cantilever beam, modifying for machining,

$$\delta = \frac{Pl^3}{3EI} = \frac{F_c l^3}{3EI} \qquad (23.10)$$

In equation 23.10, the overhang of the tool, which is represented by l, greatly affects the deflection, so it should be minimized whenever possible. In equation 23.10,

E = modulus of elasticity
I = moment of inertia of cross section of tool
$P = F_c$ = applied load or cutting force

where

$I = \pi D_1^2/64$ solid round bar
$I = \pi(D_1^4 - (D_2^4))/64$ bar with hole
D_1 = diameter of tube or bar
D_2 = inside diameter of the tube

Deflection is proportional to the fourth power of the boring bar diameter and the third power of the bar overhang. Select the largest bar diameter and minimize the overhang. Use carbide shank boring bars ($E \cong 80,000,000$ psi), and select tool geometries that direct cutting forces into the feed direction to minimize chatter. The reduction of the feed or depth of cut reduces the forces operating on the tools. The cutting speed usually controls the occurrence of chatter and vibration. See the dynamics of machining discussed in Chapter 21.

Any imbalance in the cutting forces will deflect the tool to the side, resulting in loss of accuracy in cutoff lengths. At the outset, the forces will be balanced if there is no side rake on the tool. As the cutoff tool reaches the axis of the rotating part, the tool will be deflected away from the spindle, resulting in a change in the length of the part.

Precision Boring

Sometimes bored holes are slightly bell mouthed because the tool deflects out of the work as it progresses into the hole. This often occurs in castings and forgings where the holes have draft angles so that the depth of cut increases as the tool progresses down the bore. This problem can usually be corrected by repeating the cut with the same tool setting; however, the total cutting time for the part is increased. Alternately, a more robust setup can be used. Large holes may be precision bored using the setup shown in **Figure 23.6**, where a pilot bushing is placed in the lathe spindle to mate with the hardened ground pilot of the boring bar. This setup eliminates the cantilever problems common to boring.

Because the rotational relationship between the work and the tool is a simple one and is employed on several types of machine tools—such as lathes, drilling machines, and milling machines—boring is very frequently done on such machines. However, several machine tools have been developed primarily for boring, especially in cases involving large workpieces or for large-volume boring of smaller parts. Such machines as these are also capable of performing other operations, such as milling and turning. Because boring frequently follows drilling, many boring machines can also do drilling, permitting both operations to be done with a single setup of the work.

Drilling

Drilling, discussed in detail in Chapter 25, can be done on lathes with the drill mounted in the tailstock quill of engine lathes or the turret on turret lathes and fed against a rotating workpiece. Straight-shank drills can be held in Jacobs chucks, or drills with taper shanks mounted directly in the quill hole can drill holes online (center of rotation). Drills can also be mounted in the turrets of modern turret centers and fed automatically on the rotational axis of the workpiece or off-axis with power heads. It is also possible to drill on a lathe with the drill bit mounted and rotated in the spindle while the work remains stationary, supported on the tailstock or the carriage of the lathe.

The usual speeds used for drilling should be selected for lathe work. Because the feed may be manually controlled, care must be exercised, particularly in drilling small holes. Coolants should be used where required. In drilling deep holes, the drill should be withdrawn occasionally to clear chips from the hole and to aid in getting coolant to the cutting edges. This is called *peck drilling*. See Chapter 25 for further discussion on drilling.

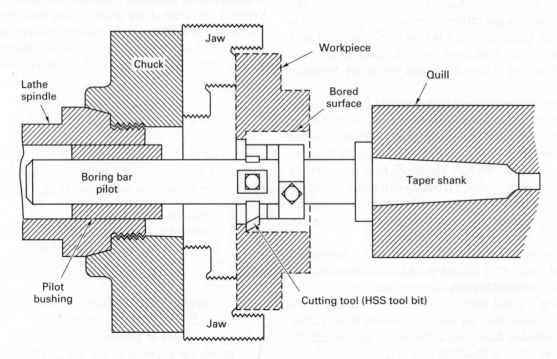

FIGURE 23.6 Pilot boring bar mounted in tailstock of lathe for precision boring large hole in casting. The size of the hole is controlled by the rotation diameter of the cutting tool.

Reaming

Reaming on a lathe involves no special precautions. Reamers are held in the tailstock quill, taper-shank types being mounted directly and straight-shank types by means of a drill chuck. Rose-chucking reamers are usually used (see Chapter 25). Fluted-chucking reamers may also be used, but these should be held in some type of holder that will permit the reamer to float (i.e., have some compliance) in the hole and conform to the geometry created by the boring process.

Knurling

Knurling produces a regularly shaped, roughened surface on a workpiece. Although knurling also can be done on other machine tools, even on flat surfaces, in most cases it is done on external cylindrical surfaces using lathes. Knurling is a chipless, cold-forming process. See **Figures 23.3** and **23.7** for examples. The two hardened rolls are pressed against the rotating workpiece with sufficient force to cause a slight outward and lateral displacement of the metal to form the knurl in a raised, diamond pattern. Another type of knurling tool produces the knurled pattern by cutting chips. Because it involves less pressure and thus does not tend to bend the workpiece, this method is often preferred for workpieces of small diameter and for use on automatic or semiautomatic machines.

Special Attachments

For engine lathes, taper turning and **milling** can be done on a lathe but require special attachments. The *milling attachment* is a special vise that attaches to the cross slide to hold work. The milling cutter is mounted and rotated by the spindle. The work is fed by means of the cross-slide screw. *Tool-post grinders* are often used to permit grinding to be done on a lathe. *Duplicating attachments* are available that, guided by a template, will automatically control the tool movements for turning irregularly shaped parts. In some cases, the first piece, produced in the normal manner, may serve as the template for duplicate parts. To a large extent, duplicating lathes using templates have been replaced by numerically controlled lathes and milling is done with power tools in numerical control turret lathes.

23.3 Lathe Design and Terminology

Knowing the terminology of a machine tool is fundamental to understanding how it performs the basic operations, how the workholding devices are interchanged, and how the cutting tools are mounted and interfaced to the work. **Lathes** are machine tools designed primarily to do turning, facing, and boring. Very little turning is done on other types of machine tools, and none can do it with equal facility. Lathes also can do facing, drilling, and reaming. Modern turning centers permit milling and drilling operations using live (also called powered) spindles in multiple-tool turrets, so their versatility permits multiple operations to be done with a single setup of the workpiece. Consequently, the lathe is probably the most common machine tool, along with milling machines.

Lathes in various forms have existed for more than 2000 years, but modern lathes date from about 1797, when Henry Maudsley developed one with a leadscrew, providing controlled, mechanical feed of the tool. This ingenious Englishman also developed a change-gear system that could connect the motions of the spindle and leadscrew and thus enable threads to be cut.

Lathe Design

The essential components of an **engine lathe** (**Figure 23.8**) are the bed, headstock assembly (which includes the spindle), tailstock assembly, carriage assembly, quick-change gearbox, and the leadscrew and feed rod. The **bed** is the base and backbone of a lathe. The bed is usually made of well-normalized or aged gray or nodular cast iron and provides a heavy, rigid frame on which all the other basic components are mounted. Two sets of parallel, longitudinal **ways**, inner and outer, are contained on

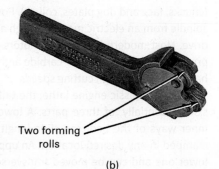

Two forming rolls

FIGURE 23.7 (a) Knurling in a lathe, using a forming-type tool, and showing the resulting pattern on the workpiece; (b) knurling tool with forming rolls. (Courtesy of Armstrong Industrial Hand Tools)

(a)   (b)

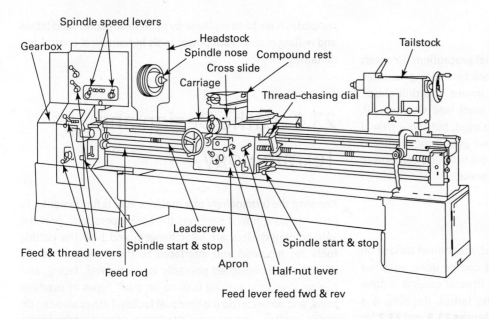

FIGURE 23.8 Schematic of an engine lathe, a versatile machine tool that is easy to set up and operate and is intended for short-run jobs.

the bed. On modern lathes, the ways are surface hardened and precision machined, and care should be taken to ensure that the ways are not damaged. Any inaccuracy in them usually means that the accuracy of the entire lathe is destroyed.

The **headstock**, mounted in a fixed position on the inner ways, provides powered means to rotate the work at various rpm values. Essentially, it consists of a hollow **spindle**, mounted in accurate bearings, and a set of transmission gears—similar to a truck transmission—through which the spindle can be rotated at a number of speeds. Most lathes provide from 8 to 18 choices of rpm. On modern lathes, all the rpm rates can be obtained merely by moving from two to four levers or the lathe has a continuously variable spindle rpm using electrical or mechanical drives.

The accuracy of a lathe is greatly dependent on the spindle. It carries the workholders and is mounted in heavy bearings, usually preloaded tapered roller or ball types. The spindle has a hole extending through its length through which long bar stock can be fed. The size of this hole is an important dimension of a lathe because it determines the maximum size of bar stock that can be machined when the materials must be fed through the spindle.

The spindle protrudes from the gearbox and contains means for mounting various types of workholding devices (chucks, face and dog plates, collets). Power is supplied to the spindle from an electric motor through a V-belt or silent-chain drive. Most modern lathes have motors of from 5 to 25 hp to provide adequate power for carbide and ceramic tools cutting hard materials at high cutting speeds.

For the classic engine lathe, the **tailstock** assembly consists, essentially, of three parts. A lower casting fits on the inner ways of the bed, can slide longitudinally, and can be clamped in any desired location. An upper casting fits on the lower one and can be moved transversely upon it, on some

type of keyed ways, to permit aligning the tailstock and headstock spindles (for turning tapers). The third major component of the assembly is the tailstock **quill**. This is a hollow steel cylinder, usually about 2 to 3 in. diameter, that can be moved longitudinally in and out of the upper casting by means of a handwheel and screw. The open end of the quill hole has a Morse taper. Cutting tools or a **lathe center** are held in the quill. A graduated scale is usually engraved on the outside of the quill to aid in controlling its motion in and out of the upper casting. A locking device permits clamping the quill in any desired position. In recent years, dual-spindle numerical control (NC) turning centers have emerged, where a subspindle replaces the tailstock assembly. Parts can be automatically transferred from the spindle to the subspindle for turning the back end of the part. See Chapter 26 for more on NC machining and turning centers.

The **carriage assembly**, together with the apron, provides the means for mounting and moving cutting tools. The **carriage**, a relatively flat H-shaped casting, rides on the outer set of ways on the bed. The **cross slide** is mounted on the carriage and can be moved by means of a feed screw that is controlled by a small handwheel and a graduated dial. The cross slide thus provides a means for moving the lathe tool in the facing or cutoff direction.

On most lathes, the tool post is mounted on a **compound rest** (see **Figure 23.9**). The compound rest can rotate and translate with respect to the cross slide, permitting further

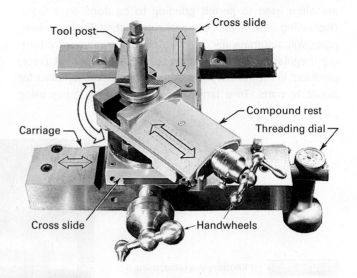

FIGURE 23.9 The cutting tools for lathe work are held in the tool post on the compound rest, which can translate and swivel. (Courtesy J T. Black)

positioning of the tool with respect to the work. The **apron**, attached to the front of the carriage, has the controls for providing manual and powered motion for the carriage and powered motion for the cross slide. The carriage is moved parallel to the ways by turning a handwheel on the front of the apron, which is geared to a pinion on the back side. This pinion engages a rack that is attached beneath the upper front edge of the bed in an inverted position.

Powered movement of the carriage and cross slide is provided by a rotating **feed rod**. The feed rod, which contains a keyway, passes through the two reversing bevel pinions and is keyed to them. Either pinion can be activated by means of the feed reverse lever, thus providing "forward" or "reverse" power to the carriage. Suitable clutches connect either the rack pinion or the cross-slide screw to provide longitudinal motion of the carriage or transverse motion of the cross slide.

For cutting threads, a **leadscrew** is used. When a friction clutch is used to drive the carriage, motion through the leadscrew is by a direct, mechanical connection between the apron and the leadscrew. A **split nut** is closed around the leadscrew by means of a lever on the front of the apron directly driving the carriage without any slippage.

Modern lathes have **quick-change gearboxes**, driven by the spindle, that connect the feed rod and leadscrew. Thus, when the spindle turns a revolution, the tool (mounted on the carriage) translates (longitudinally or transversely) a specific distance in inches—that is, inches per revolution (ipr). This revolutions per minute, rpm or N_s, times the feed, f_r, gives the feed rate, f, in inches per minute (ipm) that the tool is moving. In this way, the calculations for turning rpm and feed in ipr are "mechanically related." Typical lathes may provide as many as 48 feeds, ranging from 0.002 to 0.118 in. (0.05 to 3 mm) per revolution of the spindle, and, through the leadscrew, leads up to 92 threads per inch.

Size Designation of Lathes

The size of a lathe is designated by two dimensions. The first is known as the **swing**. This is the maximum diameter of work that can be rotated on a lathe. Swing is approximately twice the distance between the line connecting the lathe centers and the nearest point on the ways. The maximum diameter of a workpiece that can be mounted between centers is somewhat less than the swing diameter because the workpiece must clear the carriage assembly as well as the ways. The second size dimension is the *maximum distance between centers*. The swing thus indicates the maximum workpiece diameter that can be turned in the lathe, while the distance between centers indicates the maximum length of workpiece that can be mounted between centers.

Types of Lathes

Lathes used in manufacturing can be classified as speed, engine, toolroom, turret, automatics, tracer, and numerical control turning centers. Speed lathes usually have only a head-

stock, a tailstock, and a simple tool post mounted on a light bed. They ordinarily have only three or four speeds and are used primarily for wood turning, polishing, or metal spinning. Spindle speeds up to about 4000 rpm are common.

Engine lathes are the type most frequently used in manufacturing. Figures 23.2 and 23.8 are examples of this type. They are heavy-duty machine tools with all the components described previously and have power drive for all tool movements except on the compound rest. They commonly range in size from 12- to 24-in. swing and from 24- to 48-in. center distances, but swings up to 50 in. and center distances up to 12 ft are not uncommon. Very large engine lathes (36- to 60-ft-long beds) are therefore capable of performing roughing cuts in iron and steel at depths of cut of $\frac{1}{2}$ to 2 in. and at cutting speeds at 50 to 200 sfpm with WC tools run at 0.010 to 0.100 in./rev. To perform such heavy cuts requires rigidity in the machine tool, the cutting tools, the workholder, and the workpiece (using steady rests and other supports) and large horsepower (50 to 100 hp).

Most engine lathes are equipped with chip pans and a built-in coolant circulating system. Smaller engine lathes, with swings usually not greater than 13 in., are also available in *bench type*, designed for the bed to be mounted on a bench or table.

Toolroom lathes have somewhat greater accuracy and, usually, a wider range of speeds and feeds than ordinary engine lathes. Designed to have greater versatility to meet the requirements of tool and die work, they often have a continuously variable spindle speed range and shorter beds than ordinary engine lathes of comparable swing because they are generally used for machining relatively small parts. They may be either bench or pedestal type.

Several types of special-purpose lathes are made to accommodate specific types of work. On a *gap-bed lathe*, a section of the bed adjacent to the headstock can be removed to permit work of unusually large diameter to be swung. Another example is the *wheel lathe*, which is designed to permit the turning of railroad-car wheel-and-axle assemblies.

Figure 23.10 shows a vertical turning lathe. Vertical lathes are an excellent alternative to large horizontal lathes. Gravity-aided seating of large/heavy workpieces allows a high degree of process repeatability. A smaller footprint, lower initial cost, and increased productivity are all advantages when compared to traditional horizontal lathes.

Although engine lathes are versatile and very useful, the time required for changing and setting tools and for making measurements on the workpiece is often a large percentage of the cycle time. Often, the actual chip-production time is less than 30% of the total cycle time. Methods to reduce setup and tool-change time are now well known, reducing setups to minutes and unload/load steps to seconds. The placement of single-cycle machine tools into interim or lean manufacturing cells will increase the productivity of the workers because they can run more than one machine. Turret lathes, screw machines, and other types of semiautomatic and automatic lathes have been highly developed and are widely used in manufacturing as another means to improve cutting productivity.

Turret Lathes

The basic components of a **turret lathe** are depicted in **Figure 23.11**. Basically, a longitudinally feedable, hexagon turret replaces the tailstock. The turret, on which six tools can be mounted, can be rotated about a vertical axis to bring each tool into operating position, and the entire unit can be translated parallel to the ways, either manually or by power, to provide feed for the tools. When the turret assembly is backed away from the spindle by means of a capstan wheel, the turret indexes automatically at the end of its movement, thus bringing each of the six tools into operating position in sequence.

The square turret on the cross slide can be rotated manually about a vertical axis to bring each of the four tools into operating position. On most machines, the turret can be moved transversely, either manually or by power, by means of the cross slide, and longitudinally through power or manual operation of the carriage. In most cases, a fixed tool holder also is added to the back end of the cross slide; this often carries a parting tool.

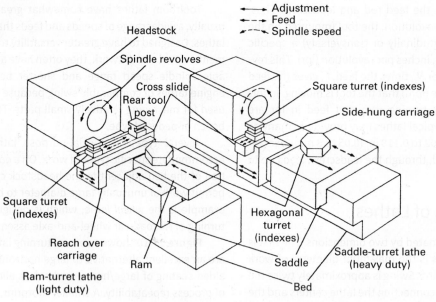

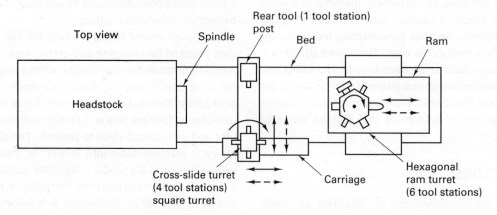

FIGURE 23.11 Block diagrams of ram- and saddle-turret lathe.

Through these basic features of a turret lathe, a number of tools can be set up on the machine and then quickly be brought successively into working position. In this way, a complete part can be machined without the necessity for further adjusting, changing tools, or making measurements.

The two basic types of turret lathes are the ram-type and the saddle-type. In the *ram-type turret lathe*, the ram and turret are moved up to the cutting position by means of the capstan wheel, and the power feed is then engaged. As the ram is moved toward the headstock, the turret is automatically locked into position so that rigid tool support is obtained. Rotary stop-screws control the forward travel of the ram, one stop being provided for each face on the turret. The proper stop is brought into operating position automatically when the turret is indexed. A similar set of stops is usually provided to limit movement of the cross slide.

The *saddle-type turret lathe* provides a more rugged mounting for the hexagon turret than can be obtained by the ram-type mounting. In saddle-type lathes, the main turret is mounted directly on the saddle, and the entire saddle and turret assembly reciprocates. Larger turret lathes usually have this type of mounting. However, because the saddle-turret assembly is rather heavy, this type of mounting provides less rapid turret reciprocation. When such lathes are used with heavy tooling for making heavy or multiple cuts, a **pilot arm** attached to the headstock engages a pilot hole attached to one or more faces of the turret to give additional rigidity. Turret-lathe headstocks can shift rapidly between spindle speeds and brake rapidly to stop the spindle very quickly. They also have automatic stock-feeding for feeding bar stock through the spindle hole. If the work is to be held in a chuck, some type of air-operated chuck or a special clamping fixture is often employed to reduce the time required for part loading and unloading.

Single-Spindle Automatic Screw Machines

There are two common types of **single-spindle screw machines**. One, an American development and commonly called the turret type (Brown & Sharpe), is shown in **Figure 23.12**. The other is of Swiss origin and is referred to as the Swiss type. The *Brown & Sharpe screw machine* is essentially a small automatic turret lathe, designed for bar stock, with the main turret mounted in a vertical plane on a ram. Front and rear tool holders can be mounted on the cross slide. All motions of the turret, cross slide, spindle, chuck, and stock-feed mechanism are controlled by cams. The turret cam is essentially a program that defines the movement of the turret during a cycle. These machines are usually equipped with an automatic rod-feeding magazine that feeds a new length of bar stock into the collet (the workholding device) as soon as one rod is completely used.

Often, screw machines of the Brown & Sharpe type are equipped with a transfer or "picking" attachment. This device picks up the workpiece from the spindle as it is cut off and carries it to a position where a secondary operation is performed by a small, auxiliary power head. In this manner, screwdriver slots are put in screw heads, small flats are milled parallel with the axis of the workpiece, or holes are drilled normal to the axis.

On the *Swiss-type automatic screw machine*, the cutting tools are held and moved in radial slides (**Figure 23.13**). Disk cams move the tools into cutting position and provide feed into the work in a radial direction only; they provide any required longitudinal feed by reciprocating the headstock.

Most machining on Swiss-type screw machines is done with single-point cutting tools. Because they are located close to the spindle collet, the workpiece is not subjected to much deflection. Consequently, these machines are particularly well-suited for machining very small parts.

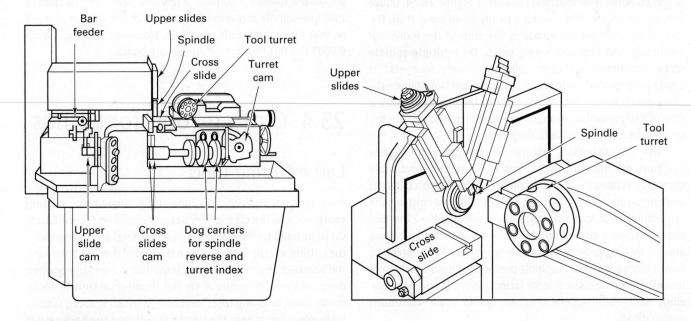

FIGURE 23.12 On the turret-type single-spindle automatic, the tools must take turns to make cuts.

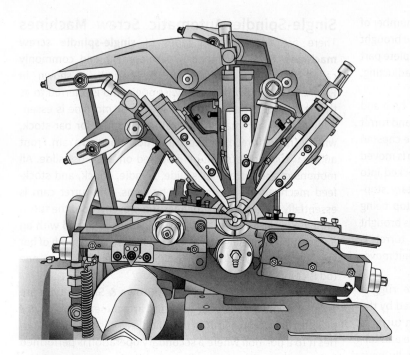

FIGURE 23.13 Close-up view of a Swiss-type screw machine, showing the tooling and radial tool sides, actuated by rocker arms controlled by a disk cam, shown in lower left. (Courtesy J T. Black)

Both types of single-spindle screw machines can produce work to close tolerances, the Swiss-type probably being somewhat superior for very small work. Tolerances of 0.0002 to 0.0005 in. are not uncommon. The time required for setting up the machine is usually one or two hours and can be much less. One person can tend many machines, after they are properly tooled. They have short cycle times, frequently less than 30 s/piece.

Multiple-Spindle Automatic Screw Machines

Single-spindle screw machines shown in **Figure 23.14**, utilize only one or two tooling positions at any given time. Thus, the total cycle time per workpiece is the sum of the individual machining and tool-positioning times. On **multiple-spindle screw machines**, sufficient spindles—usually four, six, or eight—are provided so that many tools can cut simultaneously. Thus, the cycle time per piece is equal to the maximum cutting time of a single tool position plus the time required to index the spindles from one position to the next.

The two distinctive features of multiple-spindle screw machines are the six spindles that are carried in a rotatable drum that indexes in order to bring each spindle into a different working position and a nonrotating tool slide that contains the same number of tool holders as there are spindles. The tool slides position a cutting tool (or tools) for each spindle. Tools are fed by longitudinal reciprocating motion. Most machines have a cross slide at each spindle position so that an additional tool can be fed from the side for facing, grooving, knurling, beveling, and cutoff operations. All motions are controlled automatically.

Starting with the sixth position, follow the sequence of processing steps on the tooling sheet for making a part shown in Figure 23.14. With a tool position available on the end tool slide for each spindle (except for a stock-feed stop at position 6), when the slide moves forward, these tools cut essentially simultaneously. At the same time, the tools in the cross slides move inward and make their cuts. When the forward cutting motion of the end tool slide is completed, it moves away from the work, accompanied by the outward movement of the radial slides. The spindles are indexed one position, by rotation of the spindle carrier, to position each part for the next operation. At spindle position 5, finished pieces are cut off. Bar stock 1 in. in diameter is fed to correct length for the beginning of the next operation. Thus, a piece is completed each time the tool slide moves forward and back. Multiple-spindle screw machines are made in a considerable range of sizes, determined by the diameter of the stock that can be accommodated in the spindles. There may be four, five, six, or eight spindles. The operating cycle of the end tool slide is determined by the operation that requires the longest time.

Once a multiple-spindle screw machine is set up, it requires only that the bar stock feed rack be supplied and the finished products checked periodically to make sure that they are within desired tolerances. One operator usually services many machines.

Most multiple-spindle screw machines use cams to control the motions. Setting up the cams and the tooling for a given job may require from 2 to 20 hr. However, once such a machine is set up, the processing time per part is very short. Often, a piece may be completed every 10 s. Typically, a minimum of 2000 to 5000 parts are required in a lot to justify setting up and tooling a multiple-spindle automatic screw machine. The precision of multiple-spindle screw machines is good, but seldom as good as that of single-spindle machines. However, tolerances from 0.0005 to 0.001 in. on the diameter are typical.

23.4 Cutting Tools for Lathes

Lathe Cutting Tools

Most lathe operations are done using single-point **cutting tools**, such as the classic tool designs shown in **Figure 23.15**. On right-hand turning (and left-hand turning) and facing tools, the cutting usually takes place on the side of the tool; therefore, the side rake angle is of primary importance, particularly when deep cuts are being made. On the round-nose turning tools, cutoff tools, finishing tools, and some threading tools, cutting takes place on or near the tip of the tool, and the back rake is

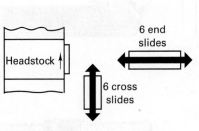

6 end slides

6 cross slides

Headstock

All spindles on multiple-spindle automatic have the same tool path

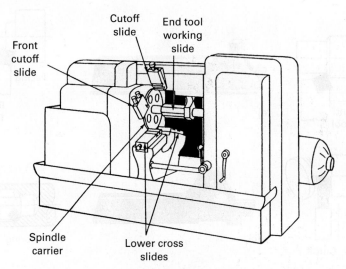

Cutoff slide

End tool working slide

Front cutoff slide

Spindle carrier

Lower cross slides

The six-spindle automatic

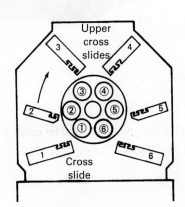

Upper cross slides

Cross slide

Spindle arrangement for six-spindle automatic. The bar stock is usually fed to a stop at position 6. The cutoff position is the one preceding the bar feed position.

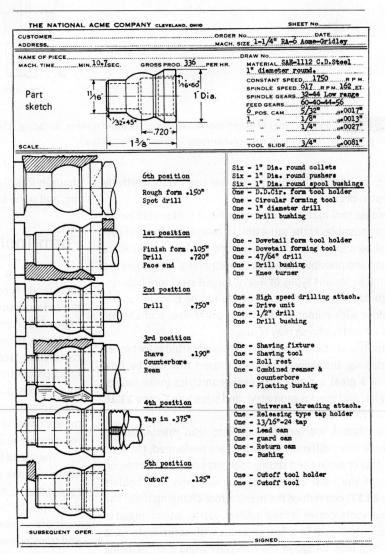

Tooling sheet for making a part on a six-spindle.

FIGURE 23.14 The multiple-spindle, automatic screw machine makes all cuts simultaneously and then performs the noncutting functions (tool withdrawal, index, and bar feed) at high speed.

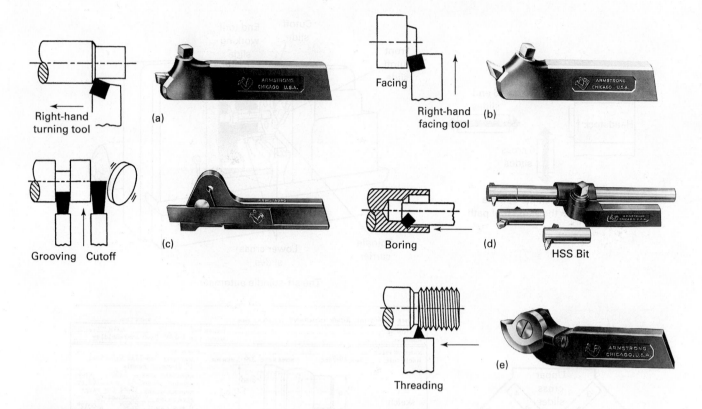

FIGURE 23.15 Common types of forged tool holders: (a) right-hand turning, (b) facing, (c) grooving cutoff, (d) boring, (e) threading. (Courtesy of Armstrong Industrial Hand Tools)

therefore of importance. Such tools are used with relatively light depths of cut.

Because tool materials are expensive, it is desirable to use as little as possible. At the same time, it is essential that the cutting tool be supported in a strong, rigid manner to minimize deflection and possible vibration. Consequently, lathe tools are supported in various types of heavy, forged steel tool holders. The high-speed steel (HSS) tool bit should be clamped in the tool holder with minimum overhang; otherwise, tool chatter and a poor surface finish may result.

In the use of carbide, ceramic, or coated carbides for higher speed cutting, throwaway inserts are used that can be purchased in a great variety of shapes, geometrics (nose radius, tool angles, and groove geometry), and sizes (see **Figure 23.16** for some examples).

When lathes are incorporated into lean manufacturing cells where many different operations are performed, the time required for changing and setting tools may constitute as much as 50% of the total cycle time. Quick-change tool holders (**Figure 23.17**) can reduce the manual tool-changing time. The individual tools, preset in their holders, can be interchanged in the special tool post in a few seconds. With some systems, a second tool may be set in the tool post while a cut is being made with the first tool and can then be brought into proper position by rotating the post.

In lathe work, the nose of the tool should be set exactly at the same height as the axis of rotation of the work. However, because any setting below the axis causes the work to tend to "climb" up on the tool, most machinists set their tools a few thousandths of an inch above the axis, except for cutoff, threading, and some facing operations.

Form Tools

In Figure 23.14, the use of form tools was shown in automatic lathe work. Form tools are made by grinding the inverse of the desired work contour into a block of HSS or tool steel. A **threading** tool is often a form tool. Although form tools are relatively expensive to manufacture, it is possible to machine a fairly complex surface with a single inward feeding of one tool. For mass-production work, adjustable form tools of either flat or rotary types, such as shown in **Figure 23.18**, are used. These are expensive to make initially but can be resharpened by merely grinding a small amount off the face and then raising or rotating the cutting edge to the correct position.

The use of form tools is limited by the difficulty of grinding adequate rake angles for all points along the cutting edge. A rigid setup is needed to resist the large cutting forces that develop with these tools. Light feeds with sharp, coated HSS tools are used on multiple-spindle automatics, turret lathes, and transfer line machines.

Turret-Lathe Tools

In turret lathes, the work is generally held in collets and the correct amount of bar stock is fed into the machine to make one part. The tools are arranged in sequence at the tool stations

Insert shape	Available cutting edges	Typical insert holder
Round	4–10 on a side 8–20 total	15° Square insert
80°/100° diamond	4 on a side 8 total	
Square	4 on a side 8 total	Triangular insert
Triangle	3 on a side 6 total	0°
55° diamond	2 on a side 4 total	35° diamond
35° diamond	2 on a side 4 total	5°

FIGURE 23.16 Typical insert shapes, available cutting edges per insert, and insert holders for throwaway insert cutting tools. (Adapted from *Turning Handbook of High Efficiency Metal Cutting*, Courtesy Armstrong Industrial Hand Tools)

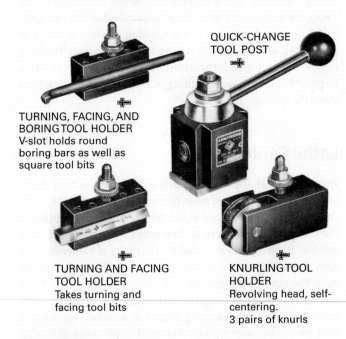

QUICK-CHANGE TOOL POST

TURNING, FACING, AND BORING TOOL HOLDER
V-slot holds round boring bars as well as square tool bits

TURNING AND FACING TOOL HOLDER
Takes turning and facing tool bits

KNURLING TOOL HOLDER
Revolving head, self-centering. 3 pairs of knurls

FIGURE 23.17 Quick-change tool post and accompanying tool holders. (Courtesy of Armstrong Industrial Hand Tools)

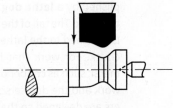

Form turning

FIGURE 23.18 Circular and block types of form tools are widely used on the lathe operations. (Courtesy J T. Black)

with depths of cut all preset. The following factors should be considered when setting up a turret lathe.

1. *Setup time:* time required to install and set the tooling and set the stops. Standard tool holders and tools should be used as much as possible to minimize setup time. Setup time can be greatly reduced by eliminating adjustment in the setup.

2. *Workholding time:* time to load and unload parts and/or stock.

3. *Machine-controlling time:* time required to manipulate the turrets. Can be reduced by combining operations where possible. Dependent on the sequence of operations established by the design of the setup.

4. *Cutting time:* time during which chips are being produced. Should be as short as is economically practical and represent the greatest percentage of the total cycle time possible.

5. *Cost:* cost of the tool, setup labor cost, lathe operator labor cost, and the number of pieces to be made.

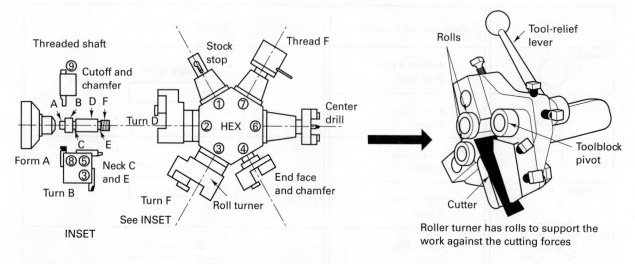

Roller turner has rolls to support the
work against the cutting forces

FIGURE 23.19 Turret-lathe tooling setup for producing part shown. Numbers in circles indicate the sequence of the operations from 1 to 9. The letters A through F refer to the surfaces being machined. Operation 3 is a combined operation. The roll turner is turning surface F, while tool 3 on the square post is turning surface B. The first operation stops the stock at the right length. The last operation cuts the finished bar off and puts a chamfer on the bar, which will next be advanced to the stock stop.

There are essentially 11 tooling stations, as shown in **Figure 23.19**, with six in the turret, four in the indexable tool post, and one in the rear tool post. The tooling is more rugged in turret lathes because heavy, simultaneous cuts are often made. Tools mounted in the hex turret that are used for turning are often equipped with pressure rollers set on the opposite side of the rotating workpiece from the tool to counter the cutting forces.

The setup times are usually 1 or 2 hr, so turret lathes are most economical in producing lots too large for engine lathes but too small for automatic screw machines or automatic lathes. In recent years, much of this work has been assumed by numerical control lathes, turning centers, or lean manufacturing cells. For example, the component (threaded shaft) shown in Figure 23.19 could have also been made on an NC turret lathe with some savings in cycle time or in a manufacturing cell with further savings in cycle time.

23.5 Workholding in Lathes

Workholding Devices for Lathes

Five methods are commonly used for supporting workpieces in lathes:

1. Held between centers.
2. Held in a chuck.
3. Held in a collet.
4. Mounted on a faceplate.
5. Mounted on the carriage.

In the first four of these methods, the workpiece is rotated during machining. In the fifth method, which is not used extensively, the tool rotates while the workpiece is fed into the tool.

A general discussion of workholding devices is found in Chapter 41, and the student involved in designing work holding devices should study the reference materials under "Tool Design." For lathes, **workholding** is a matter of selecting from standard tooling.

Lathe Centers

Workpieces that are relatively long with respect to their diameters are usually held between centers (see **Figure 23.20**). Two lathe centers are used, one in the spindle hole and the other in the hole in the tailstock quill. Two types are used, called *dead* and *live*. Dead centers are *solid*, that is, made of hardened steel with a Morse taper on one end so that it will fit into the spindle hole. The other end is ground to a taper. Sometimes the tip of this taper is made of tungsten carbide to provide better wear resistance. Before a center is placed in position, the spindle hole should be carefully wiped clean. The presence of foreign material will prevent the center from seating properly, and it will not be aligned accurately.

A mechanical connection must be provided between the spindle and the workpiece to provide rotation. This is accomplished by a **lathe dog** and **dog plate**. The dog is clamped to the work. The tail of the dog enters a slot in the dog plate, which is attached to the lathe spindle in the same manner as a lathe chuck. For work that has a finished surface, a piece of soft metal, such as copper or aluminum, can be placed between the work and the dog setscrew clamp to avoid marring. Live centers are designed so that the end that fits into the workpiece is mounted on ball or roller bearings. It is free to rotate. No

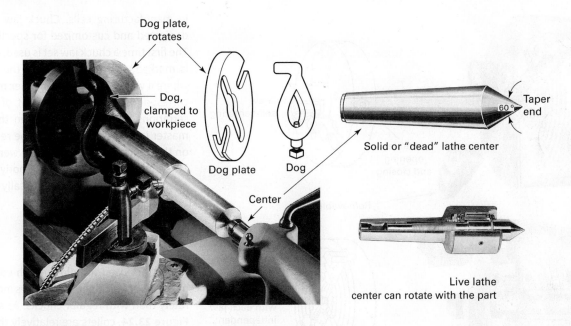

FIGURE 23.20 Work being turned between centers in a lathe, showing the use of a dog and dog plate. (Courtesy of South Bend Lathe)

lubrication is required. Live centers may not be as accurate as the solid type and therefore are not often used for precision work.

Before a workpiece can be mounted between lathe centers, a center hole must be drilled in each end. This is typically done in a drill press or on the lathe with a tool held in the rear turret. A combination center drill and countersink ordinarily is used, with care taken that the center hole is deep enough so that it will not be machined away in any facing operation and yet is not drilled to the full depth of the tapered portion of the center drill (see Chapter 25).

Because the work and the center of the headstock end rotate together, no lubricant is needed in the center hole at this end. The center in the tailstock quill does not rotate; adequate lubrication must be provided. A mixture of white lead and oil is often used. Failure to provide proper lubrication at all times will result in scoring of the workpiece center hole and the center, and inaccuracy and serious damage may occur. Live centers are often used in the tailstock to overcome these problems.

The workpiece must rotate freely, yet no looseness should exist. Looseness will usually be manifested in chattering of the workpiece during cutting. The setting of the centers should be checked after cutting for a short time. Heating and thermal expansion of the workpiece will reduce the clearances in the setup.

Mandrels

Workpieces that must be machined on both ends or are disk-shaped are often mounted on **mandrels** for turning between centers. Three common types of mandrels are shown in **Figure 23.21**. *Solid mandrels*

usually vary from 4 to 12 in. in length and are accurately ground with a 1:2000 taper (0.006 in./ft). After the workpiece is drilled and/or bored, it is pressed on the mandrel. The mandrel should be mounted between centers so that the cutting force tends to tighten the work on the mandrel taper. Solid mandrels permit the work to be machined on both ends as well as on the cylindrical surface. They are available in stock sizes but can be made to any desired size.

Gang (or disk) mandrels are used for production work because the workpieces do not have to be pressed on and thus can be put in position and removed more rapidly. However, only the cylindrical surface of the workpiece can be machined when this type of mandrel is used. *Cone mandrels* have the advantage that they can be used to center workpieces having a range of hole sizes.

Lathe Chucks

Lathe **chucks** are used to support a wider variety of workpiece shapes and to permit more operations to be performed than can be accomplished when the work is held between centers.

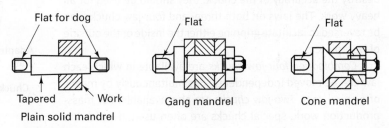

FIGURE 23.21 Three types of mandrels, which are mounted between centers for lathe work.

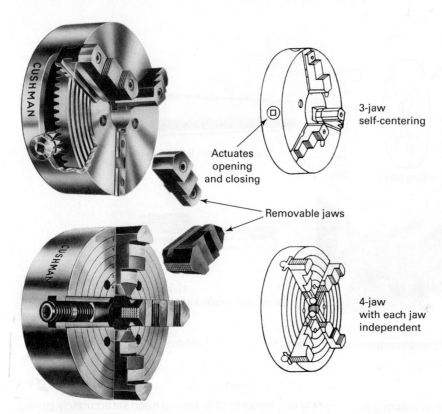

3-jaw self-centering

Actuates opening and closing

Removable jaws

4-jaw with each jaw independent

FIGURE 23.22 The jaws on chucks for lathes (four-jaw independent or three-jaw self-centering) can be removed and reversed.

Two basic types of chucks are used, three-jaw and four-jaw (**Figure 23.22**).

Three-jaw self-centering chucks are used for work that has a round or hexagonal cross section. The three jaws are moved inward or outward simultaneously by the rotation of a spiral cam, which is operated by means of a special wrench through a bevel gear. If they are not abused, these chucks will provide automatic centering to within about 0.001 in. However, they can be damaged through use and will then be considerably less accurate.

Each jaw in a *four-jaw independent chuck* can be moved inward and outward independently of the others by means of a chuck wrench. Thus, they can be used to support a wide variety of work shapes. A series of concentric circles engraved on the chuck face aid in adjusting the jaws to fit a given workpiece. Four-jaw chucks are heavier and more rugged than the three-jaw type, and because undue pressure on one jaw does not destroy the accuracy of the chuck, they should be used for all heavy work. The jaws on both three- and four-jaw chucks can be reversed to facilitate gripping either the inside or the outside of workpieces.

Combination four-jaw chucks are available in which each jaw can be moved independently or simultaneously by means of a spiral cam. *Two-jaw chucks* are also available. For mass-production work, special chucks are often used in which the jaws are actuated by air or hydraulic pressure, permitting very rapid clamping of the work. See **Figure 23.23** for a schematic. The rapid exchange of tooling is a key manufacturing strategy

in manufacturing cells. Chuck jaw sets are dedicated and customized for specific parts. The first time a chuck jaw set is used, each jaw is marked with the number of the jaw slot where it was installed and an index mark that corresponds with the alignment of the jaw serrations and the first tooth on the chuck master jaw. The jaws can now be reinstalled on the chuck exactly where they were bored. The adjustability of the chuck body lets the operator dial in part concentrically without resetting the jaw.

Collets

Collets are used to hold smooth cold-rolled bar stock or machined workpieces more accurately than with regular chucks. As shown in **Figure 23.24**, collets are relatively thin tubular steel bushings that are split into three longitudinal segments over about two-thirds of their length. At the split end, the smooth internal surface is shaped to fit the piece of stock that is to be held. The external surface of the collet is a taper that mates with an internal taper of a collet sleeve that fits into

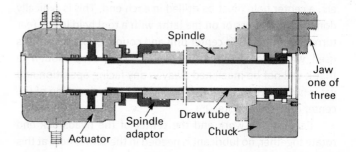

Spindle

Jaw one of three

Draw tube

Spindle adaptor

Chuck

Actuator

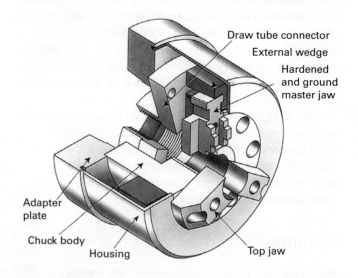

Draw tube connector

External wedge

Hardened and ground master jaw

Adapter plate

Chuck body

Housing

Top jaw

FIGURE 23.23 Hydraulically actuated through-hole three-jaw power chuck shown in section view to left and in the spindle of the lathe above connected to the actuator.

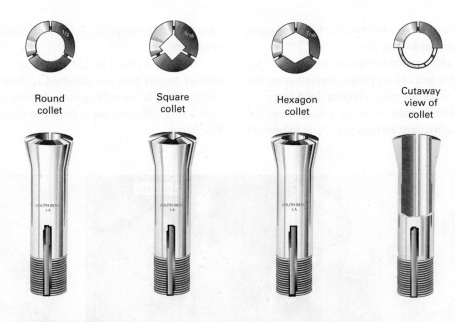

Round collet   Square collet   Hexagon collet   Cutaway view of collet

FIGURE 23.24   Several types of lathe collets. (Courtesy of South Bend Lathe)

the lathe spindle. When the collet is pulled inward into the spindle (by means of the draw bar), the action of the two mating tapers squeezes the collet segments together, causing them to grip the workpiece (see **Figure 23.25**).

Collets are made to fit a variety of symmetrical shapes. If the stock surface is smooth and accurate, good collets will provide very accurate centering, with runout less than 0.0005 in. However, the work should be no more than 0.002 in. larger or 0.005 in. smaller than the nominal size of the collet. Consequently, collets are used only on drill-rod, cold-rolled, extruded, or previously machined stock.

Collets that can open automatically and feed bar stock forward to a stop mechanism are commonly used on automatic lathes and turret lathes. Another type of collet similar to a Jacobs drill chuck has a greater size range than ordinary collets; therefore, fewer are required.

Faceplates

Faceplates are used to support irregularly shaped work that cannot be gripped easily in chucks or collets. The work can be bolted or clamped directly on the faceplate or can be supported on an auxiliary fixture that is attached to the faceplate. The latter procedure is time saving when identical pieces are to be machined.

Mounting Work on the Carriage

When no other means are available, boring is occasionally done on a lathe by mounting the work on the carriage, with the boring bar mounted between centers and driven by means of a dog.

Steady and Follow Rests

If one attempts to turn a long, slender piece between centers, the radial force exerted by the cutting tool or the weight of the workpiece itself may cause it to be deflected out of line. **Steady rests** and **follow rests** (**Figure 23.26**) provide a means for supporting such work between the headstock and the tailstock. The steady rest is clamped to the lathe ways and has three movable fingers that are adjusted to contact the work and

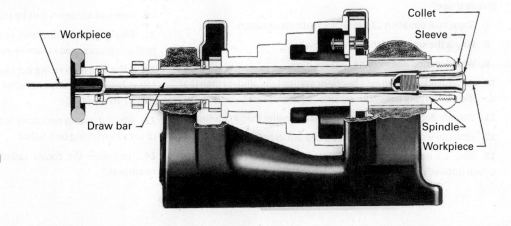

Workpiece

Draw bar

Collet

Sleeve

Spindle

Workpiece

FIGURE 23.25   A draw bar is used to close the collets, providing clamping force on the work piece. (Courtesy of South Bend Lathe)

align it. A light cut should be taken before adjusting the fingers to provide a smooth contact-surface area.

A steady rest can also be used in place of the tailstock as a means of supporting the end of long pieces, pieces having too large an internal hole to permit using a regular dead center, or work where the end must be open for boring. In such cases, the headstock end of the work must be held in a chuck to prevent

longitudinal movement. Tool feed should be toward the headstock.

The follow rest is bolted to the lathe carriage. It has two contact fingers that are adjusted to bear against the work-piece, opposite the cutting tool, in order to prevent the work from being deflected away from the cutting tool by the cut-ting forces.

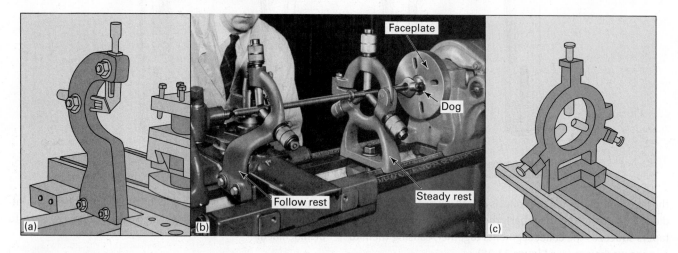

FIGURE 23.26 Cutting a thread on a long, slender workpiece, using a follow rest (a) and a steady rest (c) on an engine lathe. (b) Note the use of a dog and faceplate to drive the workpiece. (Courtesy of South Bend Lathe)

Review Questions

1. How is the tool–work relationship in turning different from that in facing?

2. What different kinds of surfaces can be produced by turning versus facing?

3. How does form turning differ from ordinary turning?

4. What is the basic difference between facing and a cutoff operation?

5. Which operations shown in Figure 23.3 do not form a chip?

6. Why is it difficult to make heavy cuts if a form turning tool is complex in shape?

7. Show how equation 23.6 is an approximate equation.

8. Why is the spindle of the lathe hollow?

9. What function does a lathe carriage have?

10. Why is feed specified for a boring operation typically less than that specified for turning if the MRR equations are the same?

11. Why are depths of cut in boring usually smaller than in turning?

12. How can work be held and supported in a lathe?

13. How is a workpiece that is mounted between centers on a lathe driven (rotated)?

14. What will happen to the workpiece when turned, if held between centers, and the centers are not exactly in line?

15. Why is it not advisable to hold hot-rolled steel stock in a collet?

16. How does a steady rest differ from a follow rest?

17. What are the advantages and disadvantages of a four-jaw independent chuck versus a three-jaw chuck?

18. Why should the distance the cutting tool overhangs from the tool holder be minimized?

19. What is the difference between a ram- and a saddle-turret lathe?

20. How can a tapered part be turned on a lathe?

21. Why might it be desirable to use a heavy depth of cut and a light feed at a given speed in turning rather than the opposite?

22. If the rpm for a facing cut (assuming given work and tool materials) is being held constant, what is happening during the cut to the speed? To the feed?

23. Why is it usually necessary to take relatively light feeds and depths of cut when boring on a lathe?

24. How does the corner radius of the tool influence the surface roughness?

25. What effect does a BUE have on the diameter of the workpiece in turning?

26. How does the multiple-spindle screw machine differ from the single-spindle machine?

27. Why does boring ensure concentricity between the hole axis and the axis of rotation of the workpiece (for boring tool), whereas drilling does not?

28. Why are vertical spindle machines better suited for machining large workpieces than horizontal lathes?

29. In which figures in this chapter is a workpiece being held in a three-jaw chuck?

30. How is the workpiece in Figure 23.13 being held?

31. In which figures in this chapter is a dead center shown?

32. In which figures in this chapter is a live center shown?

33. In which figures in this chapter showing setups do you find the following being used as a workholding device?

a. Three-jaw chuck

b. Collet

c. Faceplate

d. Four-jaw chuck

34. How many form tools are being utilized in the process shown in Figure 23.14 to machine the part?

35. From the information given in Figure 23.14, start with a piece of round bar stock and show how it progresses, operation by operation, into a finished part—a threaded shaft.

Problems

1. Select the speed, feed, and depth of cut for turning wrought, low-carbon steel (hardness of 200 BHN) on a lathe with AISI tool material of HSS M2 or M3. (*Hint:* Refer to Chapter 21 for recommended parameters.)

2. Calculate the rpm (N_s) to run the spindle on a lathe to generate a cutting speed of 80 sfpm if the outside diameter of the workpiece is 8 in.

3. The lathe in problem 2 has rpm settings of 20, 30, and 35 that are near the calculated value. What rpm would you pick, and what will actual cutting speed be?

4. Calculate the cutting time if the length of cut is 24 in., the feed rate is 0.030 ipr, and the cutting speed is 80 fpm. The allowance is 0.5 in. and the diameter is 8 in.

5. Calculate the metal removal rate for machining at a speed of 80 fpm, feed of 0.030 ipr, at a depth of 0.625 in. Use any data from Problem 4 that you need.

6. Determine the speed, feed, and depth of cut when boring wrought, low-carbon steel (hardness of 120 BHN) on a lathe with AISI tool material of HSS M2 or M3.

7. At a speed of 90 fpm, feed of 0.030 ipr, and depth at 0.625, calculate the input parameters for a lathe if the diameter to be machined (bored) is 6 in. Find the rpm (N_s) to run the spindle to generate the cutting speed.

8. Calculate the cutting time for a 4-in. length of cut, given that the feed rate is 0.030 ipr at a speed of 90 fpm.

9. For a boring operation at $V = 90$, $f_r = 0.030$, and $d = 0.625$, calculate the MMR.

10. A cutting speed of 100 sfpm has been selected for a turning cut. The workpiece is 8 in. (203.2 mm) long and a feed of 0.020 in. (0.51 mm) per revolution is used. Using cutting speeds and feeds, calculate the machining time to turn the bar.

11. The following data apply for machining a part on a turret lathe and on an engine lathe:

	Engine Lathe	Turret Lathe
Times, in minutes, to machine part	30 min	5 min
Cost of special tooling	0	$300
Time to set up the machine	30 min	3 hr
Labor rates	$8/hr	$8/hr
Machine rates (overhead)	$10/hr	$12/hr

One of the jobs for the engineer is to estimate the run time for a sequence of machining processes. For example:

a. How many pieces would have to be made for the cost of the engine lathe to just equal the cost of the turret lathe? This is the BEQ.

b. What is the cost per unit at the BEQ?

12. A finish cut for a length of 10 in. on a diameter of 3 in. is to be taken in 1020 steel with a speed of 100 fpm and a feed of 0.005 ipr. What is the machining time?

13. A workpiece 10 in. in diameter is to be faced down to a diameter of 2 in. on the right end. The lathe maintain the cutting speed at 100 fpm throughout the cut by changing the rpm. What should be the time for the cut? Now suppose the spindle rpm for the workpiece is set to give a speed of 100 fpm for the 10-in. diameter and is not changed during the cut. What is the machining time for the cut now? The feed rate is 0.005 ipr.

14. A hole 89 mm in diameter is to be drilled and bored through a piece of 1340 steel that is 200 mm long, using a horizontal boring, drilling, and milling machine. High-speed tools will be used. The sequence of operations will be center drilling; drilling with an 18-mm drill followed by a 76-mm drill; then boring to size in one cut, using a feed of 0.50 mm/rev. Drilling feeds will be 0.25 mm/rev for the smaller drill and 0.64 mm/rev for the larger drill. The center drilling operation requires 0.5 min. To set or change any given tool and set the proper machine speed and feed requires 1 min. Select the initial cutting speeds, and compute the total time required for doing the job. (Neglect setup time for the fixture.) This is often referred to as the run time or the cycle time. (*Hint:* Check Chapter 21 for recommended speeds for turning.)

CHAPTER 24

Milling

24.1 Introduction

Milling is a basic machining process by which a surface is generated by progressive chip removal. The workpiece is fed into a rotating cutting tool or the workpiece remains stationary, and the cutter is fed to the work. In nearly all cases, a multiple-tooth cutter is used so that the material removal rate is high. Often the desired surface is obtained in a single pass of the cutter or work. Because very good surface finish can be obtained, milling is particularly well-suited and widely used for mass-production work. Many types of milling machines are used, ranging from relatively simple and versatile machines that are used for general-purpose machining in job shops and tool and die work (these are NC or CNC machines) to highly specialized machines for mass production. Unquestionably, more flat surfaces are produced by milling than by any other machining process.

The cutting tool used in milling is known as a **milling cutter**. Equally spaced peripheral teeth will intermittently engage and machine the workpiece. This is called **interrupted cutting**. The workpieces are typically held in fixtures, as described in Chapter 41.

24.2 Fundamentals of Milling Processes

Milling operations can be classified into broad categories called peripheral milling, end milling, and face milling. Each has many variations. In **peripheral milling**, the surface is generated by teeth located on the periphery of the cutter body (**Figure 24.1**). The surface is parallel with the axis of rotation of the cutter. Both flat and formed surfaces can be produced by this method, the cross section of the resulting surface corresponding to the axial contour of the cutter. This process, often called **slab milling**, is usually performed on horizontal spindle milling machines. In slab milling, the tool rotates (mills) at some rpm (N_s) while the work feeds past the tool at a table feed rate, f_m, in

inches per minute (ipm), which depends on the feed per tooth, f_t.

As in the other processes, the cutting speed, V, and feed per tooth are selected by the engineer or the machine tool operator. As before, these variables depend on the work material, the tool material, and the specific process. The cutting velocity is that which occurs at the cutting edges of the teeth in the milling center. The rpm of the spindle is determined from the surface cutting speed, where D is the cutter diameter in inches according to

$$N_s = \frac{12V}{\pi D} \tag{24.1}$$

The depth of cut, called DOC or d in Figure 24.1, is simply the distance between the old and new machined surface.

The width of cut is the width of the cutter or the work, in inches, and is given the symbol W. The length of the cut, L, is the length of the work plus some allowance, L_A, for approach and overtravel. The feed of the table, f_m, in inches per minute, is related to the amount of metal each tooth removes during a revolution, the feed per tooth, f_t, according to

$$f_m = f_t N_s n \tag{24.2}$$

where n is the number of teeth in the cutter (teeth/revolution).

The **cutting time** is

$$T_m = \frac{L + L_A}{f_m} \tag{24.3}$$

The length of approach is

$$L_A = \sqrt{\frac{D^2}{4} - \left(\frac{D}{2} - \text{DOC}\right)^2} = \sqrt{d(D-d)} \tag{24.4}$$

The **metal removal rate (MRR)** is

$$\text{MRR} = \frac{\text{Volume}}{T_m} = \frac{LWd}{T_m} = Wf_m d \ \text{in.}^3/\text{min} \tag{24.5}$$

ignoring L_A. Values for recommended cutting speeds, in feet per minute, and feeds, in inches per tooth, are given in **Table 24.1**.

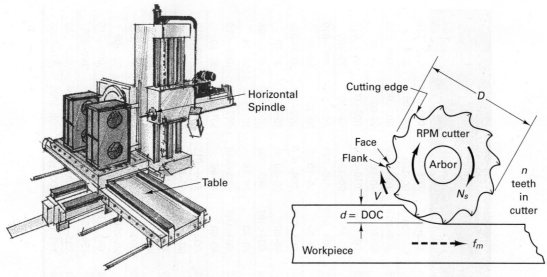

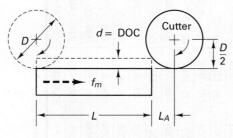

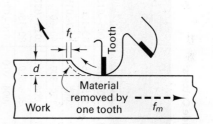

(a) Horizontal-spindle milling machine

(b) Slab milling—multiple tooth

(c) Allowances for cutter approach

(d) Feed per tooth, f_t

FIGURE 24.1 Peripheral milling can be performed on a horizontal-spindle milling machine. The cutter rotates at rpm N_s, removing metal at cutting speed, V. The allowance for starting and finishing the cut depends on the cutter diameter and depth of cut, d. The feed per tooth, f_t, and cutting speed are selected by the operator or process planner. (Courtesy J T. Black)

Face Milling and End Milling

In face milling and end milling, the generated surface is at right angles to the cutter axis (**Figure 24.2**). Most of the cutting is done by the peripheral portions of the teeth, with the face portions providing some finishing action. **Face milling** is done on both horizontal- and vertical-spindle machines.

The tool rotates (face mills) at some rpm (N_s) while the work feeds past the tool. The rpm is related to the surface cutting speed, V, and the cutting tool diameter, D, according to equation . The depth of cut is d, in inches, as shown in Figure 24.2b. The width of cut is W, in inches, and may be width of the workpiece or width of the cutter, depending on the setup. The length of cut is the length of the workpiece, L, plus an allowance, L_A, for approach and overtravel, L_o, in inches. The feed rate of the table, f_m, in inches per minute, is related to the amount of metal each tooth removes during a pass over the

work, called the feed per tooth, f_t. As before $f_m = f_t N_s n$, where the number of teeth in the cutter is n. The cutting time is

$$T_m = \frac{L + L_A + L_o}{f_m} \text{min} \qquad (24.6)$$

The metal removal rate is

$$\text{MRR} - \frac{\text{Volume}}{T_m} = \frac{LWd}{T_m} = f_m Wd \text{ in.}^3/\text{min}$$

When calculating the MRR, ignore L_o and L_A. The length of approach is usually equal to the length of overtravel, which usually equals $D/2$ in. For a setup where the tool does not completely pass over the workpiece,

$$L_o = L_A = \sqrt{W(D-W)} \text{ for } W < \frac{D}{2} \qquad (24.7)$$

TABLE 24.1   Suggested Starting Feeds and Speeds Using High-Speed Steel and Carbide Cutters[a]

Material	Feed (in./tooth) Speed (fpm)	Carbide Cutters						High-speed-Steel Cutters					
		Face Mills	Slab Mills	End Mills	Full and Half-Side Mills	Saws	Form Mills	Face Mills	Slab Mills	End Mills	Full and Half-Side Mills	Saws	Form Mills
Malleable iron	Feed per tooth	.005–.015	.005–.015	.005–.010	.005–.010	.003–.004	.005–.010	.005–.015	.005–.015	.003–.015	.006–.012	.003–.006	.005–.010
Soft/hard	Speed, fpm	200–300	200–300	200–350	200–300	200–350	175–275	60–100	60–90	60–100	60–100	60–100	60–80
Cast steel	Feed per tooth	.008–.015	.005–.015	.003–.010	.005–.010	.002–.004	.005–.010	.010–.015	.010–.015	.005–.010	.005–.010	.002–.005	.008–.012
Soft/hard	Speed, fpm	150–350	150–350	150–350	150–350	150–300	150–300	40–60	40–60	40–60	40–60	40–60	40–60
100–150	Feed per tooth	.010–.015	.008–.015	.005–.010	.008–.012	.003–.006	.004–.010	.015–.030	.008–.015	.003–.010	.010–.020	.003–.006	.008–.010
BHN steel	Speed, fpm	450–800	450–600	450–600	450–800	350–600	350–600	80–130	80–130	80–140	80–130	70–100	70–100
150–250	Feed per tooth	.010–.015	.008–.015	.005–.010	.007–.012	.003–.006	.004–.010	.010–.020	.008–.015	.003–.010	.010–.015	.003–.006	.006–.010
BHN steel	Speed, fpm	300–450	300–450	300–450	300–450	300–450	300–450	50–70	50–70	60–80	50–70	50–70	50–70
250–350	Feed per tooth	.008–.015	.007–.012	.005–.010	.005–.012	.002–.005	.003–.008	.005–.010	.005–.010	.003–.010	.005–.010	.002–.005	.005–.010
BHN steel	Speed, fpm	180–300	150–300	150–300	160–300	150–300	150–300	35–60	35–50	40–60	35–50	35–50	35–50
350–450	Feed per tooth	.008–.015	.007–.012	.004–.008	.005–.012	.001–.004	.003–.008	.003–.008	.005–.008	.003–.101	.003–.008	.001–.004	.003–.008
BHN steel	Speed, fpm	125–180	100–150	100–150	125–180	100–150	100–150	20–35	20–35	20–40	20–35	20–35	20–35
Cast iron, hard	Feed per tooth	.005–.010	.005–.010	.003–.008	.003–.010	.002–.003	.005–.010	.005–.012	.005–.010	.003–.008	.005–.010	.002–.004	.005–.010
BHN 180–225	Speed, fpm	125–200	100–175	125–200	125–200	125–200	100–175	40–60	35–50	40–60	40–60	35–60	35–50
Cast iron, medium	Feed per tooth	.008–.015	.008–.015	.005–.010	.005–.012	.003–.004	.006–.012	.010–.020	.008–.015	.003–.010	.008–.015	.003–.005	.008–.012
BHN 180–225	Speed, fpm	200–275	175–250	200–275	200–275	200–250	175–250	60–80	50–70	60–90	60–80	60–70	50–60
Cast iron, soft	Feed per tooth	.015–.025	.010–.020	.005–.012	.008–.015	.003–.004	.008–.015	.015–.030	.010–.025	.004–.010	.010–.020	.002–.005	.010–.015
BHN 150–180	Speed, fpm	275–400	250–350	275–400	275–400	250–350	250–350	80–120	70–110	80–120	80–120	70–110	60–80
Bronze	Feed per tooth	.010–.020	.010–.020	.005–.010	.008–.012	.003–.004	.008–.015	.010–.025	.008–.020	.003–.010	.008–.015	.003–.005	.008–.015
Soft/hard	Speed, fpm	300–100	300–800	300–1000	300–1000	300–1000	200–800	50–225	50–200	50–250	50–225	50–250	50–200
Brass	Feed per tooth	.010–.020	.010–.020	.005–.010	.008–.012	.003–.004	.008–.015	.010–.025	.008–.020	.005–.015	.008–.015	.003–.005	.008–.015
Soft/hard	Speed, fpm	500–1500	500–1500	500–1500	500–1500	500–1500	500–1500	150–300	100–300	150–350	150–350	150–300	100–300
Aluminum alloy	Feed per tooth	.010–.040	.010–.030	.003–.015	.008–.025	.003–.006	.008–.015	.010–.040	.015–.040	.015–.040	.010–.030	.004–.008	.010–.020
Soft/hard	Speed, fpm	2000 UP	2000 UP	2000 UP	2000 UP	2000 UP	2000 UP	300–1200	300–1200	300–1200	300–1200	300–1200	300–1200

[a]Generally lower end of range used for inserted blade cutters, higher end of range for indexable insert cutters.

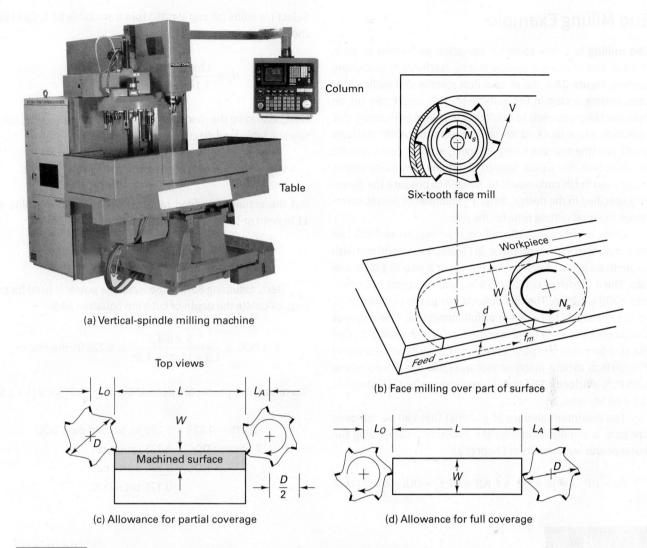

Column

Table

(a) Vertical-spindle milling machine

Six-tooth face mill

(b) Face milling over part of surface

Top views

(c) Allowance for partial coverage

(d) Allowance for full coverage

FIGURE 24.2 Face milling is often performed on a vertical spindle milling machine using a multiple-tooth cutter ($n = 6$ teeth) rotating N_s at rpm to produce cutting speed, V. The workpiece feeds at rate f_m, in inches per minute past the tool. The allowance depends on the tool diameter and the width of cut. (Courtesy J T. Black)

$$L_o = L_A = \frac{D}{2} \text{ for } W \geq \frac{D}{2} \qquad (24.8)$$

Here is an example of a face milling calculation. A 4-in.-diameter, six-tooth face mill is selected, using carbide inserts (**Figure 24.3**). The material being machined is low-alloy steel, annealed. Using cutting data recommendations, the cutting speed chosen is 400 sfpm with a feed of 0.008 in./tooth at a d of 0.12 in. Determining rpm at the spindle,

$$N_s = \frac{12V}{\pi D} = \frac{12 \times 400}{3.14 \times 4} = 392 \text{ rpm}$$

Determining the feed rate of the table, $fm = nN_s f_t$,

$$f_m = 0.008 \times 6 \times 392 = 19 \text{ in.} / \text{min}$$

If slab or side milling were being performed, with the same parameters selected as given earlier, the setup would be differ-

ent but the spindle rpm and table feed rate would be the same. The cutting time would be different because the allowances for face milling are greater than for slab milling. In milling, power consumption is usually the limiting factor. A thick chip is more power efficient than a thin chip.

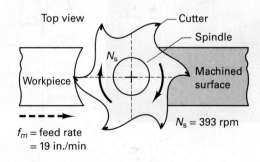

Top view — Cutter
— Spindle

N_s

Workpiece

Machined surface

f_m = feed rate = 19 in./min

N_s = 393 rpm

FIGURE 24.3 Face milling with 6 tooth cutter viewed from above with vertical spindle-machine.

End Milling Example

End milling is a very common operation performed on both vertical- and horizontal-spindle milling machines or machining centers. **Figure 24.4** shows a vertical spindle end milling process, cutting a step in the workpiece. This cutter can cut on both the sides and ends of the tool. If you were performing this operation on a block of metal (for example, 430F stainless steel), you (the manufacturing engineer) would select a specific machine tool. You would have to determine how many passes (rough and finish cuts) would be needed to produce the geometry specified in the design. Why? The number of passes determines the total cutting time for the job.

Using a vertical-spindle milling machine, an end mill can produce a step in the workpiece. In Figure 24.4, an end mill with six teeth on a 2-in. diameter is used to cut a step in 430F stainless. The d (depth of cut) is 0.375 in., and the depth of immersion (DOI) is 1.25 in. The tool deflects due to the cutting forces, so the cut needs to be made at full immersion. The engineer should check to see if there is enough power for a full DOC. Can the step be cut in one pass, or will multiple cuts be necessary? The vertical milling machine tool available has a 5-hp motor with 80% efficiency. The specific horsepower for 430F stainless is 1.3 hp/in.³/min.

The maximum amount of material that can be removed per pass is usually limited by the available power. Using the **horsepower** equation from Chapter 21,

$$hp = HP_s \times MRR = HP_s \times f_m WD = HP_s f_m \times DOI \times d \quad (24.9)$$

Select $f_t = 0.005$ ipt and $V = 250$ fpm from Table 24.1. Calculate the spindle rpm:

$$N_s = \frac{12 \times 250}{3.14 \times 2} = 477 \text{ rpm of cutter}$$

Next, assuming the machine tool has this rpm available, calculate the table feed rate:

$$f_m = f_t \times n \times N_s = 0.005 \times 6 \times 477 = 14.31 \text{ in./min}$$

But the actual table feed rates for the selected machine are 11 in./min or 16 in./min, so, being conservative, select

$$f_m = \text{table feed rate} = 11.00 \text{ in./min}$$

Next, assuming 80% of the available power is used for cutting, calculate the depth of cut from equation 24.9:

$$d = DOC \cong \frac{5 \times 0.8}{1.3 \times 11.00 \times 1.25} \cong 0.225 \text{ in. maximum}$$

Therefore, two passes are needed because (0.375/0.225 = 1.6):

$$0.375 - 0.225 = 0.150 \text{ in. second pass DOC}$$

$$2 \text{ passes: DOC} = 0.225 \text{ rough cut}$$

$$DOC = \underline{0.150} \text{ finish cut}$$

$$0.375 \text{ total DOC}$$

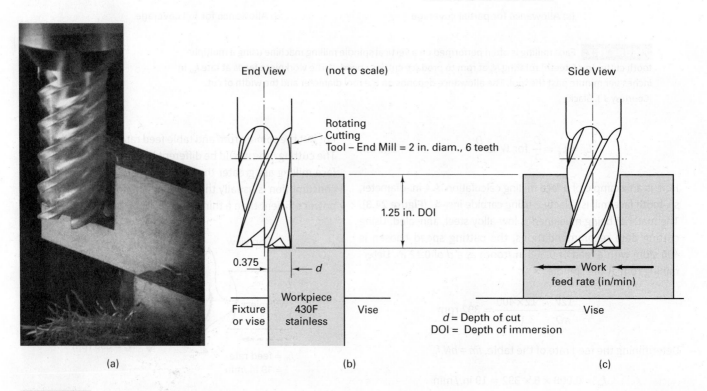

End View   (not to scale)   Side View

Rotating Cutting Tool – End Mill = 2 in. diam., 6 teeth

1.25 in. DOI

0.375   d

Workpiece 430F stainless

Fixture or vise   Vise

Work feed rate (in/min)

d = Depth of cut
DOI = Depth of immersion

Vise

(a)   (b)   (c)

FIGURE 24.4 (a) End milling a step feature in a block using a flat-bottomed, end mill cutter in a vertical spindle-milling machine. (b) End view, table moving the block into the cutter. (c) Side view, workpiece feeding right to left into tool. (Courtesy J T. Black)

Note that for $d = 0.150$, the feed per tooth would be only slightly increased to 0.0051 ipt:

$$f_t = \frac{0.5 \times 0.8}{1.3 \times 6 \times 477 \times 0.150 \times 1.25} = 0.0051 \text{ in.} / \text{tooth}$$

You may want to change f_t to improve the surface finish. With a smaller feed per tooth, a better surface finish is usually obtained. However, there are other factors to consider, like machining time.

In general (for face, slab, or end milling), if machine power is lacking, the following actions may help.

1. Use a cutter with a positive rake as this can be more efficient than one with a negative rake.
2. Use a cutter with a coarser pitch (fewer teeth).
3. Use a smaller cutter and take several passes (reduce d or DOI).

Up Versus Down Milling

One of the subtle aspects of milling concerns the direction of rotation of the cutter with respect to the movement of the workpiece. Surfaces can be generated by two distinctly different methods (**Figure 24.5**). **Up milling** is the traditional way to mill and is also called **conventional milling**. The cutter rotates against the direction of feed of the workpiece. In **climb** or **down milling**, the cutter rotation is in the same direction as the feed rate. The method of chip formation is completely different in the two cases.

In up milling, the chip is very thin at the beginning, where the tooth first contacts the work; then it increases in thickness, becoming a maximum where the tooth leaves the work. The cutter tends to push the work along and lift it upward from the table. This action tends to eliminate any effect of looseness in the feed screw and nut of the milling machine table and results in a smooth cut. However, the action tends to loosen the work from the fixture. Therefore, greater clamping forces must be employed, with the danger of distorting the part. In addition, the smoothness of the generated surface depends greatly on the sharpness of the cutting edges. In up milling, chips can be carried into the newly machined surface, causing the surface finish to be poorer (rougher) than in down milling and causing damage to the insert.

In down milling, maximum chip thickness occurs close to the point at which the tooth contacts the work. Because the relative motion tends to pull the workpiece into the cutter, any possibility of looseness in the table feed screw must be eliminated if down milling is to be used. It should never be attempted on machines that are not designed for this type of milling. Virtually all modern milling machines are capable of down milling, and it is a most favorable application for carbide cutting edges. Because the material yields in approximately a tangential direction at the end of the tooth engagement, there is less tendency (than when up milling is used) for the machined surface to show toothmarks, and the cutting process is smoother, with less chatter. Another advantage of down milling is that the cutting force tends to hold the work against the machine table, permitting lower clamping forces. However, the fact that the cutter teeth strike against the surface of the work at the beginning of each chip can be a disadvantage if the workpiece has a hard surface, as castings sometimes do. This may cause the teeth to dull rapidly. Metals that readily work-harden should be down milled, and many toolmakers recommend that down milling should always be the first choice.

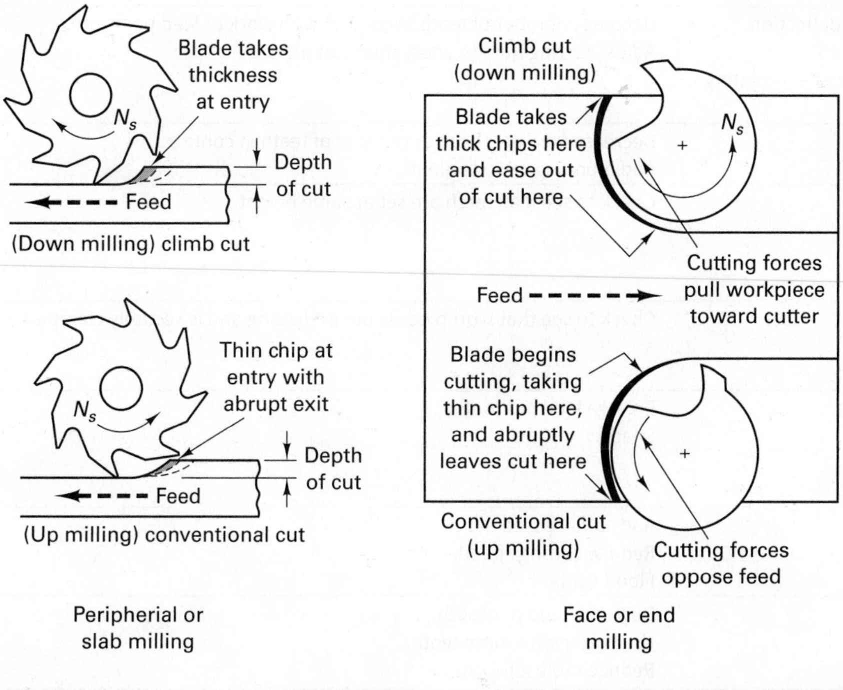

Milling Surface Finish

The average surface finishes that can be expected on free-machining materials range from 60 to 150 μin. Conditions exist, however, that can produce wide variations on either side of these ranges. For example, some inserts are designed with wiper flats (short parallel surface behind the tool tip). If the feed per revolution (Feed per tooth × Number of teeth) of the cutter is smaller than the length of the wiper flat (the land on the tool), then the surface finish on the workpiece will be generated by the highest insert. In finishing cuts, keeping the depth of cut small will limit the axial cutting force, reducing vibrations and producing a superior finish. See Chapter 31 for discussions on measuring surface finish.

Milling is an interrupted cutting process. The individual teeth enter and leave the cut and subject the tool to impact loading, cyclic heating, and cycle cutting

FIGURE 24.5 Climb cut or down milling versus conventional cut or up milling for slab or face or end milling.

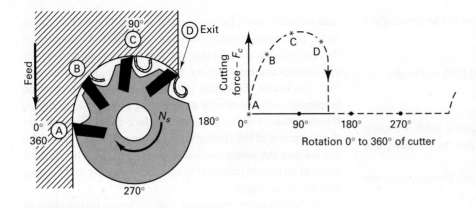

FIGURE 24.6 Conventional face milling (left) with cutting force diagram for F_c (right) showing the interrupted nature of the process. (From *Metal Cutting Principles*, 2nd ed., Ingersoll Cutting Tool Company)

forces. As shown in **Figure 24.6**, in upmilling the cutting force, F_c, builds rapidly as the tool enters the work at A and progresses to B, peaks as the blade crosses the direction of feed at C, decreases to D, and then drops to zero abruptly

upon exit. Down milling produces impact forces upon tool entry. The diagram does not indicate the impulse loads caused by impacts. The interrupted-cut phenomenon explains in large part why milling cutter teeth are designed to have small positive or negative rakes, particularly when the tool material is carbide or ceramic. These brittle materials tend to be very strong in compression, and negative rake results in the cutting edges being placed in compression by the cutting forces rather than tension. Cutters made from high-speed steel (HSS) are made with positive rakes, in the main, but must be run at lower speeds. Positive rake tends to lift the workpiece, while negative rakes compress the workpiece and allow heavier cuts to be made. **Table 24.2** summarizes some additional milling problems.

TABLE 24.2	Probable Causes of Milling Problems	
Problem	**Probable Cause**	**Cures**
Chatter (vibration)	1. Lack of rigidity in machine, fixtures, arbor, or workpiece 2. Cutting load too great 3. Dull cutter 4. Poor lubrication 5. Straight-tooth cutter 6. Radial relief too great 7. Rubbing, insufficient clearance	Use larger arbors. Change rpm (cutting speed). Decrease feed per tooth or number of teeth in contact with work. Sharpen or replace inserts. Flood coolant. Use helical cutter. Check tool angles.
Loss of accuracy (cannot hold size)	1. High cutting load causing deflection 2. Chip packing, between teeth 3. Chips not cleaned away before mounting new piece of work	Decrease number of teeth in contact with work or feed per tooth. Adjust cutting fluid to wash chips out of teeth.
Cutter rapidly dulls	1. Cutting load too great 2. Insufficient coolant	Decrease feed per tooth or number of teeth in contact. Add blending oil to coolant.
Poor surface finish	1. Feed too high 2. Tool dull 3. Speed too low 4. Not enough cutter teeth	Check to see if all teeth are set at same height.
Cutter digs in (hogs into work)	1. Radial relief too great 2. Rake angle too large 3. Improper speed	Check to see that workpiece is not deflecting and is securely clamped.
Work burnishing	1. Cut is too light 2. Tool edge worn 3. Insufficient radial relief 4. Land too wide	Enlarge feed per tooth. Sharpen cutter.
Cutter burns	1. Not enough lubricant 2. Speed too high	Add sulfur-based oil. Reduce cutting speed. Flood coolant.
Teeth breaking	1. Feed too high 2. Depth of cut too large	Decrease feed per tooth. Use cutter with more teeth. Reduce table feed rate.

Adapted from *Cutting Tool Engineering*, October 1990, p. 90, by Peter Liebhold, museum specialist, Division of Engineering and Industry, the Smithsonian Institute, Washington, DC.

24.3 Milling Tools and Cutters

Most milling work today is done with face mills and end mills. The face mills use indexable carbide insert tooling, while the end mills are either solid HSS or insert tooling (**Figure 24.7**).

Basically, *mills* are shank-type cutters having teeth on the circumferential surface and one end. They can thus be used for facing, profiling, and end milling. The teeth may be either straight or helical, but the latter is more common. Small end mills have straight shanks, whereas taper shanks are used on larger sizes (**Figure 24.8**).

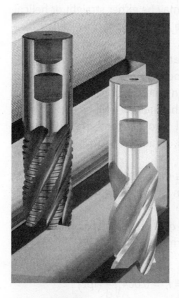

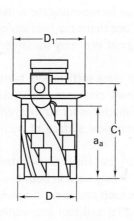
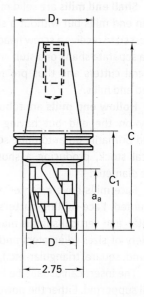

FIGURE 24.7 Solid end mills are often coated. Insert tooling end mills come in a variety of sizes and are mounted on taper shanks. (Courtesy J T. Black)

Face mill

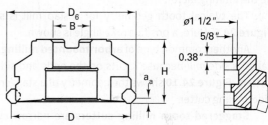

FIGURE 24.8 Face mills come in many different designs using many different insert geometries and mounting arbors.

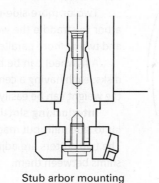

Stub arbor mounting

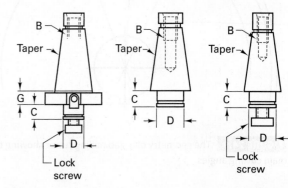

Lock screw

Lock screw

Plain end mills have multiple teeth that extend only about halfway toward the center on the end. They are used in milling slots, profiling, and facing narrow surfaces. **Two-lip mills** have two straight or helical teeth that extend to the center. Thus, they may be sunk into material, like a drill, and then fed lengthwise to form a groove, a slot, or a pocket.

Shell end mills are solid multiple-tooth cutters, similar to plain end mills but without a shank. The center of the face is recessed to receive a screw head or nut for mounting the cutter on a separate shank or a stub arbor. One shank can hold any of several cutters and thus provides great economy for larger-sized end mills.

Hollow end mills are tubular in cross section, with teeth only on the end but having internal clearance. They are used primarily on automatic screw machines for sizing cylindrical stock, producing a short cylindrical surface of accurate diameter.

Face mills have a center hole so that they can be arbor mounted. **Face-milling cutters** are widely used in both horizontal- and vertical-spindle machine tools and come in a wide variety of sizes (diameters and heights) and tool geometries (round, square, triangular, etc.), as shown in Figure 24.8.

The insert can usually be indexed four times and must be well supported. Either the power or the rigidity of the machine tool will be the limiting factor, although sometimes setup can be the limiting factor.

The typical tooth geometry for a slab mill is shown in **Figure 24.9**. Here, a positive rake angle is shown.

Another common type of **arbor-mounted milling cutter** is called a *side mill* because it cuts on the ends and sides of the cutters. **Figure 24.10** shows the geometry of a staggered-tooth side-milling cutter.

Staggered-tooth milling cutters are narrow cylindrical cutters having staggered teeth, and with alternate teeth having opposite helix angles. They are ground to cut only on the

periphery, but each tooth also has chip clearance ground on the protruding side. These cutters have a free cutting action that makes them particularly effective in milling deep slots. Staggered-tooth cutters are really special **side-milling cutters**, which are similar to plain milling cutters except that the teeth extend radially part way across one or both ends of the cylinder toward the center. The teeth may be either straight or helical. Frequently, these cutters are relatively narrow, being disk-like in shape.

The side-milling cutter can cut on sides and ends of the teeth, so it can cut slots or grooves. However, only a few teeth are engaged at any one point in time, causing heaving torsional vibrations. The average chip thickness, h_i, will be less than the feed per tooth, f_t. The actual feed per tooth, f_a, will be less than feed per tooth selected, f_t, according to $f_a = h_i\sqrt{\frac{D}{d}}$. See Figure 24.10.

For example, a thickness (h_i) of 0.004 in. corresponds to 0.012 in. feed per tooth in most side- and face-milling operations.

If the radial depth of cut, d, is very small compared to the cutter diameter, D, use this formula:

$$\text{Feed per tooth} = f_t = 0.004\sqrt{\frac{D}{d}}\ (\text{ipt})$$

For calculating the table feed, use half the number of inserts in a full side and face mill to arrive at the effective number of teeth. Thus, Table feed rate (ipm) = rpm × Number of effective teeth feed per tooth.

In **Figure 24.11**, insert-tooth side mills are arranged in a gang-milling setup to cut three slots in the workpiece simultaneously. Thus, the desired part geometry is repeatedly produced by the setup as the position of the cutters is fixed. However, in side- and face-milling operations, only a few teeth are engaged at any point in time, resulting in heavy torsional vibrations detrimental to the resulting machined product. A flywheel can solve this problem and in many cases be the key to improved productivity.

For the **gang milling** as shown in Figure 24.11, the diameter of the flywheel should be as large as possible. (The moment of inertia increases with the square of the radius.) The best position of the flywheel is inboard on the arbor at A, but depending on the setup, this may not be possible, so then position B should be chosen. It is important that the distance between the cutters and flywheel be as small as possible.

Two or more side-milling cutters often are spaced on an arbor to straddle the workpiece, also called **straddle milling**, and two or more parallel surfaces are machined at once.

A flywheel can be built up from a number of carbon steel disks, each having a center hole and keyway to fit the arbor, so the weight can be easily varied.

Interlocking slotting cutters consist of two cutters similar to side mills but made to operate as a unit for milling slots. The two cutters are adjusted to the desired width by inserting shims between them.

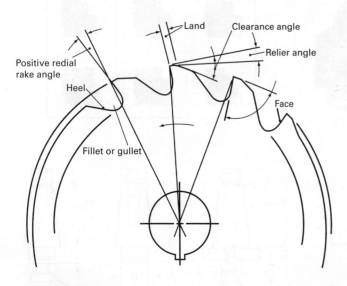

FIGURE 24.9 The geometry of a slab milling cutter showing the main cutting angles.

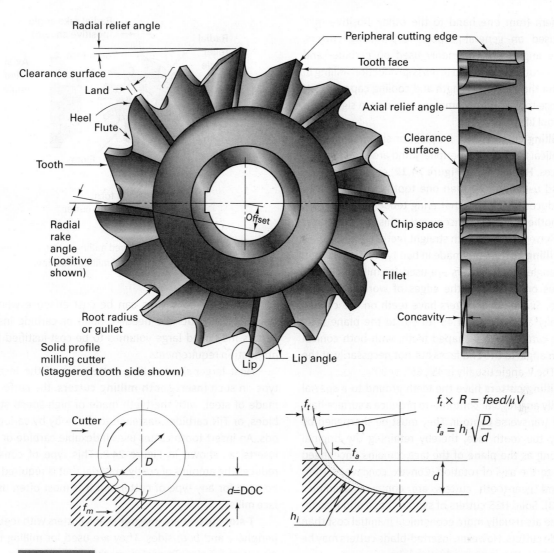

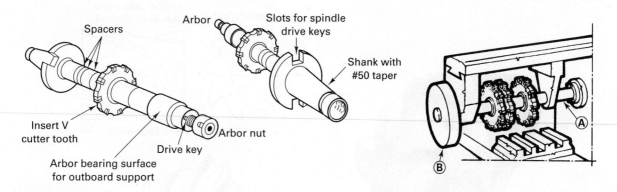

FIGURE 24.10 The **staggered tooth side mill** can cut on both the ends and sides of the cutter so it can cut slots or grooves.

FIGURE 24.11 Arbor (two views) used on a horizontal-spindle milling machine on left. On right, a gang-milling setup showing three side-milling cutters mounted on an arbor (A) with an outboard flywheel (B).

Slitting saws are thin, plain milling cutters, usually from $\frac{1}{32}$ to $\frac{3}{16}$ in. thick, which have their sides slightly "dished" to provide clearance and prevent binding. They usually have more teeth per unit of diameter than ordinary plain milling cutters and are used for milling deep narrow slots and cutting-off operations.

Another method of classification for face and end mill cutters relates to the direction of rotation. A **right-hand cutter** must rotate counterclockwise when viewed from the front end of the machine spindle. Similarly, a **left-hand cutter** must rotate clockwise. All other cutters can be reversed on the arbor to change them from one hand to the other. Positive rake angles are used on general-purpose HSS milling cutters. Negative rake angles are commonly used on carbide- and ceramic-tipped cutters employed in mass-production milling in order to obtain the greater strength and cooling capacity. TiN coating of these tools is quite common, resulting in significant increases in tool life.

Plain milling cutters used for plain or slab milling have straight or helical teeth on the periphery and are used for milling flat surfaces. **Helical mills** (**Figure 24.12**) engage the work gradually, and usually more than one tooth cuts at a given time. This reduces shock and chattering tendencies and promotes a smoother surface. Consequently, this type of cutter usually is preferred over one with straight teeth.

Angle milling cutters are made in two types: single angle and double angle. Angle cutters are used for milling slots of various angles or for milling the edges of workpieces to a desired angle. *Single-angle cutters* have teeth on the conical surface, usually at an angle of 45° to 60° to the plane face. *Double-angle cutters* have V-shaped teeth, with both conical surfaces at an angle to the end faces but not necessarily at the same angle. The V-angle usually is 45°, 60°, or 90°.

Form milling cutters have the teeth ground to a special shape—usually an irregular contour—to produce a surface having a desired transverse contour. They must be sharpened by grinding only the tooth face, thereby retaining the original contour as long as the plane of the face remains unchanged with respect to the axis of rotation. Convex, concave, corner-rounding, and gear-tooth cutters are common examples (**Figure 24.13**). Solid HSS cutters of simple shape and reasonably small size are usually more economical in initial cost than inserted-blade cutters. However, inserted-blade cutters may be lower in overall cost on large production jobs.

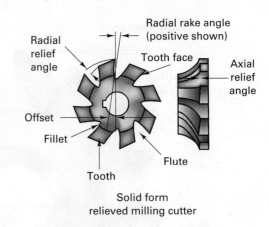

Solid form
relieved milling cutter

FIGURE 24.13 Solid form relieved milling cutter, which would be mounted on an arbor in a horizontal milling machine.

Form-relieved cutters can be cost effective where intricately shaped cuts are needed. Solid or carbide insert tool cutters may need large volumes to be cost-justified by high-production requirements.

Most larger-sized milling cutters are of the insert-tooth type. In such **insert-tooth milling cutters**, the cutter body is made of steel, with the teeth made of high-speed steel, carbides, or TiN carbides, fastened to the body by various methods. An insert-tooth cutter uses indexable carbide or ceramic inserts, as shown in Figure 24.8. This type of construction reduces the amount of costly material that is required and can be used for any type of cutter, but it is most often used with face mills.

T-slot cutters are integral-shank cutters with teeth on the periphery and *both* sides. They are used for milling the wide groove of a T-slot. To use them, the vertical groove must first

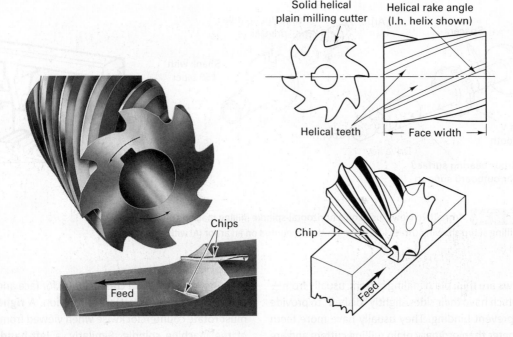

FIGURE 24.12 The chips are formed progressively by the teeth of a plain helical-tooth milling cutter during up milling.

be made with a slotting mill or an end mill to provide clearance for the shank. Because the T-slot cutter cuts on five surfaces simultaneously, it must be fed with care.

Woodruff keyseat cutters are made for the single purpose of milling the semicylindrical seats required in shafts for Woodruff keys. They come in standard sizes corresponding to Woodruff key sizes. Those less than 2 in. in diameter have integral shanks; the larger sizes may be arbor mounted.

Occasionally, **fly cutters** may be used for face milling or boring. Both operations may be done with a single tool at one setup. A single-point cutting tool is attached to a special shank, usually with provision for adjusting the effective radius of the cutting tool with respect to the axis of rotation. The cutting edge can be made in any desired shape and, because it is a single-point tool, is very easy to grind.

24.4 Machines for Milling

The four most common types of manually controlled milling machines are listed here in order of increasing power (and therefore metal removal capability):

1. Ram-type milling machines.
2. Column-and-knee-type milling machines:
   a. Horizontal spindle.
   b. Vertical spindle.
3. Fixed-bed-type milling machines.
4. Planer-type milling machines.

Milling machines whose motions are electronically (computer) controlled are listed in order of increasing production capacity and decreasing flexibility:

1. Manual data input milling machines.
2. Programmable CNC milling machines.
3. Machining centers (tool changer and pallet exchange capability).
4. Flexible manufacturing cell and flexible manufacturing system.
5. Transfer lines.

Basic Milling Machine Construction

Most basic milling machines are of **column-and-knee construction,** employing the components and motions shown in **Figure 24.14**. The column, mounted on the base, is the main supporting frame for all the other parts and contains the spindle with its driving mechanism. This construction provides controlled motion of the worktable in three mutually perpendicular directions: (1) through the *knee*, moving vertically on ways on the front of the column; (2) through the *saddle*, moving transversely on ways on the knee; and (3) through the *table*, moving longitudinally on ways on the saddle. All these motions can be imparted by either manual or powered means. In most cases, a powered rapid traverse is provided in addition to the regular feed rates for use in setting up work and in returning the table at the end of a cut.

The **ram-type milling machine** is one of the most versatile and popular milling machines, using the knee-and-column design. Ram-type machines have a head equipped with a

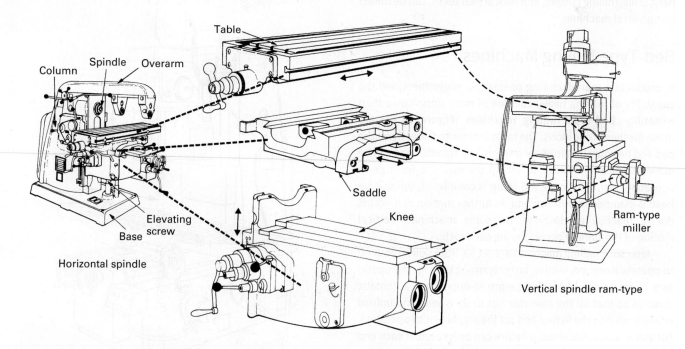

FIGURE 24.14 Major components of a plain column-and-knee-type milling machine, which can have horizontal spindle (shown on the left) or a turret type machine with a vertical spindle (shown on the right). The workpiece and workholder on the table can be translated in X, Y, and Z directions with respect to the tool.

motor-stopped pulley and belt drive as well as a spindle. The ram, mounted on horizontal ways at the top of the column, supports the head and permits positioning of the spindle with respect to the table. Ram-type milling machines are normally 10 hp or less and are suitable for light-duty milling, drilling, reaming, and so on (Figure 24.14).

Milling machines having only the three mutually perpendicular table motions are called **plain column-and-knee milling machines**. These are available with both horizontal and vertical spindles (Figure 24.14). On the older, horizontal-spindle-type machines, an adjustable overarm provides an outboard bearing support for the end of the cutter arbor, which is shown in Figure 24.11 and 24.14. These machines are well suited for slab, side, or straddle milling.

In some vertical-spindle machines, the spindle can be fed up and down, either by power or by hand. Vertical-spindle machines are especially well suited for face and end milling operations. They also are very useful for drilling and boring, particularly where holes must be spaced accurately in a horizontal plane, because of the controlled table motion.

Turret-type column-and-knee milling machines have dual heads that can be swiveled about a horizontal axis on the end of a horizontally adjustable ram. This permits milling to be done horizontally, vertically, or at any angle. This added flexibility is advantageous when a variety of work has to be done, as in tool and die or experimental shops. They are available with either plain or universal tables.

Universal column-and-knee milling machines differ from plain column-and-knee machines in that the table is mounted on a housing that can be swiveled in a horizontal plane, thereby increasing its flexibility. Helices, as found in twist drills, milling cutters, and helical gear teeth, can be milled on universal machines.

Bed-Type Milling Machines

In production manufacturing operations, ruggedness and the capability of making heavy cuts are of more importance than versatility. **Bed-type milling machines (Figure 24.15)** are made for these conditions. The table is mounted directly on the bed and has only longitudinal motion. The spindle head can be moved vertically in order to set up the machine for a given operation. Normally, once the setup is completed, the spindle head is clamped in position and no further motion of it occurs during machining. However, on some machines, vertical motion of the spindle occurs during each cycle.

After such milling machines are set up, little skill is required to operate them, permitting faster learning time for the operators. Some machines of this type are equipped with automatic controls so that all the operator has to do is load and unload workpieces into the fixture and set the machine into operation. For stand-alone machines, a fixture can be located at each end of the table so that one workpiece can be loaded while another is being machined.

Bed-type milling machines with single spindles are sometimes called *simplex milling machines;* they are made with both horizontal and vertical spindles. Bed-type machines also are made in *duplex* and *triplex* types, having two or three spindles respectively, permitting the simultaneous milling of two or three surfaces at a single pass.

Planer-Type Milling Machines

Planer-type milling machines (Figure 24.16) utilize several milling heads, which can remove large amounts of metal while permitting the table and workpiece to feed quite slowly. Often only a single pass of the workpiece past the cutters is required. Through the use of different types of milling heads and cutters, a wide variety of surfaces can be machined with a single setup of the workpiece. This is an advantage when heavy workpieces are involved.

Rotary-Table Milling Machines

Some types of face milling in mass-production manufacturing are often done on **rotary-table milling machines**. Roughing and finishing cuts can be made in succession as the workpieces are moved past the several milling cutters while held in fixtures on the rotating table. The operator can load and unload the work without stopping the machine.

Profilers and Duplicators

Milling machines that can duplicate external or internal geometries in two dimensions are called **profilers** or tracer-controlled machines. A tracing probe follows a two-dimensional

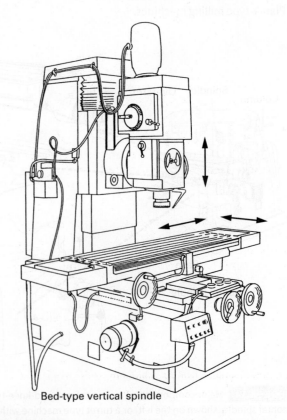

Bed-type vertical spindle

FIGURE 24.15 Bed-type vertical-spindle heavy-duty production machine tools for milling usually have three axes of motion.

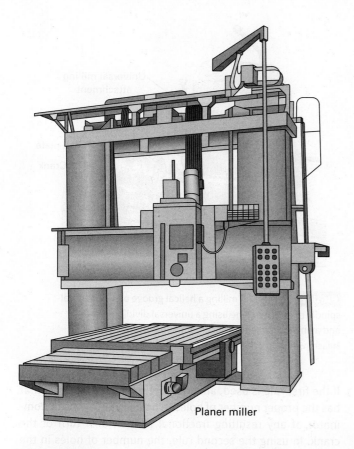
Planer miller

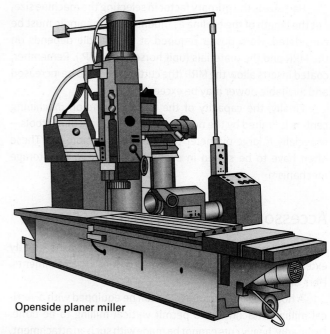
Openside planer miller

FIGURE 24.16 Large planer-type milling machines share their basic design with that of a planer with the planing tool replaced by a powered milling head (not shown). (American Machinist Special Report to Fundamentals of Milling, J Jablonowski, Feb, 1978)

pattern or template and, through electronic or hydraulic air-actuated mechanisms, controls the cutting spindles in two mutually perpendicular directions.

All hydraulic tracers work basically the same way, in that they utilize a stylus connected to a precision servomechanism for each axis of control. The servos are connected to hydraulic actuators on the machine slides. As the stylus traces a template, the servos control the motion of the slides so that the milling cutter duplicates the template shape onto the workpiece.

Duplicators produce forms in three dimensions and are widely used to machine molds and dies. Sometimes these machines are called **die-sinking machines**. They are used extensively in the aerospace industry to machine parts from wrought plate or bar stock as substitutes for forgings when the small number of parts required would make the cost of forging dies uneconomical. Many of these kinds of jobs are now done on NC and CNC-type machines. Applications are discussed in Chapter 26.

Milling Machine Selection

When purchasing or using a milling machine, consider the following issues:

1. Spindle orientation and rpm.
2. Machine capability (accuracy and precision). See Chapter 12.
3. Machine capacity (size of workpieces).
4. Horsepower available at spindle (usually 70% of machine horsepower).
5. Automatic tool changing.

The choice of spindle orientation, horizontal or vertical, depends on the parts to be machined. Relatively flat parts are usually done on vertical machines. Cubic parts are usually done on a horizontal machine, where chips tend to fall free of the part. Operations like slotting and side milling are best done on horizontal machines with outboard supports for the arbor. Use the largest-diameter arbor possible to reduce twist and deflection due to cutting forces. Machine capability refers to the tolerances, while machine refers to the size of parts and the power available.

As with all tooling applications, the tolerances that can be maintained in milling depend on the rigidity of the workpiece, the accuracy and rigidity of the machine spindle, the precision and accuracy of the workholding device, and the quality of the cutting tool itself. Milling produces forces that contribute to chatter and vibration because of the intermittent cutting action. Soft materials tend to adhere to the cutter teeth and make it more difficult to hold tolerances. Materials such as cast iron and aluminum are easy to mill.

Within these criteria, properly maintained cutters used in rigid spindles on properly fixtured workpieces can expect to machine within tolerances with surface flatness tolerances of

0.001 in./ft. Such tolerances are also possible on "slotting" operations with milling cutters, but 0.001 to 0.002 in. is more probable. Flatness specifications are more difficult to maintain in steel and easier to maintain in some types of aluminum, cast iron, and other nonferrous material.

Part size is the primary factor in selecting the machine size, but the length of the tooling as mounted in the spindle must be considered. Horsepower required at the spindle depends on the MRR and the materials (unit horsepower, HP_s). Remember, coated inserts allow the MRR (the cutting speed) to be increased and available power may be exceeded.

Finally, the capacity of the tool changers on machining centers is limited by the number, size, and weight of the tools—especially if large-diameter tools are being employed. These often have to be stored in every other space in the storage mechanism.

Accessories for Milling Machines

The usefulness of ordinary milling machines can be greatly extended by employing various accessories or attachments. Here are some examples.

A horizontal milling machine can be equipped with a vertical milling attachment to permit vertical milling to be done. Ordinarily, heavy cuts cannot be made with such an attachment.

The **universal milling attachment** (**Figure 24.17**) is similar to the vertical attachment but can be swiveled about both the axis of the milling machine spindle and a second, perpendicular axis to permit milling to be done at any angle.

The **universal dividing head** is by far the most widely used milling machine accessory, providing a means for holding and indexing work through any desired arc of rotation. The work may be mounted between centers (Figure 24.17) or held in a chuck that is mounted in the spindle hole of the dividing head. The spindle can be tilted from about 5° below horizontal to beyond the vertical position.

Basically, a dividing head is a rugged, accurate, 40:1 worm-gear reduction unit. The spindle of the dividing head is rotated one revolution by turning the input crank 40 turns. An index plate mounted beneath the crank contains a number of holes, arranged in concentric circles and equally spaced, with each circle having a different number of holes. A plunger pin on the crank handle can be adjusted to engage the holes of any circle. This permits the crank to be turned an accurate, fractional part of a complete circle as represented by the increment between any two holes of a given circle on the index plate. Utilizing the 40:1 gear ratio and the proper hole circle on the index plate, the spindle can be rotated a precise amount by the application of either of the following rules:

$$\text{Number of turns of crank} = \frac{40}{\text{Cuts per revolution of work}}$$

$$\text{Holes to be indexed} = \frac{40 \times \text{Holes index circle}}{\text{Cuts per revolution of work}}$$

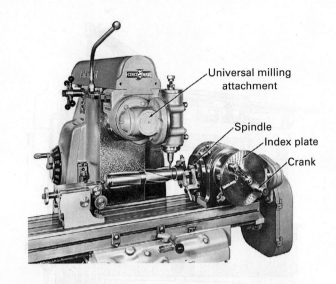

FIGURE 24.17 End milling a helical groove on a horizontal-spindle milling machine using a universal dividing head and a universal milling attachment. (Courtesy of Cincinnati Milacron, Inc.)

If the first rule is used, an index circle must be selected that has the proper number of holes to be divisible by the denominator of any resulting fractional portion of a turn of the crank. In using the second rule, the number of holes in the index circle must be such that the numerator of the fraction is an even multiple of the denominator. For example, if 24 cuts are to be taken about the circumference of a workpiece, the number of turns of the crank required would be $1\frac{2}{3}$. An index circle having 12 holes could be used with one full turn plus 8 additional holes. The second rule would give the same result. Adjustable **sector arms** are provided on the index plate that can be set to a desired number of holes, less than a full turn, so that fractional turns can be made readily without the necessity for counting holes each time. Dividing heads are made having ratios other than 40:1. The ratio should be checked before using.

Because each full turn of the crank on a standard dividing head represents 360/40, or 9° of rotation of the spindle, indexing to a fraction of a degree can be obtained. Indexing can be done in three ways. **Plain indexing** is done solely by the use of the 40:1 ratio in the dividing head. In **compound indexing**, the index plate is moved forward or backward a number of hole spaces each time the crank handle is advanced. For **differential indexing**, the spindle and the index plate are connected by suitable gearing so that as the spindle is turned by means of the crank, the index plate is rotated a proportional amount.

The dividing head can also be connected to the feed screw of the milling machine table by means of gearing. This procedure is used to provide a definite rotation of the workpiece with respect to the longitudinal movement of the table, as in cutting helical gears. This procedure is illustrated in Chapter 30.

Review Questions

1. Suppose you wanted to machine cast iron with BHN of 275. The process to be used is face milling and an HSS cutter is going to be used. What feed and speed values would you select?

2. How is the feed per tooth related to the feed rate of the table in in/sec?

3. Why must the number of teeth on the cutter be known when calculating milling machine table feed, in in./min?

4. Why is the question of up or down milling more critical in horizontal slab milling than in vertical-spindle (end or face) milling?

5. For producing flat surfaces in mass-production machining, how does face milling basically differ from peripheral milling?

6. Milling has a higher metal removal rate than planing. Why?

7. Which type of milling (up or down) is being done in Figures 24.1b, 24.1d, 24.2b, 24.10, and 24.12?

8. Why does down milling dull the cutter more rapidly than up milling when machining sand castings?

9. What parameters do you need to specify in order to calculate MRR in milling?

10. In Figure 24.2b, the tool material is carbide. Would you change the process to climb milling?

11. What is the advantage of a helical-tooth cutter over a straight-tooth cutter for slab milling?

12. What would the cutting force diagram for F_c look like if the cutter were performing climb milling?

13. Could the stub arbor–mounted face mill shown in Figure 24.8 be used to machine a T-slot? Why or why not?

14. In a typical solid arbor milling cutter shown in Figure 24.10, why are the teeth staggered? (Review the discussion of dynamics in Chapter 21.)

15. Make some sketches to show how you would you set up a plain column-and-knee milling machine to make it suitable for milling the top and sides of a large block.

16. Make some sketches to show how you would set up a horizontal milling machine to cut both sides of a block of metal simultaneously.

17. Explain how controlled movements of the work in three mutually perpendicular directions are obtained in column-and-knee-type milling machines.

18. What is the basic principle of a universal dividing head?

19. What is the purpose of the hole-circle plate on a universal dividing head?

Problems

1. You have selected a feed per tooth and a cutting speed for a face milling process, using Table 24.1. Reasonable values for feed and speed are 0.010 in. per tooth and 200 sfpm. The cutter is 8 in. in diameter, as shown in Figure 24.10. Compute the input values for the machine tool.

2. How much time will be required for a milling machine to face mill an AISI 1020 steel surface BHN 150, that is 12 in. long and 5 in. wide, using a 6-in.-diameter, eight-tooth tungsten carbide inserted-tooth face mill cutter? Select values of feed per tooth and cutting speed from Table 24.1.

3. If the depth of cut is 0.35 in., what is the metal removal rate in Problem 2?

4. Estimate the power required for the operation of Problem 3. Do not forget to consider Figure 24.6.

5. Calculate the spindle rpm and table feed (ipm) for a face milling machine after the speed and feed per tooth are selected. Use Figure 24.3 for reference. Work material-bronze.

6. A gray cast iron surface 6 in. wide and 18 in. long may be machined on either a vertical milling machine, using an 8-in.-diameter face mill having eight inserted HSS teeth, or on a horizontal milling machine using an HSS slab mill with eight teeth on a 4-in. diameter. Which machine has the faster cutting time?

7. An operation is to be performed to machine three grooves on a number of parts shown in Figure 24.11. Setup time is 40 min on a shaper (not shown) and 30 min on the horizontal milling machine. The direct time to machine each piece on the shaper is 14 min and

on the miller is 6 min. Labor costs $10/hr. The charge for the use of the shaper is $10/hr and for the milling machine $20/hr. What is the breakeven quantity, below which the shaper is more economical than the mill?

8. In Figure 24.12, the feed is 0.006 in. per tooth. The cutter is rotating at an rpm that will produce the desired surface cutting speed of 125 sfpm. The cutter diameter is 3 in. The depth of cut is 0.5 in. The block is 2 in. wide.

   a. What is the feed rate, in inches per minute, of the milling machine table?

   b. What is the MRR for this situation?

   c. What is horsepower (HP) consumed by this process, assuming an 80% efficiency and a HP_s value for this material of 1.8?

9. Suppose you want to do the job described in Problem 6 by slab milling. You have selected a 6-in.-diameter cutter with eight TiN-coated carbide teeth. The cutting speed will be 500 sfpm and the feed per tooth will be 0.010 in. per tooth. Determine the input parameters for the machine (rpm of arbor and table feed), then calculate the Tm and MRR. Compare these answers with what you got for slab milling the block with HSS teeth.

10. The Bridgeport vertical-spindle milling machine is perhaps the single most popular machine tool. Virtually every factory (or shop) that does machining has one or more of these type machines. Go to your nearest machine shop and find a Bridgeport, make a sketch to show how it works, and explain what makes it so popular.

CHAPTER 25

Drilling and Related Hole-Making Processes

25.1 Introduction

In manufacturing it is probable that more holes are produced than any other shape, and a large proportion of these are made by **drilling**. Of all the machining processes performed, drilling makes up about 25%. Consequently, drilling is a very important process. Although drilling appears to be a relatively simple process, it is really a complex one. Most drilling is done with a cutting tool called a *twist drill* that has two cutting edges, or *lips,* as shown in **Figure 25.1**. The twist drill has the most common drill geometry. The cutting edges are at the end of a relatively flexible tool. Cutting action takes place inside the workpiece. The only exit for

the chips is the hole that is mostly filled by the drill. Friction between the margin and the hole wall produces heat that is additional to that due to chip formation. The counterflow of the chips in the flutes makes lubrication and cooling difficult. There are four major actions taking place at the point of a drill:

1. A small hole is formed by the web—chips are not cut here in the normal sense.

2. Chips are formed by the rotating lips.

3. Chips are removed from the hole by the screw action of the helical flutes.

4. The drill is guided by lands or margins that rub against the walls of the hole.

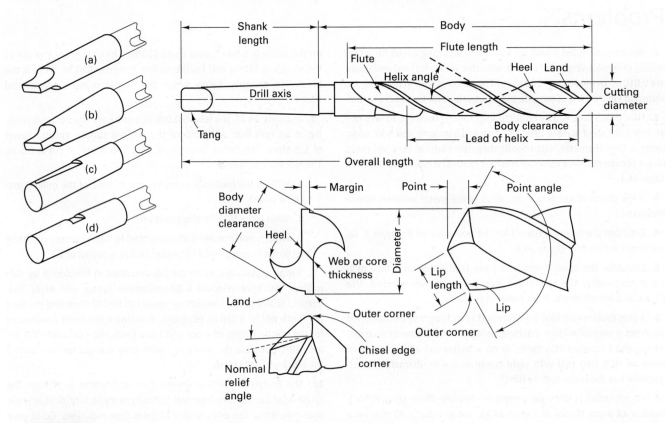

FIGURE 25.1 Nomenclature and geometry of conventional twist drill. Shank style depends on the method used to hold the drill. Tangs or notches prevent slippage. (a) Straight shank with tang; (b) tapered shank with tang; (c) straight shank with whistle notch; (d) straight shank with flat notch.

In recent years, new drill-point geometries and TiN coatings have resulted in improved hole accuracy, longer life, self-centering action, and increased feed-rate capabilities. However, the great majority of drills manufactured are twist drills. One estimate has U.S. manufacturing companies consuming 250 million twist drills per year.

When high-speed steel (HSS) drills wear out, the drill can be reground to restore its original geometry. However, if regrinding is not done properly, the original drill geometry may be lost, and so will drill accuracy and precision. Drill performance also depends on the machine tool used for drilling, the workholding device, the drill holder, and the surface of the workpiece. Poor surface conditions (sand pockets and/or chilled hard spots on castings, or hard oxide scale on hot-rolled metal) can accelerate early tool failure and degrade the hole-drilling process.

25.2 Fundamentals of the Drilling Process

The process of drilling creates two chips. A conventional two-flute drill, with drill of diameter D, has two principal cutting edges rotating at an rpm rate of N and feeding axially. Using a table like **Table 25.1**, the starting cutting speeds and feed rates are obtained, depending on the type of drill, the tool material,

and the material being machined. The rpm of the drill is established by the selected cutting speed

$$N_s = \frac{12V}{\pi D} \tag{25.1}$$

where V is in surface feet per minute and D is in inches (mm), using equation (25.1). This equation assumes that V is the cutting speed at the outer corner of the cutting lip (point X in **Figure 25.2**). The velocity is very small near the center of the chisel end of the drill.

The feed, f_r, is given in inches per revolution (ipr). The depth of cut in drilling is equal to half the feed rate, or $t = f_r/2$ (see section A–A in Figure 25.2). The feed rate in inches per minute (ipm), f_m, is $f_r N_s$. The length of cut in drilling equals the depth of the hole, L, plus an allowance for approach and for the tip of drill, usually $A = D/2$.

For drilling, cutting time is given in equation 25.2:

$$T_m = \frac{(L+A)}{f_r N_s} = \frac{L+A}{f_m} \tag{25.2}$$

The metal removal rate is

$$
\begin{aligned}
MRR &= \frac{\text{Volume}}{T_m} \\
&= \frac{\pi D^2 L/4}{L/f_r N_s}(\text{omitting allowances})
\end{aligned}
\tag{25.3}
$$

which reduces to

$$MRR = \left(\pi D^2 / 4\right) f_r N_s \text{ in.}^3 \tag{25.4}$$

TABLE 25.1 **Recommended Speeds and Feed Rates for HSS Twist Drills**

Speeds for HSS twist drills				Feeds for HSS twist drills	
Material	**Speed, sfm**	**Material**	**Speed**	**Diameter, in.**	**Feed, ipr**
Aluminum alloys	200–300	High-tensile steel,			
Brass, bronze	150–300	heat treated to 35–40 Rc	30–40	Under $\frac{1}{8}$	0.001–0.002
High-tensile bronze	70–150	heat treated to 40–45 Rc	25–35	$\frac{1}{8}-\frac{1}{4}$	0.002–0.004
Zinc-base diecastings	300–400	heat treated to 45–50 Rc	15–25	$\frac{1}{4}-\frac{1}{2}$	0.004–0.007
High-temp alloys, solution-		heat treated to 50–55 Rc	7–15	$\frac{1}{2}-1$	0.007–0.015
treated & aged	7–20	Maraging steel, heat treated	7–20	1 and over	0.015–0.025
Cast iron, soft	75–125	annealed	40–55		
medium hard	50–100	Stainless steel,			
hard chilled	10–20	free machining	30–100		
malleable	80–90	Cr-Ni, nonhardenable	20–60		
Magnesium alloys	250–400	Straight-Cr, martensitic	10–30		
Monel or high-Ni steel	30–50	Titanium, commercially pure	50–60		
Bakelite and similar	100–300	6Al-4V, annealed	25–35		
Steel, 0.2%–0.3% C	80–100	6Al-4V, solution-trt & aged	15–20		
0.4%–0.5% C	70–80	Wood	300–400		
tool, 1.2% C	50–60				
forgings	40–50				
alloy, 300–400 Bhn	20–30				

Source: Cleveland Twist Drill

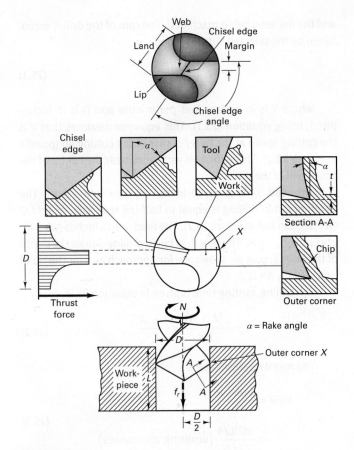

FIGURE 25.2 Conventional drill geometry viewed from the point showing how the rake angle varies from the chisel edge to the outer corner along the lip. The thrust force increases as the web is approached.

Substituting for N_s with equation 25.1, we obtain an approximate form

$$MRR \cong 3DVf_r \qquad (25.5)$$

Example of Drilling

An aluminum plate 2 in. thick needs 1-in. diameter holes drilled in it. An HSS twist drill has been selected. Looking at Table 25.1, cutting speed of 200 fpm and a feed of 0.005 ipr is selected by the engineer.

$$\text{Spindle rpm} = 12V/\pi D = 12 \times 200/3.14 \times 1 = 764 \text{ rpm}$$

What if the machine does not have this specific rpm? Pick the closest value: say, 750 rpm:

$$\text{Penetration rate or feed rate(in./min)} = \text{Feed(ipr)}$$
$$\text{rpm} = 0.005 \times 750$$
$$= 3.75 \text{ in./min}$$

$$\text{Maximum chip load} = \text{feed(ipr)}/2$$
$$= 0.005/2$$
$$= 0.0025 \text{ in./rev}$$

What if the machine does not have this specific feed rate? Pick the next lowest value as a starting value, say, 3.5 in./min:

$$\text{Material removal rate(in.}^3/\text{min}) = (\pi/4) \times (D)^2$$
$$\times \text{Feed(ipr)} \times \text{rpm}$$
$$= (\pi/4)1^2 \times \text{Feed rate}$$
$$= 3.14/4 \times 1^2 \times 3.50$$
$$= 2.75 \text{ in.}^3/\text{min}$$

The MRR can be used with the unit power for aluminum (see Chapter 21) to estimate the horsepower needed to drill the hole. Let $HP_s = 0.33$ for this aluminum:

$$HP = HP_s \times MRR = 0.33 \times 2.75 = 0.90$$

This value would typically represent 80% of the total motor horsepower (HP) needed, so in this case, a horsepower motor greater than 1.5 or 2 would be sufficient.

In estimating the cost of a job, it is often necessary to determine the time to drill a hole:

$$\text{drill time/hole} = \frac{\text{length drilled + allowance}}{\text{feed rate(in./min)}}$$
$$+ \frac{\text{rapid traverse length of withdrawal}}{\text{rapid traverse rate}}$$
$$+ \text{prorated downtime to change drills per hole}$$

The last term prorated downtime is

$$\frac{\text{drill change downtime}}{\text{holes drilled per drill (tool life)}}$$

And the cost/hole is

$$\text{Drilling time/holes} \times (\text{Labor + Machine rate})$$
$$+ \text{Prorated cost of drill/hole}$$

25.3 Types of Drills

The most common types of drills are **twist drills**. These have three basic parts: the *body*, the *point*, and the *shank*, shown in Figures 25.1 and 25.2. The body contains two or more spiral or helical grooves, called **flutes**, separated by **lands**. To reduce the friction between the drill and the hole, each land is reduced in diameter except at the leading edge, leaving a narrow margin of full diameter to aid in supporting and guiding the drill and thus aiding in obtaining an accurate hole. The lands terminate in the point, with the leading edge of each land forming a cutting edge. The flutes serve as channels through which the chips are withdrawn from the hole and coolant gets to the cutting edges. Although most drills have two flutes, some, as shown in **Figure 25.3**, have three, and some have only one.

FIGURE 25.3 Types of twist drills and shanks. Bottom to top: straight-shank, three-flute core drill; straight-shank; taper-shank; bit-shank; straight-shank, high-helix angle; straight-shank, straight-flute; taper-shank, subland drill. (Courtesy J T. Black)

The principal rake angles behind the cutting edges are formed by the relation of the flute **helix angle** to the work. This means that the rake angle of a drill varies along the cutting edges (or **lips**), being negative close to the point and equal to the helix angle out at the lip. Because the helix angle is built into the twist drill, the primary rake angle cannot be changed by normal grinding. The helix angle of most drills is 24 degrees, but drills with larger helix angles—often more than 30 degrees—are used for materials that can be drilled very rapidly, resulting in a large volume of chips. Helix angles ranging from 0 to 20 degrees are used for soft materials, such as plastics and copper. Straight-flute drills (zero helix and rake angles) are also used for drilling thin sheets of soft materials. It is possible to change the rake angle adjacent to the cutting edge by a special grinding procedure called **dubbing**.

The cone-shaped point on a drill contains the cutting edges and the various clearance angles. This cone angle affects the direction of flow of the chips across the tool face and into the flute. The 118-degree cone angle that is used most often has been found to provide good cutting conditions and reasonable tool life when drilling mild steel, thus making it suitable for much general purpose drilling. Smaller cone angles—from 90 to 118 degrees—are sometimes used for drilling more brittle materials, such as gray cast iron and magnesium alloys. Cone angles from 118 to 135 degrees are often used for the more ductile materials, such as aluminum alloys. Cone angles less than 90 degrees are frequently used for drilling plastics. Many methods of grinding drills have been developed that produce point angles other than 118 degrees.

The drill produces a **thrust force**, T, and a **torque**, M. Drill torque increases with feed (in./rev) and drill diameter, while the thrust force is influenced greatly by the web or chisel end design, as shown in **Figure 25.4**.

The relatively thin **web** between the flutes forms a metal column or backbone. If a plain conical point is ground on the drill, the intersection of the web and the cone produces a straight-line **chisel end**, which can be seen in the end view of Figure 25.2 and 25.4. The chisel point, which also must act as a cutting edge, forms a 56-degree negative rake angle with the conical surface. Such a large negative rake angle does not cut efficiently, causing excessive deformation of the metal. This results in high thrust forces and excessive heat being developed at the point. In addition, the cutting speed at the drill center is low, approaching zero. As a consequence, drill failure on a standard drill occurs both at the center, where the cutting speed is lowest, and at the outer tips of the cutting edges, where the speed is highest.

When the rotating, straight-line chisel point comes in contact with the workpiece, it has a tendency to slide or "walk" along the surface, thus moving the drill away from the desired location. The conventional point drill, when used on machining centers or high-speed automatics, will require additional supporting operations like center drilling, burr removal, and tool change—all of which increase total production time and reduce productivity.

Many special methods of grinding drill points have been developed to eliminate or minimize the difficulties caused by the chisel point and to obtain better cutting action and tool life (see Figure 25.4 for some examples).

The center core or slot-point drill shown in **Figure 25.5** has twin carbide tips brazed on a steel shank and a hole (or slot) in the center. The work material in the slot is not machined but, rather, fractured away. The center core drill has a self-centering action and greatly relieves the thrust force produced by the chisel edge of conventional twist drills. This drill operates at about 30 to 50% less thrust than that of conventional drills. All rake angles of the cutting edge are positive, which further reduces the cutting force.

The conventional point also has a tendency to produce a burr on the exit side of a hole. Some type of chip breaker is

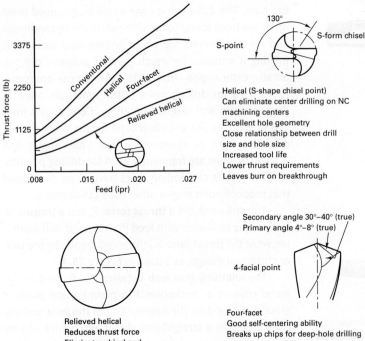

S-point

Helical (S-shape chisel point)
Can eliminate center drilling on NC
machining centers
Excellent hole geometry
Close relationship between drill
size and hole size
Increased tool life
Lower thrust requirements
Leaves burr on breakthrough

Secondary angle 30°–40° (true)
Primary angle 4°–8° (true)

4-facial point

Relieved helical
Reduces thrust force
Eliminates chisel end
Equal, rake angle

Four-facet
Good self-centering ability
Breaks up chips for deep-hole drilling
Can be generated in a single grinding
operation: reduces thrust.
Eliminates center drilling in NC

Bickford
Combination of helical and Racon
point features
Self-centering and reduced burrs
Excellent hole geometry
Increased tool life

Racon (radiused conventional point)
Increased feed rates
Increased tool life (8–10 times in C.I.)
Reduced burrs at breakthrough
Not self-centering

FIGURE 25.4 As the drill advances, it produces a thrust force. Variations in the drill-point geometry are aimed at reducing the thrust force.

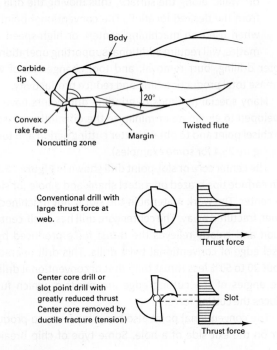

FIGURE 25.5 Center core drills can greatly reduce the thrust force.

often incorporated into drills. One procedure is to grind a small groove in the rake face, parallel with and a short distance back from the cutting edge. Drills with a special chip-breaker rib as an integral part of the flute are available. The rib interrupts the flow of the chip, causing it to break into short lengths.

The split-point drill is a form of **web thinning** to shorten the chisel edge. This design reduces thrust and allows for higher feed rates. Web thinning uses a narrow grinding wheel to remove a portion of the web near the point of the drill. Such methods have had varying degrees of success, and they require special drill-grinding equipment.

Also shown in Figure 25.4 is a four-facet self-centering point that works well in tougher materials. The facets refer to the number of edges on the clearance surfaces exposed to the cutting action. The self-centering drill lasts longer and saves machining time on numerical control (NC) machining centers as they can eliminate the need for center drills.

A common aspect in drill-point terminology is total indicator runout (TIR). This is a measure of the cutting lips' relative side-to-side accuracy. The original drill point produced by the manufacturer lasts only until the first regrind; thereafter, performance and life depend on the quality of regrind. Proper regrinding (reconditioning) of a drill is a complex and important operation. If satisfactory cutting and hole size are to be achieved, it is essential that the point angle, lip clearance, lip length, and web thinning be correct. As illustrated in **Figure 25.6**, incorrect sharpening often results in unbalanced cutting forces at the tip, causing misalignment and oversized holes. Drills, even small drills, should always be machine ground, never hand ground. Drill grinders, often computer controlled, should be used to ensure exact reproduction of the geometry established by the manufacturer of the drill. This is extremely important when drills are used on mass-production or numerically controlled machines. Companies invest huge sums in NC machining centers but overlook the value of a top-quality drill-grinding machine.

Drill shanks are made in several types. The two most common types are the straight and the taper. **Straight-shank** drills are usually used for sizes up to $\frac{3}{8}$-in. diameter and must be held in some type of drill chuck. **Taper shanks** are available on larger drills and are common on drills above 1 in. Morse tapers are used on taper-shank drills, ranging from a number 1 taper to a number 6.

Taper-shank drills are held in a female taper in the end of the machine tool spindle. If the taper on the drill is different from the spindle taper, adapter sleeves are available. The taper assures the drill's being accurately centered in the spindle. The **tang** at the end of the taper shank fits loosely in a slot at the end of the tapered hole in the spindle. The drill may be loosened for removal by driving a metal wedge, called a **drift**, through a hole in the side of the spindle and

FIGURE 25.6 Typical causes of drilling problems.

(a) Angle unequal (b) Length unequal

Outer corners break down: Cutting speed too high; hard spots in material; no cutting compound at drill point; flutes clogged with chips

Cutting lips chip: Too much feed; lip relief too great

Checks or cracks in cutting lips: Overheated or too quickly cooled while sharpening or drilling

Chipped margin: Oversize jig bushing

Drill breaks: Point improperly ground; feed too heavy; spring or backlash in drill press, fixture, or work; drill is dull; flutes clogged with chips

Tang breaks: Imperfect fit between taper shank and socket caused by dirt or chips or by burred or badly worn sockets

Drill breaks when drilling brass or wood: Wrong type drill; flutes clogged with chips

Drill spilts up center: Lip relief too small; too much feed

Drill will not enter work: Drill is dull; web too heavy; lip relief too small

Hole rough: Point improperly ground or dull; no cutting compounds at drill point; improper cutting compound; feed too great; fixture not rigid

Hole oversize: Unequal angle of the cutting edges; unequal length of the cutting edges; see part (a)

Chip shape changes while drilling: Dull drill or cutting lips chipped

Large chip coming from one flute, small chip from the other: Point improperly ground, one lip doing all the cutting

against the end of the tang. It also acts as a safety device to prevent the drill from rotating in the spindle hole under heavy loads. However, if the tapers on the drill and in the spindle are proper, no slipping should occur. The driving force to the drill is carried by the friction between the two tapered members. Standard drills are available in four size series, the size indicating the diameter of the drill body:

- *Millimeter series:* 0.01- to 0.50-mm increments, according to size, in diameters from 0.015 mm.
- *Numerical series:* no. 80 to no. 1 (0.0135 to 0.228 in.).
- *Lettered series:* A to Z (0.234 to 0.413 in.).
- *Factional series:* to 4 in. (and over) by 64ths.

TiN coating of conventional drills greatly improves drilling performance. The increase in tool life of TiN-coated drills over uncoated drills in machining steel is more than 200 to 1000%.

Gundrills and Deep-Hole Drilling

The depth of the hole to be drilled divided by the diameter of the drill is the depth-to-diameter ratio. Most machinists consider a ratio of 3:1 to be deep-hole drilling, after which hole accuracy (location), drilling speed, and tool life will be reduced. The bores of rifle barrels were once drilled using conventional drills. Today, **deep-hole drills**, or **gundrills**, are used when deep holes are to be drilled.

The oldest of these deep-hole techniques is gundrilling. The original gundrills were very likely half-round drills, drilled axially with a coolant hole to deliver cutting fluids to the cutting edge (see **Figure 25.7**). Modern gundrills typically consist

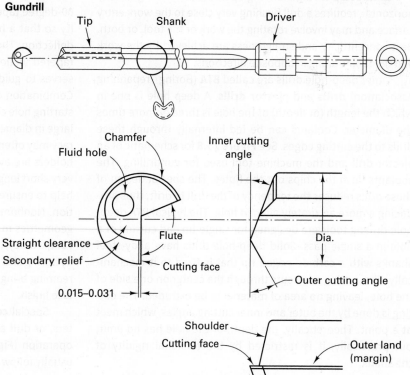

FIGURE 25.7 The gundrill geometry is very different from that of conventional drills.

of an alloy-steel-tubing shank with a solid carbide or carbide-edged tip brazed or mechanically fixed to it. Guide pads following the cutting edge by about 90 to 180 degrees are also standard.

The gundrill is a single-lipped tool, and its major feature is the delivery of coolant through the tool at extremely high pressures—typically from 300 to 1800 psi, depending on diameter—to force chips back down the flute. Successful application of a gundrill depends almost entirely on the formation of small chips that can be effectively extracted by the flow of cutting fluids.

Standard gundrills are made in diameters from 0.0078 in. (2 mm) to 2 in. or more. Depth-to-diameter ratios of 100:1 or more are possible.

In gundrilling tolerances for diameters of drilled holes under 1 in. can be held to 0.0005-in. total tolerance, and, should not exceed 0.001 in. over all. According to one source, "roundness accuracies of 0.00008 in. can be attained." Because of the burnishing effect of the guide pads, excellent surface finishes can be produced.

Hole straightness is affected by a number of variables, such as diameter, depth, uniformity of workpiece material, condition of the machine, sharpness of the gundrill, feeds and speeds used, and the specific technique used (rotation of the tool, of the work, or both), but deviation should not exceed about 0.002 in. TIR in a 4-in. depth at any diameter, and it can be held to 0.002 in./ft.

Basic setup for a gundrilling operation, which is generally horizontal, requires a drill bushing very close to the work entry surface and may involve rotating the work or the tool, or both. Best concentricity and straightness are achieved by the work and the tool rotating in opposite directions.

Other deep-hole drills are called **BTA** (Boring Trepanning Association) **drills** and **ejector drills**. A deep hole is one in which the length (or depth) of the hole is three or more times the diameter. Coolants can be fed internally through these drills to the cutting edges. See **Figure 25.8** for schematic of an ejector drill and the machine tool used for gundrilling. The coolants flush the chips out the flutes. The special design of these drills reduces the tendency of the drill to drift, thus producing a more accurately aligned hole. The typical BTA deep-hole drilling tools are designed for single-lip end cutting of a hole in a single pass. Solid deep-hole drills have alloy-steel shanks with a carbide-edged tip that is fixed to it mechanically. The cutting edge cuts through the center on one side of the hole, leaving no area of material to be extruded. The cutting is done by the outer and inner cutting angles, which meet at a point. Theoretically, the depth of the hole has no limit, but practically, it is restricted by the torsional rigidity of the shank.

Gundrills have a single-lip cutting action. Bearing areas and lifting forces generated by the coolant pressure counteract the radial and tangential loads. The single-lip construction forces the edge to cut in a true circular pattern. The tip thus follows the direction of its own axis. The **trepanning gundrill** leaves a solid core.

Hole straightness is affected by variables such as diameter, depth, uniformity of the workpiece material, condition of the machine, sharpness of cutting edges, feeds and speeds used, and whether the tool or the workpiece is rotated or counter-rotated.

Two-flute drills are available that have holes extending throughout the length of each land to permit coolant to be supplied, under pressure, to the point adjacent to each cutting edge. These are helpful in providing cooling and also in promoting chip removal from the hole in drilling to moderate depths. They require special fittings through which the coolant can be supplied to the rotating drill, and they are used primarily on automatic and semiautomatic machines. See **Table 25.2** for comparison of drilling processes.

Other Drill Designs

Larger holes in thin material may be made with a **hole cutter** (**Figure 25.9**), where the large hole is produced by the thin-walled, multiple-tooth cutter with saw teeth and the metal hole with a twist drill. Hole cutters are often called **hole saws**.

When starting to drill a hole, a drill can deflect rather easily because of the "walking" action of the chisel point. Hole location accuracy is lost. Consequently, to ensure that a hole is started accurately, a **center drill** (**Figure 25.10**) is used prior to a regular chisel-point twist drill. The center drill and countersink tool have a short, straight drill section extending beyond a 60-degree taper portion. The heavy, short body provides rigidity so that a hole can be started with little possibility of tool deflection. The hole should be drilled only partway up on the tapered section of countersink. The conical portion of the hole serves to guide the drill being used to make the main hole. Combination center drills are made in four sizes to provide a starting hole of proper size for any drill. If the drill is sufficiently large in diameter, or if it is sufficiently short, satisfactory accuracy may often be obtained without center drilling. Special drill holders are available that permit drills to be held with only a very short length protruding. The use of a center (start) drill will help to ensure that a drill will start drilling at the desired location. Nonhomogeneities in the workpiece and imperfect drill geometries may also cause the hole to be oversize or off-line. For accuracy, it is necessary to follow center drilling and drilling by boring and reaming. Boring corrects the hole alignment, and reaming brings the hole to accurate size and improves the surface finish.

Special **combination drills** can drill two or more diameters, or drill and countersink and/or counterbore, in a single operation (**Figure 25.11**). Countersinking and counterboring usually follow drilling. A **step drill** has a single set of flutes and is ground to two or more diameters. **Subland drills** have a separate set of flutes on a single body for each diameter or operation; they provide better chip flow, and the cutting edges can be ground to give proper cutting conditions for each operation. Combination drills are expensive and may be difficult to regrind, but they can be economical for production-type

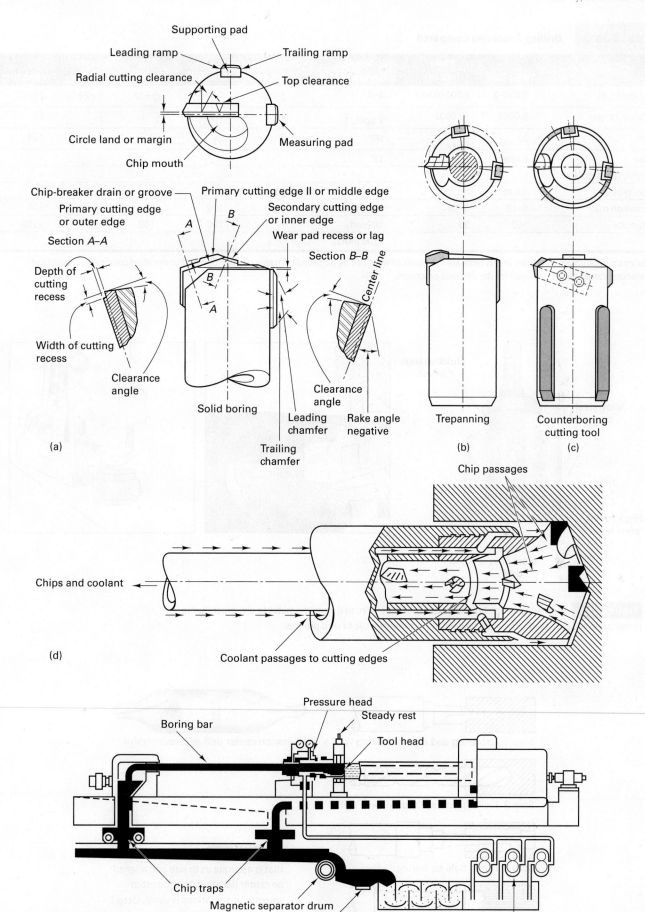

FIGURE 25.8 BTA drills for (a) boring, (b) trepanning, (c) counterboring, (d) deep-hole drilling with ejector drill, (e) horizontal deep-hole-drilling machine. (S. Azad and S. Chandeashekar, *Mechanical Engineering*, September 1985, pp. 62, 63)

TABLE 25.2	Drilling Processes Compared							
	Twist Drill	Pivot (micro)	Spade (inserted blade)	Indexable-Insert Drill	Gundrill	BTA System	Ejector Drill	Trepanning
Diameter, in.	0.020–2	0.001–0.020	1–6	$\frac{5}{8}$–3	0.078–1	$\frac{7}{16}$–18	$\frac{3}{4}$–2$\frac{1}{2}$	1$\frac{3}{4}$–10
Typical range	0.0059	<0.0001	$\frac{5}{8}$ spec, 1	$\frac{5}{8}$	0.039	$\frac{3}{4}$	$\frac{3}{4}$	1$\frac{3}{4}$
Min	3$\frac{1}{2}$ std	$\frac{1}{8}$	std	3	2$\frac{1}{2}$	12	7	>24
Max	6 spec		18					
Depth/Diameter Ratio								
Min. practical	No min.	No min.	No min	<1	1	1	1	10
Common max[a]	5–10	3–10	>40 (horiz)	2–3	100	100	50	100
Ultimate	>50	20	10 (vert)	—	200	>100	>50	>100
			>100 (horz)					

[a]Maximum depth/diameter ratios in this table are estimates of what can be achieved with special attention and under ideal conditions. Equality of tolerances should not be assumed for the different processes.

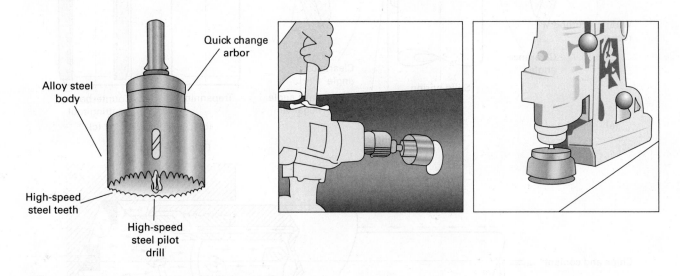

FIGURE 25.9 High-speed edge hole saws can cut holes $\frac{9}{16}$ in. to 6 in. in diameter in any machinable material up to 1$\frac{1}{8}$ in. thick. They can be used in portable electric or drill presses.

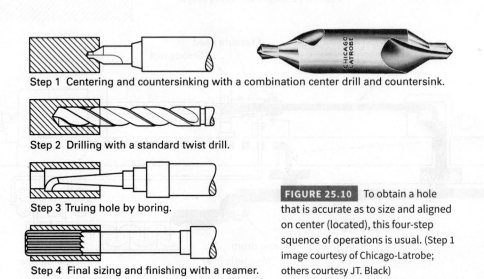

Step 1 Centering and countersinking with a combination center drill and countersink.

Step 2 Drilling with a standard twist drill.

Step 3 Truing hole by boring.

Step 4 Final sizing and finishing with a reamer.

FIGURE 25.10 To obtain a hole that is accurate as to size and aligned on center (located), this four-step sequence of operations is usual. (Step 1 image courtesy of Chicago-Latrobe; others courtesy JT. Black)

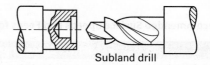

Subland drill

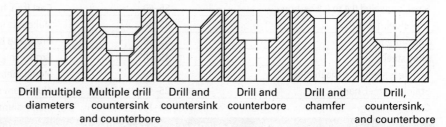

Drill multiple diameters | Multiple drill countersink and counterbore | Drill and countersink | Drill and counterbore | Drill and chamfer | Drill, countersink, and counterbore

FIGURE 25.11 Special-purpose subland drill (above) and some of the operations possible with other combination drills (below).

operations if they reduce work handling, set-ups, or separate machines and operations.

Spade drills (**Figure 25.12**) are widely used for making holes 1 in. or larger in diameter at low speeds or with high feeds (**Table 25.3**). The workpiece usually has an existing hole, but a spade drill can drill deep holes in solids or stacked materials. Spade drills are less expensive because the long supporting bar can be made of ordinary steel. The drill point can be ground with a minimum chisel point. The main body can be made more rigid because no flutes are required, and it can have a central hole through which a fluid can be circulated to aid in cooling and in chip removal. The cutting blade is easier to sharpen; only the blades need to be TiN-coated.

Spade drills are often used to machine a shallow locating cone for a subsequent smaller drill and at the same time to provide a small bevel around the hole to facilitate later tapping or assembly operations. Such a bevel also frequently eliminates the need for deburring. This practice is particularly useful on mass-production and numerically controlled machines.

Carbide-tipped drills and drills with indexable inserts are also available (see **Figure 25.13**) with one- and two-piece inserts for drilling shallow holes in solid workpieces. **Indexable insert drills** can produce a hole four times faster than a spade drill because they run at high speeds/low feeds and are really more of a boring operation than a drilling process.

However, to use indexable drills, you must have an extremely rigid machine tool and setup, adequate horsepower and lots of cutting fluid. Indexable drills are roughing tools generating surface finishes of 250 rpm or greater. The tool is designed for the inboard insert to cut past the centerline of the tool so the inboard tool is positioned radially below the center. See **Table 25.4** for an indexable drilling troubleshooting guide.

A high-pressure, pulsating coolant system can generate pressures up to 300 psi and works well with indexable drilling.

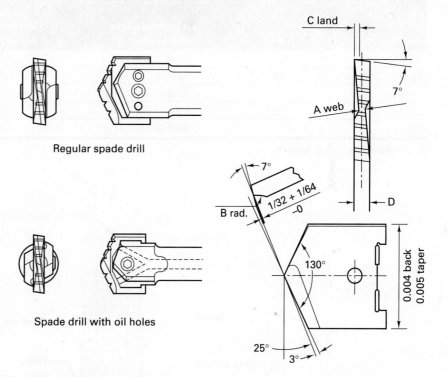

Regular spade drill

Spade drill with oil holes

FIGURE 25.12 (Top) Regular spade drill; (middle) spade drill with oil holes; (bottom) spade drill geometry, nomenclature.

It can have disadvantages, however. High pressure with pulsating action can decrease chip control and cause drill deflection. A high-pressure coolant stream can flatten chips at the point of forming, forcing them into the cut, causing recutting, insert chipping, and poor hole finishes. The pressure can force chips between the drill body and hole diameter, wrapping them around the drill. Heat from friction then will weld the chips to the tool body or hole.

The diameter of the hole and the length/diameter ratio usually determine what kind of drill to use. **Figure 25.14** explores how drill selection depends on the depth of the hole and the diameter of the drill. Section A shows the drilling areas of relatively shallow holes and small diameters. About half of all the drilling process falls within the category of this section. It is the section for which the majority of the work is done by twist drills and a very few cemented carbide drills. Section B is the

TABLE 25.3   Recommended Surface Speeds and Feeds for High-Speed Steel Spade Drills for Various Materials

Material	Surface Speed (ft per min)	Material	Speed
Mild machinery steel 0.2 and 0.3 carbon	65–100	Cast iron, medium hard	55–100
Steel, annealed 0.4 to 0.5 carbon	55–80	Cast iron, hard, chilled	25–40
Tool steel, 1.2 carbon	45–60	Malleable iron	79–90
Steel forging	35–50	Brass and bronze, ordinary	200–300
Alloy steel	45–70	Bronze, high tensile	70–150
Stainless steel, free machining	50–70	Monel metal	35–50
Stainless steel, hard	25–40	Aluminum and its alloys	200–300
Cast iron, soft	80–150	Magnesium and its alloys	250–400

Feed Rates for Spade Drilling (inches per Revolution)			
Drill Size (inches)	Cast Iron Malleable Iron Brass Bronze	Medium Steel Stainless Steel Monel Metal Drop-Forged Alloys Tool Steel (Annealed)	Tough Steel Drop Forging Aluminum
1 to $1\frac{1}{4}$	0.010–0.020	0.008–0.014	0.006–0.012
$1\frac{1}{4}$ to $\frac{3}{4}$	0.010–0.024	0.008–0.018	0.008–0.017
$1\frac{3}{4}$ to $2\frac{1}{2}$	0.010–0.030	0.010–0.024	0.010–0.017
$2\frac{1}{2}$ to 4	0.012–0.032	0.012–0.030	0.010–0.017
4 to 6	0.012–0.032	0.010–0.024	0.008–0.017

Source: Waukesha Cutting Tools, Inc.

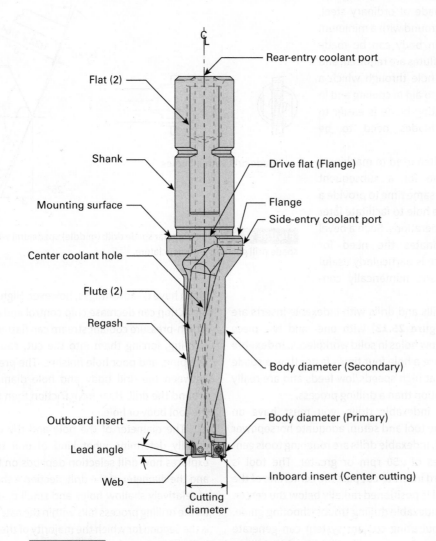

FIGURE 25.13 Design of an indexable insert drill with two inserts, the most common style.

TABLE 25.4   **Indexable Drilling Troubleshooting Guide** [a]

Problem	Source	Solution
Insert chipping or breakage [b]	Off-center drill, caused by misalignment	Maintain proper alignment.
		Concentricity not to exceed ± 0.005 TIR.
	Improper seating of tool in tool holder, spindle, or turret	Check tool shank and socket for nicks and dirt. Check parting line between tool shank and socket with feeler gage. Check to see if tool is locked tightly.
	Deflection because of too much overhand and lack of rigidity	With indicator, check if tool can be moved by hand. Check if tool can be held shorter.
	Improper seating of inserts in pockets	Clean pockets whenever indexing or changing inserts.
		Check pockets for nicks and burrs.
		Check if inserts rest completely on pocket bottoms.
	Damaged insert screws	Check head and thread for nicks and burns. Do not overtorque screws.
	Improper speeds and feeds	Check recommended guidelines for given materials.
	Insufficient coolant supply	Check coolant flow.
	Improper carbide grade in inboard station	Recommend straight grade for multiple-insert drills.
Grooving on back stroke; drill body rubbing hole wall; over- or undersize holes	Off-center drill	Maintain proper alignment and concentricity. Check bottom of hole or disk for center stub.
	Deflection	Check setup rigidity. Check speed and feed guidelines.
Poor hole surface finish	Vibrations	Check setup and part rigidity. Check seat in spindle or tool holder.
		Check speeds and feeds.
	Insufficient coolant pressure and volume	Increase coolant pressure and flow. Is coolant flow constant?
		Make sure coolant reaches inserts at all times.
	Recutting chips, causing drill to jump	Increase coolant flow. Add coolant grooves.
	Poor chip control; chips trapped in hole	Mostly speed or feed.
	Chatter	Mostly feed rate.
Very short, thick, flat chips	Feed rate too high in relation to cutting speed	Lower feed or increase speed.
Long and stringy chips	Feed rate too low in relation to cutting speed	Increase feed rate or decrease speed.
		Use dimple inserts.
Unable to loosen insert locking screws	Seized threads, caused by coolant or heat	Apply water and heat-resistant lubricant to threads.

[a] *Source:* "Fundamentals of Indexable Drilling, " K.L. Anderson, *Machining Technology*, vol. 2, no. 3, 1991.
[b] If constant chipping occurs, especially on an inner insert, and conditions are optimum, try an uncoated-carbide insert or a grade with higher transverse rapture strength.

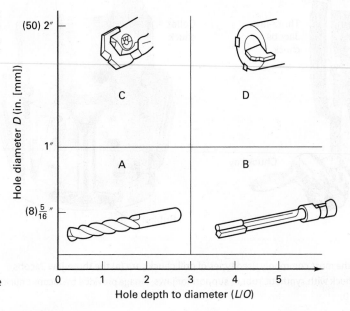

Sector	Typical Drill Types
A	Twist drill (HSS) Center core drill
B	Twist drill Gundrill BTA Ejector drill
C	Twist drill Indexable insert drill Spade drill Center core drill
D	BTA Ejector drill

FIGURE 25.14 Drill selection depends on hole diameter and hole depth.

drilling of deep holes for which cemented carbide gundrills are used. Section C is that of shallow holes having large diameters, for which spade drills are used. Section D is that of deep holes having large diameters, for which BTA tools are used.

Microdrilling

As the term suggests, **microdrilling** involves very small diameter cutting tools, including drills, end mills, routers, and other special tools. Drills from 0.002 in. (0.05 mm) and mills to 0.005 in. in diameter are used to produce geometries involving dimensions at which many workpiece materials no longer exhibit uniformity and homogeneity. Grain borders, inclusions, alloy or carbide segregates, and microscopic voids are problems in microdrilling, where holes of 0.02 to 0.0001 in. have been drilled using pivot drills, as shown in **Figure 25.15**.

Pivot drills are two-lipped (two-fluted), end-cutting tools of relatively simple geometry. Web thickness tapers toward the point, and a generous back-taper is incorporated. For softer workpiece materials, point angles are typically 118 degrees and lip clearance is 15 degrees. For steels and harder metals, 135-degree points and 8-degree clearance are recommended. The chisel edge is similar to that of a twist drill. Pivot drills are made of tungsten-alloy tool steel in standard sizes from 0.0001 to 0.125 in. and of sintered tungsten carbide from 0.001 to 0.125 in.

Small drills easily deflect, and getting accurate and precise holes requires a machine with a high quality spindle and very sensitive feeding pressure. In the medical components field, much of this machining work is performed on computer numerical control (CNC) Swiss-type turning machines. Speeds and feeds are greatly reduced with frequent pecking to clear the chips. Light, lard-based, sulfurized cutting oil is used.

Microdrill

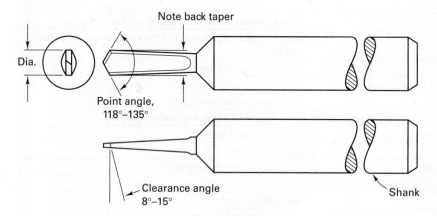

FIGURE 25.15 Pivot microdrill for drilling very small diameter holes.

25.4 Tool Holders for Drills

Straight-shank drills must be held in some type of drill **chuck** (**Figure 25.16**). Chucks are adjustable over a considerable size range and have radial steel fingers. When the chuck is tightened by means of a chuck key, these fingers are forced inward against the drill. On smaller drill presses, the chuck often is permanently attached to the machine spindle, whereas on larger drilling machines the chucks have a tapered shank that fits into the female

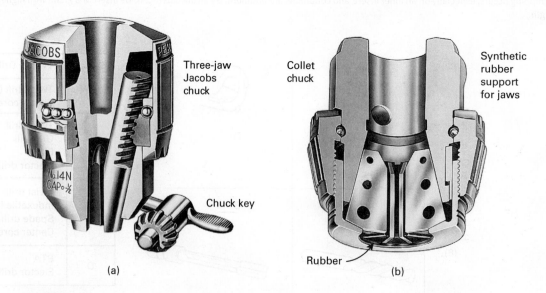

FIGURE 25.16 Two of the most commonly used types of drill chucks are (a) the three-jaw Jacobs chuck and (b) the collet chuck with synthetic rubber support for jaws. (Image provided by Jacobs Chuck, Apex Tool Group, Sparks, MD)

Morse taper of the machine spindle. Special types of chucks in semiautomatic or fully automatic machines permit quite a wide range of sizes of drills to be held in a single chuck.

Chucks using chuck keys require that the machine spindle be stopped in order to change a drill. To reduce the downtime when drills must be changed frequently, **quick-change chucks** are used. Each drill is fastened in a simple round collet that can be inserted into the chuck hole while it is turning by merely raising and lowering a ring on the chuck body. With the use of this type of chuck, center drills, drills, counterbores, reamers, and so on, can be manually changed in quick succession. For carbide drills, collet-type holders with thrust bearings are recommended (**Figure 25.17**). For drills using an internal coolant supply, a very rigid chuck with either an inducer or through-spindle coolant source is recommended.

Conventional holders such as keyless chucks cannot be used because the gripping strength is limited. Collet holders should be cleaned periodically with oil to remove small chips.

The entire flute length must protrude from the chuck. At maximum hole depth, the length of flute protruding from the hole must be at least one to one-and-a-half times the drill diameter. Radial runout at the drill tip must not exceed 0.001 in.

25.5 Workholding for Drilling

Work that is to be drilled is ordinarily held in a vise or in specially designed workholders called **jigs**. Workholding devices are the subject of Chapter 41, where the design of workholding

devices is discussed. Examples of drill jigs are shown along with some problems covering the economics of using jigs.

With regard to safety, the work should not be held on the table by hand unless adequate leverage is available, even in light drilling operations. This is a dangerous practice and can lead to serious accidents, because the drill has a tendency to catch on the workpiece and cause it to rotate, especially when the drill exits the workpiece. Work that is too large to be held in a jig can be clamped directly to the machine table using suitable bolts and clamps and the slots or holes in the table. Jigs and workholding devices on indexing machines must be free from play and firmly seated.

25.6 Machine Tools for Drilling

The basic work and tool motions required for drilling—relative rotation between the workpiece and the tool, with relative longitudinal feeding—occur in a wide variety of machine tools. Thus, drilling can be done on a variety of machine tools such as lathes, horizontal and vertical milling machines, boring machines, and machining centers. This section will focus on those machines that are designed, constructed, and used primarily for drilling.

First, the machine tools must have sufficient power (torque) and thrust to perform the cut. It is the task of the

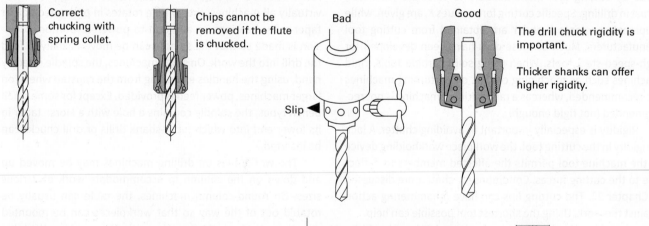

FIGURE 25.17 Here are some suggestions for correct chucking of carbide drills.

Material	Brinell Hardness	Feed (ipr)							
		0.004	0.005	0.006	0.008	0.010	0.012	0.016	0.020
	E	0.35	0.39	0.47	0.60	0.70	0.80	0.90	1.08
	C	3.3	3.5	3.8	4.4	4.4	4.4	4.0	3.6
Plain-Carbon Steel	140–220	444230	435510	431150	426790	418070	409350	391910	374460
	220–300	493590	483900	479060	474210	464520	454830	435450	416070
Free-Machining Steels	120–180	296150	290340	287440	284530	278710	272900	261270	249640
	180–260	345510	338730	335340	331950	325160	318380	304820	291250
Alloy Steels	260–340	493590	483900	479060	474210	464520	454830	435450	416070
Stainless Steels	150–200	370190	362930	359300	355660	348390	341120	326590	312050
	200–300	444230	435510	431150	426790	418070	409350	391910	374460
Cast Iron	180–250	345510	338730	335340	331950	325160	318380	304820	291250
Aluminum		148080	145170	143720	142260	139360	136450	130640	124820
Titanium		320830	314530	311390	308240	301940	295640	283040	270450
High-Temperature Alloys		542950	532290	526970	521630	510970	500310	478990	457680

F_v = Axial thrust
$$= D^{1.15} \times K_s \times f_r^{0.8}$$
where
D = Drill diameter (inches)
K_s = Specific cutting energy from table (in-lb/in²)
f_r = Feed (in./rev)
$X = 1.15$
$Y = 0.8$

← K_s Values in (in-lb/in²)
See chapter 22

Values in in.-lb/in².

FIGURE 25.18 Example of kinds of tables used for estimating the axial thrust force, F_v, in drilling using, K_s values are in in-lb/in². (Waukesha Cutting Tools)

engineer to select the correct machine or select the cutting parameters (speed and feed) based on the type and size of the drill, drill material, and the work material (hardness). Because of the complex geometry of the drill, empirical equations are widely used. **Figure 25.18** shows the type of information provided by cutting tool manufacturers to calculate (estimate) thrust in drilling. Specific cutting force values K_s are given, while empirical constants X and Y are obtained from cutting tool manufacturers. Much of these data have been developed for high-speed-steel tools. When using solid carbide tools, rigid machines such as machining centers or NC turning machines are recommended, whereas a radial drilling machine is not recommended (not rigid enough).

Rigidity is especially important in avoiding chatter. A lack of rigidity in the cutting tool, the workpiece workholding device or the machine tool permits the affected members to deflect due to the cutting forces. Conditions for chatter are discussed in Chapter 22. The cutting lips can have a hammering action against the work. Using the shortest tool possible can help.

In addition, backlash in the feed mechanism should be kept at a minimum to reduce strain on the drill when it breaks through the bottom of the hole.

The common name for the machine tool used for drilling is the **drill press**. Drill presses consist of a *base*, a *column* that supports a *powerhead*, a *spindle*, and a *worktable*. On small machines, the base rests on a workbench, whereas on larger machines it rests on the floor (**Figure 25.19**). The column may be either round or of box-type construction—the latter being used on larger, heavy-duty machines, except in radial types. The powerhead contains an electric motor and means for driving the spindle in rotation at several speeds. On small drilling

machines, this may be accomplished by shifting a belt on a step-cone pulley, but on larger machines a geared transmission is used.

The heart of any drilling machine is its **spindle**. In order to drill satisfactorily, the spindle must rotate accurately and also resist whatever side forces result from the drilling process. In virtually all machines, the spindle rotates in preloaded ball or taper-roller bearings. In addition to powered rotation, provision is made so that the spindle can be moved axially to feed the drill into the work. On small machines, the spindle is fed by hand, using the handles extending from the capstan wheel; on larger machines, power feed is provided. Except for some small bench types, the spindle contains a hole with a Morse taper in its lower end into which taper-shank drills or drill chucks can be inserted.

The worktables on drilling machines may be moved up and down on the column to accommodate work of various sizes. On round-column machines, the table can usually be rotated out of the way so that workpieces can be mounted directly on the base. On some box-column machines, the table is mounted on a subbase so that it can be moved in two directions in a horizontal plane by means of feed screws.

Figure 25.19 shows examples of common types of drilling machines used in production environments. Drilling machines are usually classified as bench, upright with single spindle, turret or NC turret, gang, multispindle, deep-hole, and transfer.

With bench drill presses, holes up to $\frac{1}{2}$ in. in diameter can be drilled. The same type of machine can be obtained with a long column so that it can stand on the floor. The size of bench and upright drilling machines is designated by *twice* the distance from the centerline of the spindle to the nearest point on

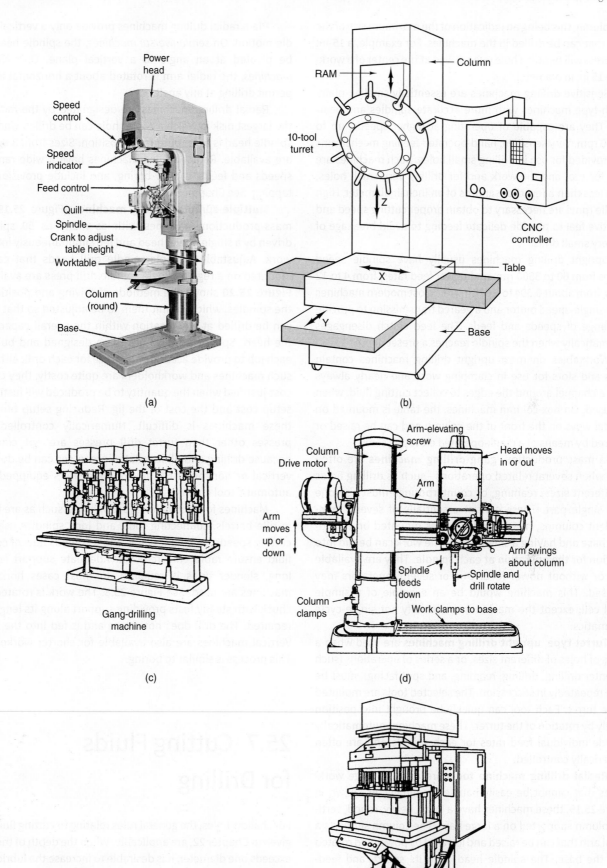

FIGURE 25.19 Examples of drilling machines: (a) small vertical spindle drilling machine, hand feed; (b) upright vertical drilling machine; (c) gang-drilling machine; (d) radial drill press; (e) multiple-spindle drilling machine. (Courtesy J T. Black)

(a) Power head / Speed control / Speed indicator / Feed control / Quill / Spindle / Crank to adjust table height / Worktable / Column (round) / Base

(b) RAM / Column / 10-tool turret / Z / CNC controller / Table / Base / X / Y

(c) Gang-drilling machine

(d) Arm-elevating screw / Column / Drive motor / Head moves in or out / Arm / Arm moves up or down / Arm swings about column / Spindle feeds down / Spindle and drill rotate / Column clamps / Work clamps to base

(e) Multiple-spindle drilling machine

the column, this being an indication of the maximum size of the work that can be drilled in the machines. For example, a 15-in. drill press will permit a hole to be drilled at the center of a workpiece 15 in. in diameter.

Sensitive drilling machines are essentially smaller, plain, bench-type machines with more accurate spindles and bearings. They are capable of operating at higher speeds, up to 30,000 rpm. Very sensitive, hand-operated feeding mechanisms are provided for use in drilling small holes. Such machines are used for tool and die work and for drilling very small holes, often less than a few thousandths of an inch in diameter. High spindle rpms are necessary to obtain proper cutting speed and sensitive feel to provide delicate feeding to avoid breakage of the very small drills.

Upright drilling machines usually have spindle speed ranges from 60 to 3500 rpm and power feed rates, from 4 to 12 steps, from about 0.004 to 0.025 in./rev. Most modern machines use a single-speed motor and a geared transmission to provide the range of speeds and feeds. The feed clutch disengages automatically when the spindle reaches a preset depth.

Worktables on most upright drilling machines contain holes and slots for use in clamping work and nearly always have a channel around the edges to collect cutting fluid, when it is used. On box-column machines, the table is mounted on vertical ways on the front of the column and can be raised or lowered by means of a crank-operated elevating screw.

In mass production, **gang-drilling machines** are often used when several related operations—such as drilling holes of different sizes, reaming, or counterboring—must be done on a single part. These consist essentially of several independent columns, heads, and spindles mounted on a common base and having a single table. The work can be slid into position for the operation at each spindle. They are available with or without power feed. One or several operators may be used. This machine would be an example of a simple small cell, except the machines are usually not single-cycle automatics.

Turret-type, **upright drilling machines** are used when a series of holes of different sizes, or a series of operations (such as center drilling, drilling, reaming, and spot facing), must be done repeatedly in succession. The selected tools are mounted in the turret. Each tool can quickly be brought into position merely by rotation of the turret. These machines automatically provide individual feed rates for each spindle and are often numerically controlled.

Radial drilling machine tools are used on large workpieces that cannot be easily handled manually. As shown in Figure 25.19, these machines have a large, heavy, round, vertical column supported on a large base. The column supports a radial arm that can be raised and lowered by power and rotated over the base. The spindle head, with its speed- and feed-changing mechanism, is mounted on the radial arm. It can be moved horizontally to any desired position on the arm. Thus, the spindle can be quickly positioned properly for drilling holes at any point on a large workpiece mounted either on the base of the machine or even sitting on the floor.

Plain radial drilling machines provide only a vertical spindle motion. On *semiuniversal machines,* the spindle head can be pivoted at an angle to a vertical plane. On *universal machines,* the radial arm is rotated about a horizontal axis to permit drilling at any angle.

Radial drilling machines are designated by the radius of the largest disk in which a center hole can be drilled when the spindle head is at its outermost position. Sizes from 3 to 12 ft are available. Radial drilling machines have a wide range of speeds and feeds, can do boring, and include provisions for tapping. See Chapter 31.

Multiple-spindle drilling machines (Figure 25.19) are mass-production machines with as many as 50 spindles driven by a single power head and fed simultaneously into the work. Adjustable multiple-spindle drill heads that can be mounted on a regular single spindle drill press are available. **Figure 25.20** shows the methods of driving and positioning the spindles, which permit them to be adjusted so that holes can be drilled at any location within the overall capacity of the head. Special drill jigs are often designed and built for each job to provide accurate guidance for each drill. Although such machines and workholders are quite costly, they can be cost-justified when the quantity to be produced will justify the setup cost and the cost of the jig. Reducing setup times on these machines is difficult. Numerically controlled drill presses other than turret drill presses are not common because drilling and all its related processes can be done on vertical or horizontal NC machining centers equipped with automatic tool changers.

Machines for drilling long (deep) holes, such as are found in rifle barrels, connecting rods, and long spindles use high cutting speeds, very light feeds, and a copious flow of cutting fluid ensure rapid chip removal. Adequate support for the long, slender drills is required. In most cases horizontal machines are used. See Figure 25.8. The work is rotated in a chuck with steady rests providing support along its length, as required. The drill does not rotate and is fed into the work. Vertical machines are also available for shorter workpieces. This process is similar to boring.

25.7 Cutting Fluids for Drilling

For shallow holes, the general rules relating to cutting fluids, as given in Chapter 22, are applicable. When the depth of the hole exceeds one diameter, it is desirable to increase the lubricating quality of the fluid because of the rubbing between the drill margins and the wall of the hole. The effectiveness of a cutting fluid as a coolant is quite variable in drilling. While the rapid exit of the chips is a primary factor in heat removal, this action also tends to restrict entry of the cutting fluid. This is of

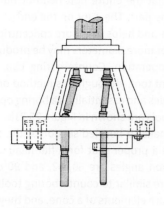

Adjustable drill head
Spindle: 6 Production: 50 pieces

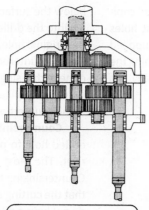

Geared drill head
Spindle: 8 Production: 80,000 pieces

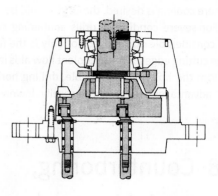

Gearless drill head
Spindle: 16 Production: 30,000 pieces

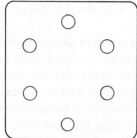

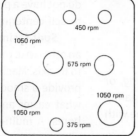

450 rpm
1050 rpm
575 rpm
1050 rpm
1050 rpm
375 rpm

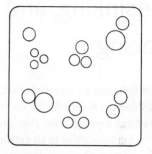

An adjustable drill head should be considered for low-production jobs. However, many short-run jobs such as this would be required to justify a multiple-spindle head.

A geared drill head is most appropriate in this situation, where there is a large difference in sizes and a high daily production.

Only a gearless head can perform this operation in one pass, due to the close proximity of the spindle centers.

FIGURE 25.20 Three basic types of multiple-spindle drill heads: (left) adjustable; (middle) geared; (right) gearless. (Courtesy of Zagar Incorporated)

TABLE 25.5 **Cutting Fluids for Drilling**

Work Material	Cutting Fluid
Aluminum and its alloys	Soluble oil, kerosene, and lard-oil compounds; light, nonviscous neutral oil; kerosene and soluble oil mixtures
Brass	Dry or a soluble oil; kerosene and lard-oil compounds; light, nonviscous neutral oil
Copper	Soluble oil, strained lard oil, oleic-acid compounds
Cast iron	Dry or with a jet of compressed air for cooling
Malleable iron	Soluble oil, nonviscous neutral oil
Monel metal	Soluble oil, sulfurized mineral oil
Stainless steel	Soluble oil, sulfurized mineral oil
Steel, ordinary	Soluble oil, sulfurized oil, high extreme-pressure-value mineral oil
Steel, very hard	Soluble oil, sulfurized oil, turpentine
Wrought iron	Soluble oil, sulfurized oil, mineral-animal oil compound

- Neat oil can be used effectively with the solid carbide drills for low-speed drilling (up to 130 sfpm).
- If the work surface becomes hard or blue in color, decrease the rpm and use neat oil.
- For heavy-duty cutting, emulsion-type oil containing some extreme pressure additive is recommended.
- A volume of 3.0 gal/min at a pressure of 37–62 lb/in.² is recommended.
- A double stream supply of fluid is recommended.

particular importance in drilling materials that have poor heat conductivity. Recommendations for cutting fluids for drilling are given in **Table 25.5**.

If the hole depth exceeds two or three diameters, it is usually advantageous to withdraw the drill each time it has drilled about one diameter of depth to clear chips from the hole. Some machines are equipped to provide this "pecking" action automatically.

Where cooling is desired, the fluid should be applied copiously. For severe conditions, drills containing coolant holes have a considerable advantage. Not only is the fluid supplied near the cutting edges, but the coolant flow aids in flushing the chips from the hole. Where feasible, drilling horizontally has distinct advantages over drilling vertically downward.

25.8 Counterboring, Countersinking, and Spot Facing

Drilling is often followed by *counterboring, countersinking,* or *spot facing.* As shown in **Figure 25.21**, each provides a bearing surface at one end of a drilled hole. They are usually done with a special tool having from three to six cutting edges.

Counterboring provides an enlarged cylindrical hole with a flat bottom so that a bolt head, or a nut, will have a smooth bearing surface that is normal to the axis of the hole; the depth may be sufficient so that the entire bolt head or nut will be below the surface of the part. The pilot on the end of the tool fits into the drilled hole and helps to ensure concentricity with the original hole. Two or more diameters may be produced in a single counterboring operation. Counterboring can also be done with a single-point tool, although this method ordinarily is used only on large holes and essentially is a boring operation. Some counterboring tools are shown in Figure 25.21b.

Countersinking makes a beveled section at the end of a drilled hole to provide a proper seat for a flat-head screw or rivet. The most common angles are 60, 82, and 90 degrees. Countersinking tools are similar to counterboring tools except that the cutting edges are elements of a cone, and they usually do not have a pilot because the bevel of the tool causes them to be self-centering.

Spot facing is done to provide a smooth bearing area on an otherwise rough surface at the opening of a hole and normal to its axis. Machining is limited to the minimum depth that will provide a smooth, uniform surface. Spot faces thus are somewhat easier and more economical to produce than counterbores. A multi-edged end-cutting tool that does not have a pilot or a counterboring tools are frequently used.

25.9 Reaming

Reaming removes a small amount of material from the surface of holes. It is done for two purposes: to bring holes to a more exact size and to improve the finish of an existing hole. Multi-edge cutting tools are used, as shown in **Figure 25.22**. No special machines are built for reaming. The same machine that was employed for drilling the hole can be used for reaming by changing the cutting tool.

To obtain proper results, only a minimum amount of materials should be left for removal by reaming. As little as 0.005 in. is desirable, and in no case should the amount exceed 0.015 in. A properly reamed hole will be within 0.001 in. of correct size and have a fine finish.

The principal types of reamers are shown in **Figures 25.22** and **25.23**. **Hand reamers** are intended to be turned and fed by hand and to remove only a few thousandths of an inch of metal. They have a straight shank with a square tang for a wrench. They can have straight or spiral flutes and can be solid or expandable. The teeth have relief along their edges and thus may cut along their entire length. However, the reamer is tapered from 0.005 to 0.010 in. in the first third of its length to assist in starting it in the hole, and most of the cutting therefore takes place in this portion.

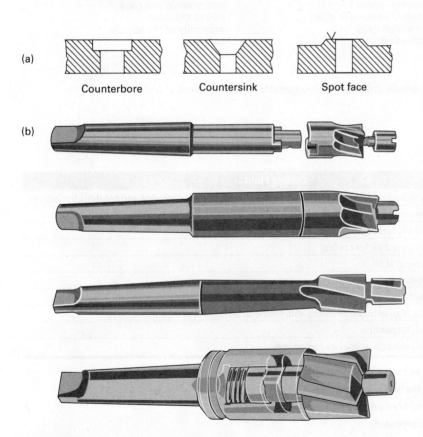

FIGURE 25.21 (a) Surfaces produced by counterboring, countersinking, and spot facing. (b) Counterboring tools: (bottom to top) interchangeable counterbore; solid, taper-shank counterbore with integral pilot; replaceable counterbore and pilot; replaceable counterbore, disassembled. (Courtesy of Ex-Cell-O Corporation and Chicago Latrobe Twist Drill Works)

(a) Counterbore   Countersink   Spot face

(b)

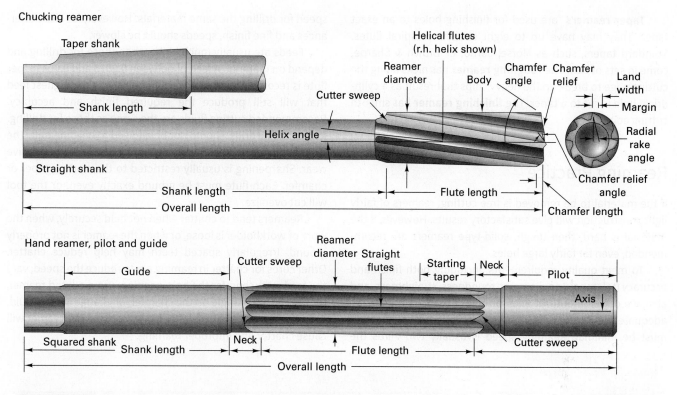

Chucking reamer

Taper shank

Shank length

Straight shank

Shank length

Overall length

Cutter sweep

Helix angle

Helical flutes
(r.h. helix shown)

Reamer
diameter

Body

Chamfer
angle

Chamfer
relief

Land
width

Margin

Radial
rake
angle

Chamfer relief
angle

Flute length

Chamfer length

Hand reamer, pilot and guide

Guide

Cutter sweep

Reamer
diameter

Straight
flutes

Starting
taper

Neck

Pilot

Axis

Squared shank

Shank length

Neck

Flute length

Cutter sweep

Overall length

FIGURE 25.22 Standard nomenclature for hand and chucking reamers. (Courtesy J T. Black)

Machine or **chucking reamers** are for use with various machine tools at slow speeds. The best feed is usually two to three times the drilling feed. Machine reamers have chamfers on the front end of the cutting edges. The chamfer causes the reamer to seat firmly and concentrically in the drilled hole, allowing the reamer to cut at full diameter. The longitudinal cutting edges do little or no cutting. Chamfer angles are usually 45 degrees. Reamers have straight or tapered shanks and straight or spiral flutes. **Rose-chucking reamers** are ground cylindrical and have no relief behind the outer edges of the teeth. All cutting is done on the beveled ends of the teeth. **Fluted-chucking reamers**, on the other hand, have relief behind the edges of the teeth as well as beveled ends. They can, therefore, cut on all portions of the teeth. Their flutes are relatively short, and they are intended for light finishing cuts. For best results, they should not be held rigidly but permitted to float and be aligned by the hole.

Shell reamers often are used for larger sizes in order to save cutting tool material. The shell, made of tool steel for smaller sizes and with carbide edges for larger sizes or for mass-production work, is held on an arbor that is made of ordinary steel. One arbor may be used with any number of shells. Only the shell is subject to wear and needs to be replaced when worn. They may be ground as rose or fluted reamers.

Expansion reamers can be adjusted over a few thousandths of an inch to compensate for wear or to permit some variation in hole size to be obtained. They are available in both hand and machine types.

Adjustable reamers have cutting edges in the form of blades that are locked in a body. The blades can be adjusted over a greater range than expansion reamers. This permits adjustment for size and to compensate for regrinding. When the blades become too small from regrinding, they can be replaced. Both tool steel and carbide blades are used.

FIGURE 25.23 Types of reamers (top to bottom): straight-fluted rose reamer, straight-fluted chucking reamer, straight-fluted taper reamer, straight-fluted hand reamer, expansion reamer, shell reamer, adjustable insert-blade reamer. (Courtesy J T. Black)

Taper reamers are used for finishing holes to an exact taper. They may have up to eight straight or spiral flutes. Standard tapers, such as Morse, Jarno, or Brown & Sharpe, come in sets of two. The **roughing reamer** has nicks along the cutting edge to break up the heavy chips that result as a cylindrical hole is cut to a taper. The **finishing reamer** has smooth cutting edges.

Reaming Practice

If the material to be removed is free cutting, reamers of fairly light construction will give satisfactory results. However, if the material is hard, then tough, solid-type reamers are recommended, even for fairly large holes.

To meet quality requirements, including both finish and accuracy (tolerances on diameter, roundness, straightness, and absence of bell-mouth at ends of holes), reamers must have adequate support for the cutting edges, and reamer deflection must be minimal. Reaming speed is usually two-thirds the speed for drilling the same materials. However, for close tolerances and fine finish, speeds should be slower.

Feeds are usually much higher than those for drilling and depend on material. A feed of between 0.0015 and 0.004 in. per flute is recommended as a starting point. Use the highest feed that will still produce the required finish and accuracy. Recommended cutting fluids are the same as those for drilling. Reamers, like drills, should not be allowed to become dull. The chamfer must be reground long before it exhibits excessive wear. Sharpening is usually restricted to the starting taper or chamfer. Each flute must be ground exactly even, or the tool will cut oversize.

Reamers tend to chatter when not held securely, when the work or workholder is loose, or when the reamer is not properly ground. Irregularly spaced teeth may help reduce chatter. Other cures for chatter in reaming are to reduce the speed, vary the feed rate, chamfer the hole opening, use a piloted reamer, reduce the relief angle on the chamfer, or change cutting fluid. Any misalignment between the workpiece and the reamer will cause chatter and improper reaming.

Review Questions

1. What functions are performed by the flutes on a standard twist drill?

2. What determines the rake angle of a drill? See Figure 25.2.

3. Basically, what determines what helix angle a drill should have?

4. When a large-diameter hole is to be drilled, why is a smaller-diameter hole often drilled first?

5. Equation 25.4 for the MRR for drilling can be thought of as ___ × ___, where $f_r N_s$ is the feed rate of the drill bit.

6. Are the recommended surface speeds for spade drills given in Table 25.3 typically higher or lower than those recommended for twist drills? How about the feeds? Why?

7. What can happen when an improperly ground drill is used to drill a hole?

8. Why are many drilled holes oversize with respect to the nominally specified diameter?

9. What are the two primary functions of a combination center drill?

10. What is the function of the margins on a twist drill?

11. What factors tend to cause a drill to "drift" off the centerline of a hole?

12. The drill shown in Figure 25.13 has coolant passages in the flutes. What is the purpose of these holes?

13. In drilling, the deeper the hole, the greater the torque. Why?

14. Why do cutting fluids for drilling usually have more lubricating qualities than those for most other machining operations?

15. How does a gang-drilling machine differ from a multiple-spindle drilling machine?

16. How does a multiple-spindle drilling machine differ from a numerical control (NC) drilling machine with a tool changer that would hold all the drills found in the multiple-spindle machine? See Chapter 26 for discussion on NC.

17. How does the thrust force vary with feed? Why?

18. Holding the workpiece by hand when drilling is not a good idea. Why?

19. What is the rationale behind the operation sequence shown in Figure 25.10?

20. In terms of thrust, what is unusual about the slot-point drill compared to other drills?

21. What is the purpose of spot facing?

22. How does the purpose of counterboring differ from that of spot facing?

23. What are the primary purposes of reaming?

24. What are the advantages of shell reamers?

25. A drill that operated satisfactorily for drilling cast iron gave very short life when used for drilling a plastic. What might be the reason for this?

26. What precautionary procedures should be used when drilling a deep, vertical hole in mild steel when using an ordinary twist drill?

27. What is the advantage of a spade drill? Is it really a drill?

28. What is a "pecking" action in drilling?

29. Why does drill feed increase with drill size?

30. Suppose you specified a drilling feed rate that was too large. What kinds of problems do you think this might cause? See Figure 25.6 and Table 25.4 for help.

Problems

1. Suppose you wanted to drill a 1.5-in.-diameter hole through a piece of 1020 cold-rolled steel that is 2 in. thick, using an HSS twist drill. What values of feed and cutting speed will you specify, along with an appropriate allowance. What other drill types could be used?

2. How much time will be required to drill the hole in Problem 1 using the HSS drill?

3. What is the metal removal rate when a 1.5-in.-diameter hole, 2 in. deep, is drilled in 1020 steel at a cutting speed of 100 fpm with a feed of 0.020 ipr? What is the cutting time?

4. If the specific horsepower for the steel in Problem 3 is 0.9, what horsepower would be required, assuming 80% efficiency in the machine tool?

5. If the specific power of an AISI 1020 steel of 0.9, and 80% of the output of the 1.0-kW motor of a drilling machine is available at the tool, what is the maximum feed that can be used in drilling a 1-in.-diameter hole with a HSS drill? (Use the cutting speed suggested in Problem 3.)

6. Show how the approximate equation 25.5 for MRR in drilling was obtained. What assumption was needed?

7. Assume that you are drilling eight holes, equally spaced in a bolt-hole circle. That is, there would be holes at 12, 3, 6, and 9 o'clock and four more holes equally spaced between them. The diameter of the bolt-hole circle is 6 in. The designer says that the holes must be 45 degrees ± 1 degree from each other around the circle.

a. Compute the tolerance between hole centers.

b. Do you think a typical multiple-spindle drill setup could be used to make this bolt circle—using eight drills all at once? Why or why not?

c. Do you think that the use of a jig may help improve the situation?

d. Do you think a CNC drilling process could do the holes best?

CHAPTER 26

CNC Processes and Adaptive Control: A(4) and A(5) Levels of Automation

26.1 Introduction

The first **numerically controlled (NC)** machine tool was developed in 1952 at the Massachusetts Institute of Technology (MIT). It had three-axis positional feedback control and is generally recognized as the first NC machine tool. By 1958, the first NC **machining center** was being marketed by Kearney and Trecker. A machining center was a compilation of many machine tools capable of performing many processes (milling, drilling, tapping, and boring), as shown in **Figure 26.1**. This NC machine had automatic positioning capability. Almost from the start, computers were needed to help program these machines. Within 10 years, NC machine tools had become **computer numerical control (CNC)** machine tools with onboard microprocessors and could be programmed directly, though the term NC is still used to refer to the foundational technology.

With the advent of the NC type of machine (and, more recently, programmable robots), two types of **automation** were defined. (See Chapter 34 for discussion on levels of automation.) **Hard** or **fixed automation** is exemplified by transfer machines or automatic screw machines controlled by a mechanical cam. **Flexible** or **programmable automation** is typified by CNC machines or robots that can be taught or programmed externally by means of computers. The control is in computer software rather than mechanical hardware.

NC uses a processing language to control the movement of the cutting tool or workpiece or both. The programs specify the path or end positions which the cutting tool or other equipment must follow, along with the machining parameters (speeds, feeds, and depth of cut) necessary to make a desired part. CNC machines can control complex geometries and replicate consecutive parts with precision. Parts made at a later date will be the same as those one made today.

Repeatability and quality are improved over conventional (job shop) machines. Workholding devices can be made more universal, and setup time can be reduced, along with tool-change time, thus making programmable machines economical for producing small lots or even a single piece. When combined with the managerial and organizational strategies of lean (cellular) manufacturing, programmable machines lead to tremendous improvements in quality and productivity. The creation of families of parts made in machining cells containing CNC level machines, plus the compatibility of the components (similarity in process and sequences of processes), greatly enhances the productivity (utility) of the programmable equipment.

A side result of CNC is the decrease in the non-chip-producing time of machine tools. The operator is relieved of the jobs of changing speeds and feeds and locating the tool relative to the work. Even simple forms of NC and digital readout equipment have provided both greater productivity and increased accuracy. Most early NC machine tools were developed for special types of work where accuracies of as much as .00005 in. were required, and many CNC machines are built to provide accuracies of at least 0.0001 in., regardless of whether it is needed. This forced many machine tool builders to redesign their machines and improve their quality because the operator was not available to compensate for machine (positioning) error. While most CNC machine tools today will provide greater accuracy than is required for most jobs, the trend is toward greater accuracy and precision (i.e., better quality) but at no increase in cost. Therefore, CNC/NC machines will continue to be the very backbone of the machine tool business.

26.2 Basic Principles of Numerical Control

As the name implies, *numerical control* is a method of controlling the motion of machine components by means of numbers or coded instructions.

For example, assume that three 1-in. holes in the part shown in **Figure 26.2** are to be drilled and bored on a

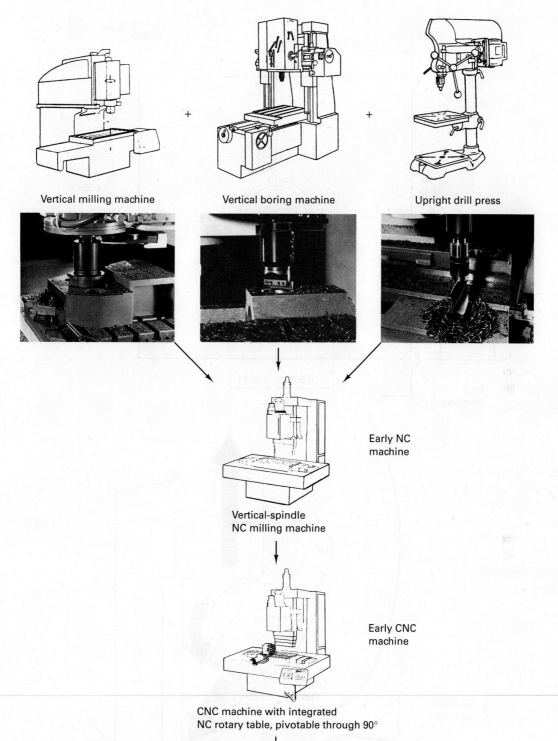

Vertical milling machine
+
Vertical boring machine
+
Upright drill press

Early NC machine

Vertical-spindle NC milling machine

Early CNC machine

CNC machine with integrated NC rotary table, pivotable through 90°

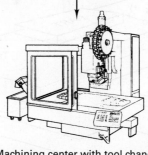

Early CNC machining center

Machining center with tool changer (20 tools), chip conveyor, pallet changer, and anti-splash booth

FIGURE 26.1 Early NC machine tools were controlled by paper tape. Soon, onboard computers were added, followed by tool changers and pallet changers. (Courtesy J T. Black)

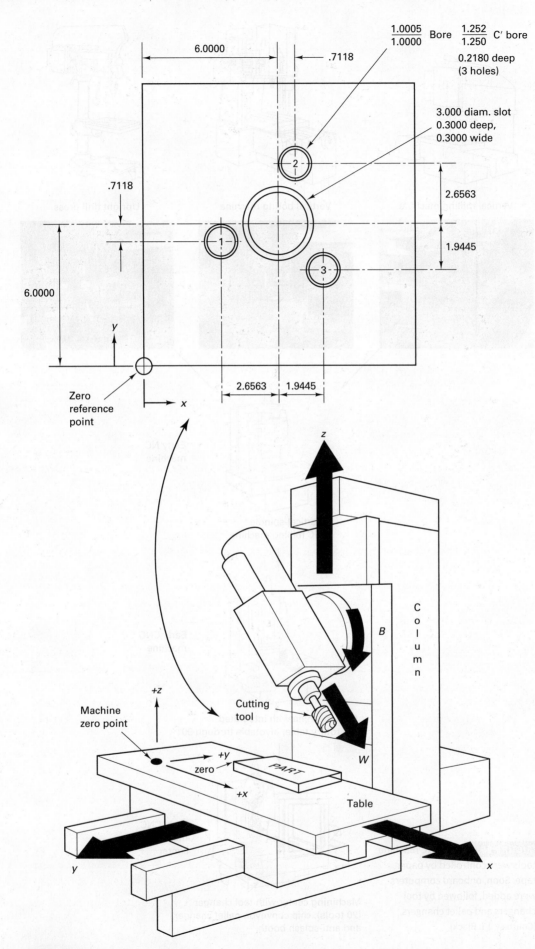

FIGURE 26.2 The part (above) to be machined on the NC machine (below) has a zero reference point. The machine also has a reference point.

vertical-spindle machine. The centers of these holes must be located relative to each other and with respect to the left-hand edge (x direction) and the bottom edge (y direction) of the workpiece. The depth of the hole will be controlled by the z- (or w-) axis. For this part, this is the zero reference point (or part reference zero) for the part.

The holes will be produced by center drilling, hole drilling, boring, reaming, and counterboring (five tool changes). If this were done conventionally or in manned cells, three or four different machines might be required. On the CNC machining center, all the tools are held in a tool magazine and are changed automatically. The movements of the table are controlled by servo mechanisms for each direction.

A software program instructs the table to move with respect to the axis of the spindle to bring the holes to the correct location for machining. The machine shown in Figure 26.2 is called a five-axis machine because it has five movements (shown by the dark arrows) under numerical control. Although this example does not show a workholding fixture, typically one would be used to obtain quick and repeatable location of the part on the table.

How CNC Machines Work

Controlling a machine tool using variable input via a computer program is known as *numerical control* and is defined by the Electronic Industries Association (EIA) as "a system in which actions are controlled by direct insertion of numerical data at some point. The system must automatically interpret at least some of the data." (The "numerical data" refers to the CNC part program.)

Traditionally, NC machine tool has a **machine control unit (MCU sometimes called the controller)** which is further divided into two elements: the **data-processing unit (DPU)** and the **control-loops unit (CLU)**. The DPU processes the coded data that are read from the tape or some other input medium that it gives to the CPU, specifically the position of each axis, its direction of motion feed, and its auxiliary-function control signals. The CLU operates the drive mechanisms of the machine.

Most CNC machines employ the concept of **feedback** control, where some aspect (usually **position**) of the process is measured using a detection device **(sensor).** This information is fed back to an electronic comparator, housed in the **MCU,** which makes comparisons with the desired position. If the output and input are not equal, an error signal is generated and the table is adjusted to reduce the error.

Using a milling machine table as an example, **Figure 26.3** shows the difference between **open-loop machine** and a **closed-loop machine**, with feedback provided on the location of the table and the part with respect to the axis of the spindle of the cutting tool. Three position control schemes are shown.

Simple CNC machines sometimes employ open-loop control. The use of a stepper motor to achieve the desired axis displacement by sending a controlled number of DC pulses is an example of **open-loop** CNC control. Open-loop control is economical, but it is not able to check that the desired motion actually was achieved and to correct for error.

Closed-loop control requires a transducer or other sensing device to detect machine table position and transmit it to the MCU as axis-position feedback, and velocity feedback if contour path control is needed. **Figure 26.4** depicts a rotary encoder or resolver which applies a pulse disk to convert the analog rotation of the screw drive to digital pulses which are used to calculate table position. A linear glass scale encoder is a common alternative to directly track linear positioning.

Figure 26.5 provides a schematic of the control feedback system for one axis of CNC table motion. In this case, in addition to the resolver or encoder, a tachometer has been at the drive motor to provide velocity feedback.

The MCU compares current status (position) with the desired state. If they are different, the control unit produces a signal to the drive motors to move the table, reducing the error signal and ultimately moving the table to the desired position at the desired velocity. At this point, the command counter reaches zero, meaning that the correct number of pulses has been sent to move the table to the desired position. In a closed-loop system, a comparator is used to compare feedback pulses with the original value, generating an error signal. Thus, when the machine control unit receives a signal to execute a command, the table is moved to the specified location, with the actual position being monitored by the feedback transducer. Table motion ceases when the error signal has been reduced to zero. At this point, that motion command is completed, and another function can begin (such as drilling a hole or starting the tool on a different cutting path).

Closed-loop systems tend to have greater accuracy than open-loop systems and respond faster to input signals. However, it is possible for them to exhibit stability problems not found in open-loop systems, oscillating about a desired value instead of achieving it.

The transducers used to track feed screw or actual movement of the machine components may provide either **digital** or **analog** signals (information). The resolver in Figure 26.5 is an example of conversion of the analog ball screw movement into digital pulses, which are then used to calculate table movement.

Two basic types of digital transducers are used. One supplies **incremental information** and tells how much motion of the input shaft or table has occurred since the last time. The information supplied is similar to telling newspaper carriers that papers are to be delivered to the first, fourth, and eighth houses from a given corner on one side of the block. To follow the instructions, the carriers would need a means of counting the houses (pulses) as they passed them. They would deliver papers as they counted 1, 4, and 8. The second type of digital information is **absolute** in character, with each pulse corresponding to a specific location of the machine components. To continue the carrier analogy, this would correspond to telling the carriers to deliver papers to the houses having house numbers 2400, 2406, and 2414. In this case, it would be necessary only for the carriers (machine component) to be able to read the house numbers (addresses) and stop to deliver a paper when arriving at a proper address. This **address system** is a common one in numerical control systems because it provides absolute location information relative to a **machine zero point**.

Open loop

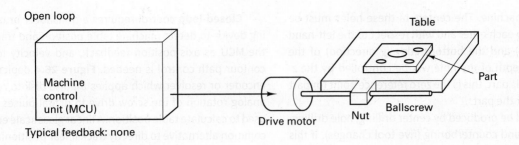

Machine control unit (MCU)

Typical feedback: none

Drive motor

Nut

Ballscrew

Part

Table

Closed loop–motor feedback

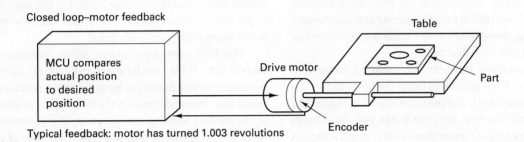

MCU compares actual position to desired position

Typical feedback: motor has turned 1.003 revolutions

Drive motor

Encoder

Part

Table

Closed loop–ball-screw feedback

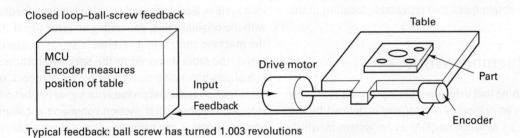

MCU Encoder measures position of table

Input

Feedback

Typical feedback: ball screw has turned 1.003 revolutions

Drive motor

Encoder

Part

Table

Closed loop–worktable position feedback

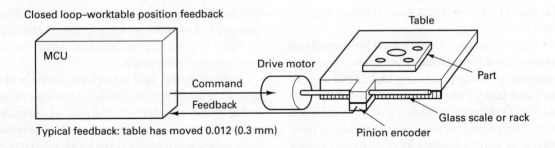

MCU

Command

Feedback

Typical feedback: table has moved 0.012 (0.3 mm)

Drive motor

Glass scale or rack

Pinion encoder

Part

Table

FIGURE 26.3 Open-loop control compared to three schemes for NC and CNC closed-loop.

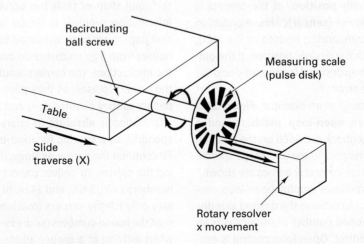

Recirculating ball screw

Measuring scale (pulse disk)

Table

Slide traverse (X)

Rotary resolver x movement

FIGURE 26.4 The table of this CNC is located using a resolver or encoder attached to the ball screw. (See Figure 26.5.)

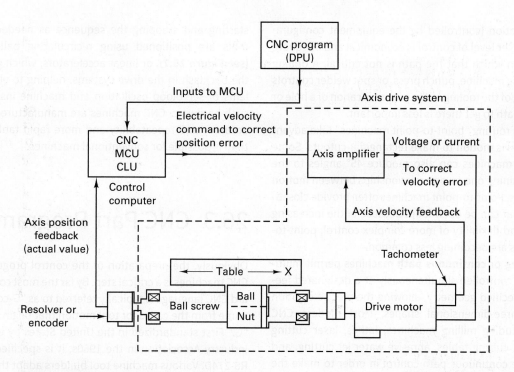

FIGURE 26.5 The table of the machine is translated with a screw mechanism and its location is detected with a resolver. The schematic shows how position and velocity feedback loops contribute to control of the table motion of the part under the cutting tool.

When analog information is used, the signal is usually in the form of an electric voltage that varies as the input shaft is rotated or the machine component is moved, the variable output being a function of movement. The movement is evaluated by measuring, or matching, the voltage or by measuring the ratios between the applied and feedback voltages; this eliminates the effect of supply-voltage variations.

The input information (i.e., the location of the holes) is given in the form of a CNC program. Early NC machines relied on a **punched paper tape** or magnetic tape to input the sequence of alphanumeric characters of the program. CNC has adapted with computer file transfer technologies and today receives programs as text files from USB memory devices, RS-232 serial or network communication, or wireless technologies. Programs can also be entered by the operator at the con-

trol panel. Regardless of the input technology, the machine control unit receives the binary form of the program character stream and converts it into a series of pulses, which in turn drives the **servomotor** or stepper motor.

Alternating-current servomotors are favored over direct-current motors on modern CNC machine tools due to better reliability, better performance-to-weight ratio, and lower power consumption.

Control and Hardware Options

NC and CNC machines can be subdivided into two types, shown in **Figure 26.6**. **Point-to-point** (or positioning) machines move to a programmed point, but do not provide control along the path between points, except when the path follows only

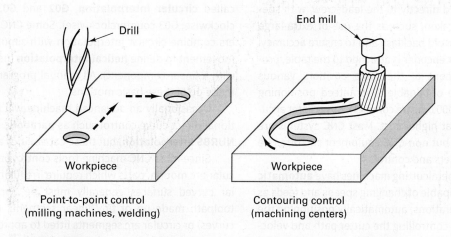

FIGURE 26.6 CNC systems are subdivided into two basic categories: point-to-point controls and contouring controls.

one axis of motion (controlled by the equipment configuration). This simpler level of control is economical and adequate for processes in which that the path is not critical. A drilling machine, boring machine, punch press, or spot welder controls the positioning of the tooling to the precise location of a hole or weld, but the path to get there is less important.

CNC wood routing "point-to-point machines" take advantage of single-axis motion to make rectangular cutouts. Some point-to-point machines can also produce 45° angles to the axes by maintaining one-to-one relationships between motion of multiple axes. Point-to-point machines often provide closed-loop control, but can be open-loop. Because of the increasing affordability and flexibility of more complex control, point-to-point machines are becoming less common.

Contouring or **continuous path** machines permit multiple axes to be controlled simultaneously at a designated feedrate along specified geometry, allowing the precise creation of two- or three-dimensional shapes. Most modern CNC machines, including milling machines, lathes, laser cutting tables, plasma cutting tables, abrasive water jet cutting, and wire EDM have continuous path control in order to make the variety of shapes generally expected by these processes.

Continuous path machines still employ point-to-point control for fast positioning moves (often called rapid traverse) that do not require the careful control of cutting motion. Continuous path machines generally provide **closed loop control**.

In contouring, it is usually necessary to control multiple paths between points by interpolation of intermediate coordinate positions. As many of these systems as desired can be combined to provide control in several axes. Two- and three-axis control are most common, but some machines have as many as seven. (Interpolation is discussed further in section 26.4.)

The components required for such numerical control systems are now standardized items of hardware. In most cases, the drive motor is electric, but hydraulic systems are also used. They are usually capable of moving the machine elements, such as tables, at high rates of speed, up to 200 in./min being common. Thus, exact positioning can be achieved more rapidly than by manual means. The transducer can be placed on the drive motor or connected directly to the leadscrew, with special precautions being taken, such as the use of extra-large screws and ball nuts to avoid backlash and to ensure accuracy. In other systems, a linear encoder is attached to the table, providing direct measurement of the table position. Various degrees of accuracy are obtainable. Guaranteed positioning accuracies of 0.001 or 0.0001 in. are common, but greater accuracies can be obtained at higher cost. Most CNC systems are built into the machines, but non-CNC equipment can often be retrofitted with transducers and controls.

Most modern CNC metalcutting machines have **automatic toolchangers** and are capable of changing speeds and feeds as needed for different operations, automatically positioning the work relative to tooling, controlling the cutter path and velocity, repositioning the tool rapidly between operations, and

starting and stopping the sequence as needed. Tables and tools are positioned using recirculating ball-screw drives (see **Figure 26.7**), or linear accelerators, which greatly reduce the backlash in the drive systems, helping to eliminate problems of servoloop oscillation and machine instability. Using such hardware, CNC machines are manufactured with greater accuracy and repeatability and more rapid table movements than is possible for conventional machines.

26.3 CNC Part Programming

Obviously, the preparation of the control program for use in CNC machines is a critical step. By far the most common, standard CNC language is typically referred to as "G-code," taking its name from the codes for tool motion that begin with the letter "G." First standardized in the United States by the Electronics Industry Association in the 1960s, it is specified by standard RS-274D. Various machine tool builders adapt the code to particular needs and preferences, but much of the basic code is standardized.

Table 26.1 lists some of the most common motion, tool, and machine commands for CNC machining programs. Each command letter-number combination is called an **NC word**. **Table 26.2** defines categories of NC words (instructions) based on their first letters.

Program lines were originally referred to as "blocks" of code. Some early NC and CNC systems required each line or block be numbered, but today N-words are generally optional within G-Code, necessary only for certain lines which may need identification. One particular use of the line numbers is the designation of the beginning of a subroutine within the program.

Types of Programmed Motion

Notice that there are very few types of basic motion in CNC: rapid traverse positioning, G00; controlled linear cutting motion, called **linear interpolation**, G01; and circular arcs, called **circular interpolation**, G02 and G03 (where G02 is clockwise, G03 counterclockwise). Some CNC machining centers combine circular interpolation with simultaneous vertical movement to define **helical interpolation** (usually G02 or G03 with a linear component). All manual programming would be made up of these basic motions.

Occasionally an advanced machine will provide an additional direct curve control such as **parabolic interpolation** or **NURBS interpolation**, but these capabilities are rare.

Since most CNC machine tools control only linear and circular arc motion, parts which require irregular curves or irregular curved surfaces generally must be approximated by a toolpath made up of tiny linear segments (**linearization** of curves) or circular arc segments fitted to approximate the original irregular curve geometry. The latter strategy is known as

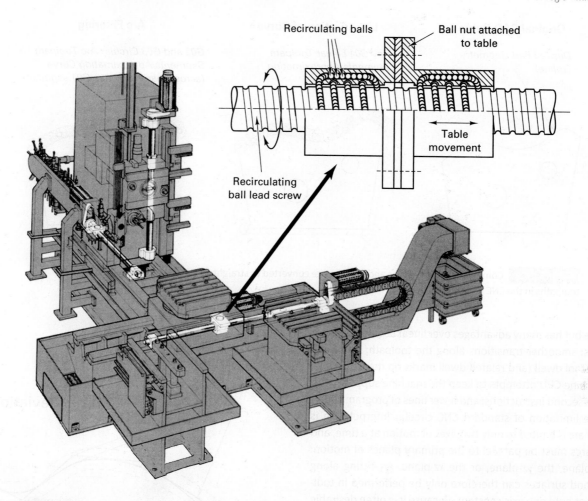

FIGURE 26.7 The ball leadscrew shown in detail provides great accuracy and position to CNC machine tools.

TABLE 26.1	Common CNC Command Words.
G90, G91	Absolute coordinates, incremental coordinates
G20, G21	Inch coordinate system, metric coordinate system
G00	Rapid traverse (positioning motion)
G01	Linear interpolation
G02, G03	Circular interpolation (clockwise, counterclockwise)
G82, G83, G80	Spot drill, peck drill, cancel canned cycle
G28	Machine zero return through reference point (specified axes)
M03, M04, M05	Spindle on clockwise, spindle on counterclockwise, spindle off
M08, M09	Coolant on, coolant off
G43 + H_; G49	Tool length offset (TLO); cancel TLO
G41,G42 + D_; G40	Cutter radius compensation left, cutter radius compensation right; cancel cutter radius compensation

TABLE 26.2	Definitions of Common CNC Word Prefixes	
NC Word	**Use**	
N	*Sequence number:* identifies the block (or line) of information	
G	*Preparatory function:* requests different control functions, including preprogrammed machining routines	
X, Y, Z, A, B, C	*Dimensional coordinate data:* linear and angular motion commands for the axes of the machine	
F	*Feed function:* sets feed rate for this operation in standard units (in/min or mm/min)	
S	*Spindle RPM:* sets spindle rotation speed, which results in an appropriate cutting speed.	
T	*Tool function:* tells the machine the location of the tool in the tool holder or tool turret	
M	*Miscellaneous function:* turns coolant on or off, opens spindle, reverses spindle, tool change, etc.	
EOB	*End of block:* Indicates to the MCU that a full block of information has been transmitted and the block can be executed. (Generally interchangeable with the "ENTER" or line-return key, though the "EOB" button remains on some CNC control panels.)	

arc-filtering, or sometimes **arc-fitting. Figure 26.8** provides a simplified example of a spline curve that is approximated by linearization, and then by arc filtering. Arc filtering requires more sophisticated CAM **Computer-Aided Manufacturing (CAM)**

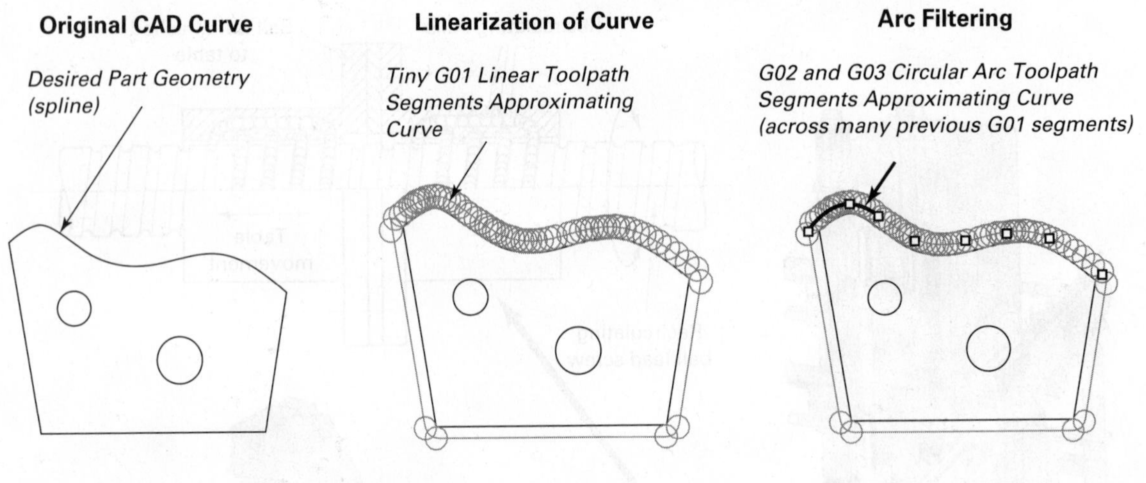

Original CAD Curve

Desired Part Geometry (spline)

Linearization of Curve

Tiny G01 Linear Toolpath Segments Approximating Curve

Arc Filtering

G02 and G03 Circular Arc Toolpath Segments Approximating Curve (across many previous G01 segments)

FIGURE 26.8 Complex, irregular curves and surfaces are converted to straight-line or tangent-arc toolpaths in the CNC code.

software but has many advantages over linear segment approximations: smoother transitions along the toolpath; less likelihood of tool dwell (and related dwell marks on the part) while the machine CPU attempts to keep the machine supplied with the split-second instructions; and fewer lines of program code.

One limitation of standard CNC circular interpolation is that the arc is limited to only two axes of motion at a time, and so the arcs must be parallel to the primary planes of motion: the *xy*-plane, the *yz*-plane, or the *xz*-plane. Arc-fitting along curves and surfaces can therefore only be performed in toolpaths that hold one axis constant. Because it is often desirable to control curved surface cutting tool motion in all three directions simultaneously, it is still common to see CAD/CAM-generated toolpaths incorporating streams of tiny G01 straight-line segments of cutting motion. Today's high-speed processors are typically able to handle the fast stream of commands without the tool dwell problems experienced by earlier CNC machine tools.

Resolution of linearized or arc-fitted toolpaths is controlled by settings in the CAM software, generally specifying **maximum chordal deviation** or a "**linearization tolerance.**" **Figure 26.9** provides an enlarged view of a linear approximation over an irregular spline curve. Chordal deviation is the distance the programmed segments deviate from the intended irregular curve.

Rapid traverse motion (G00) is used for fast positioning away from the workpiece. Interpolation motion (G01, G02, and G03) are used when cutting, approaching, and sometimes exiting the workpiece. It is also sometimes used for controlled motion and speed for positioning close to the workpiece or workholding.

For standard industrial CNC, rapid traverse motion is not necessarily a straight line to the desired point. Instead, the tool moves all axes at top speed, with no need to check intermediate positions. For a two- or three-axis move where all axes may move at the same speed, the path will follow an approximate 45° angle until it reaches the desired position of one of the axes.

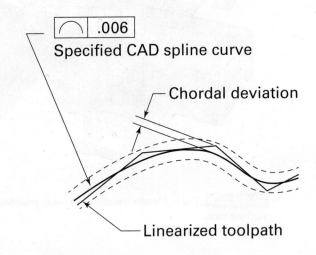

.006

Specified CAD spline curve

Chordal deviation

Linearized toolpath

FIGURE 26.9 An enlarged view of linearized toolpaths approximating a spline curve and the accompanying chordal deviation.

It then stops that axis of motion but continues on with the remaining axis or axes of motion. An example of this type of motion is shown in **Figure 26.10** (c) in the path between points P10 and P11.

Aside from the basic motion commands of G00, G01, G02, and G03, most CNC machines have built-in **canned cycles** that allow a common routine that combines rapid and interpolation motion in one line of code. **Drilling canned cycles**, for example, allow the user to specify a hole location, total depth, retract plane, and peck-drilling routine if desired. Additional holes of the same depth and routine can be specified on subsequent lines with only the new location coordinates; the routine will automatically retract and manage motion to the next hole and begin the drilling process. **Figure 26.11** depicts peck drilling motion and other examples of canned cycles for milling such as **tapping** and **pocketing** cycles. Many CNC machine tool manufacturers offer **text engraving** canned cycles for simple text and serial number engraving.

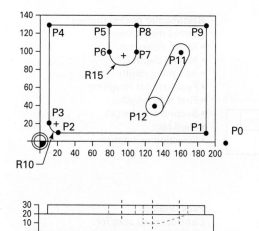

Figure 26.10 (A) Workpiece drawing (top and front views)

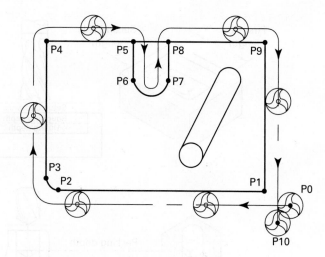

Figure 26.10 (B) Programming the outer contour

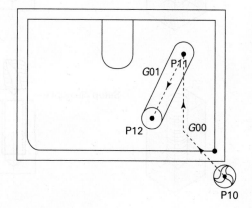

FIGURE 26.10 Example of programming of a part in a vertical-spindle CNC machine. This part requires a 20 mm flat endmill.

Figure 26.10 (C) Programming the slot

Turning centers make frequent use of **rough and finish cycles for turning** down a shape. By specifying a maximum depth of cut for rough and finish passes, desired feed rates, and the final path contour (even a complex one), the MCU will compute a roughing and finishing strategy—all with one line of code.

A variety of canned cycle capabilities are offered by CNC machine tool developers, and these are often likely to vary in the programming code. The golden rule in CNC is to always consult the manual of your own machine tool for specific programming needs.

Subroutines offer the programmer options for repeating sections of the code for cutting the same geometry at different depths or different locations on the part. Separate **subprograms** can be stored and reference to incorporate geometry such as corporate logos or other patterns that may be desired for use on several products.

Basic Steps of Programming and Setup

The basic steps of programming are demonstrated by the example part and its desired milling toolpaths in Figure 26.10 and the accompanying program provided in **Figure 26.12**.

The first step is to establish the x and y axis directions and the **part reference zero**, sometimes called the **zero reference point**, the **work coordinate system**, or the **part home**. Ideally, it is chosen at a part datum (a measurement reference), which should be related to a fixed jaw or hole location on the setup. Often the part reference zero is placed for convenience of the setup operator; for prismatic (rectangular) parts, it is often a corner along a fixed jaw of the workholding, since the movable jaw location will shift slightly with repeated parts of slight variation in size.

The location of the part reference zero should be noted on setup instructions for the operator, which are usually recorded in reference notes at the top of the program, as well as in a separate **setup sheet** with drawings of the intended workholding and part reference zero location.

For ease in establishing cutter motion endpoints, the origin of the part drawing can be shifted and ordinate dimensioning applied. **Table 26.3** demonstrates a **coordinate sheet** that records locations of each endpoint of endmill motion for the example milling program.

During setup, the operator will locate the part with respect to the cutting tool and set the differing lengths of tools when multiple tools are used. This information is stored in the

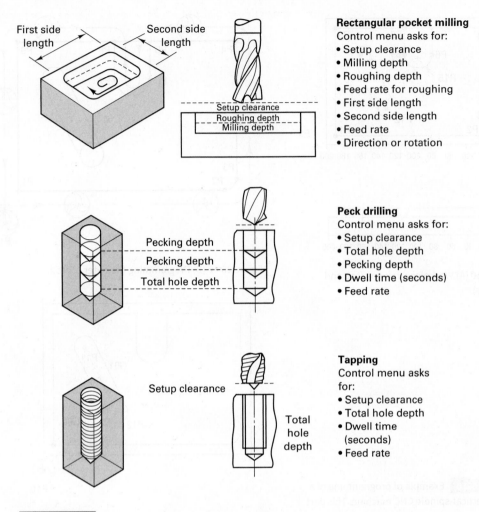

Rectangular pocket milling
Control menu asks for:
- Setup clearance
- Milling depth
- Roughing depth
- Feed rate for roughing
- First side length
- Second side length
- Feed rate
- Direction or rotation

Peck drilling
Control menu asks for:
- Setup clearance
- Total hole depth
- Pecking depth
- Dwell time (seconds)
- Feed rate

Tapping
Control menu asks for:
- Setup clearance
- Total hole depth
- Dwell time (seconds)
- Feed rate

FIGURE 26.11 Canned or preprogrammed machining routines greatly simplify programming CNC machines. (Courtesy of Heidenhain Corporation, Elk Grove Village, IL).

machine controller under the **workpiece offset** and **tool offset** menus (or other related names). The machine has its own master coordinate system called the **machine coordinate system**. Similar to the longitude and latitude of the globe, the location of the machine coordinate system never changes, and so the location of the part reference zero is stored in the MCU using the machine coordinates, similar to the location of a geographic point being specified in latitude and longitude. Tool lengths are set up with respect either to a master tool or to a reference point established on the machine bed.

The method of **locating the workpiece** involves "touching off" on the workpiece or workholding datums with a master tool, edge finder, touch probe, or other appropriate technology.

The convention of CNC programming and machine setup motion is to command motion *as if the tool is doing the motion*, though in actuality, in many cases the workpiece is actually moving with respect to the tool.

The three main sections of the program are highlighted in Figure 26.12: The program header, the tool motion section, and the end-of-program code. Many programs will require more than one tool, in which case, the tool motion section repeats in similar fashion for each tool.

It is good, standard practice to program each tool following a similar standard format: select the tool (line N020), move the tool to the first x-y position while turning on the spindle (line N030), and then approach the workpiece while simultaneously adjusting for **tool length offset (TLO)** (N040).

TLO is an adjustment for the difference in length between tools in the automatic tool changer, a length established and programmed into the controller's **offset tables** (in internal memory) during setup. Note that the TLO is established with G43 and an H-number in the first Z-axis move after changing tools. (H refers to the location where the tool length is stored in the offset tables. Usually the H-number is related to the tool number.)

Rapid motion should be halted a short distance from the workpiece in order to handle variation between workpieces, avoid damage to the tool and workpiece, and sometimes to provide operator reaction time on a first-run part, in case of setup or program error. In the case of milling, the point at which rapid motion is changed to controlled plunging motion is called the **rapid plane**, or sometimes the **feed plane**. In the example program in Figure 26.12, block N040 commands rapid motion all the way to the cutting depth of Z20.0 because the tool

```
%
o12345 (Milled Workpiece Program)
(----Setup Instructions-------------------------------)
(UNITS:  Millimeters                                   )
(WORKPIECE MAT'L: Aluminum 6061-T6                      )
(Workpiece: 200 x 140 x 30mm plate                      )
(PRZ Location G54:                                      )      Program Header
(    XY 0,0 - Lower Left of Fixture                     )
(    TOP PF PART Z=30.0                                 )
(Tool List:                                             )
(    T04 = 20mm 4 FLUTE FLAT ENDMILL - CENTER CUTTING   )
(-------------------------------------------------------)
N010 G00 G90 G17 G21 G49 G40 G80 G54

(MILL PERIPHERY CONTOUR)
N020 M06 T04 (20MM 4-FLUTE END MILL)
N030 S1900 M03 G00 X212. Y0. (P0)
N040 G43 H04 Z20. (RAPID DOWN - TOOL LENGTH OFFSET)
N050 M08 (COOLANT ON)
N060 G01 G41 D04 X190 Y10. F600. (PI - ENTER CUTTER COMP)
N070 X20. Y10. (P2)
N080 G02 X10. Y20. R10. (P3)
N090 G01 Y130. (P4)                                            Tool Motion
N100 G01 X80.(P5)
N110 G01 X80. Y100. (P6)
N120 G03 X110. Y100. R15. (P7)
N130 G01 Y130. (P8)
N140 G01 X190. (P9)
N150 G01 Y10. (P0)
N160 G01 G40 X200. Y-12. (P10 - EXIT CUTTER COMP)
N170 M01

(MILL SLOT)
N180 G00 Z35. (LIFT TOOL TO RAPID PLANE)
N190 G00 X160. Y100. (P11)
N200 G01 Z20. F200.
N210 G01 X130. Y40. Z10. F300. (P12)
N220 G00 Z35.

N230 M09 (COOLANT OFF)
N240 M05 (SPINDLE OFF)
N250 G91 G28 G49 Z0 (RAISE TOOL MAXIMUM Z)                     End-of-Program Motion
N260 G28 Y0 (BRING WORK TOWARD OPERATOR)
N270 G90
N280 M30
%
```

FIGURE 26.12 Example of programming of a part (shown in Figure 26.10) to be machined in a vertical-spindle CNC machine.

TABLE 26.3   **Coordinate Sheet for the Part Shown in Figure 26.10 and Programmed in Figure 26.12.**

	P0	P1	P2	P3	P4	P5	P6	P7	P8	P9	P10	P11	P12
x	212	190	20	10	10	80	80	130	130	190	200	160	130
y	0	10	10	20	130	130	100	100	130	130	−12	100	40
z	20	20	20	20	20	20	20	20	20	20	20	10	10

approaches off the edge of the part and has no need to slow down. However, when positioned 5 mm above point P11 at block N200, the programmer has appropriately selected slow plunging toward and into the work.

CNC codes are typically **modal**, which means that a code sets a mode that will remain valid until another code comes along with the power to make that code invalid. For example, in line N030, the G00 code sets the motion mode to rapid traverse. In the next line, the motion to the Z20.0 coordinate is a rapid move, since G00 is still the motion mode in operation. The other codes in that line merely adjust the TLO as the tool moves downward.

Another example of modal codes is line N070. The motion mode is still G01, first commanded in line N060, and the Y coordinate remains the same as the previous line; only the X coordinate changes. Note that in line N070, it was not necessary to repeat the X10.0 coordinate from the previous line, but it is still valid to include it.

In the case of milling, drilling, and many other tools, the tool is programmed from the center tip of the tool. If adjustment needs to be made for side-cutting toolpaths, the programmer can compute the toolpath centerline that is the distance of the tool radius from the desired part edge, or the programmer can take advantage of the **cutter radius compensation** function. Cutter radius compensation allows the programmer to program using the part edge coordinates, allowing the controller or CAM software to compute the offset centerline of the toolpath. An additional advantage is that the offset gives the operator the ability to shift the offset amount slightly to further compensate for tool deflection or tool wear.

The example program in Figure 26.12 provides an example of the application of cutter radius compensation applied to the cutting of the outer profile in lines N060 through N160. The G41 command begins compensation to the side of the toolpath, and the G40 command ends compensation. Note that preliminary position points must be programmed at least the length of the tool radius away from the compensation edge in order to give the tool space to move in and out of its compensation path.

Turning operations have a similar adjustment called **tool nose radius (TNR) compensation**. TNR is not necessary for end-of-part facing and straight longitudinal cuts, but adjusts for error in taper cuts and circular arcs as the actual cutter contact point moves away from the theoretical tool tip.

It is common practice to use G28 or similar code at the end of the program to move the tool up out of the way and the part toward the operator for unloading. Used by itself, G28 moves all three axes to the machine home position—their most positive positions. This is good for the y and z axes, as it lifts the tool and brings the work forward, but the x axis machine reference point carries the part away from the operator somewhat out of reach. Coupling G28 with incremental mode and an axis command specifies that only the designated axis should be moved. In the case of line N250, only the z-axis is specified, so the tool is lifted fully. In line N260, only the y-axis is specified, so it is as if the tool moves back fully as far as it can which means the work is brought straight forward to the operator as desired.

After the program has been written, it is verified, which means the steps in making the part are graphically simulated. The **verification** step can use the computer monitor, which simulates the part being made by tracing all the toolwork paths as they would occur on the machine tool. Sometimes a sample part is machined in plastic or machinable wax for checking the part specifications from the drawing against the real part.

Various techniques are employed at the machine tool to ensure that the first-piece run is safe and successful. Most modern machine controllers have a graphical mode that allows the operator to catch visible motion errors as well as code errors that prevent the program from fully running. Controllers are also equipped with modes and controls to assist the operator in making a **dry run** in the air, either without the part in place or above the part as a means of observing if the programmed motion will be as desired. Feeds, speeds, and rapid motion can be sped up or slowed down according to dry run and **first-piece run** needs. The operator can also pause the program as needed to more carefully check progress on the machined part before continuing.

Automated Part Programming

Because of the complexity of some parts and the need to offset the tool—sometimes from three dimensions—it is convenient to let a computer do the work of generating toolpath geometry. Early widespread application of automating geometric toolpath computation was a system called **APT** (Automatically Programmed Tool), which allowed the programmer to program part and tool geometry directly, though still using lines of code. This was particularly useful for automating the computation of offset toolpath endpoints in the days before the MCU could compute them with cutter radius compensation.

CAD/CAM (Computer-Aided Manufacturing) programs allow the programmer to specify toolpath needs directly from the CAD geometry already entered by the designer. Graphical interaction and a menu-driven interface assist the process designer in inputting toolpath specifications. The CAM system turns the specifications into CNC toolpath code. The advantages are elimination of wasted time re-computing and entering geometry that the CAD system already has defined, and lack of error between the CAD geometry specification and programmer interpretation of that geometry. Even more importantly, the CAM system can quickly recognize and generate the complex surface geometries and intricate curves, as well as very long programs, all which would be impractical for humans to compute and generate.

An additional advantage is the capability of the CAM software to generate different types of code for machines with different programming conventions. The CAM software saves the processing instructions and geometry provided by the CAM operator/programmer in a "neutral" format. When ready for a program, the operator selects for the program to be generated

CAD/CAM

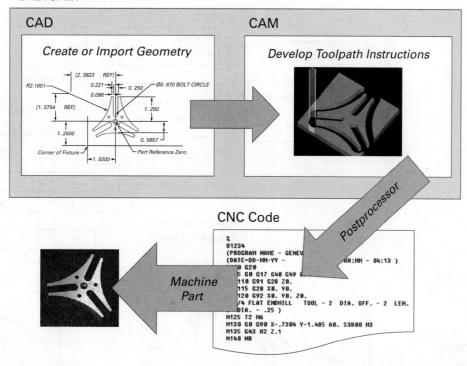

FIGURE 26.13 Part geometry and toolpath information are specified and stored at the CAM interface. The postprocessor turns the specified toolpath data into code for the selected CNC machine tool.

by the **post processor** for a particular machine tool. (This process is often called "**posting**" the code.) The post processor translates the CAM instructions into the code format most appropriate for the designated machine. This process is depicted in **Figure 26.13**. If the CAM program needs to be moved to another machine tool, the saved CAM data can be post-processed again for the different machine.

Often simple **2-D sheet cutting** programs, such as for plasma cutting, laser cutting, or waterjet cutting, can generate the necessary G-code nearly automatically from the 2-D CAD file of the cut part outline (generally DXF or DWG file formats). Operator intervention is needed to specify the cut entry position and approach specifications.

Conversational programming is the name given to programming through a series of menus, generally at the machine tool operation panel, which lead the operator to input the desired tool motion information through a series of questions (thus, the "conversation"). Programs are built by specifying location, size, and depth of standard features: holes to be drilled, slots to be milled, rectangular or circular pockets to be milled, rough or finish diameters to be turned, etc. The conversational programming system draws from information input into its user interface to build the G-code program without the user having to know or specify the code. It is particularly useful for simpler programs to be entered by the operator at the controller.

26.4 Interpolation and Adaptive Control

The toolpath interpolations developed within the program—linear interpolation, circular interpolation, or helical—must be converted by the MCU into coordinated servocontrolled motion for the multiple axes of the CNC machine tool. This discussion, highlighted by **Figure 26.14**, is limited to two dimensions, even though today's machines can involve three, four, or five axes of control.

How Interpolation Works

The commanded linear or circular interpolation path commands go to the positional x and y servomechanisms of the machine, which include motors and transmission devices driving the saddle and the table of the machine. Referring to the example in Figure 26.14, the command for linear interpolation from the starting point S, where $x_s = 20$ and $y_s = 20$, to the end point E, with $x_e = 60$ and $y_e = 50$, are dimensioned in millimeters or inches. Assume the time of the command generation cycle is $\Delta t = 0.005$ s and the commanded feed rate (velocity of motion) $f_n = 5$ mm/s. The total motions are $L_s = 40$ mm and x

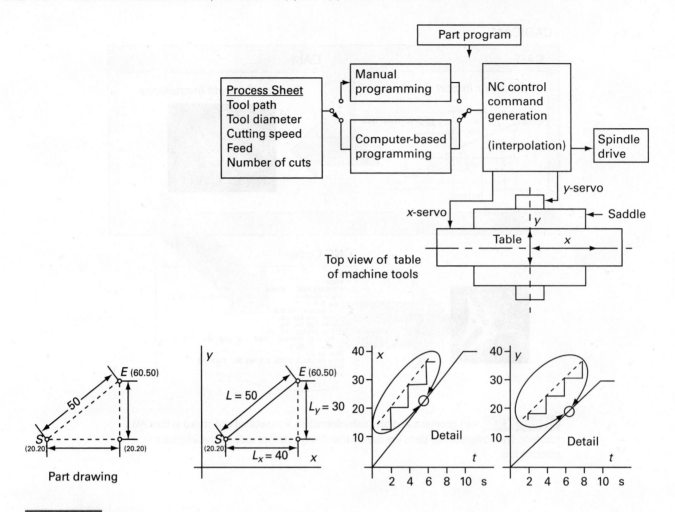

FIGURE 26.14 The 50-in. cut requires linear interpolation to move the cutter from point S to point E. The process sheet has the information on the part needed to develop the part program, either manually or computer-aided.

and L_y = 30 mm for y; the length of the motion between points S and E is L = 50 mm.

The cutter moves in small x and y steps to produce the 50 mm cut.

The control as well as the positional servos work digitally with numbers that comprise a **basic-length unit (BLU)**. Assume a basic length unit is BLU = 0.001 mm.

The total time for this motion is $T = L/f$ = 50 mm/(5 mm/s) = 10 s; the component velocities are v_x = 40 mm/10 s = 4 mm/s, and v_y = 30 mm/10 s = 3 mm/s. Correspondingly, the commands will be issued every 5 ms in increments of $\Delta x \Delta y$ represented by the following numbers:

$$\Delta x = 4000 \text{ BLU}/\text{s} \times 0.005 \text{ s} = 20 \text{ (BLU)}$$
$$\Delta y = 3000 \text{ BLU}/\text{s} \times 0.005 \text{ s} = 15 \text{ (BLU)}$$

The total numbers issued will be counted up to 40,000 in x and 30,000 in y.

The detail of motion commands for x and y versus time are shown in Figure 26.14. The increments Δx = 20 and Δy = 15 are issued discretely, as "chunks" of the travel commands per every Δt = 5 ms. In reality, however, the servos cannot execute these

small steps, so the actual motions are rather smooth along both axes. In any case, the geometric relationship of diagram would hold even if the incremental commands could be executed as quickly as they were issued.

The commands generated by the interpolator are executed by the positional servomechanisms, shown in **Figure 26.15**. The increments Δx from the interpolator are being accumulated in the counter x_{com}. Depending on the desired direction of the table, Δx numbers may be positive or negative, so x_{com} is correspondingly increasing or decreasing. Another counter, x_{act}, accumulates the feedback pulses Δx_{act}. The difference of the two registers ($x_{com} - x_{act}$), as obtained in the positional discriminator PD, is the error of position e_x, and it is interpreted by the servo as a command to move the table. These functions may be executed as a software routine in the microprocessor. The error e_x is then output through a digital/analog converter and becomes the control signal for the velocity servo.

In the velocity discriminator, VD, this command signal is compared with the **tachogenerator** feedback, and the velocity error, e_y, is fed into the power amplifier with gain, G_o, and further into the servomotor. The servomotor drives the table of the machine through a leadscrew and nut transmission of the

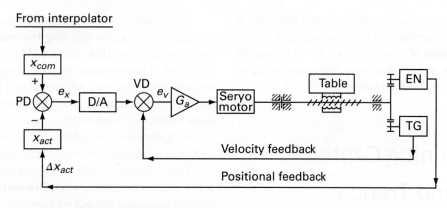

FIGURE 26.15 The positional servomechanism requires feedback from both the encoder (detects position) and the tachogenerator (defects velocity) to drive the servomotor that moves the table to the desired location.

recirculating ball type, shown in Figure 26.7. The nut consists of two halves preloaded axially against each other so as to eliminate any backlash. High-precision gears transmit the rotation of the leadscrew to a tachogenerator, TG, for velocity feedback and also to an encoder, EN, for positional feedback (see Figure 26.15). The encoder emits one pulse for every BLU of the table motion. Correspondingly, the positional feedback has the form of a train of pulses; their frequency corresponds to the velocity of the motion, and their total number to the total distance traveled.

Adaptive Control

Adaptive control (A/C), discussed in Chapter 34, can deal with the problems caused by variations in the size of the uncut workpiece, which may be a casting or a forging. The CNC program was prepared with assumptions about the amount of material to be removed. The program may have to adapt to variations in the size—that is, the depth of cut. Also, CNC programs are prepared under certain assumptions regarding the tool-wear rate for the cutting speeds that were selected, but the actual wear rate may be different from that which was assumed. Tool wear can also change the depth of cut. Consequently, the cutting speed may have to be modified to reduce the tool-wear rate. Other phenomena that are difficult to predict, such as chatter vibrations, may occur, and the cutting conditions need to be changed to stabilize the cut.

In all these instances, it is possible to use sensors to measure the significant parameters of the actual cutting process and to change the CNC program—that is, change the feed rate, f_r, the spindle rpm or the depth of cut so as to improve the cutting process. This is shown schematically in **Figure 26.16**, which depicts as an example a pocket end-milling operation (see Figure 26.5). The process parameters to be measured may be the spindle torque on the spindle, the cutting force, vibrations, and tool wear (by measuring the change in size of the part). This information is fed back into the CNC controller, which contains in its software the corresponding adaptive control strategies in the form of algorithms to modify cutting conditions. A simple A/C system can be conceived wherein the actual cutting force, F_{act}, is measured and compared with the nominal force, F_{nom}, that the cutter can safely maintain. Their relative difference establishes the force error, e_f. For $F_{act} < F_{nom}$, e_f is positive, and for $F_{act} > F_{nom}$, e_f is negative. Changing the feed rate of the table can eliminate the force error. So if $F_{act} > F_{nom}$, it is necessary to reduce the feed rate, f_r. (See Chapter 21 for relation between feed rate and cutting force.) The CNC controller will correspondingly start changing the feed rate of the commanded travel x_{com} until $F_{act} = F_{nom}$. In this way, the cutter will move rapidly when there is little

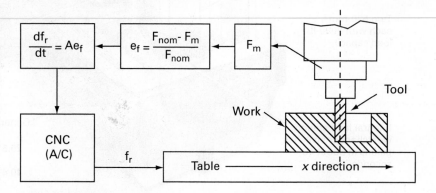

FIGURE 26.16 Adaptive control can be used to produce a constant milling force. The measured cutting force F_m is processed in the CNC (A/C) that controls the feed rate f_r as to maintain the force at a desired level.

material to cut and slowly when the depth of cut is large. In the design of CNC control systems with adaptive control, it is necessary to solve a number of problems that are mainly due to changes in the depth of cut. Thus, very few machines have adaptive control programs in their software.

26.5 Machining Center Features and Trends

Computer numerical control is used on a wide variety of machine tools. These range from single-spindle drilling machines, which often have only two-axis control and can be obtained for about $10,000, to machining centers, such as shown in **Figure 26.17**. The machining center can do drilling, boring, milling, tapping, and so on, with four- or five-axis control. It can automatically select and change 40 to 180 preset tools. The table can move left/right and in/out, and the spindle can move up/down and in/out, with positioning accuracy in the range 0.00012 in. with repeatability to ±0.00004 in. over 40 in. of travel. Beyond accuracy and repeatability and the number of controlled axes, CNC machines are also categorized by spindle speed, horsepower, and the size of the workpiece.

The accuracy and repeatability (precision) result from a combination of factors including the resolution of the control instrumentation and the accuracy of the hardware. The control resolution is the minimum length distinguishable by the control unit. It is called the basic-length unit (BLU) and is mainly a factor of the axis transducer and the quality of the leadscrew or linear translator. There are many sources of error in a machine, including wear in the machine sliding elements, machine tool assembly errors, spindle runout (wear), and leadscrew backlash.

Tool deflection due to the cutting forces produces dimensional error and chatter marks. **Thermal error**, caused by the thermal expansion of machine elements, is not uniform and is normally the greatest source of machine error. Methods used to remove heat from a machine include cutting fluids, locating drive motors away from the center of a machine, reducing friction from the ways and bearings, and spraying cooling control element of the machine.

Most modern machining centers have **automatic toolchangers** and automatic work-transfer capability using **automatic pallet changers**, so that workpieces can be loaded and unloaded while machining is in process. Such a machine can cost more than $200,000. Between these extremes are numerous machine tools that do less varied work than the highly sophisticated machining centers but that combine high output and minimum setup time with remarkable flexibility (large number of tool motions provided).

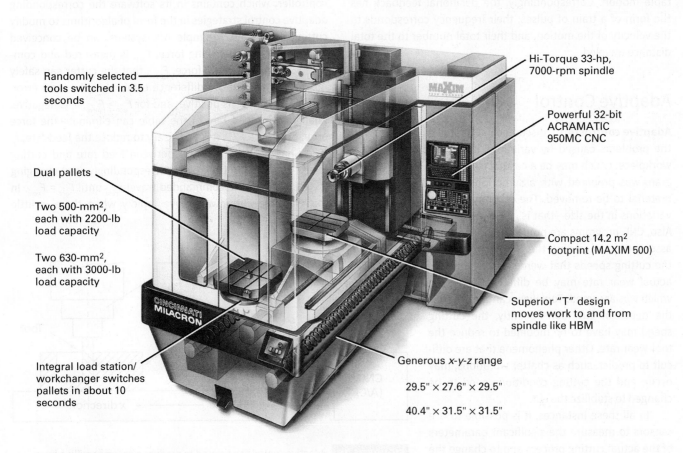

Randomly selected tools switched in 3.5 seconds

Dual pallets

Two 500-mm², each with 2200-lb load capacity

Two 630-mm², each with 3000-lb load capacity

Integral load station/ workchanger switches pallets in about 10 seconds

Hi-Torque 33-hp, 7000-rpm spindle

Powerful 32-bit ACRAMATIC 950MC CNC

Compact 14.2 m² footprint (MAXIM 500)

Superior "T" design moves work to and from spindle like HBM

Generous *x-y-z* range

29.5" × 27.6" × 29.5"

40.4" × 31.5" × 31.5"

FIGURE 26.17 Modern machining centers will typically have horizontal spindles up to 15,000 rpm, dual pallets, and cutting-tool magazines holding 40 to 100 tools. (Courtesy J T. Black)

CNC Turning Centers

The modern lathe is called a **turning center** and has CNC control and tools mounted in turrets on slanted beds. The tailstock has been replaced by a live, powered spindle and chuck. On some lathes, the concept of automatic tool changing has been implemented. The tools are held on a rotating tool magazine, and a gantry-type tool changer is used to change the tools. Each magazine holds one type of cutting tool. This is an example of the trend of providing greater versatility along with high productivity in lathes. The versatility is being further increased by combining both rotary-work and rotary-tool operations—turning and milling in a single machine. Live, powered, or driven tools, called **live tooling**, replace regular tools in the turrets and perform milling and drilling operations when the spindle is stopped. **Figure 26.18** demonstrates ways in which live tooling enhances the versatility and economy of turning center operations and equipment.

Other CNC Machines

Numerical control has been applied to a wide variety of other production processes. CNC turret punches with *x-y* control on the table, CNC wire EDM machines, laser and water-jet abrasive machining, flame cutters, CNC press brakes, tubing and wire benders, and many other machines are readily available. Some new trends are being observed in the development of machining centers, such as smaller, compact machining centers with higher spindle speeds. Machines with four- and five-axis capability are readily available. Modern machining centers have contributed significantly to improved productivity in many companies. They have eliminated the time lost in moving workpieces from machine to machine and the time needed for workpiece loading and unloading for separate operations. In addition, they have minimized the time lost in changing tools, carrying out gaging operations, and aligning workpieces on the machine.

The latest generation of machining centers is aimed at further improving utilization by reducing the time when machines are stopped, either during pauses in a shift or between shifts. Delays are caused by tool breakage, unforeseen tool wear, a limited number of tools, or an inadequate number of available workpieces. Machines are fitted with tool breakage monitors, tool-wear compensating devices, and means for increasing the number of tools and workpieces available.

Touch probes on CNC machines can greatly improve the process capability of the machine tool. There is a big difference between the claimed program resolution for a CNC machine and the accuracy and precision (the process capability) of the actual parts. As shown in **Figure 26.19**, true positioning accuracy and precision are affected by machine alignment, machine and fixture setup, variations in the workholding device, raw material variations, workpiece location in the fixture variations, and cutting-tool tolerances. Thus, the finished workpiece may be unacceptable even though the machine is more than capable of producing the part to the design specifications. The **part program** has no assurance that the part is properly located

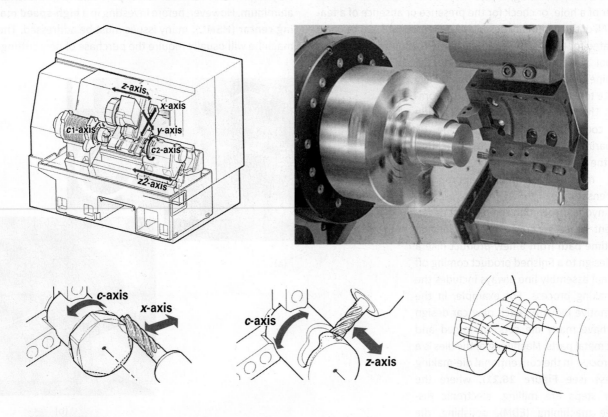

FIGURE 26.18 This CNC turning center has a multiple-axis capability with two spindles and a 12-tool turret with *x*, *y*, and *z* control as well as axis control of the spindles. (Courtesy J T. Black)

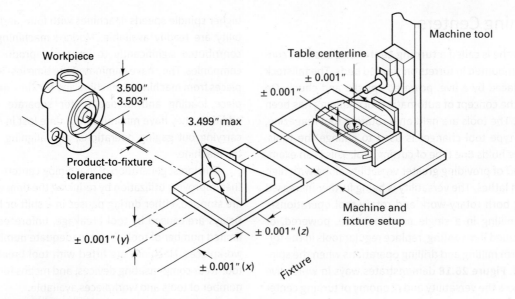

Workpiece

3.500"
3.503"

3.499" max

Product-to-fixture tolerance

± 0.001" (y)

± 0.001" (z)

± 0.001" (x)

Fixture

Machine tool

Table centerline

± 0.001"

± 0.001"

Machine and fixture setup

FIGURE 26.19 Process capability in CNC machines is affected by many factors.

in the fixture or that the fixture is properly located on the table of the machine. However, a probe—carried in the tool storage magazine and mounted when needed in the spindle like a cutting tool—can establish the location of the surface features relative to each other and to the spindle axis within 0.0005 in. (**Figure 26.20**). The machine controller, using the probe data, will then shift the program reference data accordingly.

The probe can be used to determine the amount of material on a rough casting, locate a corner of a part, define the center of a hole, or check for the presence or absence of a feature. All of the variability described in Figure 26.19 can be compensated for except for variations in the cutting-tool geometry or tool wear. A probe mounted on the machine tool can be used to automatically update tool-offset data in the control computer. Thus, the machine tool can function as a coordinate measuring machine. By comparing the actual touched location with the programmed location, the measuring routine determines appropriate compensation.

Anyone involved in the product development process knows that the longest lead-time path from a new-product (like a car) design to a finished product coming off the final assembly line always includes the die-making process. For example, in the automotive industry, each new car design may have many die sets for forged and sheet metal parts. Machining of the dies is a key process in the conventional die-making process (see **Figure 26.21**), where the major steps are milling, electronic discharge machining (EDM), polishing, die assembly, tryout, and modification.

Ultra-high-speed machining centers (UHSMCs) can shorten the die-making

process lead time. The new machining centers have highly accurate, high-speed spindles capable of 30,000 to 50,000 rpm, cutting feed rates of 60 m/min, and cutting feed accelerations of 9.8 m/s². The machine requires a spindle with high stiffness utilizing ceramic ball bearings, with a constant-pressure preload mechanism and jet lubrication for superior performance (see **Figure 26.22**).

Robust machining centers of this sort are capable of very high metal removal rates, particularly in materials like aluminum. However, before investing in a **high-speed machining center (HSMC),** many issues must be addressed. The new machine will usually require the purchase of new cutting tools

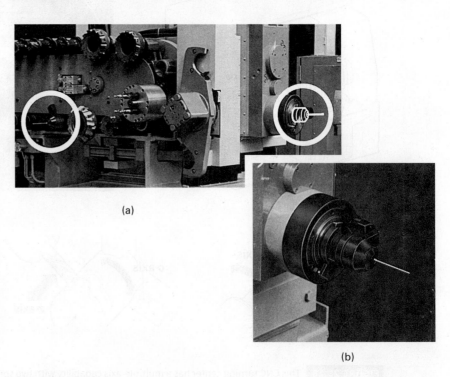

(a)

(b)

FIGURE 26.20 (a) Probe carried in the tool changer can be mounted in the spindle (b) for checking the location of part features accurately. (Courtesy J T. Black)

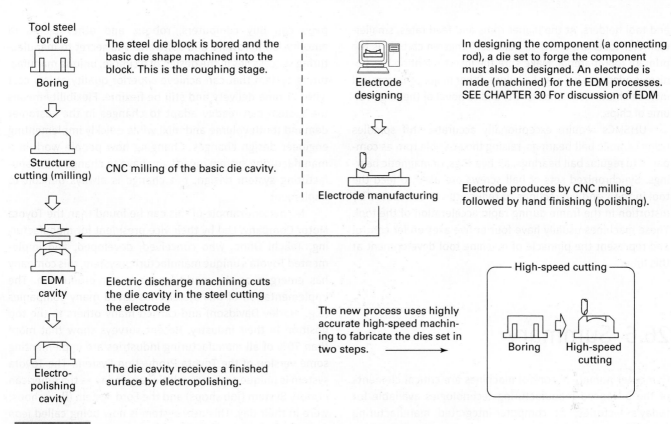

Tool steel for die

Boring

The steel die block is bored and the basic die shape machined into the block. This is the roughing stage.

Structure cutting (milling)

CNC milling of the basic die cavity.

EDM cavity

Electric discharge machining cuts the die cavity in the steel cutting the electrode

Electro-polishing cavity

The die cavity receives a finished surface by electropolishing.

Electrode designing

In designing the component (a connecting rod), a die set to forge the component must also be designed. An electrode is made (machined) for the EDM processes. SEE CHAPTER 30 For discussion of EDM

Electrode manufacturing

Electrode produces by CNC milling followed by hand finishing (polishing).

The new process uses highly accurate high-speed machining to fabricate the dies set in two steps.

High-speed cutting

Boring   High-speed cutting

FIGURE 26.21 The sequence of operations to manufacture dies for forging processes is shown. Using ultra-high-speed machining centers reduces the sequence to two steps. (See Figure 26.22 for UHSMC details.)

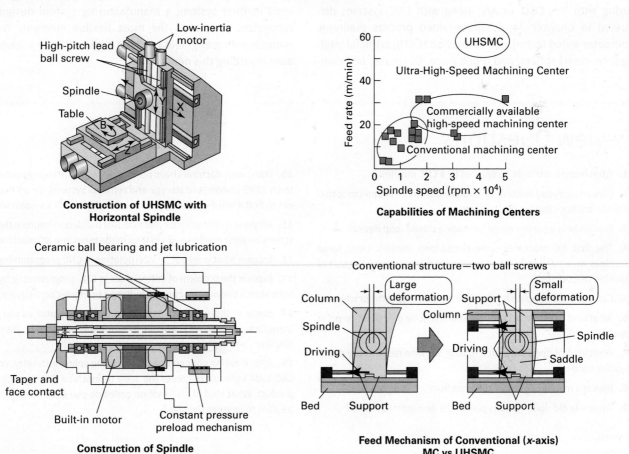

High-pitch lead ball screw

Low-inertia motor

Spindle

Table

Construction of UHSMC with Horizontal Spindle

Feed rate (m/min)

Spindle speed (rpm × 10^4)

UHSMC

Ultra-High-Speed Machining Center

Commercially available high-speed machining center

Conventional machining center

Capabilities of Machining Centers

Ceramic ball bearing and jet lubrication

Taper and face contact

Built-in motor

Constant pressure preload mechanism

Construction of Spindle

Conventional structure—two ball screws

Large deformation

Small deformation

Column

Spindle

Driving

Bed   Support

Support

Column

Driving

Bed   Support

Spindle

Saddle

Feed Mechanism of Conventional (x-axis) MC vs UHSMC

FIGURE 26.22 Ultra-high-speed machining centers (UHSMCs) employ ceramic ball bearings in the spindles and synchronized ball screws on the x-axis to reduce distortion (due to inertia) in the moving components.

and tool holders. At the higher rpms and feed rates, smaller-diameter tools are easier to balance. Tungsten carbide is the primary tool material. Make sure the insert retention mechanism is adequate in case of a failure. Other major problems will include chatter (see Chapter 21) and removal of the large volume of chips.

UHSMCs require exceptionally accurate, stiff spindles using ceramic ball bearings, raising the possible rpm as compared to regular ball bearings, air bearings, or magnetic bearings. Synchronized sets of ball screws are used to feed the tool in the x and y directions to reduce the errors caused by distortion in the frame during rapid acceleration of the tool. These machines usually have four or five axes under control and represent the pinnacle of machine tool development at this time.

26.6 Summary

Computer numerical control machines are critical elements of the advanced manufacturing technologies available for today's factories, as computer-integrated manufacturing (CIM) becomes closer to reality. Computer technology abounds: computer-aided design; computer-aided manufacturing with NC, CNC, or A/C along with DNC systems discussed in Chapter 34; computer-aided process planning; computer-aided testing and inspection (CATI); artificial intelligence; smart robots; and much more. Of course, any company can buy computers, robots, and other pieces of automation hardware and software. The secret to manufacturing success lies in the design of a simple, unique manufacturing system that can achieve superior quality at low cost with on-time delivery and still be flexible. Flexibility means the system can readily adapt to changes in the customer demand (both volume and mix) while quickly implementing engineer design changes. Changing how people work in a manufacturing system means you have to change the manufacturing system design, but change is always difficult to implement.

No better example of this can be found than the Toyota Motor Company. Led by their vice president for manufacturing, Taiichi Ohno, who conceived, developed, and implemented Toyota's unique manufacturing system, this company has emerged as the world leader in car production. The implementation of this system has saved many companies (e.g., Harley Davidson) and carried many others to the top position in their industry. Recent surveys show that more than 70% of all manufacturing industries are implementing some version of the Toyota Production System. The Toyota system is unique and as revolutionary today as the American Armory System (job shops) and the Ford system (flow shops) were in their day. This new system is now being called *lean production* (to contrast it to mass production). Toyota does not use the term *CIM* because the computer is only a tool used in their system, a manufacturing system design that recognizes people as the most flexible element. Today's manufacturing engineer is a lean engineer, fully knowledgeable regarding this new system.

Review Questions

1. What human attribute is replaced by a CNC machine?

2. Give an everyday example of a household device or appliance that exhibits feedback in its control system.

3. Explain how a toaster could be made a closed-loop device.

4. The first NC machines were closed-loop control. Later some machines were open loop. What change did this require on the part of machine tool builders?

5. Can a continuous-path NC machine be open loop? Why or why not?

6. What kind of CNC machine operations require only point-to-point control?

7. What kind of operations on a milling machine require only point-to-point control?

8. How is a machining center different from a milling machine?

9. What role did John Parsons play in the development of NC?

10. Many manufacturers have purchased large machining centers tied to an ASRS (automated storage and retrieval system). Go on the Internet to find a video of an ASRS, and explain how such a system works.

11. Why was it necessary for machine tool builders to improve the lead-screws on their machines when they made them into NC machines?

12. Explain what is meant by interpolation in CNC programming.

13. Explain the problem of cutter offset in CNC programming by making a sketch showing an end mill cutting a perimeter on a square plate.

14. Some of the functions performed by the operator in piece-part manufacturing are very difficult to automate completely. Name the functions and explain why.

15. After creating your company logo using spline geometry, you use CAD/CAM software to create the code to machine the logo into your product. What kind of CNC motion codes do you expect will primarily be used to engrave the logo?

16. What are the three basic closed-loop feedback A(4) schemes used on CNC machines?

17. What benefits would you expect to see when selecting an arc-filtering (or arc-fitting) approach in your CAM software?

18. Which method for positional feedback do you think is the most accurate? Why?

19. Draw the path that a CNC tool will most likely take between these two positions using G00 Rapid motion: (P1) X0 Y0 and (P2) X1.0 Y2.0 How would a G01 linear interpolation move between these points be different?

20. What is the difference between the work reference point and the machine zero point?

21. Name and distinguish at least three types of "offsets" or compensation discussed in this chapter.

22. What is an encoder?

23. What does a post processor do?

24. What is a peck-drilling subroutine for a CNC machine?

25. What is pocket milling, and what kind of milling cutter is usually used to perform it?

26. How are touch probes used in CNC machines to improve process capability?

27. What are G words used for in CNC?

28. What major structural changes are introduced into a UHSMCs?

29. What process steps are eliminated by the UHSMC in die production?

30. Draw a large circle and imagine you are (the tool) moving from point A to B on a semicircular segment. What is different about the increments and velocities here compared to linear interpolation?

Problems

1. What are the x and y dimensions for the center position of holes 1, 2, and 3 in the part shown in Figure 26.2?

2. A program to make the part depicted in **Figure 26.A** has been created, presented in **Figure 26.B**, but some information still needs to be filled in. Compute the tool locations, depths, and other missing information to present a completed program. (Hint: You may have to look up geometry for the center drill and standard 0.500 in. twist drill to know the required depth to drill.)

3. Suppose that the plate shown in **Figure 26.C** was to be profile milled around the periphery with a 1-in.-diameter end milling cutter. The dashed line is the cutter path. The programmer must calculate an offset path to allow for cutter diameter. Because the programmed points are followed by the cutter centerline and the profile is made at the tool's periphery, the programmer called for a ½-in. cutter offset. Working with computer assistance, the programmer would describe the part profile to be machined and specify the cutter. The computer would generate the cutter path. Complete the following table to specify the cutter path, starting with the origin at the zero reference point. Move the tool around the plate counterclockwise.

Programmed Point Locations		
PT	x	y
1		
2		
3		
4		
5		
1		

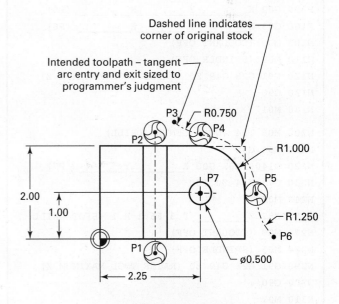

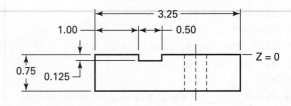

FIGURE 26.A

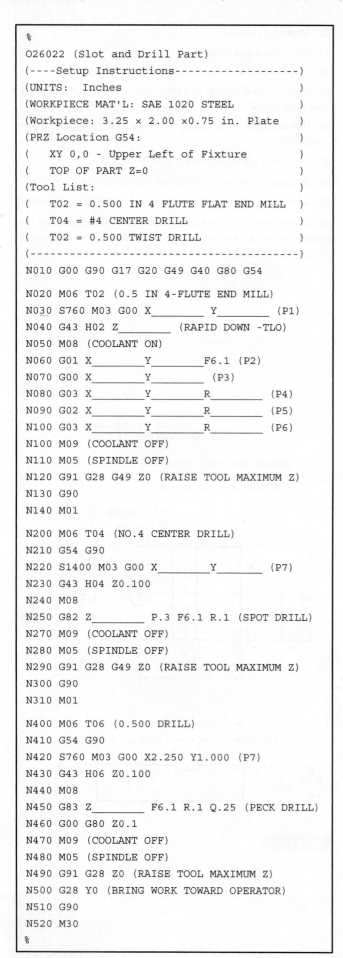

```
%
O26022 (Slot and Drill Part)
(----Setup Instructions------------------)
(UNITS:  Inches                           )
(WORKPIECE MAT'L: SAE 1020 STEEL          )
(Workpiece: 3.25 × 2.00 ×0.75 in. Plate   )
(PRZ Location G54:                        )
(    XY 0,0 - Upper Left of Fixture       )
(    TOP OF PART Z=0                       )
(Tool List:                               )
(    T02 = 0.500 IN 4 FLUTE FLAT END MILL )
(    T04 = #4 CENTER DRILL                )
(    T02 = 0.500 TWIST DRILL              )
(------------------------------------------)

N010 G00 G90 G17 G20 G49 G40 G80 G54

N020 M06 T02 (0.5 IN 4-FLUTE END MILL)
N030 S760 M03 G00 X_____ Y_____ (P1)
N040 G43 H02 Z_____ (RAPID DOWN -TLO)
N050 M08 (COOLANT ON)
N060 G01 X_____Y_____F6.1 (P2)
N070 G00 X_____Y_____ (P3)
N080 G03 X_____Y_____R_____ (P4)
N090 G02 X_____Y_____R_____ (P5)
N100 G03 X_____Y_____R_____ (P6)
N100 M09 (COOLANT OFF)
N110 M05 (SPINDLE OFF)
N120 G91 G28 G49 Z0 (RAISE TOOL MAXIMUM Z)
N130 G90
N140 M01

N200 M06 T04 (NO.4 CENTER DRILL)
N210 G54 G90
N220 S1400 M03 G00 X_____ Y_____ (P7)
N230 G43 H04 Z0.100
N240 M08
N250 G82 Z_____ P.3 F6.1 R.1 (SPOT DRILL)
N270 M09 (COOLANT OFF)
N280 M05 (SPINDLE OFF)
N290 G91 G28 G49 Z0 (RAISE TOOL MAXIMUM Z)
N300 G90
N310 M01

N400 M06 T06 (0.500 DRILL)
N410 G54 G90
N420 S760 M03 G00 X2.250 Y1.000 (P7)
N430 G43 H06 Z0.100
N440 M08
N450 G83 Z_____ F6.1 R.1 Q.25 (PECK DRILL)
N460 G00 G80 Z0.1
N470 M09 (COOLANT OFF)
N480 M05 (SPINDLE OFF)
N490 G91 G28 Z0 (RAISE TOOL MAXIMUM Z)
N500 G28 Y0 (BRING WORK TOWARD OPERATOR)
N510 G90
N520 M30
%
```

FIGURE 26.B

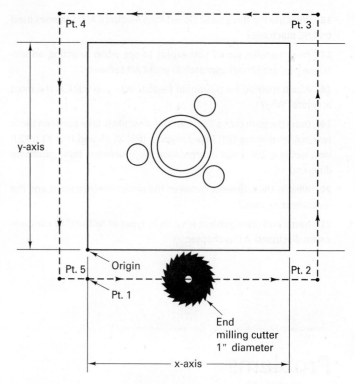

FIGURE 26.C

4. Suppose that surface finish is very important for the profile milling job described in Problem 3. Thus down milling is going to be used. Rewrite the CNC program points to accommodate this requirement. Show the new path on a sketch such as Figure 26.B. (Up versus down milling is discussed in Chapter 24.)

5. Sketch a design for a part that could be made on a turning center with one end mill installed as live tooling (in addition to typical lathe process tooling). Create a process plan depicting the tools and steps necessary to make the part.

CHAPTER 27

Sawing, Broaching, Shaping, and Filing Machining Processes

27.1 Introduction

While milling, drilling, and turning make up the bulk of the machining processes, there are many other chipmaking (metal removal) processes. This chapter will cover sawing, broaching, shaping, planing, and filing.

27.2 Introduction to Sawing

Sawing is a basic machining process in which chips are produced by a succession of small cutting edges, or *teeth*, arranged in a narrow line on a saw "blade." As shown in **Figure 27.1**, each tooth forms a chip progressively as it passes through the workpiece. The chips are contained within the spaces between successive teeth until the teeth pass from the work. Because sections of considerable size can be severed from the workpiece with the removal of only a small amount of the material in the form of chips, sawing is probably the most economical of the basic machining processes with respect to the waste of material and power consumption, and in many cases with respect to labor.

In recent years, vast improvements have been made in saw blades (design and materials) and sawing machines, resulting in improved accuracy and precision of the process. Most sawing is done to sever bar stock and shapes into desired lengths for use in other operations. There are many cases in which sawing is used to produce desired shapes. Frequently, and especially for producing only a few parts, contour sawing may be more economical than any other machining process.

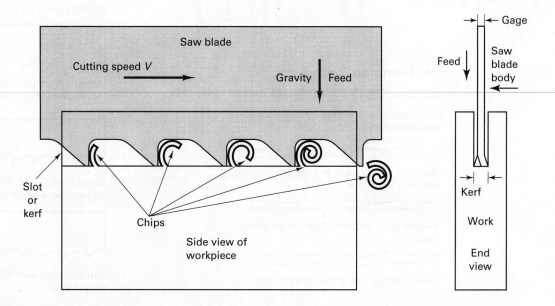

FIGURE 27.1 Progressive formation of chips in sawing.

Saw Blades

Saw blades are made in three basic configurations. The first type, commonly called a **hacksaw** blade, is straight, relatively rigid, and of limited length, with teeth on one edge. The second type, called a **bandsaw**, is sufficiently flexible that a long length can be formed into a continuous band with teeth on one edge. The third form is a rigid disk having teeth on the periphery;

these are called **circular saws** or **cold saws**. **Figure 27.2** gives the standard nomenclature for the widely used saw blades.

All saw blades have certain common and basic features: (1) material, (2) tooth form, (3) tooth spacing, (4) tooth set, and (5) blade thickness or gage. Small hacksaw blades are usually made entirely of tungsten or molybdenum high-speed steel. Blades for power-operated hacksaws are often made with teeth cut from a strip of high-speed steel that has been

MATRIX MODIFIED MIX-TOOTH	M-42 COBALT WELDED-EDGE	M-2 HIGH-SPEED WELDED-EDGE	HARD BACK CARBON	FLEXIBLE BACK CARBON
The best all-purpose welded-edge blade for sawing varying sizes, shapes, and cross sections. Cobalt-tough for cutting wide range of materials. Welded to length and coil stock.	For high-production cutting of solids, superalloys, tool steels, high-temperature alloys. Welded to length and coil stock.	The original and widely used welded-edge band blade for general-purpose sawing. Welded to length and coil stock.	Hardened back provides greater beam strength for more accurate sawing. Welded to length and coil stock.	Recommended for contour saws running over 3000 SFPM. Welded to length and coil stock.

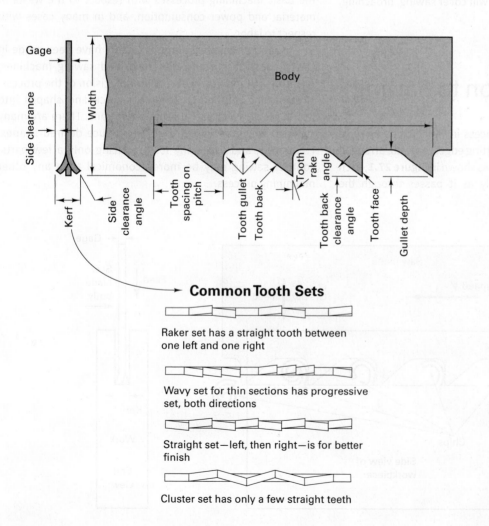

Common Tooth Sets

Raker set has a straight tooth between one left and one right

Wavy set for thin sections has progressive set, both directions

Straight set—left, then right—is for better finish

Cluster set has only a few straight teeth

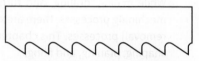

Standard design has zero rake for general purpose sawing

Skip-tooth blade clears chips, cuts non-ferrous, non-metalics

Hook tooth with 10° rake for large sections good for cast iron

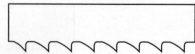

Variable pitch can change by section or individually, improves blade life

Variable pitch with 5° rake is more aggressive, smooth cutting

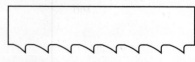

Variable pitch with 10° rake sheds chips better

FIGURE 27.2 Bandsaw blade designs and nomenclature (above). Tooth set patterns (left) and tooth designs (right).

electron-beam-welded to the heavy main portion of the blade, which is made from a tougher and cheaper alloy steel (see Figure 27.2). Bandsaw blades are frequently made with this same type of construction but with the main portion of the blade made of relatively thin, high-tensile-strength alloy steel to provide the required flexibility. Bandsaw blades are also available with tungsten carbide teeth and TiN coatings. The three most common **tooth forms** are regular, skip tooth, and hook. **Tooth spacing**, abbreviated as TPI for teeth per inch, is very important in all sawing because it determines three factors. First, it controls the size of the teeth. From the viewpoint of strength, large teeth are desirable. Second, tooth spacing determines the space (**gullet**) available to contain the chip that is formed. The chip cannot drop from this space until it emerges from the slot cut in the workpiece, called the **kerf**. **Tooth set** refers to the manner in which the teeth are offset from the centerline in order to make the kerf wider than the gage (the thickness of the back) of the blade. This allows the saw to move more freely in the kerf, reducing rubbing, friction, and heating. The kerf–gullet space must be such that there is no crowding of the chip. Chips should not become wedged between the teeth and not drop out of the gullet when the saw emerges from the cut.

Third, tooth spacing determines how many teeth will engage the workpiece at any point during the cut. This is very important in cutting thin material, such as tubing. At least two teeth should be in contact with the work at all times. If the teeth are too coarse, only one tooth rests on the work at a given time, permitting the saw to rock, and the teeth may be stripped from the saw.

Raker-tooth saws are used in cutting most steel and iron. **Straight-set teeth** are used for sawing brass, copper, and plastics. Saws with **wave-set teeth** are used primarily for cutting thin sheets and thin-walled tubing.

The gage or **blade thickness** of nearly all hand hacksaw blades is 0.025 in. Saw blades for power hacksaws vary in thickness from 0.050 to 0.100 in. Hand hacksaw blades come in two standard lengths, 10 and 12 in. All are $\frac{1}{2}$ in. wide. Blades for power hacksaws vary in length from 12 to 24 in. and in width from 1 to 2 in. Wider and thicker blades are desirable for heavy-duty work. As a general rule, in hacksawing the blade should be at least twice as long as the maximum length of cut that is to be made.

Bandsaw blades are available in straight, raker, wave, or combination sets. In order to reduce the noise from high-speed bandsawing, it is becoming increasingly common to use blades that have more than one pitch, size of teeth, and type of set. Blade width is very important in bandsawing because it determines the minimum radius that can be cut. The most common widths are from $\frac{1}{16}$ to $\frac{1}{2}$ in., although wider blades can be obtained. Because wider blades are stronger, select the widest blade possible. However, cutting small radii requires a narrower, more flexible blade. Bandsaw blades come in tooth spacings from 2 to 32 teeth per inch (TPI).

Hand hacksaw blades have 14 to 32 TPI. In order to make it easier to start a cut, some hand hacksaw blades are made with a short section at the forward end having teeth of a special form with negative rake angles. Tooth spacing for power hacksaw blades ranges from 4 to 18 TPI.

Circular Saws

Circular saws for cutting metal are often called cold saws to distinguish them from friction-type disk saws. Friction saws do not make chips but rather heat the metal to the melting temperature at the point of metal removal. Cold saws cut rapidly and produce chips like a milling cutter while producing surfaces that are comparable in smoothness and accuracy with surfaces made by slitting saws in a milling machine or by a cutoff tool in a lathe.

Disk, or circular **saws** necessarily differ somewhat from straight-blade forms. The sizes up to about 18 in. in diameter have an integral-tooth design with teeth cut directly into the disk (**Figure 27.3**). Larger saws use either *segmented* or *inserted* teeth. The teeth are made of high-speed steel or tungsten carbide. The remainder of the disk is made of ordinary, less expensive, and tougher steel. *Segmental* blades are composed of segments mounted around the periphery of the disk, usually fitted with a tongue and groove and fastened by means of screws or rivets. Each segment contains several teeth. If a single tooth is broken, only one segment needs be replaced to restore the saw to an operating condition.

As shown in Figure 27.3, circular saw teeth are usually *beveled*. A common tooth form has every other tooth beveled on both sides; that is, the first tooth is beveled on the left side, the second tooth on both sides, the third tooth on the right side, the fourth tooth on both sides, and so forth. Another method is to bevel the opposite sides of successive teeth. Beveling is done to produce a smoother cut. Precision circular saws made from carbide, which are becoming available, are very thin (0.03 in.) and have high cutting-off accuracy, around ±0.00008 in., with negligible burrs.

Types of Sawing Machines

Metal-sawing machines may be classified as follows:

1. Reciprocating saw
   a. Manual hacksaw.
   b. Power hacksaw (**Figure 27.4**).
   c. Abrasive disc.
2. Bandsaw (Figures 27.5, 27.6, 27.7)
   a. Vertical cutoff.
   b. Horizontal cutoff.
   c. Combination cutoff and contour.
   d. Friction.
3. Circular saw (Figure 27.3)
   a. Cold saw.
   b. Steel friction disk.

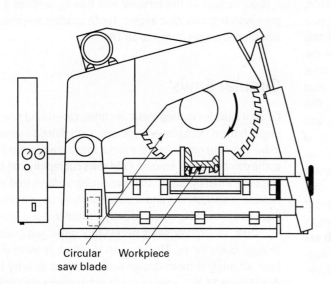

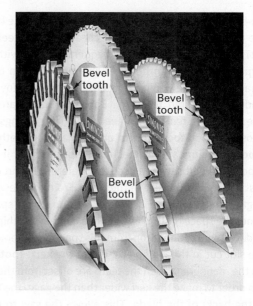

FIGURE 27.3 Circular sawing a structural shape on the left. Examples of circular saws (left to right): an insert tooth, a segmental tooth, and an integral-tooth circular saw blades on the right. (Courtesy J T. Black)

Power Hacksaws

As the name implies, *power hacksaws* are machines that mechanically reciprocate a large hacksaw blade (Figure 27.4). These machines consist of a bed, a workholding frame, a power mechanism for **reciprocating** the saw frame, and some type of feeding mechanism. Because of the inherent inefficiency of cutting in only one stroke direction, they have often been replaced by more efficient, horizontal bandsawing machines.

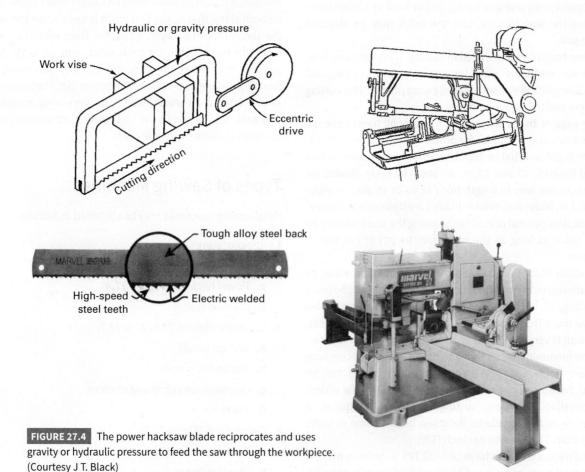

FIGURE 27.4 The power hacksaw blade reciprocates and uses gravity or hydraulic pressure to feed the saw through the workpiece. (Courtesy J T. Black)

Bandsawing Machines

The earliest metalcutting bandsawing machines were direct adaptations from woodcutting bandsaws. Modern machines of this type are much more sophisticated and versatile and have been developed specifically for metal cutting. To a large degree they were made possible by the development of vastly better and more flexible bandsaw blades and simple flash-welding equipment, which can weld the two ends of a strip of bandsaw blade together to form a band of any desired length. Three basic types of bandsawing machines are in common use.

Horizontal metal-cutting bandsawing machines were developed to combine the flexibility of reciprocating power hacksaws and the continuous cutting action of vertical bandsaws. See **Figure 27.5**. These heavy-duty automatic bandsaws feed the saw vertically by a hydraulic mechanism and have automatic stock feed that can be set to feed the stock laterally any desired distance after a cut is completed and automatically clamp it for the next cut. Such machines can be arranged to hold, clamp, and cut several bars of material simultaneously. Computer numerical control (CNC) bandsaws are available with automatic storage and retrieval systems for the bar stock. Smaller and less expensive types have swing-frame construction, with the bandsaw head mounted in a pivot on the rear of the machine. Feed is accomplished by gravity through rotation of the head about the pivot point. Because of their continuous cutting action, horizontal bandsawing machines are very efficient.

Vertical, cutoff bandsawing machines are designed primarily for cutoff work on single stationary workpieces that can be held on a table. On many machines the blade mechanism can be tilted to about 45 degrees, as shown in **Figure 27.6**, to permit cutting at an angle. They usually have automatic power feed of the blade into the work, automatic stops, and provision for supplying coolant.

Combination cutoff and contour bandsawing machines (**Figure 27.7**) can be used not only for cutoff work but also for contour sawing. They are widely used for cutting irregular shapes in connection with making dies and the production of small numbers of parts and are often equipped with rotary tables. Additional features on these machines include a table that pivots so that it can be tilted to any angle up to 45 degrees. Usually, these machines have a small flash butt welder on the vertical column, so that a straight length of bandsaw blade can be welded quickly into a continuous band. A small grinding wheel is located beneath the welder so that the flash can be ground from the weld to provide a smooth joint that will pass through the saw guides. This welding and grinding unit makes it possible to cut internal openings in a part by first drilling a hole, inserting one end of the saw blade through the hole, and then butt welding the two ends together. When the cut is finished, the band is again cut apart and removed from the opening. The cutting speed of the saw blade can be varied continuously over a wide range to provide correct operating conditions for any material. A method of power feeding the work is provided, sometimes gravity-actuated.

Contour-sawing machines are made in a wide range of sizes, the principal size dimension being the throat depth. Sizes from 12 to 72 in. are available. Cutting speeds on most machines range from about 50 to 2000 ft/min. Modern horizontal bandsaws are accurate to ±0.002 in. per vertical inch of cut but have feeding accuracy of only ±0.005 in., subject to the size of the stock and the feed rate. Repeatability from one feed to the next may be ±0.010 to ±0.020 in.

CNC-controlled sawing centers with microprocessor controls have opened up new automation aspects for sawing. Such control systems can improve accuracy to within ±0.005 in. over entire cuts by controlling saw speed, blade feed pressure, and feed rate.

Special bandsawing machines are available with very high speed ranges, up to 14,000 ft/min. These are known as **friction bandsawing** machines. Material is not cut by chip formation. Instead, the friction between the rapidly moving saw blade and the work is sufficient to raise the temperature of the material at the end of the kerf to or just below the melting point, where its strength is very low. The saw blade then pulls the molten, or weakened, material out of the kerf. Consequently, the blades do not need to be sharp; they frequently have no teeth—only occasional notches in the blade to aid in removing the metal.

Almost any material, including ceramics, can be cut by friction sawing. Because only a small portion of the blade is in contact with the work for an instant and then is cooled by its passage through the air, it remains cool. Usually, the major portion of the work, away from contact with the saw blade, also remains quite cool. The metal adjacent to the kerf is heat affected, recast, and sometimes harder than the bulk metal. It is also a very rapid method for trimming the flash from sheet metal parts, castings, and forgings.

Cutting Fluids

Cutting fluids should be used for all bandsawing, with the exception that cast iron is always cut dry. Commercially available oils or light cutting oils will give good results in cutting ferrous materials. Beeswax or paraffin are common lubricants for cutting aluminum and aluminum alloys.

Feeds and Speeds

Because of the many different types of feed involved in bandsawing, it is not practical to provide tabular feed or pressure data. Under general conditions, however, an even pressure, without forcing the work, gives best results. A nicely curled chip

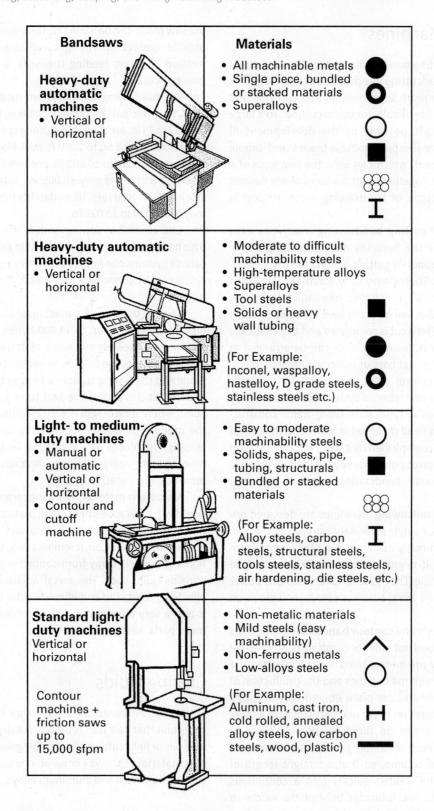

Bandsaws	Materials
Heavy-duty automatic machines • Vertical or horizontal	• All machinable metals • Single piece, bundled or stacked materials • Superalloys
Heavy-duty automatic machines • Vertical or horizontal	• Moderate to difficult machinability steels • High-temperature alloys • Superalloys • Tool steels • Solids or heavy wall tubing (For Example: Inconel, waspalloy, hastelloy, D grade steels, stainless steels etc.)
Light- to medium-duty machines • Manual or automatic • Vertical or horizontal • Contour and cutoff machine	• Easy to moderate machinability steels • Solids, shapes, pipe, tubing, structurals • Bundled or stacked materials (For Example: Alloy steels, carbon steels, structural steels, tools steels, stainless steels, air hardening, die steels, etc.)
Standard light-duty machines Vertical or horizontal Contour machines + friction saws up to 15,000 sfpm	• Non-metalic materials • Mild steels (easy machinability) • Non-ferrous metals • Low-alloys steels (For Example: Aluminum, cast iron, cold rolled, annealed alloy steels, low carbon steels, wood, plastic)

FIGURE 27.5 Bandsawing machines can be vertical or horizontal, contour and cutoff types. A wide variety of materials and shapes can be bandsawed.

usually indicates an ideal feed pressure. Burned or discolored chips indicate excessive pressure, which can cause tooth breakage and premature wear.

Most bandsaws provide recommended cutting speed information right on the machine, depending on the material being sawed. In general, HSS blades are run at 200 to 300 ft/min when cutting 1-in.-thick, low- and medium-carbon steels. For high-carbon steels, alloy steels, and tool and die steels, the range is from 150 to 225 ft/min, and most stainless steels are cut at 100 to 125 ft/min.

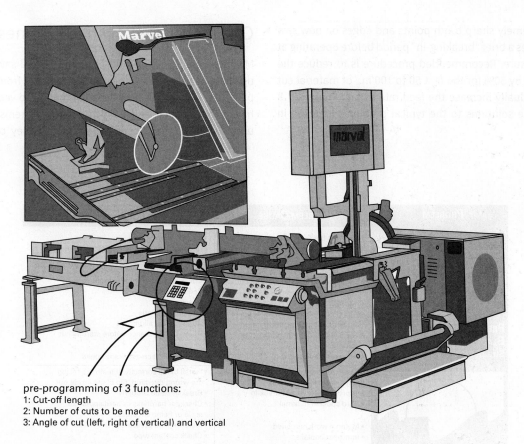

pre-programming of 3 functions:
1: Cut-off length
2: Number of cuts to be made
3: Angle of cut (left, right of vertical) and vertical

FIGURE 27.6 Vertical bandsaw setup to cut pipe. Inset shows the head tilted to 45 degrees.

FIGURE 27.7 Contour bandsawing on vertical bandsawing machine, shown in inset. (Courtesy J T. Black)

The extremely sharp tooth points and edges on new saw blades requires a brief "breaking-in" period before operating at full feed pressure. Recommended procedure is to reduce the feed pressure by 50% for the first 50 to 100 in.^{2} of material cut and then gradually increase the feed rate to full. **Figure 27.8** provides some solutions to the typical problems incurred in bandsawing.

Circular-Blade Sawing Machines

Machines employing rotating circular or cold saw blades are used exclusively for cutoff work. These range from small, simple types, in which the saw is fed manually, to very large saws having power feed and built-in coolant systems, commonly used for cutting off hot-rolled shapes as they come from a

PROBLEM	PROBLEM CAUSE	SOLUTION
 **PREMATURE BLADE BREAKAGE** Straight Break Indicates fatigue	• Incorrect blade - teeth too coarse • Blade tension too high • Side guides too tight • Damaged or misadjusted blade guides • Excessive feed • Incorrect cutting fluid • Wheel diameter too small for blade • Blade rubbing on wheel flanges • Teeth in contact with work before starting saw • Incorrect blade speed	• Use finer tooth pitch • Reduce blade tension (see machine manual) • Check side guide clearance (see machine manual) • Check all guides for alignment/damage • Reduce feed pressure • Check coolant • Use thinner blade • Adjust wheel alignment • Allow 1/2" clearance before starting cut • Increase or decrease blade speed
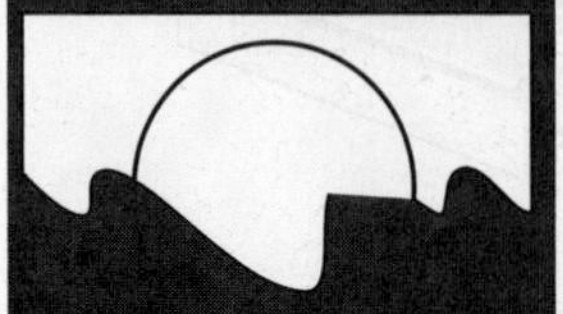 **PREMATURE DULLING OF TEETH**	• Teeth pointing in wrong direction / blade mounted backwards • Improper or no blade break-in • Hard spots in material • Material work hardened • Improper coolant • Improper coolant concentration • Speed too high • Feed too light • Teeth too small	• Install blade correctly. If teeth are facing the wrong direction, flip blade inside out • Break in blade properly • Check for hardness or hard spots like scale or flame cut areas • Increase feed pressure • Check coolant type • Check coolant mixture • Check recommended blade speed (Page 10-11) • Increase feed pressure • Increase tooth size
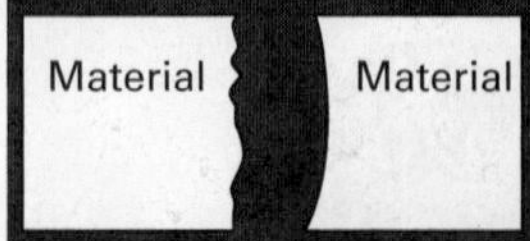 **INACCURATE CUT**	• Tooth set damage • Excessive feed pressure • Improper tooth size • Cutting fluid not applied evenly • Guides worn or loose • Insufficient blade tension	• Check for worn set on one side of blade • Reduce feed pressure • Check tooth size chart • Check coolant nozzles • Tighten or replace guides, check for proper alignment • Adjust to recommended tension
BAND LEADING IN CUT 	• Over-feed • Insufficient blade tension • Tooth set damage • Guide arms loose or set too far apart • Chips not being cleaned from gullets • Teeth too small	• Reduce feed force • Adjust recommended tension • Check material for hard inclusions • Position arms as close to work as possible. Tighten arms. • Check chip brush • Increase tooth size
 **CHIP WELDING**	• Insufficient coolant flow • Wrong coolant concentration • Excessive speed and/or pressure • Tooth size too small • Chip brush not working	• Check coolant level and flow • Check coolant ratio • Reduce speed and/or pressure • Use coarser tooth pitch • Repair or replace chip brush
TEETH FRACTURE  Back of tooth indicates work spinning in clamps	• Incorrect speed and/or feed • Incorrect blade pitch • Saw guides not adjusted properly • Chip brush not working • Work spinning or moving in vise	• Check cutting chart • Check tooth size chart • Adjust or replace saw guides • Repair or replace chip brush • Check bundle configuration/adjust vise pressure
 **IRREGULAR BREAK** Indicates material movement	• Indexing out of sequence • Material loose in vise	• Check proper machine movement • Check vise or clamp

FIGURE 27.8 Typical bandsawing problems, causes, and solutions.

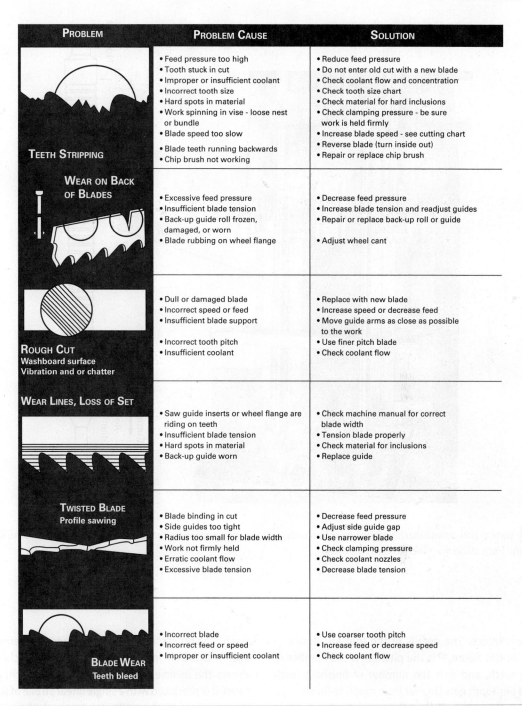

PROBLEM	PROBLEM CAUSE	SOLUTION
TEETH STRIPPING	• Feed pressure too high • Tooth stuck in cut • Improper or insufficient coolant • Incorrect tooth size • Hard spots in material • Work spinning in vise - loose nest or bundle • Blade speed too slow • Blade teeth running backwards • Chip brush not working	• Reduce feed pressure • Do not enter old cut with a new blade • Check coolant flow and concentration • Check tooth size chart • Check material for hard inclusions • Check clamping pressure - be sure work is held firmly • Increase blade speed - see cutting chart • Reverse blade (turn inside out) • Repair or replace chip brush
WEAR ON BACK OF BLADES	• Excessive feed pressure • Insufficient blade tension • Back-up guide roll frozen, damaged, or worn • Blade rubbing on wheel flange	• Decrease feed pressure • Increase blade tension and readjust guides • Repair or replace back-up roll or guide • Adjust wheel cant
ROUGH CUT Washboard surface Vibration and or chatter	• Dull or damaged blade • Incorrect speed or feed • Insufficient blade support • Incorrect tooth pitch • Insufficient coolant	• Replace with new blade • Increase speed or decrease feed • Move guide arms as close as possible to the work • Use finer pitch blade • Check coolant flow
WEAR LINES, LOSS OF SET	• Saw guide inserts or wheel flange are riding on teeth • Insufficient blade tension • Hard spots in material • Back-up guide worn	• Check machine manual for correct blade width • Tension blade properly • Check material for inclusions • Replace guide
TWISTED BLADE Profile sawing	• Blade binding in cut • Side guides too tight • Radius too small for blade width • Work not firmly held • Erratic coolant flow • Excessive blade tension	• Decrease feed pressure • Adjust side guide gap • Use narrower blade • Check clamping pressure • Check coolant nozzles • Decrease blade tension
BLADE WEAR Teeth bleed	• Incorrect blade • Incorrect feed or speed • Improper or insufficient coolant	• Use coarser tooth pitch • Increase feed or decrease speed • Check coolant flow

FIGURE 27.8 (Continued)

rolling mill. In some cases, friction saws are used for this purpose, having disks up to 6 ft in diameter and operating at surface speeds up to 25,000 ft/min. Steel sections up to 24 in. can be cut in less than 1 min by this technique.

Although technically not a sawing operation, cutoff work up to about 6 in. is often done utilizing thin, *abrasive* disks. The equipment used is the same as for sawing. It has the advantage that very hard materials that would be very difficult to saw can be cut readily. A thin rubber- or resinoid-bonded abrasive wheel is used. Usually, a somewhat smoother surface is produced.

27.3 Introduction to Broaching

The process of **broaching** is one of the most productive of the basic machining processes. The machine tool is called a *broaching machine* or a **broach**, and the cutting tool is also called the broach. **Figure 27.9** shows the basic shape of a conventional pull broach and the machine tool used to pull the cutting tool

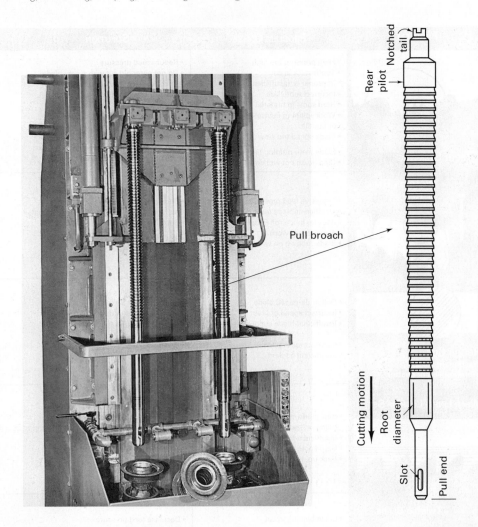

through the workpiece. The details of the tool are shown in **Figure 27.10**. In this figure, P is the pitch, n_s is the number of semiroughing teeth, and n_f is the number of finishing teeth where the rise per tooth gets smaller from rough to finish.

Figure 27.9 shows a **pull broach** in a vertical pull-down broaching machine. The pull end of the broach is passed through the part, and a key mates to the slot. The broach is pulled through the part. The broach is retracted (pulled up) out of the part. The part is transferred from the left fixture to the right fixture. One finished part is completed in every manufacturing cycle.

The feed per tooth in broaching is the change in height of successive teeth. This is called the rise per tooth (RPT, or t_r). Broaching looks similar to sawing except that the saw makes many passes through the cut, whereas the broach produces a finished part in one pass. The heart of this process lies in the broaching tool, in which roughing, semifinishing, and finishing teeth are combined into one tool, as shown in **Figure 27.11**. Broaching is unique in that it is the only one of the basic

machining processes in which **feed**, which determines the chip thickness, is built into the cutting tool. The machined surface is always the inverse of the profile of the broach, and, in most cases, it is produced with a single linear stroke of the tool across the workpiece (or the workpiece across the broach).

Broaching competes economically with milling and boring and is capable of producing precision-machined surfaces. The broach finishes an entire surface in a single pass. Broaches are used in production to finish holes, splines, and flat surfaces. Typical workpieces include small to medium-sized castings, forgings, screw-machine parts, and stampings.

This **rise per tooth (RPT)**, also known as **step** or the **feed per tooth**, determines the amount of material removed, see Figure 27.11. No feeding of the broaching tool is required. The frontal contour of the teeth determines the shape of the resulting machined surface. Because these conditions built into the tool, no complex motion of the tool relative to the workpiece is required and the need for highly skilled machine operators is minimized.

Broach Design (The Cutting Tool)

Broaches commonly are classified by the following design features:

Purpose	Motion	Construction	Function
Single	Push	Solid	Roughing
Combination	Pull	Built-up	Sizing
	Stationary		Burnishing

Figure 27.10 shows the principal components of a pull broach and the shape and arrangement of the teeth. Each tooth is essentially a single-edge cutting tool, arranged much like the teeth on a saw except for the step that determines the depth cut by each tooth. The rise per tooth, which determines the chip load, varies from about 0.006 in. for roughing teeth in machining free-cutting steel to a minimum 0.001 in. for finishing teeth. Typically, the RPT is 0.003 to 0.006 in. in surface broaching and 0.0012 to 0.0025 in. on the diameter for internal broaching. The exact amount depends on several factors.

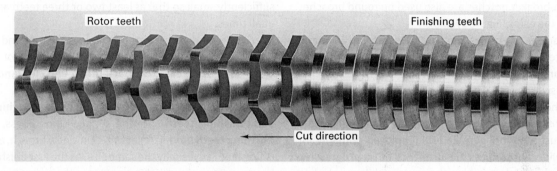

(a) Rotor- or jump-tooth broach design.

(b) Round, push-type broach with chip-breaking notches on alternate teeth except at the finishing end.

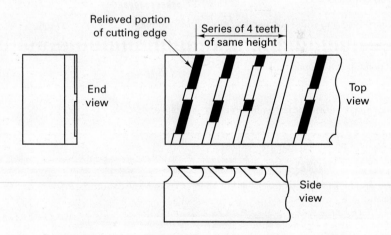

(c) Notched tooth, flat broach.

(d) Progressive surface broach.

FIGURE 27.12 Methods to decrease force or break up chip rings in broaches: (a) rotor or jump tooth; (b) notched tooth, round; (c) notched tooth, flat design (overlapping teeth permit large RPTs without increasing chip load); (d) progressive tooth design for flat broach. (Federal Broach Holdings LLC 1961 Sullivan Drive, Harrison, MI 48625)

Cuts that are too large impose undue stresses on the teeth and the work; too-small cuts result in rubbing rather than cutting action. The strength and ductility of the metal being cut are the primary factors.

Where it is desirable for each tooth to take a deep cut, as in broaching castings or forgings that have a hard, abrasive surface layer, **rotor-cut** or **jump-cut tooth design** may be used (**Figure 27.12**). In this design, two or three teeth in succession have the same diameter, or height, but each tooth of the group is notched or cut away so that it cuts only a portion of the circumference or width. This permits deeper but narrower cuts by each tooth without increasing the total load per tooth. This tooth design also reduces the forces and the power requirements. Chip-breaker notches are also used on round broaches to break up the chips (Figure 27.12b).

A similar idea can be used for flat surfaces. Tooth loads and cutting forces also can be reduced by using the **double-cut construction**, shown in Figure 27.12c. Four consecutive teeth get progressively wider. The teeth remove metal over only a portion of their width until the fourth tooth completes the cut.

Another technique for reducing tooth loads utilizes the principle illustrated in Figure 27.12d. Employed primarily for broaching wide, flat surfaces, the first few teeth in **progressive broaches** completely machine the center, while succeeding teeth are offset in two groups to complete the remainder of the surface. Rotor, double-cut, and progressive designs require the broach to be made longer than if normal teeth were used, and they therefore can be used only on a machine having adequate stroke length.

The cutting edges of the teeth on **surface broaches** may be either normal to the direction of motion or at an angle of from 5 to 20 degrees. The latter, **shear-cut broaches**, provide smoother cutting action with less tendency to vibrate. Other shapes that can be broached are shown in **Figure 27.13** along with push- or pull-type broaches used for the job.

The pitch of the teeth and the gullet between them must be sufficient to provide ample room for the chips. All chips produced by a given tooth during its passage over the full length of the workpiece must be contained in the space between successive teeth. At the same time, it is desirable to have the pitch sufficiently small so that at least two or three teeth are cutting at all times.

The **hook** determines the primary rake angle and is a function of the material being cut. It is 15 to 20 degrees for steel and 6 to 8 degrees for cast iron. **Back-off** or **end clearance angles** are from 1 to 3 degrees to prevent rubbing.

Most of the metal removal is done by the **roughing teeth**. **Semifinishing teeth** provide surface smoothness, whereas **finishing teeth** produce exact size. On a new broach, all the finishing teeth are usually the same size. As the first finishing teeth become worn, those behind continue the sizing function. On some round broaches, **burnishing teeth** are provided for finishing. These teeth have no cutting edges but are rounded disks of hard steel or carbide that are from 0.001 to 0.003 in.

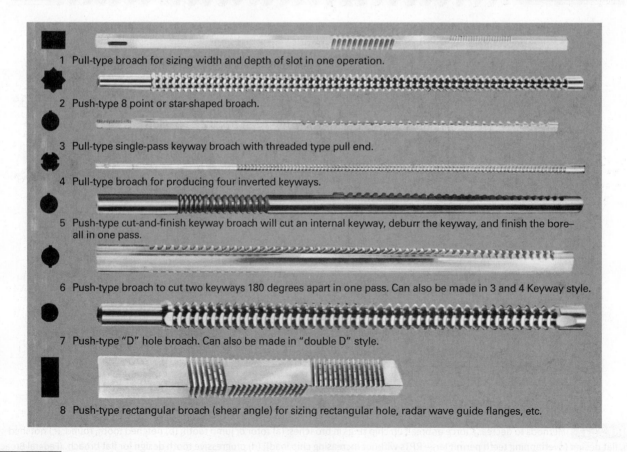

1 Pull-type broach for sizing width and depth of slot in one operation.

2 Push-type 8 point or star-shaped broach.

3 Pull-type single-pass keyway broach with threaded type pull end.

4 Pull-type broach for producing four inverted keyways.

5 Push-type cut-and-finish keyway broach will cut an internal keyway, deburr the keyway, and finish the bore—all in one pass.

6 Push-type broach to cut two keyways 180 degrees apart in one pass. Can also be made in 3 and 4 Keyway style.

7 Push-type "D" hole broach. Can also be made in "double D" style.

8 Push-type rectangular broach (shear angle) for sizing rectangular hole, radar wave guide flanges, etc.

FIGURE 27.13 Examples of push- or pull-type broaches. (Courtesy The duMONT Company Corporation)

larger than the size of the hole. The resulting rubbing action smoothes and sizes the hole. They are used primarily on cast iron and non-ferrous metals.

The pull end of a broach provides a means of quickly attaching the broach to the pulling mechanism. The **front pilot** aligns the broach in the hole before it begins to cut, and the **rear pilot** keeps the tool square with the finished hole as it leaves the workpiece. **Shank length** must be sufficient to permit the broach to pass through the workpiece and be attached to the puller before the first roughing tooth engages the work. If a broach is to be used on a vertical machine that has a tool-handling mechanism, a tail is necessary.

A broach should not be used to remove a greater depth of metal than that for which it is designed—the sum of the steps of all the teeth. In designing workpieces, a minimum of 0.020 in. should be provided on surfaces that are to be broached, and about 0.025 in. is the practical maximum.

Broaching Speeds, Accuracy, and Finish

Depending on the metal being cut, cutting speeds for broaching range from low (25 to 20 sfpm) to high while completing the surface in a single stroke, so the productivity is high. A complete cycle usually requires only from 5 to 30 s, with most of that time being taken up by the return stroke, broach handling, and workpiece loading and unloading. Such cutting conditions facilitate cooling and lubrication and result in very low tool wear rates, which reduce the necessity for frequent resharpening and prolong the life of the expensive broaching tool.

For a given cutting speed and material, the force required to pull or push a broach is a function of the tooth width, the step, and the number of teeth cutting. Consequently, it is necessary to design or specify a broach within the stroke length and power limitations of the machine on which it is to be used. The average machining precision is typically ±0.001-in.-(±0.02-mm) tolerance with surface finish 120 to 60 root mean square (RMS) or better. Burrs are minimal on the exit side of cuts.

Broaching Materials and Construction

Because of the low cutting speeds employed, most broaches are made of alloy or high-speed tool steel. Carbide-tipped broaches are seldom used for machining steel parts or forgings because the cutting edges tend to chip on the first stroke, probably due to a lack of rigidity in the combination of machine tool and cutting tool. TiN coating of high-speed-steel (HSS) broaches is becoming more common, greatly prolonging the life of broaches. When they are used in continuous mass-production machines, particularly in surface broaching of cast iron, tungsten carbide teeth may be used, permitting the broach to be used for long periods of time without resharpening.

Internal broaches are usually solid but may be made of **shells** mounted on an arbor (**Figure 27.14**). When the broach (or a section of it) is subject to rapid wear, a single shell can be replaced. This will be much cheaper than replacing an entire solid broach. Shell construction, however, is initially more expensive than a solid broach of comparable size.

Small-surface broaches may be of solid construction, but larger ones usually use modular construction (**Figure 27.15**). Building in sections makes the broach easier and cheaper to construct and sharpen. It also often provides some degree of interchangeability of the sections for different parts, bringing down the tool cost significantly.

Sharpening Broaches

Most broaches are resharpened by grinding the hook faces of the teeth. The lands of internal broaches must not be reground because this would change the size of the broach. Lands of flat-surface broaches are sometimes ground, in which case all of them must be ground to maintain their proper relationship.

FIGURE 27.14   Shell construction for a pull broach. (Courtesy J T. Black)

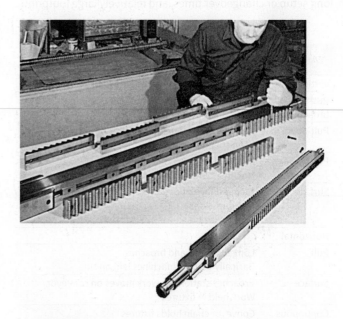

FIGURE 27.15   A modularly constructed broach is cheaper to build and can be sharpened in sections. (Courtesy J T. Black)

27.5 Broaching Machines

Because all the factors that determine the shape of the machined surface and all the cutting conditions except speed are built into the broaching tool, broaching machines are relatively simple. Their primary functions are to impart plain reciprocating motion to the broach and to provide a means for handling the broach automatically.

Most broaching machines are driven hydraulically, although mechanical drives are used in a few special types. The major classification relates to whether the motion of the broach is vertical or horizontal, as given in **Table 27.1**.

The choice between vertical and horizontal machines is determined primarily by the length of the stroke required and the available floor space. Vertical machines seldom have strokes greater than 60 in. because of height limitations. Horizontal machines can have almost any length of stroke, but they require greater floor space. The most common machine is the vertical pull-down machine shown in Figure 27.9. The worktable, usually having a spherical-seated workholder, sits below the broach elevator, with a pulling mechanism below the table. When the elevator raises the broach above the table, the work can be placed into position. The elevator then lowers the pilot end of the broach through the hole in the workpiece, where it is engaged by the puller. The elevator then releases the upper end of the broach, and it is pulled through the workpiece. The workpieces are removed from the table, and the broach is raised upward to be engaged by the elevator mechanism. In some machines with two rams, one broach is being pulled down while the work is being unloaded and the broach raised at the other station. The part is being broached in two passes, first on the left, then on the right.

Most broaching machines are found in job shops, have long setup or changeover times, and relatively large footprints.

TABLE 27.1 Broaching Machines

Vertical	
Push-broaching	Arbor press with guided ram, 5- to 50-ton capacity Internal broaching
Pull-down	Double-ram design most common Long changeover times
Pull-up	Ram above table pulling broach up Machines with multiple rams common
Surface	No handling of broach Multiple slides Short cycle times
Horizontal	
Pull	Longer strokes and broaches Basically vertical machines laid on side
Surface	Broaches stationary, work moves on conveyor Work held in fixtures
Continuous	Conveyor chain holds fixtures
Rotary	Rotary broach stationary, work translates beneath tool Work held in fixtures

In Chapter 43, a rotary broach is shown, designed for manufacturing cells and featuring rapid changeover and a narrow footprint.

27.6 Introduction to Shaping and Planing

The processes of **shaping** and **planing** are among the oldest single-point machining processes. Shaping has largely been replaced by milling and broaching as a production process, while planing still has applications in producing long flat cuts, like those in the ways of machine tools. From a consideration of the relative motions between the tool and the workpiece, shaping and planing both use a straight-line cutting motion with a single-point cutting tool to generate a flat surface.

In shaping, the workpiece is fed at right angles to the cutting motion between successive strokes of the tool, as shown in **Figure 27.16**, where f_c is the feed per stroke, V is the cutting speed, and d is the depth of cut (DOC). (In planing, discussed next, the workpiece is reciprocated and the tool is fed at right angles to the cutting motion.) For either shaping or planing, the tool is held in a clapper box, which prevents the cutting edge from being damaged on the return stroke of the tool. In addition to plain flat surfaces, the shapes most commonly produced on the shaper and planer are those illustrated in **Figure 27.17**. Relatively skilled workers are required to operate shapers and planers, and most of the shapes that can be produced on them can also be made by much more productive processes, such as milling, broaching, or grinding. Consequently, except for certain special types, planers that will do only planing have become obsolete. Today, shapers are used mainly in tool and die work, in very low volume production, or in the manufacture of gear teeth.

In shaping, the cutting tool is held in the tool post located in the ram, which reciprocates over the work with a forward stroke, cutting at velocity V and a quick return stroke at velocity, V_R. The rpm of drive crank (N_s) drives the ram and determines the velocity of the operation (see Figure 27.16d). The stroke ratio R_s is

$$R_s = \frac{\text{Cutting stroke angle}}{360°} = \frac{200°}{360°} = \frac{5}{9} \qquad (27.10)$$

The tool is advancing 55% of the time. The number of strokes per minute is determined by the rpm of the drive crank. Feed is in inches per stroke and is at right angles to the cutting direction. As in other machining processes, speed and feed are selected by the operator.

The length of cut, L, is the length of the workpiece. The length of stroke, l, must be greater than the length of the workpiece. Because velocity is position variant, let l = twice the length of the block being cut, or $2L$. The cutting velocity, V, is assumed to be twice the average forward velocity, V, of the ram.

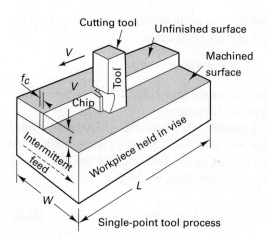

Cutting tool

Unfinished surface

Machined surface

V

f_c

Tool

V

Chip

Intermittent feed

t

Workpiece held in vise

L

W

Single-point tool process

(a) Basic geometry for shaping and planing

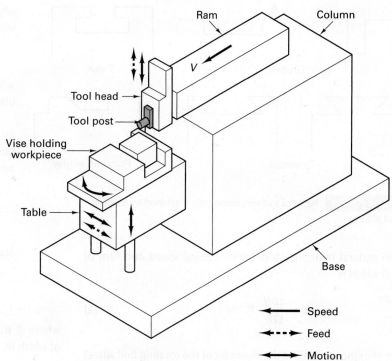

Ram

Column

V

Tool head

Tool post

Vise holding workpiece

Table

Base

→ Speed

⤙⤚ Feed

↔ Motion

(b) Shaper machine tool

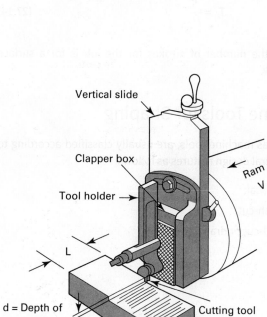

Vertical slide

Clapper box

Tool holder

Ram

V

L

d = Depth of cut DOC

Cutting tool

Workpiece

(c) Shaper tool holder, clapper box, and workpiece

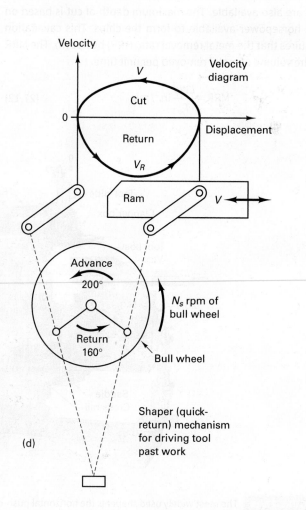

Velocity

V

Cut

0

Return

V_R

Velocity diagram

Displacement

Ram

V

Advance 200°

N_s rpm of bull wheel

Return 160°

Bull wheel

Shaper (quick-return) mechanism for driving tool past work

(d)

FIGURE 27.16 Basics of shaping and planing. (a) The cutting speed, V, and feed per stroke f_c. (b) Block diagram of the machine tool. (c) The cutting tool is held in a clapper box so the tool does not damage the workpiece on the return stroke. (d) The ram of the shaper carries the cutting tool at cutting velocity V and reciprocates at velocity V_R by the rotation of a bull wheel turning at rpm N_s.

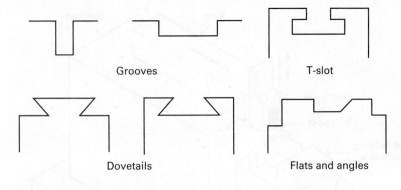

Grooves   T-slot

Dovetails   Flats and angles

FIGURE 27.17 Types of surfaces commonly machined by shaping and planing.

The general relationship between cutting speed and rpm of bull wheel is

$$V = \frac{\pi D N_s}{12 R_s} \text{ ft/min} \tag{27.11}$$

For shaping, where D is the diameter of the rotating bull wheel in inches. Once, the cutting speed, V, is selected, the rpm of the machine can be calculated. Tables for suggested feed values, f_c, are in inches per stroke (or cycle), and recommended depths of cut are also available. The maximum depth of cut is based on the horsepower available to form the chips. This calculation requires that the metal removal rate (MRR) be known. The MRR is the volume of metal removed per unit time

$$\text{MRR} = \frac{LWd}{T_m} \text{ in.}^3/\text{min} \tag{27.12}$$

where W is the width of block being cut and L is the length of block being cut, so

$$\text{Volume of cut} = WLd$$

where d is the depth of cut and T_m is the time in minutes to cut that volume.

In general, T_m is the total length of the cut divided by the feed rate. For shaping, T_m is the width of the block divided by the feed rate f_c of the tool moving across the width. Thus, for shaping,

$$T_m = \frac{W}{N_s \times f_c} \tag{27.13}$$

Also,

$$T_m = \frac{S}{N_s} \tag{27.14}$$

where S the number of strokes for the job is for a surface of width W.

Machine Tools for Shaping

Shapers, as machine tools, are usually classified according to their general design features as follows:

1. Horizontal:
   a. Push-cut.
   b. Pull-cut or draw-cut shaper.

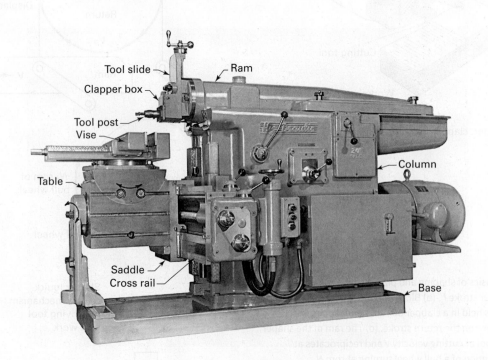

Tool slide — Ram
Clapper box —
Tool post —
Vise —
                                    Column
Table —
Saddle —
Cross rail —
                                    Base

FIGURE 27.18 The most widely used shaper is the horizontal push-cut machine tool, shown here with no tool in the tool post. (Courtesy J T. Black)

2. Vertical:

    a. Regular or slotters.

    b. Keyseaters.

3. Special.

They are also classified as to the type of drive employed: *mechanical drive* or *hydraulic drive*. Most shapers are of the **horizontal push-cut** type (**Figure 27.18**), where cutting occurs as the ram *pushes* the tool across the work.

On horizontal push-cut shapers, the work is usually held in a heavy vise mounted on the top side of the table. Shaper vises have a very heavy movable jaw, because the vise must often be turned so that the cutting forces are directed against this jaw.

In clamping the workpiece in a shaper vise, care must be exercised to make sure that it rests solidly against the bottom of the vise (on parallel bars) so that it will not be deflected by the cutting force and so that it is held securely yet not distorted by the clamping pressure.

Most shaping is done with simple high-speed steel- or carbide-tipped cutting tool bits held in a heavy, forged tool holder. Although shapers are versatile tools, the precision of the work done on them is greatly dependent on the operator. Feed dials on shapers are nearly always graduated in 0.001-in. divisions, and work is seldom done to greater precision than this. A tolerance of 0.002 to 0.003 in. is desirable on parts that are to be machined on a shaper, because this gives some provision for variations due to clamping, possible looseness or deflection of the table, and deflection of the tool and ram during cutting.

Planing Machines

Planing can be used to produce horizontal, vertical, or inclined flat surfaces on large workpieces (too large for shapers). However, planing is much less efficient than other basic machining processes, such as milling, that will produce such surfaces. Consequently, planing and planers have largely been replaced by planer milling machines or machines that can do both milling and planing.

Figure 27.19 shows the basic components and motions of planers—the most common designs being double-housing and single-housing types. In most planing, the action is opposite to that of shaping. The work is moved past one or more stationary single-point cutting tools. Because a large and heavy workpiece and table must be reciprocated at relatively low speeds, several tool heads are provided, often with multiple tools in each head. In addition, many planers are provided with tool heads arranged so that cuts occur on both directions of the

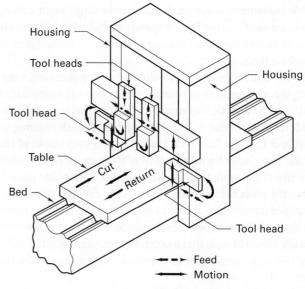

Block diagram
showing the basic components
of a double-housing planer

(a)

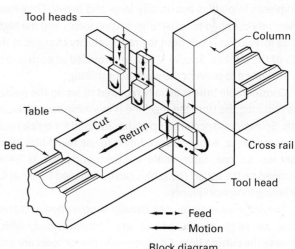

Block diagram
of an open-side planer

(b)

(c)   Four planer tools in tool holder

FIGURE 27.19 Schematic of planers: (a) double-housing planer with multiple tool heads (4) and a large reciprocating table; (b) single-housing or open-sided planer; (c) interchangeable multiple tool holder for use in planers. (Courtesy J T. Black)

table movement. However, because only single-point cutting tools are used and the cutting speeds are quite low, planers are low in productivity as compared with some other types of machine tools.

The double housing has a closed-housing structure, spanning the reciprocating worktable, with a cross rail supported at each end on a vertical column and carrying two tool heads. An additional tool head is usually mounted on each column, so that four tools (or four sets) can cut during each stroke of the table. The closed-frame structure of this type of planer limits the size of the work that can be machined. Open-side planers have the cross rail supported on a single column. This design provides unrestricted access to one side of the table to permit wider workpieces to be accommodated. Some open-side planers are convertible, in that a second column can be attached to the bed when desired so as to provide added support for the cross rail.

Workholding and Setup on Planers

Workpieces in planers are usually large and heavy. They must be securely clamped to resist large cutting forces and the high-inertia forces that result from the rapid velocity changes at the ends of the strokes. Special stops are provided at each end of the workpiece to prevent the work from shifting.

Considerable time is usually required to set up the planer, thus reducing the time the machine is available for producing chips. Sometimes special setup plates are used for quick setup of the workpiece. Another procedure is to use two tables. Work is set up on one table while another workpiece is being machined on the other. The tables can be fastened together for machining long workpieces.

The large workpieces can usually support heavy cutting forces, so large depths of cut are recommended, which decrease the cutting time. Consequently, planer tools are usually quite massive and can sustain the large cutting forces. Usually, the main shank of the tools is made of plain-carbon steel, with tips made from high-speed steel or carbide. Chip breakers should be used to avoid long and dangerous chips in ductile materials.

Theoretically, planers have about the same precision as shapers. The feed and other dimension-controlling dials are usually graduated in 0.001-in. divisions. However, because larger and heavier workpieces are usually involved, with much longer beds and tables, the working tolerances for planer work are somewhat greater than for shaping.

27.7 Introduction to Filing

Basically, the metal-removing action in **filing** is the same as in sawing, in that chips are removed by cutting teeth that are arranged in succession along the same plane on the surface of

a tool, called a **file**. There are two differences: (1) the chips are very small, and therefore the cutting action is slow and easily controlled, and (2) the cutting teeth are much wider. Consequently, fine and accurate work can be done.

Files are classified according to the following:

1. The type, or *cut*, of the teeth.
2. The degree of coarseness of the teeth.
3. Construction:
   a. Single solid units for hand use or in die-filing machines.
   b. Band segments, for use in band-filing machines.
   c. Disks, for use in disk-filing machines.

Four types of cuts are available as shown in **Figure 27.20**. **Single-cut files** have rows of parallel teeth that extend across the entire width of the file at the angle of from 65 to 85 degrees. **Double-cut files** have two series of parallel teeth that extend across the width of the file. One series is cut at an angle of 40 to 45 degrees. The other series is coarser and is cut at an opposite angle that varies from about 10 to 80 degrees. A **vixen-cut file** has a series of parallel curved teeth, each extending across the file face. On **a rasp-cut file**, each tooth is short and is raised out of the surface by means of a punch.

The coarseness of files is designated by the following terms, arranged in order of increasing coarseness: *dead smooth*, *smooth*, *second cut*, *bastard*, *coarse*, and *rough*. There is also a series of finer Swiss pattern files, designated by numbers from 00 to 8.

Reproduced by permission of Apex Tool Group, LLC, manufacturer of NICHOLSON files.

FIGURE 27.20 Four types of teeth (cuts) used in files. Left to right: single, double, rasp, and curved (vixen). (Courtesy of Nicholson File Company)

Files are available in a number of cross-sectional shapes: *flat*, *round*, *square*, *triangular*, and *half-round*. Flat files can be obtained with no teeth on one or both narrow edges, known as **safe edges**. Safe edges prevent material from being removed from a surface that is normal to the one being filed. Most files for hand filing are from 10 to 14 in. in length and have a pointed *tang* at one end on which a wood or metal handle can be fitted for easy grasping.

Filing Machines

An experienced operator can do very accurate work by hand filing, but it can be a difficult task. Therefore, three types of filing machines have been developed that permit quite accurate results to be obtained rapidly and with much less effort. **Die-filing machines** hold and reciprocate a file that extends upward through the worktable. The file rides against a roller guide at its upper end, and cutting occurs on the downward stroke; therefore, the cutting force tends to hold the work against the table. The table can be tilted to any desired angle. Such machines operate at from 300 to 500 strokes per minute, and the resulting surface tends to be at a uniform angle with respect to the table. Quite accurate work can be done. Because of the reciprocating action, approximately 50% of the operating time is non-productive.

Band-filing machines provide continuous cutting action. Most band filing is done on contour bandsawing machines by means of a special band file that is substituted for the usual bandsaw blade. The principle of a band file is shown in **Figure 27.21**. Rigid, straight file segments, about 3 in. long, are riveted to a flexible steel band near their leading ends. One end of the steel band contains a slot that can be hooked over a pin in the other end to form a continuous band. As the band passes over the drive and idler wheels of the machine, it flexes so that the ends of adjacent file segments move apart. When the band becomes straight, the ends of adjacent segments move together and interlock to form a continuous straight file. Where the file passes through the worktable, it is guided and supported by a grooved guide, which provides the necessary support to resist the pressure of the work against the file. Band files are available in most of the standard cuts and in several widths and shapes. Operating speeds range from about 50 to 250 ft/min.

Although band filing is considerably more rapid than can be done on a die-filing machine, it usually is not quite as accurate. Frequently, band filing may be followed by some finish filing on a die-filing machine.

Some **disk-filing machines** have files in the form of disks (**Figure 27.22**). These are even simpler than die-filing machines and provide continuous cutting action. However, it is difficult to obtain accurate results by their use.

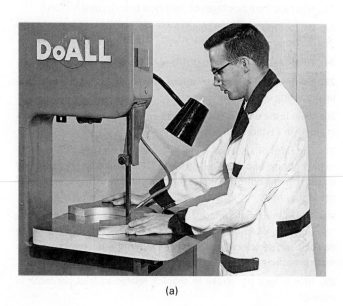

(a)

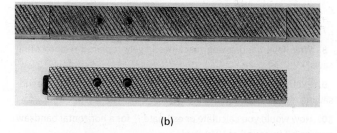

(b)

(c)

FIGURE 27.21 A bandfiling machine (a) uses segments (b) that are riveted to a steel band (c) to provide a continuous file for finishing wack. (Courtesy of DoAll Co.)

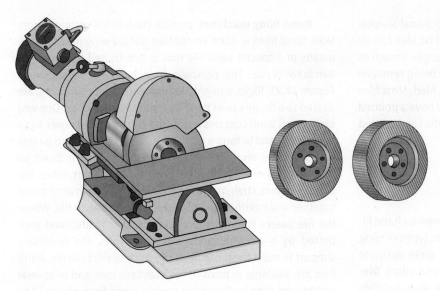

FIGURE 27.22 Disk-type filing machine and some of the available types of disk files. (Based on original drawing by Jersey Manufacturing Company)

Review Questions

1. Why is sawing one of the most efficient of the chip-forming processes?

2. Explain why tooth spacing (pitch) is important in sawing.

3. What is the tooth gullet used for on a saw blade?

4. Explain what is meant by the "set" of the teeth on a saw blade.

5. How is tooth set related to saw kerf?

6. Why can a bandsaw blade not be hardened throughout the entire width of the band?

7. What are the advantages of using circular saws?

8. Why have bandsawing machines largely replaced reciprocating saws?

9. Explain how the hole in Figure 27.7 is made on a contour band-sawing machine.

10. How would you calculate or estimate T_m for a horizontal bandsaw cutting a 3-in. round of 1040 steel.

11. What is the disadvantage of using gravity to feed a saw in cutting round bar stock?

12. What is unique about the broaching process compared to the other basic machining processes?

13. Can a thick saw blade be used as a broach? Why or why not?

14. Broaching machines are simpler in a basic design than most other machine tools. Why is this?

15. Why is broaching particularly well-suited for mass production?

16. In designing a broach, what would be the first thing you have to calculate?

17. Why is it necessary to relate the design of a broach to the specific workpiece that is to be machined?

18. What two methods can be utilized to reduce the force and power requirements for a particular broaching cut?

19. For a given job, how would a broach having rotor-tooth design compare in length with one having regular, full-width teeth?

20. Why are the pitch and radius of the gullet between teeth on a broach of importance?

21. Why are broaching speeds usually relatively low, as compared with other machining operations?

22. What are the advantages of shell-type broach construction?

23. Why are most broaches made from alloy or high-speed steel rather than from tungsten carbide?

24. What are the advantages of TiN-coated broaching tools?

25. For mass-production operations, which process is preferred, pull-up broaching or pull-down broaching?

26. What is the difference between the roughing teeth and the finishing teeth in a typical pull broach?

27. The sides of a square, blind hole must be machined all the way to the bottom (who designed this part?). The hole is first drilled to full depth, then the bottom is milled flat. Is it possible to machine the hole square by broaching? Why or why not?

28. The interior, flat surfaces of socket wrenches, which have one "closed" end, often are finished to size by broaching. By examining one of these, determine what design modification was incorporated to make broaching possible.

29. How is feed per stroke in shaping related to feed per tooth in milling?

30. What are some ways to improve the efficiency of a planer? Do any of these apply to the shaper?

31. To what extent is filing different from sawing?

32. What is a safe edge on a file?

33. How does a rasp-cut file differ from other types of files?

34. How does the process of shaping differ from planing?

Problems

1. A surface 12 in. long is to be machined with a flat, solid broach that has a rise per tooth of 0.0047 in. What is the minimum cross-sectional area that must be provided in the chip gullet between adjacent teeth?

2. The pitch of the teeth on a simple surface broach can be determined by equation 27.1. If a broach is to remove 0.25 in. of material from a gray iron casting that is 3 in. wide and 17.75 in. long, and if each tooth has a rise per tooth of 0.004 in., what will be the length of the roughing section of the broach?

3. Estimate the (approximate maximum) horsepower needed to accomplish the operation described in Problem 2 at a cutting speed of 10 m/min. (*Hint:* First find the HP used per tooth and determine the maximum number of teeth engaged at any time. What are those units?)

4. Estimate the approximate force acting in the forward direction during cutting for the conditions stated in Problems 2 and 3.

5. In cutting a 6-in.-long slot in a piece of AISI 1020 cold-rolled steel that is 1 in. thick, the material is fed to a bandsaw blade with teeth having a pitch of 1.27 mm (20 pitch) at the rate of 0.0001 in. per tooth. Estimate the cutting time for the cut.

6. The strength of a pull broach is determined by its minimum cross section, which usually occurs either at the root of the first tooth or at the pull end. Suppose the minimum root diameter is D_r, the pull end diameter is D_p, and the width of the pull slot is W. Write an equation for the allowable pull, in psi, using 200,000 as the yield strength for the broach material.

7. Suppose you want to shape a block of metal 7 in. wide and 4 in. long ($L = 4$ in.) using a shaper as set up in Figure 27.16. You have determined for this metal that the cutting speed should be 25 sfpm, the depth of cut needed here for roughing is 0.25 in., and the feed will be 0.1 in. per stroke. Determine the approximate crank rpm, and then estimate the cutting time and the MRR.

8. Could you have saved any time in Problem 7 by cutting the block in the 7-in. direction? Redo with $L = 7$ and $W = 4$ in.

9. Derive the equation for shaping cutting speed.

10. How many strokes per minute would be required to obtain a cutting speed of 36.6 m/min (120 ft/min) on a typical mechanical drive shaper if a 254-mm (10-in.) stroke is used?

11. How much time would be required to shape a flat surface 254 mm (1 in.) wide and 203 mm (8 in.) long on a hydraulic drive shaper, using a cutting speed of 45.7 m (150 ft) per minute, a feed of 0.51 mm (0.020 in.) per stroke, and an overrun of 12.7 mm ($\frac{1}{2}$ in.) at each end of the cut?

12. What is the metal removal rate in Problem 11 if the depth of cut is 6.35 mm ($\frac{1}{4}$ in.)?

13. Suppose you decide to mill the flat surface described in Problems 11 and 12. The work will be done on a vertical milling machine using a 1.25-in.-diameter end mill (four teeth) (HSS) cutting at 150 sfpm with a feed per tooth of 0.005 in. per tooth cutting at $d = 0.25$ in. Compare the milling time and MRR to that of shaping.

14. A planer has a 10-hp motor, and 75% of the motor output is available at the cutting tool. The specific power for cutting cast iron metal is 0.03 W/mm³, or 0.67 hp/in.³/min. What is the maximum depth of cut that can be taken in shaping a surface in this material if the surface is 305 × 305 mm (12 × 12), the feed is 0.25 in. per stroke, and the cutting speed is 54.9 mm/min (180 ft/min)?

15. Calculate the T_m for planing the block of cast iron in Problem 14, and then estimate T_m for milling the same surface. You will have to determine which milling process to use and select speeds and feeds for an HSS cutter.

CHAPTER 28

Abrasive Machining Processes

28.1 Introduction

Abrasive machining is a material removal process that involves the interaction of abrasive grits with the workpiece at high cutting speeds and shallow penetration depths. The chips that are formed resemble those formed by other machining processes. Unquestionably, abrasive machining is the oldest of the basic machining processes. Museums abound with examples of utensils, tools, and weapons that ancient peoples produced by rubbing hard stones against softer materials to abrade away unwanted portions, leaving desired shapes. For centuries, only natural abrasives were available for grinding, while other more modern basic machining processes were developed using superior cutting materials. However, the development of manufactured abrasives and a better fundamental understanding of the abrasive machining process have resulted in placing abrasive machining and its variations among the most important of all the basic machining processes.

The results that can be obtained by abrasive machining range from the finest and smoothest surfaces produced by any machining process, in which very little material is removed, to rough, coarse surfaces that accompany high material removal rates. The abrasive particles may be (1) free; (2) mounted in resin on a belt, called **coated product**; or, most commonly (3) close packed into wheels or stones, with abrasive grits held together by bonding material, called **bonded product** or a grinding wheel.

Figure 28.1 shows a surface grinding process using a grinding wheel. The depth of cut d is determined by the infeed and is usually very small, 0.002 to 0.005 in., so the arc of contact (and the chips) is small. The table reciprocates back and forth beneath the rotating wheel. The work feeds into the wheel in the cross-feed direction. After the work is clear of the wheel, the wheel is lowered and another pass is made, again removing a couple of thousandths of inches of metal. The metal removal process is basically the same in all abrasive machining processes but with important differences due to spacing of the grains in contact with the work (active grains) and the rigidity and degree of fixation of the grains. **Table 28.1** summarizes the primary abrasive processes. The term *abrasive machining* applied to one particular form of the grinding process is unfortunate, because all these process are machining with abrasives.

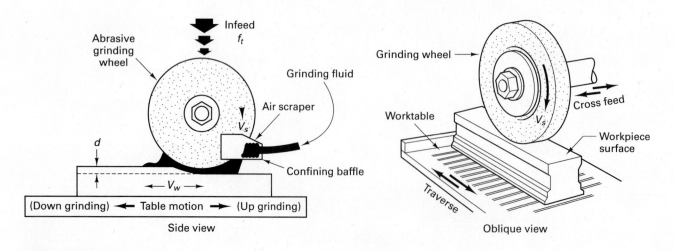

FIGURE 28.1 Schematic of surface grinding, showing infeed and cross-feed motions along with cutting speeds, V_s, and workpiece velocity, V_w.

TABLE 28.1   **Abrasive Machining Processes**

Process	Particle Mounting	Features
Grinding	Bonded	Uses wheels, accurate sizing, finishing, low MRR*; can be done at high speeds (>12,000 sfpm)*
Creep feed grinding	Bonded open, soft	Uses wheels with long cutting arc, very slow feed rate, and large depth of cut
Abrasive machining	Bonded	High MRR*, to obtain desired shapes and approximate sizes
Snagging	Bonded belted	High MRR*, rough rapid technique to clean up and deburr castings, forgings
Honing	Bonded	"Stones" containing fine abrasives; primarily a hole-finishing process
Lapping	Free	Fine particles embedded in soft metal or cloth; primarily a surface-finishing process
Abrasive waterjet	Free in jet	Water jets with velocities up to 3000 sfpm* carry abrasive particles (silica and garnet)
Ultrasonic	Free in liquid	Vibrating tool impacts abrasives at high velocity
Abrasive flow	Free in gel	Abrasives in gel flow over surface-edge finishing
Abrasive jet	Free in air	A focused jet of abrasives in an inert gas at high velocity

*sfpm = surface feet per minute; this is the cutting velocity; MRR = Metal removal rate.

Compared to normal machining, abrasive machining processes have three unique characteristics. First, each cutting edge is very small, and many of these edges can cut simultaneously. When suitable machine tools are employed, very fine cuts are possible, and fine surfaces and close dimensional control can be obtained. Second, because extremely hard abrasive grits, including diamonds, are employed as cutting tool materials, very hard materials, such as hardened steel, glass, carbides, and ceramics, can readily be machined. As a result, the abrasive machining processes are not only important as manufacturing processes, they are indeed essential. Many of our modern products, such as modern machine tools, automobiles, space vehicles, and aircraft, could not be manufactured without these processes. Third, in grinding, you have no control over the actual tool geometry (rake angles, cutting edge radius) or all the cutting parameters (depth of cut). As a result of these parameters and variables, grinding is a complex process.

To get a handle on the complexity, **Table 28.2** presents the primary grinding parameters, grouped by their independence or dependence. Independent variables are those that are controllable, by the machine operator, while the dependent variables are the resultant effects of those inputs. Not listed in the table is workpiece hardness, which has a significant effect on all the resulting effects. Workpiece hardness will be an input factor but it is not usually controllable.

TABLE 28.2   **Grinding Parameters**

Independent Parameters/Controllable	Dependent Variables/Resulting Effects
Grinding wheel selection	Forces per unit width of wheel
Abrasive type	Normal
Grain size	Tangential
Hardness grade	Surface finish
Openness of structure	Material removal rate (MRR)
Bonding media	Wheel wear (G, or grinding ratio)
Dressing of wheel	Thermal effects
Type of dressing tool	Wheel surface changes
Feed and depth of cut	Chemical effects
Sharpness of dressing tool	Horsepower
Machine settings	
Wheel speed	
Infeed rate (depth of cut)	
Cross-feed rate	
Workpiece speed	
Rigidity of setup	
Type and quality of machine	
Grinding fluid	
Type	
Cleanliness	
Method of application	

28.2 Abrasives

An **abrasive** is a hard material that can cut or abrade other substances. Natural abrasives have existed from the earliest times. For example, sandstone was used by ancient peoples to sharpen tools and weapons. Early grinding wheels were cut from slabs of sandstone, but because they were not uniform in structure throughout, they wore unevenly and did not produce consistent results. **Emery**, a mixture of alumina (Al_2O_3) and magnetite (Fe_3O_4), is another natural abrasive still in use today and is used on coated paper and cloth (emery paper). **Corundum** (natural Al_2O_3) and diamonds are other naturally occurring abrasive materials. Today, the only natural abrasives that have commercial importance are quartz, sand, garnets, and diamonds. For example, **quartz** is used primarily in coated abrasives and in air blasting, but artificial abrasives are also making inroads in these applications. The development of artificial abrasives having known uniform properties has permitted abrasive processes to become precision manufacturing processes.

Hardness, the ability to resist penetration, is the key property for an abrasive. **Table 28.3** lists the primary abrasives and their approximate Knoop hardness (kg/mm^2). The particles must be able to decompose at elevated temperatures. Two other properties are significant in abrasive grits—attrition and friability. **Attrition** refers to the abrasive wear action of the grits resulting in dulled edges, grit flattening, and wheel glazing. **Friability** refers to the fracture of the grits and is the opposite of toughness. In grinding, it is important that grits be able to fracture to expose new, sharp edges.

Artificial abrasives date from 1891, when E. G. Acheson, while attempting to produce precious gems, discovered how to make **silicon carbide** (SiC). Silicon carbide is made by charging an electric furnace with silica sand, petroleum coke, salt, and sawdust. By passing large amounts of current through the charge, a temperature of greater than 4000°F is maintained for several hours, and a solid mass of silicon carbide crystals results. After the furnace has cooled, the mass of crystals is removed, crushed, and graded or sorted into various desired sizes. As can be seen in **Figure 28.2**, the resulting grits, or grains, are irregular in shape, with cutting edges having every possible rake angle. Silicon carbide crystals are very hard (Knoop 2480), friable, and rather brittle. This limits their use. Silicon carbide is sold under the trade names Carborundum and Crystolon.

TABLE 28.3 **Knoop Hardness Values for Common Abrasives**

Abrasive Material	Year of Discovery	Hardness (Knoop)	Temperature of Decomposition in Oxygen (°C)	Comments and Uses
Quartz	?	320		Sand blasting
Aluminum oxide	1893	1600–2100	1700–2400	Softer and tougher than silicon carbide; used on steel, iron, brass, silicon
Carbide	1891	2200–2800	1500–2000	Used for brass, bronze, aluminum, and stainless and cast iron
Borazon [cubic boron nitride (CBN)]	1957	4200–5400	1200–1400	For grinding hard, tough tool steels, stainless steel, cobalt and nickel-based superalloys, and hard coatings
Diamond (synthetic)	1955	6000–9000	700–800	Used to grind nonferrous materials, tungsten carbide, and ceramics

FIGURE 28.2 Loose' abrasive grains of silicon carbide at high magnification, showing their irregular, sharp cutting edges. (Courtesy of Norton Abrasives/Saint Gobain)

Aluminum oxide (Al_2O_3) is the most widely used artificial abrasive. Also produced in an arc furnace from bauxite, iron filings, and small amounts of coke, it contains aluminum hydroxide, ferric oxide, silica, and other impurities. The mass of aluminum oxide that is formed is crushed, and the particles are graded to size. Common trade names for aluminum oxide abrasives are Alundum and Aloxite. Although aluminum oxide is softer (Knoop 2100) than silicon carbide, it is considerably tougher. Consequently, it is a better general-purpose abrasive.

Diamonds are the hardest of all materials. Those that are used for abrasives are either natural, off-color stones, called **garnets**, that are not suitable for gems, or small, synthetic stones that are produced specifically for abrasive purposes. Manufactured stones appear to be somewhat more friable and thus tend to cut faster and cooler. They do not perform as satisfactorily in metal-bonded wheels. Diamond abrasive wheels are used extensively for sharpening carbide and ceramic cutting tools. Diamonds also are used for truing and dressing other types of abrasive wheels. Diamonds are usually used only when cheaper abrasives will not produce the desired results. Garnets are used primarily in the form of very finely crushed and graded powders for fine polishing.

Cubic boron nitride (CBN) is not found in nature. It is produced by a combination of intensive heat and pressure in the presence of a catalyst. CBN is extremely hard, registering at 4700 on the Knoop scale. It is the second-hardest substance and is often referred to, along with diamonds, as a superabrasive. Hardness, however, is not everything.

CBN far surpasses diamond in the important characteristic of thermal resistance. At temperatures of 650°C, at which diamond may begin to revert to plain carbon dioxide, CBN continues to maintain its hardness and chemical integrity. When the temperature of 1400°C is reached, CBN changes from its cubic form to a hexagonal form and loses hardness. CBN can be used successfully in grinding iron, steel, and alloys of iron, nickel-based alloys, and other materials. CBN works very effectively on hardened materials (R_c 50 or higher), having long wheel life, high G ratio, good surface quality, no burn or chatter, low scrap rate, and overall increase in parts/shift. It can also be used for soft steel in selected situations. CBN does well at conventional grinding speeds (6000 to 12,000 ft/min), resulting in lower total grinding costs in conventional equipment. CBN can also perform well at high grinding speeds (12,000 ft/min and higher) and will enhance the benefits from future machine tools. CBN can solve difficult-to-grind jobs and can generate cost benefits in many production grinding operations despite its higher cost. CBN is manufactured by the General Electric Company under the trade name of Borazon.

Abrasive Grain Size and Geometry

To enhance the process capability of grinding, abrasive grains are sorted into sizes by mechanical sieving machines. The number of openings per linear inch in a sieve, or screen, through which most of the particles of a particular size can pass determines the grain size (**Figure 28.3**).

A no. 24 grit would pass through a standard screen having 24 openings per inch but would not pass through one having 30 openings per inch. These numbers have since been specified in terms of millimeters and micrometers (see ANSI B74.12 for details). Commercial practice commonly designates grain sizes from 4 to 24, inclusive, as *coarse;* 30 to 60, inclusive, as *medium;* and 70 to 600, inclusive, as *fine.* Grains smaller than 220 are usually termed *powders.* Silicon carbide is obtainable in grit sizes ranging from 2 to 240, and aluminum oxide in sizes from 4 to 240. Superabrasive grit sizes normally range from 120 grit for CBN to 400 grit for diamond. Sizes from 240 to 600 are designated as *flour* sizes. These are used primarily for lapping, or in fine-honing stones for fine finishing tasks.

The grain size is closely related to the surface finish and metal removal rate. In grinding wheels and belts, coarse grains cut faster (higher metal removal rate) while fine grains provide better finish, as shown in **Figure 28.4**.

The grain diameter can be estimated from the screen number (*S*), which corresponds to the number of openings per inch.

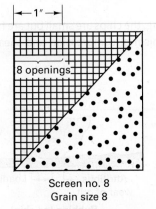
Screen no. 8
Grain size 8

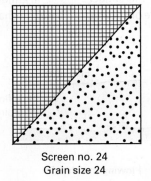
8 openings
Screen no. 24
Grain size 24

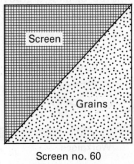
Screen
Grains
Screen no. 60
Grain size 60

FIGURE 28.3 Typical screens for sifting abrasives into sizes. The larger the screen number (of opening per linear inch), the smaller the grain size. (Courtesy of Carborundum Company)

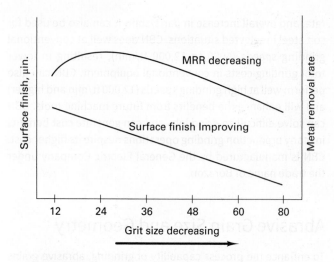

FIGURE 28.4 MRR and surface finish versus grit size.

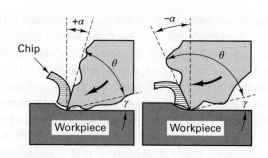

FIGURE 28.5 The rake angle of abrasive particles can be positive, zero, or negative.

The mean diameter of the grain (g) is related to the screen number by $g \cong 0.7/S$.

Regardless of the size of the grain, only a small percentage (2% to 5%) of the surface of the grain is operative at any one time. That is, the depth of cut for an individual grain (the actual feed per grit) with respect to the grain diameter is very small. Thus, the chips are small. As the grain diameter decreases, the number of active grains per unit area increases and the cuts become finer because grain size is the controlling factor for surface finish (roughness). Of course, the MRR also decreases.

The grain shape is also important, because it determines the tool geometry—that is, the back rake angle and the clearance angle at the cutting edge of the grit as shown in (**Figure 28.5**).

In this figure, γ is the clearance angle, θ is the wedge angle, and α is the rake angle. The cavities between the grits provide space for the chips, as shown in **Figure 28.6**. The volume of the cavities must be greater than the volume of the chips generated during the cut.

Obviously, there is no specific rake angle but rather a distribution of angles. Thus a grinding wheel can present to the surface a range of rake angles from +45 to −60° or greater. Grits with large negative rake angles or rounded cutting edges do not form chips but will rub or *plow* a groove in the surface (**Figure 28.7**). Thus abrasive machining is a mixture of **cutting**,

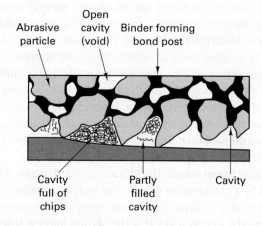

FIGURE 28.6 The cavities, or voids, between the grains must be large enough to hold all the chips during the cut.

plowing, and **rubbing**, with the percentage of each being highly dependent on the geometry of the grit. As the grits are continuously abraded, fractured, or dislodged from the bond, new grits are exposed and the mixture of cutting, plowing, and rubbing is continuously changing. A high percentage of the energy used for rubbing and plowing goes into the workpiece, but when chips are found, 95% to 98% of the energy (the heat) goes into the chip. **Figure 28.8** shows a scanning electron microscope (SEM) micrograph of a ground surface with a plowing track.

In grinding, the chips are small but are formed by the same basic mechanism of compression and shear as regular metal

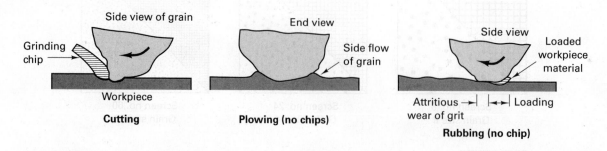

FIGURE 28.7 The grits interact with the surface in three ways: cutting, plowing, and rubbing.

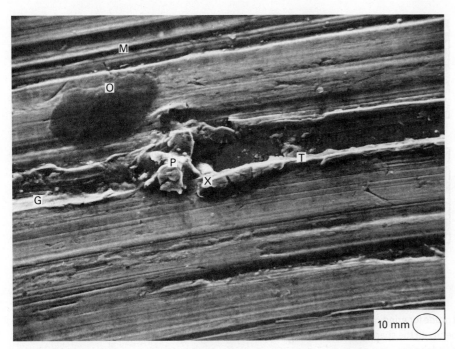

FIGURE 28.8 SEM micrograph of a ground steel surface showing a plowed track (T) in the middle and a machined track (M) above. The grit fractured, leaving a portion of the grit in the surface (X), a prow formation (P), and a groove (G) where the fractured portion was pushed farther across the surface. The area marked (O) is an oil deposit. (Courtesy J T. Black)

10 mm

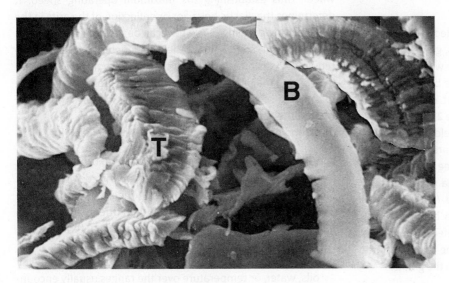

FIGURE 28.9 SEM micrograph of stainless steel chips from a grinding process. The tops (T) of the chips have the typical shear-front-lamella structure while the bottoms (B) are smooth where they slide over the grit 4800×. (Courtesy J T. Black)

cutting discussed in Chapter 21. **Figure 28.9** shows a SEM micrography of steel chips from a grinding process at high magnification. They show the same structure as chips from other machining processes. Chips flying in the air from a grinding process often have sufficient heat energy to burn or melt in the atmosphere. Sparks observed during grinding steel with no cutting fluid are really burning chips.

The feeds and depths of cut in grinding are small, but the cutting speeds are high, resulting in high specific horsepower numbers. Because cutting is obviously more efficient than plowing or rubbing, grain fracture, and grain pullout are natural phenomena used to keep the grains sharp. As the grains become dull, cutting forces increase, and there is an increased tendency for the grains to fracture or break free from the bonding material.

28.3 Grinding Wheel Structure and Grade

Grinding, wherein the abrasives are bonded together into a wheel, is the most common abrasive machining process. The performance of grinding wheels is greatly affected by the bonding material and the spatial arrangement of the particles' grits.

The spacing of the abrasive particles with respect to each other is called **structure**. Close-packed grains have dense structure; open structure means widely spaced grains. Open-structure wheels have larger chip cavities but fewer cutting edges per unit area (**Figure 28.10a**).

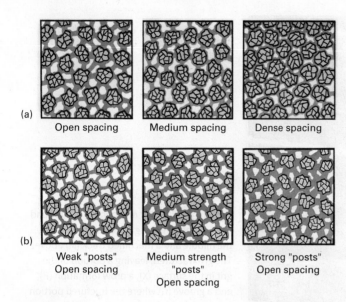

(a) Open spacing | Medium spacing | Dense spacing

(b) Weak "posts" Open spacing | Medium strength "posts" Open spacing | Strong "posts" Open spacing

FIGURE 28.10 Meaning of terms *structure* and *grade* for grinding wheels. (a) The structure of a grinding wheel depends on the spacing of the grits. (b) The grade of a grinding wheel depends on the amount of bonding agent (posts) holding abrasive grains in the wheel.

The fracturing of the grits is controlled by the bond strength, which is known as the **grade**. Thus, grade is a measure of how strongly the grains are held in the wheel. It is really dependent on two factors: the strength of the bonding materials, and the amount of the bonding agent connecting the grains. The latter factor is illustrated in **Figure 28.10b**. Abrasive wheels are really porous. The grains are held together with "posts" of bonding material. If these posts are large in cross section, the force required to break a grain free from the wheel is greater than when the posts are small. If a high dislodging force is required, the bond is said to be *hard*. If only a small force is required, the bond is said to be *soft*. Wheels are commonly referred to as hard or soft, referring to the net strength of the bond, resulting from both the strength of the bonding material and its disposition between the grains.

G Ratio

The loss of grains from the wheel means that the wheel is changing size. The grinding ratio, or **G ratio**, is defined as the cubic inches of stock removed divided by the cubic inches of wheel lost. In conventional grinding, the *G* ratio is in the range 20:1 to 80:1. The *G* ratio is a measure of grinding production and reflects the amount of work a wheel can do during its useful life. As the wheel loses material, it must be reset or repositioned to maintain workpiece size.

A typical vitrified grinding wheel will consist of 50 vol% abrasive particles, 10 vol% bond, and 40 vol% cavities; that is, the wheels have porosity. The manner in which the wheel performs is influenced by the following factors:

1. The mean force required to dislodge a grain from the surface (the grade of the wheel).

2. The cavity size and distribution of the porosity (the structure).

3. The mean spacing of active grains in the wheel surface (grain size and structure).

4. The properties of the grain (hardness, attrition, and friability).

5. The geometry of the cutting edges of the grains (rake angles and cutting-edge radius compared to depth of cut).

6. The process parameters (speeds, feeds, cutting fluids) and type of grinding (surface or cylindrical).

7. The material being ground (the workpiece material).

It is easy to see why grinding is a complex process, difficult to control.

Bonding Materials for Grinding Wheels

Bonding material is a very important factor to be considered in selecting a grinding wheel. It determines the strength of the wheel, thus establishing the maximum operating speed. It determines the elastic behavior or deflection of the grits in the wheel during grinding. The wheel can be hard or rigid, or it can be flexible or soft. Finally, the bond determines the force required to dislodge an abrasive particle from the wheel and thus plays a major role in the cutting action. Bond materials are formulated so that the ratio of bond wear matches the rate of wear of the abrasive grits.

Bonding materials in common use are the following:

1. **Vitrified bonds** are composed of clays and other ceramic substances. The abrasive particles are mixed with the wet clays so that each grain is coated. Wheels are formed from the mix, usually by pressing, and then dried. They are then fired in a kiln, which results in the bonding material becoming hard and strong, having properties similar to glass. Vitrified wheels are porous, strong, rigid, and unaffected by oils, water, or temperature over the ranges usually encountered in metal cutting. The operating speed range in most cases is 5500 to 6500 ft/min, but some wheels now operate at surface speeds up to 16, 000 ft/min.

2. **Resinoid bonds**, or phenolic resins, can be used. Because plastics can be compounded to have a wide range of properties, such wheels can be obtained to cover a variety of work conditions. They have, to a considerable extent, replaced shellac and rubber wheels. Composite materials are being used in rubber-bonded or resinoid-bonded wheels that need to have some degree of flexibility or receive considerable abuse and side loading. Various natural and synthetic fabrics and fibers, glass fibers, and nonferrous wire mesh are used for this purpose.

3. **Silicate bond** wheels use silicate of soda or waterglass, as the bond material. The wheels are formed and then baked at about 500°F for a day or more. Because they are more brittle and not so strong as vitrified wheels, the abrasive

grains are released more readily. Consequently, they machine at lower surface temperatures than vitrified wheels and are useful in grinding tools when heat must be kept to a minimum.

4. **Shellac-bonded** wheels are made by mixing the abrasive grains with shellac in a heated mixture, pressing or rolling into the desired shapes, and baking for several hours at about 300°F. This type of bond is used primarily for strong, thin wheels having some elasticity. They tend to produce a high polish and thus have been used in grinding such parts as camshafts and mill rolls.

5. **Rubber bonding** is used to produce wheels that can operate at high speeds but must have a considerable degree of flexibility so as to resist side thrust. Rubber, sulfur, and other vulcanizing agents are mixed with the abrasive grains. The mixture is then rolled out into sheets of the desired thickness, and the wheels are cut from these sheets and vulcanized. Rubber-bonded wheels can be operated at speeds up to 16,000 ft/min. They are commonly used for snagging work in foundries and for thin cutoff wheels.

6. **Superabrasive bond** wheels are either electroplated (single layer of superabrasive plated to outside diameter of a steel blank) or a thin-segmented drum of vitrified CBN surrounds a steel core. The steel core provides dimensional accuracy, and the replaceable segments provide durability, homogeneity, and repeatability while increasing wheel life. The latter type of wheels can use resin, metal, or vitrified bonding. Selection of bond grade and structure, also called abrasive concentration, is critical.

For the electroplated wheels, nickel is used to attach a single layer of CBN (or diamond) to the outside diameter (OD) of an accurately ground or turned steel blank. For the vitrified wheel, superabrasives are mixed with bonding media and

molded (or preformed and sintered) into segments or a ring. The ring is mounted on a split steel body. Porosity is varied (to alter structure) by varying preform pressure or by using "pore-forming" additives to the bond material that are vaporized during the sintering cycle. The steel-cored segmented design can rotate at 40,000 sfpm (200 m/s) whereas a plain vitrified wheel can burst at 20,000 fpm.

Abrasive Machining Versus Conventional Grinding Versus Low-Stress Grinding

The condition wherein very rapid metal removal can be achieved by grinding is the one to which some have applied the term *abrasive machining*. The metal removal rates are compared with, or exceed, those obtainable by milling or turning or broaching, and the size tolerances are comparable. It is obviously just a special type of grinding, using abrasive grains as cutting tools, as do all other types of abrasive machining. Abrasive grinding done in an aggressive way can produce sufficient localized plastic deformation and heat in the surface so as to develop tensile residual stresses, layers of overtempered martensite (in steels), and even microcracks, because this process is quite abusive. See **Figure 28.11** for a discussion of residual stresses produced by various surface-grinding processes, and conditions.

Conventional grinding can be replaced by procedures that develop lower surface stresses when service failures due to fatigue or stress corrosion are possible. Low-stress grinding (LSG) is accomplished by employing softer grades of grinding wheels, reducing the grinding speeds and infeed rates, using chemically active cutting fluids (e.g., highly sulfurized oil or KNO_2 in water), as outlined in the table of grinding conditions in Figure 28.11.

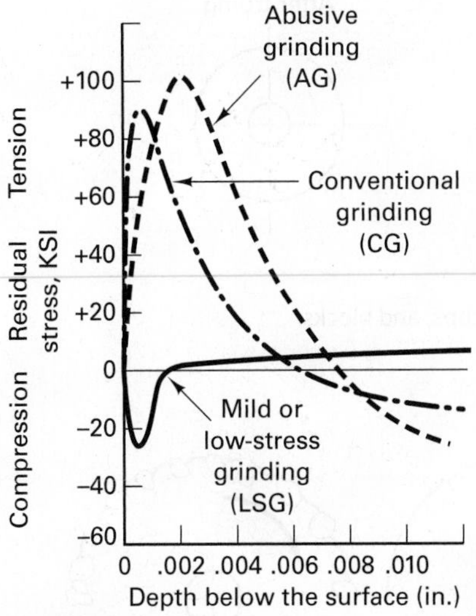

Grinding conditions	Abusive AG	Conventional CG	Low-stress LSG
Wheel	A46MV	A46KV	A46HV or A60IV
Wheel speed ft/min	6000–18,000	4500–6500	2500–3000
Down feed in./pass	.002–.004	.001–.003	.0002–.005
Cross feed in./pass	.004–.060	.040–.060	.040–.060
Table speed ft/min	40–100	40–100	40–100
Fluid	Dry	Sol oil (1 : 20)	Sulfurized oil

FIGURE 28.11 Typical residual stress distributions produced by surface grinding with different grinding conditions for abusive, conventional, and low-stress grinding. Material is 4340 steel. (From M. Field and W. P. Kosher, "Surface Integrity in Grinding, " in *New Developments in Grinding*, Carnegie-Mellon University Press, Pittsburgh, 1972, p. 666)

These procedures may require the addition of a variable-speed drive to the grinding machine. Generally, only about 0.005 to 0.010 in. of surface stock needs to be finish ground in this way, as the depth of the surface damage due to conventional grinding or abusive grinding is 0.005 to 0.007 in. High-strength steels, high-temperature nickel, and cobalt-based alloys and titanium alloys are particularly sensitive to surface deformation and cracking problems from grinding. Other post-processing processes, such as polishing, honing, and chemical milling plus peening, can be used to remove the deformed layers in critically stressed parts. It is strongly recommended, however, that testing programs be used along with service experience on critical parts before these procedures are employed in production.

In the casting and forging industries, the term often used for abrasive machining is *snagging*. **Snagging** is a type of rough manual grinding that is done to remove fins, gates, risers, and rough spots from castings or flash from forgings, preparatory to further machining. The primary objective is to remove substantial amounts of metal rapidly without much regard for accuracy, so this is a form of abrasive machining except that pedestal-type or **swing grinders** ordinarily are used. Portable electric or hand air grinders are also used for this purpose and for miscellaneous grinding in connection with welding.

Truing and Dressing

Grinding wheels lose their geometry during use. **Truing** restores the wheel to its original shape. A single-point diamond tool can be used to *true* the wheel while fracturing abrasive grains to expose new grains and new cutting edges on worn, glazed grains (**Figure 28.12**). Truing can also be accomplished by grinding the grinding wheel with a controlled-path or powered rotary device using conventional abrasive wheels. The precision in generating a trued wheel surface by these methods is poorer than by truing with a nib.

As the wheel is used, there is a tendency for the wheel to become *loaded* (metal chips become lodged in the cavities between the grains). Also, the grains dull or glaze (grits wear, flatten, and polish). Unless the wheel is cleaned and sharpened (or dressed), the wheel will not cut as well and will tend to plow and rub more. **Figure 28.13** shows an arrangement for **stick dressing** a grinding wheel.

The dulled grains cause the cutting forces on the grains to increase, ideally resulting in the grains' fracturing or being pulled out of the bond, thus providing a continuous exposure of sharp cutting edges. Such a continuous action ordinarily will not occur for light feeds and depths of cut. For heavier cuts, grinding wheels do become somewhat

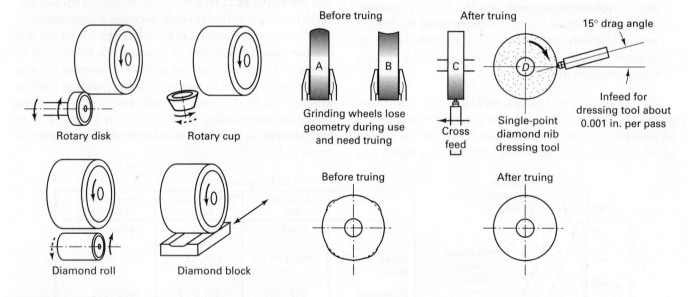

FIGURE 28.12 Truing methods for restoring grinding geometry include nibs, rolls, disks, cups, and blocks.

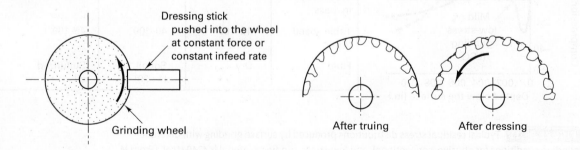

FIGURE 28.13 Schematic arrangement of stick dressing versus truing.

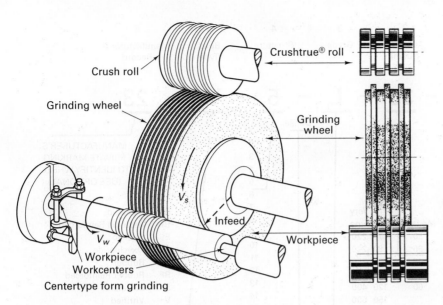

FIGURE 28.14 Continuous crush roll dressing and truing of a grinding wheel that if performing plunge-cut grinding on a cylinder held between centers. Form - truing and dressing are performed throughout the process rather than between cycles.

self-dressing, but the workpiece may become overheated and turn a bluish temper color (called **burn**) before the wheel reaches a fully dressed condition. A burned surface, the consequence of an oxide layer formation, may result in the scrapping of several workpieces before parts of good quality are ground.

Resin-bonded wheels can be trued by grinding with hard ceramics such as tungsten carbide. The procedure for truing and dressing a CBN wheel in a surface grinder might be as follows. Use 0.0002-in. downfeed per pass and cross feed slightly more than half the wheel thickness at moderate table speeds. The wheel speed is the same as the grinding speed. The grinding power will gradually increase, as the wheel is getting dull while it is being trued. When the power exceeds normal power drawn during workpiece grinding, stop the truing operation. Dress the wheel face open using a J-grade stick, with abrasive one grit size smaller than CBN. Continue the truing. Repeat this cycle until the wheel is completely trued.

Modern grinding machines are equipped so that the wheel can be dressed and/or trued continuously or intermittently while grinding continues. A common way to do this is by **crush dressing** (**Figure 28.14**). Crush dressing consists of forcing a hard roll (tungsten carbide or high speed steel) having the same contour as the part to be ground against the grinding wheel while it is revolving, usually quite slowly. A water-based coolant is used to flood the dressing zone at 5 to 10 gal/min. The crushing action fractures and dislodges some of the abrasive grains, exposing fresh sharp edges, allowing free cutting for faster infeed rates. This procedure is usually employed to produce and maintain a special contour to the abrasive wheel. Crush dressing is also called wheel profiling. Crush dressing is a very rapid method of dressing grinding wheels, and because it fractures abrasive grains, it results in free cutting and somewhat cooler grinding. The resulting surfaces of the grinding wheel may be slightly rougher than when diamond dressing is used.

28.4 Grinding Wheel Identification

Most grinding wheels are identified by a standard marking system that has been established by the American National Standards Institute (ANSI). This system is illustrated and explained in **Figure 28.15**. The first and last symbols in the marking are left to the discretion of the manufacturer.

Grinding Wheel Geometry

The shape and size of the wheel are critical selection factors. Obviously, the shape must permit proper contact between the wheel and all of the surface that must be ground. Grinding wheel shapes have been standardized, and eight of the most commonly used types are shown in **Figure 28.16**. Types 1, 2, and 5 are used primarily for grinding external or internal cylindrical surfaces and for plain surface grinding. Type 2 can be mounted for grinding on either the periphery or the side of the wheel. Type 4 is used with tapered safety flanges so that if the wheel breaks during rough grinding, such as snagging, these flanges will prevent the pieces of the wheel from flying and causing damage. Type 6, the straight cup, is used primarily for surface grinding but can also be used for certain types of off-hand grinding. The flaring-cup type of wheel is used for tool grinding. Dish-type wheels are used for grinding tools and saws.

Type 1, the straight grinding wheels, can be obtained with a variety of standard faces. Some of these are shown in **Figure 28.17**.

The size of the wheel to be used is determined primarily by the spindle rpm values available on the grinding machine and the proper cutting speed for the wheel, as dictated by the type of bond. For most grinding operations the cutting speed is about 2500 to 6500 ft/min. Different types and grades of bond often justify considerable deviation from these speeds. For certain types

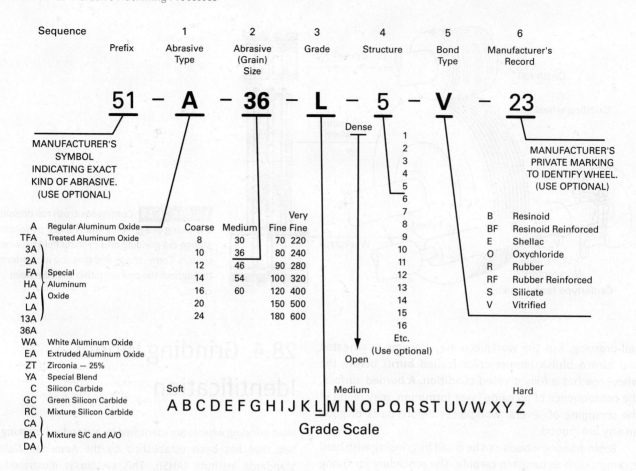

Standard bonded-abrasive wheel-marking system (ANSI *Standard* B74.13-1977).

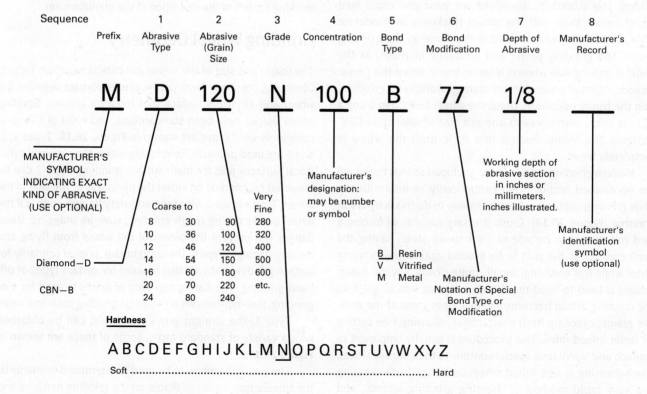

Wheel-marking system for diamond and cubic boron nitride wheels (ANSI *Standard* B74.13-1977).

FIGURE 28.15 Standard marking systems for grinding wheels (ANSI standard B74. 13-1977).

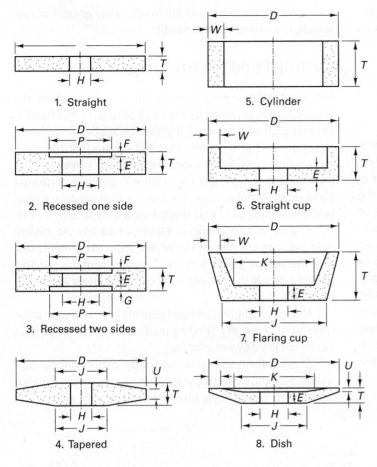

1. Straight

5. Cylinder

2. Recessed one side

6. Straight cup

3. Recessed two sides

7. Flaring cup

4. Tapered

8. Dish

FIGURE 28.16 Standard grinding wheel shapes commonly used. (Courtesy of Carborundum Company)

1. *Cutting off:* for slicing and slotting parts; use thin wheel, organic bond.

2. *Cylindrical between centers:* for grinding outside diameters of cylindrical workpieces.

3. *Cylindrical, centerless:* for grinding outside diameters with work rotated by regulating wheel.

4. *Internal cylindrical:* for grinding bores and large holes.

5. *Snagging:* for removing large amounts of metal without regard to surface finish or tolerances.

6. *Surface grinding:* for grinding flat workpieces.

7. *Tool grinding:* for grinding cutting edges on tools such as drills, milling cutters, taps, reamers, and single-point high-speed-steel tools.

8. *Offhand grinding:* work or the grinding tool is handheld.

In many cases, the classification of processes coincides with the classification of machines that do the process. Other factors that will influence the choice of wheel to be selected include the workpiece material, the amount of stock to be removed, the shape of the workpiece, and the accuracy and surface finish desired.

Workpiece material has a great impact on choice of the wheel. Hard, high-strength metals, such as tool steels or alloy steels are generally ground with aluminum oxide wheels or cubic boron nitride wheels. Silicon carbide and CBN are employed in grinding brittle materials, such as cast iron and ceramics, as well as softer, low-strength metals such as aluminum, brass, copper, and bronze. Diamonds have taken over the cutting of tungsten carbides, and CBN is used for precision grinding of tool and die steel, alloy steels, stainless steel, and other very hard materials. There are so many factors that affect the cutting action that there are no hard and fast rules with regard to abrasive selection.

Selection of grain size is determined by whether coarse or fine cutting and finish are desired. Coarse grains take larger depths of cut and cut more rapidly. Hard wheels with fine grains

of work using special wheels and machines, as in thread grinding and "abrasive machining, " much higher speeds are used.

Grinding Operations

The operation for which the abrasive wheel is intended will also influence the wheel shape and size. The major use categories are the following:

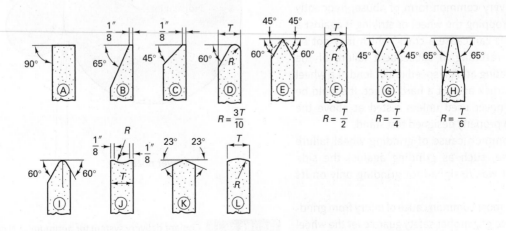

$$R = \frac{3T}{10}$$ $$R = \frac{T}{2}$$ $$R = \frac{T}{4}$$ $$R = \frac{T}{6}$$

FIGURE 28.17 Standard face contours for straight grinding wheels. (Courtesy of Carborundum Company)

leave smaller tracks and therefore are usually selected for finishing cuts. If there is a tendency for the work material to load the wheel, larger grains with a more open structure may be used for finishing.

Balancing Grinding Wheels

Because of the high rotation speeds involved, grinding wheels must never be used unless they are in good balance. A slight imbalance will produce vibrations that will cause waviness in the work surface. It may also cause a wheel to break, with the probability of serious damage and injury. The wheel should be mounted with proper bushings so that it fits snugly on the spindle of the machine. Rings of blotting paper should be placed between the wheel and the flanges to ensure that the clamping pressure is evenly distributed. Most grinding wheels will run in good balance if they are mounted properly and trued. Most machines have provisions for compensating for a small amount of wheel imbalance by attaching weights to one mounting flange. Some machines have provisions for semiautomatic balancing with weights that are permanently attached to the machine spindle.

Safety in Grinding

Because the rotational speeds are quite high, and the strength of grinding wheels is usually much less than that of the materials being ground, serious accidents occur much too frequently in the use of grinding wheels. Virtually all such accidents could be avoided and are due to one or a combination of four causes.

First, grinding wheels are occasionally operated at unsafe and improper speeds. All grinding wheels are clearly marked with the maximum rpm value at which they should be rotated. They are all tested to considerably above the designated rpm and are safe at the specified speed *unless abused. They should never, under any condition, be operated above the rated speed.*

Second, a very common form of abuse, frequently accidental, is dropping the wheel or striking it against a hard object. This can cause a crack, which may not be readily visible, resulting in subsequent failure of the wheel while rotating at high speed under load. If a wheel is dropped or struck against a hard object, it should be discarded and never used unless tested at above the rated speed in a properly designed test stand.

Third, a common cause of grinding wheel failure is improper use, such as grinding against the side of a wheel that was designed for grinding only on its periphery.

Fourth and most common cause of injury from grinding is the absence of a proper safety guard over the wheel and/or over the eyes or face of the operator. The frequency with which operators will remove safety guards

from grinding equipment, or fail to use safety goggles or face shields, is amazing and inexcusable.

Cutting Fluids in Grinding

Because grinding involves cutting, the selection and use of a cutting fluid is governed by the basic principles discussed in Chapter 22. If a fluid is used, it should be applied in sufficient quantities and in a manner that will ensure that the chips are washed away, not trapped between the wheel and the work. This is of particular importance in grinding horizontal surfaces. In hardened steel, the use of a fluid can help to prevent fine microcracks that result from highly localized heating. The air scraper shown in **Figure 28.18** lets the cutting fluid get onto the face of the wheel. Metal air scrapers disrupt the airflow. Upper and lower nozzles cool the grinding zone, while a high-pressure scrubber helps deter loading of the wheel.

Much snagging and off-hand grinding is done dry. On some types of material, dry grinding produces a better finish than can be obtained by wet grinding.

Grinding fluids strongly influence the performance of CBN wheels. Straight, sulfurized, or sulfochlorinated oils can enhance performance considerably when used with straight oils.

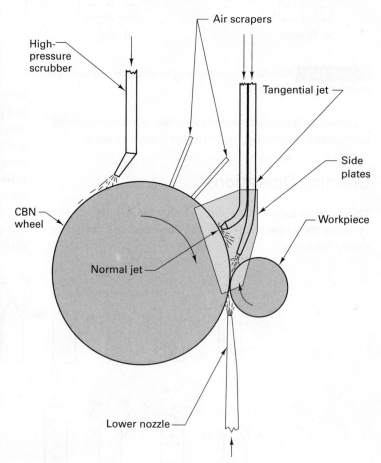

FIGURE 28.18 Coolant delivery system for optimum CBN grinding. (M. P. Hitchiner, "Production Grinding with CBN, " *Machining Technology*, Vol. 2, no. 2, 1991)

28.5 Grinding Machines

Grinding machines commonly are classified according to the type of surface they produce. **Table 28.4** presents such a classification, with further subdivision to indicate characteristic features of different types of machines within each classification.

Grinding on all machines is done in one of three ways. In the first, the depth of cut (d_t) is obtained by **infeed**—moving the wheel down into the work or the work up into the wheel (**Figure 28.19**). The desired surface is then produced by traversing

TABLE 28.4	Classification of Grinding Machines	
Type of Machine	**Type of Surface**	**Specific Types or Features**

Cylindrical external	External surface on rotating, usually cylindrical parts	Work rotated between centers
		    Centerless
		    Chucking
		    Tool post
		    Crankshaft, cam, etc.

Cylindrical internal	Internal diameters of holes	Chucking Planetary (work stationary) Centerless
Surface conventional	Flat surfaces	Reciprocating table or rotating table Horizontal or vertical spindle
Creep feed	Deep slots, profiles in hard steels, carbides, and ceramics using CBN and diamond	Rigid, chatter-free, creep feed rate Continuous dressing Heavy coolant flows NC or CNC control Variable speed wheel
Tool grinders	Tool angles and geometries	Universal Special
Other	Special or any of the above	Disk, contour, thread, flexible shaft, swing frame, snag, pedestal, bench

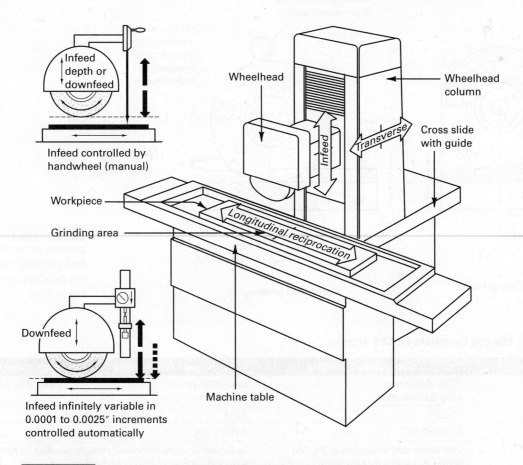

FIGURE 28.19 Horizontal-spindle surface grinder, with insets showing movements of wheelhead.

the wheel across the workpiece (cross feed), or vice versa. In the second method, known as **plunge-cut grinding**, the basic movement is of the wheel being fed radially into the work while the latter revolves on centers. It is similar to form cutting on a lathe; usually a formed grinding wheel is used (Figure 28.14). In the third method, the work is fed very slowly past the wheel and the total downfeed or depth (d) is accomplished in a single pass (**Figure 28.20**). This is called **creep feed grinding (CFG)**.

The CFG method, often done in the surface grinding mode, is markedly different from conventional surface grinding. The depth of cut is increased 1000 to 10,000 times, and the work feed ratio is decreased in the same proportion; hence the name *creep feed grinding*. The long arc of contact between the wheel and the work increases the cutting forces and the power required. Therefore, the machine tools performing this type of grinding must be specially designed with high static and dynamic stability, stick-slip-free ways, adequate damping, increased horsepower, infinitely variable spindle speed, variable but extremely consistent table feed (especially in the low ranges), high-pressure cooling systems, integrated devices for dressing the grinding wheels, and specially designed (soft with open structure) grinding wheels. The process is mainly applied when grinding deep slots with straight parallel sides or when grinding complex profiles in difficult-to-grind materials. The process is capable of producing extreme precision at relatively high metal removal rates (MRRs). Because the process can operate at relatively low surface temperatures, the surface integrity of the metals being ground is good.

However, in CFG, the grinding wheels must maintain their initial profile much longer, so continuous dressing is used that is form-truing and dressing the grinding wheel throughout the process rather than between cycles. Continuous crush dressing results in higher MRRs, improved dimensional accuracy and form tolerance, reduced grinding forces (and power), and reduced thermal effects while sacrificing wheel wear. Creep feed grinding eliminates preparatory operations such as milling or broaching, because profiles are ground into the solid workpiece. This can result in significant savings in unit part costs. (**Table 28.5** compares CFG to conventional and high-speed grinding for CBN applications.)

Grinding machines that are used for precision work have certain important characteristics that permit them to produce parts having close dimensional tolerances. They are constructed very accurately, with heavy, rigid frames to ensure permanency of alignment. Rotating parts are accurately balanced to avoid vibration. Spindles are mounted in very accurate bearings, usually of the preloaded ball-bearing type. Controls are provided so that all movements that determine dimensions of the workpiece can be made with accuracy—usually to 0.001 or 0.00001 in.

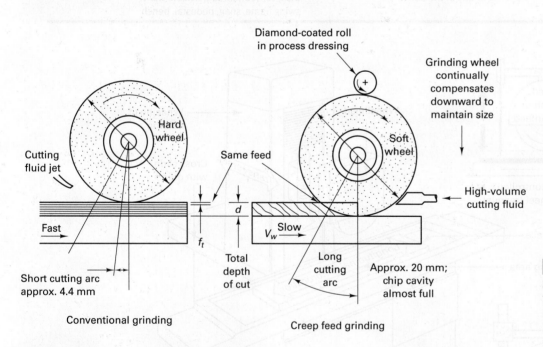

FIGURE 28.20 Conventional grinding contrasted to creep feed grinding. Note that crush roll dressing is used here; see Figure 28.14.

TABLE 28.5	Starting Conditions for CBN Grinding		
Grinding Variable	**Conventional Grinding**	**Creep Feed Grinding**	**High-speed Grinding**
Wheel speed (fpm)	5500–9500 versus 4500–6500 vitrified	5000–9000 versus 3000–5000 vitrified	12000–2500
Table speed (fpm)	80–150	0.5–5	5–20
Feed (f_t) in./pass	0.0005–0.0015	0.100–0.250	250–500
Grinding fluids	10% heavy-duty soluble oil or 3%–5% light-duty soluble for light feeds	Sulfurized or sulfochlorinated straight grinding oil applied at 80 to 100 gal/min at 100 psi or more	

The abrasive dust that results from grinding must be prevented from entering between moving parts. All ways and bearings must be fully covered or protected by seals. If this is not done, the abrasive dust between moving parts becomes embedded in the softer of the two, causing it to act as lap and abrade the harder of the two surfaces, resulting in permanent loss of accuracy.

These special characteristics add considerably to the cost of these machines and require that they be operated by trained personnel. Production-type grinders are more fully automated and have higher metal removal rates and excellent dimensional accuracy. Fine surface finish can be obtained very economically.

Cylindrical Grinding

Center-type cylindrical grinding is commonly used for producing external cylindrical surfaces. Figures 28.14 and **Figure 28.21** show the basic principles and motions of this process. The grinding wheel revolves at an ordinary cutting speed, and the workpiece rotates on centers at a much slower speed, usually from 75 to 125 ft/min. The grinding wheel and the workpiece move in opposite directions at their point of contact. The depth of cut is determined by infeed of the wheel. Because this motion also determines the finished diameter of the workpiece, accurate control of this movement is required. Provision is made to traverse the workpiece with the wheel, or the work can be reciprocated past the wheel. In very large grinders, the wheel is reciprocated because of the massiveness of the work. For form or plunge grinding, the detail of the wheel is maintained by periodic crush roll dressing.

Figure 28.22 shows a top view of a plain center-type cylindrical grinding process. The work is mounted between headstock and tailstock centers. Solid dead centers are always used in the tailstock, and provision is usually made so that the headstock center can be operated either dead or alive.

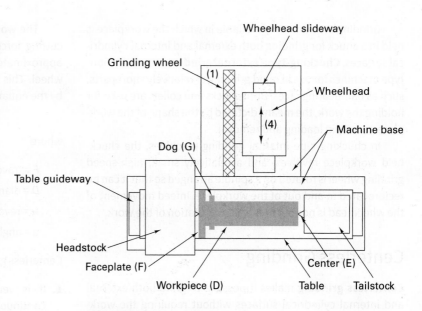

FIGURE 28.22 Schematic of a cylindrical center grinder from the overhead view. (1) Wheel (4) Infeed.

High-precision work is usually ground with a dead headstock center, because this eliminates any possibility that the workpiece will run out of round due to any eccentricity in the headstock.

The table assembly can be reciprocated in most cases, by using a hydraulic drive. The speed can be varied, and the length of the movement can be controlled by means of adjustable trip dogs.

Infeed is provided by movement of the wheelhead at right angles to the longitudinal axis of the table. The spindle is driven by an electric motor that is also mounted on the wheelhead. If the infeed movement is controlled manually by some type of vernier drive to provide control to 0.001 in. or less, the machine is usually equipped with digital readout equipment to show the exact size being produced. Most production-type grinders have automatic infeed with retraction when the desired size has been obtained. Such machines are usually equipped with an automatic diamond wheel-truing device that dresses the wheel and resets the measuring element before grinding is started on each piece.

The longitudinal traverse should be about one-fourth to three-fourths of the wheel width for each revolution of the work. For light machines and fine finishes, it should be held to the smaller end of this range. The depth of cut (infeed) varies with the purpose of the grinding operation and the finish desired. When grinding is done to obtain accurate size, infeeds of 0.002 to 0.004 in. are commonly used for roughing cuts. For finishing, the infeed is reduced to 0.00025 to 0.0005 in. The design allowance for grinding should be from 0.005 to 0.010 in. on short parts and on parts that are not to be hardened. On long or large parts and on work that is to be hardened, a grinding allowance of from 0.015 to 0.030 in. is desirable. When grinding is used primarily for metal removal (called abrasive machining), infeeds are much higher, 0.020 to 0.040 in. being common. Continuous downfeed is often used, with rates up to 0.100 in./min being common.

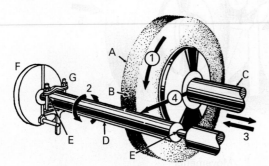

Movements
1. Wheel speed   2. Work (rotates) rpm
3. Traverse feed   4. Infeed

FIGURE 28.21 Cylindrical grinding between centers. A = edge of wheel, B = face of wheel, C = shaft for wheel, D = workpiece, E = centers, F = faceplate, G = dog.

Grinding machines are available in which the workpiece is held in a chuck for grinding both external and internal cylindrical surfaces. **Chucking-type external grinders** are production-type machines for use in rapid grinding of relatively short parts, such as ball-bearing races. Both chucks and collets are used for holding the work, the means dictated by the shape of the workpiece and rapid loading and removal.

In chucking-type internal grinding machines, the chuck-held workpiece revolves, and a relatively small, high-speed grinding wheel is rotated on a spindle arranged so that it can be reciprocated in and out of the workpiece. Infeed movement of the wheelhead is normal to the axis of rotation of the work.

Centerless Grinding

Centerless grinding makes it possible to grind both external and internal cylindrical surfaces without requiring the workpiece to be mounted between centers or in a chuck. This eliminates the requirement of center holes in some workpieces and the necessity for mounting the workpiece, thereby reducing the processing time.

The principle of centerless *external* grinding is illustrated in **Figure 28.23**. Two wheels are used. The larger one operates at regular grinding speeds and does the actual grinding. The smaller wheel is the **regulating wheel**. It is mounted at an angle to the plane of the grinding wheel. Revolving at a much slower surface speed—usually 50 to 200 ft/min—the regulating wheel controls the rotation and longitudinal motion of the workpiece (the axial feed). This wheel is a usually a plastic- or rubber-bonded wheel with a fairly wide face.

The workpiece is held against the work-rest blade by the cutting forces exerted by the grinding wheel and rotates at approximately the same surface speed as that of the regulating wheel. This traverse or axial feed is calculated approximately by the equation

$$F = ND \sin\upsilon \qquad (28.1)$$

where

F = traverse feed (mm/min or in./min)

D = diameter of the regulating wheel (mm or in.)

N = revolutions per minute of the regulating wheel

υ = angle of tilt of the regulating wheel

Centerless grinding has several important advantages:

1. It is very rapid; infeed centerless grinding is almost continuous.
2. Requires very little skill of the operator.
3. It can often be made automatic (single-cycle automatic).
4. Where the cutting occurs, the work is fully supported by the work rest and the regulating wheel. This permits heavy cuts to be made.
5. Because there is no distortion of the workpiece, accurate size control is easily achieved.
6. Large grinding wheels can be used, minimizing wheel wear.

Thus, centerless grinding is ideally suited to certain types of mass-production operations.

A. Grinding wheel
B. Grinding face
C. Regulating wheel
D. Workpiece
E. Work rest blade

υ = Angle of tilt of regulating wheel

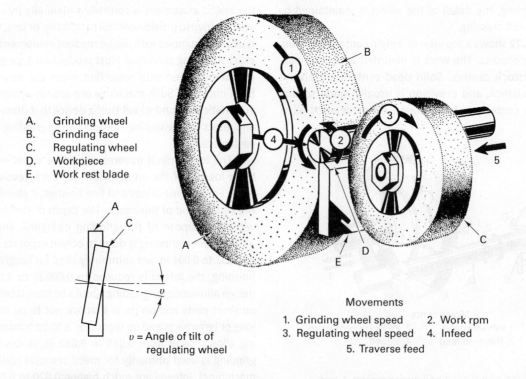

Movements
1. Grinding wheel speed
2. Work rpm
3. Regulating wheel speed
4. Infeed
5. Traverse feed

FIGURE 28.23 Centerless grinding showing the relationship among the grinding wheel, the regulating wheel, and the workpiece in centerless method. (Courtesy of Carborundum Company)

The major disadvantages are as follows:

1. Special machines are required that can do no other type of work.
2. The work must be round—no flats, such as keyways, can be present.
3. Useage on work having more than one diameter or on curved parts is limited.
4. In grinding tubes, there is no guarantee that the OD and inside diameter (ID) are concentric.

Special centerless grinding machines are available for grinding balls and tapered workpieces. The centerless grinding principle can also be applied to internal grinding, but the external surface of the cylinder must be finished accurately before the internal operation is started. However, it ensures that the internal and external surfaces will be concentric. The operation is easily mechanized for many applications.

Surface Grinding Machines

Surface grinding machines are used primarily to grind flat surfaces. However formed, irregular surfaces can be produced on some types of surface grinders by use of a formed wheel. There are four basic types of surface grinding machines. They differ in the movement of their tables and the orientation of the grinding wheel spindles (**Figure 28.24**):

1. Horizontal spindle and reciprocating table.
2. Vertical spindle and reciprocating table.
3. Horizontal spindle and rotary table.
4. Vertical spindle and rotary table.

The most common type of surface grinding machine has a reciprocating table and horizontal spindle (Figures 28.19). The table can be reciprocated longitudinally either by hand-wheel or by hydraulic power. The wheelhead is given traverse or cross-feed motion at the end of each table motion, again either by handwheel or by hydraulic power feed. Both the longitudinal and transverse motions can be controlled by limit switches. Infeed or downfeed on such grinders is controlled by hand-wheels or automatically. The size of such machines is determined by the size of the surface that can be ground.

In using such machines, the wheel should overtravel the work at both ends of the table reciprocation, so as to prevent the wheel from grinding in one spot while the table is being reversed. The traverse or cross-feed motion should be one-fourth to three-fourths of the wheel width between each stroke.

Vertical-spindle reciprocating-table surface grinders differ basically from those with horizontal spindles only in that their spindles are vertical and that the wheel diameter must exceed the width of the surface to be ground. Usually, no traverse motion of either the table or the wheelhead is provided. Such machines can produce very flat surfaces.

Rotary-table surface grinders can have either vertical or horizontal spindles, but those with horizontal spindles are limited in the type of work they will accommodate and therefore are not used to a great extent. Vertical-spindle rotary-table surface grinders are primarily production-type machines. They frequently have two or more grinding heads and therefore, both rough grinding and finish grinding are accomplished in one rotation of the workpiece. The work can be held either on a magnetic chuck or in special fixtures attached to the table.

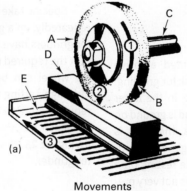

(a)

Movements
1. Wheel 2. Infeed
3. Work table traverse

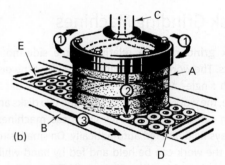

(b)

A. Grinding wheel
B. Grinding face
C. Shaft
D. Workpiece
E. Magnetic chuck on table

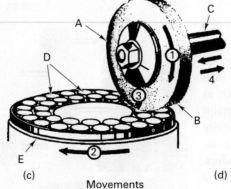

(c)

Movements
1. Wheel 2. Work table rotation
3. Infeed 4. Cross feed

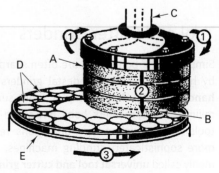

(d)

Movements
1. Wheel 2. Infeed
3. Work table rotation

FIGURE 28.24 Surface grinding: (a) horizontal surface grinding and reciprocating table; (b) vertical spindle with reciprocating table; (c) and (d) both horizontal- and vertical-spindle machines can have rotary tables. (Courtesy of Carborundum Company)

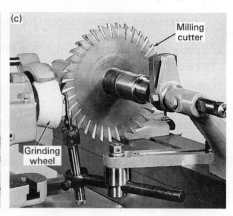

FIGURE 28.25 Three typical setups for grinding single- and multiple-edge tools on a universal tool and cutter grinder. (a) Single-point tool is held in a device that permits all possible angles to be ground. (b) Edges of a large hand reamer are being ground. (c) Milling cutter is sharpened with a cupped grinding wheel. (Courtesy J T. Black)

By using special rotary feeding mechanisms, machines of this type often are made automatic. Parts are dumped on the rotary feeding table and fed automatically onto workholding devices and moved past the grinding wheels. After they pass the last grinding head, they are automatically unloaded.

Disk-Grinding Machines

Disk grinders have relatively large side-mounted abrasive disks. The work is held against one side of the disk for grinding. Both single- and double-disk grinders are used; in the latter type, the work is passed between the two disks and is ground on both sides simultaneously. On these machines, the work is always held and fed automatically. On small, single-disk grinders, the work can be held and fed by hand while resting on a supporting table. Although manual disk grinding is not very precise, flat surfaces can be obtained quite rapidly with little or no tooling cost. On specialized, production-type machines, excellent accuracy can be obtained very economically.

Tool and Cutter Grinders

Simple, single-point tools are often sharpened by hand on bench or pedestal grinders (**off-hand grinding**). More complex tools, such as milling cutters, reamers, hobs, and single-point tools for production-type operations require more sophisticated grinding machines, commonly called **universal tool and cutter grinders**. These machines are similar to small universal cylindrical center-type grinders, but they differ in four important respects:

1. The headstock is not motorized.
2. The headstock can be swiveled about a horizontal as well as a vertical axis.

3. The wheelhead can be raised and lowered and can be swiveled through at 360° rotation about a vertical axis.

4. All table motions are manual. No power feeds being provided.

Specific rake and clearance angles must be created, often repeatedly, on a given tool or on duplicate tools. Tool and cutter grinders have a high degree of flexibility built into them so that the required relationships between the tool and the grinding wheel can be established for almost any type of tool. Although setting up such a grinder is quite complicated and requires a highly skilled worker, after the setup is made for a particular job, the actual grinding is accomplished rather easily. **Figure 28.25** shows several typical setups on a tool and cutter grinder.

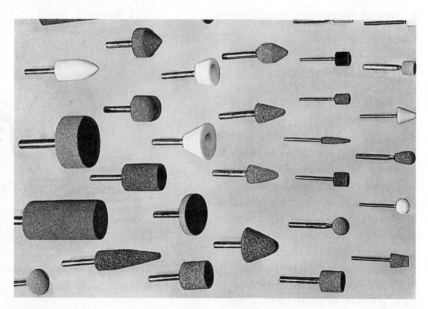

FIGURE 28.26 Examples of mounted abrasive wheels and points. (Courtesy of Norton Abrasives/Saint Gobain)

Hand-ground cutting tools are not accurate enough for automated machining processes. Many numerically controlled (NC) machine tools have been sold on the premise that they can position work to very close tolerances—within ±0.0001 to 0.0002 in.—only to have the initial workpieces produced by those machines out of tolerance by as much as 0.015 to 0.020 in. In most instances, the culprit was a poorly ground tool. For example, a twist drill with a point ground 0.005 in. off-center can "walk" as much as 0.015 in., thus causing poor hole location. Many companies are turning to computer numerical control (CNC) grinders to handle the regrinding of their cutting tools. A six-axis CNC grinder is capable of restoring the proper tool angles (rake and clearance), concentricity, cutting edges, and dimensional size.

Mounted Wheels and Points

Mounted wheels and points are small grinding wheels of various shapes that are permanently attached to metal shanks that can be inserted in the chucks of portable, high-speed electric or air motors. They are operated at speeds up to 100,000 rpm, depending on their diameters, and are used primarily for deburring and finishing in molds and dies. Several types are shown in **Figure 28.26**.

Coated Abrasives

Coated abrasives are used for finishing both metal and non-metal products. These are made by gluing abrasive grains onto a cloth or paper backing (**Figure 28.27**). Synthetic abrasives

Belt composition

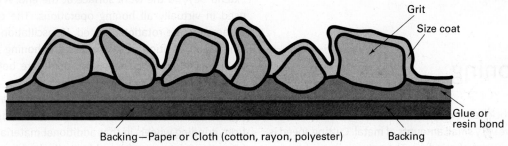

Grit
Size coat
Glue or resin bond
Backing—Paper or Cloth (cotton, rayon, polyester)
Backing

Grit Size—grade

vs	Approx.	Finish (rms)
		μin.
24	300	"
36	250	"
50	140	"
80	125	"
120	60–80	"
150	40–60	"

Bonds

Name	Make coat	Size coat	Backing
Glue bond	Glue	Glue	Non WP
Modified glue	Mod. glue	Mod. glue	"
Resin over glue	Glue	Resin	"
Resin over resin	Resin	Resin	"
Waterproof	Resin	Resin	WP

WP = waterproof

Platen grinder

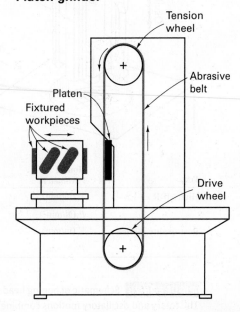

Tension wheel
Abrasive belt
Platen
Fixtured workpieces
Drive wheel

FIGURE 28.27 Belt composition for coated abrasives (top). Platen grinder (right) and examples of belts and disks for abrasive machining. (Courtesy J T. Black)

such as aluminum oxide, silicon carbide, aluminum, zirconia, CBN, and diamond are used most commonly. Some natural abrasives such as sand, flint, garnet, and emery also are employed. Various types of glues are utilized to attach the abrasive grains to the backing, usually compounded to allow the finished product to have some flexibility.

Coated abrasives are available in sheets, rolls, endless belts, and disks of various sizes. Although the cutting action of coated abrasives basically is the same as with grinding wheels, there is one major difference: they have little capability to be self-sharpened when dull grains are pulled from the backing. Consequently, when the abrasive particles become dull or the belt loaded, the belt must be replaced. Finer grades result in finer first cuts but slower material removal rates. This versatile process is now widely used for rapid stock removal as well as fine surface finishing.

28.6 Honing

Honing is a stock removal process that uses fine abrasive stones to remove very small amounts of metal. Cutting speed is much lower than that of grinding. The process is used to size and finish bored holes, remove common errors left by boring

(taper, waviness, and tool marks), or remove the tool marks left by grinding. The amount of metal removed is typically about 0.005 in. or less. Although honing is occasionally done by hand, as in finishing the face of a cutting tool, it usually is done with special equipment. Most honing is done on internal cylindrical surfaces, such as cylinder walls in engine blocks. The honing stones are usually held in a honing head, with the stones being held against the work with controlled light pressure. The honing head is not guided externally but, instead, *floats* in the hole, being guided by the work surface (**Figure 28.28**).

The stones are given a complex motion so as to prevent a single grit from repeating its path over the work surface. Rotation is combined with an oscillatory axial motion. For external and flat surfaces, varying oscillatory motions are used. The length of the motions should be such that the stones extend beyond the work surface at the end. A cutting fluid is used in virtually all honing operations. The critical process parameters are rotational speed, V_r, oscillation speed, V_o, the length and position of stroke, and the honing stick pressure. Note that V_c and the inclination angle are both products of V_o and V_r.

Virtually all honing is done with stones made by bonding together various fine artificial abrasives. **Honing stones** differ from grinding wheels in that additional materials, such as sulfur, resin, or wax, are often added to the bonding agent to modify the cutting action. The abrasive grains range in size from

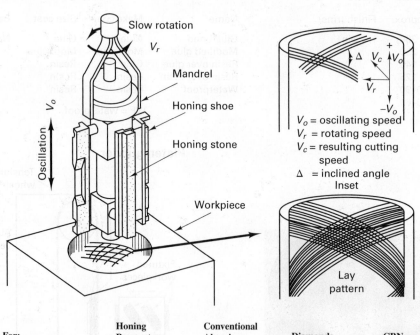

For:	Honing Parameters	Conventional Abrasives	Diamonds	CBN
High MRR	V_c(m/min)	20–30	40–70	35–90
	P_s(N/min²)	1–2	2–8	2–4
Best-quality service	V_c(m/min)	5–30	40–70	20–60
	P_s(N/min²)	0.5–1.5	1.0–3.0	1.0–2.0

FIGURE 28.28 Schematic of honing head showing the manner in which the stones are held. The rotary and oscillatory motions combine to produce a cross-hatched lay pattern. Typical values for V_c and P_s are given below.

80 to 600 grit. The stones are equally spaced about the periphery of the tool. Reference values for V_c and honing stick pressure, P_s, for various abrasives are shown in **Figure 28.28**.

Single- and multiple-spindle honing machines are available in both horizontal and vertical types. Some are equipped with special sensitive measuring devices that collapse the honing head when the desired size has been reached.

For honing single, small, internal cylindrical surfaces, a procedure is often used wherein the workpiece is manually held and reciprocated over a rotating hone. If the volume of work is sufficient, honing is a fairly inexpensive process. A complete honing cycle, including loading and unloading the work, is often less than 1 minute. Size control within 0.0003 in. is achieved routinely.

28.7 Superfinishing

Superfinishing is a variation of honing that is typically used on flat surfaces.

1. The process uses very light, controlled pressure, 10 to 40 psi.
2. Rapid (more than 400 cycles per minute), short strokes, less than ¼ in., are used.
3. Stroke paths are controlled so that a single grit never traverses the same path twice.
4. Copious amounts of low-viscosity lubricant-coolant are flooded over the work surface.

This procedure, illustrated in **Figure 28.29**, results in surfaces of very uniform, repeatable smoothness.

Superfinishing is based on the phenomenon that a lubricant of a given viscosity will establish and maintain a separating, lubricating film between two mating surfaces if their roughness does not exceed a certain value and if a certain critical pressure, holding them apart, is not exceeded. Consequently, as the minute peaks on a surface are cut away by the honing stone, applied with a controlled pressure, a certain degree of smoothness is achieved. The lubricant establishes a continuous

film between the stone and the workpiece and separates them so that no further cutting action occurs. Thus, with a given pressure, lubricant, and honing stone, each workpiece is honed to the same degree of smoothness.

Superfinishing is applied to both cylindrical and plane surfaces. The amount of metal removed usually is less than 0.002 in., most of it being the peaks of the surface roughness. Copious amounts of lubricant-coolant maintain the work at a uniform temperature and wash away all abraded metal particles to prevent scratching.

Lapping

Lapping is an abrasive surface finishing process wherein fine abrasive particles are embedded or charged into a soft material, called a **lap**. The material of the lap may range from cloth to cast iron or copper, but it is always softer than the material to be finished, being only a holder for the hard abrasive particles. Lapping is applied to both metals and nonmetals.

As the charged lap is rubbed against a surface, the abrasive particles in the surface of the lap remove small amounts of material from the surface to be machined. Thus, the abrasive does the cutting, and the soft lap is not worn away because the abrasive particles become embedded in its surface instead of moving across it. This action always occurs when two materials rub together in the presence of a fine abrasive: the softer one forms a lap, and the harder one is abraded away.

In lapping, the abrasive is usually carried between the lap and the work surface in some sort of a vehicle, such as grease, oil, or water. The abrasive particles are from 120 grit up to the finest powder sizes. As a result, only very small amounts of metal are removed, usually considerably less than 0.001 in. Because it is such a slow metal removing process, lapping is used only to remove scratch marks left by grinding or honing or to obtain very flat or smooth surfaces, such as are required on gage blocks or for liquid-tight seals where high pressures are involved.

Materials of almost any hardness can be lapped. However, it is difficult to lap soft materials because the abrasive tends to become embedded. The most common lap material is fine-grained cast iron. Copper is used quite often and is the common material for lapping diamonds. For lapping hardened metals for metallographic examination, cloth laps are used.

Lapping can be done either by hand or by special machines. In hand lapping, the lap is flat, similar to a surface plate. Grooves are usually cut across the surface of a lap to collect the excess abrasive and chips. The work is moved across the surface of the lap, using an irregular, rotary motion, and is turned frequently to obtain a uniform cutting action.

In lapping machines for obtaining flat surfaces, workpieces are placed loosely in holders and are held against the rotating lap by means of floating heads. The holders, rotating slowly, move the workpieces in

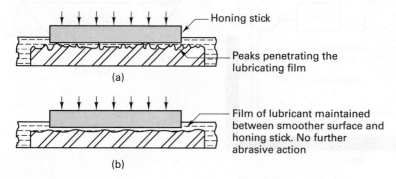

FIGURE 28.29 In superfinishing and honing, a film of lubricant is established between the work and the abrasive stone as the work becomes smoother.

an irregular path. When two parallel surfaces are to be produced, two laps may be employed—one rotating below and the other above the workpieces.

Various types of lapping machines are available for lapping round surfaces. A special type of centerless lapping machine is used for lapping small cylindrical parts, such as piston pins and ball-bearing races.

Because the demand for surfaces having only a few micrometers of roughness on hardened materials has become quite common, the use of lapping has increased greatly. However, it is a very slow method of removing metal, obviously costly compared with other methods, and should not be specified unless such a surface is absolutely necessary.

28.8 Free Abrasives

Ultrasonic Machining

Ultrasonic machining (USM), sometimes called **ultrasonic impact grinding**, employs an ultrasonically vibrating tool to impel the abrasives in a slurry at high velocity against the workpiece. The tool is fed into the part as it vibrates along an axis parallel to the tool feed at an amplitude on the order of several thousandths of an inch and a frequency of 20 kHz. As the tool is fed into the workpiece, a negative of the tool geometry is

machined into the workpiece. The cutting action is performed by the abrasives in the slurry, which is continuously flooded under the tool. The slurry is loaded up to 60 wt% with abrasive particles. Lighter abrasive loadings are used to facilitate the flow of the slurry for deep drilling (up to 2 in. deep). Boron carbide, aluminum oxide, and silicon carbide are the most commonly used abrasives in grit sizes ranging from 400 to 2000. The amplitude of the vibration should be set approximately to the size of the grit. The process can use shaped tools to cut virtually any material but is most effective on materials with hardnesses greater than R_c 40, including brittle and nonconductive materials such as glass. **Figure 28.30** shows a simple schematic of this process.

USM uses piezoelectric or magnetostrictive transducers to impart high-frequency vibrations to the tool holder and tool. Abrasive particles in the slurry are accelerated to great speed by the vibrating tool. The tool materials are usually brass, carbide, mild steel, or tool steel and will vary in tool wear depending on their hardness. Wear ratios (workpiece material removed versus tool material lost) from 1:1 (for tool steel) to 100:1 (for glass) are possible. Because of the high number of cyclic loads, the tool must be strong enough to resist fatigue failure.

The cut will be oversize by about twice the size of the abrasive particles being used, and holes will be tapered, usually limiting the hole depth-to-diameter ratio to about 3:1. Surface roughness is controlled by the size of the abrasive particles (finer finish with smaller particles). Holes, slots, or

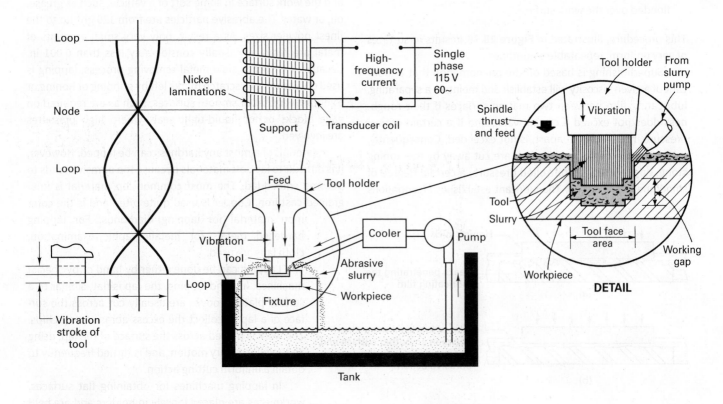

FIGURE 28.30 Sinking a hole in a workpiece with an ultrasonically vibrating tool driving an abrasive slurry.

shaped cavities can be readily eroded in any hard material, conductive or nonconductive, metallic, ceramic, or composite. One advantage of the process is that it is one of the few machining methods capable of machining glass. Also, it is the safest machining method. Skin is impervious to the process because of its ductility. High-pitched noise can be a problem due to secondary vibrations. In addition to machining, ultrasonic energy has also been employed for coining, lapping, deburring, broaching, and cleaning. Plastics can be welded using ultrasonic energy.

Waterjet Cutting and Abrasive Waterjet Machining

Waterjet cutting (WJC), also known as **waterjet machining** or **hydrodynamic machining**, uses a high-velocity fluid jet impinging on the workpiece to perform a slitting operation (**Figure 28.31**). Water is ejected from a nozzle orifice at high pressure (up to 60,000 psi). The jet is typically 0.003 to 0.020 in. in diameter and exits the orifice at velocities up to 3000 ft/s. Key process parameters include water pressure, orifice diameter, water flow rate, and working distance (distance between the workpiece and the nozzle).

Nozzle materials include synthetic sapphire, due to its resistance to wear. Tool life on the order of several hundred hours is typical. Mechanisms for tool failure include chipping from contaminants or constriction due to mineral deposits. This emphasizes the need for high levels of filtration prior to pressure intensification. In the past, long-chain polymers were added to the water to make the jet more coherent (i.e., not come out of the jet dispersed). However, with proper nozzle design, a tight, coherent waterjet can be produced without additives.

The advantages of WJC include the ability to cut materials without burning or crushing the material being cut. The mechanism for material removal is simply the impinging pressure of the water exceeding the compressive strength of the material. This limits the materials that can be cut by the process to leather, plastics, and other soft nonmetals, which is the major disadvantage of the process. Alternative fluids (alcohol, glycerine, and cooking oils) have been used in processing meats, baked goods, and frozen foods. Other disadvantages include that the process is noisy and requires operators to have hearing protection.

The majority of the metalworking applications for waterjet cutting require the addition of abrasives. This process is known as **abrasive waterjet cutting (AWC)**. A full range of materials, including metals, plastics, rubber, glass, ceramics, and composites, can be machined by AWC. Cutting feed rates vary from 20 in./min for acoustic tile to 50 in./min for epoxies and 500 in./min for paper products.

Abrasives are added to the waterjet in a mixing chamber on the downstream side of the waterjet orifice. A single, central waterjet with side feeding of abrasives into a mixing chamber is shown in **Figure 28.31**. In the mixing chamber, the momentum of the water is transferred to the abrasive particles, and the water and particles are forced out through the AWC nozzle orifice, also called the mixing tube. This design can be made quite compact; however, it also experiences rapid wear in the mixing tube. An alternate configuration is to feed the abrasives from the center of the nozzle with a converging set of angled waterjets imparting momentum to the abrasives. This nozzle design produces better mixing of the water and abrasives as well as increased nozzle life. The inside diameter of the mixing tube is normally from 0.04 to 0.125 in. in diameter. These tubes are usually made of carbide.

Generally, the kerf of the cut is about 0.001 in. greater than the nozzle orifice. AWC requires control of additional process parameters over waterjet machining, including abrasive material (density, hardness, and shape), abrasive size or grit, abrasive flow rate (pounds per minute), abrasive feed mechanism (pressurized or suction), and AWC nozzle (design, orifice diameter, and material). Typical AWC systems operate under the following conditions: water pressures of 30,000 to 50,000 psi, water orifice diameters from 0.01 to 0.022 in., and working distances of 0.02 to 0.06 in. Working distances are much smaller than in WJC to minimize the dispersion of the abrasive waterjet prior to entering the material. Abrasive materials include garnet, silica, silicon carbide, and aluminum oxide. Abrasive grit sizes range from 60 to 120 and abrasive flow rates from 0.5 to 3 lb/min. For many applications, the AWC tool is combined with a CNC-controlled *X–Y* table, which permits contouring and surface engraving.

AWC can be used to cut any material through the appropriate choice of the abrasive, waterjet pressure, and feed rate. **Table 28.6** gives cutting speeds for various metals. The ability of the abrasive waterjet to cut through thick materials (up to 8 in.) is attributed to the reentrainment of abrasive particles in the jet by the workpiece material. AWC is particularly suited for composites because the cutting rates are reasonable and they do not delaminate the layered material. In particular, AWC is used in the airplane industry to cut carbon-fiber composite sections of the airplane after autoclaving.

Abrasive Jet Machining

One of the least expensive of the nontraditional processes is **abrasive jet machining (AJM)**. AJM removes material by a focused jet of abrasives and is similar in many respects to AWC, with the exception that momentum is transferred to the abrasive particles by a jet of inert gas. Abrasive velocities on the order of 1000 ft/s are possible with AJM. The small mass of the abrasive particles produces a microscale chipping action on the workpiece material. This makes AJM ideal for processing hard, brittle materials, including glass, silicon, tungsten, and ceramics. It is not compatible with soft, elastic materials.

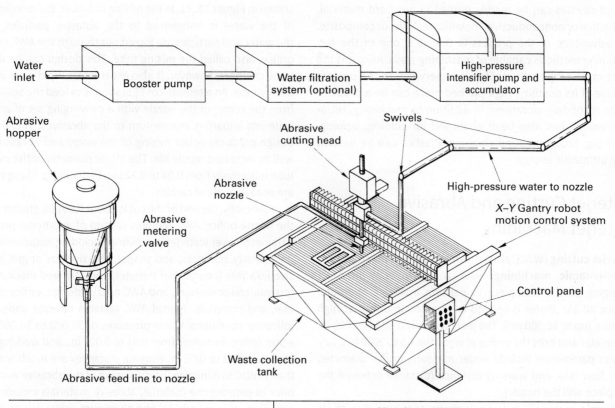

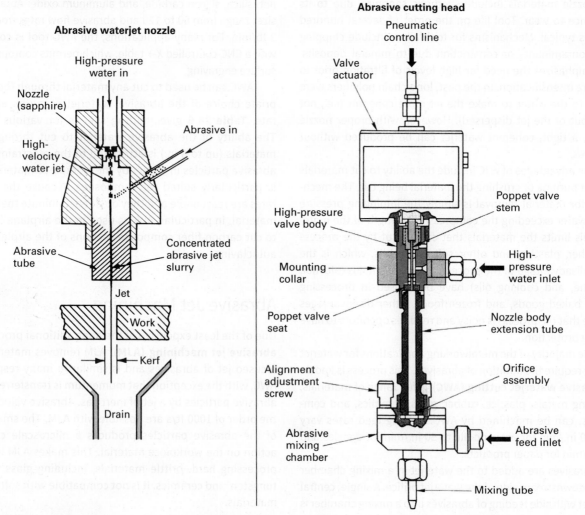

FIGURE 28.31 Schematics of hydrodynamic jet machining systems. The intensifier elevates the fluid to the desired nozzle pressure, while the accumulator smoothes out the pulses in the fluid jet. Schematic of an abrasive waterjet machining nozzle is shown on the left.

TABLE 28.6	Typical Values for Through-cutting Speeds for Simple Waterjet and Abrasive Waterjet of Machining Metals and Nonmetals

Cutting Speeds with Abrasive Waterjet

Material	Thickness (in.)	Nozzle Speed (in./min)	Edge Quality (comments)	Material	Thickness (in.)	Nozzle Speed (in./min)	Edge Quality (Comments)
Aluminum	0.130	20–40	good	Titanium	2.0	0.5–1.0	125 RMS
Aluminum tube	0.220	50	burred	Tool steel	0.250	3–15	125 RMS
Aluminum casting	0.400	15		Tool steel	1.0	2–5	
Aluminum	0.500	6–10		**Nonmetals**			
Aluminum	3.0	0.5–5		Acrylic	0.375	15–50	good to fair
Aluminum	4.0	0.2–2		C-glass	0.125	100–200	shape dependent
Brass	0.125	18–20	good or small burr	Carbon/carbon comp.	0.125	50–75	good
Brass	0.500	4–5		Carbon/carbon comp.	0.500	10–20	good
Brass	0.75	0.75–3	striations at 1 +	Epoxy/glass composite	0.125	100–250	good
Bronze	1.100	1.0	good	Fiberglass	0.100	150–300	good
Copper	0.125	22	good	Fiberglass	0.250	100–150	good
Copper-nickel	0.125	12–14	fair edge	Glass (plate)	0.063	40–150	good
Copper-nickel	2.0	1.5–4.0	fair edge	Glass (plate)	0.75	10–20	125 RMS
Lead	0.25	10–50	good to striated	Graphite/epoxy	0.250	15–70	good to practical
Lead	2.0	3–8	slower = better	Graphite/epoxy	1.0	3–5	good
Magnesium	0.375	5–15	good	Kevlar (steel reinf.)	0.125	30–50	good
Armor plate	0.200	1.5–15	good	Kevlar	0.375/0.580	10–25	good
Carbon steel	0.250	10–12	good	Kevlar	1.0	3–5	good
Carbon steel	0.750	4–8	good to bad edge	Lexan	0.5	10	good
Carbon steel	3.0	0.4	good w. sm. nozzle	Phenolic	0.25–0.50	10–15	good
4130 carbon steel	0.5	3.0		Plexiglass	0.175/0.50	25	
Mild steel	7.5	0.017–0.05		Rubber belting	0.300	200	good
High-strength steel	3.0	0.38		**Ceramic matrix composites**			
Cast iron	1.5	1.0	good edge	Toughened zirconia	0.250	1.5	
Stainless steel	0.1	10–15	good to striated	SiC fiber in SiC	0.125	1.5	
Stainless steel	0.25	4–12	good to striated	Al_2O_3/CoCrAly (60%/40%)	0.125	2	
Stainless steel	1.0	1.0	65–150 RMS	SiC./TiB_2 (15%)	0.250	0.35	
15-5 PH stainless	4.0	0.3	striated	**Metal matrix composites**			
Inconel 718	1.25	0.5–1.0	good				
Inconel	0.250	8–12	good to striated	Mg/B_4C (15%)	0.125	35	fair
Inconel	2–2.5	0.2	good to fair	Al/SiC (15%)	0.500	8–12	good to fair
Titanium	0.025–0.050	5–50	good	Al/Al_2O_3 (15%)	0.250	15–20	good to fair
Titanium	0.500	1–5	65–150 RMS				

Cutting Speeds with Simple Waterjet

Material	Thickness (in.)	Nozzle Speed (in./min)	Edge Quality (Comments)	Material	Thickness (in.)	Nozzle Speed (in./min)	Edge Quality (Comments)
ABS plastic	0.087	20–50	100% separation	Lead	0.125	10	good, slight burr
Aluminum	0.050	2–5	burr	Plexiglass	0.118	30–35	fair
Cardboard	0.055	240–600	slits very well	Printed circuit bd	0.050–0.125	50–5	good
Delrin	0.500	2–5	good to stringers	PVC	0.250	10–20	good to fair
Fiberglass	0.100	40–150	good to raggy	Rubber	0.050	2400–3600	good
Formica	0.040	1450		Vinyl	0.040	2000–2400	good
Graphite composite	0.060	25		Wood	0.125	40	fair
Kevlar	0.040–0.250	50–53	fair, some furring				

Comment on these tables: In trying to provide data on waterjet and abrasive waterjet cutting, we have collected material from diverse sources. But we must note that most of the data present are not from uniform tests. Also, note that in many cases data were largely absent on such parameters as pump horsepower, waterjet pressure, abrasive-particle rate of flow or type or size, and standoff distance. So these cutting rates vary widely in value—from laboratory control to shop floor ballpark estimates. Many of the top speeds cited either represent cuts made to illustrate speed alone, without regard to surface quality, or may reflect data from machines with very high power output. (American Machinist, *October 1989*.)

Key process parameters include working distance, abrasive flow rate, gas pressure, and abrasive type. Working distance and feed rate are controlled by hand. If necessary, a hard mask can be placed on the workpiece to control dimensions. Abrasives are typically smaller than those used in AWC. Abrasives are typically not recycled, since the abrasives are cheap and are used only on the order of several hundred grams per hour. To minimize particulate contamination of the work environment, a dust-collection hood should be used in concert with the AJM system.

Original design of base plate

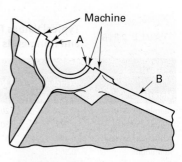

Original design of crankshaft bearing bracket

28.9 Design Considerations in Grinding

Almost any shape and size of work can be finished on modern grinding equipment, including flat surfaces, straight or tapered cylinders, irregular external and internal surfaces, cams, anti-friction-bearing races, threads, and gears. For example, the most accurate threads are formed from solid cylindrical blanks on special thread-grinding machines. Gears that must operate without play are hardened and then finish ground to close tolerances. Two important design recommendations are to reduce the area to be ground and to keep all surfaces that are to be ground in the same or parallel planes is shown in (**Figure 28.32**), an example of **design for manufacturing (DFM)**.

Abrasive machining can remove scale as well as parent metal. Large allowances of material, needed to permit conventional metal-cutting tools to cut below hard or abrasive

Redesigned to reduce weight and grinding time

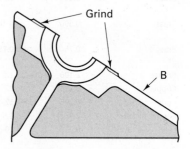

Redesign eliminated shoulders and made part suitable for grinding in single setup

FIGURE 28.32 Reducing area to be ground and keeping all surface to be ground in the same or parallel planes are two important design recommendations. (From *Machine Design*, June 1, 1972, p. 87)

inclusions, are not necessary for abrasive machining. An allowance of 0.015 in. is adequate, assuming, of course, that the part is not warped or out of round. This small allowance requirement results in savings in machining time, in material (often 60% less metal is removed), and in shipping of unfinished parts.

Review Questions

1. What are machining processes that use abrasive particles for cutting tools called?

2. What is attrition in an abrasive grit?

3. Why is friability an important grit property?

4. Explain the relationship between grit size and surface finish.

5. Why is aluminum oxide used more frequently than silicon carbide as an abrasive?

6. Why is CBN superior to silicon carbide as an abrasive in some applications?

7. What materials commonly are used as bonding agents in grinding wheels?

8. Why is the grade of a bond in a grinding wheel important?

9. How does grade differ from structure in a grinding wheel?

10. What is crush dressing?

11. How does loading differ from glazing?

12. What is meant by the statement that grinding is a mixture of processes?

13. What is accomplished in dressing a grinding wheel?

14. How does abrasive machining differ from ordinary grinding?

15. What is a grinding ratio or *G* ratio?

16. How is the feed of the workpiece controlled in centerless grinding?

17. Why is grain spacing important in grinding wheels?

18. Why should a cutting fluid be used in copious quantities when doing wet grinding?

19. How does plunge-cut grinding compare to cylindrical grinding?

20. If grinding machines are placed among other machine tools, what precautions must be taken?

21. What is the purpose of low-stress grinding?

22. How is low-stress grinding done compared to conventional grinding?

23. The number of grains per square inch that actively contact and cut a surface decreases with increasing grain diameter. Why is this so?

24. Why are centerless grinders so popular in industry compared to center-type grinders?

25. Explain how an SEM micrograph is made. Check the Internet or the library to find the answer.

26. Why are vacuum chucks and magnetic chucks widely used in surface grinding but not in milling?

27. How does creep feed grinding differ from conventional surface grinding?

28. Why does a lap not wear, even though it is softer than the material being lapped?

29. How do honing stones differ from grinding wheels?

30. What is meant by "charging" a lap?

31. Why is a honing head permitted to float in a hole that has been bored?

32. How does a coated abrasive differ from an abrasive wheel?

33. Why are the bottoms of chips shown in Figure 28.9 are so smooth. The magnification of the micrograph is 4800X. How thick are these chips?

34. What is the inclined angle in honing, and what determines it?

35. What are the common causes of grinding accidents?

36. What other machine tool does a surface grinder resemble?

37. Figure 28.11 showed residual stress distributions produced by surface grinding. What is a residual stress?

38. In grinding, what is infeed versus cross feed?

39. One of the problems with waterjet cutting is that the process is very noisy. Why?

40. In AWC, what keeps the abrasive jet from machining the orifice?

Problems

1. Perhaps you have observed the following wear phenomena: A set of marble or wooden steps shows wear on the treads in the regions where people step when they climb (or descend) the steps. The higher up the steps, the less the wear on the tread. Given that soles of shoes (leather, rubber) are far softer than marble or granite, explain:

    a. Why and how the steps wear.

    b. Why the lower steps are more worn than the upper steps.

2. Explain why it is that a small particle of a material can be used to abrade a surface made of the same material (i.e., why does the small particle act harder or stronger than the bulk material)?

3. In grinding, both the wheel and workpiece are moving (or rotating). Using the data in Figure 28.11 and assuming that you are doing surface grinding (see Figure 28.1), what are some typical MRR values? How do these compare to MRR values for other machining processes, such as milling? What is the significance of this?

You will find Chapters 29 and 30, "Nano and Micro-Manufacturing Processes" and "Nontraditional Manufacturing Processes," online at www.wiley.com/college/black. Chapter 31, "Thread and Gear Manufacturing," begins on page 678.

Thread and Gear Manufacturing

31.1 Introduction

Screw threads and gears are important machine elements. **Threading**, **thread cutting**, or **thread rolling** refers to the manufacture of threads on external diameters. **Tapping** refers to machining threads in (drilled) holes. Without these processes, our current technological society would come to a grinding halt. More screw threads are made each year than any other machined element. They range in size from those used in small watches to threaded shafts 10 in. in diameter. They are made in quantities ranging from one to several million duplicate threads. Their precision varies from that of inexpensive hardware screws to that of lead screws for the most precise machine tools. Consequently, it is not surprising that many very different procedures have been developed for making screw threads and that the production cost by the various methods varies greatly. Fortunately, some of the most economical methods can provide very accurate results. However, as in the design of most products, the designer can greatly affect the ease and cost of producing specified screw threads. Thus, understanding thread-making processes permits the designer to specify and incorporate screw threads into designs while avoiding needless and excessive cost.

Gears transmit power or motion mechanically among parallel, intersecting, or nonintersecting shafts. Although usually hidden from sight, gears are among the most important mechanical elements in our civilization, possibly even surpassing the wheel, because most wheels would not be turning were power not being applied to them through gears. They operate at almost unlimited speeds under a wide variety of conditions. Millions are produced each year in sizes from a few millimeters up to more than 30 ft in diameter. Often the requirements that must be, and are routinely, met in their manufacture are amazingly precise. Consequently, many special machines and processes have been developed for producing gears. To understand thread and gear cutting processes, we need to have an introductory understanding about the product before we can understand the processes, let's start with threads.

Screw-Thread Standardization and Nomenclature

A **screw thread** is a ridge of uniform section in the form of a helix on the external or internal surface of a cylinder, or in the form of a conical spiral on the external or internal surface of a frustrum of a cone. These are called straight or tapered threads, respectively. **Tapered threads** are used on pipe joints or other applications where liquid-tight joints are required. **Straight threads**, on the other hand, are used in a wide variety of applications, most commonly on fastening devices, such as bolts, screws, and nuts, and as integral elements on parts that are to be fastened together. But, as mentioned previously, they find very important applications in transmitting controlled motion, as in leadscrews and precision measuring equipment.

The standard nomenclature for screw-thread components is illustrated in **Figure 31.1**. The symbol P, the **pitch**, refers to the distance from a point on one screw thread to the corresponding point on the next thread, measured parallel to the length axis of the part. In 1948, representatives of the United States, Canada, and Great Britain adopted the Unified and American Screw Thread Standards, based on the form shown in **Figure 31.2**. In 1968, the International Organization for

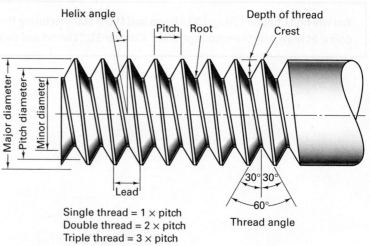

Single thread = 1 × pitch
Double thread = 2 × pitch
Triple thread = 3 × pitch

FIGURE 31.1 Standard nomenclature for a screw thread is based on pitch (the number of threads per inch), major and minor diameters, and pitch diameter.

(a) External thread

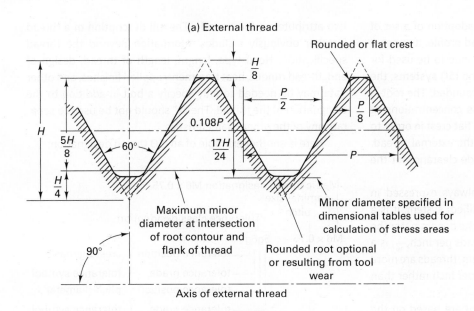

Rounded or flat crest

$\frac{H}{8}$

$\frac{P}{2}$

$\frac{P}{8}$

$0.108P$

$\frac{17H}{24}$

P

H

$\frac{5H}{8}$

$60°$

$\frac{H}{4}$

$90°$

Maximum minor diameter at intersection of root contour and flank of thread

Minor diameter specified in dimensional tables used for calculation of stress areas

Rounded root optional or resulting from tool wear

Axis of external thread

(b) Internal thread

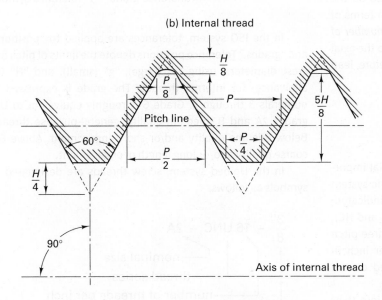

$\frac{H}{8}$

$\frac{P}{8}$

P

Pitch line

$5H/8$

$60°$

$\frac{P}{2}$

$\frac{P}{4}$

$\frac{H}{4}$

$90°$

Axis of internal thread

(c) Metric ISO/R68-1969

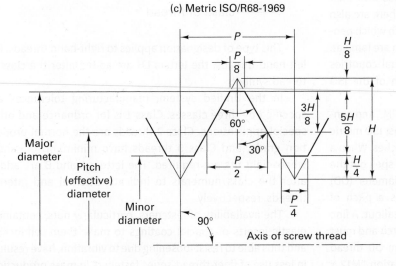

P

$\frac{P}{8}$

$\frac{H}{8}$

$\frac{3H}{8}$

$\frac{5H}{8}$

H

$60°$

$30°$

$\frac{P}{2}$

$\frac{H}{4}$

$\frac{P}{4}$

Major diameter

Pitch (effective) diameter

Minor diameter

$90°$

Axis of screw thread

$D =$ Major diameter of thread of nut ⎫ Normal
$d =$ Major diameter of screw ⎬ diameter
$D_1 =$ Minor diameter of thread of nut
$d_1 =$ Minor diameter of screw
$D_2 =$ Pitch diameter of thread of nut
$d_2 =$ Pitch diameter of screw
$H =$ Height of the complete theoretical thread profile
$H_1 =$ Engagement
$P =$ Pitch

$$H = \frac{\sqrt{3}}{2}P$$

FIGURE 31.2   Basic profiles of Unified and American screw threads: (a) external; (b) internal; (c) metric.

Standardization (ISO) recommended the adoption of a set of metric standards based on the basic thread profile. It appears likely that both types of threads will continue to be used for some time to come. In both the **Unified** and **ISO systems**, the crests of external threads may be flat or rounded. The root is usually made rounded to minimize stress concentrations at this critical area. The internal thread has a flat crest in order to mate with either a rounded or V-root of the external thread. A small round is used at the root to provide clearance for the flat crest of the external thread.

In the metric system, the pitch is always expressed in millimeters, whereas in the American (Unified) system, it is a fraction having as the numerator 1 and as the denominator the number of threads per inch. Thus, 16 threads per inch, $\frac{1}{16}$, is a 16 pitch. Consequently, in the Unified system, threads are more commonly described in terms of threads per inch rather than by the pitch.

While all elements of the thread form are based on the **pitch diameter**, screw-thread sizes are expressed in terms of the **outside**, or **major**, **diameter** and the *pitch,* or *number of threads per inch.* In threaded elements, **lead** refers to the axial advance of the element during one revolution; therefore, lead equals pitch on a single-thread screw.

Types of Screw Threads

Eleven types, or series, of threads are of commercial importance, several having equivalent series in the metric system and Unified systems. See **Figure 31.3**. As has been indicated, the Unified threads are available in a coarse (UNC and NC), fine (UNF and NF), extra-fine (UNEF and NEF), and three-pitch (8, 12, and 16) series, the number of threads per inch in accordance with an arbitrary determination based on the major diameter.

Many nations have now adopted ISO threads into their national standards. Besides metric ISO threads, there are also inch-based ISO threads, namely, the UN series with which people in the United States, Canada, and Great Britain are familiar. ISO offers a wide range of metric sizes. Individual countries have the choice of accepting all or a selection of the ISO offerings.

The size listings of metric threads start with "M" and continue with the outside diameter in millimeters. Most ISO metric thread sizes come in coarse, medium, and fine pitches. When a coarse thread is designated, it is not necessary to spell out the pitch. For example, a coarse 10-mm outside diameter (OD) thread is called out as "M12." This thread has a pitch of 1.75 mm, but the pitch may be omitted from the callout. A fine 12-mm OD thread is available. It has a 1.25 mm pitch and must be designated "M12 x 1.25"; an extra-fine 12-mm OD thread having 0.75 mm-pitch would receive the designation "M12 x 0.75." Note that the symbol "x" is not employed as a multiplication symbol in metric practice; rather, it is used to relate these

two attributes of the threads. The full description of a thread fastener obviously includes information beyond the thread specification. Head type, length, length of thread, design of end, thread runout, heat treatment, applied finishes, and other data may be needed to fully specify a bolt in addition to the designation of the thread. The "x" should not be used to separate any of the other characteristics.

Here is another example of an ISO thread designation:

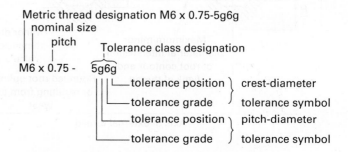

In the ISO system, tolerances are applied to "positions" and "grades." Tolerance positions denote the limits of pitch and crest diameters, using "e" (large), "g" (small), and "H" (no allowance) for internal threads. The grade is expressed by numerals 3 through 9. Grade 6 is roughly equivalent to U.S. grades 2A and B, medium-quality, general-purpose threads. Below 6 is fine quality and/or short engagement. Above 6 is coarse quality and/or long length of engagement.

In the Unified system, screw threads are designated by symbols as follows:

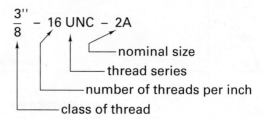

This type of designation applies to right-hand threads. For left-hand threads, the letters LH are added after the class of thread symbol.

In the Unified system, manufacturing tolerances are specified by three classes. Class 1 is for ordnance and other special applications. Class 2 threads are the normal production grade, and Class 3 threads have minimum tolerances where tight fits are required. The letters A and B are added after the class numerals to indicate external and internal threads, respectively.

The availability of fasteners, particularly nuts, containing plastic inserts or special coatings to make them self-locking and thus able to resist loosening due to vibration, have resulted in less use of finer-thread-series fasteners in mass production. Coarser-thread fasteners are easier to assemble and less subject to cross-threading (binding).

Types of Screw Threads

1. *Coarse-thread series* (UNC and NC). For general use where not subjected to vibration.

2. *Fine-thread series* (UNF and NF). For most automotive and aircraft work.

3. *Extra-fine-thread series* (UNEF and NEF). For use with thin-walled material or where a maximum number of threads are required in a given length.

4. *Eight-thread series* (8UN and 8N). Eight threads per inch for all diameters from 1 to 6 in. Used primarily for bolts on pipe flanges and cylinder-head studs where an initial tension must be set up to resist steam or air pressures.

5. *Twelve-thread series* (12UN and 12N). Twelve threads per inch for diameters from $\frac{1}{2}$ through 6 in. Not used extensively.

6. *Sixteen-thread series* (16UN and 16N). Sixteen threads per inch for diameters from $\frac{3}{4}$ through 6 in. Used for a wide variety of applications that require a fine thread.

7. *American Acme thread.* This thread and the following three are used primarily in transmitting power and motion.

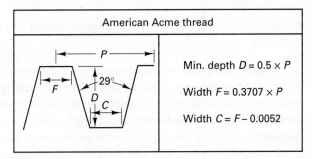

American Acme thread

Min. depth $D = 0.5 \times P$

Width $F = 0.3707 \times P$

Width $C = F - 0.0052$

8. *Buttress thread.*

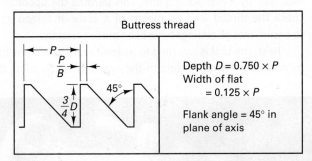

Buttress thread

Depth $D = 0.750 \times P$
Width of flat
    $= 0.125 \times P$

Flank angle = 45° in plane of axis

9. *Square thread.*

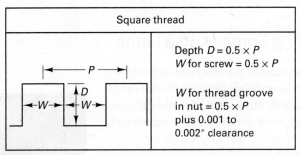

Square thread

Depth $D = 0.5 \times P$
W for screw $= 0.5 \times P$

W for thread groove
in nut $= 0.5 \times P$
plus 0.001 to
0.002″ clearance

10. *29° Worm thread.*

11. *American, standard pipe thread.* This thread is the standard tapered thread used on pipe joints in this country. The taper on all pipe threads is $\frac{3}{4}$ in./ft.

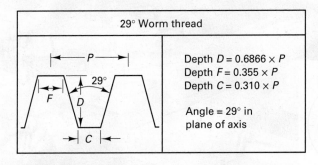

29° Worm thread

Depth $D = 0.6866 \times P$
Depth $F = 0.355 \times P$
Depth $C = 0.310 \times P$

Angle = 29° in plane of axis

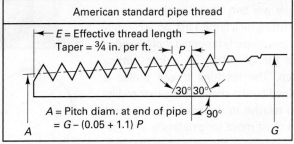

American standard pipe thread

E = Effective thread length
Taper = $\frac{3}{4}$ in. per ft.

A = Pitch diam. at end of pipe
    $= G - (0.05 + 1.1) P$

FIGURE 31.3 Types of screw threads.

31.2 Thread Making

Four basic methods are used to manufacture threads: **cutting**, **rolling**, **additive manufacture (AM),** and **casting**. Although both external and internal threads can be cast, relatively few are made in this manner, primarily in connection with die casting, investment casting, or the molding of plastics. Today, by far the largest number of threads are made by rolling. Both external and internal threads can be made by rolling, but the material must be ductile. Because rolling is a less-flexible process than thread cutting, it is restricted essentially to standardized and simple parts. Consequently, large numbers of external and internal threads still are made by cutting processes, including grinding and tapping.

External Thread Cutting Methods	Internal Thread Cutting Methods
Threading on an engine lathe	Threading (on an engine lathe or NC lathe)
Threading on an NC lathe	With a tap and holder (manual NC, machine, semiautomatic, or automatic)
With a die held in a stock (manual)	
With an automatic die (turret lathe or screw machine) or NC lathe	With a collapsible tap (turret lathe, screw machine, or special threading machine)
By milling or by grinding. See Chapter 33 for discussion on 3D printing (AM)	By milling

Cutting Threads on a Lathe

Lathes provided the first method for cutting threads by machine. Although threads can be produced by other methods, lathes still provide the most versatile and fundamentally simple method. See Chapter 23 for discussion about lathes. Consequently, they often are used for cutting threads on special workpieces where the configuration or nonstandard size does not permit them to be made by less costly methods.

There are two basic requirements for thread cutting on a lathe. First, an accurately shaped and properly mounted tool is needed because thread cutting is a form-cutting operation. The resulting thread profile is determined by the shape of the tool and its position relative to the workpiece. Second, the tool must move longitudinally in a specific relationship to the rotation of the workpiece, because this determines the lead of the thread. This requirement is met through the use of the **leadscrew** and the **split nut**, which provide positive motion of the carriage relative to the rotation of the spindle.

To cut a thread, it is also essential that constant positional relationships be maintained among the workpiece, the cutting tool, and the leadscrew. If this is not done, the tool will not be positioned correctly in the thread space on successive cuts. Correct relationship is obtained by means of a **threading dial** (**Figure 31.4**), which is driven directly by the leadscrew through a worm gear. Because the workpiece and the leadscrew are directly connected, the threading dial provides a means for establishing the desired positional relationship between the workpiece and the cutting tool. The threading dial is graduated into an even number of major and half divisions. If the feed mechanism is engaged in accordance with the following rules, correct positioning of the tool will result:

1. *For even-number threads:* at any line on the dial.
2. *For odd-number threads:* at any numbered line on the dial.
3. *For threads involving ½ numbers:* at any odd-numbered line on the dial.
4. *For ¼ or ⅛ threads:* return to the original starting line on the dial.

To start cutting a thread, the tool usually is fed inward until it just scratches the work, and the cross-slide dial reading is then noted or set at zero. The split nut is engaged and the tool permitted to run over the desired thread length. When the tool reaches the end of the thread, it is quickly withdrawn by means of the cross-slide control. The split nut is then disengaged and the carriage returned to the starting position, where the tool is clear of the workpiece. At this point, the future thread will be indicated by a fine scratch line. This permits the operator to check the thread lead by means of a scale or thread gage to ensure that all settings have been made correctly.

Next, the tool is returned to its initial zero depth position by returning the cross slide to the zero setting. By using the

FIGURE 31.4 Cutting a screw thread on a lathe, showing the method of supporting the work and the relationship of the tool to the work with the compound swiveled. Inset shows face of threading dial. (Photo Courtesy of South Bend Lathe Co.)

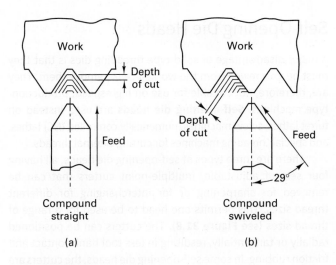

Two ways to feed the threading tool into the workpiece.

compound rest, the tool can be moved inward the proper depth for the first cut. A depth of 0.010 to 0.025 in. is usually used for the first cut and smaller amounts on each successive cut, until the final cut is made with a depth of only 0.001 to 0.003 in. to produce a good finish. When the thread has been cut nearly to its full depth, it is checked for size by means of a mating nut or thread gage. Cutting is continued until a proper fit is obtained.

Figure 31.5 illustrates two methods of feeding the tool into the work. If the tool is fed radially, cutting takes place simultaneously on both sides of the tool. With this true form-cutting procedure, no rake should be ground on the tool, and the top of the tool must be horizontal and be set exactly in line with the axis of rotation of the work. Otherwise, the resulting thread profile will not be correct. An obvious disadvantage of this method is that the absence of side and back rake results in poor cutting (except on cast iron or brass). The surface finish on steel will usually be poor. Consequently, the second method commonly is used, with the compound swiveled 29 degrees. The cutting then occurs primarily on the left-hand edge of the tool, and some side rake can be provided.

Proper speed ratio between the spindle and the leadscrew is set by means of the gear-change box. Modern industrial lathes have ranges of ratios available so that nearly all standard threads can be cut merely by setting the proper levers on the quick-change gearbox.

Cutting screw threads on a lathe is a slow, repetitious process that requires considerable operator skill. The cutting speeds usually employed are from one-third to one-half of regular speeds to enable the operator to have time to manipulate the controls and to ensure better cutting. The cost per part can be high, which explains why other methods are used whenever possible.

Cutting Threads on a CNC Lathe

Computer numerical control (CNC) lathes and turning centers can be programmed to machine straight, tapered, or scroll threads. Threads

machined using the same type of special tool have the thread shape shown in Figure 31.5. The tool is positioned at a specific starting distance from the end of the work (**Figure 31.6**). This distance will vary from machine to machine. Its value can be found in the machine's programming manual. The CNC software will have a set of preprogrammed machining routines (called G codes) specifically for threading. Beginning at the start point, the tool accelerates to the feed rate required to cut the threads. The tool creates the thread shape by repeatedly following the same path as axial infeed is applied. For standard V-threads, the infeed can be applied along a 0- or 29-degree angle. The depth of cut for the first pass is the largest. The cutting depth is then decreased for each successive pass until the required thread depth is achieved. A final finishing pass is then made with the tool set at the thread depth. More discussion about CNC is found in Chapter 26.

Cutting Threads with Dies

Straight and tapered external threads up to about 1½ in. in diameter can be manually cut quickly by means of threading dies (**Figure 31.7a**). Basically, these dies are similar to hardened, threaded nuts with multiple cutting edges. The cutting edges at the starting end are beveled to aid in starting the dies on the workpiece. As a consequence, a few threads at the inner end of the workpiece are not cut to full depth. Such threading dies are made of carbon or high-speed tool steel.

Solid-type dies are seldom used in manufacturing because they have no provision for compensating for wear. The solid-adjustable type (**Figure 31.7b**) is split and can be adjusted over a small range by means of a screw to compensate for wear or to provide a variation in the fit of the resulting screw thread. These types of threading dies are usually held in a **stock** for hand rotation (see **Figure 31.7c**). A suitable lubricant is desirable to produce a smoother thread and to prolong the life of the die because there is extensive friction during the cutting process.

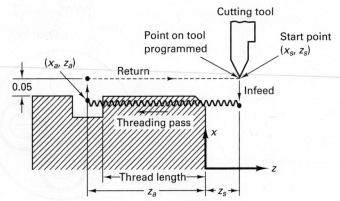

x_a	Specifies the absolute x coordinate of the tool after axial infeed.
G32	Initiates the single-pass threading cycle.
z_a	Specifies the absolute z coordinate of the tool after the threading pass.
Fn	n specifies the feed rate
$x_s\ z_s$	Specifies the absolute x and z coordinates of the start point.

Canned subroutines called G codes are used on CNC lathes to produce threads. (See Chapter 26 for CNC discussion.)

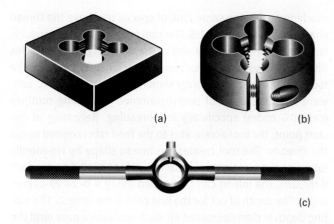

FIGURE 31.7 (a) Solid threading die; (b) solid adjustable threading die; (c) threading-die stock for round die (die removed). (Courtesy of TRW-Greenfield Tap & Die Kennametal)

Self-Opening Die Heads

A major disadvantage of solid-type threading dies is that they must be unscrewed from the workpiece to remove them. They are, therefore, not suitable for use on high-speed, production-type machines. **Self-opening die heads** are used instead on turret lathes, screw machines, numerically controlled (NC) lathes, and special threading machines for cutting external threads.

There are three types of self-opening die heads, all having four sets of adjustable, multiple-point cutters that can be removed for sharpening or for interchanging for different thread sizes. This permits one head to be used for a range of thread sizes (see **Figure 31.8**). The cutters can be positioned radially or tangentially, resulting in less tool flank contact and friction rubbing. In some self-opening die heads, the cutters are circular, with an interruption in the circular form to provide an

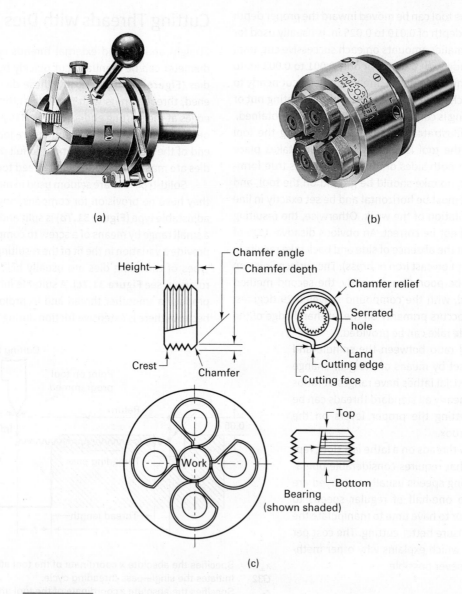

FIGURE 31.8 Self-opening die heads, with (a) radial cutter, (b) tangential cutters, (c) circular cutters. [(a) Greenfield Industries Geometric Tool; (b) Warner & Sawsey Company; (c) Bourn & Hoch Inc., Owner of the Acme product line]

easily sharpened cutting face. The cutters are mounted on the holder at an angle equal to the helix angle of the thread.

As the name implies, the cutters in self-opening die heads are arranged to open automatically when the thread has been cut to the desired length, thereby permitting the die head to be quickly withdrawn from the workpiece. On die heads used on turret lathes, the operator must usually reset the cutters in the closed position before making the next thread. The die heads used on screw machines and automatic threading machines are provided with a mechanism that automatically closes the cutters after the heads are withdrawn.

Cutting threads by means of self-opening die heads is frequently called **thread chasing**. However, some people apply this term to other methods of thread cutting, even to cutting a thread in a lathe.

31.3 Internal Thread Cutting–Tapping

The cutting of an internal thread by means of a multiple-point tool is called **thread tapping**, and the tool is called a **tap**. A hole of diameter slightly larger than the minor diameter of the thread must already exist, made by drilling/reaming, boring, or die casting. For small holes, solid **hand taps** are usually used. The flutes create cutting edges on the thread profile and provide space for the chips and the passage of cutting fluid. Such taps are made of either carbon or high-speed steel and are routinely coated with TiN. The flutes can be straight, helical, or spiral.

Hand taps have square shanks and are usually made in sets of three. The **taper tap** has a tapered end that will enter the hole a sufficient distance to help align the tap. In addition, the threads increase gradually to full depth; therefore, this type of tap requires less torque to use. However, only a through-hole can be threaded completely with a taper tap because it cuts to full depth only behind the tapered portion. A blind hole can be threaded to the bottom using three types of taps in succession. After the taper tap has the thread started in proper alignment, a **plug tap**, which has only a few tapered threads to provide gradual cutting of the threads to depth, is used to cut the threads as deep into the hole as its shape will permit. A **bottoming tap**, having no tapered threads, is used to finish the few remaining threads at the bottom of the hole to full depth. Obviously, producing threads to the full depth of a blind hole is time-consuming, and it also frequently results in broken taps and defective workpieces. Such configurations usually can be avoided if designers will give reasonable thought to the matter.

Taps operate under very severe conditions because of the heavy friction (high torque) involved and the difficulty of chip removal. Also, taps are relatively fragile. **Spiral-fluted taps** provide better removal of chips from a hole, particularly in tapping materials that produce long, curling chips. They are also helpful in tapping holes where the cutting action is interrupted

by slots or keyways. The **spiral point** cuts the thread with a shearing action that pushes the chips ahead of the tap so that they do not interfere with the cutting action and the flow of cutting fluid into the hole.

Collapsing Taps

Collapsing taps are similar to self-opening die heads in that the cutting elements collapse inward automatically when the thread is completed. This permits withdrawing the tap from the workpiece without the necessity of unscrewing it from the thread. They can either be self-setting, for use on automatic machines, or require manual setting for each cycle.

Hole Preparation

Drilling is the most common method of preparing holes for tapping, and when close control over hole size is required, **reaming** may also be necessary. The drill size determines the final thread contour and the drilling torque. Unless otherwise specified, the tap drill size for most materials should produce approximately 75% thread—that is, 75% of full thread depth.

Tapping in Machine Tools

Solid taps also are used in tapping operations on machine tools, such as lathes, drill presses, and special tapping machines. In tapping on a drill press, a tapping attachment is often used. These devices rotate the tap slowly when the drill press spindle is fed downward against the work. When the tapping is completed and the spindle raised, the tap is automatically driven in the reverse direction at a higher speed to reduce the time required to back the tap out of the hole. Some modern machine tools provide for extremely fast spindle reversal for backing taps out of holes.

When solid taps are used on a screw machine or turret, the tap is prevented from turning while it is being fed into the work. As the tap reaches the end of the hole, the tap is free to rotate with the work. The work is then reversed and the tap, again prevented from rotating, is backed out of the hole.

The machine should have adequate power, rigidity, speed and feed ranges, cutting fluid supply, and positive drive action. Chucks, tap holders, and collets should be checked regularly for signs of wear or damage. Accurate alignments of the tap holder, machine spindle, and workpiece are vital to avoid broken taps or bell-mouthed, tapered, or oversized holes.

Tapping Cutting Time

The equation to calculate the cutting time for tapping is (approximately)

$$T_m = \frac{Ln}{N} = \frac{\pi DLn}{12V} + A_L + A_R \qquad (31.1)$$

where

N = spindle rpm

T_m = cutting time (min)

D = tap diameter (in.)

L = depth of tapped hole or length of cut (in.)

n = number of threads per inch (tpi) (feed rate)

V = cutting speed (sfpm)

A_L = allowance to start the tap (min)

A_R = allowance to withdraw the tap (min)

Special Threading and Tapping Machines

Special machines are available for production threading and tapping. Threading machines usually have one or more spindles on which a self-opening die head is mounted, with suitable means for clamping and feeding the workpiece. Special tapping machines using self-collapsing taps substituted for the threading dies are also available. More commonly, tapping machines resemble drill presses, modified to provide spindle feeds both upward and downward, with the speed and feed more rapid on the upward motion.

Common Tapping Problems

Tap overloading is often caused by poor lubrication, lands that are too wide, chips packed in the flutes, or tap wear. Surface roughness in the threads has many causes. A negative grind on the heel will prevent the tap from tearing the threads when backing out.

When a tap loses speed or needs more power, it generally indicates that the tap is dull (or improperly ground) or the chips are packed in flutes (loaded). The flutes may be too shallow or the lands too deep. When tapping soft ductile metals, loading can usually be overcome by polishing the tap before usage.

Improper hole size due to drill wear increases the percentage of threads being cut. Dull tools can also produce a rough finish or work-harden the hole surface and cause the tap to dull more quickly. Check to see that the axis of the hole and tap are aligned. If the tap cuts when backing out, check to see if the hole is oversized.

Tapping High-Strength Materials

High-strength, thermal-resistant materials, sometimes called **exotics**, cause special problems in tapping. A variety of materials are classified as exotics: stainless steel, precipitation-hardened stainless steel, high-alloy steels, iron-based superalloys, titanium, Inconcel, Hastelloy, Monel, and Waspalloy. Their most important attribute is their high strength-to-weight ratio.

Each material presents different problems to efficient tapping, but they all share certain similarities. Toughness and general abrasiveness top the list. It is also difficult to impart a good surface finish to exotics; heat tends to localize in the shear zone, and exotics tend to work-harden and grab the tool.

A tap's chamfer and first full thread do virtually all the cutting. The remaining ground threads serve merely as chasers. Because of this, taps for threading exotics are increasingly manufactured with short threads and reduced necks. They diminish problems caused by material closure and provide more space for coolant and chip ejection.

When tapping exotics, the largest tap core diameter possible should be applied. Cutting 75% or 65% threads places less stress on the taps, lengthens tool life, and reduces breakage. To cut threads in exotic alloys successfully, taps must combine geometries specifically tailored for those materials and be made of premium tool steels subjected to precisely controlled heat-treatment processes.

Stainless steel is known to work-harden and to have slow heat-dissipation characteristics; stainless steel requires a tap geometry with a positive 6- to 9-degree rake, preventing work-hardening and reducing torque. Grinding an appropriate eccentric thread and back-taper relief onto the tap will reduce friction. A surface treatment promotes lubricity. That is, the tool should be made from high-vanadium, high-cobalt tool steel and have a surface treatment so that coolant adheres to it. Stainless steel generates long chips and requires a tap with a 38-degree helix angle and adequate flute depth to promote chip evacuation. Proper hook and radial relief guarantee accurate thread-hole size and long tool life.

When tapping a titanium alloy, the material's tendency to concentrate heat in a small contact area must be considered. Concentrated heat often leads to excessive cutting-edge wear. Titanium generates average-to-short chips, is abrasive, and is prone to chip welding and high friction. These characteristics degrade tap performance and shorten tool life.

Taps for threading titanium are constructed of premium tool steel, nitrided for hardness. Titanium nitride (TiN) coatings cannot be used because they react chemically with the workpiece material, causing rapid tap failure. For tapping through-holes, a 3- to 5-degree rake and short thread design with high eccentric relief will promote efficient chip evacuation.

Nickel-based Inconel, Monel, Waspalloy, and Hastelloy present severe tapping problems. Among their machining characteristics are toughness, work-hardening, heat retention, and built-up-edge (BUE). Taps designed to thread these materials need tremendous stability and a strong cross-sectional construction. The most popular tap materials for nickel alloys are high-vanadium, high-cobalt tool steel, or powdered metal (PM) tool steel. Tapping blind holes in these alloys requires a 3- to 5-degree rake to shear and deflect the cutting forces downward toward the tap's root, its strongest area. A 26-degree helix angle promotes chip evacuation, and a nitride or TiN coating reduces friction and tool wear. Because of nickel alloys' toughness, taps to cut them should have the longest taper possible. This allows

TABLE 31.1	Cutting Fluids for Tapping (HSS Tools)
Work Material	**Cutting Fluid**
Aluminum	Kerosene and lard oil; kerosene and light-base oil
Brass	Soluble oil or light-base oil
Naval brass; Manganese bronze	Mineral oil with lard or light-base oil
Phosphor bronze; Copper	Mineral oil with lard or light-base oil
Iron, cast malleable	Dry or soluble oil or sulfur-base oil
Magnesium	Light-base oil diluted with kerosene
Monel metal	Sulfur-base oil
Steel, Up to 0.25 carbon; Free machining	Sulfur-base or soluble oil
Steel, 0.30–0.60 carbon, annealed	Chlorinated sulfur-base oil
Steel, 0.30–0.60 carbon, heat treated	Sulfur-base oil
Steel, Tool, high-carbon, HSS, Stainless	Chlorinated sulfur-base oil
Titanium	Chlorinated sulfur-base oil
Zinc die castings	Kerosene and lard oil

the cutting edges to progressively gain thread height before the first full thread begins its cut, distributing the load over a wider area. Spiral-pointed, straight-flute taps have a four- to five-thread taper. For tapping blind holes, the first two or three threads—more if possible—should be tapered.

Cutting Fluid for Tapping

Cutting fluids should be kept as clean as possible and should be supplied in copious quantities to reduce heat and friction and to aid in chip removal. Long tap life has been reported to result from routing high-pressure coolants through the tap to flush out the chips and cool the cutting edges. Recommended cutting fluids are listed in **Table 31.1**.

31.4 Thread Milling

Highly accurate threads, particularly in larger sizes, are often form milled. Either a single- or a multiple-form cutter may be used. A single-form cutter having a single annular row of teeth is tilted at an angle equal to the helix angle of the thread and is fed inward radially to full depth while the work is stationary. The workpiece then is rotated slowly, and the cutter is simultaneously moved longitudinally, parallel with the axis of the work (or vice versa), by means of a leadscrew, until the thread is completed. The thread can be completed in a single cut, or roughing and finish cuts can be used. This process is used primarily for large-lead or multiple-lead threads.

Some threads can be milled more quickly by using a multiple-form cutter having multiple rows of teeth set perpen-dicular to the cutter axis (the rows having no lead). The cutter must be slightly longer than the thread to be cut. It is set parallel with the axis of the workpiece and fed inward to full-thread depth while the work is stationary. The work then is rotated slowly for a little more than one revolution, and the rotating cutter is simultaneously moved longitudinally with respect to the workpiece (or vice versa) according to the thread lead. When the work has completed one revolution, the thread is complete. This process cannot be used on threads having a helix angle greater than about 3 degrees, because clearance between the sides of the threads and the cutter depends on the cutter diameter's being substantially less than that of the workpiece. Thus, although the process is rapid, its use is restricted to threads of substantial diameter and not more than about 2 in. long.

As shown in **Figure 31.9**, advances in CNC computer controls have led to thread milling on three-axis machines. Today's CNC can helically interpolate the axial feed controlling the thread pitch with circular feed controlling the circumference of the thread. The cutter has teeth shaped like the desired thread form. The cutter rotates at high speeds while its axis slowly moves around the part in a planetary arc just over 360 degrees. The cutter advances axially a distance equal to one pitch to generate the helical path. Thread milling has advantages in diameters greater than 1.5 in., including better surface finish and concentricity and the ability to produce right- or left-hand threads with the same tool.

Thrilling (drilling plus threading) produces threaded holes by combining short-hole drilling with **thread milling** using a combination tool with a drill point and a thread mill body. The details of the process are shown in **Figure 31.10** and can be done on any CNC machining center. Compared to tapping, the process eliminates two tools, two tool holders, and two tool change cycles because the single tool combines the drill ream and tap functions into one tool. Threaded-hole depths are limited to about three hole diameters.

31.5 Thread Grinding

Grinding can produce very accurate threads, and it also permits threads to be produced in hardened materials. Three basic methods are used. **Center-type grinding with axial feed** is the most common method, being similar to cutting a thread on a lathe. A shaped grinding wheel replaces the single-point tool. Usually, a single-ribbed grinding wheel is employed, but multiple-ribbed wheels are used occasionally. The grinding wheels are shaped by special diamond dressers or by crush dressing and must be inclined to the helix angle of the thread. Wheel speeds are in the high range. Several passes are usually required to complete the thread.

Center-type infeed thread grinding is similar to multiple-form milling in that a multiple-ribbed wheel, as wide as the length of the desired thread, is used. The wheel is fed inward

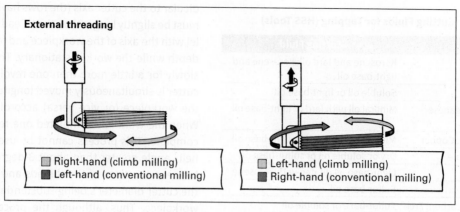

External threading

■ Right-hand (climb milling)
■ Left-hand (conventional milling)

■ Left-hand (climb milling)
■ Right-hand (conventional milling)

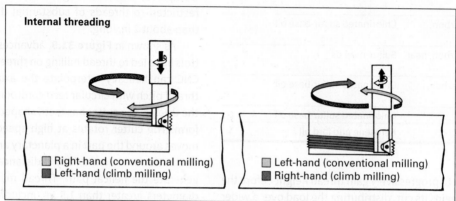

Internal threading

■ Right-hand (conventional milling)
■ Left-hand (climb milling)

■ Left-hand (conventional milling)
■ Right-hand (climb milling)

Thread milling can be done on machining centers with multitooth indexable-carbide-insert cutters. The cutter can produce a finished thread in one helical pass.

FIGURE 31.9   Thread milling on a three-axis NC machine can produce a complete thread in a single feed revolution. (Copyright © 2011 Penton Media, Inc.)

radially to full thread depth, and the thread blank is then turned through about 1½ turns as the grinding wheel is fed axially a little more than the width of one thread. **Centerless thread grinding** is used for making headless setscrews. The blanks are hopper fed to the regulating wheel, which causes them to traverse the grinding wheel face, from which they emerge in completed form. Production rates of 60 to 70 screws of 1½-in. length per minute are possible.

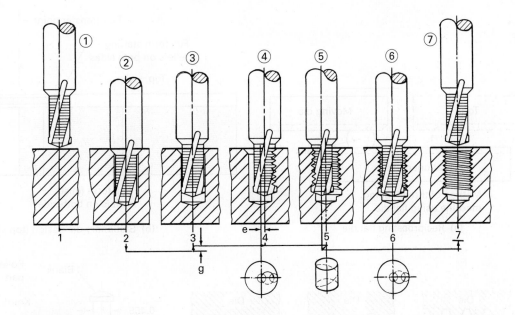

FIGURE 31.10 The process of high-speed thrilling (drilling plus threading) a hole includes (1) approach, (2) drill plus chamfer, (3) retract one thread pitch, (4) radially ramp to the major thread diameter, (5) thread-mill with helical interpolation, (6) return the tool to the centerline of the hole, and (7) retract from the finished hole. At 20,000 rpm, a hole can be thrilled in aluminum in >2 s. (Fred Mason, *American Machinist*, November, 1988)

31.6 Thread Rolling

Thread rolling is used to produce threads in substantial quantities. This is a cold-forming process operation in which the threads are formed by rolling a thread blank between hardened dies that cause the metal to flow radially into the desired shape. Because no metal is removed in the form of chips, less material is required, resulting in substantial savings. In addition, because of cold working, the threads have greater strength than cut threads, and a smoother, harder, and more wear-resistant surface is obtained. In addition, the process is fast, with production rates of one per second being common. The quality of cold-rolled products is consistently good. Chipless operations are cleaner and there is a savings in material (15% to 20% savings in blank stock weight is typical).

Thread rolling is done by four basic methods. The simplest of these employs one fixed and one movable flat rolling die (**Figure 31.11**). After the blank is placed in position on the stationary die, movement of the moving die causes the blank to be rolled between the two dies and the metal in the blank is displaced to form the threads. As the blank rolls, it moves across the die parallel with its longitudinal axis. Prior to the end of the stroke of the moving die, the blank rolls off the end of the stationary die, its thread being completed.

One obvious characteristic of a rolled thread is that its major diameter always is greater than the diameter of the blank. When an accurate class of fit is desired, the diameter of the blank is made about 0.002 in. larger than the thread-pitch diameter. If it is desired to have the body of a bolt larger than the outside diameter of the rolled thread, the blank for the thread is made smaller than the body.

Thread rolling can also be done with cylindrical dies, using the three-roll method commonly employed on turret lathes and screw machines. Two variations are used. In one, the rolls are retracted while the blank is placed in position. The rolls then move inward radially, while rotating, to form the thread. More commonly, the three rolls are contained in a self-opening die head similar to the conventional type used for cutting external threads. The die head is fed onto the blank longitudinally and forms the thread progressively as the blank rotates. With this procedure, as in the case of cut threads, the innermost one-and-a-half to two threads are not formed to full depth because of the progressive action of the rollers.

The two-roll method is commonly employed for automatically producing large quantities of externally threaded parts up to 6 in. in diameter and 20 in. long. The planetary-type machine is for mass production of rolled threads on diameters up to 1 in. Not only is thread rolling very economical, the threads are excellent as to form and strength. The cold working contributes to increased strength, particularly at the critical root areas. There is less likelihood of surface defects (produced by machining), which can act as stress raisers.

Large numbers of threads are rolled on thin, tubular products. In this case, external and internal rolls are used. The threads on electric lamp bases and sockets are examples of this type of thread.

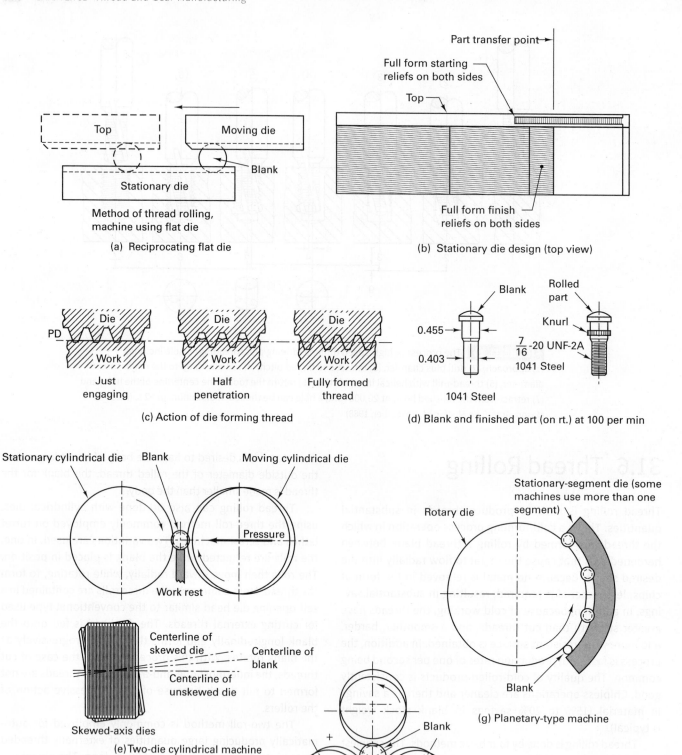

Part transfer point →

Full form starting reliefs on both sides

Top

Top

Moving die

Blank

Stationary die

Method of thread rolling, machine using flat die

Full form finish reliefs on both sides

(a) Reciprocating flat die

(b) Stationary die design (top view)

PD

Die
Work
Just engaging

Die
Work
Half penetration

Die
Work
Fully formed thread

(c) Action of die forming thread

Blank

Rolled part

Knurl

0.455

0.403

$\frac{7}{16}$-20 UNF-2A
1041 Steel

1041 Steel

(d) Blank and finished part (on rt.) at 100 per min

Stationary cylindrical die   Blank   Moving cylindrical die

Pressure

Work rest

Centerline of skewed die

Centerline of blank

Centerline of unskewed die

Skewed-axis dies

(e) Two-die cylindrical machine

Rotary die

Stationary-segment die (some machines use more than one segment)

Blank

(g) Planetary-type machine

Blank

Pressure

Three cylindrical dies in rolling position

Cylindrical dies in retracted position

(f) Three-die cylindrical machine

FIGURE 31.11 Roll forming threads using flat die thread rolling process shown in (a) and (b). The threads forming action is shown in (c) and the product in (d). Three variations of cylindrical rolling are shown in (e), (f), and (g).

Chipless Tapping

Unfortunately, most internal threads cannot be made by rolling; there is insufficient space within the hole to permit the required rolls to be arranged and supported, and the required forces are too high. However, many internal threads, up to about 1½ in. in diameter, are cold formed in holes in ductile metals by means of **fluteless taps**.

In fluteless taps, the forming action is essentially the same as in rolling external threads. Because of the forming involved and the high friction, the torque required is about double that for cutting taps. Also, the hole diameter must be controlled carefully to obtain full thread depth without excessive torque. However, fluteless taps produce somewhat better accuracy than cutting taps, and tap life is often greater than that of high-speed-steel (HSS) machine taps. A lubricating fluid should be used; water-soluble oils are quite effective. Fluteless taps are especially suitable for forming threads in dead-end holes because no chips are produced. They come in both plug and bottoming types.

Machining Versus Rolling Threads

Threads are machined or cut when full thread depth is needed (more than one pass necessary) for short production runs, when the blanks are not very accurate, when proximity to the shoulder in end threading is needed, for tapered threads, or when the workpiece material is not adaptable for rolling.

31.7 Gear Theory and Terminology

Basically, gears are modifications of wheels, with **gear teeth** added to prevent slipping and to ensure that their relative motions are constant. However, it should be noted that the relative surface velocities of the wheels (and shafts) are determined by the diameters of the wheels. Although wooden teeth or pegs were attached to disks to make gears in ancient times, the teeth of modern gears are produced by machining or forming teeth on the outer portion of the wheel. The **pitch circle** (**Figures 31.12** and **31.13**) corresponds to the diameter of the wheel. Thus, the angular velocity of a gear is determined by the diameter of this imaginary pitch circle. All design calculations relating to gear performance are based on the pitch-circle diameter or, more simply, the *pitch diameter* (PD).

For two gears to operate properly, their pitch circles must be tangential to each other. The point at which the two pitch circles are tangent, at which they intersect the centerline connecting their centers of rotation, is called the **pitch point**. The common

normal at the point of contact of mating teeth must pass through the pitch point. This condition is illustrated in Figure 31.13.

To provide uniform pressure and motion and to minimize friction and wear, gears are designed to have rolling motion between mating teeth rather than sliding motion. To achieve this condition, most gears utilize a tooth form that is based on an **involute curve**. This is the curve that is generated by a *point* on a straight line when the line rolls around a *base circle*. A somewhat simpler method of developing an involute curve is that shown in **Figure 31.14**. By unwinding a tautly held string from a point on the base circle, point *A*, an involute curve is generated.

There are three other reasons for using the involute form for gear teeth. First, such a tooth form provides the desired pure rolling action. Second, even if a pair of involute gears is operated with the distance between the centers slightly too large or too small, the common normal at the point of contact between mating teeth will always pass through the pitch point. Obviously, the theoretical pitch circles in such cases will be increased or decreased slightly. Third, the **line of action** or **path of contact**—that is, the locus of the points of contact of

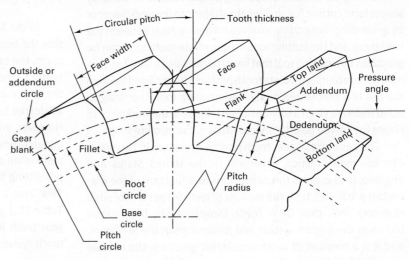

FIGURE 31.12 Gear-tooth nomenclature.

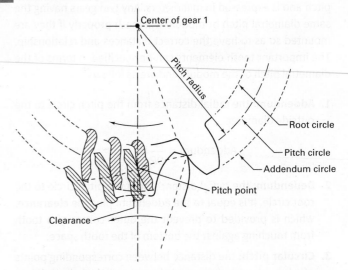

FIGURE 31.13 Tangent pitch circles between two gears produce a pitch point.

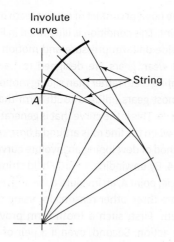

FIGURE 31.14 Method of generating an involute curve by unwinding a string from a cylinder.

mating teeth—is a straight line that passes through the pitch point and is tangent to the base circles of the two gears.

Cutting an involute shape in gear blanks can be done by simple form cutting (i.e., milling the shape into the workpiece) or by generating. Generating involves relative motion between the workpiece and the cutting tool. True involute tooth form can be produced by a cutting tool that has straight-sided teeth. This permits a very accurate involute tooth profile to be obtained through the use of a simple and easily made cutting tool. The straight-sided teeth are given a rolling motion relative to the workpiece to create the curved gear-tooth face—that is, the involute shape.

The basic size of gear teeth may be expressed in two ways. The common practice, especially in the United States and England, is to express the dimensions as a function of the **diametral pitch** (DP). DP is the number of teeth (N) per unit of pitch diameter (PD); thus DP = N/PD. Dimensionally, DP involves inches in the English system and millimeters in the SI system, and it is a measure of tooth size. Metric gears use the module system (M), defined as the pitch diameter divided by the number of teeth, or M = PD/N. It thus is the reciprocal of diametral pitch and is expressed in millimeters. Any two gears having the same diametral pitch or module will mesh properly if they are mounted so as to have the correct distances and relationship. The important tooth elements can be specified in terms of the diametral pitch or the module and are as follows:

1. **Addendum:** the radial distance from the pitch circle to the outside diameter.

$$\text{Addendum} = \frac{1}{DP}\text{in.}$$

2. **Dedendum:** the radial distance from the pitch circle to the root circle. It is equal to the addendum plus the **clearance**, which is provided to prevent the outer corner of a tooth from touching against the bottom of the tooth space.

3. **Circular pitch:** the distance between corresponding points of adjacent teeth, measured along the pitch circle:

$$\pi/\text{Diametral pitch}$$

4. **Tooth thickness:** the thickness of a tooth, measured along the pitch circle. When tooth thickness and the corresponding **tooth space** are equal, no **backlash** exists in a pair of mating gears.

5. **Face width:** the length of the gear teeth in an axial plane.

6. **Tooth face:** the mating surface between the pitch circle and the addendum circle.

7. **Tooth flank:** the mating surface between the pitch circle and the root circle.

8. **Pressure angle:** the angle between a tangent to the tooth profile and a line perpendicular to the pitch surface.

Four shapes of involute gear teeth are used in the United States:

1. $14\frac{1}{2}$ degree pressure angle, full depth (used most frequently).
2. $14\frac{1}{2}$ degree pressure angle, composite (seldom used).
3. 20-degree pressure angle, full depth (seldom used).
4. 20-degree pressure angle, stub tooth (second most common).

In the 14½ degree full-depth system, the tooth profile outside the base circle is an involute curve. Inward from the base circle, the profile is a straight radial line that is joined with the bottom land by a small fillet. With this system, the teeth of the basic rack have straight sides. The composite system and the 20-degree full-depth system provide somewhat stronger teeth. However, with the 20-degree full-depth system, considerable undercutting occurs in the dedendum area; therefore, stub teeth often are used. The addendum is shortened by 20%, thus permitting the dedendum to be shortened a similar amount. This results in very strong teeth without undercutting. **Table 31.2** gives the formulas for computing the dimensions of gear teeth in the 14½ degree full-depth and 20-degree stub-tooth systems.

TABLE 31.2	**Formula for Calculating the Standard Dimensions for Involute Gear Teeth**	
	$14\frac{1}{2}$ **degree Full Depth**	**20 degree, Stub Tooth**
Pitch diameter (PD)	$\dfrac{N}{DP}$	$\dfrac{N}{DP}$
Addendum	$\dfrac{1}{DP}$	$\dfrac{0.8}{DP}$
Dedendum	$\dfrac{1.157}{DP}$	$\dfrac{1}{DP}$
Outside diameter	$\dfrac{N+2}{DP}$	$\dfrac{N+1.6}{DP}$
Clearance	$\dfrac{0.157}{DP}$	$\dfrac{0.2}{DP}$
Tooth thickness	$\dfrac{1.508}{DP}$	$\dfrac{1.508}{DP}$

DP = Number of teeth (N) per unit of pitch diameter (PD).

Physical Requirements of Gears

A consideration of gear theory leads to five requirements that must be met in order for gears to operate satisfactorily:

1. The actual tooth profile must be the same as the theoretical profile.

2. Tooth spacing must be uniform and correct.

3. The actual and theoretical pitch circles must be coincident and must be concentric with the axis of rotation of the gear.

4. The face and flank surfaces must be smooth and sufficiently hard to resist wear and prevent noisy operation.

5. Adequate shafts and bearings must be provided so that desired center-to-center distances are retained under operational loads.

The first four of these requirements are determined by the material selection and manufacturing process. The various methods of manufacture that are used represent attempts to meet these requirements to varying degrees with minimum cost, and their effectiveness must be measured in terms of the extent to which the resulting gears embody these requirements.

Before looking at the ways to manufacture gears, let's look at some examples of gears.

31.8 Gear Types

The most common types of gears are shown in **Figure 31.15**. **Spur gears** have straight teeth and are used to connect parallel shafts. They are the most easily made and the cheapest of all types.

The teeth on **helical gears** lie along a helix, the angle of the helix being the angle between the helix and a pitch cylinder element parallel with the gear shaft. Helical gears can connect either parallel or nonparallel nonintersecting shafts. Such gears are stronger and quieter than spur gears because the contact between mating teeth increases more gradually and more teeth are in contact at a given time. Although they usually are slightly more expensive to make than spur gears, they can be manufactured in several ways and are produced in large numbers.

Helical gears have one disadvantage. When they are in use, a side thrust is created that must be absorbed in the bearings. **Herringbone gears** neutralize this side thrust by having, in effect, two helical-gear halves, one having a right-hand and the other a left-hand helix. The continuous herringbone type is rather difficult to machine but is very strong. A modified herringbone type is made by machining a groove, or gap, around the gear blank where the two sets of teeth would come together. This

provides a runout space for the cutting tool in making each set of teeth.

A **rack** or rack bar is a gear having teeth that lie on a straight line on a bar or a plane. The teeth may be normal to the axis of the rack or helical so as to mate with spur or helical gears, respectively. This gear has an infinite radius.

A **worm** is similar to a screw. It may have one or more threads, the multiple-thread type being very common. Worms usually are used in conjunction with a **worm gear**. High gear ratios are easily obtainable with this combination. The axes of the worm and worm gear are nonintersecting and are usually at right angles. If the worm has a small helix angle, it cannot be driven by the mating worm gear. This principle is frequently employed to obtain nonreversible drives. Worm gears are usually made with the top land concave to permit greater area of contact between the worm and the gear. A similar effect can be achieved by using a **conical worm**, in which the helical teeth are cut on a double-conical blank, thus producing a worm that has an hourglass shape.

Bevel gears, teeth on a cone, are used to transmit motion between intersecting shafts. The teeth are cut on the surface of a truncated cone. Several types of bevel gears are made, the types varying as to whether the teeth are straight or curved and whether the axes of the mating gears intersect. On *straight-tooth* bevel gears, the teeth are straight, and if extended all would pass through a common apex. *Spiral-tooth* bevel gears, have teeth that are segments of spirals. Like helical gears, this design provides tooth overlap so that more teeth are engaged at a given time and the engagement is progressive. *Hypoid* bevel gears also have a curved-tooth shape but are designed to operate with nonintersecting axes. Rear-drive automobiles used hypoid gears in the rear axle so that the drive-shaft axis can be below the axis of the axle and thus permit a lower floor height. *Zerol* bevel gears have teeth that are circular arcs, providing somewhat stronger teeth than can be obtained in a comparable straight-tooth gear. They are not used extensively. When a pair of bevel gears are the same size and have their shafts at right angles, they are termed **miter gears**.

A **crown gear** is a special form of bevel gear having a 180-degree cone apex angle. In effect, it is a disk with the teeth on the side of the disk. It may also be thought of as a rack that has been bent into a circle so that its teeth lie in a plane. The teeth may be straight or curved. On straight-tooth crown gears, the teeth are radial. Crown gears are seldom used, but they have the important quality that they will mesh properly with a bevel gear of any cone angle, provided that the bevel gear has the same tooth form and diametral pitch. This important principle is incorporated in the design and operation of two very important types of gear-generating machines that will be discussed later.

Most gears are of the external type, the teeth forming the outer periphery of the gear. Internal gears have the teeth on the inside of a solid ring, pointing toward the center of the gear.

Different types of gears

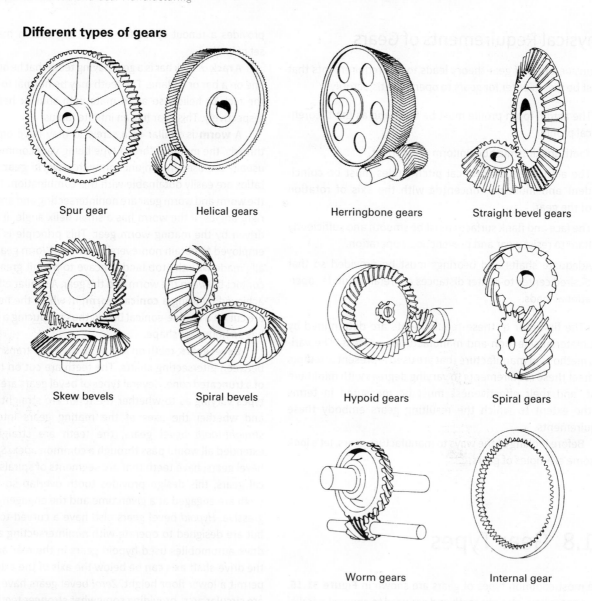

Spur gear   Helical gears   Herringbone gears   Straight bevel gears

Skew bevels   Spiral bevels   Hypoid gears   Spiral gears

Worm gears   Internal gear

FIGURE 31.15 There are many different types of gears for transmitting motion, resulting in a wide variety of gear-making processes. (Courtesy of TRW-Greenfield Tap & Die Kennametal)

31.9 Gear Manufacturing

Whether produced in large or small quantities, in manufacturing cells, or in job shop batches, the standard sequence of operations for gear manufacturing requires four sets of operations (see **Figure 31.16**):

1. Blanking (turning).
2. Gear cutting (milling, hobbing, shaping, shaving, broaching, and CNC).
3. Heat treatment.
4. Grinding.

Blanking refers to the initial forming or machining operations that produce a semifinished part ready for gear cutting, starting from a piece of raw material. Turning on chucker lathes

or CNC turning centers, followed by facing and centering of shafts, milling, and sometimes grinding fall into this category of operations. Good-quality blanks are essential in precision gear manufacturing.

Milling, hobbing, shaping, broaching, and shaving machines are the most frequently used machines for gear cutting, producing gears for automotive, truck, agricultural, and construction equipment. Other processes used in industrial gear production include CNC gearcutting, rolling, grinding, and shaving. The process selected depends on finding a cost-effective application based on quality specification, production volumes, and economic conditions.

The gear-cutting or -machining operations can be divided into operations executed prior to heat treatment, when the material is still soft and easily machinable, and after heat treatment, performed on parts that have acquired high hardness and strength.

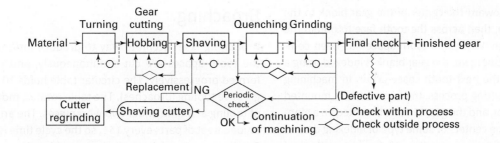

FIGURE 31.16 The sequence of operations for gearcutting (hobbing) typically starts with producing a blank, cutting the feath, heat threating to harden, and finishing by grinding.

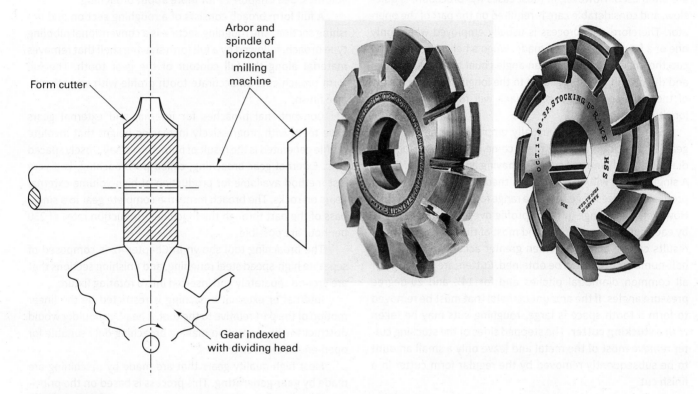

FIGURE 31.17 The basic method of machining a gear uses a form cutter (left) to mill out the space between the teeth using the form cutter (middle) or the stocking cutter (right) to machine the gear. (Image Courtesy of Hexagon Metrology, Inc.)

Heat treatment gives the material the strength and durability to withstand high loads and wear but results in a reduction in dimensional and geometrical accuracy. The metallurgical transformations that occur during hardening, quenching, and tempering cause a general quality deterioration in the gears. Therefore, precision grinding operations are used on external and internal bearing diameters, critical length dimensions, and fine surface finishes after heat treatment. Cylindrical grinders, angle-head grinders, internal grinders, and surface grinders are commonly used.

Gears are made in very large numbers by cold-roll forming. In addition, significant quantities are made by extrusion, blanking, and casting and some by powder metallurgy or a forging process. However, it is only by machining that all types of gears can be made in all sizes, and although roll-formed gears can be

made with accuracy sufficient for most applications, even for automobile transmissions, machining still is unsurpassed for gears that must have very high accuracy. Also, roll forming can be used only on ductile metals.

31.10 Machining of Gears

Form Milling

Form cutting or **form milling** on a horizontal milling machine is illustrated in **Figure 31.17**. The multiple-tooth form cutter has the same form as the *space* between adjacent teeth. The tool is fed radially toward the center of the gear blank to the desired tooth depth, then across the tooth face to obtain the required tooth width. When one tooth space has been completed, the tool is withdrawn, the gear blank is indexed using a dividing head, and the next tooth space is cut. In machining gears by the form-cutting process, the form cutter is mounted on the machine arbor, and the gear blank is mounted on a mandrel held between the centers of some type of indexing device. Basically, form cutting is a simple and flexible method of machining gears. The equipment and cutters required are relatively simple, and standard machine tools (milling machines) are often used. However, in most cases the procedure is quite slow, and considerable care is required on the part of the operator. Therefore, this process is usually employed where only one or a few gears are to be made. When a helical gear is to be cut, the table must be set at an angle equal to the helix angle, and the dividing head is geared to the longitudinal feed screw of the table so that the gear blank will rotate as it moves longitudinally.

Standard cutters are usually employed in form-cutting gears. In the United States, these come in eight sizes for each diametral pitch and will cut gears having any number of teeth. A single cutter will not produce a theoretically perfect tooth profile for all sizes of gears in the range for which it is intended. However, the change in tooth profile over the range covered by each cutter is very slight, and most of the time satisfactory results can be achieved. When greater accuracy is required, half-number cutters can be obtained. Cutters are available for all common diametral pitches and for 14½ and 20-degree pressure angles. If the amount of metal that must be removed to form a tooth space is large, roughing cuts may be taken with a **stocking cutter**. The stepped sides of the stocking cutter remove most of the metal and leave only a small amount to be subsequently removed by the regular form cutter in a finish cut.

Straight-tooth bevel gears can be form cut on a milling machine, but this is seldom done. Because the tooth profile in bevel gears varies from one end of the tooth to the other, after one cut is taken to form the correct tooth profile at the smaller end, the relationship between the cutter and the blank must be altered. Shaving cuts are then taken on the side of each tooth to form the correct profile throughout the entire tooth length.

Although the form cutting of gears on a milling machine is a flexible process and is suitable for gears that are not to be operated at high speeds or that need not operate with extreme quietness, the process is slow and requires a skilled operator. Semiautomatic machines are available for making gears by the form-cutting process. The procedure is essentially the same as on a milling machine, except that, after setup, the various operations are completed automatically. Gears made on such machines are no more accurate than those produced on a milling machine, but the possibility of error is less, and they are much cheaper because of reduced labor requirements. For large quantities, however, form cutting is not used.

Broaching

Broaching is an accurate way to produce internal gears. All the tooth spaces are cut simultaneously, and the tooth is formed progressively. The circular table holds 10 sets of progressive tooling (broaches). The table rotates, moving one set of tooling at a time under two workpieces. The arms load and unload a set of parts every 15 s, so the cycle time is very quick. Excellent gears can be made by broaching. However, a separate broach must be provided for each size of gear. The tooling tends to be expensive, restricting this method to large volumes. See Chapter 27 for more about broaching.

A full form broach consists of a roughing section and finishing section. The roughing section is a conventional nibbling type broach, followed by a full form shaving shell that removes material along the full contour of the gear tooth. The full form broach cuts an accurate tooth profile with smooth surface finish.

Conventional broaches for internal and external gears have teeth with progressively increasing height that involute profile generated is the result of numerous small closely spaced cuts. External gear broaching, called potbroaching, is a very fast method available for production of high-volume external gears on rocks. The broach forms the complete gear in a single pass of the part through the machine. Production rates of 250 per hour are possible.

The broaching tool shown in Chapter 42 is composed of separate high-speed steel roughing, and finishing sections that are ground separately and inserted into a rotating fixture.

Internal or external broaching is restricted by the linear motion of the part relative to the tool. A nearby shoulder would obstruct the linear motion making broaching only suitable for open-ended tooth structures.

Most high-quality gears that are made by machining are made by **gear generating**. This process is based on the principle that any two involute gears, or any gear and a rack, of the same diametral pitch will mesh together properly. Utilizing this principle, one of the gears (or the rack) is made into a cutter by proper sharpening. It can be used to cut into a mating gear blank and thus generate teeth on the blank. The two principal methods for gear generating are *shaping* and *hobbing*.

Shaping   To carry out the **gear shaping** process, the cutter and the gear blank must be attached rigidly to their respective shafts, and the two shafts must be interconnected by suitable gearing so that the cutter and the blank rotate positively with respect to each other and have the same pitch-line velocities. To start cutting the gear, the cutter is reciprocated vertically and is fed radially into the blank between successive strokes. When the desired tooth depth has been obtained, the cutter and blank are then slightly indexed after each cutting stroke. The resulting generating action is indicated schematically in **Figure 31.18** and shown in the cutting of an actual gear tooth in **Figure 31.19** along with a machine called a gear shaper. **Gear shapers** generate gears by a reciprocating tool motion. The gear blank is mounted on the rotating table (or

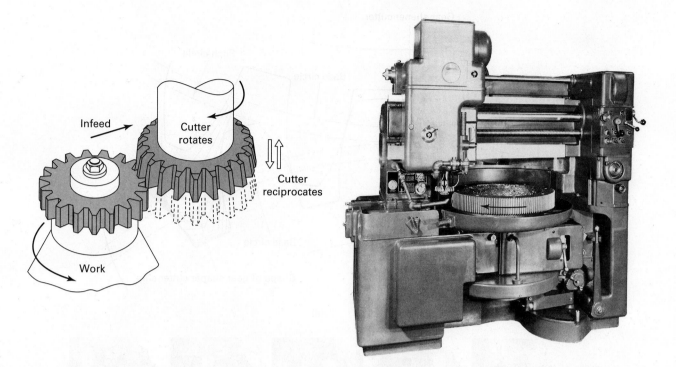

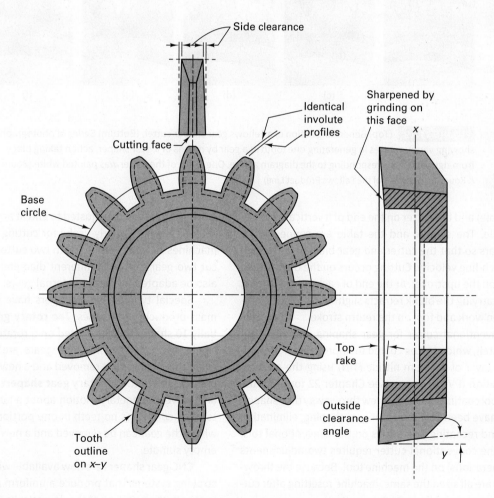

Infeed

Cutter rotates

Cutter reciprocates

Work

Side clearance

Cutting face

Identical involute profiles

Sharpened by grinding on this face

x

Base circle

Top rake

Outside clearance angle

y

Tooth outline on *x–y*

FIGURE 31.18 This machine tool is a gear shaper. The blank is rotating while the cutter is reciprocating vertically, as shown in the inset. The tool is very complex and is shown in detail below. (Bourn & Kouch Inc., Owners of the Fellows Product Line)

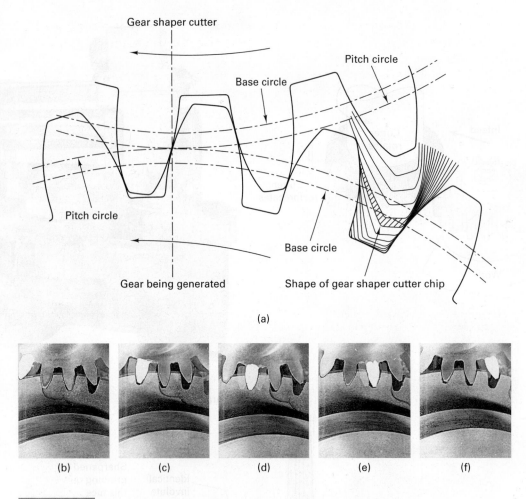

Gear shaper cutter

Pitch circle

Base circle

Pitch circle

Pitch circle

Base circle

Gear being generated

Shape of gear shaper cutter chip

(a)

(b) (c) (d) (e) (f)

FIGURE 31.19 (Top) Generating action of a Fellows gear shaper cutter. (Bottom) Series of photographs showing various stages in generating one tooth in a gear by means of a gear shaper, action taking place from right to left, corresponding to the diagram above. One tooth of the cutter was painted white.(Bourn & Kouch Inc., Owners of the Fellows Product Line)

vertical spindle) and the cutter on the end of a vertical, reciprocating spindle. The spindle and the table are connected by means of gears so that the cutter and gear blank revolve with the same pitch-line velocity. Cutting occurs on the downstroke (sometimes on the upstroke). At the end of each cutting stroke, the spindle carrying the blank retracts slightly to provide clearance between work and tool on the return stroke.

The conventional cutter for gear shaping is made from high-speed stell, which can be coated to improve wear life with a superhard layer of titanium nitride (TiN) using the physical vapor deposition (PVD) process. See Chapter 22 for a discussion about tool coatings. Recently, new throwaway disk-shaped insert tools have been developed for gear shaping, eliminating regrinding and recoating operations on the conventional tols. Regrinding the conventional cutter requires two adjustments (resetting operations) on the machine tool. Because the throwaway blades are all sized the same, machine resetting after cutting tool changeover is eliminated.

Either straight- or helical-tooth gears can be cut on gear shapers. To cut helical teeth, both the cutter and the blank are given an oscillating rotational motion during each stroke of the cutter, turning in one direction during the cutting stroke and in the opposite direction during the return stroke. Because the

cutting stroke can be adjusted to end at any desired point, gear shapers are particularly useful for cutting cluster gears. Some machines can be equipped with two cutters simultaneously to cut two gears, often of different diameters. Gear shapers can also be adapted for cutting internal gears.

Special types of gear shapers have been developed for mass-production purposes. The **rotary gear shaper** is essentially 10 shaper units mounted on a rotating base that have a single drive mechanism. Nine gears are cut simultaneously while a finished gear is removed and a new blank is put in place on the tenth unit. **Planetary gear shapers** hold six gear blanks and move in planetary motion about a large, central gear cutter. The cutter has no teeth in one portion to provide a space where the gear can be removed and a new blank placed on the empty spindle.

CNC gear shapers are now available with hydromechanical stroking systems that produce a uniform cutting velocity during the cutting portion of the downstroke. These machines can operate at 500 to 1700 rpm and use TiN-coated cutters to enhance tool life. **Vertical shapers** for gear generating have a vertical ram and a round table that can be rotated in a horizontal plane by either manual or power feed. These machine tools are sometimes called **slotters**. Usually, the ram is pivoted near the top so that it can be swung outward from the column through an arc of about 10 degrees.

Because one circular and two straight-line motions and feeds are available, vertical shapers are very versatile tools and thus find considerable use in one-of-a-kind manufacturing. Not only can vertical and inclined flat surfaces be machined, but external and internal cylindrical surfaces can be generated by circular feeding of the table between strokes. This may be cheaper than turning or boring for very small lot sizes. A vertical shaper can be used for generating gears or machining curved surfaces, interior surfaces, and arcs by using a stationary tool and rotating the workpiece. A **keyseater** is a special type of vertical shaper designed and used exclusively for machining keyways on the inside of wheel and gear hubs.

For machining continuous herringbone gears, a *Sykes gear-generating machine* is used.

Hobbing

Involute gear teeth could be generated by a cutter that has the form of a rack. Such a cutter would be simple to make but has two major disadvantages. First, the cutter (or the blank) would have to reciprocate, with cutting occurring only during one stroke direction. Second, because the rack would have to move longitudinally as the blank rotated, the rack would need to be very long (or the gear very small) or the two would not be in mesh after a few teeth were cut. A **hob** overcomes the preceding two difficulties. As shown in **Figure 31.20**, a hob can be thought of, basically, as one long rack tooth that has been wrapped around a cylinder in the form of a helix and

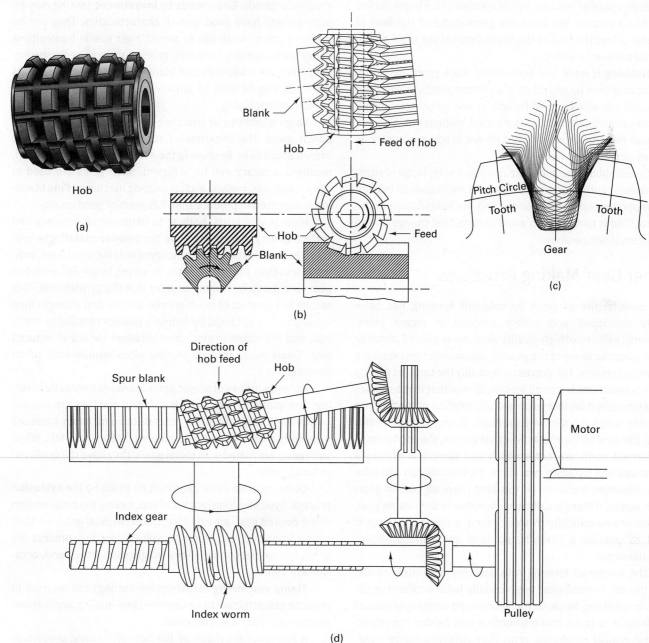

FIGURE 31.20 Hobs are rotary cutters with helped rows of teeth sharpened on these faces, with straight tooth sides that can generate involutes. (a) Hob. (b) Relationship of a hob and the gear blank in machining a spur gear. (c) The action of a hob as its teeth progress through blank to form the gear tooth. (d) Schematic of the gear hobbing mechanism.

fluted at intervals to provide a number of cutting edges. Relief is provided behind each of the teeth. The cross section of each tooth, normal to the helix, is the same as that of a rack tooth. (A hob can also be thought of as a gashed worm gear.)

The action of a **gear hobbing** machine cutting a spur gear is also illustrated in Figure 31.20. To cut a spur gear, the axis of the hob must be set off from the normal to the rotational axis of the blank by the helix angle of the hob. In cutting helical gears, the hob must be set over an additional amount equal to the helix angle of the gear. The cutting of a gear by means of a hob is a continuous action. The hob and the blank are connected by proper gearing so that they rotate in mesh. To start cutting a gear, the rotating hob is fed inward until the proper setting for tooth depth is obtained. The hob is then fed in a direction parallel with the axis of rotation of the blank. As the gear blank rotates, the teeth are generated and the feed of the hob across the face of the blank extends the teeth to the desired tooth-face width.

Hobbing is rapid and economical. More gears are cut by this process than by any other. The process produces excellent gears and can also be used for splines and sprockets. Single-, double-, and triple-thread hobs are used. Multiple-thread types increase the production rate but do not produce accuracy as high as single-thread hobs.

Gear-hobbing machines are made in a wide range of sizes. Machines for cutting accurate large gears are frequently housed in temperature-controlled rooms, and the temperature of the cutting fluid is controlled to avoid dimensional change due to variations in temperature.

Other Gear-Making Processes

The manufacture of gears by **cold-roll forming** has been highly developed and widely adopted in recent years. Currently, millions of high-quality gears are produced annually by this process; many of the gears in automobile transmissions are made this way. The process is basically the same as that by which screw threads are roll formed, except that in most cases the teeth cannot be formed in a single rotation of the forming rolls; the rolls are fed inward gradually during several revolutions. Because of the metal flow that occurs, the top lands of roll-formed teeth are not smooth and perfect in shape; a depressed line between two slight protrusions can often be seen. However, because the top land plays no part in gear-tooth action, if there is sufficient clearance in the mating gear, this causes no difficulty. Where desired, a light turning cut is used to provide a smooth top land and correct addendum diameter.

The hardened forming rolls are very accurately made, and the roll-formed gear teeth usually have excellent accuracy. In addition, because the severe cold working produces tooth faces that are much smoother and harder than those on the typical machined gear, they seldom require hardening or further finishing, and they have excellent wear characteristics.

The process is rapid (up to 50 times faster than gear machining) and easily mechanized. No chips are made, and thus, less material is needed. Less skilled labor is required. Small gears are often made by rolling a length of shaft and then slicing off the individual gear blanks. Usually, soft steel is required and 4 to 5 in. in diameter is about the limit, with fewer than six teeth, coarser than 12 diametral pitch, and no pressure angle less than 20 degrees.

Gears can be made by the various casting processes. **Sand-cast gears** have rough surfaces and are not accurate dimensionally. They are used only for services where the gear moves slowly and where noise and inaccuracy of motion can be tolerated. Gears made by **die casting** are more accurate and have fair surface finish. They can be used to transmit light loads at moderate speeds. Gears made by **investment casting** may be accurate and have good surface characteristics. They can be made of strong materials to permit their use in transmitting heavy loads. In many instances, gears that are to be finished by machining are made from cast blanks, and in some larger gears the teeth can be cast to approximate shape to reduce the amount of machining.

Large quantities of gears are produced by blanking in a punch press. The thickness of such gears usually does not exceed about $\frac{1}{16}$ in. By shaving the gears after they are blanked, excellent accuracy can be achieved. Such gears are used in clocks, watches, meters, and calculating machines. **Fine blanking** is also used to produce thin, flat gears of good quality.

High-quality gears, both as to dimensional accuracy and surface quality, can be made by the **powder metallurgy** process. Usually, this process is employed only for small sizes, ordinarily less than 1 in. in diameter. However, larger and excellent gears are made by forging powder metallurgy preforms. This results in a product of much greater density and strength than usually can be obtained by ordinary powder metallurgy methods, and the resulting gears give excellent service at reduced cost. Gears made by this process often require little or no finishing.

Large quantities of plastic gears are made by **plastic molding**. The quality of such gears is only fair, and they are suitable only for light loads. Accurate gears suitable for heavy loads are frequently machined out of laminated plastic materials. When such gears are mated with metal gears, they have the quality of reducing noise.

Quite accurate small gears can be made by the **extrusion** process. Typically, long lengths of rod, having the cross section of the desired gear, are extruded. The individual gears are then sawed from this rod. Materials suitable for this process are brass, bronze, aluminum alloys, magnesium alloys, and, occasionally, steel.

Flame machining (oxyacetylene cutting) can be used to produce gears that are to be used for slow-moving applications wherein accuracy is not required.

A few gears are made by the hot-roll-forming process. In this process, a cold master gear is pressed into a hot blank as the two are rolled together.

31.12 Gear Finishing

To operate efficiently and have satisfactory life, gears must have accurate tooth profiles, and the faces of the teeth must be smooth and hard. These qualities are particularly important when gears must operate quietly at high speeds. When they are produced rapidly and economically by most processes except cold-roll forming, the tooth profiles may not be as accurate as desired, and the surfaces are somewhat rough and subject to rapid wear. Also, it is difficult to cut gear teeth in a hardened gear blank, and therefore economy dictates that the gear be cut in a relatively soft blank and subsequently be heat treated to obtain greater hardness. Such heat treatment usually results in some slight distortion and surface roughness. Although most roll-formed gears have sufficiently accurate profiles, and the tooth faces are adequately smooth and frequently have sufficient hardness, this process is feasible only for relatively small gears. Consequently, a large proportion of high-quality gears are given some type of finishing operation after they have received primary machining or after heat treatment. Most of these **gear-finishing** operations can be done quite economically because only minute amounts of metal are removed, and they are fast and often automatic.

Gear shaving is the most commonly used method for finishing spur- and helical-gear teeth prior to hardening. The gear is run, at high speed, in contact with a shaving tool, usually of the type shown in **Figure 31.21**. Such a tool is a very accurate, hardened, and ground gear that contains a number of peripheral serrations, thus forming a series of sharp cutting edges on each tooth. The gear and shaving cutter are run in mesh with their axes crossed at a small angle, usually about 10 degrees. As they rotate, the gear is reciprocated longitudinally across the shaving tool (or vice versa). During this action, which usually requires less than 1 min, very fine chips are *shaved* from the gear-tooth faces, thus eliminating any high spots and producing a very accurate tooth profile.

Rack shaving cutters are sometimes used for shaving small gears, the cutter reciprocating lengthwise, thus causing the gear to roll along it as it is moved sideways across the cutter and fed inward.

Although shaving cutters are costly, they have a relatively long life because only a very small amount of metal is removed, usually 0.001 to 0.004 in. Some gear-shaving machines produce a slight crown on the gear teeth during shaving. Some gears are not hardened prior to shaving, although it is possible to remove very small amounts of metal from hardened gears if they are not too hard. However, modern heat-treating equipment makes it possible to harden gears after shaving without harmful effects, and therefore this practice is followed if possible.

Roll finishing is a cold-forming process that is used to finish helical gears. The unhardened gear is rolled with two hardened, accurately formed rolling dies. See **Figure 31.22**. The center distance between the dies is reduced so as to cold work the surfaces and produce highly accurate tooth forms. High points on the unhardened gear are plastically deformed so that a smoother surface and more accurate tooth form are achieved. Because the operation is one of localized cold working, some undesirable effects may accrue, such as localized residual stresses and nonuniform surface characteristics. Surface finishes of 6 to 8 μm have been achieved. If roll finishing is to be used, attention must be paid to the prerolled geometry. Designers should consult the manufacturers of gear-rolling machines for specific recommendations. See Chapter 19 for discussion about cold rolling.

Gear grinding is used to obtain very accurate teeth on hardened gears. Two methods are used. One employs a formed grinding wheel that is trued to the exact form of a tooth by means of diamonds mounted on a special holder and guided by a large template. The other method is involute-generation

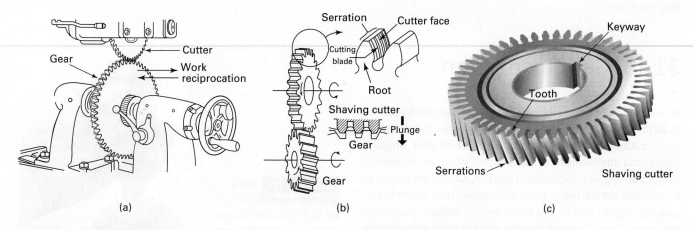

(a)   (b)   (c)

FIGURE 31.21 Gear shaving (on the left) uses a gear shaving tool (shown on the right) to accurately finish a gear.

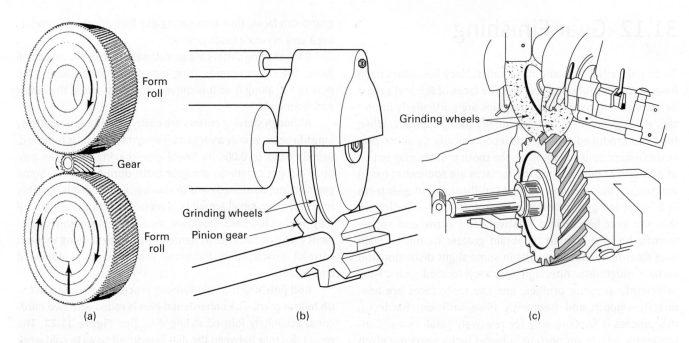

FIGURE 31.22 Two methods of gear finishing (other than gear shaving) are (a) gear roll finishing and gear grinding using (b) formed grinding wheels and (c) involve-generation grinding using straight sided grinding wheels.

grinding, which uses straight-sided grinding wheels that simulate one side of a rack tooth. The surface of the gear tooth is ground as the gear rolls (and reciprocates) past the grinding wheels. Grinding produces very accurate gears, but because it is slow and expensive, it is used only on the highest-quality, hardened gears. See Chapter 28 for additional discussion about grinding.

Lapping can also be used for finishing hardened gears. The gear to be finished is run in contact with one or more cast iron lapping gears under a flow of very fine abrasive in oil. Because lapping removes only a very small amount of metal, it is usually employed on gears that have previously been shaved and hardened. This combination of processes produces gears that are nearly equal to ground gears in quality, but at considerably lower cost.

Gear-tooth vernier calipers can be used to measure the thickness of gear teeth on the pitch circle (**Figure 31.23**). CNC gear inspection machines (**Figure 31.24**) can quickly check several factors, including variations in circular pitch, involute profile, lead, tooth spacing, and variations in pressure angle. The gear is usually mounted between centers. The probe is moved to the gear through x, y, and z translations. The gear is rotated between measurements. The inset in Figure 31.24 shows a typical display for an involute profile.

Because noise level is important in many applications—not only from the viewpoint of noise pollution but also as an indicator of probable gear life—special equipment for noise measurement is quite widely used, sometimes integrated into mass-production assembly lines.

31.13 Gear Inspection

As with all manufactured products, gears must be checked to determine whether the resulting product meets the design specifications and requirements. Because of their irregular shape and the number of factors that must be measured, inspection of gears is somewhat difficult. Among the factors to be checked are the linear tooth dimensions such as thickness, spacing, depth, and so on; tooth profile; surface roughness; and noise. Several special devices, most of them automatic or semiautomatic, are used for such inspection.

FIGURE 31.23 Using gear-tooth vernier calipers to check the tooth thickness at the pitch circle. (Courtesy E. Paul DeGarmo)

FIGURE 31.24 A CNC gear inspection machine has *x*, *y*, *z* motions plus a rotary table.
(Courtesy of Magnaflux a division of Illinois Tool Works, Inc.)

Review Questions

1. How does the pitch diameter differ from the major diameter for a standard screw thread?

2. For what types of threads are the pitch and the lead the same?

3. What is the helix angle of a screw thread?

4. Why are pipe threads tapered?

5. What are three basic methods by which external threads are produced?

6. Explain the meaning of ¼"-20 UNC-3A.

7. What is meant by the designation M20 x 2.5 6g6g? (What does the "x" mean?)

8. What are two reasons fine-series threads are being used less now than in former years?

9. In cutting a thread on a lathe, how is the pitch controlled?

10. What is the function of a threading dial on a lathe?

11. In this chapter, which figure(s) showed lapping threading, or a threading of deal?

12. What controls the lead of a thread when it is cut by a threading die?

13. What is the basic purpose of a self-opening die head?

14. What is the reason for using a taper tap before a plug tap in tapping a hole?

15. What difficulties are encountered if full threads are specified to the bottom of the dead-end hole?

16. Can a fluteless tap be used for threading a hole in gray cast iron? Why or why not?

17. What provisions should a designer make so that a dead-end hole can be threaded?

18. What is the major advantage of a spiral-point tap?

19. Can a fluteless tap be used for threading to the bottom of a dead-end hole? Why or why not?

20. Is it desirable for a tapping fluid to have lubricating qualities? Why?

21. How does thread milling differ from thread turning?

22. What are the advantages of making threads by grinding?

23. Why has thread rolling become the most commonly used method for making threads?

24. How can you determine whether a thread has been produced by rolling rather than by cutting?

25. What is a fluteless tap?

26. Why is the involute form used for gear teeth?

27. What is the diametral pitch of a gear?

28. What is the relationship between the diametral pitch and the module of a gear?

29. On a sketch of a gear, indicate the pitch circle, addendum circle, dedendum circle, and circular pitch.

30. What five requirements must be met for gears to operate satisfactorily? Which of these are determined by the manufacturing process?

31. What are the advantages of helical gears compared with spur gears?

32. What is the principal disadvantage of helical gears?

33. What difficulty would be encountered in hobbing a herringbone gear?

34. What is the only type of machine on which full-herringbone gears can be cut?

35. What modification in design is made to herringbone gears to permit them to be cut by hobbing?

36. Why aren't more gears made by broaching?

37. What is the most important property of a crown gear?

38. What are three basic processes for machining gears?

39. Which basic gear-machining process is utilized in a Fellows machine?

40. When a helical gear is machined on a milling machine, the table leadscrew and the universal dividing head have to be connected by a gear train. Why?

41. Why is a gear-hobbing machine much more productive than a gear shaper?

42. In gear shaping and gear hobbing, the tooth profiles are generated. What does that mean?

43. What are the advantages of cold-roll forming for making gears?

44. Why is cold-roll forming not suitable for making gray cast iron gears?

45. Under what conditions should shaving not be used for finishing gears?

46. What inherent property accrues from cold-roll forming of gears that may result in improved gear life?

47. Can lapping be used to finish cast iron gears?

48. What factors are usually checked in inspecting gears?

49. What are the basic methods for gear finishing?

Problems

1. A hob that has a pitch diameter of 76.2 mm is used to cut a gear having six teeth. If a cutting speed of 27.4 m/min is used, what will be the rpm value of the gear blank?

2. In Figure 31.17, a form-milling operation and cutters are shown. The gear is to be made from 4340 steel, R_c 50 prior to heat treat and final grind. Select the proper speeds and feeds for the job (the cutter is 4 in. in diameter) and compute the cutting time to mill this gear. Would you use up milling versus down milling?

3. A gear-broaching machine can complete the gear in 15 s (about 240 part/hr). How many additional gears per year are needed to

cover the broaching tooling cost if each broach on the machine cost $250? Do you think the broach tool life is sufficient to handle that number of parts? What about TiN coating the broaches (cost $10.00/broach)?

4. Assume that 10, 000 spur gears, 1⅛ in. in diameter and 1 in. thick, are to be made of 70-30 brass. What manufacturing method would you consider?

5. If only three gears described in Problem 4 were to be made, what process would you select?

CHAPTER 32

Surface Integrity and Finishing Processes

32.1 Introduction

Surface finishing processes are used to tailor the properties of the surfaces of manufactured components so that their function and serviceability can be improved. Processes include solidification treatments such as hot-dip coatings, weld-overlay coatings, and thermal spray surfaces; heat treatment coatings such as diffusion coatings and surface hardening; deposition surface treatments such as electrodeposition, chemical vapor deposition, and physical vapor deposition; and Electroplating means the electrodeposition of an adherent metallic coating onto an object that serves as the cathode in an electrochemical reaction. The resulting surface provides wear resistance, corrosion resistance, high-temperature resistance, or electrical properties different from those in the bulk material.

Many manufacturing processes influence surface properties, which in turn may significantly affect the way the component functions in service. The demands for greater strength and longer life in components often depend on changes in the surface properties rather than the bulk properties. These changes may be mechanical, thermal, chemical, and/or physical and therefore are difficult to describe in general terms.

Many metal-cutting processes specified by the manufacturing engineer to produce a specific geometry can produce alterations in the surface material of the component, which, in turn, produces changes in performance.

32.2 Surface Integrity

The term **surface integrity** was coined by Field and Kahles in 1964 in reference to the nature of the surface condition that is produced by the manufacturing process. If we view the process as having five main components (workpiece, tool, machine tool, environment, and process variables) as outlined in

Table 32.1 we see that surface properties can be altered by all of these parameters by producing the following:

- High temperatures involved in the machining process.
- Plastic deformation of the work material (residual stress).
- Surface geometry (roughness, waviness, cracks, distortions).
- Chemical reactions, particularly between the tool and the workpiece.

More specifically, surface integrity refers to the impaired or enhanced surface condition of a component or specimen that influences its performance in service. Surface integrity has two aspects: topography characteristics and surface-layer characteristics. **Topography** is made up of surface roughness, waviness, errors of form, and flaws (**Figure 32.1**).

A typical roughness profile includes the peaks and valleys that are considered separately from waviness. Flaws also add to texture but should be measured independent of it. Changes in the surface layer, as a result of processing, include plastic deformation, **residual stresses**, cracks, and other metallurgical changes including hardness, overaging, phase changes, recrystallization, intergranular attack, and hydrogen embrittlement. The surface layer will always contain local surface deformation due to any machining passes.

The material removal processes generate a wide variety of surface textures, generally referred to as **surface finish**. The cutting processes leave a wide variety of surface patterns on the materials. **Lay** is the term used to designate the direction of the predominant surface pattern produced by the machining process see **Figure 32.2**. In addition, certain other terms and symbols have been developed and standardized for specifying the surface quality. The most important terms are *surface roughness*, *waviness*, and *lay*. For machined surfaces, **Roughness** refers to the finely spaced surface irregularities. It resulting from specific machining operations. **Waviness** is surface irregularity of greater spacing than in roughness. It may be the result of warping, vibration, or the work being deflected during machining.

A variety of instruments are available for measuring surface roughness and surface profiles. The majority of these

TABLE 32.1   **Characteristics of Manufacturing Processes That Affect Surface Integrity**

Workpiece–Tool–Machine–Environment–Process Variables	
Workpiece characteristics	**Tool characteristics**
Geometry	Tool body
Shape	Type of tool
Dimensions	Size
Material	Shape
Type	Number of cutting edges
Route of manufacture	Cutting edge
Mechanical properties	Shape (angles)
Elastic constants	Nose geometry/topography
Plastic constants	Microgeometry
Physical properties	Wear
Melting point	Material
Thermal diffusivity, conductivity, capacity	Type
Coefficient of thermal expansion	Coating
Phase transformations	Type
Chemical properties	Thickness
Chemical composition	Number and kind of layers
Chemical affinity to tool material and environment	Mechanical properties
Metallurgical properties	Elastic constants
Structure	Plastic properties
Grain size	Physical properties
Hardness	Thermal diffusivity, conductivity, capacity
	Coefficient of thermal expansion
	Chemical properties
Environment characteristics	Chemical composition
Type of medium (gas, fluid, mist)	Chemical affinity to tool material
Lubricity	Metallurgical properties
Cooling ability	Structure
Flow rate	Grain size
Temperature	
Chemical composition	**Process variables**
	Speed
Machine tool characteristics	Feed
Error motions	Depth of cut

Source: Advanced Manufacturing Engineering, Vol. 1, July 1989.

devices use a diamond stylus that is moved at a constant rate across the surface, perpendicular to the lay pattern. The rise and fall of the stylus is detected electronically [often by a linear-variable differential transformer (LVDT)], amplified and recorded on a strip-chart, or processed electronically to produce average or root-mean-square readings for a meter (**Figure 32.3**). The unit containing the stylus and the driving motor may be handheld or supported by skids that ride on the workpiece or some other supporting surface.

Roughness is measured by the height of the irregularities with respect to an average line. These measurements are usually expressed in micrometers or microinches. In most cases, the arithmetic average, R_A, is used. In terms of the measurements, the R_A would be as follows:

$$R_A = \frac{\sum_{i=1}^{n} y_i}{n} \tag{32.1}$$

Cutoff refers to the sampling length used for the calculation of the roughness height. When it is not specified, a value of 0.030 in. (0.8 mm) is assumed. In the previous equation, y_i is a vertical distance from the centerline and n is the total number of vertical measurements taken within a specified cutoff distance. This average roughness value is also called **arithmetic average (AA)**. Occasionally, the **root-mean-square (rms)** value, R_q, is used which is defined as

$$R_q = \text{rms} = \sqrt{\frac{\sum_{i=1}^{n} y_i^2}{n}} \tag{32.2}$$

The resolution of stylus profile devices is determined by the radius or the diameter of the tip of the stylus. When the magnitude of the geometric surface features begins to approach the magnitude of the tip of the stylus, great caution should be used in interpreting the output from these devices. As a case in

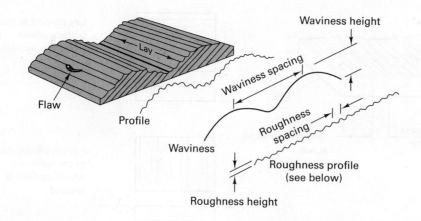

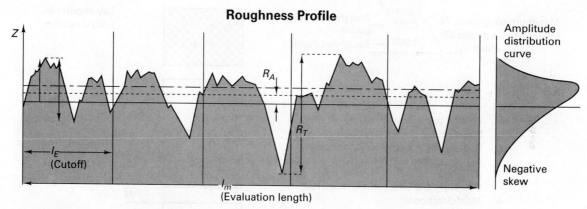

Roughness Profile

R_T = Maximum roughness depth (peak to valley) along l_m

R_A = Arithmetic roughness average

FIGURE 32.1 Machining processes produce surface flaws, waviness, and roughness that can influence the performance of the component.

point, **Figure 32.4** shows a scanning electron micrograph of a face-milled surface on which has been superimposed (photographically) a scanning electron micrograph of the tip of a diamond stylus (tip radius of 0.0005 in.). Both micrographs have the same final magnification. Surface flaws of the same general size as the roughness created by the machining process are difficult to resolve with the stylus-type device, where both these features are about the same size as the stylus tip.

This example points out the difference between *resolution* and *detection*. Stylus tracing devices can often *detect* the presence of a surface crack, step, or ridge on the part but cannot *resolve* the geometry of the defect when the defect is of the same order of magnitude as the stylus tip or smaller.

Another problem with these devices is that they produce a reading (a line on the chart) where the stylus tip is not touching the surface. This is demonstrated in **Figure 32.5a**, which shows the *S* from the word *TRUST* on a U.S. dime. The scanning electron microscope (SEM) micrograph was made after the topographical map of **Figure 32.5b** had been made. Both figures are at about the same magnification. The tracks produced by the stylus tip are easily seen in the micrograph. Notice the difference between the features shown in the micrograph and the trace, indicating that the stylus tip was not in contact with the

surface many times during its passage over the surface (left no track in the surface), yet the trace itself is continuous.

Surface integrity has become the subject of intense interest because the traditional, nontraditional, and posttreatment methods used to manufacture hardware can change the material's properties. Although the consequence of these changes becomes a design problem, the preservation of properties is a manufacturing consideration. Designs that require a high degree of surface integrity are the ones that display the following qualities:

• Are highly stressed.

• Employ low safety factors.

• Operate in severe environments.

• Must have prime reliability.

• Have a high surface areas–to–volume ratio.

• Are made with alloys that are sensitive to processing.

Surface integrity should be a joint concern of manufacturing and engineering. Manufacturing must balance cost and producibility with design requirements. It bears repeating to say that engineering must design components with knowledge of

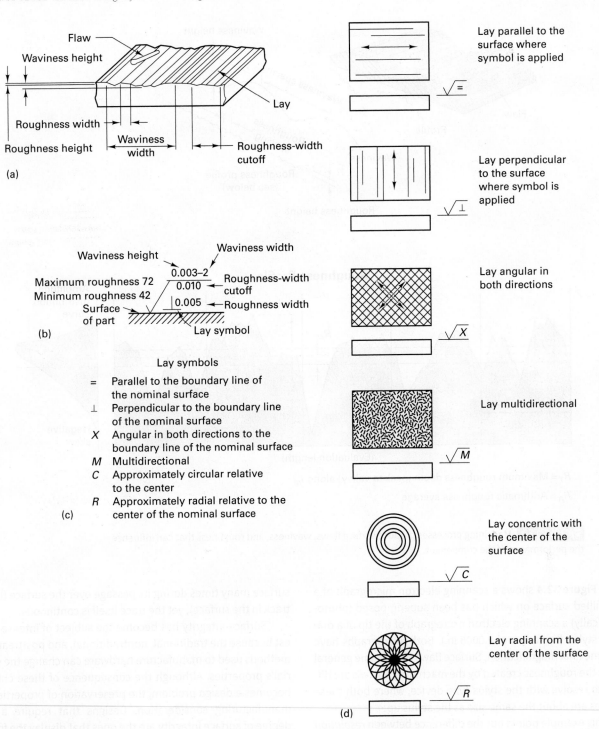

FIGURE 32.2 (a) Terminology used in specifying and measuring surface quality; (b) symbols used on drawing by part designers, with definitions of symbols; (c) lay symbols; (d) lay symbols applied on drawings.

manufacturing processes. A reduction in fatigue life resulting from processing can be reversed with a posttreatment. This is another example of design for manufacturing.

The range of surface roughnesses that are typically produced by various manufacturing processes is indicated in **Figure 32.6**, which is a very general picture of typical ranges associated with these processes. However, one can usually count on its being more expensive to generate a fine finish (low

roughness). To aid designers, metal samples with various levels of surface roughness are available.

All of the processes used to manufacture components are important if their effects are present in the finished part. It is convenient to divide processes that are used to manufacture parts into three categories: traditional, nontraditional, and finishing treatments. In **traditional processes**, the tool contacts the workpiece. Examples are grinding, milling, and turning.

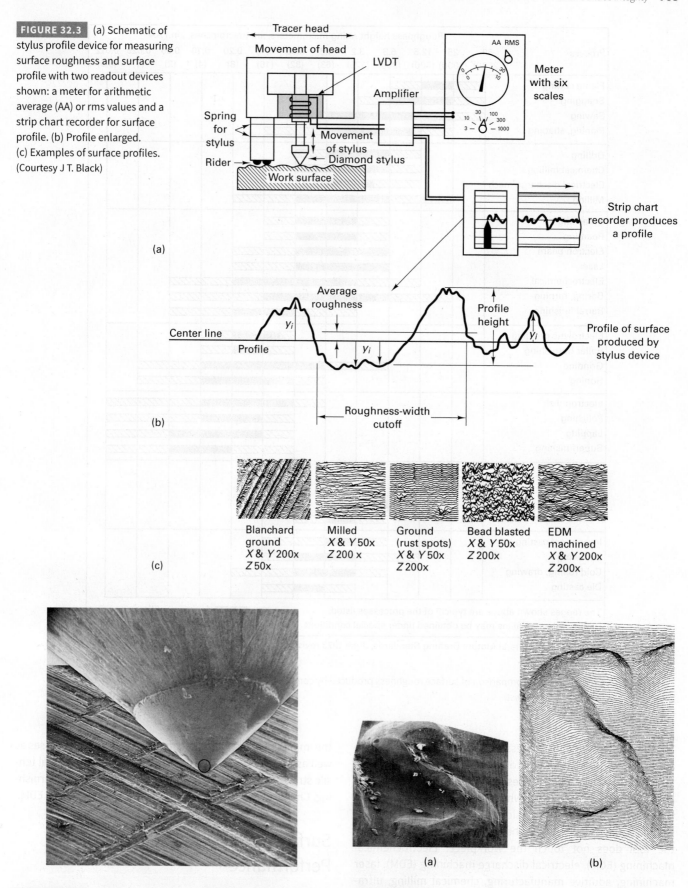

FIGURE 32.3 (a) Schematic of stylus profile device for measuring surface roughness and surface profile with two readout devices shown: a meter for arithmetic average (AA) or rms values and a strip chart recorder for surface profile. (b) Profile enlarged. (c) Examples of surface profiles. (Courtesy J T. Black)

Tracer head
Movement of head
LVDT
Amplifier
AA RMS
Meter with six scales
Spring for stylus
Movement of stylus
Rider
Diamond stylus
Work surface
Strip chart recorder produces a profile

(a)

Average roughness
y_i
Center line
Profile
y_i
Profile height
y_i
Profile of surface produced by stylus device
Roughness-width cutoff

(b)

Blanchard ground
X & Y 200x
Z 50x

Milled
X & Y 50x
Z 200 x

Ground (rust spots)
X & Y 50x
Z 200x

Bead blasted
X & Y 50x
Z 200x

EDM machined
X & Y 200x
Z 200x

(c)

FIGURE 32.4 Typical machined steel surface as created by face milling and examined in the SEM. A micrograph (same magnification) of a 0.00005-in. stylus tip has been superimposed at the top. (Courtesy J T. Black)

FIGURE 32.5 (a) SEM micrograph of a U.S. dime, showing the S in the word *TRUST* after the region has been traced by a stylus-type machine. (b) Topographical map of the S region of the word *TRUST* from a U.S. dime (compare to part a). (Courtesy J T. Black)

(a)

(b)

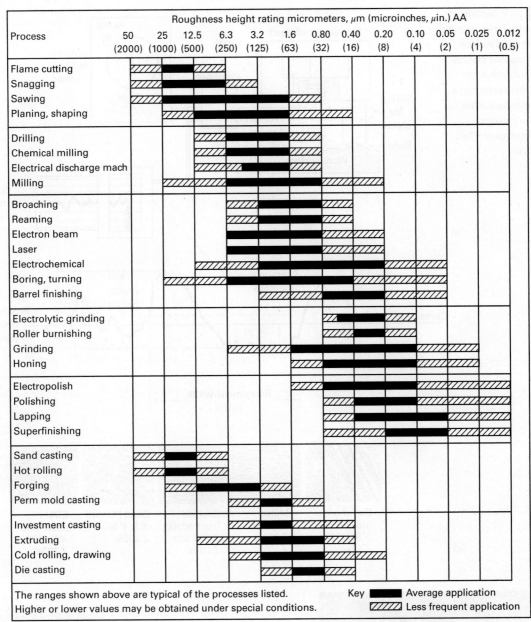

Process	Roughness height rating micrometers, μm (microinches, μin.) AA												
	50 (2000)	25 (1000)	12.5 (500)	6.3 (250)	3.2 (125)	1.6 (63)	0.80 (32)	0.40 (16)	0.20 (8)	0.10 (4)	0.05 (2)	0.025 (1)	0.012 (0.5)
Flame cutting													
Snagging													
Sawing													
Planing, shaping													
Drilling													
Chemical milling													
Electrical discharge mach													
Milling													
Broaching													
Reaming													
Electron beam													
Laser													
Electrochemical													
Boring, turning													
Barrel finishing													
Electrolytic grinding													
Roller burnishing													
Grinding													
Honing													
Electropolish													
Polishing													
Lapping													
Superfinishing													
Sand casting													
Hot rolling													
Forging													
Perm mold casting													
Investment casting													
Extruding													
Cold rolling, drawing													
Die casting													

The ranges shown above are typical of the processes listed. Higher or lower values may be obtained under special conditions.

Key ▮ Average application ▨ Less frequent application

Extracted from General Motors Drafting Standards, June 1973 revision

FIGURE 32.6 Comparison of surface roughness produced by common production processes. (Courtesy of *American Machinist*)

These material removal processes will inflict damage to the surface if improper parameters are used. Examples of improper parameters are dull tools, excessive infeed, inadequate coolant, and improper grinding wheel hardness. The **nontraditional processes** have intrinsic characteristics that, even if well controlled, will change the surface. In these processes, the tool does not touch the workpiece. Electrochemical machining (ECM), electrical discharge machining (EDM), laser machining, additive manufacturing, chemical milling, ultrasonic grinding, and abrasive water-jet machining are examples of these kinds of processes. Such methods can leave stress-free surfaces, remelted layers, and excessive surface roughness. **Finishing treatments** can be used to negate or remove the impact of both traditional and nontraditional processes as well as provide good surface finish. For example, residual tensile stresses can be removed by shot peening or roller burnishing. Chemical milling can remove the recast layer left by EDM.

Surface Properties and Product Performance

It is important to understand that the various manufacturing and surface-finishing processes each impart distinct properties to the materials that will influence the performance of the product. The achievement of satisfactory product performance

obviously depends on a good design, high-quality manufacturing (including surface treatment), and proper assembly. The failure of parts in service, however, is usually the result of a combination of factors.

The various machining processes will each produce characteristic surface textures (roughness, waviness, and lay) on the workpieces. In addition, the various processes tend to produce changes in the chemical, physical, mechanical, and metallurgical properties on or near the surfaces that are created. For the most part, these changes are limited to a depth of 0.005 to 0.050 in. below the surface. The effects can be beneficial or detrimental, depending on the process, material, and function of the product.

Machining processes (both chip-forming and chipless) induce plastic deformation into the surface layer, as shown in **Figure 32.7**. The cut surfaces are generally left with tensile residual stresses, microcracks, and a hardness that is different from the bulk material.

Consider **Figure 32.8**, which shows the depth of "surface damage" due to machining as a function of the rake angle of the tool. To increase the cutting speed and thereby increase the rate of production, an engineer might change from a high-speed tool steel cutter with a large rake angle (such as 30 degrees) to a carbide tool with a zero rake. While the resulting surface finish might look the same, the depth of surface damage might doubled. The increase in cutting speed may increase the heat going into the surface. If sufficient heat is generated, phase transformations can occur in the surface and subsurface regions. Failures may occur in service, whereas previous parts had performed quite admirably.

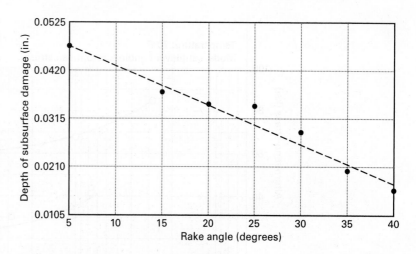

FIGURE 32.8 The depth of damage to the surface of a machined part increases with decreasing rake angle of the cutting tool.

Processes such as EDM and laser machining leave a layer of hard, recast metal on the surface that usually contains microcracks. Ground surfaces can have either residual tension or residual compression, depending on the mix between chip formation and plowing or rubbing during the grinding operation. **Figure 32.9** shows the effect on fatigue life of a component finished by EDM versus grinding. Look how much improvement occurred with gentle grinding versus EDM. Additional information about surface finish versus fatigue life is found in the case studies.

Processes such as **roller burnishing** and **shot peening** produce a smooth surface with compressive residual stresses. Shot peening (and tumbling) can increase the hardness in the surface and introduce a residual compressive stress, as shown in **Figure 32.10** and **32.11**. Welding processes produce tensile residual stresses as the deposited material shrinks upon cooling. Similar shrinkage occurs in castings, but the resulting stresses may be complex due to the variation of shrinkage or the lack of restraint. Tensile stresses on the surface can often be offset by a subsequent shot peening or tumbling.

The objectives of the surface-modification processes can be quite varied. Some are designed to clean surfaces and remove the kinds of defects that occur during processing or handling (such as scratches, pores, burrs, fins, and blemishes). Others further improve or modify the products' appearance, providing features such as smoothness, texture, or color. Numerous techniques are available to improve resistance to wear or corrosion or to reduce friction or adhesion to other materials. Scarce or costly materials can be conserved by making the interior of a product from a cheaper, more common material and then coating or plating the product surface.

As with all other processes, surface treatment requires time, labor, equipment, and material handling—and all of these have an associated cost. Efficiencies can be realized through process optimization and the integration of surface treatment into the entire manufacturing system. Design modifications can often facilitate automated or bulk finishing, eliminating the need for labor-intensive or single-part operations.

Illustration removed to comply with copyright law

FIGURE 32.7 A machining process can leave the machined surface with a plastically deformed surface, as shown here metallographically. (B.W. Kruszynski and C.W. Cuttervelt, *Advanced Manufacturing Engineering*, Vol. 1, 1989)

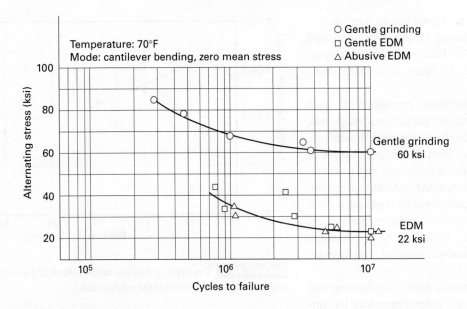

FIGURE 32.9 Fatigue strength of Inconel 718 components after surface finishing by grinding or EDM. (Field and Kahles, 1971)

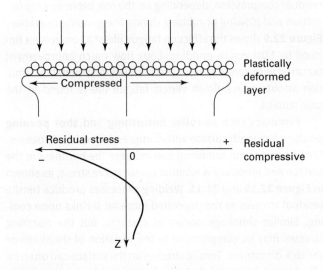

FIGURE 32.10 Mechanism for formation of residual compressive stresses in surface by cold plastic deformation (shot peening).

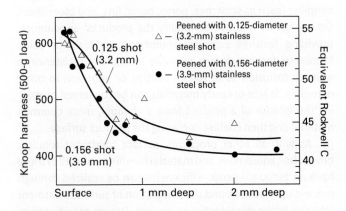

FIGURE 32.11 Hardness increased in surface due to shot peening.

Process selection should further consider the size of the part, the shape of the part, the quantity to be processed, the temperatures required for processing, the temperatures encountered during subsequent use, and any dimensional changes that might occur due to the surface treatment. Through knowledge of the available processes and their relative advantages and limitations, finishing costs can often be reduced or eliminated while maintaining or improving the quality of the product.

In recent years, the field of surface finishing has undergone another significant change. Many chemicals that were once "standard" to the field—such as cyanide, cadmium, chromium, and chlorinated solvents—have now come under strict government regulation. Wastewater treatment and waste disposal have also become significant concerns. As a result, processes may have to be modified or replacement processes may have to be used.

Because of their similarity to other processes, many surface finishing techniques have been presented elsewhere in the book. The **case hardening** techniques, both selective heating (flame, induction, and laser hardening) and altered surface chemistry (diffusion methods such as carburizing, nitriding, and carbonitriding), are presented in Chapter 5 as variations of heat treating. Shot peening and roller burnishing are presented in Chapter 19 as cold-working processes. Roll bonding and explosive bonding are discussed in Chapter 8 as means of producing laminar composites. Hard facing and metal spraying are included in Chapter 38 as adaptations of welding techniques. Chemical vapor deposition and physical vapor deposition are discussed in Chapters 22 and 29. In this chapter, we focus on techniques for cleaning and surface preparation as well as the remaining methods of surface finishing or surface modification.

32.3 Abrasive Cleaning and Finishing

It is not uncommon for the various manufacturing processes to produce certain types of surface contamination. Sand from the molds and cores used in casting often adheres to product surfaces. **Scale** (metal oxide) can be produced whenever metal is processed at elevated temperatures. **Oxides** such as rust can form if material is stored between operations. These and other contaminants must be removed before decorative or protective surfaces can be produced. While vibratory shaking can be useful, some form of **blast cleaning** is usually required to remove the foreign material. Blast cleaning uses a media (abrasive) propelled into the surface using air, water, or even a wheel (wheel blasting uses a high-rpm blocked wheel to deliver the media). The bulk of the work is done by kinetic energy of the impacting media; $KE = \frac{1}{2} MV^2$, where M = mass of the median and V = the velocity. Abrasives, steel grit, metal shot, fine glass shot, plastic beads, and even CO_2 are mechanically impelled against the surface to be cleaned. When sand is used, it should be clean, sharp-edged silica sand. Steel grit tends to clean more rapidly and generates much less dust, but it is more expensive.

When the parts are large, it may be easier to bring the cleaner to the part rather than the part to the cleaner. A common technique for such applications is **sand blasting** or **shot blasting**, where the abrasive particles are carried by a high-velocity blast of air emerging from a nozzle with about a $\frac{3}{8}$-in. opening. Air pressures between 60 and 100 psi, producing particle speeds of 400 mph, are common when cleaning ferrous metals, and 10 to 60 psi is common for nonferrous metals. The abrasive may be sand or shot, or materials such as walnut shells, dry-ice pellets, or even baking soda. Pressurized water can also be used as a carrier medium.

When production quantities are large or the parts are small, the operation can be conducted in an enclosed hood, with the parts traveling past stationary nozzles. For large parts or small quantities, the blast may be delivered manually. Protective clothing and breathing apparatus must be provided and precautions taken to control the spread of the resulting dust. The process may even require a dedicated room or booth that is equipped with integrated air pollution control devices.

From a manufacturing perspective, these processes are limited to surfaces that can be reached by the moving abrasive (line-of-sight) and cannot be used when sharp edges or corners must be maintained (because the abrasive tends to round the edges).

Barrel Finishing or Tumbling

Barrel finishing or **tumbling** is an effective means of finishing large numbers of small parts. In the Middle Ages, wooden casks were filled with abrasive stones and metal parts and were rolled about until the desired finish was obtained. Today, modifications of this technique can be used to deburr, radius (remove sharp corner on part), descale, remove rust, polish, brighten, surface-harden, or prepare parts for further finishing or assembly. The amount of stock removal can vary from as little as 0.0001 to as much as 0.005 in.

In the typical operation, the parts are loaded into a special barrel or drum until a predetermined level is reached. Occasionally, no other additions are made, and the parts are simply tumbled against one another. In most cases, however, additional media of metal slugs or abrasives (such as sand, granite chips, slag, or ceramic pellets) are added. Rotation of the barrel causes the material to rise until gravity causes the uppermost layer to cascade downward in a "landslide" movement, as depicted in **Figure 32.12**. The sliding produces abrasive cutting that can effectively remove fins, flash, scale, and

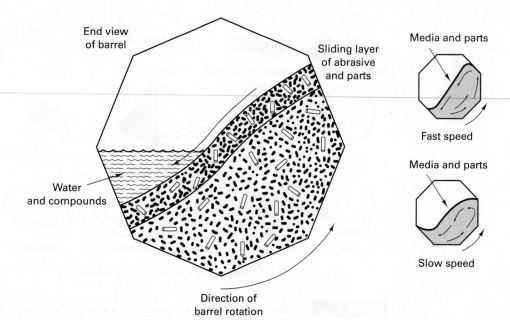

FIGURE 32.12 Schematic of the blow of material in tumbling or barrel finishing. The parts and media mass typically account for 50% to 60% of capacity.

adhered sand. Because only a small portion of the load is exposed to the abrasive action, long times may be required to process the entire contents.

Increasing the speed of rotation adds centrifugal forces that cause the material to rise higher in the barrel. The enhanced action can often accelerate the process, provided that the speed is not so great as to destroy the cascading action and that the additional action does not damage the workpiece. By a suitable selection of abrasives, filler, barrel size, ratio of workpieces to abrasive, fill level, and speed, a wide range of parts can be tumbled successfully. Delicate parts may have to be attached to racks within the barrel to reduce their movement while permitting the media to flow around them.

Natural and synthetic abrasives are available in a wide range of sizes and shapes, including those depicted in **Figure 32.13**, that enable the finishing of complex parts with irregular openings. The various media are often mixed in a given load so that some will reach into all sections and corners to be cleaned.

Tumbling is usually done dry, but it can also be performed with an aqueous solution in the barrel. Chemical compounds can be added to the media to assist in cleaning, or descaling, or to provide features such as rust inhibition. Support equipment is used to assist with loading and unloading the barrels as well as with the separation of the workpieces from the abrasive media. The latter operation often uses mesh screens with selected size openings.

Barrel tumbling can be a very inexpensive way to finish large quantities of small parts and produce rounded edges and corners. Unfortunately, the abrasive action occurs on all surfaces and cannot be limited to selected areas. The cycle time is often long, and the process can be quite noisy.

In the **barrel burnishing** process, no cutting action is desired. Instead, the parts are tumbled against themselves or with media such as steel balls, shot, rounded-end pins, or ballcones. If the original material is free of visible scratches and pits, the combination of peening and rubbing will reduce minute irregularities and produce a smooth, uniform surface.

Barrel burnishing is normally done wet, using a solution of water and lubricating or cleaning agents, such as soap or cream of tartar. Because the rubbing action between the work and the media is very important, the barrel should not be loaded more than half full, and the volume ratio of media to work should be about 2:1 so the workpieces rub against the media, not each other. The speed of rotation should be set to maintain the cascading action and not fling the workpieces free of the tumbling mass.

Centrifugal barrel tumbling places the tumbling barrel at the end of a rotating arm. This adds centrifugal force to the weight of the parts in the barrel and can accelerate the process by as much as 25 to 50 times.

In **spindle finishing**, the workpieces are attached to rotating shafts, and the assembly is immersed in media moving in a direction opposite to part rotation. This process is commonly applied to cylindrical parts and avoids the impingement of workpieces on one another. The abrasive action is accelerated, but time is required for fixturing and removal of the parts.

Vibratory Finishing

Vibratory finishing is a versatile process widely used for deburring, radiusing, descaling, burnishing, cleaning, brightening, and fine finishing. In contrast to the barrel processing, vibratory finishing is performed in open containers which allows for direct observation of the process. As illustrated in **Figure 32.14**, tubs or bowls are loaded with workpieces and media and are vibrated at frequencies between 900 and 3600 cycle/min. The specific frequency and amplitude are determined by the size, shape, weight, and material of the pan, as

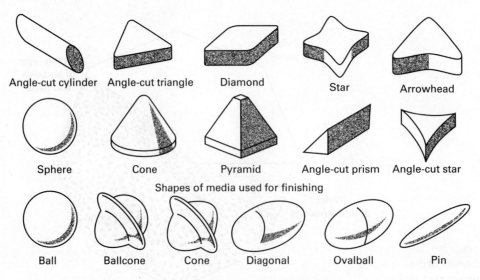

Angle-cut cylinder Angle-cut triangle Diamond Star Arrowhead

Sphere Cone Pyramid Angle-cut prism Angle-cut star

Shapes of media used for finishing

Ball Ballcone Cone Diagonal Ovalball Pin

Steel media shapes used for burnishing

FIGURE 32.13 Synthetic abrasive media are available in a wide variety of sizes and shapes. Through proper selection, the media can be tailored to the product being cleaned.

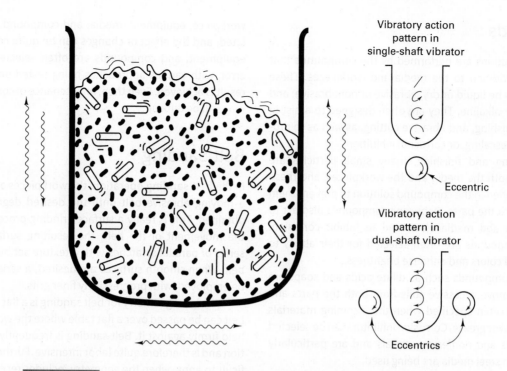

FIGURE 32.14 Schematic of a vibratory-finishing tub loaded with parts and media. The single eccentric shaft drive provides maximum motion at the bottom, which decreases as one moves upward. The dual-shaft design produces more uniform motion of the tub and reduces processing time.

well as the media and compound. Because the entire load is under constant agitation, cycle times are less than with barrel operations. The process is less noisy and is easily controlled and automated. In addition, the process can also deburr or smooth internal recesses or holes.

Media

The success of any of the mass-finishing processes depends greatly on **media** selection and the ratio of media to parts, as presented in **Table 32.2**. The media may prevent the parts from impinging upon one another as they simultaneously clean and finish. Fillers such as scrap punchings, minerals, leather scraps, and sawdust are often added to provide additional bulk and cushioning.

Natural abrasives include slag, cinders, sand, corundum, granite chips, limestone, and hardwood shapes such as pegs, cylinders, and cubes. Synthetic media typically contain 50 to 70 wt% of abrasives, such as alumina (Al_2O_3), emery, flint, and silicon carbide. This material is embedded in a matrix of ceramic, polyester, or resin plastic, which is softer than the abrasive and erodes, allowing the exposed abrasive to perform the work. The synthetics are generally produced by some form of casting operation, so their sizes and shapes are consistent and reproducible as opposed to the random sizes and shapes of the natural media. Steel media with no added abrasive are frequently specified for burnishing and light deburring.

Media selection should also be correlated with part geometry, because the abrasives should be able to contact all critical surfaces without becoming lodged in recesses or holes.

This requirement has resulted in a wide variety of sizes and shapes, including those presented in Figure 32.13. The different abrasives, sizes, and shapes can be selected or combined to perform tasks ranging from light deburring with a very fine finish to heavy cutting with a rough surface.

TABLE 32.2	Typical Media-to-Part Ratios for Mass Finishing
Media/Part Ratio by Volume	**Typical Application**
0:1	Part-on-part processing or burr removal without media
1:1	Produces very rough surfaces and is suitable for parts in which part-on-part damage is not a problem
2:1	Somewhat less severe part-on-part damage, but more action from less media
3:1	May be acceptable for very small parts and very small media. Part-on-part contact is likely on larger and heavier parts
4:1	In general, a good average ratio for many parts; a good ratio for evaluating a new deburring process
5:1	Better for nonferrous parts subject to part-on-part damage
6:1	Suitable for nonferrous parts, especially preplate surfaces on zinc parts with resin-bonded media
8:1	For improved preplate surfaces with resin-bonded media
10:1	Produces very fine finishes

Source: American Machinist, August 1983.

Compounds

A variety of functions are performed by the **compounds** that are added in addition to the media and workpieces. These compounds can be liquid or dry, abrasive or nonabrasive, and acid, neutral, or alkaline. They are often designed to assist in deburring, burnishing, and abrasive cutting, as well as to provide cleaning, descaling, or corrosion inhibition.

In deburring and finishing, many small particles are abraded from both the media and the workpieces, and these must be suspended in the compound solution to prevent them from adhering to the parts. Deburring compounds also act to keep the parts and media clean and to inhibit corrosion. Burnishing compounds are often selected for their ability to develop desired colors and enhance brightness.

Cleaning compounds such as dilute acids and soaps are designed to remove excessive soils from both the parts and media and are often specified when the incoming materials contain heavy oil or grease. Corrosion inhibitors can be selected for both ferrous and nonferrous metals and are particularly important when steel media are being used.

Another function of the compounds may be to condition the water when aqueous solutions are being used. Consistent water quality, in terms of "hardness" and metal ion content, is important to ensure uniform and repeatable finishing results. Liquid compounds may also provide cooling to both the workpieces and the media.

Summary of Mass-Finishing Methods

The barrel and vibratory finishing processes are really quite simple and economical and can process large numbers of parts in a batch procedure. Soft, nonferrous parts can be finished in as little as 10 min, while harder steel parts may require 2 hr or more. Sometimes the operations are sequenced, using progressively finer abrasives. **Figure 32.15** shows a variety of parts before and after the mass-finishing operation, using the triangular abrasive shown with each component.

Despite the high volume and apparent success, these processes may still be as much art as science. The key factors of workpiece, equipment, media, and compound are all interrelated, and the effect of changes can be quite complex. Media, equipment, and compounds are often selected by trial and error, with various approaches being tested until the desired result is achieved. Even then, maintenance of consistent results may still be difficult.

Belt Sanding

In the **belt sanding** operation, the workpieces are held against a moving abrasive belt until the desired degree of finish is obtained. This is really a surface grinding process. Because of the movement of the belt, the resulting surface contains a series of parallel scratches with a texture set by the grit of the belt. When smooth surfaces are desired, a series of belts may be employed, with progressively finer grits.

The ideal geometry for belt sanding is a flat surface, for the belt can be passed over a flat table where the workpiece can be held firmly against it. Belt sanding is frequently a hand operation and is therefore quite labor intensive. Furthermore, it is difficult to apply when the geometry includes recesses or interior corners. As a result, belt sanding is usually employed when the number of parts is small and the geometry is relatively simple See Chapter 28 for discussion about grinding.

Wire Brushing

High-speed rotary **wire brushing** is sometimes used to clean surfaces and can also impart some small degree of material removal or smoothing. The resulting surface consists of a series of uniform curved scratches. For many applications, this *may* be an acceptable final finish. If not, the scratches can easily be removed by barrel finishing or buffing.

Wire brushing is often performed by hand application of a small workpiece to the brush or the brush to a larger workpiece. Automatic machines can also be used where the parts are moved past a series of rotating brushes. In another modification, the brushes are replaced with plastic or fiber wheels that are loaded with abrasive.

Buffing

Buffing is a polishing operation in which the workpiece is brought into contact with a revolving cloth wheel that has been charged with a fine abrasive, such as polishing rouge. The "wheels," which are made of disks of linen, cotton, broadcloth, or canvas, achieve the desired degree of firmness through the amount of stitching used to fasten the layers of cloth together. When the operation calls for very soft polishing or polishing into interior corners, the stitching may be totally omitted, the centrifugal force of the wheel

FIGURE 32.15 A variety of parts before and after barrel finishing with triangular-shaped media. (Courtesy of Norton Abrasives/Saint Gobain)

rotation being sufficient to keep the layers in the proper position. Various types of polishing compounds are also available, with many consisting of ferric oxide particles in some form of binder or carrier.

The buffing operation is very similar to the lapping process that was discussed in Chapter 28. In buffing, however, the abrasive removes only minute amounts of metal from the workpiece. Fine scratch marks can be eliminated and oxide tarnish can be removed. A smooth, reflective surface is produced. When soft metals are buffed, a small amount of metal flow may occur, which further helps reduce high spots and produce a high polish.

In manual buffing, the workpiece is held against the rotating wheel and manipulated to provide contact with all critical surfaces. Once again, the labor costs can be quite extensive. If the workpieces are not too complex, semiautomatic machines can be used, where the workpieces are held in fixtures and move past a series of individual buffing wheels. By designing the part with buffing in mind, good results can be obtained quite economically.

Electropolishing

Electropolishing is the reverse of *electroplating* (discussed later in this chapter) because material is removed from the surface rather than being deposited. A DC electrolytic circuit is constructed with the workpiece as the anode. As current is applied, material is stripped from the surface, with material removal occurring preferentially from any raised location. Unfortunately, it is not economical to remove more than about 0.001 in. of material from any surface. However, if the initial surface is sufficiently smooth (less than 8 in. rms), and the grain size is small, the result will be a smooth polish with irregularities of less than 2 μin.—a mirrorlike finish.

Electropolishing was originally used to prepare metallurgical specimens for examination under the microscope. It was later adopted as a means of polishing stainless steel sheets and other stainless products. It is particularly useful for polishing irregular shapes that would be difficult to buff.

32.4 Chemical Cleaning

Chemical cleaning operations are effective means of removing oil, dirt, scale, or other foreign material that may adhere to the surface of a product, as a preparation for subsequent painting or plating. Because of environmental, health, and safety concerns, however, many processes that were once the industrial standard have now been eliminated or substantially modified. While the major concern with the mechanical methods has usually been airborne particles, the chemical methods often require the disposal of spent or contaminated solutions, and they occasionally use hazardous, toxic, or environmentally unfriendly materials. Chlorofluorocarbons (CFCs) and carbon

tetrachloride, for example, have been identified as ozone-depleting chemicals and have been phased out of commercial use. Process changes to comply with added regulations can significantly shift process economics. Manufacturers must now ask themselves if a part really has to be cleaned, what soils have to be removed, how clean the surfaces have to be, and how much they are willing to pay to accomplish that goal. Selection of the cleaning method will depend on cost of the equipment, power, cleaning materials, maintenance and labor, plus the cost of recycling and disposal of materials. Specific processes will depend on the quantity of parts to be processed (parts per hour), part configuration, part material, desired surface finish, temperature of the process, and flexibility. Manufacturers want machines they can integrate with manufacturing cells so changes in products can be quickly handled.

Alkaline Cleaning

Alkaline cleaning is basically the "soap and water" approach to parts cleaning and is a commonly used method for removing a wide variety of soils including oils, grease, wax, fine particles of metal, and dirt from the surfaces of metals. The cleaners are usually complex solutions of alkaline salts, additives to enhance cleaning or surface modification, and surfactants or soaps that are selected to reduce surface tension and displace, emulsify, and disperse the insoluble soils. The actual cleaning occurs as a result of one or more of the following mechanisms: (1) saponification, the chemical reaction of fats and other organic compounds with the alkaline salts; (2) displacement, where soil particles are lifted from the surface; (3) dispersion or emulsification of insoluble liquids; and (4) dissolution of metal oxides.

Alkaline cleaners can be applied by immersion or spraying, and they are usually heated to accelerate the cleaning action. The cleaning is then followed by a water rinse to remove all residue of the cleaning solution as well as to flush away some small amounts of remaining soil. A drying operation may also be required because the aqueous cleaners do not evaporate quickly, and some form of corrosion inhibitor (or rust preventer) may be required, depending on subsequent use.

Environmental issues relating to alkaline cleaning include (1) reducing or eliminating phosphate effluent, (2) reducing toxicity and increasing biodegradability, and (3) recycling the cleaners to extend their life and reduce the volume of discard.

Solvent Cleaning

In **solvent cleaning**, oils, grease, fats, and other surface contaminants are removed by dissolving them in organic solvents derived from coal or petroleum, usually at room temperature. The common solvents include petroleum distillates (such as kerosene, naphtha, and mineral spirits); chlorinated hydrocarbons (such as methylene chloride and trichloroethylene); and liquids such as acetone, benzene, toluene, and the various alcohols. Small parts are generally cleaned by immersion, with or without assisting agitation, or by spraying. Products that are too large to immerse can be cleaned by spraying or wiping. The process is quite simple, and capital equipment costs are rather low. Drying is usually accomplished by simple evaporation.

Solvent cleaning is an attractive means of cleaning large parts, heat-sensitive products, materials that might react with alkaline solutions (such as aluminum, lead, and zinc), and products with organic contaminants (such as soldering flux or marking crayon). Virtually all common industrial metals can be cleaned, and the size and shape of the workpiece are rarely limitations. Insoluble contaminants—such as metal oxides, sand, scale, and the inorganic fluxes used in welding, brazing, and soldering—cannot be removed by solvents. In addition, resoiling can occur as the solvent becomes contaminated. As a result, solvent cleaning is often used for preliminary cleaning.

Many of the common solvents have been restricted because of health, safety, and environmental concerns. Fire and excessive exposure are common hazards. Adequate ventilation is critical. Workers should use respiratory devices to prevent inhalation of vapors and wear protective clothing to minimize direct contact with skin. In addition, solvent wastes are often considered to be hazardous materials and may be subject to high disposal cost.

Vapor Degreasing

In **vapor degreasing**, the vapors of a chlorinated or fluorinated solvent are used to remove oil, grease, and wax from metal products. A nonflammable solvent, such as trichloroethylene, is heated to its boiling point, and the parts to be cleaned are suspended in its vapors. The vapor condenses on the work and washes the soluble contaminants back into the liquid solvent. Although the bath becomes dirty, the contaminants rarely volatilize at the boiling temperature of the solvent. Therefore, vapor degreasing tends to be more effective than cold solvent cleaning because the surfaces always come into contact with clean solvent. Because the surfaces become heated by the condensing solvent, they dry almost instantly when they are withdrawn from the vapor.

Vapor degreasing is a rapid, flexible process that has almost no visible effect on the surface being cleaned. It can be applied to all common industrial metals, but the solvents may attack rubber, plastics, and organic dyes that might be present in product assemblies. A major limitation is the inability to remove insoluble soils, forcing the process to be coupled with another technique, such as mechanical or alkaline cleaning. Because hot solvent is present in the system, the process is often accelerated by coupling the vapor cleaning with an immersion or spray using the hot liquid.

Unfortunately, environmental issues have forced the almost complete demise of the process. While the vapor degreasing solvents are chemically stable, have low toxicity, are nonflammable, evaporate quickly, and can be recovered for reuse, the CFC materials have been identified as ozone-depleting compounds and have essentially been banned from use. Solvents that can be used in the same process, or in a

replacement process that offers the necessary cleaning qualities, include chlorinated solvents (methylene chloride, perchloroethylene, and trichloroethylene). Most manufacturers have converted to some form of water-based process using alkaline, neutral, or acid cleaners or to a process using chlorine-free, hydrocarbon-based solvents. Sealed chamber machines use non-**volatile organic compounds (VOCs)** non-chlorinated solvents that are continuously recycled.

Ultrasonic Cleaning

When high-quality cleaning is required for small parts, **ultrasonic cleaning** may be preferred. Here, the parts are suspended or placed in wire-mesh baskets that are then immersed in a liquid cleaning bath, often a water-based detergent. The bath contains an ultrasonic transducer that operates at a frequency that causes **cavitation** in the liquid. The bubbles that form and implode provide the majority of the cleaning action, and if gross dirt, grease, and oil are removed prior to the immersion, excellent results can usually be obtained in 60 to 200 s. Most systems operate at between 10 and 40 kHz. Because of the ability to use water-based solutions, ultrasonic cleaning has replaced many of the environmentally unfriendly solvent processes.

Acid Pickling

In the **acid-pickling** process, metal parts are first cleaned to remove oils and other contaminants and then dipped into dilute acid solutions to remove oxides and dirt that are left on the surface by the previous processing operations. The most common solution is a 10% sulfuric acid bath at an elevated temperature between 150° and 185°F. Muriatic acid is also used, either cold or hot. As the temperature increases, the solutions can become more dilute.

After the parts are removed from the pickling bath, they should be rinsed to flush the acid residue from the surface and then dipped in an alkaline bath to prevent rusting. When it will not interfere with further processing, an immersion in a cold milk of lime solution is often used. Caution should be used to avoid overpickling because the acid attack can result in a roughened surface.

32.5 Coatings

Each of the surface finishing methods previously presented has been a material removal process designed to clean, smooth, and otherwise reduce the size of the part. Many other techniques have been developed to add material to the surface of a part. If the material is deposited as a liquid or organic gas (or deposited from a liquid or gas medium), the process is called **coating**. If the added material is a solid during deposition, the process is known as **cladding**.

Painting, Wet or Liquid

Paints and **enamels** are by far the most widely used finishes on manufactured products, and a great variety is available to meet the wide range of product requirements. Most of today's commercial paints are synthetic organic compounds that contain pigments and dry by polymerization or by a combination of polymerization and adsorption of oxygen. Water is the most common carrying vehicle for the pigments. Heat can be used to accelerate the drying, but many of the synthetic paints and enamels will dry in less than an hour without the use of additional heat. The older oil-based materials have a long drying time and require excessive environmental protection measures. For these reasons, they are seldom used in manufacturing applications.

Paints are used for a variety of reasons, usually to provide protection and decoration but also to fill or conceal surface irregularities, change the surface friction, or modify the light or heat absorption or radiation characteristics. **Table 32.3** provides a list of some of the more commonly used organic finishes, along with their significant characteristics. *Nitrocellulose lacquers* consist of thermoplastic polymers dissolved in organic solvent. Although fast drying (by the evaporation of the solvent) and capable of producing very beautiful finishes, they are not sufficiently durable for most commercial applications. The *alkyds* are a general-purpose paint but are not adequate for hard-service applications. *Acrylic enamels* are widely used for automotive finishes and may require catalytic or oven curing. *Asphaltic paints*, solutions of asphalt in a solvent, are used extensively in the electrical industry where resistance to corrosion is required and appearance is not of prime importance.

When considering a painted finish, the temptation is to focus on the outermost coat, to the exclusion of the underlayers. In reality, painting is a complex system that includes the substrate material, cleaning and other pretreatments (such as anodizing, phosphating, and various conversion coatings), priming, and possible intermediate layers. The method of application is another integral feature to be considered.

Paint Application Methods

In manufacturing, almost all painting is done by one of four methods: *dipping, hand spraying, automatic spraying,* or *electrostatic spraying or electrocoating*. In most cases, at least two coats are required. The first coat (or **prime coat**) serves to (1) ensure adhesion, (2) provide a leveling effect by filling in minor porosity and other surface blemishes, and (3) improve corrosion resistance and thus prevent later coatings from being dislodged in service. These properties are less easily attainable in the more highly pigmented paints that are used in the final coats to promote color and appearance. When using multiple coats, however, it is important that the carrying vehicles for the final coats do not unduly soften the underlayers.

Dipping is a simple and economical means of paint application used when all surfaces of the part are to be coated. The products can be manually immersed into a paint bath or passed through the bath while on or attached to a conveyor. Dipping is attractive for applying prime coats and for painting small parts where spray painting would result in a significant waste due to overspray. Conversely, the process is unattractive where only some of the surfaces require painting or where a very thin, uniform coating would be adequate, as on automobile bodies. Other difficulties are associated with the tendency of paint to run, producing both a wavy surface and a final drop of paint attached to the lowest drip point. Good-quality dipping requires that the paint be stirred at all times and be of uniform viscosity.

Spray painting is probably the most widely used paint application process because of its versatility and the economy in the use of paint. In the conventional technique, the paint is atomized and transported by the flow of compressed air. In a variation known as **airless spraying**, mechanical pressure forces the paint through an orifice at pressures between 500 and 4500 psi. This provides sufficient velocity to produce atomization and also propel the particles to the workpiece. Because no air pressure is used for atomization, there is less spray loss Paint efficiency may be as high as 99% with less generation of gaseous fumes.

Hand spraying is probably the most versatile means of application but can be quite costly in terms of labor and production time. When air or mechanical means provide the atomization, workers must exercise considerable skill to obtain the proper coverage without allowing the paint to "run" or "drape." Only a very thin film can be deposited at one time, usually less than 0.001 in. As a result, several coats may be required with intervening time for drying.

One means of applying thicker layers in a single application is known as **hot spraying**. Special solvents are used that

Material	Durability (Scale of 1–10)	Relative Cost (Scale of 1–10)	Characteristics
Nitrocellulose lacquers	1	2	Fast drying; low durability
Epoxy esters	1	2	Good chemical resistance
Akyd-amine	2	1	Versatile; low adhesion
Acrylic lacquers	4	1.7	Good color retention; low adhesion
Acrylic enamels	4	1.3	Good color retention; tough high baking temperature
Vinyl solutions	4	2	Flexible; good chemical resistance; low solids
Silicones	4–7	5	Good gloss retention; low flexibility
Flouropolymers	10	10	Excellent durability; difficult to apply

TABLE 32.3   **Commonly Used Organic Finishes and Their Qualities**

reduce the viscosity of the material when heated. Upon atomization, the faster-evaporating solvents are removed, and the drop in temperature produces a more viscous, run-resistant material that can be deposited in thicker layers.

When producing large quantities of similar or identical parts, some form of **automatic spraying** system is usually employed. The simplest automatic equipment consists of some form of parts conveyor that transports the parts past a series of stationary spray heads. While the concept is simple, the results may be unsatisfactory. A large amount of paint is wasted, and it is difficult to get uniform coverage.

Industrial robots are widely used to move the spray heads in a manner that mimics the movements of a human painter, maintaining uniform separation distance and minimizing waste. This is an excellent application for the robot because a monotonous and repetitious process can be performed with consistent results. In addition, use of a robot removes the human from an unpleasant, and possibly unhealthy, environment. Cars are painted almost exclusively with robots.

Both manual and automatic spray painting can benefit from the use of **electrostatic deposition**. A DC electrostatic potential is applied between the atomizer and the workpiece. The atomized paint particles assume the same charge as the atomizer and are therefore repelled. The oppositely charged workpiece then attracts the particles, with the actual path of the particle being a combination of the kinetic trajectory and the electrostatic attraction. The higher the DC voltage, the greater the electrostatic attraction. Overspraying can be reduced by as much as 60% to 80%, as can the generation of airborne particles and other emissions. Unfortunately, part edges and holes receive a heavier coating than flat surfaces due to the concentration of electrostatic lines of force on any sharp edge. Recessed areas will receive a reduced amount of paint, and a manual touch-up may be required using conventional spray techniques. Despite these limitations, electrostatic spraying is an extremely attractive means of painting complex-shaped products where the geometry would tend to create large amounts of overspray.

In an electrostatic variation of airless spraying, the paint is fed onto the surface of a rapidly rotating cone or disk that is also one electrode of the electrostatic circuit. Centrifugal force causes the thin film of paint to flow toward the edge, where charged particles are spun off without the need for air assist. The particles are then attracted to the workpiece, which serves as the other electrode of the electrostatic circuit. Because of the effectiveness of the centrifugal force, paints can be used with high-solids content, reducing the amount of volatile emissions and enabling a thicker layer to be deposited in a single application.

Electrocoating or **electrodeposition** applies paint in a manner similar to the electroplating of metals. As shown schematically in **Figure 32.16**, the paint particles are suspended in an aqueous solution and are given an electrostatic charge by applying a DC voltage between the tank (cathode) and the workpiece (anode). As the electrically conductive workpiece enters and passes through the tank, the paint particles are attracted to it and deposit on the surface, creating a uniform,

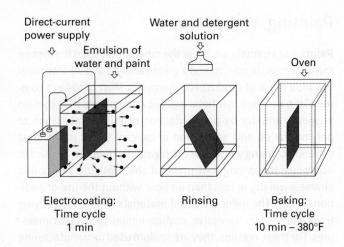

FIGURE 32.16 Basic steps in the electrocoating process.

thin coating that is more than 90% resin and pigment. When the coating reaches a desired thickness, determined by the bath conditions, no more paint is deposited. The workpiece is then removed from the tank, rinsed in a water spray, and baked at a time and temperature that depends on the particular type of paint. Baking of 10 to 20 min at 375°F is somewhat typical.

Electrocoating combines the economy of ordinary dip painting with the ability to produce thinner, more uniform coatings. The process is particularly attractive for applying the prime coat to complex structures, such as automobile bodies, where good corrosion resistance is a requirement. Hard-to-reach areas and recesses can be effectively coated. Because the solvent is water, no fire hazard exists as there is with many solvents, and air and water pollution are reduced significantly. In addition, the process can be readily adapted to conveyor-line production.

Drying

Most paints and enamels used in manufacturing require from 2 to 24 hr to dry at normal room temperature. This time can be reduced to between 10 min and 1 hr if the temperature can be raised to between 275° and 450°F. As a result, elevated-temperature drying is often preferred. Parts can be batch processed in ovens or continuously passed through heated tunnels or under panels of infrared heat lamps.

Elevated-temperature drying is rarely a problem with metal parts, but other materials can be damaged by exposure to moderate temperatures. For example, when wood is heated, the gases, moisture, and residual sap are expanded and driven to the surface beneath the hardening paint. Small bubbles tend to form that roughen the surface, or break, producing small holes in the paint.

Powder Coating

Powder coating is yet another variation of electrostatic spraying, but here the particles are solid rather than liquid. Several coats, such as primer and finish, can be applied and then followed by a single baking, in contrast to the baking after each coat that is required in conventional spray processes. In addition, the overspray powder can often be collected and reused. While volatilized solvents are no longer a concern, operators must now address the possibility of powder explosion, as well as the health hazards of airborne particles.

Modern powder technology can produce a high-quality finish with superior surface properties and usually at a lower cost than liquid painting. Powder painting is more efficient in the use of materials with lower energy requirements. The economic advantages must be weighed against the limitations of powder coating. Dry systems have a longer color change time than wet systems. The process is not good for large objects (massive tanks) or heat-sensitive objects. It is not easy to produce film thickness less than 1 mil (0.03 mm).

Table 32.4 provides details about powders that are used in powder coatings. Thermoplastics can also be used, but thermosetting powders are most common. The elements of a powder coating system are shown in **Figure 32.17**. When using the powder coating process, the following aspects must be considered:

- Types of guns—corona charged or tribo charged.
- Number of guns—depends on many factors, such as parts per hour, size of parts, line speed, and powder types.
- Color change time/frequency.
- Safety.
- Curing oven—coated parts put in ovens to melt, flow, and cure the powder.

Hot-Dip Coatings

Large quantities of metal products are given corrosion-resistant coatings by direct immersion into a bath of molten metal. The most common coating materials are zinc, tin, and aluminum. Terne, an alloy of lead and tin was used but has been discontinued.

Hot-dip galvanizing is the most widely used method of imparting corrosion resistance to steel. After the products, or sheets, have been cleaned to remove oil, grease, scale, and rust, they are fluxed by dipping into a solution of zinc ammonium chloride and dried. Next, the article is completely immersed in a bath of molten zinc. The zinc and iron react metallurgically to produce a coating that consists of a series of zinc–iron compounds and a surface layer of nearly pure zinc. (The zinc acts as a sacrificial anode, protecting the underlying iron.)

The coating thickness is usually specified in terms of weight per unit area. Values between 0.5 and 3.0 oz/ft^2 are typical, with the specific value depending on the time of immersion and speed of withdrawal. Thinner layers can be produced by incorporating some form of air-jet or mechanical wiping as the product is withdrawn. Because the corrosion resistance is provided through the sacrificial action of the zinc, the thin layers do not provide long-lasting protection. Extremely heavy coatings, on the other hand, may tend to crack and peel. The appearance of the coating can be varied through both the process conditions and alloy additions of tin, antimony, lead, and aluminum. When the coatings are properly applied, bending or forming can often follow galvanizing without damage to the integrity of the coating. Zinc-galvanized sheets can be heat treated by annealing (galvannealing) with a zinc–iron alloy coating. The 10% iron content adds strength and makes for good corrosion and pitting/chipping resistance. In auto applications, galvannealing beats out pure zinc on several counts: spot weldability, pretreatability, and ease of painting. Electrogalvanized zinc–nickel coatings that contain 10% to 15% nickel can be used in thinner layers (5 to 6 μm) and are easier to form and spot weld.

The primary limitations to hot-dip galvanizing are the size of the product, which is limited to the size of the tank holding the molten zinc, and the "damage" that might occur when a metal is exposed to the temperatures of the molten zinc

TABLE 32.4 Thermosetting Powder Coatings (Dry Painting) Have a Wide Variety of Properties and Applications

Properties	Epoxy	Epoxy/Polyester Hybrid	TGIC Polyester	Polyester Urethane	Acrylic Urethane
Application thickness	0.5–20 mils[a]	0.5–10 mils	0.5–10 mils	0.5–10 mils	0.5–10 mils
Cure cycle (metal temperatures)[b]	450 °F—3 min 250°F—30 min	450 °F—3 min; 325 °F—25 min	400 °F—7 min; 310 °F—20 min	400 °F—7 min; 350 °F—17 min	400 °F—7 min; 360 °F—25 min
Outdoor weatherability	Poor	Poor	Very good	Very good	Excellent
Pencil hardness	HB-5H	HB-2H	HB-2H	HB-3H	H-3H
Direct impact resistance, in lb[c]	80–160	80–160	80–160	80–160	20–60
Chemical resistance	Excellent	Very good	Good	Good	Very good
		Least expensive			Most expensive
Cost (relative)	2	1	3	4	5
Applications	Furniture, cars, ovens, appliances	Water heaters, radiators, office furniture	Architectural aluminum, outdoor furniture, farm equipment	Car wheels/rims, playground equipment	Washing machines, refrigerators, ovens

[a] Thickness up to 150 mils can be applied via multiple coats in a fluidized bed.
[b] Time and temperature can be reduced, by utilizing accelerated curing mechanisms, while maintaining the same general properties.
[c] Tested at a coating thickness of 2.0 mils.

Powder application equipment

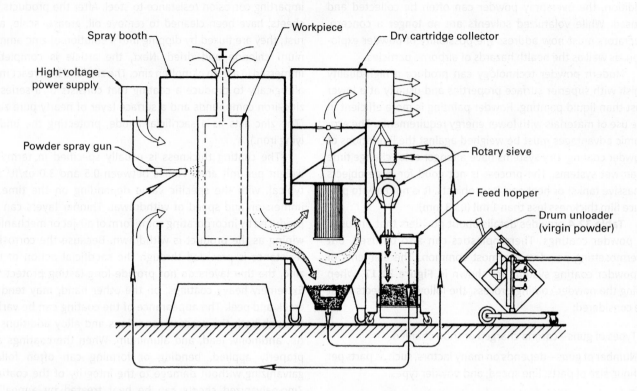

FIGURE 32.17 A schematic of a powder coating system. The wheels on the color modules permit it to be exchanged with a spare module to obtain the next color. (Source: Nordson Corp).

material, approximately 850°F, **Tin coatings** can also be applied by immersing in a bath of molten tin with a covering of flux material. Because of the high cost of tin and the relatively thick coatings applied by hot dipping, most tin coatings are now applied by electroplating.

Chemical Conversion Coatings

In **chemical conversion coating**, the surface of the metal is chemically treated to produce a nonmetallic, nonconductive surface that can impart a range of desirable properties. The most popular types of conversion coatings are chromate and phosphate. Aluminum, magnesium, zinc, and copper (as well as cadmium and silver) can all be treated by a **chromate** conversion process that usually involves immersion in a chemical bath. The surface of the metal is convened into a layer of complex chromium compounds that can impart colors ranging from bright clear through blue, yellow, brown, olive drab, and black. Most of the films are soft and gelatinous when they are formed but harden upon drying. They can be used to (1) impart exceptionally good corrosion resistance; (2) act as an intermediate bonding layer for paint, lacquer, or other organic finishes; or (3) provide specific colors by adding dyes to the coating when it is in its soft condition.

Phosphate coatings are formed by immersing metals (usually steel or zinc) in baths where metal phosphates (iron,

zinc, and manganese phosphates are all common) have been dissolved in solutions of phosphoric acid. The resultant coatings can be used to precondition surfaces to receive and retain paint or enhance the subsequent bonding with rubber or plastic. In addition, phosphate coatings are usually rough and can provide an excellent surface for holding oils and lubricants. This feature can be used in manufacturing, where the coating holds the lubricants that assist in forming, or in the finished product, as with black-color bolts and fasteners, whose corrosion resistance is provided by a phosphate layer impregnated with wax or oil.

Blackening or Coloring Metals

Many steel parts are treated to produce a black, iron oxide coating—a lustrous surface that is resistant to rusting when handled. Because this type of oxide forms at elevated temperatures, the parts are usually heated in some form of special environment, such as spent carburizing compound or special blackening salts.

Chemical solutions can also be used to blacken, blue, and even "brown" steels. Brown, black, and blue colors can also be imparted to tin, zinc, cadmium, and aluminum through chemical bath immersions or wipes. The surfaces of copper and brass can be made to be black, blue, green, or brown, with a full range of tints in between.

Electroplating

Large quantities of metal *and plastic* parts are electroplated to produce a metal coating that imparts corrosion or wear resistance, improves appearance (through color or luster), or increases the overall dimensions. Virtually all commercial metals can be plated, including aluminum, copper, brass, steel, and zinc-based die castings. Plastics can be electroplated, provided that they are first coated with an electrically conductive material.

Figure 32.18 depicts the typical **electroplating** process. A DC voltage is applied between the parts to be plated (which is made the cathode) and an anode material that is either the metal to be plated or an inert electrode. Both of these components are immersed in a conductive electrolyte, which may also contain dissolved salts of the metal to be plated as well as additions to increase or control conductivity. In response to the applied voltage, metal ions migrate to the cathode, lose their charge, and deposit on the surface. While the process is simple in its basic concept, the production of a high-quality plating requires selection and control of a number of variables, including the electrolyte and the concentrations of the various dissolved components, the temperature of the bath, and the electrical voltage and current. The interrelation of these features adds to the complexity and makes process control an extremely challenging problem.

The surfaces to be plated must also be prepared properly if satisfactory results are to be obtained. Pinholes, scratches, and

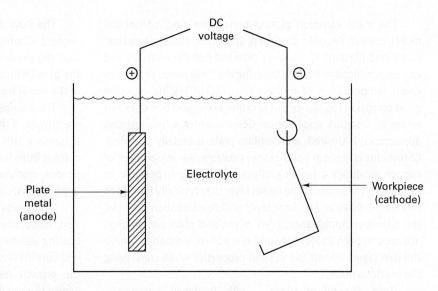

FIGURE 32.18 Basic circuit for an electroplating operation, showing the anode, cathode (workpiece), and electrolyte (conductive solution).

other surface defects must be removed if a smooth, lustrous finish is desired. Combinations of degreasing, cleaning, and pickling are used to ensure a chemically clean surface, one to which the plating material can adhere.

As shown in **Figure 32.19**, the plated metal tends to be preferentially attracted to corners and protrusions. This makes it particularly difficult to apply a uniform plating to irregular shapes, especially ones containing recesses, corners, and edges. Design features can be incorporated to promote plating uniformity, and improved results can often be obtained through the use of multiple spaced anodes or anodes whose shape resembles that of the workpiece.

Convex surfaces: Plate uniformly especially if edges are rounded

Concave recesses: Platability depends on dimensions

Flat surfaces: Not desirable. Use slight crown to hide undulations.

Slots: Narrow slots and holes should have rounded corners.

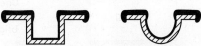
Blind holes: Must be exempted from minimum thickness requirements. Require vent hole at blind end.

V-shaped grooves: Difficult to plate. Should be avoided.

Sharply angled edges: Plating is thinner in center areas. Round all areas.

Fins: Increase plating time and costs. Reduce durability of finish.

FIGURE 32.19 Design recommendations for electroplating operations.

The most common plating metals are zinc, chromium, nickel, copper, tin, gold, platinum, and silver. The electrogalvanized zinc platings are thinner than the hot-dip coatings and can be produced without subjecting the base metal to the elevated temperatures of molten zinc. Nickel plating provides good corrosion resistance but is rather expensive and does not retain its lustrous appearance. Consequently, when lustrous appearance is desired, a chromium plate is usually specified. Chromium is seldom used alone, however. An initial layer of copper produces a leveling effect and makes it possible to reduce the thickness of the nickel layer that typically follows to less than 0.0006 in. The final layer of chromium then provides the attractive appearance. Gold, silver, and platinum platings are used in both the jewelry and electronics industries, where the thin layers impart the desired properties while conserving the precious metals.

Hard chromium plate, with Rockwell hardnesses between 66 and 70, can be used to build up worn parts to larger dimensions and to coat tools and other products that need reduced surface friction and good resistance to both wear and corrosion. Hard chrome coatings are always applied directly to the base material and are usually much thicker than the decorative treatments, typically ranging from 0.003 to 0.010 in. thick. Even thicker layers are used in applications such as diesel cylinder liners. Because hard chrome plate does not have a leveling effect, defects or roughness in the base surface will be amplified. If smooth surfaces are desired, subsequent grinding and polishing may be necessary.

Electroplating is frequently performed as a continuous process, where the individual parts to be plated are hung from conveyors. As they pass through the process, they are lowered into successive plating, washing, and fixing tanks. Ordinarily, only one type of workpiece is plated at a time, because the details of solutions, immersion times, and current densities are usually changed with changes in workpiece size and shape.

In the **electroforming** process, the coating becomes the final product. Metal is electroplated onto a mandrel, or mold, to a desired thickness and is then stripped free to produce small quantities of molds or other intricate-shaped sheet-metal-type products.

Anodizing

Anodizing is an electrochemical process, somewhat the reverse of electroplating, that produces a conversion-type coating on aluminum that can improve corrosion and wear resistance and impart a variety of decorative effects. If the workpiece is made the anode of an electrolytic cell, instead of a plating layer being deposited on the surface, a reaction progresses inward, increasing the thickness of the hard hexagonal aluminum oxide crystals on the surface. The hardness depends on thickness, density, and porosity of the coating, which are controlled by the cycle time and applied currents along with the chemistry, concentration, and temperature of the electrolyte. The surface texture very nearly duplicates the prefinishing texture, so a buffing prefinish produces a smooth, lustrous coating, while sand blasting produces a grainy or satiny coating.

The flow diagram in **Figure 32.20** shows the anodizing process. Coating thicknesses range from 0.1 to 0.25 mil. Note that the product dimensions will increase, however, because the aluminum oxide coating occupies about twice the volume of the metal from which it formed.

The nature of the developed coating is controlled by the electrolyte. If the oxide coating is not soluble in the anodizing solution, it will grow until the resistance of the oxide prevents current from flowing. The resultant coating, which is thin, non-porous, and nonconducting, is used in a variety of electrical applications.

If the oxide coating is slightly soluble in the anodizing solution, dissolution competes with oxide growth and a porous coating will be produced, where the pores provide for continued current flow to the metal surface. As the coating thickens, the growth rate decreases until it achieves steady state, where the growth rate is equal to the rate of dissolution. This condition is determined by the specific conditions of the process, including voltage, current density, electrolyte concentration, and electrolyte temperature. Sulfuric, chromic, oxalic, and phosphoric acids all produce electrolytes that dissolve oxide, with a sulfuric acid solution being the most common.

In a process variation known as **color anodizing**, a sulfuric acid bath is used to produce a layer of microscopically porous oxide that is transparent on pure aluminum and somewhat opaque on alloys. When this material is immersed in a dye solution, capillary action pulls the dye into the pores. The dye is then trapped in place by a sealing operation, usually performed simply by immersing the anodized metal in a bath of hot water. The aluminum oxide coating is converted to a monohydrate, with accompanying increase in volume. The pores close and become resistant to further staining or the leaching out of the dye.

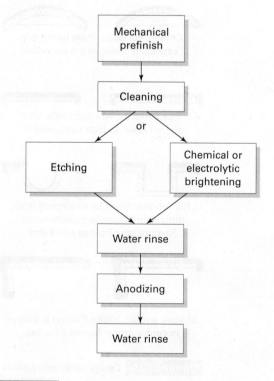

FIGURE 32.20 The anodizing process has many steps.

While most people are familiar with the variety of colors in aluminum athletic goods, such as softball bats, the actual applications range from giftware through automotive trim to architectural use. Aluminum can be made to look like gold, copper, or brass, or it can take on a variety of colors with a combined metallic luster that cannot be duplicated by other methods.

If PTFE (Teflon) is introduced into the pores, coatings can be produced that couple high hardness and low friction. The porous oxide layer can also be used to enhance the adhesion of an additional layer of material, such as paint, or carry lubricant during a subsequent forming operation. Because the coating is integral to the part, subsequent operations can often be performed without destroying its integrity or reducing its protective qualities.

Anodizing can also be performed on other metals, such as magnesium, and the process is similar to the passivation of stainless steel.

Electroless Plating

When using electroplating, it is almost impossible to obtain a uniform plating thickness on even moderately complex shapes: the platings cannot be applied to nonconductors, and a large amount of energy is required. For these reasons, a substantial effort has been directed toward the development of plating techniques that do not require an external source of electricity. These methods are known as **electroless** (or **autocatalytic**) **plating**. Considerable success has been achieved with nickel, but copper and cobalt, as well as some of the precious metals, can also be deposited.

In the electroless process, complex plating solutions are brought into contact with a substrate surface that acts as a catalyst or has been pretreated with catalytic material (The solution contains metal salts, reducing agents, complexing agents, pH adjusters, and stabilizers). The metallic ion in the plating solution is reduced to metal and deposits on the surface. Because the deposition is purely a chemical process, the coatings are uniform in thickness, independent of part geometry. Unfortunately, the rate of deposition is considerably slower than with electroplating.

Probably, the most popular of the electroless coatings is electroless nickel, and various methods exist for its deposition using both acid and alkaline solutions. The coatings offer good corrosion resistance, as well as hardnesses between Rockwell C 49 and 55. In addition, the hardness can be increased further to as high as Rockwell C 80 by subsequent heat treatment.

Electroless Composite Plating

A very useful adaptation of the electroless process has been developed wherein minute particles are co-deposited along with the electroless metal to produce composite-material coatings; the process is called **electroless composite plating**. Finely divided solid particles, with diameters between 1 and 10 μ in., are added to the plating bath and deposit up to 50 vol% with the matrix. While it may appear that a large variety of materials could be co-deposited, commercial applications have largely been limited to diamond, silicon carbide, aluminum oxide, and Teflon (PTFE).

Figure 32.21 shows a deposit of silicon carbide particles in a nickel–alloy matrix, where the particles constitute about 25 percent/volume. The coating offers the same corrosion resistance as nickel, but the high hardness of the silicon carbide particles (about 4500 on the Vickers scale, where tungsten carbide is 1300 and hardened steel is about 900) contributes outstanding resistance to wear and abrasion. Because the deposition is electroless, the thickness of the coating is not affected by the

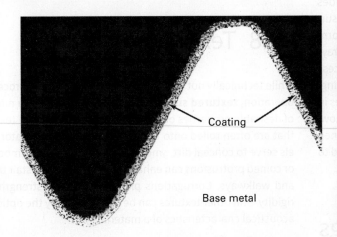

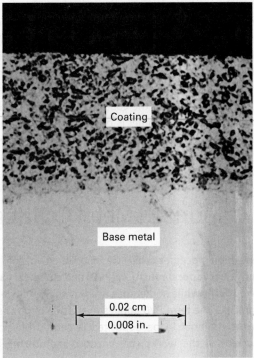

FIGURE 32.21 (Left) Photomicrograph of nickel carbide plating produced by electroless deposition. Notice the uniform thickness coating on the irregularly shaped product. (Right) High-magnification cross section through the coating. (Images Courtesy of The L. S. Starrett, Co.)

shape of the part. Applications include the coating of plastic-molding dies, for use where the polymer resin contains significant amounts of abrasive filler.

Mechanical Plating

Mechanical plating, also known as **peen plating** or **impact plating**, is an adaptation of barrel finishing in which coatings are produced by cold-welding soft, malleable metal powder onto the substrate. Numerous small products are first cleaned and may be given a thin galvanic coating of either copper or tin. They are then placed in a tumbling barrel, along with a water slurry of the metal powder to be plated, glass or ceramic tumbling media, and chemical promoters or accelerators. The media particles peen the metal powder onto the surface, producing uniform-thickness deposits (possibly a bit thinner on edges and thicker in recesses—the opposite of electroplating). Any metal that can be made into fine powder can be deposited, but the best results are obtained for soft materials, such as cadmium, tin, and zinc. Because the material is deposited mechanically, the coatings can be layered or involve mixtures with bulk chemistries that would be chemically impossible due to solubility limits. The fact that the coatings are deposited at room temperature, and in an environment that does not induce hydrogen embrittlement, makes mechanical plating an attractive means of coating hardened steels.

Porcelain Enameling

Metals can also be coated with a variety of specially formulated, glass fused to metal. These inorganic material impart resistance to corrosion and abrasion, decorative color, electrical insulation, or the ability to function in high-temperature environments in a process known as **porcelain enameling**. In the process known as fritting, the first or ground coat provides adhesion to the substrate and the cover coat provides the surface characteristics. The material is usually applied in the form of a multicomponent suspension or slurry (by dipping or spraying), which is then dried and fired. An alternative dry process uses electrostatic spraying of powder and subsequent firing. During the firing operation, which may require temperatures in the range of 900° to 1600°F, the coating materials melt, flow, and resolidify. Porcelain enamel is often found on the inner, perforated tubs of many washing machines and may be used to impart the decorative exterior on cookpots and frying pans.

32.6 Vaporized Metal Coatings

Vapor deposition processes can be classified into two main categories: **physical vapor deposition (PVD)** and **chemical vapor deposition (CVD)**. While sometimes used as though it were a specific process, the term *PVD* applies to a group of processes in which the material to be deposited is carried physically to the surface of the workpiece. **Vacuum metallizing** and

sputtering are key PVD processes, as are complex variations, such as ion plating. All are carried out in some form of vacuum, and most are line-of-sight processes in which the target surfaces must be positioned relative to the source. In contrast, the CVD processes deposit material through chemical reactions and generally require significantly higher temperatures. Tool steels treated by CVD may have to be heat treated again, while most PVD processes can be conducted below normal tempering temperatures. See Chapter 22 for additional discussions about PVD and CVD processes.

32.7 Clad Materials

Clad materials are actually a form of composite in which the components are joined as solids, using techniques such as roll bonding, explosive welding, and extrusion. The most common form is a laminate, where the surface layer provides properties such as corrosion resistance, wear resistance, electrical conductivity, thermal conductivity, or improved appearance, while the substrate layer provides strength or reduces overall cost. Alclad aluminum is a typical example. Here, surface layers of weaker but more corrosion-resistant single-phase aluminum alloys are applied to a base of high-strength but less corrosion-resistant, age-hardenable material. Aluminum-clad steel meets the same objective but with a heavier substrate. Stainless steel can be used to clad steels, reducing the need for nickel- and chromium-alloy additions throughout.

Wires and rods can also be made as claddings. Here, the surface layer often imparts conductivity, while the core provides strength or rigidity. Copper-clad steel rods that can be driven into the ground to provide electrical grounding for lightning rod systems are one example.

32.8 Textured Surfaces

While technically not the result of a surface finishing process or operation, **textured surfaces** can be used to impart a number of desirable properties or characteristics. The types of textures that are often rolled onto the sheets used for refrigerator panels serve to conceal dirt, smudges, and fingerprints. Embossed or coined protrusions can enhance the grip of metal stair treads and walkways. Corrugations provide enhanced strength and rigidity. Still other textures can be used to modify the optical or acoustical characteristics of a material.

32.9 Coil-Coated Sheets

Traditionally, sheet metal components, such as panels for appliance cabinets, have been fabricated from bare-metal sheets. Pans are blanked and shaped by the traditional metal-forming operations, and the shaped panels are then finished on an individual basis. This requires individual handling and the painting or plating of geometries that contain holes, bends, and contours. In addition, there is the time required to harden, dry, or cure the applied surface finish.

An alternative approach is to apply the finish to the sheet material after rolling but before coiling. Coatings can be applied continuously to one or both sides of the material while it is in the form of a flat sheet. Thus, the coiled material is effectively prefinished, and efforts need to be taken to protect the surface during the blanking and forming operations used to produce the final shape. Various paints have been applied successfully, as well as a full spectrum of metal coatings and platings. The sheared edges will not be coated, but if this feature can be tolerated, the additional measures to protect the surface may be an attractive alternative to the finishing of individual components. A second sequence that has some advantages takes the coils of steel that have been cut to length and stamps the holes and notches into them to create blanks. The blanks are pretreated, dried, powder coated, cured, and restacked. Then they are postformed to shape them into the back, side, and front panels of appliance cabinets.

The manufacturers call this **blank coating**. The coating thickness is about 1.5 mil ± 0.2 mil versus 2 mil ± 0.5 mil (less powder, better quality). Rusting at the corners of the holes is eliminated.

32.10 Edge Finishing and Burr Removal

Burrs are the small, sometimes flexible projections of material that adhere to the edges of workpieces that are formed by machining, like the exit-side burrs formed in the milled slot of **Figure 32.22**. Dimensionally, they are typically only 0.003 in. thick and 0.001 to 0.005 in. in height, but if not removed, they can lead to assembly failures, short circuits, injuries to workers, or even fatigue failures.

The most basic way to detect a burr is to run your finger or fingernail over the edges of the part. Probes and visual inspection techniques (microscopes) are used to find burrs as well.

Many different processes have been used for **burr removal**, including some discussed previously in this chapter and others presented as special types of machining. These include grinding, chamfering, barrel tumbling, vibratory finishing, centrifugal and spindle finishing, abrasive-jet machining, water-jet cutting, wire brushing, belt sanding, chemical machining, electropolishing, buffing, electrochemical machining, filing, ultrasonic machining, and abrasive flow machining.

Other burr removal methods may be quite specialized, such as thermal-energy deburring, where parts are loaded into a chamber, that is then filled with a combustible gas mixture. When the gas is ignited, the short-duration wavefront heats the small burrs to as much as 6000°F, while the remainder of the

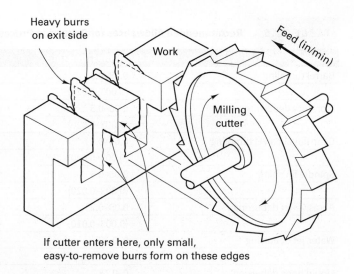

FIGURE 32.22 Schematic showing the formation of heavy burrs on the exit side of an up milled slot. (From L. X. Gillespie, *American Machinist*, November 1985)

workpiece rarely exceeds 300°F. The burrs are vaporized in less than 20 ms, including those in inaccessible or difficult-to-reach locations. Because the process does not use abrasive media, there is no change to any of the product dimensions. The product surfaces are rarely affected by the generated heat, and the cycle (including loading and unloading) can be repeated as many as 100 times an hour. Unfortunately, there is a thin recast layer and heat-affected zone that forms where the burrs were removed. This region is usually less than 0.001 in. thick, but it may be objectionable in hardened steels and highly stressed parts.

Of all of the burr removal methods, tumbling and vibratory finishing are usually the most economical, typically costing about a few cents per part. Because most of the common methods also remove metal from exposed surfaces and produce a radius on all edges, it is important that the parts be designed for deburring. **Table 32.5** provides a listing of the various deburring processes, as well as the edge radius, stock loss, and surface finish that would result from removal of a "typical burr" that is 0.003 in. thick.

By knowing how and where burrs are likely to *form,* the design engineer may be able to design parts to make the burrs easy to remove or even eliminate them. As shown in **Figure 32.23**, extra recesses or grooves can eliminate the need for deburring, because the burr produced by a cutoff tool or slot milling cutter will now lie below the surface. In this approach, one must determine whether it is cheaper to perform another machining operation (undercutting or grooving) or to remove the resulting burr.

Chamfers on sharp corners can also eliminate the need to deburr. The chamfering tool removes the large burrs formed by facing, turning, or boring and produces a relief for mating parts. The small burr formed during chamfering may be allowable or can easily be removed. Often, it may be preferable to give the manufacturer the freedom to use either a chamfer (produced by machining) or an edge radius (formed during the deburring operation) on all exposed corners or edges.

TABLE 32.5	Recommended Allowances for Deburring Processes[a]		
Process	**Edge Radius, mm (in.)**	**Stock Loss, mm (in.)**	**Surface Finish, μmAA (μin.AA[b])**
Barrel tumbling	0.08–0.5	0.0025	1.5–0.5
	(0.003–0.020)	(0–0.001)	(60–20)
Vibratory deburring	0.08–0.5	0–0.025	1.8–0.9
	(0.003–0.020)	(0–0.001)	(70–35)
Centrifugal barrel tumbling	0.08–0.5	0–0.025	1.8–0.5
	(0.003–0.020)	(0–0.001)	(70–20)
Spindle finishing	0.08–0.5	0–0.025	1.8–0.5
	(0.003–0.020)	(0–0.001)	(70–20)
Abrasive-jet deburring	0.08–0.25	0–0.05	0.8–1.3
	(0.003–0.010)	(0–0.002)[c]	(30–50)
Water-jet deburring	0–0.13	0(p)	
	(0–0.005)(p)		
Liquid hone deburring	0–0.13	0–0.013	
	(0–0.005)	(0–0.0005)	
Abrasive-flow deburring	0.025–0.5	0.025–0.13	1.8–0.5
	(0.001–0.020)	(0.001–0.005)[d]	(70–20)
Chemical deburring	0–0.5	0–0.025	1.3–0.5
	(0–0.002)	(0–0.001)	(50–20)
Ultrasonic deburring	0–0.05	0–0.025	0.5–0.4
	(0–0.002)	(0–0.001)	(20–15)
Electrochemical deburring	0.05–0.25	0.025–0.08	
	(0.002–0.010)	(0.001–0.003)[e]	
Electropolish deburring	0–0.25	0.025–0.08	0.8–0.4
	(0–0.010)	(0.001–0.003)[e]	(30–15)
Thermal-energy deburring	0.05–0.5	0	1.5–1.3(p)
	(0.002–0.020)		(60–50)
Power brushing	0.08–0.5	0–0.013	
	(0.003–0.020)	(0–0.0005)	
Power sanding	0.08–0.8	0.013–0.08	1.0–08
	(0.003–0.030)[f]	(0.0005–0.003)	(40–30)
Mechanical deburring	0.08–1.5		
	(0.003–0.060)		
Manual deburring	0.05–0.4		
	(0.002–0.015)[g]		

[a] Based on a burr 0.08 mm (0.003 in.) thick and 0.13 mm (0.005 in.) high in steel. Thinner burrs can generally be removed much more rapidly. Values shown are typical. Stock-loss values are for overall thickness or diameter. Location A implies that loss occurs over external surfaces, B that loss occurs over all surfaces, and C that loss occurs only near edge. (p) indicates best estimate.
[b] Values shown indicate typical before and after measurements in a deburring cycle.
[c] Abrasive is assumed to contact all surfaces.
[d] Stock loss occurs only at surfaces over which medium flows.
[e] Some additional stray etching occurs on some surfaces.
[f] Flat sanding produces a small burr and no radius.
[g] Chamfer is generally produced with a small burr.
Source: L. X. Gillespie, *American Machinist*, November 1985.

Recess eliminates deburring

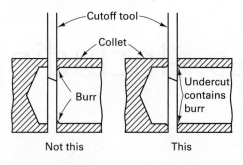

Groove minimizes milling burr

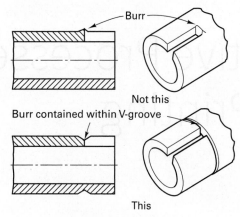

FIGURE 32.23 Designing extra recesses and grooves into a part may eliminate the need to deburr. (From L. X. Gillespie, *American Machinist*, November 1985)

Review Questions

1. Why are surface processes so important?

2. What are some of the factors that should be considered when selecting a surface-modification process?

3. How is the surface and its integrity altered by the process of metal cutting?

4. Two surfaces can have the same microinch roughness but be different in appearance. Explain.

5. What limits the resolution of a stylus-type surface-measuring device

6. What is the general relationship between surface roughness and tolerance?

7. What is the relationship between tolerance and cost to produce the surface and/or tolerance?

8. What are some common abrasive media used in blasting or abrasive cleaning operations?

9. What types of quantities and part sizes are most attractive for barrel finishing operations?

10. Why might there be an optimum fill level in barrel finishing?

11. Using statistical methods, how might you find the optimum rotational speed?

12. Describe the primary differences between barrel finishing and vibratory finishing.

13. What are some of the possible functions of the compounds that are used in abrasive finishing operations?

14. What are some of the mechanisms of alkaline cleaning, and what types of soils can be removed?

15. What types of surface contaminants cannot be removed by solvent cleaning?

16. In view of its many attractive features, why has vapor degreasing become an unattractive process?

17. What is the primary type of surface contaminant removed by acid pickling?

18. What is the difference between coating and cladding operations?

19. What are some of the reasons that paints may be specified for manufactured items?

20. What are some of the functions of a prime coat in a painting operation? What features are desired in the final coat?

21. What produces atomization and propulsion in airless spraying?

22. What features make industrial robots attractive for spray-painting automobiles?

23. What are some of the attractive features of electrostatic spraying?

24. Why would it be difficult to apply electrostatic spray painting to products made from wood or plastic?

25. What are some of the metal coatings that can be applied by the hot-dip process?

26. What are the two most common types of chemical conversion coatings?

27. How can nonconductive materials such as plastic be coated by electroplating?

28. What are the attractive properties of hard chrome plate?

29. What are some of the common process variables in electroplating?

30. Why is it difficult to mix parts of differing size and shape in an automated electroplating system?

31. How is electroforming different from electroplating?

32. When anodizing aluminum, what features determine the thickness of the resulting oxide when the oxide is not soluble in the electrolyte? When it is partially soluble?

33. What produces the various colors in the color anodizing process?

34. What are some of the attractive features of electroless plating?

35. What types of particulate composites can be deposited by electroless plating?

36. What is mechanical plating?

37. What are some of the attractive properties of a porcelain enamel coating?

38. How and why are burrs made by the milling process? How can they be removed?

39. What deburring processes are available that were not described in this chapter?

CHAPTER 33

Additive Processes—Including 3-D Printing

33.1 Introduction

The beginning of Chapter 13 introduced five families of processes by which useful shaped products are made. Liquids assume the shape of their container, and **casting processes** produce shaped containers, fill them with liquid, and allow the material to solidify while being held in that shape. **Deformation processes** exploit the plasticity of certain materials and use mechanical forces to rearrange solids into a more desirable shape. **Material removal processes** begin with an oversized solid and progressively remove unwanted segments to create the desired product. Discrete pieces of material are joined together by the **consolidation processes**. Each of these families, as well as the various processes within them, has distinct assets and limitations. Deformation processes, for example, frequently require strong, part-specific tooling, and the cost of this tooling is usually distributed over a large number of identical products. In contrast, basic machine tools, in the hands of a skilled operator, can produce millions of different shapes, one-of-a-kind if desired, and with extremely high dimensional precision. The material must be clamped or fixtured for each of the material removal operations, and the removed material, often in the form of cuttings, turnings or chips, must be discarded or recycled, producing added cost. Liquids can flow around inserts, enabling castings to easily incorporate internal holes and passages, but most liquids also shrink when they solidify, causing significant concern for the casting processes.

It was not long after the development of the computer that its capabilities were put to use in manufacturing, and these applications have continued to grow to include computer-numerically-controlled (CNC) machines, industrial robots, and computer-aided design/computer-aided manufacturing processes (CAD/CAM) to name just a few. CNC machining processes are now common throughout the manufacturing industry. They begin with a three-dimensional computer model that is used to generate tool paths that progressively subtract or remove material. This chapter will present a relatively new set of processes that create a desired shape by the incremental addition of material in a layer-by-layer fashion, and are known as **additive processes**. They begin with the same three-dimensional computer model, which is then sectioned into a large number of individual layers. Liquids, powders, plasticized solids or sheets of material are then used to deposit the layers and build products.

The first patent for an additive process was granted in March 1986, and the earliest application was the manufacture of prototype products that could be produced with extremely short lead times. As design concepts are transitioned from engineering drawings or CAD models to working products, **prototypes** are often produced to test and verify a part's design and provide a physical representation of the product that can be used to assess form, fit, and function. While form and fit can be evaluated by a prototype made from cheap, weak and easy-to-fabricate material, a fully functional prototype generally requires a material that is the same or sufficiently similar to the desired product. In either case, one-of-a-kind or small quantity prototypes often have high unit cost and fabrication delays because of the skilled labor required or the need for specialized tooling (e.g. patterns or molds for casting, dies for forming, or fixtures for machining). Based on prototype evaluation, design changes might be required, which may require another iteration of tooling. Cheaper and faster prototype production, therefore, can provide an attractive competitive edge.

The past three decades have been ones of extensive activity in the area of additive manufacturing. A number of new approaches have been developed and the spectrum of build materials continues to broaden, now including plastics, metals, ceramics, composites, organics, and even living tissue. The array of applications has expanded to include: 1) **rapid prototyping (RP)**: the production of scale models to enhance visualization or for purposes such as wind-tunnel testing; 2) **rapid tooling (RT)**: the production of tooling to be used in another process, such as foundry patterns, cores or molds, or jigs and fixtures for machining or joining processes; and 3) **direct-digital manufacturing (DDM)**: the manufacture of finished products directly from a computer file. With no required tooling, one-of-a-kind and small-batch production can become quite economical, provided material cost, build time, surface finish and resulting precision are all considered to be acceptable.

Along with the evolving technology, the terminology has also evolved. The **rapid prototyping** term was commonly applied. Because some of the original machines were quite compact[1] and produced small products, they were often classified as **desktop manufacturing**. Other general terms included *freeform fabrication, rapid manufacturing, and layered manufacturing.* While the **additive manufacturing** name encompasses all processes using the layer-build approach, and 3-D printing is only one member of this family, the term **3-D printing** has come to be associated with the entire family of processes, and is currently the term most frequently encountered by the general public.

33.2 Layerwise Manufacturing

The additive processes build the shape progressively through the accumulation of thin layers. They are **free-form fabrication** processes, capable of producing any required geometry, providing that the dimensions fit within the working envelope of the equipment. They are also **tool-less manufacturing** processes, able to make any product geometry without requiring part-specific tooling, such as dies, molds, or fixtures. Complex, free-form surfaces and contours can be produced as quickly as simple planar and cylindrical features. When the geometry is simple, subtractive machining processes, using general-purpose tools, will usually produce a quicker and cheaper product. As the shape complexity increases, the additive processes begin to excel.

As shown in **Figure 33.1**, the first step in all of the additive processes is **preprocessing**, the conversion of the three-dimensional **computer-aided design (CAD)** into the set of instructions that will be used to control and direct the manufacturing tool. Using **computer-aided manufacturing (CAM)** software, the part is mathematically sectioned into a succession of layers, starting at the bottom and moving up.[2] **Bases** and **supports** can be added to the part during this stage to facilitate separation from the machine tool after fabrication and to support free-standing geometric features that would distort during the process, such as cantilevers. Each layer is then reduced to a set of **tool paths** that will guide the deposition of energy or material during fabrication. In many ways, the computerized preprocessing for additive manufacturing is similar to the reduction of CAD geometries to the series of tool paths used in the subtractive process of computer numerical control (CNC) machining.

Figures 33.1 and **33.2** both illustrate the manufacture of an additive product, using the process of stereolithography (to be presented in more detail later) as an example. For any given layer, a laser follows the programmed tool path, converting light-curable resin (liquid photopolymer) into polymerized solid. When the layer is completed, the build platform descends one layer of thickness, and uncured material flows across the surface. The laser then follows the path for the next successive layer, bonding it to the previously cured solid. The cycle then repeats, building the part layer-by-layer. Each layer is essentially a two-dimensional X-Y raster scan, but by processing each layer on top of the preceding, a three-dimensional shape is produced.

Postprocessing operations might also be necessary. These might include the removal of bases and supports, removal of excess uncured resin or unbounded material, firing or sintering to increase strength, mechanical finishing to remove surface artifacts (such as *stairstepping*), or additional curing operations.

During the past three decades, a number of additive manufacturing processes have emerged using various approaches to the layer-build principle. While these processes have attained different levels of development and process maturity, they typically have low material addition rates compared to the material removal rates in machining, poorer dimensional accuracy and surface finish, and smaller work envelopes. The types and range of materials are often limited, and material cost is typically higher. The primary advantage is the ability to produce complex geometries with a minimal lead time and no part-specific tooling. Each individual process has its own set of advantages and limitations, but each builds the product in a layer-by-layer fashion, and they share common features and a common terminology.

They can be best classified by the nature of material deposition as:

1. **Liquid-based processes**
2. **Powder-based processes**: powder bed
3. **Deposition-based processes**: including powder feed

These families and the key members within them will be presented in later sections of this chapter.

Computer-Aided Design (CAD) to Computer-Aided Manufacturing (CAM)

Because each of the additive manufacturing processes is highly automated, computer planning and control software is essential to their operation. The manufacturing sequence begins with the conversion of the CAD solid model into the data needed to drive and control the additive machine tool. The three-dimensional geometric solid is sliced into a series of two-dimensional cross-sectional layers, with a thickness specified by the process. Each layer is then reduced to a series of linear tool paths that will be used to guide the deposition of material or the application of energy used to form that specific layer.

When the manufacturing objective is to replicate an existing product, as with **reverse engineering**, the initial geometric

---

[1]The original 10 cubic-inch build chambers were selected based on the sizes of typical injection-molded parts. The first stereolithography machines used this size build chamber, which then became the default standard for other entry-level or first-generation process platforms.

[2]ASTM Standard F2915 presents the standard three-dimensional file interchange called AMF (additive manufacturing file format). It replaces the former STL (stereolithography) format, and provides for easy exchange of printable object files and a seamless transition from design to physical printed object.

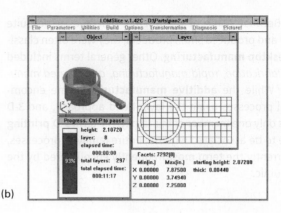

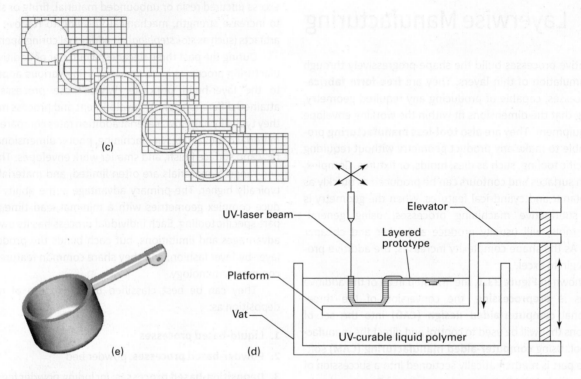

FIGURE 33.1 Conceptual framework for additive processes: (a) development of a virtual model in CAD; (b) model is converted to STL or other additive manufacturing file format and loaded into CAM software; (c) CAM software slices the model to generate tool paths for the laser; (d) product is produced layer-by-layer in machine tool, (e) final prototype or product.

input can be obtained from devices such as three-dimensional laser scanners or coordinate measuring machines. The manufacture of medical components often begins with data from CT (computed tomography) or MRI (magnetic resonance imaging) scans.

Thickness Control, Stairstepping, and Surface Finish

When a product is produced by creating thousands of stacked layers, it is important to control the thickness of the individual layers. Some processes have excellent control and permit fabrication by simple deposition of layer upon layer. Others can require intermediate attention, such as a surface milling after each or a series of depositions, or an intermediate measurement

of build height followed by a modification of slice data to compensate for any inaccuracies. The maximum thickness of a given layer is typically set by the depth of solid that can be formed by the deposition process, but considerations of surface finish and dimensional accuracy may dictate layer thickness and set the required number of layers.

The various additive manufacturing processes create products with a range of surface finishes and surface textures, but all suffer from a phenomenon known as **stairstepping**, illustrated in **Figure 33.3**. All surfaces not perpendicular to the slice plane (i.e. curved surfaces, rounded corners, and tapered sides) are approximated by the edges of stacked, uniform-thickness layers. Surface geometry, therefore, will tend to be rougher in the z-direction than around the perimeter of the x-y planes. Surfaces that have shallow inclination angles with respect to the x-y plane will have more obvious stairstepping

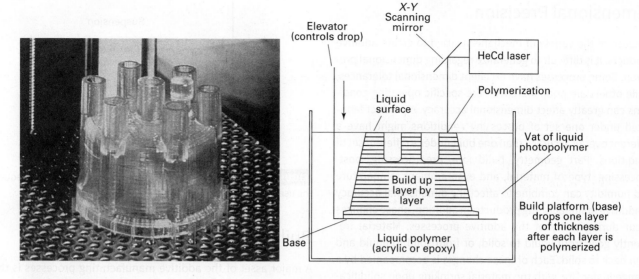

FIGURE 33.2 The stereolithography apparatus (SLA) (right) can build a part like the one shown in plastic (left) layer-by-layer, using a laser to polymerize liquid photopolymer (Arcam AB)

artifacts because the distance between the edges of the z-height layers will be greater. If finish is critical in the z-direction, it can be improved by reorienting the part so that this surface is closer to being perpendicular to the build plane, or by reducing the layer thickness during the build cycle. Decreasing the layer thickness, however, typically results in an increase in the time required to build the part. For certain applications, such as the manufacture of investment casting patterns or scale models for wind tunnel testing, the as-deposited surface finish might not be adequate. Secondary grinding, sanding, or polishing might be required, with these operations adding to both the time and the cost. Processes with relatively thick layers (some metal processes can have layers up to 2 mm thick) will also have a meniscus feature on the top (called a crown) and on the edge of each layer that will generally have to be removed for fit-up, elimination of stress concentrations, and weight minimization. In some process/application combinations, this can require light machining of 100% of the surfaces.

Coordinates and Voxel Geometry

Figure 33.4 shows a generic beam scanning process, and presents the coordinate system used with all of the additive manufacturing techniques. The x-y plane corresponds to the material layers, with the z-axis being the build direction. For processes that involve scanning, the x-y plane contains both the direction of scan travel and the width of the scanned line.

Voxel geometry is a useful concept in understanding the scanning-types of layerwise deposition. A voxel is a volume element that describes the depth, width and breadth of material that is instantaneously being altered by the deposition process. As such, the voxel determines the thickness of layers, the distance between adjacent scans, and ultimately the number of layers and scans needed to complete a part, as well as the resultant surface finish. To understand the capabilities of various processes, it is important to understand how the voxel geometry is affected by changes in the material being deposited and process parameters such as beam focus and beam intensity.

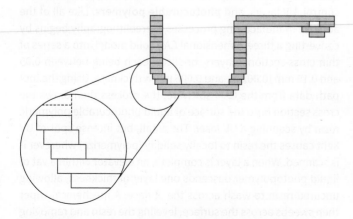

FIGURE 33.3 Building by layers results in stairstepping of curved surfaces and rounded corners.

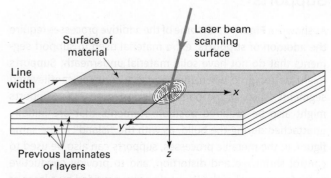

FIGURE 33.4 Schematic of a generic laser-based additive process showing the traditional coordinate system.

Dimensional Precision

Because of the variety of mechanisms used to create additive products, it is difficult to generalize regarding dimensional precision. Some processes have excellent dimensional tolerances, while others are poor. Moreover, the specific operating conditions can greatly affect dimensional accuracy. A product fabricated under one set of processing conditions might have a different overall accuracy than one built under a different set of conditions. Part geometry, build rate, preprocessing, postprocessing, type of material, and even ambient temperature and humidity can combine to affect the dimensional accuracy of additive manufacturing products. In addition, phase changes occur during many of the additive processes. Material frequently goes from liquid to solid, or from solid to liquid and then back to solid. Each of these changes is accompanied by a change in density, with the material shrinking upon solidification as it is deposited layer by layer and voxel by voxel.

When layers are produced by scanning techniques, it is common for the initial scans to trace the perimeter of the particular cross section, followed a filling in of the cross sectional area, a process known as **hatching** or **rastoring**. If the amount of shrinkage is small, hatching could occur in a straightforward manner, such as a simple sweep across the area. If the shrinkage is large, alternative methods of hatching can be used to minimize the effects of shrinkage on dimensional accuracy. The space between scan lines can be increased to retain small pockets of uncured resin that is then cured during postprocessing. If the product use requires only external surfaces of the correct size and shape, hatching can be modified to create internal honeycomb structures, with the uncured resin being drained prior to final postprocessing.

On a more macroscopic scale, the interior temperatures of a solid part tend to exceed the surface temperatures during the build process, leading to the creation of residual stresses within the material as it cools and possible part warpage. Because the additive processes are generally used with complex geometries, and complex geometries cool in complex ways, the resultant residual stresses and associated dimensional changes are often quite complex.

Supports

As shown in **Figure 33.5**, some of the additive processes require the addition of **supports**, extra material used to support segments that do not have solid material underneath. Supports might be required when the geometry contains cantilevered segments, such as the pan handle in Figure 33.5. Supports might also be required position segments that are initially unattached during the build, as with the "island" in the same figure. In the metallic processes, supports can also be used to control shrinkage and distortion, and to provide conductive paths for removal of the thermal energy provided by a laser or electron beam. The supports must be designed into the part before processing, and must be removed after the build, possibly negating some of the advantages of tool-less processing.

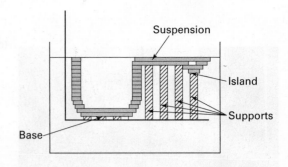

FIGURE 33.5 Many of the additive manufacturing processes require the use of bases and supports.

Build Time

A major asset of the additive manufacturing processes is the significant reduction in the time required to produce a product. The **build time** for these processes consists of three major components: preprocessing, fabrication, and postprocessing. Preprocessing involves the creation of the CAM software necessary to drive the machinery. Fabrication, often the largest component of both time and cost, varies depending on the specific process, part size, part complexity, and batch size. Because all of the additive processes permit the manufacture of multiple products within the working envelope, there is the possibility for reduction in both time and cost. Postprocessing involves cleaning and finishing the product after fabrication with details varying greatly depending upon the process being used.

33.3 Liquid-Based Processes

Stereolithography (SLA)

The first of the additive manufacturing processes, **stereolithography**, already presented in Figures 33.1 and 33.2, was made possible by a series of technological advances, including high-speed computers, computer-aided design (CAD), precise motion control, UV lasers, and **photocurable polymers**. Like all of the additive manufacturing processes, stereolithography begins by converting a three-dimensional CAD solid model into a series of thin cross-sectional layers, or slices, each being between 0.05 and 0.15 mm (0.002 in. and 0.006 in.) in thickness. Using the tool path data from the CAM software, the process then draws the cross section onto the surface of liquid photocurable polymeric resin by scanning a UV laser. The small, but intense spot of UV light causes the resin to locally solidify (polymerize) wherever it is scanned. When a layer is complete, an elevator within a vat of liquid photopolymer descends one layer of thickness, allowing uncured resin to wash across the surface. A mechanical wiper then sweeps across the surface, leveling the resin and removing excess material. The uniform layer of liquid resin is then ready for the next curing scan. By repeating these steps over and over, the desired three-dimensional geometry is created with typical build times ranging from an hour to more than a day. The completed part is then removed from the resin bath and excess partially cured resin is removed from surfaces with the use of a solvent. Supports are mechanically removed while they are still soft, and a postcuring operation in an ultraviolet oven is used to achieve more thorough polymerization and enhanced material properties.

Advantages of stereolithography (SLA) include the manufacture of extremely complex shapes with high dimensional accuracy and good surface finish. Tolerances can be as tight as +/−0.4 mm (0.015 in.) or 0.002 cm/cm or in/in. Part size is limited by the dimensions of the resin bath, but large parts can be produced as joined segments by breaking the original CAD model and adding joint features that might include dowel holes or gaps for adhesives. Wax investment casting patterns can be replaced by hollow SLA shells supported by an internal lattice network as illustrated in **Figure 33.6** showing an SLA investment casting pattern and the resulting aluminum/silicon carbide (metal matrix composite) casting. Ceramic products can be produced by loading a high volume fraction of nano- or micro-sized ceramic powder into the photocure resin to produce a starting paste. After the stereolithography build, the products are cleaned of excess resin, and then heat treated through debinding and sintering to eliminate the cured resin and produce a near-theoretical-density ceramic part. Hydroxyapatite can be processed into biocompatible bone replacements and implants. Other ceramic materials that have been loaded into the resin include alumina, zirconia, aluminum nitride, mullite, and cordierite.

The major disadvantage to the process, however, is the use of expensive acrylic, epoxy or vinyl ether photopolymer resins.

The most common acrylic resins offer good strength, but lower than engineering-grade resins, and are rather brittle and aggressively hygroscopic. Because they are transparent to UV radiation, full-volume postprocessing cures are common, but the material will continue to cure in sunlight and become more brittle over time. Unfortunately, many of the resins have toxicity concerns ranging from possible skin irritation through being possible carcinogens. Because of the necessary safety precautions, SLA is generally not performed in an office environment. In addition, the process might not be attractive for parts with multiple free-hanging segments, because each will require support.

Stereolithography was the first of the additive processes to be commercialized, and it continues to be the most-used and most-studied of the techniques. Additional scanned laser polymerization (SLP) processes have been developed by modifying certain features of the basic SLA technique, each having their own advantages and limitations. In the **masked-lamp descending-platform** approach, photomasks are created for each layer and are positioned above the resin surface. A single photoexposure forms the entire layer as opposed to the serial laser scan, thereby decreasing the required build time. Moreover, the processing time for a single layer is independent of its complexity or the number of single components in the build. In the **ascending surface** approach, the base and partially built solid remain fixed and a premeasured amount of uncured resin is added to the chamber to create each new layer. Yet another approach inverts the process, placing the laser beneath a transparent window. As the part is built, the partial product is raised by incremental amounts, allowing a new layer of resin to flow between the window and the product.

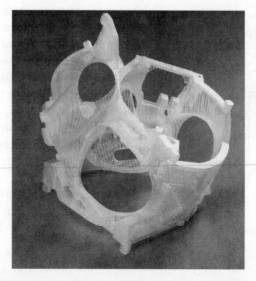
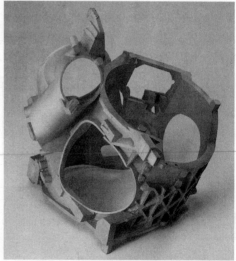

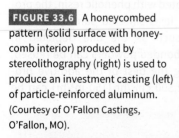

FIGURE 33.6 A honeycombed pattern (solid surface with honeycomb interior) produced by stereolithography (right) is used to produce an investment casting (left) of particle-reinforced aluminum. (Courtesy of O'Fallon Castings, O'Fallon, MO).

Stereolithography in the Medical Field

Stereolithography has been linked with image scanning systems such as computed tomography (CT) scanners and magnetic resonance imaging (MRI) machines to produce accurate recreations of an individual's anatomical components, such as broken, shattered, or deformed bones. Using these solid, hand-held objects, surgeons can more precisely identify obstacles and plan necessary steps prior to the actual surgery.

Inkjet Deposition (ID) or Droplet Deposition Manufacturing (DDM)

Using similar layer-sectioning software and tool path geometries, the **inkjet deposition (ID)** or **droplet deposition manufacturing** systems create the individual layers by selectively depositing molten material (wax, thermoplastic polymers or low-melting-point metals) onto a substrate as a series of uniformly spaced, uniform size, micro-droplets, often at rates of several thousand per second. As depicted in **Figure 33.7**, these droplets then adhere to the substrate, forming the new layer, and ultimately a solid mass.

By using multiple jet heads, build time can be reduced or different materials can be deposited simultaneously, enabling the production of multicolored products, products with varying properties, or more easily removable supports. Supports are often deposited in a perforated pattern to further facilitate easy removal, and alternative hatching patterns can be used to reduce build time when solid products are not required. If multi-axis robots are used that can deposit material at non-vertical angles, the need for supports can often be reduced or even eliminated. If the build material is liquid photopolymer, exposure to UV light is required as each layer is deposited. Ultrathin layers, on the order of 16 μm (0.0006 in.), provide exceptionally fine details and smooth surfaces. Dimensional accuracy is good, typically within 0.13 mm (0.005 in.) or 0.1%.

The most common applications of inkjet deposition are the production of prototype parts and complex wax patterns for investment casting. Fully functional metal products can be made by the inkjet approach through the selective deposition of a binder onto powdered materials, a process known as 3-D printing that will be presented later in the chapter.

33.4 Powder-Based Processes

Selective Laser Sintering (SLS)

Like most of the other processes, **selective laser sintering (SLS)** begins with a sliced CAD file and a build chamber containing a vertically positionable elevator. As depicted in **Figure 33.8**, the process begins by spreading a layer of heat-fusible powder across the part build chamber. A CO_2 or YAG laser is then scanned over the layer to selectively fuse those areas defined by cross-section geometry, joining the powder particles to one another and also to the layer below. The unfused material remains in place as the support structure for subsequent layers. The elevator then lowers, and another layer of powdered material is deposited, leveled, and laser sintered, with the process repeating until the part is complete. After processing, the part is removed from the build chamber and the unfused powder is removed for reuse.

Selective laser sintering uses one of the broadest range of build materials, including plastics, waxes (such as those used in investment casting), metals, ceramics, and particulate composites. The variety of polymeric materials includes nylons, polyamides, polycarbonates, polyvinylchloride, elastomers, and acrylic styrene. These materials are less expensive than the photosensitive resins used in stereolithography and are also nontoxic. Strength, toughness and elongation values are among the best of the additive manufacturing processes, and are often near-isotropic. Glass beads and other fillers can easily be dry-blended with the polymers to enhance stiffness and reduce shrinkage. Steels, stainless steels, tool steels, aluminum, titanium, and even superalloys have been used. When the starting material is aluminum silicate sand or quartz sand that has been coated with phenolic resin, the process can be used to create laser-sintered sand cores and molds for metal casting, such as the one depicted in **Figure 33.9**. Because the unbonded powders can be reused, material utilization is among the best of the additive processes. Build speeds up to 2,500 cm³/hr (150 in³/hr) can be achieved.

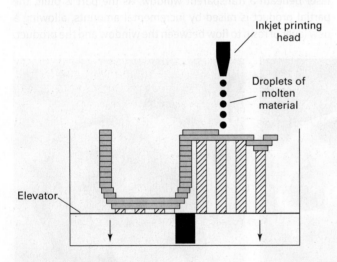

Inkjet printing head

Droplets of molten material

Elevator

FIGURE 33.7 Inkjet deposition uses the ballistic particle concept to build solid objects.

Expanded Applications

By expanding the range of available "inks" to include electrically conductive materials and even biological tissue, researchers have printed products as diverse as electrodes and other microbattery components, sensors, a bionic ear, and retinal eye tissue. All of the inks can be printed at room temperature and must be formulated to flow through the nozzles under pressure, but retain their form after they are in place.

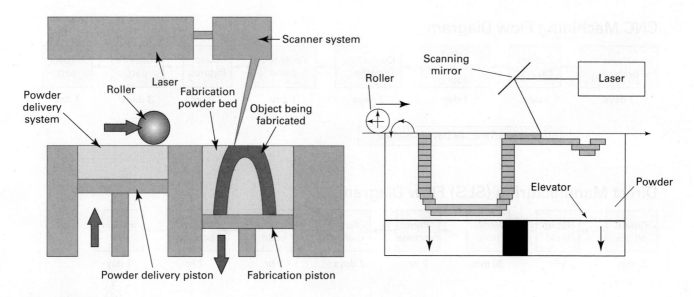

FIGURE 33.8 Selective laser sintering (SLS) uses a laser to scan and sinter powdered material into solid shapes in a layer-by-layer manner.

Depending upon the material and product objectives, the power of the laser can be adjusted to induce viscous flow of thermally softened (but not melted) surface material, produce partial melting (surface only), or produce full-melting. When the particles simply bind at points of contact, porosity can be as great as 40% by volume. When the particles are completely melted, fully dense parts can be produced.

Because the bonding of metal and ceramic powders requires more energy than the fusing of polymers, metal and ceramic powders can be coated with a thin layer of polymer binder. During the build, the laser heats the polymer coating, fusing it to the coatings on adjacent particles. The part is then transferred to a separate sintering operation, where the polymer binder is burned off, leaving a porous metal or ceramic product. The temperature is then raised to sintering temperatures to complete bonding and densification. In some cases, a full-density product is created by infiltrating a lower-melting

temperature metal, such as bronze into tool steel, using capillary action to pull it into the loosely sintered product. Because of its ability to fabricate metals and ceramics, selective laser sintering was among the first of the additive processes to offer direct fabrication of metal tooling for processes such as injection molding and die casting.

To minimize the required laser power and increase process speed, radiant heaters are often used to bring the powder to temperatures just below the fusing point. This also helps to decrease the thermal residual stresses and material distortion that occur during processing. Because of the high temperatures created by the laser, the build chamber is typically filled with nitrogen or argon to avoid oxygen contamination of the bonding surfaces, and to reduce the possibility of combustion or explosion because of the high surface energy of powder particles.

Because the excess powder acts as a **natural support**, there is no need for the design, fabrication, and removal of support structures. Optimal packing densities can be achieved within a build because additional geometries can be produced within the open volume of a part. Dimensional accuracy and surface roughness tend to be poorer than for other additive processes. Dimensional stability is largely a function of the volumetric shrinkage of the materials during solidification and cooling. Dimensions that require the greatest accuracy should be built in the x-y plane if possible. The particles used in the SLS process are approximately spherical and range in diameter from 50 to 125 μm (0.002 to 0.005 in.). If they fuse only at their respective contact points, significant surface roughness can result. Mechanical properties tend to be very isotropic in both the thermoplastic materials and metals, and, at full density, are often equivalent to those produced by alternative processes. Aerospace components made from titanium-6Al-4V regularly achieve performance equivalent to parts produced by investment casting.

FIGURE 33.9 Silica sand casting mold produced by selective laser sintering. (Courtesy of Mr. C. Huskamp, Charleston, SC.)

CNC Machining Flow Diagram

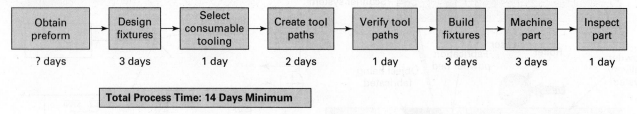

Obtain preform	Design fixtures	Select consumable tooling	Create tool paths	Verify tool paths	Build fixtures	Machine part	Inspect part
? days	3 days	1 day	2 days	1 day	3 days	3 days	1 day

Total Process Time: 14 Days Minimum

Direct Manufacturing (SLS) Flow Diagram

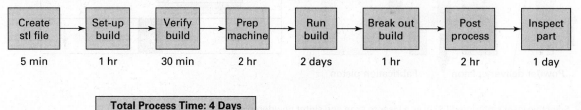

Create stl file	Set-up build	Verify build	Prep machine	Run build	Break out build	Post process	Inspect part
5 min	1 hr	30 min	2 hr	2 days	1 hr	2 hr	1 day

Total Process Time: 4 Days

FIGURE 33.10 Comparison of metal prototype manufacture by CNC machining and direct-digital manufacture using selective laser sintering. Note the significant reduction in total process time. (Courtesy of Mr. C. Huskamp, Charleston, SC.)

Figure 33.10 compares the required steps and associated time for the manufacture of a metal prototype by CNC machining and direct-digital manufacturing using selective laser sintering. Note the significant reduction in total process time, which could be further reduced by advances in the additive process.

Selective Laser Melting (SLM) and Direct Metal Laser Sintering

The processes of selective laser sintering (described above), **selective laser melting (SLM)** and **direct metal laser sintering (DMLS)** are sufficiently similar that the terms are often used interchangeably. Selective laser sintering typically uses polymeric powders, or powders of other materials that have been coated with a thin polymeric film. Selective laser melting and direct metal laser sintering employ high power lasers to sinter or fully melt powders of almost any metal alloy, including aluminum alloys, bronze, steel, tool steel, stainless steel, maraging steel, titanium, cobalt-chrome, and nickel-based superalloys. Products can have surface finish equivalent to a fine investment casting, full density, and mechanical properties equivalent to conventional processes. They can be heat-treated or surface-treated just like conventional parts. Ideal products would be small, fully functional, metal parts of highly complex geometry that would be difficult to machine, too expensive to investment cast, are needed in small quantities, and have a short lead time. Materials with high operating temperatures are commonly fabricated. Geometric features can include internal features, cooling passages and lattice structures.

Electron Beam Melting (EBM)

Electron beam melting (EBM) is another powder-based, scanning-beam process, where the higher power of an electron beam can be used to produce full-density metal products from layers of metal powder. A vacuum is now required, however, to permit the electron beam to travel within the build chamber. The vacuum also provides a clean environment, leading to the production of superior-quality products with excellent mechanical properties. A high-speed diffuse beam is first scanned over the topmost layer to preheat the material and reduce thermal stresses. A new powder layer is then spread and a high-power focused-beam then scans to melt the selected regions. Build rates are on the order of 80 cm³/hr (5 in³/hr), and product dimensions are typically within 0.2 mm (0.008 in.).

Electron-beam melting has been used to produce fully functional products, especially in areas where strength and elevated temperature performance is required. Build materials must be electrically conductive, but include reactive metals, such as titanium, high-temperature metals, such as tantalum, cobalt-chrome, tool steels, and the superalloys, and intermetallic compounds, like titanium aluminide. Compared to selective laser melting, surface finish is typically rougher and the minimum feature size is larger. The mechanical properties are greater, however, because of the lack of oxygen in the chamber (vacuum), and the high energy of the electron beam fully consolidates the powder. Applications include injection molding dies, die-casting tooling, and stamping tools, as well as finished parts for direct service, such as aerospace parts and medical implants. **Figure 33.11** presents the sequence of electron beam melting manufacture, and **Figure 33.12** shows a 120 mm (4.72 in.) diameter, 30 mm (1.2 in.) thick impeller made from titanium (Ti-6Al-4V).

Candy Manufacture

Intricately shaped candy can be produced in a similar manner using a bed of sugar and a directed stream of heated air.

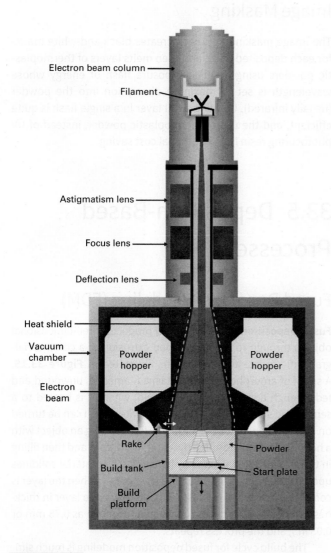

Electron beam column

Filament

Astigmatism lens

Focus lens

Deflection lens

Heat shield

Vacuum chamber

Powder hopper

Powder hopper

Electron beam

Rake

Build tank

Build platform

Powder

Start plate

FIGURE 33.11 Schematic of the electron beam manufacturing process. (Courtesy of Arcam AB, Molandal, Sweden)

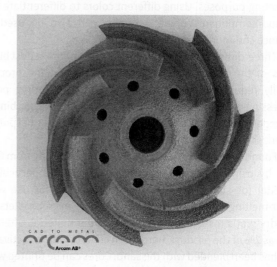

FIGURE 33.12 Small metal impeller made from titanium (Ti-6Al-4V). (Courtesy of Arcam AB, Molandal, Sweden)

The primary assets of three-dimensional printing are the variety of engineering materials that can be used, (virtually any material that can be obtained in particulate form, including metals, ceramics, cermets and polymers), and the elimination of thermal residual stresses and their related problems because of the absence of thermal processing during the build. Inorganic binders, such as colloidal silica, are often used with ceramic powders. Rather than being evaporated away or removed like organic binders, these inorganic binders fuse together with the

Three-Dimensional Printing (3DP)

In yet another approach, ink-jet printing has been combined with powdered materials by using the ink-jet to deposit a low-viscosity liquid binder. As shown in the sequence of **Figure 33.13**, **three-dimensional printing (3DP)** [also known as *selective inkjet binding* to differentiate it from the inkjet deposition process presented earlier] begins by loosely depositing a thin layer of powder for processing. An inkjet printing head then scans the powder surface and selectively injects a binder material (such as polymeric resin or colloidal silica), joining the powder together in those areas defined by the cross-section geometry. The unbounded powder then remains in place, surrounding the product and serving as supports.

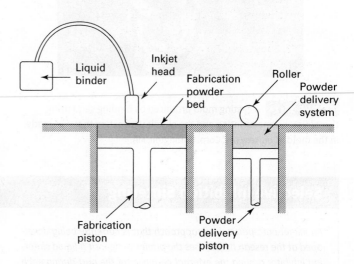

Liquid binder

Inkjet head

Fabrication powder bed

Roller

Powder delivery system

Fabrication piston

Powder delivery piston

FIGURE 33.13 3-D printing builds components by selectively binding powdered material in a layer-by-layer manner. Metal parts are then sintered and possibly infiltrated to produce strong, high-density products.

ceramic powder and reduce the amount of volumetric shrinkage. When product strength is not required and low material cost is preferred, plaster powder can be used along with deposited drops of water.

By using colored binders and multiple printheads, multicolored products can be produced. These are often used for marketing purposes. Using different colors to differentiate the various components of an assembly can also be used to enhance communication and understanding.

Three-dimensional printing offers some of the fastest build times of the additive processes. Unfortunately, the process usually requires multiple processing steps, including postprocessing in a sintering furnace to remove the binder (if needed) and to fuse and densify the product. Residual porosity tends to compromise material strength.

Foundry molds and cores can be created directly from CAD data by using quartz or specialty sands as the particulate material (or starch- or plaster-based powders for use with lower-melting-point metals) and injecting microdroplets of foundry-grade resin as a binder. The binder material, often as little as 2% of the product, hardens in contact with an activator that has been blended with the sand. Cores can be printed with a hollow interior to allow for better venting. **Figure 33.14** shows a printed mold and the resulting aluminum casting.

In another variation, radiation absorbing material is selectively deposited in place of the usual binder. After each layer is

printed, infrared light is flashed over the surface. The deposited material absorbs the energy, causing the selected material to sinter. By sintering as each layer is deposited, subsequent processing can often be eliminated.

Image Masking

The image masking approach creates black-and-white masks for each deposited layer, and then melts layers of thermoplastic powders using a single exposure flash of energy whose wavelength is set to maximize absorption into the powder (usually infrared). Creating a full layer in a single flash is quite efficient, and the use of thermoplastic powder, instead of UV photocuring resin, is an additional cost saving.

33.5 Deposition-Based Processes

Fused Deposition Modeling (FDM)

Fused Deposition Modeling (FDM) produces three-dimensional objects through robotically guided extrusion of a commercial-grade thermoplastic material, as illustrated in **Figure 33.15**. A spool of amorphous thermoplastic filament is unwound and fed through a robotic extruding head, where it is heated to a semi-liquid state. The emerging extrudate, which can be turned on and off, then builds the x-y plane, like building an object with a hot glue gun, first forming the external contour and then filling in the cross-sectional areas by hatching. The material solidifies upon contact, and bonds with the layer below. When the layer is completed, the build platform indexes down one layer in thickness (as fine as 0.05 mm or 0.002 in. or as much as 0.75 mm or 0.03 in.), and the process repeats.

The build cycle for fused deposition modeling is much simpler than other additive manufacturing processes, and the equipment is compact and relatively low cost, ideal for applications in an engineering office environment. In addition, the process uses nontoxic thermoplastic materials that do not require venting or postprocessing. Desktop units are ideal for concept-modeling applications early in product development

FIGURE 33.14 A casting mold produced by binding sand using three-dimensional printing and the aluminum casting that was made in the mold. (Courtesy of Z Corp, Burlington, MA.)

Selective Inhibition Sintering

An ink-jet-onto-particulates approach that is currently being developed at the research level uses the printer to deposit a liquid sintering inhibitor around the external periphery of the part during each layer of the build. When the build is completed, the entire volume is then bulk sintered, and the product is removed by separating it along the sinter-inhibited surfaces. The sintered powder external to the part simply becomes a sacrificial mold. Compared to existing processes, speed is increased, because only the perimeter of the part is treated and not the entire cross-sectioned surface. Because no binder has been deposited into the volume of the product, the sintering characteristics and resultant properties are superior to those of bindered powders.

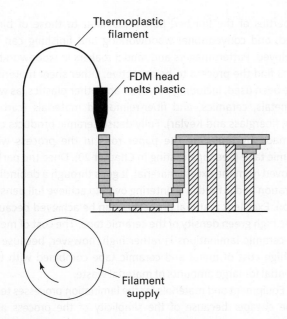

Thermoplastic filament

FDM head melts plastic

Filament supply

FIGURE 33.15 Schematic of fused deposition modeling (FDM). This process can be "office friendly"—i.e. small in size and using nontoxic materials.

cycles. Colors can be incorporated in the build material and are easily changed for the various regions of a product.

Because the material does not cool fast enough to maintain rigidity and dimensionality during deposition, the build must begin with a firm base that can later be easily separated from the finished product. If supports are needed, a dual extrusion head is used and a different material is used. Depending on the material selected for this application, supports can be removed by dissolving in a water-based solution or by breaking them free of the targeted part.

Common applications include concept models, functional prototypes, low-volume end-use products, and manufacturing tooling, such as jigs, fixtures, and tooling masters. Disadvantages of the process include the limited selection of materials, which includes machinable wax, investment-casting wax, polycarbonate, polyphenolsulfone, high density polyethylene, ABS, and low-melting-point eutectic metals. Most have poor mechanical properties as well as anisotropic behavior in the z-direction. While parts do not require postprocessing, the build time is the longest among the additive processes, and the material cost of the starting filaments is among the highest per unit volume. Common defects include gaps between contour perimeter traces and the rastored fill traces, voids between rastored fill traces when the extrusion carriage makes a 180° direction reversal, and visible hatch lines from the extrusion of round profile material. If the surface roughness or stairstepping effect is objectionable, a heated tool can be used to smooth the surface, the surface can be sanded, or a coating can be applied, provided the necessary tolerances can be maintained.

Laser Engineered Net Shaping (LENS); Direct Metal Deposition (DMD); and Similar Material-Feed Deposition Processes

In a process known as **direct metal deposition (DMD),** lasers and powders have been combined to enable the direct fusing of uncoated metal powders into near-net shape fully dense functional metal products. Powder delivery nozzles direct streams of metal powder into the focal point of a high-power Nd:YAG laser. The powder particles melt and fuse with the underlying layer as the part is scanned on an x-y positioning stage beneath the laser. Upon completion of the layer, the laser and powder delivery nozzle index upward and begin patterning the next layer. By adjusting the focal point of the laser (as small as 0.2 mm or 0.1 in. to as large as 2.5 mm or 0.1 in.) the material can be deposited as either fine detail or bulk build. In **laser engineered net shaping (LENS)** or **laser metal deposition (LMD),** the high-power laser first produces a tiny molten pool on the surface of a build, and a nozzle disperses a small amount of metal powder into the pool to increase its volume. In yet another alternative, the laser melts and fuses a solid wire with a flat cross-sectional profile. In essence, these processes are microscopic, computer-controlled versions of laser cladding performed in a controlled-atmosphere chamber, typically argon, to prevent undesirable oxidation.

A variety of metals, alloys, and nonmetallic materials can be deposited by these processes, including alloys of aluminum, copper and titanium, tool steels, stainless steels, nickel and cobalt superalloys, cermets, and other high-performing materials, such as tungsten and the metallic glasses. By using multiple delivery systems, different materials can be deposited simultaneously or the composition can be varied during a build to create products with composition gradients. These technologies were initially targeted for the rapid manufacture of metal molds for the injection molding of plastic parts, but have expanded into the production of molds, dies, and low-volume finished products with full density, fine grain structure, and mechanical and metallurgical properties that are equivalent or superior to

Deposition with a Multi-Axis Robot

By placing the deposition head on the end of a multi-axis robot, builds no longer need to be made in the z-axis direction. Material can be deposited onto angled, sloped or curved surfaces, or even onto vertical walls with considerable precision and accuracy. Both polymers and metals can be deposited, and the absence of an x-y printing bed enables the production of exceptionally large structures, limited only by the range of the robot.

forged materials. By inputting both current and desired geometries, the process can also be used to add material to existing components to repair damage, restore worn products to useful life, or reconfigure or add features to an existing part. Wear or corrosion-resistant coatings or claddings can also be applied.

An electron-beam equivalent has also been developed in which a metal wire is fed into a molten pool created by the beam. The beam focus and power output is adjustable, allowing for bulk deposition or fine detail. The vacuum environment allows for the deposition of reactive materials, and the high-energy density of the electron beam allows for the deposition of refractory materials, including tantalum and tungsten. This process can have very high build rates, but also has a large puddle size, meniscus and crown. Other wire feed processes use a laser or plasma torch.

Laminated-Object Manufacturing (LOM)

Lamination-based processes involve the sequential bonding and patterning of solid material sheets or laminae. A process known as **laminated object manufacturing (LOM)** uses sheets of paper with a thermally activated organic adhesive coating one side. As shown in **Figure 33.16**, the process begins by delivering a sheet of paper onto the working surface either by roller or some other means. Next, a heated roller is passed over the surface to bond the paper sheet to the layer below it. Once bonded, a laser or cutter is guided to cut the perimeter of that particular cross section, cutting only the uppermost layer. The build platform descends, and these steps are repeated.

A unique characteristic of the lamination processes is the method for extracting the product by removing the excess paper mass that accumulates around the periphery of the part. This is accomplished by drawing cross-hatches in the excess material regions during patterning. Over many layers, these cross-hatches form the boundaries of small blocks, which can later be removed from around the finished part. By remaining in place during the build, the excess solid material provides a natural support of the part, permitting complex cantilevered geometries to be fabricated as easily as any other structure.

Any material that can be made into a foil and cut with a laser is a candidate for this process. With paper laminates, the properties of the finished part are similar to those of birch wood, and conventional woodworking and finishing can be employed. Patternmakers and model makers in woodworking shops find the process to be attractive. Other sheet materials have been used, including polyester and other plastics, as well as metals, ceramics, and fiber-reinforced materials (carbon fiber, fiberglass and Kevlar). Fully dense ceramic products can be made by replacing the paper rolls in the process with ceramic tape (See tape casting in Chapter 20). Once the part is removed from the excess material, it goes through a debinding operation and is fired in a sintering oven to achieve full densification. Excellent material properties can be achieved because of the high green density of the ceramic tape. The cost of metal and ceramic laminations is rather high, however, because of the high cost of metal and ceramic tape combined with the potential for large amounts of material waste.

Equipment and materials for the lamination processes tend to be cheaper because of the simplicity of the process and reduced maintenance requirements. Large components can be produced with sizes up to about 0.5 x 0.5 x 0.75 (20 x 20 x 30 in.). Because the material does not undergo a phase change (e.g., liquid to solid), residual stresses are low, resulting in less warpage and better dimensional stability. Because paper is hygroscopic, laminated parts must be surface treated with water-resistant resins to prevent swelling. Combining multiple parts into a single build is often difficult, because the removal of excess material often prohibits parts from being spaced closely together. Layer thickness is fixed at the thickness of the lamina, and z-axis precision tends to be worse than for the other additive processes.

A variation of the process described above, **selective deposition lamination (SDL)** uses standard copier paper. Adhesive is selectively applied to the top of the previous layer, based on the specific cross section of the subsequent sheet—higher adhesive density to the area that will become the part and much lower density to the surrounding region. Heat and pressure is applied, and a tungsten carbide blade cuts the uppermost sheet along the product perimeter. Adhesive is applied for the next layer, and the process repeats. Geometric complexity is less than other additive processes, and build size is limited to the size of the paper feedstock.

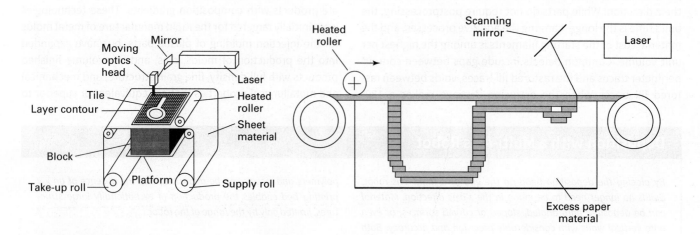

FIGURE 33.16 Schematic of the laminated object manufacturing process in which solid sheets are used to create the layers.

Other Deposition-Based Processes

In a modification of the lamination approach, **ultrasonic consolidation** (i.e. ultrasonic welding) is used to bond layers of metal foil into full-density products. Parts have been made from foils of aluminum, copper, stainless steel and titanium. True metallurgical bonds are produced with modest pressure and temperatures that do not exceed half of the melting point. A periodic machining pass can be incorporated to maintain flatness and dimensional tolerance. Features include the ability to produce extremely complex shapes, as well as laminate dissimilar metals or produce metal matrix composites. Because low temperatures are maintained, material properties are not altered, and electronics, sensors, and fibers can be embedded in fully dense metal products.

33.6 Uses and Applications

In just three decades, additive manufacturing processes have evolved from a method of making low-strength, three-dimensional, plastic design prototypes, to a family of processes that is changing the way organizations design and manufacture products. Starting with a CAD drawing, computer tomography (CT scan), magnetic resonance imaging (MRI), or three-dimensional digitizing, this data can be quickly converted to a fully functional prototype, scale model, manufacturing tool, or finished product, facilitating considerable savings in both time and money. The processes have become faster, more accurate and less expensive; build volumes have increased, and the range of available materials has expanded considerably.

Prototype Manufacture

A large number of additive products are used produce both **conceptual models** and **functional models** to assess aspects of form, fit, and function. Converting a design concept to a physical representation offers the following advantages:

1. Ability to evaluate concepts and seek input from others during the design phase.

2. Ability to detect and correct design errors before the costly tooling stage.

3. Ability to have multiple iterations of a design before manufacture, allowing comparison of different shapes and styles.

4. Ability to assess assembly concerns and ensure proper alignment and interaction with companion parts.

5. Can possibly allow for functional testing under real-life or simulated conditions, including wind-tunnel. testing, assessment of fluid-flow through orifice designs, and evaluation of stress distribution and strength under load.

6. Ability to serve as communication models between: management, design, manufacturing, sales and marketing, and potential customers.

7. Can serve as sales and marketing models or samples.

8. Can serve as a visual aid during bidding, tooling design, and other activities.

9. Often enables a significant reduction in product development time.

Figure 33.17 presents a small scale model of a vehicular space frame that was produced by additive manufacturing. The model enables meaningful discussion about the proposed design, as well as methods and sequences of assembly, and the possible use of subassemblies and/or welding jigs and fixtures. **Figure 33.18** shows a half-size model of a Hellfire air-to-surface missile with the computer screens in the background showing the originating CAD representations.

Rapid Tooling

Rapid tooling is an excellent application for the additive manufacturing processes for several reasons:

1. Tooling is usually a low-production-volume or one-of-a-kind commodity.

2. Tooling typically has an associated long lead time.

3. Tooling typically requires the high cost of skilled labor.

4. Tooling is usually specific to one or a few part geometries.

FIGURE 33.17 Small scale model of a space frame for a dragster produced by additive manufacturing (shown with and without partial body work.) (Courtesy of Mikael Madsen/MagDragster)

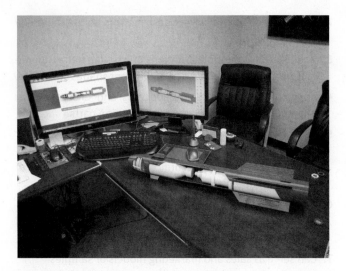

FIGURE 33.18 Half-size scale model of a Hellfire air-to-surface missile with CAD representation showing on computer screens. (Rapid Processing Solutions, Inc., Wichita, KS.)

5. The high cost of tooling often requires its use in producing a large number of identical products.

6. Tooling is a requirement of most manufacturing processes, so there are a number of potential applications.

As a rapid source of cheap yet effective tooling, additive manufacturing technology can lower fixed tooling costs. Manufacturers could possibly justify die casting, stamping, or injection molding runs of a few hundred or a few thousand parts or produce those parts with a much shorter lead time.

The simplest of the rapid tooling approaches is the direct production of patterns or molds for use in metal casting. Almost all of the additive techniques can be used to produce patterns for the sand casting processes, where the direct conversion of design to pattern can significantly reduce lead time while requiring no change in standard foundry practices or procedures. Patterns must be abrasion-resistant and able to withstand the ramming forces of sand compaction, and many additive process materials are adequate. The choice of CNC machining or additive manufacturing will depend upon a number of factors, including size of the casting, geometric complexity, number to be produced, and required precision.

The additive manufacturing techniques that deposit binder onto sand or fuse polymer-coated sand can be used to directly produce both molds and cores, eliminating the patternmaking step and enabling the production of small-quantity casting orders without tooling. Numerous small molds and cores can often be manufactured side-by-side in a single build. Molds can be produced with no draft. Undercuts are possible, and molds can have complex parting surfaces. Cores can be hollow, providing excellent venting and permeability, or possibly produced as an integral part of the sand mold, improving product precision. The approach offers low-volume production with short lead time, ease of design change, and the manufacture of parts whose complexity would be difficult for conventional methods. Several examples of this application have been presented in the previous figures of this chapter.

Investment casting (or lost wax casting) is frequently used to manufacture intricate parts that are difficult, if not impossible, to machine, forge, or cast by other methods. Each casting requires an expendable wax or thermoplastic pattern, a replica of the desired shape, usually produced in an injection mold. Designs often include internal passages and ports (as in valve bodies), curved surfaces (as in the vanes of impellers), and internal cooling channels (such as those in turbine blades)—features that would be extremely difficult or impossible to produce by injecting wax into a rigid die. Many of the additive manufacturing processes, including stereolithography (SLA), selective laser sintering (SLS), and fused deposition modeling (FDM), can now produce expendable patterns directly from design data, bypassing the costly and time-consuming manufacture of injection mold tooling. There is no increase in cost as part complexity increases. Hollow patterns, or patterns with a honeycomb or scaffold interior, can be produced, significantly reducing the possibility of shell cracking when the pattern material is melted out of the ceramic shell.

Another investment casting approach bypasses the pattern entirely and builds the ceramic mold directly from digital data. The mold is created by selectively exposing a mixture of photosensitive resin and ceramic particles to a beam of ultraviolet light and building layer-by-layer, as in stereolithography. The resulting solid then undergoes a binder burnout to remove the resin, followed by furnace sintering of the ceramic.

Die casting dies and die inserts have been produced by the additive techniques that can create full-density metal products. They can be made with internal cooling channels that conform to cavity contours or are otherwise are positioned for optimum performance. Cycle times have been reduced by 15% to 45% and part quality has been improved. Molds for plastic injection molding can be economically produced, and these molds can be produced in a modular form that reduces cost and enables simple replacement of worn or broken segments. Rubber or epoxy molds for vacuum casting of polymers or spin casting of low-temperature metal alloys can be produced by pouring the mold material over a pattern that has been produced by additive manufacturing and allowing it to cure.

Molds have been produced to perform the blow molding of prototype and production polyethylene and PET containers. Mold manufacturing time can be reduced from several weeks to several days, and cost to less than half of a typical aluminum tool. Both male and female molds and form tools have been produced for the thermoforming of thermoplastic sheets into parts that include consumer packaging, automotive interior and body panels, aerospace panels and ducts, and even boat hulls. In a similar manner, the molds and form tools have been used in the manufacture of fiber-reinforced lay-ups, and lightweight, honeycombed or semi-hollow additive products have been used as enclosed cores for fiber-reinforced products.

More durable metal molds and dies can be produced by first creating a pattern using one of the additive processes. A metal deposition operation, such as thermal spray or electroforming, is then employed to coat the pattern and build a metal shell of the desired thickness. The resulting metal shell is then separated from the pattern and is reinforced with a backing

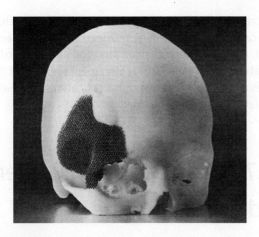

FIGURE 33.19 Titanium skull patch produced by the electron beam melting process beginning with data obtained from a CT scan of an injured person. This is an example of a one-of-a-kind, or custom, product. (Courtesy of Arcam AB, Molandal, Sweden)

material such as a metal-reinforced epoxy or chemically bonded ceramic. Production runs in excess of 5000 have been made on a plastics compression molding machine using a pure Ni shell backed with chemically bonded ceramic.

The additive metal deposition techniques can be used to repair cracked or worn tooling, apply hard facing coatings, or deposit variable materials to different areas of die surfaces to tailor friction and wear characteristics. It is now possible to produce durable tooling for metal-forming processes, including the sheet-forming processes of hydroforming, rubber-pad forming, and conventional punch-and-die forming. These tools can withstand high forming pressures and have good dimensional precision. Cost and access time are both reduced, and rapid-response design changes are possible. Difficult-to-machine geometries are easily produced.

There are also a number of tools that are used within the manufacturing system, including jigs, fixtures, templates, and gauges. They are used to align, assemble, clamp, hold, test, and calibrate components and subassemblies. Much like any other manufactured product, they too go through the operations of design and production, and have the same concerns of lead time and cost. These tools can be made quickly, with little effort and expense. Damaged or lost tools are easily replaced.

Direct-Digital Manufacture

With the numerous additive manufacturing processes and the developments that have occurred over the past several decades, what was once considered to be the fabrication of low-strength, plastic prototypes has expanded to include the direct production of metal, ceramic, polymer, and composite material components, many with properties equivalent to standard-production-process equivalents. By producing finished parts directly from digital data, eliminating all tooling, a number of applications have emerged, including:

1. The production of small batch sizes that could not be produced economically by traditional methods that required dedicated tooling.

2. The production of parts with intricate geometries or internal features that could not be made by conventional processes or without extensive assembly operations.

3. The production of one-of-a-kind custom or personalized products, such as bone replacements, dental implants, orthodontics, custom-fit hearing aids, and other items that must be tailored to a specific individual. **Figure 33.19** shows such an application where CT-scan data from a skull injury was used to create a custom titanium patch using the electron beam melting process.

4. The production of replacement parts on demand, enabling reduced inventory costs.

5. The repair or reworking of worn, distorted or damaged parts, restoring to original dimensions and mechanical properties comparable to the original product, thereby enabling reuse as opposed to costly replacement. This is especially important when replacement parts are not available.

6. Manufacture of products from hard-to-fabricate or hard-to-machine materials.

7. The production of near-net parts in costly materials, like titanium, where conventional processing would be more costly because of poor material utilization.

8. The manufacture of products within pressing time constraints.

Designing for direct-digital manufacture is different from traditional design. Much of traditional design is design for manufacture. For example, castings must consider the various features of solidification and shrinkage, with preference for uniform wall thickness, and parting lines and draft incorporated to facilitate pattern removal or part extraction. Consideration must be given to core placement and support, and aspects such as minimum wall thickness, gate placement, risering, and others. Machining operations require accessibility of the cutting tool and chip removal. Forging and the other deformation processes have similar considerations. A part designed for forging would not make an ideal casting, and a part designed for casting would not make an ideal forging.

Direct-digital manufacture, a form of **design-driven manufacturing**, breaks most of the conventional rules. Simply stated, if a part can be modeled on a computer in 3-D, it can be sliced and printed, layer-by-layer, on an additive manufacturing system. Functionality is now the basis of design. Because cost and time are no longer a function of complexity, parts can be as intricate, complex or detailed as they need to be. Internal features and/or sculptured surfaces do not add to cost. Continued iterations and refinements are encouraged. There is no penalty for changes late in the product development cycle. Design changes can be made or variations of a product can be offered even after production has started. There is no need to simplify, because additional features do not increase cost. Wall thickness can vary as needed, and weight can be reduced by making some walls or features with an internal lattice, or even completely hollow. Part consolidation should be considered to eliminate assembly operations, provide better dimensional precision, and reduce parts inventory. Conversely, because there is no cost of part-specific tooling, breaking a product into subcomponents to better facilitate serviceability or reduce replacement costs does not add to the manufacturing expense. One should note, however, that for performance-critical applications, such as aerospace, the requirements for control and qualification can impose restrictions on frequent or rapid changes in geometry or processing.

Direct digital manufacturing is the fastest-growing segment of the additive industry, comprising more than one-third of the market in 2015 (up from 3.9% in 2003). Application areas span a broad spectrum from costume jewelry, awards and trophies, through a variety of consumer products (including furniture and entertainment products), to automotive and aerospace components.

Aerospace: Building with titanium, superalloys, and cobalt-chrome, existing parts can be duplicated or designs can be modified to utilize the unique additive capabilities. Weight can be reduced, parts can be consolidated, and design features such as curved internal cooling channels are possible. Parts made from ceramic composites offer a significant reduction in weight and can endure much higher temperatures than metal alloys. The low-volume, high-value aspects of aerospace components make them ideal for additive manufacture, and parts can be manufactured when they are needed.

Medical/dental: The greatst demand for additive manufactured products is likely to be in the medical/dental area where one-of-a-kind, patient-specific items can be made from biocompatible materials and be cost-effective. CT and MRI scan data can provide the starting 3-D image. The plastics-based additive methods can produce surgical aids and models, hearing aid cases, and polymeric implants, while the metal-based methods can produce tailored implants of titanium, cobalt-chrome, and other alloys. Surfaces can be created with specified pore size, pore geometry, overall density, and roughness, optimized for the infusion of living bone (osseointegration), and these surfaces can be integral to a solid interior. The deposition of binder material onto a bed of particles enables the manufacture of ceramic products, including bone grafts and dental implants and crowns. Bioprinting, building with tissue-engineered and biological materials, has numerous medical

applications and is the subject of intensive research. Skin, bone, cartilege, blood vessels, and heart tissue have been generated, along with replacement ears. The ultimate goal is the generation of fully functional replacement organs and tissues, such as aortic heart valves. Pharmaceutical inks can enable the preparation of custom-dose medications.

33.7 Pros, Cons and Current and Future Trends

Additive manufacturing has grown from a concept to a diverse and fully developed manufacturing family in less than three decades. **Table 33.1** summarizes the primary features of some of the more popular processes. What began as a means of producing plastic models directly from computer data has expanded to the production of a multitude of components, including direct-usage products. Available materials have expanded from low-strength photopolymers to now include engineering plastics and elastomers, ceramics, metals, and even composites. Properties include good tensile and flexural strength, good impact resistance, and reduced anisotropy. Metals have moved to full density with properties equivalent to wrought and cast products, and now range from low-strength, low-melting-temperature metals through titanium, tool steels, and superalloys. Candidate applications are most attractive when complexity is high, lead times are short, and volumes are low, often as low as a single part.

Assets of the additive processes include:

1. Outstanding design freedom. There is no penalty for added complexity. Shapes can be produced that are not possible with conventional methods. A range of wall thicknesses can be used on a single part—permitting increased strength where needed, weight savings at other locations, and lower material cost.

2. Lead times or time to market can be drastically reduced.

TABLE 33.1 Comparison of Additive Processes

Process	Starting Material	Layering Mechanism	Possible Materials
Stereolithography (SLA)	Liquid photopolymer	Polymer curing by UV light	Photopolymers
Inkjet deposition (ID)	Molten liquid	Solidification of droplets	Thermoplastic polymers, waxes, low melting-point metals
Selective laser sintering (SLS)	Powders	Sintering or melting	Thermoplastic polymers, waxes, metals, binder coated sands
Electron beam melting (EBM)	Powders	Melting	Metals (titanium, tool steels, superalloys)
Three-dimensional printing	Powders	Deposited molten binder	Metals, ceramics, polymers, cermets, sand
Fused deposition modeling (FDM)	Semi-liquid polymer	Solidification of extruded strand	Thermoplastic polymers
Laminated-object manufacturing (LOM)	Solid sheets	Fusing of sheet material	Paper, polymers, ceramics, metals, composites

3. One-of-a-kind or short production runs can be made economically. Products can be customized to the customer/user.

4. Part consolidation is often possible because there is no penalty for added complexity.

5. Design modifications are quick and easy, and can be made at any stage of manufacture. Multiple iterations of a design can be quickly and inexpensively produced.

6. Scrap can be substantially reduced compared to conventional subtractive processes. This is very attractive for expensive materials.

7. Unused powder materials can be readily recycled.

8. Doesn't require dies, molds or other tooling.

9. Can make hollow parts for weight savings.

10. Product properties are generally consistent and independent of orientation and location.

11. Just-in-time manufacturing is possible. Products are often suitable for direct use or application. Parts can be made as needed, considerable reducing inventory. Parts can be made at the location of need, eliminating shipping and associated delays.

12. Parts can be "reverse engineered" by dimensionally scanning an existing part.

13. Damaged, worn or corroded parts can be repaired or remanufactured, thereby extending their in-service life.

14. Existing parts can be easily reconfigured or acquire added features.

Current problems, limitations, and concerns include:

1. Equipment cost might be prohibitive for some companies.

2. Build rates (time to produce a part) are slow compared to the production rates of conventional processes making large production runs uneconomical. Improvement efforts are being made, such as the use of multiple lasers in a single build.

3. Raw material costs are often high. Some decreases are expected, however, as process use grows.

4. The range of available materials is still limited, especially within certain processes.

5. The material property data are often poor, especially as they relate to products of a particular process. The long-term durability of products, especially as it relates to fatigue and fracture, is not well documented.

6. Process optimization guidelines are needed—layer thickness, build speed, temperature control, time between layers, etc. Will product properties vary with deposition variables?

7. Build space is often limited. Need larger build space for larger products.

8. Postprocessing is often required, including postcure, cleaning, surface finishing, stress relief, and support removal.

9. In some cases, inconsistent quality (structure and properties) has been observed machine-to-machine (even though nominally identical), operator-to-operator, and material batch-to-material batch.

10. Some processes have rough surface finish and porosity is often common. It might be necessary to begin with less demanding applications (start conservatively) and build confidence. When porosity is present, the mechanical properties are often less than those of products made by alternative processes using similar materials. For some materials and processes, hot isostatic pressing can be used to eliminate internal porosity or lack of fusion.

11. Certain additive systems use toxic resins that require high-capacity venting or cleaning solvents.

12. Supports and anchors might be needed during the build.

13. Not appropriate for high-volume, low-complexity parts.

Current trends include:

1. Overall reduction in cost accompanying rapid growth in software, hardware, techniques, and materials.

2. The development of high productivity machines with large build volumes. Single large parts could be produced. Many small parts in a single build would allow 3-D printing to compete with injection molding for high-volume products.

3. The development of small, low-cost machines, aimed at desktop users and educational institutions.

4. The development of machines for targeted applications, such as dental, jewelry, etc.

5. Increased use in the direct-digital manufacturing of complex-shaped, low-volume products.

6. Improvements in dimensional precision and surface finish.

LEAP Engine Fuel Nozzles

A jet engine fuel nozzle is currently being made by additive manufacturing for the LEAP engine, produced by CFM International (a joint venture of GE and the French company Snecma). The additive process enables manufacture from a more advanced material than traditional methods. Cobalt chromium can withstand temperatures of 1650°C/3000°F. The single piece product replaces a multipiece assembly (one part replaces 20), offering weight reduction and the absence of brazes and weldments. The additive process also enables the incorporation of built-in cooling channels. There was a 20% reduction in cost compared to the conventional process, largely because of reduced labor in multipiece assembly and inspection. By 2017, GE Aviation will be producing 30,000 of these fuel nozzles a year.

7. Expansion of candidate materials:

   a. High temperature polymers

   b. Ceramics and cermets

   c. Elastomers

   d. Carbon-fiber composite (chopped fiber in polymer)

   e. Amorphous alloys (twice the strength and 10 times the elasticity of steel and corrosion resistant)

8. Ability to produce multicolored products.

9. Ability to build with dissimilar materials (multiple materials in a single component for tailored properties or functional gradients).

Evolving and future trends:

1. Broader expansion of composite materials.

   a. Incorporation of graphite for thermal conductivity.

   b. Incorporation of copper and aluminum microspheres.

2. Ability to produce multi-functional products with components—such as conductors, insulators, dielectrics, microfluidics, photovoltaics or sensors—printed directly on or embedded inside a part.

3. Ability to print electronics and battery components that conform to the shape of a product. Also microbatteries.

4. Fabrication of nano-scale objects.

5. Ability to produce parts in space.

6. Availability of combined additive/subtractive hybrid machines. Examples include an additive powder process combined with EDM or multi-axis CNC milling. Attractive features: can integrate cut-off and finishing into one machine, can perform machining during the build when surfaces are most accessible to cutters, can combine build and repair into one machine for remanufacturing.

7. Expansion of biomaterial applications, including skin grafts, bone repair, and functional implants.

8. Could significantly affect the way certain parts are manufactured:

   a. Potential for mass customization of products, tailored to the individual customer.

   b. Potential for distributed manufacturing – making the parts where they are to be used, multiple locations of production.

   c. Part-specific molds or dies would be replaced by part-specific software instructions.

Table 33.2 lists the primary applications of the additive manufacturing processes for a recent year in decreasing order of use. Over the past decade, direct part production has risen from near bottom to its current dominant position.

TABLE 33.2	Primary Uses of Additive Manufacturing—2014 (In decreasing order of use)
1.	Functional parts
2.	Assess fit and function
3.	Patterns for prototype tooling
4.	Patterns for metal casting
5.	Visual aids
6.	Presentation models
7.	Education and research
8.	Tooling components
9.	Other

Data taken from Wohlers Report—2014, Wohlers Associates, Inc, Fort Collins, CO.

33.8 Economic Considerations

When deciding to use additive manufacturing processes, one must determine which process to choose and whether to buy equipment for in-house manufacture or to outsource work to a service provider. It is important to first define the purpose for using the additive manufacturing technology. There are many different applications, including concept modeling, form/fit verification, marketing demonstrations, functional testing, prototype tooling, production tooling, and direct manufacture of finished products. Processes should be considered with respect to size of the build envelope, build rate, layer thickness, surface finish, dimensional tolerances, residual stresses, the need for subsequent heat treatment, resultant material properties, and dimensional stability over time. The batch size of products to be fabricated should be considered, as well as the possible need for trained, experienced personnel.

Potential benefits resulting from the use of additive manufacturing, in addition to those previously cited, can include:

1. Reduction of current operating costs including labor.

2. Improved sales and marketing because of the ability to respond to customer requests for bids with actual models or the delivery of small quantities of actual parts.

3. Improved product development, including improved customer satisfaction and improved product manufacturability.

4. Improved process development as lead times and costs are reduced, leading to more iterations.

5. No operator observation is required during the build.

Review Questions

1. What are the four traditional families of shape-production processes?

2. What are some of the assets and limitations of the traditional process families?

3. What are some of the manufacturing changes or new approaches that have been enabled by the computer?

4. Describe the basic approach of the additive processes.

5. What is a prototype and how is it used?

6. What are the four areas of application for the additive manufacturing processes?

7. What are some of the other terms that have been applied to additive manufacturing? What is the basis of these terms?

8. What are the attractive benefits of "free-form fabrication" and "tool-less manufacturing"?

9. What is involved in preprocessing?

10. What is the role of bases and supports?

11. What are some possible postprocessing operations?

12. What are some common limitations of the additive processes?

13. What is the key, somewhat universal, advantage of the additive manufacturing processes?

14. What are the three families of layerwise manufacturing processes based on the nature of material deposition?

15. What is reverse engineering and how is the input data obtained?

16. What are some of the ways of controlling the thickness of a multi-layered part?

17. What factors influence or determine the thickness of a single layer?

18. What is stairstepping?

19. Describe the coordinate system that has become common for all of the additive processes.

20. What is a voxel? Why is it significant in the additive processes?

21. What are some of the factors that affect the dimensional accuracy in additive processes?

22. What is hatching and how might it be affected by dimensional changes, such as those occurring during phase transformations?

23. Why are the residual stresses in an additive product often complex?

24. Of the three components of build time (preprocessing, fabrication, and postprocessing), which typically requires the greatest amount of time?

25. What technical advances were necessary for the stereolithography process?

26. How are the various layers produced during stereolithography?

27. What postprocessing is required for SLA products?

28. What are some of the advantages of stereolithography? Disadvantages?

29. What are the attractive features of the "masked-lamp descending platform" approach compared to traditional stereolithography?

30. How are the various layers produced during inkjet deposition (ID)?

31. What materials are commonly deopsited by the inkjet process?

32. What are the advantages of inkjet deposition (ID) processes? Disadvantages?

33. How are the various layers produced during selective laser sintering (SLS)?

34. How is the laser of selective laser sintering different from the laser of stereolithography?

35. What are some of the build materials that can be used in selective laser sintering (SLS)?

36. How might the products of selective laser sintering differ as the power of the laser is increased?

37. What are some potential benefits of preheating the powder in selective laser sintering?

38. Why might some form of protective atmosphere be required when using a laser and powdered material?

39. Why are designed supports unnecessary in selective laser sintering?

40. What is the difference between selective laser sintering and selective laser melting?

41. Describe the ideal product for manufacture by selective laser melting.

42. What are some of the advantages of using an electron beam as opposed to a laser beam?

43. How are the various layers produced during 3-D printing (or selective inkjet binding)?

44. What types of materials can be used in 3-D printing?

45. How can 3-D printing produce multicolord products?

46. What type of postprocessing is required for 3-D printing processes?

47. How can foundry molds and cores be produced by 3-D printing?

48. What are some of the potential benefits of selective inhibition sintering compared to 3-D printing?

49. What are some of the potential benefits of the image masking approach?

50. How are the various layers produced during fused deposition modeling (FDM)?

51. What are some of the advantages and disadvantages of fused deposition modeling (FDM)?

52. How are the various layers produced during direct metal deposition (DMD) and laser engineered net shaping (LENS)?

53. What types of materials can be used in the laser engineered net shaping (LENS) and direct metal deposition (DMD) processes?

54. How can the direct metal deposition (DMD) and laser engineered net shaping (LENS) processes be used to restore or modify existing parts?

55. How are the various layers produced during laminated-object manufacturing (LOM)?

56. How are products extracted after a lamination-process build?

57. What are some of the cited advantages and limitations of the lamination approach?

58. How are the various layers produced during ultrasonic consolidation?

59. What lamination materials are used in the ultrasonic consolidation process?

60. What are some of the advantages of a physical prototype or model?

61. What are some of the attractive features of being able to produce rapid tooling?

62. How can the additive manufacturing processes be employed in the various metal casting processes?

63. How might the investment casting patterns produced by additive manufacture be superior to ones produced using conventional molds?

64. How can additive manufacturing be coupled with thermal spray or electroforming to create shaped metal shells?

65. What are some of the benefits of creating jigs, fixtures, templates, and gauges by additive manufacturing techniques?

66. What are some of the areas where direct-digital manufacture has been applied?

67. How is design for direct-digital manufacture different from traditional design?

68. What are some of the medical/dental applications that have emerged for direct-digital manufacture?

69. What are some of the most cited assets of the additive manufacturing processes?

70. What are some of the commonly cited limitations or concerns relating to the additive manufacturing processes?

71. What significant changes or trends are currently occurring in the additive manufacturing area?

72. What might we expect in the future with respect to additive manufacturing?

73. What are some of the considerations when deciding whether to buy additive manufacturing equipment or outsource work to vendors?

74. What are some potential benefits for a business to invest in the additive processes?

Problems

1. a. Which of the additive manufacturing processes can produce full-density metal products from the higher-strength, higher-melting-temperature metals (such as titanium, tool steels, and nickel superalloys)?

b. Which of the additive manufacturing processes can produce parts from ceramic materials?

c. Which of the additive manufacturing processes can directly produce foundry cores and molds?

2. Perform a Web survey of additive manufacturing service providers and determine which of the various processes appear to be in most common use? Would your answer differ if it were to be focused on prototype products? Rapid tooling? Direct-digital manufacturing?

3. Describe several biomaterials-type applications where the manufacture of products tailored to a specific individual would be attractive and desirable.

4. The ability to restore worn parts to original dimensions or repair damaged components has been cited in the chapter. For what types of products would this ability be most attractive?

5. Looking to the future, people have cited the ability to build parts on demand from computer files and the potential to significantly reduce inventory of spare parts in areas as diverse as automotive, aerospace, and military. What do you see as potentially limiting concerns to this future projection?

6. Three-D printing has been performed on the International Space Station to demonstrate its capability to produce parts in space. This has been presented as a precursor to the production of needed parts and products for long-range space missions and future space colonies. Briefly discuss what you see as both assets and concerns.

7. Identify a product currently being made by additive manufacturing in three or more of the following areas: automotive; aerospace; sporting goods; dental; orthopedic or prosthetic; production tooling; direct consumer product; repair, rework or remanufacture; and agriculture or food processing.

8. How might the additive manufacturing processes be used within a household, i.e., coupled to a home computer?

CHAPTER 34

Manufacturing Automation and Industrial Robots

34.1 Introduction

The term **automation** has many definitions. Apparently, it was first used in the early 1950s to mean automatic handling of materials, particularly equipment used to unload and load stamping equipment. It has now become a general term referring to services performed, products manufactured and inspected, information handling, materials handling, and assembly—all done automatically (i.e., as an automatic operation without human involvement).

In 1962, Amber and Amber presented their *Yardstick for Automation*, which is based on the concept that all work requires energy and information and that certain functions must be provided by workers or machines. Whenever a machine function replaces a human function or attribute, it is considered to have taken a step up in an "order" of automaticity. The chart that they developed is shown in **Table 34.1** with updates in bold type to account for modern terms. Notice that each order of automation is tied to the human attribute that is being replaced (mechanized or automated) by the machine. Therefore, the A(0) level of automation, in which no human attribute was mechanized, covers the Stone Age through the Iron Age. Two of the earliest machine tools were the crude lathes the Etruscans used for making wooden bowls around 700 B.C. and the windlass-powered broach for machining grooves into rifle barrels used over 300 years ago.

The A(1) level and first industrial revolution are tied to the development of powered machine tools, dating from 1775 in England. The energetic, "iron-mad" John Wilkinson constructed a horizontal boring machine for machining internal cylindrical surfaces, such as piston-type pumps. In Wilkinson's machine, a model of which is shown in **Figure 34.1**, the boring bar extended through the casting to be machined and was supported at its outer ends by bearings. Modern boring machines still employ this basic design. Wilkinson reported that his machine could bore a 57-in.-diameter cylinder to such accuracy that nothing greater than an English shilling (about 1.59 mm) could be inserted between the piston and the cylinder. This machine tool made James Watt's steam engine a reality. At the time of his death, Wilkinson's industrial complex James was the largest in the world.

Another early machine tool was developed in 1794 by Henry Maudsley. It was an engine lathe with a practical slide tool rest. This machine tool, also shown in Figure 34.1, was the forerunner of the modern engine lathe. The lead screw and change-gear mechanism, which enabled threads to be cut, were added about 1800. The first planer was developed in 1817 by Richard Roberts in Manchester, England. Roberts was a student of Maudsley, who also had a hand in the career of Joseph Whitworth, the designer of screw threads. Roberts also added back gears and other improvements to the lathe. The first horizontal milling machine is credited to Eli Whitney in 1818 in New Haven, Connecticut. The development of machine tools that not only could make specific products but could also produce other machines to make other products was fundamental to the industrial revolution.

While early work in machine tools and precision measurement was done in England, the earliest attempts at interchangeable-parts manufacturing apparently occurred almost simultaneously in Europe and the United States with the development of *filing jigs*, with which duplicate parts could be hand filed to substantially identical dimensions. In 1798, Eli Whitney, using this technique, was able to obtain and eventually fulfill a contract from the U.S. government to produce 10,000 army muskets, the parts of each being interchangeable. However, this truly remarkable achievement was accomplished primarily by painstaking handwork, not by specified machines.

Joseph Whitworth, starting about 1830, accelerated the use of Wilkinson's and Maudsley's machine tools by developing precision measuring methods. He developed a measuring machine using a large micrometer screw and later worked toward establishing thread standards and made plug-and-ring gages. His work was valuable because precise methods of measurement were the prerequisite for developing interchangeable parts, a requirement for later mass production.

Perhaps the next significant machine tool developed was the drill press with automatic feed, by John Nasmyth, another student of Maudsley's, in 1840 in Manchester, England. Surface grinding machines came along in England about Maudsley's

751

TABLE 34.1   Yardstick for Automation

Order of Automation	Human Attribute Replaced	Examples
A(0)	*None:* lever, screw, pulley, wedge	Hand tools, manual machine
A(1)	*Energy:* muscles replaced with power	Powered machines and tools, Whitney's milling machine
A(2)	*Dexterity:* self-feeding	Single-cycle automatics; self-feeding
A(3)	*Diligence:* no feedback but repeats cycle automatically	Repeats cycle; **open-loop numerical control** or automatic screw; transfer lines
A(4)	*Judgment:* positional feedback	Closed loop; **numerical control;** self-measuring and adjusting; CAD, CAM, CNC
A(5)	*Evaluation:* adaptive control; deductive	Computer control; model of process required for analysis; feedback from the process analysis and optimization with data from sensors
A(6)	*Learning:* from experience	Limited self-programming; **some artificial intelligence** (A1); **expert systems**
A(7)	*Reasoning:* exhibits intuition; relates causes and effects	Inductive reasoning; advanced A1 in control software
A(8)	*Creativeness:* performs design work unaided	Originality **creates new process programs,** neural networks; fuzzy logic
A(9)	*Dominance:* supermachine, commands others	Machine is master (Hal in *2001, A Space Odyssey*)

Source: G. Amber & P. Amber, *Anatomy of Automation*, Prentice Hall, Englewood Cliffs, NJ, 1962. Used by permission, modified by Black.

1880, and the era was completed with the development of the bandsaw blades that could cut metal. In total, there were eight basic machine tools in the first industrial revolution for machining: lathe, milling machine, drill press, broach, boring mill, planer (shaper), grinder, and saw. The first factories were developed so that power could be added to drive the machines. This is the A(1) level of automation. Early on, water power was used and, later, steam engines and, still later, large electric motors.

A(2) Level Is Single-Cycle Automatics

The A(2) level of automation was clearly delineated when machine tools became single-cycle, self-feeding machines displaying *dexterity*. Many examples of this level of machine are given in this text. They exist in great numbers in many factories today as milling, drilling, and turning machines. The A(2) level of machine can be loaded with a part and the cycle is initiated by the worker. The machine completes the processing cycle and stops automatically. **Mechanization** refers to the first and second orders of automaticity, which includes semiautomatic machines. Virtually all of the machine tools described in previous chapters are A(2) machines. The machines used in manned manufacturing cells used in lean manufacturing are basically A(2) or A(3) machines. Some examples of A(2) machines are shown in Figure 34.1. Most of the machines found in the home (kitchen) are A(1) or A(2) level machines.

A(3) Is Repeat-Cycle Automatics

The A(3) level of automation requires that the machine be *diligent*, or repeat cycles automatically. These machines are open loop, meaning that they do not have *feedback* and are controlled by either an internal fixed program, such as a cam, or are externally programmed with a tape, **programmable logic**

controller (PLC), or computer. Figure 34.1 provides some examples of A(3) machines that include transfer lines. **Figures 34.2** and **34.3** show typical linear, or in-line, and rotary transfer lines. The automatic ice cube maker in a refrigerator is an example of the A(3) level.

When people talk about mass production, they are often referring to a *transfer line*, which is really an automated flow line. Workpieces are automatically transferred from station to station, from one machine to another. Operations are performed sequentially. Ideally, workstations perform the operation(s) simultaneously on separate workpieces, with the number of parts equal to the number of stations. Each time the machine cycles, a part is completed.

Transferring is usually accomplished by one or more of four methods. Frequently, the work is pulled along supporting rails by means of an endless chain that moves intermittently as required. Alternately, the work is pushed along on continuous rails by air or hydraulic pistons. A third method, restricted to lighter workpieces, is to move them by an overhead chain conveyor that may lift and deposit the work at the machining stations.

A fourth method is often employed when a relatively small number of operations, usually 3 to 10, are to be performed. The machining heads are arranged radially around a rotary indexing table, which contains fixtures in which the workpieces are mounted (Figure 34.3). The table movement may be continuous or intermittent. Face-milling operations are sometimes performed by moving the workpieces past one or more vertical axis heads at the required feed rate. Such circular configurations have the advantage of being compact, permitting the two pieces to be loaded and unloaded at a single station without having to interrupt the machining.

Means must be provided for positioning (locating) the workpieces correctly as they are transferred to the various stations. One method is to attach the work to carrier pallets or

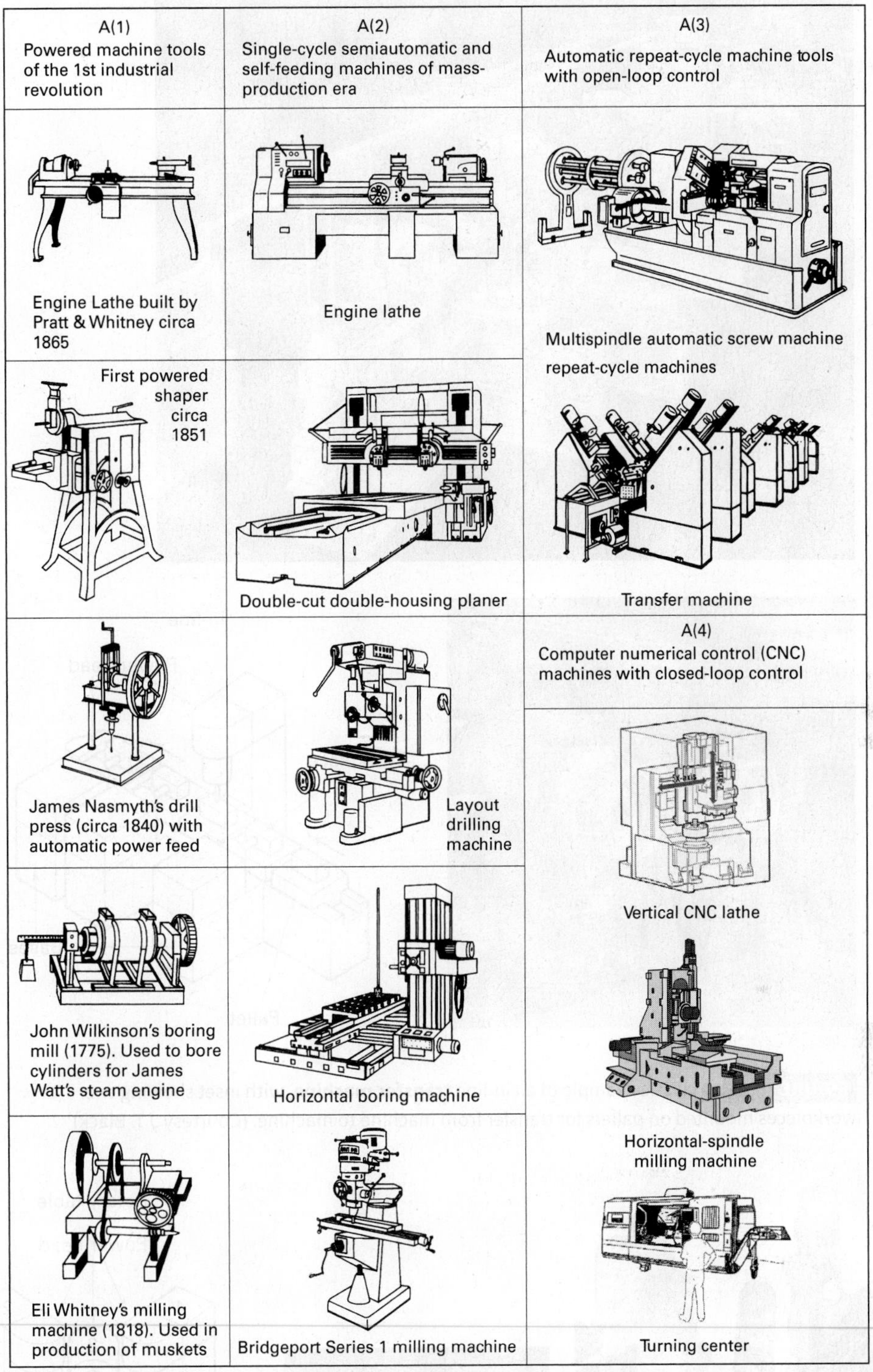

A(1) Powered machine tools of the 1st industrial revolution	A(2) Single-cycle semiautomatic and self-feeding machines of mass-production era	A(3) Automatic repeat-cycle machine tools with open-loop control

FIGURE 34.1 Machine tools of the first industrial revolution A(1) and the mass production era A(2), with examples of A(3) open-loop level and A(4) closed-loop level.

fixtures that contain locating holes or points that mate with retracting pins or fingers at each workstation. Excellent station-to-station precision is obtained as the fixtures thus are located and then clamped in the proper positions. However, a set of carrier fixtures is costly and each one must be identical. When it is possible, they are eliminated, and the workpiece is transported between the machines on rails, which locate the parts by self-contained holes or surfaces. This procedure eliminates

the pallets and saves the labor required for fastening the work-pieces to the carrier pallets.

In the design of large transfer machines, the matter of the geometric arrangement of the various production units must be considered. Whether transfer fixtures or pallets must be used is an important factor. These fixtures and pallets are usually quite heavy. Consequently, when they are used, a closed rectangular arrangement is often employed so that the fixtures

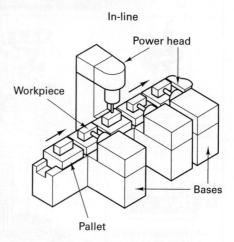

In-line

Power head

Workpiece

Bases

Pallet

FIGURE 34.2 A classic example of an in-line **transfer machine**, with inset showing two workpieces mounted on pallets for transfer from machine to machine. (Courtesy J T. Black)

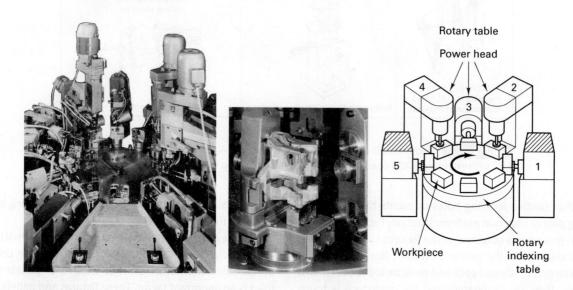

Rotary table

Power head

Workpiece

Rotary indexing table

FIGURE 34.3 Rotary transfer lines, also called dial-index machines, typically come in 6–10 stations. The power heads can be vertical or horizontal and can hold multiple tools. The workpiece on the right is mounted in a specially designed fixture (1 of 10, which must be identical). (Courtesy J T. Black)

are returned to the loading point automatically. If no fixtures or pallets are required, straight-line configurations can be employed. Whether pallets or fixtures must be used depends primarily on the degree of precision needed as well as on the size, rigidity, and design of the workpieces. If no transfer pallet or fixture is to be used, locating bosses or points must be designed or machined into the workpiece.

Cutting tool wear and replacement (of worn tools) is a significant problem when a large number of operations are incorporated in a single production unit. Tools must be replaced before they become worn and produce defective parts. Transfer machines often have more than 100 cutting tools. If the entire complex machine had to be shut down each time a single tool became dull and had to be replaced, overall productivity would be very low. This is avoided by designing the tooling so that certain groups have similar tool lives, monitoring tool thrust and torque, and shutting down the machine before the tooling has deteriorated. All the tools in the affected group can be changed so that repeated shutdowns are not necessary. If the transfer machine is equipped with AC drives, providing diagnostic feedback information to individual processes is easily accomplished. Programmable drives replace feedboxes, limit switches, and hydraulic cylinders and eliminate changing belts, pulleys, or gears to change feed rates and depths of cut, thus making the system much more flexible.

Methods have been developed for accurately presetting tools and for changing them rapidly. Tools can be changed in a few minutes, thereby reducing machine downtime. Increasingly, tools are preset in standard quick-change holders with excellent accuracy, often to within 0.0002 in. (0.005 mm).

In large transfer lines, to prevent entire machines from being shut down when one or two stations become inoperative, the individual machines are grouped in sections with 10 to 12 stations per section. A small amount of buffer storage (*banking* of workpieces) is provided between the sections. This permits production to continue on all remaining sections for a short time while one section is shut down for tool changing or repair.

The most significant problem involves designing the line itself so that it operates efficiently as a whole. Transfer machines or, for that matter, any system wherein a number of processes are connected sequentially will require that the line be *balanced*. *Line balancing* means that the processing time at each station must be the same, with the total nonproductive time for all other stations minimized. Practically speaking, there will be one station that will have the longest time, and this station will control the cycle time for all the stations. Computer algorithms have been developed to deal with line balancing, which, when the line is very large, becomes a rather complex problem. Ford Motor has a single transfer line comprising 15 transfer machines, two assembly machines, part washers, gages, and inspection equipment. One of the transfer machines has 30 stations just for performing rough milling on cylinder heads. The machines perform a variety of drilling, milling, rough and finish boring, reaming, and end-milling operations. This line produces over 200,000 engine blocks and cylinder heads per year (100 cylinder heads per hour).

Transfer machines are usually A(3)-level machines but can be A(4)- or even A(5)-level machines, depending on whether they have the built-in capacity for sensing when corrective action is required and how such corrections are made. Sensing and feedback control systems are essential requirements for the fourth level and all higher levels of automation.

A(3), A(4), and A(5) levels are basically superimposed on A(2)-level machines, which must be A(1) by definition. The A(3) level also includes robots, numerical control (NC) machines that have no feedback, and many special-purpose machine tools.

Thus, automation as we know it today begins with the A(3) level. In recent years, this level has taken on two forms: *hard* (or *fixed-position*) *automation* and *soft* (or *flexible* or *programmable*) *automation* (see **Table 34.2**). Instructions to the machine, telling it what to do, how to do it, and when to do it, are called the *program*. In hard-automation transfer lines, the programming consists of cams, stops, slides, and hard-wired electronic circuits using relay logic. A widely used example of the A(3) level of automation is the automatic screw machine. If the machine is programmed with a tape, a programmable controller (PC), a handheld control box, a microprocessor, or a computer, control instructions are easily changed, with the software making the system or device much more adaptable.

In order to improve their reliability, transfer machines are equipped with automated inspection stations or probing heads that determine whether the operation was performed correctly and detect whether any tool breakage has occurred that might cause damage in subsequent operations. For example, after drilling, a hole is checked to make certain that it is clear prior to tapping.

Automation and transfer principles are also used very successfully for assembly operations. In addition to saving labor, automatic testing and inspection can be incorporated into such machines at as many points as desired. Such in-process

TABLE 34.2 **Fixed Automation vs. Flexible Automation**

Hard-wired Automation	Programmable Automation
• Machine control using mechanical and electrical components hardwired to perform a single function, or multiple functions on a part.	• Software-driven manufacturing systems are typically multipurpose systems designed to produce a family of parts. Soft-wired machines are standard machine tools selected for their ability to fabricate a required product. The basis for soft-wired machining is the *numerically controlled* (NC) machine.
• Advantages • Each machine is specifically designed to perform a specific task. Therefore, the task can be optimized in the design. • Location and allocation of process needs can be optimized.	• NC can be defined as "a system in which movements are controlled by a program containing numerical data and may employ feedback to compare the desired action to the actual movement." See Chapter 26 for details.
• Disadvantages • High setup cost • Low flexibility	

inspection should be used to prevent defects from being made rather than finding defects after they are made. This ensures superior quality. When defective assemblies are simply discovered and removed for rework or scrapped, the cause of the problem is not necessarily corrected.

In many cases, some manual operations are combined with some automatic operations. For example, one transfer machine for assembling steering knuckle, front-wheel hub, and disk-brake assemblies has 16 automatic and five manual workstations. As with manufacturing operations, automatic assembly can often be greatly improved through proper part design. This is called design for assembly.

34.2 The A(4) Level of Automation

A(4) Level Has Feedback

The A(4) level of automation required that *human judgment* be replaced by a capability in the machine to measure and compare results with desired position or size and adjustments to minimize errors. This is *feedback*, or closed-loop, control. The first numerically controlled machine tool with positional feedback control was developed in 1952 and is generally recognized as the first A(4) machine tool. By 1958, the first NC **machining center** was being marketed by Kearney and Trecker. A machining center was a compilation of many machine tools capable of performing many processes: in this case, milling, drilling, tapping, and boring (see **Figure 34.4**). The machining center can automatically change tools to give it greater flexibility. Almost from the start, computers were needed to help program these machines. Within 10 years, NC machine tools became **computer numerical control (CNC)** machine tools with an onboard microprocessor and could be programmed directly. See Chapter 26 for further discussion of CNC.

In the concept of feedback control, a simple A(4) level is one in which some aspect (usually position) of the process is measured using a detection device (sensor). This information is fed back to an electronic comparator, housed in the **machine control unit (MCU)**, which makes comparisons with the desired level of operation. If the output and input are not equal, an error signal is generated and the table is adjusted to reduce the error.

Figure 34.5 shows the difference between an open-loop A(3) machine and a closed-loop A(4) machine, with feedback provided on the location of the table with respect to the axis of the spindle of the cutting tool. Three position control methods are commonly used:

1. Transducer on table itself
2. Transducer (encoder) on the motor, as shown in Figure 34.5
3. Transducer (encoder) on the drive motor.

In CNC turning machines, the feedback is on the tool tip with respect to the rotating part creating tool paths. **Figure 34.6** shows how a part can be turned (machined) from a round bar in a CNC lathe. A program is written that directs the machine to execute the necessary roughing and finishing passes.

$$\text{number of rough passes} = \frac{\text{stock diameter} - \text{minimum diameter} + \text{finish}}{\text{depth of cut} \times 2}$$

In this case, eight roughing passes and one finishing cut were specified to get the part to size.

A Brief History of Numerical Control

The advent and wide-scale adoption of numerically (tape- and computer-) controlled machine tools has been the most significant development in machine tools in the past 50 years. These machines raised automation to a new level, A(4), by providing positional feedback as well as programmable flexibility. Numerical control of machine tools created entirely new concepts in manufacturing. Certain operations are now routine that previously were very difficult if not impossible to accomplish.

In earlier years, highly trained NC programmers were required. The development of low-cost, solid-state microprocessing chips has resulted in machines that can be quickly programmed by skilled machinists after only a few hours of training, using only simple machine shop language. As a consequence, there are few manufacturing facilities today, from the largest factories to the smallest job shops, that do not have one or more numerically controlled machine tools in routine use.

NC came into being to fill a need. The U.S. Air Force (USAF) and the airframe industry were seeking a means to manufacture complex contoured aircraft components to close tolerances on a highly repeatable basis. John Parsons of the Parsons Corporation of Traverse, Michigan, had been working on a project for developing equipment that would machine templates to be used for inspecting helicopter blades. He conceived of a machine controlled by numerical data to make these templates and took his proposal to the USAF. Parsons convinced the USAF to fund the development of a machine. The Massachusetts Institute of Technology (MIT) was subcontracted to build a prototype machine. The prototype was a conventional two-axis tracer mill retrofitted with servomechanisms. As luck would have it, the servomechanism lab was located next to a lab where one of the very first digital computers (Whirl-wind) was being developed. This computer generated the digital numerical data for the servomechanisms.

By 1962, NC machines accounted for about 10% of total dollar shipments in machine tools. Today, about three-fourths of the $35 billion spent for machine tools (drill presses, milling machines, lathes, and machining centers) goes for CNC equipment.

Early on, NC machines were continuous-path or contouring machines where the entire path of the tool was controlled with close accuracy in regard to position and velocity. Today,

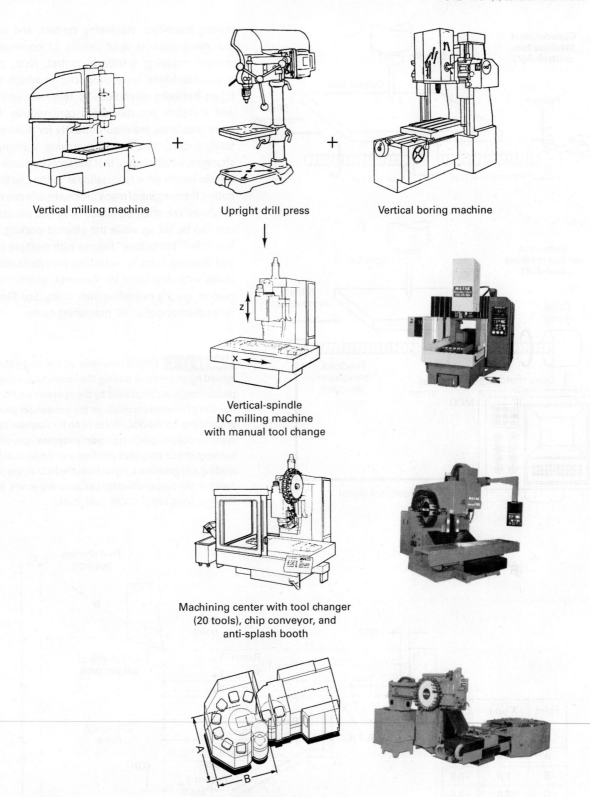

Vertical milling machine

Upright drill press

Vertical boring machine

Vertical-spindle
NC milling machine
with manual tool change

Machining center with tool changer
(20 tools), chip conveyor, and
anti-splash booth

Machining center with pallet changer

FIGURE 34.4 The NC machine combined the capability of a milling machine, a drilling machine, and a boring machine into one machine with programmable control of the movement of the work with respect to the tool. It became a machining center when a tool changer was added. Later, a pallet changer was added. (Courtesy JT. Black)

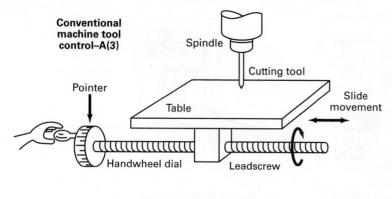

Conventional machine tool control–A(3)

Spindle

Cutting tool

Pointer

Table

Slide movement

Handwheel dial

Leadscrew

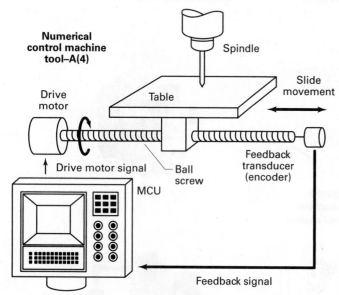

Numerical control machine tool–A(4)

Spindle

Drive motor

Table

Slide movement

Drive motor signal

Ball screw

Feedback transducer (encoder)

MCU

Feedback signal

milling machines, machining centers, and lathes are the most popular applications of continuous-path control requiring feedback control. Next, **point-to-point machines** were produced in which the path taken between operations is relatively unimportant and therefore not monitored continuously. Point-to-point machines are used primarily for drilling, milling straight cuts, cutoff, and punching. Automatic tool changers, which require that the tools be precisely set to a given length prior to installation in the machines, permitted the merging of many processes into one machine. Machines are often equipped with two pallets so that one can be set up while the other is working. Two- or four-sided "tombstone" fixtures with multiple mounting and locating holes for attaching part-dedicated fixture plates were developed for horizontal-spindle machining centers, greatly extending their utility. See **Figure 34.7** for a schematic of a CNC machining center.

FIGURE 34.5 (Top) A conventional machine's slide is moved by an operator turning the handwheel. Accurate positioning is accomplished by the operator counting the number of revolutions made on the handwheel plus the graduations on the dial. (Bottom) An NC machine takes the commanded position from a **part program**—any difference between the commanded position and the feedback signal reading will generate a signal from the MCU to run the drive motor in the proper direction to cancel any errors. (Modern Machine Shop 1991 NC/CIM Guidebook).

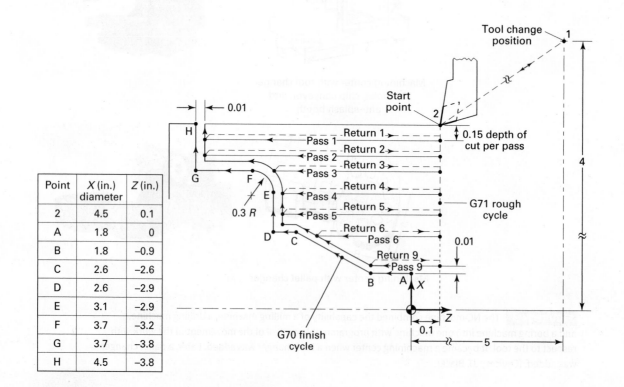

Point	X (in.) diameter	Z (in.)
2	4.5	0.1
A	1.8	0
B	1.8	−0.9
C	2.6	−2.6
D	2.6	−2.9
E	3.1	−2.9
F	3.7	−3.2
G	3.7	−3.8
H	4.5	−3.8

FIGURE 34.6 The tool paths necessary to rough and finish turn a part in a CNC lathe are computer generated.

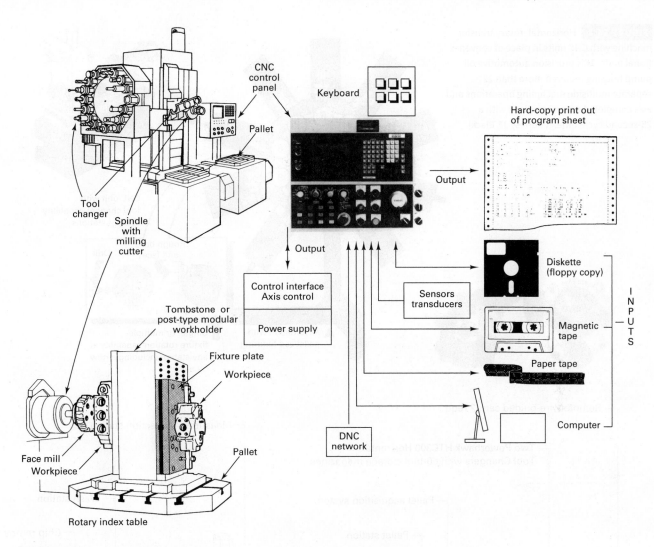

FIGURE 34.7 Horizontal-spindle CNC machining center with four axes of control (*x, y, z, r* table), tool changer, and multiple pallets. It receives inputs to the control panel from many sources.

Today CNC machines are very common throughout the manufacturing world as the technology has been applied to machines for sheet metal cutting and forming using lasers, wire electrical discharge machines (EDM), and plasma arc (flame) cutting, and additive manufacturing, as well as metal and wood machining. It is hard to imagine a manufacturing world without CNC.

Flexible Manufacturing Systems (FMSs)

In the 1960s, companies began combining the repeatability and productivity of the transfer line with the programmable flexibility of the NC machine so that a variety of parts could be produced on the same set of machines.

Figure 34.8 shows a dial-indexing transfer machine that has CNC machines replacing slide units. The CNC units have tool changers, multiaxis fixture positioning, palletized automation, in-process gaging, size control, fault diagnostics, and excellent process capability, so this machine has A(4) capability.

In the United States, the first systems were called variable mission or **flexible manufacturing systems (FMSs)**. In the late 1960s, Sundstrand installed a system for machining aircraft speed drive housings that was used for more than 30 years. Overall, however, very few of these systems were sold until the late 1970s and early 1980s, when a worldwide FMS movement began. But even today international trade in FMS is not significant, and there are fewer than 2000 systems in the world (less than 0.1% of the machine tool population). There is also some evidence that the market for these large, expensive systems became saturated around the mid-1980s.

The FMS permits (schedules) the products to take random paths through the machines. This system is fundamentally an automated, conveyorized, computerized job shop making the system complex to schedule. Because the machining time for different parts varies greatly, the FMS is difficult to link to an integrated system and often remains an island of very expensive automation.

About 60% to 70% of FMS implementations are for components consisting of nonrotational (prismatic) parts such as crankcases and transmission housings. **Figure 34.9** shows an

FIGURE 34.8 Horizontal, rotary transfer machine with CNC units in place of conventional units. This precision automotive oil pump housing required more than 22 separate multiside machining operations and was completed on nine stations with a 25-second cycle time. (Courtesy JT. Black)

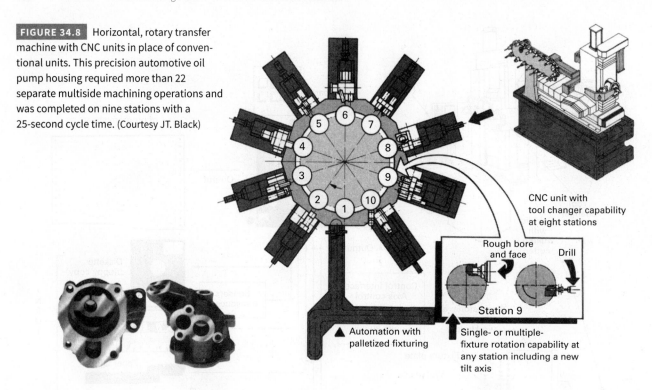

CNC unit with tool changer capability at eight stations

Rough bore and face

Drill

Station 9

▲ Automation with palletized fixturing

Single- or multiple-fixture rotation capability at any station including a new tilt axis

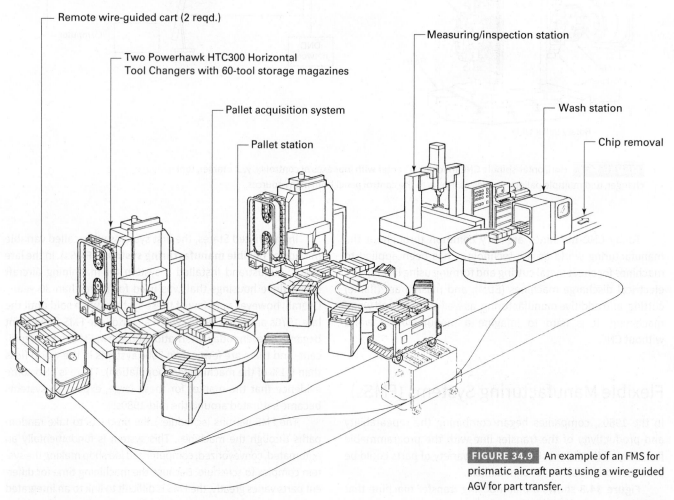

Remote wire-guided cart (2 reqd.)

Two Powerhawk HTC300 Horizontal Tool Changers with 60-tool storage magazines

Pallet acquisition system

Pallet station

Measuring/inspection station

Wash station

Chip removal

FIGURE 34.9 An example of an FMS for prismatic aircraft parts using a wire-guided AGV for part transfer.

Application:

An aircraft parts manufacturer needed parts transfer mobility, in/out parts queue, cutting tool library, and quality control management for production of high-technology parts.

Wire-guided vehicles offer interdepartment transfer capability as well as in-cell transport. The Q.C. center manages the machining accuracy for continous flow of acceptable parts. Parts are scheduled in batch and/or random, controlled by a management computer.

The machines are equipped with telemetry probes, adaptive control, bulk tool storage, and completed tool management.

FMS with two machine tools serviced by a pallet system and an **automated guided vehicle (AGV)**. The balance of FMS installations are for rotational parts or a mixture of both types of parts.

The number of machine tools in an FMS varies from 2 to 10, with 3 or 4 being typical. Annual production volumes for the systems are usually in the range of 3000 to 10,000 parts, the number of different parts ranging from 2 to 20, with 8 being typical. The lot sizes are typically 20 to 100 parts, and the typical part has a machining time of about 30 minutes, with a range of 6–90 minutes per part. Each part typically needs two or three chucking or locating positions and 30 or 40 machining operations. NC machining centers were used on older FMSs. In recent years, CNC machine tools have been favored, leading to a considerable number of systems being operated under direct numerical control (DNC). The machining centers always have tool changers. To overcome the limitation of a single spindle, some systems are being built with head changers. These are sometimes referred to as modular machining centers.

Common features of FMSs (see **Table 34.3**) are **pallet changers**, underfloor conveyor systems for the collection of chips (not shown), and a conveyor system that delivers parts to the machine. This is also an expensive part of the system, as the conveyor systems are either powered rollers, mechanical pallet transfer conveyors, or AGVs operating on underground towlines or buried guidance cables. Carts are more flexible than the conveyors. The AGVs also serve to connect the islands of automation, operating between FMSs, replacing human guided vehicles (forklift trucks).

Pallets are a significant cost item for the FMS because the part must be accurately located on the pallet and the pallet accurately located in the machine. Since many pallets are required for each different component, a lot of pallets are needed; they typically represent anywhere from 15% to 20% of the total system cost. FMSs cost about $1 million per machine tool. Thus, the seven-machine FMS costs $6 million for hardware and software, with the transporter costing over $1 million.

The computer control system for an FMS system usually has three levels. The master control monitors the entire system for tool failures or machine breakdowns, schedules the work, and routes the parts to the appropriate machine. The DNC computer distributes programs to the CNC machines and supervises their operations, selecting the required programs and transmitting them at the appropriate time. It also keeps track of the completion of the cutting programs and sends this information to the master computer. The bottom level of computer control is at the machines themselves.

It is difficult to design an FMS because it is, in fact, a very complex assembly of elements that must work together. Designing the FMS to be flexible is difficult. Many companies have found that between the time they ordered their system and had it installed and operational, design changes had eliminated a number of parts from the FMS. That is, the system was not as flexible as they thought. **Figure 34.10** shows some typical FMS designs. If the system has only one or two machines, it is often called a flexible manufacturing cell (FMC).

FMSs are, in fact, classic examples of supermachines. Such large, expensive systems must be examined with careful and complete planning. It is important to remember that even though they are often marketed and sold as a *turnkey* installation (the buyer pays a lump sum and receives a system that can be turned on and run), this is only rarely possible with a system that has so many elements that must work together reliably. Taken in the context of integrated manufacturing systems, large FMSs may be difficult to synchronize to the rest of the system. The flexibility of the FMS requires variable speeds and cycles, numerical control, and a supervisory computer to coordinate cell operation. In the long run, smaller manned or unmanned cells may well be the better solution, in terms of system flexibility. Perhaps a better name for these systems is *variable mission* or *random-path manufacturing systems*.

The use of computer coding and classification systems, a group technology technique, to identify the initial family of parts around which the FMS is designed greatly improves the FMS design. One might say that FMSs were developed before their time, since they are being more readily accepted since group technology has been used (at least conceptually) to identify families of parts for the system to produce.

As an FMS generally needs about three or four workers per shift to load and unload parts, change tools, and perform general maintenance, it cannot really be said to be self-operating. FMS systems are rarely left untended, as in third-shift operations. Other than the personnel doing the loading and unloading, the workers in the FMS are usually highly skilled and trained in NC and CNC. Most installations run fairly reliably (once they are debugged) over three shifts, with uptime ranging from 70% to 80%, and many are able to run one shift untended.

TABLE 34.3	Common Features of Flexible Manufacturing Systems
Pallet changers	
Multiple machine tools: NC or CNC	
Automated material handling system (to deliver parts to machines)	
Computer control for system: DNC	
Multiple parts: Medium-sized lots (200–10,000) with families of parts	
Random sequencing of parts to machines (optional)	
Automatic tool changing	
In-process inspection	
Parts washing (optional)	
Automated storage and retrieval (optional, to deliver parts to system)	

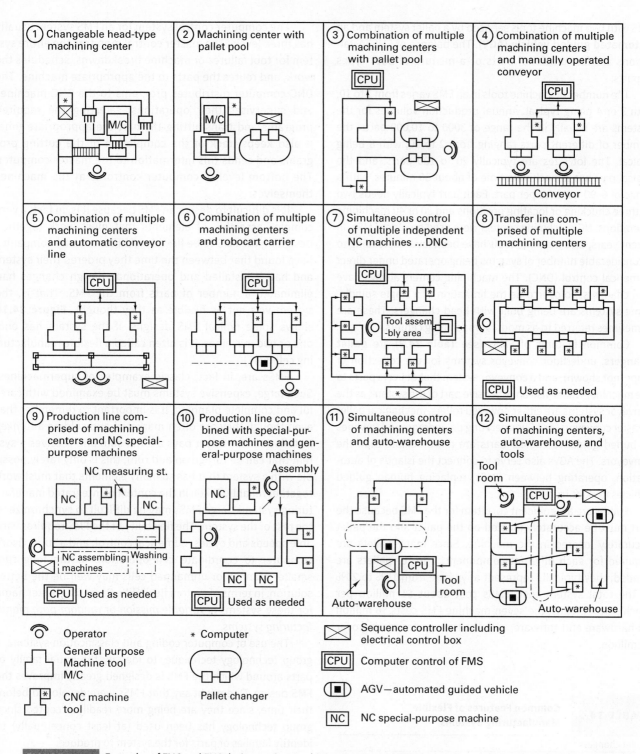

FIGURE 34.10 Examples of FMC and FMS designs, using machining centers and AGVs.

34.3 A(5) Level of Automation Requires Evaluation

In the yardstick for automation, the NC or CNC machine represents the A(4) level. The next level of automation, A(5), requires that the control system perform an evaluation function of the process. **Figure 34.11** compares block diagrams for A(4) closed-loop control (with encoder) to A(5) adaptive control. A(5) requires a feedback loop from the process or the product and seeks to either constrain or optimize the process. Note how this loop lies inside the position feedback loop. The multiple loops can cause conflicts in the controller's decision-making software. CNC machines, with their onboard computers, are capable of the A(5) level of automation. In the standard versions of NC and CNC machines in use today, speed and feed are fixed in the program unless the operator overrides them at the

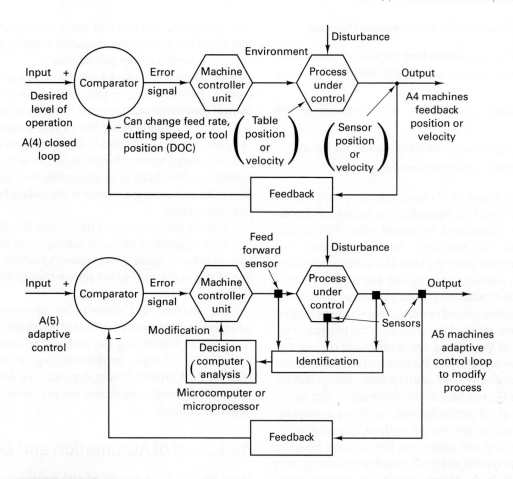

FIGURE 34.11 Block diagrams of A(4) and A(5) level of control: A(4), closed-loop NC; A(5) closed-loop NC with adaptive control loop.

control panel of the machine. If either speed or feed is too high, the result can be rapid tool failure, poor surface quality, or damaged parts. If the speed and feed are too low, production time is greater than desired for best productivity. An **adaptive control (A/C)** system that can sense deflection, force, heat, and changes in geometry or torque will use these measurements to make decisions about how the input parameters might be altered to *constrain* or *optimize* the behavior of the process. This means that the computer must have *mathematical models* in its software that describe how this process behaves and mathematical functions that state what is to be constrained or optimized (surface quality, cutting force, metal removal rate, power consumed, and so on).

The A(5) level requires that machines perform *evaluation* of the process itself. Thus, the machine must be cognizant of the multiple factors on which the process performance is predicted, evaluate the current setting of the input parameters versus the outputs from the process, and then determine how to alter the inputs to optimize the process. A(5) level, typified by *adaptive control*, emulates human evaluation.

In summary then, A(5) machines are capable of adapting the process itself so as to optimize it in some way, so this level of automation requires that the system have a computer pro-

grammed with models (mathematical equations) that describe how this process behaves, how this behavior is bounded, and what aspects of the process or system are to be constrained or optimized. *This modeling obviously requires that the process be sufficiently well understood theoretically so that equations (models) can be written that describe how the real process behaves.*

This level of automation has been achieved for continuous processes (e.g., oil refineries) where the theory (of heat transfer and fluid dynamics) of the process is well understood and parameters are easy to measure. Unfortunately, the theory of metal forming and metal removal is less well understood and parameters are usually difficult to measure. These processes have resisted adequate theoretical modeling, and as a consequence there are few A(5) machines on the shop floor. A/C applications are important where the components vary in size (like castings), which alters the depth of cut and changes the cutting forces, perhaps producing chatter. Similarly, changes in the hardness of the workpiece, perhaps produced by improper heat treating, prior nonuniform rolling, or surface processing, can produce significant variations in the cutting forces, again leading to dynamic instability or quality problems.

The basic elements of the adaptive control loop are:

1. *Identification*: measurements from the process itself, its output, or process inputs using sensors to determine operating conditions

2. *Decision analysis*: optimization of the process in the computer

3. *Modification*: signal to the **controller** to alter the inputs

4. *Monitoring*: continuous control over the process.

As shown in Figure 34.11, many machines have built-in **feedforward** devices. This means that the system takes information from the input side of the process rather than the output side and uses that information to alter the process. For example, the temperature of the billet as it enters a hot-rolling or hot-extrusion process can be sensed and used as feedforward information to alter the process parameters. This would be an A(5) or adaptive control example. Sensors can be located in three positions: ahead of the process, in the process, or on the output side of the process. The feedforward concept can also be applied at the A(4) level. Suppose that you have a transfer line that processes flywheel housings from castings that are similar in shape but will have slight differences in size, which will alter depth of cut during different machining operations. When the housings are fed into the machine, a sensing device contacts the casting and determines the variation from the nominal size. The sensing and feedforward system then selects the proper tooling for the housing and adjusts cutting parameters accordingly.

Here is an example of what is required to raise the automatic CNC grinder to the A(5) level, assuming that it is a cylindrical center-type grinder on which part deflection and part size are measured (see **Figure 34.12**). The position of the grinding wheel with respect to the part is controlled numerically in X–Y coordinates. However, the cutting forces tend to deflect the part more as the grinding wheel gets farther from the centers. The adaptive control software program would have equations that relate deflection to grinding forces and infeed rates. The infeed would be decreased to reduce the force and minimize deflection. Notice that the overall system would still have to compensate for the grit dulling and grit attrition that accompanies wheel wear and that also alters the grinding forces. The A(5) level reflects deductive reasoning whereby particular outcomes are predicted from general theory.

A/C systems that place a constraint on a process variable (such as forces, torque, or temperature) are called **adaptive control constraint (ACC)** systems. Thus, if the thrust force and the cutting force (and hence the torque) increase excessively (e.g., because of the presence of a hard region in a casting), the adaptive control system changes the depth of cut or the feed to lower the cutting force to an acceptable level. Note that altering the feed may cause a change in the surface finish that may be unacceptable.

Without adaptive control (or without the direct intervention of the operator), excessive cutting forces may cause the tools to break or cause the workpiece to deflect, resulting in a loss of size in the part, so the system tries to maintain a constant force.

A/C systems that optimize an operation are called **adaptive control optimization (ACO)** systems. Optimization may involve maximizing the material removal rate (MRR) between tool changes (or resharpening) or improving surface finish. Currently, most systems are based on ACC, because the development and the proper implementation of ACO is complex.

A(6) Level of Automation and Beyond

There are very few examples of A(5) machines on the factory floor and fewer at the A(6) level, wherein the machine control has expert systems capability. *Expert systems* try to infuse the software with the deductive decision-making capability of the human brain by having the system get smarter *through experience*. The software is designed to emulate human learning by configuring *neural networks* where the computer relates inputs to outputs. *Artificial intelligence* (AI) carries this step higher by *infecting* the control software with programs that exhibit the ability to reason inductively.

The A(6) level tries to *relate cause to effect*. Suppose that the effect was tool failure. At the A(5) level, the system would detect the increase in deflection due to increased forces (due to the tool's dulling) and reduce the feed to reduce the force.

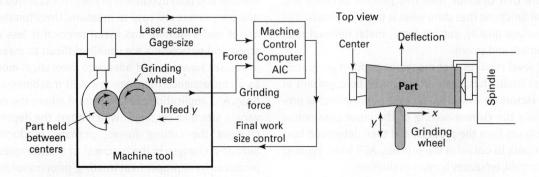

Developing adaptive control systems for metalcutting machines usually requires that the part be measured as well as the cutting forces.

However, it might simultaneously try to increase the speed to maintain the MRR constant, which would increase the rate of tool wear. This was not the desired result. At the A(6) level, multiple factors are evaluated so that the system can recognize the need to change the tool rather than reduce the cutting feed. That is, the system learned by experience that feed reduction or speed increase was the wrong decision, and that it was tool wear that caused the increase in cutting forces. Such systems are often called *expert systems* when the software contains the collective experience of human experts or prior replications of the same process. The use of neural networks will allow the process to learn (i.e., the expert system will become more *intelligent* about the process).

An alternative approach is being evaluated using *fuzzy logic* control. The behavior of the process is described with linguistic terms. Fuzzy logic control statements replace closed-form mathematical models. The decision analysis element depends on a set of if–then statements rather than specific equations.

The A(6) level reflects the beginnings of artificial intelligence, in which the control software is infected with elements (subroutines) that permit some thinking on the part of the software. Few, if any, such systems exist on the factory floor. **Figure 34.13** presents a summary of the A(3) through A(6) levels of automation.

The A(7) level reflects the next level of AI, whereby inductive reasoning is used. The system software can determine a general principle (the theory) based on the particular facts (the database) collected. Levels A(6) and A(7) are the subjects of intensive worldwide research efforts. Levels A(8) and A(9) are left to your imagination.

Automation involves machines, or integrated groups of machines, that automatically perform required machining, forming, assembly, handling, and inspection operations. Through sensing and feedback devices, these systems automatically make necessary corrective adjustments. That is, *human thinking must be automated*. There are relatively few completely automated systems, but there are numerous examples

ORDER OF AUTOMATICITY	HUMAN ATTRIBUTE MECHANIZED	DISCUSSION	EXAMPLES
A (3) Automatic repeat cycle or Open loop control	DILIGENCE Carries out routine instructions without aid of man Open end or nonfeedback	All automatic machines Loads, processes, unloads, repeats System assumed to be doing okay; probability of malfunctions negligible Obeys fixed internal commands or external program	• Record player with changer • Automatic screw machine • Bottling machines • Clock works • Doughnut maker • Spot welder • Engine production lines • Casting lines • Newspaper printing machines • Transfer machines
A (4) Self-measuring and adjusting feedback or closed loop systems	JUDGMENT Measures and compares result (output) to desired size or position (input) and adjusts to minimize any error	Self-adjusting devices Feedback from product position, size, velocity, etc. 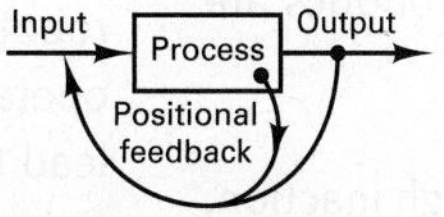 Input → Process → Output Positional feedback Multiple loops are possible	• Product control • Can filling • NC machine tool with position control • Self-adjusting grinders • Windmills • Thermostats • Waterclock • Fly ball governor on steam engine
A (5) Adaptive or computer control: Automatic cognition; Fuzzy logic control	EVALUATION Senses multiple factors on process performance, evaluates and reconciles them Uses mathematical algorithms or fuzzy logic	Process performance must be expressed as equation 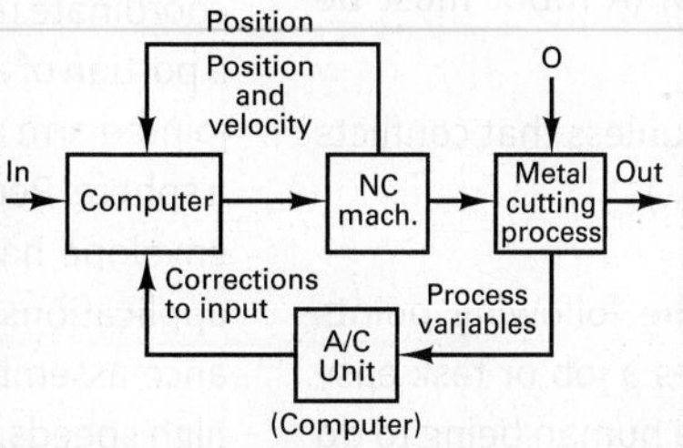 Position Position and velocity O In → Computer → NC mach. → Metal cutting process → Out Corrections to input Process variables A/C Unit (Computer)	CNC machine with A/C capability Maintaining pH level Turbine fuel control Maintaining constant cutting force
A (6) Expert systems or neural networks or limited self-programming	LEARNING BY EXPERIENCE	The subroutines in the AC computer are using a form of limited self-programming Trial-and-error sequencing develops history of usage	Phone circuits Elevator dispatching

FIGURE 34.13 Yardstick for automation: Levels A(3) to A(6) are compared.

of highly mechanized machines. While the potential advantages of a completely automated plant are tremendous, in practice, step-by-step automation of individual operations is required. Therefore, it is important to have a piecewise plan to first convert the classical job shops to a simplified, integrated factory that can be automated step by step. Perhaps the most serious limitation in automation is available capital, as the initial investments for automation equipment and installations are large. Because proper engineering economics analysis must be employed to evaluate these investments, students who anticipate a career in manufacturing should consider a course in this area, a firm requirement.

34.4 Industrial Robotics

Robots, or steel-collared workers, are typically A(3)-, A(4)-, or A(5)-level machines. As defined by the Robot Institute of America, "A *robot* is a reprogrammable, multifunctional manipulator designed to handle material, parts, tools, or specialized devices through variable programmed motions for the performance of a variety of tasks." The word *robot* was coined in 1921 by Karel Capek in his play *R.U.R.* (Rossum's Universal Robots). The term is derived from the Czech word for worker. The principal inventor of the underlying control technology for robots was George Devol, who worked for Remington-Rand in the 1950s. His patents were purchased by CONDEC, Inc., which developed the first commercial robot, called *Unimate*. Much of the work for the next 20 years was spearheaded by Joseph Engelberger, who came to be called the father of robotics. A famous author, Isaac Asimov, depicted robots in many of his stories and proposed three laws that hold quite well for industrial robotic applications. Asimov's three laws of robotics are the following:

1. A robot may not injure a human being or, through inaction, allow a human being to be harmed. (Safety first.)

2. A robot must obey orders given by human beings except when that conflicts with the First Law. (A robot must be programmable.)

3. A robot must protect its own existence unless that conflicts with the First or Second Law. (Reliability.)

In considering the use of a robot, the following points should be considered. Anything that makes a job or task easy for the robot to do makes the job easy for a human being to do as well. Why? The robot is a handicapped worker and is less flexible than a human being. Robots cannot think or solve problems on the plant floor. The big advantage of the robot is that it will do a job in an exact cycle time, and do it every time, whereas a human being often cannot. This is an important feature in robotic assembly lines and manufacturing and assembly cells.

For our purposes, if a machine is programmable, capable of automatic repeat cycles, and can perform manipulations in an industrial environment, it is an industrial robot.

Most industrial robots have the following basic components (**Figure 34.14**):

1. *Manipulators*: The mechanical unit, often called the *arm*, that does the actual work of the robot. It is composed of mechanical linkages and joints with actuators to drive the mechanism directly or indirectly through gears, chains, or ball screws.

2. *End effector*: The *hand* or *gripper* portion of the robot, which attaches the end of the arm and performs the operations of the robot.

3. *Controller*: The brains of the system that direct the movements of the manipulator. In higher-level robots, computers are used for controllers. The functions of the controller are to initiate and terminate motion, store data for position and motion sequence, and interface with the outside world, meaning other machines and human beings. *Teach pendants* are often used to "program" the robot. The operator manipulates the arm and gripper and the robot remembers the taught path.

4. *Feedback devices*: Transducers that sense the positions of various linkages and joints and transmit this information to the controllers in either digital or analog form on A(4)- or A(5)-level robots.

5. *Power supply* (**not shown**): Electric, pneumatic, and hydraulic power supplies provide and regulate the energy needed for their manipulator's actuators.

Early robots were programmed using analog rather than digital control technology. The next generation of robots was more accurate and precise, with reliable electric (digital) controls. Many robot manufacturers made it hard to integrate their robots with other CNC equipment by retaining proprietary operating systems. Today, Japan, China, and the United States lead the world in the implementation of robots, with most of them going into automative or electronics factories.

Most commercially available robots have one of the mechanical configurations shown in **Figure 34.15**. Cylindrical coordinate robots have a work envelope (shaded region), that is, a portion of a sphere. Jointed-spherical coordinate robots have a jointed arm and a work envelope that approximates a portion of a sphere. Rectangular coordinate robots with a rectangular work envelope have been developed for high precision in assembly applications. Another design, called SCARA (selective compliance assembly robot arm), also has good positioning accuracy, high speeds, and low cost for a robot with three or four controllable axes, which are usually adequate for assembly tasks.

Jointed-arm robots may have two or three additional minor axes of motion at the end of the arm (commonly called the *wrist*). These three movements are *pitch* (vertical movement), *yaw* (horizontal motion), and *roll* (wrist rotation). The hand (or *gripper* or *end effector*), which is usually custom-made by the user, attaches to the wrist.

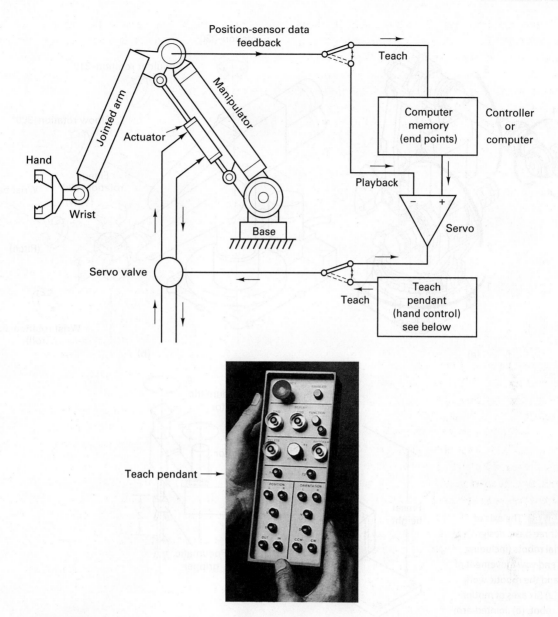

FIGURE 34.14 Major elements of an industrial robot. (Courtesy JT. Black)

Tactile and Visual Sensing in Robots

Many of the industrial robots in use at this time are A(4) point-to-point machines. Most robots operate in systems wherein the items to be handled or processed are placed in precise locations with respect to the robots. Even robots with computer control that can follow a moving auto conveyor line while performing spot-welding operations have point-to-point feedback information, but this is satisfactory for most industrial applications.

To expand the capability of this handicapped worker made of steel, sensors are used to obtain information regarding position and component status. Tactile sensors provide information about force distributions in the joints and in the hand of the robot during manipulations. This information is then used to control movement rates. Visual sensors collect data on spatial dimensions by means of image recording and analysis.

Visual sensors are used to identify workpieces; determine their position and orientation; check position, orientation, geometry, or speed of parts; determine the correct welding path or point; and so on.

To provide a robot with tactile or visual capability, powerful computers and sophisticated software are required, but this is the most logical manner to raise the robot to the A(5) or A(6) levels of automation, at which it can adapt to variations in its environment. Vision systems can locate parts moving past a robot on a conveyor, identify those parts that should be removed from the conveyor, and communicate this information to the robot. The robot tracks the moving part, orients its gripper, picks up the part, and moves it to the desired work station. See **Table 34.4** for a summary on sensors.

Industrial robots used in industry today fill three main functions: material handling, assembly, and material processing. They have, for the most part, very primitive motor and

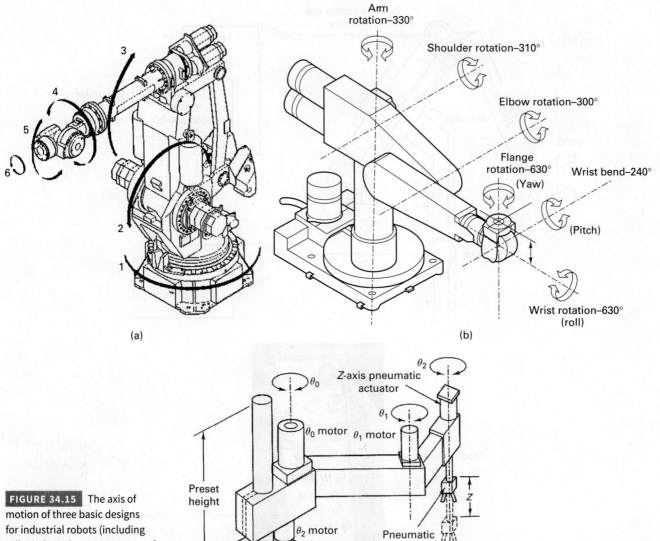

FIGURE 34.15 The axis of motion of three basic designs for industrial robots (including roll, pitch, and yaw movement of the wrist) and the robotic work envelope. (a) Six axes of motion of a Fanuc robot. (b) Jointed-arm robot. (c) Scara robot.

TABLE 34.4 **Sensors Used on Robots with Some Typical Application**

Sensor Type	Design	Application
Visual	• Video pickup tubes (TV camera)	• Position detection and part inspection
	• Position detection and part inspection	• Parts detection, identification, and sorting
	• Semiconductor sensors (lasers)	• Consistency testing (e.g., in manipulation, welding, and assembly)
	• Fiber optics	• Guidance and control
Tactile	• Feelers	• Position detection
	• Pin matrix	• Tool monitoring (e.g., in casting, cleaning, grinding, manipulation, and assembly)
	• Load cells (piezo and capacitance)	
	• Conductive elastometers	• Force sensing and part identification
	• Silicon	• Object recognition and pressure
Electrical (inductive capacitative)	• Shunt (current determination)	• Position detection
	• Capacitor	• Status determination (e.g., in manipulation and welding)
	• Coil manipulation and welding	

intelligence capabilities, because most robots are A(3)-level machines. The sensory interactive control, decision making, and artificial intelligence capabilities of robots are still inferior to those of human beings at this time. With regard to short-comings in performance, the robot's major stumbling blocks are its accuracy and repeatability (i.e., process capability or dexterity), but robot capabilities are progressing steadily, with improvements in controls, ease of programming, operating speeds, and precision.

Robots are having a strong impact in certain industrial environments, designed for doing jobs that are unhealthy, hazardous, extremely tedious, and unpleasant. Robots perform well doing paint spraying, loading and unloading small forgings or die-casting machines, spot welding, and so on.

The A(3)-level robot, usually called a *pick-and-place machine*, is capable of performing only the simplest repeat-cycle movements, on a point-to-point basis, being controlled by an electronic or pneumatic control with manipulatory movement controlled by end stops. A(3) robots are usually small robots with relatively high-speed movements, good repeatability (0.010 in.), and low cost. They are simple to program, operate, and maintain but have limited flexibility in terms of program capacity and positioning capability.

To raise the robot to an A(4)-level machine, sensory devices must be installed in the joints of the arm(s) to provide positional feedback and error signals to the servo-mechanisms, just as was the case for NC machines. The addition of an electronic memory and digital control circuitry allows this level of robot to be programmed by a human being guiding the robot through the desired operations and movements using a hand controller (**Figure 34.16**). The handheld control box has rate-control buttons for each axis of motion of the robot arm. When the arm is in the desired position, the record or program button is pushed to enter that position or operation into the

memory. This is similar to point-to-point NC machines, as the path of the robot arm movement is defined by selected end-points when the program is played back. The electronic memory can usually store multiple programs and randomly access the required one, depending on the job to be done. This allows for a product mix to be handled without stopping to reprogram the machine. The addition of a computer, usually a mini-computer, makes it possible to program the robot to move its hand or gripper in straight lines or other geometric paths between given points, but the robot is still essentially a point-to-point machine. Using computer simulations, robots can be programmed offline to do assembly and processing tasks, but they must usually be taught the final location of the points by the operator.

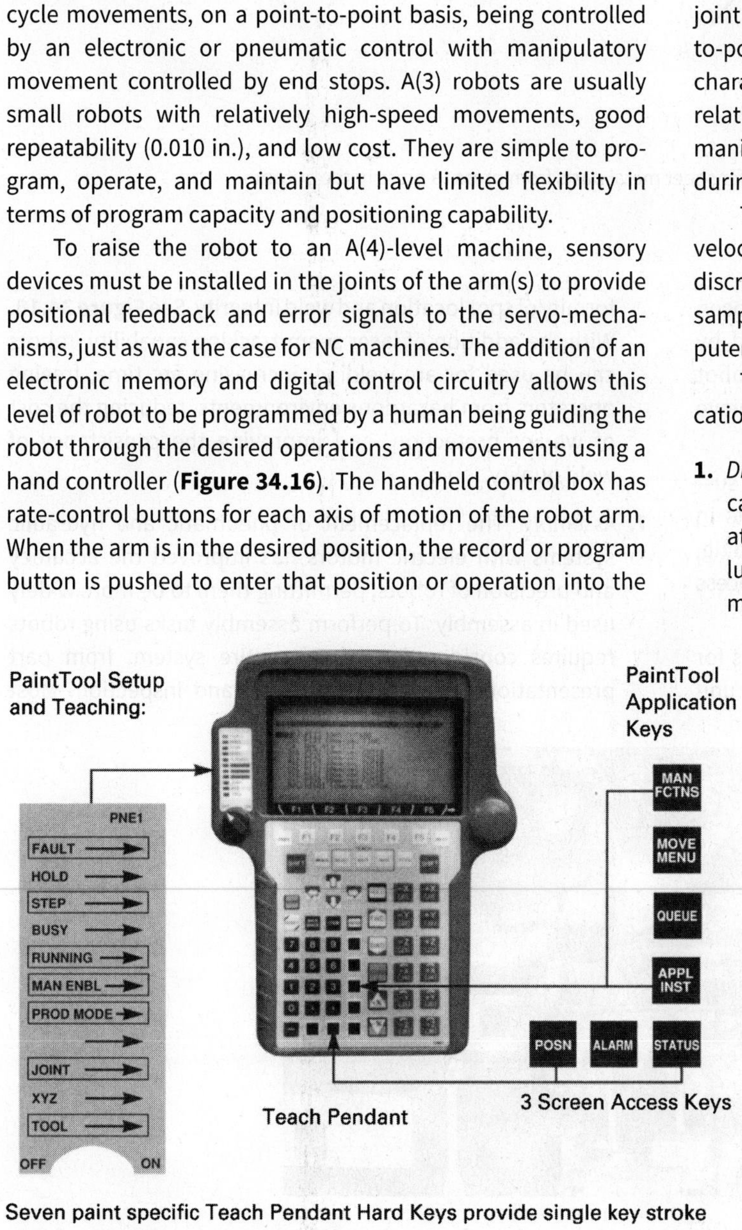

There are three ways of controlling point-to-point motion independently: (1) sequential joint control, (2) noncoordinate joint control, and (3) terminal coordinate joint control. Point-to-point servocontrolled robots have the following common characteristics: high load capacity, large working range, and relatively easy programming, but the path followed by the manipulator during operation may not be the path followed during teaching.

To make an A(4) robot continuous-path, position, and velocity data must be sampled on a time basis rather than as discretely determined points in space. Due to the high rate of sampling, many spatial positions must be stored in the computer memory, thus requiring a mass storage system.

The current generation of industrial robots is finding applications in the following areas:

1. *Die casting.* In single- or multishift operations, custom or captive shops, robots unload machines, quench parts, operate trim presses, load inserts, ladle metal, and perform die lubrication. Die life is increased because die-casting machines can be operated without breaks or shutdowns. Die temperatures remain stable and better controlled, with uniform cycle times.

2. *Press transfer.* Robots in sheet metal press transfer lines guarantee consistent throughput shift after shift. Large and unwieldy parts can be handled at piece rates as high as 400 per hour, with no change in cycle time due to fatigue. Robots are adaptable for long- or short-run operations. Programming for new part sizes can be accomplished in minutes (**Figure 34.17**).

3. *Material handling.* Strength, dexterity, and a versatile memory allow robots to pack goods in complex palletized arrays or to transfer workpieces to and from moving or indexing conveyors or to and from machines. These are boring, labor-intensive operations. Operating costs are reduced when robots feed forge presses and upsetters. They work continuously without fatigue or the need for relief in the hot, hostile environments commonly found in forging factories. Robots can easily manipulate the hot parts in the presses.

PaintTool Setup and Teaching:

PNE1

FAULT
HOLD
STEP
BUSY
RUNNING
MAN ENBL
PROD MODE

JOINT
XYZ
TOOL

OFF   ON

Teach Pendant

PaintTool Application Keys

MAN FCTNS
MOVE MENU
QUEUE
APPL INST

POSN   ALARM   STATUS

3 Screen Access Keys

Seven paint specific Teach Pendant Hard Keys provide single key stroke

FIGURE 34.16 This teach pendant allows the operator to teach the robot paths for painting. (Courtesy of Fanuc Robotics.)

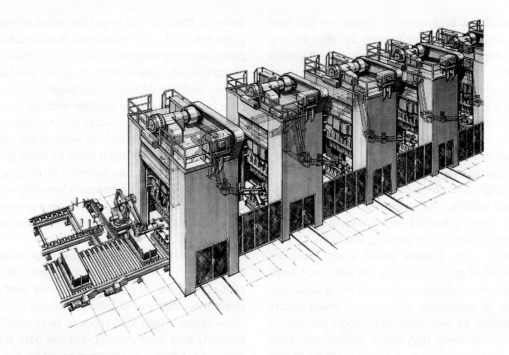

4. *Investment casting.* Scrap rates as high as 85% have been reduced to less than 5% when molds are produced by robots. The smooth, controlled motions of the robot provide consistent mold quality impossible to achieve manually.

5. *Material processing.* Product quality is improved and sustained with point-to-point, or continuous-path robots in jobs such as routing, flame cutting, mold drying, polishing, and grinding. Once programmed, the robot will process each part with the same high quality.

6. *Welding and cutting.* Robots spot weld cars and trucks for almost every major manufacturer in the world, with uni-

formity of spot location and weld integrity. See **Figure 34.18**. With the addition of laser seam-tracking capability, robots can be used for arc welding, increasing arc time, freeing operators from hazardous environments, reducing the cost of worker protection, and improving the consistency of weld quality.

7. *Assembly.* The replacement of pneumatic and hydraulic systems with electric motors has improved the accuracy and precision of robots, permitting them to be more widely used in assembly. To perform assembly tasks using robots requires consideration of the entire system, from part presentation through joining, test, and inspection. Most

FIGURE 34.18 Robots are widely employed for welding auto bodies. (Courtesy J T. Black)

important, the parts must be designed for robotic assembly. Flexible assembly, as this is now referred to, usually addresses mid-volume products. As opposed to hard automation that uses special-purpose fixtures, part feeders, and work heads, flexible (or robotic) assembly uses general-purpose and programmable equipment and combinations of visual and tactile sensing so that a variety of parts can be assembled using the same equipment. This requires multiple degrees of freedom and general-purpose grippers, which usually decreases the accuracy and precision of the robot, so vision systems are needed to compensate. See **Figure 34.19**.

8. *Painting.* The automobile painting process uses robots to apply paint to thicknesses of about 0.1 mm using repeated painting and drying cycles. The painting process is the highest energy user in the entire automobile manufacturing and assembly process. Each car gets about 100 liters of paint. The system, shown in **Figure 34.20**, can paint about 100,000 cars per year at a rate of 30 cars per hour, using an air volume of 860,000 cubic inches. This is a huge volume of air recirculated every 30 minutes. The figure shows how the typical paint booth was redesigned to reduce energy consumption while improving quality, using robots to apply the materials.

The integration of robots into cellular arrangements with machine tools to process families of component parts where the robot performs tasks right along with one or more human beings is very efficient. The robot can perform part loading and unloading, as well as material processing (like joining), when machines are grouped properly in a machining or forming cell.

Totally unmanned cells can facilitate maximum automation and productivity while maintaining programmable flexibility in producing small to medium-sized production lots of parts from compatible parts families. The robots can also change tools in the machines and even the workholding devices, thereby adding more flexibility to the cell. Unmanned manufacturing cells help to achieve maximum machine tool utilization by greatly increasing the percentage of time the machines spend cutting, which in turn increases the output for the same investment.

Here are some things to consider when selecting a robot:

1. Economic analysis: Can this robotic application be cost-justified?

2. Process capability: Can this robot do the job—is it accurate and precise enough?

3. Changing product designs: Can this robot handle new product designs?

4. Changing existing machine and tooling (either cutting tools or workholders): Will this change the product?

5. Doing simple, hazardous, harmful, or fatiguing jobs first: Is this a good application for a robot?

6. Does the supplier have a training program and will the supplier help during initial installation?

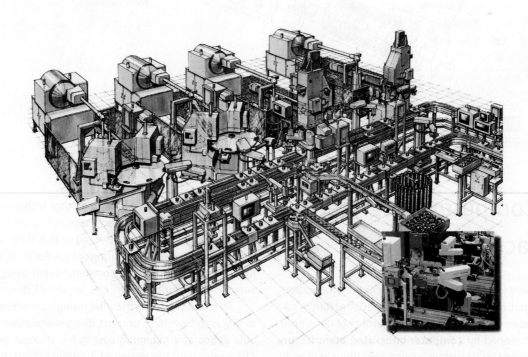

FIGURE 34.19 Robots are used to assemble an airbag inflator, providing safety and process control while improving quality and reducing cost. (Courtesy of Fanuc Robotics)

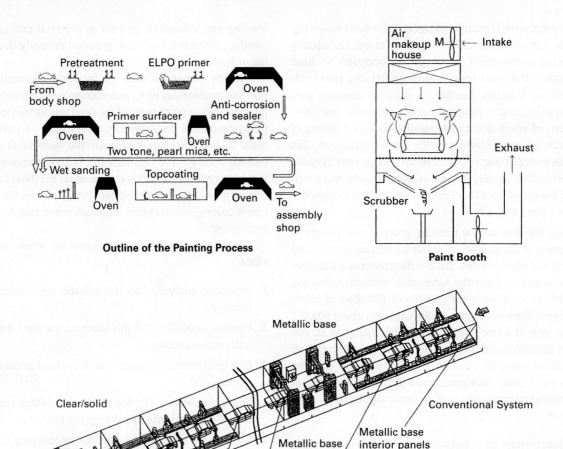

Outline of the Painting Process

Paint Booth

FIGURE 34.20 The painting process for automobiles uses many robots to consistently apply the primer and topcoating to the car bodies. (Source: *Toyota Technical Review,* vol. 48, No. 2, March 1999)

34.5 Computer-Integrated Manufacturing (CIM)

When many people in manufacturing think about automation, they think about computers, and no three-letter acronym has been more widely discussed than CIM. A number of definitions have been developed for **computer-integrated manufacturing (CIM)**. However, a CIM system is commonly thought of as an integrated system that encompasses all the activities in the production system, from the planning and design of a product through the manufacturing system, including control. CIM attempts to combine existing computer technologies to manage and control the entire business. CIM is an approach that very few companies have adopted at this time, since surveys show that few US manufacturing companies have implemented full-scale use of FMS and computer-aided design/computer-aided manufacturing (CAD/CAM), let alone CIM systems.

As with traditional manufacturing approaches, the vision of CIM is to transform product designs and materials into salable goods at a minimum cost in the shortest possible time. CIM begins with the design of a product (CAD) and ends with the computer-aided manufacture of that product (CAM). With CIM, the customary split between the design and manufacturing functions is (supposed to be) eliminated.

CIM differs from the traditional job shop manufacturing system in the role the computer plays in the manufacturing process. CIM systems are basically a network of computer systems tied together by a single integrated database. Here the word *integrated* means everyone can access (via computers) the same database. Using the information in the database, a CIM system can direct (not perform) manufacturing activities, record results, and maintain accurate data. CIM is the computerization of design, process planning (how to make it), manufacturing, distribution, and financial functions into one coherent system.

Figure 34.21 presents a block diagram illustrating the functions and their relationship in CIM. These functions are identical to those found in a traditional production (planning and control) system for a job shop–flow shop manufacturing system. With the introduction of computers, changes have occurred in the organization and execution of production planning and control through the implementation of such systems as material requirements planning (mrp), capacity planning, inventory management, shop floor control, and cost planning and control.

Engineering and manufacturing databases contain all the information needed to fabricate the components and assemble the products. The design engineering and process planning functions provide the inputs for the engineering and manufacturing database. This database includes all the data on the product generated during design, such as part geometric data, parts lists, and material specifications. The bill of materials is shown separately, but it is a key part of the database. **Figure 34.22** shows how CAD/CAM is related to the design and manufacturing activities. Included in the CAM is a CAPP (*computer-aided process planning*) module, which acts as the interface between CAD and CAM.

Capacity planning is concerned with determining what labor and equipment capacity is required to meet the current master production schedule as well as the long-term future production needs of the firm. Capacity planning is typically performed in terms of labor and/or machine hours available. The master schedule is transformed into material and component requirements using mrp. These requirements are then compared with available plant capacity over the planning

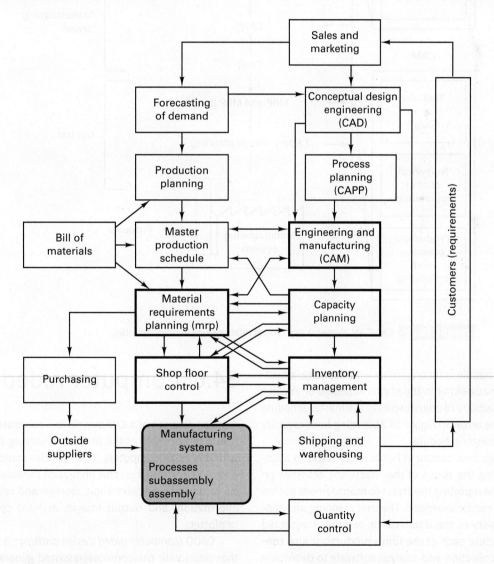

FIGURE 34.21 Cycles of activities in a computer-integrated manufacturing system, in the darker box.

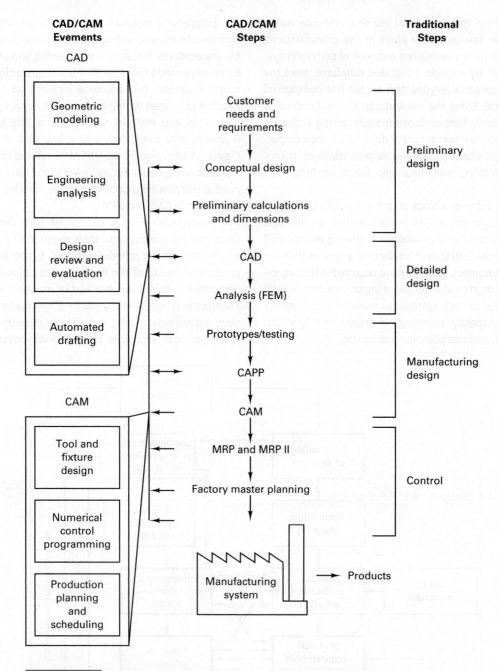

CAD/CAM Evements

CAD

- Geometric modeling
- Engineering analysis
- Design review and evaluation
- Automated drafting

CAM

- Tool and fixture design
- Numerical control programming
- Production planning and scheduling

CAD/CAM Steps

- Customer needs and requirements
- Conceptual design
- Preliminary calculations and dimensions
- CAD
- Analysis (FEM)
- Prototypes/testing
- CAPP
- CAM
- MRP and MRP II
- Factory master planning
- Manufacturing system → Products

Traditional Steps

- Preliminary design
- Detailed design
- Manufacturing design
- Control

FIGURE 34.22 CAD/CAM elements and typical relationships in manufacturing.

horizon. If the schedule is incompatible with capacity, adjustments must be made either in the master schedule or in plant capacity. The possibility of adjustments in the master schedule is indicated by the arrow in Figure 34.21 leading from capacity planning to the master schedule.

The term *shop floor control* in Figure 34.21 refers to a system for monitoring the status of manufacturing activities on the plant floor and reporting the status to management so that effective control can be exercised. The cost planning and control system consists of the database to determine expected costs to manufacture each of the firm's products. It also consists of the cost collection and analysis software to determine what the actual costs of manufacturing are and how these costs compare with the expected costs.

34.6 Computer-Aided Design

A major element of a CIM system is a **computer-aided design (CAD)** system. CAD is the process of solving design problems with the aid of computers. This includes computer generation and modification of graphic images on the video display, analysis of design data, electronic storage and retrieval of design information, and output images as hard copy to a printer or plotter.

CADD (*computer-aided design drafting*) is a subset of CAD that deals with the computer-assisted generation of working drawings and other engineering documents. Most good design engineers agree that the computer does not change the nature of the design process and consider it simply a tool to improve efficiency and productivity. The designer provides knowledge, creativity, and control while the computer assists with accurate, easily modified geometric models and the capacity to perform complex design analysis and recall design information with great speed.

So, CADD offers the advantages of being able to do the following:

- visualize and correct the drawings
- store and retrieve design data
- integrate computation design analysis (such as stress analysis)
- simulate and test designs
- ultimately improve accuracy (fewer errors).

As shown in Figure 34.22, *geometric modeling* corresponds to the synthesis phase of the design process in the CAD/CAM part of CIM. To use geometric modeling, the designer constructs the graphical image of the object on the computer screen. The object can be represented using several different methods. The simplest systems use wire frames to represent the object. The wire-frame geometric modeling is classified as one of the following:

- Two-dimensional representation, used for flat objects
- Three-dimensional modeling, used for more complex geometries.

Other enhancements to wire-frame geometric modeling are:

- Color graphics
- Dashed lines to portray rear edges that would be invisible from the front
- Removal of hidden lines
- Surface representation that makes the object appear solid to the viewer.

The more advanced (intelligent) CAD work uses solid geometry shapes called *primitives* to construct objects—the object is a collection of geometric shapes that can be understood by both the CAD and the CAM software. To tie the geometric shapes to the processing, *features* are used. Features are higher-order abstract geometric forms that allow the designer to consider both design and manufacturing activities when describing an object to be manufactured.

Engineering analysis is required in the formulation of any engineering design project. The analysis may involve stress calculations, finite element analysis for heat-transfer computations, or differential equations to describe the dynamic behavior of the system. Generally, commercially available general-purpose programs are used to perform these analyses.

Design review and evaluation techniques check the accuracy of the CAD design. Semiautomatic dimensioning and tolerancing routines that assign size specification to surfaces help to reduce the possibility of dimensioning errors. The designer can zoom in on the part design details for close scrutiny. Many systems have a layering feature that involves overlaying the geometric image of the final shape of a machined part on top of the image of a rough part. Other features are interference checking and animation, which enhance the designer's visualization of the operation of the mechanism and help to ensure against interferences.

Automated drafting involves the creating of hard-copy engineering drawings from the CAD database. Typical features of CAD systems include automatic dimensioning, generation of cross-hatched areas, scaling of drawings, the ability to develop sectional views and enlarged views of particular part details, and the ability to rotate the part and perform transformations such as oblique, isometric, and perspective views.

Using the CAD database, a group technology (GT) *coding and classification* system can also be developed. Parts coding involves assigning letters or numbers to parts to define their geometry or manufacturing process sequence. The codes are used to classify or group similar parts into families for manufacturing using groups of machines. Designers can use the classification and coding system to retrieve existing part designs rather than design new parts.

Computer-Aided Process Planning

Process planning is responsible for the conversion of design data to work instructions, through the specification of the process parameters to be used as well as those machines (cells) capable of performing these processes in order to convert the piece part from its initial state to final form. The output of the planning includes the specification of machine and tooling to be used, the sequence of operations, machining parameters, and time estimates. Doing all this with computer-aided assistance is called **computer-aided process planning (CAPP)**. CAPP uses computer software to determine how a part is to be made. If group technology is used, parts are grouped into part families according to how they are to be manufactured. For each part family, a standard process plan is established. That is, each part in the family is a variation of the same theme. The standard process plan is stored in computer files and then retrieved for new parts that belong to that family.

Figure 34.23 explains the CAPP process. The designer can initiate the program by entering the GT part code. The CAPP program then searches the part family file to determine whether a match exists. If the file contains an identical code number, the standard machine routing and operation sequence are retrieved for display to the user. The standard operation sequence is examined by the user to permit any necessary editing of the plan to make it compatible with the new part design. This is *variant* CAPP. After editing, the process plan formatter prepares the paper documents (the route sheet and operation sheets) for the job shop.

If an exact match cannot be found, a new plan has to be *generated. Generative process planning* synthesizes process

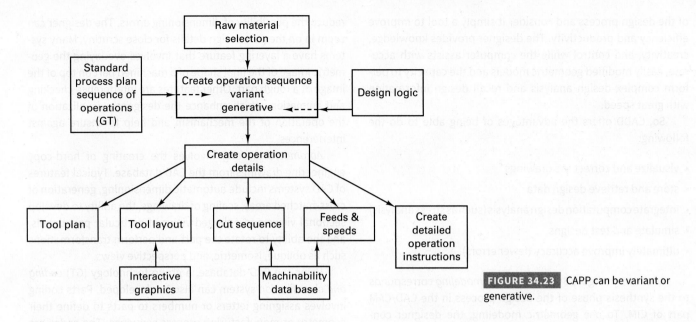

FIGURE 34.23 CAPP can be variant or generative.

information in order to create a process plan for a new component. Design logic and machinability data are required for generative process planning. The system automatically retrieves the part code, operation code, and tool code from the feature database. These codes are compared to the status database in order to establish a process plan. The machinability data utilized are in the form of *mathematical* models. These models relate cutting stability and forces as a function of speed, feed, and depth of cut using parameters such as surface finish and tool life to constrain the selected machining parameters. Given these relationships, the system may generate (compute) process parameters that result in minimum cost of time and do not violate set constraints. Once the process plan for a new part code number has been entered and verified, it becomes the standard process for future parts of the same GT classification. The process plan formatter may include software to compute machining conditions, layouts, and other detailed operation information for the job shop.

34.7 Computer-Aided Manufacturing

Another major element of CIM is **computer-aided manufacturing (CAM).** See Figure 34.22. An important reason for using a CAD system is that it provides a database for manufacturing the product. However, not all CAD databases are compatible with manufacturing software (CNC machine tool software).

The tasks performed by a CAM system are:

- Numerical control or CNC programming
- Production planning and scheduling
- Tool and fixture design.

Numerical control can use special computer languages. APT and COMPACT II are two of the early language-based computer-assisted programming systems. These systems take the CAD data and adapt them to the particular machine control unit/machine tool combination used to make the part. The development of computer-aided design in the 1980s led to an attractive alternative method for NC programming. However, virtually all CAD graphics systems were developed as part (or product) design tools, with NC capability added as an afterthought. CAD programs depict the final part and do not usually deal with how the part is processed from raw material to finished goods, that is, the sequence of operations. The major advantage of NC graphics is visualization of a part. You can see how to machine it. Even so, it takes considerable training (months) to achieve proficiency in the use of CADNC, or, as it is commonly called, CAD/CAM.

APT was designed in exactly the opposite way to CADNC. APT starts with the cutter and proceeds to "machine" the raw material into the final part shape. However, APT users must be able to visualize or imagine the part and "see" it being machined in their mind's eye, so APT also requires considerable training to become expert. For modern CAM applications for complex parts, the software simulates the machining steps necessary to generate the part. However, the CAD capability to design a complex part may still exceed the CAM capability of the CAD system. Chapter 26 has additional information on CNC programming, Chapter 41 discusses production planning and scheduling, Chapter 22 presents the fundamentals of cutting-tool design, and Chapter 40 presents the basics of jig and fixture design.

34.8 Summary

The yardstick for automation developed in 1962 still serves us well today, with only some minor additions. Clearly, numerical control machines, robots, FMSs, and computers are seen as critical elements of the advanced manufacturing technologies available for the next decade, which is being touted as the time when computer-integrated manufacturing (CIM) will become a widespread reality. Clearly computer technology and digital manufacturing abound, often labeled intelligent or integrated manufacturing; computer-aided design; computer-aided manufacturing with NC, CNC, or A/C and DNC; computer-aided process planning; computer-aided testing and inspection (CATI); artificial intelligence; smart robots; and much more. But a word of caution: Any company can buy computers, robots, and other pieces of automation hardware and software. The secret to manufacturing success lies in the design of a unique manufacturing system that can produce superior-quality goods at the lowest unit cost with on-time delivery and still be flexible. Flexibility means the system can readily adapt to changes in the customer demand (both volume and mix) while quickly implementing design changes as well as new designs. The flexibility characteristic is often difficult to couple with automation and requires a visionary management team and a change in culture on the factory floor (i.e., an empowered and involved workforce). Changing how people work in a manufacturing system means the manufacturing system must be redesigned. No better example of this can be found than at the Toyota Motor Company. Led by their vice president for manufacturing, Taiichi Ohno, who conceived, developed, and implemented Toyota's unique manufacturing system, this company has emerged as the world leader in car production. The implementation of this system has saved many companies (e.g., Harley Davidson) and carried many others to the top position in their industry. The Toyota system is unique and as revolutionary today as the American Armory System (job shops) and the Ford system (flow shops) for mass production were in their day. It is significant that virtually every manufacturing technology cited in this chapter is practiced at Toyota. This new system is now being called *lean production*, to contrast it to mass production. Toyota does not use the word *CIM* because the computer is viewed as just a tool used in its system, a system that recognizes people as the most flexible element. This new system is discussed in Chapters 42 and 43. Students of industrial, mechanical, and manufacturing engineering are well advised to be knowledgeable about this unique system.

Review Questions

1. What human attribute is replaced by a machine capability for the first five levels of automation?

2. Give an everyday example of a household device or appliance that exhibits automation levels A(1), A(2), A(3), and A(4).

3. Explain how a windmill is an A(4) device or machine (i.e., self-adjusting).

4. What device in the typical modern bathroom has a feedback control device in it?

5. What level of automation is the modern automobile?

6. In terms of a transfer line, explain what is meant by *line balancing*. Discuss using the rotary transfer line shown in Figure 34.3.

7. How is a machining center different from a milling machine?

8. How does an adaptive control system differ from a numerical control system?

9. What makes an NC machine a machining center?

10. At what level of automation is your typical kitchen toaster?

11. Figure 34.6 shows tool paths to rough and finish turn a part. What is the difference between a rough and a finish turn?

12. In Figure 34.6, what is a G70 finish cycle (see Chapter 26)?

13. What are some ways an NC machine can receive inputs (information)?

14. What is an AGV?

15. What has prevented widespread applications of FMS?

16. In FMS configurations, what is the CPU?

17. What is the difference between an FMS and an FMC?

18. What does the A(5) level of automation require beyond the A(4) level?

19. What is feedforward compared to feedback?

20. Name an A(5)-level machine or device in your home.

21. What is AI? (Ask a computer science friend for an example of an expert system.)

22. Does your cell phone have AI?

23. What is a PLC?

24. Flexible manufacturing systems use CNC machines for processing and AGVs, robots, or conveyors to transport parts. What differentiates the FMS from a transfer line?

25. Why are transfer lines replacing relay logic with more programmable controls?

26. The A(6) level tries to relate causes to effect. How is this done?

27. A standard piece of CAD analysis is called finite element analysis (FEA). What does FEA do?

28. What are the basic components of all robots?

29. What are work envelopes for industrial robots?

30. How is positional feedback obtained in robots?

31. Compare a rotary transfer machine with an unmanned **robotic cell.**

32. Compare a human worker in a manned cell to the robotic worker in an unmanned cell. What are the advantages of one over the other?

33. The vast majority of robots are used in automobile fabrication for welding and painting. Why?

34. What is a teach pendant?

35. What is CIM versus CAD and CAM?

Problems

1. Who was John Parsons and what role did he play in the development of numerical control?

2. In Figure 34.13, what is a fly ball governor and how does it work?

3. What is tactile sensing? Name some examples of tactile sensing in the home.

CHAPTER 35

Fundamentals of Joining

35.1 Introduction to Consolidation Processes

Large products, products with a high degree of shape complexity, or products with a wide variation in required properties are often manufactured as joined assemblies of two or more component pieces. These pieces can be smaller and therefore easier to handle, simpler shapes that are easier to manufacture, or segments that have been made from different materials. Assembly is an important part of the manufacturing process, and a wide variety of **consolidation processes** have been developed to meet the various needs.

Each of the methods has its own distinctive characteristics, strengths, and weaknesses. The metallurgical processes of fusion welding, brazing, and soldering are usually used to join metals and involve the solidification of molten material. The use of discrete fasteners (such as bolts and nuts, screws, and rivets) requires the creation of aligned holes and produces stress localization. While the holes might affect performance, disassembly and reassembly can often be performed with relative ease. Adhesive bonding has grown with new developments in polymeric materials and is being used extensively in automotive and aircraft production. Any material can be joined to any other material, and the low-temperature joining is particularly attractive for composite materials. Production rates are often low, however, because of the time required for the adhesive to develop full strength. Lesser-known joining techniques include shrink fits, slots and tabs, and a wide variety of other mechanical methods. From a technical viewpoint, powder metallurgy can be considered to be a consolidation process, because the end product is built up by the joining of a multitude of individual particles. Additive manufacturing or 3-D printing is another technical candidate because the product is built by the sequential deposition of multiple layers.

Welding is the dominant method of joining in manufacturing, and a large fraction of products would have to be drastically modified if welding were not available. Our survey of consolidation or joining processes, therefore, begins with a spectrum of techniques known by the generic term of **welding**—the permanent joining of two materials, usually metals, by **coalescence**, which is induced by a combination of temperature, pressure, and metallurgical conditions. The particular combination of these variables can range from high temperature with no pressure to high pressure with no increase in temperature. Because welding can be accomplished under such a wide variety of conditions, a number of different processes have been developed.

Coalescence requires sufficient proximity and activity between the two pieces so as to create the formation of common crystals or common structure. The ideal metallurgical bond, for which there would be no noticeable or detectable interface, would require (1) perfectly smooth, flat, or matching surfaces; (2) surfaces that are clean and free from oxides, absorbed gases, grease, and other contaminants; and (3) the joining of crystals with identical crystallographic structure and orientation alignment. These conditions would be difficult to obtain under laboratory conditions, and are virtually impossible to achieve in normal production. Consequently, the various joining methods have been designed to overcome or compensate for the common deficiencies.

Surface roughness can be overcome either by force, causing plastic deformation and flattening of the high points, or by melting the two surfaces so that fusion occurs. The various processes also employ different approaches to cleaning the metal surfaces prior to welding and preventing further oxidation or contamination during the joining process. In solid-state welding, contaminated layers are generally removed by mechanical or chemical cleaning prior to welding, or by causing sufficient metal flow along the interface so that new surface is created and existing impurities are displaced from the joint. In **fusion welding**, where molten material is produced and high temperatures accelerate the reactions between the metal and its surroundings, contaminants are often removed from the pool of molten metal through the use of fluxing agents. Protective atmospheres prevent the formation of new contaminants. If welding is performed in a vacuum, the contaminants are removed much more easily, and coalescence is easier to achieve. In the vacuum of outer space, mating parts can weld under extremely light loads, even when welding is not intended.

When the process requires heat, the structure of the metal can be significantly altered. Melting and resolidification will certainly change structure. Even when no melting occurs, the heating and cooling of the welding process can affect the metallurgical structure and quality of both the weld and the adjacent material. Because many of the changes are detrimental, the possible consequences of heating and cooling should be a major consideration when selecting a joining process.

To produce a high-quality weld, we will need: (1) a source of satisfactory heat and/or pressure, (2) a means of cleaning and protecting the metals to be joined, and (3) caution to avoid, or compensate for, harmful metallurgical effects. These aspects will be developed in the sections that follow.

35.2 Classification of Welding and Thermal Cutting Processes

Wherever possible, this text will utilize the nomenclature of the American Welding Society (AWS). The various welding processes have been classified in the manner presented in **Figure 35.1**, and letter symbols have been assigned to facilitate process designation. The variety of processes provide multiple ways of achieving coalescence, and make it possible to produce effective and economical welds in nearly all metals and combinations of metals. Chapter 36 will present the gas and arc welding processes; Chapter 37 will cover resistance and solid-state welding; and Chapter 38 will present a variety of other processes, including brazing and soldering.

For many years, welding equipment (such as oxyfuel torches and electric arc units) has also been used to cut metal sheets and plates. Developed originally for salvage and repair work, then used to prepare plates for welding, this type of

equipment is now widely used to cut sheets and plates into desired shapes for a variety of uses and operations. Laser and electron-beam equipment can now cut both metals and non-metals at speeds up to 25 m/min (1000 in./min), with accuracies of up to 0.25 mm (0.01 in.). **Figure 35.2** summarizes the commonly used **thermal cutting** processes, and provides their AWS designations. Because cutting is often an adaptation of welding, the welding and cutting capabilities will be presented together as the individual processes are discussed.

35.3 Some Common Concerns

Many of the problems that are inherent to welding and joining can be avoided by selecting the proper process with further consideration of both general and process-specific characteristics and requirements. Proper design of the joint is extremely critical. Heating, melting, and resolidification can produce drastic changes in the properties of base and filler materials. **Weld metal** properties can also be changed by dilution of the filler by melted base metal, vaporization of various alloy elements, and gas-metal reactions.

Various types of weld defects can also be produced. These include cracks in various forms, cavities (both gas and shrinkage), inclusions (slag, flux, and oxides), **incomplete fusion** between the weld and base metals, **incomplete**

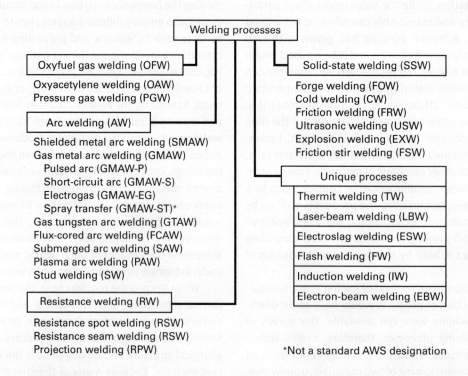

Welding processes

- Oxyfuel gas welding (OFW)
  - Oxyacetylene welding (OAW)
  - Pressure gas welding (PGW)
- Arc welding (AW)
  - Shielded metal arc welding (SMAW)
  - Gas metal arc welding (GMAW)
    - Pulsed arc (GMAW-P)
    - Short-circuit arc (GMAW-S)
    - Electrogas (GMAW-EG)
    - Spray transfer (GMAW-ST)*
  - Gas tungsten arc welding (GTAW)
  - Flux-cored arc welding (FCAW)
  - Submerged arc welding (SAW)
  - Plasma arc welding (PAW)
  - Stud welding (SW)
- Resistance welding (RW)
  - Resistance spot welding (RSW)
  - Resistance seam welding (RSW)
  - Projection welding (RPW)

- Solid-state welding (SSW)
  - Forge welding (FOW)
  - Cold welding (CW)
  - Friction welding (FRW)
  - Ultrasonic welding (USW)
  - Explosion welding (EXW)
  - Friction stir welding (FSW)
- Unique processes
  - Thermit welding (TW)
  - Laser-beam welding (LBW)
  - Electroslag welding (ESW)
  - Flash welding (FW)
  - Induction welding (IW)
  - Electron-beam welding (EBW)

*Not a standard AWS designation

FIGURE 35.1 Classification of common welding processes along with their AWS (American Welding Society) designations.

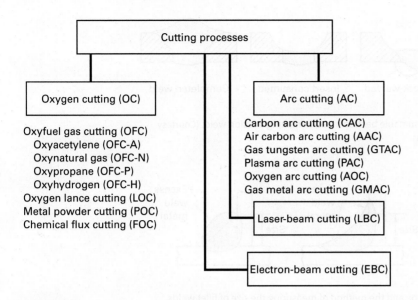

FIGURE 35.2 Classification of thermal cutting processes along with their AWS (American Welding Society) designations.

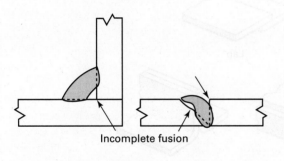

Incomplete fusion

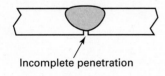

Incomplete penetration

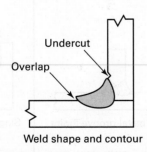

Weld shape and contour

FIGURE 35.3 Some common welding defects.

penetration (insufficient weld depth), unacceptable weld shape or contour, arc strikes, spatter, undesirable metallurgical changes (aging, grain growth, or transformations), and excessive distortion. **Figure 35.3** depicts several of these defects.

35.4 Types of Fusion Welds and Types of Joints

Figure 35.4 illustrates four basic types of fusion welds. **Bead welds**, or surfacing welds, are made directly onto a flat surface and therefore require no edge preparation. Because the penetration depth is limited, bead welds are used primarily for joining thin sheets of metal, building up surfaces, and depositing hard-facing (wear-resistant) materials.

Groove welds are used when full-thickness strength is desired on thicker material. Some sort of edge preparation is required to form a groove between the abutting edges. V, double-V (top and bottom), U, and J (one-sided V) configurations are most common and are often produced by oxyacetylene flame cutting. The specific type of groove usually depends upon the thickness of the joint, the welding process to be employed, and the position of the work. The objective is to obtain a sound weld throughout the full thickness with a minimum amount of additional weld metal. If possible, single-pass welding is preferred, but multiple passes may be required, depending upon the thickness of the material and the welding process being used. As shown in **Figure 35.5**, special consumable **inserts** can be used to ensure proper spacing between the mating edges and good quality in the root pass. These inserts are particularly useful in pipeline welding and other applications where welding must be performed from only one side of the work.

Fillet welds are used for tee, lap, and corner joints, and require no special edge preparation. The size of the fillet is measured by the leg of the largest 45-degree right triangle that can be inscribed within the contour of the weld cross section. This is shown in **Figure 35.6**, which also depicts the proper shape for fillet welds to avoid excess metal deposition and reduce stress concentration.

Plug welds attach one part on top of another and are often used to replace rivets or bolts. A hole is made in the top plate and welding is started at the bottom of this hole.

Figure 35.7 shows five basic types of joints (**joint configurations**) that can be made with the use of bead, groove, and fillet welds, and **Figure 35.8** shows some of the methods to construct these joints. In selecting the type of joint to be used,

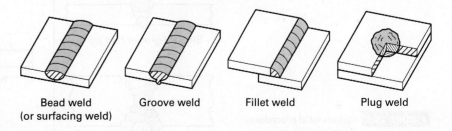

Bead weld (or surfacing weld) Groove weld Fillet weld Plug weld

FIGURE 35.4 Four basic types of fusion welds.

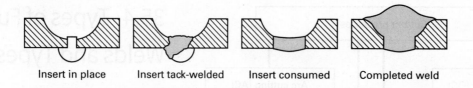

Insert in place   Insert tack-welded   Insert consumed   Completed weld

FIGURE 35.5 The use of a consumable backup insert in making a fusion weld. (Courtesy Arcos Industries, Mount Carmel, PA)

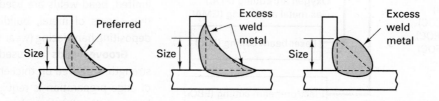

Preferred   Excess weld metal   Excess weld metal

Size   Size   Size

FIGURE 35.6 Preferred shape and the method of measuring the size of fillet welds.

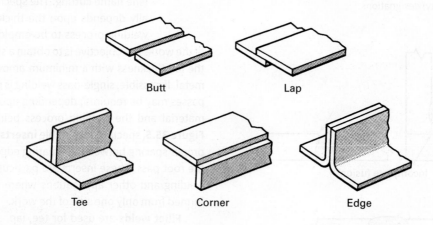

Butt   Lap

Tee   Corner   Edge

FIGURE 35.7 Five basic joint designs for fusion welding.

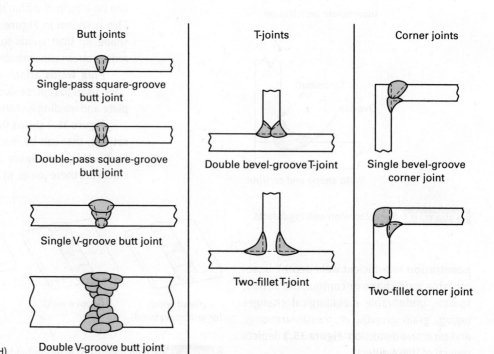

Butt joints

Single-pass square-groove butt joint

Double-pass square-groove butt joint

Single V-groove butt joint

Double V-groove butt joint

T-joints

Double bevel-groove T-joint

Two-fillet T-joint

Corner joints

Single bevel-groove corner joint

Two-fillet corner joint

FIGURE 35.8 Various weld procedures used to produce welded joints. (Courtesy Republic Steel Corporation, Youngstown, OH)

a primary consideration should be the type of loading that will be applied. A large portion of what are erroneously called "welding failures" can more accurately be attributed to inadequate consideration of loading. Cost and accessibility for welding are other important factors when specifying joint design, but should be viewed as secondary to loading. Cost is affected by the amount of required edge preparation, the amount of weld metal that must be deposited, the type of equipment that must be used, and the speed and ease with which the welding can be accomplished.

35.5 Design Considerations

Welding is a unique process that cannot be directly substituted for other methods of joining without proper consideration of its particular characteristics and requirements. Unfortunately, welding is also easy and convenient, and the considerations of proper design and implementation are often overlooked.

One very important fact is that welding produces **monolithic**, or one-piece, structures. When two pieces are welded together, they become one continuous piece. This can cause significant complications. For example, a crack in one piece of a multipiece structure may not be catastrophic, because it will seldom progress beyond the single piece in which it occurs. However, when a large structure, such as a ship hull, pipeline, storage tank, or pressure vessel, consists of many pieces welded together, a crack that starts in a single plate or weld can propagate for a great distance and cause complete failure. Obviously, this kind of failure is not the fault of the welding process itself, but is simply a reflection of the monolithic nature of the product.

It is also important to note that a given material in small pieces might not behave as it does in a larger size. This feature is clearly illustrated in **Figure 35.9**, which shows the

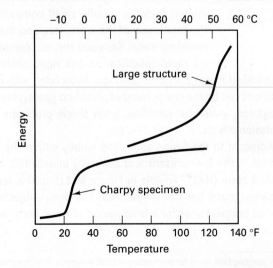

FIGURE 35.9 Effect of size on the transition temperature and energy-absorbing ability of a certain steel. While the larger structure absorbs more energy because of its size, it becomes brittle at a much higher temperature.

relationship between the energy required to fracture and temperature for the same steel tested as a small Charpy impact specimen (see Chapter 2 and Figure 2-20) and as a large, welded structure. In the form of a small Charpy bar, the material exhibits ductile behavior and good energy absorption at temperatures down to −4°C (25°F). When welded into a large structure, however, brittle behavior is observed at temperatures as high as 43°C (110°F). More than one welded structure has failed because the designer overlooked the effect of size on the notch-ductility of metal.

Another common error is to make welded structures too rigid, thereby restricting their ability to redistribute high stresses and avoid failure. Considerable thought may be required to design structures and joints that provide sufficient flexibility, but the multitude of successful welded structures attests to the fact that such designs are indeed possible.

Accessibility, welding position, component match-up, and the specific nature of a joint are other important considerations in welding design.

35.6 Heat Effects

Welding Metallurgy

Heating and cooling are essential and integral components of almost all welding processes, and tend to produce metallurgical changes that are often undesirable. In fusion welding, the heat is sufficient to melt some of the **base metal** (the material being welded), and this is often followed by a rapid cooling. Thermal effects tend to be most pronounced for this type of welding, but also exist to a lesser degree in processes where the heating–cooling cycle is less severe. If the thermal effects are properly considered, adverse results can usually be avoided or minimized, and excellent service performance can be obtained. If they are overlooked, however, the results can be disastrous.

Because such a wide range of metals are welded and a variety of processes are used, welding metallurgy is an extensive subject, and the material presented here serves only as an introduction. In fusion welding, a pool of molten metal is created, with the molten metal coming from either the parent plate alone (**autogenous welding**) or a mixture of parent and filler material. **Figure 35.10** shows a butt weld between plates of material A and material B. A backing strip of material C is used with filler metal of material D. In this situation, the molten pool is actually a complex alloy of all four materials. The molten material is held in place by a metal "mold" formed by the surrounding solids. Because the molten pool is usually small compared to the surrounding metal, fusion welding can often be viewed as *a small metal casting in a large metal mold*. The resultant structure and its properties can be best understood by first analyzing the casting and then considering the effects of the associated heat on the adjacent base material.

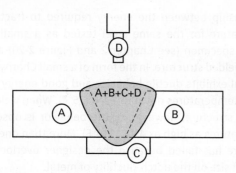

FIGURE 35.10 Schematic of a butt weld between a plate of metal A and a plate of metal B, with a backing plate of metal C and filler of metal D. The resulting weld nugget becomes a complex alloy of all four metals.

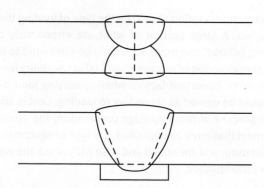

FIGURE 35.12 Comparison of two butt-weld designs. In the top weld, a large percentage of the weld pool is base metal. In the bottom weld, most of the weld pool is filler metal.

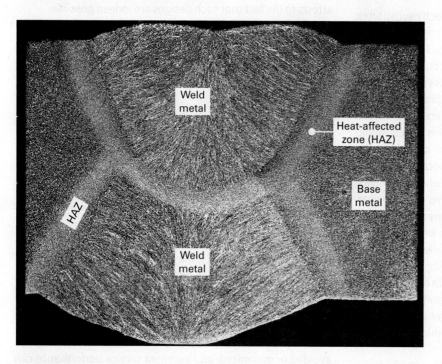

FIGURE 35.11 Grain structure and various zones in a fusion weld. (Courtesy Ronald Kohser)

Figure 35.11 shows a typical microstructure produced by a fusion weld. In the center of the weld is a region composed of metal that has solidified from the molten state. The material in this **weld pool**, or **fusion zone**, is actually a mixture of parent metal and electrode or filler metal, with the ratio depending upon the particular process, the type of joint, and the edge preparation. **Figure 35.12** compares two butt weld designs where the weld pool in the upper design would contain a large percentage of base metal and the weld pool in the lower design would be largely filler material. The metal in the fusion zone is cast material with a microstructure reflecting the cooling rate of the weld. This region cannot be expected to have the same properties and characteristics as the wrought material being welded because their processing histories and resulting structures are usually different. Adequate mechanical properties, therefore, can only be achieved by selecting filler rods or

electrodes, which have properties *in their as-deposited condition* that equal or exceed those of the wrought parent metal. It is not uncommon, therefore, for the filler metal to have a slightly different chemistry than the metal being welded[1]. The grain structure in the fusion zone can be fine or coarse, equiaxed or dendritic, depending upon the type and volume of weld metal and the rate of cooling, but most electrode and filler rod compositions tend to produce fine, equiaxed grains. The matching or exceeding of base metal strength, in the as-solidified condition, is the basis for several AWS specifications for electrodes and filler rods.

The pool of molten metal created by fusion welding is prone to all of the problems and defects associated with metal casting, such as gas porosity, inclusions, blowholes, cracks, and shrinkage. Because the amount of molten metal is usually small compared to the total mass of the workpiece and the surrounding metal has good thermal conductivity, rapid solidification and rapid cooling of the solidified metal are quite common. Associated with these conditions can be the entrapment of dissolved gases, chemical segregation, grain-size variation, grain shape problems, and orientation effects.

Adjacent to the fusion zone, and wholly within the base material, is the ever-present and generally undesirable **heat-affected zone (HAZ)**—(visible in Figure 35.11). In this region, the parent metal has not melted, but has been subjected to elevated temperatures for a brief period of time. Because the

---

[1] When selecting filler metal for the welding of steel, a number of factors should be considered. Low carbon steels have good weldability and can normally be welded with many different fillers and welding procedures. Medium and high carbon steels have poorer weldability and produce more crack sensitive welds. The minimum tensile strength of the filler should generally be at least as high as that of the metal being welded. Low hydrogen filler metals should be specified when welding high strength steels to reduce the susceptibility to cracking.

temperature and its duration vary widely with location, fusion welding might be more appropriately described as "a metal casting in a metal mold, coupled with an abnormal and widely varying heat treatment." The adjacent metal might experience sufficient heat to bring about structure and property changes, such as phase transformations, recrystallization, grain growth, precipitation or precipitate coarsening, embrittlement, or even cracking. The variation in thermal history can produce a variety of microstructures and a range of properties. In steels, the structures can range from hard, brittle martensite all the way through the much softer and weaker coarse pearlite and ferrite.

Because of its altered structure, the heat-affected zone might no longer possess the desirable properties of the parent material, and because it was not melted, it cannot assume the properties of the solidified weld metal. Consequently, this is often the weakest area in the as-welded joint. Except where there are obvious defects in the weld deposit, most welding failures originate in the heat-affected zone. This region extends outward from the weld to the location where the base metal has experienced too little heat to be affected or altered by the welding process. **Figure 35.13** presents a schematic of a fusion weld in steel, and uses standard terminology for the various regions and interfaces. Part of the heat-affected zone has been heated above the A_1 transformation temperature, and could assume a totally new structure through phase transformation. The lower temperature portion of the heat-affected zone (peak temperatures below the A_1 value) can experience diffusion-induced changes within the original structure.

Because of the melting, solidification, and exposure to a range of high temperatures, the structure and properties of welds can be extremely complex and varied. Through proper concern, however, associated problems can often be reduced or totally eliminated. Consideration should first be given to the thermal characteristics of the various processes. **Table 35.1** classifies some of the more common welding processes with regard to their **rate of heat input**. Processes with low rates of

TABLE 35.1	Classification of Common Welding Processes by Rate of Heat Input	
Low Rate of Heat Input		**High Rate of Heat Input**
Oxyfuel welding (OFW)		Plasma arc welding (PAW)
Electroslag welding (ESW)		Electron-beam welding (EBW)
Flash welding (FW)		Laser welding (LBW)
		Spot and seam resistance welding (RW)
		Percussion welding
Moderate Rate of Heat Input		
Shielded metal arc welding (SMAW)		
Flux cored arc welding (FCAW)		
Gas metal arc welding (GMAW)		
Submerged arc welding (SAW)		
Gas tungsten arc welding (GTAW)		

heat input (slow heating) tend to produce large total heat content within the metal, slow cooling rates, large heat-affected zones, and resultant structures with lower strength and hardness, but higher ductility. High-heat-input processes, on the other hand, have low total heats, fast cooling rates, and small heat-affected zones. The size of the heat-affected zone will also increase with increased starting temperature, decreased welding speed, increased thermal conductivity of the base metal, and a decrease in base metal thickness. Weld geometry is also important, with fillet welds producing smaller heat-affected zones than butt welds.

If the as-welded properties are unacceptable, the entire welded assembly might be heat-treated after welding. Structure variations can be reduced or eliminated, but the results are restricted to those that can be produced through heat treatment. The structures and properties associated with cold working, for example, could not be achieved. In addition, problems may be encountered in trying to achieve controlled heating and cooling within the large, complex-shaped structures commonly produced by welding. Moreover, furnaces,

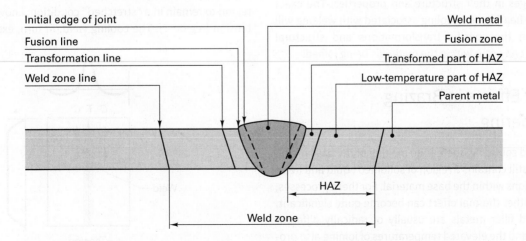

FIGURE 35.13 Schematic of a fusion weld in steel, presenting proper terminology for the various regions and interfaces. Part of the heat-affected zone has been heated above the transformation temperature and will form a new structure upon cooling. The remaining segment of the heat-affected zone experiences heat alteration of the initial structure. (Courtesy Sandvik AB, Sandviken, Sweden)

quench tanks, and related equipment might not be available to handle the full size of welded assemblies.

An alternative technique to reduce microstructural variation, or the sharpness of that variation, is to **preheat** either the entire base metal or material at least 10 centimeters (4 in.) on either side of the joint just prior to welding. This heating serves to reduce the cooling rate of both the weld deposit and the immediately adjacent metal in the heat-affected zone. The slower cooling produces a softer, more ductile structure and provides more time for the out-diffusion of harmful dissolved hydrogen. The welding stresses are distributed over a larger area, reducing the amount of weld distortion and the possibility of cracking.

Preheating is more common with higher carbon and alloy steels, and is particularly important with the high-thermal-conductivity metals, such as copper and aluminum, where the cooling rate would otherwise be extremely rapid. If the carbon content of plain carbon steels is greater than about 0.3%, the cooling rates encountered in normal welding can be sufficient to produce hard, untempered martensite, with an accompanying loss of ductility. Because alloy steels possess higher hardenability, the likelihood of martensite formation will be even greater with these materials. Special pre- and postwelding heat cycles (**preheat** and **postheat**) might be required when welding the higher carbon and alloy steels. For plain carbon steels, a preheat temperature of 100 to 200°C (200 to 400°F) is usually adequate. Because they can be welded without the need for preheating or postheating, low-carbon, low-alloy steels are extremely attractive for welding applications.

In joining processes where little or no melting occurs, considerable pressure is often applied to the heated metal (as in forge or resistance welding). The weld region experiences deformation, and the resultant structure exhibits the characteristics of a wrought material.

Because steel is the primary metal that is welded, our discussion of metallurgical effects has largely focused on steel. It should be noted, however, that other metals also exhibit heat-related changes in their structure and properties. The exact effects of the heating and cooling associated with welding will depend upon the specific transformations and structural changes that can occur within the materials being joined.

Thermal Effects in Brazing and Soldering

In brazing and soldering, there is no melting of the base metal, but the joint still contains a region of solidified liquid and heat-affected sections within the base material. For these processes, however, another thermal effect can become quite significant. The base and filler metals are usually of radically different chemistries, and the elevated temperatures of joining also promote interdiffusion. Intermetallic phases can form at interfaces and alter the properties of the joint. If present in small amounts, they can enhance bonding and provide strength reinforcement. Most **intermetallic compounds**, however, are extremely

brittle. Too much intermetallic material can result in the formation of continuous layers with significant loss of both strength and ductility.

Thermal-Induced Residual Stresses

Another effect of heating and cooling is the introduction of **residual stresses**. In welding, these can be of two types and are most pronounced in fusion welding, where the greatest amount of heating occurs. Their effects can be observed in the form of dimensional changes, distortion, and cracking.

Residual welding stresses are the result of restraint to thermal expansion and contraction by the pieces being welded. Consider a rectangular bar of metal that is uniformly heated and cooled. When heated, the material expands and becomes larger in length, width, and thickness. Upon cooling, the material contracts, and each dimension returns to its original value. Now insert the bar between rigid restraints so that lengthwise expansion cannot take place and repeat the thermal cycle. Upon heating, all of the expansion is restricted to the width and thickness, but the contraction upon cooling will still occur uniformly. The resulting rectangle will be shorter, thicker, and wider than the original specimen.

Now consider a weld being made between two flat plates, as illustrated in **Figure 35.14**. As the weld is produced, the liquid region conforms to the shape of the "mold," and the adjacent material becomes hot and expands (the heat-affected zone). The molten pool can absorb expansion of the plate perpendicular to the weld line, but expansion along the length of the weld tends to be restrained by the adjacent plate material that has remained cooler and stronger. This resistance or restraint is often sufficient to induce deformation of the hot, weak, and thermally expanding heat-affected zone, which now becomes thicker instead of longer.

After the weld pool solidifies, both the weld metal and adjacent heat-affected region cool and contract. The surrounding metal, however, resists this contraction, forcing the weld region to remain in a "stretched" condition, known as residual tension (region T). The cooling weld, in turn, exerts forces on

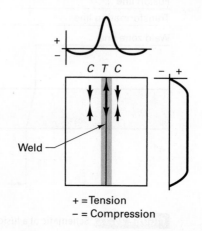

+ = Tension
− = Compression

FIGURE 35.14 Schematic of the longitudinal residual stresses in a fusion-welded butt joint.

the adjacent material, producing regions of residual compression (regions C). While the net force must remain at zero (in keeping with the equilibrium laws of physics and mechanics), the localized tensions and compressions can be substantial. As Figure 35.14 depicts, a high longitudinal residual tension is observed in the weld metal, which becomes longitudinal compression and then returns to zero as one moves away from the weld centerline. The magnitude of the residual tension will be relatively uniform along the weld line, except at the ends where the stresses can be relieved by a pulling in of the edges.

During cooling, the thermal contractions occur both parallel (longitudinal) and perpendicular (transverse) to the weld line. Lateral movement of the material being welded can often compensate for the transverse contractions. The width of the welded assembly simply becomes less than that of the positioned components at the time of welding. **Figure 35.15a** depicts this reduction in width, while **Figure 35.15b** illustrates the longitudinal contractions that generate the complex stresses of Figure 35.14.

Components being joined during fabrication typically have considerable freedom of movement, but welds made on nearly completed structures or repair welds often join components that are somewhat restrained. If the welded plates in Figure 35.15a are restrained from horizontal movement, *additional stresses will be induced*. These residual stresses are known as **reaction stresses**, and they can cause cracking of the hot weld or heat-affected material, or contribute to failure during subsequent use. Their magnitude will be an inverse function of the length between the weld joint and the point of restraint, and can be as high as the yield strength of the parent metal (because yielding would occur to relieve any higher stresses).

Effects of Thermal Stresses

Distortion or warping can easily result from the nonuniform temperatures and thermal stresses induced by welding. **Figure 35.16** depicts some of the distortions that can occur during various welding configurations. Because welding conditions can vary widely, no fixed rules can be provided to ensure the absence of warping. The following suggestions can help, however.

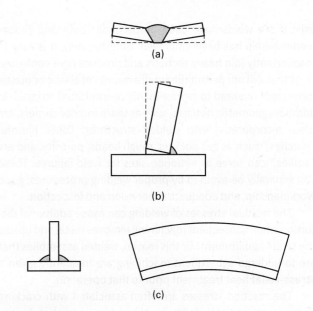

FIGURE 35.16 Distortions or warpage that can occur as a result of welding operations: (a) V-groove butt weld where the top of the joint contracts more than the bottom; (b) one-side fillet weld in a T-joint; (c) two-fillet weld T-joint with a high vertical web.

Total heat input to the weld should be minimized. Welds should be made with the least amount of weld metal necessary to form the joint. Overwelding is not an asset, because it actually increases residual stresses and distortion. Faster welding speeds reduce the welding time and also reduce the volume of metal that is heated. Welding sequences should be designed to use as few passes as possible, and the base material should be permitted to have a high freedom of movement. When constructing a multiweld assembly, it is beneficial to weld toward the point of greatest freedom, such as from the center to the edge.

The initial components can also be oriented out of position, so that the subsequent distortion will move them to the desired final shape. Another common procedure is to completely restrain the components during welding, thereby forcing some plastic flow in the joint and surrounding material. This procedure is used most effectively on small weldments where the reaction stresses will not be high enough to cause cracking.

Still another procedure is to balance the resulting thermal stresses by depositing the weld metal in a specified pattern, such as short lengths along a joint (intermittent or stitch welding) or on alternating sides of a plate (such as a double-V weld on thick plate). Warping can also be reduced by the use of **peening**. As the weld bead surface is hammered with the peening tool or material, the metal is flattened and tries to spread. Being held back by the underlying material, the surface becomes compressed or squeezed. Surface rolling of the weld-bead area can have the same effect. In both processes, the compressive stresses induced by the surface deformation serve to offset the tensile stresses induced by welding.

The presence of notches or high rigidity in structures can both be extremely harmful. These two conditions should not

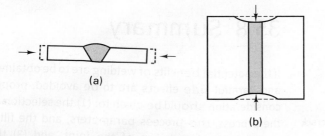

FIGURE 35.15 Shrinkage of a typical butt weld in the (a) transverse and (b) longitudinal directions as the material responds to the induced stresses. Note that restricting transverse motion will place the entire weld in transverse tension.

exist if the welds have been properly designed and proper workmanship has been employed. Unfortunately, it is easy to inadvertently join heavy sections and produce rigid configurations that will not permit the small amounts of elastic or plastic movement required to reduce highly concentrated stresses. In addition, geometric notches, such as sharp interior corners, are often incorporated into welded structures. Other harmful "notches," such as gas pockets, rough beads, porosity, and arc "strikes," can serve as initiation sites for weld failures. These can generally be avoided by proper welding procedures, good workmanship, and adequate supervision and inspection.

The residual stresses of welding can cause additional distortion when subsequent machining removes metal and upsets the stress equilibrium. For this reason, welded assemblies that are to undergo subsequent machining are frequently given a **stress-relief** heat treatment prior to that operation.

The reaction stresses are often associated with cracking during or immediately following the welding operation (as the weld is cooling). This cracking is most likely to occur when there is great restraint to the shrinkage transverse to the direction of welding. When a multipass weld is being made, cracking tends to occur in the early beads where there is insufficient weld metal to withstand the shrinkage stresses. These cracks can be quite serious if they go undetected and are not chipped out and repaired, or melted and resolidified during subsequent passes. **Figure 35.17** shows the various forms of cracking that can occur as a result of welding.

To minimize the possibility of cracking, welded joints should be designed to keep restraint to a minimum. The metals and alloys of the structure should be selected with welding in mind (more problems exist with higher-carbon steels, higher-alloy steels, and high-strength materials), and special consideration should be given when welding thicker materials. Crack-prevention efforts can also include maintaining the proper size and shape of the weld bead. While a concave fillet is desirable when machining, a concave weld profile has a greater tendency to crack upon cooling because contraction actually increases the length of the surface. With a convex profile, the length of the surface will contract simultaneously with the volume, reducing the possibility of surface tension and cracking. Weld beads with high penetration (high depth/width ratio) are also more prone to cracking.

Still other methods to suppress cracking focus on reducing the stresses by making the cooling more uniform or relaxing the stresses by promoting plasticity in the metals being welded. The metals to be welded can be preheated and additional heat can be applied between the welding passes to retard cooling. Some welding codes also require the inclusion of a thermal stress relief (post-weld heat treatment) after welding but prior to use. Hydrogen dissolved in the molten weld metal can also induce cracking. Slower welding and cooling or a post-weld heating will allow any hydrogen to escape, and the use of low-hydrogen electrodes and low-moisture fluxes will reduce the likelihood of hydrogen being present. Welding stresses can also be relieved through a post-weld vibratory stress relief.

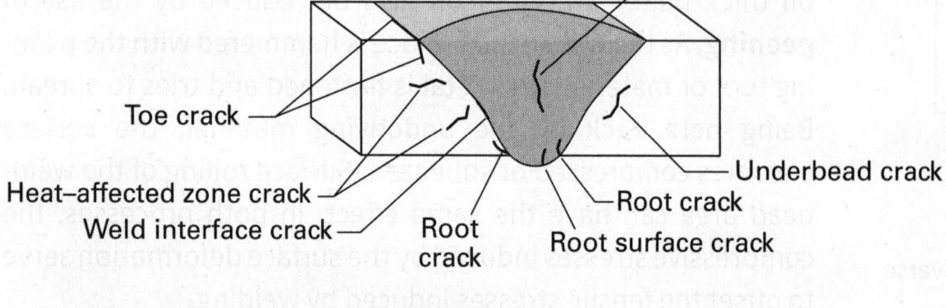

35.7 Weldability or Joinability

It is important to note that not all joining processes are compatible with all engineering materials. While the terms **weldability** or **joinability** imply a reliable measure of a material's ability to be welded or joined, they are actually quite nebulous. One process might produce excellent results when applied to a given material, whereas another may produce a dismal failure. Within a given process, the quality of results can vary greatly with differences in the process parameters, such as electrode material, shielding gas, welding speed, and cooling rate.

Table 35.2 presents the compatibility of various joining processes with some of the major classes of engineering materials. In each case, the process is classified as recommended (R), commonly performed (C), performed with some difficulty (D), seldom used (S), and not used (N). It should be noted, however, that the classifications are generalizations, and exceptions often exist within both the family of materials (such as the various types of stainless steels) and the types of processes (arc welding here encompasses a large variety of specific processes). Nevertheless, the table can serve as a guideline to assist in process selection.

35.8 Summary

If the potential benefits of welding are to be obtained and harmful side effects are to be avoided, proper consideration should be given to: (1) the selection of the process, the process parameters, and the filler material, (2) the design of the joint, and (3) the effects of heating and cooling on both the weld and parent material. Joint design should consider manufacturability, durability, fatigue resistance, corrosion resistance, and safety. Weldi... ...urgy helps

Weld metal crack
Toe crack
Heat–affected zone crack
Weld interface crack
Root crack
Underbead crack
Root crack
Root surface crack

FIGURE 35.17 Various types and locations of cracking that can occur as a result of welding.

TABLE 35.2   **Weldability or Joinability of Various Engineering Materials**[a]

Material	Arc Welding	Oxyacetylene Welding	Electron-Beam Welding	Resistance Welding	Brazing	Soldering	Adhesive Bonding
Cast iron	C	R	N	S	D	N	C
Carbon and low-alloy steel	R	R	C	R	R	D	C
Stainless steel	R	C	C	R	R	D	C
Aluminum and magnesium	C	C	C	C	C	C	C
Copper and copper alloys	C	C	C	C	C	S	R
Nickel and nickel alloys	R	C	C	R	R	R	C
Titanium	C	N	C	C	R	C	C
Lead and zinc	C	C	N	D	D	S	C
Thermoplastics	Heated tool R	Hot gas R	N	Induction C	N	R	R
Thermosets	N	N	N	N	N	N	C
Elastomers	N	N	N	N	N	N	C
Ceramics	N	S	C	N	N	N	R
Dissimilar metals	D	D	C	D	D/C	R	R

[a] R, recommended (easily performed with excellent results); C, commonly performed; D, difficult; S, seldom used; N, not used.

determine the structure and properties across the joint, and the need for additional thermal treatments. Further attention can be required to control or minimize residual stresses and distortion.

Parallel considerations apply to brazing and soldering operations, with additional attention to the effects of interdiffusion between the filler and base metals. Flame and arc-cutting operations involve localized heating and cooling, and they also create altered structures in the heat-affected zone. Because many thermally-cut products undergo further welding or machining, however, the regions of undesirable structure might not be retained in the final product.

Review Questions

1. What types of design features favor manufacture as a joined assembly?

2. What types of manufacturing processes fall under the classifica-

4. What conditions are required to produce an ideal metallurgical bond?

5. What are some of the ways in which welding processes compensate for the inability to meet the conditions of an ideal bond?

6. What are some possible problems associated with the high temperatures that are commonly used in welding?

7. What are the three primary aspects required to produce a high-quality weld?

8. How are welding processes identified by the American Welding Society?

9. What is thermal cutting?

10. What are some of the common types of weld defects?

11. What are the four basic types of fusion welds?

12. What are some of the common edge configurations used in preparation for groove welding?

13. What is the role of an insert in welding?

14. What types of weld joints commonly employ fillet welds?

15. What are the five basic joint types for fusion welding?

16. What are some of the factors that influence the cost of making a weldment?

17. Why is it important to consider welded products as monolithic structures?

18. How does the fracture resistance and temperature sensitivity of a steel vary with changes in material thickness?

19. How might excessive rigidity actually be a liability in a welded structure?

20. What is autogenous welding?

21. In what way is the weld-pool segment of a fusion weld like a small metal casting?

22. Why is it possible for the fusion zone to have a chemistry that is different from the filler metal?

23. Why is it not uncommon for the selected filler metal to have a chemical composition that is different from the material being welded?

24. What are some of the defects or problems that can occur in the molten metal region of a fusion weld?

25. Why can the material properties vary widely within welding heat-affected zones?

26. What are some of the structure and property modifications that can occur in welding heat-affected zones?

27. Why do most welding failures occur in the heat-affected zone?

28. Discuss the various regions within the heat-affected zone of a fusion weld in steel.

29. What are some of the characteristics and consequences of welding with processes that have low rates of heat input?

30. What process features can increase the size of the heat-affected zone?

31. What are some of the difficulties or limitations encountered in heat treating large, complex structures after welding?

32. What is the purpose of pre- and postheating in welding operations?

33. What heat-related metallurgical effects can produce adverse results when brazing or soldering?

34. What causes weld-induced residual stresses?

35. What is the cause of reaction-type residual stresses?

36. How are reaction stresses affected by the distance between the weld and the point of fixed constraint?

37. What are some of the techniques that can reduce the amount of distortion in a welded structure?

38. How can the surfaces of weldments be put into residual compression?

39. In what ways might welding create geometric notches in a welded structure?

40. Why might a welded structure warp if the structure is machined after welding?

41. Why might a stress relief heat treatment be performed prior to machining?

42. What are some of the techniques that can be employed to reduce the likelihood of cracking in a welded structure?

43. Why are the terms *weldability* and *joinability* somewhat nebulous?

Problems

1. Through the 1940s, the hulls of ocean-going freighters were constructed by riveting plates of steel together. When the defense efforts of World War II demanded accelerated production of freighters to supply US troops overseas, construction of the hulls was converted to welding. The resulting Liberty Ships proved quite successful but also drew considerable attention when minor or moderate impacts (usually under low-temperature conditions) produced cracks of lengths sufficient to scuttle the ship, often up to 50 feet or more. Because the material was essentially the same and the only significant process change had been the conversion from riveting to welding, the welding process was blamed for the failures.

    a. Is this a fair assessment?

    b. What do you think might have contributed to the problem?

    c. What evidence might you want to gather to support your beliefs?

2. Two pieces of AISI 1025 steel are being shielded-metal-arc welded with E6012 electrodes. Some difficulty is being experienced with cracking in the weld beads and in the heat-affected zones. What possible corrective measures might you suggest?

3. Figure 35.A schematically depicts the design of a go-cart frame with cross bars and seat support. The assembly is to be constructed from hot-rolled, low-carbon, box-channel material with miter, butt, and fillet welds at the 12 numbered joints. Because of the solidification shrinkage and subsequent thermal contraction of the joint material, the welds are best made when one or more of the sections are unrestrained. If the structure is too rigid at the time of welding, the associated dimensional changes are restricted, causing the generation of residual stresses that can lead to distortion, cracking, or tears.

    a. Consider the 12 welds in the proposed structure and recommend a welding sequence that would minimize the possibility of hot tears and cracks because of the welding of a restrained joint.

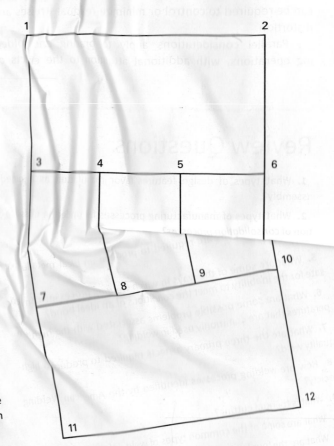

FIGURE 35.A

b. Your company is developing a computer-assisted design program. Suggest one or more rules that could be programmed to aid in the selection of an acceptable weld sequence.

4. Investigate and summarize some of the important metallurgical considerations when attempting to join each of the following materials?

  a. Medium- or high-carbon plain carbon steels

  b. Advanced high-strength steels

  c. Aluminum

  d. Titanium

  e. Dissimilar metals

5. What do you foresee as the major difficulties when attempting to join the following materials?

  a. Thermoplastic polymers

  b. Thermoset polymers

  c. Ceramics

  d. Composites (consider polymer-matrix and metal-matrix, as well as particulate and fiber reinforcements)

CHAPTER 36

Gas Flame and Arc Processes

36.1 Oxyfuel-Gas Welding

Oxyfuel-Gas Welding Processes

Oxyfuel-gas welding (OFW) refers to a group of welding processes that use the flame produced by the combustion of a fuel gas and oxygen as the source of heat. It was the development of a practical **torch** to burn acetylene and oxygen, shortly after 1900, that brought welding out of the blacksmith's shop, demonstrated its potential, and started its development as a manufacturing process. Other processes have largely replaced gas-flame welding in large-scale manufacturing, but the process is still popular for small-scale and repair operations because of its portability, versatility (most ferrous and nonferrous metals can be welded), and the low capital investment required. Acetylene is still the principal fuel gas because of its high temperature of combustion and ease of production and transportation.

The combustion of oxygen and **acetylene** (C_2H_2) by means of a welding torch of the type shown in **Figure 36.1** produces a temperature of about 3250°C (5850°F) in a two-stage reaction. In the first stage, the supplied oxygen and acetylene react to produce carbon monoxide and hydrogen:

$$C_2H_2 + O_2 \rightarrow 2CO + H_2 + heat$$

This reaction occurs near the tip of the torch and generates intense heat. The second stage of the reaction involves the combustion of the CO and H_2 and occurs just beyond the first combustion zone. The specific reactions of the second stage are:

$$2CO + O_2 \rightarrow 2CO_2 + heat$$
$$H_2 + \tfrac{1}{2}O_2 \rightarrow H_2O + heat$$

The oxygen for these secondary reactions is generally obtained from the surrounding atmosphere.

The two-stage combustion process produces a flame having two distinct regions—a light blue inner cone and a darker blue outer cone. As shown in **Figure 36.2**, the maximum temperature occurs near the end of the inner cone, where the first stage of combustion is complete. Most welding should be performed with the torch positioned so that this point of maximum temperature is just above the metal being welded. The outer envelope of the flame serves to preheat the metal and, at the same time, provides shielding from oxidation, because oxygen from the surrounding air is consumed in the secondary combustion.

Three different types of flames can be obtained by varying the oxygen-to-acetylene (or oxygen-to-fuel gas) ratio. If the ratio is between 1:1 and 1.15:1, all reactions are carried to completion and a **neutral flame** is produced. Most welding is done with a neutral flame, because it will have the least chemical effect on the heated metal.

A higher ratio, such as 1.5:1, produces an **oxidizing flame**, which is hotter than the neutral flame (about 3600°C or 6000°F) but similar in appearance. It is often accompanied by a crackling sound. Such flames are used when welding copper and copper alloys but are generally considered harmful when welding steel because the excess oxygen reacts with the carbon in the steel, lowering the carbon in the region around the weld.

Excess fuel, on the other hand, produces a **carburizing flame**. The excess fuel decomposes to carbon and hydrogen, and the flame temperature is not as great (about 3050°C or 5500°F). Flames with a slight excess of fuel are reducing flames and are characterized by

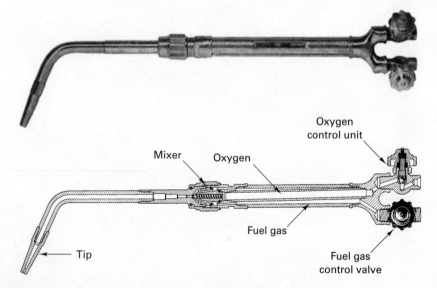

Oxygen control unit

Mixer
Oxygen

Fuel gas

Tip

Fuel gas control valve

FIGURE 36.1 Typical oxyacetylene welding torch and cross-sectional schematic. (Courtesy of Thermadyne Industries, Inc., St. Louis, MO)

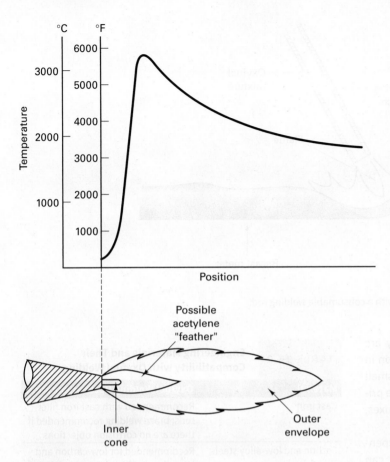

FIGURE 36.2 Typical oxyacetylene flame and the associated temperature distribution.

three flame zones—the inner cone, an "acetylene feather," and the outer cone. Flames of this type are used in welding Monel (a nickel–copper alloy), high-carbon steels, and some alloy steels, and for applying some types of hard-facing material.

For welding purposes, oxygen is usually supplied from pressurized tanks in a relatively pure form, but, in rare cases, air (20% oxygen) can also be used. The acetylene is usually obtained in portable storage tanks that hold up to 8.5 m³ (300 ft³) at 1.7 MPa (250 psi) pressure. Because acetylene is not safe when stored as a gas at pressures above 0.1 MPa (15 psi), it is usually dissolved in acetone. The storage cylinders are filled with a porous filler. Acetone is absorbed into the voids in the filler material and serves as a medium for dissolving the acetylene.

Acetylene is the hottest and most versatile of the fuel gases. Alternative fuels include propane, propylene, and stabilized **methyl-acetylene-propadiene**, best known by the trade name of **MAPP** gas. MAPP is the second-hottest gas, with a flame temperature between 2875° and 3000°C (5200° and 5400°F). **Propylene** is actually a generic name for a variety of mixed gases, often consisting of propane and ethylene or other hotter-burning chemicals. The flame temperature is usually between 2650° and 2925°C (4800° and 5300°F). Propane has a flame temperature between 2480° and 2540°C (4500° and 4600°F). While the flame temperature is slightly lower, these gases can be safely stored in ordinary pressure tanks. Three to

four times as much gas can be stored in a given volume, and cost per cubic foot can be less than acetylene. Butane, natural gas, and hydrogen have also been used in combination with air or oxygen for brazing and to weld the low-melting-temperature, nonferrous metals. They are generally not suited to the ferrous metals because the flame atmosphere is oxidizing and the heat output is too low.

The pressures used in gas-flame welding range from 0.006 to 0.1MPa (1 to 15 psi) and are controlled by pressure regulators on each tank. Because mixtures of acetylene and oxygen or air are highly explosive, precautions must be taken to avoid mixing the gases improperly or by accident. All acetylene fittings have left-hand threads, while those for oxygen are equipped with right-hand threads. This prevents improper connections.

The tip size, or orifice diameter of the torch, can be varied to control the shape of the inner cone and the flow rate of the gases. Larger tips permit greater flow of gases, resulting in greater heat input without the higher gas velocities that might blow the molten metal from the weld puddle. Larger torch tips are used for the welding of thicker metal.

Uses, Advantages, and Limitations

Almost all oxyfuel-gas welding is **fusion welding**. The metals to be joined are simply melted where a weld is desired and no pressure is required. Because a slight gap often exists between the pieces being joined, **filler metal** can be added in the form of a solid metal wire or rod. Welding rods come in standard sizes, with diameters from 1.5 to 9.5 mm ($^1/_{16}$ to $^3/_8$ in.) and lengths from 0.6 to 0.9 m (24 to 36 in.). They are available in standard grades that provide specified minimum tensile strengths or in compositions that match the base metal. **Figure 36.3** shows a schematic of oxyfuel-gas welding using a consumable welding rod.

To promote the formation of a better bond, **fluxes** can be used to clean the surfaces and remove contaminating oxide. In addition, the gaseous shield produced by vaporizing flux can prevent further oxidation during the welding process, and the slag produced by solidifying flux can protect the weld pool as it cools. Flux can be added as a powder, the welding rod can be dipped in a flux paste, or the rods can be precoated.

The oxyfuel-gas welding (OFW) processes can produce good-quality welds if proper caution is exercised. Welding can be performed in all positions, the temperature of the work can be easily controlled, and the puddle is visible to the welder. However, exposure of the heated and molten metal to the various gases in the flame and atmosphere makes it difficult to prevent contamination. Because the heat source is not concentrated, heating is rather slow. As the heat spreads, a large volume of metal is heated, and distortion is likely to occur. The thickness of the material being joined is usually less than 6.5 mm (¼ in.). In production applications, therefore, the

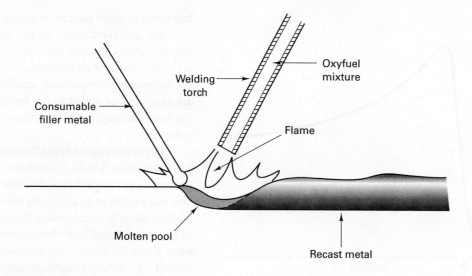

FIGURE 36.3 Oxyfuel-gas welding with a consumable welding rod.

flame-welding processes have largely been replaced by arc welding. Nevertheless, flame welding is still quite common in field work, in maintenance and repairs, and in fabricating small quantities of specialized products. Low-carbon steel is the primary material being welded, but, with the use of proper fluxes, a variety of ferrous and nonferrous metals can be joined.

Oxyfuel equipment is quite portable, relatively inexpensive, and extremely versatile. A single set of equipment can be used for welding, brazing, and soldering, and as a heat source for bending, forming, straightening, and hardening. With the modifications to be discussed shortly, it can also perform flame cutting. **Table 36.1** summarizes some of the key features of oxyfuel-gas welding. **Table 36.2** shows its compatibility with some common engineering materials.

Pressure Gas Welding

Pressure gas welding (PGW) is a process that uses equipment similar to the oxyfuel-gas process to produce butt joints between the ends of objects such as pipe and railroad rail. The ends are heated with a gas flame to a temperature below the

TABLE 36.1	Process Summary: Oxyfuel Gas Welding (OFW)
Heat source	Fuel gas—oxygen combustion
Protection	Gases produced by combustion
Electrode	None
Material joined	Best for steel and other ferrous metals
Rate of heat input	Low
Weld profile (Depth/Width)	$\frac{1}{3}$
Maximum penetration	3 mm
Assets	Cheap, simple equipment, portable and versatile
Limitations	Large HAZ, slow and skill required

TABLE 36.2	Engineering Materials and Their Compatibility with Oxyfuel Welding
Material	**Oxyfuel Welding Recommendation**
Cast iron	Recommended with cast iron filler rods; braze welding recommended if there are no corrosion objections
Carbon and low-alloy steels	Recommended for low-carbon and low-alloy steels, using rods of the same material; more difficult for higher carbon
Stainless steel	Common for thinner material; more difficult for thicker
Aluminum and magnesium*	Common for aluminum thinner than 1 in.; difficult for magnesium alloys
Copper and copper alloys*	Common for most alloys; more difficult for some types of bronzes
Nickel and nickel alloys	Common for nickel, Monels, and Inconels
Titanium	Not recommended
Lead and zinc	Recommended
Thermoplastics, thermosets, and elastomers	Hot-gas welding used for thermoplastics, not used with thermosets and elastomers
Ceramics and glass	Seldom used with ceramics, but common with glass
Dissimilar metals	Difficult; best if melting points are within 50°F; concern for galvanic corrosion
Metals to nonmetals	Not recommended
Dissimilar nonmetals	Difficult

*Due to the high thermal conductivity, a large volume of metal can reach the melting temperature and it might be difficult to control the size of the weld pool.

melting point, and the soft metal is then forced together under pressure. Pressure gas welding, therefore, is actually a form of solid-state welding where the gas flame simply softens the metal, and coalescence is produced by pressure.

36.2 Oxygen Torch Cutting

Processes

Oxyfuel-gas cutting (OFC), commonly called *flame cutting*, is the most common **thermal cutting** process. In some cases, the metal is merely melted by the flame of the oxyfuel-gas torch and is blown away to form a gap, or **kerf**, as illustrated in **Figure 36.4**. When ferrous metal is cut, however, the process becomes one where the iron actually burns (or oxidizes) at high temperatures according to one or more of the following reactions:

$$Fe + O \rightarrow FeO + heat$$
$$3Fe + 2O_2 \rightarrow Fe_3O_4 + heat$$
$$4Fe + 3O_2 \rightarrow 2Fe_2O_3 + heat$$

Because these reactions do not occur until the metal is above 815°C (1500°F), the oxyfuel flame is first used to raise the metal to the temperature where burning can be initiated. Then, a stream of pure oxygen is added to the torch (or the oxygen content of the oxyfuel mixture is increased) to oxidize the iron. The liquid iron oxide and any unoxidized molten iron are then expelled from the joint by the kinetic energy of the oxygen-gas stream. Because of the low rate of heat input and the need for preheating ahead of the cut, oxyfuel cutting produces a relatively large heat-affected zone and associated distortion compared to competing techniques. Therefore, the process is best used where the edge finish or tolerance is not critical, and the edge material will either be subsequently welded or removed by machining. Cutting speeds are relatively slow, but the low cost of both the required equipment and its operation make the process attractive for many applications.

Theoretically, the heat supplied by the oxidation will be sufficient to keep the cut progressing, but additional heat is often necessary to compensate for losses to the atmosphere and the surrounding metal. If the workpiece is already hot from other processing, such as solidification or hot working, no supplemental heating is required, and a supply of oxygen through a small pipe is all that is needed to initiate and continue a cut. This is known as **oxygen lance cutting (LOC)**. A workpiece temperature of about 1200°C (2200°F) is required to sustain continuous cutting.

Oxyfuel-gas cutting works best on metals that oxidize readily but do not have high thermal conductivities. Carbon and low-alloy steels can be readily cut in thicknesses from 5 mm to in excess of 75 cm (30 in.). Stainless steels contain oxidation-resistant ingredients and are difficult to cut, as are aluminum and copper alloys.

Fuel Gases for Oxyfuel-Gas Cutting

Acetylene is by far the most common fuel used in oxyfuel-gas cutting, and the process is often referred to as **oxyacetylene cutting (OFC-A)**. **Figure 36.5** shows a typical cutting torch. The tip contains a circular array of small holes through which the oxygen–acetylene mixture is supplied to form the heating flame. A larger hole in the center supplies a stream of oxygen and is controlled by a lever valve. The rapid flow of the cutting oxygen not only oxidizes the hot metal, but also blows the formed oxides from the cut.

If the torch is adjusted and manipulated properly, it is possible to produce a relatively smooth cut. Cut quality, however, depends upon careful selection of the process variables, including preheat conditions, oxygen flow rate, and cutting speed. Oxygen purities higher than 99.5% are required for the most efficient cutting. If the purity drops to 98.5%, cutting speed will be reduced by 15%, oxygen consumption will increase by 25%, and the quality of the cut will diminish.

Cutting torches are often manipulated manually. However, when the process is applied to manufacturing, the desired path is usually controlled by mechanical or programmable means. Specialized equipment has been designed to produce straight cuts in flat stock and square-cut ends on pipe. The marriage of computer numerically controlled (CNC) machines and cutting torches has also proven to be quite popular. This approach, along with the use of robot-mounted torches, provides great flexibility along with good precision and control.

Fuel gases other than acetylene can also be used for oxyfuel-gas cutting, the most common being *natural gas (OFC-N)* and *propane (OFC-P)*. While their flame temperatures are lower than acetylene, their use is generally a matter of economics and gas availability. For certain special work, *hydrogen* can also be used (*OFC-H*).

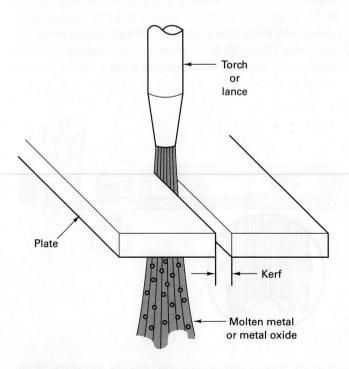

FIGURE 36.4 Flame cutting of a metal plate.

Torch or lance

Plate

Kerf

Molten metal or metal oxide

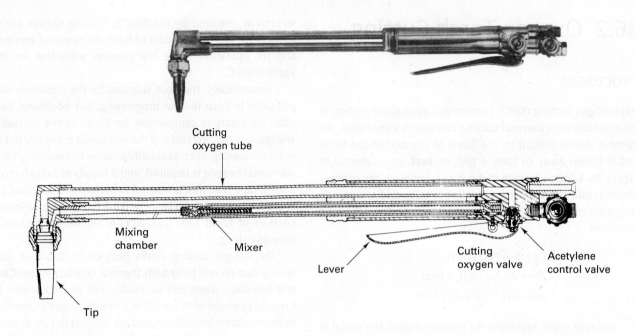

FIGURE 36.5 Oxyacetylene cutting torch and cross-sectional schematic. (Courtesy of Thermadyne Industries, Inc., St. Louis, MO)

Stack Cutting

When a modest number of duplicate parts are to be cut from thin sheet, but not enough to justify the cost of a blanking die, **stack cutting** may be the answer. The sheets should be flat, smooth, and free of scale, and they should be stacked and clamped together tightly so that there are no intervening gaps that could interrupt uniform oxidation or permit slag or molten metal to be entrapped. The resulting cut, however, will be less accurate than one produced by a blanking die.

Metal Powder Cutting, Chemical Flux Cutting, and Other Thermal Methods

When cutting hard-to-cut materials, modified torch techniques may be required. Metal powder cutting (POC) injects iron or aluminum powder into the flame to raise its cutting temperature. Chemical flux cutting (FOC) adds a fine stream of special flux to the cutting oxygen to increase the fluidity of the high-melting-point oxides. Both of these methods, however, have largely been replaced by plasma arc cutting (PAC), which is discussed as an extension of plasma arc welding later in this chapter. Laser- and electron-beam cutting will be presented with their welding parallels in a future chapter.

Underwater Torch Cutting

The thermal cutting of materials underwater presents a special challenge. A specially designed torch, such as the one shown in **Figure 36.6**, is used to cut steel. An auxiliary skirt surrounds the main tip, and an additional set of gas passages conducts a flow of compressed air that provides secondary oxygen for the oxyacetylene flame and expels water from the zone where the burning of metal occurs. The torch is either ignited in the usual manner before descent or by an electric spark device after being submerged. Acetylene gas is used for depths up to about 7.5 m (25 ft). For greater depths, hydrogen is used, because the environmental pressure is too great for the safe use of acetylene.

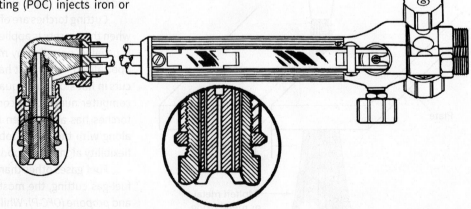

FIGURE 36.6 Underwater cutting torch. Note the extra set of gas openings in the nozzle to permit the flow of compressed air and the extra control valve. (Courtesy of Bastian-Blessing Company, Chicago, IL)

36.3 Flame Straightening

Flame straightening uses controlled, localized **upsetting** as a means of straightening warped or buckled material. **Figure 36.7** illustrates the theory of the process. If a straight piece of metal is heated in a localized area, such as the shaded area of the upper diagram, the metal on side *b* will be upset (i.e., plastically deformed) as it softens and tries to expand against the cooler restraining metal. When the upset portion cools, it will contract, and the resulting piece will be shorter on side *b*, forcing it to bend to the shape in the lower diagram.

If the starting material is bent or warped, as in the lower segment of Figure 36.7, the upper surface can be heated. Upsetting and subsequent thermal contraction will shorten the upper surface at a′, bringing the plate back to a straight or flat configuration. This type of procedure can be used to restore structures that have been bent in an accident, such as automobile frames.

A similar process can be used to flatten metal plates that have become dished. Localized spots about 50 mm (2 in.) in diameter are quickly heated to the upsetting temperature while the surrounding metal remains cool. Cool water is then sprayed onto the plate, and the contraction of the upset spot brings the buckle into an improved degree of flatness. To remove large buckles, the process might have to be repeated at several spots within the buckled area.

Several cautions should be noted. When straightening steel, consideration should be given to the possible phase transformations that could occur during the heating and cooling. Because rapid cooling is used and martensite can form, a subsequent tempering operation might be required. In addition, one should also consider the residual stresses that are induced and their effect on subsequent cracking, stress-corrosion cracking, and other modes of failure. The effects of phase transformations and residual stresses have been discussed more fully in Chapter 35.

Also, flame straightening should not be attempted with thin material. For the process to work, the metal adjacent to the heated area must have sufficient strength and rigidity to induce upsetting. If the material is too thin, localized heating and cooling will simply transfer the buckle from one area to another.

36.4 Arc Welding

With the development of commercial electricity in the late 19th century, it was soon recognized that an **arc** between two electrodes was a concentrated heat source that could produce temperatures approaching 4000°C (7000°F). As early as 1881, various attempts were made to use an arc as the heat source for fusion welding. A carbon rod was selected as one **electrode** and the metal workpiece became the other. **Figure 36.8** depicts the basic electrical circuit. If needed, filler metal was provided by a wire or rod that was independently fed into the arc. As the process developed, the filler metal replaced the carbon rod as the upper electrode. The metal wire not only carried the welding current, but as it melted in the arc, it also supplied the necessary filler.

The results of these early efforts were extremely uncertain. Because of the instability of the arc, a great amount of skill was required to maintain it, and contamination of the weld resulted from the exposure of hot metal to the atmosphere. There was little or no understanding of the metallurgical effects and requirements of arc welding. Consequently, while the great potential was recognized, very little use was made of the process until after World War I. Shielded metal electrodes were developed around 1920. These electrodes enhanced the stability of the arc by shielding it from the atmosphere and provided a fluxing action to the molten pool. The major problems of arc welding were overcome, and the process began to expand rapidly. It is estimated that 90% of all industrial welding is now performed with arc welding.

All **arc-welding** processes employ the basic circuit depicted in Figure 36.8. Welding currents vary from 1 to 4000 A (amps), with the range from 100 to 1000 being most typical. Voltages are generally in the range of 20 to 50 V. If direct current is used and the electrode is made negative, the condition is known as **DCEN**, for **direct-current electrode-negative**, or **straight polarity** (SPDC). Electrons are attracted to the positive workpiece, while ionized atoms in the arc column are accelerated toward the negative electrode. DCEN processes are characterized by fast melting of the electrode (high metal deposition rates) and a shallow molten pool on the workpiece (weld penetration). If the work is made negative and the electrode positive, the condition is known as **DCEP**, for

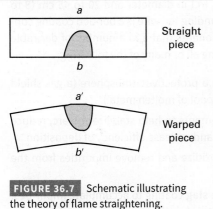

FIGURE 36.7 Schematic illustrating the theory of flame straightening.

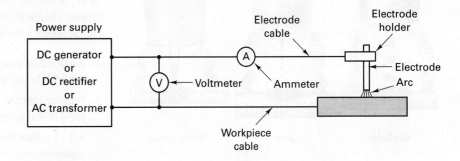

FIGURE 36.8 The basic electrical circuit for arc welding.

direct-current electrode-positive, or **reverse polarity (RPDC)**. The positive ions impinge on the workpiece, breaking up any oxide films and giving deeper penetration. Metal deposition rate is lower, however. Sinusoidal **alternating current** provides a 50–50 average of the preceding two modes and is a popular alternative to the direct current conditions. **Variable polarity** power supplies also alternate between DCEP and DCEN conditions, using rectangular waveforms to vary the fraction of time in each mode, as well as the frequency of switching. With these power supplies, weld characteristics can now be varied over a continuous range between DCEN and DCEP conditions.

In one group of arc-welding processes, the electrode is consumed (**consumable electrode processes**) and thus supplies the metal needed to fill the joint. Consumable electrodes have a melting temperature below the temperature of the arc. Small droplets are melted from the end of the electrode and pass to the workpiece. The size of these droplets varies greatly, and the transfer mechanism depends on the type of electrode, welding current, and other process parameters. **Figure 36.9** depicts metal transfer by the globular, spray, and short-circuit transfer modes. As the electrode melts, the arc length and the electrical resistance of the arc path will vary. To maintain a stable arc and satisfactory welding conditions, the electrode must be moved toward the work at a controlled rate. Manual arc welding is almost always performed with shielded (covered) electrodes, where the coating provides molten slag or isolating gas that protects the hot weld metal from oxidation and contamination. Continuous bare-metal wire can be used as the electrode in automatic or semiautomatic arc welding, but this is always in conjunction with some form of shielding and arc-stabilizing medium and automatic feed control devices that maintain the proper arc length.

The second group of arc-welding processes employs a tungsten (or carbon) electrode, which is not consumed by the arc, except by relatively slow vaporization. In these **nonconsumable-electrode processes**, a separate metal wire is required to supply the filler metal.

Because of the wide variety of processes available, arc welding has become a widely used means of joining material.

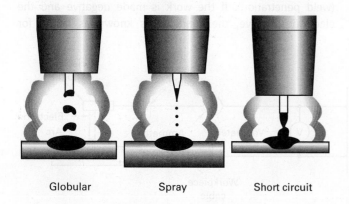

| Globular | Spray | Short circuit |

FIGURE 36.9 Three modes of metal transfer during arc welding. (Courtesy of Republic Steel Corporation, Youngstown, OH)

Each process and application, however, requires the selection or specification of the welding voltage, welding current, arc polarity (DCEN, DCEP, AC, or variable), arc length, welding speed (how fast the electrode is moved across the workpiece), arc atmosphere, electrode or filler material, and flux. Filler materials must be selected to match the base metal with respect to properties and/or alloy content (chemistry). For many of the processes, the quality of the weld also depends on the skill of the operator. Automation and robotics are reducing this dependence, but the selection and training of welding personnel are still of great importance.

36.5 Consumable-Electrode Arc Welding

Four processes make up the bulk of consumable-electrode arc welding:

1. Shielded metal arc welding (SMAW)
2. Flux-cored arc welding (FCAW)
3. Gas metal arc welding (GMAW)
4. Submerged arc welding (SAW)

These processes all have a medium rate of heat input and produce a fusion zone whose depth is approximately equal to its width. Because the fusion zone is composed of metal from both of the pieces being joined plus melted filler (i.e., electrode), the electrode must be of the same material as that being welded. Dissimilar metals or ceramics cannot be joined by these processes.

Shielded Metal Arc Welding

Shielded metal arc welding (SMAW), also called **stick welding** or covered-electrode welding, is among the most widely used welding processes because of its versatility and because it requires only low-cost equipment. The key to the process is a finite-length electrode that consists of metal wire, usually from 1.5 to 8 mm ($\frac{1}{16}$ to $\frac{5}{16}$ in.) in diameter and 20 to 45 cm (8 to 18 in.) in length. Surrounding the wire is a bonded coating containing chemical components that add a number of desirable characteristics, including all or many of the following:

1. Vaporize to provide a protective atmosphere (a gas shield around the arc and pool of molten metal).
2. Provide ionizing elements to help stabilize the arc, reduce weld metal spatter, and increase efficiency of deposition.
3. Act as a **flux** to deoxidize and remove impurities from the molten metal.
4. Provide a protective **slag** coating to accumulate impurities, prevent oxidation, and slow the cooling of the weld.

5. Add alloying elements that often enhance the ductility or strength of the weld.

6. Add additional filler metal.

7. Affect arc **penetration** (the depth of melting in the workpiece).

8. Influence the shape of the weld bead.

The coated electrodes are classified by the tensile strength of the deposited weld metal, the welding position in which they can be used, the preferred type of current and polarity (if direct current), and the type of coating. For mild-steel electrodes, a four- or five-digit system of designation has been adopted by the American Welding Society (AWS classification A5.1) and is presented in **Figure 36.10**. As an example, type E7016 is a low-alloy steel electrode that will provide a deposit with a minimum tensile strength of 70,000 psi (485 MPa) in the non-stress-relieved condition; it can be used in all positions, with either alternating or reverse-polarity direct current; and it has a low-hydrogen plus potassium coating. To assist in identification, all electrodes are marked with colors in accordance with a standard established by the National Electrical Manufacturers Association. Electrode selection consists of determining the electrode coating, coating thickness, electrode composition, and electrode diameter. The current type and polarity are matched to the electrode.

A variety of electrode coatings have been developed, and can be classified as cellulosic, rutile (titanium oxide), and basic. The cellulosic and rutile coatings contain variable amounts of SiO_2, TiO_2, FeO, MgO, Na_2O, and volatile matter. Upon decomposition, the volatile matter can release hydrogen, which can dissolve in the weld metal and lead to embrittlement or cracking in the joint. The basic coatings contain large amounts of calcium carbonate (limestone) and calcium fluoride (fluorspar), and produce low hydrogen weld metal. Because many of the electrode coatings can absorb moisture, and this is another source of undesirable hydrogen, the coated low-hydrogen electrodes are often baked at temperatures between 200° and

300°C (400° and 600°F) just prior to use and stored in an oven at 110° to 150°C (225° to 300°F).

To initiate a weld, the operator briefly touches the tip of the electrode to the workpiece and quickly raises it to a distance that will maintain a stable arc. The intense heat quickly melts the tip of the electrode wire, the coating, and portions of the adjacent base metal. As part of the electrode coating melts and vaporizes, it forms a protective atmosphere of CO, CO_2, and other gases that stabilizes the arc and protects the molten and hot metal from contamination. Other coating components surround the metal droplets with a layer of liquid flux and slag. The fluxing constituents unite with any impurities in the molten metal and float them to the surface to be entrapped in the slag coating that forms over the weld. The slag coating then protects the cooling metal from oxidation and slows the cooling rate to prevent the formation of hard, brittle structures. The glassy slag is easily chipped from the weld when it has cooled. **Figure 36.11** illustrates the shielded metal arc welding process, and **Figure 36.12** provides a schematic of metal deposition from a shielded electrode.

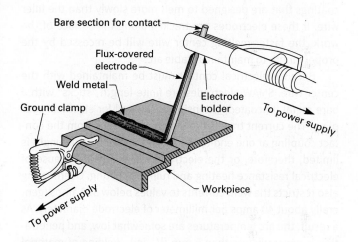

FIGURE 36.11 A shielded metal arc welding (SMAW) system.

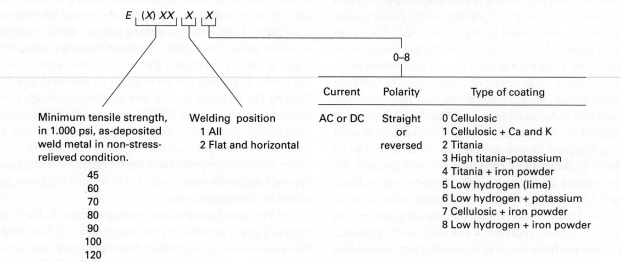

		Current	Polarity	Type of coating
Minimum tensile strength, in 1.000 psi, as-deposited weld metal in non-stress-relieved condition.	Welding position 1 All 2 Flat and horizontal	AC or DC	Straight or reversed	0 Cellulosic 1 Cellulosic + Ca and K 2 Titania 3 High titania–potassium 4 Titania + iron powder 5 Low hydrogen (lime) 6 Low hydrogen + potassium 7 Cellulosic + iron powder 8 Low hydrogen + iron powder
45 60 70 80 90 100 120				

$E \ (X) \ XX \ X \ X$

0–8

FIGURE 36.10 Designation system for arc-welding electrodes.

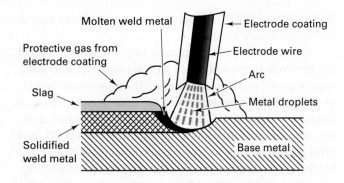

FIGURE 36.12 Schematic diagram of shielded metal arc welding (SMAW). (Courtesy of American Iron and Steel Institute, Washington, DC)

TABLE 36.3	Process Summary: Shielded Metal Arc Welding (SMAW)
Heat source	Electric arc
Protection	Slag from flux and gas from vaporized coating material
Electrode	Discontinuous, consumable
Material joined	Best for steel
Rate of heat input	Medium
Weld profile (D/W)	1
Current	<300 amps
Maximum penetration	3–6 mm
Assets	Cheap, simple equipment, all positions
Limitations	Discontinuous, shallow welds; requires slag removal, manual operation (slow and requires some skill)

Metal powder (usually iron) can be added to the electrode coating to significantly increase the amount of weld metal that can be deposited with a given size electrode wire and current—but with a noticeable decrease in penetration depth. Alloy elements can also be incorporated into the coating to adjust the chemistry of the weld. Special contact or drag electrodes utilize coatings that are designed to melt more slowly than the filler wire. If these electrodes are tracked along the surface of the work, the faster melting center wire will be recessed by the proper length to maintain a stable arc.

Because electrical contact must be maintained with the center wire, SMAW electrodes are finite-length "sticks" with a bare metal (uncoated) extension at the end for electrical coupling. The current travels the length of the stick from the contact coupling at one end to the arc at the other. Stick length is limited, therefore, or the electrode will overheat because of electrical resistance heating and ruin the coating. Overheating also restricts the weld currents to values below 300 amps (generally about 40 amps per millimeter of electrode diameter). As a result, the arc temperatures are somewhat low, and penetration is generally less than 5 mm ($^3/_{16}$ in.). Welding of material thicker than 5 mm will require multiple passes, and the slag coating must be removed between each pass.

Shielded metal arc welding is a simple, inexpensive, and versatile process, requiring only a power supply, power cables, electrode holder, and a small variety of electrodes. The equipment is portable and can even be powered by gasoline or diesel generators. Therefore, it is a popular process in job shops and is used extensively in repair operations. The process is best used for welding ferrous metals; carbon steels, alloy steels, stainless steels, and cast irons can all be welded. The electrode provides and regulates its own flux, and there is less sensitivity to wind and drafts than with the gas-shielded processes. Welds can be made in all positions. DCEP conditions are used to obtain the deepest possible penetration, with alternate modes being employed when welding thin sheets. The mode of metal transfer is either globular or short circuit. Unfortunately, the process is discontinuous (only short lengths of weld can be produced before a new electrode needs to be inserted into the holder), produces shallow welds, and requires slag removal after each welding pass. **Table 36.3** presents a process summary for shielded metal arc welding.

Flux-Cored Arc Welding

Flux-cored arc welding (FCAW) overcomes some of the limitations of the shielded metal arc process by moving the powdered flux to the interior of a continuous tubular electrode (**Figure 36.13**). When the arc is established, the vaporizing flux again produces a protective atmosphere and molten material forms an isolating slag layer over the weld pool that will require subsequent removal. Alloy additions (metal powders) can be blended into the flux to create a wide variety of filler metal chemistries. Compared to the stick electrodes of the shielded metal arc process, the flux-cored electrode is both continuous and less bulky, because binders are no longer required to hold the flux in place. The American Welding Society provides a designation system for the tubular electrodes in its standard AWS A5.20.

The continuous electrode is fed automatically through a welding gun, with electrical contact being maintained through the bare metal exterior of the wire at a position near the exit of the gun. [Note: By positioning the contact a short distance from the arc, some electrical resistance heating of the wire will occur prior to melting in the arc.] Overheating of the electrode is no longer a problem, and welding currents can be increased to about 500 A. The higher heat input increases penetration depth to about 1 cm ($^3/_8$ in.). The process is best used for welding steels, and welds can be made in all positions (the fast-freezing flux enabling vertical and overhead welding). Direct current electrode positive (DCEP) conditions are almost always used for the enhanced penetration. High deposition rates are possible, but the equipment cost is greater than SMAW because of the need for a controlled wire feeder and more costly power supply. Good ventilation is required to remove the fumes generated by the vaporizing flux.

In the basic flux-cored arc welding process (FCAW-S), the shielding gas is provided by the vaporization of flux components (self-shielding electrodes). Better protection and cleaner welds can be produced by a process variation (FCAW-G) that combines the electrode flux with a flow of externally supplied shielding gas, such as CO_2.

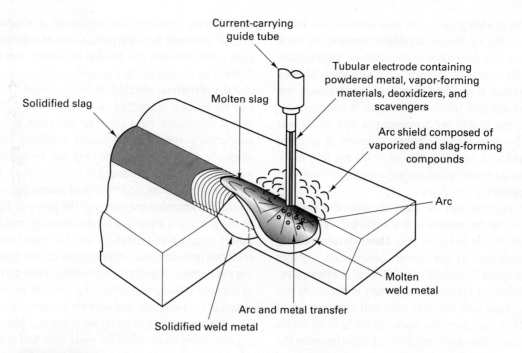

FIGURE 36.13 The flux-cored arc welding (FCAW) process. (Courtesy of American Welding Society, New York, NY.)

TABLE 36.4	Process Summary: Flux-Cored Arc Welding (FCAW)
Heat source	Electric arc
Protection	Slag and gas from flux (optional secondary gas shield)
Electrode	Continuous and consumable
Material joined	Best for steel
Rate of heat input	Medium
Weld profile (D/W)	1
Current	<500 amps
Maximum penetration	6–10 mm
Assets	Continuous electrode, all positions
Limitations	Requires slag removal, limited to ferrous and nickel-based metals

Table 36.4 presents a process summary of flux-cored arc welding.

Gas Metal Arc Welding

If the supplemental shielding gas flowing through the torch in the FCAW-G process becomes the primary protection for the arc and molten metal, there is no longer a need for the volatilizing flux. The consumable electrode can now become a continuous, solid, uncoated metal wire or a continuous hollow tube with powdered alloy additions in the center, known as a metal-cored electrode. The resulting process, shown in **Figure 36.14**, was formerly called **metal inert-gas (MIG) welding**, and is now known as **gas metal arc welding (GMAW)**. The arc is still maintained between the workpiece and the automatically fed bare-wire electrode, which continues to provide the

necessary filler metal. Electrode diameters range from 0.6 to 6.4 mm (0.02 to ¼ in.). The welding current, penetration depth, and process cost are all similar to the flux-cored process.

Because shielding is provided by the flow of gas, and fluxing and slag-forming agents are no longer required, the gas metal arc process can be applied to all metals. Carbon and stainless steels, and alloys of aluminum, magnesium, copper and nickel are the most common. Argon, helium, and mixtures of the two are the primary shielding gases. When welding steel, some O_2 or CO_2 is usually added to improve the arc stability and reduce weld spatter. The cheaper CO_2 can also be used alone when welding steel, provided that a deoxidizing electrode wire is employed. Nitrogen and hydrogen can also be added to modify arc characteristics. Because these shielding gases only provide protection and do not remove existing contamination, starting cleanliness is critical to the production of a good weld.

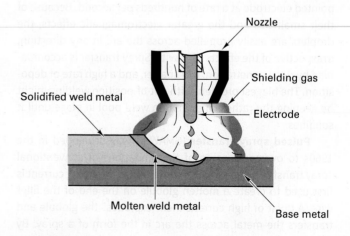

FIGURE 36.14 Schematic diagram of gas metal arc welding (GMAW). (Courtesy of American Iron and Steel Institute, Washington, DC)

The specific shielding gases can have considerable effect on the stability of the arc, the metal transfer from the electrode to the work, and also the heat transfer behavior, penetration, and tendency for undercutting (weld pool extending laterally beneath the surface of the base metal). Helium produces the hottest arc and deepest penetration. Argon is intermediate, and CO_2 yields the lowest arc temperatures and shallowest penetrations. Because argon is heavier than air, it tends to blanket the weld area, enabling the use of low gas flow rates. The lighter-than-air helium generally requires higher flow rates than either argon or carbon dioxide.

Electronic controls can be used to alter the welding current, enabling further control of the metal transfer mechanism, shown previously in Figure 36.9. **Short-circuit transfer** (GMAW-S) is promoted by the lowest currents and voltages (14 to 21 V), the use of CO_2 shield gas, and small-diameter electrodes. The advancing electrode (or molten metal on its tip) makes direct contact with the weld pool, and the short circuit causes a rapid rise in current. Big molten globs form on the tip of the electrode and then separate, forming a gap between the electrode and workpiece. This gap reinitiates a brief period of arcing, but the rate of electrode advancement exceeds the rate of melting in the arc, and another short circuit occurs. The power conditions oscillate between arcing and short circuiting at a rate of 20 to 200 cycles per second. Short-circuit transfer is preferred when joining thin materials and can be used in all welding positions. The deposition rate is relatively low.

If the voltage and amperage are increased, the mode becomes one of **globular transfer**. The electrode melts from the heat of the arc, and metal drops form with a diameter equal to or greater than the diameter of the electrode wire. Gravity and electromagnetic forces then transfer the drops to the workpiece at a rate of several per second. Because gravity plays a role in metal transfer, there is a definite limitation on the positions of welding. This is the least desirable mode of transfer because the arc is loud and erratic, with lots of splashing or spatter.

Spray transfer (GMAW-ST) occurs with even higher currents and voltages (25 to 32 V and about 200 A), argon gas shielding, and DCEP conditions. Small droplets emerge from a pointed electrode at a rate of hundreds per second. Because of their small size and the greater electromagnetic effects, the droplets are easily propelled across the arc in any direction, irrespective of the effects of gravity. Spray transfer is accompanied by deep penetration, low spatter, and a high rate of deposition. The biggest problem with out-of-position welding might be keeping the rather large molten weld pool in place until it solidifies.

Pulsed spray transfer (GMAW-P) was developed in the 1960s to overcome some of the limitations of conventional spray transfer. In the **pulsed arc** mode, a low welding current is first used to create a molten globule on the end of the filler wire. A burst of high current then "explodes" the globule and transfers the metal across the arc in the form of a spray. By alternating low and high currents at a rate of 60 to 600 times per second, the filler metal is transferred in a succession of rapid bursts, similar to the emissions of a rapidly squeezed aerosol atomizer. With the pulsed form of deposition, the weld pool cools between the periods of molten metal deposition. There is less heat input to the weld, and the weld temperatures and size of the heat-affected zone are reduced. Thinner material can be welded, distortion is reduced, workpiece discoloration is minimized, heat-sensitive parts can be welded, high-conductivity metals can be joined, electrode life is extended, electrode cooling techniques may not be required, and fine microstructures are produced in the weld pool. Welds can be made in all positions, and the use of pulsed power lowers spattering and improves the safety of the process. The high speed of the process is attractive for productivity, and the energy or power required to produce a weld is lower than with other methods (reduced cost). Adjustments can be made to the pulsing parameters of background current, peak current, duration of the peak pulse, and number of pulses per second to alter the shape of the weld pool and vary the penetration.

In general, the gas metal arc process is fast and economical and currently accounts for more than half of all weld metal deposition. There is no frequent change of electrodes as with the shielded metal arc process. No flux is required, and no slag forms over the weld. Thus, multiple-pass welds can be made without the need for intermediate cleaning. The process can be readily automated, and the lightweight, compact welding unit lends itself to robotic manipulation. A direct-current electrode-positive (DCEP) arc is generally used because of its deep penetration, spray transfer, and the ability to produce smooth welds with good profile. Process variables include type of current, current magnitude, shielding gas, electrode diameter, electrode composition, electrode stickout (extension beyond the gun), welding speed, welding voltage, and arc length. **Table 36.5** provides a process summary for gas metal arc welding.

Several process modifications have recently emerged for GMAW. While nearly all GMAW welding is performed in the direct-current electrode-positive (DCEP) mode because of the features cited above, newer power supplies allow a variable combination of DCEP and DCEN. Weld penetration can be varied, enabling this **variable-polarity GMAW** to weld thin steel

TABLE 36.5	Process Summary: Gas-Metal Arc Welding (GMAW)
Heat source	Electric arc
Protection	Externally supplied shielding gas
Electrode	Continuous and consumable
Material joined	All common metals
Rate of heat input	Medium
Weld profile (D/W)	1
Current	<500 amps
Maximum penetration	6–10 mm
Assets	No slag to remove, all positions, easily automated and minimal post-weld cleaning
Limitations	More costly equipment than SMAW or FCAW, torch and cables may restrict accessibility

and aluminum sheet metal and tubing. In a process modification known as *advanced gas metal arc welding* (AGMAW), a second power source is used to preheat the filler wire before it emerges from the welding torch. Less arc heating is needed to produce a weld, so less base metal is melted, producing less dilution of the filler metal and less penetration. In *hybrid induction arc welding* (HIAW), an induction coil immediately before the GMAW torch preheats the joint edges to near the melting point. The welding arc then melts the filler metal, with the combination yielding a two to four times increase in welding speed with reduced weld distortion. The ability to weld square butt edges reduces the cost of weld preparation. *Twin-wire GMAW* increases the deposition rate by feeding two wires through the same torch. *Cold metal transfer* provides a lower-temperature short-circuit process, by oscillating the tip of the electrode forward and back, the withdrawal acting to assist in molten metal transfer. The time of high current flow during the short circuit is reduced, lowering the heat input and enabling the welding of thinner or more heat-sensitive materials.

Another recent modification is the use of flat electrode wire, typically having a rectangular cross section of about 4 mm by 0.5 mm. (0.15 in. by 0.02 in.). By having a larger surface area participating in the arc, deposition rate is similar to a two-wire feed with only a single wire delivery system. The arc is also

asymmetric. Orienting the wire perpendicular to the weld seam produces a wide, shallow weld pool, suitable for bridging gaps and often eliminating the need to weave during deposition. A narrower, deeper weld pool results when the wire is parallel to the weld. Varying the angle between parallel and perpendicular generates a spectrum of weld pool geometries.

Similar penetration variation has been achieved by rotating the emerging tip of a round electrode wire (between 2000 and 4000 rpm) with the ability to vary both the spin diameter and speed. As the spin diameter increases, the droplets that are deposited around the circumference of rotation produce a wider, shallower weld. A slower, narrower rotation increases penetration.

Submerged Arc Welding

No shielding gas is used in the **submerged arc welding (SAW)** process, depicted in **Figure 36.15**. Instead, a thick layer of granular flux is deposited just ahead of a solid bare-wire consumable electrode, and the arc is maintained beneath the blanket of flux with only a small amount of smoke or flames being visible. A portion of the flux melts and acts to remove impurities from the rather large pool of molten metal. The molten flux solidifies into a glasslike covering over the weld,

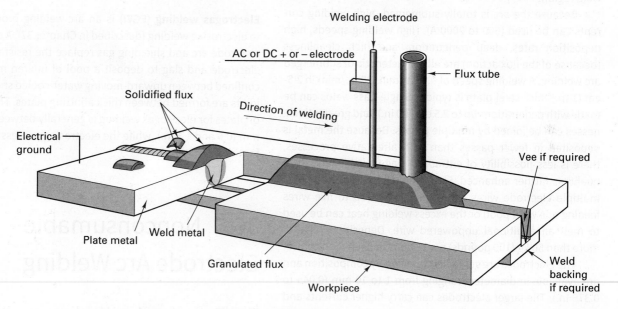

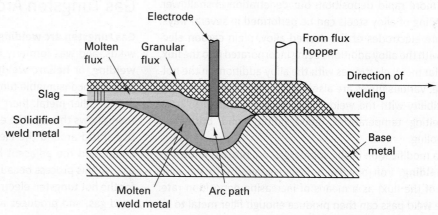

FIGURE 36.15 (Top) Basic features of submerged arc welding (SAW). (Courtesy of Linde Division, Union Carbide Corporation, Houston, TX) (Bottom) Cutaway schematic of submerged arc welding. (Courtesy of American Iron and Steel Institute, Washington, DC)

and, upon further cooling, cracks loose from the weld (because of differences in thermal contraction) and is easily removed. The unmelted granular flux is recovered by a vacuum system and is reused. Both layers of flux provide shielding and thermal insulation to the weld pool, where slow cooling produces soft, ductile welds.

Submerged arc welding is most suitable for making flat-butt or fillet welds in low-carbon steels (less than 0.3% carbon). With some preheat and postheat precautions, medium-carbon and alloy steels and some cast irons, stainless steels, copper alloys, and nickel alloys can also be welded. Welding is restricted to the horizontal position because the large, high-fluidity weld pool, flux and slag are all held in place by gravity. For circumferential joints, as when joining pipe, the workpiece is rotated under a fixed welding head so that all welding takes place in the flat position. Other process limitations include the need for extensive flux handling, possible contamination of the flux by moisture (leading to porosity in the weld), the large volume of slag that must be removed, and shrinkage problems because of the large weld pool. The high heat inputs can produce large grain-size structures, and the slow cooling rate can enable segregation and possible hydrogen or hot cracking. In addition, chemical control is quite important, because the electrode material often contributes more than 70% of the molten weld region.

Because the arc is totally submerged, high welding currents can be used (600 to 2000 A). High welding speeds, high deposition rates, deep penetration, and high cleanliness (because of the flux action) are all characteristic of submerged arc welding. A welding speed of 0.75 m/min (25 in./min) in 2.5-cm (1-in.) thick steel plate is typical. Single-pass welds can be made with penetrations up to 2.5 cm (1.0 in.), and greater thicknesses can be joined by multiple passes. Because the metal is deposited in fewer passes than with alternative processes, there is less possibility of entrapped slag or voids, and weld quality is further enhanced. For even higher deposition rates, multiple electrode wires can be employed (up to five wires feeding one weld pool) or the excess welding heat can be used to melt an additional unpowered wire. Deposition rates of more than 50 kg (100 pounds) per hour have been reported.

The electrodes are generally classified by composition and are available in diameters ranging from 1 to 10 mm (0.045 to 0.375 in.). The larger electrodes can carry higher currents and enable more rapid deposition, but penetration is shallower. The welding of alloy steels can be performed in several ways: solid wire electrodes of the desired alloy, plain carbon electrodes with the alloy additions being incorporated into the flux, or tubular metal electrodes with the alloy additions in the hollow core. Various fluxes are also available and are selected for compatibility with the weld metal. All are designed to have low melting temperatures, good fluidity, and brittleness after cooling.

In a modification of the submerged arc process known as **bulk welding**, iron powder is first deposited into the joint (ahead of the flux) as a means of increasing deposition rate. A single weld pass can then produce enough filler metal to be

TABLE 36.6	Process Summary: Submerged Arc Welding (SAW)
Heat source	Electric arc
Protection	Granular flux provides slag and an isolation blanket
Electrode	Continuous and consumable
Material joined	Best for steel
Rate of heat input	Medium
Weld profile (D/W)	1
Current	<1000 amps
Maximum penetration	25 mm
Assets	High-quality welds, high deposition rates and deep penetration
Limitations	Requires slag removal, difficult for overhead and out-of-position welding and joints often require backing plates

equivalent to seven or eight conventional submerged arc passes.

Table 36.6 provides a summary of the submerged arc welding process.

Electrogas Welding

Electrogas welding (EGW) is an arc-welding process similar to electroslag welding (described in Chapter 37). A consumable electrode arc and shielding gas replace the resistance-melted electrode and slag to deposit a pool of molten metal that is confined between upward-moving water-cooled shoes. Vertical joints are formed between thick abutting plates. The thickness of plates for electrogas welding is generally between 13 and 38 mm (0.5 and 1.5 in.), while the electroslag process is preferred for thicker material.

36.6 Nonconsumable Electrode Arc Welding

Gas Tungsten Arc Welding

Gas tungsten arc welding (GTAW) produces very high quality welds, and was formerly known as **tungsten inert-gas (TIG) welding**, or **heliarc welding** when helium was the shielding gas. A nonconsumable tungsten electrode provides the arc but not the filler metal. Inert gas (argon, helium, or a mixture of them) flows through the electrode holder to provide a protective shield around the electrode, the arc, the pool of molten metal, and the adjacent heated areas. (*Note*: CO_2 cannot be used in this process because it provides inadequate protection for the hot tungsten electrode.) While argon is the most widely used gas, and produces a smoother, more stable arc, helium can be added to increase the heat input (higher welding speeds and deeper penetration). Helium alone might be preferred for overhead welding because it is lighter than air and flows upward. Hydrogen is a reactive gas and is sometimes added to the shielding mixture to prevent the formation of undesirable oxides in the molten weld metal.

The composition, diameter, length, and tip geometry (balled, pointed, or truncated cone and the angle of the point or cone) of the tungsten or tungsten alloy electrode are selected based on the material being welded, the thickness of the material, and the type of current being used. The tungsten is often alloyed with 1% to 4% thorium oxide, zirconium oxide, cerium oxide, or lanthanum oxide to provide better current-carrying and electron-emission characteristics and longer electrode life. Because tungsten is not consumed at the temperatures of the arc, the arc length remains constant, and the arc is very stable and easy to maintain. **Figure 36.16** shows a typical GTAW torch with cables and passages for gas flow, power, and cooling water.

In applications where there is a close fit between the pieces being joined, no filler metal might be needed. When filler metal is required, it is usually supplied as a separate rod or wire as illustrated in **Figure 36.17**. The filler metal is generally selected to match the chemistry and/or tensile strength of the metal being welded. When high deposition rates are desired, a separate resistance heating circuit can be provided to preheat the filler wire. As shown in **Figure 36.18**, the deposition rate of heated wire can be several times that of a cold wire. By oscillating the filler wire from side to side while making a weld pass, the deposition rate can be further increased.

GTAW is considered to be the most versatile of the arc-welding processes. With skilled operators, it can produce

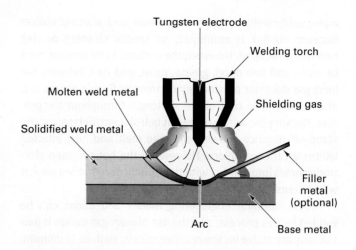

FIGURE 36.17 Diagram of gas tungsten arc welding (GTAW). (Courtesy of American Iron and Steel Institute, Washington, DC)

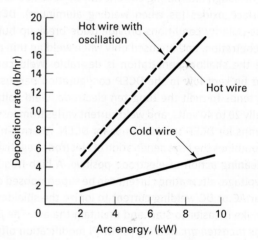

FIGURE 36.18 Comparison of the metal deposition rates in GTAW with cold, hot, and oscillating-hot filler wire. (Courtesy of Welding Journal)

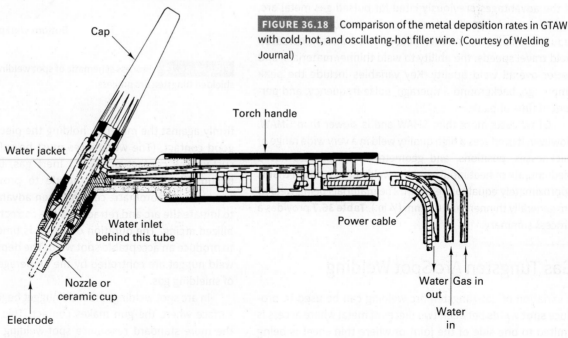

FIGURE 36.16 Welding torch used in nonconsumable electrode, gas tungsten arc welding (GTAW), showing feed lines for power, cooling water, and inert gas flow. (Courtesy of Linde Division, Union Carbide Corporation, Houston, TX)

high-quality welds that are very clean and scarcely visible. Because no flux is employed, no special cleaning or slag removal is required. However, the surfaces to be welded must be clean and free of oil, grease, paint, and rust, because the inert gas does not provide any cleaning or fluxing action. It is also important to control the arc length throughout the process. Because the arc is somewhat bell-shaped, decreasing the stand-off distance will decrease the melt and heat-affected widths on the workpiece. However, if the hot tungsten electrode comes into contact with the workpiece or molten pool, it will contaminate the electrode.

Most common engineering metals and alloys can be welded by this process, and the use of inert gas makes it particularly attractive for the reactive metals, such as aluminum, magnesium, and titanium, as well as the high-temperature superalloys and refractory metals. Maximum penetration is obtained with direct current electrode negative (DCEN) conditions, although alternating current can be specified to break up surface oxides (as when welding aluminum). DCEP or reverse-polarity conditions provide oxide break-up but with low penetration, and are used only when welding thin pieces where the shallow penetration is desirable. Weld currents should be kept low in the DCEP configuration because this mode tends to melt the tungsten electrode. Weld voltage is typically 20 to 40 volts, and weld current varies from less than 125 amps for DCEP to 1000 amps for DCEN. Alternating current combines the high penetration of electrode negative with the cleaning action of electrode positive. A high-frequency, high-voltage, alternating current can be superimposed on the regular AC or DC welding current to ionize the shielding gas and make it easier to start and maintain the arc. The *pulsed arc gas tungsten arc welding (GTAW-P)* modification offers all of the advantages previously cited for pulsed gas metal arc, including reduced bead width, increased penetration, reduced heat input, smaller heat-affected zones, increased weld travel speeds, the ability to weld thinner materials, and better overall weld quality. Key variables include the peak amperage, background amperage, pulse frequency, and percent of time at peak.

GTAW costs more than SMAW and is slower than GMAW. However, it produces a high quality weld in a very wide range of thicknesses, positions, and geometries. The process has a medium rate of heat input, and the welds have a depth that is approximately equal to the width. The materials being welded are generally thinner than 6.5 mm (¼ in.). **Table 36.7** provides a process summary.

Gas Tungsten Arc Spot Welding

A variation of gas tungsten arc welding can be used to produce **spot welds** between two pieces of metal where access is limited to one side of the joint or where thin sheet is being attached to heavier material. The basic procedure is illustrated in **Figure 36.19**. A modified tungsten inert-gas gun is used with a vented nozzle on the end. The nozzle is pressed

TABLE 36.7	Process Summary: Gas Tungsten Arc Welding (GTAW)
Heat source	Electric arc
Protection	Externally supplied shielding gas
Electrode	Nonconsumable
Material joined	All common metals
Rate of heat input	Medium
Weld profile (D/W)	1
Current	<500 amps
Maximum penetration	3 mm
Assets	High-quality welds, no slag to be removed, all positions
Limitations	Slower than consumable electrode GMAW, sensitive to cleanliness

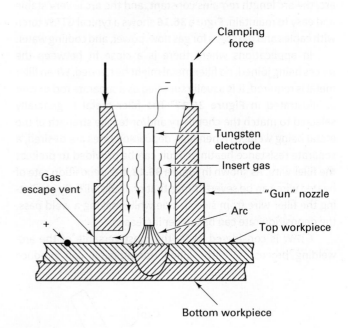

FIGURE 36.19 Process schematic of spot welding by the inert-gas-shielded tungsten arc process.

firmly against the material, holding the pieces in reasonably good contact. (The workpieces must be sufficiently rigid to sustain the contact pressure.) Inert gas, usually argon or helium, flows through the nozzle to provide a shielding atmosphere. Automatic controls then advance the electrode to initiate the arc and retract it to the correct distance for stabilized arcing. The duration of arcing is timed automatically to produce an acceptable spot weld. The depth and size of the weld nugget are controlled by the amperage, time, and type of shielding gas.

In arc spot welding, the weld nugget begins to form at the surface where the gun makes contact. This is in contrast to the more standard resistance spot-welding methods, where the weld nugget forms at the interface between the two members. Each technique has its characteristic advantages and disadvantages.

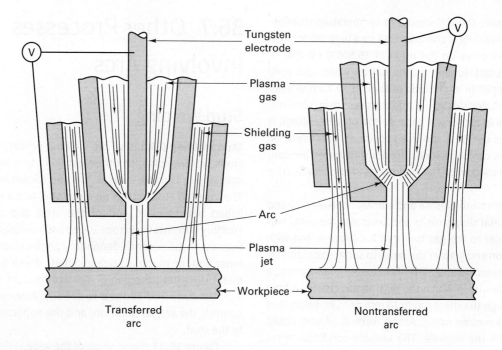

FIGURE 36.20 Two types of plasma arc torches: (left) transferred arc; (right) nontransferred arc.

Plasma Arc Welding

In **plasma arc welding (PAW),** the arc is maintained between a nonconsumable electrode and either the welding gun (**nontransferred arc**) or the workpiece (**transferred arc**) as illustrated in **Figure 36.20**. The nonconsumable tungsten electrode is set back within the "torch" in such a way as to force the arc to pass through or be contained within a small-diameter nozzle. An inert gas (usually argon) is forced through this constricted arc, where it is heated to a high temperature and forms a hot, fast-moving **plasma**. The emerging stream of superheated ionized gas, called the **orifice gas**, then transfers its heat to the workpiece and melts the metal. A second flow of inert gas surrounds the plasma column and provides shielding to the weld pool. When filler metal is needed, it is provided by an external feed.

Figure 36.21 presents a comparison of the nonconstricted arc of the GTAW process and the constricted arc of plasma arc

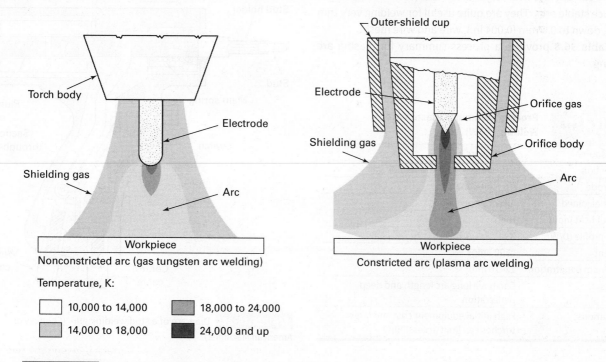

FIGURE 36.21 Comparison of the nonconstricted arc of gas tungsten arc welding and the constricted arc of the plasma arc process. Note the level and distribution of temperature. (Courtesy ASM International, Materials Park, OH)

welding, and shows the differences in temperature distribution. Plasma arc welding is characterized by a high rate of heat input and temperatures on the order of 16,500°C (30,000°F). This in turn offers fast welding speeds, narrow welds with deep penetration (a depth-to-width ratio of about 3), a narrow heat-affected zone, reduced distortion, and a process that is insensitive to variations in arc length because the plasma column is cylindrical. Welds can be made in all positions, and nearly all metals and alloys can be welded. The nontransferred arc configuration can be used when the workpieces are nonconductive materials.

With a low-pressure plasma and currents between 20 and 100 amps, the metal simply melts and flows into the joint. This condition is similar to the gas tungsten arc process, but with higher penetration and greater tolerance to surface contamination. At higher pressures and currents in excess of 100 amps, a **keyhole effect** occurs in which the plasma gas creates a hole completely through the sheet (up to 10 mm or $3/8$ in. thick) that is surrounded by molten metal. As the torch is moved, liquid metal flows to fill the keyhole. The keyhole condition offers deep penetration and high welding speeds. If the gas pressure is increased even further, the molten metal is expelled from the region, and the process becomes one of plasma cutting, which is discussed later in this chapter.

Many plasma torches employ a small, nontransferred arc within the torch to heat the orifice gas and ionize it (a pilot arc). The ionized gas then forms a good conductive path for the main transferred arc. This dual-arc technique permits instant ignition of a low-current arc, which is more stable than that of an ordinary plasma torch. Separate DC power supplies are used for the pilot and main arcs. **Microplasma,** or **needle arc,** torches can operate with very low currents (0.1 to 20 A) and still produce stable arcs. They are quite useful for welding very thin sheet, down to 0.1 mm (0.004 in.), wire and wire mesh.

Table 36.8 provides a process summary for plasma arc welding.

TABLE 36.8	**Process Summary: Plasma Arc Welding (PAW)**
Heat source	Plasma arc
Protection	Externally supplied shielding gas
Electrode	Nonconsumable
Material joined	All common metals
Rate of heat input	High
Weld profile (D/W)	3
Current	<500 amps
Maximum penetration	12–18 mm
Assets	Can have long arc length and deep penetration
Limitations	High initial equipment cost and large torches can limit accessibility

36.7   Other Processes Involving Arcs

Stud Welding

Stud welding (SW) is an arc-welding process used to attach studs, screws, pins, or other fasteners to a metal surface. A special gun is used, such as the one shown in **Figure 36.22**. The inserted stud acts as an electrode, and a DC arc is established between the end of the stud and the workpiece. Welding currents vary from 200 to 2400 amps, and weld times from 0.1 to 1.5 seconds depending on the diameter of the fastener and the materials being joined. After a small amount of metal is melted, the two pieces are brought together under light pressure and allowed to solidify. Automatic equipment controls the arc, its duration, and the application of pressure to the stud.

Figure 36.23 shows some of the wide variety of studs that are specially made for this process. Many contain a recessed end that is filled with flux. A ceramic ferrule, such as the one shown in the center photo of Figure 36.23, can be placed over the end of the stud before it is positioned in the gun. During the arc, the ferrule serves to concentrate the heat and isolate the hot metal from the atmosphere. It also confines the molten or softened metal and shapes it around the base of the stud, as shown in the photo on the right of Figure 36.23. After the weld has cooled, the brittle ceramic is broken free and removed.

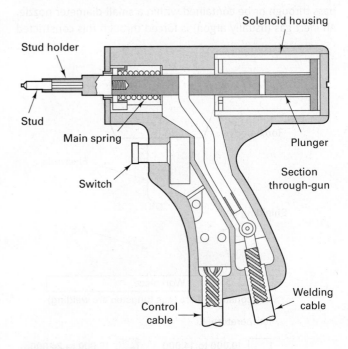

FIGURE 36.22   Diagram of a stud welding gun. (Courtesy of American Machinist)

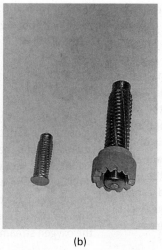

(a)　　　　　　　　　(b)　　　　　　　　　(c)

FIGURE 36.23　(a) Threaded studs being welded utilizing the drawn arc-welding process; (b) (left) Flanged stud (right) stud with flux and ceramic ferrule; (c) cross section of an internally tapped stud weld. Note that the fusion zone is composed of metal from both the stud and the plate. (Courtesy of Nelson Stud Welding Co., Elyria, OH)

Because burn-off or melting reduces the length of the stud, the original dimensions should be selected to compensate.

Stud welding requires almost no skill on the part of the operator. Once the stud and ferrule are placed in the gun and the gun positioned on the work, all the operator has to do is pull the trigger. Thus, the process is well suited to manufacturing and can be used to eliminate the drilling and tapping of many special holes. Production-type stud welders can produce more than 1000 welds per hour.

Flash and Percussion Welding

Flash welding (FW) is a process used to produce butt welds between similar or dissimilar metals in solid or tubular form. The two pieces of metal are first secured in current-carrying grips and lightly touched together. An electric current can be passed through the joint to provide optional preheat, after which the pieces are withdrawn slightly. An intense flashing arc forms across the gap, which melts material on both surfaces. The pieces are then forced together under high pressure (on the order of 70 MPa, or 10,000 psi), expelling the liquid and oxides, and upsetting the softened metal.

The electric current is turned off, and the force is maintained until solidification is complete. If desired, the upset portion can then be removed by machining. **Figure 36.24** shows a schematic of the flash welding process, including both the equipment setup and the completed weld.

To produce a high-quality weld, it is important that the initial surfaces be flat and parallel so that the flashing is even across the area to be joined. The flashing action must be continued long enough to melt the interface and also soften the

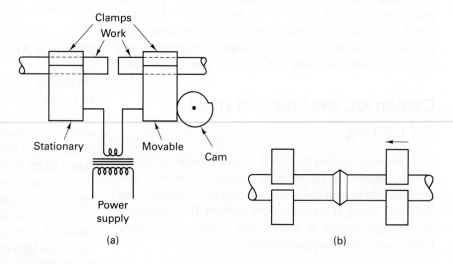

(a)　　　　　　　　　　　　　　　　(b)

FIGURE 36.24　Schematic of the flash-welding process (a) equipment and setup; (b) completed weld.

adjacent metal. Sufficient plastic deformation must occur during the upsetting to transfer the impurities and contaminants outward into the flash. The equipment required is generally large and expensive, but excellent welds can be made at high production rates. The process can be applied to any material that can be forged. Unlike friction welding, no rotational symmetry is required.

Percussion welding (PEW) is a similar process, in which a rapid discharge of stored energy from a capacitor bank produces a brief period of arcing, which is followed by the rapid application of force to expel the molten metal and produce the joint. In percussion welding, the duration of the arc is on the order of 1 to 10 ms. The heat is intense but highly concentrated. Only a small amount of molten metal is produced, little or no upsetting occurs at the joint, and the heat-affected zone is quite small. Application is generally restricted to the butt welding of bar or tubing where heat damage is a major concern.

36.8 Arc Cutting

While oxygen torch cutting has been discussed earlier in this chapter, and laser cutting will be covered in a future chapter, there are a number of arc-cutting methods. Virtually all metals can be cut by some form of electric arc. In the **arc cutting** processes, the material is melted by the intense heat of the arc and then permitted, or forced, to flow away from the region of the slit or notch (**kerf**). Most of the techniques are simply adaptations of the arc-welding procedures discussed in this chapter. Each has its inherent characteristics and capabilities, including tolerance, thickness capability, kerf width, edge squareness, size of the heat-affected zone, and cost. Selection depends upon factors such as tolerance requirements, the subsequent processes that will be performed on the cut part, and the end use of the product. The ideal cut would have a low bevel angle (less than 1°), no rounding of the top edge, no dross on the bottom edge, minimal heat-affected zone, and a smooth cut face.

Carbon Arc and Shielded Metal Arc Cutting

The carbon arc cutting (CAC) and shielded metal arc cutting (SMAC) methods use the arc from a carbon or shielded metal arc electrode to melt the metal, which is then removed from the cut by gravity or the force of the arc itself. These processes are generally limited to small shops, garages, and homes, where investment in equipment is limited.

Air Carbon Arc Cutting

In air carbon arc cutting (AAC), the arc is again maintained between a carbon electrode and the workpiece, but high-velocity jets of air are directed at the molten metal from holes in the electrode holder. While there is some oxidation, the primary function of the air is to blow the molten material from the cut. Air carbon arc cutting is particularly effective for cutting cast iron and preparing steel plates for welding. Speeds up to 0.6 m/min (2 ft/min) are possible, but the process is quite noisy, and hot metal particles tend to be blown over a substantial area.

Oxygen Arc Cutting

In oxygen arc cutting (AOC), an electric arc and a stream of oxygen are combined to make the cut. The stick electrode is a coated ferrous-metal tube. The coated metal serves to establish a stable arc, while oxygen flows through the bore and is directed on the area of incandescence. With easily oxidized metals, such as steel, the arc preheats the base metal, which then reacts with oxygen, becomes liquefied, and is expelled by the oxygen stream.

Gas Metal Arc Cutting

If the wire feed rate and other variables of gas metal arc welding (GMAW) are adjusted so that the electrode penetrates completely through the workpiece, cutting rather than welding will occur, and the process becomes gas metal arc cutting (GMAC). The wire feed rate controls the quality of the cut, and the voltage determines the width of the slit or kerf. The process can be performed continuously, but there is a high rate of electrode consumption and high current is required.

Gas Tungsten Arc Cutting

Gas tungsten arc cutting (GTAC) employs the same basic circuit and shielding gas as used in gas tungsten arc welding, with a high-velocity jet of gas added to expel the molten metal.

Plasma Arc Cutting

The torches used in plasma arc cutting (PAC) produce the highest temperatures available from any practical source. With the nontransferred type of torch, the arc column is completely within the nozzle, and a temperature of about 16,500°C (30,000°F) is obtained. With the transfer-type torch, depicted in **Figure 36.25**, the arc is maintained between the electrode and the workpiece, and temperatures can be as high as 33,000°C (60,000°F). Ionized gases flowing at these temperatures and near supersonic speeds are capable of cutting virtually any electrically conductive material simply by melting it and blowing it away from the cut. Materials up to 75 mm (3 in.) thick can be cut with plasma torches.

Early efforts to employ this technique showed that the speed, versatility, and operating cost were far superior to those of the oxyfuel cutting methods. However, the early systems could not constrict the arc sufficiently to produce the quality of cut needed to meet the demands of manufacturing. Therefore, plasma arc cutting was generally limited to those materials

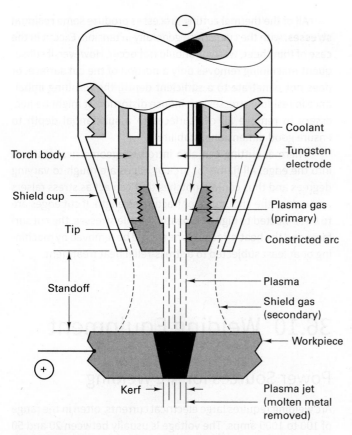

Torch body

Shield cup

Tip

Standoff

Coolant

Tungsten electrode

Plasma gas (primary)

Constricted arc

Plasma

Shield gas (secondary)

Workpiece

Kerf

Plasma jet (molten metal removed)

FIGURE 36.25 Plasma arc cutting with a transferred arc torch.

that could not be cut by the oxidation type of cutting techniques, particularly aluminum and stainless steel. In the 1970s, swirling of the plasma gas combined with radial impingement of water on the arc was found to produce the desired constriction. A steam curtain forms at the plasma-water interface, shielding the plasma, shrinking its diameter, and creating an intense, highly focused column of heat that liquefies metal and ejects it out the bottom of the column. Water-injected torches can now cut virtually any metal in any position. Magnetic fields have also been used to constrict the arc and can produce high-quality cuts without the need for water impingement.

Compared to oxyfuel cutting, plasma cutting is more economical (cost per cut is a fraction of oxyfuel), more versatile (can cut all metals as easily as mild steel), and much faster (typically, five to eight times faster than oxyfuel). The combination of the extremely high temperatures and jet-like action of the plasma produces narrow kerfs and remarkably smooth surfaces, nearly as smooth as can be obtained by sawing. Plasma-cut surfaces are often within 2° of vertical, and surface oxidation is nearly eliminated by the cooling effect of the water spray. In addition, the heat-affected zone in the metal is only one-third to one-fourth as large as that produced by oxyfuel cutting, and a preheat cycle is not required in the cutting of steel. Heat-related distortion is extremely small. Cuts can be made through scale, rust, paint, or surface primer. Plasma cutters are being progressively replaced by lasers for cutting thin material, but plasma systems can cut thicker materials faster than lasers and produce quality cuts.

Transferred-arc torches are usually used for cutting metals, while the nontransferred type are employed with the low-conductivity nonmetals. Ordinary air or inexpensive nitrogen can be used as the plasma gas for the cutting of all types of metal. Oxygen plasma systems were introduced in the 1980s and are used on carbon and low-alloy steel products with thicknesses ranging from 2 to 50 mm (up to 2 in.). As with the oxyfuel cutting systems, the oxygen gas reacts with the iron to provide additional heat, and produces a better quality cut at higher speeds. When cutting thick sections (greater than 12 mm or ½ in.), stainless steels, or nonferrous metals, an argon–hydrogen mixture might be preferred to provide a hotter, deeper-penetrating arc. A secondary flow of shielding gas (nitrogen, air, or carbon dioxide) can be used to help cool the torch, blow the molten metal away, shield the arc, and prevent oxidation of the cut surface. The arc-constricting water flow can also serve as a shielding medium.

During the 1990s, *high-density*, or **precision plasma**, systems began to appear. Various designs are used to restrict the orifice (i.e., superconstrict the plasma), producing vertical edges (less than 1° taper), close tolerances ($^1/_3$ that of conventional), and dross-free plasma cutting of thin materials. The lower-amperage torches (10 to 100 amps) are limited to cutting carbon and low-alloy steels less than 16-mm ($^5/_8$ in.) thick and higher-performance metals (such as stainless and high-strength steels, nickel alloys, titanium, and aluminum) less than 12-mm ($^1/_2$ in.) thick. The cutting speeds are slower than conventional plasma cutting, but there is no change in the size of the heat-affected zone. **Pulsed plasma arc cutting**, another recent development, can reduce heat input to the workpiece while producing cleaner edges on the cuts and kerfs that are 50% narrower.

Combining a plasma torch with CNC manipulation can provide fast, clean, and accurate cutting, like that shown in **Figure 36.26**. Because plasma torches work under water, the cutting table can be submerged as a means of reducing noise, smoke and air pollution, dust, and arc glare (dyes are placed in the water). Plasma arc torches can also be incorporated

FIGURE 36.26 Precision cutting of 6.5 mm ($^1/_4$ in.) steel with automated plasma arc. (Courtesy of FMA Communications, Inc.)

into punch presses to provide a manufacturing machine with outstanding flexibility in producing cut and punched products from a variety of materials. Plasma arc cutting is also suitable for robot application. A single robot system can be used for both cutting and welding of intricate shapes and contours. Water constriction of the manipulated arc is a problem, however, making it important to select the right process parameters and type of gas for the particular application.

Table 36.9 compares the features of oxyfuel cutting, plasma arc cutting, and laser cutting.

36.9 Metallurgical and Heat Effects in Thermal Cutting

When used for cutting, the flame and arc processes expose materials to high localized temperatures and can produce harmful metallurgical effects. If the cut edges will be subsequently welded, or if they will be removed by machining, there is little cause for concern. When the edges are retained in the finished product, however, consideration should be given to the effects of cutting heat and their interaction with the applied loads. In some cases, additional steps might be required to avoid or overcome harmful consequences.

For carbon steels with less than 0.25% carbon, thermal cutting does not produce serious metallurgical effects. However, in steels of higher carbon content, the metallurgical changes can be quite significant, and preheating and/or post-heating might be required. For alloy steels, additional consideration should be given to the effects of the various alloy elements.

Because of the low rate of heat input, oxyacetylene cutting will produce a rather large **heat-affected zone (HAZ)**. The arc-cutting methods produce intermediate effects that are quite similar to those of arc welding. Plasma arc cutting is so rapid, and the heat is so localized, that the original properties of a metal are modified only within a very short distance from the cut.

All of the thermal cutting processes produce some **residual stresses**, with the cut surface generally in tension. Except in the case of thin sheet, warping should not occur. However, if subsequent machining removes only a portion of the cut surface, or does not penetrate to a sufficient depth, the resulting imbalance in residual stresses can induce distortion. It might be necessary to remove all cut surfaces to a substantial depth to ensure good dimensional stability.

Thermal cutting can also introduce geometrical features into the edge. All flame- or arc-cut edges are rough to varying degrees and thus contain notches that can act as stress raisers and reduce the endurance or fracture strength. If cut edges are to be subjected to high or repeated tensile stresses, the cut surfaces and the heat-affected zone should be removed by machining or at least subjected to a stress-relief heat treatment.

36.10 Welding Equipment

Power Sources for Arc Welding

Arc welding requires large electrical currents, often in the range of 100 to 1000 amps. The voltage is usually between 20 and 50 volts. Both DC and AC **power supplies** are used to generate and maintain the arc, and generally employ the "drooping voltage" characteristics shown in **Figure 36.27**. These characteristics are designed to minimize changes in welding current as the welding voltage fluctuates within anticipated limits (voltage changing with variation in arc length).

In the past, most direct current units were gasoline- or diesel-powered motor-generator sets, and these are still used when welding is to be performed in remote locations. Most welding today, however, uses solid-state transformer-rectifier machines, such as the one shown in **Figure 36.28**. Operating on a three-phase electrical line, these machines can usually provide both AC and DC output.

If only AC welding is to be performed, relatively simple transformer-type power supplies can be used to convert the high-voltage, low-amperage primary power into the low-voltage, high-amperage power needed for welding. These are

TABLE 36.9	Cutting Process Comparison: Oxyfuel, Plasma Arc, and Laser		
Feature	**Oxyfuel Cutting**	**Plasma Arc Cutting**	**Laser Cutting**
Preferred Materials	Carbon steel and titanium	All electrically conductive metals	Metal, plastic, wood and textiles
Quality of Cut	Average	Similar to oxyfuel, almost as good as laser on thin material	Good quality—best for plate material $> \frac{1}{2}$ in. thick
Thickness Range			
1. Steel	$\frac{3}{16}$ to unlimited	26 ga. to 3 in.	Foil to 1 in.
2. Stainless	not used	26 ga. to 5 in.	20 ga. to $\frac{3}{4}$ in.
3. Aluminum	not used	22 ga. to 6 in.	20 ga. to $\frac{3}{4}$ in.
Cutting Speed or Time	Long preheat is required	Fast cutting	Slower than plasma, but faster than oxyfuel

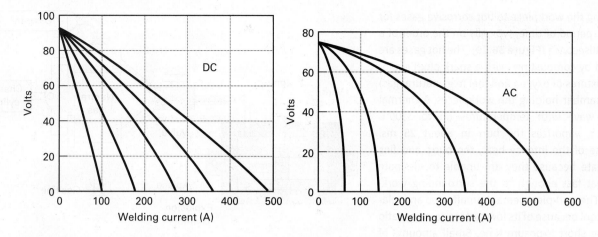

FIGURE 36.27 Drooping-voltage characteristics of typical arc-welding power supplies: (Left) Direct current; (right) alternating current.

usually single-phase devices with low power factors. When multiple machines are to be operated, as in a production shop, they are often connected to the various phases of a three-phase supply to help balance the load.

Inverter-based power supplies, introduced in the 1980s, provide great flexibility. Through solid-state electronics, these AC machines can quickly modify the shape (either sinusoidal or

FIGURE 36.28 Rectifier-type AC and DC welding power supply. (Photo used with permission of The Lincoln Electric Company, Cleveland, OH)

square) and frequency of the pulse waveform, or momentarily change the power output. The square wave technology currently being employed provides improved arc starts and more stable arcs. The percent of time in electrode negative or electrode positive can be adjusted along with the amperages in each of those conditions. Increasing the electrode negative portion narrows the weld bead and increases penetration. Reducing the electrode negative portion widens the bead, decreases penetration and produces greater cleaning action to remove adhered oxides. Output frequency can be varied from 20 to 400 Hz, with higher frequencies giving a tighter, more focused arc and a narrower weld bead. The specific settings can be tailored to the material, size, type of joint, type of shielding gas, travel speed, and so on. Through feedback and logic control, the power supply can actually adjust to compensate for changes in a number of process variables.

Jigs, Positioners, and Robots

Jigs or fixtures (also called **positioners**) are frequently used to hold the work in production welding. By positioning and manipulating the workpiece, the welding operations can often be performed in a more favorable orientation. Parts can also be mounted on numerically controlled (NC) tables that position them with respect to the welding tool.

Industrial robots have replaced humans for many welding applications. They can operate in hostile environments, and are capable of producing high quality welds in a repetitive mode.

36.11 Thermal Deburring

While not an adaptation of a welding or joining process, and not a cutting process, material can also be removed by a thermal process known as **thermal deburring**. Burrs and fins can be removed from machined, cast or forged components by exposing the workpiece to hot corrosive gases for a short period of time, typically on the order of a few milliseconds (**Figure 36.29**). The hot gases are formed by combusting (with a spark plug) explosive mixtures of oxygen and fuel (e.g., natural gas) in a chamber holding the workpieces. A thermal shock wave with temperatures up to 3300°C (6000°F), vaporizes the burr in about 25 ms. Because of the intense heat, the burrs and fins sublimate because they are unable to dissipate the heat fast enough to the surrounding workpiece. The workpiece remains unaffected and relatively cool because of its low surface-to-mass ratio and the short exposure time. Small amounts of metal are removed from all exposed surfaces, and this must be permissible if the process is to be used. The procedure is generally used for the removal of small burrs, and care must be taken with parts with thin cross sections.

Thermal deburring can remove burrs or fins from a wide range of materials, including thermosetting plastics, but not thermoplastic materials. It is particularly effective with materials of low thermal conductivity that can easily oxidize. Any workpiece of modest size requiring manual deburring or flash removal should be considered a candidate for thermal deburring. Die castings, gears, valves, rifle bolts, and similar small

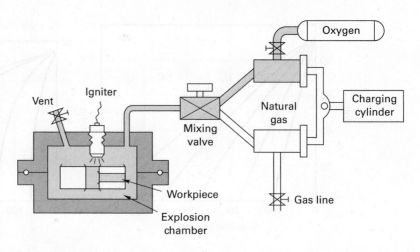

FIGURE 36.29 Thermochemical machining process for the removal of burrs and fins.

parts are deburred readily, including blind, internal, and intersecting holes in inaccessible locations. The major advantage of thermal deburring is that fine burrs are removed much more quickly and cheaply than if they were removed by hand. Uniformity of results and greater quality assurance over hand deburring are additional advantages of thermal deburring. See Chapter 32 (Surface Finishing) for additional material on deburring.

Review Questions

1. What is the principal fuel gas used in oxyfuel-gas welding?

2. Why does an oxyfuel-gas welding torch usually have a flame with two distinct regions?

3. What is the location of the maximum temperature in an oxyacetylene flame?

4. What function or functions are performed by the outer zone of the welding flame?

5. What three types of flames can be produced by varying the oxygen-fuel ratio?

6. Which type of oxyfuel flame is most commonly used?

7. What are some of the alternative fuels (other than acetylene) for oxyfuel-gas welding? What attractive features might they offer?

8. Why might a welder want to change the tip size (or orifice diameter) in an oxyacetylene torch?

9. What is filler metal, and why might it be needed to produce a joint?

10. What is the role of a welding flux?

11. Oxyfuel-gas welding has a low rate of heat input. What are some of the adverse features that result from the slow rate of heating?

12. For what types of applications is oxyfuel-gas welding commonly used?

13. What are some of the more attractive features of the oxyfuel-gas process?

14. How does pressure gas welding differ from the oxyfuel-gas process?

15. In what way does the torch cutting of ferrous metals differ from cutting nonoxidizing metals?

16. Why might it be possible to use only an oxygen lance to cut hot steel strands as they emerge from a continuous casting operation?

17. How does an oxyacetylene cutting torch differ from an oxyacetylene welding torch?

18. What are some of the ways in which cutting torches can be manipulated?

19. When might stack cutting be an attractive process?

20. What modification must be incorporated into a cutting torch to permit it to cut metal under water?

21. If a curved plate is to be straightened by flame straightening, should the heat be applied to the longer or shorter surface of the arc? Why?

22. Why does the flame-straightening process not work for thin sheets of metal?

23. What sorts of problems plagued early attempts to develop arc welding?

24. What are the three basic types of current and polarity that are used in arc welding?

25. What is the attractive feature of variable polarity power supplies?

26. What is the difference between a consumable and nonconsumable electrode? For which processes does a filler metal have to be added by a separate mechanism?

27. What are the three types of metal transfer that can occur during arc welding?

28. What are some of the process variables that must be specified when setting up an arc-welding process?

29. What are the four primary consumable-electrode arc-welding processes?

30. What are some general properties of the consumable-electrode arc-welding processes?

31. What are some of the functions of the electrode coatings used in shielded metal arc welding?

32. How are welding electrodes commonly classified, and what information does the designation usually provide?

33. Why are shielded metal arc electrodes often baked just prior to welding?

34. What are the functions of the fluxing constituents and slag coating that forms over a shielded metal arc weld?

35. What benefit can be obtained by placing iron powder in the coating of shielded metal arc electrodes that will be used to weld ferrous metals?

36. Why are shielded metal arc electrodes generally limited in length, forcing the process to be one of intermittent operation?

37. Why is the shielded metal arc welding process limited to low welding currents and shallow penetration?

38. What are some of the attractive features of the shielded metal arc welding process?

39. What is the advantage of placing the flux in the center of an electrode (flux-cored arc welding) as opposed to a coating on the outside (shielded metal arc welding)?

40. What feature enables the welding current in FCAW to be higher than in SMAW?

41. What are some of the advantages of gas metal arc welding compared to the shielded metal arc process?

42. Describe the relative performance of argon, helium, and carbon dioxide gases in creating a high-temperature arc and promoting weld penetration.

43. For what welding conditions would short circuit transfer be preferred?

44. What is the least desired mode of metal transfer?

45. Which of the metal transfer mechanisms is most used in arc welding?

46. Describe the metal transfer that occurs during pulsed arc gas metal arc welding.

47. What are some of the benefits that can be obtained by the reduced heating of the pulsed arc process?

48. What are some of the attractive features of gas metal arc welding?

49. What are some of the primary process variables in the gas metal arc welding process?

50. What is the attractive feature of advanced gas metal arc welding?

51. What other gas metal arc welding modifications have been developed to improve performance?

52. What benefits can be gained by using a rectangular crosssection electrode wire as opposed to a round one?

53. What are some of the functions of the flux in submerged arc welding?

54. What are some of the attractive features of submerged arc welding? Major limitations?

55. What is the primary goal or objective in bulk welding?

56. What is the current (or proper) designation for MIG welding? TIG welding? Heliarc welding?

57. What types of shielding gases are used in the gas tungsten arc process?

58. What are some of the features that must be specified for the tungsten electrode?

59. What can be done to increase the rate of filler metal deposition during gas tungsten arc welding?

60. What are some of the attractive features of gas tungsten arc welding?

61. For the GTAW process, what are the attractive features of the DCEN polarity? DCEP?

62. What are some of the advantages of employing a pulsed arc in the gas tungsten arc process?

63. How are the spot welds produced by gas tungsten arc spot welding different from those made by conventional resistance spot welding?

64. How is the heating of the workpiece during plasma arc welding different from the heating in other arc welding techniques?

65. What are the two different gas flows in plasma arc welding?

66. What are some of the attractive features of plasma arc welding?

67. What is the keyhole effect in plasma arc welding?

68. What is the primary difference between plasma arc welding and plasma arc cutting?

69. What is the primary objective of stud welding?

70. What is the function of the ceramic ferrule placed over the end of the stud in stud welding?

71. Describe the sequence of activity in flash welding.

72. How is percussion welding different from flash welding?

73. What is the kerf in thermal cutting operations?

74. Describe the ideal thermal cut.

75. What is the purpose of the oxygen in oxygen arc cutting?

76. Why is plasma arc cutting an attractive way of cutting high-melting-point materials?

77. What techniques can be used to constrict the arc in plasma arc cutting, producing a narrower, more controlled cut?

78. Compared to oxyfuel cutting, what are some of the attractive features of the plasma technique?

79. How can a nontransferred arc plasma torch be used to cut low-conductivity nonmetals?

80. What are some of the attractive features of precision plasma and pulsed plasma arc cutting?

81. What are some of the benefits of performing plasma cutting on a submerged (under water) cutting table?

82. Describe the relative size of the heat-affected zone for the various cutting processes: oxyfuel, arc, and plasma.

83. Why might the residual stresses induced during cutting operations be objectionable?

84. Why might it be wise to machine away the thermally cut edge and heat-affected zone of metal that will be used as a stressed machine part?

85. What are typical arc welding currents and voltages?

86. What are the attractive features or benefits of an inverter-based power supply?

87. What are jigs, fixtures or positioners, and how are they used in welding?

88. Describe thermal deburring.

Problems

1. When welds fail, it is not uncommon for the failure to occur in the material adjacent to the weld. Inexperienced welders have been known to comment that the weld was obviously good and the problem is the poor quality material that was being welded. With your knowledge of the fusion welding processes and the heat-affected zone (HAZ) that is produced adjacent to the weld pool, comment on the weld-related problems that could occur if the welding was being performed on:

   a. Heat-treated medium carbon alloy steel

   b. Age-hardened (aerospace grade) aluminum

   c. Thin pieces of high-carbon steel

   d. Plates of electrical grade copper

2. For each of the conditions in Problem 1, could some form of post-weld processing be used to restore desirable properties? What type of post-weld processing? What concerns might you have?

3. Using the Internet or technical literature, identify a "modern advance or improvement" in a gas flame or arc welding or cutting process. What are the claimed advantages of this advance? Do you see any limitations or disadvantages? If so, discuss them.

4. Using the Internet or technical literature, identify a "modern advance or improvement" in welding or thermal cutting equipment (consider power supplies, jigs, positioners, robots, torch design, electrode material, etc). What are the claimed advantages of this advance? Do you see any limitations or disadvantages? If so, discuss them.

5. Identify a new or recent EPA or OSHA regulation relating to gas flame or arc welding. Discuss the purpose of this regulation and what impact it might have on the process or the industry.

CHAPTER 37

Resistance and Solid-State Welding Processes

37.1 Introduction

As indicated in the lists of Figures 35.1 and 35.2, a number of welding and cutting processes utilize heat sources other than oxyfuel flames and electric arcs, and some use no heat source at all. We begin this chapter with a group of processes that use electrical resistance heating to form the joint. A second group of processes in this chapter, known as **solid-state welding** processes, create joints without any melting of the workpiece or filler material.

37.2 Theory of Resistance Welding

In **resistance welding**, heat and pressure are combined to induce coalescence. **Electrodes** are placed in contact with the material and electrical current is passed between them. Electrical resistance heating raises the temperature of the workpieces and the interface between them. The same electrodes that supply the current also apply pressure, which is usually varied throughout the weld cycle. A certain amount of pressure is applied initially to hold the workpieces in contact and thereby control the electrical resistance at the interface. When the proper temperature has been attained, the pressure is increased to induce coalescence and is maintained as the weld nugget cools. Because pressure is utilized, coalescence occurs at a lower temperature than that required for oxyfuel gas or arc welding. In fact, melting of the base metal does not occur in many resistance welding operations. Resistance welding processes might well be considered as a form of solid-state welding, although they are not officially classified as such by the American Welding Society.

In some resistance-welding processes, additional pressure is applied immediately after coalescence to provide a specified amount of forging action. Accompanying the deformation is a certain amount of grain refinement. Additional heating can also be employed after welding to provide tempering and/or stress relief.

The required temperature can often be attained, and coalescence can be achieved, in a few seconds or less. Resistance welding, therefore, is a very rapid and economical process, extremely well suited to automated manufacturing. No filler metal is required, and the tight contact maintained between the workpieces excludes air and eliminates the need for fluxes or shielding gases.

Heating

The heat for resistance welding is obtained by passing a large electrical current through the workpieces for a short period of time. The amount of heat input can be determined by the basic relationship:

$$H = I^2 R t$$

where:

H = total heat input in joules

I = current in amperes

R = electrical resistance of the circuit in ohms

t = length of time, in seconds, during which current is flowing

It is important to note that the workpieces actually form part of the electrical circuit, as illustrated in **Figure 37.1**, and that the total resistance between the electrodes consists of three distinct components:

1. The bulk resistance of the electrodes and workpieces—the upper electrode, upper workpiece, lower workpiece, and lower electrode.

2. The contact resistance between the electrodes and the workpieces—between the upper electrode and upper workpiece and the lower electrode and lower workpiece.

3. The resistance between the surfaces to be joined, known as the **faying surfaces**.

817

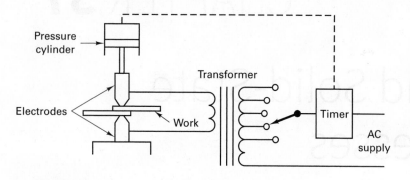

FIGURE 37.1 The basic resistance-welding circuit.

Because the maximum amount of heat is generated at the point of maximum resistance, it is desirable for this to be the location where the weld is to be made. Therefore, it is essential to keep components 1 and 2 as low as possible with respect to resistance 3. The bulk resistance of the electrodes is always quite low, and that of the workpieces is determined by the type and thickness of the metal being joined. Because of the large areas involved and the relatively high electrical conductivity of most metals, the workpiece resistances are usually much less than the contact or interface values. Resistance 2 (the resistance between the electrodes and workpieces) can be minimized by using electrode materials that are excellent electrical conductors, and by controlling the size and shape of the electrodes and the applied pressure. Any change in the pressure between the electrodes and workpieces, however, also affects the contact between the faying surfaces. Therefore, the ability to control the electrode-to-work resistance by pressure variation is quite limited.

The final resistance, that between the faying surfaces, is a function of: (1) the quality of the surfaces (surface finish or roughness); (2) the presence of nonconductive scale, dirt, or other contaminants; (3) the pressure; and (4) the contact area. These factors must all be controlled if uniform resistance welds are to be produced.

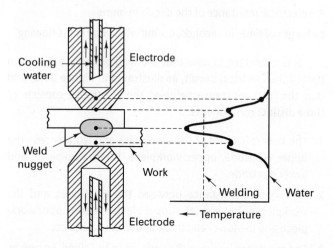

FIGURE 37.2 The desired temperature distribution across the electrodes and workpieces during resistance welding.

As indicated in **Figure 37.2**, the objective of resistance welding is to bring both of the faying surfaces to the proper temperature, while simultaneously keeping the remaining material and the electrodes relatively cool. Water cooling is usually used to keep the electrode temperature low and thereby extend its useful life.

Pressure

Because the applied pressure promotes a forging action, resistance welds can be produced at lower temperatures than welds made by other processes. Controlling both the magnitude and timing of the pressure, however, is very important. If too little pressure is used, the contact resistance will be high and surface burning or pitting of the electrodes may result. If excessive pressure is applied, molten or softened metal might be expelled from between the faying surfaces or the electrodes might indent the softened workpiece. Ideally, moderate pressure should be applied to hold the workpieces in place and establish proper resistance at the interface prior to and during the passage of the welding current. The pressure should then be increased considerably just as the proper welding temperature is attained. This completes the coalescence and forges the weld to produce a fine-grained structure.

On small, foot-operated machines, only a single spring-controlled pressure is used. On larger, production-type welders, the pressure is generally applied through controllable air or hydraulic cylinders.

Current and Current Control

Although surface conditions and pressure are important variables, the temperature achieved during resistance welding is primarily determined by the magnitude and duration of the welding current. The various resistances change as current flows and the material heats. The bulk resistances of metal increase as temperature rises, and the contact resistances decrease as the metal softens and pressure improves the contact. Because the best conditions are the initial ones, high currents and short time intervals are generally preferred. [Note also the I-squared current component in the heat input equation, making weld current extremely significant.] The weld location can attain the desired temperature while minimizing the amount of heat generated in or dissipated to the adjacent material.

In production-type welders, the magnitude, duration, and timing of both current and pressure can be programmed to follow specified cycles. **Figure 37.3** shows a relatively simple cycle for a resistance weld that includes both forging and post-heating operations. The quality of the final weld, therefore, often depends more on the development of a proper schedule and the subsequent setup, adjustment, and maintenance of equipment than it does on operator skill.

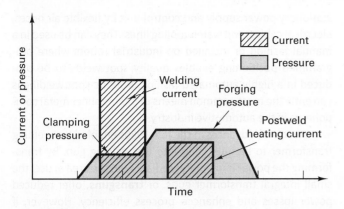

FIGURE 37.3 A typical current and pressure cycle for resistance welding. This cycle includes forging and postheating operations.

Power Supply

Because the overall resistance in the welding circuits can be quite low, high currents are generally required to produce a resistance weld. Power transformers convert the high-voltage, low-current line power to the high-current (typically about 10,000 amps, but possibly up to 100,000 amps), low-voltage (0.5 to 10 volts) power required for welding. Smaller machines can utilize single-phase circuitry, but the larger units generally operate on three-phase power. Many resistance welders use DC welding current, obtained through solid-state rectification of the three-phase power. These machines reduce the current demand per phase, give a balanced load, and produce excellent welds.

Newer power supplies can also adapt to changes in the process conditions. The resistance changes as the initial contact interface is replaced with a weld. The resistance can also change from one weld to the next with variations in part geometry, surface conditions, and part alignment. Over time, the electrodes can mushroom, changing the area of contact, and the temperature of the electrodes can vary. These and other changes can be compensated by adjustments in voltage, current, or duration of current flow.

37.3 Resistance Welding Processes

Resistance Spot Welding

Resistance spot welding (RSW) is the simplest and most widely used form of resistance welding, providing a fast, economical means of joining overlapped materials that will not require subsequent disassembly. Even with all of the advances in technology, resistance spot welding is still the dominant method for joining sheet material, and the average steel-bodied automobile contains between 2000 and 5000 spot welds. **Figure 37.4** presents a schematic of the process.

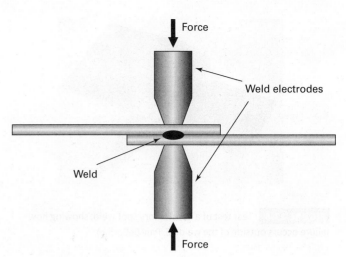

FIGURE 37.4 The arrangement of the electrodes and workpieces in resistance spot welding.

Overlapped metal sheets are positioned between water-cooled electrodes, which have reduced areas at the tips to produce welds that are usually from 1.5 to 13 mm ($^1/_{16}$ to $^1/_2$ in.) in diameter. The electrodes close on the work, and a controlled cycle of pressure and current is applied to produce a weld at the metal interface. The electrodes are then opened, and the work is removed.

A satisfactory spot weld, such as the one shown in **Figure 37.5**, consists of a **nugget** of coalesced metal formed between the faying surfaces. There should be little indentation of the metal under the electrodes. As shown in **Figure 37.6**, the strength of the weld should be such that in a tensile or tear test, the weld will remain intact while failure occurs in the heat-affected zone surrounding the nugget. Sound spot welds can be obtained with excellent consistency if proper current density and timing, electrode shape, electrode pressure, and surface conditions are maintained,

Spot-Welding Equipment
A variety of spot-welding equipment is available to meet the needs of production operations. For light-production work where complex current-pressure cycles are not required, a simple **rocker-arm machine**

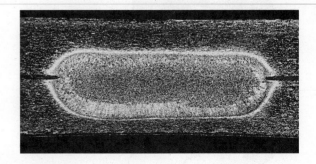

FIGURE 37.5 A spot-weld nugget between two sheets of 1.3-mm (0.05-in.) aluminum alloy. The nugget is not symmetrical because the radius of the upper electrode was greater than that of the lower electrode. (Courtesy of Lockheed Martin Corporation, Bethesda, MD)

FIGURE 37.6 Tear test of a satisfactory spot weld, showing how failure occurs outside of the weld. (E. Paul DeGarmo)

is often used. The lower electrode arm is stationary, while the upper electrode, mounted on a pivot arm, is brought down into contact with the work by means of a spring-loaded foot pedal. Rocker-arm machines are available with throat depths up to about 1.2 m (48 in.) and transformer capacities up to 50 kVa. They are used primarily on steel.

Larger spot welders, and those used at high production rates, are generally of the press type, as shown in **Figure 37.7**. On these machines, the movable electrode is controlled by an air or hydraulic cylinder, and complex pressure cycles can be programmed. Capacities up to 500 kVa with a 1.5-m (60-in.) throat depth are quite common. Special-purpose press-type welders can employ multiple welding heads to make up to 200 simultaneous spot welds in less than 60 seconds.

Quite often, the desired products are too large to be manipulated and positioned on a welding machine. Portable **spot-welding guns** have been instrumental in extending the process to such applications. The guns are connected to a

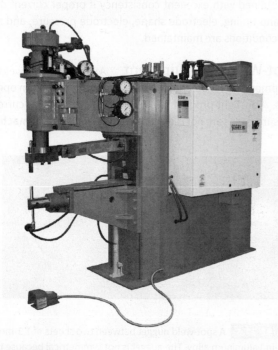

FIGURE 37.7 Three-phase, air-operated, press-type resistance welder with microprocessor control. (Courtesy of Sciaky, Inc., Chicago, IL.)

stationary power supply and control unit by flexible air hoses, electrical cables, and water-cooling lines. They can be used in a manual fashion or installed on industrial robots where programmed positioning enables quality spot welds to be produced in a highly automated fashion. Robotic spot welding is currently the most common means of joining sheet metal components in the automotive industry.

Electronic advances in the late 1980s enabled the welding transformer to be integrated into the welding gun. By transforming the power immediately adjacent to the area of use, the small integral transformer guns, or **transguns**, offer reduced power losses and enhanced process efficiency. However, if accurate positioning is required in an articulated system such as an industrial robot, the added weight of the integral transformer might become a disadvantage. Servomotors have also been incorporated into a variety of spot-welding machines to control the electrode positioning, speed of closure, level of applied torque or pressure, and the rate at which the load is applied.

Electrodes Resistance spot-welding electrodes must conduct the welding current to the work, set the current density at the weld location, apply force, and help dissipate heat during the noncurrent portions of the welding cycle. Electrical and thermal conductivity properties are important considerations for electrode selection. Hot compressive strength must be sufficient to resist electrode deformation during the application of pressure. In addition, the electrode should not melt under welding conditions, and should be of a composition that does not alloy with the material being welded—a phenomenon that promotes sticking or galling and electrode wear. The Resistance Welder Manufacturers Alliance (RWMA) has standardized various electrode geometries (size and shape) and has approved a variety of electrode materials, including copper-base alloys, refractory metals, and refractory–metal composites. While the electrode tip diameter should be suited to the thickness of the material being welded, a straight electrode with a 6.5 mm ($^1/_4$ in.) diameter flat face suits most sheet metal welding applications. A dome-shaped nose might be preferred for applications where alignment may be difficult. Offset electrodes can be used for difficult-to-reach welds. Since the tip faces progressively deteriorate with use, a regular schedule of electrode maintenance and redressing should be followed.

Spot-Weldable Metals and Geometries While steel is clearly the most common metal that is spot welded, one of the greatest advantages of the process is that virtually all of the commercial metals can be joined, and most of them can be joined to each other. In only a few cases do the welds tend to be brittle. **Table 37.1** shows some of the many combinations of metals that can be successfully spot welded.

While spot welding is primarily used to join wrought sheet material, other forms of metal can also be welded. Sheets can be attached to rolled shapes and steel castings, as well as some types of nonferrous die castings. Most metals require no special preparation, except to be sure that the surface is free of

TABLE 37.1	Metal Combinations That Can Be Spot Welded										
Metal	Aluminum	Brass	Copper	Galvanized Iron	Iron (Wrought)	Monel	Nichrome	Nickel	Nickel Silver	Steel	Tin Plate	Zinc
Aluminum	x										x	x
Brass		x	x	x	x	x	x	x	x	x	x	x
Copper		x	x	x	x	x	x	x	x	x	x	x
Galvanized Iron		x	x	x	x	x	x	x	x	x	x	
Iron (Wrought)		x	x	x	x	x	x	x	x	x	x	
Monel		x	x	x	x	x	x	x	x	x	x	
Nichrome		x	x	x	x	x	x	x	x	x	x	
Nickel		x	x	x	x	x	x	x	x	x	x	
Nickel Silver		x	x	x	x	x	x	x	x	x	x	
Steel		x	x	x	x	x	x	x	x	x	x	
Tin Plate	x	x	x	x	x	x	x	x	x	x		
Zinc	x	x	x									x

corrosion and is not badly pitted. For best results, aluminum and magnesium should be cleaned immediately prior to welding by some form of mechanical or chemical technique. Metals that have high electrical conductivity require clean surfaces to ensure that the electrode-to-metal resistance is low enough for adequate temperature to be developed within the metal itself. Silver and copper are especially difficult to weld because of their high thermal conductivity. Higher welding currents coupled with water cooling of the surrounding material might be required if adequate welding temperatures are to be obtained.

When the two pieces being joined are of the same thickness, the practical limit for spot welding is about 3 mm ($\frac{1}{8}$-in.) for each sheet. Sheets of differing thickness can also be joined, and thin pieces can be attached to material that is considerably thicker than 3 mm. When metals of different thickness or different conductivity are to be welded, however, a larger electrode or one with higher conductivity is often used against the thicker or lower-conductivity material to ensure that both workpieces will be brought to the desired temperature simultaneously.

Resistance Seam Welding

Resistance seam welding (RSEW) can be performed in two distinctly different ways. In the first process, sheet metal segments are joined to produce gas- or liquid-tight vessels, such as gas tanks, mufflers, and simple heat exchangers. The weld is made between overlapping sheets of metal, and the seam is simply a series of overlapping spot welds, such as those shown in **Figure 37.8**. The basic equipment is the same as for spot welding, except that the electrodes now assume the form of rotating disks, such as those shown schematically in **Figure 37.9**. As the metal passes between the electrodes, timed pulses of current form the overlapping welds. The timing of the welds and the movement of the work are controlled to ensure that the welds overlap, and the workpieces do not get too hot. The welding current is usually a bit higher than in conventional spot welding to compensate for the short circuit of the adjacent weld, and the workpiece is often cooled by a flow of air or water. In a variation of the process, a continuous current is passed through the rotating electrodes to produce a continuous seam.

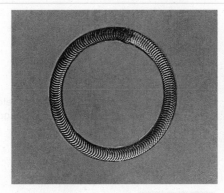 

FIGURE 37.8 Seam welds made with overlapping spots of varied spacing. (Courtesy of Taylor-Winfield Corporation, Youngstown, OH)

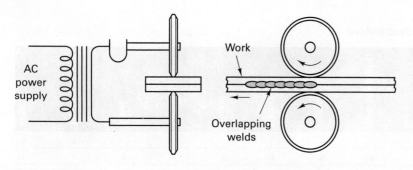

FIGURE 37.9 Schematic representation of the seam welding process.

This form of seam welding is best suited for thin materials, but metals up to 6 mm (¼ in.) can be joined. A typical welding speed is about 2 m/min (80 in./min) for thin sheet.

The second type of resistance seam welding, known as **resistance butt welding**, is used to produce butt welds between thicker metal plates. The electrical resistance of the abutting metals is still used to generate heat, but high-frequency current (up to 450 kHz) is now employed. At high frequencies, current flows only on or near the surface of conductive material—the higher the frequency, the lower the penetration depth. (*Note:* This is similar to the results obtained in the process of high-frequency induction heating.) When the abutting surfaces attain the desired temperature, they are pressed together to form a weld.

Resistance butt welding is used extensively in the manufacture of pipes and tubes, as illustrated in **Figure 37.10**, but the process is also used to construct simple structural shapes from sections of plate. Material from 0.1 mm (0.004 in.) to more than 20 mm (¾ in.) in thickness can be welded at speeds up to 80 m/min (250 fpm). The combination of high-frequency current and high welding speed produces a very narrow heat-affected zone. Almost any type or combination of metal can be welded, including difficult dissimilar metals and the high-conductivity metals, such as aluminum and copper.

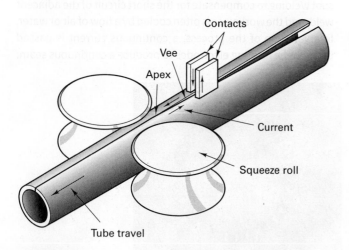

FIGURE 37.10 Using high-frequency AC current to produce a resistance seam weld in butt-welded tubing. Arrows from the contacts indicate the path of the high-frequency current.

Projection Welding

In a mass-production operation, conventional spot welding is plagued by two significant limitations. Because the small electrodes provide both the high currents and the required pressure, the electrodes generally require frequent attention to maintain their geometry. In addition, the process is designed to produce only one spot weld at a time. When increased strength is required, multiple welds are often needed, and this means multiple operations. **Projection welding (RPW)** provides a means of overcoming these limitations.

Figure 37.11 illustrates the principle of projection welding. A dimple is **embossed** into one of the workpieces at the location where a weld is desired. The two workpieces are then placed between large-area electrodes in a press machine, and pressure and current are applied as in spot welding. Because the current must flow through the points of contact (i.e., the dimples or protrusions), the heating is concentrated where the weld is desired. As the metal heats and becomes plastic, pressure causes the dimple to flatten and form a weld.

Because the projections are press-formed into the sheet, they can often be produced during previous blanking and forming operations with virtually no additional cost, and can be precisely located both on the part and with respect to one another. In addition, the dimples or projections can be made in almost any shape—round, oval, or circular ring-shaped—to produce welds of shapes that optimize a given design. It is important, however, that the shape be such that the weld will form outward from the center of the projection. Because the heating is concentrated at the points of contact, materials of widely different thicknesses can be joined.

Multiple dimples can be incorporated into a sheet, enabling multiple welds to be produced at one time. The number of projections is limited only by the ability of the machine to provide the required current and pressure, and the need to uniformly distribute both. If more than three projections are to be made at one time, however, the height of all projections should be uniform to ensure uniform contact and heating.

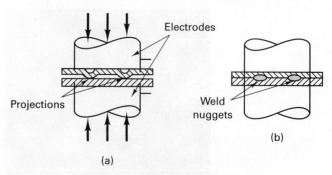

FIGURE 37.11 Principle of projection welding (a) prior to application of current and pressure and (b) after formation of the welds.

An attractive feature of projection welding is the fact that it does not require special equipment. Conventional spot-welding machines can be converted to projection welding simply by changing the size and shape of the electrodes. In addition, projection welding leaves no indentation mark on the exterior surfaces, a definite advantage over spot welding when good surface appearance is required.

In a variation of the process, projection welding can also be used to attach bolts and nuts to other metal parts. Contact is made at one or more projections that have been machined or forged onto the bolt or nut. Current is applied, and the pieces are pressed together to form a weld. In a variation known as **capacitor-discharge stud welding,** a burst of current from an electrostatic storage system melts the projection and the pieces are pushed together, all within a time of 6 to 10 milliseconds.

TABLE 37.2	Process Summary: Resistance Welding (RW)
Heat source	Electrical resistance heating with high current
Protection	None; isolation of weld site is adequate
Material joined	All common metals (steel, aluminum, and copper)
Rate of heat input	High
Weld profile (D/W)	Does not apply
Maximum penetration	Does not apply
Assets	High speed; Small HAZ; no flux, filler metal, or shielding gas required; adaptable to mass production
Limitations	Equipment is more expensive than arc welding; welds are weaker than arc welds; requires access to both sides of a joint

37.4 Advantages and Limitations of Resistance Welding

The resistance welding processes have a number of distinct advantages that account for their wide use, particularly in mass-production operations:

1. They are very rapid.
2. The equipment can often be fully automated.
3. They conserve material, because no filler metal, shielding gases, or flux is required.
4. There is minimal distortion of the parts being joined.
5. Skilled operators are not required.
6. Dissimilar metals can be easily joined.
7. A high degree of reliability and reproducibility can be achieved.

The primary limitations of resistance welding include:

1. The equipment has a high initial cost.
2. There are limitations to the thickness of material that can be joined (generally less than 6 mm or ¼ in.), and the type of joints that can be made (mostly **lap joints**). Lap joints tend to add weight and material.
3. Access to both sides of the joint is usually required to apply the proper electrode force or pressure.
4. Skilled maintenance personnel are required to service the control equipment.
5. Clean surfaces are necessary. For some materials, the surfaces must receive special preparation prior to welding. The oxide surfaces of aluminum and stainless steel, and coatings

such as zinc in galvanized steel and other platings, generally must be removed prior to welding.

6. The resistance welding machines impose a heavy electrical load, and power surges can be disruptive to nearby equipment.

The resistance welding processes are among the most common techniques for high-volume joining. The rapid heat inputs, short welding times, and rapid quenching by both the base metal and the electrodes can produce extremely high cooling rates in and around the weld. While these conditions can be quite attractive for most nonferrous metals, untempered martensite can form in steels containing more than 0.15% carbon. For these materials, some form of postweld heating is generally required to eliminate possible brittleness.

Table 37.2 provides a process summary for resistance welding.

37.5 Solid-State Welding Processes

Forge Welding

Being the most ancient of the welding processes, **forge welding (FOW)** has both historical and practical value as it helps us to understand how and why the modern welding practices were developed. The armor makers of ancient times occupied positions of prominence in their society, largely because of their ability to join pieces of metal into single, strong products. The village blacksmith was a more recent master of forge welding. With his hammer and anvil, coupled with skill and training, he could create a wide variety of useful shapes from metal.

Using a charcoal forge, the blacksmith heated the pieces to be welded to a practical forging temperature and then prepared the ends by hammering so that they could be properly fitted together. The ends were then reheated and dipped into a borax flux. Heating was continued until the blacksmith judged (by color) that the workpieces were at the proper temperature for welding. They were then withdrawn from the heat and either struck on the anvil or hit by the hammer to remove any loose scale or impurities. The ends to be joined were then overlapped on the anvil and hammered to the degree necessary to produce an acceptable weld.

As the two pieces reduced in thickness, they spread in width, resulting in the creation of new, fresh, uncontaminated metal surface. As these surfaces were being created, the hammer blows also provided the necessary pressure to produce instant coalescence. Thus, by the correct combination of heat and deformation, a competent blacksmith could produce joints that might be every bit as strong as the original metal. However, because of the crudeness of the heat source, the uncertainty of temperature, and the difficulty in maintaining metal cleanliness, a great amount of skill was required and the results were highly variable.

Forge-Seam Welding

Although forge welding has largely been replaced by other joining methods, a large amount of **forge-seam welding** is still used in the manufacture of pipe. In this process, a heated strip of steel is first formed into a cylinder, and the edges are simply pressed together in either a lap or a butt configuration. Welding is the result of pressure and deformation when the metal is pulled through a conical welding bell or passed between welding rolls.

Cold Welding

Cold welding (CW) is a variation of forge welding that uses no heating, but produces metallurgical bonds by means of room-temperature plastic deformation. The surfaces to be joined are first cleaned and placed in contact. They are then subjected to high localized pressure, sufficient to cause about 30% to 50% cold work. While some heating will occur from the severe deformation, the primary factor in producing coalescence is the high pressure acting on newly created surface material. The cold-welding process is generally confined to the joining of small parts made from soft, ductile metal, such as those encountered in various electrical connections. Many ductile metals and combinations of dissimilar metals can be joined.

Roll Welding or Roll Bonding

In the **roll-welding (ROW)** or **roll-bonding** process, shown schematically in **Figure 37.12**, two or more sheets or plates of ductile metal are joined by passing them simultaneously through a rolling mill. As the materials are reduced in thickness

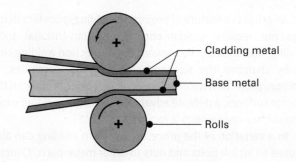

FIGURE 37.12 Schematic representation of the roll bonding process.

(generally more than 60% if performed cold and less if performed hot), the length and/or width must increase to compensate. The newly created uncontaminated interfaces are pressed together by the rolls and coalescence is produced. Roll bonding can be used to join either similar or dissimilar metals (such as the Alclad aluminums—a skin of high-corrosion-resistance aluminum over a core of high-strength aluminum—conventional steel with a stainless steel cladding, or the bimetallic strips used in thermostat controls). The resulting bond can be quite strong, as evidenced by the roll-bonded "sandwich" material used in the production of various U.S. coins (copper clad with cupro-nickel).

By pre-coating select portions of one interface surface with a material that prevents bonding, the roll-bonding process can be used to produce sheets that have both bonded and non-bonded areas. The no-bond regions are subsequently expanded to produce flow paths for gases or liquids by either heating in an oven or furnace to cause the no-bond coating to volatilize or by the direct introduction of compressed air. A common example of this technique is in the manufacture of heat exchanger panels or refrigerator freezer panels, where inexpensive sheet metal is used to produce structural panels that also serve to conduct the coolant or heat-carrying medium.

Friction Welding: Rotational Friction Welding, Inertia Welding, and Linear Friction Welding

In **friction welding (FRW)**, the heat required to produce the joint is generated by mechanical friction at the interface between a moving workpiece and a stationary component. The pieces to be joined are first prepared to have smooth, square-cut surfaces. As shown in **Figure 37.13**, one piece is mounted in a motor-driven chuck or collet and rotated against a stationary piece at high speed. A low contact pressure might be applied initially to permit cleaning of the surfaces by a burnishing action. The pressure is then increased, and contact friction quickly generates enough heat to soften both components and raise the abutting surfaces to the welding temperature. As soon as this temperature is reached, rotation is stopped and the pressure is further increased to complete the weld. The softened metal is squeezed out to the edges of the joint, forming a **flash**,

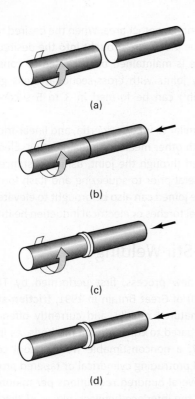

FIGURE 37.13 Sequence for making a friction weld. (a) Components with square surfaces are inserted into a machine where one part is rotated and the other is held stationary. (b) The components are pushed together with a low axial pressure to clean and prepare the surfaces. (c) The pressure is increased, causing an increase in temperature, softening, and possibly some melting. (d) Rotation is stopped and the pressure is increased rapidly, creating a forged joint with external flash.

which can be removed by subsequent machining. Clean, uncontaminated material is left on the interface, and the force creates a "forged" structure in the joint. Friction welding has been used to join steel bars up to 20 cm (8 in.) in diameter and tubes of even larger diameter. The process is also ideal for welding dissimilar metals with very different melting temperatures and physical properties, such as copper to aluminum, titanium to copper, and nickel alloys to steel. **Figure 37.14** shows a schematic of the equipment required for friction welding.

Inertia welding is a modification of friction welding where the moving piece is attached to a rotating flywheel

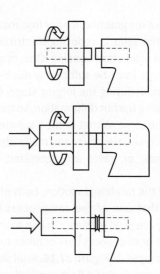

FIGURE 37.15 Schematic representation of the various steps in inertia welding. The rotating part is now attached to a large flywheel.

(**Figure 37.15**). The flywheel is brought to a specified rotational speed, storing a predetermined amount of kinetic energy, and is then separated from the driving motor. The rotating and stationary components are then pressed together, and the kinetic energy of the flywheel is converted into frictional heat at the interface between the two pieces. The weld is formed when the flywheel stops its motion and the pieces are firmly pressed together. Because the conditions of inertia welding are easily duplicated, welds of extremely consistent quality can be produced, and the process can be readily automated.

With inertia welding, the time required to form a weld can be very short, often on the order of several seconds. Because of the high rate of heat input and the limited time for heat to flow away from the joint, both the weld and heat-affected zones are usually very narrow. Oxides and other surface impurities tend to be displaced radially into the upset flash, which is generally removed after welding. Because virtually all of the energy is converted to heat, the process is very efficient. No material is melted, so joints can be formed with a wide variety of metals or combinations of metals, including some not normally considered compatible, such as aluminum to steel. Thermoplastic polymers can also be joined. Graphite-bearing cast irons, free-machining metals, and some bearing materials must be

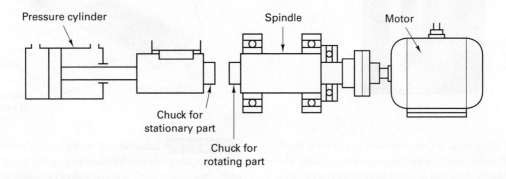

FIGURE 37.14 Schematic diagram of the equipment used for friction welding. (Courtesy of Materials Engineering)

excluded because the graphite, lead or free-machining additive smears across the surface, reducing the friction heating and preventing good solid-state bonding. One, or preferably both, of the components must be sufficiently ductile (when hot) to permit deformation during the forging stage. Grain size tends to be refined during the hot deformation, so the strength of the weld is about the same as that of the base metal. In addition, the friction processes are environmentally attractive because no smoke, fumes, or gases are generated, and no fluxes are required.

Because of the rotational motion, both of the above processes require that one of the components have rotational symmetry. They are used primarily to join round bars or tubes of the same size or to connect bars or tubes to flat surfaces as shown in the examples of **Figure 37.16**. Axial shortening occurs because of the upsetting and flash generation, and should be factored into product design.

In a variation known as **linear friction welding (LFW)**, the moving component oscillates laterally (parallel to the interface) across the surface of a second, rigidly clamped piece. The speeds are lower and the pieces are kept in constant pressure,

heating the entire contact area. When the desired temperature is reached, the pieces are brought into the desired alignment and the force is maintained or increased to consolidate the joint. Quality joints with cross-sectional areas greater than 65 cm² (10 in²) can be formed in 3 to 5 seconds with no filler metal.

Welds similar to friction, inertia, and linear-friction can be produced with other means of initial heating. Electric current can be passed through the joint, using resistance heating to soften the metal prior to squeezing and flash formation. The surfaces to be joined can also be brought to elevated temperature by oxyfuel torches or electrical induction heating.

Friction-Stir Welding

A relatively new process, first performed by The Welding Institute (TWI) of Great Britain in 1991, **friction-stir welding (FSW)** has matured rapidly, and currently offers significant benefits compared to conventional methods. As illustrated in **Figure 37.17**, a nonconsumable welding tool containing a shoulder and protruding cylindrical or tapered probe or pin is rotated at several hundred revolutions per minute. It is then lowered into the interface between pieces of rigidly clamped material, and frictional heat is generated along the top surface (under the rotating shoulder) and along the surfaces of the rotating probe. After a period of time for heating and softening (temperature rising to 60% to 80% of the melting temperature), the tool is driven along the material interface. As the probe traverses, a plasticized region is continually created. This softened material is swept along the periphery of the pin, flows to the back of the advancing probe, and coalesces to form a solid-state bond. The most common application is the formation of butt welds, usually between plates of the lower-melting-point metals or thermoplastic polymers. The photo in **Figure 37.18** shows a friction-stir weld in aluminum plate, viewed from the

(a)

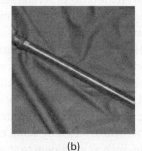
(b)

(c)

FIGURE 37.16 (a) An array of parts produced by friction or inertia welding; (b) a ball socket assembly for the automotive industry where a pre-machined casting is friction welded to a cylindrical shaft; (c) a motor coupling where a pre-machined tubular component is friction welded to a solid base. Note the flash in the weld location. (Courtesy of Spinweld, Inc., Waukesha, WI)

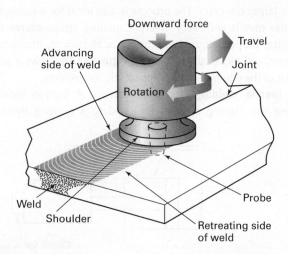

FIGURE 37.17 Schematic of the friction-stir welding process. The rotating probe generates frictional heat, while the shoulder provides additional friction heating and prevents expulsion of the softened material from the joint. (*Note*: To provide additional forging action and confine the softened material, the tool can be tilted so the trailing edge is lower than the leading segment.)

FIGURE 37.18 Top surface of a friction-stir weld joining 1.5-mm- and 1.65-mm-thick aluminum sheets with 1500-rpm pin rotation. The welding tool has traversed left-to-right and has retracted at the right side of the photo. (Courtesy Ronald Kohser)

TABLE 37.3	Attractive Features of Friction-Stir Welding

Metallurgical benefits:

Excellent weld quality

Applicable to a wide range of materials, including both wrought and cast alloys, as well as some "nonweldable" by fusion methods

Solid-state process

Low distortion of the workpiece

High joint strength and good fatigue properties

No loss of alloy elements

Fine microstructure

No cracking or porosity

Low shrinkage

Environmental benefits:

No shielding gas is required

No surface cleaning is required

No solvent degreasing is used

No fumes, gases, or smoke are produced

Post-weld finishing is often unnecessary

No arc glare or reflected laser beams

Energy benefits:

Welds produced with far less energy than other processes

Enables weight reduction in aircraft, automobiles, and ships

Process features:

Nonconsumable tool

No filler metal

Can tolerate small alignment and surface imperfections

Thin oxide layers are acceptable

Can be performed in all positions

top. Key process variables include probe geometry (diameter, depth, and profile), shoulder diameter, rotation speed, downward force, travel speed, and possible tilt to the tool.

Weld quality is excellent. The extensive plastic deformation creates a refined grain structure with no entrapped oxides or gas porosity. The strength, ductility, fatigue life, and toughness are all quite good. Welds in aircraft aluminum are 30% to 50% stronger than those formed by arc welding. Because no material is melted, both wrought and cast alloys can be joined, and they can be joined to each other. No filler material or shielding gas is required, and the process is environmentally friendly (no fumes, weld spatter, or arc glare). Because of the high energy efficiency, total heat input and associated distortion and shrinkage are all low. Joint preparation is minimal and surface oxides need not be removed. Welding can be performed in any position and requires access to only one side of the plate. Full penetration welds can be produced in materials with thicknesses between 0.5 and 65 mm (0.02 to 2.5 in.). Gaps up to 10% of the material thickness can be accommodated with no reduction in weld quality or performance. The workpieces must be rigidly clamped, however, to prevent the joint faces from being forced apart. **Table 37.3** summarizes some of the attractive features of the friction-stir process.

Friction-stir welding has been used to weld nearly all of the wrought aluminum alloys, including some that are classified as "unweldable" by fusion processes (the 2xxx and 7xxx series). Aluminum plates up to 75 mm (3 in.) thick have been successfully welded from a single side in a single pass. Copper, lead, magnesium, and zinc have all been welded with relative ease. High-strength and high-hardness tools that will withstand temperatures in the order of 650°C or 1200°F, along with process

developments, have enabled the single pass welding of titanium; ferritic, high-strength, and stainless steels; and the nickel-base superalloys in thicknesses up to 25 mm. (1 in.). Various thermoplastics, including polyethylene, polypropylene, nylon, polycarbonate, and ABS, have also been successfully welded, with several exhibiting as-welded strengths exceeding 95% of the tensile strength of the base material. Metal-matrix composites and different alloys of the same base metal can be welded. Wrought plate or forgings can be joined to castings.

Example: The Eclipse 500 Very-light Jet

As an indication of process capability, the Eclipse 500 very-light jet, a six-passenger, short-hop air taxi built by Eclipse Aviation of Albuquerque, New Mexico, contains more than 260 friction-stir welds on the aluminum fuselage and wing panel assemblies with a combined length of 135 meters (nearly 450 ft.). These welds replaced more than 7000 rivets and other fasteners with a significant reduction in cost, and produced a structure whose fatigue life is equal to or greater than a riveted assembly.

A hybrid laser—friction-stir process has been developed in which a laser is used to preheat the metal ahead of the stir tool. Because the spinning tool no longer needs to generate all of the required heat, the load on the tool can be reduced, producing a significant increase in tool life. In addition, the hybrid process enhances the ability of friction-stir to weld the high-strength or high-melting-temperature materials.

The friction-stir process has also been modified to create **friction-stir spot welding**, and thin sheets of aluminum and magnesium can now be welded. A rotating friction-stir tool is pressed against the surface of overlapped sheets, and after sufficient frictional rubbing, plunges into the heated material, with the pin penetrating through the upper sheet and part, but not all, of the lower one. Material adjacent to the pin is stirred and the tool is extracted. The resulting weld has a circular indentation in the upper surface created by the retracting pin, and an unblemished bottom surface. To overcome the hole in the upper surface, a process modification called **refill stir spot welding (RSSW)** has been developed. The rotating pin and the shoulder sleeve become separate pieces, and as the pin is retracted, the sleeve presses onto the surface, causing the hole to fill with stirred metal. No special surface preparation is required for the friction-stir spot welds, and joint strength is comparable to resistance spot welding. Dissimilar metals can be joined, such as aluminum to steel, provided the stronger material is place on the bottom.

In an extension known as **friction-stir processing**, the thermo-mechanical features of friction-stir have been used for purposes other than creating a joint. By tracking the stir tool through the material with overlapping passes, ultrafine grain size (<10 μm) can be produced that enables superplastic forming at comparatively high strain rates. While superplasticity is usually limited to thin sheets, thicker material can now be made superplastic, and by stirring only selected locations, large parts can be made from less-expensive conventional materials, with enhanced formability in only the needed locations. In other applications, key surfaces of large castings can be enhanced by the passage of the probe. The cast structure, with possible microporosity and segregation, is replaced by a fine, homogeneous, wrought microstructure. Strength, ductility, corrosion resistance, and fatigue resistance are all improved. Fusion welds can be stirred to replace the cast structure with a fine, worked structure, removing any weld defects, and enhancing properties. Reinforcement particles can be stirred into a material to create a particle-reinforced composite surface on a standard alloy substrate. Unique composition powder products can be brought to full density with attractive strength and ductility.

Ultrasonic Welding

In **ultrasonic welding (USW)**, coalescence is produced by the localized application of high-frequency (10,000 to 200,000 Hz with 20,000 Hz being most common) shear vibrations of low amplitude (usually < 100 microns or 0.004 in.) to surfaces that are held together under rather light normal pressure. Although there is some heating at the faying surfaces, the interface temperature rarely exceeds one-half of the melting point of the material on an absolute-temperature scale. Instead, it appears that the rapid reversals of stress along the contact interface deform, shear, and flatten the surface asperities and break up and disperse the oxide films and surface contaminants, allowing the clean metal surfaces to coalesce into a high-strength bond.

Figure 37.19 depicts the basic components of the ultrasonic welding process. The ultrasonic piezoelectric transducer is essentially the same as that employed in ultrasonic machining. It is coupled to a force-application system that contains a welding tip on one end, either stationary for spot welds or rotating for seams. The pieces to be welded are placed between this tip and a reflecting anvil, thereby concentrating the vibratory energy to a relatively small area, typically about 40 mm^2 (0.06 in^2). Seam welds can be produced by rolling an ultrasonically vibrated disc over the workpieces.

Ultrasonic welding is restricted to the overlapping joint welding of thin materials—sheet, foil, and wire—or the attaching of thin sheets to heavier structural members. The maximum thickness is about 2.5 mm (0.1 in.) for aluminum and 1.0 mm (0.04 in.) for harder metals. As indicated in **Table 37.4**, the process is particularly attractive because of the number of metals and dissimilar metal combinations that can be joined. It is even possible to bond metals to nonmetals, such as aluminum to ceramics or glass. Because the temperatures are low and no arcing or current flow is involved, the process is often preferred for heat-sensitive electronic components. Intermetallic compounds seldom form, and there is no contamination of the weld or surrounding area. The equipment is simple and reliable, and only moderate skill is required of the operator.

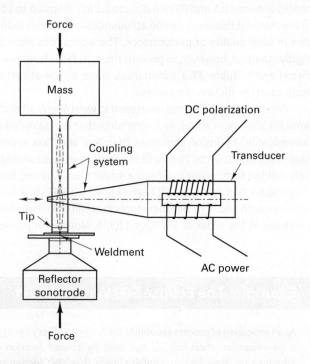

FIGURE 37.19 Diagram of the equipment used in ultrasonic welding.

TABLE 37.4 **Metal Combinations Weldable by Ultrasonic Welding**

Metal	Aluminum	Copper	Germanium	Gold	Molybdenum	Nickel	Platinum	Silicon	Steel	Zirconium
Aluminum	x	x	x	x	x	x	x	x	x	x
Copper		x		x		x	x		x	x
Germanium		x	x	x		x	x	x		
Gold				x		x	x	x		
Molybdenum					x	x			x	x
Nickel						x	x		x	x
Platinum							x		x	
Silicon										
Steel									x	x
Zirconium										x

Surface preparation is less than for most competing processes (such as resistance welding), and less energy is needed to produce a weld. Typical applications include joining the dissimilar metals in bimetallics, making microcircuit electrical contacts, welding refractory or reactive metals, bonding ultra-thin metal, and encapsulating explosives, combustible materials, or chemicals.

Ultrasonic welding has also been used to produce spot and seam welds on thin thermoplastics, and to seal foil or plastic envelopes and pouches. Compared to joining methods that employ solvents or adhesives, the ultrasonic method is considerably faster and results in products with cleaner surfaces.

Table 37.5 summarizes some of the advantages and limitations of ultrasonic welding.

TABLE 37.5 **Advantages and Limitations of Ultrasonic Welding**

Advantages

Solid-state process—no melting of materials.

Excellent for aluminum, copper, and other high-thermal-conductivity materials that are often difficult to join by fusion processes.

Can join many dissimilar material combinations.

Able to weld thin materials to thick materials (difficult for fusion processes).

Can weld through oxides and contaminants.

No filler metals or gases required.

Low energy requirement.

Fast and easily automated.

Limitations

Restricted to lap joint configuration.

Limited in joint thickness.

More difficult with high-strength and high-hardness materials.

Material may deform under the tooling.

Diffusion Welding

Diffusion welding (DFW) or **diffusion bonding** occurs when properly prepared surfaces are maintained in contact under sufficient pressure and time at elevated temperature. In contrast to the deformation welding methods, plastic flow is limited and the principal bonding mechanism is atomic diffusion. Clean, flat, well-prepared surfaces can produce an interface that can be viewed as a planar grain boundary with intervening voids and impurities. Under low pressure (enough to produce firm contact) and elevated temperature (usually between 0.5 and 0.8 times the melting temperature on an absolute temperature scale), atomic diffusion will provide the necessary void shrinkage and grain boundary migration to form a metallurgical bond. The required time can range from one minute to more than an hour, depending on the materials being bonded and the process conditions. Most diffusion bonding is performed in a vacuum or an inert gas atmosphere.

Diffusion bonding is capable of joining a wide range of metal and ceramic materials, as well as dissimilar material combinations and composite materials, in both small and large sizes. Furnaces with inert or protective atmospheres can be used to produce high-quality joints with the reactive metals, such as titanium, beryllium, and zirconium, and the high-temperature refractory metals. Because the bonding process is quite slow, multiple parts are generally loaded into a furnace, or the application is restricted to low-volume production. The quality of a diffusion weld depends on the surface condition of the materials, temperature, time at temperature, and pressure.

Modifications of the basic process often use intermediate material layers, which can either promote diffusion or prevent the formation of undesirable intermetallic compounds. Intermediate layers are often used when joining dissimilar materials, especially metals to ceramics, where they aid the bonding process and help to modify the stress distribution in the bonded product. A dissimilar material layer can also be designed to melt at a lower temperature than the materials being joined, forming a temporary liquid that significantly accelerates the rate of atom movement. While being held at the bonding temperature, the atoms in the liquid subsequently diffuse into the adjoining solids, producing a strong, solid-state bond.

Because the conditions for the diffusion bonding of titanium and titanium alloys coincide with the conditions required for superplastic forming, aerospace companies have combined the two processes into a single operation known as **superplastic forming/diffusion bonding**. The products are generally sandwich-type sheet metal structures with the bonds being used to create internal reinforcing elements.

Explosive and Magnetic-Pulse Welding

Explosive welding (EXW) is typically used to bond flat surfaces of widely different materials, particularly when large areas are involved. As shown in **Figure 37.20**, the bottom sheet or plate is positioned on a rigid base or anvil, and the top sheet is inclined to it with a small open angle between the surfaces to be joined. An explosive material, usually in the form of a sheet, is placed on top of the two layers of metal and detonated in a progressive fashion, beginning where the surfaces touch. A compressive stress wave, on the order of thousands of megapascals (hundreds of thousands of pounds per square inch), sweeps across the surface. Surface films are liquefied or scarfed off the metals and are jetted out of the interface. The clean metal surfaces are then thrust together under high contact pressure. The result is a low-temperature weld with a characteristic wave-like interface, like that shown in Figure 37.20. Because the bond strength is quite high (equal to or greater than the strength of the weaker material), explosively clad plates can be subjected to a wide variety of subsequent processing, including further reduction in thickness by rolling. Because it is a solid-state welding process, numerous combinations of dissimilar metals can be joined.

Explosive welding is best suited to large products and material combinations that involve widely different melting points or might produce undesirable intermetallic compounds. A common application is the thin cladding of corrosion-resistant or wear-resistant metal to heavier plates of base metal. Claddings of aluminum, titanium, zirconium, copper, nickel, tantalum, and stainless steel have been applied to steel, stainless steel, and aluminum substrates. Dimensions are often measured in meters or tens-of-feet. Process assets include: the flexibility in product size, the wide variety of material combinations (including those unsuitable for fusion welding), high bond strength, retention of material properties, little surface preparation required, and the quick delivery of a finished product. The material being clad must have sufficient ductility and toughness to withstand the explosive process.

Magnetic-pulse welding is another high-velocity, solid-state, joining technique used with electrically conductive metals. A sudden discharge from a bank of capacitors is passed through a conductive coil, generating a transient magnetic field and eddy currents in a nearby conductive material. Repulsion between the coil and conductor drives the conductive material at impact velocity onto an adjacent surface. Process features are similar to explosive welding—an angular gap between the materials being joined, a cleansing jet of escaping gas, and the resulting wavy interface. Tubes can be bonded to an internal rod or smaller diameter tube. Overlapping sheet metal sections can be joined. Bond quality is high; cycle times are short; and the process can be automated for rapid production.

Yet another high-velocity solid-state process that has demonstrated success in the laboratory is **vaporized foil actuator (VFA) welding**. A high-voltage capacitor bank discharges a short (millionths of a second) electrical pulse inside a thin piece of aluminum foil. The foil vaporizes, and the burst of hot plasma propels the pieces to be joined together at immense speed. The pieces do not melt, but form an interface with features that resemble explosive welding. Strong bonds have been formed between otherwise "unweldable" metals, with a major asset being the ability to successfully bond different combinations of copper, aluminum, magnesium, iron, nickel, and titanium. Producing high strength bonds between high-strength steel and aluminum is extremely attractive to the automotive industry as it seeks to decrease the weight of its vehicles. In addition, the pressures can also deform the materials, enabling forming and welding to be performed as a single operation.

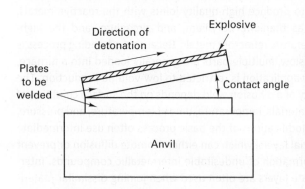

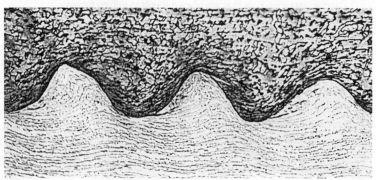

FIGURE 37.20 (Left) Schematic of the explosive welding process. (Right) Explosive weld between mild steel and stainless steel, showing the characteristic wavy interface. (Courtesy Ronald Kohser)

Review Questions

1. What are the two primary functions of the electrodes in resistance welding?

2. What are the two major roles of the applied pressure in resistance welding?

3. Why might resistance welding be considered a form of solid-state welding?

4. Why is there no need for fluxes or shielding gases in resistance welding?

5. Based on the heat input equation, which term is most significant in providing heat—current, resistance, or time?

6. What are the three components that contribute to the total resistance between the electrodes?

7. What measures can be taken to reduce the resistance between the electrodes and the workpieces?

8. What factors control the resistance between the faying surfaces?

9. What are the possible consequences of too little pressure during the resistance-welding cycle? Too much pressure?

10. What is the ideal sequence for pressure application during resistance welding?

11. Why do the resistance-welding conditions become less favorable as the material heats and softens?

12. What magnitude of current might be required to produce resistance welds?

13. What are some of the changes that can occur in process conditions as resistance welds are made?

14. What is the simplest and most widely used form of resistance welding?

15. What is the typical size of a spot-weld nugget?

16. What are the two basic types of stationary spot-welding machines?

17. What is the major advantage of spot-welding guns?

18. What are the pros and cons of a resistance spot welding transgun?

19. What are some of the properties that must be possessed by resistance-welding electrodes?

20. What is the most common metal that is spot welded?

21. What is the practical limit of the thicknesses of material that can be readily spot welded?

22. What design features can be altered to permit the joining of different thicknesses or different conductivity metals?

23. What are the two methods used to produce resistance seam welds?

24. For what products would resistance butt welding be a common approach?

25. What two limitations of spot welding can be overcome by using the projection approach?

26. What limits the number of projection welds that can be formed in a single operation?

27. What are some of the attractive features of resistance welding when viewed from a manufacturing standpoint?

28. What are some of the primary limitations to the use of resistance welding?

29. What type of metallurgical problem might be encountered when spot welding medium- or high-carbon steels?

30. What were some of the limitations that made the forge welds of a blacksmith somewhat variable in terms of quality?

31. What features promote coalescence in cold welding?

32. Describe how the roll-bonding process can be used to fabricate products that contain pressure-tight, fluid-flow channels that once required the use of metal tubing.

33. Describe the friction welding process.

34. How is inertia welding similar to friction welding? Different from friction welding?

35. How are surface impurities removed in the friction- and inertia-welding processes?

36. Why are inertia welds of more consistent quality than friction welds?

37. What are some of the geometric limitations of friction and inertia welding?

38. How does linear friction welding differ from friction welding?

39. How does friction-stir welding differ from friction welding?

40. What are the primary process variables in friction-stir welding?

41. What are some of the attractive features of friction-stir welding?

42. What are some of the materials that have been welded by the friction-stir process?

43. What is the benefit of adding a preheat laser to the friction-stir process?

44. Describe the friction-stir spot welding process.

45. What types of material or property modifications can be induced through friction-stir processing?

46. How do ultrasonic vibrations produce a weld?

47. What are some of the geometric limitations of ultrasonic welding?

48. What are some of the attractive features of ultrasonic welding?

49. What are the conditions necessary to produce high-quality diffusion welds?

50. What kinds of materials can be joined by diffusion bonding?

51. How might intermediate layers be used to enhance the diffusion bonding process?

52. How are surface contaminants removed during explosive welding?

53. If the interface of a weld is viewed in cross section, what is the distinctive geometric feature of an explosive weld?

54. What are some typical applications of explosive welding?

Problems

1. Many advanced engineering products, as well as composite materials, require the joining of dissimilar materials. Select several of the processes discussed in this chapter and investigate the capability of the process to join dissimilar materials and the associated limitations.

2. Using the Internet or technical literature, identify a "modern advance or improvement" in one of the processes described in this chapter. What are the claimed advantages of this advance? Do you see any limitations or disadvantages? If so, discuss them.

3. Friction-stir processing is an interesting extension of friction-stir welding. Can you identify other examples where a welding or joining process is currently being used for purposes other than those for which it was initially developed?

4. Investigate the various types of power supplies currently being used for resistance welding. What are some of the features being offered?

CHAPTER 38

Other Welding Processes, Brazing, and Soldering

38.1 Introduction

We have already surveyed gas-flame and arc welding (Chapter 36), as well as resistance and solid-state joining processes (Chapter 37). Other processes within the realm of welding include some that are quite old (thermit welding) and others that are relatively new (laser and electron beam). These and several others will be presented here, along with a brief section devoted to the application of welding and welding-related processes to surfacing and thermal spray coating.

There are also many joining or assembly operations where welding might not be the best choice. Perhaps the heat of welding is objectionable, the materials possess poor weldability, welding is too expensive, or the joint involves thin or dissimilar materials. In such cases, low-temperature joining methods might be preferred. These include brazing, soldering, adhesive bonding, and the use of mechanical fasteners. Brazing and soldering will be explored in this chapter, while adhesive bonding and mechanical fasteners are deferred to Chapter 39.

38.2 Other Welding and Cutting Processes

Thermit Welding

Thermit welding (TW) is an extremely old process in which superheated molten metal and slag are produced from an exothermic chemical reaction between a metal oxide and a metallic reducing agent. The name **thermit** usually refers to a mechanical mixture of about one part (by weight) finely divided aluminum and three parts iron oxide (either Fe_2O_3 or Fe_3O_4), plus possible alloy additions. When this mixture is ignited by a magnesium fuse (the ignition temperature is about 1150°C or 2100°F), it reacts according to one of the following chemical equations:

$$2\,Al + Fe_2O_3 \rightarrow 2\,Fe + Al_2O_3 + Heat$$

$$8\,Al + 3\,Fe_3O_4 \rightarrow 9\,Fe + 4\,Al_2O_3 + Heat$$

The temperature rises to higher than 2750°C (5000°F) in about 30 seconds, superheating the molten iron, which then flows by gravity into a prepared joint, providing both heat and filler metal. Runners and risers must be provided, as in a casting, to channel the molten metal and compensate for solidification shrinkage. The molten aluminum oxide slag rises to the surface, separating from the metal.

Steels and cast irons can be welded using the process described above. Copper, brass, and bronze can be joined using a starting mixture of copper oxide and aluminum. Nickel, chromium, and manganese oxides have also been used in the thermit welding of more exotic metals.

To a large degree, thermit welding has been replaced by alternative methods. Nevertheless, it is still effective and can be used to produce economical, high-quality welds in thick sections of material, particularly in remote locations or where more sophisticated welding equipment is not available. The field repair of large steel castings that have broken or cracked is one such application.

Electroslag Welding

Electroslag welding (ESW), depicted in **Figure 38.1**, is a very effective process for welding thick sections of steel plate. No arc is involved (except to start the weld), so the process is entirely different from submerged arc welding, and the electrical resistance of the metal being welded plays no part in producing the heat. Instead, heat is derived from the passage of electrical current through a pool of electrically conductive liquid slag. Resistance heating raises the temperature of the slag to around 1750°C (3200°F). The molten slag then melts the edges of the pieces that are being joined, as well as continuously fed solid or flux-cored electrodes. Multiple electrodes are often used to provide an adequate supply of filler metal and maintain the molten pool, giving the process one of the highest metal deposition rates. Under normal operating conditions, there is a 65-mm (2.5-in.)-deep layer of molten slag, which serves to protect and cleanse the underlying 12- to 20-mm (½ to ¾ in.)-deep pool of molten metal as well as the molten drops as they pass from the electrode to the pool below. The liquids are confined to the region between the materials being joined

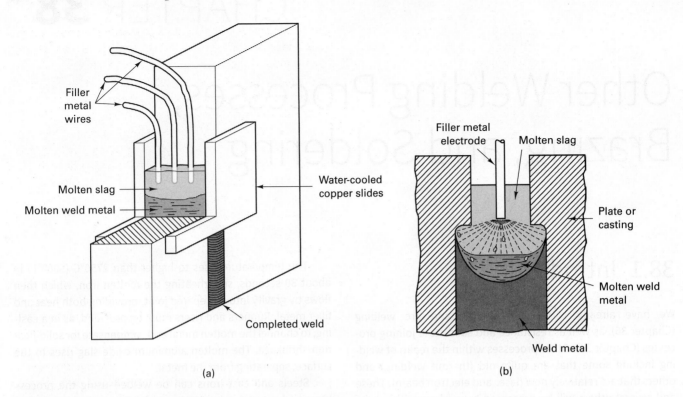

FIGURE 38.1 (a) Arrangement of equipment and workpieces for making a vertical weld by the electroslag process. (b) Cross section of an electroslag weld, looking through the water-cooled copper slide.

by means of sliding, water-cooled **molding plates** that are usually made of copper. As the weld metal solidifies at the bottom of the pool, the molding plates move upward at a rate that is typically between 12 and 40 mm/min (½ to 1½ in./min), a relatively slow welding rate.

Because a vertical joint provides the easiest geometry for maintaining a deep slag bath, the process is used most frequently in this configuration. Circumferential joints can also be produced in large pipe by using special curved slag-holder plates and rotating the pipe to maintain the welding area in a vertical position.

Because large amounts of weld metal and heat can be supplied, electroslag welding is the best of all the welding processes for making welds in thick plates. The thickness of the plates can vary from 25 to 900 mm (1 to 38 in.), and the length of the weld (amount of vertical travel) is almost unlimited. Edge preparation is minimal, requiring only squared edges separated by 25 to 35 mm (1 to 1³/₈ in.). Applications have included building construction, shipbuilding, machine manufacture, heavy pressure vessels, and the joining of large castings and forgings.

Solidification control is vitally important to obtaining a good electroslag weld because the large molten pool and slow cooling tends to produce a coarse grain structure. Cracking tendencies can be suppressed by adjusting the current, voltage, slag depth, number of electrodes, and electrode extension to produce a shallower pool of molten metal. A large heat-affected zone and extensive grain growth are common features of the process. While these are undesirable metallurgical features, the long thermal cycle does serve to minimize residual stresses, distortion, and cracking in the heat-affected zone. Reducing the gap between the plates being welded has been shown to reduce grain size and improve impact properties—a modification known as Narrow Gap Improved Electroslag Welding (NGI-ESW). Subsequent heat treatment of the welded structure, usually by normalizing, can also be used to enhance fracture toughness.

Electron-Beam Welding

In the **electron-beam welding (EBW)** process, the metal to be welded is heated by the impingement of a beam of high-velocity electrons. Originally developed for obtaining ultrahigh-purity welds in reactive and refractory metals, the unique qualities of the process have led to a much wider range of applications.

Figure 38.2 presents the electron optical system. An electric current heats a tungsten filament to about 2200°C (4000°F), causing it to emit a stream of electrons by thermal emission. By means of a control grid, accelerating voltage, and focusing coils, these electrons are collected into a concentrated beam, accelerated, and directed to a focused spot between 0.8 and 3.2 mm. (¹/₃₂ to ¹/₈ in.) in diameter. Because electrons accelerated at 150 kV achieve speeds nearly ²/₃ the speed of light, the electron beam is concentrated energy (up to 10^8 watts/cm²), capable of producing temperatures in excess of one million degrees Celsius when its kinetic energy is converted to heat.

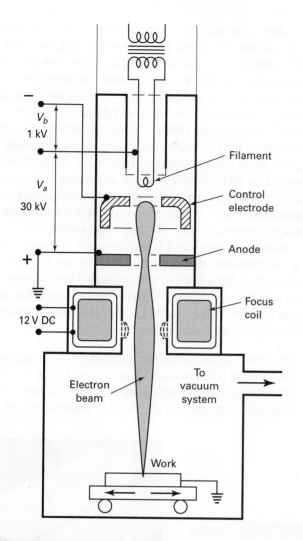

FIGURE 38.2 Schematic of the electron-beam welding process.

While these machines offer more production freedom, they do produce shallower, wider welds because the beam loses energy and diffuses as the pressure increases.

Two distinct ranges of accelerating voltage are generally employed in electron-beam welding. High-voltage equipment operates between 60 and 150 kV and produces a smaller spot size and greater penetration than does the lower-voltage type, which uses from 10 to 50 kV. Because of their high electron velocities, the high-voltage units emit considerable quantities of harmful X-rays and thus require expensive shielding and indirect viewing systems for observing the work. The X-rays produced by the low-voltage machines are sufficiently soft that the walls of the vacuum chamber absorb them, and the parts can be viewed directly through viewing ports.

Almost any metal can be welded by the electron-beam process, including those that are difficult to weld by other methods, such as zirconium, beryllium, and tungsten. Dissimilar metals, including those with extremely different melting points, can also be readily welded, because the intense beam will melt both metals simultaneously. Electron-beam welds typically exhibit a narrow profile and remarkable penetrations like those shown in **Figure 38.3**. The high power and heat concentrations can produce fusion zones with depth-to-width ratios up to 25:1. This is coupled with low total heat input, low distortion, and a very narrow heat-affected zone. Heat-sensitive materials can often be welded without damage to the base metal. Deep welds can be made in a single pass. High welding speeds are common; no shielding gas, flux, or filler metal is required; the process can be performed in all positions; and preheat or postheat is generally unnecessary.

On the negative side, the equipment is quite expensive, and extensive joint preparation is required. Because of the deep and narrow weld profile, joints must be straight and

Because the beam is composed of charged particles, it can be positioned and moved by electromagnetic lenses. Unfortunately, the electrons cannot travel well through air. To be effective as a welding heat source, the beam must be generated and focused in a very high vacuum, typically at pressures of 0.01 Pa (1×10^{-4} mm Hg) or less.

In many operations, the workpiece must also be enclosed in the high-vacuum chamber, with provision for positioning and manipulation. The vacuum then ensures degasification and decontamination of the molten weld metal, and welds of very high quality are obtained. The size of the vacuum chamber, however, tends to impose serious limitations on the size of the workpiece that can be accommodated, and the need to break and reestablish the high vacuum as pieces are inserted and removed places a considerable restriction on productivity. As a consequence, electron-beam welding machines have been developed that operate at pressures considerably higher than those required for beam generation. Some permit the workpiece to remain outside the vacuum chamber, with the beam emerging through a small orifice in the vacuum chamber to strike an adjacent surface. High-capacity vacuum pumps are required to compensate for the leakage through the orifice.

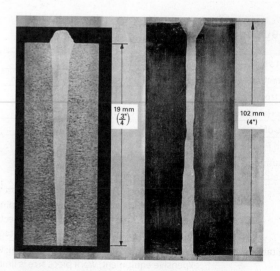

FIGURE 38.3 (Left to right) Electron-beam welds in 19-mm-thick 7079 aluminum, and 102-mm-thick stainless steel. (Courtesy of Hamilton Standard Division of United Technologies Corporation, Hartford, CT)

precisely aligned over the entire length of the weld. Machining and fixturing tolerances are often quite demanding. The vacuum requirements tend to limit production rate, and the size of the vacuum chamber can restrict the size of workpiece that can be welded.

The electron-beam process is best employed where welds of extremely high quality are required or where other processes will not produce the desired results. Electron-beam welds often exhibit joint strength 15% to 25% greater than arc welds in the same material. The unique capabilities have resulted in its routine use in a number of applications, particularly in the automotive and aerospace industries. **Table 38.1** provides a process summary for electron-beam welding.

Laser-Beam Welding

Laser beams can be used as a heat source for welding, cutting, hole making, cladding, heat treating, marking, micromachining, and additive building a wide variety of engineering materials. When used for **laser-beam welding (LBW)**, the beam of monochromatic, coherent light can be focused to a diameter of 0.1 to 1.0 mm (0.004 to 0.04 in.), providing a power density in excess of 10^6 watts/mm^2. The high-intensity beam can be used to simply melt the material at the joint, but more often, it produces a very narrow column of vaporized metal (a "keyhole") with a surrounding liquid pool. As the beam traverses, the liquid flows into the joint to produce a weld with depth-to-width ratio generally greater than 5:1. Because of the narrow weld pool geometry, high travel speed of the beam (typically several meters per minute), and low total heat input, the molten metal solidifies quickly, producing a very thin heat-affected zone and little thermal distortion. Finishing costs are quite low. Because welds require only one-side access, many different joint configurations are possible.

Laser-beam welding is most effective for simple fusion welds without filler metal (**autogenous welds**), but careful joint preparation is required to produce the narrow gap and necessary level of cleanliness. Filler metal can be added if the gap is excessive, and a low-velocity flow of inert gas (generally helium or argon) can be used to protect the weld pool from oxidation. Carbon steel, stainless steel, aluminum, titanium, nickel alloys, and thermoplastics, as well as dissimilar materials, can all be welded with laser techniques. Some of these materials are shiny and reflective, and a light-absorbing surface treatment might be required prior to the welding operation. **Figure 38.4** shows a typical laser beam weld and **Table 38.2** provides a process summary for laser-beam welding.

As shown in **Figure 38.5**, laser-beam welding and electron-beam welding both offer some of the highest power densities of the welding processes. The well-collimated beam of intense laser energy can produce deep penetration welds that are similar to electron-beam welds, but the laser-beam technique offers several distinct advantages:

1. The beam can be transmitted through air [i.e., a vacuum environment is not required]. There is no physical contact between the welding equipment and the workpiece. The originating laser can be a considerable distance removed.

FIGURE 38.4 Laser butt weld of 3-mm (0.125-in.) stainless steel, made at 1.5 m/min with a 1250-w laser. (Courtesy of Coherent, Inc., Santa Clara, CA)

TABLE 38.1	Process Summary: Electron-Beam Welding (EBW)
Heat source	High-energy electron beam
Protection	Vacuum
Electrode	None
Material joined	All common metals
Rate of heat input	High
Weld Profile (D/W)	20
Maximum penetration	175 mm (7 in.)
Assets	High precision; high quality; deep and narrow welds; small HAZ, low distortion; fast welding speed; beam is easily positioned and deflected; no filler metal, flux, or shielding gas required
Limitations	Beam and target must be within a vacuum; very expensive equipment; work piece size is limited by the vacuum chamber; significant edge preparation and alignment required; requires safety protection from X-rays, and visible radiation

TABLE 38.2	Process Summary: Laser-Beam Welding (LBW)
Heat source	Laser light
Protection	None, or externally supplied gas
Electrode	None
Material joined	All common metals, dissimilar metals, and thermoplastics
Rate of heat input	High
Weld profile (D/W)	5
Maximum penetration	25 mm (1 in.)
Assets	High heat-transfer efficiency; high rate of heat input but low total heat input; can weld any location that is light-accessible; small HAZ; low distortion; can accurately focus the beam with light optics; high welding speed easily automated
Limitations	Possible problems with reflectivity of some metals; good positioning and fit-up required

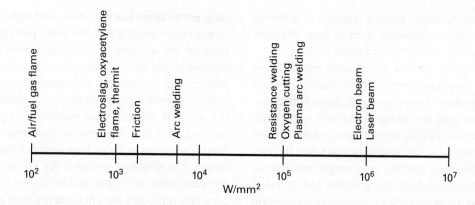

FIGURE 38.5 Comparison of the power densities of various welding processes. The high power densities of the electron-beam and laser-beam welding processes enable the production of deep, narrow welds with small heat-affected zones. Welds can be made quickly and at high travel speeds.

2. No harmful X-rays are generated.

3. The laser beam can be easily shaped, directed, and focused with both transmission and reflective optics (lenses and mirrors), and some beams can be transmitted through fiber-optic cables.

4. The only restriction on weld location is optical accessibility (if you can see it, you can weld it). Welds can be made in difficult-to-reach places, and materials can be joined within transparent containers, such as inside a vacuum tube.

5. The laser process is not electrical. It is not influenced by magnetism and is not limited to electrically conductive materials.

A laser welding system consists of an industrial laser, a means of guiding and focusing the beam, and a means of positioning and manipulating the parts to be welded. There are three types of industrial lasers, each with their own capabilities and limitations. The traditional equipment for laser welding is a CO_2 laser with a power output range of 1.5 to 10 kW (or even higher). The laser gas is actually a mixture of CO_2, nitrogen, and helium, and the rising cost of helium has become a concern. Because CO_2 lasers emit light with a far-infrared wavelength, they require mirror systems or special optical materials to focus and position the beam. They can be used to weld steel up to 25 mm (1 in.) thick at speeds ranging from 1 to 20 m/min (3 to 65 ft/min).

Nd:YAG (neodymium:yttrium–aluminum–garnet) lasers are more limited in power and capability, but operate in a near-infrared wavelength (approximately $1/10$th that of the CO_2 variety), and can utilize conventional glass lenses or delivery by flexible fiber-optic cable (as much as 3000 W of energy can be transmitted up to 150 m through a 0.6-mm-diameter fiber). Because of their limited power, these lasers are relatively slow and are typically used to weld thin materials (less than 2 mm or 0.08 in. in thickness). They are most often operated in the pulsed mode. Pulsing can enable high-power, deep-penetration bursts from an otherwise lower-power laser. By controlling

peak power, pulse duration, and pulse shape, the equipment can be tailored to the product and heat-input can be reduced and controlled, enabling the welding of heat-sensitive materials.

A relatively new type of laser, the high-power, higher-efficiency fiber-delivery laser,[1] now accounts for 20% to 25% of the industrial laser market. The beam of this laser is produced by the excitation of doped optical fiber by diode lasers, the most common dopant being ytterbium. The wavelength is near infrared, giving the delivery benefits of the Nd:YAG laser. It can be focused to finer spot sizes than the other lasers, and is absorbed well by most metallic materials, including reflective ones such as copper, brass, and aluminum. Disk and fiber-delivery direct-diode are similar types of solid-state lasers.

The industrial lasers lend themselves to automation and robotic manipulation. A single Nd:YAG laser with a fiber-optic system can distribute its beam to multiple workstations in either a simultaneous (distributed power) or time-sharing (full-power) fashion. Multiple welds can be made on a single part, thereby speeding production, or the power can be distributed to individual stations as much as 100 meters (300 feet) from the laser. By using fiber-optic cables, laser energy can be piped directly to the end of a robot arm. This eliminates the need to mount and maneuver a heavy, bulky laser and, by reducing weight, enhances the speed and accuracy of both positioning and manipulation. Cutting, drilling, welding, and heat treating can all be performed with the same unit, and multiple axes of motion can provide a high degree of mobility and accessibility.

The equipment cost for a CO_2 or Nd:YAG laser-beam welding system is quite high, but this cost can be somewhat offset by the faster welding speeds, the ability to weld without filler metal, and low distortion, which enables a reduction in post-weld straightening and machining. Caution should be used

---

[1]The fiber laser has a "wall-plug" efficiency of 25% to 30%, compared to 8% to 12% for CO_2 lasers and 3% to 4% for Nd:YAG.

with such equipment, however, because reflected or scattered laser beams can be quite dangerous, even at great distances from the welding site. Eye protection is a must.

Because laser welds do not significantly reduce sheet metal formability, they have been used to produce tailored blanks for the production of sheet products. Steels with different thicknesses, strengths, or coatings can be butt welded with no filler metal to produce single-piece products with different properties at different locations. Weight can be saved by placing thicker material where needed for strength and thinner gages elsewhere. Advanced, high-strength steel can be joined to thinner mild steel. Laser welding has made great progress in the welding of aluminum alloys and has replaced gas tungsten-arc welding or riveting in a number of applications. Lasers have also been used to join automotive powertrain components, as shown in **Figure 38.6**, where the process has enabled weight and cost reduction, coupled with enhanced performance.

With a sharply focused beam and short exposure times, laser welds can be very small and have a low total heat input, often on the order of 0.1 to 10 Joules. These conditions are ideal for use in the electronics industry, and laser welding is frequently used to connect lead wires to small electronic components. Lap, butt, tee, and cross-wire configurations can all be used. It is even possible to weld wires without removing the polyurethane insulation. The laser simply evaporates the insulation and completes the weld with the internal wire.

Lasers have also been used in **hybrid processes** that combine laser welding with arc welding (GMAW, GTAW, or PAW), with both operating in one process zone and producing one weld pool. Hybrid laser-arc welding (HLAW) combines the deep penetration, low distortion, and high-welding speed features of laser welding with the wider pool, gap-bridging capability of arc welding. The resulting weld pool is wide and shallow at the surface, transitioning to deep and narrow. In addition to the unique and flexible weld pool geometry, another benefit is the enhanced arc stabilization provided by the material that the laser evaporates. Laser power can be reduced from that required for lasers operating alone, and welds can be made faster than with just the arc-welding processes. The shielding gas from the arc-welding process protects the entire weld pool, and the added filler metal from the arc welding process adds mass to the weld pool, slowing cooling, thereby helping to retard weld cracking.

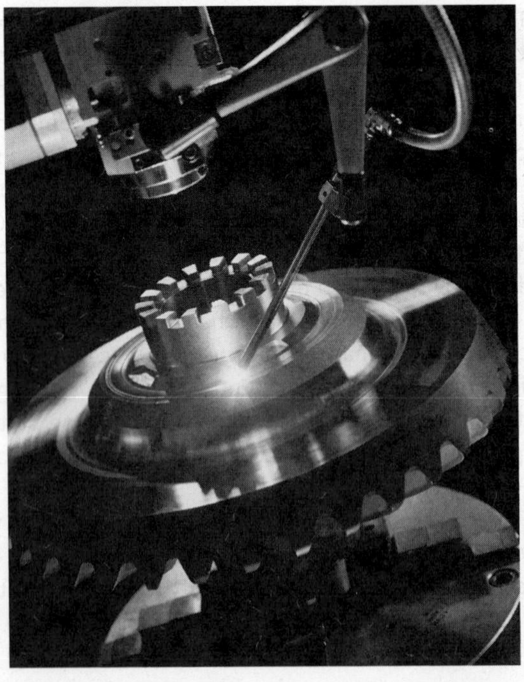

Laser-Beam Cutting

Cutting small holes, narrow slots, or closely spaced patterns in a variety of materials, or producing small quantities of complex-contoured sheet or plate, is another widely used application of industrial lasers. All metals, all plastics, glass, leather, rubber, wood, and paper can be cut. **Laser-beam cutting (LBC)** begins by "drilling" a hole through the material and then moving the beam along a programmed path. As shown in **Figure 38.7**, the intense heat from the laser is used to melt, burn, or vaporize the material being cut. A stream of **assist gas** blows the molten material through the cut, cools the workpiece, minimizes the heat-affected zone, and may participate in a combustion reaction with the material being cut.

Oxygen is the usual gas for cutting mild steel. The laser heats the metal to a temperature where the iron and oxygen

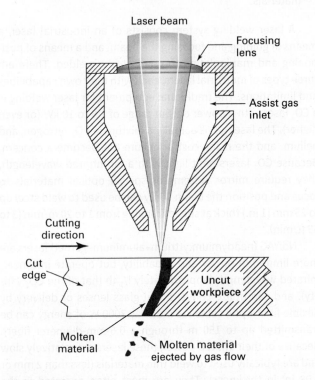

FIGURE 38.7 Schematic of laser-beam cutting. The laser provides the heat, and the flow of assist gas propels the molten droplets from the cut.

FIGURE 38.6 Laser welding of automotive differential gears reduced weight by replacing screw fasteners and increased torque capacity. (Courtesy TRUMPF, Inc., Farmington, CT)

combine in an exothermic reaction. The molten iron and iron oxide have a low viscosity and are easily blown away by the flow of assist gas. In this **exothermic cutting** process, the assist gas actually contributes additional heat, up to 50% of the energy required to cut the material. High cutting speeds are possible, the speed being limited by the rate of material burning. Nitrogen is used with stainless steel and aluminum, and, because of its high reactivity, titanium requires an inert gas, such as argon. Inert gas or compressed air is generally used to cut a wide range of nonmetallic materials. The latter are **endothermic cutting processes**, because the gas actually absorbs energy as it is heated. Cutting speed is set by the rate at which the laser can melt and/or vaporize material. Exothermic cutting produces an oxidized edge, while endothermic cutting (also called *clean cutting*) results in oxide-free surfaces.

The same three types of lasers described in laser welding are also used in laser cutting. The traditional CO_2 laser is an all-purpose cutting machine, capable of cutting steel over a range of thicknesses from 0.5 to 32 mm (0.02 to 1.25 in.). The more efficient, but lower power, solid-state lasers have become the preferred means of cutting thin sheet and highly reflective materials, such as titanium, aluminum, copper, and brass. A lower-power fiber laser can cut thin steel two to three times faster than a higher-power CO_2 laser, and is nearly three times more efficient. Cutting speeds have been reported in the range of 25 to 38 meters/min (1000 to 1500 ipm). The CO_2 laser is still the preferred equipment for cutting thicker material, generally greater than 9.5 mm or $3/8$-in.

Because of its features and recent advancements, laser cutting has replaced a number of plasma and oxyfuel cutting operations, as well as some mechanical cutting processes. Cutting speeds are higher. Clean, accurate, square-edged cuts are characteristic, and the **kerf** (typically as small as 0.25 mm or 0.01 in.) and heat-affected zone are narrower than with any other thermal cutting process. Because laser cutting is non-contact, no clamping or fixturing is required, as is common with mechanical cutting. Virtually any material can be cut, and no postcut finishing is required in many applications, even though the process does produce a thin recast surface. **Figure 38.8** shows the edge of 6-mm (0.25-in.) thick carbon steel, laser cut at 1.8 m/min with a 1250-W laser.

When used for cutting, lasers can be operated in either a continuous wave or pulsed mode. Cutting speed is greatest in the continuous mode, the preferred method for straight and mildly contoured cuts. The pulsed mode is preferred for thin

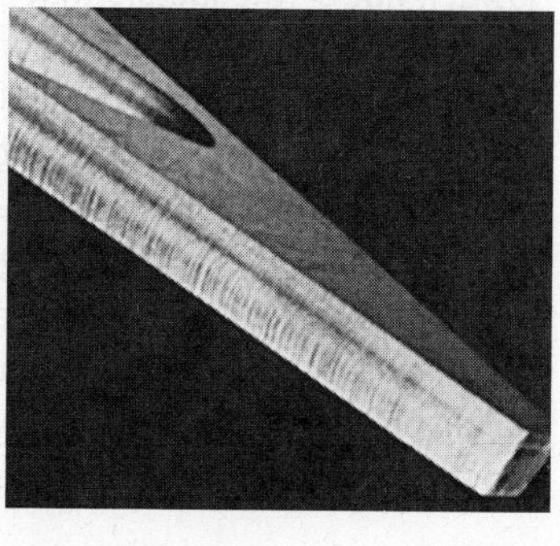

materials and enables tight corners and intricate details to be cut without excessive burning. Because the heat input is reduced, precision is improved and the size of the heat-affected zone and thickness of the recast layer are both reduced.

In addition to common robotic applications, lasers have also been mounted on CNC-type machines, or combined with traditional tools, such as punch presses, to produce extremely flexible hybrid equipment. Because no dedicated dies or tooling are required to produce a cut and there is no setup time, the laser is an economical alternative to blanking or nibbling for prototype or short-run products or for materials that are difficult to cut by conventional methods, such as plastics, wood, and composites. Lasers can drill holes as small as 0.025 mm (0.001 in.) in milliseconds in polyimide or polyester printed circuit board sheets.

Because lasers can cut a wide variety of metals and non-metals, laser cutting has become a dominant process in the cutting of composite materials. The more uniform the thermal characteristics of the components, the better the cut and the less thermal damage to the material. Kevlar-reinforced epoxy cuts easiest and gives a narrow heat-affected zone. Glass-reinforced epoxy is more difficult because of the greater thermal differences, and graphite-reinforced epoxy is even worse because of the high dissociation temperature and thermal conductivity of the graphite. By the time the graphite has absorbed sufficient cutting heat, the epoxy matrix will have decomposed to a significant depth.

Laser Spot Welding

Lasers have also been used to produce spot welds in a manner that offers unique advantages when compared to the conventional resistance methods. A small clamping force is applied to ensure contact of the workpieces, and a fine-focused beam then scans the area of the weld. Welding is performed in the keyhole mode, where the laser produces a small hole through the molten puddle. As the beam is moved, molten metal flows into the hole and solidifies, forming a fusion-type nugget.

Laser spot welding can be performed with access to only one side of the joint. It is a noncontact process and produces no indentations. No electrodes are involved, so electrode wear is no longer a production problem. Weld quality is independent of material resistance, surface resistance, and electrode condition, and no water cooling is required. The total heat input is low, so the heat-affected zone is small. Speed of welding and strength of the resulting joint are comparable to resistance spot welds.

Other Laser Applications

Lasers have also been used in a number of other applications. Laser heat treating of metal surfaces has been discussed in Chapter 5. Lasers are being used to fuse metal powders in an additive manufacturing process known as direct metal sintering. Fully functional metal components are being manufactured,

FIGURE 38.8 Surface of 6-mm-thick carbon steel cut with a 1250-w laser at 1.8 m/min. (Courtesy of Coherent, Inc., Santa Clara, CA)

as described in Chapter 33. Lasers are being used for marking and engraving products, replacing the protrusions or recesses produced by alternative techniques. **Laser peening** uses short bursts of intense laser light to create pressure pulses on metal surfaces. The shock waves travel into the metal, compressing it to generate compressive residual stresses.

In addition to the cutting operations previously described, lasers can also be used for material removal, similar to machining. UV lasers can thermally evaporate polymers or induce the breaking of bonds within the polymer chains. With low power and fine focus, the lasers can be used for micromachining operations. Cooling micro-channels can be "drilled" in turbine engine blades, and holes as small as 0.025 mm (0.001 in.) in diameter can be made in polyimide or polyester printed circuit boards. Hole depth-to-diameter ratios of 10:1 are quite common.

38.3 Surface Modification by Welding-Related Processes

Surfacing (Including Hardfacing)

Surfacing or thermal **cladding** is the process of depositing a layer of weld metal on the surface or edge of a different composition base material. The usual objectives are to obtain improved resistance to wear, abrasion, heat, or chemical attack, without having to make the entire piece from a more expensive material, one that is difficult to fabricate, or one that would not possess the desired bulk properties. The process is also used to restore the dimensions of worn surfaces on used implements as a means of extending their useful working life. Because the deposited surfaces are generally harder than the base metal, the process is often called **hardfacing**. This is not always true, however, for in some cases a softer metal (such as bronze) is applied to a harder base material.

Surfacing Materials
The materials most commonly used for surfacing include carbon and low-alloy steels; high-alloy steels and irons; cobalt-based alloys; nickel-based alloys, such as Monel, Nichrome, and Hastelloy; copper-based alloys; stainless steels; and ceramic and refractory carbides, oxides, borides, silicides, and similar compounds. Widely dissimilar materials can be effectively bonded to one another.

Surfacing Methods and Applications
Because some of the base metal melts during the deposition, surfacing is actually a variation of fusion welding and can be performed by nearly all of the gas-flame or arc-welding techniques, including oxyfuel gas, shielded metal arc, gas metal arc, gas tungsten arc, submerged arc, electroslag, and plasma arc. Arc welding is frequently used for the deposition of high-melting-point alloys. The submerged arc and electroslag processes are used when large areas are to be surfaced or a large amount of surfacing material is to be applied. The plasma arc process further extends the process capabilities because of its extreme temperatures. To obtain true fusion of the surfacing material, a transferred arc is used and the surfacing material is injected in the form of a powder. If a nontransferred arc is used, only a mechanical bond is produced, and the process becomes a form of metallizing. Lasers are also used in surfacing or thermal cladding operations where they melt powders that have been deposited on the substrate surface.

Thermal Spray Coating or Metallizing

The **thermal spray** or **metallizing** processes offer a means of applying a coating of high-performance material (metals, alloys, ceramics, intermetallics, cermets, carbides, or even plastics) to more economical and more easily fabricated base metals. A wire or rod of the coating material is fed into a gas flame or arc, where it melts and becomes atomized by a stream of gas, such as argon, nitrogen, combustion gases, or compressed air. The gas stream propels the 0.01 to 0.05 mm (0.0004 to 0.002 in.)-diameter molten or semi-molten particles toward the target surface, where they impact ("splat"), cool, and bond. Very little heat is transferred to the substrate, whose peak temperatures generally range from 100° to 250°C (212° to 500°F). As a result, thermal spraying does not induce undesirable metallurgical changes or excessive distortion, and coatings can be applied to thin or delicate targets or to heat-sensitive materials such as plastics. The applied coating can range in thickness from 0.1 to 12 mm (0.004 to 0.5 in.) and there is no limit to the size of the workpiece that can be coated. Unlike processes like plating, thermal spray is a line-of-sight process. As a result, it cannot coat features such as deep bores, slots, or hidden surfaces.

Several of the thermal spray processes use adaptations of oxyfuel welding equipment. **Figure 38.9** shows a schematic of an oxyacetylene metal spraying gun designed to utilize a solid wire feed. The flame melts the wire and a flow of compressed air disintegrates the molten material and propels it to the workpiece. An alternative type of oxyfuel gun uses material in the form of powder, which is gravity- or pressure-fed into the flame, where it is melted and carried by the flame gas onto the target. The powder feed permits the deposition of material that would be difficult to fabricate into wire, such as cermets, oxides, and carbides, and the droplet size is controlled by the size of the powder, not by the factors that control atomization. The oxyfuel processes are simple to operate and can be performed in remote areas without electricity. Unfortunately, the lower temperatures and lower particle velocities of the oxyfuel deposition methods result in coatings with high porosity and low cohesive strength. By modifying the combustion process to produce a supersonic stream of emerging hot gas, an adaptation known as **high-velocity oxyfuel spraying (HVOF)**, the molten powder particles now impact with high kinetic energies, and the resulting coating is dense and well bonded. The HVOF

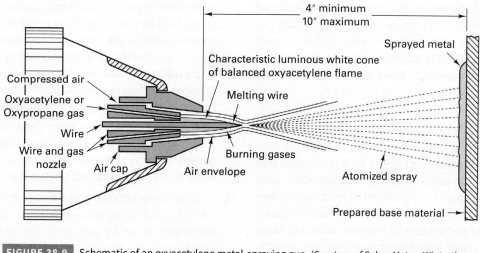

FIGURE 38.9 Schematic of an oxyacetylene metal-spraying gun. (Courtesy of Sulzer Metco, Winterthur, Switzerland)

process is widely used to apply coatings of tungsten carbide for wear resistance, chrome carbide for corrosion resistance, and quality coatings of various metals and alloys.

The simplest of the electric arc methods is wire arc or electric arc spraying. In the twin-wire arc spray process, two oppositely charged electrode wires are fed through a gun, meeting at the tip, where they form an arc with a temperature around 4000°C (7000°F). A stream of atomizing gas flows through the gun, stripping off the molten metal to produce a high-velocity spray. Wires of different chemistry or cored wires can be used to create unique coatings. Because all of the input energy is used to melt the metal, this process is extremely energy efficient and deposition rates are higher than for most other processes. Large areas can be coated economically, but the coatings typically have up to 15% porosity.

Plasma spray metallizing, illustrated in **Figure 38.10**, is a more sophisticated technique. A plasma-forming gas serves as both the heat source and propelling agent for the coating

material, which is usually fed in the form of powder. The molten particles attain high velocity and therefore produce a dense, strongly bonded coating. Because temperatures can reach 16,500°C (30,000°F), plasma spraying can be used to deposit materials with extremely high melting points. Metals, alloys, ceramic oxides and carbides, cermets, intermetallics, and plastic-based powders have all been successfully deposited, and deposition rates can be high.

Yet another means of spray deposition is detonation spray. Oxygen and fuel is fed into the barrel of a D-gun, along with a charge of powder. A spark ignites the gas mixture, and the resulting detonation heats and accelerates the powder to supersonic velocity. A pulse of nitrogen purges the barrel and the process repeats multiple times per second. The ultrahigh velocity produces a dense, strong coating with less than 1% porosity.

While thermal spraying or metallizing is similar to surfacing and is often applied for the same reasons, the coatings are

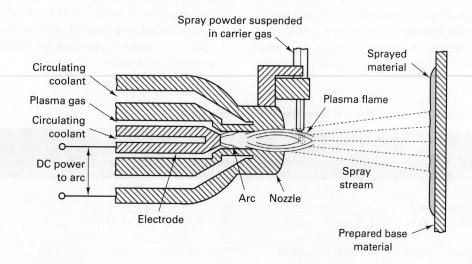

FIGURE 38.10 Diagram of a plasma-arc spray gun. (Courtesy of Sulzer Metco, Winterhur, Switzerland)

usually thinner and the process is more suitable for irregular surfaces or heat-sensitive substrates. The deposition guns can be either hand-held or machine-driven. A standoff distance of 0.15 to 0.25 m (6 to 10 in.) is usually maintained between the spray nozzle and the workpiece. **Table 38.3** compares the features of five methods of thermal spray deposition.

Surface Preparation for Metallizing

Unlike surfacing, thermal spray or metallizing does not melt the base metal. Adhesion is entirely mechanical, so it is essential that the base metal be prepared in a way that promotes good mechanical interlocking. The target material must first be clean and free of dirt, moisture, oil, and other contaminates. Techniques include dry abrasive blasting or brushing, vapor degreasing, acid pickling, oven baking, and ultrasonic cleaning. The surface is then roughened by one of a variety of methods to create minute crevices that can anchor the solidifying particles. Grit blasting with a sharp, abrasive grit is the most common technique, and a surface roughness of 2.5 to 7.5 microns is adequate for most applications.

Characteristics and Applications of Sprayed Metals

During deposition, the atomized, molten, or semi-molten particles mix with air and then cool rapidly upon impact with the base metal. The resultant coatings consist of bonded particles that span a range of size, shape, and degree of melting. Some particles become oxidized and interparticle voids can become entrapped. Compared to conventional wrought material, the coatings are harder, more porous (0.1% to 15% porosity), and more brittle. Thermal spray coatings add little, if any, additional strength to a part, because the strength of the porous coating is usually between one-third and one-half of its normal wrought strength. Applications, therefore, generally look to the coating to provide resistance to heat, wear, erosion, and/or corrosion, or to restore worn parts to original dimensions and specifications. Some typical applications include:

1. *Protective coatings.* Zinc and aluminum are sprayed on iron and steel to provide corrosion resistance—a process that can extend the lifetime of bridges, buildings, and other infrastructure items. Compared to electroplating or hot-dip immersion, there is no size limit, coating thickness can be varied from location to location as needed, and the coating can be applied on site. The interior surfaces of power boilers can be coated with high-chromium alloys to extend wall life by providing both heat resistance and corrosion resistance. Coatings are often applied to aircraft engine components to help them endure extreme temperatures and pressures.

2. *Building up worn surfaces.* Worn parts can be salvaged or their life extended by adding new metal to the depleted regions. The repair and restoration of aircraft engine components is probably the largest single use of thermal spraying.

3. *Hard surfacing.* Although metal spraying should not be compared to hard-facing deposits that are applied by welding techniques, it can be used when thin coatings are considered to provide adequate wear resistance. Typical applications might include automobile cylinder liners and piston rings; thread guides in textile plants; and critical parts within pumps, bearings, and seals. The process is often used as an alternative to hard chrome plating, which suffers from regulatory and environmental restrictions.

4. *Applying coatings of expensive metals.* Metal spraying provides a simple method for applying thin coatings of noble metals to surfaces where conventional plating would not be economical.

5. *Electrical properties.* Because metal can be deposited on almost any surface, thermal spraying can be used to apply a conductive surface to an otherwise poor conductor or nonconductor. Copper, aluminum, or silver is frequently sprayed onto glass or plastics for this purpose. Conversely, sprayed alumina (Al_2O_3) can be used to impart insulating or dielectric properties. Metal deposits can also provide electromagnetic and radiation shielding.

6. *Reflecting surfaces.* Aluminum, sprayed on the back of glass by a special fusion process, makes an excellent mirror.

7. *Decorative effects.* One of the earliest and still important uses of metal spraying was to obtain decorative effects. Because sprayed metal can be treated in a variety of ways, such as buffed, wire brushed, or left in the as-sprayed condition, it is frequently specified for finishing manufactured products and architectural materials.

TABLE 38.3   Comparison of Five Thermal Spray Deposition Techniques

Method	Source	Heat Temperature (°C)	Deposited Materials	Particle Impact Velocity (m/sec)	Adhesion Strength (Mpa)	Porosity (%)	Maximum Spray Rate (kg/hr)
Flame spray							
Wire	Oxyfuel	3000	Metals	180	14–21	10–15	9
Powder	Oxyfuel	3000	Metals, ceramics, plastics	30	14–21	10–15	7
High-velocity oxyfuel (HVOF)	Oxyfuel	>3000	Metals, carbides	600–1000	>70	<1	14
Wire arc	DC arc	4000	Metals only	250	28–42	10–15	16
Plasma spray	DC arc	5500 to 16, 500	All	250–1200	40–70	2–10	5–25

8. *Tailored surface characteristics.* Porous coatings of cobalt or titanium alloys, or certain ceramic materials, have been applied to medical implants to help promote adhesion and in-growth of bone and tissue.

38.4 Brazing

In brazing and soldering, the surfaces to be joined are first cleaned, the components assembled or fixtured, and a low-melting-point nonferrous metal is then melted, drawn into the space between the two solids by capillary action, and allowed to solidify. **Brazing** is the permanent joining of similar or dissimilar metals or ceramics (or composites based on those two materials) through the use of heat and a filler metal whose melting temperature (actually, liquidus temperature) is higher than 450°C (840°F),[2] but below the melting point (or solidus temperature) of the materials being joined. Within this definition, joint design, materials to be joined, filler material and flux selection, heating methods, and joint preparation can all vary widely. The brazing process is different from welding in a number of ways:

1. The *composition* (or chemistry) of the brazing alloy is significantly different from that of the materials being joined.

2. The *strength* of the brazing alloy is usually lower than that of the materials being joined.

3. The *melting point* of the brazing alloy is lower than that of the materials being joined, so none of those materials is melted.

4. Bonding requires **capillary action** to distribute the filler metal between the closely fitting surfaces of the joint. The specific flow is dependent upon the viscosity of the liquid, the geometry of the joint, and surface wetting characteristics.

Because of these differences, the brazing process has several distinct advantages:

1. A wide range of metallic and nonmetallic materials can be brazed. The process is ideally suited for joining dissimilar materials, such as ferrous metal to nonferrous metal, cast metal to wrought metal, metals with widely different melting points, or even metal to ceramic or ceramic to ceramic. Fiber- and dispersion-strengthened composites and porous materials can be joined. A wide array of filler materials is available.

2. Because less heating is required than for welding, the process can be performed quickly and economically.

3. The lower temperatures reduce problems associated with heat-affected zones (or other material property alteration), warping, and distortion. Thinner and more complex assemblies can be joined successfully. Thin sections can be joined

to thick. Metal as thin as 0.01 mm (0.0004 in.) and as thick as 150 mm (6 in.) can be brazed.

4. Assembly tolerances are closer than for most welding processes, and joint appearance is usually quite neat.

5. Brazing is highly adaptable to automation and performs well when mass-producing complex or delicate assemblies. Numerous joints can be brazed at the same time, or complex products requiring multiple joints can be brazed in several steps using filler metals with progressively lower melting temperatures.

6. A strong permanent joint is formed.

Successful brazing or soldering requires that the parts have relatively good fit-up (i.e., small joint clearances) to promote capillary flow of the filler metal. The parts must be thoroughly cleaned prior to joining, and many parts will require flux removal after joining. It is also important to remember that any subsequent heating of the assembly can cause inadvertent melting of the braze metal, thereby weakening or destroying the joint.

Another concern with brazed joints is their enhanced susceptibility to corrosion. Because the **filler metal** is of different composition from the materials being joined, the brazed joint is actually a localized galvanic corrosion cell. Corrosion problems can often be minimized, however, by proper selection of the filler metal.

Nature and Strength of Brazed Joints

The One-Cent Copper Insert

A young engineering student was attempting to impress his date by using a 60,000-pound press to deform a copper penny (before the current copper-plated zinc ones). He placed it between the press surfaces and ran the machine to its full capacity. When the platens separated, he found that the penny had not deformed, but was embedded into the steel press platen. How was this possible? The steel was more than four times stronger than the penny!

Consider a stack of ten pennies, and the desire to shorten the stack by half the thickness of a single penny. As the press surface descends, only a small amount of radial movement is necessary to accommodate the downward compression. Now remove nine of the pennies and perform the same compression. The required radial movement is significant, and if no lubricant is provided, the movement is extremely difficult. Even though the contact area and strength of the pennies remained the same, the required pressure would increase as pennies were progressively removed from the stack. With a single penny, it required less force to deform the thicker steel platen than the thin piece of copper. This is the so-called "thin specimen effect."

What does this have to do with brazing? Let's bond the surfaces of the penny to the press surfaces, and apply tension instead of compression—or join two steel rods with a thin copper braze. If the copper layer is thin enough, it requires less force to deform the stronger steel rod than to localize deformation and produce failure within the weaker copper layer. The strength of the brazed assembly, therefore, is significantly stronger than the strength of the braze metal, and failure typically occurs in the materials being brazed, even though they are of higher strength.

Brazing, like welding, forms a strong metallurgical bond at the interfaces. Clean surfaces, proper clearance, good wetting, and good fluidity will all enhance the bonding. The strength of the resulting joint can be quite high, certainly higher than the strength of the brazing alloy and often greater than the strength of the metal being brazed. Attainment of a high-strength joint, however, requires optimum processing and design.

Of all of the factors contributing to joint strength, **joint clearance** is the most important. If the joint is too tight, it can be difficult for the braze metal to flow into the gap (leaving unfilled voids), and flux might not be able to escape (remaining in locations that should be filled with braze material). There must be sufficient clearance for the braze metal to wet the joint and flow into it under the force of capillary action. As the gap is increased beyond an optimum value, however, the joint strength decreases rapidly, dropping off to the strength of the braze metal itself. If the gap becomes too great, the capillary forces might be unable to draw the material into the joint or hold it in place during solidification. **Figure 38.11** shows the tensile strength of a butt-joint braze as a function of joint clearance.

Proper clearance can vary considerably, depending primarily on the type of braze metal being used. The ideal clearance is usually between 0.01 and 0.04 mm (0.0005 and 0.0015 in.), an "easy-slip" fit. A press fit can even be acceptable provided that fluxes are not used and surface roughness is sufficient to ensure adequate flow of the filler metal into the joint. Clearances up to 0.075 mm (0.003 in.) can be accommodated with a more sluggish filler metal, such as nickel. When clearances range between 0.075 and 0.13 mm (0.003 and 0.005 in.), however, acceptable brazing becomes somewhat difficult, and joints with gaps in excess of 0.13 mm (0.005 in.) are almost impossible to braze. It should be noted that the specified gap should be maintained over the entire braze area—braze surfaces should be parallel.

It is also important to recognize that the dimensions cited above are the clearances that should exist *at the temperature of the brazing process*. Any effects of thermal expansion should be compensated when specifying the dimensions of the starting components. This is particularly significant when dissimilar materials are to be joined, because the joint clearance will change as one material expands at a faster rate than the other. Consider a joint between brass and steel, such as the one depicted in **Figure 38.12**. Brass expands more than steel when temperature is increased. Therefore, if the insert tube is the brass component, the initial fit should be somewhat loose. The brass will expand more than the steel as the temperature is increased, and at the brazing temperature, the gap will assume the desired dimensions. Conversely, if a steel tube is to be inserted into a brass receiver, an initial force fit might be required because the interface will widen as the brass expands more than the steel. Problems can also occur when the reverse dimensional changes occur during cooldown. Significant residual stresses can form in the new joint, and tensile stresses can induce cracking.

Wettability is a strong function of the surface tensions between the braze metal and the base alloy. Generally, the wettability is good when the surfaces are clean and the two metals can form intermediate diffused alloys. Sometimes the wettability can be improved, as is done when steel is tin plated to accept a lead-tin solder, or plated with nickel or copper to enhance brazing. **Fluidity** is a measure of the flow characteristics of the molten braze metal and is a function of the metal, its temperature, surface cleanliness, and clearance.

Design of Brazed Joints

Because the strength of a braze filler metal is generally less than that of the materials being joined, a good joint design is required if one is to obtain adequate mechanical strength. The desired load-carrying ability is usually obtained by assuring proper joint clearance and providing sufficient area for the

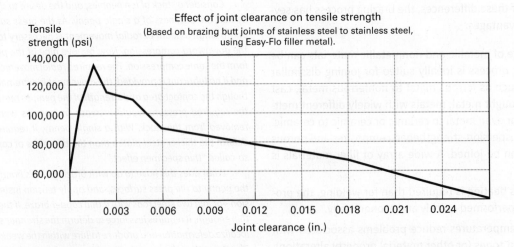

Effect of joint clearance on tensile strength
(Based on brazing butt joints of stainless steel to stainless steel, using Easy-Flo filler metal).

FIGURE 38.11   Typical variation of tensile strength with clearance in a butt-joint braze. (Courtesy of Handy & Harman, Rye, NY)

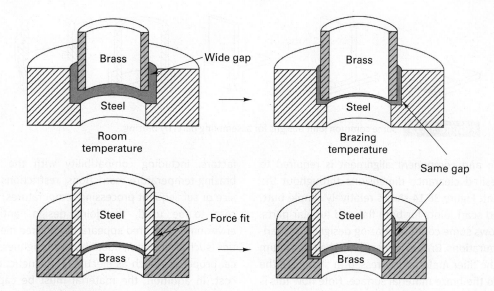

FIGURE 38.12 When brazing dissimilar metals, the initial joint clearance should be adjusted for the different thermal expansions (here brass expands more than steel). Proper brazing clearances should exist at the temperature where the filler metal flows.

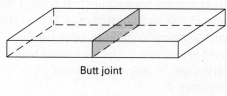

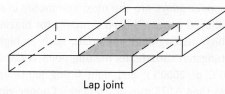

FIGURE 38.13 The two most common types of braze joints are butt and lap. Butt offers uniform thickness across the joint, whereas lap offers greater bonding area and higher strength.

bond. **Figure 38.13** depicts the two most common types of brazed joints: *butt* and *lap*. **Butt joints** do not require additional thickness in the vicinity of the joint, and are most often used where the strength requirements are not that critical. The bonding area is limited to the cross-sectional area of the thinner or smaller member. In contrast, **lap joints** can provide bonding areas that are considerably larger than the butt configuration. Hence, they are often preferred when maximum strength is desired. If the joints are made very carefully, a lap of 1 to 1¼ times the material thickness can develop strength equal to that of the joined materials. For joints that are made by routine production, it is best to use a lap of three to six times the material thickness to

ensure that failure will occur in the base material, not in the brazed joint.

Variations of the two basic joint designs include the *butt-lap* and *scarf* configurations, shown in **Figure 38.14**. The butt-lap design is an attempt to combine the advantage of a uniform thickness with a large bonding area and companion high strength. Unfortunately, it also requires a higher degree of joint preparation. The scarf joint maintains uniform thickness and increases bonding area by tilting the butt joint interface. Careful

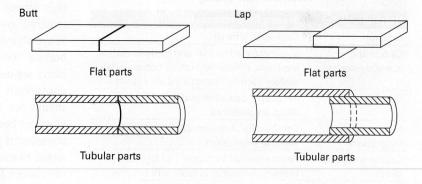

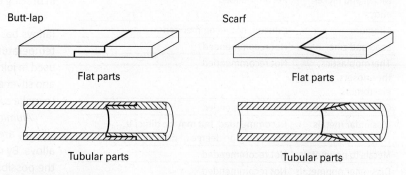

FIGURE 38.14 Variations of the butt and lap configurations include the butt-lap and scarf. The four types are shown for both flat and tubular parts.

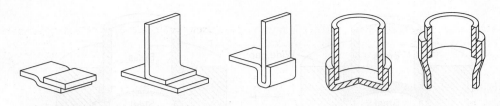

FIGURE 38.15 Some common joint designs for assembling parts by brazing.

joint preparation and component alignment is required to maintain the desired clearance dimensions throughout the length of the joint. Figure 38.14 shows relatively simple butt, lap, butt-lap, and scarf joints for both flat and tubular parts. **Figure 38.15** shows some common brazing designs for a variety of joint configurations. Brazed joints exhibit their maximum strength when the filler material is stressed in shear, i.e. the load is parallel to the braze material surface. Note how this is accomplished in some of the designs of Figure 38.15.

The materials being brazed also need to be considered when designing a brazed joint. **Table 38.4** summarizes the compatibility of various engineering materials with the brazing process.

Filler Metals

The filler metal used in brazing can be any metal that melts between 450°C (840°F) and the melting point of the material being joined. Actual selection, however, considers a variety of

TABLE 38.4	**Compatibility of Various Engineering Materials with Brazing**
Material	**Brazing Recommendation***
Cast iron	Somewhat difficult
Carbon and low-alloy steels	Recommended for low- and medium-carbon materials; difficult for high-carbon materials; seldom used for heat-treated alloy steels
Stainless steel	Recommended; silver and nickel brazing alloys are preferred
Aluminum and magnesium	Common for aluminum alloys and some alloys of magnesium
Copper and copper alloys	Recommended for copper and high-copper brasses; somewhat variable with bronzes
Nickel and nickel alloys	Recommended
Titanium	Difficult, not recommended
Lead and zinc	Not recommended
Thermoplastics, thermosets, and elastomers	Not recommended
Ceramics and glass	Not recommended
Dissimilar metals	Recommended, but may be difficult, depending on degree of dissimilarity
Metals to nonmetals	Not recommended
Dissimilar nonmetals	Not recommended

*For some of the "Not recommended" materials, brazing might be possible but other low temperature joining methods, such as adhesive or diffusion bonding might be preferred.

factors, including compatibility with the base materials, brazing-temperature restrictions, restrictions because of service or subsequent processing temperatures, the brazing process to be used, the joint design, anticipated service environment, desired appearance, desired mechanical properties (such as strength, ductility, and toughness), desired physical properties (such as electrical, magnetic, or thermal), and cost. In addition, the material must be capable of flowing through small capillaries, "wetting" the joint surfaces, and partially alloying with the metals being joined. The most commonly used brazing metals are copper and copper alloys, silver and silver alloys, and aluminum alloys. Many of the brazing metals are based on eutectic reactions (see Chapter 4) in which the material melts at a single temperature that is lower than the melting points of the individual metals in the alloy. **Table 38.5** presents some common braze metal families, the metals they are used to join, and the typical brazing temperatures. These materials can be used in the form of wires, foils, tapes, powders, or pastes.

Copper and copper alloys are the most commonly used braze metals. Unalloyed copper is used primarily for brazing steel and other high-melting-point materials, such as high-speed steel and tungsten carbide. Its melting point is rather high (about 1100°C or 2000°F), and tight fitting joints are required (gaps less than 0.075 mm or 0.003 in.). Copper–zinc alloys offer lower melting points and are used extensively for brazing steel, cast irons, and copper. Copper–phosphorus alloys are used for the fluxless-brazing of copper because the phosphorus can reduce the copper oxide film. These alloys should not be used with ferrous or nickel-based materials, however, because these metals form brittle compounds with phosphorus and the resulting joints can be brittle. A copper–nickel–titanium alloy can be used to braze titanium and some of its alloys. Manganese bronzes can also be used as filler metal in brazing operations.

Pure silver can be used for brazing titanium. **Silver solders** (alloys based on the silver–28% copper eutectic) have brazing temperatures significantly below that of pure copper and are used in joining steels, copper, brass, and nickel. Although silver and silver alloys are expensive, only a small amount is required to make a joint, so the cost per joint is still low.

Aluminum–silicon alloys, containing between 6% and 12% silicon, are used for brazing aluminum and other aluminum alloys. By using a braze metal that is similar to the base metal, the possibility of galvanic corrosion is reduced. These brazing alloys, however, have melting points of about 610°C (1130°F), and the melting temperature of commonly brazed aluminum alloys, such as 3003, is around 670°C (1290°F). Therefore,

TABLE 38.5 Some Common Braze Metal Families, Metals They Are Used to Join, and Typical Brazing Temperatures

Braze Metal Family	Materials Commonly Joined	Typical Brazing Temperature (°C)
Aluminum–silicon	Aluminum alloys	565–620
Copper and copper alloys	Various ferrous metals, as well as copper and nickel alloys and stainless steel	925–1150
Copper–phosphorus	Copper and copper alloys	700–925
Silver alloys	Ferrous and nonferrous metals, except aluminum and magnesium	620–980
Precious metals (gold-based)	Iron, nickel, and cobalt alloys	900–1100
Magnesium	Magnesium alloys	595–620
Nickel alloys	Stainless steel, nickel, and cobalt alloys	925–1200

control of the brazing temperature is critical. In brazing aluminum, proper fluxing action, surface cleaning, and/or the use of a controlled atmosphere or vacuum environment is required to ensure adequate flow of the braze metal.

Nickel- and cobalt-based alloys are attractive for joining assemblies that will be subjected to elevated-temperature service conditions and/or extremely corrosive environments. The service temperature for brazed assemblies can be as high as 1200°C (2200°F). Gold and palladium alloys offer outstanding oxidation and corrosion resistance, as well as good electrical and thermal conductivity. Magnesium alloys can be used to braze other types of magnesium.

A variety of brazing alloys are currently available in the form of amorphous foils, formed by cooling metal at extremely rapid rates. These foils are extremely thin (0.04 mm or 0.0015 in. being typical) and exhibit excellent ductility and flexibility, even when they are made from alloys whose crystalline form is quite brittle. Shaped inserts can be cut or stamped from the foil and positioned in the joint region. Because the braze material is fully dense, no shrinkage or movement occurs during the brazing operation.

One amorphous alloy, composed of nickel, chromium, iron, and boron, is used to produce assemblies that can withstand high temperatures. When the filler metal is liquid (during the brazing operation), the boron diffuses into the base metal, raising the melting point of the remaining filler. The brazed assembly can then be reheated to temperatures above the melting point of the original braze alloy, and the brazed joint will not melt.

Fluxes

In a normal atmosphere, the heat required to melt the brazing alloy would also cause the formation of surface oxides that oppose the wetting of the surface and subsequent bonding. Brazing **fluxes**, therefore, play an important part in the process by dissolving oxides that might have formed on the surfaces prior to heating; preventing the formation of new oxides during heating; and lowering the surface tension between the molten brazing metal and the surfaces to be joined, thereby promoting the flow of the molten material into the joint. Ideally, the flux will melt and become active at a temperature below the solidus of the filler metal, yet remain active throughout the entire range of temperatures encountered while making the braze.

In addition, the flux should be sufficiently fluid that it can be expelled from the joint by the capillary inflow of filler metal.

Surface cleanliness is one of the most significant factors affecting the quality and uniformity of brazed joints. Although fluxes can dissolve modest amounts of oxides, *they are not cleaners*. Before a flux is applied, dirt, grease, oil, rust, and heat-treat scale should be removed from the surfaces that are to be brazed. Cleaning operations can involve water- or solvent-based techniques; high-temperature burn-off of oils, greases, and fuel residues; acid pickling; grit blasting with selected media; other mechanical methods; or exposure to high-temperature reducing atmospheres. The less cleaning the flux has to do, the more effective it will be during the brazing operation. Because the presence of surface graphite impairs wetting, cast iron materials often require special treatment. Graphite removal by chemical etching might be required before cast iron can be brazed.

Brazing fluxes usually take the form of chemical compounds in which the most common ingredients are borates, fused borax, fluoroborates, fluorides, chlorides, acids, alkalies, wetting agents, and water. The particular flux should be selected for compatibility with the base metal being brazed and the particular process being used. Paste fluxes are used for furnace, induction, and dip brazing, and they are usually applied by brushing. Either paste or powdered fluxes can be used with the torch-brazing process, in which application is usually achieved by dipping the heated end of the filler wire into the flux material. Gaseous fluxes have also been employed.

Applying the Braze Metal

The brazing filler metal can be applied to joints in several ways. The oldest method (and still a common technique when torch brazing) uses brazing metal in the form of a rod or wire. The joint area is first heated to a temperature high enough to melt the braze alloy and ensure that it remains molten while flowing into the joint. The torch is then used to melt the braze metal, and capillary action draws it into the prepared gap.

The above method of braze metal application requires considerable labor, and care must be taken to ensure that the filler metal has flowed into the inner portions of the joint. To avoid these difficulties, the braze metal is often inserted into the joint prior to heating, usually in the form of wires, foils, shims, powders, or preformed rings, washers, disks, or slugs.

Rings or shims can also be fitted into internal grooves in the joint before the parts are assembled.

When using preloaded joints, care must be exercised to ensure that the filler metal is not drawn away from the intended surface by the capillary action of another surface of contact. Capillary action will always pull the molten braze metal into the smallest clearance, regardless of whether that was the intended location. In addition, the flow of filler metal must not be cut off by inadequate clearances or the presence of entrapped or escaping air. Fillets and grooves within the joint can also act as reservoirs and trap the filler metal.

Yet another approach is to precoat one or both of the surfaces to be joined with the brazing alloy. Simply placing the materials in contact and heating forms the desired bond. By having the braze material already in place over the full area of contact, the joining operation does not have to rely on capillary action and metal flow. More complex assemblies can be produced than with conventional methods, and the thickness of the braze material is precisely controlled to provide maximum strength to the joint.

All of the components must maintain fixed positions during the brazing operation, and some form of restraint or fixturing is often required. Alignment and clearances can often be maintained by tack welding, riveting, staking, expanding or flaring, swaging, knurling, or dimpling. Shims, wires, ribbons, and screens can also be employed to assist in locating pieces or maintaining fit. For more complex components, special brazing **jigs and fixtures** are often used to hold the components during the heating. When these are used, however, it is necessary to provide springs that will compensate for thermal expansion, particularly when dissimilar metals are being joined.

Heating Methods Used in Brazing

Because molten metal tends to flow toward the location of highest temperature, it is important that the heat sources used in brazing control both the temperature and the uniformity of that temperature throughout the joint. In specifying the heating method, a number of factors should be considered, including the size and shape of the parts being brazed, the type of material being joined, and the desired quantity and rate of production.

A common source of heat for brazing is a gas-flame torch. In the **torch-brazing** procedure, oxyacetylene, oxyhydrogen, or another gas-flame combination (natural gas, MAPP, butane, or propane) can be used. Most repair brazing and small quantity production is done in this manner because of its flexibility and simplicity, and the equipment is inexpensive, readily available, and usually portable. The process is also widely used in production applications where specially shaped torches speed the heating and reduce the amount of skill required. Local heating permits the retention of most of the original material strength and enables large components to be joined with little or no distortion. The joint area should be heated only with the outer envelope of the flame, avoiding contact with the bright inner cone where the temperature is much higher. The major

drawbacks are the difficulty in controlling the temperature and maintaining uniformity of heating, as well as meeting the cost of skilled labor. Because the heating is performed in air, a protective flux is usually required, and the flux residue must be removed after brazing.

If the filler metal can be preloaded into the joints and the part can endure uniform heating, a number of assemblies can be brazed simultaneously in controlled atmosphere or vacuum furnaces, a process known as **furnace brazing**. If the components are not likely to maintain their alignment, brazing jigs or fixtures must be used. Fortunately, for most assemblies that are to be furnace brazed, a light press fit is usually sufficient to maintain alignment. **Figure 38.16** shows some typical furnace-brazed assemblies.

Because excellent control of the furnace temperature is possible and no skilled labor is required, furnace brazing is particularly well suited for mass-production operations, with either batch- or continuous-type furnaces being used. Furnace brazing heats the entire assembly in a uniform manner and therefore produces less warpage and distortion than processes that employ localized heating. Extremely complex assemblies can be produced, with multiple joints being formed in a single heating.

A variety of furnace atmospheres can be used to reduce oxide films and prevent both the base and filler metals from oxidizing during the brazing operation. A chemical flux might no longer be needed, and the parts emerge clean and free of contaminants. When reactive materials are to be joined or the joint must meet the highest of standards, a vacuum furnace is frequently used.

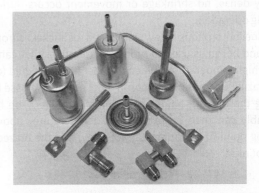

FIGURE 38.16 Two views of an array of furnace-brazed assemblies. (www.franklinbrazing.com)

Another type of heating is **salt-bath brazing**, where the joints are preloaded with filler metal and then immersed in a bath of molten salt that is maintained at a temperature slightly above the melting point of the brazing metal. This process offers three distinct advantages:

1. The salt bath acts as the brazing flux, preventing oxidation and enhancing wettability.

2. The work heats very rapidly because it is in complete contact with the heating medium.

3. Temperature can be accurately controlled so thin pieces can be attached to thicker pieces without danger of overheating. This last feature makes the process well suited for brazing aluminum, where precise temperature control is often required.

To ensure that the bath remains at the desired temperature during the immersion process, its volume must be substantially larger than that of the assemblies to be brazed. Massive parts might need to be preheated prior to immersion.

In **dip brazing**, the assemblies are immersed in a bath of molten brazing metal, often covered with a layer of molten flux. The bath thus provides both the heat and the metal for the joint. Because the braze metal will usually coat the entire workpiece, it is a somewhat wasteful process and is usually employed only for small products.

Induction brazing uses high-frequency induced currents as the source of heat and is therefore limited to the joining of electrically conductive materials. A variety of high-frequency AC power supplies is available in large and small capacities. These are coupled to a simple heating coil designed to fit around the joint. The heating coils are generally formed from copper tubing and typically carry a supply of cooling water. Although the filler metal can be added to the joint manually after it is heated, the usual practice is to use preloaded joints to speed the operation and produce more uniform bonds. Induction brazing offers the following advantages, which account for its extensive use:

1. The complete heating cycle is very rapid, usually only a few seconds in duration.

2. The operation can be made semiautomatic so that only semiskilled labor is required.

3. Heating can be confined to the specific area of the joint through use of specially designed coils, frequency control, and short heating times. This minimizes softening and distortion and reduces problems associated with scale and discoloration.

4. Uniform results are easily obtained because of the precise control of both heating rate and final temperature.

5. By making relatively simple, part-specific, heating coils, a wide variety of work can be brazed with a single power supply.

Resistance brazing can be used to produce relatively simple joints in metals with high electrical conductivity. The parts to be joined are pressed between two water-cooled electrodes and a current is passed through, as in spot welding. Unlike resistance welding, however, the carbon or graphite electrodes provide most of the resistance in resistance brazing, and the heating of the joint is primarily by conduction from the hot electrodes.

Infrared heat lamps, lasers, and electron beams can also be used to provide the heat required for brazing. Recent studies have also shown microwave energy to be an efficient heat source. Silicon carbide plates are positioned around the joint and are heated by the microwaves. Heat is then transferred to the joint by radiation.

The heating for brazing has also been combined with the heating for other metallurgical processes, such as the aging, tempering, or stress-relieving components of heat treatment.

Flux Removal and Other Postbraze Operations

Because most brazing fluxes are corrosive, the flux residue should be removed from the work as soon as brazing is completed. Rapid and complete flux removal is particularly important in the case of aluminum, where chlorides can be particularly detrimental. Fortunately, many brazing fluxes are water soluble, and an immersion in a hot-water tank for a few minutes will often provide satisfactory results. Blasting with grit or sand is another effective method of flux removal, but this procedure might not be attractive if a good surface finish is to be maintained. Fortunately, such drastic treatment is seldom necessary.

Other postbraze operations can include heat treating, cleaning, and inspection. A visual examination is probably the simplest of the inspection techniques and is most effective when both sides of a brazed joint are accessible for examination. A proof test can be performed by subjecting the joint to loads in excess of those expected during service. Leak tests or pressure tests can ensure gas- or liquid-tightness. Cracks and other flaws can be detected by dye-penetrant, magnetic particle, ultrasonic, or radiographic examination. Destructive forms of evaluation include peel tests, tension, or shear tests, and metallographic examination.

Fluxless Brazing

Because the application and removal of brazing flux involves significant costs, particularly where complex joints and assemblies are involved, a large amount of work has been devoted to the development of procedures where a flux is not required (**fluxless brazing**). Controlled furnace atmospheres can make a flux unnecessary by reducing existing oxides and preventing the formation of new ones. Vacuum furnaces can also be used to create and preserve clean brazing surfaces. Special brazing metals have been developed with alloy additions, such as phosphorus, that can also fulfill the role of a flux.

Braze Welding

Braze welding differs from straight brazing in that capillary action is not required to distribute the filler metal. The molten filler is simply deposited by gravity, as in oxyacetylene gas welding, but unlike welding, the base metal does not melt. Because relatively low temperatures are required and warping is minimized, braze welding is very effective for the repair of steel products and ferrous castings. It is also attractive for joining cast irons because the low heat does not alter the graphite shape, and the process does not require good wetting characteristics. Strength is determined by the braze metal being used and the amount applied. Considerable buildup might be required if full strength is to be restored to the repaired part.

Braze welding is almost always done with an oxyacetylene torch. The surfaces are first "tinned" with a thin coating of the brazing metal, and the remainder of the filler metal is then added. **Figure 38.17** shows a schematic of braze welding.

38.5 Soldering

Soldering is a brazing-type operation where the filler metal has a melting temperature (or liquidus temperature if the alloy has a freezing range) below 450°C (840°F). It is typically used for joining thin metals, connecting electronic components, joining metals while avoiding exposure to high elevated temperatures, and filling surface flaws and defects. The process generally involves six important steps: design of an acceptable joint; selection of the correct solder for the job; selection of the proper type of flux; cleaning the surfaces to be joined; application of flux, solder, and sufficient heat to allow the molten solder to fill the joint by capillary action and solidify; and removal of the flux residue.

Design and Strength of Soldered Joints

Soldering can be used to join a wide variety of sizes, shapes, and thicknesses, and is employed extensively to provide electrical coupling or gas- or liquid-tight seals. While the low joining temperatures are attractive for heat-sensitive materials, solders are intrinsically weak. Soldered joints seldom develop shear strengths in excess of 1.75 MPa (250 psi). Consequently, if appreciable strength is required, soldered joints should be avoided, the contact area should be large, or some form of mechanical joint, such as a rolled-seam lock, should be made prior to soldering. Butt joints should never be used, and designs where peeling action is possible should be avoided. **Figure 38.18** shows some of the more common solder joint designs, including lap, flanged butt, and interlock. Design selection should also consider the method of heating, as well as the method of introducing the flux and solder to the joint. Whenever possible, the solder should be applied from one side of the joint. The appearance of solder on the other side will then provide visual confirmation that the solder has distributed through the entire joint area.

As with brazing, there is an optimal clearance for best performance. For typical solder joints, a clearance of 0.025 to 0.13 mm (0.001 to 0.005 in.) provides for capillary flow of the solder, expulsion of the flux, and reasonable joint strength. To avoid cracks, the parts should be held firmly so that no movement can occur until the solder has cooled to well below the solidification temperature.

Metals to Be Joined

Table 38.6 summarizes the compatibility of soldering with a variety of engineering materials. Copper, silver, gold, and tin, as well as steels plated with these metals, are all easily soldered. Because aluminum has a strong, adherent oxide that makes soldering difficult, special fluxes and modified techniques can be required. Adequate joints are indeed possible, however, as evidenced by the large number of soldered aluminum radiators currently in automotive use. Ceramic materials can be joined if they have solderable metal coatings.

Soldering is used extensively in electronic assemblies where the joints provide sufficient strength while allowing the various components to expand and contract, dissipate heat, and transmit electronic signals.

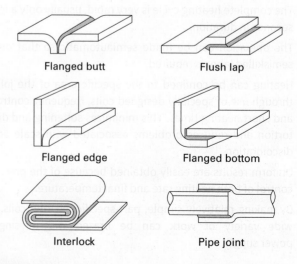

Flanged butt   Flush lap

Flanged edge   Flanged bottom

Interlock   Pipe joint

FIGURE 38.18 Some common designs for soldered joints.

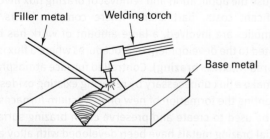

Filler metal   Welding torch

Base metal

FIGURE 38.17 Schematic of the braze-welding process.

TABLE 38.6	Compatibility of Various Engineering Materials with Soldering
Material	**Soldering Recommendation**
Cast iron	Seldom used since graphite and silicon inhibit bonding
Carbon and low-alloy steels	Difficult for low-carbon materials; seldom used for high-carbon materials
Stainless steel	Common for 300 series; difficult for 400 series
Aluminum and magnesium	Seldom used; however, special solders are available
Copper and copper alloys	Recommended for copper, brass, and bronze
Nickel and nickel alloys	Commonly performed using high-tin solders
Titanium	Seldom used
Lead and zinc	Recommended, but must use low-melting-temperature solders
Thermoplastics, thermosets, and elastomers	Not recommended
Ceramics and glass	Not recommended
Dissimilar metals	Recommended, but with consideration for galvanic corrosion
Metals to nonmetals	Not recommended
Dissimilar nometals	Not recommended

Solder Metals

Soldering alloys are generally combinations of low-melting-temperature metals, such as lead, tin, bismuth, indium, cadmium, silver, gold, and germanium. Because of their low cost, low melting temperature, acceptable mechanical and physical properties, and many years of use, the most common solders are alloys of lead and tin with the addition of small amounts of antimony, usually less than 0.5%. The three most common alloys contain 60%, 50%, and 40% tin and all melt below 240°C (465°F). Because tin is expensive, those alloys having higher proportions of tin are used only where their higher fluidity, higher strength, and lower melting temperature are desired. For wiped joints and for filling dents and seams, where the primary desire is appearance and little strength is required, solders containing only 10% to 20% tin are preferred. Joints made with a 5% tin alloy require higher temperatures to produce, but will withstand service temperatures as high as 150°C (300°F).

Other soldering alloys might be specified for special purposes or where environmental or health concerns dictate the use of lead-free joints. Lead and lead compounds can be quite toxic. Since 1988, the use of lead-containing solders in drinking water lines has been prohibited in the United States, and concern has been expressed regarding other applications and industries. Japan and the European Union have banned the use of lead-containing solders in electronic equipment. The ideal substitute solder should not only be harmless to the environment, but should also exhibit desirable characteristics in the areas of melting temperature, wettability, electrical and thermal conductivity, thermal-expansion coefficient, mechanical strength, ductility, creep resistance, thermal fatigue resistance, corrosion resistance, manufacturability, and cost. Compatible fluxes should be available and the application procedures should be the same or only slightly different. At present, none of the **lead-free solders** meets all of these requirements, and most are deficient in more than one area.

Most of the alternative solders have been proposed from other eutectic alloy systems. Tin–antimony and tin–copper alloys are useful in electrical applications and have good strength and creep resistance but high melting points. Bismuth alloys have very low melting points and good fluidity, but suffer from poor wettability. Indium alloys offer low melting points, ductility that is retained even at cryogenic temperatures, and rapid creep that allows joints between dissimilar metals to adjust to changes in temperature without generating internal stresses. Tin–indium alloys have been used to join metal-to-glass and glass-to-glass. They have very low melting points and good wettability, but they are expensive and can be somewhat brittle. Aluminum is often soldered with tin–zinc, cadmium–zinc, or aluminum–zinc alloys. Tin–silver and tin–gold offer possibilities when a somewhat higher melting point is desired (typically higher than 205°C, or 400°F) coupled with good mechanical strength and creep resistance, but both systems are limited by the high cost of their components. Lead–silver and cadmium–silver alloys can also be used for higher-temperature service.

The three-component tin–silver–copper system has emerged as the predominant lead-free solder for electrical and electronics applications. Typical compositions include 3% to 4% silver and 0.5% to 0.8% copper, with the remainder being tin. Other ternary (three-component) systems showing promise include tin–silver–bismuth and bismuth–indium–zinc. Binary tin–silver is recommended for wave soldering. All of the lead-free solders have higher melting temperatures than the lead-base alloys.

Like the filler metal used in brazing and braze welding, solders are available as wire and paste, as well as in a variety of standard and special preshaped forms. **Table 38.7** presents some of the more common solder alloys with their melting properties and typical applications.

Soldering Fluxes

As in brazing, soldering requires that the metal surfaces be clean and free of oxide so that the solder can wet the surfaces and be drawn into the joint to produce an effective bond. Soldering fluxes are used to remove surface oxides and prevent oxide formation during the soldering process, but it is essential that dirt, oil, and grease be removed before the flux is applied. This precleaning or surface preparation can be performed by a variety of chemical or mechanical means, including solvent or alkaline degreasers, acid immersion (pickling), grit blasting, sanding, wire brushing, and other mechanical abrasion techniques.

| TABLE 38.7 | Some Common Solders and Their Properties | | | |

| Composition (wt %) | Freezing Temperature (°C) | | | Applications |
	Liquidus	Solidus	Range	
Lead–tin solders				
98 Pb–2 Sn	322	316	6	Side seams in three-piece can
90 Pb–10 Sn	302	268	34	Coating and joining metals
80 Pb–20 Sn	277	183	94	Filling and seaming auto bodies
70 Pb–30 Sn	255	183	72	Torch soldering
60 Pb–40 Sn	238	183	55	Wiping solder, radiator cores and heater units
50 Pb–50 Sn	216	183	33	General purpose
40 Pb–60 Sn	190	183	7	Electronic (low temperature)
Silver solders				
97.5 Pb–1 Sn–1.5 Ag	308	308	0	Higher–temperature service
36 Pb–62 Sn–2 Ag	189	179	10	Electrical
96 Sn–4 Ag	221	221	0	Electrical
Other alloys				
45 Pb–55 Bi	124	124	0	Low temperature
43 Sn–57 Bi	138	138	0	Low temperature
95 Sn–5 Sb	240	234	6	Electrical
50 Sn–50 In	125	117	8	Metal-to-glass
37.5 Pb–25 In–37.5 Sn	138	138	0	Low temperature
95.5 Sn–3.9 Ag–0.6 Cu	217	217	0	Electrical
91.8 Sn–3.4 Ag–4.8 Bi	213	211	2	Electrical (must be lead-free)

Note: In the United States, specifications of solder alloy compositions are listed in ASTM B 32.

Soldering fluxes are generally classified as **corrosive** or **noncorrosive**. The most common noncorrosive flux is **rosin** (the residue after distilling turpentine) dissolved in alcohol. Rosin fluxes are suitable for making joints to copper and brass, and to tin-, cadmium-, and silver-plated surfaces, provided that the surfaces have been adequately cleaned prior to soldering. Aniline phosphate is a more active noncorrosive flux, but it has limited use because it emits toxic gases when heated. The wide variety of corrosive fluxes provide enhanced cleaning action, but require complete removal after the soldering operation to prevent corrosion problems during service.

The Soldering Operation

Soldering requires a source of sufficient heat and a means of transferring it to the metals being joined. Any method of heating that is suitable for brazing can be used for soldering, but furnace and salt-bath heating are seldom used. Most hand soldering is still done with localized heating—soldering irons or small oxyfuel or air-fuel (acetylene, propane, butane or MAPP) torches. Induction heating is used when large numbers of identical parts are to be soldered. For low-melting-point solders, infrared heat sources or ovens can also be employed. The joints can be preloaded with solder, or the filler metal can be supplied from a wire. The particular method of heating usually dictates which procedure is used.

Wave soldering, depicted in **Figure 38.19**, is an automated process used to solder wire ends, such as the multiple connectors that protrude through holes in electronic circuit boards. Molten solder is pumped upward through a submerged nozzle to create a wave or crest in a pool of molten metal. The circuit boards are then passed across this wave at a height where each of the protruding pins sees contact with the molten metal. Wetting and capillary action pulls solder into each joint, and numerous connections are made as each board passes across the wave. The boards are passed across the wave at a slight slant to aid in draining excess solder back into the pool.

In **vapor-phase soldering**, a product with pre-positioned solder is passed through a chamber containing hot, saturated vapors, usually of boiling fluorocarbon, which condense on the cooler product, transferring the heat of vaporization. The result is rapid and uniform heating, with excellent temperature control, combined with the possibility of an oxygen-free environment. The soldering temperature is linked to the boiling point of the fluid, with current materials operating in the range of 100° to 265°C (212° to 510°F). The vapor-phase process has also been used to cure epoxies and stress relieve metals, but its primary application is the soldering of surface-mounted components to substrate materials. Because of the precise temperature control, multi-pass soldering is possible, using up to three different solder compositions with three different melting temperatures. Because the solder is pre-positioned, this process is also known as **vapor-phase reflow soldering**.

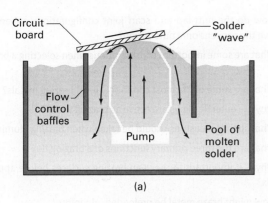

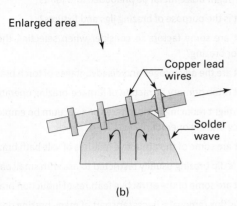

(a) Schematic of the wave soldering process; (b) Magnified view showing the contact with the molten solder.

Dip soldering, where the entire piece is immersed in molten metal, has been used to produce automobile radiators and "tinned" coatings.

Flux Removal

After soldering, the flux residues should be removed from the finished joints, either to prevent corrosion or for the sake of appearance. Flux removal is rarely difficult, provided that the type of solvent in the flux is known. Water-soluble fluxes can be removed with hot water and a brush. Alcohol will remove most rosin fluxes. However, when the flux contains some form of grease, as in most paste fluxes, a grease solvent must be used, followed by a hot water rinse. In the past, solvents containing chlorofluorocarbons (CFCs) were the cleaners of choice, but because they have been implicated in the depletion of atmospheric ozone, an alternative means of flux removal should be employed or the process converted to fluxless soldering.

Fluxless Soldering

Several **fluxless soldering** techniques have been developed using controlled atmospheres (such as hydrogen plasma), thermomechanical surface activation (such as plasma gas impingement), or protective coatings that prevent oxide formation and enhance wetting. Additional successes have been reported with both laser and ultrasonic soldering.

Review Questions

1. What are some joining conditions where welding might not be the best choice?

2. What are some of the lower-temperature methods of joining?

3. In what ways is a thermit weld similar to the production of a casting?

4. What is the source of the welding heat in thermit welding?

5. For what types of applications might thermit welding be attractive?

6. What is the source of the welding heat in electroslag welding?

7. What are some of the various functions of the slag in electroslag welding?

8. Electroslag welding would be most attractive for the joining of what types of geometries and thicknesses?

9. What is the source of heat in electron-beam welding?

10. Why is a high vacuum required in the electron-beam chamber of an electron-beam welding machine?

11. What types of production limitations are imposed by the high-vacuum requirements of electron-beam welding? What compromises are made when welding is performed on pieces outside the vacuum chamber?

12. What are the major assets and negative features of high-voltage electron-beam welding equipment?

13. What are some of the attractive features of electron-beam welding? Negative features?

14. What is unique about the fusion zone geometry of electron-beam welds?

15. Describe the weld pool geometry and size of the heat-affected zone in laser-beam welding.

16. What is an autogenous weld?

17. What might be necessary to permit the laser welding of shiny or reflective materials?

18. What are some of the ways in which laser-beam welding is more attractive than electron-beam welding?

19. What are the three common types of industrial lasers?

20. Which type of laser light can be transmitted through fiber optic cable?

21. What are some of the attractive features of a fiber-optic laser coupled with robotic manipulation?

22. Why is laser-beam welding an attractive process for producing tailored blanks for sheet metal forming? For use on small electronic components?

23. What are the attractive properties of hybrid processes that combine laser and arc welding?

24. What materials can be cut by a laser beam?

25. What is the function of the *assist gas* in laser-beam cutting?

26. What is the difference between exothermic cutting and endothermic cutting, and how do the cut edges differ?

27. Which type of laser is preferred for cutting thicker material?

28. What are the attractive features of laser-beam cutting compared to the plasma and oxyfuel processes?

29. What are the benefits of using the pulsed mode in laser beam cutting?

30. What features have made lasers a common means of cutting composite materials?

31. What are some of the attractive features of laser spot welding?

32. What are some common objectives of surfacing operations?

33. What types of materials are applied by surfacing methods?

34. What are some of the primary methods by which surfacing materials can be deposited onto a metal substrate?

35. What is the benefit of high-velocity oxyfuel spraying compared to the conventional oxyfuel process?

36. What are some of the arc or plasma techniques that can be used to apply a thermal spray coating?

37. How is thermal spraying similar to surfacing? How is it different?

38. Why is surface preparation such a critical feature of metallizing?

39. What are some of the more common applications of sprayed coatings?

40. Provide a reasonable definition of brazing?

41. What are some key differences between brazing and fusion welding?

42. What kinds of materials or combinations can be joined by brazing?

43. What advantages can be gained by the lower temperatures of the brazing process?

44. Why do brazed joints have an enhanced susceptibility to corrosion?

45. What is the most important factor influencing the strength of a brazed joint?

46. How does capillary action relate to joint clearance?

47. Why is it necessary to adjust the initial room-temperature clearance of a joint between two significantly dissimilar metals?

48. What is wettability? Fluidity? How do each relate to brazing?

49. What are the two most common types of brazed joints and the attractive features of each?

50. How do the butt-lap and scarf joint configurations enhance or improve the conventional butt design?

51. What are some important considerations when selecting a brazing alloy?

52. What are some of the most commonly used brazing metals?

53. Why are eutectic alloys attractive as brazing metal?

54. What special measures should be taken when brazing aluminum?

55. What are the three primary functions of a brazing flux?

56. Why is it important to preclean brazing surfaces before applying the flux?

57. How might braze metal be preloaded into joints?

58. What is the purpose of brazing jigs and fixtures?

59. What are some factors to consider when selecting the heating method for brazing?

60. What are the advantages and disadvantages of torch brazing?

61. What is the primary attraction of furnace-brazing operations?

62. Why might reducing atmospheres or a vacuum be employed during furnace brazing operations?

63. What are some of the attractive features of salt-bath brazing?

64. Why is dip brazing usually restricted to use with small parts?

65. What are some of the attractive features of induction brazing?

66. Why is flux removal a necessary part of many brazing operations?

67. What benefits can be achieved through fluxless brazing?

68. How does braze welding differ from traditional brazing?

69. What is the primary difference between brazing and soldering?

70. What are the six steps of a soldering operation?

71. Why is soldering unattractive if a high-strength joint is desired?

72. For many years, the most common solders have been alloys of what two base metals?

73. What is driving the conversion to lead-free solders?

74. What are some of the difficulties encountered when attempting a conversion to lead-free solder?

75. What are the two basic families of soldering flux?

76. What are some of the more common heat sources for producing a soldered joint?

77. Why is wave soldering attractive for making the multiple connections of circuit boards?

78. Describe the vapor-phase soldering process.

Problems

1. A common problem with brazed or soldered joints is galvanic corrosion, because the joint usually involves dissimilar metals in direct metal-to-metal electrical contact.

 a. For each of the various solder or braze joints described below, determine which material will act as the corroding anode.

  i. Two pieces of low-carbon steel being brazed with a copper-base brazing alloy

  ii. A copper wire being soldered to a steel sheet using lead–tin solder

  iii. Pieces of tungsten carbide being brazed into recesses in a carbon-steel plate

 b. How do the various lead-free solders compare to the conventional lead–tin solders with regard to their potential for galvanic corrosion?

c. If galvanic corrosion becomes a significant and chronic problem in a brazed assembly, what changes might you suggest that could possibly reduce or eliminate the problem?

2. When molten metal deposition is applied to a substrate (as with surfacing and thermal spray), a heat affected zone will be created in the unmelted material, just like in welding. Identify several specific applications of the deposition processes, and discuss any concerns you might have relating to the heat-affected zone.

3. It is not uncommon for surface defects in large castings (incomplete mold filling, localized shrinkage, surface cracks, etc.) or in large forgings (incomplete mold filling, cracks, etc.) to be repaired by some form of surface preparation, such as machining or grinding, followed by weld or braze metal deposition.

a. How important is it that the deposited metal match the material being repaired? Consider both type (casting alloy vs. wrought alloy) and specific chemistry.

b. How might the properties of the deposited material differ from those of the material being repaired? Why?

c. What type of concerns might you have relating to heat-affected zones or residual stresses?

d. Might any post-repair processing be required?

CHAPTER 39

Adhesive Bonding, Mechanical Fastening, and Joining of Non-Metals

39.1 Adhesive Bonding

The **ideal adhesive** bonds to any material, needs no surface preparation, cures rapidly, and maintains a high bond strength under all operating conditions. It also don't exist. However, tremendous advances have been made in the development of adhesives that are stronger, easier to use, less costly, and more reliable than many of the alternative methods of joining. The use of structural adhesives has grown rapidly from early applications, such as plywood. Adhesives are everywhere—in construction, packaging, furniture, appliances, electronics, bookbinding, product assembly, and even medical and dental applications. They are used to bond metals, ceramics, glass, plastics, rubbers, composite materials, woods, and even a variety of roofing materials. Even such quality- and durability-conscious fields as the automotive and aircraft industries now make extensive use of adhesive bonding. Adhesives now comprise 15% to 20% of all automotive joining. They have advanced from the attaching of interior and exterior trim to the joining of major components, such as door, hood, and trunk assemblies, and the installation of nonmoving front and rear windows. Adhesive bonding has become the preferred means of assembly for polymeric body panels made from sheet-molding compounds and reaction-injection-molded (RIM) materials. Moreover, because adhesive bonding has the ability to bond such a wide variety of materials, its use has grown significantly with the ever-expanding applications of plastics and composites.

Adhesive Materials and Their Properties

In **adhesive bonding**, a nonmetallic material (the **adhesive**) is used to fill the gap and create a joint between two surfaces. The actual adhesives span a wide range of material types and forms, including **thermoplastic** resins, **thermosetting** resins, artificial **elastomers**, and even some ceramics. They can be applied as drops, beads, pellets, tapes, or coatings (films) and are available in the form of liquids, pastes, gels, and solids. **Curing** can be induced by the use of heat, radiation or light (photoinitiation), moisture, activators, catalysts, multiple-component reactions, or combinations thereof. Applications can be full load bearing (structural adhesives); light-duty holding or fixturing; or simply sealing (the forming of liquid- or gas-tight joints). With such a wide range of possibilities, the selection of the best adhesive for the task at hand can often be quite challenging.

Structural adhesives are selected for their ability to effectively transmit load across the joint, and include epoxies, cyanoacrylates, anaerobics, acrylics, urethanes, silicones, high-temperature adhesives, and hot melts. Both strength and rigidity might be important, and the bond must be able to be stressed to a high percentage of its maximum load for extended periods of time without failure.

Consider some of the more important families of adhesives:

1. **Epoxies.** The thermosetting epoxies are the oldest, most common, and most diverse of the adhesive systems, and can be used to join most engineering materials, including metal, glass, and ceramic. They are strong, versatile adhesives that can be designed to offer high adhesion, good tensile and shear strength, toughness, high rigidity, creep resistance, easy curing with little shrinkage, good chemical resistance, and tolerance to elevated temperatures. Various epoxies can be used over a temperature range from −50° to +250°C (−60° to +500°F). After curing, shear strengths can be as high as 35 to 70 MPa (5,000 to 10,000 psi).

   Single-component epoxies use heat as the curing agent. Most epoxies, however, are two-component blends involving a reaction between a resin and a curing agent, with possible additives such as accelerators, plasticizers, and fillers that serve to enhance cure rate, flexibility, peel resistance, impact resistance, or other characteristics. Heat might be used to drive or accelerate the cure.

   Low peel strength and poor flexibility limit epoxy adhesives, and the bond strength can be sensitive to moisture and

Superglue Curing

Knowing that moisture on the surface promotes the curing of super-glue, can you better understand why it so effectively bonds fingertips to one another, but might not create an effective joint between clean, dry surfaces?

surface contamination. Epoxies are often brittle at low temperatures, and the rate of curing is comparatively slow. Sufficient strength for structural applications is generally achieved in 8 to 12 hours, with full strength often requiring two to seven days.

2. **Cyanoacrylates.** These are liquid monomers that polymerize when spread into a thin film between two surfaces. Trace amounts of moisture on the surfaces promote curing at amazing speeds, often in as little as two seconds. Thus, the cyanoacrylates offer a one-component adhesive system that cures at room temperature with no external impetus. Commonly known as **superglues**, this family of adhesives is now available in the form of liquids and gels of varying viscosity, toughened versions designed to overcome brittleness, and even nonfrosting varieties.

   The cyanoacrylates provide excellent tensile strength, fast curing, and good shelf life, and adhere well to most commercial plastics, metals, and rubbers. They are limited by their high cost, poor peel strength, and brittleness. Bond properties are poor at elevated temperatures and effective curing requires good component fit (gaps must be smaller than 0.25 mm or 0.010 in.).

3. **Anaerobics**. These one-component, thermosetting, polyester acrylics remain liquid when exposed to air. When confined to small spaces and shut off from oxygen, as in a joint to be bonded or along the threads of an inserted fastener, the polymer becomes unstable. In the presence of iron or copper, it polymerizes into a bonding-type resin, without the need for elevated temperature. Curing can occur across gaps as large as one millimeter (0.04 in.). Additives can reduce odor, flammability, and toxicity and can speed the curing operation. Slow-curing anaerobics require 6 to 24 hours to attain useful strength. With selected additives and heat, however, curing can be reduced to as little as 5 minutes.

   The anaerobics are extremely versatile and can bond almost anything, including oily surfaces. The joints resist vibrations and offer good sealing to moisture and other environmental influences. Unfortunately, they are somewhat brittle and are limited to service temperatures below 150°C (300°F).

4. **Acrylics.** The acrylic-based adhesives offer good strength, toughness, and versatility, and they are able to bond a variety of materials, including plastics, metals, ceramics, and composites, even oily or dirty surfaces. Most involve application systems where a catalyst primer (curing agent) is applied to one of the surfaces to be joined and the adhesive is applied to the other. The pretreated parts can be stored separately for weeks without damage. Upon assembly, the components react to produce a strong bond at room temperature. Heat can often accelerate the curing, and at least one variety cures with ultraviolet light. In comparison to other varieties of adhesives, the acrylics offer strengths comparable to the epoxies, flexible bonds, good resistance to water and humidity, and the added advantages of room temperature curing and a no-mix application system. Major limitations include poor strength at high temperatures, flammability, and an unpleasant odor when still uncured.

5. **Urethanes.** Urethane adhesives are a large and diverse family of polymers that are generally targeted for applications that involve temperatures below 65°C (150°F) and components that require great flexibility. Both one-part thermoplastic and two-part thermosetting systems are available. Urethanes cure quickly to handling strength but are slow to reach the full-cure condition. Two minutes to handling with 24 hours to complete cure is common at room temperature.

   Compared to other structural adhesives, the urethanes offer good flexibility and toughness, even at low operating temperatures. They are somewhat sensitive to moisture, degrade in many chemical environments, and can involve toxic components or curing products.

6. **Silicones.** The silicone thermosets cure from the moisture in the air or adsorbed moisture from the surfaces being joined. They form low-strength structural joints and are usually selected when considerable amounts of expansion and contraction are expected in the joint; flexibility is required (as in sheet metal parts); or good gasket, gap-filling, or sealing properties are necessary. Metals, glass, paper, plastics, and rubbers can all be joined. The adhesives are relatively expensive, and curing is slow, but the bonds that are produced can resist moisture, hot water, oxidation, and weathering, and they retain their flexibility at low temperature.

7. **High-temperature adhesives.** When strength must be retained at temperatures in excess of 300°C (500°F), high-temperature structural adhesives should be specified. These include epoxy phenolics, modified silicones or phenolics, polyamides, and some ceramics. High cost and long cure times are the major limitations for these adhesives, which see primary application in the aerospace industry.

8. **Hot melts.** Hot-melt adhesives can be used to bond dissimilar substrates, such as plastics, rubber, metals, ceramics, glass, wood, and fibrous materials such as paper, fabric, and leather. They can produce permanent or temporary bonds, seal gaps, and plug holes. While generally not considered to be true structural adhesives, the hot melts are

being used increasingly to transmit loads, especially in composite material assemblies. The joints can withstand exposure to vibration, shock, humidity, and numerous chemicals, and offer the added features of sound deadening and vibration damping.

Most hot-melt adhesives are thermoplastic resins that are solid at room temperature, but melt abruptly when heated into the range of 100° to 150°C (212° to 300°F). They are usually applied as heated liquids (between 160° and 180°C or 320° and 355°F) and form a bond as the molten adhesive cools and resolidifies. Another method of application is to position the adhesive in the joint prior to operations such as the paint bake process in automobile manufacture. During the baking, the adhesive melts, flows into seams and crevices, and seals against the entry of corrosive moisture. These adhesives contain no solvents and do not need time to cure or dry. Hot melts achieve more than 80% of their bond strength within seconds of solidification, but do soften and creep when subsequently exposed to elevated temperatures, and can become brittle when cold.

The traditional hot melts do not cross-link or form three-dimensional network structures, but retain their linear thermoplastic structure throughout their history. As a result, they are characterized by poor strength, poor heat resistance, and the tendency to creep under load. A relatively new class of material, the **reactive hot melts**, overcomes many of these limitations, positioning the hot melts as true structural adhesives. These materials are applied as liquids at elevated temperature, cool to solids at room temperature, but then react (often with moisture) to form a cross-linked or three-dimensional network thermoset polymer with enhanced performance properties. They melt at lower temperatures than the conventional hot melts, and can bond to many different surfaces. Their tensile strengths range from 14 to 24 MPa (2000 to 3500 psi), with elongations between 290% and 750%. They can also endure higher service temperatures than their conventional counterparts (100°C or 212°F for long time, and 125°C or 260°F for intermittent exposure).

Additives also play a large part in the success of industrial adhesives. They can impart or enhance properties such as toughness, joint durability, moisture resistance, adhesion, and flame retardance. Rheological additives and plasticizers control viscosity and flow. Adhesives must penetrate the surfaces to be bonded, but not flow in an uncontrollable fashion. Pigments, antioxidants, and ultraviolet light stabilizers impart still other properties. Fillers and extenders provide bulk and reduce cost.

Figure 39.1 provides the distribution of various types of adhesive and sealant products for a recent year, and **Figure 39.2** classifies adhesives by end-use markets. **Table 39.1** lists some popular structural adhesives along with their service and cur-

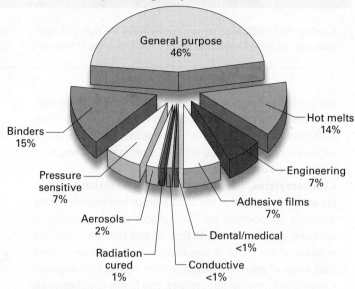

US consumption of adhesives and sealants by product, 2003 (percentages by dollar value)

General purpose 46%
Hot melts 14%
Binders 15%
Engineering 7%
Pressure sensitive 7%
Adhesive films 7%
Aerosols 2%
Dental/medical <1%
Radiation cured 1%
Conductive <1%

FIGURE 39.1 Distribution among the common types of adhesives and sealants. (Reprinted with permission from *The Rauch Guide to the US Adhesives & Sealants Industry*, Fifth Edition, 2006, Grey House Publishing, Millerton, NY).

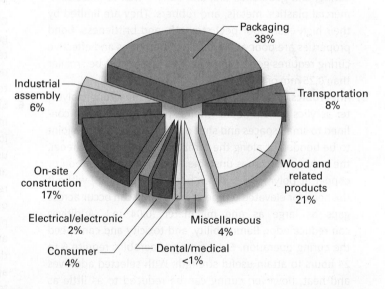

Packaging 38%
Industrial assembly 6%
Transportation 8%
On-site construction 17%
Wood and related products 21%
Electrical/electronic 2%
Miscellaneous 4%
Consumer 4%
Dental/medical <1%

FIGURE 39.2 Distribution of adhesives and sealants by end-use areas. (Reprinted with permission from *The Rauch Guide to the US Adhesives & Sealants Industry*, Fifth Edition. 2006, Grey House Publishing, Millerton, NY).

ing temperatures and expected strengths. **Table 39.2** presents the advantages and disadvantages of various curing processes.

Nonstructural and Special Adhesives

There are a number of other types of adhesives whose limited load-bearing capabilities place them in a nonstructural classification. Nevertheless, they still play roles in manufacturing through a variety of uses, such as labeling and packaging. The hot-melt adhesives are often placed in this category but can be

| TABLE 39.1 | Some Common Structural Adhesives, Their Cure Temperatures, Maximum Service Temperatures, and Strengths Under Various Types of Loading | | | |

Adhesive Type	Cure Temperature (°F)	Service Temperature (°F)	Lap Shear Strength (psi at °F) [a]	Peel Strength at Room Temperature (lb/in.)
Butyral-phenolic	275 to 350	−60 to 175	1000 at 175	10
			2500 at RT	
Epoxy				
Room-temperature cure	60 to 90	−60 to 180	1500 at 180	4
			2500 at RT	
Elevated-temperature cure	200 to 350	−60 to 350	1500 at 350	5
			2500 at RT	
Epoxy-nylon	250 to 350	−420 to 180	2000 at 180	70
			6000 at RT	
Epoxy-phenolic	250 to 350	−420 to 500	1000 at 175	10
			2500 at RT	
Neoprene-phenolic	275 to 350	−60 to 180	1000 at 180	15
			2000 at RT	
Nitrile-phenolic	275 to 350	−60 to 250	2000 at 250	60
			4000 at RT	
Polyimide	550 to 650	−420 to 1000	1000 at 1000	3
			2500 at RT	
Urethane	75 to 250	−420 to 175	1000 at 175	50
			2500 at RT	

[a]RT, room temperature.

TABLE 39.2	Advantages and Disadvantages of Various Structural Adhesive Curing Processes	
Curing Process	**Advantages**	**Disadvantages**
Mixing reactive components	Good shelf life, unlimited depth of cure and accelerated with heat	High processing costs, mix ratio critical to performance
Anaerobic cure	Single-component adhesive and good shelf life	Poor depth of cure, require primer on many surfaces and sensitive to surface contaminants
Heat cure	Unlimited depth of cure and heat can aid adhesion	Expenses for oven energy cost and heat can adversely affect some substrates
Moisture cure	Room-temperature process, one component and no curing equipment required	Long cure cycles (12–72 hour), minimum % humidity required and limited depth of cure
Light cure	Rapid cure and cure on demand	Expenses for UV light source, limited depth of cure and most allow light to reach bond
Surface-initiated cure	Rapid cure	Poor depth of cure

used for applications in both classifications. **Evaporative adhesives** use an organic solvent or water base, coupled with vinyls, acrylics, phenolics, polyurethanes, or various types of rubbers. Some common evaporative adhesives are rubber cements and floor waxes. **Pressure-sensitive adhesives** are usually based on various rubbers, compounded with additives to bond at room temperature with a brief application of pressure. No cure is involved (they bond immediately), and the tacky adhesive-coated surfaces require no activation by water, solvents, or heat. Peel-and-stick labels, cellophane tape, and Post-it notes are examples of this group of adhesives. Acrylic pressure-sensitive adhesives perform reliably at continuous temperatures up to 200°C (400°F). **Delayed-tack adhesives** are similar to the pressure sensitive systems, but are nontacky until activated by exposure to heat. Once heated, they remain tacky for several minutes to a few days to permit use or assembly.

While most adhesives are electrical and thermal insulators, **conductive adhesives** can be produced by incorporating selected fillers, such as silver, copper, aluminum, nickel, and gold in the form of flakes or powder. These conductive materials must be added in sufficient quantity so as to be in physical contact with one another—an amount that often compromises flexibility and adhesion. The resin must then provide bonding between the particles and between the adhesive and the substrates being joined. Metal particles can also be used to provide thermal conductivity. When thermal conductivity is desired without electrical conductivity, certain ceramic fillers can be used, including aluminum oxide, beryllium oxide, boron nitride, and silica.

Still another group of commercial adhesives are those designed to cure by exposure to radiation, such as visible, infrared, or ultraviolet light; microwaves; or electron beams. These **radiation-curing adhesives** offer rapid conversion from liquid to solid at room temperature and a curing mechanism that occurs throughout, rather than progressing from exposed surfaces (as with the competing low-temperature air or moisture cures). Current applications include a wide variety of dental amalgams that can fill cavities or seal surfaces while matching the color of the remaining tooth. In the manufacturing realm, heat-sensitive materials can be bonded effectively, and the rapid cure time significantly reduces the need for fixturing.

Selection and Design Considerations

The structural adhesives have been used for a wide range of applications in fields as diverse as automotive, aerospace, appliances, biomedical, electronics, construction, machinery, and sporting goods. Proper selection and use, however, requires consideration of a number of factors, including:

1. What materials are being joined? What are their surface finishes, hardnesses, and porosities? Will the thermal expansions or contractions be different?

2. How will the joined assembly be used? What type of joint is proposed, what will be the bond area, and what will be the applied stresses? How much strength is required? Will there be mechanical vibration, acoustical vibration, or impacts?

3. What temperatures might be required to affect the cure, and what temperatures might be encountered during service? Consideration should be given to the highest temperature, lowest temperature, rates of temperature change, frequency of change, duration of exposure to extremes, the properties required at the various conditions, and differential expansions or contractions.

4. Will there be subsequent exposure to solvents, water or humidity, fuels or oils, light, ultraviolet radiation, acid solutions, or general weathering?

5. What is the desired level of flexibility or stiffness? How much toughness is required?

6. Over what length of time is stability desired? What portion of this time will be under load?

7. Is appearance important?

8. How will the adhesive be applied? What equipment, labor, and skill are required?

9. Are their restrictions relating to storage or shelf life? Cure time? Disposal? Recyclability?

10. What will it cost?

Because there is such a large difference in bonding area, adhesive-bonded joints are often classified as either continuous surface or core-to-face. In **continuous-surface bonds**, both of the adhering surfaces are relatively large and are of the same size and shape. **Core-to-face bonds** have one **adherend** area that is very small compared to the other, such as when the edges of lightweight honeycomb core structures are bonded to the face sheets (See Figure 20-19).

A major design consideration for both types is the nature of the stresses that the joint will experience. As shown in **Figure 39.3**, applied stresses can subject the joint to **tension**,

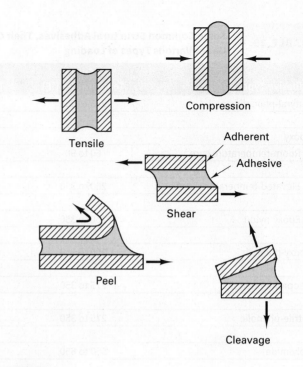

FIGURE 39.3 Types of stresses in adhesive-bonded joints.

compression, **shear**, **cleavage**, and **peel**. While it might be tempting to use joint designs intended for other methods of fastening, adhesives require specially designed joints to optimize their properties. Most of the structural adhesives are significantly weaker in peel and cleavage, where the stress is concentrated on only a very small area of the total bond, than they are in shear or tension. Therefore, adhesively bonded joints should be designed so as much of the stress as possible is in shear, tension, or compression, where all of the bonded area shares equally in bearing the load. The shear strengths of structural adhesives range from 14 to 40 MPa (2000 to 6000 psi) at room temperature, while the tensile strengths are only 4 to 8 MPa (600 to 1200 psi). The best adhesive-bonded joints, therefore, will be those that are designed to utilize the superior shear strengths. Creep, vibration and associated fatigue, thermal shock, and mechanical shocks can all induce additional stresses. When vibration or shock loading is expected, the elastomeric adhesives are quite attractive, because they can provide valuable damping. Rigid adhesives are better for shear loadings, while flexible adhesives are better in peel.

Figure 39.4 shows some commonly used joint designs and indicates their relative effectiveness. The butt joint is unsatisfactory because it offers only a minimum of bond surface area and little resistance to cleavage. Useful strength is generally obtained by increasing the bond area through the addition of straps or the conversion to some form of lap design. The scarf joint, shown previously in Figure 38-13, is also used when uniform thickness is required. **Figure 39.5** shows a few recommended designs for corner and angle joints. Adhesives can also be used in combination with welding, brazing, or mechanical fasteners. Spot welds or rivets can provide additional strength or simply prevent movement of the components when the

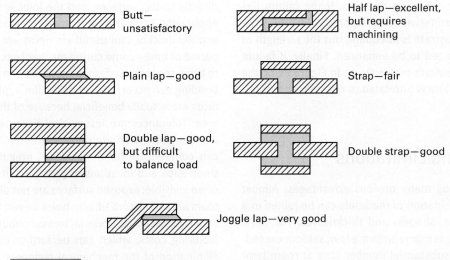

Butt—
unsatisfactory

Plain lap—good

Double lap—good,
but difficult
to balance load

Half lap—excellent,
but requires
machining

Strap—fair

Double strap—good

Joggle lap—very good

FIGURE 39.4 Possible designs of adhesive-bonded joints and a ratings of their performance
in service.

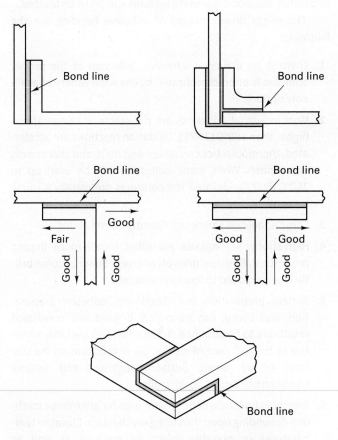

Bond line

Bond line

Bond line

Bond line

Good

Good

Good

Good

Fair

Good

Good

Good

Bond line

FIGURE 39.5 Adhesively bonded corner and angle joint designs.

To obtain satisfactory and consistent quality in adhesive-bonded joints, it is essential that the surfaces be properly prepared. Procedures vary widely, but frequently include cleaning the surfaces to be joined. Contaminants, such as oil, grease, rust, scale, or even mold-release agents, must be removed to ensure adequate wetting of the surfaces by the adhesive. Solvent or vapor cleaning is usually adequate. Chemical alteration of the surface to form a new intermediate layer, chemical etching, steam cleaning, or abrasive techniques also might be employed to further enhance wetting and bonding. While thick or loose oxide films are detrimental to adhesive bonding, a thin porous oxide or surface primer can often provide surface roughness and enhance adhesion.

The destructive testing of adhesive joints, or the examination of joint failures, can reveal much about the effectiveness of an adhesive system. If failure occurs by separation at the adhesive-substrate interface, as shown in **Figure 39.6a**, it is indica-

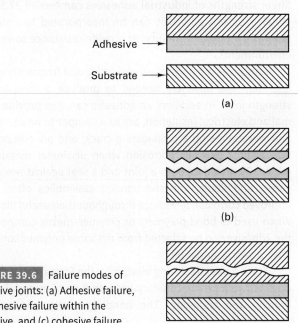

Adhesive

Substrate

(a)

(b)

(c)

FIGURE 39.6 Failure modes of adhesive joints: (a) Adhesive failure, (b) cohesive failure within the adhesive, and (c) cohesive failure within the substrate.

adhesive is not fully cured or is softened by exposure to elevated temperature.

As with brazing and soldering, the strength of an adhesive joint varies with the thickness of the adhesive layer, generally increasing with a decrease in thickness (the thin-specimen effect described in Chapter 38). When an optimum thickness is observed, it can often be achieved consistently by incorporating pre-sized glass beads into the adhesive and pressing until bead contact is made with both adherend surfaces.

tive of a bonding or adhesion problem. If the failure lies entirely within the adhesive as in Figure 39-6b, then the bonding with the substrate is adequate, but the strength of the adhesive might need to be enhanced. Finally, if failure occurs within the substrate materials, as in Figure 39-6c the joint is good, and failure is unrelated to the adhesive bonding operation.

Advantages and Limitations

Adhesive bonding has many obvious advantages. Almost any material or combination of materials can be joined in a wide variety of sizes, shapes, and thicknesses. For most adhesives, the curing temperatures are low, seldom exceeding 180°C (350°F). A substantial number cure at room temperature or slightly above and can provide adequate strength for many applications. As a result, very thin or delicate materials, such as foils, can be joined to each other or to heavier sections. Heat-sensitive materials can be joined without damage, and heat-affected zones are not present in the product. When joining dissimilar materials that experience subsequent changes in temperature, the adhesive often provides a bond that can tolerate the stresses of differential expansion and contraction. Industrial adhesives offer useful properties in the temperature range from −40° to 170°C (−40° to 340°F).

Because adhesives bond the entire joint area, good load distribution and fatigue resistance are obtained, and stress concentrations (such as those observed with screws, rivets, and spot welds) are avoided. Because of the high extension and recovery properties of flexible adhesives, the fatigue resistance can be up to 20 times that of riveted or spot-welded assemblies. The large contact areas that are usually employed provide a total joint strength that compares favorably with alternative methods of joining or attachment. Shear strengths of industrial adhesives can exceed 27.5 MPa (4000 psi), and additives can be incorporated to enhance strength, increase flexibility, or provide resistance to various environments.

Adhesives are generally inexpensive and frequently weigh less than the fasteners needed to produce a comparable-strength joint. In addition, an adhesive can also provide thermal and electrical insulation; act as a damper to noise, shock, and vibration; stop a propagating crack; and provide protection against galvanic corrosion when dissimilar metals are joined. By providing both a joint and a seal against moisture, gases, and fluids, adhesive-bonded assemblies often offer improved corrosion resistance throughout their useful lifetime. When used to bond polymers or polymer–matrix composites, the adhesive can be selected from the same polymer family to ensure good compatibility.

From a manufacturing viewpoint, the formation of a joint does not require the capillary-induced flow of material, as in brazing and soldering. The bonding adhesive is applied directly to the surfaces, and the joint is then formed by the application of heat and/or pressure. Most adhesives can be applied quickly, and useful strengths are achieved in a short period of time—some curing mechanisms take as little as two to three seconds. Surface preparation can be reduced because bonding can occur with an oxide film in place, and rough surfaces are actually beneficial because of the increased contact area. Tolerances are less critical because the adhesives are more forgiving than alternative methods of bonding. Adhesives can compensate for dimensional irregularities by filling in small gaps and locations of poor part fit. The adhesives are often invisible; exposed surfaces are not defaced; smooth contours are not disturbed; and holes do not have to be made, as with rivets or bolts. These factors contribute to reduced manufacturing costs, which can be further reduced through the elimination of the mechanical fasteners and the absence of highly skilled labor. Bonding can often be achieved at locations that would prevent the access of many types of welding apparatus. Robotic dispensing systems can often be utilized.

The major disadvantages of adhesive bonding are the following:

1. There is no universal adhesive. Selection of the proper adhesive is often complicated by the wide variety of available options.

2. Most industrial adhesives are not stable at temperatures higher than 180°C (350°F). Oxidation reactions are accelerated, thermoplastics can soften and melt, and thermosets decompose. While some adhesives can be used up to 260°C (500°F), elevated temperatures are usually a cause for concern.

3. Some adhesives shrink significantly during curing.

4. High-strength adhesives are often brittle (poor impact properties). Resilient ones often creep. Some become brittle when exposed to low temperatures.

5. Surface preparation and cleanliness, adhesive preparation, and curing can be critical if good and consistent results are to be obtained. Some adhesives are quite sensitive to the presence of grease, oil, or moisture on the surfaces to be joined. Surface roughness and wetting characteristics must be controlled.

6. Assembly times might be greater than for alternative methods depending upon the curing mechanism. Elevated temperatures or pressure might be required as well as specialized fixtures.

7. It is difficult to determine the quality of an adhesive-bonded joint by traditional nondestructive techniques, although some inspection methods have been developed that give good results for certain types of joints.

8. Some adhesives contain objectionable or toxic chemicals or solvents, or produce them upon curing.

9. Many structural adhesives deteriorate under certain operating conditions. Environments that might be particularly

hostile include heat, ultraviolet light, ozone, acid rain (low pH), water and humidity, salt, and numerous solvents. Thus, long-term durability and reliability might be questioned, and life expectancy is hard to predict.

10. Adhesively bonded joints cannot be readily disassembled.

11. Storage life might be limited or special storage conditions required.

Nevertheless, the extensive and successful use of adhesive bonding provides ample evidence that these limitations can be overcome if adequate quality-control procedures are adopted and followed.

Weld Bonding

Welding and adhesive bonding are two very different methods of joining, and are typically considered to be competing technologies. When used together, however, in a process called **weld bonding**, they often combine in a way that complements one another and cancels each other's negative features. Various forms of spot welding (resistance, laser, friction stir, ultrasonic, or other) can be combined with adhesives to join steel and aluminum. The adhesives distribute the load over large areas and provide sealing, while the welds provide high peel resistance. The welds can also act as fixtures, holding the pieces in position or alignment while the adhesives cure. In some combinations, the welds are made through the adhesives, while others employ a staggered or patterned approach.

39.2 Mechanical Fastening

Introduction and Methods

Mechanical fastening includes a wide variety of techniques and fasteners designed to suit the individual requirements of a multitude of joints and assemblies. Included within this family are integral fasteners, threaded discrete fasteners (which includes screws, bolts, studs, and inserts), nonthreaded discrete fasteners (such as rivets, pins, retaining rings, nails, staples, and wire stitches), special-purpose fasteners (such as the quick-release and tamper-resistant types), shrink and expansion

fits, press fits, seams, and others. Selection of the specific fastener or fastening method depends primarily on the materials to be joined, the function of the joint, strength and reliability requirements, weight limitations, dimensions of the components, and environmental conditions. Other considerations include cost, installation equipment and accessibility, appearance, and the need or desire for disassembly. When disassembly and reassembly are desired (as for parts replacement, maintenance, or repair), threaded fasteners, snap-fits, or other fasteners that can be removed quickly and easily should be specified. Such fasteners should not have a tendency to loosen after installation, however. If disassembly is not necessary, permanent fasteners are often preferred, or threaded fasteners can be coupled with anaerobic adhesives that cure to full strength at room temperature and "lock" the fastener in place.

A mechanical joint acquires its strength through either mechanical interlocking or interference as a result of a clamping force. No fusion or adhesion of the surfaces is required. The fasteners and fastening processes should be selected to provide the required strength and properties in view of the nature and magnitude of subsequent loading. Consider the possibility of vibrations and/or cyclic stresses that might promote loosening over a period of time. The added weight of fasteners might be a significant factor in certain applications, such as those in aerospace and automotive. The need to withstand corrosive environments, operate at high or low temperatures, or face other severe conditions might be additional constraints to the selection of fasteners or fastening processes.

The effectiveness of a mechanical fastener often depends upon the material of the fastener, the fastener design (including the load-bearing area of the head), hole preparation, and the installation procedure. The general desire is to achieve a uniform load transfer, a minimum of stress concentration, and uniformity of installation torque or interference fit. Various means are available for achieving these goals, as described in the following paragraphs.

Integral fasteners are formed areas of a component that interfere or interlock with other components of the assembly, and are most commonly found in sheet metal products. Examples include lanced or shear-formed tabs, extruded hole flanges, embossed protrusions, edge seams, and crimps. **Figure 39.7** shows some of these techniques, each of which involves some form of metal shearing and/or forming. Bare, plated, and pre-painted materials can all be easily joined.

Two-Piece Beverage Can

The common two-piece beverage can contains several integral fastener joints – an edge seam to join the top of the can to the body (as *in Figure 39-7e) an an embossed protrusion that is subsequently flattened to attach the opener tab (as in Figure 39-7d).*

Rivets in a Commercial Airliner

There are approximately 1 million rivets in the riveted fuselage of a commercial aircraft.

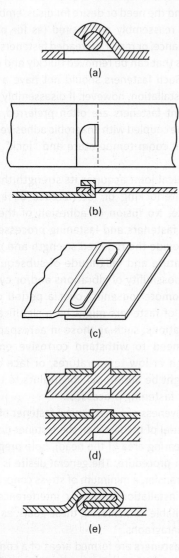

FIGURE 39.7 Several types of integral fasteners: (a) lanced tab to fasten wires or cables to sheet or plate; (b) and (c) assembly through folded tabs and slots for different types of loading; (d) use of a flattened embossed protrusion; (e) single-lock seam.

Discrete fasteners, such as those illustrated in **Figures 39.8** and **39.9**, are separate pieces whose function is to join the primary components. These include bolts and nuts (with accessory washers, etc.), screws, nails, rivets, quick-release fasteners, staples, and wire stitches. More than 200 billion discrete fasteners are consumed annually in the United States, with a variety so immense that the major challenge is usually selection of an appropriate, and hopefully optimum, fastener for the task at hand. Fastener selection is further complicated by inconsistent nomenclature and identification schemes. Some fasteners are identified by their specific product or application, while others

are classified by the material from which they are made, their size, their shape, their strength, or primary operational features.[1] The commercial availability of such a wide range of standard and special types, sizes, materials, strengths, and finishes virtually ensures that an appropriate fastener can be found for most all joining needs. Discrete fasteners are easy to install, remove, and replace. In addition, most standard varieties are interchangeable. Steel is the most common material because of its high strength and low cost. Various finishes and coatings can be applied to withstand a multitude of service conditions.

Shrink and **expansion fits** form another major class of mechanical joining. Here, a dimensional change is introduced to one or both of the components by heating or cooling (heating one part only, heating one and cooling the other, or cooling one). Assembly is then performed, and a strong interference fit is established when the temperatures return to uniformity. Joint strength can be exceptionally high. This technique can also be used to apply a corrosion-resistant cladding or lining to a less costly, but corrosion-prone, bulk material.

Press fits are similar to shrink and expansion fits, but the results are obtained through mechanical force instead of differential temperatures. Anaerobic structural adhesives can be introduced into the interfaces of shrink, expansion, and press fits to increase strength, permit some relaxation of dimensional precision, and provide sealing that prevents corrosion and/or leakage.

Reasons for Selection

Mechanical fastening offers a number of attractive features:

1. They are easy to disassemble and reassemble. The threaded fasteners are noteworthy for this feature, and semipermanent fasteners (such as rivets) can be drilled out for a major disassembly.

2. They can be used to join similar or different materials (including materials and combinations not easily welded) in a wide variety of sizes, shapes, and joint designs, including some—such as hinges and slides—that permit limited motion between the components.

3. Manufacturing cost is low. The fasteners are usually small components that cost little compared to the components being joined. They are readily available in a variety of mass-produced sizes.

---

[1]Discussion of the primary terms used in identifying discrete fasteners can be found in ANSI Standard B18.12. Both the Society of Automotive Engineers (SAE) and the American Society of Testing and Materials (ASTM) have formalized various "grades" of bolts and identified them by markings that are stamped on the bolt head. These grades often relate to allowable stress and temperatures of operation.

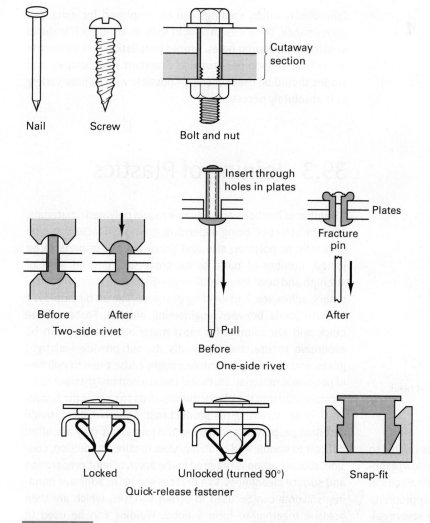

Nail

Screw

Bolt and nut

Cutaway section

Insert through holes in plates

Plates

Fracture pin

Before / After
Two-side rivet

Pull
Before
One-side rivet

After

Locked

Unlocked (turned 90°)
Quick-release fastener

Snap-fit

FIGURE 39.8 Various types of discrete fasteners, including a nail, screw, nut-and-bolt, two-side-access supported rivet, one-side-access blind rivet, quick-release fastener, and snap-fit.

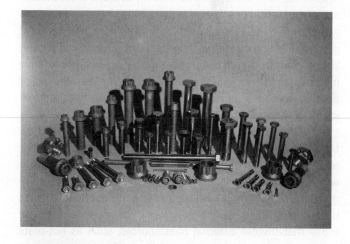

FIGURE 39.9 An array of commercially available discrete fasteners. (Courtesy of Fastener Dimensions, Ozone Park, NY).

4. Installation does not adversely affect the base materials, as is often the case with techniques involving the application of heat and/or pressure.

5. Little or no surface preparation or cleaning is required.

6. Speed of assembly is usually rather high.

Manufacturing Concerns

Many mechanical fasteners require that the components contain aligned holes. Castings, forgings, extrusions, and powder metallurgy components can be designed to include integral holes. Holes can also be produced by such techniques as punching, drilling, and electrical, chemical, or laser-beam machining. Each of these techniques produces holes with characteristic surface finish, dimensional features, and properties. Secondary operations, such as shaving, deburring, reaming, and honing, can be used to improve precision and surface finish. Hole making and the proper positioning and alignment of the holes are major considerations in mechanical fastening.

Some fasteners, such as bolts coupled with nuts, require access to both sides of an assembly during joining. In contrast, screws offer one-side joining. If a bolt can be inserted into a threaded (tapped) hole, however, the nut can be eliminated, and only one-side access is required. If the bolt or screw is sufficiently hard, the fastener can often form its own threads, thereby eliminating the need for a threaded receptacle. This self-tapping feature is particularly attractive when assembling plastic products.

Rivets offer good strength, but produce permanent or semipermanent joints. Rivet heads can be shaped by upset forming, impact riveting, or orbital forming—a technique whereby a tilted and rotating peen tool progressively shapes the head with about 80% less force than conventional methods. Orbital forming is depicted in **Figure 39.10**. Stapling is a fast way of joining thin materials and does not require prior holemaking. Snap-fits utilize the elasticity of one of the components, but the necessary elastic deformation must be possible without fracture.

Design and Selection

The design and selection of a fastening method requires numerous considerations, including the possible means of joint failure. When a product is assembled with fastened joints, the fasteners are extremely vulnerable sites. Mechanical joints generally fail because of oversight or lack of control in one of four areas: the design of the fastener itself and the manufacturing techniques used to make it; the material from which the fastener is made; joint design; or the means and details of installation. Fasteners might have insufficient strength or corrosion resistance or might be subject to stress corrosion

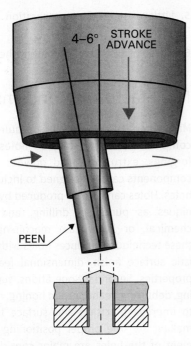

Orbital Head, Peen Holder, and Peen

FIGURE 39.10 Schematic of the orbital forming of a rivet head. (Courtesy of Orbitform Group, Jackson, MI).

cracking or hydrogen embrittlement. They might be unable to withstand the temperature extremes (both high and low) experienced by the final assembly. Metal fasteners provide electrical conductivity between the components, and an inappropriate choice of fastener or component material can cause severe galvanic corrosion. Nonmetallic fasteners (such as threaded nylon) can be used for low-strength applications where corrosion is a concern, but creep under load is a concern for these materials. Because mechanical fasteners join only at discrete points, gases or liquids can easily penetrate the joint area and further aggravate conditions.

Many failures are the result of poor joint preparation or improper fastener installation. A high percentage of the cracks in aircraft structures originate at fastener holes, and fatigue of fasteners is the largest single cause of fastener failure. Installation frequently imparts too much or too little preload (too tight or too loose). The joint surfaces might not be flat or parallel, and the area under the fastener head might be insufficient to bear the load. Vibrational loosening enhances fastener fatigue. The details of joint design should further consider stress distribution, because much of the load will be concentrated on the fasteners (in contrast to the previously discussed adhesive joints that distribute the load uniformly over the entire joint area).

Nearly all fastener failures can be avoided by proper design and fastener selection. Consideration should be given to the operating environment, required strength, and magnitude and frequency of vibration. Fastener design should incorporate a shank-to-head fillet whenever possible. Rolled threads can be specified for their superior strength and fracture resistance.

Corrosion-resistant coatings can be employed for enhanced performance. Joint design should seek to avoid such features as offset or oversized holes. Proper installation and tightening are critical to good performance. Standard sizes, shapes, and grades should be used whenever possible with as little variety as is absolutely necessary.

39.3 Joining of Plastics

A number of methods are available to join polymeric materials, with the selection being dependent upon such factors as type of plastic or polymer, size and geometry of the parts to be joined, number of parts to be processed, desired level of strength and need for a liquid- or gas-tight seal. Mechanical fasteners, adhesives, and welding processes can all be employed to form joints between engineering plastics. Fasteners are quick and are suitable for most materials, but they can be expensive to use, they generally do not provide leak-tight joints, and the localized stresses might cause them to pull free of polymeric material. Threaded metal inserts might have to be incorporated into the plastic components to receive the fasteners, further increasing the product cost. Adhesives can provide excellent properties and fully sound joints, but they are often difficult to handle and relatively slow to cure. In addition, considerable attention is required in the areas of joint preparation and surface cleanliness. In a process similar to adhesive bonding, solvents can be used to soften surfaces, which are then pressed together to form a bond. Welding can be used to produce bonded joints with mechanical properties that approach those of the parent material. Unfortunately, only the *thermoplastic polymers* (polyethylene, polypropylene, polyvinylchloride, and others) can be welded, because these materials can be melted or softened by heat without degradation and good bonds can be formed with the subsequent application of pressure. The *thermosetting polymers* do not soften with heat, tending only to char or burn, and must be joined by alternative methods, such as mechanical fasteners, adhesives, snap fits, or possible co-curing (placing the components together and curing while in contact).

Because the thermoplastics soften and melt at such low temperatures, the heat required to weld these materials is significantly less than that required to weld. The processes used to weld plastics can be divided into two groups: those that utilize mechanical movement and friction to generate heat, such as ultrasonic welding, spin welding, and vibration welding, and those that involve external heat sources, such as hot-plate welding, hot-gas welding, and resistive and inductive implant welding. In both groups, it is important to control the rate of heating. Plastics have low thermal conductivity, and it is easy to induce burning, charring, or other material degradation before softening has occurred to the desired depth.

Ultrasonic welding, using high-frequency mechanical vibrations to create the bond, is probably the most widely used process for welding. Parts are held together under pressure and are subjected to ultrasonic vibrations (20 to 40 kH frequency and 10 to 100 micron amplitude) perpendicular to the area of contact. [Note: In ultrasonic welding of metals, the vibration is parallel to the weld interface.] The high-frequency vibrations are transmitted through the workpiece to the joint area, where they generate sufficient heat through friction to produce a high-quality weld in a period of ½ to 1½ seconds. The process can be readily automated, and produces fast, strong, clean, and reliable welds. The tools are expensive, however, and large production runs are generally required. Ultrasonic welding is usually restricted to small components where relative movement is restricted and weld lengths do not to exceed a few centimeters.

In **vibration welding**, or **linear friction welding**, relative movement between the two parts is again used to generate the heat, but the direction of movement is now parallel to the interface and aligned with the longest dimension of the joint. The vibration amplitudes are significantly larger than in ultrasonic welding (1 to 4 mm), and the frequencies are considerably less (on the order of 100 to 240 Hz). When molten material is produced, the vibration is stopped, parts are aligned, and the weld region is allowed to cool and solidify. The entire process takes about 1 to 5 seconds. Long-length, complex joints can be produced at rather high production rates. Nearly all thermoplastics can be joined.

The **friction welding** of plastics (also called **spin welding**) is essentially the same as the friction welding of metals, but the continuous rotational motion now induces melting at the joint interface. High-quality welds are produced with good reproducibility, and little end preparation is required. The major limitation is that at least one of the components must exhibit circular symmetry, and the axis of rotation must be perpendicular to the mating surface. Weld strengths vary from 50% to 95% of the parent material in bonds of the same plastic. Joints between dissimilar materials generally have poorer strengths.

Friction stir welding (described in Chapter 37), can be used to produce butt welds between plates of thermoplastic material.

Hot-plate welding (or **hot-tool welding**) uses an external heat source and is probably the simplest of the mass-production techniques used to join the thermoplastic polymers. The parts to be joined are held in fixtures and pressed against the opposite sides of an electrically heated tool. Contact is maintained until the surfaces have melted and the adjacent material has softened to a specified distance from the interface. The parts separate, the tool is removed, and the two prepared surfaces are pressed together and allowed to cool. Contaminated surface material is usually displaced into a flash region. Weld times are comparatively slow, ranging from 10 seconds to several minutes. The joint strength can be equal to that of the parent material, but the joint design is usually limited to a square-butt configuration, like that encountered when joining sections of plastic pipe. If the bond interface has a nonflat profile, shaped heating tools can be employed. Heated-tool welding can also be used to produce lap seams between flexible plastic sheets. Rollers apply pressure after the material has passed over a heater.

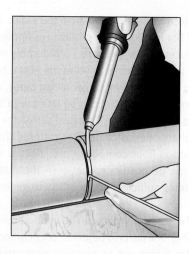

Using a hot-gas torch to make a weld in plastic pipe.

The **hot-gas welding** of plastics is similar to the oxyacetylene welding of metals, and V-groove or fillet welds are the most common joint configuration. A gas (usually compressed air or nitrogen) is heated by an electric coil as it passes through a welding gun, such as the one shown in **Figure 39.11**, emerging from a nozzle at 200° to 400°C (400° to 750°F) to impinge on the joint area. Thin rods of thermoplastic material (the same material as the parts being joined) are heated along with the workpieces and are then forced into the softened joint area, providing both the filler material and the pressure needed to produce coalescence. Because this process is usually slow and the results are generally dependent on operator skill, it is seldom used in production applications. It is, however, a popular process for the repair of thermoplastic materials, and the inexpensive equipment is extremely portable.

Extrusion welding is an established technique for joining polyethylene and polypropylene when the fabrications are large and the materials being joined are thick. A flow of hot gas is used to heat the substrate, as in the hot-gas process, but the external filler material rod is replaced by a continuously extruded stream of fully molten polymer that emerges from a heated extruder as it moves along the joint. [Note: This is similar to the electrode feed in the consumable-electrode arc welding processes presented in Chapter 36]. When welding thick materials, the hot-gas process is often used to produce tack welds and then to create a single pass along the base to ensure full root penetration. Extrusion welding is then used to fill the remainder of the joint, using the same thermoplastic material as the substrate.

Implant welding processes involve positioning metal inserts between the pieces to be joined and then heating them by means of electrical resistance or induction. The resistance method requires a continuous current-carrying path, and the implants are often wire, braid, or mesh. As the implant heats because of the passage of DC or low-frequency AC current, the surrounding thermoplastic softens and melts, and the subsequent application of pressure forms a weld. In the induction welding approach, a high-frequency AC coil is placed in proximity to the implants and the eddy currents induced in the metal produce the desired heating. Because the welds form only in the vicinity of the implants, the process resembles spot welding and produces joints that are considerably weaker than those formed by processes that bond the entire contact area. When bonding is desired over larger areas, tapes, rods, or gaskets of thermoplastic material can be laced with iron oxide or metal particles, which then provide heat across the entire interface and simultaneously provide filler material. To bond polymer-matrix carbon-fiber composite materials, a single ply of carbon fiber can be positioned between the materials to be bonded, providing sufficient conductivity to generate the necessary heat.

Sheets of certain polymers can be bonded using **radio frequency welding**. The parts to be joined are positioned between two metal bars, which act as pressure applicators, and subjected to a high frequency (13 to 100 MHz) electromagnetic field. Molecular oscillations within polar thermoplastics, including polyvinylchloride and polyurethanes, generate the heat required for the weld.

In a relatively new process, **microwave energy** is being used to weld thermoplastics. Most thermoplastics do not experience a temperature rise when irradiated with microwaves, but the insertion of a microwave-susceptible implant and the application of pressure after heating can be used to create an acceptable weld. Because the waves penetrate the entire component, complex three-dimensional joints can be produced. As with all forms of implant welding, the inserted material becomes an integral part of the final assembly.

The welding of thermoplastics using **infrared radiation welding** is another relative newcomer to the list of processes. One approach is essentially a noncontact version of hot-plate welding, where the tool is heated to higher temperatures than in the contact method. The parts to be welded are brought into close proximity to the heated plate (typically within 0.2 mm or 0.01 in.) where they heat by radiation and convection. When large surface areas are to be joined, a bank of infrared lamps might replace the heated plate. Because the heating is performed without contact, there is less possibility of contamination entering the joint.

Low-power lasers have also been used to weld thermoplastics. The radiation from a CO_2 laser is readily absorbed by plastics, and is used in a variation known as **direct laser welding**. The energy absorption restricts heat to the outer surface, limiting the depth of heating, and thereby restricting the process to the joining of thin films. The radiation from Nd:YAG, fiber and diode lasers is easily transmitted through plastics. In **transmission laser welding**, one of the pieces transmits the laser light to the bond interface where the other component or an opaque surface coating absorbs the energy and converts it to heat. Thicker materials, up to 10 mm (0.4 in.) can be welded. The attractive features of laser welding include controllable beam power and the ability to precisely focus the beam. Because the process is noncontact, it is quite attractive for hygienic applications, such as medical devices and food packaging, as well as applications where cleanliness is critical, such as microelectronics.

Thermal energy or heating is used to achieve coalescence in all of the welding techniques discussed above. In **solvent welding**, a solvent is applied to both surfaces to temporarily dissolve the polymer at room temperature and allow for molecular entanglement. The solvent then permeates through the polymer and into the environment, leaving a solid material in the joint area. Probably the most common use of this technique is the joining of PVC (polyvinylchloride) pipe and fittings in plumbing-type applications.

Mechanical fasteners are actually the most common method of joining plastics to one another, and joining metal to plastic. Here, it is important that the plastic be able to withstand the strain of fastener insertion and the localized stresses around the fastener. Conventional machine screws are rarely used, except with extremely strong plastics. Instead, there are a number of fasteners designed specifically for use with plastics. Threaded fasteners work best with thick sections. Self-tapping, thread-cutting screws are used on hard plastics, and thread-forming screws are used with softer materials. If the joint is to undergo disassembly and reassembly, threaded metal inserts can be incorporated into the part to receive the fasteners. If recycling of the plastic product is desired, however, welding might be preferred over mechanical fasteners because only similar materials are involved and fasteners do not need to be removed or separated.

Because of the low elastic modulus of plastic materials, **snap-fit** assemblies are often an attractive alternative to the use of fasteners. The fastening system can be molded into the part, and no additional inserts, fasteners, solvents, or adhesives are required. The method is well suited to high volume production, and assembly is easy and instantaneous. Access panels to the battery compartments of electronic devices are a classic example of a snap-fit assembly.

Adhesives, described previously in this chapter, provide yet another attractive means of joining plastics. Because adhesives are polymeric materials, the joint material can be selected for compatibility with the material being joined.

39.4 Joining of Ceramics and Glass

The properties of ceramic materials are significantly different from the engineering metals, and these differences restrict or limit the processes that can be used for joining. High melting temperatures can be a significant deterrent to fusion welding. More significant, however, are the effects of low thermal conductivity and brittleness. Heating and cooling will likely result in nonuniform temperatures, and the thermally induced stresses are likely to result in cracking or fracture. The lack of useful ductility virtually eliminates any form of deformation bonding. Mechanical fasteners, and their associated threads and holes, create high concentrated stresses, and these stresses often lead to material fracture. As a result, most ceramic materials are joined by some form of adhesive bonding, brazing, diffusion or sinter bonding, or ceramic cements.

Adhesives and cements are probably the most common methods of joining ceramics to ceramics, ceramics to glasses, and ceramics to metals and other materials. The inserted material (polymer adhesive, glass or glass-ceramic frit, or ceramic cement or mortar) will bond to the surfaces and bridge what are often radically different compositions and structures.

Brazing and soldering use a low-melting-point metal or lower-melting ceramic as the intermediate material. Some materials, such as indium solders directly wet ceramic surfaces. Other standard braze alloys can be made "active" by adding elements such as titanium, hafnium, or zirconium that promote wetting of the ceramic. To promote adhesion and bonding with nonwetting alloys, it might be necessary to first coat the ceramic with some form of metallized or deposited layer. These coatings bond to the ceramic, and the braze or solder material bonds to the coating.

Sinter bonding is a means of joining ceramic materials during their initial production. As the component pieces are held together and co-fired, diffusion bonds form across the interface while similar bonds form within the components. This process is best performed when the components are of identical material or materials with similar composition and structure. Various intermediate materials have been used to assist the joining of dissimilar ceramics.

The joining of glass is a much easier operation. Heating softens the two materials, which are then pressed together and cooled. A wide variety of heating methods are used, depending on the size, shape, and quantity of components to be joined.

39.5 Joining of Composites

Joining processes are often used to bond the various components within composite materials. Examples might include the bonding of honeycomb cores to the face materials in a honeycomb sandwich structure, or the bonding of the individual layers in a laminate. At various times, however, we might need to join composite materials to other materials or other composites.

The joining of composite materials can be an extremely complex subject, especially when one considers the variety of composites and the fact that the joint interface is likely to be a distinct disruption to the continuity of structure and properties. Particulate composites might have the least structural difference at an interface. Laminar composites will certainly behave differently if the joint surface is a core (or multi-layer) surface or a single-material exposed face. Fiber-reinforced composites, regardless of the type of fiber and fiber configuration, will certainly lack fiber continuity across the joint.

The usual joining methods tend to be those used with the matrix of the composite. Metal-matrix composites can be welded, brazed, or soldered, or joined with screws or bolts or any of the other techniques applied to metals. The techniques for polymer-matrix and ceramic-matrix follow those for plastics and ceramics, as discussed previously in this chapter.

Theoretically, all composites can be bonded adhesively, but there might be limits set by the applied stresses, operating temperatures, or size of the workpiece (especially if the adhesives require a thermal cure). When working with polymer-matrix composites, the adhesive is often selected to match or be compatible with the matrix polymer. In many cases, however, the adhesive is being asked to bond to already-cured polymer surfaces.

Review Questions

1. What would be some of the characteristics of an ideal adhesive?

2. What are some of the applications of adhesives in the automotive industry?

3. What are some of the types of materials that have been used as industrial adhesives?

4. What are some of the ways in which adhesives can be cured?

5. What is a structural adhesive?

6. Characterize the temperature range over which epoxies might be used, typical values of shear strength, and commonly observed curing times.

7. What are some of the limiting characteristics of epoxies?

8. What promotes the curing of cyanoacrylates? Of anaerobics?

9. Urethane adhesives might be favored when what properties are desired?

10. What features or characteristics might favor the selection of a silicone adhesive?

11. How can the use of hot-melt adhesives be combined with the paint bake operation in automotive manufacture?

12. What property enhancement characteristic is provided by the reactive hot melt adhesives? What properties are improved?

13. What features or properties are provided or enhanced by adhesive additives?

14. What are some types of nonstructural or special adhesives?

15. How can polymeric adhesives be made electrically or thermally conductive?

16. What types of radiation can be used with radiation-curing adhesives?

17. What are some of the temperature considerations that should be made when selecting an adhesive?

18. What are some of the environmental conditions that might reduce the performance or lifetime of a structural adhesive?

19. What is the difference between a continuous-surface and core-to-face bond?

20. Why is it desirable for adhesive joints to be designed so the adhesive is loaded in shear, tension, or compression?

21. Why are butt joints unattractive for adhesive bonding?

22. What types of joints provide large bonding areas?

23. How does the strength of an adhesive joint vary with the thickness of the adhesive layer?

24. What are some common techniques by which surfaces are prepared for adhesive bonding?

25. How can destructive testing and the examination of the failed joint provide useful information about the effectiveness of an adhesive system?

26. Why are the structural adhesives an attractive means of joining dissimilar metals or materials? Different sizes or thicknesses?

27. In view of the relatively low strengths of the structural adhesives, how can adhesively bonded joints attain strengths comparable to other methods of joining?

28. Why might an adhesive joint provide enhanced corrosion resistance compared to alternative joining methods?

29. What are some of the other attractive properties of structural adhesives?

30. In what ways might a structural adhesive offer manufacturing ease or reduced manufacturing cost?

31. Why are adhesive joints unattractive for applications that involve exposure to elevated temperature?

32. Describe some of the ways that structural adhesives might deteriorate over time.

33. What are some other common limitations to adhesives?

34. What is weld bonding?

35. What are some common types of mechanical fastening?

36. What factors would influence the selection of a specific type of mechanical fastener or fastening method?

37. What types of fasteners are attractive if the application requires the ability to disassemble and reassemble the product?

38. What factors determine the overall effectiveness of a mechanical fastener?

39. What is an integral fastener? Provide an example.

40. What are some of the primary types of discrete fasteners?

41. How are press fits similar to shrink or expansion fits? How do they differ?

42. What are some of the major assets of mechanical fasteners?

43. What are some of the ways that fastener holes can be made in manufactured products?

44. What is the benefit of a self-tapping fastener?

45. What are some of the common causes for failure of mechanically fastened joints?

46. What are galvanic corrosion problems common when discrete fasteners are used?

47. From a manufacturing viewpoint, why is it desirable to use standard fasteners and minimize the variety of fasteners within a given product?

48. What are some of the ways that plastics can be joined?

49. Why can the thermoplastic polymers be welded, but not the thermosetting varieties?

50. What are the two distinct approaches to providing the heat necessary to weld thermoplastic materials?

51. Describe several of the plastic joining processes that use mechanical movement or friction to generate the required heat. Which of these is most widely used?

52. What are some of the external heat sources that can be used in the welding of plastic materials?

53. How can metal inserts be used to produce welds in plastics?

54. How is infrared radiation welding of plastics similar to and different from hot plate welding?

55. What is the difference between direct laser welding and transmission laser welding? What type of laser is used with each?

56. Give an everyday example of solvent welding.

57. What material property enables snap-fits to be a common means of connecting plastic parts?

58. Why are the crystalline ceramic materials particularly difficult to join?

59. What are some of the inserted or bridging materials that can be used to join ceramics?

60. What is sinter bonding, and how does it join ceramic materials?

61. When materials are joined, interfaces are created between the various components. Describe the structural features that might result at the interface when we join: (a) particulate composites, (b) laminar composites, and (c) fiber-reinforced composites.

Problems

1. Some automakers are using adhesives and sealants that cure under the same conditions used for the paint-bake operation. Determine the conditions used for paint-bake, and identify some adhesives and sealants that could be used. What are some of the pros and cons of such an integration?

2. Select two types of additives to industrial adhesives and summarize the pros and cons of their incorporation into the adhesive.

3. Identify at least five types of adhesives that could be found in a typical American household. How are these various adhesives used?

4. A contractor has installed aluminum siding on a house with steel nails. Use the galvanic series to evaluate the corrosion properties of this assembly. (*Note*: The aluminum is exposed to air, so it should be considered to be in its passive condition.) What do you expect will be the outcome of this fastener selection? Can you recommend a better alternative?

5. Mechanical fasteners are an attractive means of joining composite materials because they avoid exposing the composite to heat and/or high pressure. Assume that the composite is a polymer-based

fiber-reinforced material with either uniaxial or woven fibers. For this particular system, what are some possible fastener-related problems? Consider joint preparation, assembly, and possible service failure.

6. The heat-resisting tiles on the U.S. space shuttle were made from heat-resisting ceramics. Determine the method or methods used to attach them to the structure. What difficulties or problems might have been encountered relating to this bonding?

7. The bicycle frames used by riders in recent Tour de France races have been single piece fiber-reinforced composites. What difficulties or property compromises might be associated with a fabrication method that uses joints? What methods might be available to join the carbon fiber–epoxy composite materials commonly used in these bicycles?

8. The processes described for the joining of plastics focused almost exclusively on the joining of thermoplastic polymers. What types of joining techniques could be applied to thermosetting polymers? To elastomeric polymers? To ceramic materials?

8. Select a small kitchen appliance, such as a toaster, hand mixer, electric griddle, or waffle maker. Locate and identify the various types of assembly being employed. What features might have led to the specific selection?

9. A number of techniques were presented in the chapter for the joining of plastics. Secure a plastic product that contains one or more assembled joints, and identify, if possible, the method that was used to produce the joint or joints. Why might this method have been selected? Are there any associated cons?

You will find Chapters 40, "Jig and Fixture Design," 41, "The Enterprise (Production Systems)," 42, "Lean Engineering," and 43, "Mixed Model Final Assembly," online at www.wiley.com/college/black.

Index

Selected References for Additional Study

J T. Black and Ronald A. Kohser

Handbooks and General References

ASM Engineered Materials Handbook, Desk Edition, ASM, Materials Park, OH (1995).

ASM Engineered Materials Reference Book, 2nd ed., ASM, Materials Park, OH (1993).

ASM Metals Reference Book. 3rd ed., ASM, Materials Park, OH (1993).

ASTM Standards (multiple volumes, published annually), ASTM, Philadelphia, PA.

Binary Alloy Phase Diagrams, 2nd ed., T. B. Massalski et al. (eds.), ASM (1990).

Desk Handbook: Phase Diagrams for Binary Alloys, 2nd ed., Hiroaki Okamoto, ASM (2010).

Engineered Materials Handbook Series, ASM, Materials Park, OH.
   Vol. 1, Composites (1987)
   Vol. 2, Engineering Plastics (1988)
   Vol. 3, Adhesives and Sealants (1990)
   Vol. 4, Ceramics and Glasses (1991)

Machinery's Handbook, 28th ed., Industrial Press (2008).

Manufacturing Engineering Handbook, Hinauyu Geng, McGraw-Hill (2004).

Mark's Standard Handbook for Mechanical Engineers, 11th ed., E. A. Avallone and T. Baumeister III (eds.), McGraw-Hill (2006).

Materials Handbook, 15th ed., George S. Brady, et al., McGraw-Hill (2002).

Materials Handbook: A Concise Reference, 2nd ed., F. Cardelli, Springer (2008).

Metals Handbook, 10th ed., ASM, Materials Park, OH
   Vol. 1, Properties and Selection: Irons, Steels and High-Performance Alloys (1990)
   Vol. 2, Properties and Selection: Nonferrous Alloys and Special-Purpose Materials (1990)
   Vol. 3, Alloy Phase Diagrams (1992)
   Vol. 4, Heat Treating (1991)
   Vol. 4A, Steel Heat Treating Fundamentals and Processes (2013)
   Vol. 4B, Steel Heat Treating Technologies (2014)
   Vol. 4C, Induction Heating and Heat Treatment (2014)
   Vol. 4D, Heat Treating of Irons and Steels (2014)
   Vol. 5, Surface Engineering (1994)
   Vol. 5A, Thermal Spray Technology (2013)
   Vol. 5B, Protective Organic Coatings (2015)
   Vol. 6, Welding, Brazing and Soldering (1993)
   Vol. 6A, Welding Fundamentals and Processes (2011)
   Vol. 7, Powder Metallurgy (2015)
   Vol. 8, Mechanical Testing and Evaluation (2000)
   Vol. 9, Metallography and Microstructures (2004)
   Vol. 10, Materials Characterization (1986)
   Vol. 11, Failure Analysis and Prevention (2002)
   Vol. 12, Fractography (1987)
   Vol. 13A, Corrosion: Fundamentals, Testing, and Protection (2003)
   Vol. 13B, Corrosion: Materials (2005)
   Vol. 13C, Corrosion: Environments and Industries (2006)
   Vol. 14A, Metalworking: Bulk Forming (2005)
   Vol. 14B, Metalworking: Sheet Forming (2006)
   Vol. 15, Casting (2008)
   Vol. 16, Machining (1989)
   Vol. 17, Nondestructive Evaluation and Quality Control (1989)
   Vol. 18, Friction, Lubrication and Wear Technology (1992)
   Vol. 19, Fatigue & Fracture (1996)
   Vol. 20, Materials Selection and Design (1997)
   Vol. 21, Composites (2001)
   Vol. 22A, Fundamentals of Modeling for Metals Processing (2009)
   Vol. 22B, Metals Process Simulation (2010)
   Vol. 23, Materials for Medical Devices (2012)

Metals Handbook, Desk Edition, 2nd ed., ASM, Materials Park, OH (1998).

MMPDS-02: Metallic Materials Properties Development and Standardization, Battelle Memorial Institute (2005) [NOTE: This is the successor to MIL-HDBK-5].
   Vol. 1, Introduction and Guidelines
   Vol. 2, Steel and Magnesium Alloys
   Vol. 3, 2000-6000 Series Aluminum Alloys
   Vol. 4, 7000 Series and Cast Aluminum Alloys
   Vol. 5, Miscellaneous Alloys & Structural Joints

SAE Handbook, 3 vols., Society of Automotive Engineers, Warrendale, PA (2005).

Smithells Metals Reference Book, 8th ed., W. F. Gale and T. C. Totemcier, Butterworths (2004).

Tool and Manufacturing Engineers Handbook—Desk Edition, Society of Manufacturing Engineers, Dearborn, MI (1988).

Tool and Manufacturing Engineers Handbook, 4th ed., Society of Manufacturing Engineers, Dearborn, MI
   Vol. 1, Machining (1983)
   Vol. 2, Forming (1984)
   Vol. 3, Materials, Finishing and Coating (1985)
   Vol. 4, Quality Control and Assembly (1987)
   Vol. 5, Manufacturing Management (1987)
   Vol. 6, Design for Manufacturability (1992)
   Vol. 7, Continuous Improvement (1994)
   Vol. 8, Plastic Part Manufacturing (1996)
   Vol. 9, Material and Part Handling in Manufacturing (1997)

Woldman's Engineering Alloys, 9th ed., J. Frick, ASM, Materials Park, OH (2000).

Basic and General Textbooks in Materials and Processes

21st Century Manufacturing, Paul Kenneth Wright, Prentice Hall (2001).

Elementary Materials Science, William F. Hosford, ASM (2013).

Elements of Metallurgy and Engineering Alloys, Flake Campbell (ed.), ASM, Materials Park, OH (2008).

Engineering Materials, 4th ed., 2 vols., M. F. Ashby and D. R. H. Jones, Butterworth-Heinemann (2012).

Engineering Materials, 9th ed., Kenneth Budinski and Michael Budinski, Prentice Hall (2009).

Essentials of Materials Science and Engineering, 3rd ed., D. R. Askeland and Wendelin Wright, Cengage Learning (2014).

Foundations of Materials Science and Engineering, 5th ed., W. F. Smith and J. Hashemi, McGraw-Hill (2010).

Fundamentals of Materials Science and Engineering: An Integrated Approach, 4th ed., W. D. Callister and D. G. Rethwisch, Wiley (2012).

Fundamentals of Manufacturing, 3rd ed., Philip Rufe (ed.), Society of Manufacturing Engineers (2011).

Fundamentals of Modern Manufacturing, 5th ed., Mikell P. Groover, Wiley (2013).

Handbook of Manufacturing Processes: How Products, Components, and Materials Are Made, James Bralla, Industrial Press (2007).

Introduction of Manufacturing Processes, 3rd ed., John A. Schey, McGraw-Hill (2000).

Introduction to Manufacturing Processes and Materials, Robert C. Creese, Marcel Dekker (1999).

Introduction to Materials Science for Engineers, 8th ed., James F. Shackelford, Prentice Hall (2015).

Lightweight Materials: Understanding the Basics, F. C. Campbell (ed.), ASM (2012).

Manufacturing Engineering and Technology, 7th ed., Serope Kalpakjian and Steven R. Schmid, Prentice Hall (2014).

Manufacturing Processes and Equipment, Jiri Tlusty, Prentice Hall (2000).

Manufacturing Processes and Materials, 5th ed., Ahmad K. Elshennawy and Gamal Weheba, Society of Manufacturing Engineers (2011).

Manufacturing Processes and Systems, 9th ed., Phillip F. Ostwald and Jairo Muñoz, Wiley (1997).

Materials: Engineering, Science, Processing and Design 2nd ed., M. Ashby et al., Elsevier/Butterworth-Heinemann (2010).

Materials Science and Engineering: An Introduction, 8th ed., W. D. Callister and D. G. Rethwisch, Wiley (2010).

Mechanical Metallurgy, 3rd ed., G. E. Dieter, McGraw-Hill, New York (1986).

Metallurgy for the Non-Metallurgist, 2nd ed., Arthur Reardon (ed.), ASM (2011).

Phase Diagrams: Understanding the Basics, F. C. Campbell, ASM (2012).

Principles of Materials Science and Engineering, 3rd ed., W. F. Smith, McGraw-Hill (1996).

Processes and Design for Manufacturing, 2nd ed., Sherif D. El Wakil, PWS Publishing Company (1998).

Structures and Properties of Engineering Alloys, 2nd ed., W. F. Smith, McGraw-Hill, New York (1993).

The Science and Design of Engineering Materials, 2nd ed., J. P. Schaffer et al., McGraw-Hill (1999).

The Science and Engineering of Materials, 7th ed., D. R. Askeland and Wendelin Wright, Cengage Learning (2016).

Ferrous Metals

Advanced High-Strength Steels: Science, Technology and Applications, Mahmoud Demeri, ASM (2013)

ASM Specialty Handbook—Carbon and Alloy Steels, ASM (1996).

ASM Specialty Handbook—Cast Irons, ASM (1996).

ASM Specialty Handbook—Stainless Steels, ASM (1994).

ASM Specialty Handbook—Tool Materials, ASM (1995).

Cast Iron: Physical and Engineering Properties, H. T. Angus, Butterworths (1976).

Ductile Iron Handbook, American Foundry Society (1992).

Engineering Properties of Steel, ASM, Materials Park, OH (1982).

Handbook of Stainless Steels, Donald Peckner and I. M. Bernstein (eds.), McGraw-Hill (1977).

Introduction to Stainless Steels, 3rd ed., J. Beddoes and J. G. Parr, ASM (1999).

Metallurgy and Heat Treatment of Tool Steels, Robert Wilson, McGraw-Hill (1976).

Stainless Steel, R. A. Lula, ASM, Materials Park, OH (1985).

Steel Metallurgy for the Non-Metallurgist, John D. Verhoeven, ASM (2007).

Steels: Metallurgy & Applications, 3rd ed., D. T. Llewellyn and R. C. Hudd, Butterworth-Heinemann (1999).

Steels: Heat Treatment and Processing Principles, G. Krauss, ASM, Materials Park, OH (1990).

Steels: Processing, Structure and Performance, 2nd ed., George Krauss, ASM, Materials Park, OH (2015).

The Making, Shaping and Treating of Steel, 11th ed., Association for Iron and Steel Technology (1998).

The Tool Steel Guide, Jim Szumera, Industrial Press (2003).

Tool Steels, 5th ed., G. A. Roberts, G. Krauss, and R. Kennedy, ASM, Materials Park, OH (1998).

Worldwide Guide to Equivalent Irons and Steels, 5th ed., ASM, Materials Park, OH (2006).

Nonferrous Metals

Aluminum, ASM, Materials Park, OH (1967).
    Vol. 1, Properties, Physical Metallurgy and Phase Diagrams
    Vol. 2, Design and Application
    Vol. 3, Fabrication and Finishing

Aluminum Standards and Data, The Aluminum Association, Washington, DC (bi-annual updates).

Aluminum: Properties and Physical Metallurgy, John E. Hatch (ed.), ASM, Materials Park, OH (1984).

ASM Specialty Handbook—Aluminum and Aluminum Alloys, ASM (1993).

ASM Specialty Handbook—Copper and Copper Alloys, ASM (2001).

ASM Specialty Handbook—Magnesium and Magnesium Alloys, ASM (1999).

ASM Specialty Handbook—Nickel, Cobalt, and Their Alloys, ASM (2000).

Beryllium Chemistry and Processing, Kenneth A. Walsh, ASM, Materials Park, OH (2009).

Engineering Properties of Zinc Alloys, International Lead Zinc Research Organization, New York (1980).

Introduction to Aluminum Alloys and Tempers, J. G. Kaufman, ASM (2000).

Materials Properties Handbook: Titanium Alloys, R. Boyer, E. W. Collings, and G. Welsch, ASM, Materials Park, OH (1994).

Properties of Aluminum Alloys: Fatigue Data and the Effects of Temperature, Product Form and Processing, J. G. Kaufman (ed.), ASM, Materials Park, OH (2008).

Properties of Aluminum Alloys: Tensile, Creep and Fatigue Data at High and Low Temperatures, J. G. Kaufman (ed.), ASM and Aluminum Association (1999).

Standards Handbook: Copper, Brass, and Bronze, 7 vols., Copper Development Association.

Superalloys: Alloying and Performance, Blaine Geddes, Hugo Leon, and Xiao Huang, ASM, Materials Park, OH (2010)

Superalloys: A Technical Guide, 2nd ed., M. J. Donachie and S. J. Donachie, ASM (2002).

The Superalloys: Fundamentals and Applications, C. Reed, Cambridge University Press (2008).

Titanium, 2nd ed., G. Lutjering and J. C. Williams, Springer (2007).

Titanium: Physical Metallurgy, Processing and Applications, F. H. Froes (ed.), ASM (2015).

Titanium: A Technical Guide, 2nd ed., M. Donachie (ed.), ASM, Materials Park, OH (2000).

Worldwide Guide to Equivalent Nonferrous Metals and Alloys, 4th ed., ASM, Materials Park, OH (2001).

Ceramics

Advanced Ceramics Technologies & Products, Springer (2012).

Advanced Structural Ceramics, B. Basu and K. Balani, Wiley/American Ceramic Society (2011).

An Introduction to the Mechanical Properties of Ceramics, D. J. Green, Cambridge University Press (1998).

Bioceramics: Properties, Characterization and Applications, J. Park, Springer (2008).

Ceramic and Glass Materials: Structure, Properties and Processing, G. F. Shackelford and R. H. Doremus, Springer (2008).

Ceramic Manufacturing Practices and Technologies, B. Hiremath et al., American Ceramic Society (1997).

Ceramic Materials: Science and Engineering, C. B. Carter and M. G. Norton, Springer (2008).

Ceramic Processing, M. N. Rahaman, CRC/Taylor & Francis (2007).

Ceramics—Applications in Manufacturing, D. W. Richerson, Society of Manufacturing Engineers (1989).

Ceramics and Composites: Processing Methods, N. P. Bansal and A. R. Boccaccini (eds.), Wiley/American Ceramic Society (2012).

Ceramics Processing and Technology, A. G. King, Noyes Publishing (2001).

Design and Properties of Glass Ceramics, 2nd ed., W. Holland and G. H. Beall, Wiley/American Chemical Society (2012).

Engineering Ceramics, M. Bengisu, Springer (2010).

Fundamentals of Ceramics, M. W. Barsoum, CRC/Taylor & Francis (2003).

Glass Ceramic Technology, 2nd ed., W. Holland and G. H. Beall, Wiley/American Ceramic Society (2012).

Glass Engineering Handbook, 3rd ed., G. McLellan and E. B. Shand, McGraw-Hill (1984).

Handbook of Advanced Ceramics: Materials, Applications, Processing and Properties, S. Somiya et al., Academic Press (2004).

Handbook of Ceramics, Glasses and Diamonds, C. A. Harper (ed.), McGraw-Hill (2001).

Introduction to Ceramics, 2nd ed., W. D. Kingery, et al., Wiley (1995).

Mechanical Properties of Ceramics, J. B. Wachtman, et al., Wiley (2009).

Modern Ceramic Engineering: Properties, Processing, and Use in Design, 3rd ed., D. W. Richerson, CRC/Taylor & Francis (2006).

Physical Ceramics, Y.-M. Chiang et al., Wiley (1997).

Principles of Ceramics Processing, 2nd ed., J. S. Reed, Wiley (1995).

Structural Ceramics, M. Schwartz, McGraw-Hill (1994).

    Vol. 1, Manufacturing Process Fundamentals

    Vol. 2, Material Selection Fundamentals & Product Design

    Vol. 3, Mold Design & Construction Fundamentals

    Vol. 4, Manufacturing Startup and Management

Plastics

Characterization and Failure Analysis of Plastics, ASM, Materials Park, OH (2003).

Concise Polymeric Materials Encyclopedia, J. C. Salamone (ed.), CRC Press (1999).

Degradable Polymers and Materials: Principles and Practice, K. Khemani and C. Scholz, American Chemical Society (2006).

Elements of Polymer Science and Engineering, 2nd ed., A. Rudin, Academic Press (1998).

Engineering Plastics and Composites, 2nd ed., W. A. Woishnis (ed.), ASM, Materials Park, OH (1993).

Engineering Plastics Handbook, J. M. Margolis, McGraw-Hill (2006).

Handbook of Elastomers, 2nd ed., A. K. Bhowmick and H. L. Stephens, CRC Press (2000).

Handbook of Plastics, Elastomers and Composites, 4th ed., C. A. Harper (ed.), McGraw-Hill (2003).

Handbook of Plastics Technologies: The Complete Guide to Processes and Performance, 2nd ed., C. Harper, McGraw-Hill (2006).

Handbook of Thermoset Plastics, S. H. Goodman, Noyes Publications (1986).

High Performance Polymers and Engineering Plastics, V. Mittal (ed.), Wiley-Scrivener (2011).

Introduction to Polymers, 3rd ed., R. J. Young and P. Lovell, CRC/Taylor & Francis (2008).

Modern Plastics Encyclopedia, published annually by Modern Plastics Magazine, a McGraw-Hill publication.

Modern Plastics Handbook, C. A. Harper, McGraw-Hill (2000).

Plastic Blow Molding Handbook, N. C. Lee (ed.), Van Nostrand Reinhold (1991).

Plastic Injection Molding, 4 vols., Society of Manufacturing Engineers (1996–1999).

Plastic Part Technology, E. A. Muccio, ASM, Materials Park, OH (1991).

Plastics: Materials and Processing, 3rd ed., A. B. Strong, Pearson Educational (2006).

Plastics Processing Technology, E. A. Muccio, ASM, Materials Park, OH (1994).

Plastics Engineering Handbook, 5th ed., M. L. Berins (ed.), Chapman & Hall (1994).

Plastics Technology Handbook, 4th ed., M. Chanda and S. K. Roy (eds.), CRC Press (2006).

Polymer Handbook, 4th ed., J. Brandrup and E. E. Immergut (eds.), Wiley (2004).

Polymer Processing, D. H. Morton-Jones, Chapman & Hall, London (2008).

Polymer Processing Principles and Design, D. G. Baird and D. I. Collias, Wiley (1998).

Principles of Polymer Engineering, 2nd ed., C. P. Buckley et al., Oxford University Press (1997).

Principles of Polymer Processing, 2nd ed., Z. Tadmore and C. G. Gogos, Wiley (2006).

Properties and Behavior of Polymers, 2 vols., Wiley (2011).

Science and Technology of Rubber, 3rd ed., J. E. Mark and B. Erman, Academic Press (2005).

Selecting Thermoplastics for Engineering Applications, 2nd ed., C. P. MacDermott and A. V. Shenoy, Marcel Dekker (1997).

Textbook of Polymer Science, 3rd ed., F. W. Billmeyer Jr., Wiley (1984).

Thermoforming: Improving Process Performance, Stanley R. Rosen, Society of Manufacturing Engineers (2002).

Composites

Advanced Polymer Composites: Principles and Applications, B. J. Jang. ASM, Materials Park, OH (1993).

Analysis and Performance of Fiber Composites, 3rd ed., B. D. Agarwal et al., Wiley (2006).

Ceramic Matrix Composites, R. Naslain and B. Harris, Elsevier Applied Science (1990).

Composite Materials Handbook, 2nd ed., M. M. Schwartz, McGraw-Hill, New York (1995).

Composite Materials: Science and Engineering, 3rd ed., K. K. Chawla, Springer-Verlag (2008).

Composite Filament Winding, Stan Peters (ed.), ASM, Materials Park, OH (2011).

Composites Engineering Handbook, P. K. Mallick, Marcel Dekker (1997).

Composites Manufacturing: Materials, Products and Process Engineering, S. K. Mazumdar, CRC Press (2001).

Composite Materials: Science and Applications, 2nd ed., D. D. L. Chung, Springer (2010).

Composite Materials: Science and Engineering, 3rd ed., K. K. Chawla, Springer-Verlag (2008).

Engineers' Guide to Composite Materials, John W. Weeton (ed.), ASM, Materials Park, OH (1986).

Fiber-Reinforced Composites: Materials, Manufacturing, and Designs, 3rd ed., P. K. Mallick, CRC/Taylor & Francis (2007).

Fundamentals of Composite Manufacturing: Materials, Methods, and Applications 2nd ed., A. B. Strong, Society of Manufacturing Engineers (2008).

Fundamentals of Metal Matrix Composites, S. Suresh (ed.), Butterworth-Heinemann (1993).

Handbook of Ceramic Composites, N. P. Bansal (ed.), Springer (2005).

Handbook of Composites, George Lubin (ed.), Van Nostrand Reinhold, New York (1982).

Manufacturing Processes for Advanced Composites, E. Campbell, Elsevier (2004).

Metal and Ceramic Matrix Composites, B. Cantor et al., Taylor & Francis (2003).

Metal Matrix Composites, N. Chawla and K. K. Chawla, Springer (2006).

Structural Composite Materials, F. C. Campbell, ASM, Materials Park, OH (2010).

Elevated Temperature Applications

ASM Specialty Handbook—Heat-Resistant Materials, ASM (1997).

Fatigue and Durability of Metals at High Temperatures, S. S. Manson and G. R. Halford, ASM, Materials Park, OH (2009).

High Temperature Property Data: Ferrous Alloys, M. F. Rothman (ed.), ASM, Materials Park, OH (1987).

Electronic and Magnetic Applications

Advanced Electronic Packaging, 2nd ed., R. K. Ulrich and W. D. Brown, IEEE Press and Wiley (2006).

ASM Ready Reference: Electrical and Magnetic Properties of Materials, ASM (2000).

Electrical Properties of Materials, I. Solymar and D. Walsh, Oxford (2004)

Electronic Materials Handbook, Vol. 1, Packaging, ASM (1989).

Electronic Materials & Processes Handbook, 3rd ed., C. A. Harper, McGraw-Hill (2009).

Electronic Packaging and Interconnection Handbook, 4th ed., C. A. Harper, McGraw-Hill, New York (2004).

Electronic Properties of Materials, 4th ed., R. E. Hummel, Springer (2011).

Fundamentals of Semiconductor Manufacturing and Process Control, G. S. May and C. J. Spanos, Wiley (2006).

Fundamentals of Semiconductor Processing Technology, B. El-Kareh Kluwer Academic Publishers, Boston (1995).

Handbook of Electronic Package Design, M. Pecht (ed.), Marcel Dekker (1991).

Handbook of Electronics Manufacturing, 3rd ed., B. S. Matisoff, Chapman & Hall (1996).

Handbook of Photomask Manufacturing Technology, S. Rizvi, CRC Press (2005).

Handbook of Semiconductor Manufacturing Technology, 2nd ed., R. Doering and Y. Nishi, CRC Press (2008).

Handbook of Semiconductor Technology, K. A. Jackson and W. Schroter, Wiley (2000).

Integrated Circuit Fabrication Technology, D. J. Elliott, McGraw-Hill, New York (1989).

Introduction to Microelectronic Fabrication, 2nd ed., R. C. Jaeger, Prentice Hall (2001).

Manufacturing Technology in the Electronics Industry, P. R. Edwards, Chapman & Hall (1991).

Microchip Fabrication, 5th ed., P. Vanzant, McGraw-Hill (2005).

Microelectronics Failure Analysis Desk Reference, 6th ed., Richard J. Ross (ed.), ASM/Electronic Device Failure Analysis Society (2011).

Printed Circuit Assembly Design, L. Marks and J. Caterina, McGraw-Hill (2000).

Printed Circuit Boards: Design, Fabrication and Assembly, R. S. Khandpur, McGraw-Hill (2005).

Printed Circuits Handbook, 6th ed., C. F. Coombs Jr. (ed.), McGraw-Hill (2006).

Semiconductor Devices: Physics and Technology, 3rd ed., S. M. Sze and M. K. Lee, Wiley (2011).

Semiconductor Manufacturing Handbook, H. Geng, McGraw-Hill (2005).

Springer Handbook of Electronic and Photonic Materials, S. Kasap and P. Capper (eds.), Springer (2006).

The Science and Engineering of Microelectronic Fabrication, S. A. Campbell, Oxford University Press (2001).

Design

Application of Fracture Mechanics—for Selection of Metallic Structural Materials, J. E. Campbell, et al., ASM, Materials Park, OH (1982).

Atlas of Stress-Strain Curves, H. E. Boyer (ed.), ASM, Materials Park, OH (1986).

Atlas of Fatigue Curves, Howard E. Boyer (ed.), ASM, Materials Park, OH (1986).

Atlas of Stress Corrosion and Corrosion Fatigue Curves, A. J. McEvily, ASM, Materials Park, OH (1990).

Axiomatic Design, Nam P. Suh, Oxford University Press (2001).

Concurrent Design of Products and Processes, J. L. Nevins and Dan Whitney, McGraw-Hill (1989).

Design for Assembly, M. Andreasen et al., Springer-Verlag (1988).

Design for Manufacturability Handbook, 2nd ed., J. G. Bralla (ed.), McGraw-Hill (1998).

Design for Manufacturing: A Structured Approach, C. Poli, Butterworth-Heinemann (2001).

Design for the Environment, 2nd ed., J. Fiksel, McGraw-Hill (2011).

Designing for Economical Production, 2nd ed., H. E. Trucks and G. Lewis, Society of Manufacturing Engineers (1987).

Die Design Handbook, 3rd ed., David A. Smith, Society of Manufacturing Engineers (1990).

Engineering Design: A Materials and Processing Approach, 3rd ed., George Dieter, McGraw-Hill (2000).

Fatigue and Durability of Structural Materials, S. S. Manson and G. R. Halford, ASM, Materials Park, OH (2006).

Fatigue and Fracture: Understanding the Basics, F. C. Campbell (ed.), ASM (2012).

Fatigue Data Book: Light Structural Alloys, ASM, Materials Park, OH (1995).

Fracture Mechanics: Fundamentals and Applications, T. L. Anderson, CRC Press (1991).

Fundamentals of Tool Design, 6th ed., J. Nee (ed.), Society of Manufacturing Engineers (2010).

Handbook of Design, Manufacturing and Automation, R. C. Dorrf and A. Kusiak (eds.), Wiley (1995).

Handbook of Materials for Product Design, C. A. Harper (ed.), McGraw-Hill (2001).

Handbook of Product Design for Manufacturing—A Practical Guide for Low-Cost Production, James G. Bralla (ed.), McGraw-Hill, New York (1986).

Integrated Product and Process Design and Development: The Product Realization Process, 2nd ed., E. B. Magrab et al., CRC Press (2009).

Introduction to Engineering Design, Andrew Samuel and John Weir, Butterworth Heinemann (1999).

Material Selection in Mechanical Design 4th ed., Michael Ashby, Elsevier (2011).

Mechanical Assemblies, Daniel E. Whitney, Oxford University Press (2004).

Process Selection: From Design to Manufacture, 2nd ed., K. G. Swift and J. D. Booker, Butterworth-Heinemann (2003).

Product Design and Development, 5th ed., K. Ulrich and S. Eppinger, McGraw-Hill (2011).

Product Design for Manufacture and Assembly, 3rd ed., G. Boothroyd, P. Dewhurst, W. Knight, CRC Press (2010).

The Principles of Design, Nam P. Suh, Oxford University Press (1990).

The Principles of Material Selection for Engineering Design, Pat L. Mangonon, Prentice Hall (1999).

Tool Design, 3rd ed., C. Donaldson, G. LeCain, and V. C. Gould, Glencoe (1993).

Casting

Aluminum Alloy Castings: Properties, Processes and Applications, J. G. Kaufman and E. L. Rooy, ASM and American Foundry Society (2004)

Aluminum Permanent Mold Handbook, American Foundry Society (2001).

Basic Principles of Gating and Risering, 2nd ed., American Foundry Society (2008).

Best Practices in Aluminum Metalcasting, Geoffrey Sigworth, American Foundry Society (2014).

Casting Copper-base Alloys, 2nd ed., American Foundry Society (2007).

Casting Design and Performance, ASM, Materials Park, OH (2009).

Casting Design Handbook, ASM (2012).

Castings, 2nd ed., John Campbell, Elsevier (2003).

Centrifugal Casting, Nathan Janco, American Foundry Society (1988).

Complete Casting Handbook: Metal Casting Processes, Techniques and Design, 2nd ed., J. Campbell, Butterworth-Heinemann (2011).

Continuous Casting, W. Schneider, Wiley (2006).

Cupola Handbook, 6th ed., American Foundry Society (1999).

Foundry Technology, P. R. Beeley, Butterworth-Heinemann (2002).

Fundamentals of Metal Casting, R. A. Flinn, American Foundry Society (1987).

Handbook on the Investment Casting Process, American Foundry Society (1993).

High-Integrity Die Casting, North America Die Casting Association (NADCA) (2008).

Introduction to Die Casting, North America Die Casting Association (NADCA) (2006).

Investment Casting Handbook, Investment Casting Institute, Chicago, IL (1997).

Iron Castings Engineering Handbook, Iron Castings Society and American Foundry Society (2003).

Lost Foam Casting Made Simple, F. Sonnenberg (ed.), American Foundry Society (2008)

Magnesium Die Casting Handbook, North America Die Casting Association (NADCA) (2006).

Metalcasting, C. W. Ammen, McGraw-Hill (1999).

Metalcasting Principles & Techniques, Yury Lerner and P. N. Rao, American Foundry Society (2014).

Mold and Core Test Handbook, 4th ed., American Foundry Society (2015).

Principles of Metal Casting, 3rd ed., Mahi Sahoo and Sam Sahu, McGraw-Hill (2014).

Principles of Sand Control, American Foundry Society (2004).

Principles of the Shell Process, American Foundry Society (2002).

Steel Castings Handbook, 6th ed., Steel Founder's Society of America, OH (1995).

The NFFS (Non-Ferrous Founders Society) Guide to Aluminum Casting Design: Sand and Permanent Mold, NFFS (1994).

The Process of Metalcasting, Cast Metals Institute and American Foundry Society (2007).

Product Design for Die Casting, 7th ed., North America Die Casting Association (NADCA) (2015).

Rapid Tooling Guidelines for Sand Castings, Wanlong Wong, et al., Springer (2010).

Science and Engineering of Casting Solidification, 2nd ed., Doru Michael Stefanescu, Springer Science (2008).

Technology for Magnesium Castings: Design, Products & Applications, American Foundry Society (2011).

Technology of Metalcasting, Frederick Schleg, American Foundry Society (2003).

Zinc Die Casting Process, The, North America Die Casting Association (NADCA) (2015).

Welding and Joining

Adhesive Bonding: Materials, Applications, and Technology, W. Brockman et al., Wiley (2009).

Adhesive Bonding: Science, Technology and Applications, R. S. Adams (ed.), CRC/Taylor & Francis (2005).

Adhesives Technology Handbook, Arthur Landrock, Noyes (1985).

Brazing Handbook, 5th ed., American Welding Society (2007).

Brazing, 2nd ed., M. M. Schwartz, ASM, Materials Park, OH (2003).

Ceramic Joining, M. M. Schwartz, ASM, Materials Park, OH (1990).

Design of Welded Structures, James F. Lincoln Foundation.

Design of Weldments, James F. Lincoln Foundation.

Friction Stir Welding and Processing, R. S. Mishra and M. W. Mahoney (eds.), ASM, Materials Park, OH (2007).

Handbook of Adhesives, 3rd ed., Irving Skiest (ed.), Chapman & Hall (1990).

Handbook of Adhesives and Sealants, 2nd ed., E. M. Petrie, McGraw-Hill (2006).

Handbook of Plastics Joining: A Practical Guide, William Andrew Inc. (1996).

Industrial Brazing Practice, P. Roberts, CRC Press (2004).

Joining: Understanding the Basics, F. C. Campbell (ed.), ASM (2011).

Joining of Composite Matrix Materials, M. M. Schwartz, ASM, Materials Park, OH (1994).

Joining of Plastics: Handbook for Designers and Engineers, J. Rotheiser, Hanser Gardner (2004).

Joining Processes: Introduction to Brazing and Diffusion Bonding, M. G. Nicholas, Chapman & Hall (1998).

Lead-free Soldering, J. Bath (ed.), Springer (2007).

Mechanical Assemblies, D. E. Whitney, Oxford University Press (2004).

Mechanical Fastening of Plastics, B. Lincoln et al., Marcel Dekker (1984).

Mechanical Fastening, Joining, and Assembly, J. A. Speck, Marcel Dekker (1997).

Modern Welding Technology, 6th ed., H. B. Cary and S. C. Helzer, Pearson/Prentice Hall, NJ (2005).

Principles of Brazing, D. M. Jacobson and G. Humpston, ASM, Materials Park, OH (2005).

Principles of Soldering, G. Humpston and D. M. Jacobson, ASM, Materials Park, OH (2004).

Principles of Welding: Processes, Physics, Chemistry, and Metallurgy, R. W. Messler Jr., Wiley (1999).

Resistance Welding: Fundamentals and Applications, 2nd ed., H. Zhang and J. Senkara, CRC Press (2011).

Soldering: Understanding the Basics, M. M. Schwartz, ASM (2014).

Soldering Manual, 2nd ed., American Welding Society (1978).

Solders and Soldering, 3rd ed., H. H. Manko, McGraw-Hill (1994).

Standard Handbook of Fastening and Joining, 3rd ed., R. O. Parmley (ed.), McGraw-Hill (1997).

Structural Adhesives: Chemistry and Technology, R. S. Hartshorn (ed.), Plenum Press (1986).

The Basics of Soldering, A. Rahn, Wiley (1993).

The Science and Practice of Welding, 10th ed., 2 vols., A. C. Davies, Cambridge University Press (1993).

Welding and Cutting: A Guide to Fusion Welding and Associated Cutting Processes, T. Houldcroft, Industrial Press (2001).

Welding: Fundamentals and Procedures, J. Galyen et al., Prentice Hall (1991).

Welding Handbook, 9th ed., American Welding Society, New York (2007).

Welding Processes Handbook, K. Weman, CRC Press (2003).

Welding: Principles and Applications, 6th ed., L. F. Jeffus, Cengage Learning (2007).

Welding Technology for Engineers, B. Raj et al., Narosa Publishing House and ASM, Materials Park, OH (2006).

Fabrication and Forming

Aluminum Extrusion Technology, P. K. Saha, ASM, Materials Park, OH (2000).

Cold and Hot Forging: Fundamentals and Applications, T. Altan et al., ASM, Materials Park, OH (2005).

Die Design Handbook, 3rd ed., David A. Smith, Society of Manufacturing Engineers (1990).

Extrusion, 2nd ed., M. Bauser et al. (eds.), ASM, Materials Park, OH (2006).

Extrusion, K. Laue and H. Stenger, ASM (1981).

Extrusion of Aluminum Alloys, T. Sheppard, Springer (2010).

Finite-Element Plasticity and Metalforming Analysts, G. W. Rowe, C. E. N. Sturgess, P. Hartley, and I. Pillingen, Cambridge University Press (1991).

Forging Handbook, T. G. Bryer (ed.), Forging Industry Association, Cleveland, OH and ASM (1985).

Fundamentals of Hydroforming, Harjinder Singh, Society of Manufacturing Engineers (2003).

Fundamentals of Metal Forming, R. H. Wagoner and J.-L. Chenot, Wiley (1997).

Fundamentals of Pressworking, David A. Smith, Society of Manufacturing Engineers (1994).

Fundamentals of Tool Design, 5th ed., David A. Smith and John Nee (eds.), Society of Manufacturing Engineers (2003).

Handbook of Fabrication Processes, O. D. Lascoe, ASM, Materials Park, OH (1988).

Handbook of Metal Forming, Kurt Lange (ed.), Society of Manufacturing Engineers (2006).

Handbook of Metalforming Processes, B. Avitzur, Wiley-Interscience, New York (1983).

Handbook of Workability and Process Design, G. E. Dieter et al. (eds.), ASM, Materials Park, OH (2003).

Mechanics of Sheet Metal Forming, D. J. Hu et al., Butterworth-Heinemann (2002).

Metal Forming: Fundamentals and Applications, T. Altan, S.-I. Oh, and H. Gegal, ASM, Materials Park, OH (1983).

Metal Forming: Mechanics and Metallurgy, 4th ed., W. F. Hosford and R. M. Caddell, Prentice Hall (2011).

Metal Forming Practice: Processes, Machines, Tools, H. Tschaetch, Springer (2007).

Metal Forming Science and Practice, J. G. Lenard, Elsevier (2002).

Metalforming and the Finite-Element Method, S. Kobayshi, S. Oh, and T. Altan, Oxford University Press (1989).

Metals Fabrication: Understanding the Basics, F. C. Campbell (ed.), ASM (2013).

Metalworking Science and Engineering, Edward M. Mielnik, McGraw-Hill (1991).

Press Brake Technology, S. Benson, Society of Manufacturing Engineers (1997).

Progressive Dies: Principles and Practices of Design and Construction, Don Peterson (ed.), Society of Manufacturing Engineers (1994).

Roll Forming Handbook, George T. Halmos, Taylor & Francis (2005)

Sheet Metal Forming, R. Pearce, Springer (2006).

Sheet Metal Forming: Fundamentals, Taylan Altan and A. Erman Tekkaya (eds.), ASM (2012).

Sheet Metal Forming: Processes and Applications, Taylan Altan and A. Erman Tekkaya (eds.), ASM (2012).

Sheet Metal Forming Processes and Die Design, Vukota Boljanovic, Industrial Press (2004).

Steel Forgings: Design, Production, Selection, Testing, and Application, E. G. Nisbett, ASTM, West Conshohocken, PA (2005).

The Metal Stamping Process, Jim Szumera, Industrial Press (2003).

Tube Forming Processes: A Comprehensive Guide, Greg Miller, Society of Manufacturing Engineers (2002).

Powder Metallurgy

A-Z of Powder Metallurgy, R. M. German, Elsevier Science (2006).

Fundamentals of Powder Metallurgy, Leander Pease and William West, Metal Powder Industries Federation (2002)

Handbook of Non-Ferrous Metal Powders: Technologies and Applications, O. D. Neikov et al., Elsevier (2009).

Handbook of Powder Metallurgy, H. H. Hausner, Chemical Publishing Co. (2007).

Handbook of Powder Science and Technology, M. E. Fayed and L. Otten (eds.), 2nd ed., Chapman & Hall (1997).

Injection Molding of Metals and Ceramics, Randall M. German and Animesh Bose, Metal Powder Industries Federation (1997).

Introduction to Powder Metallurgy, F. Thummler and R. Oberacker, The Institute of Materials (1994).

Powder Injection Molding, R. M. German, Metal Powder Industries Federation (1990).

Powder Injection Molding: Design & Applications, Randall M. German, Innovative Materials Solutions (2003).

Powder Metallurgy: Science, Materials and Technology, A. Upadhyaya and G. S. Upadhyaya, University Press (2011).

Powder Metallurgy & Particulate Materials Processing, Randall M. German, Metal Powder Industries Federation (2006).

Powder Metallurgy Design Manual, 3rd ed., Metal Powder Industries Federation (1998).

Powder Metallurgy Science, 2nd ed., R. M. German, Metal Powder Industries Federation (1994).

Powder Metallurgy, Principles and Applications, F. V. Lenel, Metal Powder Industries Federation, Princeton, NJ (1980).

Powder Metallurgy Stainless Steels: Processing, Microstructure and Properties, E. Klar and P. Samal, ASM (2007).

Machining

Abrasive Processes, J. A. Webster, et al., Marcel Dekker (1999).

Advanced Machining Technology Handbook, J. Brown, McGraw-Hill (1998).

Analysis of Material Removal Processes, W. R. DeVries, Springer-Verlag (1992).

Applied Machining Technology, H. Tschatsch, Springer (2009).

Creep Feed Grinding, C. Andrew et al., Holt, Rinehart and Winston (1985).

Cutting Tool Technology: Industrial Handbook, G. T. Smith, Springer (2008).

Fundamentals of Metal Machining and Machine Tools, 3rd ed., G. Boothroyd and Winston Knight, CRC/Taylor & Francis (2006).

Gear Materials, Properties and Manufacture, J. R. Davis (ed.), ASM International (2006).

Grinding Technology, 2nd ed., Changsheng Guo and Stephen Malkin, Industrial Press and SME (2007).

Grinding Technology: Theory and Applications of Machining with Abrasives, 2nd ed., S. Malkin and C. Guo, Industrial Press (2008).

Handbook of Machine Tools, 4 vols., M. Weck, Wiley (1984).

Handbook of Machining with Grinding Wheels, I. D. Marinescu et al., CRC Press (2006).

High-Speed Machining, Berthold Erdel, Society of Manufacturing Engineers (2003).

Machine Tool Practices, 9th ed., R. R. Kibbe, et al., Prentice Hall (2009).

Machine Tools for High Performance Machining, L. N. Lopez and A. Lamikiz (eds.), Springer (2009).

Machine Tools Handbook, P. H. Joshi, McGraw-Hill (2008).

Machining: Fundamentals and Recent Advances, J. P. Davim (ed.), Springer (2010).

Machining and Metal Working Handbook, 3rd ed., R. A. Walsh and D. Cormier, McGraw-Hill (2006).

Machining Data Handbook, 3rd ed., 2 vols., Machinability Data Center, Metcut Associates, Cincinnati, OH (1989).

Machining Dynamics: Fundamentals, Applications and Practice, K. Cheng, Springer (2008).

Machining Fundamentals and Recent Advances, J. P. Davim (ed.), Springer (2008).

Machining Technology: Machine Tools and Operations, H. A. Youssef and H. El-Hofy, CRC Press (2008).

Machining with Abrasives, M. J. Jackson and M. J. Davim, Springer (2010).

McGraw-Hill Machining and Metalworking Handbook, 3rd ed., R. A. Walsh (2006).

Metal Cutting, 4th ed., E. M. Trent and P. K. Wright, Butterworth-Heinemann (2000).

Metal Cutting Mechanics, V. P. Astakhov, CRC Press (1998)

Metal Cutting Principles, 2nd ed., M. C. Shaw, Oxford University Press (2005).

Metal Cutting Theory and Practice, 2nd ed., David A. Stephenson and John S. Agapiou, CRC/Taylor & Francis (2006).

Metal Cutting Tool Handbook, 7th ed., Industrial Press (1989).

Metal Machining: Theory and Application, T. H. C. Childs et al., Butterworth-Heinemann (2000).

Modern Grinding Process Technology, S. C. Salmon, McGraw-Hill (1992).

Modern Metal Cutting, AB Sandvik Coromant (1994).

Principles of Modern Grinding Technology, W. Rowe, Elsevier Applied Science (2009).

Technology of Machine Tools, 6th ed., S. F. Krar and A. F. Check, Macmillan/McGraw-Hill (2009).

Tribology of Metal Cutting, V. Astakhov, Elsevier (2007).

Heat Treatment

Atlas of Isothermal and Cooling Transformation Diagrams, ASM, Materials Park, OH (1977).

Atlas of Isothermal and Cooling Transformation Diagrams for Engineering Steels, ASM, Materials Park, OH (1980).

Atlas of Time-Temperature Diagrams for Nonferrous Alloys, G. VanderVoort (ed.), ASM, Materials Park, OH (1991).

Atlas of Time-Temperature Diagrams for Irons and Steels, G. VanderVoort (ed.), ASM, Materials Park, OH (1991).

Atmosphere Heat Treatment: Principles, Applications, Equipment, vol. 1, Daniel Herring, BNP Media (2014).

Handbook of Quenchants and Quenching Technology, G. E. Totten, C. E. Bates, and N. A. Clinton, ASM, Materials Park, OH (1992).

Heat Treater's Guide: Practices and Procedures for Irons and Steels, 2nd ed., ASM, Materials Park, OH (1995).

Heat Treater's Guide: Practices and Procedures for Nonferrous Alloys, H. Chandler (ed.), ASM (1996).

Heat Treatment, Structure and Properties of Nonferrous Alloys, Charles R. Brooks, ASM, Materials Park, OH (1982).

Heat Treating Data Book, 9th ed., Seco/Warwick Corporation (2006).

Practical Heat Treating, 2nd ed., J. L. Dossett and H. E. Boyer, ASM, Materials Park, OH (2006).

Practical Induction Heat Treating, R. E. Haimbaugh, ASM (2001).

Principles of Heat Treatment of Steel, G. Krauss, ASM, Materials Park, OH (1980).

Principles of Heat Treatment, M. A. Grossman and E. C. Bain, ASM, Materials Park, OH (1964).

Principles of the Heat Treatment of Plain Carbon and Low Alloy Steels, C. R. Brooks, ASM, Materials Park, OH (1996).

Steel and Its Treatment, Bofors 2nd ed., G. E. Totten and M. A. H. Howes (eds.), Marcel Dekker (2006).

Steel Heat Treatment Handbook, 2nd ed., S. S. Babu and G. E. Totten, CRC/Taylor & Francis (2006).

Surface Hardening of Steels: Understanding the Basics, J. R. Davis (ed.), ASM (2002).

Surfaces and Finishes

Advanced Machining Technology Handbook, James Brown, McGraw-Hill (1998).

Chemical Vapor Deposition, J. Park, ASM International (2001).

Coating and Surface Treatment Systems for Metals: A Comprehensive Guide to Selection, J. Edwards, ASM International (1997).

Coatings and Coating Processes for Metals, J. H. Lindsay (ed.), ASM (1998).

Coatings Technology Handbook, 2nd ed., A. A. Tracton (ed.), CRC/Taylor & Francis (2006).

Electroplating Engineering Handbook, 4th ed., Lawrence Durney (ed.), Van Nostrand Reinhold (1984).

Graham's Electroplating Engineering Handbook, 4th ed., L. J. Durney (ed.), Chapman & Hall (1996).

Guide to High Performance Powder Coating, Bob Utech, Society of Manufacturing Engineers (2002).

Handbook of Hard Coatings: Deposition Technologies, Properties and Applications, R. F. Bunshah (ed.), Noyes (2001).

Handbook of Physical Vapor Deposition (PVD) Processing, D. M. Mattox, Noyes (1999).

Handbook of Surface Treatments and Coatings, ASME Press (2003).

Handbook of Thermal Spray Technology, J. R. Davis (ed.), Thermal Spray Society and ASM, Metals Park, OH (2004).

Industrial Painting and Powder Coating: Principles and Practice, 3rd ed., Norman Roobol, Hanser-Gardner Publications (2005).

Mass Finishing Handbook, LaRoux Gillespie, Industrial Press and SME (2006).

Metal Cutting and High Speed Machining, D. Dudzinski, et al. (eds.), Kluwer Academic (2002).

Metal Finishing: Guidebook and Directory, Metals and Plastics Publications, Inc., Westwood, NJ (published annually).

Metallic and Ceramic Coatings, M. G. Hocking et al., Addison-Wesley (1989).

Metallizing of Plastics: Handbook of Theory and Practice, R. Suchentruck, Finishing Publications Ltd and ASM (1993).

Modern Electroplating, 4th ed., M. Schlesinger and M. Paunovic, Wiley (2001).

Properties of Electroplated Metals and Alloys, 2nd ed., W. H. Safranek, American Electroplaters and Finishers Society (1986).

Surface Engineering for Corrosion and Wear Resistance, J. R. Davis (ed.), IOM Communications and ASM (2001).

Surface Engineering for Wear Resistance, K. G. Budinski, Prentice Hall (1988).

Surface Engineering of Metals: Principles, Equipment, Technologies, T. Burakowski and T. Wiershon, CRC Press (1998).

Surface Finishing Systems, George Rudski, ASM, Materials Park, OH (1983).

Surface Hardening of Steels, J. R. Davis (ed.), ASM (2002).

Surface Modification Technologies, T. S. Sudarshan (ed.), ASM International (1998).

The Surface Treatment and Finishing of Aluminum and Its Alloys, 6th ed., S. Wernick, R. Pinner and P. B. Sheasby, Finishing Publications Ltd. and ASM (2001).

Corrosion

An Introduction to Metallic Corrosion, 3rd ed., Ulick R. Evans, Edward Arnold Ltd. and ASM (1981).

Corrosion and Corrosion Control, 3rd ed., H. H. Uhlig and R. Winston Revie, Wiley-Interscience, New York (1985).

Corrosion and Its Control—An Introduction to the Subject, Atkinson and Van Droffelaar, National Association of Corrosion Engineers, Houston, TX (1982).

Corrosion Engineering, 3rd ed., M. G. Fontana, McGraw-Hill, New York (1986).

Corrosion Prevention by Protective Coatings, Charles G. Munger, National Association of Corrosion Engineers (1984).

Corrosion Resistance Tables, 4th ed., P. A. Schweitzer, Marcel Dekker (1995).

Corrosion, 3rd ed.—Vol. 1: Corrosion of Metals and Alloys and Vol. 2: Corrosion Control, L. L. Shrier (ed.), Elsevier (1994).

Corrosion: Understanding the Basics, J. R. Davis (ed.), ASM (2000).

Encyclopedia of Corrosion Technology, 3rd ed., P. A. Schweitzer, Marcel Dekker (2004).

Fundamentals of Electrochemical Corrosion, E. E. Stansbury and R. A. Buchanan, ASM (2000).

Handbook of Corrosion Data, 2nd ed., B. Craig and D. Anderson (eds.), ASM, Materials Park, OH (1995).

Handbook of Corrosion Engineering, 2nd ed., P. R. Roberge, McGraw-Hill (2012).

Material Selection for Corrosion Control, S. I. Chawla and R. K. Gupta, ASM, Materials Park, OH (1993).

NACE Corrosion Engineer's Reference Book, 2nd ed., R. S. Treseder (ed.), National Association of Corrosion Engineers (NACE) (1991).

Mechanical Testing, NDT, and Metrology

Atlas of Stress-Strain Curves, 2nd ed., ASM (2002).

Dimensioning and Tolerancing, American National Standards Institute (ANSI-Y14.5M-2009) and American Society of Mechanical Engineers (2009).

Dimensioning and Tolerancing Handbook, P. J. Drake, McGraw-Hill (1999).

Geometric Dimensioning and Tolerancing for Mechanical Design, G. Cogorno, McGraw-Hill (2006).

Geometric Dimensioning and Tolerancing Handbook, J. D. Meadows, ASME (2009)

Handbook of Dimensional Measurement, 4th ed., M. Curtis, Industrial Press (2007).

Handbook of Metrology, Brown and Sharpe (1992).

Handbook of Surface and Nanometrology, 2nd ed., D. J. Whitehouse, CRC Press (2010).

Hardness Testing, 2nd ed., H. Chandler (ed.), ASM (1999).

Hardness Testing: Principles and Applications, Konrad Herrmann et al., ASM (2011).

ISO System of Limits and Fits, General Tolerances and Deviations, American National Standards Institute.

Measurement of Geometrical Tolerances in Manufacturing, J. D. Meadows, Marcel Dekker (1998).

Mechanical Behavior of Materials, 2nd ed., T. H. Courtney, Waveland Press (2005).

Mechanical Behavior of Materials, W. F. Hosford, Cambridge (2005).

Mechanical Behavior of Materials: Engineering Methods for Deformation, Fracture, and Fatigue, 3rd ed., N. E. Dowling, Prentice Hall (2006).

Metrology Handbook, The Society for Quality, J. L. Bucher (ed.) (2004).

Nondestructive Testing Handbook (11 vols.), 2nd & 3rd eds. (multiple volumes), P. O. Moore (ed.), American Society for Nondestructive Testing (2008).

Nondestructive Testing, Louis Cartz, ASM, Materials Park, OH (1995).

Practical Non-Destructive Testing, 2nd ed., B. Raj, T. Jaykumar, and M. Thavasimuthu, Narosa Publishing and ASM (2002).

Principles of Measurement Systems, 4th ed., J. P. Bentley, Prentice Hall (2005).

Springer Handbook of Materials Measurement Methods, H. Czichos et al. (eds.), Springer (2006).

Tensile Testing, 2nd ed., J. R. Davis (ed.), ASM, Materials Park, OH (2004).

The Testing of Engineering Materials, 4th ed., H. E. Davis, G. E. Troxell, and G. F. W. Hauck, McGraw-Hill, New York (1982).

Quality Control, Taguchi Methods, and Six Sigma

A Primer on the Taguchi Method, Ranjit Roy, Society of Manufacturing Engineers (2010).

Creating Quality, William J. Kolarik, McGraw-Hill (1995).

Inspection of Metals: Understanding the Basics, F. C. Campbell (ed.), ASM (2013).

Introduction to Quality Engineering, Genichi Taguchi, Kraus International Publications (1986).

Introduction to Statistical Quality Control, 7th ed., D. C. Montgomery, Wiley (2012).

Juran's Quality Handbook, 6th ed., J. De Feo and J. M. Juran, McGraw-Hill (2010).

Juran's Quality Planning & Analysis, 5th ed., F. M. Gryna, R. C. H. Chua, and J. A. DeFeo, McGraw-Hill (2006).

Lean Six Sigma, M. L. George, McGraw-Hill (2002).

Metallographer's Guide: Practices and Procedures for Irons and Steels, B. L. Bramfitt and A. O. Benscoter, ASM (2002).

Metallography: Principles and Practice, G. VanderVoort, ASM (1984).

Process Quality Control, 2nd ed., E. R. Ott and E. G. Schilling, McGraw-Hill (1990).

Quality Control, 8th ed., D. H. Besterfield, Prentice Hall (2008).

Quality Control Handbook, 4th ed., J. M. Juran and F. M. Gigna, McGraw-Hill (1988).

Quality Engineering Handbook, 2nd ed., T. Pyzdek and P. Keller, CRC/Taylor & Francis (2003).

Quality Engineering in Production Systems, G. Taguchi, E. A. Elsayed, and T. Hsiang, McGraw-Hill (1989).

Quality Planning and Analysis, 3rd ed., Juran and Gryan, McGraw-Hill (1993).

Quality, Productivity, and Competitive Position, W. Edwards Deming, MIT Press, Boston (1982).

Statistical Methods for Quality Engineering, 3rd ed., T. P. Ryan, Wiley (2011).

Statistical Quality Control, 7th ed., Eugene L. Grant and Richard S. Leavenworth, McGraw-Hill (1996).

Statistical Quality Design and Control, 2nd ed., R. E. DeVor, T. Chang, and J. W. Sutherland, Prentice Hall (2007).

Statistical Process Control and Quality Improvement, 5th ed., G. M. Smith, Prentice Hall (2004).

Taguchi Methods, A. Bendell, Springer (2007).

Taguchi Techniques for Quality Engineering, 2nd ed., P. J. Ross, McGraw-Hill (1996).

Taguchi's Quality Engineering Handbook, G. Taguchi, et al., Wiley-Interscience (2004).

The Six Sigma Black Belt Handbook, Thomas McCarty et.al., The Motorola University (2005).

The Six Sigma Handbook, 3rd ed., T. Pyzdek and P. Keller, McGraw-Hill (2009).

The Six Sigma Performance Handbook, P. Gupta, McGraw-Hill (2005).

Total Quality Management—Text, Cases and Readings, Joel E. Ross, St. Lucie Press (1993).

Work Systems and the Methods, Measurement and Management of Work, M. P. Groover, Prentice Hall (2007).

Zero Quality Control: Source Inspection and the Poka-Yoke System, Shigeo Shingo, Productivity Press (1986).

Failure Analysis and Product Liability

Analysis of Metallurgical Failures, V. J. Conangelo and F. A. Heiser, Wiley-Interscience (1974).

Case Histories in Failure Analysis, ASM, Materials Park, OH (1979).

Engineering Aspects of Product Liability, V. J. Conangelo and P. A. Thornton, ASM, Materials Park, OH (1981).

Failure Analysis of Engineering Structures: Methodology and Case Histories, V. Ramachandran et al., ASM, Materials Park, OH (2005).

Failure Analysis: Case Histories, F. K. Naumann, ASM, Materials Park, OH (1983).

Failure Analysis: The British Engine Technical Reports, F. K. Hutchings and P. M. Unterweiser, ASM, Materials Park, OH (1981).

Failure Analysis of Heat Treated Steel Components, L. C. F. Canale et al. (eds.), ASM, Materials Park, OH (2008)

Failure of Materials in Mechanical Design, J. A. Collins, Wiley-Interscience (1981).

Handbook of Case Histories in Failure Analysis: Vols. 1 and 2, K. A. Esaklul (ed.), ASM, Materials Park, OH (1992–93).

How to Organize and Run a Failure Investigation, Daniel P. Dennies, ASM, Materials Park, OH (2005).

Metal Failures: Mechanisms, Analysis, Prevention, A. J. McEvily (ed.), John Wiley & Sons, New York (2002).

Metallography in Failure Analysis, J. L. McCall and P. M. French, Plenum (1977).

Metallurgical Failure Analysis, C. R. Brooks and A. Choudhury, McGraw-Hill (1993).

Residual Stresses and Fatigue in Metals, J. O. Almen and P. H. Black, McGraw-Hill, New York (1963).

Stress Corrosion Cracking: Materials Performance and Evaluation, Russell Jones (ed.), ASM, Materials Park, OH (1992).

Tool and Die Failures Source Book, ASM, Materials Park, OH (1982).

Understanding How Components Fail, 3rd ed., Donald J. Wulpi, ASM, Materials Park, OH (2013).

Why Metals Fail, R. D. Barer and B. F. Peters, Gordon and Breach, New York (1970).

Manufacturing Systems

Computational Intelligence in Design and Manufacturing, Andrew Kusiak, Wiley (2000).

Fundamentals of Systems Engineering, C. Jotin Kristy and Jamshid Mohammadi, Prentice Hall (2001).

Intelligent Manufacturing Systems, A. Kusiak, Prentice Hall (1990).

Manufacturing Cells: A Systems Engineering View, C. Moodie, et al., Taylor & Francis (1995).

Manufacturing Intelligence, P. K. Wright and D. A. Bourne, Addison-Wesley (1988).

Manufacturing Systems: Theory and Practice, G. Chryssolouris, Springer (2006).

Manufacturing Systems Design and Analysis, 2nd ed., B. Wu, Chapman & Hall (1994).

Manufacturing Systems Engineering, S. B. Gershwin, PTR, Prentice Hall (1994).

Manufacturing Systems Engineering, 2nd ed., K. Hitomi, Taylor & Francis Ltd. (1996).

Modeling and Analysis of Manufacturing Systems, R. G. Askin and C. R. Standridge, Wiley (1993).

Performance Modeling of Automated Manufacturing Systems, N. Viswanadham and Y. Narahari, Prentice Hall (1992).

Stochastic Models of Manufacturing Systems, J. A. Buzacott and J. C. Shanthikumar, Prentice Hall (1993).

Systems Analysis and Modeling, Donald W. Boyd, Academic Press (2001).

Lean Manufacturing Systems

A Revolution in Manufacturing: The SMED System, Shigeo Shingo, Productivity Press (1985).

A Study of the Toyota Production System, Shigeo Shingo, Productivity Press (1989).

Becoming Lean, J. Liker (ed.), Productivity Press (2004).

Competitive Manufacturing Management, John Nicholas, McGraw-Hill (1998).

Creating a Lean Culture: Tools to Sustain Lean Conversions, 2nd ed., David Mann, Taylor & Francis (2010).

Design and Analysis of Lean Production Systems, R. G. Askin and J. B. Goldberg, Wiley (2002).

Design, Analysis and Control and Manufacturing Cells, PED-Vol. 53, J T. Black, B. C. Jiang, and G. J. Wiens, ASME (1991).

Design of the Factory with a Future, J T. Black, McGraw-Hill (1991).

Factory Physics, W. J. Hopp, and M. L. Spearman, McGraw-Hill (2001).

Flexible Manufacturing Systems: The Technology and Management, R. A. Maleki, Prentice Hall (1991).

Handbook of Cellular Manufacturing Systems, Shahrukh A. Irani (ed.), Wiley (1999).

How to Implement Lean Manufacturing, L. Wilson, McGraw-Hill (2009).

Inside the Mind of Toyota, Satoshi Hino, Productivity Press (2006).

Japanese Manufacturing Techniques: Nine Hidden Lessons in Simplicity, Richard J. Schonberger, Free Press (1982).

JIT Factory Revolution, H. Hirano and J T. Black, Productivity Press (1988).

Just-in-Time in American Manufacturing, William Duncan, SME (1988).

Kaikaku, The Power and Magic of Lean, N. Bodek, PCS Press (2004).

Kaizen Assembly: Designing, Constructing, and Managing a Lean Assembly Line, C. A. Ortiz, CRC Press (2006).

Kaizen for Quick Changeover, Kenichi Sekine and Keisuke Arai (translated by Bruce Talbot), Productivity Press (1992).

Kanban: Just-In-Time at Toyota, translation by D. J. Lu, Japan Management Association, Productivity Press (1986).

Kaizen Event Fieldbook: Foundation, Framework, and Standard Work for Effective Events, Mark Hamel, Society of Manufacturing Engineers (2010)

Lean Assembly, Michael Baudin, Productivity Press (2002).

Lean Engineering, J T. Black and Don T. Phillips, Virtualbookworm.com Publishing (2014)

Lean Enterprise Systems, Steve Bell, Wiley (2006).

Lean Manufacturing for the Small Shop, 2nd ed., G. Connor, Society of Manufacturing Engineers (2008).

Lean Manufacturing Implementation, D. P. Hobbs, APICS (2004).

Lean Manufacturing Systems and Cell Design, J T. Black and Steve L. Hunter, Society of Manufacturing Engineers (2003).

Lean Production: Implementing a World Class System, John Black, Industrial Press and SME (2008).

Lean Production Simplified, Pascal Dennis (2002).

Lean Supply Chain and Logistics Management, P. Myerson, McGraw-Hill (2012)

Lean Thinking, James Womack and Daniel Jones, Simon & Schuster, New York (1996).

One-Piece Flow: Cell Design for Transforming the Production Process, Kenichi Sekine, Productivity Press (1990).

The Design of the Factory with a Future, J T. Black, McGraw-Hill (1991).

The Hitchhiker's Guide to Lean: Lessons from the Road, Jamie Flinchbaugh and Andy Carlino, Society of Manufacturing Engineers (2005)

The Lean Design Guidebook, Ronald Mascitelli, Technology Publications (2004).

The Machine That Changed the World, James P. Womack, Daniel T. Jones, and Daniel Roos, Harper Perennial (1991).

The Principles of Group Technology and Cellular Manufacturing, H. Parsai et al., Wiley (2006).

The Toyota Product Development System, J. M. Morgan and J. K. Liker, Productivity Press (2006).

The Toyota Way: 14 Management Principles from the World's Greatest Manufacturer, Jeffrey Liker, McGraw-Hill (2004).

The Toyota Way to Lean Leadership, J. Liker and G. L. Convys, McGraw-Hill (2011).

Toyota Production System: Beyond Large-Scale Production, Taiichi Ohno, Productivity Press (1988).

Toyota Production System—An Integrated Approach to Just-In-Time, 3rd ed., Yasuhiro Monden. Norcross, Georgia (1998).

World Class Manufacturing Casebook: Implementing JIT and TOC, Richard J. Schonberger, The Free Press (1987).

World Class Manufacturing, Richard J. Schonberger, The Free Press (1986).

World Class Manufacturing, The Next Decade, Richard Schonberger, The Free Press (1996).

Zero Inventories, Robert W. Hall, Dow Jones-Irwin (1983).

Robotics

Handbook of Industrial Robotics, 2nd ed., Shimon Nof (ed.), Wiley (1999).

Industrial Robotics: How to Implement the Right System for Your Plant, Andrew Glaser, Industrial Press (2008)

Industrial Robotics: Technology, Programming and Applications, M. P. Groover et al., McGraw-Hill (1986).

Introduction to Robotics: Mechanics and Control, 3rd ed., J. J. Craig, Prentice Hall (2004).

Introduction to Robot Technology, P. Coiffet and M. Chizouze, McGraw-Hill (1993).

Introduction to Robots in CIM Systems, J. A. Rehg, Prentice Hall (2000).

Robot Analysis, L.-W. Tsai, Wiley (1999).

Robotics and Automation Handbook, T. R. Kurfess (ed.), CRC Press (2004).

Robotics and Manufacturing Automation, 2nd ed., C. R. Asfahl, Wiley (1992).

Automation, CNC, CAM, CAD, FMS, and CIM

An Introduction to CNC Machining and Programming, D. Gibbs and T. M. Crandall, Industrial Press (1991).

Automation, Production Systems, and Computer-Integrated Manufacturing, 3rd ed., Mikell P. Groover, Prentice Hall (2008).

Computer-Aided Manufacturing, 3rd ed., T. C. Chang, R. A. Wysk and Hsu-Pin Wang, Prentice Hall (2006).

Computer-Integrated Design and Manufacturing, D. D. Bedworth, M. R. Henderson, and P. M. Wolfe, McGraw-Hill (1991).

Computer-Integrated Manufacturing, 3rd ed., J. A. Rehg and H. W. Kraebber, Prentice Hall (2005).

Computer Control of Machines and Processes, J. G. Bollinger and N. A. Duffie, Addison-Wesley (1989).

Computer Numerical Control: Operation and Programming, 3rd ed., J. Stenerson and K. S. Curran, Prentice Hall (2006).

Engineering Robust Designs, J. X. Wang, Prentice Hall (2005).

Fundamentals of Computer-Integrated Manufacturing, A. L. Foston, et al., Prentice Hall (1991).

Getting Factory Automation Right (The First Time), Edwin H. Zimmerman, Society of Manufacturing Engineers (2001).

Handbook of Machine Vision, A. Hornberg, Wiley (2007).

Introduction to Computer Numerical Control, 5th ed., J. V. Valentino and J. Goldenberg, Prentice Hall (2012).

Industrial Automation, D. W. Pessen, Wiley (1989).

The Principles of Group Technology and Cellular Manufacturing Systems, H. Parsaei, et al., Wiley (2006).

Cost Estimating

Cost Estimator's Reference Manual, R. D. Steward and R. W. Wyskida, Wiley (1987).

Engineering Cost Estimating, 3rd ed., P. E. Ostwald, Prentice Hall (1992).

Realistic Cost Estimating for Manufacturing, 2nd ed., W. Winchell, Society of Manufacturing Engineers (1989).

Nontraditional Machining/Laser Machining Sources

Advanced Machining Processes: Nontraditional and Hybrid Machining Processes, H. El-Hofy, McGraw-Hill (2005).

Advanced Machining Processes of Metallic Materials: Theory, Modelling and Applications, W. Grzesik, Elsevier (2008).

Advanced Machining Technology Handbook, James Brown, McGraw-Hill, New York (1998).

Advanced Methods of Machining, J. A. McGeough, Chapman & Hall, New York (1988).

Advanced Machining Processes: Nontraditional and Hybrid Machining Processes, H. El-Hofy, McGraw-Hill (2005).

Electrical Discharge Machining, E. C. Jameson, Society of Manufacturing Engineers (2001).

Laser Fabrication and Machining of Metals, N. B. Dahotre and S. P. Harimkar, Springer (2007).

Laser Material Processing, 4th ed., W. M. Steen and J. Mazumder, Springer (2010).

Laser Materials Processing, L. Migliore, Marcel Dekker (1996).

Laser Machining, G. Chryssolouris and P. Sheng, Springer-Verlag (1991).

Laser Machining of Advanced Materials, N. B. Dahotre and A. Samant, CRC Press (2011).

Non-Traditional Machining Handbook, 2nd ed., C. Sommer, Advance Publishing (2009).

Nontraditional Machining Processes, 2nd ed., E. J. Weller ed., Society of Manufacturing Engineers (1984).

Nontraditional Manufacturing Processes, G. F. Bennedict, Marcel Dekker (1987).

Principles of Abrasive Water Jet Machining, A. W. Momber and R. Kovacevic, Springer (1998).

Principles of Laser Materials Processing, E. K. Asibu Jr., Wiley (2009).

The EDM Handbook, E. B. Guitrau and Hanser-Gardner, Cincinnati, OH (1997).

Theory and Practice of Electrochemical Machining, V. K. Jain and P. C. Pandey, Wiley (1993).

Ultrasonic Processes and Machines, V. K. Astashev and V. I. Babitsky, Springer (2010)

Additive Manufacturing and Rapid Prototyping

Additive Manufacturing Technologies: Rapid Prototyping to Direct Digital Manufacturing, I. Gibson, et al., Springer (2010).

Automated Fabrication: Improving Productivity in Manufacturing, M. Burns, Prentice Hall, Englewood Cliffs, NJ (1993).

Engineering Design and Rapid Prototyping, T. Grimm, Springer (2010).

Fabricated: The New World of 3D Printing, Hod Lipson and Melba Kurman, Wiley (2013).

Rapid Manufacturing: The Technologies and Applications of Rapid Prototyping and Rapid Tooling, D. T. Pham and S. S. Dimov, Springer-Verlag, London, UK (2001).

Rapid Prototyping, A. Gebhardt, Hanser Gardner (2004).

Rapid Prototyping and Other RP&M Technologies: From Rapid Prototyping to Rapid Tooling, P. F. Jacobs, SME, Dearborn, MI (1996).

Rapid Prototyping: Laser Based and Other Technologies, P. K. Venuvinod and W. Ma, Springer (2010).

Rapid Prototyping: Principles and Applications, 2nd ed., C. C. Kai et al., World Scientific Publishing Company, Singapore (2003).

Rapid Prototyping: Principles and Applications, R. I. Noorani, Wiley (2006).

Rapid Prototyping: Principles and Applications in Manufacturing, C. K. Chua and L. K. Fua, Wiley (1997).

Rapid Prototyping: Theory and Practice, A. Kamrani and E. A. Nasr (eds.), Springer (2007).

Rapid Tooling: Technologies and Industrial Applications, Peter D. Hilton and P. F. Jacobs, Marcel Dekker/CRC Press (2000).

Rapid Tooling: The Technologies and Applications of Rapid Prototyping and Rapid Tooling, D. T. Pham and S. S. Dimov, Springer-Verlag (2001).

Solid Freeform Fabrication, J. J. Beaman et al., Kluwer (1997).

User's Guide to Rapid Prototyping, Todd Grimm, Society of Manufacturing Engineers (2004).

Wohler's Report (published annually), Terry Wohlers, Wohlers Associates, CO.

Microfabrication and Nanotechnology

An Introduction to Microelectromechanical Systems Engineering, 2nd ed., N. Maluf and K. Williams, Artech House (2004).

Applications of Microfabrication and Nanotechnology, M. J. Madou, CRC Press (2009).

Basics of Nanotechnology, 3rd ed., H.-G. Rubahn, Wiley-VCH, Germany (2008).

Encyclopedia of Nanoscience and Nanotechnology, American Scientific Publishers (2004).

Foundations of MEMS, 2nd ed., C. Liu, Prentice Hall (2011).

Fundamentals of Microfabrication, M. J. Madou, CRC Press, Boca Raton, FL (2002).

Fundamentals of Nanotechnology, G. L. Hornyak et al., CRC/Taylor & Francis (2009).

Handbook of Nanoscience, Engineering and Technology, W. A. Goddard III, et al., CRC Press (2003).

Introduction to Micromachining, V. K. Jain. (ed.), Alpha Science International Ltd., Kanpur, India (2010)

Introduction to Nanoscience and Nanotechnology, C. Binns, Wiley (2010).

Introduction to Nanotechnology, C. P. Poole Jr., Wiley-Interscience (2003).

Introductory MEMS: Fabrication and Applications, T. M. Adams and R. W. Layton, Springer (2009).

Manufacturing Techniques for Microfabrication and Nanotechnology, M. Madou, CRC/Taylor & Francis (2009).

MEMS Materials and Processes Handbook, R. Ghodssi and P. Lin (eds.), Springer (2011).

MEMS/NEMS Handbook (5 vols.), C. T. Leondes, Springer (2007).

MEMS and Microsystems: Design, Manufacture, and Nanoscale Engineering, 2nd ed., T.-R. Hsu, Wiley (2008).

Micro and Nanomanufacturing, M. L. Jackson, et al., Springer (2007).

Microfabrication and Nanomanufacturing, M. J. Jackson, CRC/Taylor & Francis, Boca Raton, FL (2006).

Micromachining of Engineering Materials, J. A. McGeugh, Marcel Dekker, New York (2002)

Micromanufacturing and Nanotechnology, N. P. Mahalik, Springer Publishing, New York, NY (2010)

Micromechanical System Design, J. J. Allen, CRC Press (2006).

Nanofabrication Fundamentals and Applications, A. A. Tseng (ed.), World Scientific, Singapore (2008).

Nanofuture: What's next for nanotechnology, J. S. Hall, Amherst, NY: Prometheus Books (2005).

Nanomaterials, S. Mitura, Elsevier (2000).

Nanomaterials: An Introduction to Synthesis, Properties and Applications, D. Vollath, Wiley (2008).

Nanomaterials Handbook, Y. Gogotsi, CRC Press (2006).

The MEMS Handbook, 2nd ed. (3 vols.), M. Gad-el-Hak (ed.), CRC Press (2006).

Springer Handbook of Nanotechnology, 3rd ed., B. Bhushan (ed.), Springer (2010).